Marine Natural Products Research in South-East Asia

Lik Tong Tan · Novriyandi Hanif
Editors

Marine Natural Products Research in South-East Asia

Current Status and Perspectives

Springer

Editors
Lik Tong Tan
Department of Natural Sciences and Science Education
National Institute of Education
Nanyang Technological University
Singapore, Singapore

Novriyandi Hanif
Department of Chemistry
Faculty of Mathematics and Natural Sciences
IPB University
Bogor, Indonesia

ISBN 978-3-032-01097-1 ISBN 978-3-032-01098-8 (eBook)
https://doi.org/10.1007/978-3-032-01098-8

This Springer imprint is published by the registered company Springer Nature Switzerland AG
The registered company address is: Gewerbestrasse 11, 6330 Cham, Switzerland

Preface

The marine environments of Southeast Asia are among the most biologically diverse ecosystems on the planet, harboring a vast array of marine organisms that serve as invaluable sources of novel bioactive compounds. As the global demand for new therapeutics and biotechnological solutions intensifies, marine natural products (MNPs) have emerged as a critical frontier in drug discovery and scientific innovation. This book, *Marine Natural Products Research in South-East Asia: Current Status and Perspectives*, brings together a diverse collection of studies and perspectives that reflect the current landscape, emerging trends, and future directions of MNP research across the region.

This edited volume comprises ten chapters, each focusing on a specific geographical area or thematic aspect of marine natural products research. From Indonesia's chemical evolution of marine-derived molecules to Malaysia's extensive NMR-based structure characterizations, and from the Philippines' vast compound catalog spanning decades to Singapore's nascent integration of biotechnological tools for marine drug discovery, the contributions highlight the rich scientific tapestry of Southeast Asia's marine biota. Vietnam's coastal biodiversity and its associated invertebrate chemistry, as well as microbial metabolites from the East Sea, further exemplify the depth of underexplored marine potential. Additionally, the volume extends its scope beyond Southeast Asia, incorporating insights from Papua New Guinea—a region of comparable ecological significance. This book not only provides an overview of the state of marine biotechnology related to drug discovery in each nation but also critically examines the challenges that hinder its progress, such as uneven research infrastructure, environmental degradation, and limited international collaboration. Nevertheless, through these chapters, the authors illuminate the region's growing capacity to contribute to the global biopharmaceutical pipeline, offering a wealth of novel scaffolds, mechanisms of action, and opportunities for sustainable development.

The editors hope that this volume will serve as both a comprehensive reference for established researchers and an inspiring resource for new entrants into the field. By documenting current findings and envisioning future possibilities, this book aspires to catalyze further research, foster regional collaborations, and underscore the urgent need to preserve our marine ecosystems as we mine them for life-saving innovations.

Singapore
Bogor, Indonesia
May 2025

Lik Tong Tan
Novriyandi Hanif

Contents

Chapter 1
Status of Drug Discovery-Based Marine Biotechnology in Southeast Asia

Lik Tong Tan, Clarissa Widyantoro, and Novriyandi Hanif

1 Introduction

Southeast Asia (SEA) is a global hotspot of marine biodiversity, boasting some of the richest and most diverse coastal ecosystems on the planet. This region, comprising ten countries under the Association of Southeast Asian Nations (ASEAN) umbrella, has coastlines that collectively span tens of thousands of kilometers. These marine environments are home to a broad array of organisms, including sponges, soft corals, tunicates, echinoderms and marine algae, that have evolved complex biochemical strategies to survive. These adaptations have made marine organisms a promising source of bioactive compounds with applications in pharmaceuticals, cosmetics and nutraceuticals. Recognizing this potential, several Southeast Asian countries have embarked on bioprospecting and drug discovery initiatives grounded in marine biotechnology.

Since the early 1990s, marine natural products (MNPs) research in the region has grown steadily, driven by both regional efforts, international collaborations as well as government supported initiatives and investments in the blue economy related to marine biotechnology.

Countries, such as the Philippines, Indonesia, Vietnam and Thailand, have taken leading roles, publishing extensively on the bioactivity of marine-derived metabolites. Universities, marine institutes, and governmental agencies have launched various programs aimed at cataloging marine biodiversity, screening for bioactivity and translating findings into therapeutic applications. However, despite this

L. T. Tan (✉) · C. Widyantoro
Natural Sciences and Science Education, National Institute of Education, Nanyang Technological University, Singapore, Singapore
e-mail: liktong.tan@nie.edu.sg

N. Hanif
Department of Chemistry, Faculty of Mathematics and Natural Sciences, IPB University, Bogor, Indonesia
e-mail: nhanif@apps.ipb.ac.id

L. T. Tan and N. Hanif (eds.), *Marine Natural Products Research in South-East Asia*,
https://doi.org/10.1007/978-3-032-01098-8_1

momentum, many parts of the region remain underexplored, and the research output is disproportionately concentrated in a few countries. Issues such as lack of infrastructure, limited funding, and inadequate taxonomic expertise continue to constrain the growth of this scientific domain.

Adding to the complexity is the increasing threat to marine ecosystems posed by overfishing, habitat destruction, pollution and climate change. The ecological pressures not only endanger the marine species themselves but also limit access to yet-undiscovered compounds with potential biomedical value. To address these challenges, countries are beginning to align MNPs research with sustainable development strategies under the emerging framework of the Blue Economy. This integration of economic development and environmental stewardship offers a path forward that could amplify the region's contribution to global drug discovery while preserving its natural heritage. As this chapter outlines, advancing marine biotechnology in Southeast Asia requires a multi-pronged approach involving scientific innovation, regional collaboration and sustainable marine resource management.

This introductory chapter aims to provide an overview of the current landscape of MNPs research in Southeast Asia, with a particular focus on its applications in drug discovery. It highlights the region's rich marine biodiversity and identifies key marine organisms, such as sponges, tunicates, soft corals, echinoderms and algae, that have shown potential in yielding bioactive compounds with pharmacological uses. The chapter also examines the uneven development of MNPs research across Southeast Asian countries, drawing attention to disparities in infrastructure, funding and scientific collaboration. In addition, it addresses critical challenges facing marine biodiversity, including overfishing, pollution and climate change. By integrating insights from traditional marine medicine and emphasizing regional cooperation, this chapter sets the stage for the rest of the book, offering context and identifying knowledge gaps. Subsequent chapters highlight the chemistry and biology of marine natural products in Southeast Asian countries, including Indonesia, the Philippines, Malaysia, Vietnam and Singapore. A chapter on marine natural products research from Papua New Guinea is also included due to its biogeographic connections to Southeast Asia and ecologically linked, especially through the Coral Triangle region.

2 Overview of Coastal Marine Biodiversity in Southeast Asia

Southeast Asia is one of the most biodiverse regions in the world, with its coastal marine ecosystems playing a critical role in global biodiversity (Fig. 1) [1]. The region encompasses vast coastlines, islands, and archipelagic waters that support a wide array of marine life. Marine biodiversity represents an invaluable reservoir of chemical and biological novelty, offering immense potential for drug discovery and biomedical innovation [2]. The vast array of marine organisms, ranging from sponges and tunicates to soft corals and mollusks, produce unique secondary metabolites

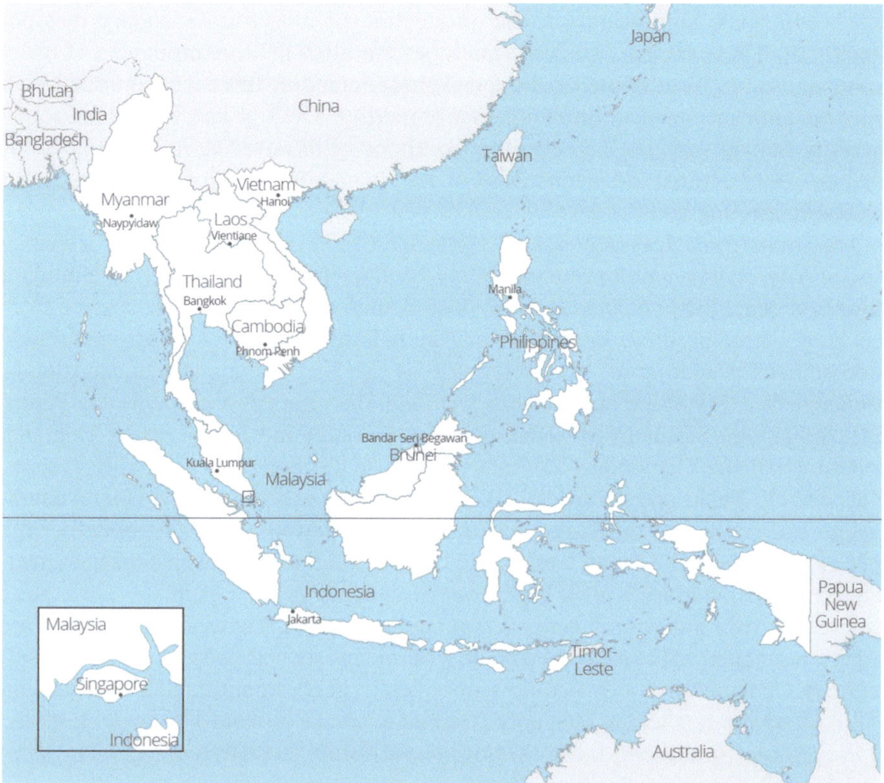

Fig. 1 Map of Southeast Asia (image from https://aseanup.com/free-maps-asean-southeastasia/)

as part of their survival strategies in complex and competitive ecosystems. These compounds often possess potent bioactivities, including antimicrobial, anticancer, anti-inflammatory and antiviral properties, making them promising leads for pharmaceutical development. In particular, the largely unexplored and richly diverse benthic marine ecosystems of Southeast Asia present a critical frontier for natural product research. The following sections provide an overview of the marine biodiversity in this region, focusing on key groups of benthic organisms, including porifera, cnidaria, echinoderms and tunicates, that have emerged as valuable sources of novel bioactive compounds [3].

2.1 Marine Sponges

Sponge biodiversity plays a crucial role in drug discovery due to the remarkable chemical diversity found within these marine organisms [4]. Sponges host a rich array of symbiotic microorganisms, sometimes accounting for up to 35% of their biomass, which contribute significantly to the production of bioactive secondary metabolites [5, 6]. These spongeassociated microbes are often the true producers of many compounds initially attributed to the sponge host, including those with antimicrobial, antiviral, anticancer and antiinflammatory properties. The high species and microbial diversity among sponges increases the likelihood of discovering novel compounds with unique structures and mechanisms of action, making them invaluable resources for bioprospecting and the development of new pharmaceuticals.

The South China Sea is recognized as one of the richest regions globally for sponge biodiversity, with scientific interest dating back to the late 1700s [7]. A landmark annotated checklist published in 2000 documented over 1500 sponge species from this area, including those known for marine natural products [7]. This vast region encompasses a wide array of habitats across the Nicobar and Andaman Islands, Andaman Sea, Myanmar and western Thailand, the west coast of the Malay Peninsula, Singapore, Straits of Malacca, northern Sumatra and Java, Gulf of Thailand, eastern Thailand, Cambodia, Vietnam (including the Paracel and Spratly Islands, Macclesfield Bank, and Tizard Atoll), Borneo, the southern coast of China, southeastern Indonesia and the Philippines including Palawan. The exceptionally high species richness in the South China Sea, comparable to other diverse marine invertebrate groups, is marked by a high level of endemism, with estimates suggesting hundreds to over a thousand endemic sponge species [7]. These endemic forms are largely associated with specialized ecological niches such as deeper coral reefs, soft sediments and trawl grounds. A follow-up checklist in 2016 confirmed the uniqueness of this biodiversity, showing that only 16 sponge species (around 4%) are widespread in the region [8]. Among these, *Xestospongia testudinaria*, *Mycale* (*Zygomycale*) *parishii* and *Cinachyrella australiensis* are the only species found continuously from the Singapore Strait to the Taiwan Strait, underscoring the biogeographic distinctiveness of sponge communities across the South China Sea.

Using Indonesia as a case study highlights the remarkable sponge biodiversity found in the South China Sea, which holds great promise for marine drug discovery. Among Southeast Asian nations, Indonesia boasts the highest recorded sponge diversity, with a total of 731 species identified to date [9]. This includes approximately 45 species, 566 species, 115 species and five species from Class Calcarea, Demospongiae, Hexactinellida and Homoscleromorpha, respectively [9]. These Poriferan species have been documented from 12 marine ecoregions across the Indonesian archipelago, as well as freshwater habitats (Spongillida), based on records spanning from 1820 to 2021. Indonesia's sponges have been the subject of scientific interest since the nineteenth century, explored through historical expeditions and international collaborations. Since the first national checklist compiled by Hooper and colleagues in 2000 [7], at least 63 new sponge species have been described from Indonesia's marine ecosystems [10–37]. Despite this progress, several ecoregions, such as Northeast Sulawesi, Papua, Southern Java and Western Sumatra, remain underexplored. Additionally, several sponge diversity studies in Indonesia are coupled with biomedical assessments of sponges as well as associated microbes [38–43].

Given Southeast Asia's extensive and complex coastline, it is not surprising that other countries in the region, including the Philippines, Vietnam, Malaysia, Thailand and Myanmar, also exhibit high sponge diversity. Even in Singapore, with a relatively short coastline of only 193 km, at least 130 sponge species have been recorded [8]. The abundance and uniqueness of sponge species, especially endemic species, in the region represent a valuable reservoir for the discovery of novel marine natural products. Therefore, establishing a strong taxonomic baseline for Porifera is essential for advancing our understanding of aquatic biodiversity and supporting drug discovery efforts in the South China Sea.

2.2 Tunicates

Tunicate diversity in Southeast Asia, particularly in the South China Sea region, is notably rich but remains underexplored. Surveys conducted across areas, such as Malaysia, Thailand, the Philippines, Vietnam and Indonesia reveal a high variety of solitary and colonial ascidian species, many of which are likely undescribed or new to science. Based on a checklist on the status of ascidian diversity in South China Sea published in 2016, about 381 species of ascidians, represented by 17 families, including Clavelinidae, Diazonidae, Didemnidae, Holozoidae, Polycitoridae, Polyclinidae, Pseudodistomidae, Stomozoidae, Ascidiidae, Cionidae, Ciallusiidae, Corellidae, Perophoridae, Plurellidae, Molgulidae, Pyuridae and Styelidae, as well as 68 genera have been recorded [44]. A significant number of the tunicate species in this checklist were recorded from Indonesia (229 species) and the Philippines (183 species). Despite the biological richness, taxonomic studies are limited and often outdated, making it difficult to assess the full extent of biodiversity.

Since the publication of the checklist in 2016, several tunicate biodiversity studies have been conducted, particularly in Indonesia. For instance, studies on tunicate biodiversity within the Sangkarang (Spermonde) Archipelago have contributed valuable baseline data for sustainable marine resource management. In the waters of Samalona Island, South Sulawesi, researchers recorded 18 tunicate species, an initial but comparatively low count relative to global diversity and even lower than species richness documented from neighboring Baranglompo Island [45]. A broader survey aimed to expand the tunicate distribution map in the region, covering Pannikiang and Badi Islands, revealed a total of 35 species, with 24 species found at Pannikiang and 18 at Badi [46]. Additionally, across the Sangkarang Archipelago, including sites such as Samalona, Barrang Caddi, Barranglompo, Laelae, Bone Batang and Badi, a total of 55 tunicate species have now been identified [47]. At Badi Island specifically, species like *Polycarpa aurata, Didemnum molle* and *Polycarpa* sp. were noted for their wide distribution [48]. In addition to these biodiversity assessments, other research in Indonesia has focused on tunicate bioprospecting, highlighting the ecological and pharmacological importance of these marine invertebrates [46].

2.3 Soft Corals

More than 100 genera of shallow-water soft corals have been described from the Indo-Pacific alone [49], underscoring the ecological significance and diversity of this group. Importantly, soft corals are also a valuable source of novel bioactive compounds: between 2010 and 2019, a remarkable 95% of all new MNPs from invertebrates derived from species within the order Alcyonacea. Nearly 20% of these new MNPs were sourced specifically from the family Alcyoniidae, highlighting this family's dominant role in marine drug discovery [3]. Moreover, the genera *Sinularia, Sarcophyton, Lobophytum* and *Dendronephthya* have yielded a wide array of secondary metabolites with promising anti-inflammatory, anticancer, antimicrobial and antiviral properties [50–53]. However, the identification of Alcyoniidae to specific level is often extremely difficult, and there is a large number of very similar species, many of which are still undescribed [54]. These findings not only highlight the therapeutic potential of soft corals but also underscore the urgency of conserving coral reef habitats and supporting taxonomic and chemical investigations in underexplored regions of Southeast Asia.

Within Southeast Asia, Vietnam stands out as one of the most comprehensively studied countries in terms of soft coral biodiversity. Hoang and Dautova recorded 200 species belonging to 45 genera and 14 families, with the Alcyoniidae family being the most diverse (124 species), followed by Nephtheidae (21 species) and Plexauridae (18 species) [55]. The genus *Sinularia* alone accounts for 66 species, followed by *Lobophytum* (26 species) and *Sarcophyton* (20 species). In Nha Trang Bay, a particularly rich site, 94 species from 15 genera and five families were documented [56], with *Sclerophytum* being the most speciose genus. Subsequent additions to Vietnam's soft coral fauna include two *Sclerophytum* species from Con Dao Islands [57] and 15 soft coral species newly recorded for the South China Sea, including *Xenia novaebritaniae* and *Heteroxenia elisabethae* [58]. Recent survey of soft corals belonging to Family Xeniidae collected from coastal fringing reefs along the coast of central Vietnam, revealed five species belonging to two genera, *Xenia* and *Heteroxenia* [59]. Of five species, two were recorded for the first time in Vietnam are *Xenia novaebritaniae* and *Heteroxenia elisabethae.*

In Thailand, soft coral diversity is unevenly studied, with a greater richness observed along the Andaman Sea coast compared to the Gulf of Thailand. A survey identified soft corals from four families and 19 genera. Alcyoniidae, including *Sinularia, Sarcophyton* and *Lobophytum*, was the dominant family in the Gulf of Thailand, while Nephtheidae (including Dendronephthya) dominated the Andaman coast [60]. The most observed genera across both regions were *Sinularia, Dendronephthya* and *Sarcophyton*, which are also known for their chemical richness and bioactivity.

In contrast, Indonesia, a megadiverse archipelago, remains relatively underexplored regarding soft coral distribution. Over the last decade, only eight studies have focused on this group [61]. A recent survey by Putra et al. identified 41 species from 14 genera and eight families across three islands in the Makassar Strait [61]. Some genera exhibited restricted distributions, such as *Acrossota* and *Dendronephthya* (only found in Barrang Caddi Island), while *Litophyton*, *Lobophytum*, *Sarcophyton* and *Sinularia* were found across all sites surveyed.

Singapore's soft coral assemblages, though limited by coastal development, still exhibit noteworthy diversity. Goh and associates reported seven genera from 10 sites, with Seah et al. (2015) later identifying 16 distinct morphotypes [62, 63]. Similarly, Mohammad et al. [64] documented 15 genera across eight families in the North Malacca Straits, while Lalas et al. [65] noted approximately 10 families in the West Philippine Sea.

2.4 Echinoderms

In a 2000 publication on checklist of echinoderm records for South China Sea revealed that close to a thousand species have been recorded, including 113 crinoids, 227 asteroids, 272 ophiuroids, 167 echinoids and 203 holothuroids [66]. A total of 178 echinoderms has their type locality in the South China Sea, and 63% of these can be considered endemic species. Among the five major subgroups of echinoderms, including Asteroidea (sea stars), Ophiuroidea (brittle stars), Echinoidea (sea urchins), Holothuroidea (sea cucumbers) and Crinoidea (feather stars), the Holothuroidea (sea cucumbers) and Asteroidea (sea stars) have been the most extensively investigated for their natural products [67]. Sea cucumbers and starfishes are rich in secondary metabolites, especially triterpene glycosides (saponins), which exhibit a wide range of bioactivities including anticancer, antimicrobial, antiviral, antiinflammatory and anticoagulant properties [68, 69].

Among the various echinoderms, there is a growing body of biodiversity research on sea cucumbers in Southeast Asia, driven by their known mega biodiversity, high commercial value, ecological importance and the need for sustainable management due to overfishing and declining populations [70]. For instance, Vietnam is home to approximately 90 species of sea cucumbers, with the families Holothuriidae (genera Holothuria and Stichopus), Cucumariidae (genera Colochirus and Cucumaria) and Sinaptidae (genus Protankyra) being the most prevalent [71]. Among these, around 10 species are highly valued for their nutritional and commercial significance, including *Holothuria martensii*, *H. atra*, *H. scabra*, *H. echinites*, *H. mauritiana*, *H. nobilis* and *H. ananas*. Notably, the waters surrounding the Phu Quoc Archipelago host at least 25 sea cucumber species, with 11 being commercially exploited [72].

The biodiversity of sea cucumbers in Thailand has been extensively studied since the early twentieth century, beginning with Dr. Theodor Mortensen's surveys along the eastern coast of the Gulf of Thailand in 1899–1900. Over the decades, both foreign and Thai researchers have significantly contributed to the taxonomy and ecological understanding of sea cucumbers across various marine habitats, including coral reefs, seagrass beds, sandy and rocky shores [73]. Notable findings include several new species and detailed checklists compiled from both the Gulf of Thailand and the Andaman Sea [73]. Currently, a total of 90 valid sea cucumber species, representing 29 genera, 8 families, and 5 orders, had been confirmed from 959 collected samples.

Of these, 33 species were exclusive to the Gulf of Thailand, and 32 were exclusive to the Andaman coast, highlighting Thailand's rich and regionally distinct sea cucumber fauna [73].

Sea cucumber biodiversity has also been reported to be significantly high in other Southeast Asian countries. Indonesia is recognized for its rich sea cucumber biodiversity where approximately 350 species have been recorded in Indonesian waters, with more than half collected from depths exceeding 3000 m. Of these, at least 55 species are known to have economic value [74]. Additionally, studies have shown that sea cucumber in Indonesia, such as *Stichopus hermanni*, *Thelenota ananas*, *T. anax*, *Holothuria fuscogilva*, *H. leucospilota*, *H. atra*, *H. scabra* and *Actinopyga mauritiana* are edible species with beneficial medical effects and low toxicity [70]. Malaysia is estimated to host over 80 species of sea cucumbers, primarily from the order Aspidochirotida, which includes genera such as *Holothuria*, *Actinopyga*, *Bohadschia*, *Stichopus* and *Thelenota* [75, 76]. Notably, the east coast of Malaysia exhibits higher sea cucumber species diversity compared to the west coast. Relatively high sea cucumber diversity has been reported in the Philippines and Singapore waters with over 170 and 50 species, respectively [77–79].

2.5 Others

In addition to the above marine taxonomic groups, South China Sea is recognized as a global hotspot for marine biodiversity, with remarkable species richness across other taxonomic groups. Molluscs, including diverse gastropods and bivalves, are especially abundant [80]. In a 2000 publication on checklist of the nudibranchs, about 193 species of nudibranchs have been documented along the coasts of South China sea, including South China, Taiwan, Hong Kong, Vietnam, Cambodia, Thailand, East and West Malaysia, Singapore, Brunei and the Philippines [81]. Additionally, bryozoans and marine algae, particularly red and green macroalgae, exhibit high diversity and endemism [82, 83]. This exceptional megadiversity provides a vast reservoir of bioactive compounds, making the region highly valuable for marine biotechnology as well as drug discovery efforts.

2.6 Threats on Marine Biodiversity

The marine biodiversity of the South China Sea and broader Southeast Asian region is under increasing threat from a combination of anthropogenic and environmental pressures [84]. Overfishing remains one of the most pressing concerns, with unsustainable fishing practices such as bottom trawling, blast fishing, and the use of fine-meshed nets leading to the depletion of fish stocks and destruction of benthic habitats [85]. Illegal, unreported, and unregulated fishing further exacerbates this issue, undermining conservation efforts and threatening the livelihoods of coastal communities.

The decline of keystone species and top predators due to overexploitation can have cascading effects on marine ecosystems, disrupting ecological balance and resilience [86].

Coastal development and land-based pollution are also major contributors to the degradation of marine ecosystems in the region [87]. Rapid urbanization, industrialization, and agriculture result in increased sedimentation, nutrient loading, and the release of pollutants into coastal waters [84]. These changes lead to habitat loss, particularly in sensitive areas such as mangroves, coral reefs and seagrass beds. Eutrophication from nutrient runoff encourages algal blooms, which can lead to hypoxia and the death of marine organisms [88]. In addition, plastic pollution, including microplastics, is an emerging concern, impacting a wide range of marine species through ingestion and entanglement [89].

Climate change further compounds existing threats by altering ocean temperatures, increasing sea levels, and causing ocean acidification [90]. Coral reefs, a cornerstone of marine biodiversity in the South China Sea, are especially vulnerable to coral bleaching events triggered by rising sea temperatures [91]. Ocean acidification threatens the calcification process in corals, molluscs and other marine organisms, weakening reef structures and affecting the marine food web [92]. As climate-related impacts intensify, they interact with local stressors to reduce the overall resilience of marine ecosystems. Additionally, impact of climate change on the chemical ecology of marine organisms, especially on the production, functionality and perception of allelochemicals, poses threats to marine biodiversity [93]. Addressing these multifaceted threats requires coordinated regional efforts, robust marine conservation policies and sustainable management practices to safeguard the rich marine biodiversity of Southeast Asia for future generations [94].

3 Marine-Based Traditional Medicines in Southeast Asia

For centuries, coastal communities across Southeast Asia have harnessed the rich biodiversity of marine environments for traditional medicine. From seaweeds and mollusks to fishes and corals, marine organisms have been integral to indigenous healing practices, often used to treat ailments ranging from infections to inflammatory diseases. These remedies, deeply rooted in local ethnomedicinal knowledge, reflect a unique interplay between cultural traditions and ecological resources. Notably, the influence of Traditional Chinese Medicine (TCM) has further enriched the region's marine-based healing practices, especially in areas with significant Chinese diaspora [95]. TCM's emphasis on the medicinal value of marine materials, such as sea cucumbers, cephalopods, seahorses, jellyfish and various types of shellfish, has contributed to a broader pharmacopeia and influenced preparation techniques and diagnostic frameworks (Fig. 2) [96]. With growing scientific interest in marine-derived bioactive compounds, many of these traditional remedies are now being investigated for their pharmacological potential, bridging the gap between ancient wisdom and modern

Fig. 2 Traditional marine-based products, including **a** dried octopus, **b** dried cuttlefish, **c** dried seahorses, **d** dried sea cucumbers, **e** sea cucumber medicinal oil and **f** red marine algae supplements

drug discovery [97, 98]. The following sections provide a brief account on the extensive range of marine organisms being used as traditional medicines, particularly for coastal communities in Southeast Asia.

3.1 Marine Macroalgae

Seaweeds have long been used as traditional medicines in Southeast Asia, where coastal communities have recognized their therapeutic properties for generations. Rich in bioactive compounds, such as polysaccharides, phenolics and bioavailable minerals, various seaweed species are incorporated into folk remedies to treat ailments ranging from digestive disorders to skin conditions [99]. In countries like Indonesia, the Philippines and Malaysia, seaweeds are consumed as herbal infusions, topical treatments, and dietary supplements to promote overall health and wellness [100]. Modern scientific studies are now validating many of these traditional uses, highlighting the potential of seaweeds as sources of novel bioactive compounds for pharmaceutical and nutraceutical applications.

Various species of Vietnamese seaweeds have long been used in traditional medicine. Of the more than 60 species of recorded macroalgae, about 20 species are used in folk medicine in Vietnam [101]. For instance, *Gracilaria* sp. is served as a natural laxative, while certain *Kappaphycus* and *Eucheuma* species are valued for their potential in reducing tumors, ulcers and headaches. *Sargassum* sp., influenced by Chinese medicinal practices, are widely used in Vietnamese folk medicine as herbal teas and traditional remedies [102]. Additionally, *Caloglossa* sp., *Codium* sp., *Dermonema* sp., *Gymnogongrus* sp. and *Hypnea* sp. are utilized as vermifuge drugs. *Ulva lactuca* and *U. reticulata* are also sold for medicinal purposes. *Laminaria* is both an ingredient in "Che" and a component of traditional medicine.

Macroalgae are often utilized in Filipino traditional medicine, especially in certain northern areas of the archipelago, such as the Ilocos Norte region, eastern Camarines Sur and eastern Sorsogon regions [103–105]. A total of 34 seaweed species have traditionally been recognized by the Ilocano people for their medicinal properties, addressing eight types of emic health concerns [105]. These algal species belong to three major groups, namely rhodophytes (16 species), ochrophytes (7 species) and chlorophytes (11 species), spanning 12 families and 20 genera. Most were utilized to treat digestive, pulmonary and glandular-related ailments. Among coastal communities, the Ilocano people of Pagudpud and Burgos were the most knowledgeable about these medicinal seaweeds. *Caulerpa racemosa* (locally known as ar-arusip) was the most widely recognized species, traditionally used to relieve cough and asthma. This species and *C. lentillifera* were claimed by locals from eastern Camarines Sur to have medicinal properties to cure digestive and glandular diseases [103].

Other medicinal uses of macroalga by the Ilocano people included treatment for sexual dysfunction, dermatological problems (e.g., pruritis, skin allergy and other skin infection), musculoskeletal pain (e.g., headache or rheumatism), digestive disorders, thyroid dysfunction and insect bites. Similar to Japanese traditional medicine, *Diginea simplex* was the only recorded algal species with an anti-helminthic usage. Interestingly, both *D, simplex* and *Chondria armata* claimed to have anti-helminthic property by locals from eastern Sorsogon in the Philippines [104]. In addition, seaweeds with medicinal value were collected from the wild by the local people, whenever required, while certain species, such as *C. racemosa* (ararusip) and *Codium intricatum* (pok-poklo) are sold regularly in the local market.

Indonesians, particularly those in coastal areas have traditionally utilized various seaweed species as both food supplements and herbal medicine. The study by Anggadiredja reported that 38 macroalgal species were traditionally utilized for medicinal purposes at Panguhalodo areas of East Sumba District at Sumba island [106]. The herbal species consisted of 7 species of green algae, 13 species of red algae and 18 species of brown algae. As a medicinal resource, seaweeds have been used as antipyretics and antiseptics, vermifuges and treatments for ailments, such as cough and asthma, hemorrhoids, nosebleeds, boils, goiter, scrofula, stomach disorders and urinary diseases. Several species possess antiseptic properties, including *Chaetomorpha linum*, *Valonia aegagropila*, *Hypnea cervicornis* and *H. musciformis.*

Seaweeds with anthelmintic properties, commonly used as vermifuges, include *Codium arabicum*, *Dictyosphaeria cavernosa*, *D. versluysii*, *Ascophyllum* sp. and *Hydroclathrus clathratus*. Antipyretic seaweeds, often prepared as decoctions or salads, include *Chaetomorpha linum*, *Ulva lactuca*, *Hypnea cervicornis*, *H. musciformis*, *Sargassum crassifolium*, *S. echinocarpum*, *S. polyphyllum*, *S. siliquosum*, *Turbinaria conoides* and *T. ornata*.

3.2 Cnidarians

Marine corals have been extensively used in traditional medicine across various Asian medical systems, including Traditional Chinese Medicine, Tibetan medicine, Mongolian medicine and Uyghur medicine [107]. In TCM, coral has been documented since ancient times, first appearing in the Newly Revised Materia Medica (659 AD) and used for treating blood stasis, corneal opacity and epilepsy. Tibetan medicine, known as Qiwuru or Pazhuma, utilizes coral for clearing liver heat, detoxifying toxins and treating encephalopathy and various fevers. Mongolian medicine, referring to coral as Shuru or Ulan Shuru, employs it to alleviate liver and lung heat, detoxify toxins and address neurological disorders. In Uyghur medicine, coral, often called Bihe Marjiang or Busai, is recognized for its ability to treat blood-related diseases, inflammation and gastrointestinal issues. These diverse applications highlight the enduring medicinal value of marine corals in these traditional healing systems. In contrast, documentation on the use of corals as traditional medicine is rather sparse in Southeast Asia. Four coral species, including *Acropora concinna*, *A. brueggemani*, *Goniopora stutchhunyi* and *G. lobata*, were listed for the treatment of diarrhoea, fever, headache, urinary disorder, indigestion and food poison in Myanmar traditional medicine [108]. Other studies by Salleh and co-workers as well as Mustarichie and Kelutur listed/mentioned "coral" used by communities at East Coast of Peninsular Malaysia for treatment of kidney stones and sea fan (*Gorgonia Mariae*) in Maluku, Indonesia for treatment of asthma [109, 110].

3.3 Molluscs and Cephalopods

Molluscs have played a significant role in traditional medicine across Southeast Asia, as well as in Chinese, Indian, South African and Middle Eastern healing practices [111–113]. Various parts of molluscs, including shells, soft tissues, basal portions, and mucilage, have been used in remedies for ailments such as cancer, inflammation and aging-related conditions. In North Borneo, the Iranun community in Kampung Rampayan Ulu, Kota Belud, has long relied on the marsh clam (*Geloina expansa*) for postnatal care, believing that its soup enhances milk production in nursing mothers due to its rich iron, zinc, and copper content [114]. Additionally, the mantle organ, which has the highest iron concentration, is applied topically to wounds to aid blood clotting and reduce bruising. Similarly, the Kinh people at Beibu Gulf, who have historically migrated from Vietnam since the sixteenth century, have an extensive ethnozoological knowledge of marine animal-based medicine, including the use of molluscs [115]. Field studies have recorded 18 molluscan species utilized by the Kinh, primarily for treating conditions, such as stomachache, gastric and duodenal ulcers, ulcerative gingivitis, dizziness, hyperacidity, gonorrhea, hematemesis and insomnia. The most commonly used molluscan parts are shells and meat, with decoction, soups and powdered forms being the preferred modes of preparation.

Cephalopods have been traditionally used in Southeast Asia for medicinal purposes, such as local communities in the East Coast of Peninsular Malaysia and Indonesia. In Malaysia, cuttlefish bone, known as Tulang sotong, is commonly used to treat gallstones and relieve itching, while squid (sotong) is believed to alleviate asthma, cough and body swelling [109, 116]. Meanwhile, the Samawa people of Sumbawa Island, Indonesia, prepare a remedy by mixing cuttlefish (*Sepia* sp.) with essential oil to effectively remove mucus from the body [117]. However, the active ingredient in cuttlefish, which possesses a strong, unpleasant odor, has a short shelf life. The medicinal effect of *Sepia* sp. is attributed to its ability to inhibit the COX-2 enzyme, thereby reducing the production of prostaglandin E2, a key inflammatory mediator involved in asthma and bronchitis [118]. Additionally, the ethnopharmacological practices of Maluku (Tepa Islands, Southeast Maluku) and Papua have incorporated octopus ink in cancer treatment, particularly for breast cancer [119]. *Octopus cyanea*, a species abundant in Eastern Indonesian waters, secretes an ink rich in melanin-based alkaloids and lysosome proteins, which have been found to kill pathogenic bacteria, such as *Staphylococcus aureus*, and exhibit potential anti-cancer properties [119]. These traditional medicinal uses of cephalopods highlight their significance in local healing practices and their potential for modern pharmacological research.

3.4 Arthropods

3.4.1 Crustacean

Several crustacean species have been identified as traditional medicines to cure various illnesses based on a survey conducted on communities from the East Coast of Peninsular Malaysia, including traditional practitioners, fishermen, villagers and members of the public [109, 116]. These crustaceans include mud crab (ketam nipah), *Thenus* sp. (slipper lobster) and Mantis shrimp (udang katok). Mud crab, when prepared as a soup, is used by the elderly in Malaysia as a traditional remedy and folk medicine to reduce the symptoms of dengue fever. Other traditional medicinal uses of mud crab include treatment for high blood pressure, malaria and chikungunya infection. Additionally, the slipper lobster (udang ketak) and mantis shrimp have been documented as tradition medicine for treatment of asthma and kidney stone [109].

3.4.2 Horseshoe Crab

Traditional medicine, particularly in East and Southeast Asia, has long utilized horseshoe crabs for their perceived healing properties. Their morphology has remained virtually unchanged for over 400 million years, earning them the title of "Living Fossils" and making them a key focus of conservation efforts [120]. They are commercially utilized in biomedical science, as amoebocyte lysates derived from their hemolymph are used to detect bacterial contamination in drugs, blood products and pharmaceutical devices [121]. However, overharvesting for traditional medicine and biomedical applications poses a significant threat to their populations, raising conservation concerns [122].

In some cultures, horseshoe crab eggs and body parts are believed to have medicinal benefits, including enhancing vitality and treating ailments like asthma and joint pain. Reports indicate that tail tips of horseshoe crabs are used in West Bengal by faith healers to treat arthritis and joint pain [96]. In Singapore, pregnant women consume the egg mass of horseshoe crabs, believing it helps immunize their fetus. Additionally, in many Asian countries, the horseshoe's carapace is boiled with mustard oil and used as a remedy for rheumatic pain [123]. A series of studies conducted on coastal communities at Peninsular Malaysia revealed the traditional medicinal use of horseshoe crab (belangkas) for the treatment of various illness, including heart diseases, cancer, muscle endurance as well as HIV [109, 116]

3.5 Echinoderms

3.5.1 Sea Cucumbers

Among marine organisms, sea cucumbers stand out as a rich natural source of bioactive compounds with potential applications in both food and biomedicine [124]. They have been consumed as food and used in traditional medicine across Asia for centuries, with species such as *Stichopus hermanni*, *Thelenota ananas*, *Thelenota anax*, *Holothuria fuccogilva* and *Actinopyga mauritiana*, being particularly prized [70]. These organisms contain high-value bioactive compounds, including triterpene glycosides, carotenoids, bioactive peptides, vitamins, minerals, fatty acids, collagens, gelatins, chondroitin sulfates and amino acids. Historical records from the Ming dynasty (1368–1644 AD) indicate that sea cucumbers were considered to have medicinal properties similar to ginseng, earning them the name "haishen," meaning "ocean ginseng".

Globally, there are approximately 1716 species of sea cucumbers, with the highest biodiversity found in the Asia–Pacific region [70]. In Indonesia, sea cucumbers are commonly known as "teripang" or "trepang". Indonesia is recognized as one of the oldest and largest exporters of sea cucumbers, with around 350 species recorded in its waters, over half of which are found at depths exceeding 3000 m [125]. Of these, at least 26 species hold significant economic value. Malaysia and the Philippines also play key roles in sea cucumber production and export. Southeast Asia remains a major hub in the global sea cucumber trade, benefiting from its remarkable marine biodiversity [126].

Many sea cucumber species are harvested for human consumption, while others are cultivated in aquaculture systems [127]. In Chinese medicine, sea cucumber is highly valued as a tonic, particularly for blood purification, "jing" nourishment, and warming properties. It is traditionally prescribed for a range of health conditions, including joint and skeletal issues associated with aging, kidney disorders, impotence, constipation, poor lipid digestion, circulatory ailments, and frequent urination. Its reputed aphrodisiac properties may be linked to its phallic appearance.

In Malaysian traditional medicine, sea cucumber, known as "gamat", is widely used in various therapeutic preparations [128]. Gamat gels and fluids are consumed as general tonics, similar to nutraceuticals, and are traditionally given after trauma or childbirth. Gamat oil and extracts, derived from the body wall and viscera of aspidochirote sea cucumbers, are formulated into topical creams/ointments (Fig. 2e) for treating wounds, muscle aches, and skin conditions. Research indicates that sea cucumber extracts are effective in wound healing, with positive outcomes reported in treating various types of wounds, including excisional, incisional, ulcerative, diabetic, and burn wounds in both rodents and humans [129].

3.5.2 Other Echinoderms

In Asia and Southeast Asia, starfish and sea urchins have been traditionally used in folk medicine for their purported healing properties. In Chinese and Vietnamese folk medicine, dried starfish are sometimes used in soups and tonics believed to treat asthma, pneumonia, diabetes and heart disease [130]. Similarly, in coastal communities of Southeast Asia, sea urchins are used as a remedy for treatment of convulsions, seasickness, asthma and heart diseases [131, 132]. Certain species of sea urchins are valued for their bioactive compounds, which are thought to have antimicrobial and immune-boosting properties [133]. While traditional practices persist, modern scientific research is increasingly exploring these marine echinoderms for their potential pharmaceutical applications.

3.6 Marine Fish

3.6.1 Seahorse

Hippocampus is predominantly found in the Indo-Pacific Ocean, inhabiting shallow waters at depths of less than 30 m across temperate, subtropical and tropical regions [134]. Its habitat includes seagrass beds, coral reefs, mangroves and seaweed, with coral reefs serving as the primary habitat for tropical species of *Hippocampus*. Of the 44 species of Hippocampus found in 159 countries, about 20 *Hippocampus* species, including *H. trimaculatus*, *H. kelloggi*, *H. histrix*, *H. kuda* and *H. japonicus*, have been documented in the literature for their medicinal uses [95]. The medicinal use of seahorses has been reported in at least 33 countries worldwide, including Cambodia, China, Hong Kong SAR, India, Indonesia, Japan, Korea, Laos, Malaysia, Philippines, Singapore, Sri Lanka, Thailand and Vietnam [135].

Seahorses have been associated with 284 therapeutic properties and applications. Among the 18 health conditions categorized by the International Statistical Classification of Diseases and Related Health Problems (ICD-10), mental and behavioral disorders were the most frequently cited, followed by genitourinary diseases. This reflects the common use of seahorses for treating sexual dysfunction (e.g., impotence) and conditions affecting the reproductive and urinary systems [136]. May and Tomoda noted that while seahorses are often regarded as sexual stimulants, they have a long history of use in aiding difficult childbirth [137]. They argued that the popularity of seahorses as medicinal remedies, particularly as sex tonics, was largely influenced by the commercial production and sale of seahorse-based pills across Asia since the 1970s.

The medicinal applications of seahorses vary by region. For example, they are used as aphrodisiacs and treatments for sexual disorders in Taiwan and Indonesia, whereas in China and Hong Kong, they are more commonly employed to address respiratory ailments (Vincent 1996). In Indonesia, Vincent (1996) documented the use of seahorse-based jamu pills (e.g., Health Ginjal Seahorse Ghenshen Kidney Pills), typically taken twice daily with warm water. Another remedy, Jamu Ramuan (Traditional) "Tangkur Jaya", is consumed three times weekly, mixed with honey, an egg, and orange juice in boiling water. In the Philippines, traditional preparation involves roasting and boiling one or two seahorses in four cups of water until reduced to one cup, grinding them, and drinking the resulting extract twice daily.

In Southeast Asia, seahorses are predominantly collected from shallow coastal waters, coral reefs, seagrass beds, and mangrove areas. Harvesting methods vary by region but often include bycatch from non-selective fishing techniques such as trawling and seine netting, as well as targeted collection using hand nets or diving. Small-scale fishers frequently capture seahorses as part of mixed fisheries, with many individuals unaware of the species' conservation status. In some areas, particularly in the Philippines, Indonesia, and Vietnam, seahorses are deliberately harvested for the traditional medicine trade, often by night divers who manually extract them from their habitats.

Dried seahorses are the most commonly traded form, as preservation extends their shelf life and facilitates transportation. Fishermen typically sun-dry the seahorses or dry them over low heat before selling them to intermediaries or local markets. While live seahorses are occasionally traded for aquarium purposes, the overwhelming demand comes from traditional medicine markets, especially in China and other East Asian countries that source supplies from Southeast Asia. The seahorse trade in Southeast Asia operates through a complex network of fishermen, middlemen, wholesalers, and international exporters. Large quantities of seahorses pass through major trade hubs such as Hong Kong, Singapore, and mainland China, where they are further processed or sold in traditional medicine shops. Hong Kong alone has historically accounted for a significant proportion of global seahorse imports, with much of its supply coming from Southeast Asian nations like the Philippines, Thailand, Malaysia and Vietnam.

Seahorse populations in Southeast Asia are under severe pressure due to overexploitation, habitat destruction, and the impacts of non-selective fishing practices. As slow-moving and low-fecundity species, seahorses are particularly vulnerable to overfishing. Furthermore, the degradation of coral reefs, seagrass beds, and mangrove ecosystems—largely due to coastal development, pollution, and destructive fishing techniques—has exacerbated population declines. Recognizing these threats, *Hippocampus* species were listed in Appendix II of the Convention on International Trade in Endangered Species of Wild Fauna and Flora (CITES) in 2002, requiring export permits to regulate international trade. Many Southeast Asian nations have since imposed restrictions or outright bans on seahorse collection and export. For example, the Philippines officially banned seahorse harvesting and trade in 2004 under the Fisheries Administrative Order No. 208. However, enforcement remains a major challenge, as illegal and unregulated trade continues in many parts of the region.

Efforts to balance conservation with local livelihoods have led to various initiatives, including seahorse aquaculture, community-based fisheries management, and marine protected areas (MPAs). Seahorse farming has been explored as an alternative to wild collection, but challenges such as high mortality rates and breeding difficulties have limited its large-scale success. Community-driven conservation programs, like Project Seahorse in the Philippines, have worked to empower local fishers through education, alternative livelihoods and habitat restoration. Moving forward, stronger enforcement of trade regulations, increased monitoring of wild populations, and greater public awareness are crucial for ensuring the long-term survival of seahorses in Southeast Asia. Collaborative efforts between governments, conservation organizations, and local communities will be key to striking a balance between traditional medicinal use and sustainable seahorse populations.

3.6.2 Other Marine Fishes

Several fish species found in Southeast Asia have been documented to have medicinal properties and are used by coastal communities. At least three species, including sea perch fish, Golden pomfret and fins of sharks, have been used in coastal areas of Peninsular Malaysia for treatment of heart diseases. These fishes are prepared in several ways, such as fried and ground to powder (sea pearch fish), eaten raw (Golden pomfret) or made into soup (shark fin) [131]. Additionally, from the east coast of Peninsular Malaysia, a range of fishes, including sea eel (belut laut), flying fish (ikan belalang), Bronze croaker (Ikan gelama selampai), Sharpnose Stingray (Ikan Tuke), Croaker fish (Ikan Gelama), Goby fish (Ikan Goby), Tembiring fish (ikan tembiring) and Dragon fish (ikan naga), have been used for treatment of illnesses/symptoms ranging from asthma, skin disease, intestinal parasites and gout [109]. In another study conducted on the use of marine-based animal drugs by Kinh people in Beibu Gulf, more than 30 fish species have been recorded for medicinal usage [115].

3.7 Sea Turtle

Marine turtles have long been used in traditional medicine across Southeast Asia, where various parts of the turtle, including their shells, bones, blood and internal organs, are believed to have medicinal properties [138]. These include treating ailments ranging from asthma and arthritis to improving vitality and fertility. Redang Island, located off the northeast coast of Peninsular Malaysia, is one of Southeast Asia's key nesting sites for green sea turtles (*Chelonia mydas*). For decades, the local community has relied on turtle eggs as both a food source and a means of livelihood [139]. Traditionally consumed raw, boiled or pickled in brine for long-term preservation, turtle eggs have held cultural, medicinal, and dietary significance. Many older community members recalled that, in the past, limited access to Western medicine led to the use of turtle eggs as an alternative remedy. The eggs were believed to possess therapeutic properties, particularly for treating asthma, muscle pain and supporting pregnant women's health. Some respondents even claimed that the first egg laid by a mother turtle had unique medicinal value for curing asthma [139].

Beyond Malaysia, similar beliefs about turtle eggs exist in Sumatra, Indonesia, where herpetofauna have long been utilized for food, medicine and other purposes [140]. In a study conducted in Padang and Mentawai, 307 medicinal uses of turtle products were reported, with 77% in Padang and 20% in Mentawai [141]. Turtle eggs were commonly believed to enhance stamina, boost immunity and maintain physical strength, appetite and blood pressure. Some respondents also attributed aphrodisiac properties to the eggs, a belief shared in parts of Latin America [142]. Additionally, in Mentawai, turtle fat was traditionally used for treating rheumatism, supporting child growth, and even as a remedy in poultry medicine [141]. These long-standing cultural and medicinal associations highlight the deep-rooted reliance on turtle-derived products, despite ongoing conservation efforts to protect marine turtle populations.

Despite the lack of scientific evidence supporting the traditional medicinal claims, the demand for marine turtle products persists, fueling illegal trade, particularly in countries like Vietnam, China and Indonesia [138]. This trade, often linked to black markets and organized wildlife trafficking, poses a significant threat to already vulnerable turtle populations, including the critically endangered hawksbill and green turtles. In response, Southeast Asian nations have implemented conservation measures, including stricter enforcement of CITES (Convention on International Trade in Endangered Species of Wild Fauna and Flora) regulations, habitat protection and community-based conservation programs [143]. However, the effectiveness of these efforts is often hindered by weak enforcement, corruption and persistent demand, making it crucial to enhance regional cooperation and public awareness to ensure the long-term survival of marine turtles.

3.8 Marine Mammals

Marine mammals have long been incorporated into traditional medicinal practices across Southeast Asia, with various species believed to possess healing and protective properties. The dugong (*Dugong dugon*), a marine mammal found in the coastal waters of Southeast Asia, has long been exploited for its meat, oil and body parts [144]. In several communities, particularly among the Ubian and Bajau people in Sabah, Malaysia, dugong parts hold significant medicinal and cultural value. These coastal communities have traditionally used dugong tusks and bones to treat various ailments, including asthma, back pain and shock [145]. A common practice involves shaving the tusks and mixing the fine powder with lukewarm water to create a medicinal concoction. Additionally, dugong tusks are believed to possess aphrodisiac properties, similar to the well-known herbal root Tongkat Ali, and are consumed with the intent of enhancing vitality and sexual health.

Beyond Malaysia, the medicinal and cultural use of dugongs extends to other parts of Southeast Asia, including Cambodia, where the species is reportedly utilized for various health treatments. Dugong-derived remedies are traditionally used to address fever in children, alleviate rheumatism and serve as a household protection against malevolent spirits [146]. In some Cambodian practices, dugong parts are also considered aphrodisiacs, reinforcing a widespread belief in the potency of these marine mammals in traditional medicine. Despite these deep-rooted practices, the hunting and trade of dugongs pose a severe threat to their populations, as they are already classified as vulnerable due to habitat loss and incidental bycatch. Conservation efforts and legal protections are crucial in mitigating the decline of this species while also promoting sustainable alternatives for communities reliant on their traditional medicinal uses.

The Irrawaddy dolphin (*Orcaella brevirostris*), found in freshwater and coastal environments of Cambodia and the Philippines, has been traditionally used to treat both human and livestock ailments [146]. In certain communities in Cambodia, parts of the dolphin, such as bones or oil, are believed to have medicinal benefits and are used in treatments for sick animals or individuals suffering from various illnesses [147]. Dolphin bones are finely ground into powder and mixed with water for sick livestock or humans to consume as a remedy. Dried dolphin flesh is hung inside chicken coops to ward off evil spirits believed to cause poultry illnesses. Dolphin teeth are inserted into small holes made in the horns of ailing buffaloes as part of a traditional healing practice. Additionally, dolphin teeth are crafted into necklaces and placed around the necks of sick children to aid in their recovery [148]. These cultural practices reflect the deep connection between local communities and their surrounding marine ecosystems, where animals like the Irrawaddy dolphin are valued beyond their ecological roles.

3.9 Seagrasses

As a coastal plant, seagrass is not only important in maintaining the stability of coastal ecosystems, but also provides many benefits in daily life, including their medicinal use by various Southeast Asian coastal communities. Rich in bioactive compounds, seagrasses are commonly utilized in folk remedies to treat wounds, skin infections, fever and digestive ailments [149]. One common seagrass, *Enhalus acoroides*, found in major seagrass regions of Southeast Asia, has been used extensively as traditional medicine. Some traditional coastal residents of Maluku, Indonesia believe that the seeds of *E. acoroides* (lamun) can strengthen the immune system [150]. In Thailand, only the rhizome of *E. acoroides* is used in traditional medicine, primarily for its carminative properties and to improve blood and lymph circulation [151]. It is also a key ingredient in several Thai traditional medicine formulations, such as "Maha-Anan-Ta-Khun", "Ya-Kae-Sama-That" and "Prab-Chom-Phu-Tha-Weep" [152].

In a study by Noor and co-workers, it was reported that the common *Enhalus acoroides* was used as traditional medicine by the local community at Pahawang Island, Indonesia, to avoid mosquito bites [153]. The seagrass, especially the leaves, are first pounded by adding small amounts of water and making it into a smooth paste before applying on body to avoid mosquito bites. Phytochemical analysis was carried out and concluded that the active components contained flavonoid, tannin, terpenoid, steroid, triterpenoid and saponin, which could be attributed to its possible antioxidative, antibiotic and antimalarial properties.

A survey carried out by Digdo and associates found complex and intertwining links between humans and seagrass in rural area in North Minahasa, North Sulawesi, Indonesia [154]. The survey found that about 9% of the local population recognize the importance of seagrass for medicinal usage beyond its role as a food source. The community uses seagrass in traditional medicine, with different parts believed to have specific healing properties. For instance, the root of *Enhalus acoroides* is considered an anti-toxin. Fresh roots of large *E. acoroides* can be cut into pieces and applied to wounds to stop bleeding. When boiled, the roots are believed to refresh the body and serve as a remedy for common colds (maso dingin). Additionally, based on local knowledge, the shoots of *E. acoroides* can be crushed and applied as an antitoxin for stingray stings. If the plant remains intact, it can be eaten raw to treat ulcers and boost immunity. Meanwhile, another seagrass, *Halophila ovalis*, is traditionally used as an anti-toxin against venomous fish, including lionfish (Scorpaenidae), rabbitfish (Siganidae), surgeonfish (Acanthuridae) and various stingrays.

A study by Zulkifli and co-workers revealed that various species of seagrasses have long been used as traditional medicine by coastal residents in South Lombok [155]. Based on interviews, four medicinal species commonly used by residents include *Enhalus acoroides*, *Syringodium isoetifolium*, *Thalassia hemprichii* and *Cymodocea cerulata.* They are used as an external medicine for itching-skin diseases and external wounds as well as internal medicine, such as heart, cancer and kidney disease.

3.9.1 Mangrove Plants

Mangrove plants have been widely used as traditional medicine across Southeast Asia, with various species providing remedies for ailments ranging from fever and infections to rheumatism and digestive disorders [156]. Certain common mangrove plant species found throughout Southeast Asia have been utilized as folk medicine by respective local communities. For instance, *Pluchea indica*, in the form of traditional medicine and food have been reported in countries of Southeast and South Asia [157]. Its leaves are used as a traditional remedy for stomachache, cough, dysentery and leucorrhoea in Indonesia and Malaysia. In Thailand and Java, *P. indica*, known locally as "khlu" among the Thai people, is valued for its astringent and antipyretic properties. The leaves and roots are commonly used as diaphoretic to induce sweating in cases of fever. Fresh leaves are also applied as poultices to treat atonic and gangrenous ulcers, while cigarettes made from the chopped stem bark are smoked to relieve sinusitis pain. In Indo-China, including Vietnam and Cambodia, crushed leaves and young shoots mixed with alcohol are applied to the back to alleviate lumbago and are also used in baths to treat scabies and rheumatic pain [156, 157].

Several surveys have been conducted on the use of mangrove plants in Thailand as traditional medicines. For instance, a survey conducted in the Sating Phra Peninsula of Songkhla Province, Thailand, documented the medicinal use of 110 plant species from mangrove and beach forests, representing 100 genera and 51 families [158]. Among these, 69 species were exclusive to beach forests, 35 were found only in mangrove forests, and six were present in both ecosystems. The Fabaceae family had the highest number of medicinal species, with many plants traditionally used to treat fever, skin diseases and gastrointestinal issues. This highlights the diversity of medicinal flora in coastal regions in Thailand and their significance in local healing practices.

In Indonesia, the use of mangrove plants as traditional medicine is widespread, particularly among coastal communities in Papua, Sumatra, Kalimantan, Java, Bawean, Rote, Timor and Halmahera [159]. A review of 12 studies identified 36 mangrove species utilized for medicinal purposes, comprising 30 true mangrove species and six mangrove associates from 17 different plant families [159]. The commonly used plant parts include leaves, stems, bark, buds, flowers, seeds, roots, exudates and fruits. These are processed in various ways, such as direct consumption, boiling, pounding, burning, slicing, grating and squeezing. Application methods range from drinking and eating to smearing, affixing, chewing and bathing, demonstrating the versatility of mangrove-based traditional remedies.

A recent survey conducted in Lampung Timur, Sumatra, Indonesia, revealed that the tradition of using mangrove plants as herbal medicine remains prevalent, particularly among the older generation [160]. Several key mangrove plant species are known for their medicinal properties: *Avicennia marina* serves as an antiseptic, anti-histamine and vitamin source; *Acanthus ilicifolius* is used for rheumatism and muscle pain relief; *Rhizophora apiculata* functions as an antipyretic; *R. mucronata* is believed to enhance stamina and vitality; and *Sonneratia caseolaris* is valued for its anti-cholesterol and antioxidant properties. In another survey, six mangrove plants, including *Lumnitzera littorea*, *Rhizophora mucronata*, *Scyphiphora hydrophyllacea*, *Sonneratia alba*, *Xylocarpus granatum* and *X. moluccensis*, are used by six ethnic groups of North Maluku Province, Indonesia, as malaria medicine [161]. These findings underscore the deep-rooted traditional knowledge of mangrove plants in Southeast Asia and their potential for further pharmacological exploration.

4 Overview of Marine Natural Products Research in Southeast Asia from 2003 to 2022

4.1 Marine Organisms Studied for New MNPs in Southeast Asia

This overview of marine natural products research in Southeast Asia is not intended to be comprehensive. Rather, it highlights key developments undertaken by scientists from SEA and their collaborators as well as research trends from the past two decades (2003–2022). It is worth noting that research activities in this field began as early as the late 1980s, initially driven by institutions from user countries, such as the United States, Canada, Japan, Australia, Germany and France, who conducted exploratory expeditions and collaborative studies in the region. For instance, in the early phases of global marine drug discovery, the National Cooperative Natural Products Drug Discovery Group (NCNPDDG) and International Cooperative Biodiversity Group (ICBG) programs, funded by the U.S. National Institutes of Health (NIH) and related agencies, played pivotal role in supporting bioprospecting efforts in several biodiversity hotspot, including Southeast Asia region. These initiatives aimed to discover novel bioactive compounds for therapeutic development while fostering biodiversity conservation and capacity building in host countries [162]. Additionally, contracts with the Coral Reef Research Foundation, Harbor Branch Oceanographic Institute, the Australian Institute of Marine Science, the University of Canterbury in New Zealand, approximately 20,000 marine invertebrates and marine algae, mainly from the Indo-Pacific region (e.g., Indonesia, Thailand, the Philippines and Papua New Guinea), were collected to support the Natural Products Branch at National Cancer Institute (NCI) efforts in marine drug discovery, especially for cancer-fighting molecules. this effort. Research groups from several academic institutions in the US, including the Scripps Institution of Oceanography, University of California at Santa Cruz, Arizona State University, University of Utah and University of Hawaii, contributed to marine drug discovery efforts from this region.

An analysis of MNPs research in Southeast Asia from 2003 to 2022, based on annual reviews of new marine natural products by Blunt et al. and Carroll et al., reveals trends in both taxonomic focus and publication output over time [163–182]. From 2003 to 2022, a total of 756 MNPs publications on new molecules were sourced from 18 different phyla across Southeast Asia. Not surprisingly, Porifera consistently

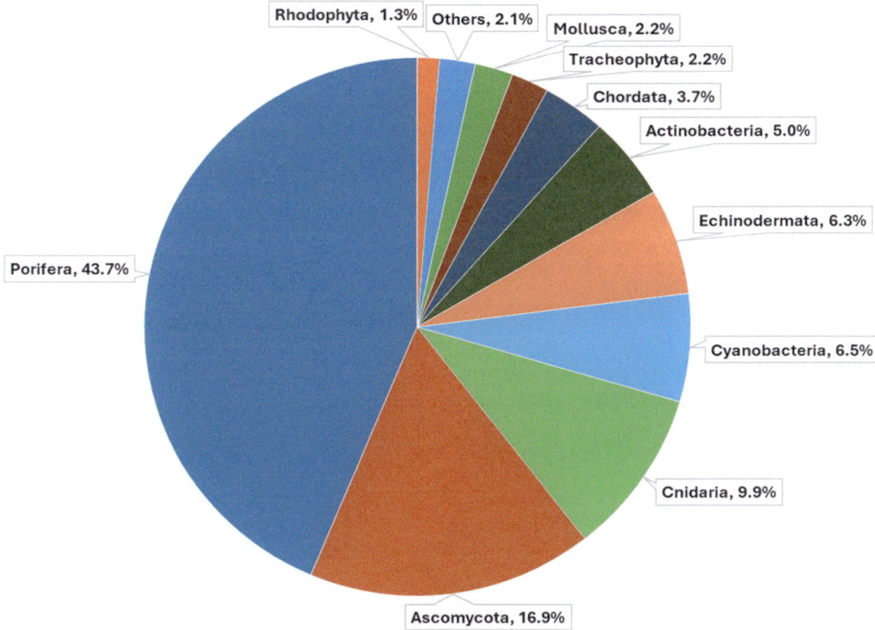

Fig. 3 Pie chart showing the breakdown of new MNPs publications (2003–2022) as percentage from various phyla reported from Southeast Asia

remained the dominant source, accounting for 43.7% of the total publications (330 publications) (Fig. 3). Ascomycota followed with 16.9% (128 publications), and Cnidaria contributed 9.9% (75 publications) (Fig. 3). A similar trend was observed on the global biodiscovery of new molecules from marine invertebrates over 1990–2019 period, where sponges and Cnidaria were the main sources of new natural products [3].

Within Southeast Asia, the number of new MNPs publications remained relatively stable between the periods of 2003–2012 and 2013–2022, with a slight decrease from 168 to 162 publications (Fig. 4). However, publications sourced from Ascomycota increased dramatically by 145.9%, from 37 to 91 publications during this time, suggesting a growing interest in fungal MNPs. Similar trends were observed in Echinodermata, which saw a 280% increase (from 10 to 38 publications), and Actinobacteria, which grew by 180% (from 9 to 28 publications). Although Porifera still held the largest share, these sharp increases in the number of publications from Ascomycota, Echinodermata and Actinobacteria reflect a shift toward more diverse sources of MNPs in Southeast Asia.

The number of new MNPs publications generally remained consistent year to year, with an overall increase of 24.3% (from 337 to 419 publications) between two decadal periods of 2003–2012 and 2013–2022 (Fig. 4). While the overall number of publications has risen, there are significant disparities between Southeast Asian

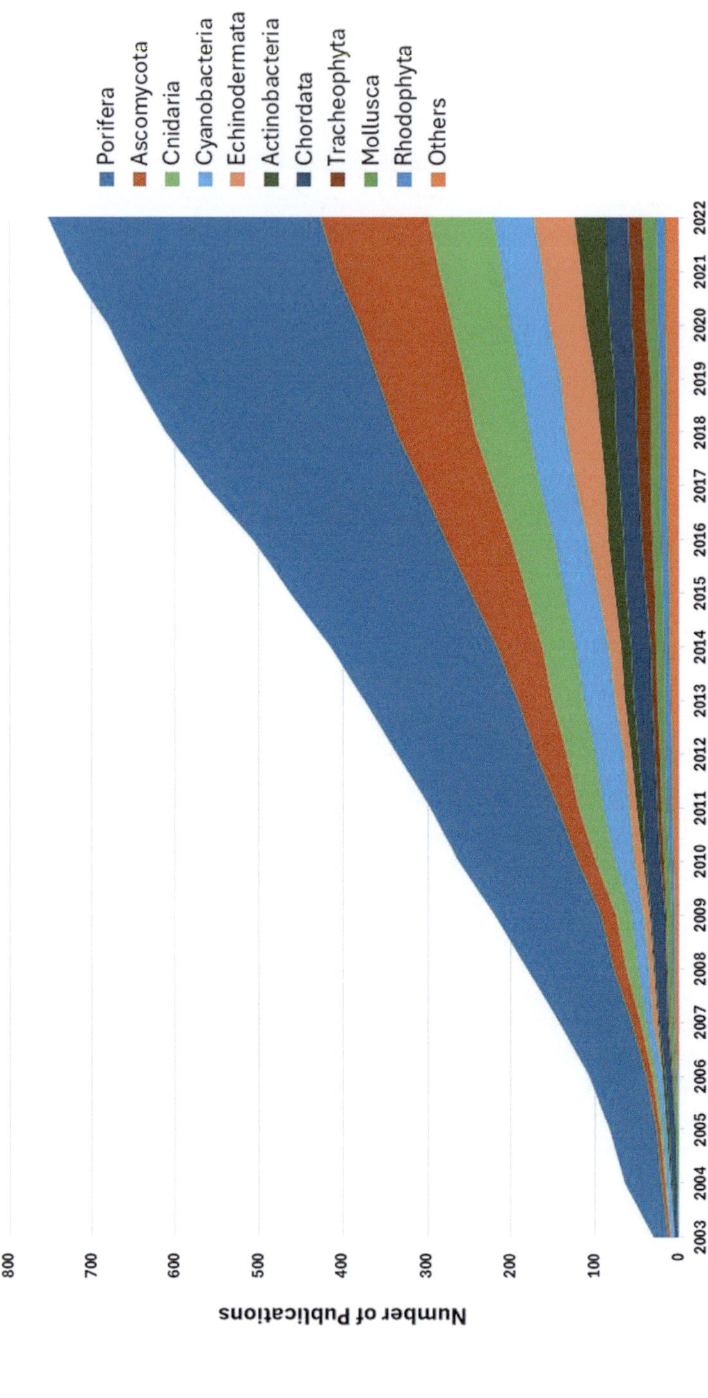

Fig. 4 Cumulative number of new MNPs publications from Southeast Asia between 2003 and 2022 based on the phylum of the source organism. The phyla Chlorophyta, Proteobacteria, Bacteroidetes, Ochrophytina, Myzozoa, Firmicutes and Basidiomycota were grouped under "Other" due to small number of publications. Data was compiled from the annual review of marine natural products published between 2003 and 2022 [163–182]

countries and other countries in terms of authorship and lead authorship in these publications. By using a 50-year dataset on the discovery of marine natural products, Leal and co-workers highlights that many nations with limited financial resources, typically in the lower 60th percentile of global GDP per capita, are underrepresented as lead authors in studies involving organisms from their own waters [183]. This pattern is particularly evident in Indonesia, which has contributed numerous marine organisms since 1993 but continues to have relatively low representation in terms of authorship and lead authorship [183]. These disparities point to the influence of a country's economic capacity in its ability to engage with and benefit from its marine biodiversity through scientific research.

4.1.1 Porifera

Class Demospongiae accounts for 94.5% of all Porifera-related publications (2003–2022) in SEA. Class Homoscleromorpha and Class Calcarea accounted for 3.1% and 2.4% respectively (Fig. 5). Although publications featured sponges from 43 different families, 69.3% were concentrated on ten families, namely: Thorectidae (15.3%) and Petrosiidae (14.7%), Halichondriidae (6.7%), Dictyonellidae (5.5%), Spongiidae (5.5%), and Ancorinidae (5.5%), Dysideidae (5.2%), Theonellidae (4.0%), Suberitidae (3.7%) and Chalinidae (3.1%) (Fig. 5). Thorectidae research was distributed across 12 genera, yet onethird of all publications focused on *Hyrtios*, which alone accounts for 4.3% of total publications from Porifera. Petrosiidae was dominated by *Xestospongia* (5.5%) and *Acanthostrongylophora* (4.6%), followed closely by *Petrosia* (2.7%).

Globally, the most studied families are: Thorectidae, Petrosiidae, Dysideidae, Plakinidae, Ancorinidae, Halichondriidae, Spongiidae, Theonellidae, Chalinidae, and Agelasidae [3]. Thorectidae and Petrosiidae are widely studied both globally and in Southeast Asia. However, Dysideidae ranks slightly lower regionally. Conversely, Halichondriidae and Dictyonellidae appear higher regionally than they do globally. Also of note is Agelasidae, which appears in global rankings but does not make the regional top ten. However, it remains close, accounting for 2.5% of publications from Porifera. Families like Dictyonellidae and Suberitidae are somewhat unique to SEA, indicating a possible ecological or sampling emphasis distinct from global trends.

Using *Xestospongia* sp. as an example illustrates diverse new natural products from samples collected within Southeast Asia. *Xestospongia*, (class Demospongiae), is one of the most extensively studied groups of sponges and is widely distributed throughout Southeast Asian waters [184]. For example, *X. testudinaria* spans the entire coastline, from the Singapore Strait to the Taiwan Strait [8]. This genus is of particular interest due to its ability to produce a diverse array of primary and secondary metabolites, which exhibit a broad spectrum of biological activities. Notably, a total of 40 different metabolites have been identified in specific species, such as *X. bergquistia*, *X. testudinaria*, *X. muta*, *X. exigua*, *X. ashmorica* and *X. vansoesti* [185]. These include alkaloids, fatty acids, steroids and quinones, all of

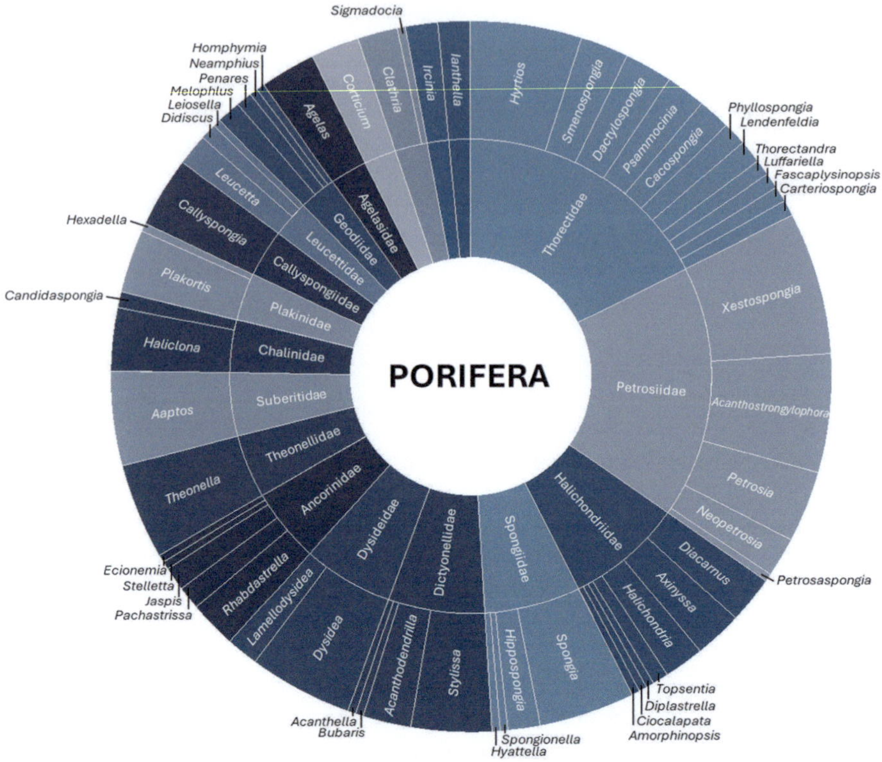

Fig. 5 Taxonomic breakdown of marine sponges from SEA investigated for new MNPs (2003–2022). The sunburst diagram displays the hierarchical distribution of organisms by family and specific levels. Each segment represents a family (inner ring) and its corresponding species (outer ring), sized proportionally by the number of publications. Only Poriferan families with four or more publications are included. Excluded Porifera families (genus) are: Aplysinellidae (*Aplysinella*), Aplysinidae (*Verongula*), Arenoscleridae (*Arenosclera*), Axinellidae (*Axinella*, *Cymbastela*), Bubaridae (*Auletta*), Clionaidae (*Cliona*), Coelosphaeridae (*Lissodendoryx*), Crambeidae (*Monanchora*), Desmacididae (*Biemna*), Desmapsammidae (*Desmapsamma*), Desmoxyidae (*Myrmekioderma*), Hymedesmiidae (*Damiria*, *Hamigera*, *Stylotella*), Iotrochotidae (*Iotrochota*), Leucosoleniidae (*Leucosolenia*), Mycalidae (*Mycale*), Niphatidae (*Gelliodes*, *Niphates*), Oscarellidae (*Oscarella*), Phloeodictyidae (*Oceanapia*), Pseudoceratinidae (*Pseudoceratina*), Scleritodermidae (*Scleritoderma*) and Tetillidae (*Cinachyrella*)

which hold significant potential for biotechnological and pharmaceutical applications (Fig. 6). Open-sea mariculture of the Philippine blue sponge *Xestospongia* sp. has been explored as a route for the production of the biomedically-important renieramycins [186]. Recent studies have also revealed microbial symbiotic relationships within *Xestospongia* species. Four distinct groups of Actinobacteria have been identified across various *Xestospongia* species, with three of these groups being present in both *X. muta* from Key Largo, Florida and *X. testudinaria* from Manado, Indonesia [187]. In addition, microbial communities in *X. testudinaria* from Karah Island,

Terengganu, Malaysia, are predominantly composed of Proteobacteria and Actinobacteriota [188]. These findings highlight the critical role of symbiotic microbes in the sponge's bioactive compound production.

4.1.2 Cnidaria

Publications of new marine natural products (2003–2022) from Cnidaria is heavily concentrated on Class Anthozoa (98.6%). Only a single study was attributed to Class Hydrozoa. Publications have featured Cnidarians from 11 families, though majority (52%) focused on Family Alcyoniidae. This was followed by Nephtheidae (21.3%), Xeniidae (8.0%), Plexauridae (5.3%), Ellisellidae (4.0%) and Gorgoniidae (2.7%) (Fig. 7). Families Subergorgiidae, Clavulariidae, Aglaopheniidae, Stichodactylidae and Pennatulidae were featured in one publication each. The prominence of Xeniidae in Southeast Asia is somewhat interesting as this family is not typically featured as prominently in global MNPs publications [3]. This could reflect either local ecological abundance or a targeted research interest in the region. Publications from Family Alcyoniidae were dominated by *Sinularia* (26.7%), followed by *Sarcophyton* (12%), *Lobophytum* (10.7%) and *Cladiella* (2.7%). Notably, *Sinularia* was also the single most-published genus under Cnidaria. Within Nephtheidae, *Nephthea* was the dominant genus, contributing 8% of all Cnidarian publications.

4.1.3 Echinodermata

Within SEA, four main classes were represented in publications of new marine natural products (2003–2022) from Echinodermata: Asteroidea (54.2%), Holothuroidea (33.3%), Crinoidea (10.4%) and Echinoidea (2.1%). Family Oreasteridae and Sclerodactylidae accounted for 18.8% and 12.5% respectively, followed by Acanthasteridae (8.3%) (Fig. 7). Family Archasteridae, Astropectinidae, Cucumariidae, Holothuriidae, Ophidiasteridae, Stichopodidae and Comatulidae each contributed to 6.3% publications from Echinodermata. Family Goniasteridae, Asterinidae, Asteropseidae, Toxopneustidae, Echinasteridae, Asterometridae, Himerometridae and Synaptidae were each featured in one publication. Oreasteridae and Sclerodactylidae are also top contributors of global MNPs publications from Echinodermata [3]. However, Acanthasteridae, which ranks highly in SEA, is less featured globally, suggesting a possible regional specificity or interest. At the generic level, *Pentaceraster* under Oreasteridae and *Neothyonidium* under Sclerodactylidae, each contribute 10.4% of total publications in Echinodermata.

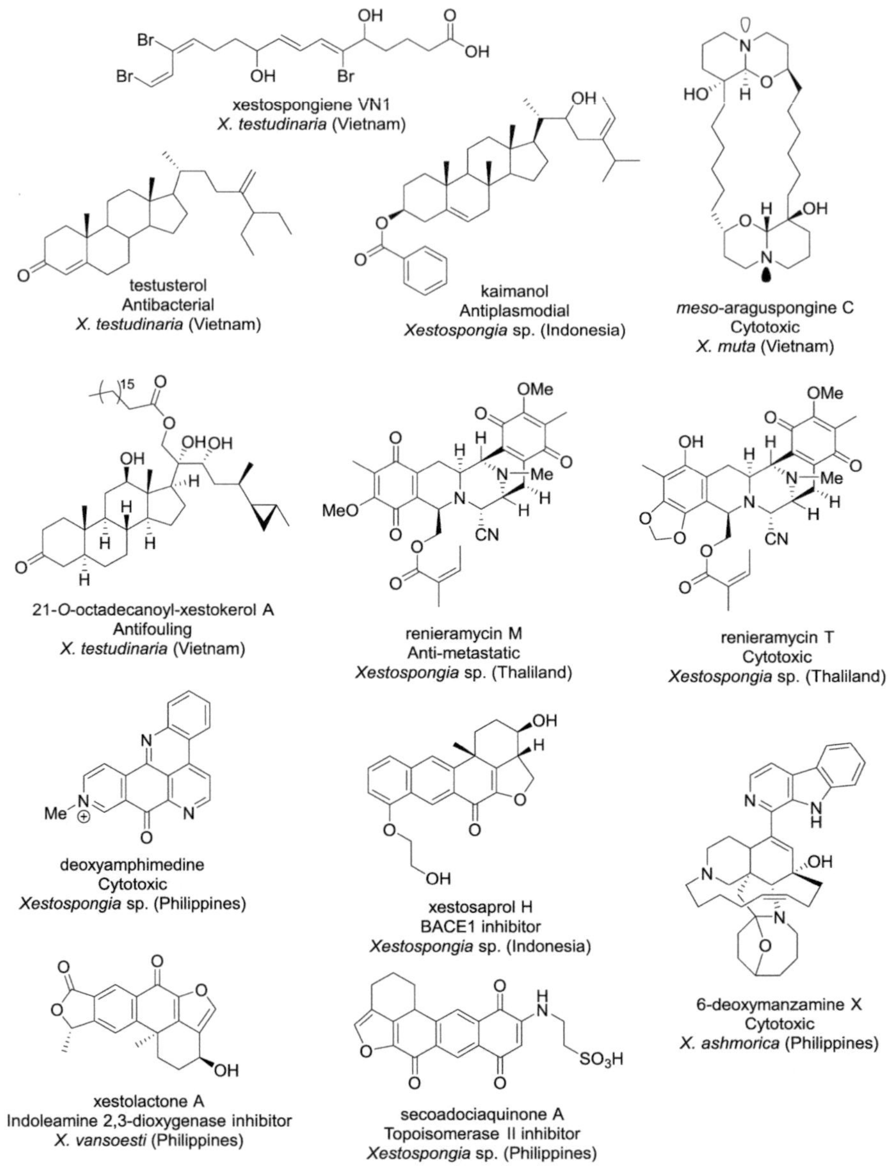

Fig. 6 Selection of diverse new bioactive natural products isolated from various *Xestospongia* species found within Southeast Asia: xestospongiene VN [189], testusterol [190], kaimanaol [191], 21-*O*-octadecanoyl-xestokerol A [192], *meso*-araguspongine C [193], renieramycins M and T [194, 195], deoxyamphimedine [196], xestolactone A [197], xestosaprol H [198], secoadociaquinone A [199] and 6-deoxymanzamine X [200]

Fig. 7 Taxonomic breakdown of cnidarians and echinoderms from SEA investigated for new MNPs (2003–2022)

4.2 *Country-Specific Investigations of Marine Organisms for New MNPs (2003–2022)*

Different countries show different patterns of taxonomic focus. Porifera accounts for the majority of new MNPs publications in Indonesia (71.6%), the Philippines (53.6%) and Papua New Guinea (48.6%) (Fig. 8). Notably, Myanmar's sole publication in this field is also based on Porifera. Thailand's MNPs publications are dominated by Ascomycota (53.1%), indicating a strong interest in fungal sources of natural products. In fact, within the past 10 years (2014–2023) most new marine fungal-derived compounds were reported by Dr. Anake Kijjoa's research group at Biomedical Sciences Institute of Abel Salazar (ICBAS), University of Porto, Portugal, in collaboration with Dr. Tida Dethoup at Kasetsart University, Thailand. Other Thai researchers working on marine fungal natural products include scientists from Prince of Songkla University (Dr. Vatcharin Rukachaisirikul), Thailand's National Center for Genetic Engineering and Biotechnology (BIOTEC) and Chulalongkorn University.

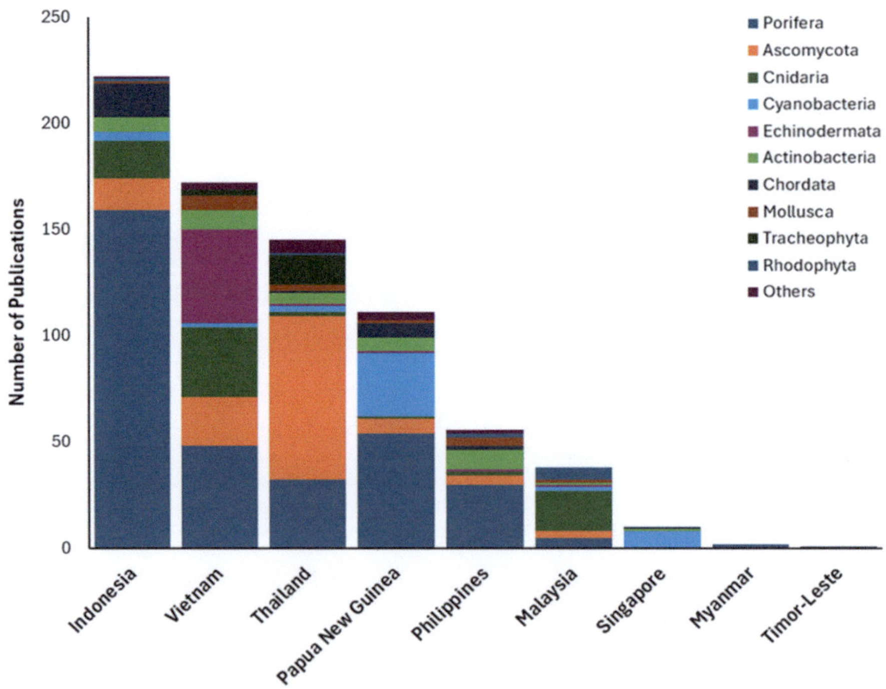

Fig. 8 Total number of publications on new MNPs (2002–2022) from SEA countries and Papua New Guinea based on the country origin of the source organism. The phyla Chlorophyta, Proteobacteria, Bacteroidetes, Ochrophytina, Myzozoa, Firmicutes and Basidiomycota were grouped under "Other" due to small number of publications. Data was compiled from the annual reviews of new marine natural products published between 2003 and 2022 [163–182]

Malaysia's output is primarily centered on Cnidaria (50.0%), while Singapore's small but focused contribution is heavily skewed toward cyanobacteria (80.0%) (Fig. 8). Interestingly, Papua New Guinea also shows a relatively high proportion of cyanobacteria-related publications (27.0%), making it a secondary focus after Porifera. Vietnam stands out for its balanced distribution across multiple phyla, with Porifera (27.9%), Echinodermata (25.6%) and Cnidaria (19.2%) each contributing significantly to its overall MNPs publications. This more distributed spread suggests a broad-based exploration of marine biodiversity and taxonomic diversity in its bioprospecting efforts.

When narrowing in on selected countries, namely Indonesia, Thailand and Vietnam, with most new MNPs publications (2003–2022) focusing solely on key marine invertebrates, such as Porifera, Cnidaria and Echinodermata, distinctive patterns emerged that shed light on the taxonomic focus of each country's MNPs research. These patterns may reflect the ecological availability of marine samples. Indonesia's MNPs research is heavily skewed toward Porifera (Fig. 8). There is a particular emphasis on Family Petrosiidae, which accounts for 17.5% of Indonesian publications from Porifera (Fig. 9). This is followed by Thorectidae (10.5%). Although the analysis is based on 20 years survey (2003–2022), this observation is interesting given that, regionally and globally, Thorectidae surpasses Petrosiidae in number of publication [3]. Within Family Petrosiidae, *Acanthostrongylophora* (8.8%) and *Xestospongia* (5.3%) are the most studied genera. On the other hand, Indonesia's Cnidarian research aligns more closely with global and regional patterns, with Alcyoniidae as the most studied family and *Sinularia* as the dominant genus.

Vietnam's new MNPs publications (2003–2022) presents a more even distribution across the three marine invertebrate phyla (Fig. 10). In Porifera, Family Suberitidae (5.6%), Spongiidae (5.6%), Ancorinidae (4.8%), Petrosiidae (4.0%) and Dysideidae (4.0%) contribute most significantly (Fig. 10). However, generic-level diversity appears limited, as many families are represented by only a single genus. Cnidaria research was centered on Family Alcyoniidae (18.4%), specifically on the genus *Sinularia* (9.6%). However, Vietnam's strength lies in its Echinodermata research, which is conducted in close collaboration of Vietnamese scientists with Russian counterpart at the G.B. Elyakov Pacific Institute of Bioorganic Chemistry, Far Eastern Branch of the Russian Academy of Science [201, 202]. There is a broad diversity of echinoderm families investigated, with Oreasteridae (7.2%) and Sclerodactylidae (4.8%) contributing most significantly. However, generic-level diversity also appears to be limited.

Like Indonesia, Thailand's new MNPs publications (2003–2022) show a strong dominance in Porifera, with Thorectidae (17.6%) and Halichondriidae (14.7%) contributing most significantly for the period 2003–2022 (Fig. 10). Thailand's Cnidaria publications are unique in the region, focusing on Family Plexauridae and Subergorgiidae which are not common in other publications from SEA. Despite being among the top three in total publications, Thailand's prominence in MNPs literature stems largely from its work Ascomycota, rather than marine invertebrates.

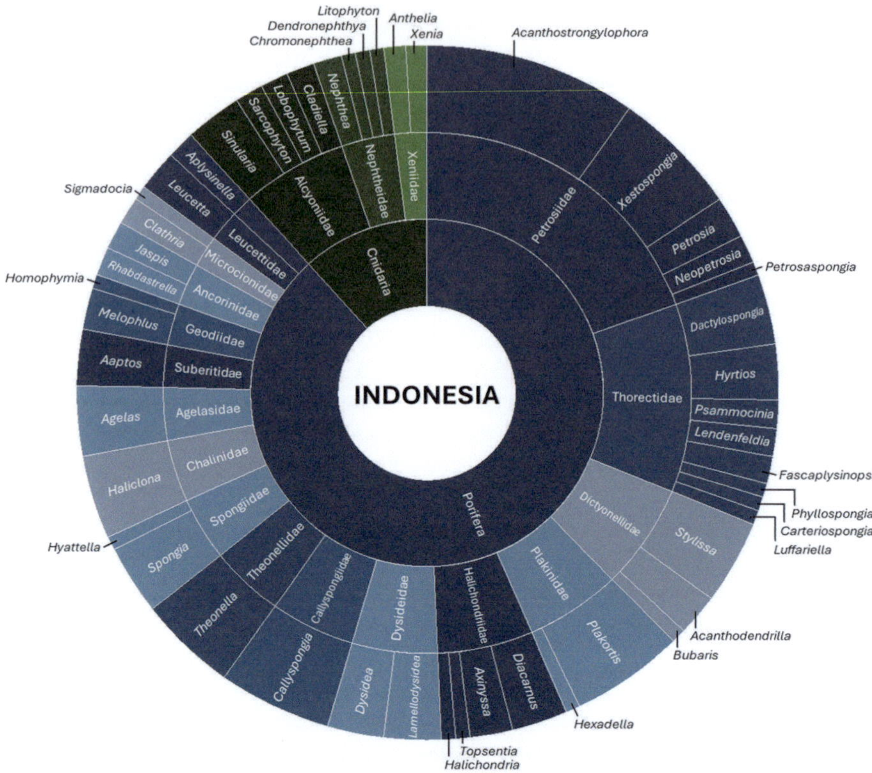

Fig. 9 Taxonomic breakdown of marine invertebrates studied for new MNPs in Indonesia (2003–2022). The sunburst diagram displays the hierarchical distribution of organisms by family and specific levels. Each segment represents a phylum (inner ring), a family (middle ring) and its corresponding species (outer ring), sized proportionally by the number of publications. Only families with three or more publications were included. Excluded Poriferan families (genus) are: Coelosphaeridae (*Lissodendoryx*), Iotrochotidae (*Iotrochota*), Placospongiidae (*Corticium*), Tetillidae (*Cinachyrella*), Axinellidae (*Axinella*), Desmacididae (*Biemna*), Desmoxyidae (*Myrmekioderma*), Ianthellidae (*Ianthella*), Irciniidae (*Ircinia*), Mycalidae (*Mycale*), Niphatidae (*Niphates*), Phloeodictyidae (*Oceanapia*); excluded Cnidarian families (genus) are: Stichodactylidae (*Heteractis*), Xeniidae (*Anthelia, Xenia*)

4.3 Contribution of Countries to the Overall Publications of New MNPs (2003–2022)

The distribution of research on new MNPs between 2003 and 2022 varies significantly across countries in the region, with contributions to the literature being highly uneven. Three countries, namely Indonesia, Vietnam and Thailand, account for 71% of all publications, with Indonesia contributing the largest share at 29.4% (Fig. 11). Indonesia maintained its leading position with the highest number of new MNPs publications consistently between 2003–2012 and 2013–2022 time frame but had a modest increase of 3.7% (from 109 to 113 publications) (Fig. 12). Vietnam saw an impressive 480% increase (from 24 to 145 publications), reflecting a significant rise in MNPs research output. Thailand also demonstrated growth of 57% (from 56 to 88 publications). Malaysia emerged as a surprising contender, showing a 92% increase from 13 to 25 publications. Myanmar and Timor-Leste, which had no publications in the first decade, managed to increase their contributions to two and one publications, respectively, in the following decade. However, Papua New Guinea experienced a concerning decline of 80% (from 94 to 18 publications). Brunei and

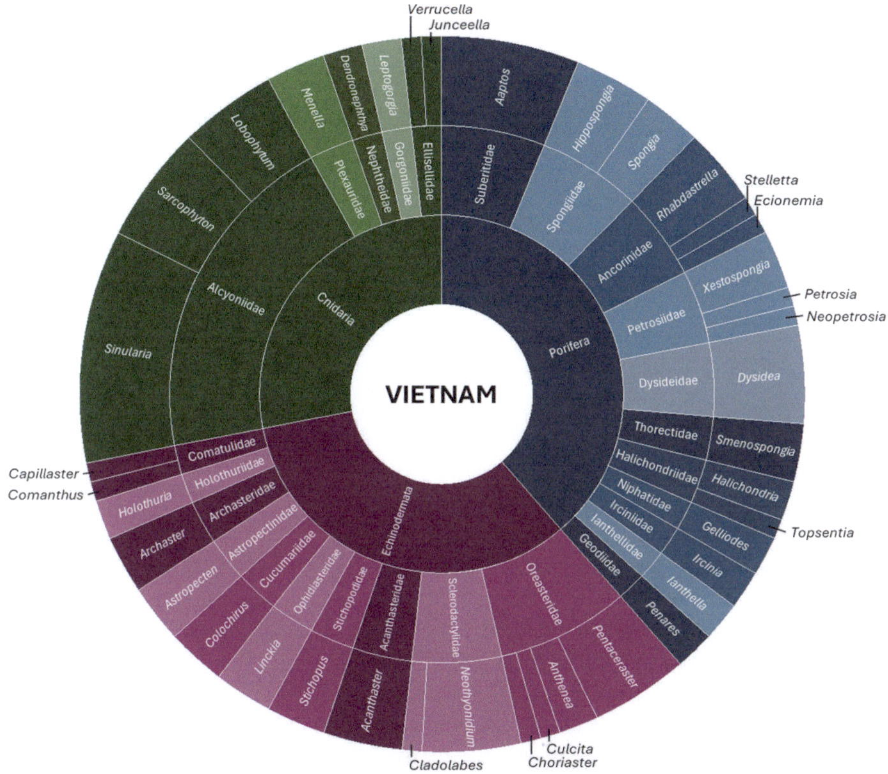

Fig. 10 Taxonomic breakdown of marine invertebrates studied for new MNPs in Vietnam and Thailand (2003–2022). The sunburst diagram displays the hierarchical distribution of organisms by family and specific levels. Each segment represents a phylum (inner ring), a family (middle ring) and its corresponding species (outer ring), sized proportionally by the number of publications. For Vietnam's sunburst diagram, only families with more than one publication were included. Excluded Poriferan families (genus) are: Arenoscleridae (*Arenosclera*), Microcionidae (*Clathria*), Desmapsammidae (*Desmapsamma*), Chalinidae (*Haliclona*); excluded Echinoderm families (genus) are: Echinasteridae (*Echinaster*), Synaptidae (*Synapta*), Toxopneustidae (*Cercodemas*), Goniasteridae (*Anthenoides*), Asterinidae (*Asterina*), Asteropseidae (*Asteropsis*); excluded Cnidarian family (genus) is: Clavulariidae (*Carijoa*)

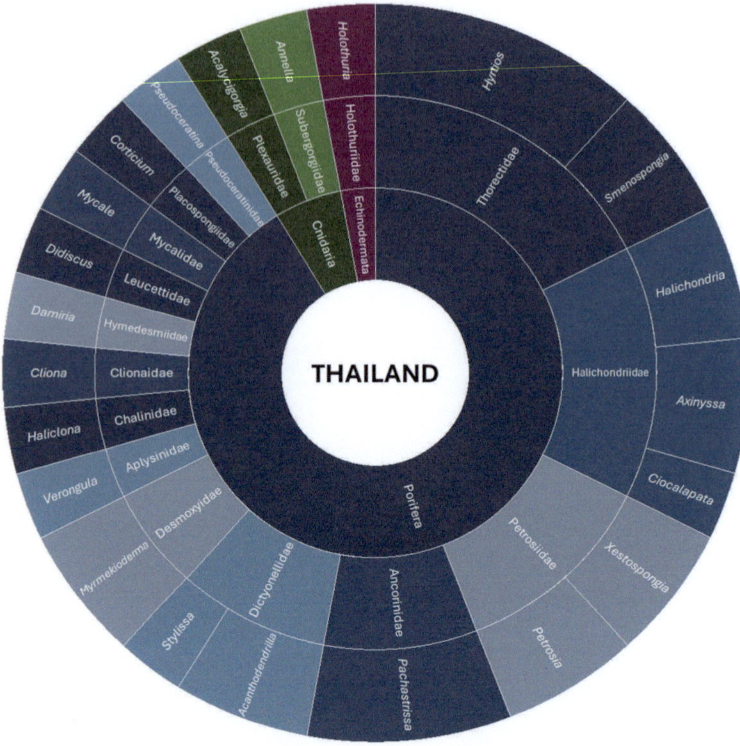

Fig. 10 (continued)

Cambodia consistently reported no MNPs publications throughout the period, highlighting a lack of research output despite their biological richness. These disparities underscore the need for greater regional collaboration, capacity-building initiatives and investments to better leverage Southeast Asia's marine biodiversity, ensuring that all countries in the region can contribute more effectively to MNPs research.

For the period of 2003–2022, Indonesia was the dominant contributor to MNPs publications from Porifera accounting for nearly half (48.2%) of Porifera publications in SEA (Fig. 13). This was followed by Papua New Guinea (16.4%) and Vietnam (14.5%) (Fig. 13). For Cnidaria, publications were mainly Vietnam (44%), Malaysia (25.3%) and Indonesia (24%). Vietnam also showed an extraordinary specialization in Echinodermata, accounting for 91.7% of all regional publications in this phylum. In terms of less commonly studied phyla, Indonesia led Chordata-related research (57.1%) while Thailand dominated Tracheophyta publications (82.4%).

New MNPs publications from Ascomycota (2003–2022) were led by Thailand, which contributed 60% of the regional total (Fig. 13). Additionally, actinobacteria studies (2003–2022) were more evenly distributed across five countries: the Philippines (23.7%), Vietnam (23.7%), Indonesia (18.4%), Papua New Guinea (15.8%) and Thailand (13.2%) (Fig. 13). If publications of marine microbes obtained from

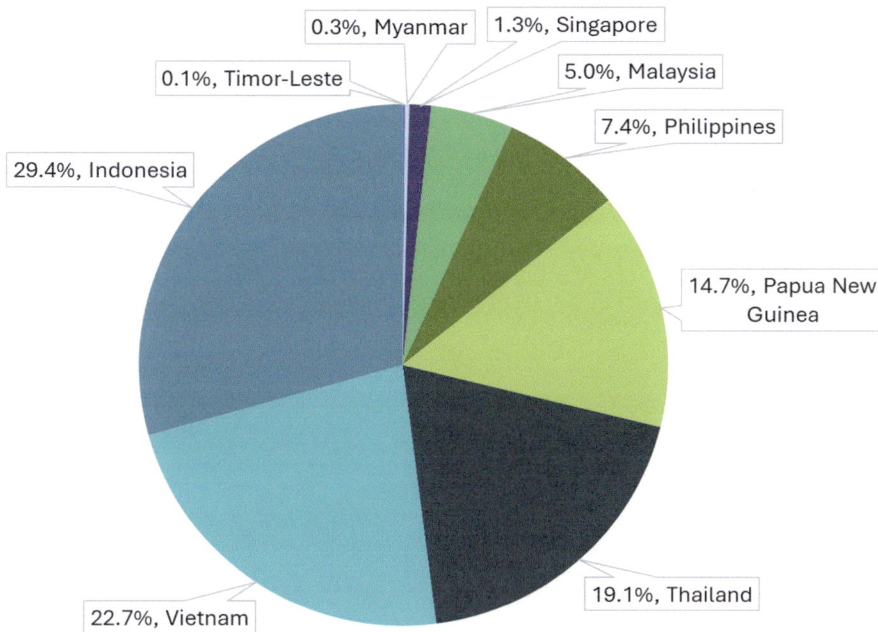

Fig. 11 Percentage contribution of publications of new MNPs (2003–2022) for countries within ASEAN and Papua New Guinea

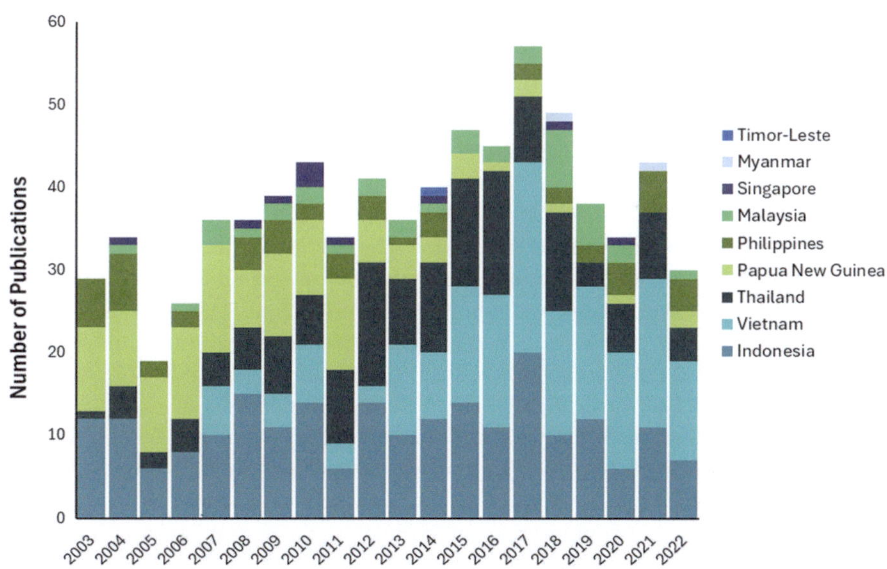

Fig. 12 Trends in the number of publications of new MNPs from SEA, including Papua New Guinea, between 2003 and 2022. Data was compiled from the annual reviews of marine natural products published between 2003 to 2022 [163–182]

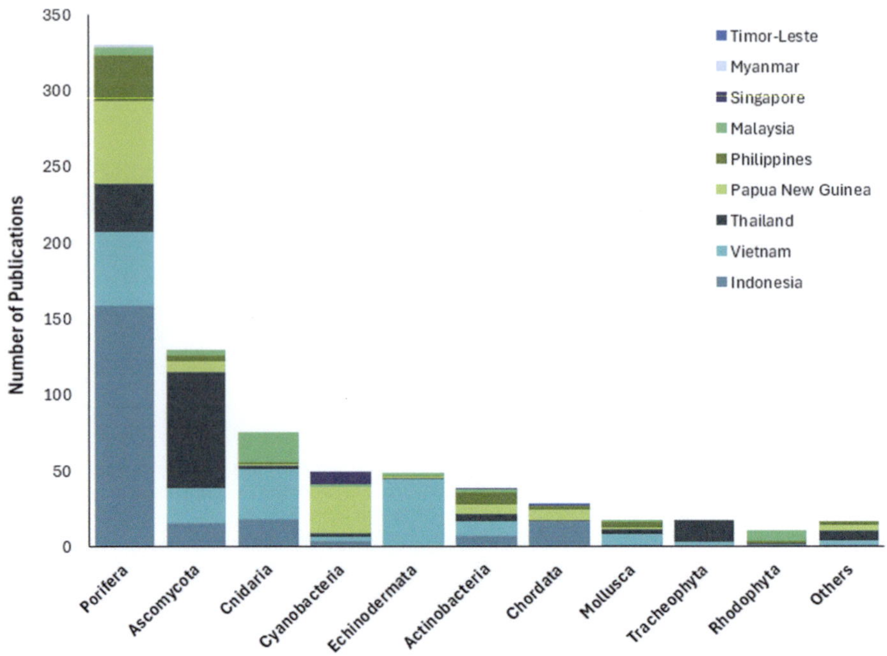

Fig. 13 Total number of publications on new MNPs based on the phylum of the source organism from SEA and Papua New Guinea (2003 and 2022). The phyla Chlorophyta, Proteobacteria, Bacteroidetes, Ochrophytina, Myzozoa, Firmicutes and Basidiomycota were grouped under "Other" due to small number of publications. Data was compiled from the annual reviews of new marine natural products published between 2003 and 2022 [163–182]

major archipelagos, such as the Spratly Islands, Paracel Islands and Pratas Islands, were to be included in the analysis, they would result in significant contribution to overall new microbial marine natural products publications in the region.

When the contribution from Papua New Guinea is included in the analysis, the country stood out for its reports on new marine cyanobacterial molecules, producing 61.2% of the region's total new MNPs publications (2003–2022) in this taxonomic group (Fig. 13). Most of these marine cyanobacterial publications were from Prof. William Gerwick's research group at the Scripps Institution of Oceanography, University of California, San Diego. A range of structurally diverse and potent secondary metabolites, including polyketides and hybrid peptide-polyketides, have been isolated from Papua New Guinean marine cyanobacteria, including *Moorena*, *Lyngbya* and *Symploca* species (Fig. 14). Several of these cyanobacterial compounds, such as the hoiamides and apratoxins, and their synthetic derivatives are in preclinical testing for treatment of various human diseases. Additionally, based on MS/MS features, marine cyanobacteria from Papua New Guinea have been identified as having high chemical diversity [203].

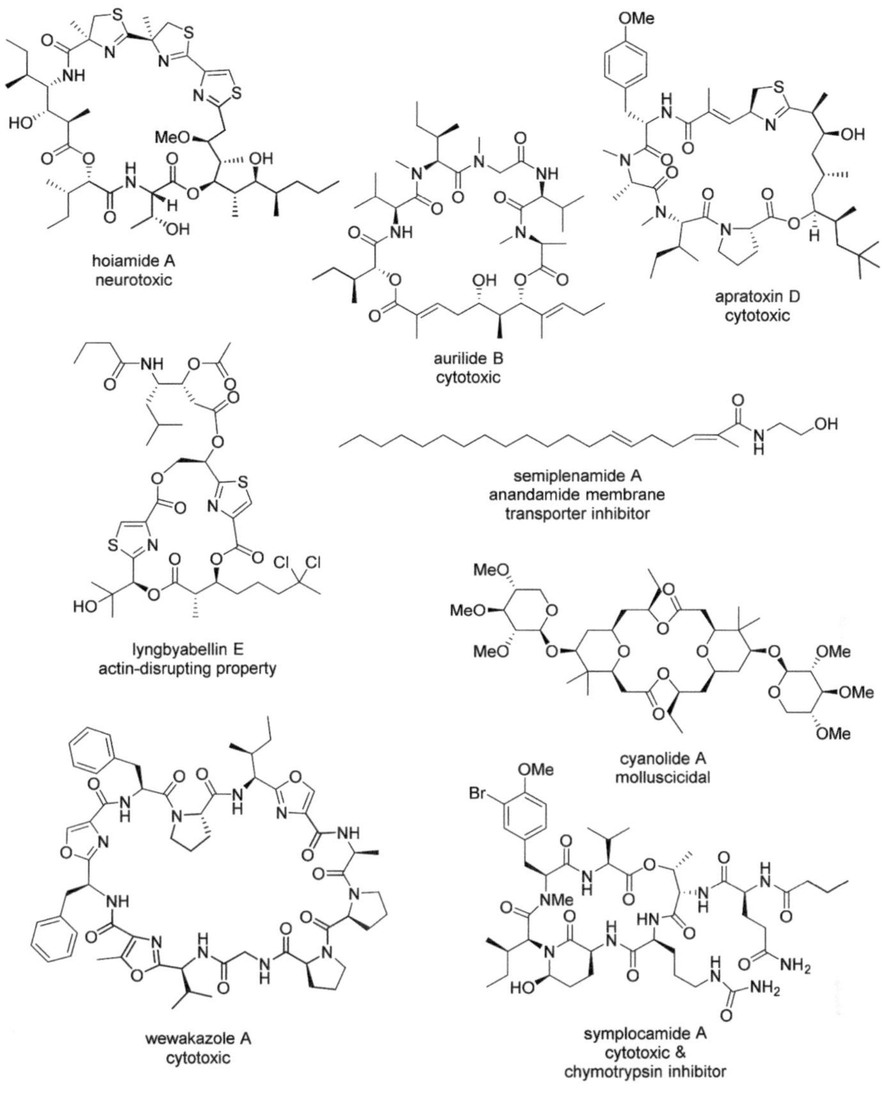

Fig. 14 A selection of bioactive natural products isolated from Papua New Guinean marine cyanobacteria: hoiamide A [204], aurilide B [205], apratoxin D [206], lyngbyabellin E [207], semiplenamide A [208], cyanolide A [209], wewakazole A [210], symplocamide A [211]

5 Marine Drug Discovery Initiatives and Programmes in Southeast Asia

The transition to the Blue Economy is a crucial development for Southeast Asia in the context MNPs research, as it aligns with the region's broader policy goals. The Blue Economy refers to the sustainable utilization of ocean resources for economic growth, improved livelihoods, and job creation while maintaining the health of marine and coastal ecosystems. The transition to the Blue Economy is particularly significant for Southeast Asia due to its potential to enhance food security, mitigate climate change impacts, create highvalue jobs and promote marine conservation through sustainable economic activities. As such, the transition to the Blue Economy in Southeast Asia has intensified development efforts in both traditional marine sectors (such as fisheries, aquaculture, seafood processing and marine tourism) and emerging industries, including marine biotechnology and bioprospecting. The role of the MNPs in drug discovery and biopharmaceuticals varies significantly across ASEAN member states. While some countries have explicitly incorporated MNPs discovery into their Blue Economy strategies, others focus more broadly on marine biotechnology while others are still in the early stages of development.

5.1 Philippines

The Philippines, a nation of 7641 islands stretching approximately 1880 km from north to south, is one of the world's largest archipelagic states. With a vast land area of 298,170 km^2, a coastline extending 37,008 km and territorial waters covering around 679,800 km^2, the country is endowed with exceptional marine biodiversity. This rich marine environment provides a valuable reservoir of biologically active compounds, making the Philippines a significant player in MNPs research and marine-based drug discovery efforts by local and foreign researchers. The earliest work on the discovery of bioactive and chemistry of diverse natural products from Philippine marine organisms started with the research conducted by National Scientist Lourdes J. Cruz and Prof. Baldomero M. Olivera on cone snails during the 1970s [212].

An earlier initiative, the National Cooperative Natural Products Drug Discovery Group (NCNPDDG) program, titled "Anticancer Agents from Unique Natural Products Sources" (CA 67786), was initiated in September 1995 with funding from the U.S. NCI. Its primary aim was to discover and develop novel anticancer agents from diverse natural product sources while promoting equitable partnerships through technology transfer, education and fair benefit-sharing with source countries [162]. Under this programme, the Marine Natural Products Laboratory of the University of the Philippines Marine Science Institute (UP-MSI) partnered with the Department of Medicinal Chemistry at the University of Utah to investigate the rich marine biodiversity of the Philippines. This collaboration focused on the collection, documentation and bioactivity-guided isolation of anticancer compounds from marine invertebrates, such as ascidians, sponges and their symbiotic microorganisms [213].

All bioprospecting activities in the Philippines were conducted with legal permits and in compliance with the country's bioprospecting regulations. At the outset, a Memorandum of Understanding (MOU) between the University of Utah and UP-MSI was executed, establishing the co-ownership of patents and equal sharing of revenues from intellectual property resulting from the research [214]. This agreement emphasized collaborative research, co-authorship and the importance of building local capacity through training and infrastructure development. UP-MSI received dedicated funding from the NCNPDDG program to support these objectives. To further ensure legal and ethical compliance, a Commercial Research Agreement (CRA) was also signed among the University of Utah, the University of the Philippines and the Philippine Department of Agriculture [214]. The Philippines thus played a significant role not just as a source of marine natural products, but as an equal research partner [214]. The success of this initiative catalyzed further programs that were built upon the foundation laid by the NCNPDDG collaboration. Notable among these are the Anti-pain Peptides from Marine Organisms Research (AMOR) program, Conus/Turrid project, PharmaSeas program and the Philippine Mollusk Symbiont–International Cooperative Biodiversity Group (PMS-ICBG) program, each contributing to the sustained advancement of marine natural products research and drug discovery in the Philippines.

A landmark initiative on marine drug discovery is the Philippine PharmaSeas Drug Discovery Program, launched in 2008 through the efforts of the Department of Science and Technology (DOST) and UP-MSI [215]. Under the leadership of Prof. Gisela Concepcion, the program aimed to discover novel anti-infective agents from sponge-associated marine microorganisms and anti-pain compounds from cone snails (conoidea). The PharmaSeas team includes local scientists, medical professionals and international collaborators, highlighting the multidisciplinary and global scope of the research. Additionally, several national institutes have played pivotal roles in supporting this initiative, including the Philippine Council for Aquatic and Marine Research and Development (PCAMRD) and the Research Institute for Tropical Medicine. Their involvement has helped bridge marine sciences with biomedical research, fostering the translation of marine natural products into potential therapeutic applications.

Currently within UP-MSI, the Marine Pharmacognosy Laboratory, led by Prof. Lilibeth A. Salvador-Reyes, has been at the forefront of marine drug discovery efforts. The lab focuses on exploring bioactive compounds from a wide array of marine organisms, including bacteria, cyanobacteria, sponges, mollusks, seaweeds as well as harmful algal bloom-causing species. Additionally, this research lab is key contributor to two major drug discovery consortia in the Philippines, namely the Discovery and Development of Health Products (DDHP) and the PMS-ICBG programmes. These consortia aim to develop new antibiotics, anticancer agents and analgesics derived from marine natural compounds. Beyond these initiatives, there is also growing interest in underexplored marine organisms. Studies have assessed the biotechnological potential of cyanobacteria and evaluated the antimalarial properties of marine-derived microorganisms. This indicates an ongoing expansion of marine biodiscovery efforts into novel ecological niches and less-studied taxa.

International collaboration has further enriched the field, with Philippine marine samples contributing to MNPs research in partnership with institutions from the United States, Canada, China, Korea, Japan, Europe and Australia. These collaborations underscore the global significance of Philippine marine biodiversity and highlight the country's role as a key contributor and regional leader to the search for new marine-derived pharmaceuticals.

5.2 *Indonesia*

Indonesia, as the world's largest archipelagic nation, comprises approximately 16,056 islands and boasts a vast coastline extending 99,093 km. Approximately 75% of the country's total territory, equivalent to 5.8 million km^2, is constituted by ocean waters. Situated within the Coral Triangle, which it shares with the Philippines, Indonesia occupies one of the most biologically diverse marine regions globally. This extraordinary marine biodiversity positions Indonesia as a key player in the global search for marine natural products, particularly bioactive compounds with potential pharmaceutical and nutraceutical applications.

Recognizing the strategic importance of its marine resources, the Government of Indonesia has initiated the Blue Economy Roadmap 2023–2045, a long-term national development plan aimed at promoting sustainable economic growth from marine-based sectors [216]. Central to this roadmap are emerging blue economy sectors, including marine biotechnology, bioeconomy, renewable energy, marine scientific research and ecosystem conservation and management. Notably, Phase II (2025–2029) and Phase III (2030–2034) of the roadmap place significant emphasis on the development and expansion of marine biotechnology industries. These phases specifically prioritize the sustainable extraction and utilization of bioactive compounds from marine organisms, underscoring Indonesia's intent to enhance its role in the global MNPs research landscape.

Indonesia's contributions to MNPs research have evolved considerably over the past two decades, primarily facilitated through sustained international collaborations. For the period of 2002–2023, Indonesian researchers from a range of universities and research institutions, including Sam Ratulangi University, University of Hasanuddin, Diponegoro University, Padjadjaran University, Lampung University, IPB University (formerly known as Bogor Agricultural University), Ganesha University of Education, Universitas Pembangunan Indonesia, Gadjah Mada University, Universitas Negeri Manado, and the Indonesian Institute of Sciences (LIPI), have been actively engaged in international collaborations aimed at exploring marine natural products from Indonesia's rich marine biodiversity, particularly marine sponges. These partnerships, involving prominent natural products researchers from overseas universities and pharmaceutical companies from Japan (the University of the Ryukyus, Osaka University, Nagoya University, Hokkaido University, Tohoku Pharmaceutical

University, Kumamoto University, Tokyo University of Marine Science and Technology), Australia (Australian National University, The University of Queensland), Germany (Heinrich Heine University), Canada (University of British Columbia), the United States (University of Mississippi), Italy (University of Naples Federico II), and Spain (PharmaMar). A crucial component of this research has been the involvement of taxonomists from the Naturalis Biodiversity Center in the Netherlands, who have contributed significantly to the accurate identification and classification of Indonesian sponge species, thereby supporting the integrity and advancement of natural products research in the region.

Despite Indonesia's rich marine biodiversity and its increasingly central role in MNPs research, the country has continued to rely predominantly on external funding and technical resources, including financial support from the Japan Society for the Promotion of Science (JSPS) and various European grant agencies. While local government support has been present and gradually increasing, it has remained secondary in comparison to international contributions. Increasing domestic investment in research infrastructure, enhancing institutional capabilities and fostering integrated national research programs will be essential to ensure that Indonesia can transition from a supporting to a leading role in the global marine biodiscovery effort.

5.3 Vietnam

Located on the eastern edge of the Indochina Peninsula, Vietnam holds a strategic position near the center of Southeast Asia. The nation's extensive coastline, stretching over 3,260 km and encompassing 28 coastal provinces, forms the backbone of its maritime economy and plays a vital role in ecological conservation and marine biodiversity management. Vietnam has long recognized the immense potential of its marine resources and has committed itself to the advancement of marine ecology and biotechnology. This commitment is particularly evident in the country's robust efforts toward the sustainable exploration and utilization of marine natural products. At the forefront of this research is the Vietnam Academy of Science and Technology (VAST), the country's leading scientific institution [217].

Under the VAST umbrella, the Institute of Marine Biochemistry (IMBC) serves as Vietnam's premier center for MNPs research. The IMBC focuses on both fundamental and applied aspects of marine natural product chemistry, with a special emphasis on marine-derived bioactive compounds sourced from diverse habitats, such as mangrove forests, coral reefs, coastal ecosystems and deep-sea environments. Between 2015 and 2020, the IMBC successfully isolated and characterized around 427 compounds from a wide range of marine organisms. Impressively, 77 of these were novel compounds that demonstrated promising cytotoxic, anti-inflammatory, and antimicrobial properties, highlighting the pharmacological potential of Vietnam's marine biodiversity.

A hallmark of Vietnam's MNPs research landscape is its strong international collaborations. From 2003 to 2023, research initiatives based on Vietnamese marine samples have involved scientific partnerships with overseas institutions, such as Russia, Korea and Japan. The collaboration with Russia has been particularly significant, involving joint projects between VAST and the G. B. Elyakov Pacific Institute of Bioorganic Chemistry of the Far Eastern Branch of the Russian Academy of Sciences. These partnerships often include joint expeditions aboard Russian research vessels, such as the 'Akademik Oparin', to conduct marine biodiversity surveys and sample collections in Vietnamese waters [202].

5.4 *Thailand*

Thailand, officially known as the Kingdom of Thailand, is a Southeast Asian nation with a land area of 513,115 km^2 and boasts a coastline of 3148 km, divided into two key maritime zones: the Gulf of Thailand (2055 km) on the east and the Andaman Sea (1093 km) on the west. This geographical positioning, coupled with its tropical climate and biodiversity-rich marine ecosystems, makes Thailand an important hub for MNPs research in the region.

Currently, MNPs research in Thailand is conducted by several leading academic institutions and government-funded research organizations, many of which are engaged in the isolation and characterization of new bioactive compounds from marine organisms, including marine microbes, marine invertebrates and mangrove-associated flora. Notable academic/research institutions engaged in MNPs research include Chulalongkorn University (Drs Khanit Suwanborirux and Khanitha Pudhom), Chulabhorn Research Institute (Drs. Hunsa Prawat and Prasat Kittakoop) and Prince of Songkla University (Dr. Vatcharin Rukachaisirikul). These universities house specialized research units and centers that contribute to marine biotechnology and natural products discovery. Chulalongkorn University, in particular, plays a significant role in MNPs research through its multiple marine-focused centers, including Center of Excellence for Natural Products and Nanoparticles and Center of Excellence for Marine Biotechnology. Additionally, at the national level, the National Center for Genetic Engineering and Biotechnology (BIOTEC), under the National Science and Technology Development Agency (NSTDA), is one of the foremost research organizations in Thailand and the broader Asian region. Within BIOTEC, the Marine Biotechnology Research Team focuses specifically on harnessing marine bioresources for applications in pharmaceuticals, agriculture and industry.

Thai institutions are also active in international collaborations, engaging in joint research with partners from countries, such as Japan (e.g., Meiji Pharmaceutical University) and Portugal (e.g., Universidade do Porto). These collaborations have strengthened Thailand's capacity for high-quality research, technology transfer and scientific publication in the field of marine natural products. In terms of funding, the Thailand Research Fund (TRF) has played a pivotal role in supporting MNPs research since its establishment in 1992. TRF provides financial support for research projects across various scientific disciplines, including marine biotechnology. Its Royal Golden Jubilee (RGJ) PhD Programme has been especially influential, offering scholarships that promote doctoral-level research in marine natural products and contribute to the training of a new generation of scientists. With a solid research infrastructure, growing international partnerships, and strong funding support, Thailand continues to make significant contributions to the discovery and development of marine natural products in Southeast Asia and beyond.

5.5 *Malaysia*

Malaysia stands out in the marine biotechnology landscape due to its unique geographic characteristic. While the country's total land area is approximately 329,750 km^2, its marine territory, including the Exclusive Economic Zone (EEZ), spans around 450,000 km^2. This vast marine expanse is complemented by an extensive coastline of about 4675 km, which borders both East and West Malaysia and includes more than 870 islands. These factors position Malaysia as a nation rich in marine biodiversity and natural resources, making it a promising frontier for MNPs research.

Research into MNPs in Malaysia is primarily driven by researchers from several key institutions, including the Institute of Marine Biotechnology (IMB), established in 2006 at the Universiti Malaysia Terengganu, Universiti Malaysia Sabah (UMS), University of Malaya and others. As a government initiative, IMB plays a leading role in advancing marine biotechnology through comprehensive research, development and commercialization efforts. The institute serves as a hub for innovation in marine bioresource utilization, including bioprospecting for novel compounds with potential pharmaceutical, nutraceutical and industrial applications. The scope of ongoing research at these institutes encompasses general screening of a wide array of marine organisms, such as sponges, algae and marine microbes, for their biological activities and bioactive compounds [218–223].

Between 1992 and 2023, the field of MNPs research, especially on the reports of new marinederived molecules, in Malaysia evolved significantly, marked by growing international collaboration, diversification of funding sources and expanded institutional partnerships. A notable trend from 2000 to 2023 is the deepening of Malaysia-Japan research ties, with UMS working closely with Japanese institutions like Kagoshima University and Hokkaido University. These collaborations have facilitated joint expeditions, shared facilities and copublications in peer-reviewed journals, strengthening the scientific foundation of MNPs discovery. Funding for MNPs research at UMS has been supported by a combination of domestic and international sources. Key local contributors include the Sabah Biodiversity Centre and the Ministry of Higher Education, Malaysia. International funding, especially from Japanese government agencies and private foundations, has played a crucial role in enabling cross-border collaborative projects and advanced chemical analysis.

5.6 Singapore

Singapore is a small, highly urbanized island-state at the southern tip of the Malay Peninsula, comprising one main island and over 60 islets, as well as a 209 km coastline, strategically located between the Western and Central Indo-Pacific ecoregions near the Coral Triangle. Marine natural products research in Singapore has evolved gradually, with early initiatives beginning in the 1990s through institutions like the Bioscience Centre at the National University of Singapore and the Institute of Molecular and Cell Biology (IMCB), which collaborated with Glaxo Wellcome to establish the Centre for Natural Product Research in 1993 [224, 225]. This centre later became Merlion Pharmaceuticals in 2002, which maintained extensive natural product libraries, including marine-derived sources, now housed at the A*Star's Bioinformatics Institute [226]. Despite Singapore's strategic location between major marine ecoregions and its strong infrastructure in marine biotechnology, dedicated MNPs discovery remains limited. Notable efforts include the National Research Foundation (NRF)'s Marine Science R&D Programme (MSRDP, 2015–2021) and a dedicated MNPs lab at the National Institute of Education (NIE), which also leads collaborations with local and international partners and has been supported by local funding bodies, such as the NIEAcademic Research Fund (AcRF) and NRF MSRDP.

5.7 Others

Myanmar possesses one of the longest coastlines in mainland Southeast Asia, stretching approximately 3000 km. The country has outlined a vision for a Blue Economy that encompasses key sectors, such as fisheries, tourism and energy. However, the sustainable development of these sectors is hindered by persistent challenges, including socio-political conflicts in coastal regions and the withdrawal of foreign investment. At present, marine biotechnology is not recognized as a national priority, and there is limited evidence of active research in marine-derived drug discovery. In fact, between 2003 and 2022, Myanmar contributed only two publications in the field of MNPs chemistry, both in collaboration with Japanese researchers and funded by Japanese government and Japanese private foundations [227, 228].

Brunei's marine bioprospecting status is in its early stages, with more focus on biodiversity research and mariculture rather than direct bioprospecting activities for pharmaceuticals or biotechnology. While there's a recognition of the potential of Brunei's marine resources, there is no strong emphasis on developing the marine biotechnology sector. Given Brunei's geographical position, the country has access to an abundance of unique flora and fauna, both on land and in the surrounding marine ecosystems. This makes it a prime location for bioprospecting, as these natural resources may yield bioactive natural products that could have significant commercial value.

6 Challenges and Opportunities

Marine natural products research in Southeast Asia is uniquely positioned at the intersection of immense biodiversity and urgent conservation needs. The region, home to globally significant marine ecosystems, holds tremendous potential for drug discovery and biotechnological innovation. However, Southeast Asian countries face a complex set of challenges, including underexplored ecosystems, limited infrastructure and fragmented research efforts, that must be addressed to unlock this potential. At the same time, key opportunities exist in the form of rich endemic species, growing academic interest, technological advances and collaborative frameworks for conservation and discovery. This section explores several critical areas that define both the promise and the hurdles of marine natural product research in Southeast Asia.

6.1 Rich Biodiversity, Endemic Marine Species and Novel Compounds

Southeast Asia, with its extensive coastlines stretching across archipelagic nations like Indonesia, the Philippines and Malaysia, encompasses some of the most ecologically rich marine regions in the world. This area is part of the Coral Triangle, a global biodiversity hotspot known for its exceptional variety of marine species and complex ecosystems, including coral reefs, mangroves and seagrass beds. Despite this ecological wealth, many of these marine habitats remain poorly studied, particularly in terms of their chemical and biological diversity. For instance, marine organisms from Myanmar have not been studied extensively for natural products. Despite Myanmar's long coastline along the Bay of Bengal and the Andaman Sea, the country remains largely underrepresented in marine natural products research. Political instability, limited infrastructure for marine science and a lack of coordinated bioprospecting efforts have contributed to the scarcity of published studies from this region. However, this also means that Myanmar's marine ecosystems represent an untapped reservoir of chemical diversity, offering promising opportunities for future natural products discovery.

Certain marine ecosystems in Indonesian coral triangle areas are also understudied, including areas around North Maluku, Maluku (Aru, Arafura Sea, Banda Sea), Papua (Cenderawasih Bay, Raja Ampat Archipelago) and Lesser Sunda Islands (East and West Nusa Tenggara). Among the 114 sponge genera explored from Indonesia for their marine natural products, several sponge genera, such as *Homophymia*, *Daedolopelta*, *Siliquariaspongia*, *Histodermella* and *Pachypellina* are rare with limited distribution in Indonesia. Additionally, members of *Minabea*, a rare genus of octocorals in Indonesia are underexplored for their natural products. To initiate their chemical investigation, information on the distribution and occurrence of these underexplored marine organisms can be obtained from Global Biodiversity Information Facility (GBIF) (https://www.gbif.org/), World Register of Marine

Species (WORMS) (https://www.marinespecies.org/) and Ocean Biodiversity Information System (OBIS) (https://obis.org/). These understudied ecosystems present significant opportunities for the discovery of novel marine natural products with pharmaceutical potential. The vast and varied marine biota found in Southeast Asia holds promise for the development of new drugs, particularly in the search for antimicrobial, anticancer, and antiinflammatory agents, making the region a vital frontier for marine drug discovery and bioprospecting efforts.

While a precise, quantifiable "rate" of new marine species discoveries in Southeast Asia is difficult to pinpoint, the region is a biodiversity hotspot with ongoing discoveries, including 234 new species in the Greater Mekong region in 2023 alone [229]. Since 2000, more than 60 new sponge species, including one species (*Svenzea devoogdae*) belonging to a new genus *Svenzea* [11], have been described in Indonesian waters. Such new discoveries are usually described by taxonomists, without information on their chemistry. One way to evaluate chemical diversity and detection of possible novel compounds from newly described sponge species is through the strategic use of museum-preserved specimens. By utilizing these archival samples, researchers can bypass the logistical, economic, and ecological costs associated with fresh collection expeditions. A recent study exemplified this approach by applying a dual mass spectrometry/NMR-guided workflow to analyze these "ready-made" marine invertebrate extracts, maintained at the Smithsonian National Museum of Natural History, using SIRIUS 5 for structure prediction from mass spectral data, followed by selective NMR experiments for structural confirmation [230]. This method successfully led to the isolation of a previously unreported heteroaromatic molecule from the octocoral, *Carijoa riisei*, collected in 1950 from the Gulf of Mexico. This technique not only advances the discovery of novel natural products but opens new avenues in chemical ecology and chemotaxonomy, demonstrating that historical museum samples remain an underutilized source for exploring chemical diversity.

6.2 Natural Product-Driven Marine Conservation Efforts

Marine biodiversity in Southeast Asia and other regions faces numerous threats, including overfishing, habitat destruction, pollution, coastal development and the growing impacts of climate change. These pressures have led to the degradation of coral reefs, loss of mangroves, and declining populations of ecologically and economically important marine species. In response, natural product-driven conservation efforts offer a promising and pragmatic approach to biodiversity preservation. By highlighting the immense pharmaceutical and biotechnological value of marine organisms, natural products research provides tangible incentives for protecting marine ecosystems, including coral reefs and deep-sea habitats. These incentives can also be invested in lifelong learning opportunities for the public, including students, families and social groups, through interactive media, marine field trips and educational resources, such as illustrated nature guidebooks. Such efforts help foster awareness, especially among children, and encourage a deeper appreciation and long-term stewardship of marine biodiversity [231]. Realizing the economic and medicinal potential of marine biodiversity can strengthen public support for conservation and provide compelling arguments to policymakers.

These arguments have been made towards marine conservation efforts in at least two marine areas, including island of Coiba, Panama and the Caribbean island of Curacao [232]. For instance, the discovery of coibamide A, a potent anticancer compound from a cyanobacterium, *Leptolyngbya* sp., near Coiba Island, highlighted the island's rich marine biodiversity and its potential for pharmaceutical development [233]. This finding played a key role in conservation efforts, contributing to Coiba's designation as a UNESCO World Heritage Site in 2005, by demonstrating the tangible value of preserving its natural resources. It is therefore imperative that conservationists and natural product researchers collaborate closely to protect and study biologically rich marine habitats. With sufficient funding, strong policy backing from regional governments and international research partnerships, Southeast Asia can take significant strides toward conserving its marine biodiversity, ensuring both ecological sustainability and continued discovery of novel bioactive compounds.

6.3 Marine Microbial Natural Products Research

There is a pressing need to strengthen marine microbial natural products research within Southeast Asia, particularly considering the region's rich marine biodiversity and untapped microbial resources. Despite this potential, Southeast Asia continues to lag significantly behind countries like China, which produced 897 publications on bioactive compounds from marine-derived bacteria and fungi between 2009 and 2018 alone [234]. Within Southeast Asia, Thailand and Vietnam are currently the most active in this field, with 221 and 88 new compounds reported respectively from 2014 to 2023, based on annual marine natural products reviews by Blunt et al. and Carroll et al. [174–182, 235]. Indonesia, despite its vast marine ecosystems, reported about 53 new marine microbial molecules during the same period. Overall, only about 414 new marine microbial molecules were reported from the entire Southeast Asian region over the past decade (2014–2023), averaging about 41 new molecules/year, a number that pales in comparison to the output of more research-intensive countries/regions. Of the total number of marine microbial compounds, about 76.8%, 17.1% and 4.3% were from marine fungi, bacteria (e.g., actinobacteria) and cyanobacteria, respectively. These figures highlight a significant underutilization of marine microbial resources in the region and underscore the urgent need for increased investment, capacity building and regional collaboration to accelerate natural product discovery and development from marine microorganisms.

6.4 Traditional Knowledge Integration

The integration of traditional marine medicine with modern bioprospecting efforts presents a valuable opportunity for drug discovery in Southeast Asia, where coastal communities have long relied on marine resources for their traditional health practices. These ethnomedicinal traditions, passed down through generations, reflect a deep, empirical knowledge of marine biodiversity and its therapeutic potential [236]. Validating the medicinal uses of marine organisms through scientific research not only helps preserve this cultural heritage but also provides a strategic framework for identifying novel bioactive compounds. By systematically documenting traditional marine remedies and evaluating their pharmacological properties, researchers can prioritize species with proven local usage for targeted bioprospecting. To ensure equitable benefit-sharing, bioprospecting initiatives should actively involve local communities, particularly when traditional knowledge originates from them, by allocating a fair share of royalties to support education, employment and capacity building in the collection areas. Moreover, these activities should promote sustainability and conservation incentives, ensuring that sample collection is both environmentally responsible and socially inclusive. This integrative approach not only enhances the efficiency of drug discovery but also fosters community engagement and conservation awareness, ensuring that the benefits of marine natural product research are shared with the indigenous knowledge holders of Southeast Asia [237].

6.5 Technological Availability, Advancements and Regional/International Collaborations

Technological advancements in the identification and sustainable production of bioactive marine natural products have significantly expanded the potential of marine bioprospecting, particularly from invertebrates and marine microbes. Innovations such as genome mining, metabolomics, metagenomics and synthetic biology now enable the discovery and optimization of bioactive compounds with greater precision and efficiency [238, 239]. However, many research institutions in Southeast Asia still lack access to advanced instrumentation and technologies necessary for such sophisticated work [240]. To effectively carry out marine natural products research at academic institutions, it is essential to equip laboratories with a core set of specialized instruments and equipment [241]. Fundamental items include evaporators paired with a water chiller and diaphragm pump for efficient solvent removal, and a full suite of chromatographic tools, such as high-performance liquid chromatography (HPLC) and column chromatography, for compound separation and purification. Given the high cost and maintenance demands of advanced analytical instruments, equipment such as nuclear magnetic resonance spectrometers, mass spectrometers, X-ray diffractometers and electronic circular dichroism spectrometers should be housed within a centralized, shared analytical facility. Additionally, there are no close collaboration and coordination among the universities, research institutes and industry on marine natural products research within countries, such as Indonesia [231].

To bridge this gap, regional and international collaborations are essential. ASEAN-wide cooperation offers a strategic platform to pool resources, share expertise, strengthen human capital and establish standardized research frameworks and benefit-sharing mechanisms. International cooperation is especially critical, as many marine biodiversity hotspots are in developing countries that face limitations in research funding, technology access and industrial infrastructure for drug development. Partnerships with technologically advanced nations or institutions, under fair and transparent benefit-sharing agreements, can help build local capacity while ensuring ethical and equitable outcomes [242]. These efforts are crucial not only for advancing science but also for ensuring that marine bioprospecting aligns with conservation goals and respects the cultural and ecological values of biodiversity-rich nations. Such international partnerships have been highly productive and include ongoing and past collaborations between Southeast Asian countries and nations such as Japan, Russia and the United States.

An effective regional and international partnership is illustrated by a recent quadripartite collaboration between scientists from Indonesia, Singapore, Scotland and Netherlands which led to the discovery of potentially novel anti-quorum sensing and anti-biofilm molecules from marine samples collected in the underexplored marine areas of Indonesia, including Raja Ampat [243]. In this partnership, HR-MS/MS support was provided by Prof. Marcel Jaspar's team at the University of Aberdeen, Scotland, while extensive bioassay was conducted by Prof. Staffan Kjelleberg's and Dr. Tan's lab at the Singapore Centre for Environmental Life Sciences Engineering and National Institute of Education, Nanyang Technological University, Singapore, respectively. Unique marine samples were collected, and their extracts prepared by Dr. Hanif's lab at IPB University, Indonesia, while the taxonomy of sponge samples was provided by Prof. de Voogd at Naturalis Biodiversity Center, Netherlands. Additionally, as part of the collaborative efforts, a short training video was created by scientists at Prof. Jaspar's group on the use of MS/MS data for compound dereplication as well as elucidation of possible chemical structures. Such training videos are valuable tools for knowledge sharing and for building the capacity of all collaborative partners involved. Furthermore, research grants that support multi-party research groups can play a key role in strengthening international collaboration, particularly in efforts to unlock the medicinal value of unique marine organisms from resource-rich countries.

6.6 Youth and Academic Engagement

Youth and academic engagement represent a vital opportunity for advancing marine natural products research within Southeast Asia. With the region's rich marine biodiversity and growing recognition of its biotechnological potential, nurturing the next generation of researchers is essential for sustainable scientific progress. Universities across Southeast Asia could incorporate marine biotechnology into their curricula, offering research projects and hands-on training that inspire and equip students to explore the field. Such academic integration not only fosters local talent but also builds a foundation for homegrown innovation and long-term research capacity. Additionally, academics at universities in Southeast Asia could co-develop marine biotechnology-related courses, with grants from universities, for their students. For instance, the Nanyang Technological University, Singapore, provides Collaborative Online International Learning Grant to support faculty to collaborate with partners from overseas institutions to develop undergraduate courses.

The involvement of STEM (Science, Technology, Engineering, and Mathematics) education at both secondary and tertiary levels plays a key role in sparking early interest and awareness of marine science and biotechnology. By engaging youth through research internships, science fairs, mentorship programs and collaborative projects, institutions can cultivate a generation of scientists committed to exploring and conserving marine resources. For example, the Marine Science Research and Development Program, in partnership with the St John's Island National Marine Laboratory, National University of Singapore, had an Explore programme for young marine scientists. The objectives of the student programme include naturing of multi-disciplinary problem-solving skills and interest in marine science as well as encouraging interest in marine science and maritime careers.

Enhancing academic networks and fostering collaboration within ASEAN and beyond is vital for advancing marine natural products research. Strengthening regional academic partnerships and student exchange programs can significantly promote knowledge-sharing, stimulate cooperation and cultivate a resilient community of marine biotechnologists committed to sustainable development and natural product discovery. An excellent example of such efforts is the series of International Summer Course on Marine Natural Products in Indonesia, spearheaded by Dr. Novriyandi Hanif of IPB University. This program offers a dynamic platform for engaging young scientists from both ASEAN and international institutions, combining lectures by global experts in marine natural products, hands-on laboratory training and field trips for marine organism collection at Pramuka Island. Such initiatives not only enhance technical skills and scientific understanding but also foster crosscultural collaboration and long-term partnerships crucial for the future of marine biotechnology in the region.

7 Conclusion

Marine natural products research in Southeast Asia stands at a pivotal juncture, with immense potential grounded in the region's unparalleled marine biodiversity. Countries like the Philippines, Indonesia, Vietnam and Thailand, are making notable contributions, yet there remains a significant need for broader participation, capacity building and infrastructure development across the region. The integration of traditional marine knowledge with modern biotechnological approaches presents a unique opportunity to enhance drug discovery efforts while preserving cultural heritage. However, to truly harness the full potential of marine biotechnology, Southeast Asia must confront environmental threats, promote regional cooperation, and embed sustainability at the core of its scientific and economic strategies. Through a cohesive and collaborative approach, the region can solidify its position as a global leader in marine-based drug discovery while ensuring the sustainable use of its vast marine wealth.

References

1. Sodhi NS, Posa MRC, Lee TM, Bickford D, Koh LP, Brook BW (2010) The state and conservation of Southeast Asian biodiversity. Biodivers Conserv 19:317328
2. La Barre S (2014) Marine biodiversity and chemodiversity—the treasure troves of the future. Biodiversity-the dynamic balance of the planet 1–26
3. Calado R, Mamede R, Cruz S, Leal MC (2022) Updated trends on the biodiscovery of new marine natural products from invertebrates. Mar Drugs 20:389
4. Hong L-L, Ding Y-F, Zhang W, Lin H-W (2022) Chemical and biological diversity of new natural products from marine sponges: a review (2009–2018). Marine Life Sci Technol 4:356–372
5. Brinkmann CM, Marker A, Kurtböke Dİ (2017) An overview on marine sponge-symbiotic bacteria as unexhausted sources for natural product discovery. Diversity 9:40
6. Webster Nicole S, Thomas T (2016) The sponge hologenome. mBio 7. https://doi.org/10.1128/mbio.00135-00116
7. Hooper J, Kennedy J, Van Soest R (2000) Annotated checklist of sponges (Porifera) of the South China Sea region. Raffles Bull Zool Suppl 8:125–207
8. Lim S-C, Putchakarn S, Thai M-Q, Wang D (2016) Inventory of sponge fauna from the Singapore Strait to Taiwan Strait along the western coastline of the South China Sea. Raffles Bull Zool Suppl
9. Putra SA, Ambo-Rappe R, Jompa J, de Voogd NJ (2023) Two centuries of sponges (phylum Porifera) taxonomic studies in Indonesia (1820–2021): checklist and bibliography. Zootaxa 5298:1–74
10. de Weerdt W, van Soest R (2001) *Haliclona* (*Halichoclona*) *vanderlandi* spec. nov. (Porifera: Demospongiae: Haplosclerida) from Indonesia. Zoologische Verhandelingen 334:189–194
11. Alvarez B, van Soest RWM, Rützler K (2002) Svenzea, a new genus of Dictyonellidae (Porifera: Demospongiae) from tropical reef environments, with description of two new species. Contrib. Zool. 71:171–176
12. de Voogd N, van Soest R (2002) Indonesian sponges of the genus Petrosia Vosmaer (Demospongiae: Haplosclerida). Zoologische Mededelingen 76:193–209
13. de Voogd N (2003) *Amphimedon denhartogi* spec. nov. (Porifera: Haplosclerida) from deep reef habitats in Indonesia. Zoologische Verh 345:413–418
14. de Voogd NJ (2004) *Callyspongia* (*Euplacella*) *biru* spec. nov. (Porifera: Demospongiae: Haplosclerida) from Indonesia. Zoologische Mededelingen 78, 1828:477–483
15. Tabachnick KR, Janussen D (2004) Description of a new species and subspecies of *Fieldingia*, erection of a new family Fieldingidae and a new order Fieldingida (Porifera; Hexactinellida; Hexasterophora). Boll Mus Ist Biol Univ Genova 68:623–637
16. Calcinai B, Bavestrello G, Cerrano C (2005) Excavating sponge species from the Indo-Pacific Ocean. Zool Stud 44:5–18

17. Calcinai B, Cerrano C, Totti C, Romagnoli T, Bavestrello G (2006) Symbiosis of Mycale (Mycale) vansoesti sp. nov. (Porifera, Demospongiae) with a coralline alga from North Sulawesi (Indonesia). Invertebr Biol 125:195–204
18. Azzini F, Calcinai B, Pansini M (2007) A new species of Coelocarteria (Porifera: Demospongiae) from Sulawesi, Indonesia. J Mar Biol Assoc UK 87:1349–1353
19. de Voogd NJ, Van Soest RW (2007) *Acanthotetilla celebensis* sp. nov., a new species from North Sulawesi, Indonesia (Porifera: Demospongiae: Spirophorida: Tetillidae). Zootaxa 1397:25–28
20. de Voogd NJ, Parra-Velandia FJ, van Soest RWM (2008) A new *Agelas* (Demospongiae: Agelasida: Agelasidae) from the Thousand Islands, West Java, Indonesia. Zool Med Leiden 82:235–243
21. Becking L, Lim S-C (2009) A new *Suberites* (Demospongiae: Hadromerida: Suberitidae) from the tropical Indo-West Pacific. Zoologische Mededelingen 83:853862
22. Calcinai B, Bastari A, Makapedua DM, Cerrano C (2017) Mangrove sponges from Bangka Island (North Sulawesi, Indonesia) with the description of a new species. J Mar Biol Assoc UK 97:1417–1422
23. Muricy G (2011) Diversity of Indo-Australian Plakortis (Demospongiae: Plakinidae), with description of four new species. J Mar Biol Assoc UK 91:303–319
24. De Voogd NJ (2012) On sand-bearing myxillid sponges, with a description of *Psammochela tutiae* sp. nov. (Poecilosclerida, Myxillina) from the northern Moluccas, Indonesia. Zootaxa 3155:21–28–21–28
25. Becking LE (2013) Revision of the genus *Placospongia* (Porifera, Demospongiae, Hadromerida, Placospongiidae) in the Indo-West Pacific. Zookeys 39–76
26. Cavalcanti FF, Rapp HT, Klautau M (2013) Taxonomic revision of Leucascus Dendy, 1892 (Porifera: Calcarea) with revalidation of Ascoleucetta Dendy & Frederick, 1924 and description of three new species. Zootaxa 3619:275–314
27. Calcinai B, Bavestrello G, Bertolino M, Pica D, Wagner D, Cerrano C (2013) Sponges associated with octocorals in the Indo-Pacific, with the description of four new species. Zootaxa 3617:1–61
28. van Soest RW, de Voogd NJ (2015) Calcareous sponges of Indonesia. Zootaxa 3951:1–105
29. Alvarez B, Voogd NJ, Soest RW (2016) Sponges of the family Axinellidae (Porifera: Demospongiae) in Indonesia. Zootaxa 4137:451–477
30. Van Soest RWM (2017) *Flagellia*, a new subgenus of *Haliclona* (Porifera, Haplosclerida). Eur J Taxonomy 351:1–48
31. Calcinai B, Bastari A, Bavestrello G, Bertolino M, Horcajadas SB, Pansini M, Makapedua DM, Cerrano C (2017) Demosponge diversity from North Sulawesi, with the description of six new species. Zookeys 105–150
32. Sanamyan K, Sanamyan N, Martynov A, Korshunova T (2019) A new species of *Ernstia* (Porifera: Calcarea) described from marine aquarium. Zootaxa 4603:192–200
33. Marlow J, Bell JJ, Shaffer M, Haris A, Schönberg CHl (2021) Bioeroding sponge species from the Wakatobi region of southeast Sulawesi, Indonesia. Zootaxa 4996:148
34. Johan O, de Voogd NJ, Budianto A, Peters S, Aljbour SM, Kunzmann A (2024) Epizoic yellow sponge (Poecilosclerida, Demospongiae) expansion on living Scleractinian corals in Bintan, Riau Archipelago, Indonesia. Mar Ecol 45:e12796
35. Lim S-C, Setiawan E (2021) A new *Semperella* (Hexactinellida: Amphidiscosida: Pheronematidae) from Indonesia, Indian Ocean. Raffles Bull Zool 36:2328
36. Bertolino M, Cerrano C, Bavestrello G, Thung DC, Núñez-Pons L, Rispo F, Efremova J, Mazzella V, Makapedua DM, Calcinai B (2023) New insight into the genus *Cladocroce* (Porifera, Demospongiae) based on morphological and molecular data, with the description of two new species. J Marine Sci Eng 11:1240
37. Putra SA, Ambo-Rappe R, Jompa J, de Voogd NJ (2024) Preliminary study of marine sponges (Porifera) in the littoral of Spermonde Archipelago, Indonesia. Zookeys 1208:275–313
38. Abdillah S, Nurhayati APD, Nurhatika S, Setiawan E, Heffen WL (2013) Cytotoxic and antioxidant activities of marine sponge diversity at Pecaron Bay Pasir Putih Situbondo East Java, Indonesia. J Pharm Res 6:685–689

39. Trianto A, Radjasa OK, Sabdono A, Muchlissin SI, Afriyanto R, Sulistiowati S, Radjasa SK, Crews P, Mccauley E (2019) Exploration culturable bacterial symbionts of sponges from Ternate Islands, Indonesia. Biodiversitas J Biol Divers 20:776–782
40. Julianti E, Singgih M, Ikram M, Naufal A, Putra M, Hadi T (2019) Antimicrobial activity of fungi isolated from the marine sponges collected from Sekotong Beach Lombok, Indonesia. IOP Conf Series Earth Environ Sci 339:012058
41. Sahidin I, Sabandar CW, Hamsidi R, Mardikasari SA, Zubaydah WO, Sadarun B, Musnina WOS, Darmawan A, Sundowo A (2020) Investigation of compounds and biological activity of selected Indonesian marine sponges. Nat Prod J 10:312–321
42. Retnowati Y, Katili AS (2023) Antibacterial activity of sponge-associated bacteria from Torosiaje marine area, Gorontalo, Indonesia. Biodiversitas J Biol Divers 24:1151–1156
43. Prabowo B, Fahlevy K, Subhan B, Santoso P, Hadi TA, Atmaja HE, Hudha F, Andriani Y, Arafat D, Bashari MH (2023) Variation in species diversity and abundance of sponge communities near the human settlement and their bioprospect in Pramuka Island, Jakarta, Indonesia. Aquac Aquarium Conserv Legislation 16:1186–1198
44. Lee SS, Chan JY, LM Teo S, Lambert G (2016) State of knowledge of ascidian diversity in South China Sea and new records for Singapore. Raffles Bull Zool 718–743
45. Litaay M, Santosa S, Johannes E, Agus R, Moka W, Tanjung JDD (2018) Biodiversity of marine tunicates in Samalona waters, Sangkarang Archipelago, Indonesia. Spermonde 4:25–30
46. Litaay M, Piri R, Jabir NB, Priosambodo D, Putra AW (2023) Diversity of marine tunicate from waters of Pannikiang Island and Badi Island of South Sulawesi, Indonesia. Biodiversitas J Biol Divers 24:1431–1437
47. Litaay M (2018) Marine tunicates from Sangkarang Archipelago Indonesia: recent finding and bio-prospecting. In: Journal of physics: conference series. IOP Publishing
48. Piri R, Litaay M, Priosambodo D (2022) Biodiversity of Tunicate (Ascidiacea) in the waters of Badi Island, Pangkajene regency of South Sulawesi. Intl J Appl Biol 6:149155
49. Fabricius K, Alderslade P (2001) Soft corals and sea fans: a comprehensive guide to the tropical shallow water genera of the central-west Pacific, the Indian Ocean and the Red Sea
50. Yan X, Liu J, Leng X, Ouyang H (2021) Chemical diversity and biological activity of secondary metabolites from soft coral genus *Sinularia* since 2013. Mar Drugs 19:335
51. Rodrigues IG, Miguel MG, Mnif W (2019) A brief review on new naturally occurring Cembranoid Diterpene derivatives from the soft corals of the Genera *Sarcophyton*, *Sinularia*, and *Lobophytum* Since 2016. Molecules 24:781
52. Elkhawas YA, Elissawy AM, Elnaggar MS, Mostafa NM, Al-Sayed E, Bishr MM, Singab ANB, Salama OM (2020) Chemical diversity in species belonging to soft coral genus *Sacrophyton* and its impact on biological activity: a review. Mar Drugs 18:41
53. Chen Y-H, Chang Y-C, Chen Y-H, Zheng L-G, Huang P-C, Huynh T-H, Peng B-R, Chen Y-Y, Wu Y-J, Fang L-S, Su J-H, Hsu C-M, Sung P-J (2020) Natural products from octocorals of the genus *Dendronephthya* (Family Nephtheidae). Molecules 25:5957
54. Dautova TN (2011) Biodiversity of soft corals Alcyoniidae (Cnidaria: Octocorallia) and their taxonomy problem. In: Lutaenko KA (eds) Proceedings of the workshop: coastal marine biodiversity and bioresources of Vietnam and adjacent areas to the South China Sea. Nha Trang, Vietnam, Asia-Pacific network for global change research
55. Hoang XB, Dautova TN (2010) Diversity of soft corals (Alcyonacea) in Vietnam. International conference marine biodiversity of East Asian Sea: status, challenges and sustainable development 23:82–87
56. Tixier-Durivault A (1970) Les Octocoralliaires de Nhatrang. Cah Pac 14:115–236
57. Malyutin A (1990) Two new species of *Sinularia* (Octocorallia: Alcyonacea) from south Vietnam. Asian Mar Biol 7:9–14
58. Dautova T, van Ofwegen L, Savinkin O (2010) New species of the genus *Sinularia* (Octocorallia: Alcyonacea) from Nha Trang Bay, South China Sea, Vietnam. Zoologische Meded 84:47–91

59. Hoang XB (2024) The family Xeniidae (Octocorallia: Malacalcyonacea) of Vietnam with two new records. Acad J Biol 46:103–111
60. Chanmethakul T, Chansang H, Watanasit S (2010) Soft coral (Cnidaria: Alcyonacea) distribution patterns in Thai waters. Zool Stud 49:72–84
61. Putra AW, Priawandiputra W, Litaay M, Atmowidi T (2022) Comparison of shallow water soft coral (Octocorallia) diversity and distribution among three islands in Makassar Strait, Indonesia. Biodiversitas J Biol Divers 23:59515961
62. Goh BPL, Tan GE, Tan LT (2009) Diversity, distribution and biological activity of soft corals (Octocorallia, Alcyonacea) in Singapore. J Coast Develop 12:89–98
63. Seah JZS, Yap NWL, Tan LT, Goh BPL (2015) Distribution and abundance of octocoral (Octocorallia, alcyonacea) communities at three Southern Islands of Singapore. Ocean Sci J 50:299–306
64. Mohammad M, Apandi ZNi, Marican HAW, Kamphol N, Mubin NAAA, Salleh S, Ismail MN (2016) The identification of Octocorals from the northern region of straits of Malacca. Trop Life Sci Res 27:87–93
65. Lalas JAA, Lim RTS, Cabasan JP, Segumalian CS, Luciano RMA, Valino DAM, Jacinto MR, Arceo HO, Baria-Rodriguez MV (2022) Spatial and short-term temporal patterns of octocoral assemblages in the West Philippine Sea. Front Mar Sci 8
66. Lane D, Marsh L, Vandenspiegel D, Rowe F (2000) Echinoderm fauna of the South China Sea: an inventory and analysis of distribution patterns. Raffles Bull Zool Suppl 8:459–493
67. Kalinin VI, Silchenko AS (2022) Echinoderms metabolites: structure, functions and biomedical perspectives II. Mar Drugs 20:492
68. Khotimchenko Y (2018) Pharmacological potential of sea cucumbers. Int J Mol Sci 19:1342
69. Dong G, Xu T, Yang B, Lin X, Zhou X, Yang X, Liu Y (2011) Chemical constituents and bioactivities of starfish. Chem Biodivers 8:740–791
70. Pangestuti R, Arifin Z (2018) Medicinal and health benefit effects of functional sea cucumbers. J Tradit Complement Med 8:341–351
71. Hung NQ (2007) Sea cucumber fisheries, utilization, and trade in Vietnam. Report of the regional study on sea cucumber fisheries, utilization and trade in Southeast Asia 2008:113–126
72. Otero MDM, Vu U (2007) Sea cucumber fisheries around Phu Quoc Archipelago: a cross-border issue between South Vietnam and Cambodia. SPC Beche-de-mer Inf Bull 25:32–36
73. Putchakarn S, Sonchaeng P (2004) Echinoderm fauna of Thailand: history and inventory reviews. Sci Asia 30:417–428
74. Setyastuti A, Wirawati I, My I (2018) Identification and distribution of sea cucumber exploited in Lampung, Indonesia. Biodiversitas J Biol Divers 19:676–682
75. Kamarudin KR, Usup G, Hashim R, Rehan MM (2015) Sea Cucumber (Echinodermata: Holothuroidea) species richness at selected localities in Malaysia. Pertanika J Trop Agric Sci 38:7–32
76. Kamarudin KR, Hashim R, Usup G (2010) Phylogeny of sea cucumber (Echinodermata: Holothuroidea) as inferred from 16S mitochondrial rRNA gene sequences. Sains Malaysiana 39:209–218
77. Olavides RDD, Edullantes CMA, Menez MAJ (2010) Assessment of the sea cucumber resource and fishery in the Bolinao-Anda reef system. Sci Diliman 22:1–12
78. Ong JY, Wong HP-S (2015) Sea cucumbers (Echinodermata: Holothuroidea) from the Johor Straits, Singapore. Raffles Bull Zool 31:273–291
79. Ong JY, Wirawati I, Pei-San Wong H (2016) Sea cucumbers (Echinodermata: Holothuroidea) collected from the Singapore Strait. Raffles Bull Zool 34:666–717
80. Lutaenko KA (2016) Biodiversity of bivalve mollusks in the western South China Sea: an overview. In: Adrianov AV, Lutaenko KA (eds) Biodiversity of the western part of the South China Sea. Vladivostok, Dalnauka, p 315–384
81. Willan R, Chou L (2000) Checklist of the nudibranchs (Opisthobranchia: Nudibranchia) of the South China Sea. Raffles Bull Zool Suppl 7:513537
82. Gordon DP (2016) Bryozoa of the South China Sea—an overview. Raffles Bull Zool Suppl 604–618

83. Phang S-M, Yeong H-Y, Ganzon-Fortes ET, Lewmanomont K, Prathep A, Gerung GS, Tan KS (2016) Marine algae of the South China Sea bordered by Indonesia, Malaysia, Philippines, Singapore, Thailand and Vietnam. Raffles Bull Zool 34:13–59
84. Todd PA, Ong X, Chou LM (2010) Impacts of pollution on marine life in Southeast Asia. Biodivers Conserv 19:1063–1082
85. Teh LS, Witter A, Cheung WW, Sumaila UR, Yin X (2017) What is at stake? Status and threats to South China Sea marine fisheries. Ambio 46:57–72
86. Baum JK, Worm B (2009) Cascading top-down effects of changing oceanic predator abundances. J Anim Ecol 78:699–714
87. Vo ST, Pernetta JC, Paterson CJ (2013) Lessons learned in coastal habitat and landbased pollution management in the South China Sea. Ocean Coast Manag 85:230–243
88. Nguyen TTN, Némery J, Gratiot N, Strady E, Tran VQ, Nguyen AT, Aimé J, Peyne A (2019) Nutrient dynamics and eutrophication assessment in the tropical river system of Saigon—Dongnai (southern Vietnam). Sci Total Environ 653:370383
89. Omeyer LCM, Duncan EM, Aiemsomboon K, Beaumont N, Bureekul S, Cao B, Carrasco LR, Chavanich S, Clark JR, Cordova MR, Couceiro F, Cragg SM, Dickson N, Failler P, Ferraro G, Fletcher S, Fong J, Ford AT, Gutierrez T, Shahul Hamid F, Hiddink JG, Hoa PT, Holland SI, Jones L, Jones NH, Koldewey H, Lauro FM, Lee C, Lewis M, Marks D, Matallana-Surget S, Mayorga-Adame CG, McGeehan J, Messer LF, Michie L, Miller MA, Mohamad ZF, Nor NHM, Müller M, Neill SP, Nelms SE, Onda DFL, Ong JJL, Pariatamby A, Phang SC, Quilliam R, Robins PE, Salta M, Sartimbul A, Shakuto S, Skov MW, Taboada EB, Todd PA, Toh TC, Valiyaveettil S, Viyakarn V, Wonnapinij P, Wood LE, Yong CLX, Godley BJ (2022) Priorities to inform research on marine plastic pollution in Southeast Asia. Sci Total Environ 841:156704
90. Kay S, Avillanosa AL, Cheung VV, Dao HN, Gonzales BJ, Palla HP, Praptiwi RA, Queirós AM, Sailley SF, Sumeldan JDC, Syazwan WM, Then AY-H, Wee HB (2023) Projected effects of climate change on marine ecosystems in Southeast Asian seas. Front Mar Sci 10:1082170
91. Sobha TR, Vibija CP, Fahima P (2023) Coral reef: a hot spot of marine biodiversity. In: Sukumaran ST, Keerthi TR (eds) Conservation and sustainable utilization of bioresources. Springer Nature Singapore, Singapore, pp 171–194
92. Lam VWY, Chavanich S, Djoundourian S, Dupont S, Gaill F, Holzer G, Isensee K, Katua S, Mars F, Metian M, Hall-Spencer JM (2019) Dealing with the effects of ocean acidification on coral reefs in the Indian Ocean and Asia. Reg Stud Marine Sci 28:100560
93. Tan LT (2023) Impact of marine chemical ecology research on the discovery and development of new pharmaceuticals. Mar Drugs 21:174
94. Hoi NC, Dang VH (2015) Building a regional network and management regime of marine protected areas in the South China Sea for sustainable development. J Int Wildlife Law Policy 18:128–138
95. Fu X-M, Zhang M-Q, Shao C-L, Li G-Q, Bai H, Dai G-L, Chen Q-W, Kong W, Fu X-J, Wang C-Y (2016) Chinese marine Materia Medica resources: status and potential. Mar Drugs 14:46
96. Chatterji A, Kassim Z, Hassan A, Therwath A, Shaharom F (2010) Marine living resources in the practice of traditional medicine. J Coast Environ 1:4152
97. Sun S, Gao Z, Wang M, Chen S, Guo W, Li X (2024) Chemical constituents of the marine traditional Chinese medicine of *Pegasus laternarius* Cuvier (Hai-E Yu). Pharm Fronts 6:e149–e154
98. Sadeer NB, Gokhan Z, Mahomoodally MF (2023) Biotechnological applications of mangrove plants and their isolated compounds in medicine-a mechanistic overview. Crit Rev Biotechnol 43:393–414
99. El-Beltagi HS, Mohamed AA, Mohamed HI, Ramadan KMA, Barqawi AA, Mansour AT (2022) Phytochemical and potential properties of seaweeds and their recent applications: a review. Mar Drugs 20:342
100. Kalasariya HS, Pereira L (2025) Role of seaweed as a functional ingredients in nutraceuticals, pharmaceuticals, cosmetics, and edible salts. In: Trivedi N, Reddy CRK, Critchley AT (eds) Recent advances in seaweed biotechnology: biomass, emerging applications and bioeconomy. Springer Nature Singapore, Singapore, pp 347–390

101. Titlyanov EA, Titlyanova TV, Pham VH (2012) Stocks and the use of economic marine macrophytes of Vietnam. Russ J Mar Biol 38:285–298
102. Hong DD, Hien HM, Son PN (2007) Seaweeds from Vietnam used for functional food, medicine and biofertilizer. J Appl Phycol 19:817–826
103. Sindo EL, Clores MA, Bigcas EJJ, Dumilag RV (2025) Seaweed ethnobotany of Eastern Camarines Sur, the Philippines. Philippine J Sci 154:27–35
104. Dumilag RV, Belgica THR, Mendoza LC, Hibay JM, Arevalo AE Jr, Malto MAD, Orgela EG, Longavela MR, Corral LEH, Olipany RD (2022) Seaweed ethnobotany of eastern Sorsogon, Philippines. Algae 37:227–237
105. Dumilag RV, Javier RF (2022) Ethnobotany of medicinal seaweeds of Ilocos Norte, Philippines. Philippine J Sci 151
106. Anggadiredja JT (2009) Ethnobotany study of seaweed diversity and its utilization in Warambadi, Panguhalodo areas of East Sumba District. Jurnal Teknologi Lingkungan 10:297–310
107. Han M, Wang Z, Li Y, Song Y, Wang Z (2024) The application and sustainable development of coral in traditional medicine and its chemical composition, pharmacology, toxicology, and clinical research. Front Pharmacol 14:1230608
108. Awale S, Linn TZ, Than MM, Swe T, Saiki I, Kadota S (2006) The healing art of traditional medicines in Myanmar. J Tradit Med 23:47–68
109. Salleh H, Mohamed WN, Mat NHN, Yusof Y, Syamimi N (2020) Traditional medicines from marine resources understanding the consumer's knowledge and perceptions. Int J Adv Appl Sci 7:110–118
110. Mustarichie R, Kelutur FJ (2021) In silico studies: virtual screening of the compound of sea fan (Gorgonia Mariae) as antiasthmatic. Asian J Pharm Res Develop 9:16–23
111. Ngandjui YAT, Kereeditse TT, Kamika I, Madikizela LM, Msagati TAM (2024) Nutraceutical and medicinal importance of marine molluscs. Mar Drugs 22:201
112. Benkendorff K, Rudd D, Nongmaithem BD, Liu L, Young F, Edwards V, Avila C, Abbott CA (2015) Are the traditional medical uses of Muricidae molluscs substantiated by their pharmacological properties and bioactive compounds? Mar Drugs 13:5237–5275
113. Ahmad TB, Liu L, Kotiw M, Benkendorff K (2018) Review of anti-inflammatory, immune-modulatory and wound healing properties of molluscs. J Ethnopharmacol 210:156–178
114. Hamdan DM, Shah JM, Gumpulan F, Foo J, Lukman KA (2020) The North Borneo Iranun's community's ethnomedicine knowledge on marsh clam (Geloina expansa). Asian J Ethnobiol 3
115. Luo B, Nong Y, Zhang T, Zhang S, Hu R (2023) The use and sustainable development of marine animal drugs by the Kinh people in Beibu Gulf. Front Marine Sci 10
116. Arzaman AFM, Salleh HS, Mat NHN, Wan WIA (2023) Traditional knowledge of marine resources and its impact on the well-being of coastal communities in Peninsular Malaysia. Int J Des Nat Ecodyn 18:693–703
117. Hakim A, Burhanuddin B, Siahaan J, Irawan J, Fitri ZN (2024) Development of an ointment to medicate medicine for dosage caused by mucus in babies with the active compound sotong (*Sepia* sp.) and essential oils *Cymbopogon citratus*. Jurnal Pijar Mipa 19:1063–1066
118. Sastre B, del Pozo V (2012) Role of PGE2 in asthma and nonasthmatic eosinophilic bronchitis. Mediators Inflamm 2012:645383
119. Kalor JD, Simaremare ES, Futwembun A, Wabiser G, Gunawan E, Yabansabra YR (2019) Cytotoxic test of *Octopus cyanea* ink extract. J Ecol Eng 20:144–152
120. Rudkin DM, Young GA (2009) Horseshoe crabs—an ancient ancestry revealed. In: Tanacredi JT, Botton ML, Smith D (eds) Biology and conservation of horseshoe crabs. Springer, US, Boston, MA, pp 25–44
121. Ashrafuzzaman M, Razu MH, Showva N-N, Bondhon TA, Moniruzzaman M, Rahman SAR, Rabby MRI, Akter F, Khan M (2022) Biomolecules of the horseshoe crab's hemolymph: components of an ancient defensive mechanism and its impact on the pharmaceutical and biomedical industry. Cell Microbiol 2022:3381162

122. John BA, Nelson BR, Sheikh HI, Cheung SG, Wardiatno Y, Dash BP, Tsuchiya K, Iwasaki Y, Pati S (2018) A review on fisheries and conservation status of Asian horseshoe crabs. Biodivers Conserv 27:3573–3598
123. Mikkelsen T (1988) The secret in the blue blood. Science Press, 124 pp
124. Kiew PL, Don MM (2012) Jewel of the seabed: sea cucumbers as nutritional and drug candidates. Int J Food Sci Nutr 63:616–636
125. Wirawati I, Pratiwi R, Widyastuti E, Ibrahim PS (2021) Commercial sea cucumber trading status in Indonesia. Aquac Aquarium Conserv Legislation 14:3204–3216
126. Rahman MA, Yusoff FM (2017) Sea cucumber fisheries: market potential, trade, utilization and challenges for expanding the production in the South-East Asia. Int J Adv Chem Eng Biol Sci 4:26–30
127. Slater M (2015) Use and exploitation of sea cucumbers. In: Echinoderm aquaculture, pp 57–73
128. Mazlan NB, Abd Rahman NNB, Shukhairi SSB, Nazahuddin MNAB (2023) Sea cucumbers: source of nutritional, medicinal, and cosmeceutical products. In: Shah MD, Ransangan J, Venmathi Maran BA (eds) Marine biotechnology: applications in food, drugs and energy. Springer Nature Singapore, Singapore, pp 171–188
129. Poh Yuen Wen A, Halim AS, Mat Saad AZ, Mohd Nor F, Wan Sulaiman WA (2018) A prospective study evaluating wound healing with sea cucumber gel compared with hydrogel in treatment of skin graft donor sites. Complement Ther Med 41:261–266
130. Thao NP, Cuong NX, Luyen BT, Quang TH, Hanh TT, Kim S, Koh YS, Nam NH, Van Kiem P, Van Minh C, Kim YH (2013) Anti-inflammatory components of the starfish *Astropecten polyacanthus*. Mar Drugs 11:2917–2926
131. Halim M, Salleh H, Mohamed W, Mat N, Yusof Y (2017) Malaysian traditional medicine: the usage of marine resources as a treatment and complementary medicine for heart disease. J Fundam Appl Sci 9:816–827
132. Wulandari DA, Warsito MF (2022) Nutritional value and health benefit of Sea Urchin. Omni-Akuatika 18:101–111
133. Sibiya A, Jeyavani J, Sivakamavalli J, Ravi C, Divya M, Vaseeharan B (2021) Bioactive compounds from various types of sea urchin and their therapeutic effects—a review. Reg Stud Mar Sci 44:101760
134. Foster SJ, Vincent ACJ (2004) Life history and ecology of seahorses: implications for conservation and management. J Fish Biol 65:1–61
135. Rosa IL, Defavari GR, Alves RRN, Oliveira TPR (2013) Seahorses in traditional medicines: a global overview. In: Alves RRN, Rosa IL (eds) Animals in traditional folk medicine: implications for conservation. Springer, Berlin, Heidelberg, pp 207–240
136. Kumaravel K, SR, TB, Sonneschein L (2012) Seahorses—a source of traditional medicine. Nat Prod Res 26:2330–2334
137. May B, Tomoda T (2002) Seahorses in the Ben cao gang mu and contemporary Chinese medicine. J Aust Chin Med Educ Res Counc 7:2–13
138. Chen T-H, Chang H-C, Lue K-Y (2009) Unregulated trade in turtle shells for Chinese traditional medicine in East and Southeast Asia: the case of Taiwan. Chelonian Conserv Biol 8:11–18
139. Poti M, Long SL, Rusli MU, Jani JM, Hugé J, Dahdouh-Guebas F (2021) Changing trends and perceptions of sea turtle egg consumption in Redang Island, Malaysia. Ecol Soc 26:14
140. Mardiastuti A, Masy'ud B, Ginoga L, Sastranegara H (2021) Traditional uses of herpetofauna practiced by local people in the island of Sumatra, Indonesia: implications for conservation. IOP Conf Series Earth Environ Sci 762:012003
141. Hemelikova A, Chajma P, Ferasyi TR, Awaluddin A, Fadli N, Sari W, Hof C, Riskas K, Vojar J, Sea turtle exploitation in Sumatra, Indonesia: disentangling the roles of religion, culture and socio-demographics to support effective conservation. Indonesia: disentangling the roles of religion, culture and socio-demographics to support effective conservation
142. Campbell LM (1998) Use them or lose them? Conservation and the consumptive use of marine turtle eggs at Ostional, Costa Rica. Environ Conserv 25:305319

143. Strang K, Rusli N (2021) The challenges of conserving biodiversity: a spotlight on Southeast Asia. In: Underkoffler SC, Adams HR (eds) Wildlife biodiversity conservation: multidisciplinary and forensic approaches. Springer, Cham, pp 47–66
144. Panyawai J, Prathep A (2022) A systematic review of the status, knowledge, and research gaps of dugong in Southeast Asia. Aquat Mamm 48:203–222
145. Rajamani L, Cabanban AS, Rahman RA (2006) Indigenous use and trade of Dugong (*Dugong dugon*) in Sabah, Malaysia. AMBIO: A J Hum Environ 35:266–268
146. Alves RRN, Souto WMS, Oliveira REMCC, Barboza RRD, Rosa IL (2013) Aquatic mammals used in traditional folk medicine: a global analysis. In: Alves RRN, Rosa IL (eds) Animals in traditional folk medicine: implications for conservation. Springer, Berlin, Heidelberg, pp 241–261
147. Ashwell D, Walston N (2008) An overview of the use and trade of plants and animals in traditional medicine systems in Cambodia, 1st edn. TRAFFIC Southeast Asia, Greater Mekong Programme, Ha Noi, Vietnam
148. Beasley I, Davidson P, Somany P, Samanth L (2001) Abundance, distribution and conservation management of marine mammals in Cambodia's coastal waters. Wildlife Conservation Society, Phnom Penh, Cambodia
149. Gono CMP, Ahmadi P, Hertiani T, Septiana E, Putra MY, Chianese G (2022) A comprehensive update on the bioactive compounds from seagrasses. Mar Drugs 20:406
150. Nindatu M, Noya F, Seimahuira T, Kaya E, Wakano D, Leasa M (2016) Potential of Lamun (*Enhalus acoroides*) seeds from West Seram Coastal area as natural antioxidant. Res J Pharm Biol Chem Sci 7:66–71
151. Wuthidhammvej W (1997) Herbal encyclopedia and a theory of Thai traditional pharmacy. OA Printing, Bangkok
152. Klangprapun S, Buranrat B, Caichompoo W, Nualkaew S (2018) Pharmacognostical and physicochemical studies of *Enhalus acoroides* (L.F.) Royle (Rhizome). Pharmacognosy J 10:s89–s94
153. Noor NM, Febriani D, Ali M (2022) Seagrass of enhalus acoroides as a traditional body scrubs in preventing Malarial bites by Pahawang Island community in Indonesia. IOP Conf Series Earth Environ Sci 1012:012037
154. Digdo AA, Astari E, Arinda BR, Arendege J (2023) Humans and seagrass: a complex and intertwining links—an illustration from North Sulawesi, Indonesia. IOP Conf Series Earth Environ Sci 1220:012027
155. Zulkifli L, Muksin YD, Hartanto P, Desimarlina Y, Idrus AA, Syukur A (2021) Phytochemical profiles and ethnomedicine preliminary studies on seagrass species in the Southern Coast of Lombok Island Indonesia. IOP Conf Series Earth Environ Sci 913:012102
156. Bandaranayake WM (1998) Traditional and medicinal uses of mangroves. Mangrove Salt Marshes 2:133–148
157. Chan E, Ng YK, Wong S, Chan H (2022) *Pluchea indica*: an updated review of its botany, uses, bioactive compounds and pharmacological properties. Pharm Sci Asia 49:77–85
158. Neamsuvan O, Singdam P, Yingcharoen K, Sengnon N (2012) A survey of medicinal plants in mangrove and beach forests from sating Phra Peninsula, Songkhla Province, Thailand. J Med Plants Res 6:2421–2437
159. Herningtyas W, Rahmila YI, Pujiono E (2023) Mangrove plants as traditional medicine by local coastal communities in Indonesia. IOP Conf Series Earth Environ Sci 1266:012006
160. Duryat D, Yuwono SB, Riniarti M, Hidayat KF, Hidayat W, Rodiani R, Damai AA (2024) Mangrove species diversity and its use as medicinal plant by coastal communities of Lampung Timur Indonesia. J Multidisc Appl Nat Sci 4:304–318
161. Tamalene MN, Uday UK, Bhakat RK, Vianti E, Bahtiar S (2021) Utilization of mangrove plants as a source of malaria medicine in north Maluku province, Indonesia. Asian J Ethnobiol 4:86–92
162. Suffness M, GCG, MGM, FGF, GJ, RMJA, SSS, JVJ, Wolpert M (1995) The national cooperative natural products drug discovery group (NCNPDDG) and international cooperative biodiversity group (ICBG) programs. Int J Pharm 33:5–16

163. Blunt JW, Copp BR, Munro MH, Northcote PT, Prinsep MR (2005) Marine natural products. Nat Prod Rep 22:15–61
164. Blunt JW, Copp BR, Munro MH, Northcote PT, Prinsep MR (2006) Marine natural products. Nat Prod Rep 23:26–78
165. Blunt JW, Copp BR, Hu W-P, Munro MH, Northcote PT, Prinsep MR (2007) Marine natural products. Nat Prod Rep 24:31–86
166. Blunt JW, Copp BR, Hu W-P, Munro MHG, Northcote PT, Prinsep MR (2008) Marine natural products. Nat Prod Rep 25:35–94
167. Blunt JW, Copp BR, Hu W-P, Munro MH, Northcote PT, Prinsep MR (2009) Marine natural products. Nat Prod Rep 26:170–244
168. Blunt JW, Copp BR, Munro MH, Northcote PT, Prinsep MR (2010) Marine natural products. Nat Prod Rep 27:165–237
169. Blunt JW, Copp BR, Munro MH, Northcote PT, Prinsep MR (2011) Marine natural products. Nat Prod Rep 28:196–268
170. Blunt JW, Copp BR, Keyzers RA, Munro MH, Prinsep MR (2012) Marine natural products. Nat Prod Rep 29:144–222
171. Blunt JW, Copp BR, Keyzers RA, Munro MH, Prinsep MR (2013) Marine natural products. Nat Prod Rep 30:237–323
172. Blunt JW, Copp BR, Keyzers RA, Munro MH, Prinsep MR (2014) Marine natural products. Nat Prod Rep 31:160–258
173. Blunt JW, Copp BR, Keyzers RA, Munro MH, Prinsep MR (2015) Marine natural products. Nat Prod Rep 32:116–211
174. Blunt JW, Copp BR, Keyzers RA, Munro MH, Prinsep MR (2016) Marine natural products. Nat Prod Rep 33:382–431
175. Blunt JW, Copp BR, Keyzers RA, Munro MH, Prinsep MR (2017) Marine natural products. Nat Prod Rep 34:235–294
176. Blunt JW, Carroll AR, Copp BR, Davis RA, Keyzers RA, Prinsep MR (2018) Marine natural products. Nat Prod Rep 35:8–53
177. Carroll AR, Copp BR, Davis RA, Keyzers RA, Prinsep MR (2019) Marine natural products. Nat Prod Rep 36:122–173
178. Carroll AR, Copp BR, Davis RA, Keyzers RA, Prinsep MR (2020) Marine natural products. Nat Prod Rep 37:175–223
179. Carroll AR, Copp BR, Davis RA, Keyzers RA, Prinsep MR (2021) Marine natural products. Nat Prod Rep 38:362–413
180. Carroll AR, Copp BR, Davis RA, Keyzers RA, Prinsep MR (2022) Marine natural products. Nat Prod Rep 39:1122–1171
181. Carroll AR, Copp BR, Davis RA, Keyzers RA, Prinsep MR (2023) Marine natural products. Nat Prod Rep 40:275–325
182. Carroll AR, Copp BR, Grkovic T, Keyzers RA, Prinsep MR (2024) Marine natural products. Nat Prod Rep 41:162–207
183. Leal MC, Anaya-Rojas JM, Munro MHG, Blunt JW, Melian CJ, Calado R, Lürig MD (2020) Fifty years of capacity building in the search for new marine natural products. Proceed Nat Acad Sci 117:24165–24172
184. Zhou X, Xu T, Yang XW, Huang R, Yang B, Tang L, Liu Y (2010) Chemical and biological aspects of marine sponges of the genus *Xestospongia*. Chem Biodivers 7:2201–2227
185. Khodzori FA, Mazlan NB, Chong WS, Ong KH, Palaniveloo K, Shah MD (2023) Metabolites and bioactivity of the marine *Xestospongia* sponges (Porifera, Demospongiae, Haplosclerida) of Southeast Asian waters. Biomolecules 13:484
186. Santiago VS, Manzano GG, Yu CC, Aliño PM, Salvador-Reyes LA (2019) Mariculture potential of renieramycin-producing Philippine blue sponge *Xestospongia* sp. (Porifera: Haplosclerida). Aquaculture 502:356–364
187. Montalvo NF, Mohamed NM, Enticknap JJ, Hill RT (2005) Novel actinobacteria from marine sponges. Antonie Van Leeuwenhoek 87:29–36

188. Hayami Y, Ambalavanan L, Zainathan SC, Danish-Daniel M, Sharifah NE, Iehata S (2023) Diversity and functional roles of the symbiotic microbiome associated to marine sponges off Karah Island, Terengganu, Malaysia. Reg Stud Mar Sci 62:102967
189. Tai BH, Hang DTT, Thung DC, Trang DT, Yen PH, Yen DTH, Huong PTT, Dung DT, Cuong NC, Thao DT, Nhiem NX, Van Kiem P (2022) Xestospongiene VN1, a new brominated polyunsaturated lipid from marine sponge *Xestospongia testudinaria* (Lamark, 1815). Vietnam J Chem 60:540–545
190. Vu Luu P, Hien MN, Phuc MP, An DV, Ton-Nu HL (2024) Testusterol, a new sterol of the sponge species *Xestospongia testudinaria* from Phu Quoc island, Vietnam. Nat Prod Res 39:1–9
191. Murtihapsari M, Supriatno S, Dikdik K, Darwati D, Kadarusman K, Fauzi AF, Tati H, Hafiz HM, Khalijah A, Yoshihito S, Nurul AM, Supratman U (2021) A new antiplasmodial sterol from Indonesian marine sponge, *Xestospongia* sp. Nat Prod Res 35:937–944
192. Nguyen XC, Longeon A, Pham VC, Urvois F, Bressy C, Trinh TTV, Nguyen HN, Phan VK, Chau VM, Briand J-F, Bourguet-Kondracki M-L (2013) Antifouling 26,27cyclosterols from the Vietnamese marine sponge *Xestospongia testudinaria*. J Nat Prod 76:1313–1318
193. Dung DT, Thuy HDT, Hai YP, Hong QT, Xuan NN, Huu TB, Van MC, Youn-Chul K, Cheol KD, Hyuncheol O, Kiem PV (2019) Macrocyclic bis-quinolizidine alkaloids from *Xestospongia muta*. Nat Prod Res 33:400–406
194. Daikuhara N, Tada Y, Yamaki S, Charupant K, Amnuoypol S, Suwanborirux K, Saito N (2009) Chemistry of renieramycins. Part 7: renieramycins T and U, novel renieramycin–ecteinascidin hybrid marine natural products from Thai sponge *Xestospongia* sp. Tetrahedron Lett 50:4276–4278
195. Suwanborirux K, Amnuoypol S, Plubrukarn A, Pummangura S, Kubo A, Tanaka C, Saito N (2003) Chemistry of renieramycins. Part 3. Isolation and structure of stabilized renieramycin type derivatives possessing antitumor activity from Thai sponge *Xestospongia* species, pretreated with potassium cyanide. J Nat Prod 66:1441–1446
196. Tasdemir D, Marshall KM, Mangalindan GC, Concepción GP, Barrows LR, Harper MK, Ireland CM (2001) Deoxyamphimedine, a new pyridoacridine alkaloid from two tropical *Xestospongia* sponges. J Org Chem 66:3246–3248
197. Centko RM, Steinø A, Rosell FI, Patrick BO, de Voogd N, Mauk AG, Andersen RJ (2014) Indoleamine 2,3-dioxygenase inhibitors isolated from the sponge *Xestospongia vansoesti*: structure elucidation, analogue synthesis, and biological activity. Org Lett 16:6480–6483
198. Dai J, Sorribas A, Yoshida WY, Kelly M, Williams PG (2010) Xestosaprols from the Indonesian marine sponge *Xestospongia* sp. J Nat Prod 73:11881191
199. Concepción GP, Foderaro TA, Eldredge GS, Lobkovsky E, Clardy J, Barrows LR, Ireland CM (1995) Topoisomerase II-mediated DNA cleavage by adocia- and xestoquinones from the Philippine sponge *Xestospongia* sp. J Med Chem 38:45034507
200. Edrada RA, Proksch P, Wray V, Witte L, Müller WEG, Van Soest RWM (1996) Four new bioactive manzamine-type alkaloids from the Philippine marine sponge *Xestospongia ashmorica*. J Nat Prod 59:1056–1060
201. Kalinin VI, Silchenko AS, Avilov SA, Stonik VA (2021) Progress in the studies of triterpene glycosides from Sea Cucumbers (Holothuroidea, Echinodermata) Between 2017 and 2021. Nat Prod Commun 16:1934578X211053934
202. Yurchenko EA, Yurchenko AN, Van Minh C, Aminin DL (2019) Achievements in the study of marine low-molecular weight biologically active metabolites from the Vietnamese territorial waters as a result of expeditions aboard the research vessel 'Akademik Oparin' (2004–2017). Chem Biodivers 16:e1800654
203. Luzzatto-Knaan T, Garg N, Wang M, Glukhov E, Peng Y, Ackermann G, Amir A, Duggan BM, Ryazanov S, Gerwick L, Knight R, Alexandrov T, Bandeira N, Gerwick WH, Dorrestein PC (2017) Digitizing mass spectrometry data to explore the chemical diversity and distribution of marine cyanobacteria and algae. eLife 6:e24214
204. Pereira A, Cao Z, Murray TF, Gerwick WH (2009) Hoiamide A, a Sodium channel activator of unusual architecture from a consortium of two Papua New Guinea cyanobacteria. Chem Biol 16:893–906

205. Han B, Gross H, Goeger DE, Mooberry SL, Gerwick WH (2006) Aurilides B and C, cancer cell toxins from a Papua New Guinea collection of the marine cyanobacterium *Lyngbya majuscula*. J Nat Prod 69:572–575
206. Gutiérrez M, Suyama TL, Engene N, Wingerd JS, Matainaho T, Gerwick WH (2008) Apratoxin D, a potent cytotoxic cyclodepsipeptide from Papua New Guinea collections of the marine cyanobacteria *Lyngbya majuscula* and *Lyngbya sordida*. J Nat Prod 71:1099–1103
207. Han B, McPhail KL, Gross H, Goeger DE, Mooberry SL, Gerwick WH (2005) Isolation and structure of five lyngbyabellin derivatives from a Papua New Guinea collection of the marine cyanobacterium *Lyngbya majuscula*. Tetrahedron 61:1172311729
208. Han B, McPhail KL, Ligresti A, Di Marzo V, Gerwick WH (2003) Semiplenamides A−G, fatty acid amides from a Papua New Guinea collection of the marine cyanobacterium *Lyngbya semiplena*. J Nat Prod 66:1364–1368
209. Pereira AR, McCue CF, Gerwick WH (2010) Cyanolide A, a glycosidic macrolide with potent molluscicidal activity from the Papua New Guinea cyanobacterium *Lyngbya bouillonii*. J Nat Prod 73:217–220
210. Nogle LM, Marquez BL, Gerwick WH (2003) Wewakazole, a novel cyclic dodecapeptide from a Papua New Guinea *Lyngbya majuscula*. Org Lett 5:3–6
211. Linington RG, Edwards DJ, Shuman CF, McPhail KL, Matainaho T, Gerwick WH (2008) Symplocamide A, a potent cytotoxin and chymotrypsin inhibitor from the Marine Cyanobacterium Symploca sp. J Nat Prod 71:22–27
212. Olivera BM, Cruz LJ (2001) Conotoxins, in retrospect. Toxicon 39:7–14
213. Concepcion GP, Anas ARJ, Azcuna MA (2014) Anticancer compounds from Philippine marine organisms act on major pathways in cancer. Philippine Sci Lett 7:207–227
214. Ireland C, William A, Raymond A, Semiramis A-K, Roberto B, Valerie B, Guy C, Alice C, Jon C, Gisela C, E. DS, Carolyn D, Tito F, Philip F, Donna G, Lee G, Michael G, Kay HM, Robert M, Frank L, Maria N, Marianne P, Zask A (2003) Anticancer agents from unique natural products sources. Pharm Biol 41:15–38
215. Concepcion GP (2008) Marine biodiversity and marine biotechnology: a Philippine competitive advantage. Selected essays on science and technology for securing a better Philippines. Department of Science and Technology
216. Sambodo LAAT (2023) Indonesia blue economy roadmap. Ministry of National Development Planning/National Development Planning Agency: Indonesia, 233 p
217. Baark E (2020) Basic research and national goals: the Vietnam Academy of Science and technology in an emerging innovation system. J Asian Dev Stud 9:1937
218. Jamaludin NA, Bakar K, Saidin J (2023) In vitro biological activity of three marine sponges from *Theonella* and *Haliclona* Genera collected from Bidong Island, Terengganu, Malaysia. Malays Appl Biol 52:51–59
219. Gazali M, Nur Izzah Alissa Wan Rosli W, Sarah Mohd Zahidi S, Amira Jamaludin N, Bakar K, Saidin J (2024) Investigation of antibiofilm activities from Marine Sponges' crude extract of Pulau Bidong, Terengganu, Malaysia. BIO Web Conf 147:01008
220. Rahman NIA, Ramzi MM, Rawi NN, Siong JYF, Bakar K, Bhubalan K, Ariffin F, Saidin J, Azemi AK, Ismail N (2024) Characterization of antibiofilm compound from marine sponge *Stylissa carteri*. Environ Sci Pollut Res 31:37552–37563
221. Shah MD, Venmathi Maran BA, Shaleh SRM, Zuldin WH, Gnanaraj C, Yong YS (2022) Therapeutic potential and nutraceutical profiling of North Bornean seaweeds: a review. Mar Drugs 20:101
222. Zainuddin N, Alias SA, Lee CW, Ebel R, Othman NA, Mukhtar MR, Awang K (2010) Antimicrobial activities of marine fungi from Malaysia. Bot Mar 53:507–513
223. Paliany AS, Sivasothy Y, Awang K, Rizman-Idid M, Alias SA (2014) Marine derived fungi of peninsular Malaysia—a biochemical perspective. Chiang Mai J Sci 41:894909
224. Lim LPL, Gregory MJ (2004) Singapore's biomedical science sector development strategy: Is it sustainable? J Commer Biotechnol 10:352–362
225. Lim TM, Chua KL (1994) Marine biotechnology in Singapore. Australas Biotechnol 4:226–228

226. Ng SB, Kanagasundaram Y, Fan H, Arumugam P, Eisenhaber B, Eisenhaber F (2018) The 160K Natural Organism Library, a unique resource for natural products research. Nat Biotechnol 36:570–573
227. Woo S-Y, Win NN, Wong CP, Ito T, Hoshino S, Ngwe H, Aye AA, Han NM, Zhang H, Hayashi F, Abe I, Morita H (2018) Two new pyrrolo-2-aminoimidazoles from a Myanmarese marine sponge, *Clathria prolifera*. J Nat Med 72:803807
228. Win NN, Kodama T, Aye AA, Lae KZW, Ngwe H, Han NM, Abe I, Morita H (2021) Pyrrolactams from marine sponge *Stylissa massa* collected from Myanmar and their anti-Vpr activities. Chem Pharm Bull 69:702–705
229. Jeffries B, Signs M (2024) New species discoveries in the Greater Mekong
230. Netjes J, Afoullouss S, Rocca J, Quattrini A, Keel W, Baker BJ (2023) Old samples, new tricks: modern natural product methodologies for century-old museum collections. In: XVII international symposium on marine natural products & XIII European conference on marine natural products. Granada, Spain
231. Dewi AS, Tarman K, Uria AR, Greifswald E (2008) Marine natural products: prospects and impacts on the sustainable development in Indonesia. In: Proceeding of Indonesian students' scientific meeting, Delft, The Netherlands
232. Naman CB, Leber CA, Gerwick WH (2017) Modern natural products drug discovery and its relevance to biodiversity conservation. In: Kurtböke I (ed) Microbial resources: from functional existence in nature to applications. Academic Press, London, UK, pp 103–120
233. Medina RA, Goeger DE, Hills P, Mooberry SL, Huang N, Romero LI, Ortega-Barría E, Gerwick WH, McPhail KL (2008) Coibamide A, a potent antiproliferative cyclic depsipeptide from the panamanian marine cyanobacterium *Leptolyngbya* sp. J Am Chem Soc 130:6324–6325
234. Sun W, Wu W, Liu X, Zaleta-Pinet DA, Clark BR (2019) Bioactive Compounds isolated from marine-derived microbes in China: 2009–2018. Mar Drugs 17:339
235. Carroll AR, Copp BR, Grkovic T, Keyzers RA, Prinsep MR (2025) Marine natural products. Nat Prod Rep 42:257–297
236. Demunshi Y, Chugh A (2010) Role of traditional knowledge in marine bioprospecting. Biodivers Conserv 19:3015–3033
237. Bhatia P, Chugh A (2015) Role of marine bioprospecting contracts in developing access and benefit sharing mechanism for marine traditional knowledge holders in the pharmaceutical industry. Glob Ecol Conserv 3:176–187
238. Ragozzino C, Casella V, Coppola A, Scarpato S, Buonocore C, Consiglio A, Palma Esposito F, Galasso C, Tedesco P, Della Sala G, de Pascale D, Vitale L, Coppola D (2025) Last decade insights in exploiting marine microorganisms as sources of new bioactive natural products. Mar Drugs 23:116
239. Sweeney D, Bogdanov A, Chase AB, Castro-Falcón G, Trinidad-Javier A, Dahesh S, Nizet V, Jensen PR (2024) Pattern-based genome mining guides discovery of the antibiotic indanopyrrole A from a marine *Streptomycete*. J Nat Prod 87:2768–2778
240. Tapilatu YH (2015) Status of drug discovery research based on marine organisms from Eastern Indonesia. Procedia Chem 14:484–492
241. Tanaka J (2020) How can we develop marine natural products chemistry in Indonesia? J Phys Conf Ser 1460:012079
242. Mustafa S, Estim A, Shaleh SRM (2020) A call for open access for marine bioprospecting. Environ Policy Law 49:232–236
243. Hanif N, Miftah JA, Yanti HD, Oluwabusola ET, Zahra VA, Salleh NF, Kundukad B, Tan LT, Voogd NJ, Rachmania N, Jaspars M, Kjelleberg S, Noviendri D, Murni A, Tanaka J (2025) Integrated biological and chemical investigation of Indonesian marine organisms targeting anti-quorum-sensing, anti-biofilm, anti-biofouling, and anti-biocorrosion activities. Molecules 30:1202

Chapter 2
Chemical Evolution of Indonesian Marine Natural Products

Novriyandi Hanif
and Muhammad Naufal Khotmi Ramadhan Ashari Jaya Sudrajat

1 Introduction

Indonesia is located at one of the centers of biodiversity hotspots [1], endowed with a coral reef triangle, and is an epicenter of marine biodiversity covering 10.36 million km^2 of ocean and coastal waters surrounding Indonesia, Malaysia, Papua New Guinea, the Philippines, Timor Leste, and Australia [2]. The country consists of 38 provinces (Fig. 1) [3] and surrounding water areas encompassing 5.79 million km^2, of which 1.08% or 62,600 km^2 is protected as national parks [4]. All the province has the coastline except for highland Papua (HPA) indicating maritime potency of each province. The features of this area include not only species richness, endemism and habitat diversity, but their relatively pristine condition. Because of high biodiversity and hence high competition for survival for million years within special environment, marine natural products (MNPs) exhibit unique chemical structures and biological activities different from those of traditionally investigated terrestrial natural products. These molecules have been biosynthesized by a vast range of enzymatic chemical reactions giving unique structures including building blocks, skeletons, scaffolds, functional groups, and chiral properties. As a consequence, more than 850 new molecules have been discovered from Indonesian waters since 1972. The source of Indonesian MNPs is diverse ranging from sponges, cnidarians, tunicates, molluscs, algae, marine-sourced bacteria, marine-sourced fungi, cyanobacteria, dinoflagellates, ciliates, and mangroves. The biosynthesized MNPs are mainly used for chemical defense and/or chemical communication by marine organisms, while human beings can utilize MNPs for drugs, chemical probes, and other scientific developments in the

N. Hanif (✉) · M. N. K. R. A. J. Sudrajat
Department of Chemistry, Faculty of Mathematics and Natural Sciences, IPB University, Bogor, Indonesia
e-mail: nhanif@apps.ipb.ac.id

M. N. K. R. A. J. Sudrajat
e-mail: naufalkrajsudrajat@apps.ipb.ac.id

L. T. Tan and N. Hanif (eds.), *Marine Natural Products Research in South-East Asia*,
https://doi.org/10.1007/978-3-032-01098-8_2

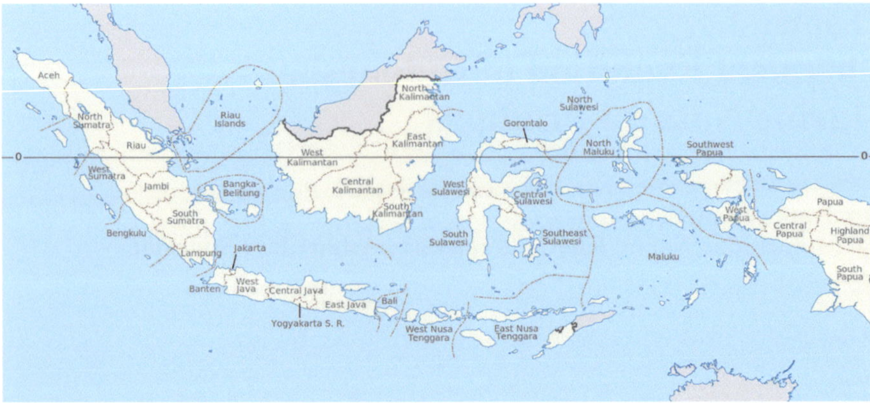

Fig. 1 Map of Indonesia and its 38 provinces [3] [Aceh (ACH), North Sumatra (NST), Riau (RAU), Riau Islands (RUI), West Sumatra (WST), Jambi (JMI), Bengkulu (BKU), South Sumatra (SST), Bangka-Belitung Islands (BBI), Lampung (LPG), Banten (BTN), Jakarta Special Capital Region (JSCR), West Java (WJV), Central Java (CJV), Special Region of Yogyakarta (SRY), East Java (EJV), Bali (BLI), WNT (West Nusa Tenggara), ENT (East Nusa Tenggara), West Kalimantan (WKM), Central Kalimantan (CKM), East Kalimantan (EKM), South Kalimantan (SKM), North Kalimantan (NKM), North Sulawesi (NSW), Gorontalo (GTO), West Sulawesi (WSW), Central Sulawesi (CSW), South Sulawesi (SSW), Southeast Sulawesi (SES), North Maluku (NMU), Maluku (MLU), Southwest Papua (SWP), West Papua (WPA), Central Papua (CPA), Papua (PUA), Highland Papua (HPA), and South Papua (SPA). The code of UEP (unidentified exact province) was used when the collection place was not defined or concluded]

field of maritime, agricultural, medicine, material, energy, biology, and environment. Considering the number of MNP discovered and novel sources of marine organisms, the contribution of new Indonesian MNPs is about 2% of total new MNPs discovered to date [5–12] indicating the underexplored research in marine natural products in Indonesia.

To provide a current status of Indonesian MNPs, we update the literature review [13] of new chemical structures found in Indonesian waters in the perspective of chemistry: isolation, structure determination, and potent biological activities from marine macro and microorganisms published in the period of January 1972–December 2024. Moreover, supply challenge of talented MNPs is briefly identified. Classification of MNPs is updated based on their biosynthetic pathways either from experimental or plausible data. This is included for revisited terpenoid classification [14, 15]. As a result, distribution of Indonesian MNPs based on structural types can be classified into 6 different structural types: 338 terpenoids, 329 alkaloid derived amino acid, 85 peptides, 11 fatty acids, 133 polyketides, and 15 miscellaneous unique structures. The present work covers a total of 911 molecules in which approximately 879 molecules are counted as natural products. The chemical structures of Indonesian MNPs are disclosed with a variety of spectroscopic methods (NMR, MS, IR, UV, ECD, X-ray diffraction, optical rotation), chemical transformations (derivatization or selective degradation, total synthesis), molecular modelling, DFT-based NMR, ECD, optical rotation calculations, molecular networking, biosynthetic consideration, and analysis of microbial genomes. Disclosing of absolute configurations (ACs) and certain relative configurations (RCs) using *J*-based configurational analysis (JBCA) and DFT-based calculation are pointed out in the current review. The source of Indonesian MNPs is approximately from 108 species, 116 genera, 76 families, 37 orders, 17 classes, 12 phyla, and 5 kingdoms. The potent bioactivity criteria [16] of Indonesian MNPs (reported as IC_{50}, EC_{50}, or GI_{50} < 0.1 μM; MIC < 1.0 μg/mL) is also highlighted. Consequently, around 37 MNPs are classified as potent activities, which consisted of cytotoxic (7 molecules), antiproliferative (9 molecules), anticancer (1 molecule), antibacterial (18 molecules), antiparasite (1 molecule), and immunomodulator (1 molecule) activities.

2 Terpenoid

The organization of Indonesian marine terpenoid is updated based on the analysis of their plausible biosynthetic pathways. Structurally, terpenoids are composed of C_5 (isoprene) units. Terpenoid with a mixed biosynthetic pathway is classified as meroterpenoid [14, 15]. Meroterpenoid class can be a mixed biosynthesis of polyketide-terpenoid, non-ribosomal peptide (NRPS)-terpenoid, terpenoid-alkaloid, and others (C_1 or C_2 unit-terpenoid). A total of 338 marine terpenoids are isolated from Indonesian marine organisms consisting of 26 sesquiterpenoids, 35 diterpenoids, 24 sesterterpenoids, 10 triterpenoids and steroids as well as 243 meroterpenoids. The meroterpenoid class can be further divided into five subclasses: merosesquiterpenoid (61 molecules), meroditerpenoid (70 molecules), merosestert-erpenoid (37 molecules), merotriterpenoid (27 molecules), and merosteroid (48 molecules).

Among the 92 skeletal types (Fig. 2) encountered in the terpenoid metabolites, approximately 27 are unprecedented skeletons covering 4 new skeletons C_{15} (lamellodysidane, citreane, dysideane, lamellolane), 9 new skeletons C_{20} (trinordolabellane, seco-norisocopalane, litosetoane, spirolitosetoane, niphatane, dinorcembrane, norphytane, butonane/prenyllingshuiane, carnusane), 11 new skeletons C_{25} (acanthodendrane, acanthane, 4,24-cylcogeranylfarnesane, 3-nor-14,19-cyclogeranylfarnesane, psammociniane, nonanorprenylhalimane, habdosane/norprenylclerodane, 21-homoscalarane, 20,24-bishomoscalarane/honulactane, nor-20,24-bishomoscalarane/honuane, 22,24-bishomoscalarane), and 3 new skeleton C_{30} (haliclonane, vannusane, cortistane/*abeo*-9(10–19) cholestane) (Fig. 2). Of the 92 skeletal types, 57 skeletons are only discovered in marine life forms, 2 skeletons are derived from marine environment, and the rest are found in terrestrial and marine organisms.

From the analysis of carbon skeletons, the structural complexity of the Indonesian marine terpenoids is derived from relatively simple biosynthetic roots: regular and irregular coupling reactions of C_5 precursors giving a variety of molecular structures as linear and or cyclic (fused, bridged, spiro) entities. Moreover, rearrangement is

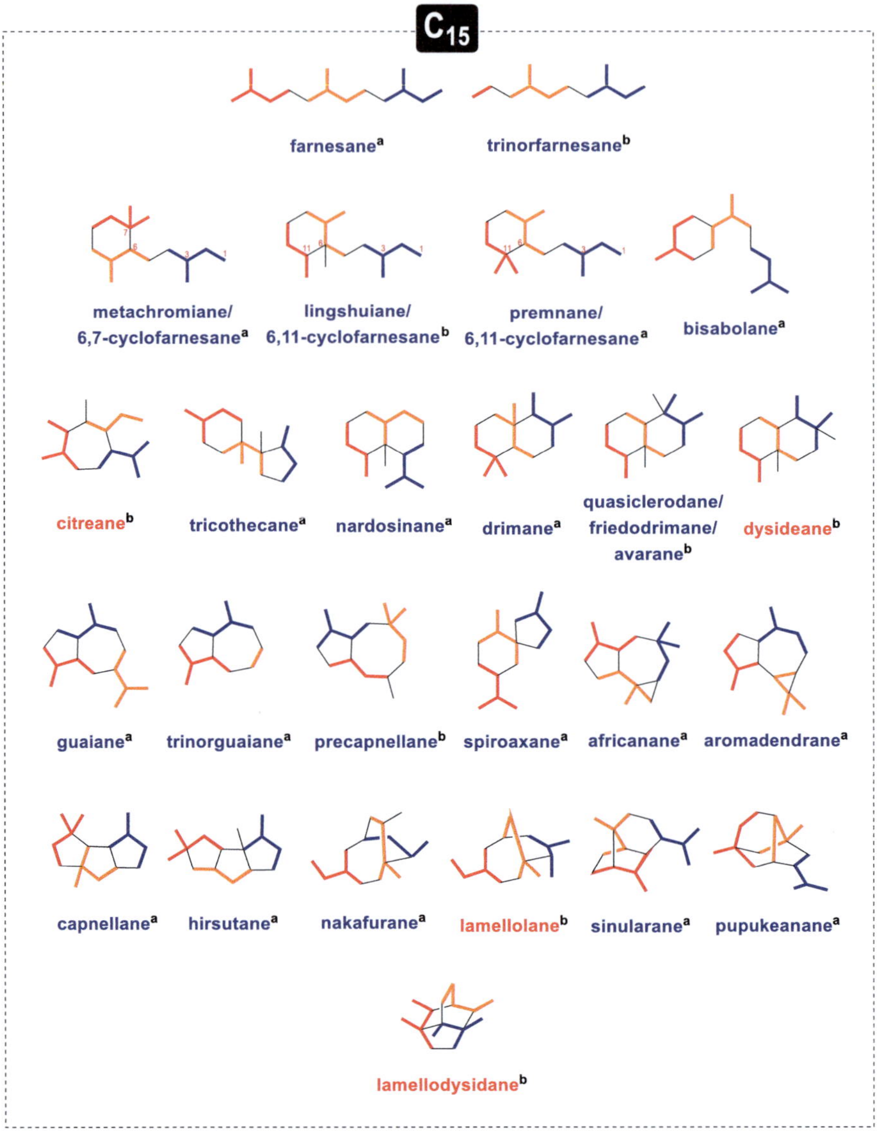

Fig. 2 Terpenoid (C_{15}, C_{20}, C_{25}, and C_{30}) skeletons found in Indonesian marine organisms. *Note* [a] is a skeleton found in marine and terrestrial organisms, [b] is a skeleton found only in marine organisms, [c] is a skeleton derived from marine environment. Isoprene unit or its degraded form is given in different color. Skeleton names in blue were already reported in the literature, those in red were new skeleton and named in this review. Skeleton names in green were known and named in this review

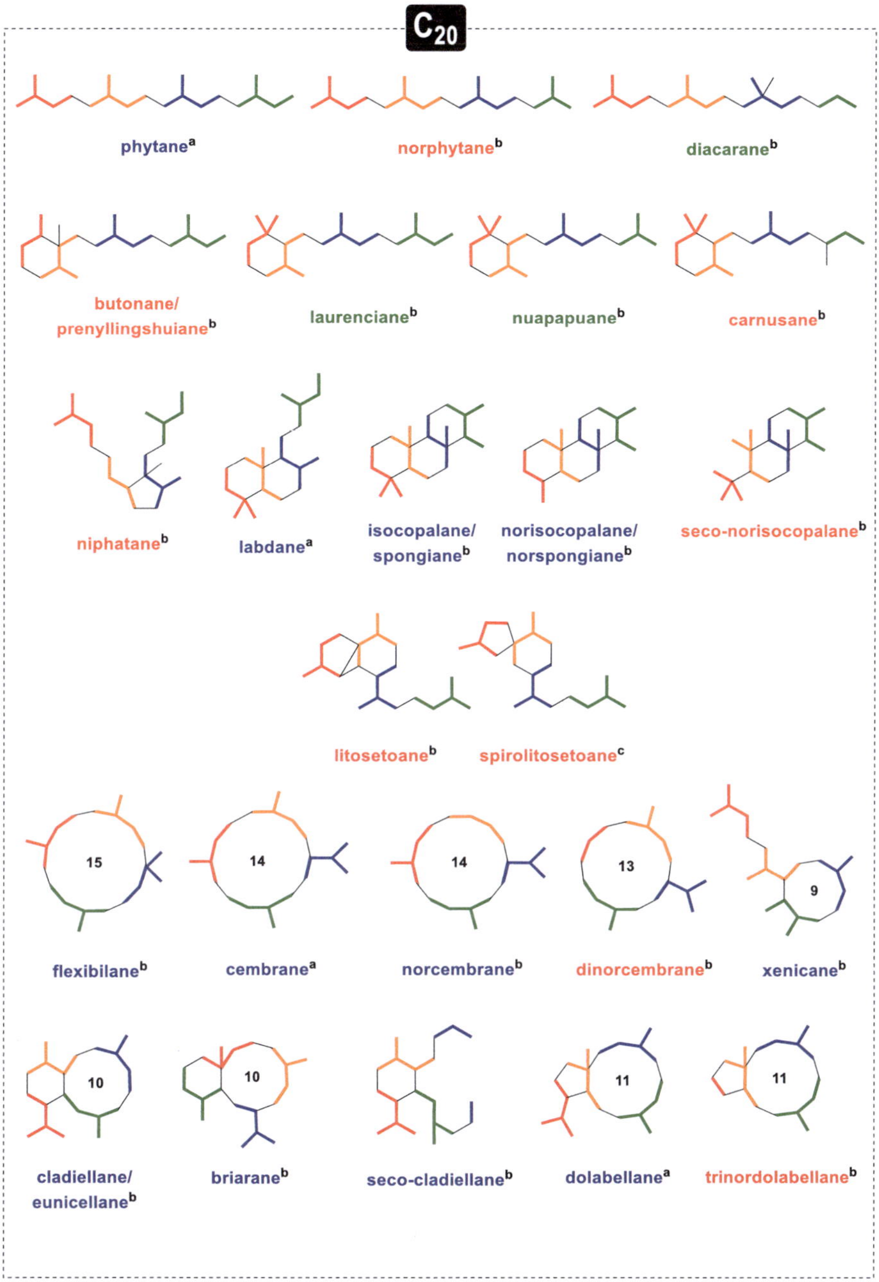

Fig. 2 (continued)

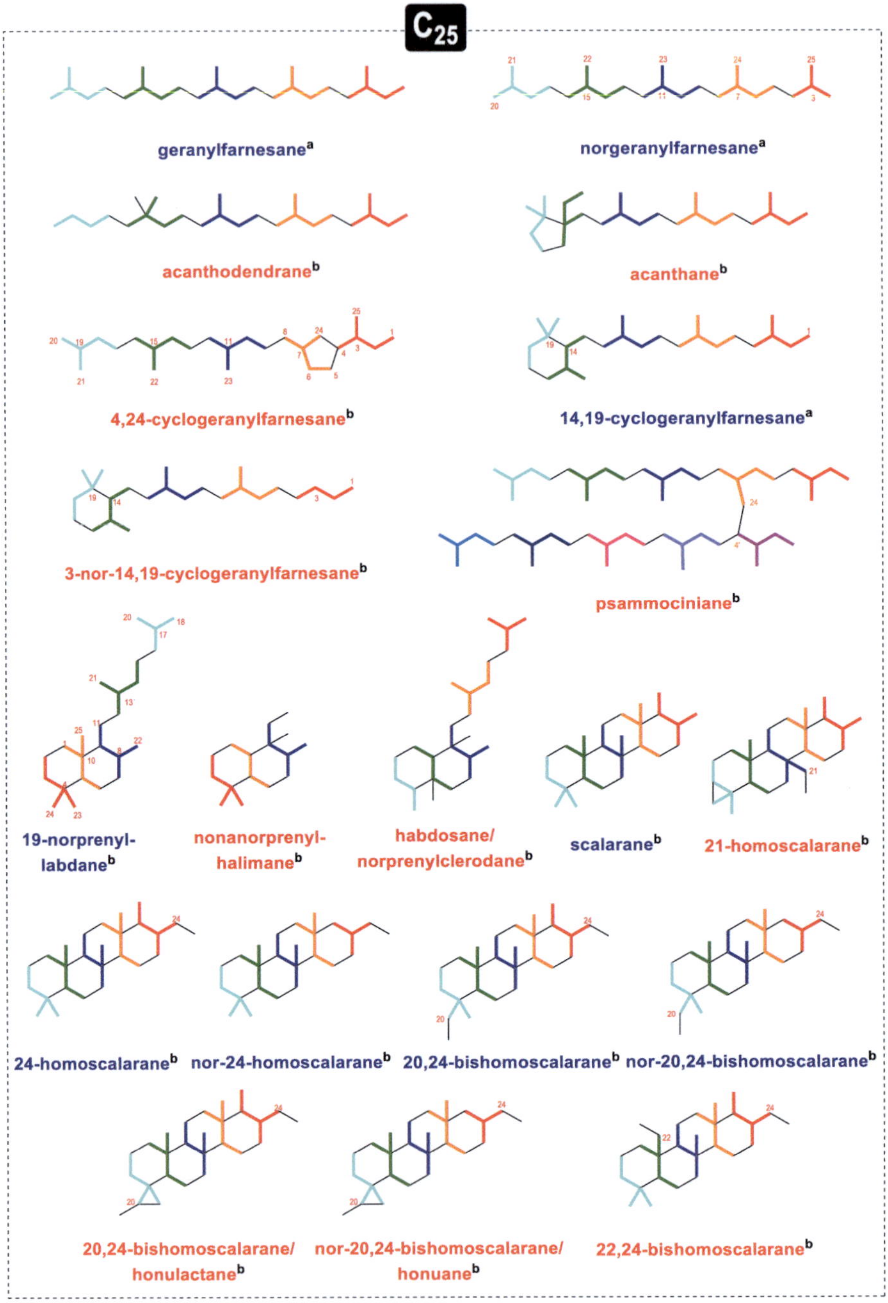

Fig. 2 (continued)

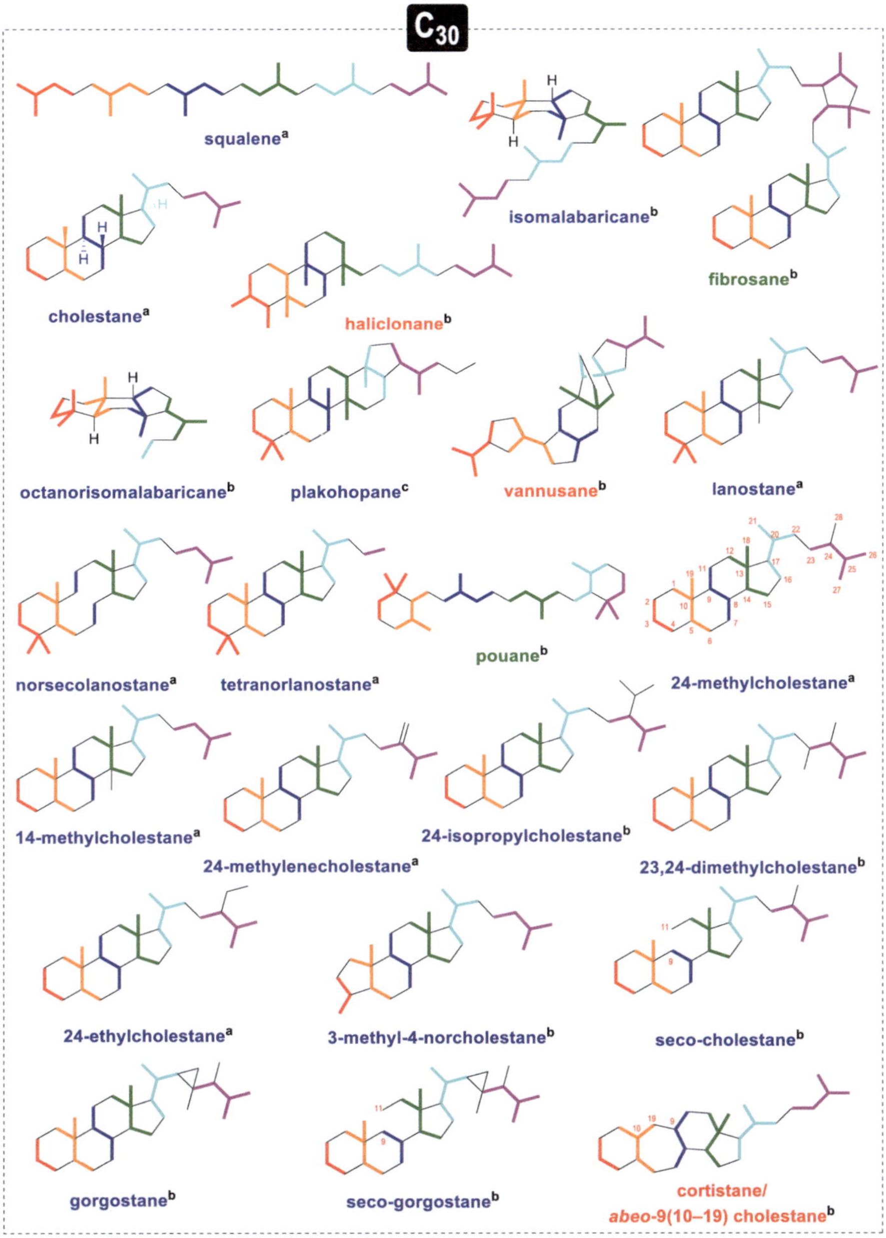

Fig. 2 (continued)

the most reaction catalyzed by a terpenoid synthase followed by other reactions such as degradation, cleavage, dimerization, degradation-rearrangement, degradation-cleavage, and degradation-dimerization. Thus, unique enzymatic chemical reactions are involved in the biosynthesis of Indonesian marine terpenoids.

2.1 Sesquiterpenoid

Three new nardosinane-type molecules (**1–3**) were isolated from MLU collections of *Lemnalia* and *Paralemnalia* species [17, 18]. The ACs of (−)-lemnacarnol (**2**) and (−)-2-deoxy-12-oxolemnacarnol (**3**) were elucidated by X-ray crystallographic analysis [19, 20]. An investigation of the soft coral *Capnella imbricata* collected from NMU and MLU led the isolation of 7 new capnellane-type molecules (**4–10**) with a variety of hydroxylations [21–25]. The ACs of (+)-$\Delta^{9(12)}$-capnellene-8β,10α-diol (**6**), (+)-$\Delta^{9(12)}$-capnellene-3β,8β,10α-triol (**7**), and (+)-$\Delta^{9(12)}$-capnellene-3β,8β,10α,14β-tetraol (**10**) were supported by X-ray diffraction analysis [25–27], while the AC of (−)-$\Delta^{9(12)}$-capnellene (**4**) was secured by asymmetric total synthesis featured with Heck reaction [28]. From GTO province, hirsutanols A–C, F (**11**, **12a**, **13a**, **14**) were isolated from an unidentified fungus derived a sponge *Haliclona* sp. [29]. X-ray study established the ACs of (−)-**11**, (+)-**13a**, and (−)-**14** [30, 31]. With the aid of DFT-based NMR calculation [32], the initial configuration of hirsutanol B (**12a**) was revised to be *epi*-1,9-hirsutanol B (**12b**). Moreover, the application of Mosher's method was also used to determine the ACs of (−)-hirsutanols A (**11**) and C (**13b**) [33]. The optical rotation sign of (+)-hirsutanol C (**13a**) was revised to (−)-hirsutanol C (**13b**) based on the isolation work of terrestrial mushrooms *Gloestereum incarnatum*. In 1974, the carbon skeleton of africanane (5/7/3 tricyclic skeleton) as in (+)-africanol (**15**) was discovered from the soft coral *Lemnalia africana* (MLU) [34]. X-ray diffraction analysis confirmed its decahydro-1*H*-cyclopropa[e]azulene core structure and determined its AC as depicted in Fig. 3 [35]. Interestingly, a plausible biosynthetic pathway analysis indicated a series of cyclizations (1,11; 1,3; 1,5) and a sequence of 1,2 or 1,3-hydride shifts to form the stable africanyl cation featured with cyclopropylcarbinyl carbocation were likely occurred. Two new aromadendrane-type molecules, (+)-**16** and (−)-**17** were identified from the soft coral *Sinularia mayi* collected at NST [36]. The AC of (+)-**16** was determined by chemical correlation, synthesis, and optical rotation, while the AC of (−)-**17** was disclosed via total synthesis featured with a stereodivergent gold (I)-catalyzed cascade reaction [37]. From WST province, (+)-hydroxycolorenone (**18**) possessing the guaiane skeleton was isolated from the marine soft coral *Nephthea chabrolii* [38]. In addition to the skeleton, a marine soft coral *Anthelia* sp. collected from BTN province gave also a new trinorguaiane-type molecule, (−)-antheliol (**19a**) [39]. The stereochemistry at C7 was revised from *R*, (−)-**19a** to *S*, (−)-**19b** based on the DFT-based NMR calculation [40].

An investigation of the soft coral *Capnella imbricata* collected from MLU and *Dendronephthya rubeola* from BLI gave two new precapnellane-type molecules,

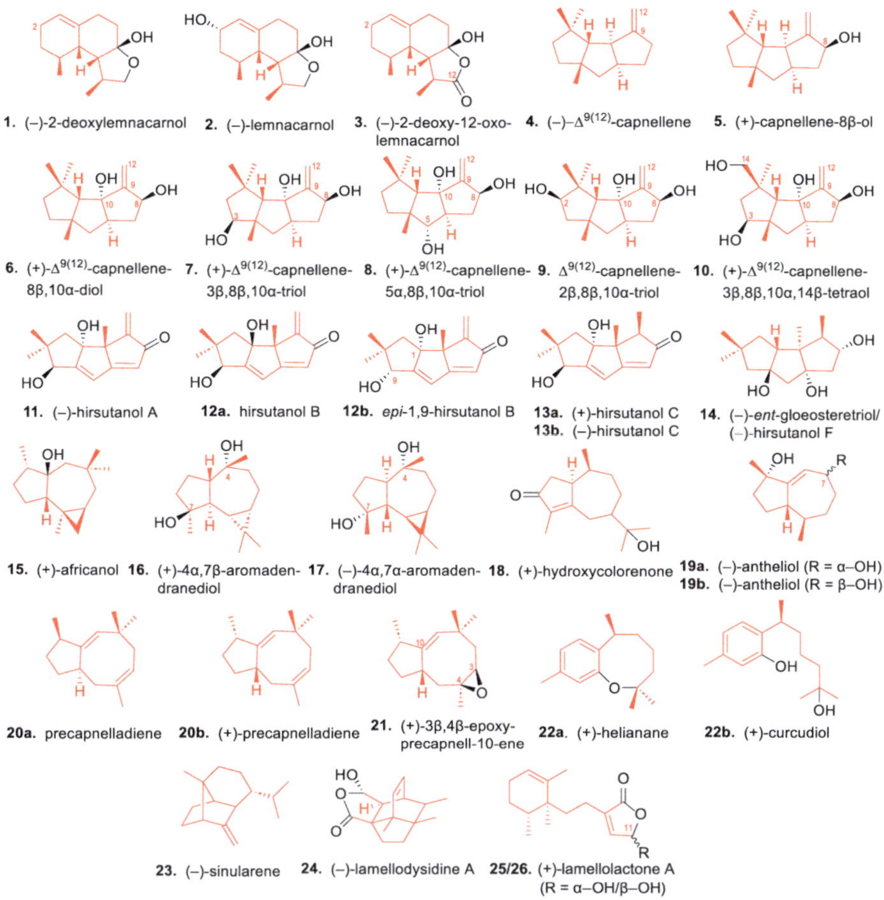

Fig. 3 Structures of marine sesquiterpenoids (**1–26**) from Indonesian waters. The marine sesquiterpene skeletons were proposed with red color

precapnelladiene (**20a**) [41] and (+)-3β,4β-epoxyprecapnell-10-ene (**21**), respectively [42]. The AC of **20a** was revised to be (+)-**20b** by enantioselective formal synthesis using chiral copper-catalyzed asymmetric [2 + 2]-cycloaddition reaction [43]. (+)-Helianane (**22a**) possessing the bisabolane-type molecule was isolated from CSW collection of the marine sponge *Haliclona ?fascigera* [44]. The true structure of (+)-helianane (**22a**) was revealed as (+)-curcudiol (**22b**) after total synthesis [45] and DFT-based NMR calculation [46]. The sinularane, the 5/5/6 tricyclic skeleton as in (−)-sinularene (**23**) was discovered from the NST collection of *Sinularia mayi* in 1977 [47]. The AC of (−)-**23** was secured by X-ray analysis. (−)-Lamellodysidine A (**24**) possessing a new tricyclic 6/6/5 skeleton was obtained from the NSW collection of *Lamellodysidea herbacea* [48]. Comparison of calculated and experimental ECD, total synthesis, and X-ray analysis [49] determined the AC of (−)-**24**. The carbon skeleton of lingshuaine was exemplified by a mixture of epimers, (+)-10*R*/10*S*-lamellolactones A (**25/26**) isolated from the marine sponge *Lamellodysidea* cf. *herbacea* collected from NSW [50].

2.2 Diterpenoid

Extraction of the soft coral *Sinularia flexibilis* (MLU) led to the isolation of a rare 15-membered macrocyclic flexibilane-based diterpenoid, flexibilene (**27**) [51]. The presence of 15-membered ring (Fig. 4) was confirmed by total synthesis [52]. The soft coral *Sarcophyton ehrenbergi* collected from BLI gave a new (+)-(3*E*, 11*E*)-cembra-3,8(19),11,15-tetraene-7α-ol (**28**) [53]. A newly isolated 15-hydroxycembrenene (**29**) obtained from the soft coral *Nepthea* sp. (JSCR) was discovered for the first time in nature and also found as in synthetic compound [54]. Reinvestigation of the extract soft coral *Sinularia flexibilis* from MLU furnished (+)-3,4,11,12-diepoxycembrane A (**30**) [55]. (−)-2-Hydroxynephthenol (**31**) was isolated from the MLU collection of *Litophyton viridis* [56, 57]. In 1974, the investigation of the soft coral *Lobophytum cristagalli* collected from MLU led the isolation of new (+)-lobophytolide (**32**) [58]. The AC of (+)-**32** was determined through X-ray diffraction study [59]. A cembranolide, 2-hydroxycrassocolide E (**33**) was also isolated from *Sarcophyton* sp. (NSW) [60]. The genus *Lobophytum* collected from NSW was also the source of new (−)-decaryiols B (**34**) and D (**35**) [61]. A combination of techniques including Mosher's method and chemical transformation was used to establish the ACs of (−)-**34** and (−)-**35**.

In 1975, the isolation work of *Sinularia flexibilis* collected from MLU gave a new (+)-sinulariolide (**36**) possessing the cembrane skeleton [62]. The AC of (+)-**36** was assigned by X-ray diffraction analysis [62, 63]. Following the discovery of (+)-**36**, two related molecules 6ξ-hydroxysinulariolide (**37**) and (+)-11-dehydrosinulariolide (**38**) were also isolated from the same species [55, 64]. The latter was secured its AC through X-ray analysis [65, 66]. Two new norcembrane-type molecules, (+)-chloroscabrolide A (**39**) and (−)-prescabrolide C (**40**) were obtained from a soft coral *Sinularia* sp. (NSW) [67]. The RC of (+)-**39** was elucidated by DFT-based NMR calculation. Two diterpenoids possessing the xenicane carbon skeleton, (+)-xeniolide F (**41**) and (+)-9-hydroxyxeniolide F (**42**) were characterized from a marine soft coral *Xenia* sp. collected from CSW province [68]. A soft coral *Cladiella* sp. (UEP) yielded three new-cladiellane-type molecules assigned as (−)-cladieunicellin F (**43**) [69], (−)-6-*epi*-cladieunicellin F (**44**) [70], and (−)-solenopodin C (**45**) [69]. (−)-Solenopodin C (**45**) was found as a new enantiomer confirmed through its specific optical rotation. Another skeleton, seco-cladiellane was found as in (+)-cladielloide D (**46**) [71] obtained from the same species. Its RC was established by a combination of molecular modelling dan NMR analysis. A cascade of 1,15 and 1,14 cyclizations as well as 1,3- and 1,5-hydride shifts reactions were probably taken place in the biosynthesis of cladiellane (eunicellane) based on deuterium labeling studies and quantum chemical calculations [72].

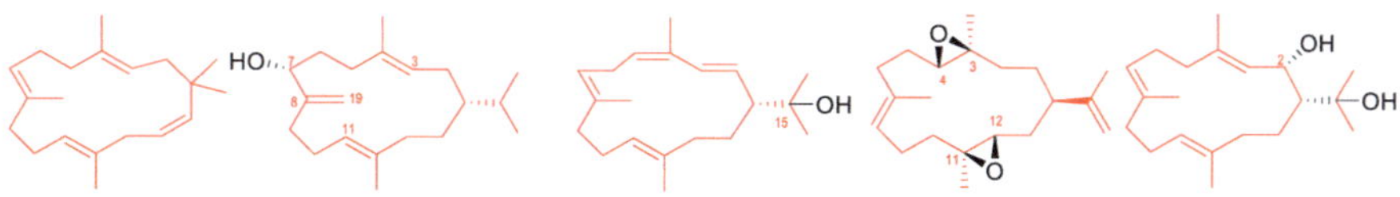

27. flexibilene **28.** (+)-(3*E*,11*E*)-cembra-3,8(19),11,15-tetraene-7α-ol **29.** 15-hydroxy-cembrenene **30.** (+)-3,4,11,12-diepoxycembrane A **31.** (−)-2-hydroxy-nephthenol

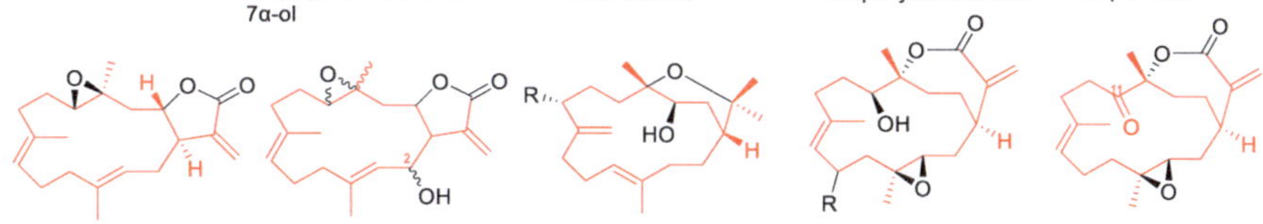

32. (+)-lobophytolide **33.** 2-hydroxy-crassocolide E **34.** (−)-decaryiol B (R = OH) **35.** (−)-decaryiol D (R = OOH) **36.** (+)-sinulariolide (R = H) **37.** 6ξ-hydroxysinula-riolide (R = OH) **38.** (+)-11-dehydro-sinulariolide

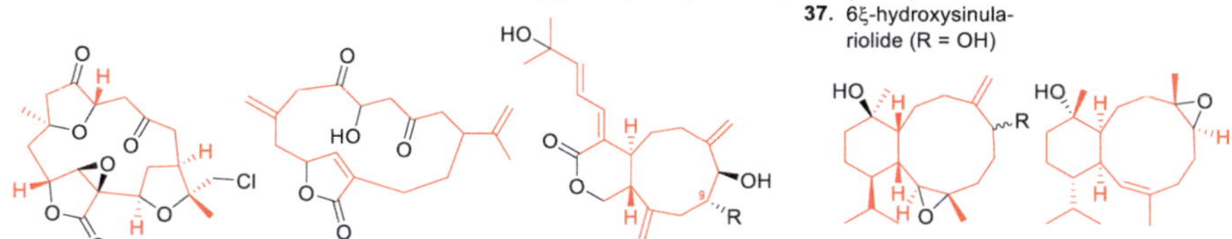

39. (+)-chloroscabrolide A **40.** (−)-prescabrolide C **41.** (+)-xeniolide F (R = H) **42.** (+)-9-hydroxyxeniolide F (R = OH) **43.** (−)-cladieunicellin F (R = α–OH) **44.** (−)-6-*epi*-cladieuni-cellin F (R = β–OH) **45.** (−)-solenopodin C

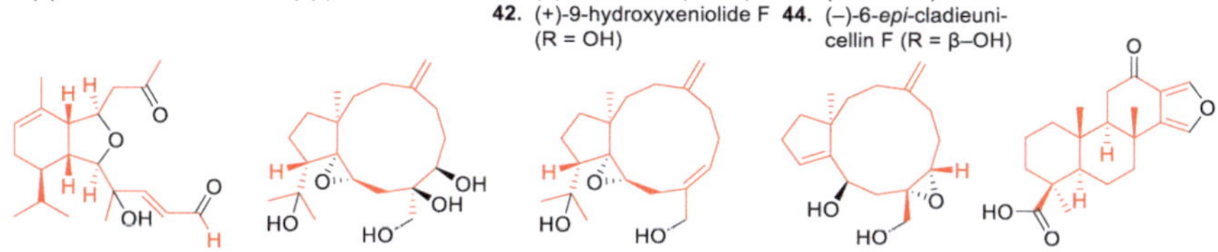

46. (+)-cladielloide D **47.** (+)-sangiangol B **48.** (−)-sangiangol C **49.** (+)-sangiangol A **50.** (+)-ceylonin G

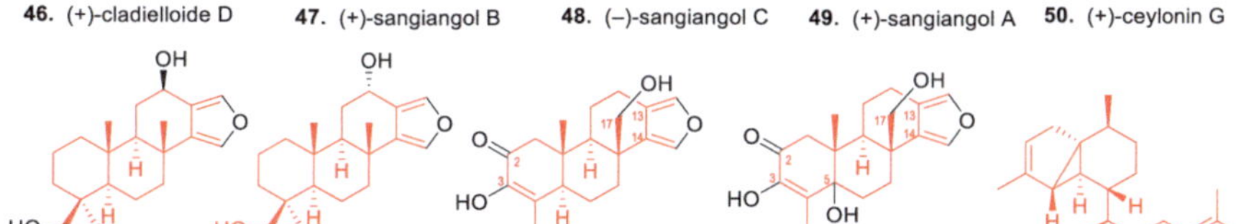

51. (+)-ceylonin H **52.** (−)-ceylonin I **53.** (−)-18-*nor*-3,17-dihydroxy-spongia-3,13(16),14-trien-2-one **54.** (−)-18-*nor*-3,5,17-trihydroxy-spongia-3,13(16),14-trien-2-one **55.** (+)-litosetoenin A

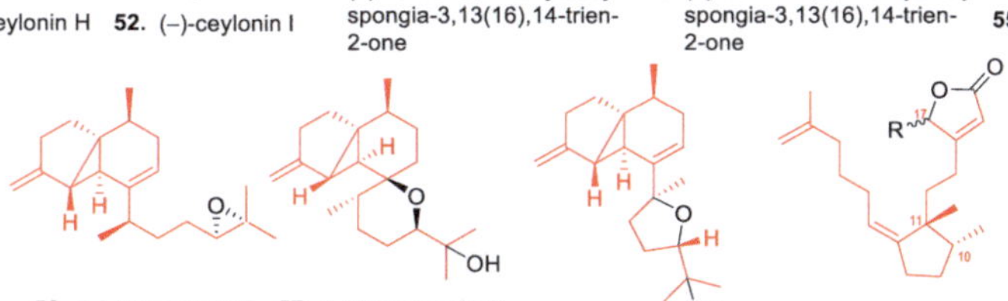

56. (−)-litosetoenin B **57.** (−)-litosetoenin C **58.** (−)-litosetoenin E **59/60.** (+)-niphateolide A (R = β–OH/α–OH)

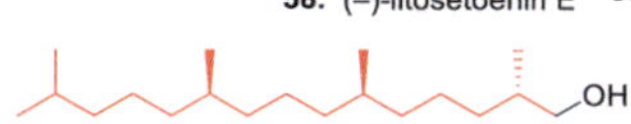

61. (+)-lamellodysideol

Fig. 4 Structures of marine diterpenoids (**27–61**) from Indonesian waters. The marine diterpene skeletons were proposed with red color

An investigation of a soft coral *Anthelia* sp. collected from BTN waters gave (+)-sangiangol B (**47**) [73] and (−)-sangiangol C (**48**) [74] possessing the dolabellane skeleton, as well as (+)-sangiangol A (**49**) [73] having a new trinordolabellane skeleton. The RC of (+)-**47** was elucidated by DFT-based NMR calculation [40]. (+)-Ceylonins G (**50**), H (**51**), and (−)-ceylonin I (**52**) possessing the spongiane (isocopalane) skeleton were obtained from NSW collection of the marine sponge *Spongia ceylonensis* [75]. The ACs of (+)-**50**, (+)-**51**, and (−)-**52** were ascertained by ECD studies. Two other norspongiane compounds assigned as (−)-18-nor-3,17-dihydroxyspongia-3,13(16),14-trien-2-one (**53**) and (−)-18-nor-3,5,17-trihydroxyspongia-3,13(16),14-trien-2-one (**54**) were isolated from a marine sponge *Spongia* sp. (NSW) [76]. Four new diterpenoids having a new skeleton litosetoane, (+)-litosetoenin A (**55**), (−)-litosetoenins B (**56**), C (**57**), and E (**58**) (Fig. 4) were isolated for the first time from the soft coral *Litophyton setoensis* (BLI) [77] featuring tricyclo[3.0.4]decane core. Their RCs were elucidated using DFT-based NMR calculations. A mixture of epimers, (+)-17*R*/17*S*-niphateolides A (**59/60**) possessing a new niphatane skeleton was elucidated from the marine sponge *Niphates olemda* (NSW) [78]. The C10 and C11 stereochemistry of (+)-**59/60** were elucidated by comparison of the calculated and measured ECD spectra. The marine sponge *Lamellodysidea herbacea* was also the source of a norditerpene having a new norphytane skeleton named as (+)-lamellodysideol (**61**) [79]. The RC of (+)-**61** was mainly elucidated using Hehre's protocol [80].

2.3 Sesterterpenoid

The early report of Indonesian marine sesterterpenoid was in 2002 by the isolation work of the genus *Hippospongia* sp. from SSW waters giving monocarbocylic sesterterpenoids containing 6-membered ring. The sesterterpenoids were (+)-barangcadoic acid A (**62**), rhopaloic acids D/E (**63/64**), rhopaloic acids F (**65**), and G (**66**) [81]. A geranylfarnesane skeleton was identified as in **62**, while **63/64**, **65**, and **66** contained a norgeranylfarnesane framework. (−)-Acantholides C (**67**), D (**68**), E (**69**) were secondary metabolites isolated from a marine sponge *Acanthodendrilla* sp. (SSW) [82]. A new acyclic rearrangement skeleton named acanthodendrane was shown by (−)-**67**, while a new rearrangement skeleton identified as a rare 5-membered ring system, the acanthane was observed in (−)-**68** and **69**.

An investigation of the extract sponge *Psammocinia* sp. from SSW gave (−)-sulawesins A and B as a mixture of stereoisomers (**70–73**) and (**74–77**) [83] with a new 4,24-cyclogeranylfarnesane skeleton. Although the ACs of stereoisomers (**70–73**) and (**74–77**) were difficult to disclose, the RCs of (−)-sulawesins A and B were revealed by NMR and DFT-based NMR calculation. Two new (−)-acantholides A (**78**) and B (**79**) possessing a 14,19-cyclogeranylfarnesane skeleton were isolated from a marine sponge *Acanthodendrilla* sp. (SSW) [82]. One new sesterterpenoid containing a 3-nor-14,19-cyclogeranylfarnesane skeleton assigned as (−)-diacardiol

A (**80**) was isolated from SSW collection of the sponge *Diacarnus megaspinorhabdosa* [84]. The sponge was also the source of a 19-norprenyllabdane skeleton as in the three new molecules (−)-diacarperoxides D (**81**), E (**82**), and (+)-diacarperoxide F (**83**) [84]. The new carbon skeleton of nonanorprenylhalimane was observed in (−)-euplectellodiol (**84**) isolated from *Mycale euplectelloides* (NSW) [85]. Further study from *Psammocinia* sp. (SSW) identified the present of (−)-sulawesin C characterized as a mixture of stereoisomers [83] having a new psammociniane skeleton (Fig. 5).

2.4 *Triterpenoid and Steroid*

About ten Indonesian MNPs were classified as triterpenoid and steroid. In 1999, the first Indonesian triterpenoid assigned as a linear squalene (−)-**86** was discovered from a collection of *Hyrtios erectus* species (SSW) [86]. The AC of (−)-**86** was unequivocally established by an asymmetric total synthesis [87]. The sponge *Rhabdastrella globostellata* was found to have isomalabaricane metabolites including (−)-globostellatic acid F (**87**), (+)-globostelletin (**88**), and a mixture of globostellatic acids M (**89**)/L (**90**) collected from SSW waters [88]. From NSW province, a soft coral *Minabea* sp. was the source of (+)-25*S*-3-oxocholesta-1,4-dien-26-oic acid (**91**) which was the first natural product isolated. Previously, the compound isolated as one of degradation products [89]. Two new-14-methylcholestane-type molecules assigned as (+)-lembehsterols A (**92**) and B (**93**) obtained from the marine sponge *Petrosia strongylata* (NSW) [90]. The AC of (+)-**92** was determined by Mosher's chemistry. Rare steroid dimers, (+)-manadosterols A (**94**) and B (**95**) were obtained from *Lissodendoryx fibrosa* (NSW) [91] owning the fibrosane skeleton (Fig. 6). Apparently, [4 + 2] cycloaddition was involved in the biosynthesis of (+)-**94** and **95**. The presence of [4 + 2] cycloaddition dimeric steroids was also observed in fibrosterol sulfates A–C isolated from the Philippines collection of the same species [92].

2.5 *Meroterpenoid*

2.5.1 Merosesquiterpenoid

Fourteen farnesane-type molecules (**96–109**) were isolated marine from sponges, nudibranchs, tunicates, and fungus-associated with tunicate. (−)-Aaptocarbamates A, C (**96**, **98**) and (+)-aaptocarbamate B (**97**) were furnished from a marine sponge *Aaptos* sp. (NSW) [93]. The marine fungus *Penicillium verruculosum* TPU1311 obtained from *Polycarpa aurata* (NSW) was the source of (−)-verruculide B (**99**) [94]. The chirality of (−)-**99** in polyketide part was partially determined by CD data compared with those of (*R*)-6-hydroxymellein. Chemical investigation of the

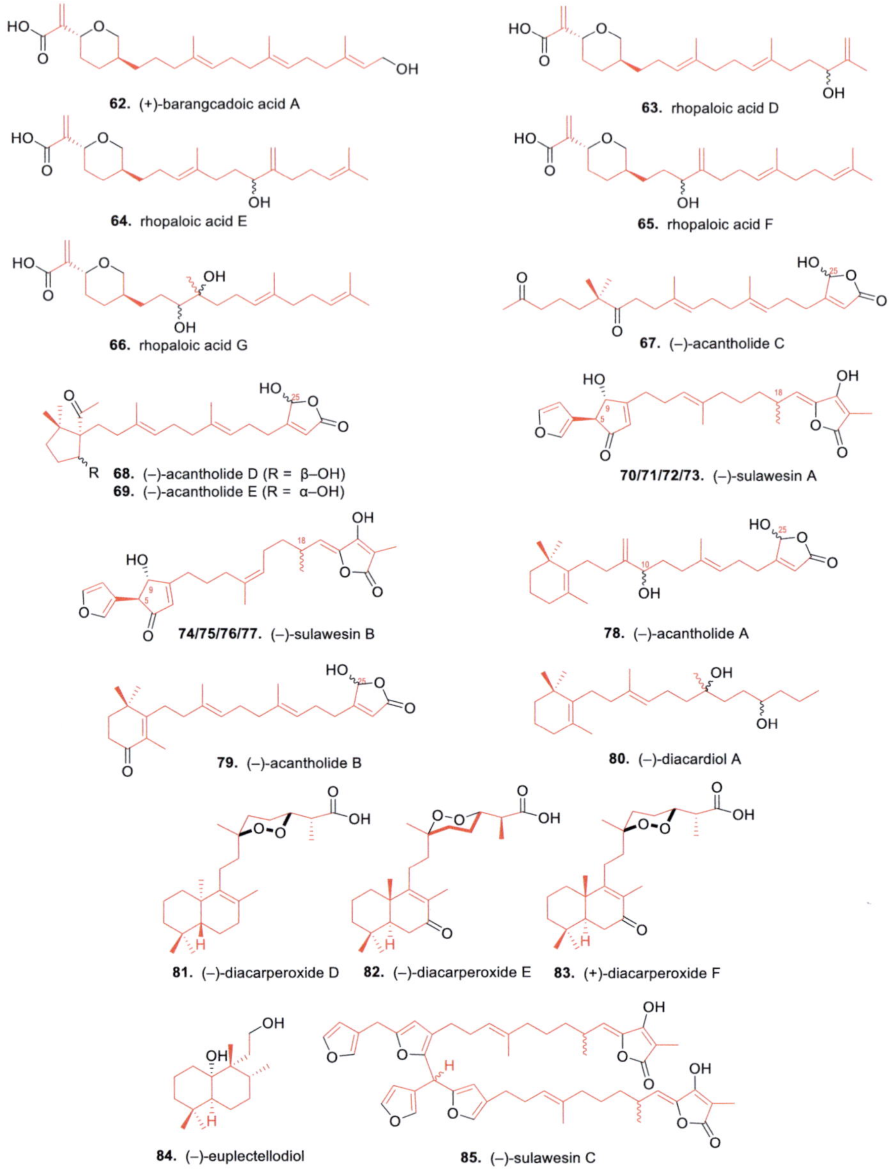

Fig. 5 Structures of marine sesterterpenoids (**62–85**) from Indonesian waters. The marine sesterterpene skeletons were proposed with red color

(–)-**86**

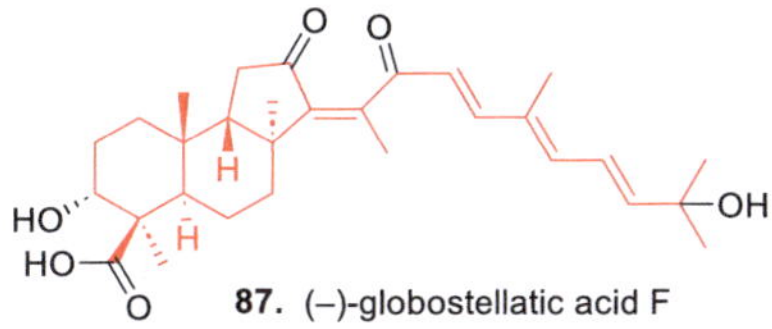

87. (–)-globostellatic acid F

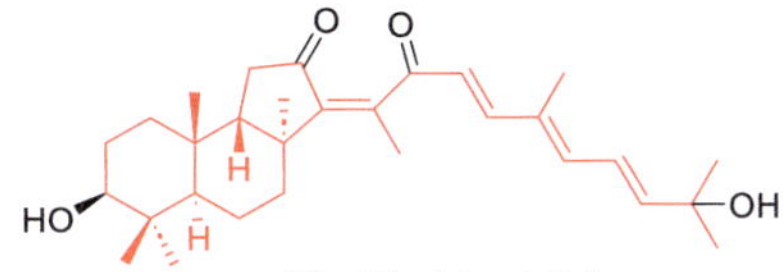

88. (+)-globostelletin

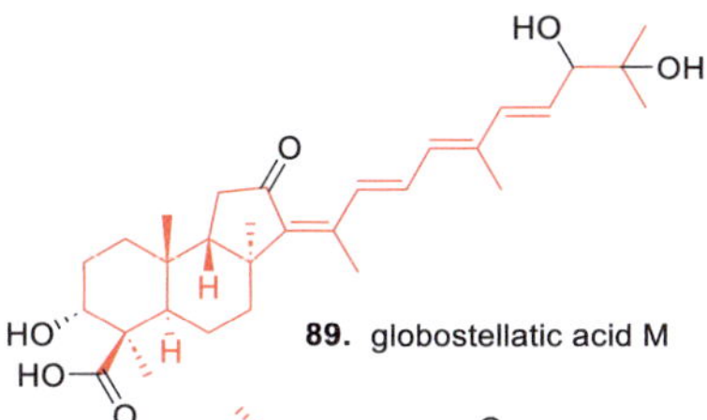

89. globostellatic acid M

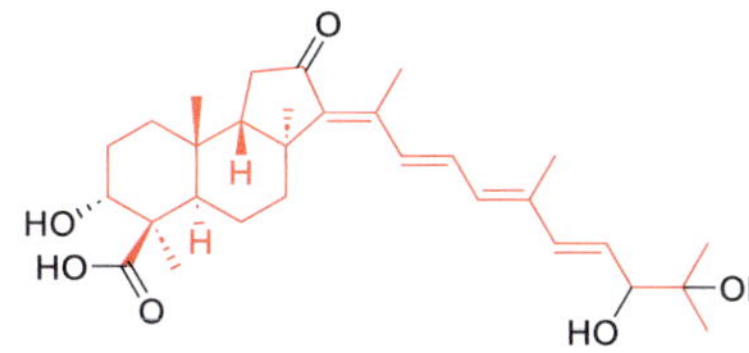

90. globostellatic acid L

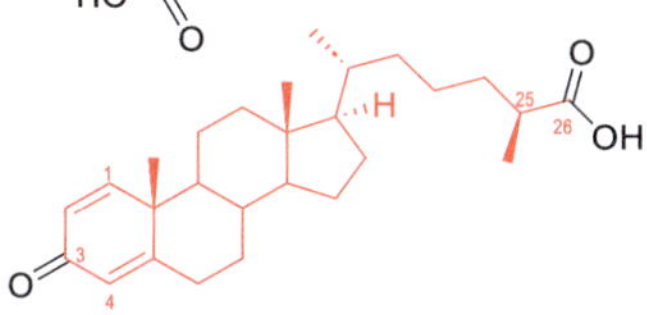

91. (+)-25*S*-3-oxocholesta-1,4-dien-26-oic acid

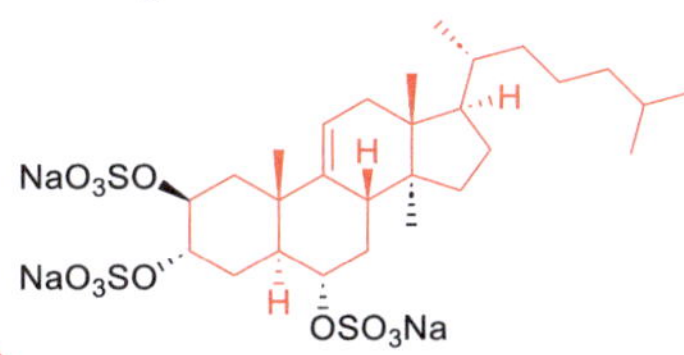

92. (+)-lembehsterol A

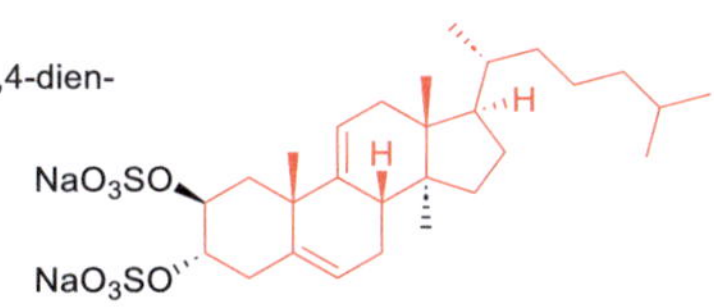

93. (+)-lembehsterol B

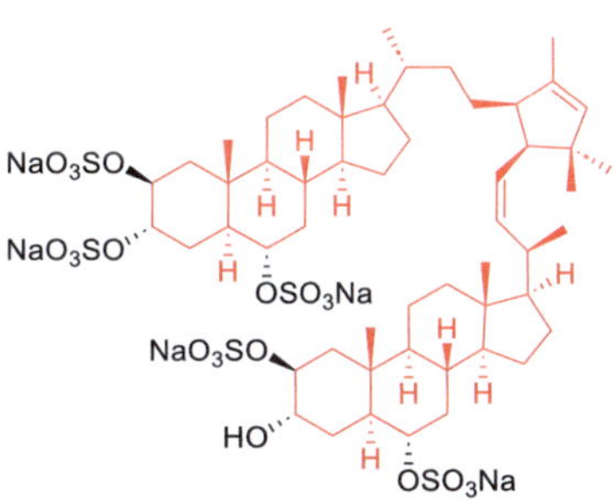

94. (+)-manadosterol A

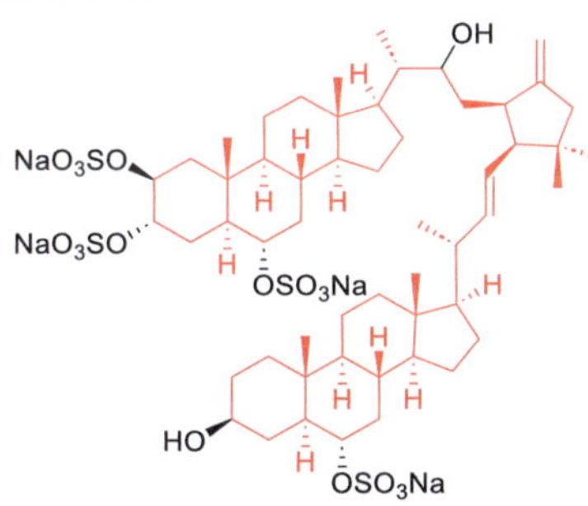

95. (+)-manadosterol B

Fig. 6 Structures of marine triterpenoids and steroids (**86–95**) from Indonesian waters. The marine triterpene/steroid skeletons were proposed with red color

nudibranch *Phyllidiella pustulosa* collected from NSW waters using molecular networking analysis revealed the presence of rare dichloroimidic sesquiterpene (−)-**100** and its derivative (−)-**101** [95]. The AC of dihydrofuranone ring in (−)-**101** was assigned by CD spectra and compared with those of ascofuranone. A marine sponge *Aaptos* sp. (NSW) was also the source of three linear merosesquiterpenoids containing tetrahydrofuran ring assigned as (+)-aaptocarbamates D (**102**), E (**103**), and (−)-aaptocarbamate F (**104**) [93]. Molecular modelling was also used to determine the RCs of compounds (+)-**102**, (+)-**103**, and (−)-**104**.

A mixed biosynthesis of the farnesane skeleton and 3 × C_2 building blocks as in 3-[(3*E*,7*E*)-4,8-dimethylpentadeca-3,7-dienyl] furan (**105**) was obtained from a marine sponge *Lendenfeldia* sp. (UEP) [96]. The carbon skeleton of trinorfarnesane was part of (−)-aaptocarbamate G (**106**) structure isolated from *Aaptos* sp. (NSW) [93]. Chemical investigation of a marine tunicate *Aplidium* sp. (ENT) led to the isolation of three new merosesquiterpenoids, (+)-floresolides A–C (**107–109**) [97]. A combination of total synthesis and X-ray analysis established the ACs of (+)-**108** [98, 99] as shown in Fig. 7. In addition to the MNPs, four new merosesquiterpenoids possessing the lingshuiane (6,11-cyclofarnesane) skeleton were reported as a mixture of epimers as in *O*,*O*-dimethyl lingshuiolide A (**110/111**) [48] and (−)-lamellolactone B1 (**112**)/(+)-lamellolactone B2 (**113**) [50] isolated from NSW collection of the marine sponge *Lamellodysidea herbacea*. The ACs of (−)-**112**/(+)-**113** were partially elucidated by a comparison of the calculated and measured ECD spectra. A mixed biosynthesis of terpenoid and alkaloid yielded two new merosesquiterpenoids, fasciospyrinadinone (**114**) and fasciospyrinadinol (**115**) possessing the metachromiane (6,7-cyclofarnesane) skeleton [100]. The natural products were obtained from a Papuan marine sponge *Petrosaspongia* sp. (PUA). The planar structure of **114** was also confirmed through total synthesis. The nudibranch *Phyllidiella pustulosa* collected from NSW waters was the source of five new merosesquiterpenoids having the premnane (6,11-cyclofarnesane) skeleton, (+)-**116**, **117–119**, and (+)-**120** [95]. Compound **116** was carboimidic dichloride sesquiterpene, while the rest possessed carbamate functional group. Compounds **117/119** and **118/120** were also found as a mixture. The RCs of **117–119** were disclosed by a molecular modelling study while the AC of (+)-**120** at C5 was revealed by comparison between experimental and calculated ECD data. The unstable carboimidic dichloride functional group [94] may be the true functional group of (−)-**96**, **98**, **101**, **104**, **106**, **117–119** and (+)-**97**, **102**, **103**, **120**. The molecular networking including LC–MS/MS analysis is a useful strategy to probe such artifact formation [101]. The source of C_1N in sponge could be derived from an amino acid [102].

Two new NRPS-terpenoids, (+)-halichonic acid B (**121**) [103] and (+)-halichonic acid (**122**) [104] owned the bisabolane skeleton were discovered from a marine sponge *Axinyssa* sp. (NSW) and *Halichondria* sp. (NSW), respectively. Both ECD analysis and total synthesis were used to reveal the ACs of (+)-**121** and (+)-**122** [103, 105]. The marine-derived fungus, *Penicillium citreonigrum* obtained from the marine sponge *Pseudoceratina purpurea* (BLI) gave a new oxygenated (−)-citreonigrin A (**123**) [106] having a new skeleton citreane. The AC of (−)-**123** was revealed by X-ray structure analysis. A mixed biosynthesis of polyketide-terpenoid possessing

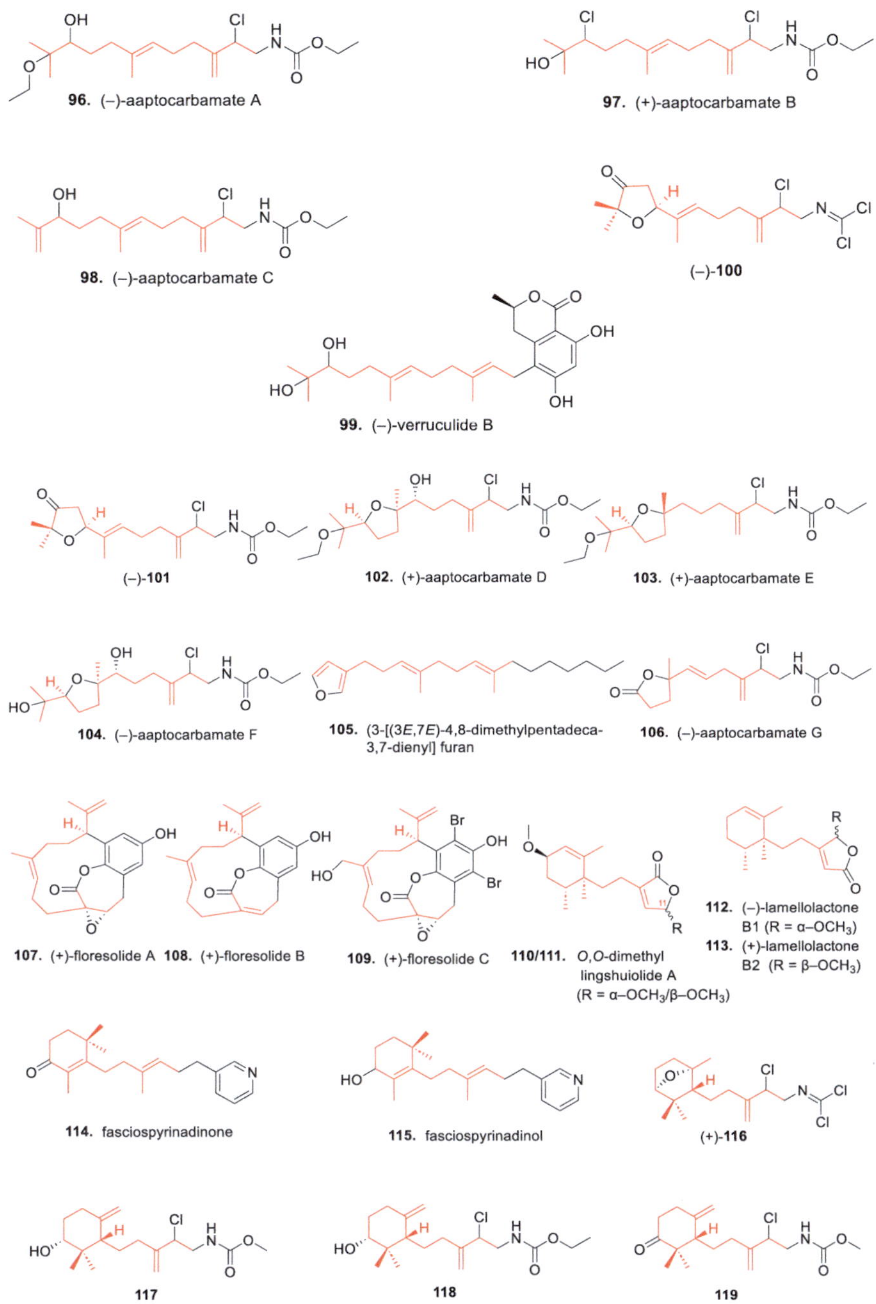

Fig. 7 Structures of marine merosesquiterpenoids (**96–156**) from Indonesian waters. The marine sesquiterpene skeletons were proposed with red color

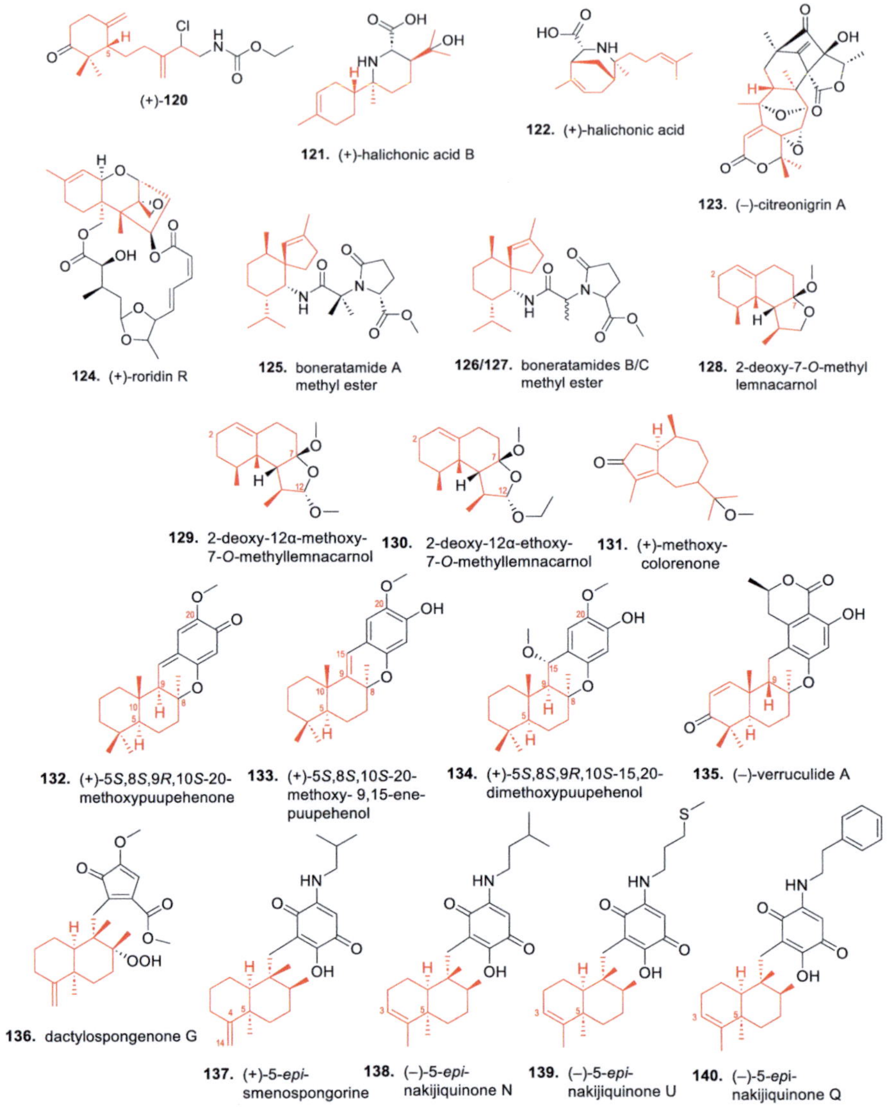

Fig. 7 (continued)

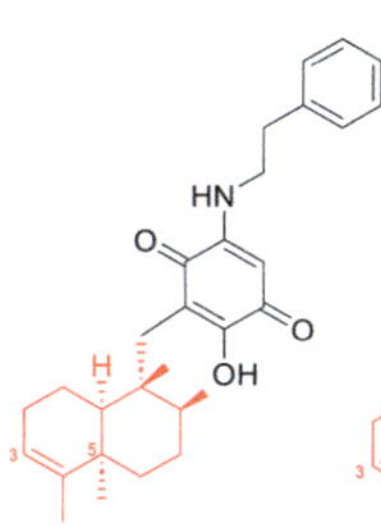
141. (−)-5-*epi*-nakijiquinone S

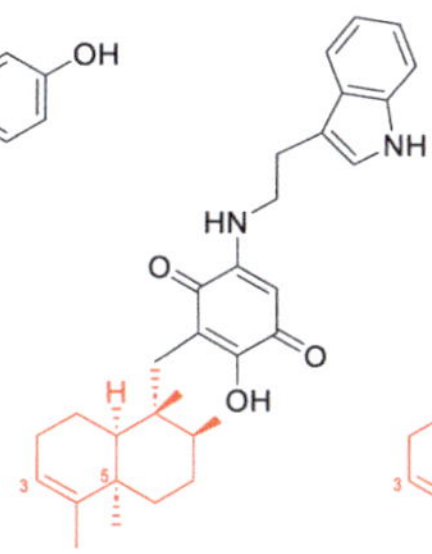
142. (−)-5-*epi*-nakijiquinone T

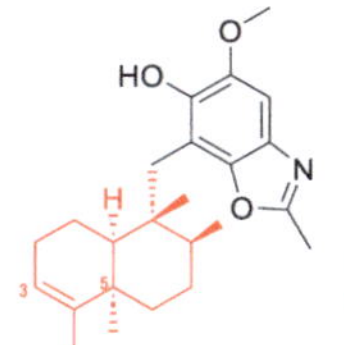
143. (−)-5-*epi*-nakijinol C

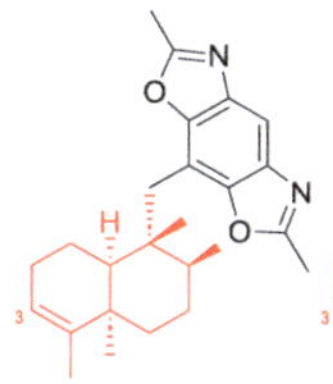
144. (−)-5-*epi*-nakijinol D

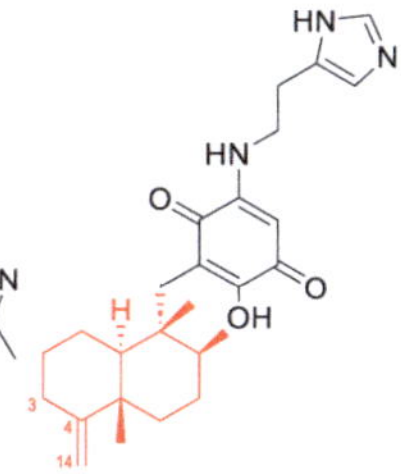
145. (+)-nakijiquinone V

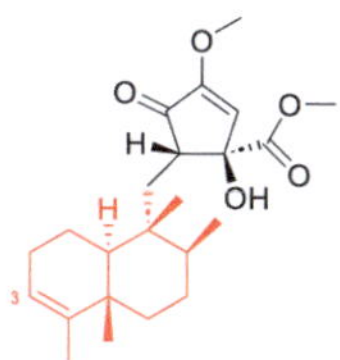
146. dactylospongenone H

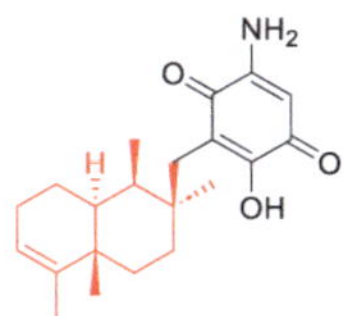
147. (−)-dysideamine

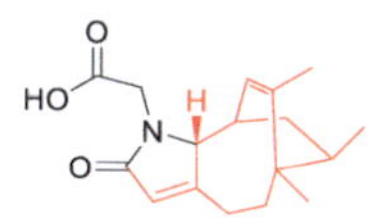
148. (−)-lamellodysidine B

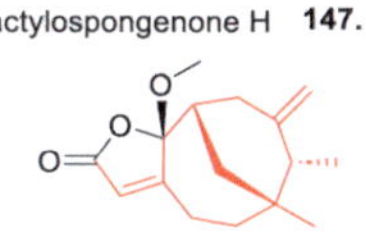
149. (−)-bicyclolamellolactone A

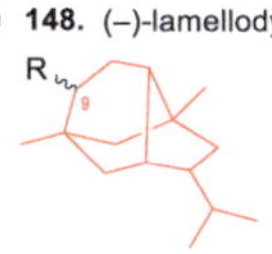
150/151. 9-thiocyanatopupukeanane (R = α–SCN/β–SCN)

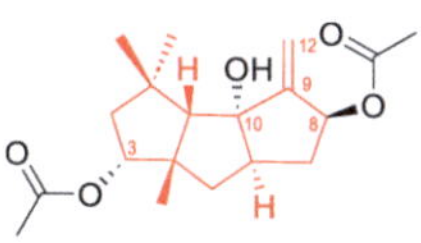
152. (+)-3α,8β-diacetoxycapnell-9(12)-ene-10α-ol

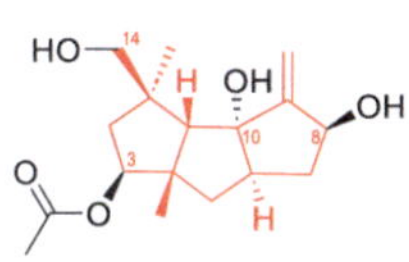
153. (+)-3β-acetoxycapnellene-8β,10α,14-triol

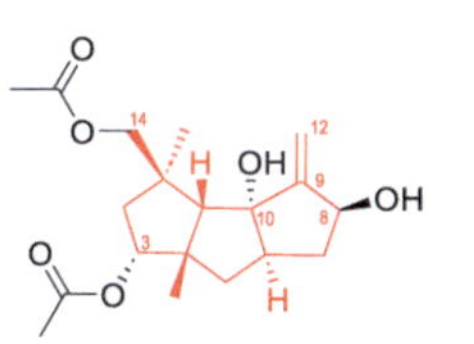
154. (+)-3α,14-diacetoxy-capnell-9(12)-ene-8β,10α-diol

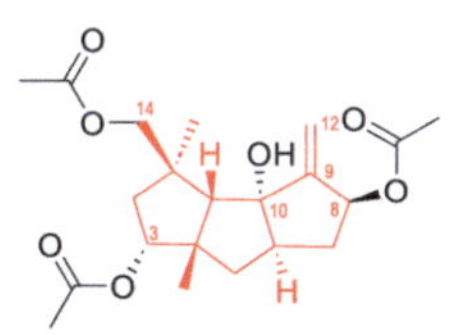
155. (+)-3α,8β,14-triacetoxy-capnell-9(12)-ene-10α-ol

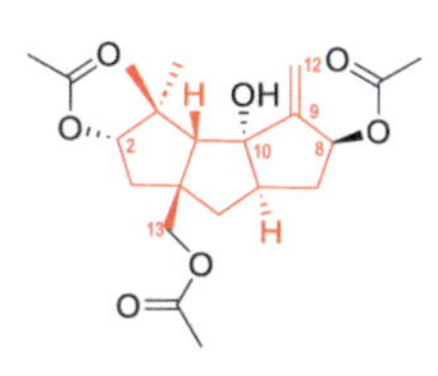
156. (−)-2α,8β,13-triacetoxy-capnell-9(12)-ene-10α-ol

Fig. 7 (continued)

the tricothecane skeleton, (+)-roridin R (**124**) was isolated from a marine fungus *Myrothecium* sp. TUF 02F6 associated with a sponge (BLI) [107]. From the SSW marine sponge *Axinyssa aplysinoides*, three merosesquiterpenoids, boneratamides A methyl ester (**125**), boneratamides B/C methyl ester (**126/127**) were isolated after modification of the compounds using TMS-CH_2N_2 [108]. The AC of **125** was supported by X-ray analysis and total synthesis on enantiomer of **125** [109]. A mixed biosynthesis of terpenoid and C_1 building block exemplified by **128–130** was found in a marine soft coral *Nephthea* sp. (NSW) owning the nardosinane skeleton [110]. In 1997, (+)-methoxycolorenone (**131**) possessing a guaiane skeleton with additional OMe group (C_1 building block) was obtained from the WST marine sponge *Nephthea chabrolii* [38].

Three merosesquiterpenoids assigned as (+)-5*S*,8*S*,9*R*,10*S*-20-methoxypuupehenone (**132**) and (+)-5*S*,8*S*,10*S*-20-methoxy-9,15-ene-puupehenol (**133**), (+)-5*S*,8*S*,9*R*,10*S*-15,20-dimethoxypuupehenol owning drimane skeleton were obtained from GTO collection of a sponge *Hyrtios* sp. [111]. The ACs of (+)-**132–134** were revealed by a degradation study [111], comparison of experimental and calculated ECD data [112]. Compounds (+)-**132** and **134** were likely obtained as artifacts. A merosesquiterpenoid with an unusual drimane skeleton with *trans-syn-trans* fusion assigned as (−)-verruculide A (**135**) was also isolated from the marine fungus *Penicillium verruculosum* TPU1311 obtained from *Polycarpa aurata* (NSW) [94]. Chemoenzymatic total synthesis established the AC of (−)-**135** [113]. Two *cis*-avarane $\Delta^{4(14)}$ MNPs, dactylospongenones G (**136**) [114] and (+)-5-*epi*-smenospongorine (**137**) [115] were isolated from *Dactylospongia elegans.* Compound **136** was obtained as a mixture stereoisomer with dactylospongenone H (**146**) possessing *trans*-avarane Δ^3 skeleton [114]. Another modification of *cis*-avarane skeleton, *cis*-avarane Δ^3 was observed in the marine sponge *Dactylospongia metachromia* represented by (−)-5-*epi*-nakijiquinones N (**138**), U (**139**), Q (**140**), S (**141**), T (**142**), (−)-5-*epi*-nakijinols C (**143**), and D (**144**) [116]. Revisited *Dactylospongia elegans* from NSW collection afforded a new (+)-nakijiquinone V (**145**), a *trans*-avarane $\Delta^{4(14)}$ marine natural product [117].

A new merosesquiterpenoid possessing a new skeleton dysideane, (−)-dysideamine (**147**) was isolated from a Papuan marine sponge *Dysidea* sp. (PUA) [118]. In addition to (−)-lamellodysidine A (**24**), (−)-lamellodysidine B (**148**) was also isolated as a mixed biosynthesis of nakafurane skeleton and glycine amino acid from the same marine organism [48]. Its RC was proposed by DFT based NMR calculation, while the AC of (−)-**148** was assigned by comparison of experimental and calculated ECD data. An unusual skeleton, lamellolane was identified from (−)-bicyclolamellolactone A (**149**) (NSW) isolated from *Lamellodysidea* cf. *herbacea* [119]. Its AC was established by comparison of experimental and calculated ECD data. Lamellolane and nakafurane skeletons could be connected by a remarkable sequence of 1,2-methide shifts. A mixture of epimers, 9-thiocyanatopupukeanane (**150/151**) possessing the pupukeanane skeleton was obtained from *Phyllidia varicosa* and its prey *Axinyssa aculeata* (JSCR) [120]. Moreover, five capnellane molecules with -OAc groups (C_2 building block), (+)-**152**, (+)-**154**, (+)-**155**, (−)-**156** were isolated from *Dendronephthya rubeola* (BLI) [42], while (+)-**153** was obtained from the marine soft coral *Capnella imbricata* (NMU) [22].

2.5.2 Meroditerpenoid

A mixed biosynthesis of a phytane skeleton and a cysteine derivative gave two meroditerpenoids assigned as (+)-sinulasulfoxide (**157**) and (+)-sinulasulfone (**158**) isolated from a soft coral *Sinularia* sp. (NSW) [121]. The ACs of (+)-**157** and **158** were secured via analysis of ECD and chemical correlation. Extraction of the sponge *Diacarnus megaspinorhabdosa* collected from SSW yielded two new molecules, (−)-diacarperoxides B (**159**) and C (**160**) possessing the diacarane skeleton [84]. In 2011, (+)-halioxepine (**161a**) was discovered from a marine sponge *Haliclona* sp. from SES waters [122] possessing a new butonane skeleton. DFT-based NMR calculation of halioxepine stereoisomers suggested a revision of configuration from (+)-**161a** to (+)-**161b1** [123]. Subsequently, total synthesis of (+)-halioxepine indicated a revision of its relative configuration at C10, (+)-**161b2** [124]. Another metabolite possessing the butonane skeleton, (−)-halioxepine C (**162**), was isolated from a marine sponge *Haliclona* (*Reneira*) sp. (NSW) [123]. Its AC was established by a combination of NMR calculation and Mosher's analysis. Likely, (−)-**162** was an artifact. Another type of halioxepine skeleton was the laurenciane framework identified as in (+)-halioxepine B (**163**) from the same organism [123]. NMR calculation and JBCA analysis established its RC. A meroditerpenoid, (+)-diacarperoxide A (**164**) having the nuapapuane skeleton was a mixed biosynthesis of terpenoid and C_1 building block isolated from *Diacarnus megaspinorhabdosa* (SSW) [84]. Compound **165** classified as meroditerpenoid containing a new carnusane skeleton was isolated from the same marine organism [85]. A terpenoid-polyketide metabolite named as (+)-subersic acid (**166**) was a new enantiomer obtained from a marine sponge *Acanthodendrilla* sp. (SSW) [125]. The marine sponge *Agelas nakamurai* collected from JSCR was the source of (−)-agelasine D (**167**) and (−)-ageloxime D (**168**) [126] having the labdane skeleton. Compound (−)-**167** was a new enantiomer. Another *Acanthodendrilla* metabolite (SSW), (+)-makassaric acid (**169**) owning the spongiane skeleton was coisolated with compound (+)-**166** [125]. The AC of (+)-**169** was secured by total synthesis [127]. Spongiane is the major skeleton for marine meroditerpenoids (+)-**170, 183, 184** and (−)-**171–182, 185–187**. (+)-Coelodiol (**170**) was isolated from the marine sponge *Coelocarteria* cfr. *singaporensis* (NSW) [128]. The AC of (+)-**170** was determined through the application of the exciton chirality CD method. The marine sponge *Spongia ceylonensis* (NSW) [129, 130] was the source of compounds (−)-**171**, **172**, **175**, **176**, **178–182**, **185**, **186** and (+)-**183**, **184**, while *Spongia* sp. (PUA) [76, 131] contained (−)-**173**, **174**, **177**, **187**. The RCs of (−)-**171**, **180–182**, **185**, **186** and (+)-**183**, **184** were elucidated by molecular modelling [129, 130]. Moreover, the AC of (−)-ceylonin A (**180**) was established by comparison of calculated and experimental ECD [130]. (−)-**171** was likely isolated as an artifact.

Another *Coelocarteria* cfr. *singaporensis* metabolite, (−)-coeloic acid (**188**) [128] possessed a new skeleton, seco-norisocopalane was obtained from NSW. Chemical investigation of the soft coral *Litophyton setoensis* collected from BLI yielded meroditerpenoid, (+)-litosetoenin D (**189**), (−)-spirolitosetoenin A (**190**), (+)-isospirolitosetoenin A (**191**) [77], possessing a new litosetoane and spirolitosetoane skeletons, respectively (Fig. 8). The RCs of (+)-**189**, **191** and (−)-**190** were elucidated by DFT-based NMR calculation. Compounds (−)-**190** and (+)-**191** were artifacts. Meroditerpenoids with a mixed biosynthesis of the cembrane skeleton and C_2 unit exemplified by 3,4-epoxynephthenol acetate (**192**) [54], (+)-decaryiol C (**193**) [61], (−)-crassolide (**194**) [132], (−)-11-*epi*-sinulariolide acetate (**195**) [64], and (−)-danielid (**196**) [133] were isolated from marine soft corals: *Nephthea* sp., *Lobophytum* sp., *Lobophytum crissum*, *Sinularia flexibilis*, and *Sinularia asterolobata*, respectively. The AC of (+)-**193** was confirmed by chemical transformation [61], while the ACs of (−)-**194** and (−)-**195** were established by X-ray analysis [132, 134]. In addition (−)-chloroscabrolide B (**197**) [67] was one example of meroditerpenoids having a norcembrane skeleton and C_1 unit isolated from *Sinularia* sp. (NSW).

Two new (−)-sarcofuranocembrenolide B (**198**) and (+)-sarcofuranocembrenolide A (**199**) were meroditerpenoids consisting of a new dinorcembrane skeleton and C_1 or C_2 building block isolated from a marine soft coral *Sarcophyton* sp. (NSW) [135]. A series of xenicane-C_1 metabolites, (−)-xenimanadin A (**200**), (+)-xenimanadins B–D (**201–203**) [136] were isolated a marine soft coral *Xenia* sp. (NSW). The AC of (−)-**200** was confirmed by the application of Mosher's method [136]. A total of nineteen briarane metabolites with a variety of C_1, C_2 building blocks [(+)-**204–206**, **208–210**, **212**, **213**, **216**, **217**, **219**, **222**; (−)-**207**, **214**, **215**, **218**, **220**, **221**; and **211**] were isolated from five species of octocorals (*Veretillum malayense*, *Briareum excavatum*, *Briareum* sp., *Pteroeides* sp., and *Junceela fragilis*). (+)-Malayenolides A–D (**204**, **205**, **209**, **210**) were obtained from NSW collection of *Veretillum malayense* [137]. The octocoral *Briareum excavatum* was the source of three new molecules (+)-briantheins A–C (**206**, **212**, **217**) collected from Sulawesi Island (UEP) [138] possessing a mixed biosynthesis of briarane skeleton and C_2 building block. The RC of (+)-brianthein A (**206**) was elucidated by molecular modelling, while its AC was determined by Mosher's analysis. Two new metabolites, (−)-13-dehydroxystecholide J (**207**) and (−)-2,9-diacetyl-2-debutyrylstecholide H (**221**) and one newly isolated 2β-acetoxy-2-(debutyryloxy)-stecholide E acetate (**211**) were obtained from *Briareum* sp. from CSW province [139]. Molecular modelling was use to probe the RC of (−)-**206**. Four meroditerpenoids, (+)-**208**, **213**, **222**, and (−)-**214** were isolated from ENT collection of *Pteroeides* sp. [140]. Meroditerpenoids with briarane skeleton as in (−)-4-deacetyljunceellolide D (**215**), (+)-junceellolide A (**216**), (−)-11α,20α-epoxy-4-deacetyljunceellolide D (**218**), (+)-11α,20α-epoxyjunceellolide D (**219**), (−)-11α,20α-epoxy-4-deacetoxyjunceellolide D (**220**) were isolated from the octocoral *Junceela fragilis* (NMU) [141]. The ACs of (−)-**215** was determined by Mosher's chemistry, while the ACs of (−)-**218**, **219** were elucidated by chemical transformation. (+)-Junceellolide A (**216**) was a new enantiomer confirmed by its specific optical rotation. Chemical investigation of *Cladiella* sp. (UEP) gave three meroditerpenoids possessing the cladiellane skeleton assigned as (−)-cladielloides A, B (**223a**, **224a**) and (+)-cladielloide C (**225**) [71, 142]. The AC of (−)-**223a** was determined by Mosher's method. In 2011, (−)-**223a** and (−)-**224a** was revised to (−)-**223b** and (−)-**224b**, respectively by reinvestigation of their 2D NMR (HMBC) [69]. (−)-Cladieunicellin G (**226**) derived from the cladiellane skeleton and C_2 unit

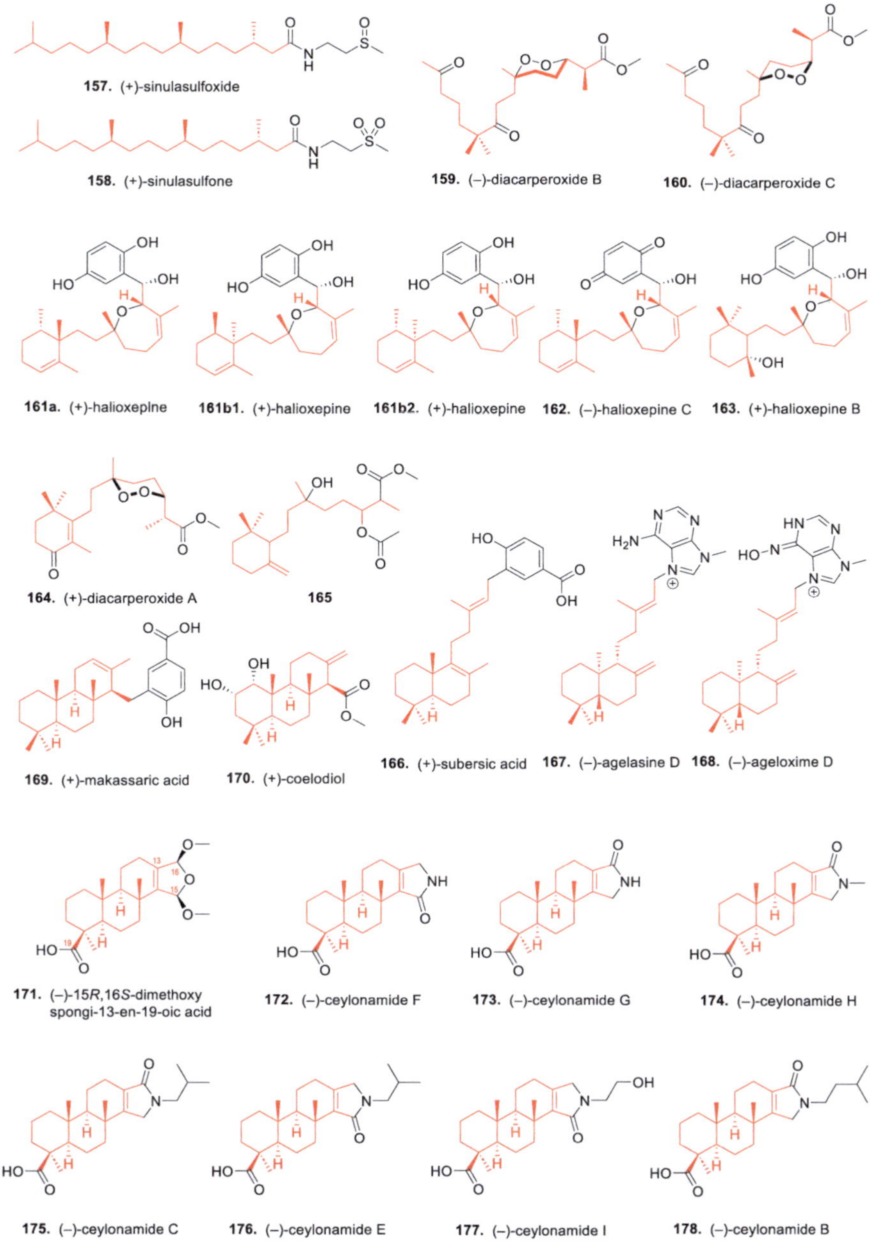

Fig. 8 Structures of marine merodíterpenoids (**156–226**) from Indonesian waters. The marine diterpene skeletons were proposed with red color.

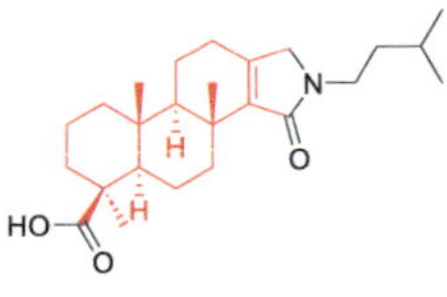

179. (–)-ceylonamide D

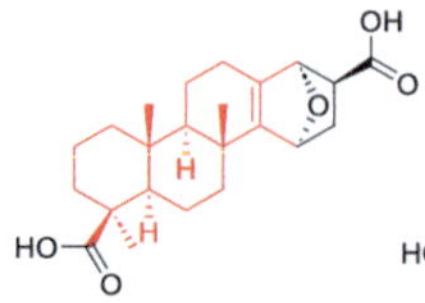

180. (–)-ceylonin A

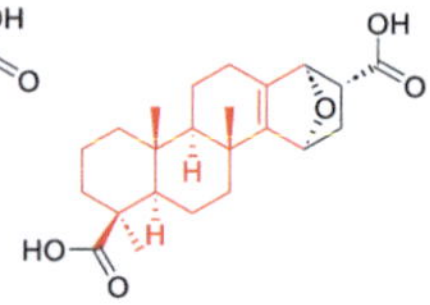

181. (–)-ceylonin C

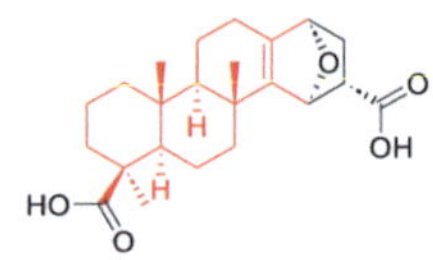

182. (–)-ceylonin E

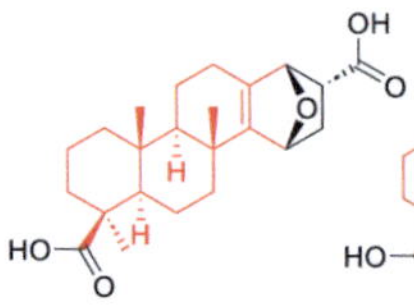

183. (+)-ceylonin B

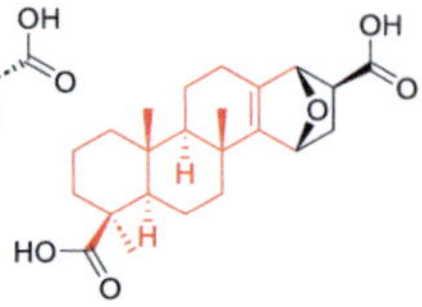

184. (+)-ceylonin D

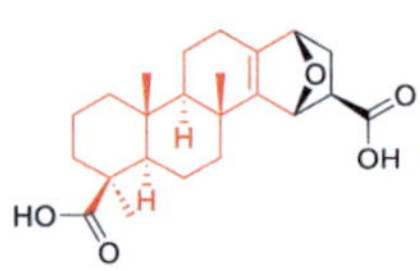

185. (–)-ceylonin F

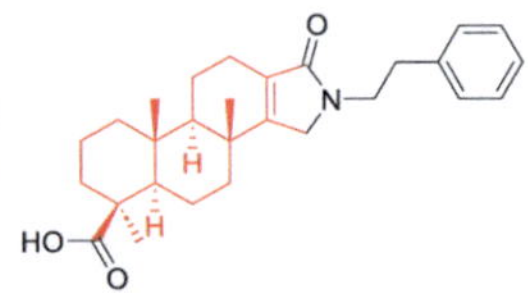

186. (–)-ceylonamide A

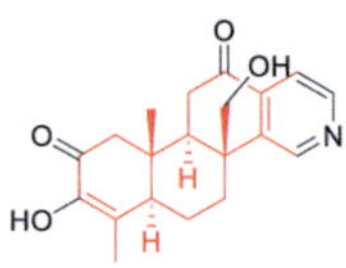

187. (–)-spongiapyridine

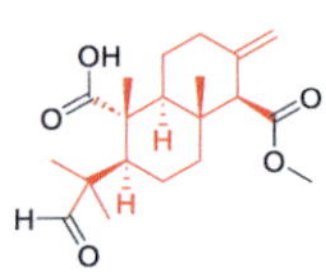

188. (–)-coeloic acid

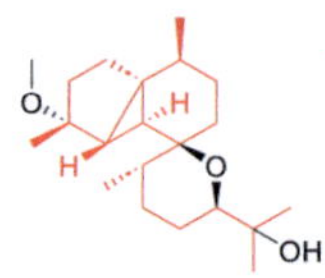

189. (+)-litosetoenin D

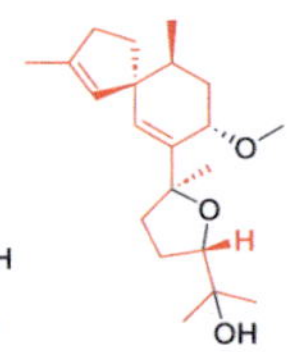

190. (–)-spirolito-setoenin A

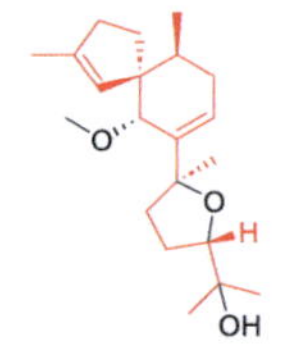

191. (+)-isospirolito-setoenin A

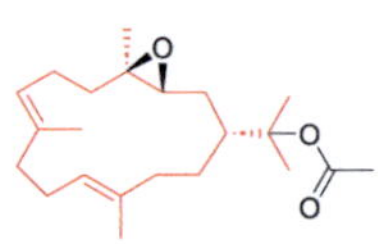

192. 3,4-epoxynephthenol acetate

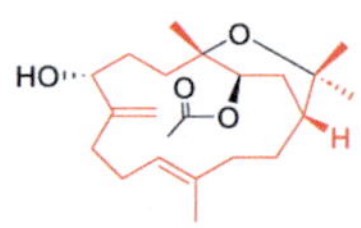

193. (+)-decaryiol C

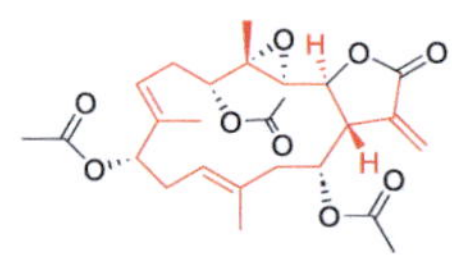

194. (–)-crassolide

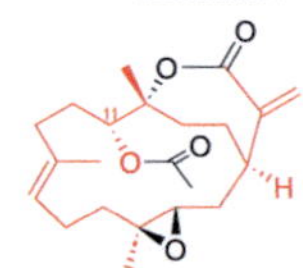

195. (–)-11-*epi*-sinulari-olide acetate

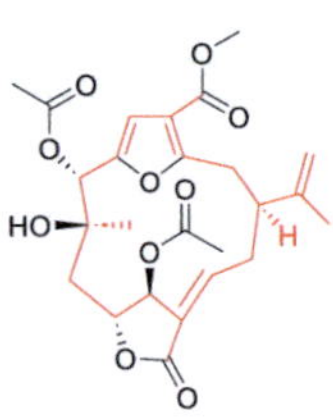

196. (–)-danielid

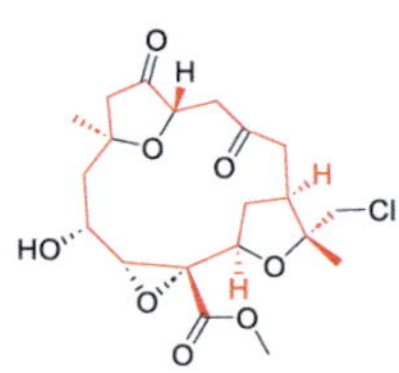

197. (–)-chloroscabrolide B

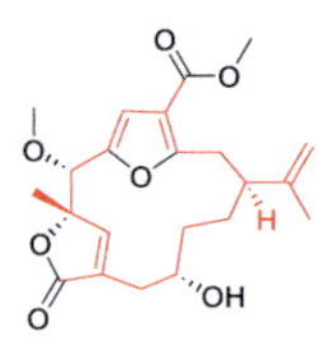

198. (–)-sarcofurano-cembrenolide B

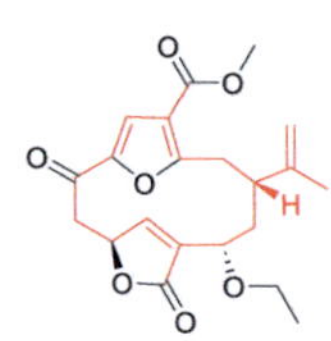

199. (+)-sarcofurano-cembrenolide A

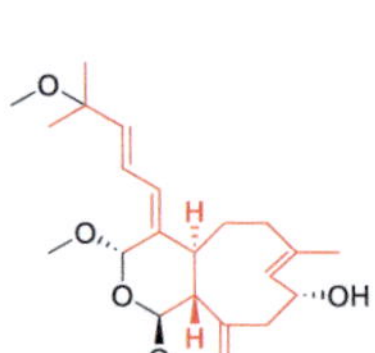

200. (–)-xenimanadin A

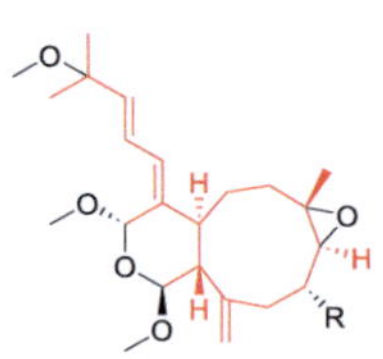

201. (+)-xenimanadin B (R = OH)
202. (+)-xenimanadin C (R = H)

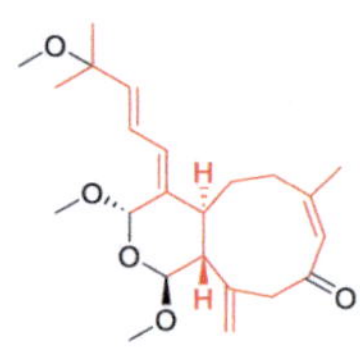

203. (+)-xenimanadin D

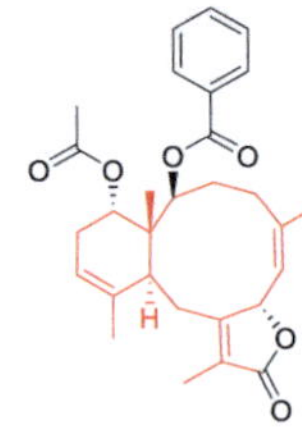

204. (+)-malayenolide A

Fig. 8 (continued)

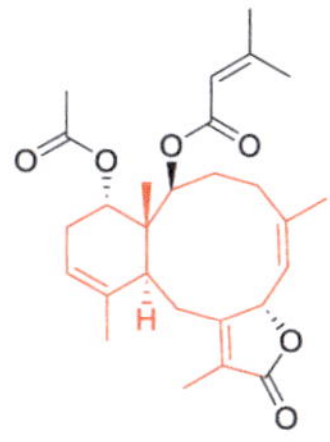

205. (+)-malayenolide C

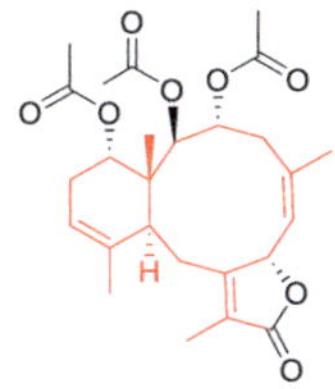

206. (+)-brianthein A

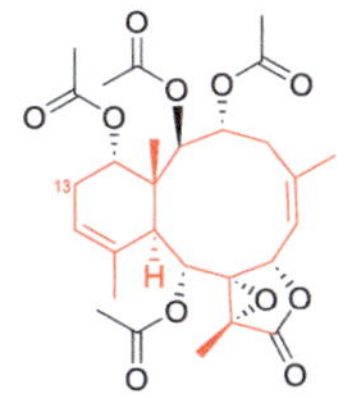

207. (–)-13-dehydroxy-stecholide J

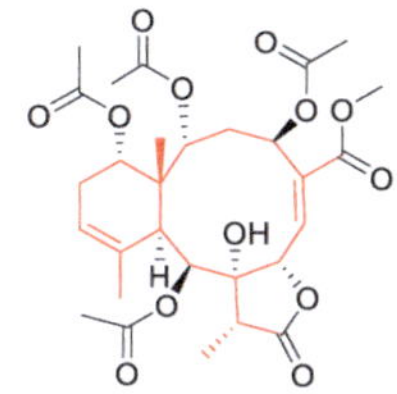

(+)-**208**

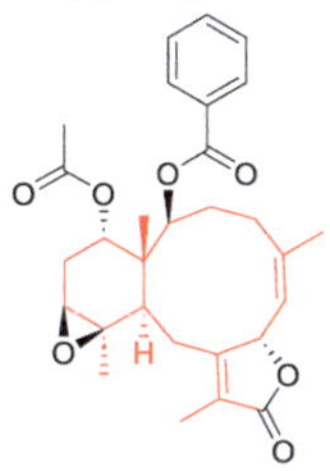

209. (+)-malayenolide B

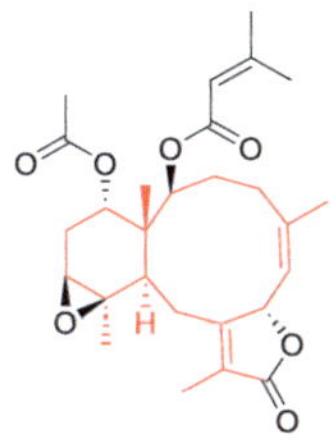

210. (+)-malayenolide D

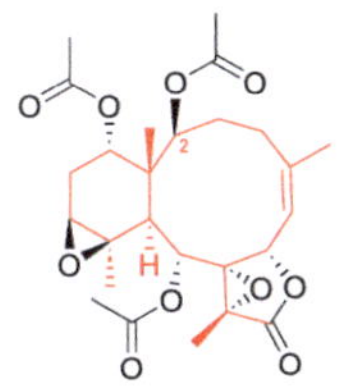

211. 2β-acetoxy-2-(debutyryloxy)-stecholide E acetate

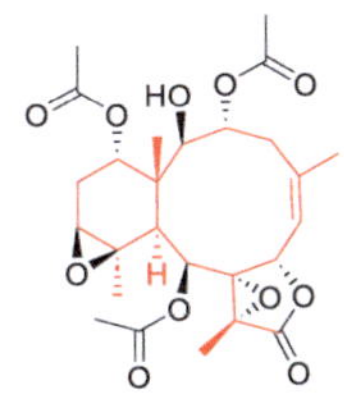

212. (+)-brianthein C

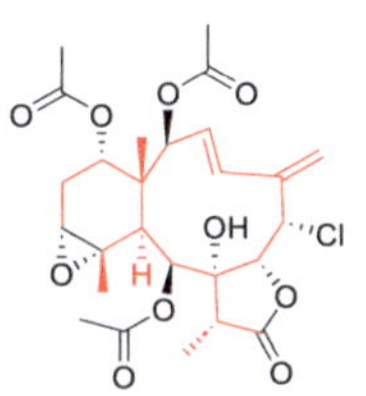

(+)-**213**

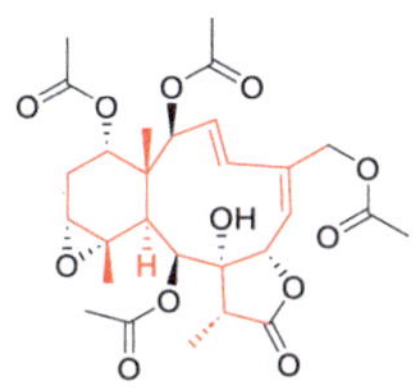

(–)-214

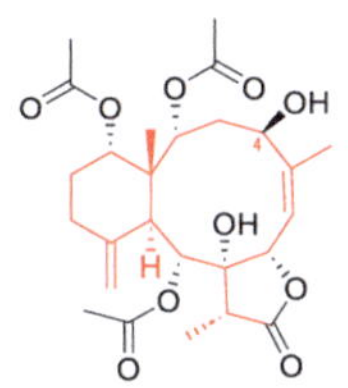

215. (–)-4-deacetyl-junceellolide D

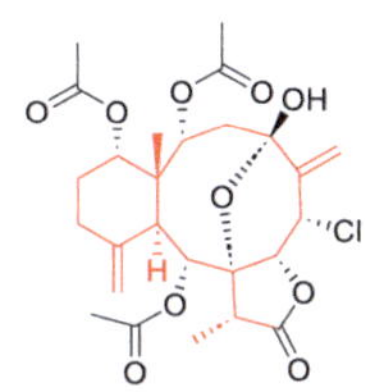

216. (+)-junceellolide A

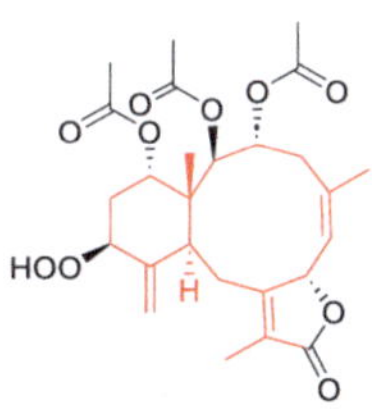

217. (+)-brianthein B

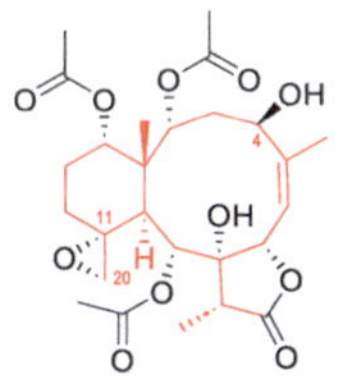

218. (–)-11α,20α-epoxy-4-deacetyljunceellolide D

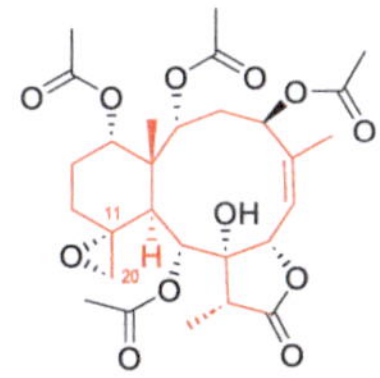

219. (+)-11α,20α-epoxy-junceellolide D

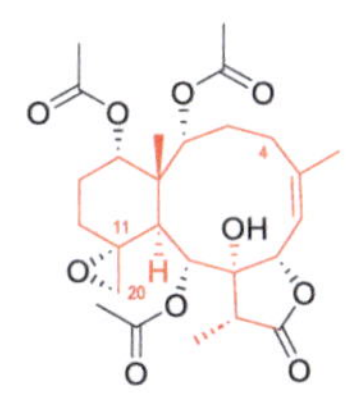

220. (–)-11α,20α-epoxy-4-deacetoxyjunceell-olide D

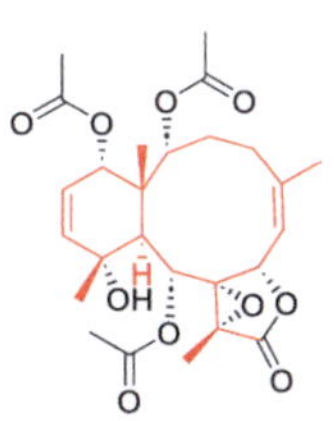

221. (–)-2,9-diacetyl-2-debutyrylstecholide H

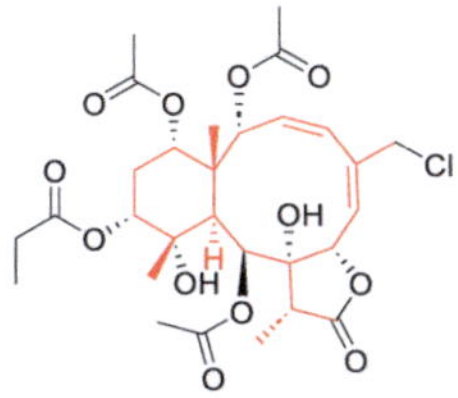

(+)-**222**

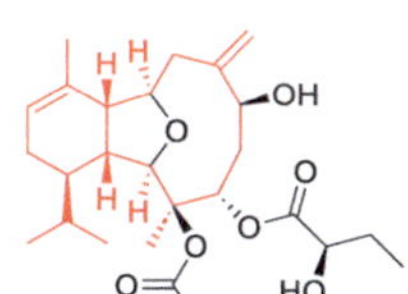

223a. (–)-cladielloide A

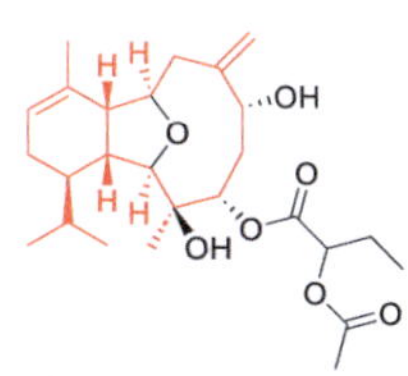

223b. (–)-cladielloide A

Fig. 8 (continued)

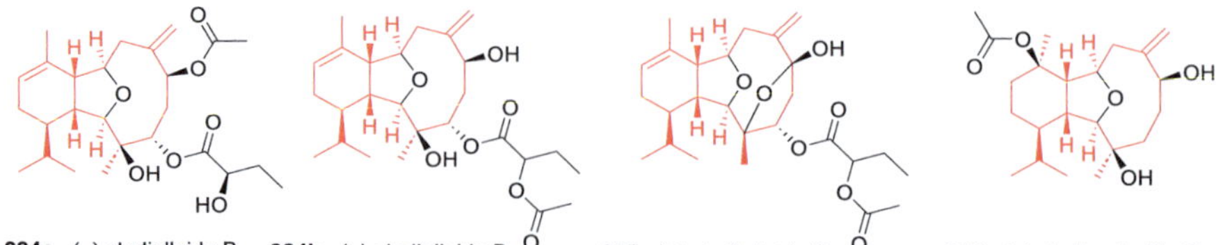

224a. (–)-cladielloide B **224b.** (–)-cladielloide B **225.** (+)-cladielloide C **226.** (–)-cladieunicellin G

Fig. 8 (continued)

was isolated from a marine soft coral *Cladiella* sp. [70]. Its RC was determined by molecular modelling.

2.5.3 Merosesterterpenoid

A 19-norprenyllabdane and C_1 unit constructed a merosesterterpenoid molecule, (+)-diacarperoxide G (**227**) through several enzymatic reactions. Compound (+)-**227** was isolated from SSW collection of the marine sponge *Diacarnus megaspinorhabdosa* [84]. (−)-Diacarperoxide S (megaspinoxide A) (**228**) was also the constituent of the marine sponge *Diacarnus megaspinorhabdosa* possessing a new habdosane (norprenylclerodane) skeleton [143, 144]. The AC of (+)-**227** was assigned by Mosher's method. The skeleton was formed by the sequence of 1,2-hydride and 1,2-methide shifts.

Two MNPs having scalarane skeleton, (−)-**229** and (+)-**230** were isolated from BLI collection of nudibranch *Glossodoris hikuerensis* [145]. Investigations on the biochemical relationship between *Doriprismatica stellata* nudibranchs, their egg ribbons, and the associated dietary sponge *Spongia* cf. *agaricina* (NSW) led to the isolation of the structurally new scalarane-type merosesterterpenoids (+)-12-deacetoxy-4-demethyl-11,24-diacetoxy-3,4-methylenedeoxoscalarin (**231**) possessing a new 21-homoscalarane skeleton and C_2 unit [146]. A marine sponge *Phyllospongia* sp. (SSW) was the source of four 24-homoscalarane metabolites with C_2 units, (+)-**232–235** [147]. A combination of X-ray analysis and Mosher's chemistry secured the AC of (+)-**234**. Furthermore, the RC of (+)-**235** was assigned via a chemical correlation. Compound (+)-**233** was likely an artifact. Additionally, two nor-24-homoscalarane metabolites, (+)-**236** and (+)-**237** were also isolated from the same organism [147].

Four marine sponges, *Strepsichordaia aliena* (EKM) [148, 149], *Phyllospongia* sp. (SSW) [147], *Phyllospongia papyrecea* (NSW) [150], and *Carteriospongia foliascens* (SSW) [151] furnished (+)-phyllofolactones H (**238**), I (**240**), J (**241**), K (**242**), (+)-honulactones C (**246**), D (**247**), I (**248**), J (**249**), K (**250**), L (**251**) [148, 149], (+)-**239** [147], (+)-phyllactone H (**243**) [150], **244/245** [151], respectively possessing a 20,24-bishomoscalarane skeleton. The AC of (+)-honulactone D (**247**) was established by X-ray analysis except for its flexible side chain [149]. Three additional merosesterterpenoids having a nor-20,24-bishomoscalarane framework,

(+)-phyllofenone C (**252**) [148], **253**, and **254** [151] were isolated from *Strepsichordaia aliena* (EKM) and *Carteriospongia foliascens* (SSW), respectively. From the same province, the marine sponge *Strepsichordaia aliena* (EKM) was also origin of a new skeleton honulactane (20,24-bishomoscalarane) as in (+)-honulactones A (**255**), B (**256**), E (**257**), F–H (**258–260**). Partial AC of (+)-**256** was elucidated by X-ray crystallography [149]. Moreover, a nor-20,24-bishomoscalarane (honuane) framework as in (+)-honu'enone (**261**, Fig. 9) was also obtained from the same marine organism [148]. Only 7 molecules possessed unique a functionalized spiro [2, 5] octane as in (+)-**255–261**. A marine sponge *Hyattella* sp. contained two new 22,24-bishomoscalarane metabolites as in (+)-hyattellactones A (**262**) and B (**263**) [152]. The AC of the two compounds was assigned based on comparison of calculated and experimental ECD.

2.5.4 Merotriterpenoid

Two merotriterpenoids, (−)-haliclotriols A (**264**) and B (**265**) possessing a new haliclonane skeleton and ketide part were characterized from a marine sponge *Haliclona* sp. collected from NMU waters [153]. Isomalabaricane skeleton is a signature of marine triterpene metabolites. The marine sponge of *Rhabdastrella globostellata* was the source of thirteen MNPs including (+)-13*Z*, 17*E*-globostellatic acid X methyl ester (**266**) (UEP) [154], globostellatic acid G (**267**) (SSW) [88], globostellatic acid I (**268**) (SSW) [88], globostellatic acid K (**269**) (SSW) [88], (−)-3-*O*-deacetyl-13*Z*-stelliferin riboside (**270**) (SSW) [88], (−)-13*Z*, 17*Z*-globostellatic acid X methyl ester (**271**) (UEP) [154], globostellatic acid J (**272**) (SSW) [88], (−)-13*E*, 17*E*-globostellatic acid X methyl ester (**273**) (UEP) [154], (−)-globostellatic acid F methyl ester (**274**) (UEP) [154], (−)-13*E*-globostellatic acid B methyl ester (**275**) (UEP) [154], globostellatic acid H (**276**) (SSW) [88], (+)-13*E*-stelliferin riboside (**277**) (SSW) [88], (−)-13*E*, 17*Z*-globostellatic acid X methyl ester (**278**) (UEP) [154], possessing isomalabaricane framework. Compounds (+)-**266**, (−)-**271**, (−)-**273–275**, **278** were modified by TMS-CH_2N_2 [154], while compounds **267/276**, **268/276**, **269/272** were obtained as a mixture except for compounds **270** and **277** [88]. Moreover, one octanorisomalabaricane skeleton as in (−)-acetyljaspiferal E (**279**) was also co-isolated with compounds (+)-**266**, (−)-**271,** (−)-**273–275**, and **278** [154]. (+)-Plakohopanoid peracetate (**280**, Fig. 10) owning the plakohopane framework was isolated from the marine sponge *Plakortis* cf. *lita* after the fraction of the sponge extract was modified by Ac_2O [155]. The AC of (+)-**280** was determined by chemical transformation and CD analysis. The tropical marine ciliate *Euplotes vannus* collected from NSW furnished merotriterpenoid, (−)-vannusals A (**281a**) and B (**282a**) [156] possessing an unprecedented vannusane skeleton. The RCs of (−)-**281a** and **282a** were established by molecular modelling [156]. Later, total syntheses [157–162] required stereochemical revision of (−)-**281a** and **282a** to (−)-**281b** and **282b**, respectively establishing the correct AC of the two compounds. Probably, the oldest marine merotriterpenoid in Indonesia was (−)-23ξ-acetoxy-17-deoxy-7,8-dihydroholothurinogenin (**283**) isolated from *Stichopus chloronotus*

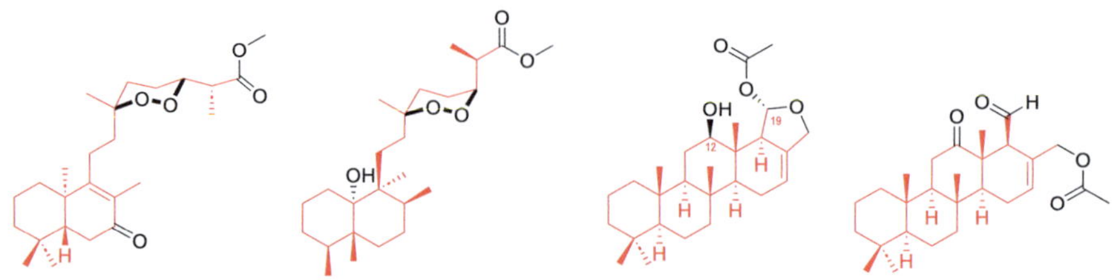

227. (+)-diacarperoxide G **228.** (−)-diacarperoxide S (megaspinoxide A) **229.** (−)-19-acetyl-12-deacetyl-12-*epi*-deoxoscalarin (+)-**230**

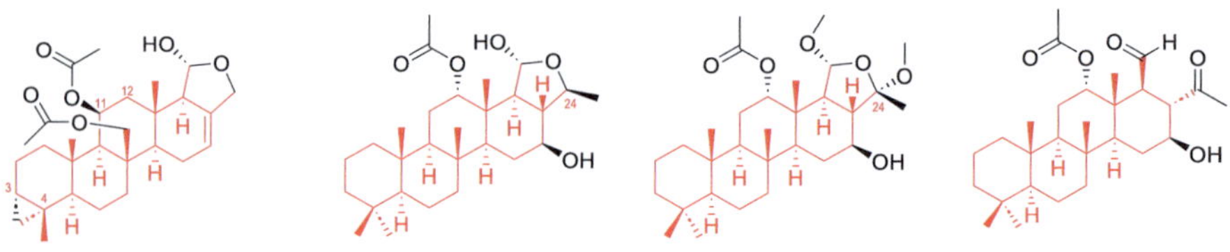

231. (+)-12-deacetoxy-4-demethyl-11,24-diacetoxy-3,4-methylenedeoxoscalarin (+)-**232** (+)-**233** (+)-**234**

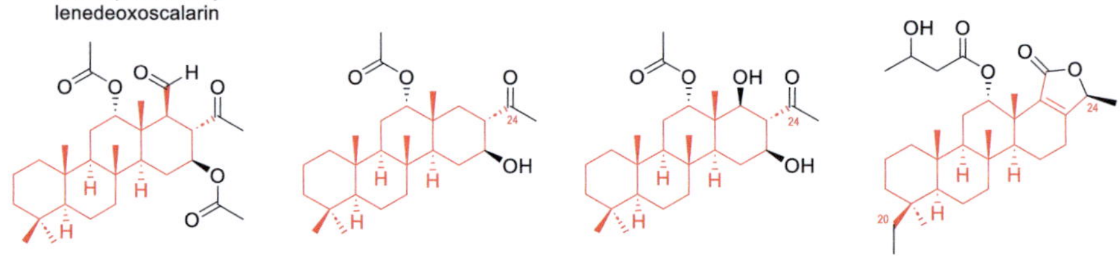

(+)-**235** (+)-**236** (+)-**237** **238.** (+)-phyllofolactone H

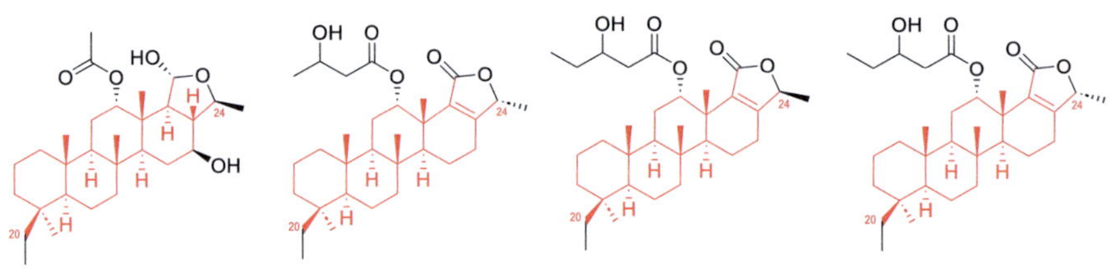

(+)-**239** **240.** (+)-phyllofolactone I **241.** (+)-phyllofolactone J **242.** (+)-phyllofolactone K

Fig. 9 Structures of marine merosesterterpenoids (**227–263**) from Indonesian waters. The marine sesterterpene skeletons were proposed with red color

Brandt (NST) in 1973 [163]. Its RC was elucidated by a combination of techniques including NMR, ORD, CD, and chemical transformation. Three (−)-sarasinosides J–L (**284–286**) [164] and two (−)-sarasinosides M (**287**) [164], S (**288**) [165] were isolated as merotriterpenoid from the marine sponge *Melophlus sarasinorum* and *Petrosia* sp., (NSW) respectively. The RC of (−)-**284** was assigned by chemical transformation. (−)-Sarasinoside M (**287**) possessed a 11-oxabicyclo [4,4,1] system with anti-Bredt and bridgehead double bonds [166]. Moreover, the marine sponge *Melophlus sarasinorum* was also source of (−)-melophluosides A (**289**) and B (**290**)

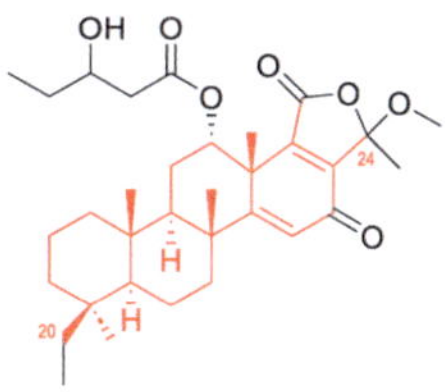

243. (+)-phyllactone H

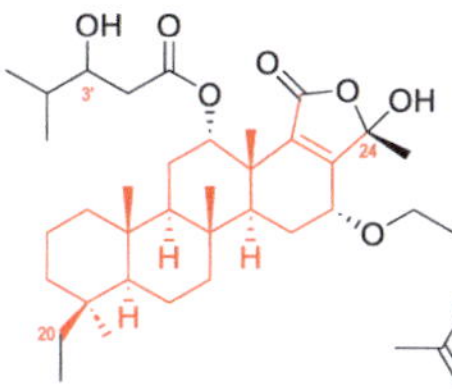

244/245

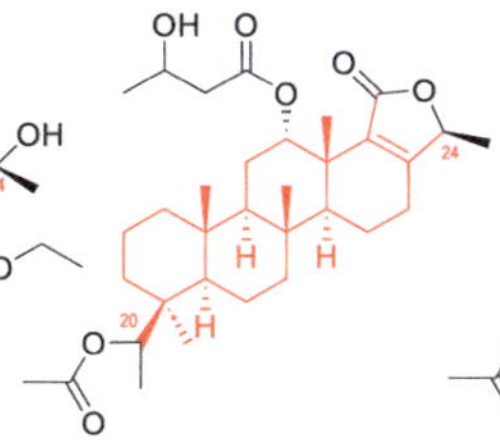

246. (+)-honulactone C

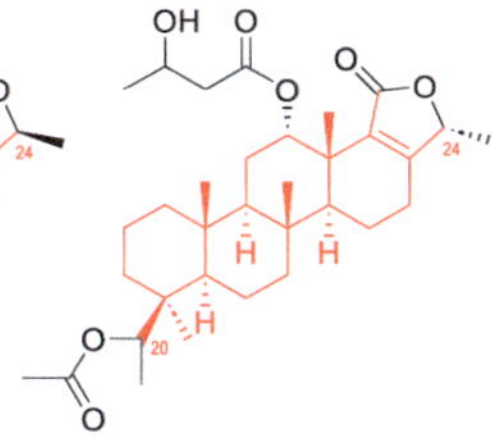

247. (+)-honulactone D

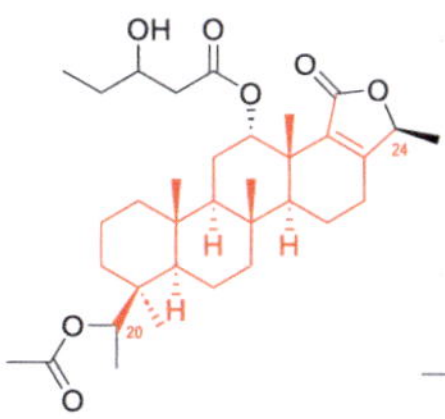

248. (+)-honulactone I

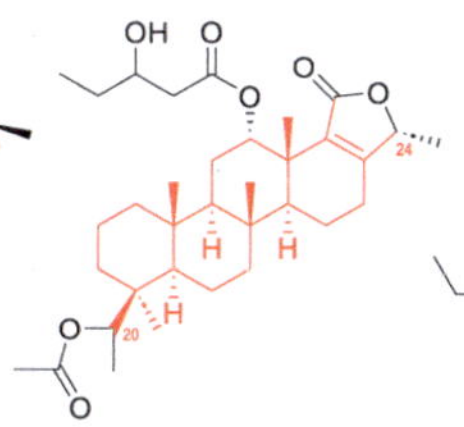

249. (+)-honulactone J

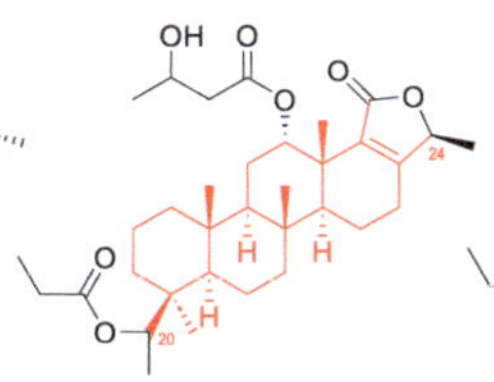

250. (+)-honulactone K

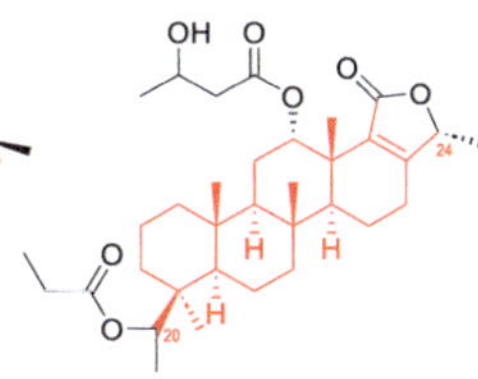

251. (+)-honulactone L

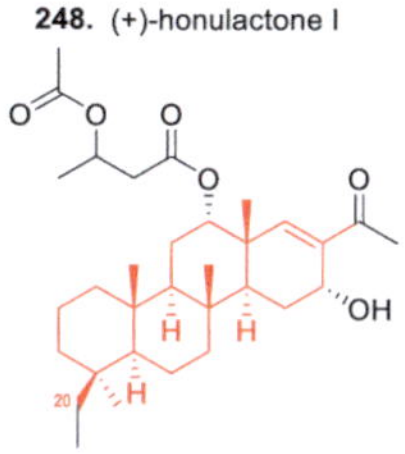

252. (+)-phyllofenone C

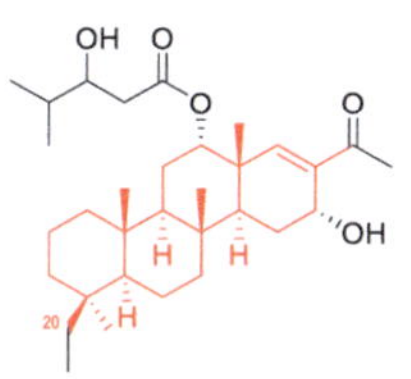

253

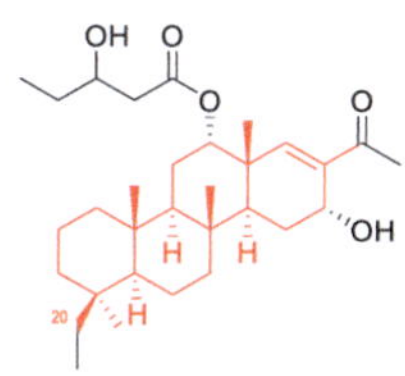

254

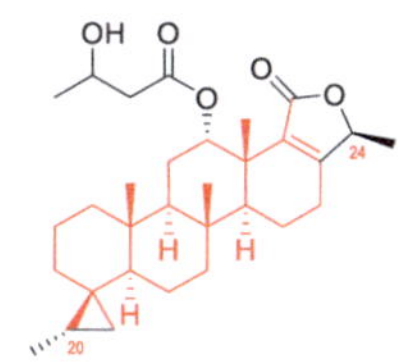

255. (+)-honulactone A

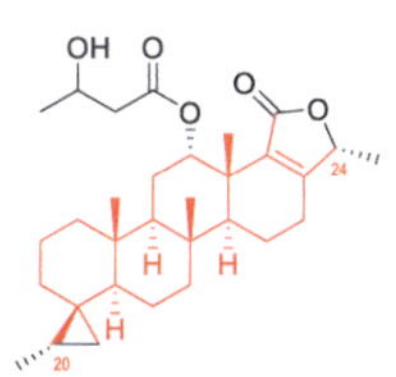

256. (+)-honulactone B

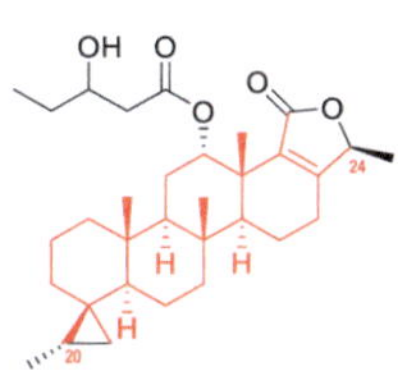

257. (+)-honulactone E

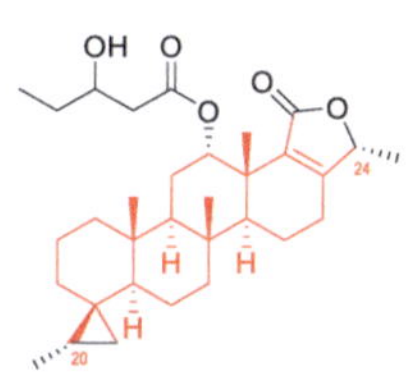

258. (+)-honulactone F

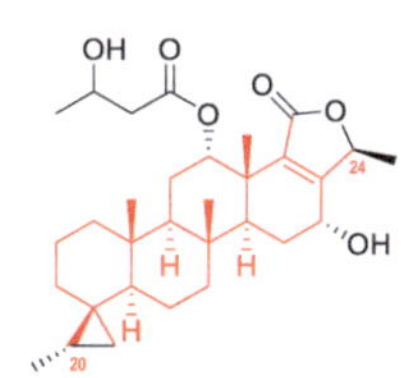

259. (+)-honulactone G

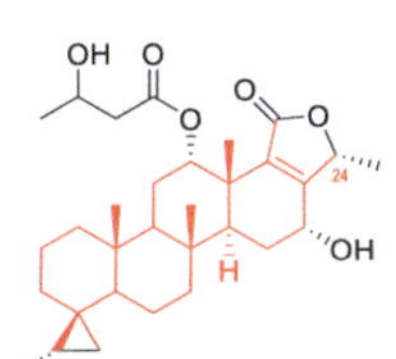

260. (+)-honulactone H

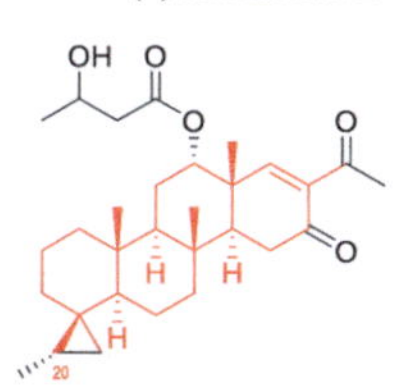

261. (+)-honu'enone

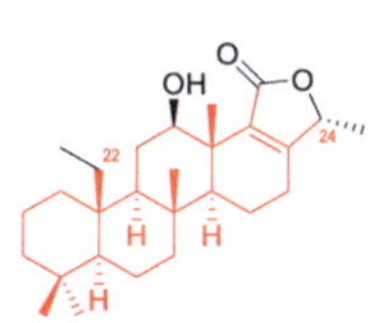

262. (+)-hyattellactone A

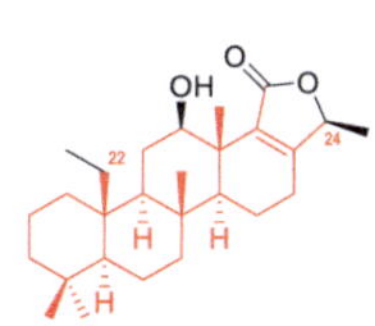

263. (+)-hyattellactone B

Fig. 9 (continued)

[167] possessing pouane skeleton with irregular coupling of C_5 units. The ACs of (−)-**289** and **290** were determined via comparison of calculated and experimental ECD. The skeleton was named after the discovery of pouosides A–E from a marine sponge *Asteropus* sp. collected from Pou Bay, Micronesia [168].

2.5.5 Merosteroid

(+)-25*S*-18-Acetoxy-3-oxocholesta-1,4-dien-26-oic acid (**291**) was isolated from a soft coral *Minabea* sp. (NSW) having the cholestane skeleton and the C_2 building block [89]. Its AC was elucidated by comparative ^{1}H NMR study using chiral anisotropic reagents [(*S*)- and (*R*)-phenylglycine methyl esters]. At least, fifteen MNPs, (−)-25-hydroxy-24ξ-methylcholesterol (**292**) [169], 24ξ-methylcholestane-3β,5α,6β-triol (**293**), (−)-24ξ-methylcholestane-3β,5α,6β-triol-6-monoacetate (**294**) [170], (−)-24*S*-methylcholestan-1α,3β-diol (**295**), (−)-24*S*-methylcholestan-1α,3β,25-triol (**296**), (−)-24*S*-methylcholestan-11-acetoxy-1α,3β,5α,6β-tetraol (**297**) [171], (+)-hippuristerol D (**298**), (+)-hippuristerone D (**299**), (−)-11-dehydroxy-22-*epi*-hippuristanol (**300**) [172], (−)-3-acetyl-22-*epi*-hippuristanol (**301**), (−)-11-dehydroxy-22-*epi*-hippuristan-3-one (**302**), (+)-orthohippurinsterols A (**303**), B (**304**), (−)-orthohippurinsterone A (**305**) [172], and (−)-desulfohaplosamate (**306**) [173] possessing a 24-methylcholestane skeleton were distributed into four species of marine soft corals (*Nepthea* sp. [169], *Sinularia dissecta* [170], *Sinularia* sp. [171], *Isis hippuris* [172]) and one species of marine sponge (*Dasychalina* sp.) [173], respectively. The RCs of (−)-**292–294** were established by chemical correlations [169, 170]. Six merosteroids 12α-hydroxy-24-methylenecholesterol (**307**, MLU) [64], (−)-24-methylenecholestane-3β,5α,6β-triol-6-monoacetate (**308**, MLU) [170], (+)-24-methylenecholest-5-en-3β,7β,19-triol (**309**, MLU) [174], (+)-4α-methyl-3β,8β-dihydroxy-5α-ergost-24(28)-en-23-one (**310**, UEP) [175], (+)-dehydroconicasterol (**311**, NSW) [176], 24-methylenecholestane-3β,5α,6β-triol (**312**, MLU) [170] having a 24-methylenecholestane skeleton were obtained from marine soft corals and a sponge. Compounds **307**, (+)-**309**, and (+)-**310** were isolated from *Litophyton viridis*, while compounds (−)-**308** and **312** were obtained from *Sinularia dissecta.* Compound (+)-**311** was found in *Theonella swinhoei*. X-ray studies established the ACs of (+)-**309** and **310** [175, 177]. Compounds (−)-**308** and **312** were obtained as a mixture [170].

24-Isopropylcholestane skeleton was a signature of marine merosteroid as demonstrated by (−)-topsentinol K (**313**), (+)-topsentinol K trisulfate (**314**), and (+)-topsentinol L (**315**) [178]. They were isolated from *Topsentia* sp. collected from EKM province. The octocoral *Isis hippuris* (UEP) was also the source of 23,24-dimethylcholestane metabolites like (+)-hippuristerols B (**316**), A (**317**), C (**318**), (+)-hippuristerones B (**319**) and C (**320**) [172]. Three merosteroids, (+)-lokysterolamine A (**321**), (−)-lokysterolamine B (**322**) [179] and (−)-kaimanol (**323**) [180] having a 24-ethylcholestane skeleton were characterized from a marine sponge *Corticium* sp. (NSW) [179] and *Xestospongia* sp. (SWP) [180], respectively. A mixed biosynthesis

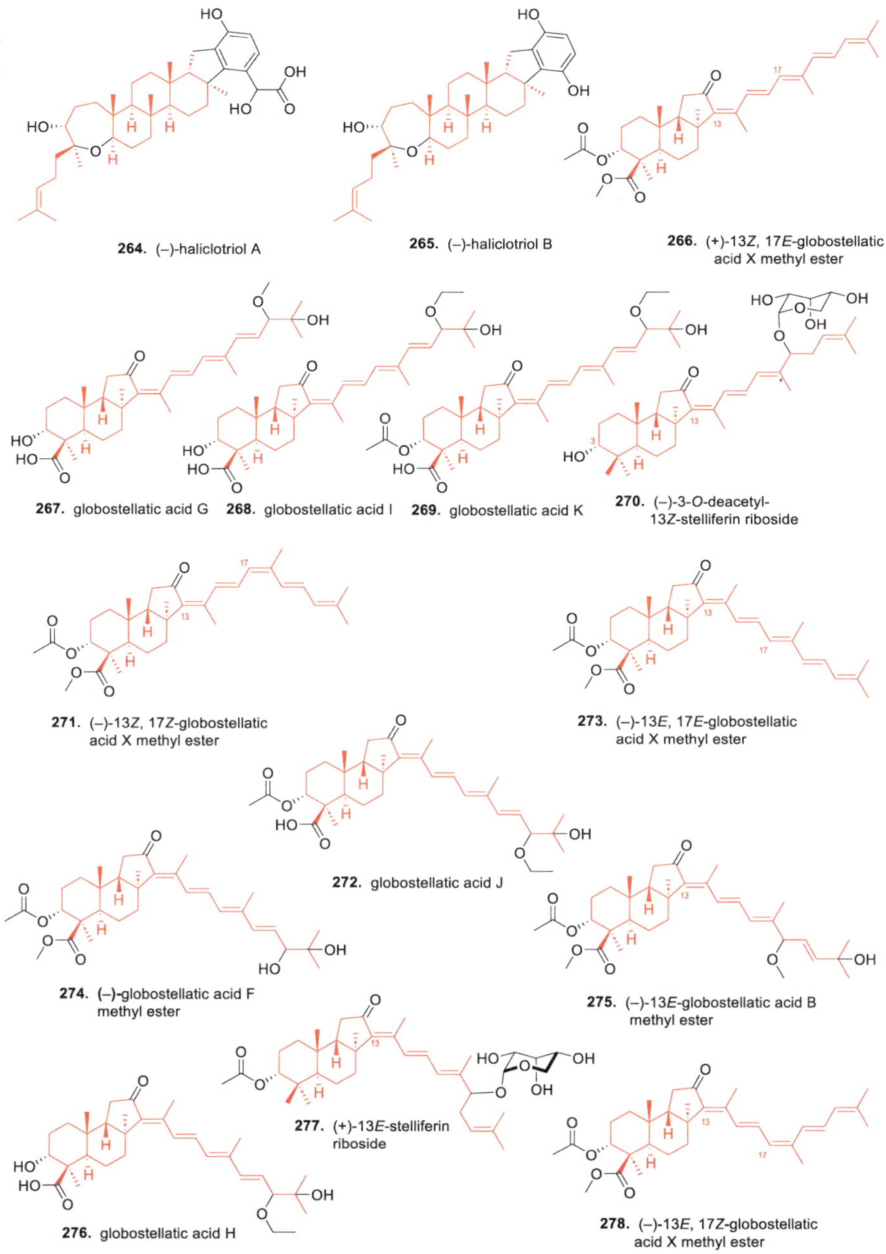

Fig. 10 Structures of marine merotriterpenoids (**264–290**) from Indonesian waters. The marine triterpene skeletons were proposed with red color

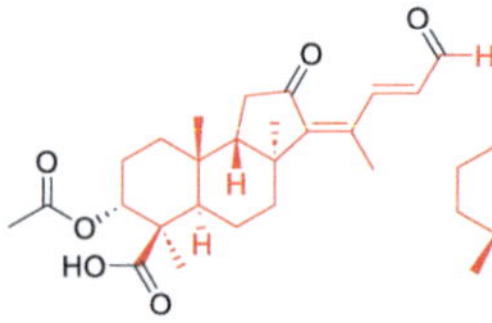

279. (–)-acetyljaspiferal E

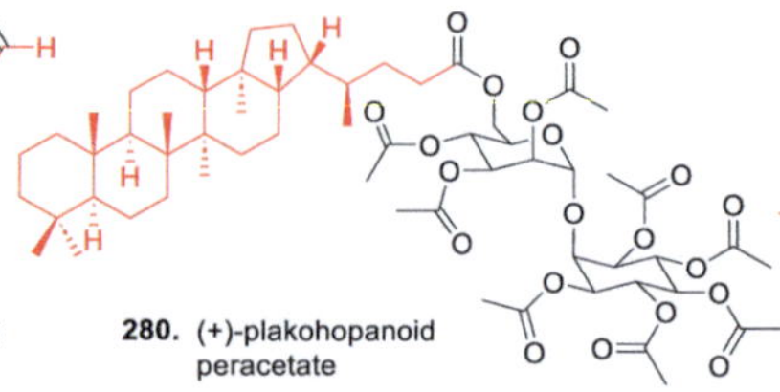

280. (+)-plakohopanoid peracetate

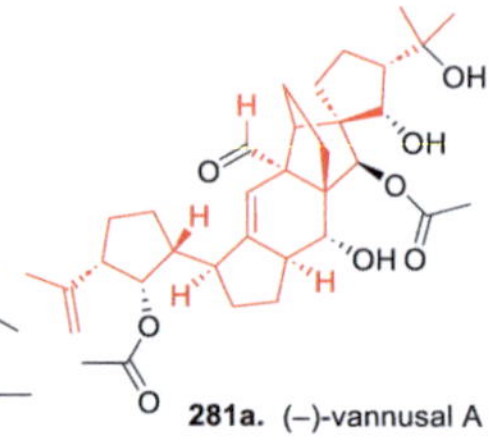

281a. (–)-vannusal A

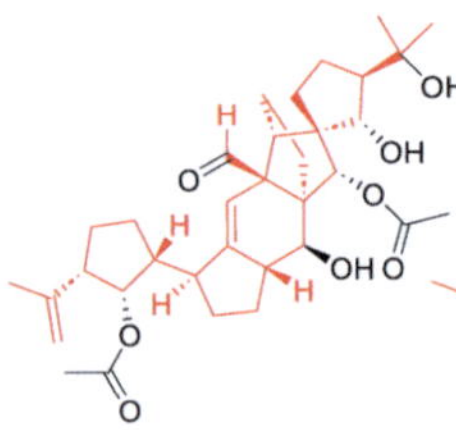

281b. (–)-vannusal A

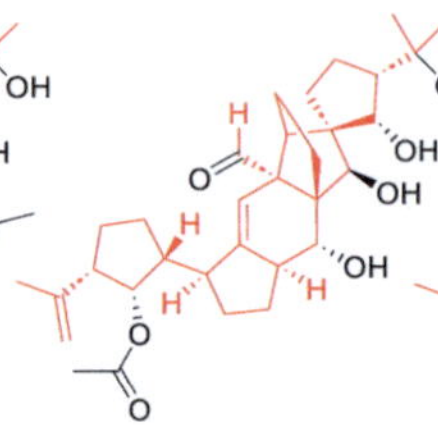

282a. (–)-vannusal B

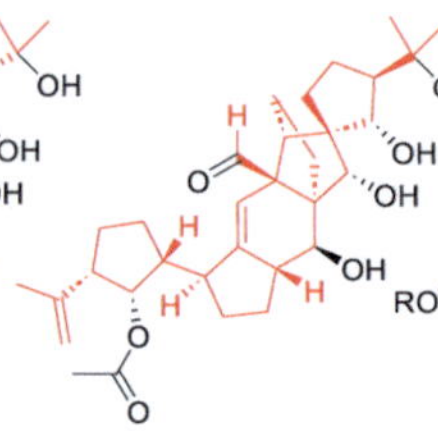

282b. (–)-vannusal B

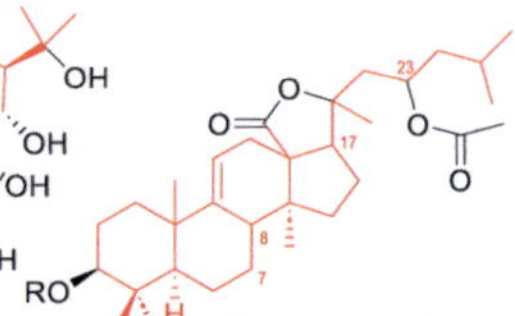

283. (–)-23ξ-acetoxy-17-deoxy-7,8-dihydroholothurinogenin

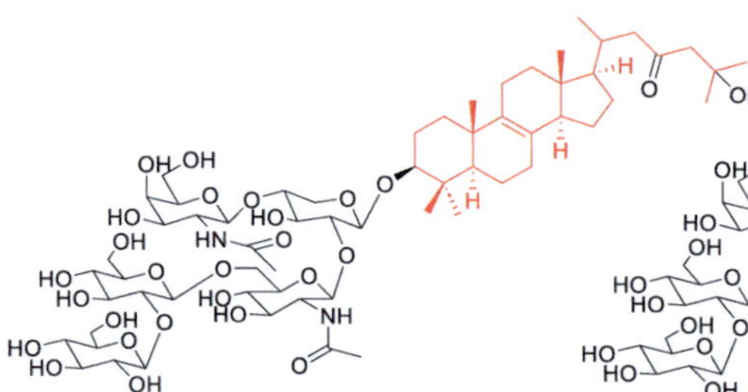

284. (–)-sarasinoside J

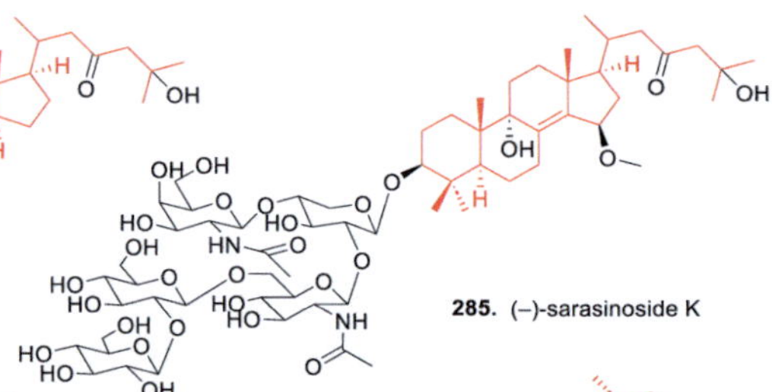

285. (–)-sarasinoside K

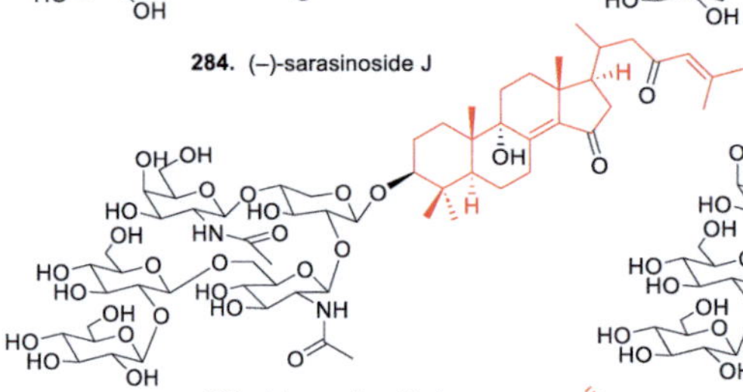

286. (–)-sarasinoside L

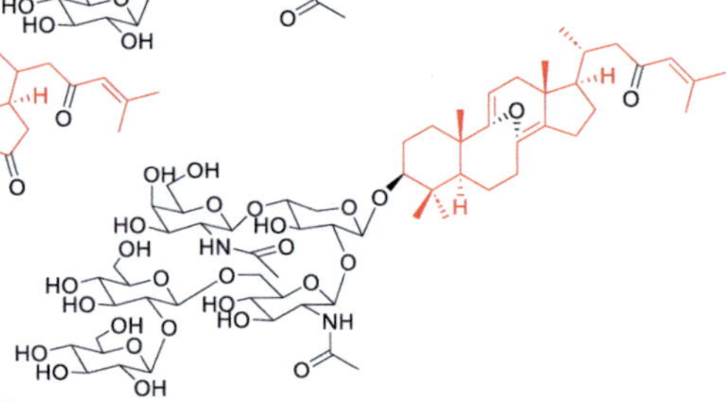

287. (–)-sarasinoside M

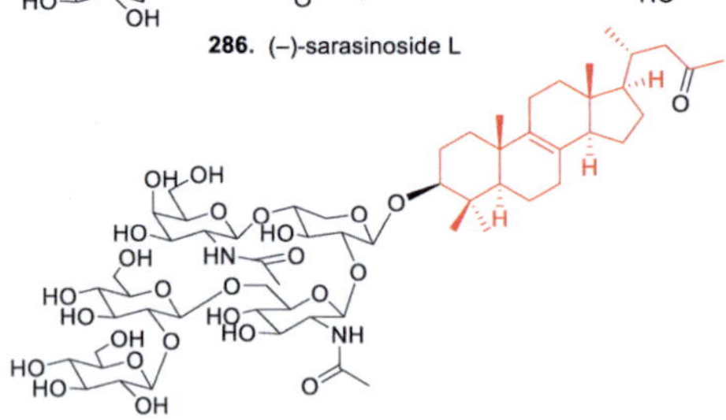

288. (–)-sarasinoside S

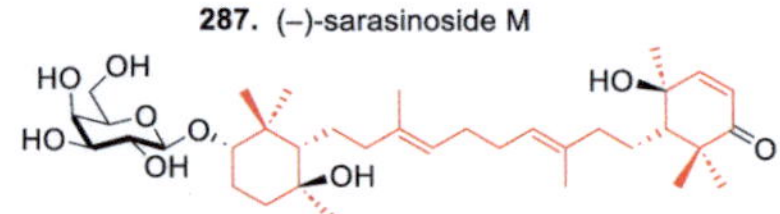

289. (–)-melophluoside A

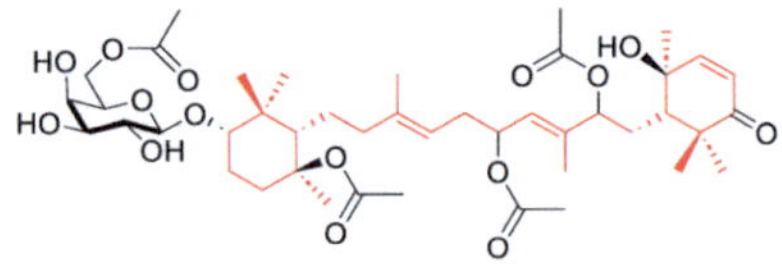

290. (–)-melophluoside B

Fig. 10 (continued)

of 3-methyl-4-norcholestane and butanoic acid gave a merosteroid named clathurohoate (**324**) collected from SES [181]. (−)-24*S**,3β,11-Dihydroxy-5β,6β-epoxy-24-methyl-9,11-secocholestan-9-one (**325**) possessing a seco-cholestane skeleton was found in the CSW collection of the marine soft coral *Pachyclavularia violacea* [182]. Gorgostane, a signature skeleton of MNPs, was isolated from *Sinularia* sp. (NSW) represented by (−)-3α,5β-dihydroxygorgostan-6-one (**326**) [171]. A marine soft coral *Lobophytum* sp. (NMU) was the origin of (+)-3β,7β,11-trihydroxy-5α,6α-epoxy-9,11-secogorgostan-9-one (**327**) [183] possessing the seco-gorgostane skeleton. Eleven merosteroids with a new cortistane or *abeo*-9(10–19) cholestane skeleton (Fig. 11), (+)-cortistatins A (**328**), B (**329**) [184], F (**338**) [185] and (−)-cortistatins C (**330**), D (**331**) [184], J (**332**), K (**333**), L (**334**) [186], G (**335**), H (**336**), E (**337**) [185] were discovered from the marine sponge *Corticium simplex* (ENT). The AC of (+)-**328** was determined by X-ray and CD analysis [184] as well as total synthesis [187]. In addition to (+)-**328**, the AC of (−)-**332** was secured by total synthesis [187].

3 Alkaloid Derived Amino Acid

The classification of alkaloid can be approached by analysis of amino acids building blocks. At least thirteen regular amino acids including L-tryptophan (Trp), L-tyrosine (Tyr), L-lysine (Lys), L-arginine (Arg), L-serine (Ser), glycine (Gly), L-alanine (Ala), L-aspartic acid (Asp), L-histidine (His), L-phenylalanine (Phe), L-valine (Val), L-asparagine (Asn), and L-cysteine (Cys) were involved in the biosynthesis of Indonesian marine alkaloids. Moreover, eight additional building blocks can be identified from the Indonesian marine alkaloids including tryptamine, degraded tryptamine, kynuramine, degraded kynuramine, anthranilic acid, tyramine, dopamine, and L-ornithine. Therefore, the alkaloids derived amino acid were composed of 329 molecules. All the building blocks are commonly found in terrestrial and marine organism metabolites except for degraded tryptamine and degraded kynuramine. Furthermore, the 329 molecules can approximately be classified into 7 subclasses consisted of 51 alkaloid derived tryptophan, 82 alkaloid derived tyrosine, 38 alkaloid derived lysine, 67 alkaloid derived arginine, 27 alkaloid derived serine, 26 alkaloid derived glycine, 24 alkaloid derived alanine, and 14 alkaloid derived miscellaneous amino acids. The building blocks of Indonesian marine alkaloid derived amino acids are presented in Fig. 12.

3.1 *Alkaloid Derived Tryptophan*

Six building blocks derived tryptophan can be identified from Indonesian MNPs including L-tryptophan, tryptamine, degraded tryptamine, kynuramine, degraded kynuramine and anthranilic acid. Eleven MNPs possessing L-tryptophan

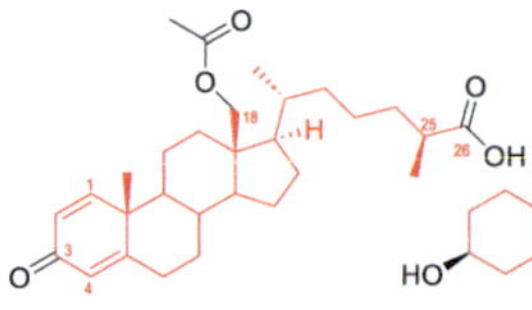

291. (+)-25*S*-18-acetoxy-3-oxocholesta-1,4-dien-26-oic acid

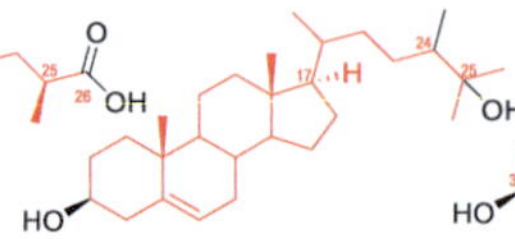

292. (–)-25-hydroxy-24ξ-methylcholesterol

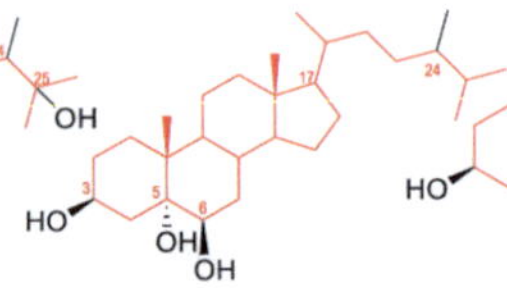

293. 24ξ-methylcholestane-3β,5α,6β-triol

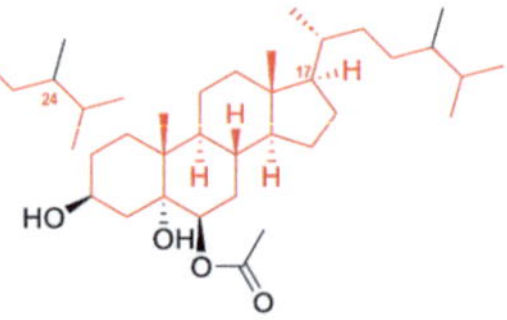

294. (–)-24ξ-methylcholestane-3β,5α,6β-triol-6-monoacetate

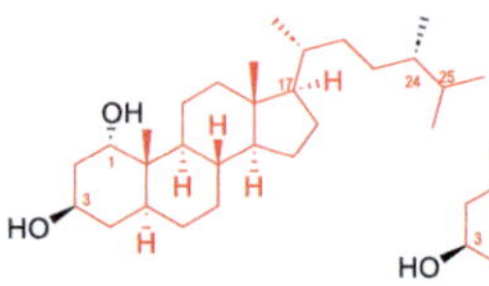

295. (–)-24*S*-methylcholestan-1α,3β-diol

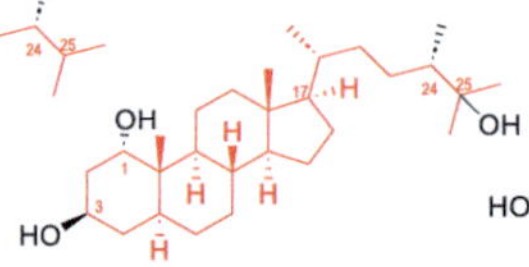

296. (–)-24*S*-methylcholestan-1α,3β-25-triol

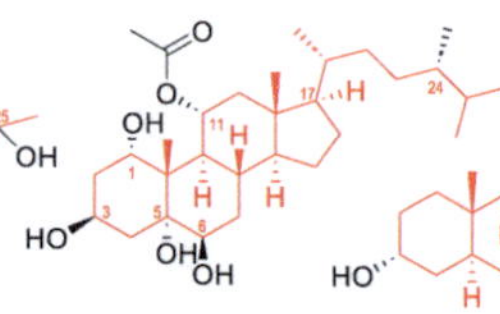

297. (–)-24*S*-methylcholestan-11-acetoxy-1α,3β,5α,6β-tetraol

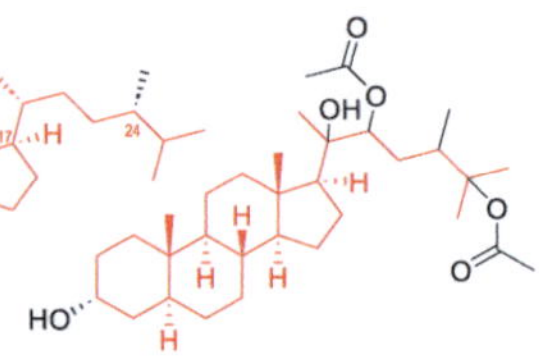

298. (+)-hippuristerol D

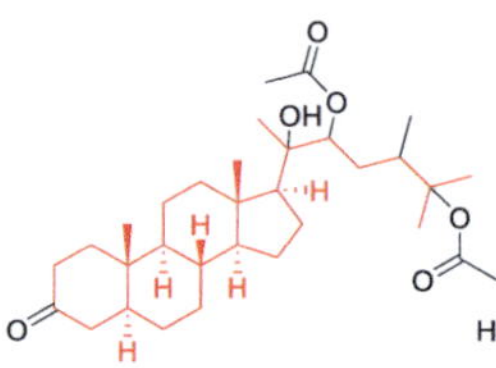

299. (+)-hippuristerone D

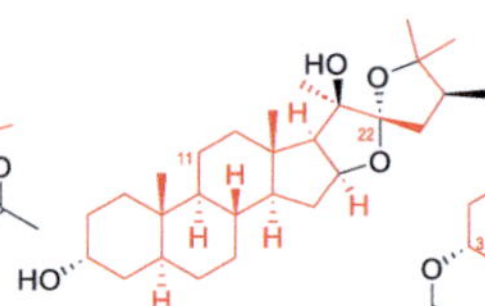

300. (–)-11-dehydroxy-22-*epi*-hippuristanol

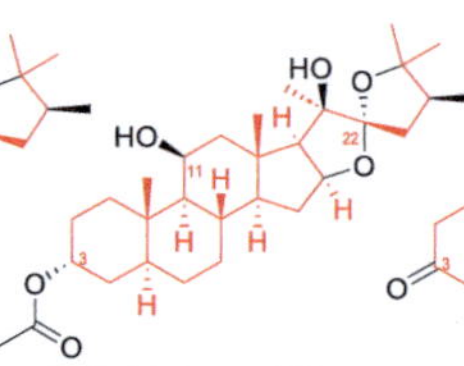

301. (–)-3-acetyl-22-*epi*-hippuristanol

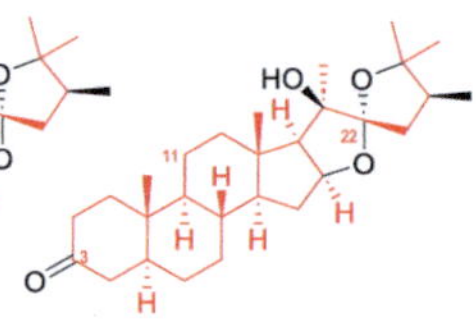

302. (–)-11-dehydroxy-22-*epi*-hippuristan-3-one

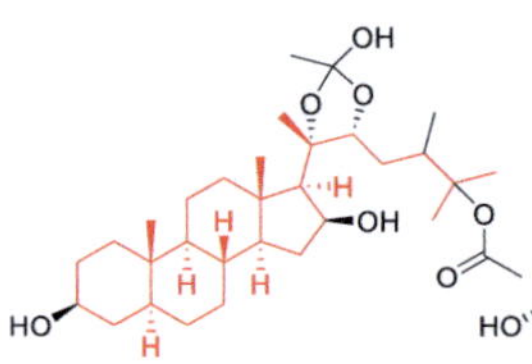

303. (+)-orthohippurinsterol A

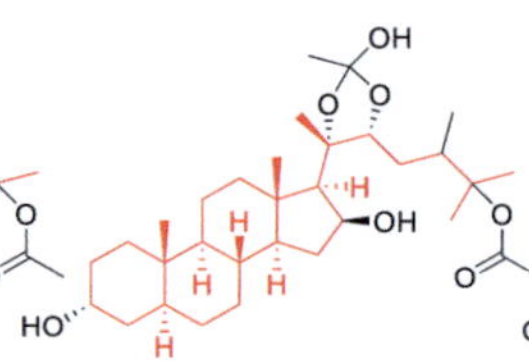

304. (+)-orthohippurinsterol B

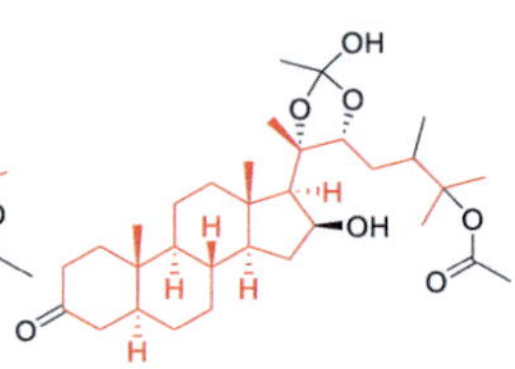

305. (–)-orthohippurinsterone A

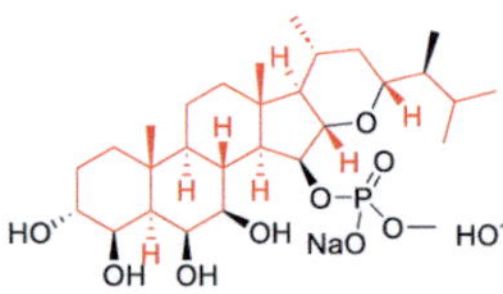

306. (–)-desulfohaplosamate

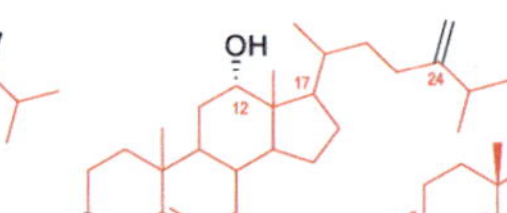

307. 12α-hydroxy-24-methylenecholesterol

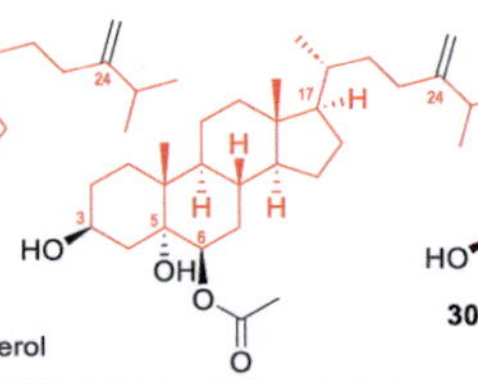

308. (–)-24-methylenecholestane-3β,5α,6β-triol-6 monoacetate

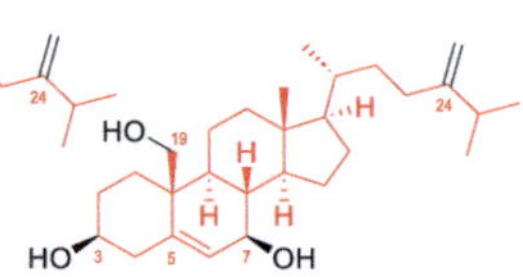

309. (+)-24-methylenecholest-5-en-3β,7β,19-triol

Fig. 11 Structures of marine merosteroids (**291–338**) from Indonesian waters. The marine steroid skeletons were proposed with red color

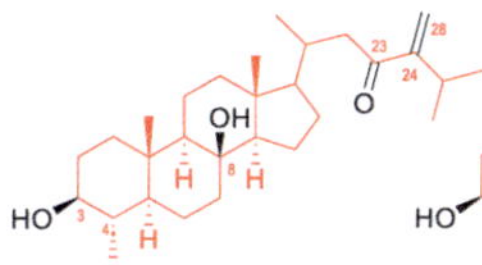
310. (+)-4α-methyl-3β,8β-dihydroxy-5α-ergost-24(28)-en-23-one

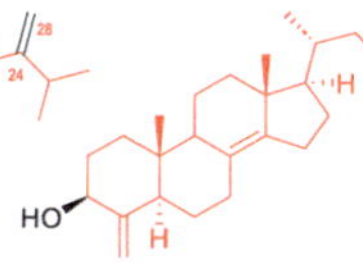
311. (+)-dehydroconicasterol

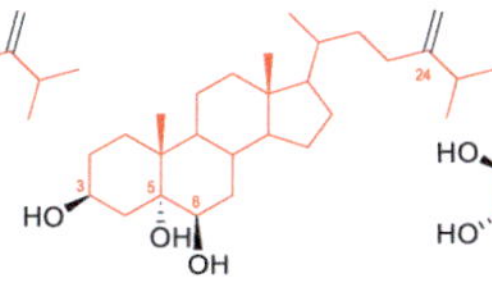
312. 24-methylenecholestane-3β,5α,6β-triol

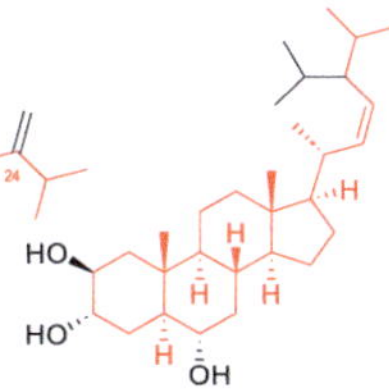
313. (−)-topsentinol K

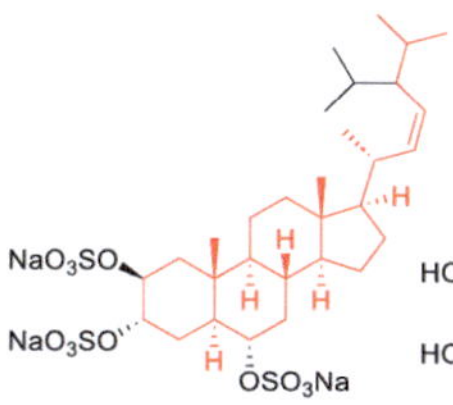
314. (+)-topsentinol K trisulfate

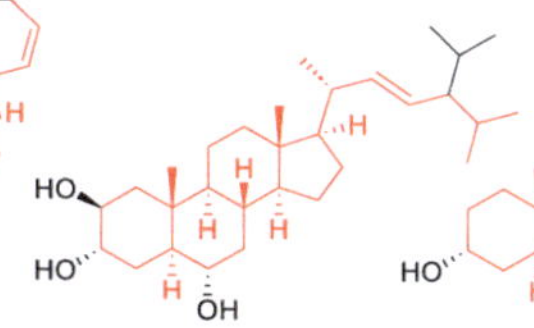
315. (+)-topsentinol L

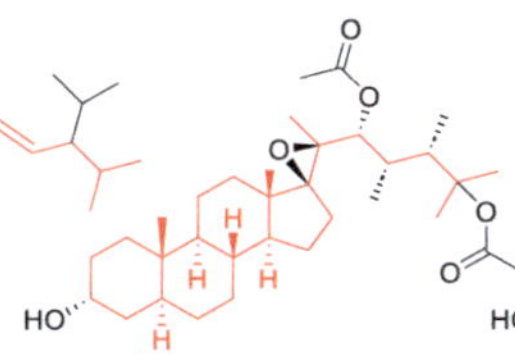
316. (+)-hippuristerol B

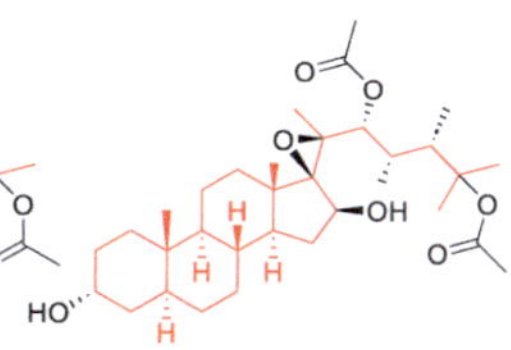
317. (+)-hippuristerol A

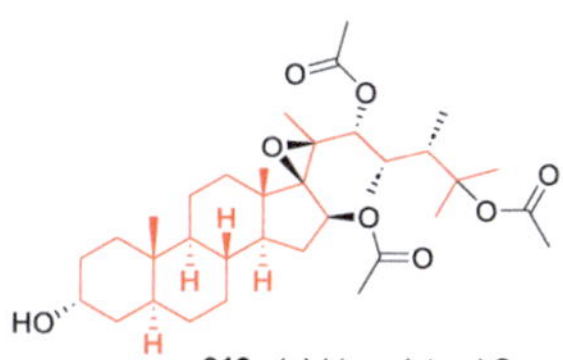
318. (+)-hippuristerol C

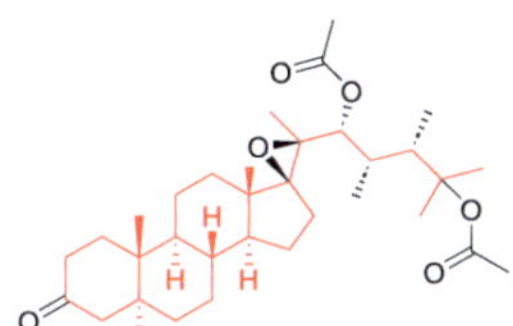
319. (+)-hippuristerone B

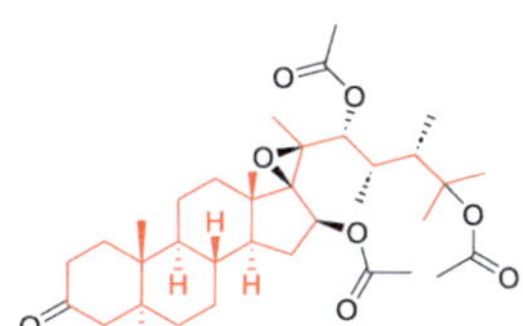
320. (+)-hippuristerone C

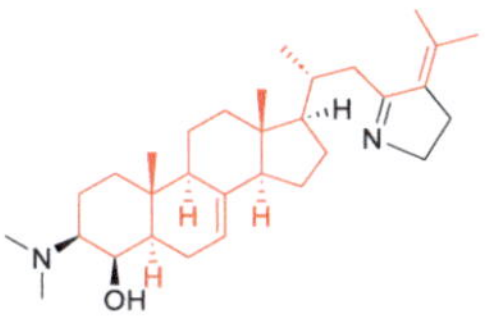
321. (+)-lokysterolamine A

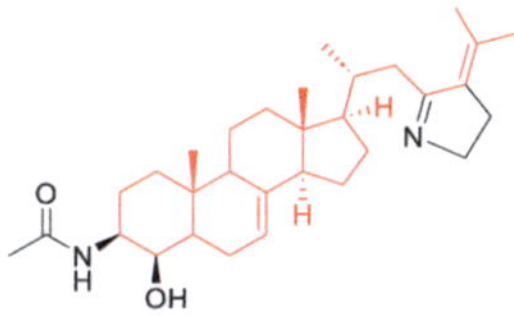
322. (−)-lokysterolamine B

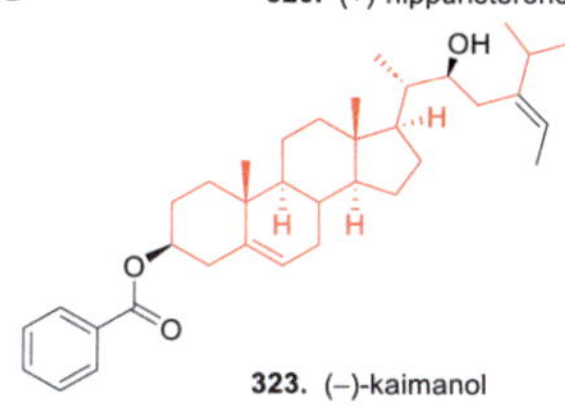
323. (−)-kaimanol

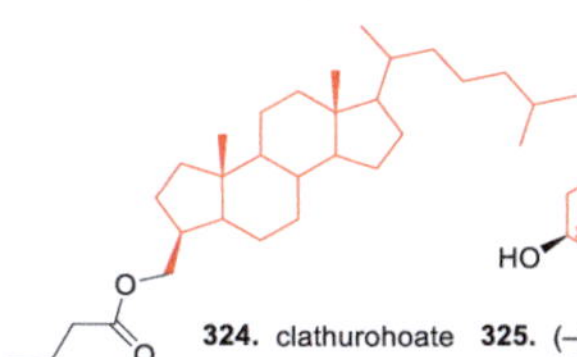
324. clathurohoate

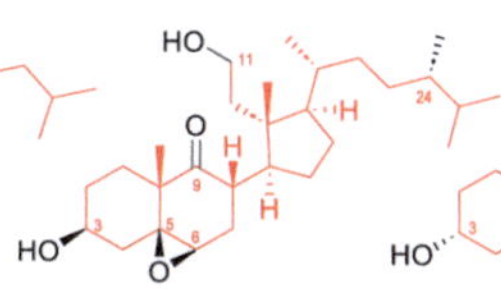
325. (−)-24*S**,3β,11-dihydroxy-5β,6β-epoxy-24-methyl-9,11-secocholestan-9-one

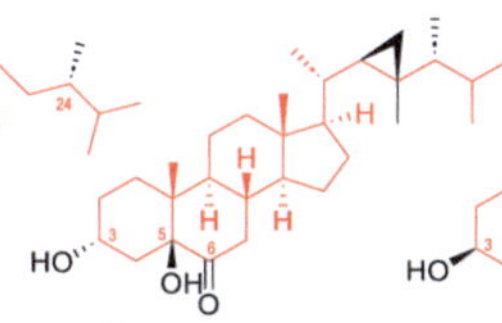
326. (−)-3α,5β-dihydroxy gorgostan-6-one

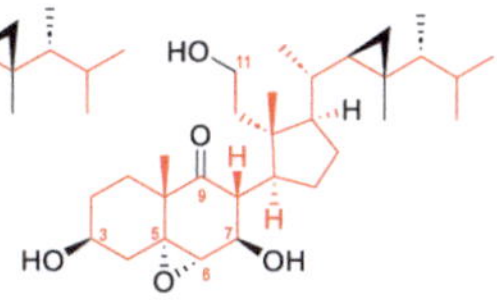
327. (+)-3β,7β,11-trihydroxy-5α,6α-epoxy-9,11-seco-gorgostan-9-one

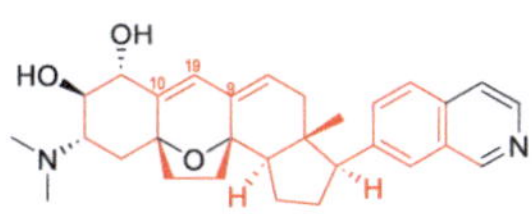
328. (+)-cortistatin A

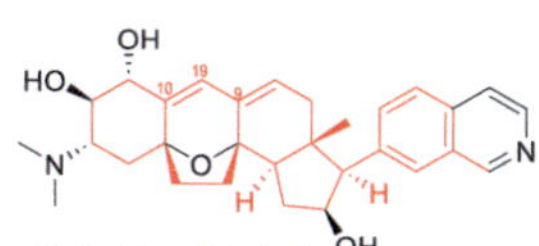
329. (+)-cortistatin B

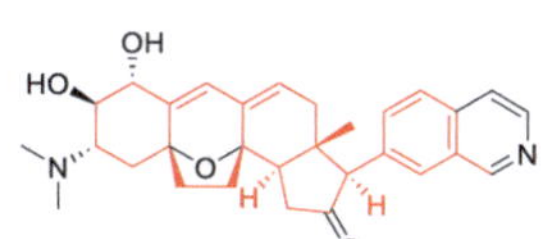
330. (−)-cortistatin C

Fig. 11 (continued)

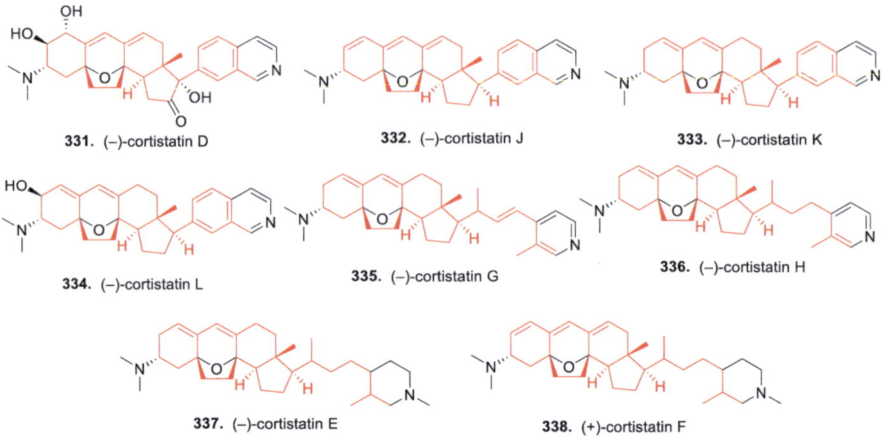

Fig. 11 (continued)

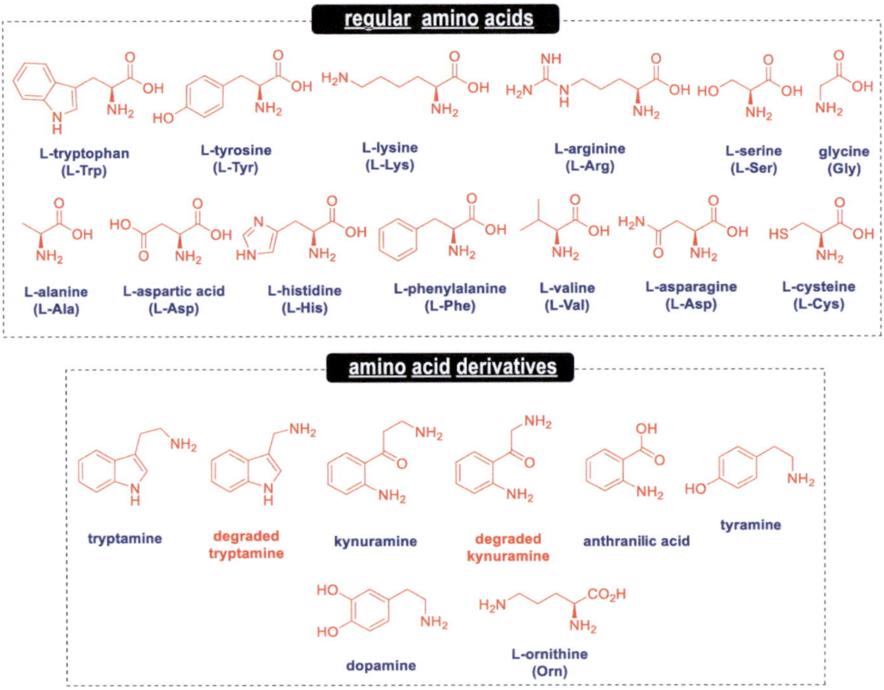

Fig. 12 Structure of amino acids and their derivative building blocks for Indonesian alkaloid derived amino acids. Skeleton names in blue were already reported in the literature, those in red were proposed as new building blocks

building block were distributed into three marine macroorganisms: the sea anemone *Heteractis aurora* [(+)-2-amino-1,5-dihydro-5-(1*H*-indol-3-ylmethyl)-4*H*-imidazol-4-one (**339**, BLI), (+)-2-amino-5-[(6-bromo-1*H*-indol-3-yl)methyl]-3,5-dihydro-3-methyl-4*H*-imidazol-4-one (**340**, BLI)] [188], the marine sponges *Hyrtios reticulatus* [(+)-hyrtioreticulin C (**341a**, NSW), (−)-hyrtioreticulins D (**342a**, NSW), E (**343**, NSW), (+)-hyrtioreticulin A (**345**, NSW), (−)-hyrtioreticulin B (**346**, NSW) [189], (−)-hyrtioreticulatin F (**344**, NSW) [190], *Luffariella variabilis* [variabines B (**347**, NSW), A (**348**, NSW)] [191] and one marine microorganism: the marine fungus *Aspergillus clavatus* associated with a tunicate [(−)-12*S*-deoxynortryptoquivaline (**349**, WST)] [192]. Two compounds (+)-**341a** and (−)-**342a** were revised their optical rotations to (−)-**341b** and (+)-**342b**, respectively by total synthesis [193]. A trace amount of trifluoracetic acid (TFA) during isolation work induced a racemization. The ACs of (+)-**339**, **340**, (−)-**341b**, (+)-**342b** were determined by total synthesis [188, 193] while the AC of (−)-12*S*-deoxynortryptoquivaline (**349**) was elucidated by comparison of calculated and experimental ECD after revealing its RC by residual dipolar coupling (RDC) analysis and NMR calculation [192].

Tryptamine building block was identified from twenty-eight Indonesian MNPs as in **350–377** (Fig. 13). A marine sponge *Oceanapia* sp. collected from NSW waters was the source of 6-Br-8-keto-conicamin A (**350**) [194]. Its planar structure was confirmed through synthesis. Two type Pictet-Spengler-enzymatic reactions were involved in the marine metabolites through cyclizations at C4 giving azepinoindole as in **351** and C2 affording β-carboline as in **352–377**. A mixed biosynthesis of makaluvamine G (**351**) was discovered from a marine sponge *Histodermella* sp. (NSW) [195]. 1,6-Dihydroxy-1,2,3,4-tetrahydro-β-carboline was predicted to present in a marine sponge *Hyrtios* sp. [196]. An unidentified sponge from family *Petrosiidae* furnished 1,2,3,4-tetrahydronorharman-1-one (**352**) and des-*N*-methylxestomanzamine A (**357**) [197]. Both were obtained from NSW province. Compound **352** was the first report as natural compound. The marine sponge *Acanthostrongylophora ingens* was the source of (+)-ingenine F (**353**, UEP) [198], (−)-acanthomine A (**356**, NSW) [199], (+)-ingenine A (**358**, NSW) [200], (−)-ingenines C (**359**, UEP), D (**360**, UEP) [201], and ingenine E (**361**, UEP) [202]. Two marine alkaloids assigned as 1-carboxytrypargine (**354**) and trypargimine (**355**) [203] were isolated from a marine tunicate *Eudiostoma* sp. (SSW). The RC of **354** was elucidated by experimental and literature ECD of its derivative [203]. Hyrtiosulawesine (**362**) was isolated from a marine sponge *Hyrtios* sp. (SSW) [196]. Its planar structure was confirmed by total synthesis [204, 205]. Another marine tunicate, *Didemnum* sp. was the source of fascaplysin metabolites exemplified by (+)-homofascaplysate A (**370**, CSW), 3-bromohomofascaplysin B (**371**, CSW) [206], homofascaplysin B-1 (**372**, CSW) [206], 3-bromohomofascaplysin B-1 (**373**, CSW) [206], 3-bromohomofascaplysin C (**374**, CSW) [206], 3-bromosecofascaplysin B (**375**, CSW) [206]. Moreover, the marine sponge *Fascaplysinopsis reticulata* was also the source of fascaplysin compounds including 14-bromoreticulatol (**363**, CSW) [206], 14-bromoreticulatate (**364**, UEP), 14-bromoreticulatine (**365**, UEP) [207], 7,14-dibromoreticulatine (**366**, CSW) [206], 3-bromofascaplysin (**367**, UEP) [207], 10-bromofascaplysin (**368**, CSW) [206], 3,10-dibromofascaplysin (**369**, CSW)

[206], and 3-bromosecofascaplysin A (**376**) (CSW) [206]. Compounds **364**, **365**, **367–369**, **371–373** were confirmed their planar structures by total syntheses [208–210]. (−)-Ingenine B (**377**) was the only Indonesian MNPs possessing γ-carboline moiety isolated from the marine sponge *Acanthostrongylophora ingens* (SSW) [200]. Its planar structure was secured by total synthesis [211]. The compound possessed a degraded tryptamine building block.

Kynuramine building block was found in nine marine acridine metabolites (**378–386**) (Fig. 13). Styelsamines A–D (**378–381**) were isolated from the marine tunicate *Eusynstyela latericius* (SSW) [212]. The planar structures of **379–381** were confirmed by total syntheses [213–215]. *Biemna fortis* was the source of labuanine (**382a**) [216] which was revised by reisolation work from *Biemna* sp. (Japan) to give ecionine A (**382b**) [217] which was previously isolated from the Australian marine sponge, *Ecionemia geodides* [218]. Sagitol C (**383**) was obtained from a marine sponge *Oceanapia* sp. (MLU) [219]. The marine sponge *Xestospongia* cf. *carbonaria* (UEP) was the source of three acridine metabolites, 5-methoxyneoamphimedine (**384**), neoamphimedine Y (**385**), and neoamphimedine Z (**386**) [220]. Two new degraded kynuramine building blocks were obtained as in plakinidine D (**387**) [221] and alpkinidine (**388**) [220]. The former was found from *Didemnum* sp. (SSW), while the latter was identified from *Xestospongia* cf. *carbonaria* (UEP). Aaptoline (**389**) was characterized from *Aaptos suberitoides* possessing anthranilate building block [222]. Although synthesis of **389** was performed, the spectral data of the synthesized aaptoline A (**389**) were not consistent with those of previous studies [223].

3.2 Alkaloid Derived Tyrosine

Indonesian alkaloids derived tyrosine can be subclassified on the basis of their building blocks including tyrosine (**390–395**), tyramine (**396**), and dopamine (**397–431**). The combination of the building blocks and their modifications such as tyrosine-tyramine (**432–433**), tyrosine-tyramine-3-amino propanol (**434**), tyrosine-3-aminopropanol-tyramine (**435–459**), tyrosine-3-amino propanol-tyramine-oxidative tyramine (**460**), tyrosine-3-amino propanol-degraded tyramine (**461–463**), oxidative tyramine-3-aminopropanol-tyramine (**464–468**), tyramine-tyrosine-tyrosine-tyramine (**469**, **470**), and tyramine-tyrosine-tyrosine-tyramine-tyrosine-tyramine (**471**) are also described. Mixed biosynthesis metabolites between tyrosine-arginine (e.g., **390**, **391**), tyrosine-arginine-C_1 unit (e.g., **395**), tyrosine-aspartic acid as in aaptamine related molecules (e.g., **397–408**) were observed in the marine alkaloid derived tyrosine.

Six alkaloids lissodendrins A, B (**390**, **391**) [224], polycarpaurines A–C (**392–394**) [225], and methyldorimidazole (**395**) [226] possessing a tyrosine building block were isolated from *Lissodendoryx* (*Achantodoryx*) *fibrosa* (MLU) [224], *Polycarpa aurata* (NSW) [225], and *Leucetta chagosensis* (SSW) [226], respectively. The planar structure of **391** and **392** were confirmed by total synthesis [227, 228]. A marine

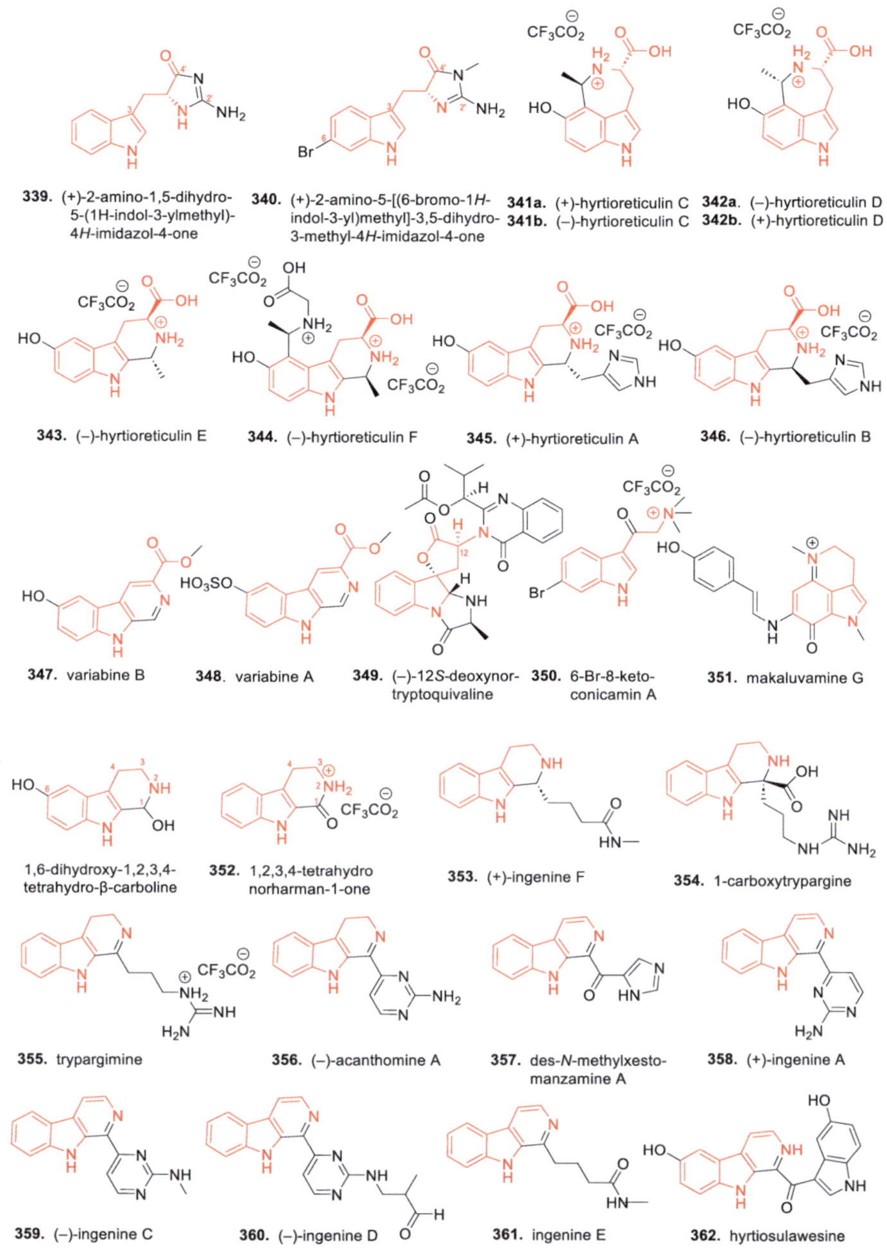

Fig. 13 Structures of marine alkaloid derived tryptophan (**339–389**) from Indonesian. The marine alkaloid derived tryptophan building blocks were proposed with red color

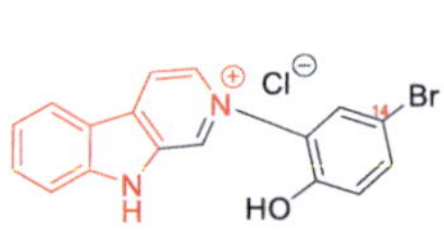

363. 14-bromoreticulatol

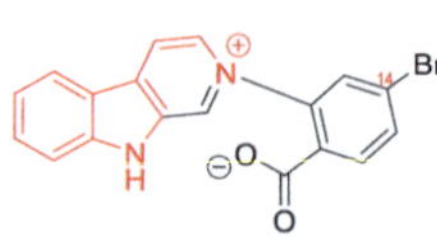

364. 14-bromoreticulatate

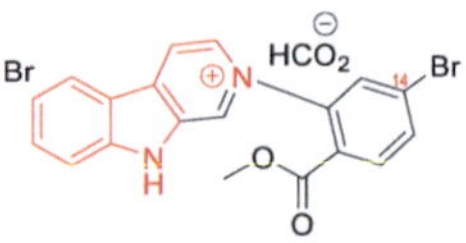

365. 14-bromoreticulatine

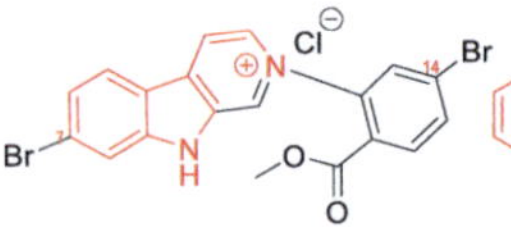

366. 7,14-dibromoreticulatine

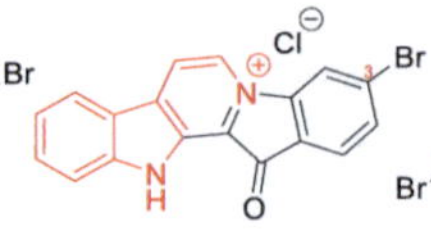

367. 3-bromofascaplysin

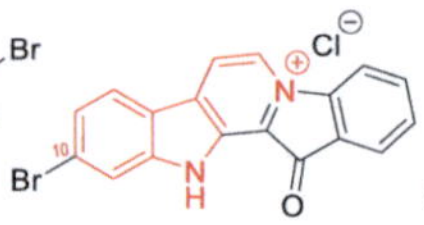

368. 10-bromofascaplysin

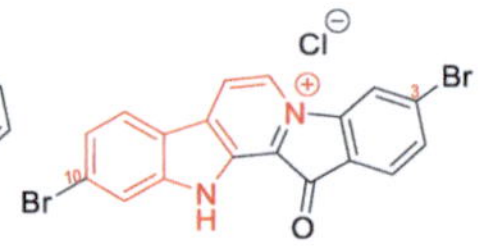

369. 3,10-dibromofascaplysin

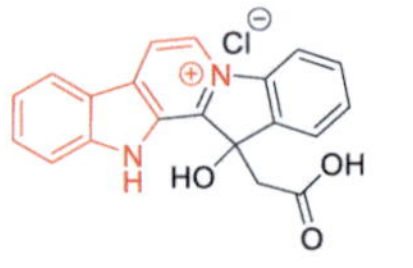

370. (+)-homofascaplysate A

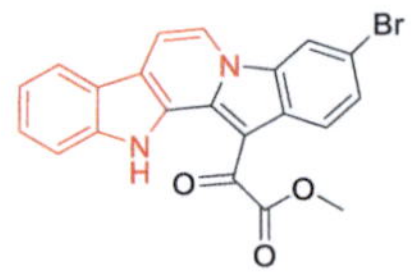

371. 3-bromohomofascaplysin B

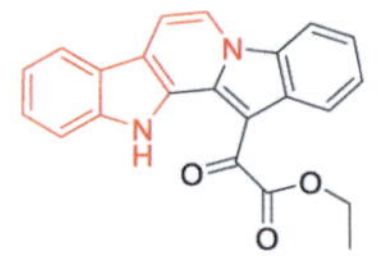

372. homofascaplysin B-1

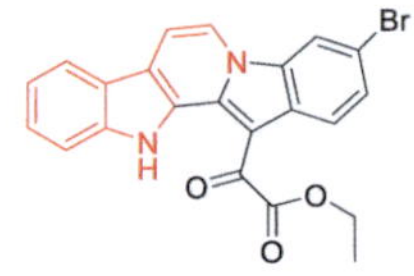

373. 3-bromohomofascaplysin B-1

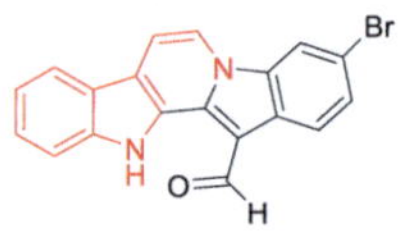

374. 3-bromohomofascaplysin C

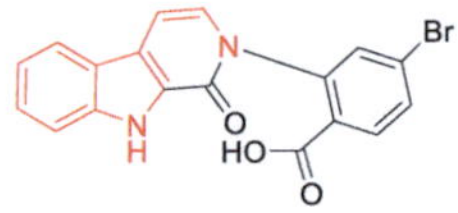

375. 3-bromosecofascaplysin B

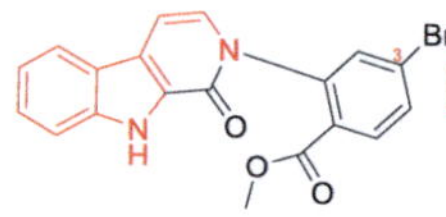

376. 3-bromosecofascaplysin A

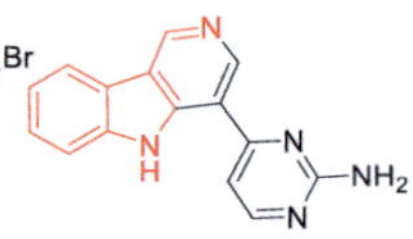

377. (–)-ingenine B

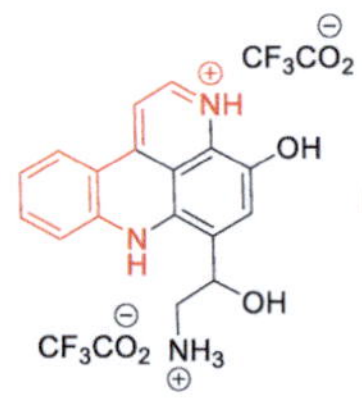

378. styelsamine A

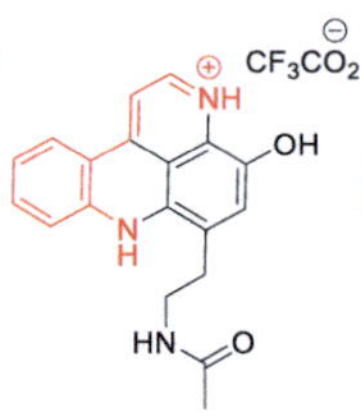

379. styelsamine B

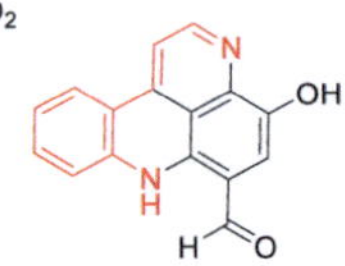

380. styelsamine C

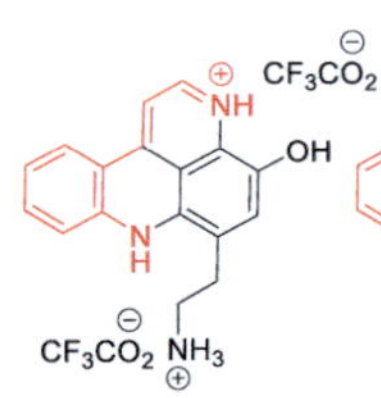

381. styelsamine D

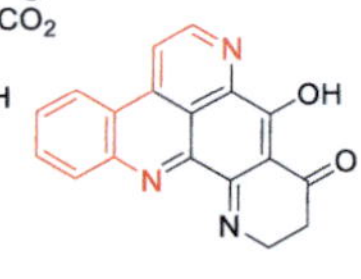

382a. labuanine

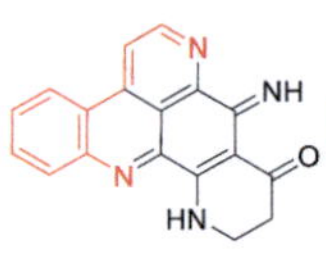

382b. ecionine A

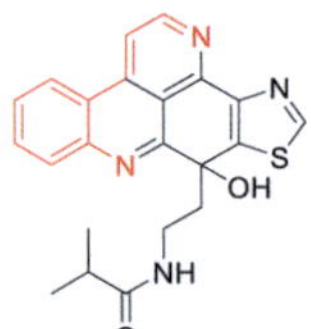

383. sagitol C

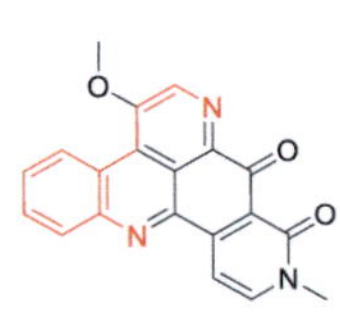

384. 5-methoxyneo-amphimedine

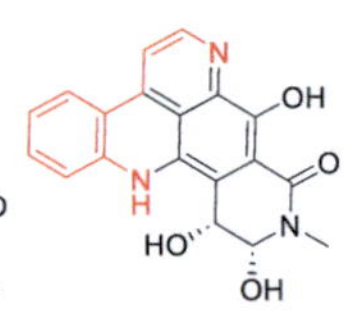

385. neoamphimedine Y

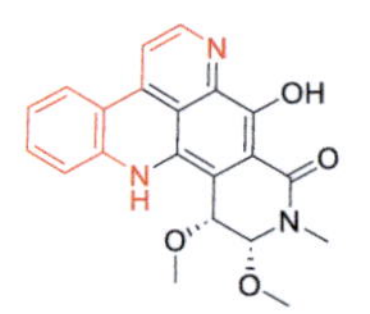

386. neoamphimedine Z

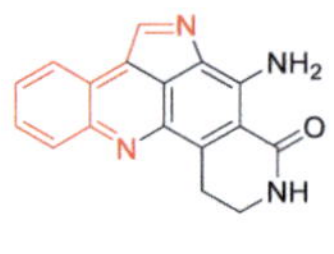

387. plakinidine D

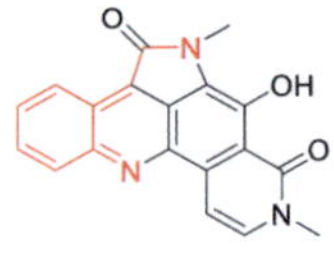

388. alpkinidine

389. aaptoline

Fig. 13 (continued)

tunicate *Eudistoma* (SSW) yielded a new tyramine metabolite, 3′,5′-dibromo-4′-methoxyphenethylamine (**396**) [203] although its structure had been reported [229]. Thirty-five Indonesian MNPs (**397–431**) were further distributed into isoquinoline (**397–408**), polysulfur aromatic (**409–422**), and imidazole alkaloids (**423–431**). A variety of isoquinoline alkaloids were biosynthesized by *Xestospongia* sp. (JSCR) (**397**, **398**, **401**, **402**) [230], *Aaptos suberitoides* (MLU) (**399**, **400**, **405**, **406**) [231], *Aaptos* sp. (NSW, ENT) (**403**, **404**, **407**) [232, 233], and *Aaptos lobata* (NSW) (**408**) [234]. The isoquinoline skeleton may be derived from a biochemical Pictet–Spengler condensation between L-dopamine (tyrosine as a precursor) and β-alanine aldehyde (L-aspartic acid as precursor). The marine tunicate *Lissoclinum* cf. *badium* (NSW) was the source of lissoclibadins (**409–422**) [235–237]. Molecular modelling established the planar structure of (−)-lissoclibadins 1 (**419**) [235], 8 (**422**), and 9 (**421**) [237]. Nine imidazole alkaloids possessing tyrosine building block, (+)-spironaamidine (**423**) [238], naamines F (**424**), G (**425**), kealiinines A–C (**426–428**) [239], naamidines H, I (**429**, **430**) [240], and preclathridine B (**431**) [226] were exclusively found in the marine sponge *Leucetta chagosensis* (SSW and NSW). Synthesis was used to confirm the planar structures of **424** [241], **425** [242], **426** [243, 244], while X-ray crystallography was employed to establish the planar structures of **427** and **428** [245]. During structure confirmation of **426** via synthesis, it should be noted that the NMR comparison between synthetic material and natural products required certain conditions such as TFA salt form and specific NMR solvent i.e., CD_3OD [243]. Kealiinines B (**427**) and C (**428**) were obtained as an inseparable mixture during synthesis.

Two tyrosine-tyramine metabolites were isolated from the marine sponge *Iotrochota* cf. *iota* (CSW) as in (+)-1-*O*-methylhemibastadinols 2 (**432**), 4 (**433**) [246]. Moreover, (+)-aplyzanzine B (**434**) was obtained from the two-sponge association *Jaspis* sp. and *Bubaris* sp. (SSW) showing a tyrosine-tyramine-3-amino propanol scaffold [247]. In addition to this MNPs, twenty-five tyrosine-3-amino propanol-tyramine natural products found in (+)-anomoian B (**435**, *Hexadella* cf. *indica*, NSW) [247], purpuramines M, N (**436**, **437**, respectively, Aplysinellidae, EKM), (−)-araplysillins VII, X, XI (**438**, **439**, **440** respectively, Aplysinellidae, EKM) [248], (−)-psammaplysins L, M, Q, O, P, N, S, T, U, V, W (**442**, **443**, **444**, **446**, **447**, **449**, **451**, **453**, **455**, **457**, **458**, respectively, *Aplysinella strongylata*, BLI), its 19-hydroxy molecules obtained as a mixture of *R*/*S* stereoisomers as in (−)-19-hydroxypsammaplysin E, Q, P, S, T, U, W, psammaplysin R (**441**, **445**, **448**, **452**, **454**, **456**, **459**, **450** respectively, *Aplysinella strongylata*, BLI) [249]. Compounds **437–439** have a spirocyclohexadienyloxazoline, while compounds **441–459** contain spirooxepinisoxazoline framework. Apparently, fatty acid synthase was also involved in the biosynthesis of metabolites **444–459**. (−)-Araplysillin VIII (**460**, *Aplysinellidae*, EKM) was an example of a tyrosine-3-amino propanol-tyramine-oxidative tyramine metabolite, while (−)-araplysillin IX (**461**) was a tyrosine-3-amino propanol-degraded tyramine natural product obtained from the same organism [248]. Both possessed spirocyclohexadienyloxazoline moieties. Two other tyrosine-3-amino propanol-degraded tyramine metabolites, (−)-psammaplysin K (**462**) and its dimethoxy acetal (**463**) were also isolated from *Aplysinella strongylata* (BLI) possessing spirooxepinisoxazoline moiety [249]. Compound **463** was an artifact.

The marine sponge *Iotrochota* cf. *iota* was also the source of enisorines A–E (**464–468**) [246] possessing oxidative tyramine-3-amino propanol-tyramine units. The planar structure of enisorine D (**467**) was also confirmed by total synthesis [250]. Bastadins 16 (**469**) and 17 (**470**) were isolated from the marine sponge *Ianthella basta* (NSW) [251]. The planar structure of **469** was confirmed by total synthesis [252]. The two compounds (**469, 470**) possessed tyramine-tyrosine-tyrosine-tyramine scaffold. Compound **471** assigned as sesquibastadin 1 was obtained from the same organism collected from MLU province having tyramine-tyrosine-tyrosine-tyramine-tyrosine-tyramine units [253]. Compound **471** has a unique coupling between biphenyl ring. The structures of Indonesian alkaloid derived tyrosine are presented in Fig. 14.

3.3 Alkaloid Derived Lysine

Thirty-eight Indonesian marine alkaloids (**472–509**) are biosynthetically subclassified into alkaloid derived lysine compounds possessing ornithine (**472–474**), ornithine-3-aminopropanoic acid (**475, 476**), ornithine-4-aminobutanoic acid (**477–480, 486**), ornithine-taurine (**481**), pyrrole-imidazole (**482, 485, 491a–500, 506, 507**), pyrrole-imidazole-taurine (**483a, 484**), pyrrole-3-aminopropanol-tyramine (**487–490**), pyrrole-3-aminopyrimidine (**501–505**), indolizidine (**508**), and quinolizidine (**508**) building blocks or scaffolds. Pyrrole and pyrrole-imidazole as well as their derivatives could be derived from lysine amino acid through several interesting enzymatic reaction which were supported by metabolomic and genomic works [254, 255].

Two ornithine-derived molecules, 3,5-dibromo-1*H*-pyrrole-2-carboxylic acid (**472**) and methyl 3,4-dibromo-1*H*-pyrrole-2-carboxylate (**473**) were isolated from a marine sponge *Acanthostylotella* sp. (BLI) [256] while ethyl 3,4-dibromo-1*H*-pyrrole-2-carboxylate (**474**) was obtained from the marine sponge *Stylissa massa* (PUA) [257]. Ornithine-3-aminopropanoic acid scaffold was found in a marine sponge *Acanthostylotella* sp. as in acanthamides B (**475**) and C (**476**) [256]. Moreover, a series of ornithine-4-aminobutanoic acid as in acanthamides A (**477**), D (**478**) [256], 4-(4′,5′-dibromo-1′-methylpyrrole-2′-carboxamido) butanoic acid (**479**), (−)-agelanin B (**480**), and (−)-longamide C (**486**) [126] was also discovered in the same species except for the last three obtained from *Agelas linnaei* (JSCR). Compounds **473–478, 486** were likely artifacts. A mixed biosynthesis of two amino acids, lysine and cysteine was observed in the metabolite mauritamide D (**481**) isolated from *Agelas linnaei* (JSCR) [126]. Pyrrole-imidazole metabolites were distributed into *Stylissa massa* (**482**, PUA) [257], *Stylissa carteri* (**494, 495**, MLU) [258], *Stylissa* sp. (**496, 497**, EKM) [259], *Agelas linnaei* (**485, 498/499**, JSCR) [126], *Agelas nakamurai* (**491a, 492**, MLU) [260], *Agelas* sp. [(−)-**500** [261], **506/507** [262], NSW], and *Axinella carteri* (**493**, UEP) [263]. Structure confirmation of (−)-nakamuric acid (**491a**) gave a revision of its ^{13}C NMR data as in (−)-**491b1** by total synthesis

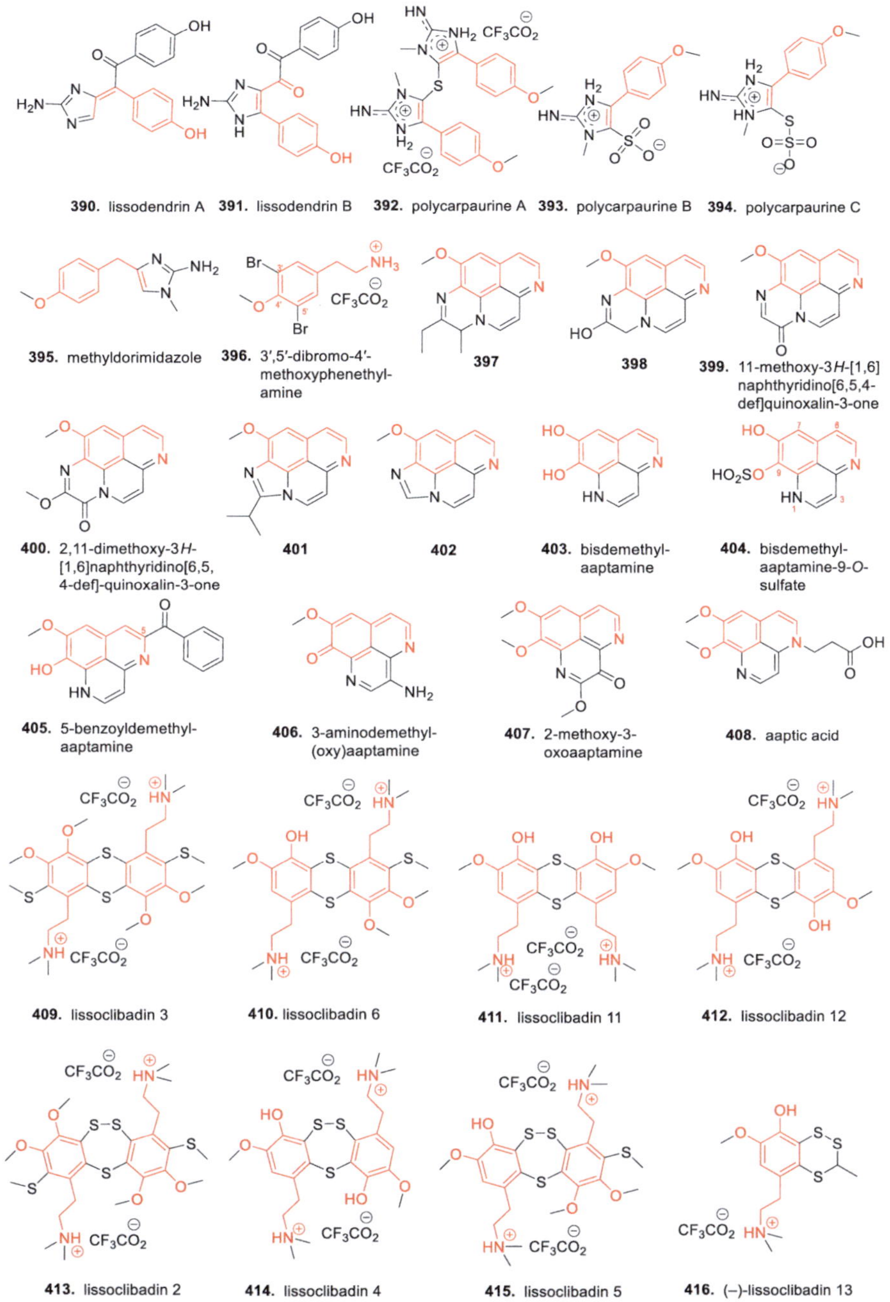

Fig. 14 Structures of marine alkaloid derived tyrosine (**390–471**) from Indonesian waters. The marine alkaloid derived tyrosine building blocks were proposed with red and blue colors

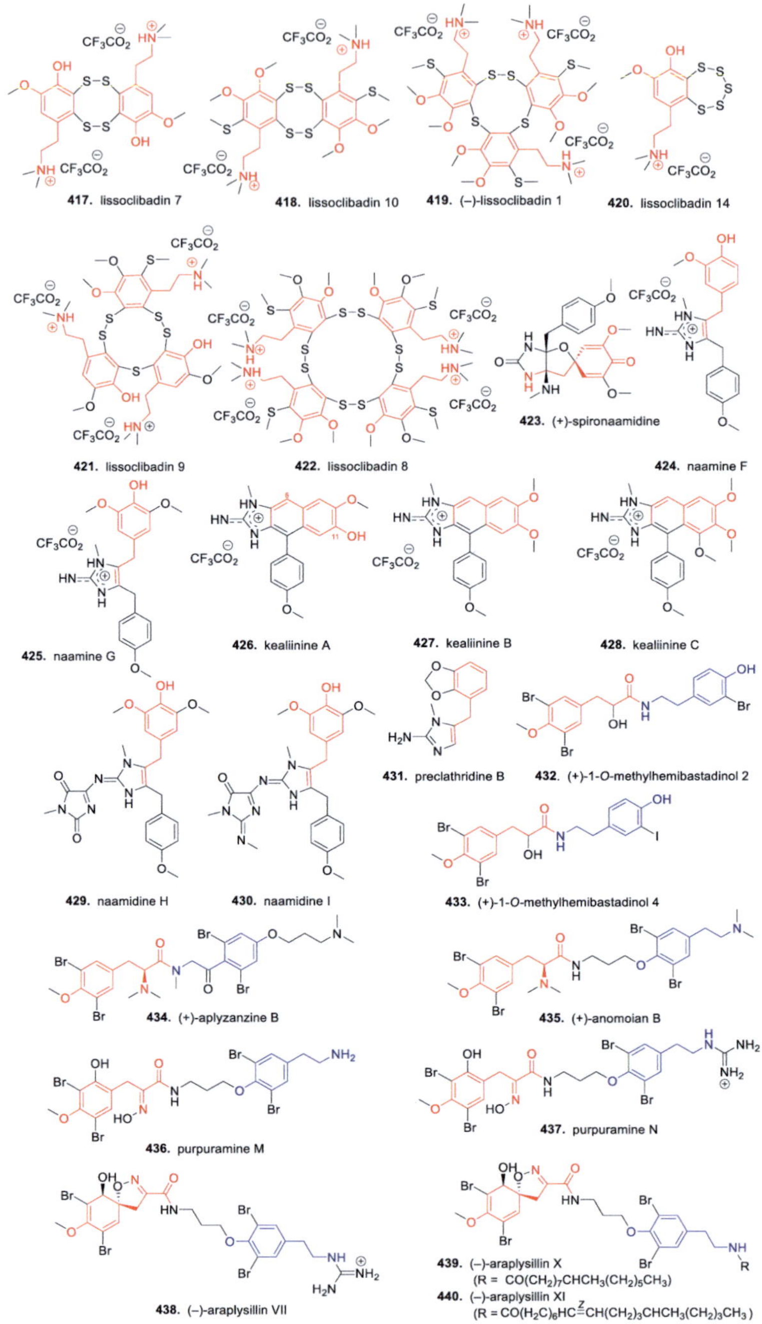

Fig. 14 (continued)

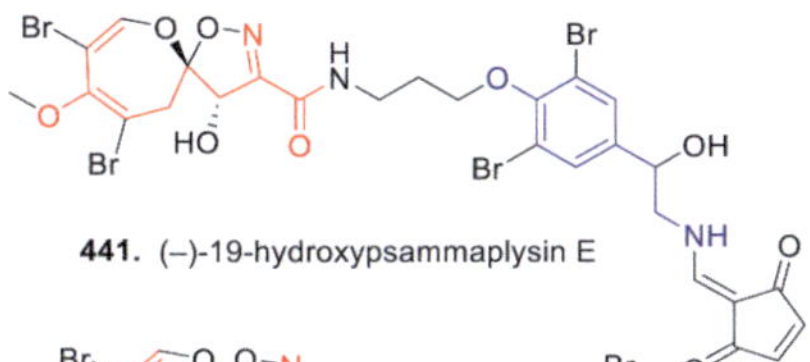

441. (–)-19-hydroxypsammaplysin E

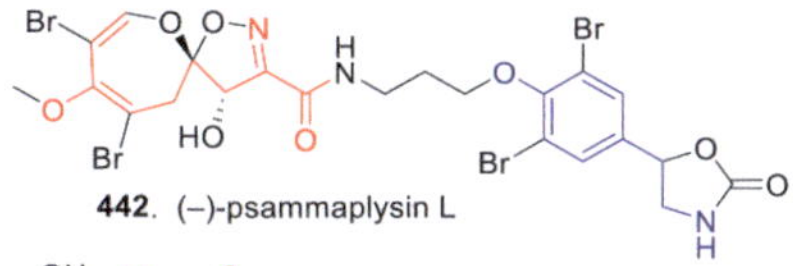

442. (–)-psammaplysin L

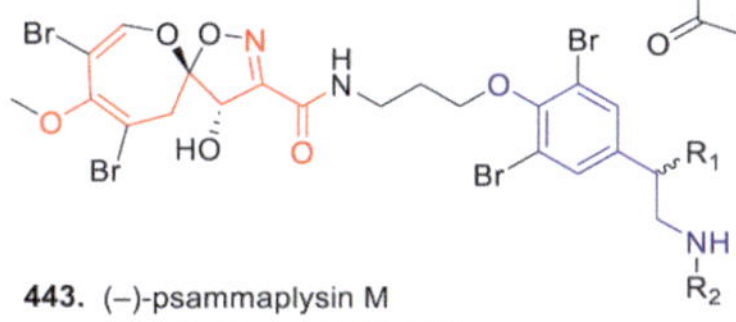

443. (–)-psammaplysin M (R_1 = H, R_2 = $COCH_2OH$)
444. (–)-psammaplysin Q (R_1 = H, R_2 = $CO(CH_2)_{12}CH_3$)
445. (–)-19-hydroxypsammaplysin Q (R_1 = OH, R_2 = $CO(CH_2)_{12}CH_3$)
446. (–)-psammaplysin O (R_1 = H, R_2 = $CO(CH_2)_{14}CH_3$)
447. (–)-psammaplysin P (R_1 = H, R_2 = $CO(CH_2)_{16}CH_3$)
448. (–)-19-hydroxypsammaplysin P (R_1 = OH, R_2 = $CO(CH_2)_{16}CH_3$)
449. (–)-psammaplysin N (R_1 = H, R_2 = $CO(CH_2)_{12}CH(CH_3)_2$)
450. (–)-psammaplysin R (R_1 = OH, R_2 = $CO(CH_2)_{11}CHCH_3CH_2CH_3$)
451. (–)-psammaplysin S (R_1 = H, R_2 = $CO(CH_2)_{12}CHCH_3CH_2CH_3$)
452. (–)-19-hydroxypsammaplysin S (R_1 = OH, R_2 = $CO(CH_2)_{12}CHCH_3CH_2CH_3$)
453. (–)-psammaplysin T (R_1 = H, R_2 = $CO(CH_2)_{14}CHCH_3CH_2CH_3$)
454. (–)-19-hydroxypsammaplysin T (R_1 = OH, R_2 = $CO(CH_2)_{14}CHCH_3CH_2CH_3$)
455. (–)-psammaplysin U (R_1 = H, R_2 = $CO(H_2C)_7HC\overset{Z}{=}CH(CH_2)_3CH(CH_3)_2$)
456. (–)-19-hydroxypsammaplysin U (R_1 = OH, R_2 = $CO(H_2C)_7HC\overset{Z}{=}CH(CH_2)_3CH(CH_3)_2$)
457. (–)-psammaplysin V (R_1 = H, R_2 = $CO(H_2C)_8HC\overset{Z}{=}CH(CH_2)_4CH_3$)
458. (–)-psammaplysin W (R_1 = H, R_2 = $CO(H_2C)_{11}HC\overset{Z}{=}CH(CH_2)_4CH_3$)
459. (–)-19-hydroxypsammaplysin W (R_1 = OH, R_2 = $CO(H_2C)_{11}HC\overset{Z}{=}CH(CH_2)_4CH_3$)

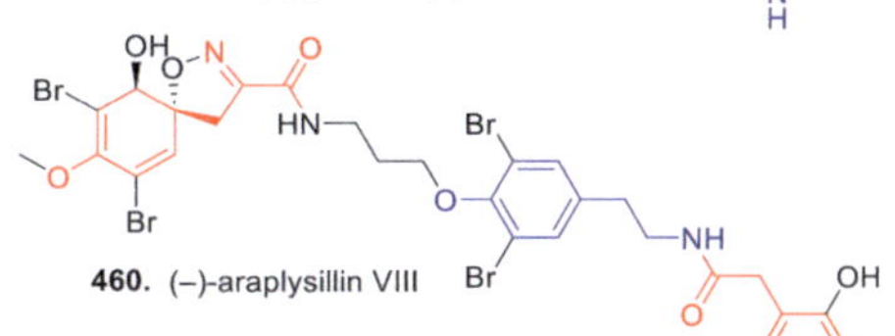

460. (–)-araplysillin VIII

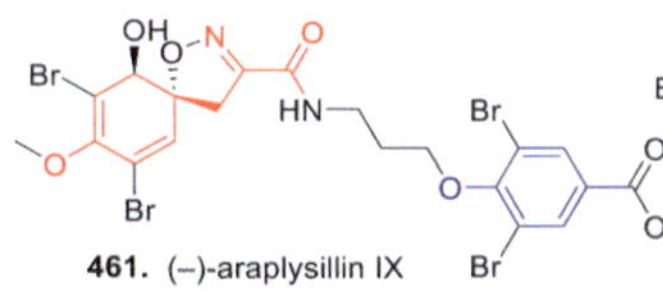

461. (–)-araplysillin IX

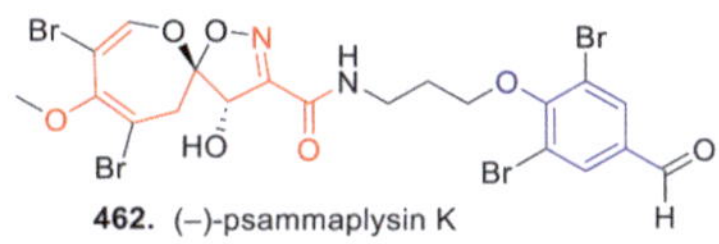

462. (–)-psammaplysin K

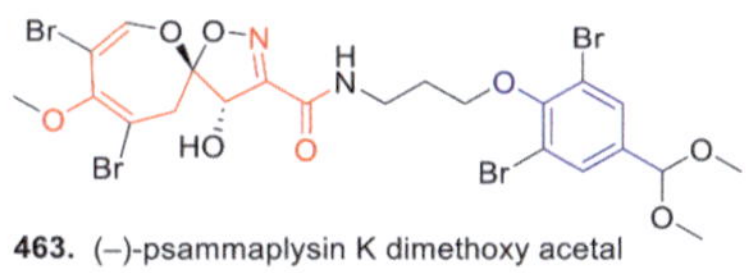

463. (–)-psammaplysin K dimethoxy acetal

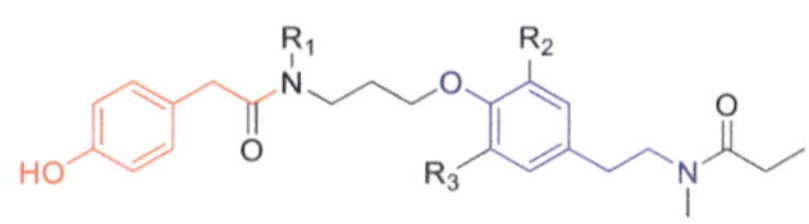

464. enisorine A (R_1 = R_3 = H, R_2 = Br)
465. enisorine B (R_1 = R_3 = H, R_2 = I)
466. enisorine C (R_1 = CH_3, R_2 = Br, R_3 = H)
467. enisorine D (R_1 = H, R_2 = R_3 = Br)
468. enisorine E (R_1 = H, R_2 = I, R_3 = Br)

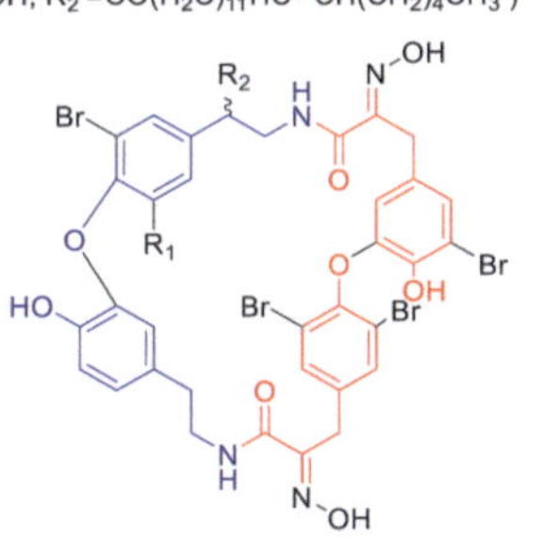

469. bastadin 16 (R_1 = Br, R_2 = H)
470. bastadin 17 (R_1 = H, R_2 = OH)

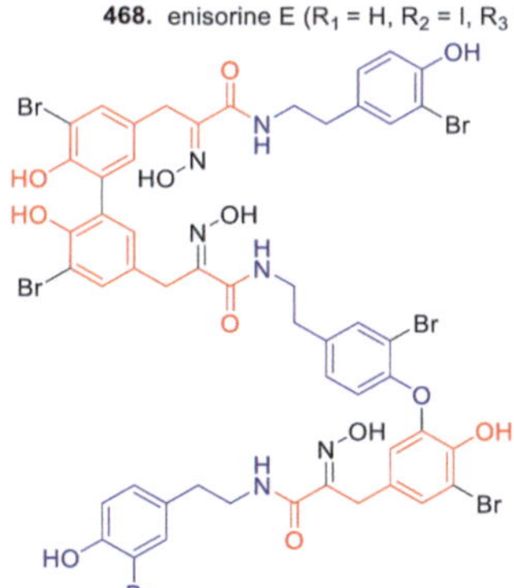

471. sesquibastadin 1

Fig. 14 (continued)

[264]. In 2017, reisolation of (−)-nakamuric acid from Chinese *Agelas* sp. established its AC by comparison of calculated and measured ECD spectra and hence its stereochemistry was revised as in (−)-**491b2** [265]. Compound **492** was likely an artifact. The planar structure of **495** was confirmed by CD [258] and total synthesis [266]. Compounds **488/489** were obtained as a mixture of enantiomers. The ACs of **506/507**, an inseparable mixture was assigned by comparison of calculated and experimental ECD [262].

Two pyrrole-imidazole-taurine metabolites, (−)-mauritamides B (**483a**) and C (**484**) were obtained from the marine sponge *Agelas linnaei* (JSCR) [126]. The original structure of mauritamide B (**483a**) was revised by total synthesis to be **483b** [267]. A series of pyrrole-3-aminopropanol-tyramine metabolites were exemplified by agelanesins A–D (**487–490**) discovered from the same organism [126]. The genus *Stylissa* collected from SSW and EKM waters was the source of six pyrrole-3-aminopyrimidine given by latonduine A (**501**) [268], 3-debromolatonduine A (**502**) [259], latonduine B methyl ester (**503d1**) [268], latonduine B ethyl ester (**503d2**) [268], latonduine B (**504**) [268], and 3-debromolatonduine B methyl ester (**505**) [261]. The planar structure of **501** was established by total synthesis [268]. Compounds **503d1**, **503d2**, and **505** were possible artifacts. Two marine tricyclic alkaloids bearing perhydropyrrolo- and pyrido[2,1-*j*] framework, (+)-polycitorol A (**508**) and (−)-polycitorol B (**509**) were isolated from a tunicate (ENT) [269]. Compound (+)-**508** contained indolizidine framework while (−)-**509** possessed quinolizidine scaffold. The structure confirmation (+)-polycitorol A (**508**) and (−)-polycitorol B (**509**) remained elusive as the NMR data of both synthetic samples (**508** and **509**) were not identical with those of natural ones [270]. The structures of marine alkaloid derived lysine are depicted as in Fig. 15.

3.4 *Alkaloid Derived Arginine*

Sixty-seven MNPs (**510–576**) were considered to involve L-arginine in their biosynthesis. The classification can be further divided into six subclasses including 3-alkyl pyridine (**510–519**), halicyclamine [(−)-**520**, (−)-**524**, (−)-**525** and (+)-**521–523**, (+)-**526–528**], neopetrocylamine [(−)-**529**, **530**], manzamine [(+)-**531–538**, (+)-**540–546**, (+)-**550–557**, (+)-**559**, **560**, (+)-**562–565**, (+)-**568**, **569** and (−)-**539**, (−)-**547–549**, (−)-**561**, (−)-**566**, (−)-**567**, as well as **558**], crambescidin [(−)-**570–572**], thiadiazole/chorismic acid (**573**, **574**, **576**), and chorismic acid (**575**). A mixed biosynthesis of fatty acid and nitrogenous compounds (e.g., arginine, tryptophan) through several enzymatic reactions could be observed in five subclasses including 3-alkyl pyridine, halicyclamine, neopetrocyclamine, manzamine, and crambescidin [271–274]. Meanwhile, a thiadiazole subclass is assumed to be derived from a mixed biosynthesis of chorismic acid and arginine [275].

3-Alkyl pyridine metabolites were found in *Haliclona* sp. (**510**, UEP [276], **513–515**, NSW [277]), *Xestospongia* sp. (**511**, UEP) [278], *Echinochalina* sp. ((−)-**512g1**, EKM) [279], and *Neopetrosia chaliniformis* (**516–519**, NSW) [280]. The

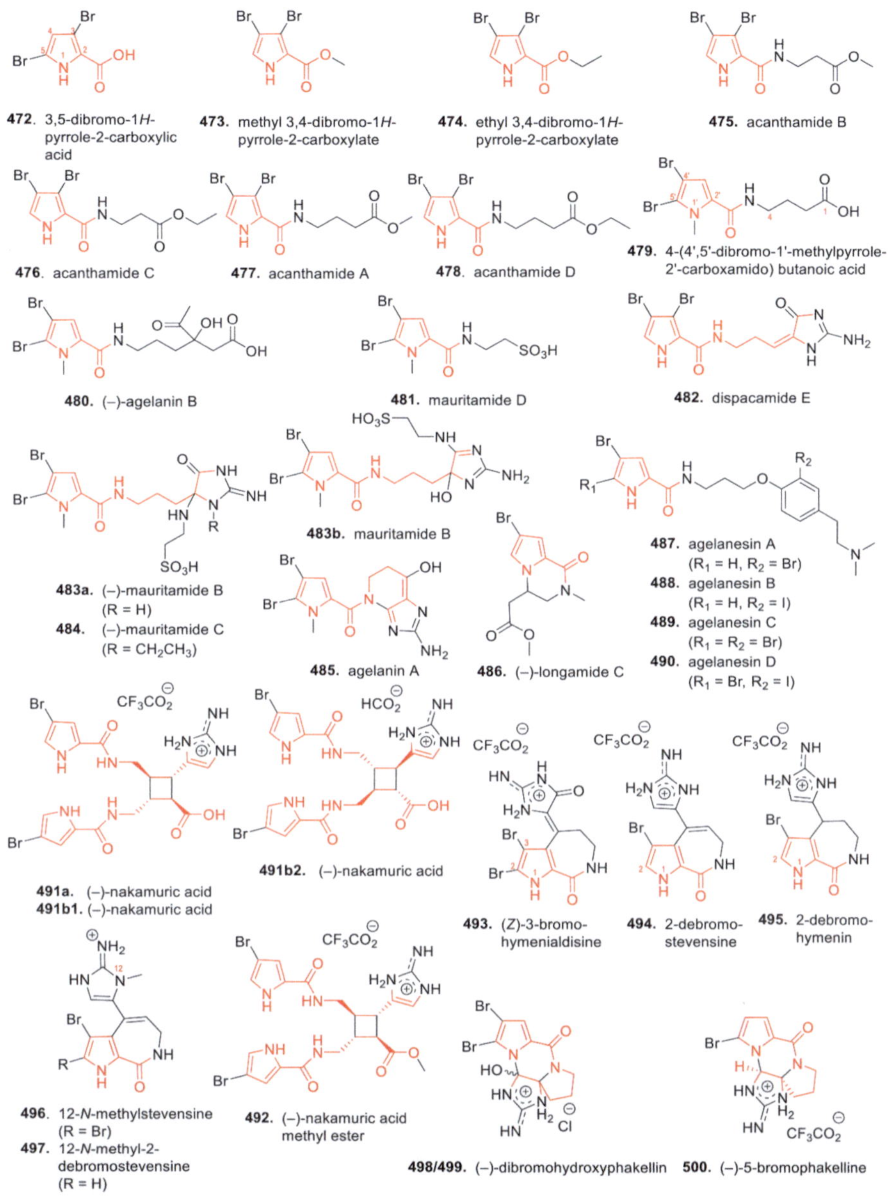

Fig. 15 Structures of marine alkaloid derived lysine (**472–509**) from Indonesian waters. The marine alkaloid derived lysine building blocks were proposed with red color

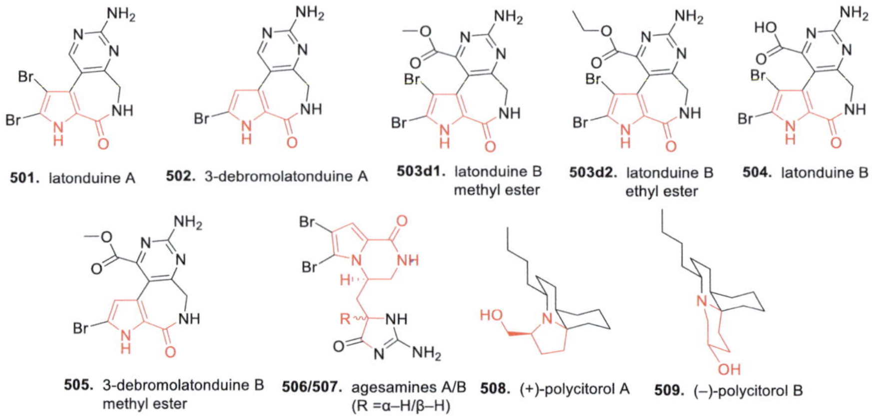

Fig. 15 (continued)

planar structure of *N*-methylniphatyne A (**511**) was established by MS/MS analysis and total synthesis [278]. Total synthesis was also used to confirm the structure of (−)-upenamide, but the NMR data of synthetic materials both diastereomers as in (**512g1**) and (**512g2**) [281, 282] were not matched to the natural materials indicating the structure of upenamide was remained elusive (Fig. 16). MS/MS study was also utilized to disclose a planar structure possessing long carbon chain as in haliclocyclamines A–C (**514–516**) and neopetrosidines A–D (**516–519**). In addition to MS/MS study, chemical degradation was also employed to reveal the planar structures **516–519**. Compounds **518** and **519** were obtained as a mixture. Halicyclamine skeleton was isolated for the first time as in (−)-halicyclamine A (**520**) from a marine sponge *Haliclona* sp. [283]. Since then, eight molecules (**521–528**) related to **520** have been discovered. In addition to *Haliclona* sp., *Halichondria* sp. was the source of (+)-tetradehydrohaliclonacyclamine A (**521**), (+)-2-*epi*-tetradehydrohaliclonacyclamine (**522**), (+)-tetradehydrohaliclonacyclamine A mono-*N*-oxide (**523**) [284], while (−)-acanthocyclamine A (**524**, SES) [285], (−)-chloromethylhalicyclamine B (**525**, SSW) [286], (+)-tetradehydrohalicyclamine B (**527**, NSW) [287], (+)-*epi*-tetradehydrohalicyclamine B (**528**, NSW) [288] were isolated from *Acanthostrongylophora ingens*. (+)-22-Hydroxyhaliclonacyclamine B (**526**) was characterized from a marine sponge *Haliclona* sp. [289]. The ACs of (+)-**521** and (−)-**524** were established by X-ray analysis [284, 285]. Moreover, chemical correlation was also applied to determine the AC of (+)-**521** [284]. A comparison of calculated and experimental ECD was assigned to determine the AC of (−)-**525** [286] and (+)-**527** [287]. The RC of (+)-**528** was initially proposed by molecular modelling, while its AC was confirmed by total synthesis [290]. Natural compound (+)-**528** was obtained as a mixture. Compound (−)-**525** was likely an artifact. Another family of alkaloid derived arginine was (−)-neopetrocyclamines A and B (**529**, **530**) [291] possessing neopetrocyclamine skeleton isolated from the marine sponge *Neopetrosia* cf. *exigua* (EKM). Compound (−)-**529** contained a formamidinium moiety. The RCs of (−)-**529** and **530** were disclosed by molecular modelling.

The family *Petrosiidae* (NSW) including genus *Acanthostrongylophora* (NSW, JSCR) was the major source of manzamine alkaloids consisting of thirty-seven new, two new form of natural products, and one artifact as in (+)-**531–538** [197, 292–295], **540–546** [292, 296, 297], **550–565** [292, 294, 296–301] and (−)-**539** [302], (−)-**547–549** [293, 302], (−)-**566–567** [303] except for (+)-kauluamine (**568**) [304] discovered from a marine sponge *Prianos* sp. The AC of (+)-31-hydroxymanzamine A (**531**) was determined with the application of Mosher ester [292]. A combination of X-ray diffraction analysis and Mosher's methodology assigned the AC of (+)-32,33-dihydro-31-hydroxymanzamine A (**532**) [293]. Molecular modelling tactic was applied to determine the RCs of (+)-12,28-oxamanzamine A (**536**) [294], (+)-acantholactone (**554**) [298], (−)-manadomanzamine A (**566**) [303], and (+)-*neo*-kauluamine (**569**) [299, 302]. Moreover, DFT-based NMR calculation was also utilized to (+)-manzamine J N-oxide-HCl (**552**), (+)-3,4-dihydromanzamine J N-oxide-HCl (**553**) to assign their RCs [292]. The ACs of (+)-acantholactone (**554**) [298], (−)-manadomanzamines A, B (**566**, **567**) [303] were assigned by CD or ECD analysis. Moreover, specific optical rotation $[\alpha]_D$ experiment confirmed the presence of new enantiomer of **538** as in (−)-8-hydroxymanzamine A (**539**) [302], (−)-12,34-oxamanzamine E (**547**) [293], (−)-manzamine F (**548**) [302], and (−)-12,34-oxamanzamine F (**549**) [293]. (+)-*neo*-Kauluamine (**569**) was apparently an artifact.

Three marine alkaloids assigned as (−)-crambescidins 345, 361, 373 (**570–572**) [305] possessing crambescidin skeleton were isolated from the SSW collection of *Clathria bulbotoxa.* The building block chorismic acid and arginine could be identified from polycarpathiamines A, B (**573**, **574**) [306], polyaurines A, B (**575**, **576**) [307]. The planar structure of **573** and **574** was also elucidated by chemical transformation, while the planar structure of **576** were confirmed by DFT-based NMR calculation.

3.5 *Alkaloid Derived Serine*

About twenty-seven MNPs were estimated derived from serine amino acid as in (+)-**577–593**, **595–598** and (−)-**594**, **599–601** as well as **602** and **603**. Furthermore, the class can be classified into at least four subclasses including serinolipids [(+)-**577–593**, **595** and (−)-**594**], ceramides [(+)-**596–598**, (−)-**599**, **600**), azirine [(−)-**601**], and oxazoles (**602**, **603**). Family *Didemnidae* or genus *Didemnum* sp. was the source of those serinolipids so far. Three (+)-didemniserinolipids A–C (**577**, **578a**, **579a**, respectively) were the first example Indonesian marine serinolipids contained 6,8-dioxabicyclo[3.2.1]octanes (6,8-DOBCOs) moiety obtained from NMU waters [308]. The serinolipids were possibly biosynthesized from C28 very-long-chain highly unsaturated fatty acids (VLC-HUFA) and serine. The C28 VLC-HUFA could be derived from C18-long chain (LC)-polyunsaturated fatty acid

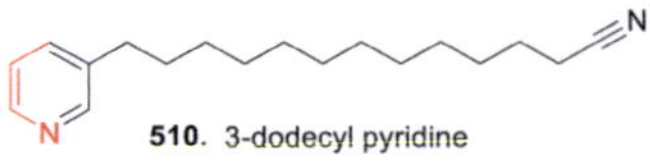

510. 3-dodecyl pyridine

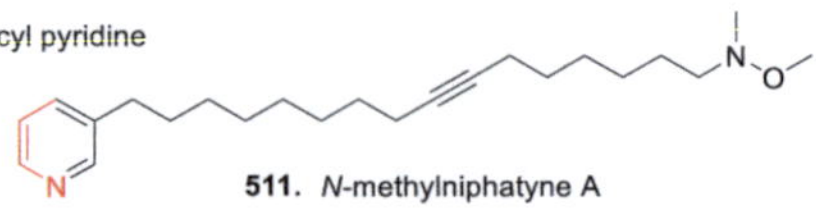

511. *N*-methylniphatyne A

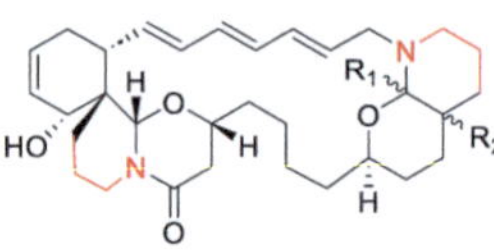

512g1. (–)-upenamide ($R_1 = R_2 = \beta$–H)
512g2. (–)-upenamide ($R_1 = R_2 = \alpha$–H)

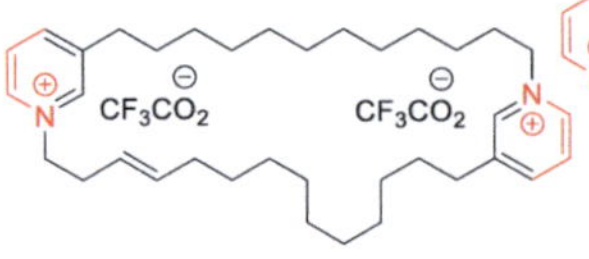

513. haliclocyclamine A

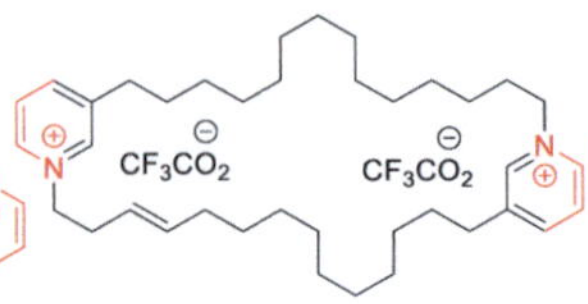

514. haliclocyclamine B

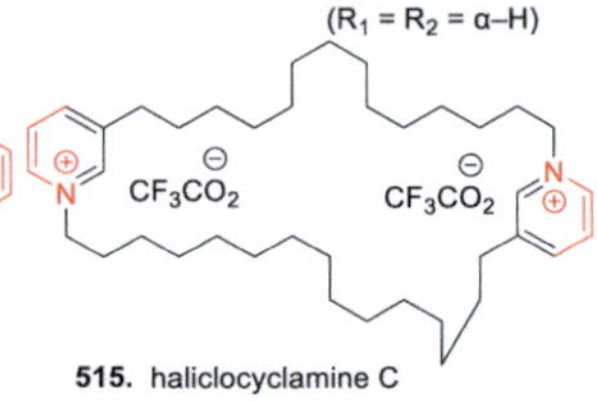

515. haliclocyclamine C

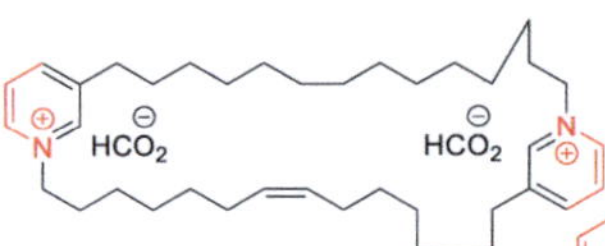

516. neopetrosidine A

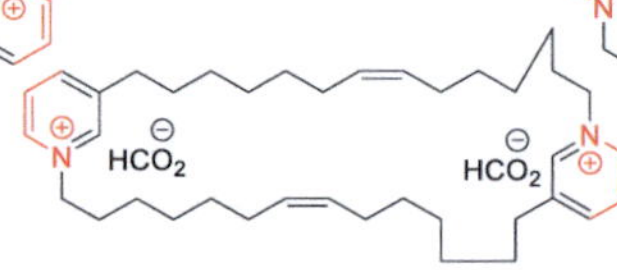

517. neopetrosidine B

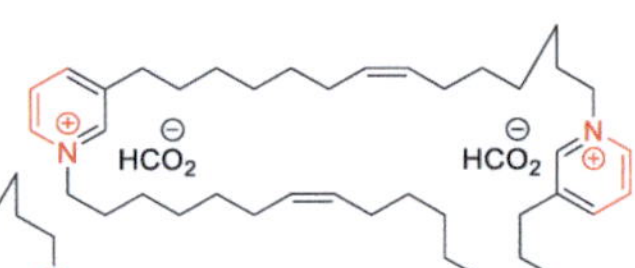

518. neopetrosidine C

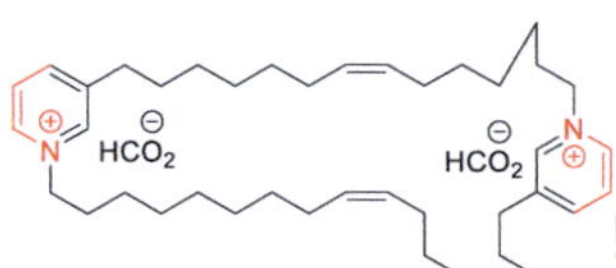

519. neopetrosidine D

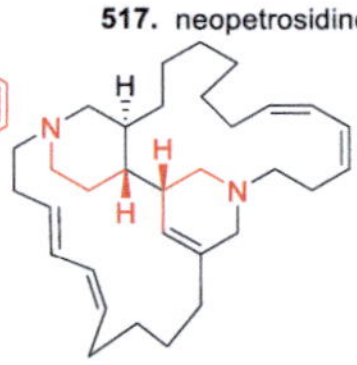

520. (–)-halicyclamine A

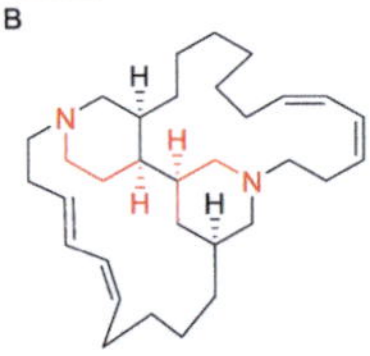

521. (+)-tetradehydro-haliclonacyclamine A

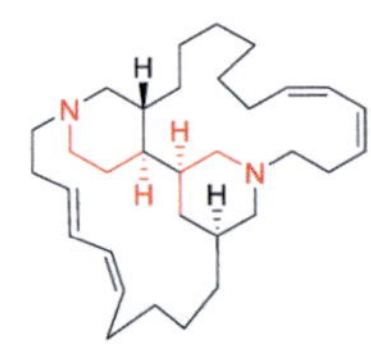

522. (+)-2-*epi*-tetradehydro-haliclonacyclamine

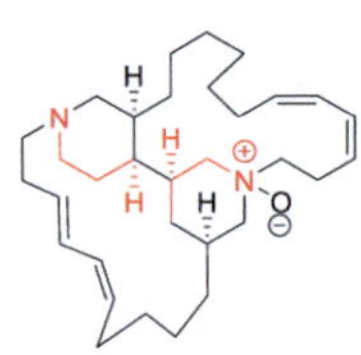

523. (+)-tetradehydro-haliclonacyclamine A mono-*N*-oxide

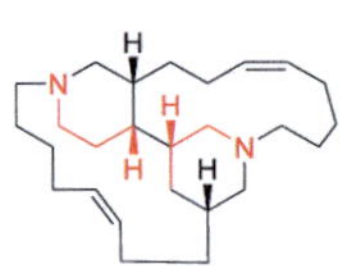

524. (–)-acantho-cyclamine A

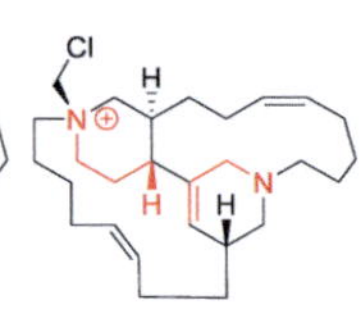

525. (–)-chloromethyl-halicyclamine B

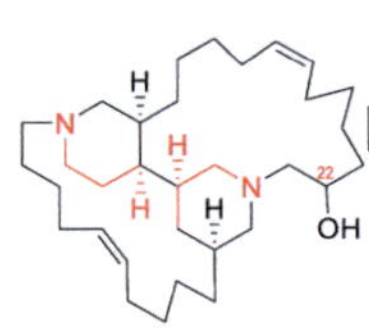

526. (+)-22-hydroxy-haliclonacyclamine B

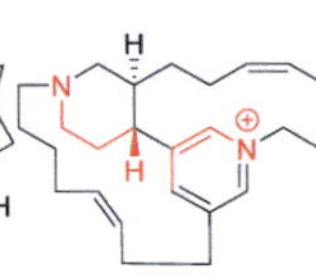

527. (+)-tetradehydro-halicyclamine B

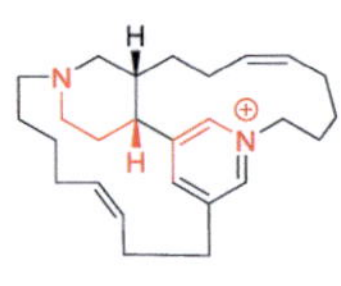

528. (+)-*epi*-tetradehydro halicyclamine B

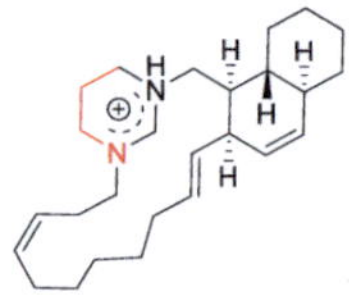

529. (–)-neopetro-cyclamine A

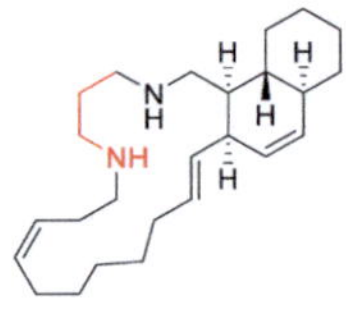

530. (–)-neopetro-cyclamine B

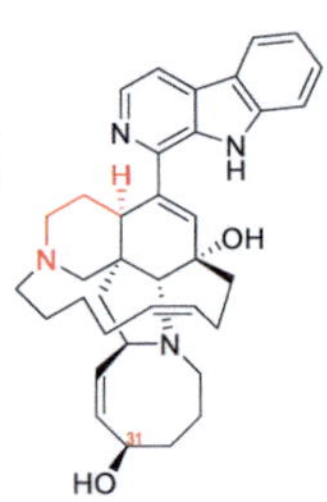

531. (+)-31-hydroxy-manzamine A

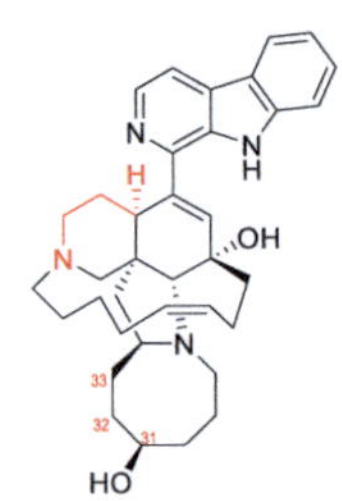

532. (+)-32,33-dihydro-31-hydroxymanzamine A

Fig. 16 Structures of marine alkaloid derived arginine (**510–576**) from Indonesian waters. The marine alkaloid derived arginine building blocks were proposed with red color

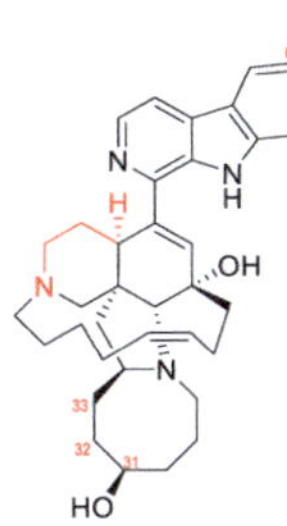

533. (+)-32,33-dihydro-6,31-dihydroxymanzamine A

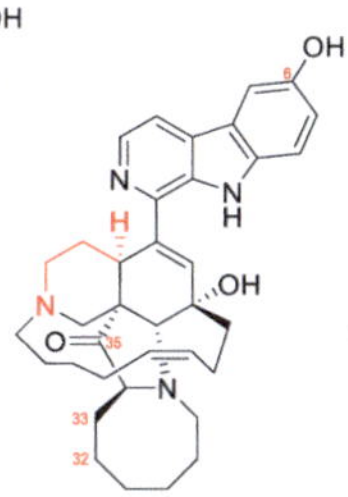

534. (+)-32,33-dihydro-6-hydroxymanzamine A-35-one

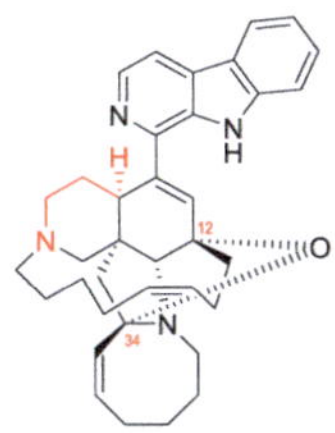

535. (+)-12,34-oxamanzamine A

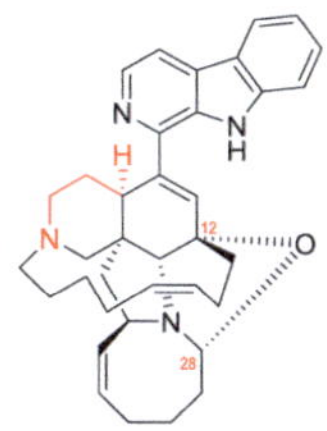

536. (+)-12,28-oxamanzamine A

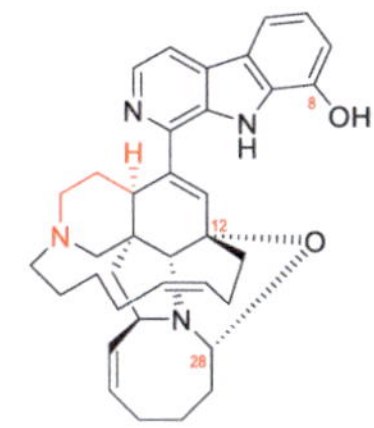

537. (+)-12,28-oxa-8-hydroxymanzamine A

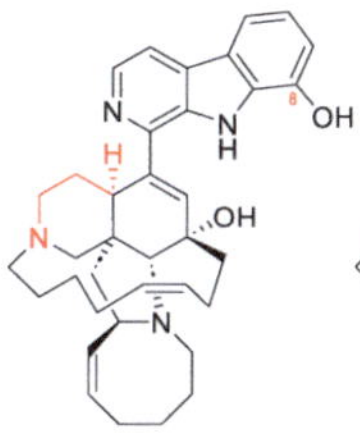

538. (+)-8-hydroxymanzamine A

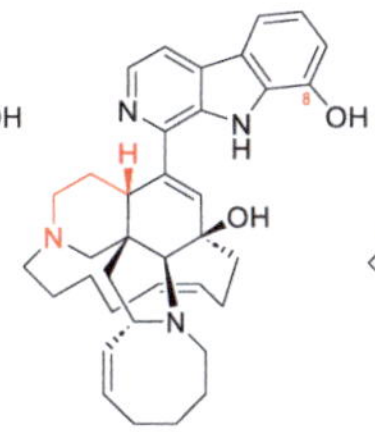

539. (–)-8-hydroxymanzamine A

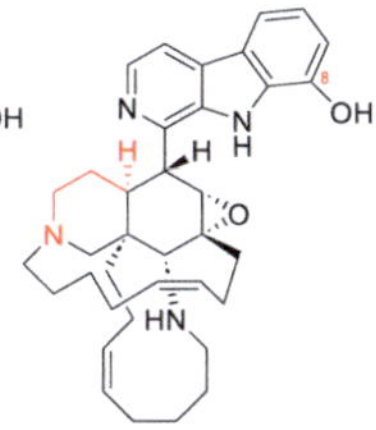

540. (+)-8-hydroxymanzamine B

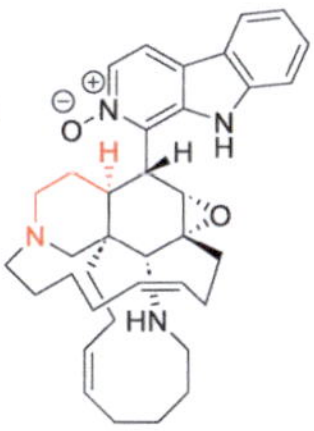

541. (+)-manzamine B N-oxide

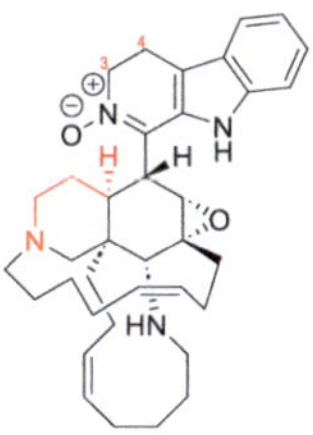

542. (+)-3,4-dihydromanzamine B *N*-oxide

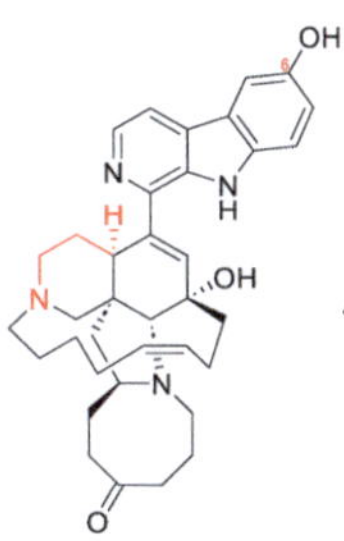

543. (+)-6-hydroxymanzamine E

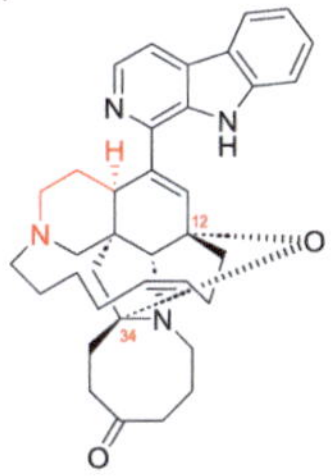

544. (+)-12,34-oxamanzamine E

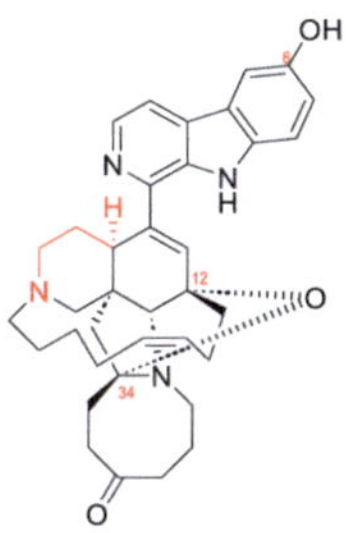

545. (+)-12,34-oxa-6-hydroxymanzamine E

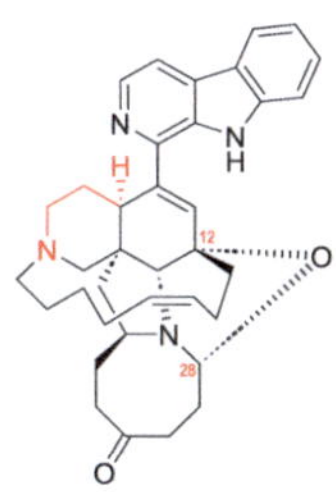

546. (+)-12,28-oxamanzamine E

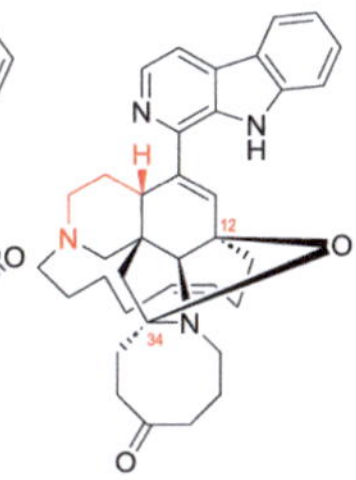

547. (–)-12,34-oxamanzamine E

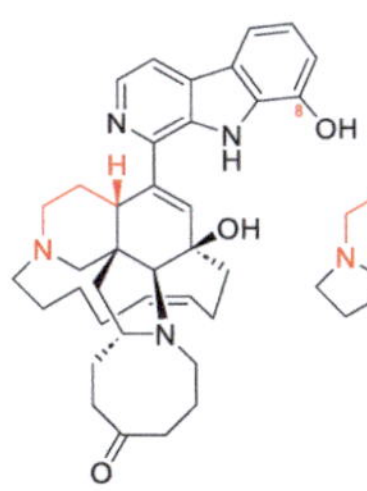

548. (–)-manzamine F

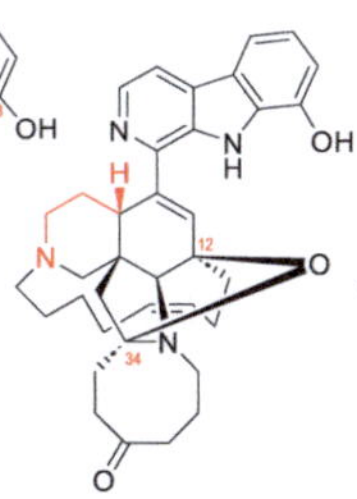

549. (–)-12,34-oxamanzamine F

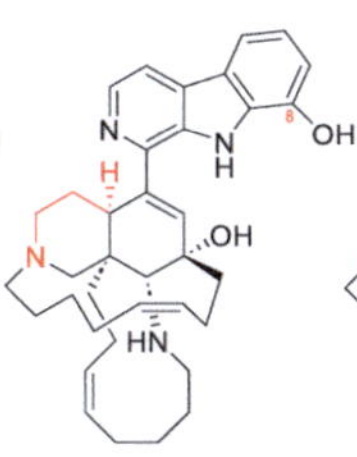

550. (+)-8-hydroxymanzamine J

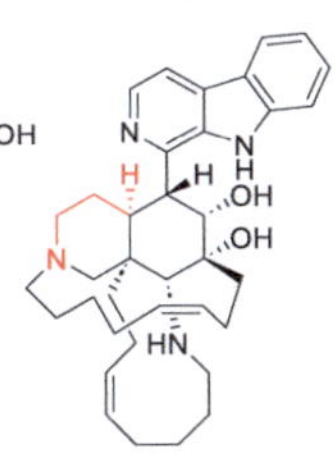

551. (+)-11-hydroxymanzamine J

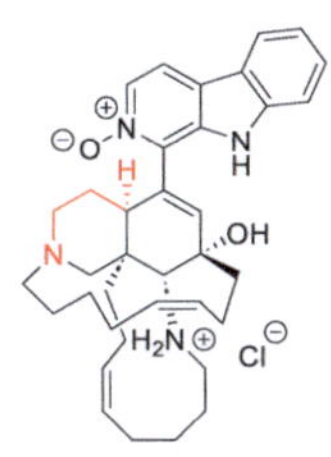

552. (+)-manzamine J N-oxide-HCl

Fig. 16 (continued)

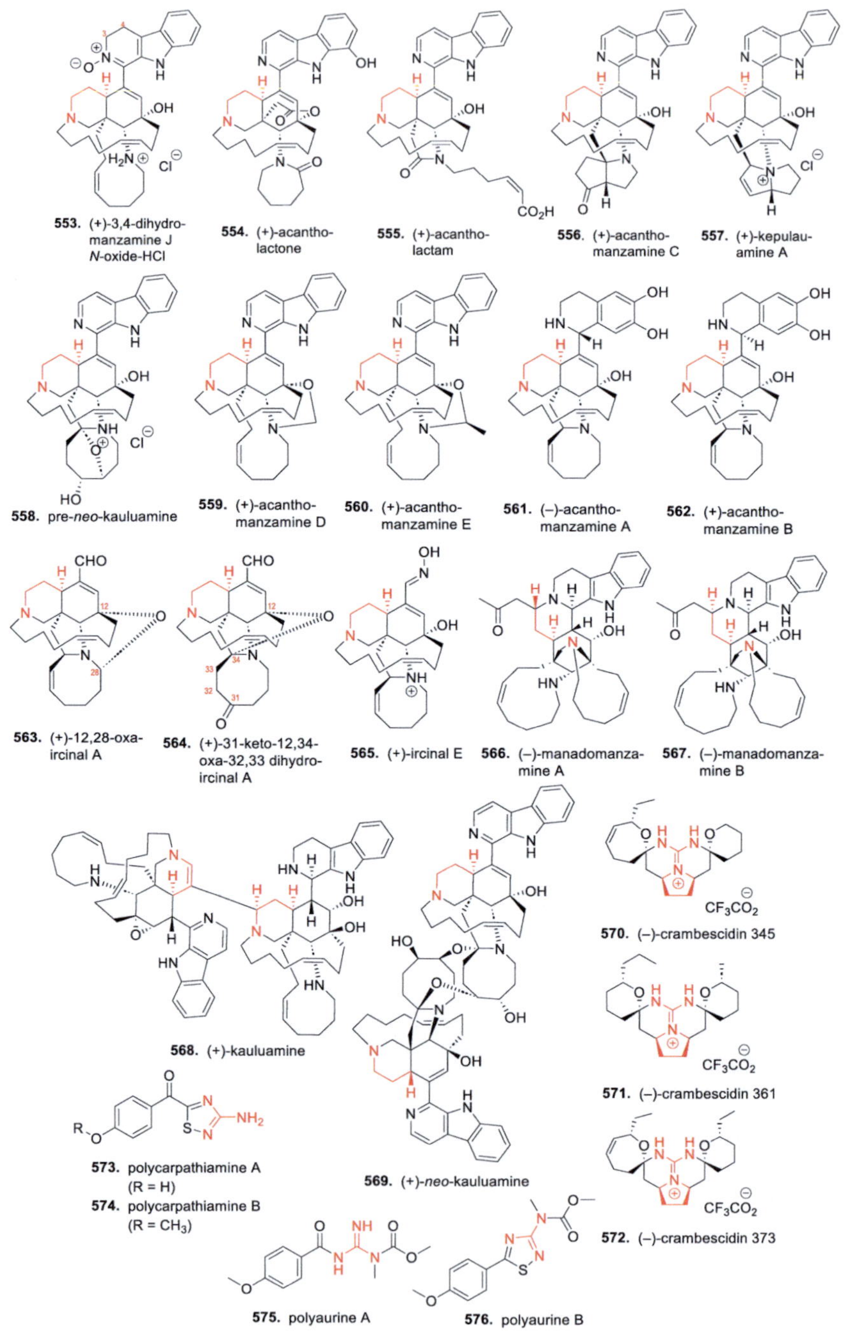

Fig. 16 (continued)

(PUFA) [309, 310]. The AC of (+)-**578b** was determined by total synthesis [311, 312] leading to stereochemical revision of (+)-**578a**. In case of (+)-**579a**, the NMR spectra for the synthetic material were not consistent with those reported for natural didemniserinolipid C, (+)-**579a** indicated that the correct structure might be as in (+)-**579b**. Sixteen siladenoserinols A–L [(+)-**580–591**] [313], M, N [(+)-**592**, **593**], O [(−)-**594**], P [(+)-**595**] [314] were isolated from *Didemnidae* tunicates obtained from NSW waters. The AC of (+)-**580** was established by combination of Mosher and ECD analyses [313] and confirmed by total synthesis [315]. Mosher ester was used to determine the AC of (+)-**592** (sulfonated serinol part). Compound (+)-**595** could be an artifact. Three ceramides, (+)-strepsiamides A–C (**596–598**) [316] were isolated from the marine sponge *Strepsichordaia lendenfeldi* (UEP). Two additional ceramides related molecules, (−)-iotrochotamides I (**599**), II (**600**) were also isolated from *Iotrochota purpurea* (SSW) [317]. Another mixed biosynthesis of serine, fatty acid, and C_1 molecules was (−)-debromoantazirine (**601**) possessing azirine skeleton isolated from a marine sponge *Dysidea* sp. (PUA) [318]. In addition to this work, a new marine sponge species *Homophymia* sp. from GTO was the source of hybrid polyketides-NRPS possessing oxazole moiety, enigmazoles D and E (**602**, **603**) [319]. The AC of **601** was determined by the application of Mosher's method. Moreover, **603** was likely an artifact. The structures of marine alkaloid derived serine are presented in Fig. 17.

3.6 Alkaloid Derived Glycine

About twenty-six alkaloids derived glycine (Fig. 18) consisted of melophlins A (**604/605**) [320], D (**606/607**), E (**608/609**), (−)-melophlin F (**610/611/612/613**), melophlins G (**614/615**), H (**616/617**), (−)-melophlin I (**618/619/620/621**), K (**622/623**) [321], (+)-lobozoanthamine (**624**) [322], (+)-kumusine (**625**) [323], 1,3,O^7-trimethylisoxanthopterin (**626**) [324], nocarimidazoles C (**627/628**), and D (**629**) [325] were discovered from Indonesian waters. The marine sponge *Melophlus sarassinorum* was a rich source of melophlin MNPs collected from SSW waters. Melophlins (**604–628**) were tetramic acids possessing a mixture of tautomers and diastereoisomers. A mixed of biosynthesis alkaloid and polyketide, (+)-lobozoanthamine (**624**), was isolated from a marine soft coral *Lobophytum* sp. (NSW). The AC of (+)-**624** was elucidated by Mosher ester. A marine sponge *Theonella* sp. (NSW) was the source of (+)-kumusine (**625**) possessing purine base. The AC of (+)-**625** was determined by CD analysis from carbamate derivative indicating D-configuration for the ribose. An example pterine natural product, **626** was isolated from *Eudistoma* sp. (SSW). Three degraded purines as in **627/628**, **629** were obtained from a symbiotic soft coral-bacteria *Mycedium* sp.-*Kocuria* sp. (CJV). A mixture of enantiomer *R* (73%) and *S* (27%) were observed after chemical transformation. Melophlin natural products are assumed to be derived from a mixed biosynthesis of glycine or alanine and carboxylic acid of β-keto acid [326]. Moreover, zoanthamine alkaloid illustrated by (+)-lobozoanthamine (**624**) is likely consisted of

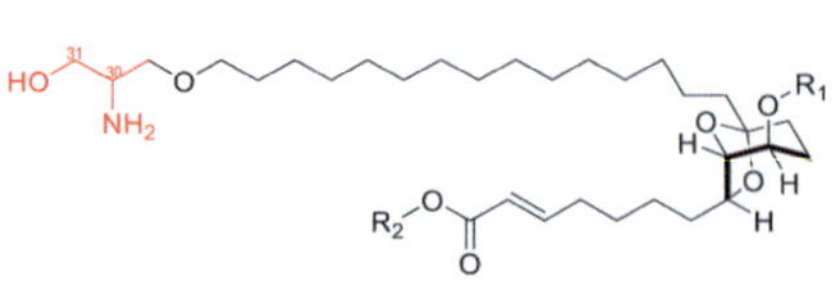

577. (+)-didemniserinolipid A (R_1 = Ac, R_2 = H)
578a. (+)-didemniserinolipid B (R_1 = H, R_2 = CH_2CH_3)
579a. (+)-didemniserinolipid C (R_1 = R_2 = H)

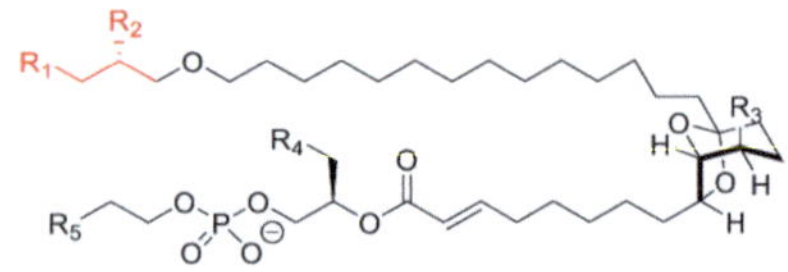

580. (+)-siladenoserinol A (R_1 = OH, R_2 = $NHSO_3H$, R_3 = R_4 = OAc, R_5 = $\overset{\oplus}{N}(CH_3)_3$)
581. (+)-siladenoserinol B (R_1 = OH, R_2 = $NHSO_3H$, R_3 = R_4 = OAc, R_5 = $\overset{\oplus}{N}H_3$)
582. (+)-siladenoserinol C (R_1 = OSO_3H, R_2 = NH_2, R_3 = OH, R_4 = OAc, R_5 = $\overset{\oplus}{N}H_3$)
583. (+)-siladenoserinol D (R_1 = OH, R_2 = $NHSO_3H$, R_3 = OAc, R_4 = OH, R_5 = $\overset{\oplus}{N}(CH_3)_3$)
584. (+)-siladenoserinol E (R_1 = OSO_3H, R_2 = NH_2, R_3 = OH, R_4 = OAc, R_5 = $\overset{\oplus}{N}(CH_3)_3$)
585. (+)-siladenoserinol F (R_1 = OH, R_2 = $NHSO_3H$, R_3 = OH, R_4 = OAc, R_5 = $\overset{\oplus}{N}(CH_3)_3$)
586. (+)-siladenoserinol G (R_1 = OSO_3H, R_2 = NH_2, R_3 = R_4 = OAc, R_5 = $\overset{\oplus}{N}(CH_3)_3$)

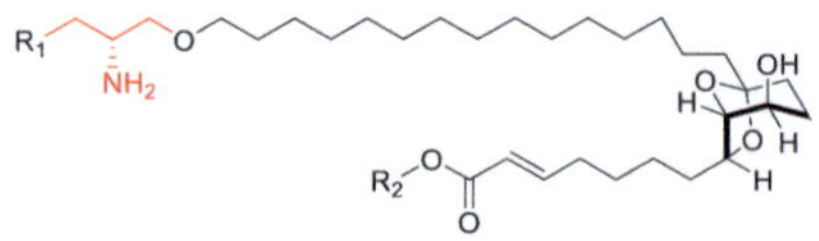

578b. (+)-didemniserinolipid B (R_1 = OSO_3Na, R_2 = CH_2CH_3)
579b. (+)-didemniserinolipid C (R_1 = OH, R_2 = H)

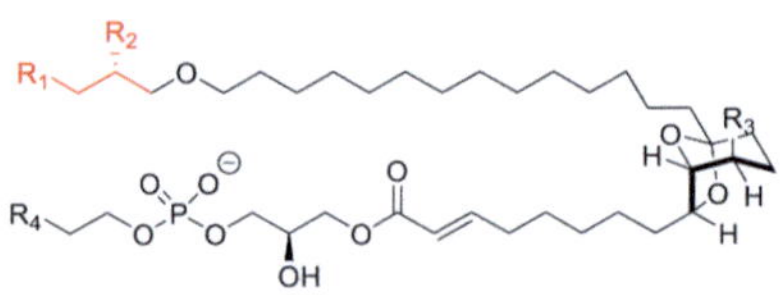

587. (+)-siladenoserinol H (R_1 = OH, R_2 = $NHSO_3H$, R_3 = OAc, R_4 = $\overset{\oplus}{N}(CH_3)_3$)
588. (+)-siladenoserinol I (R_1 = OSO_3H, R_2 = NH_2, R_3 = OAc, R_4 = $\overset{\oplus}{N}(CH_3)_3$)
589. (+)-siladenoserinol J (R_1 = OSO_3H, R_2 = NH_2, R_3 = OH, R_4 = $\overset{\oplus}{N}H_3$)
590. (+)-siladenoserinol K (R_1 = OH, R_2 = $NHSO_3H$, R_3 = OH, R_4 = $\overset{\oplus}{N}(CH_3)_3$)
591. (+)-siladenoserinol L (R_1 = OSO_3H, R_2 = NH_2, R_3 = OH, R_4 = $\overset{\oplus}{N}(CH_3)_3$)

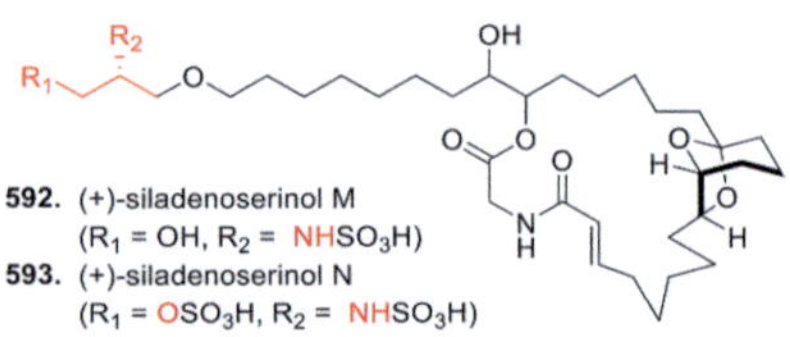

592. (+)-siladenoserinol M (R_1 = OH, R_2 = $NHSO_3H$)
593. (+)-siladenoserinol N (R_1 = OSO_3H, R_2 = $NHSO_3H$)

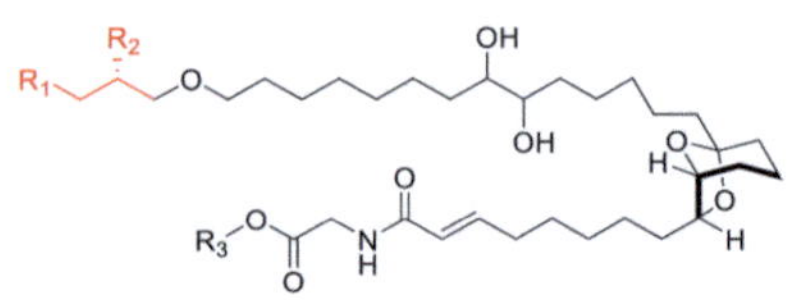

594. (–)-siladenoserinol O (R_1 = OH, R_2 = $NHSO_3H$, R_3 = H)
595. (+)-siladenoserinol P (R_1 = OSO_3H, R_2 = $NHSO_3H$, R_3 = CH_3)

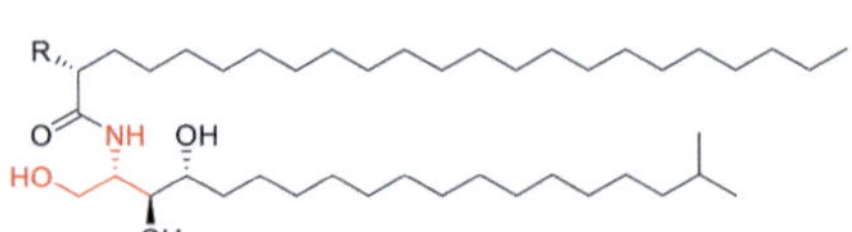

596. (+)-strepsiamide A (R = H)
597. (+)-strepsiamide B (R = OH)

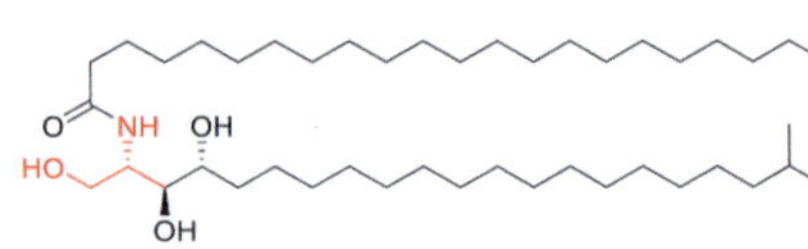

598. (+)-strepsiamide C

Fig. 17 Structures of marine alkaloid derived serine (**577–603**) from Indonesian waters. The marine alkaloid derived serine building blocks were proposed with red color

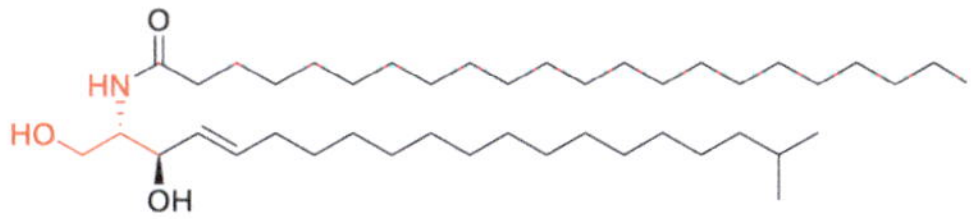

599. (–)-iotrochotamide I

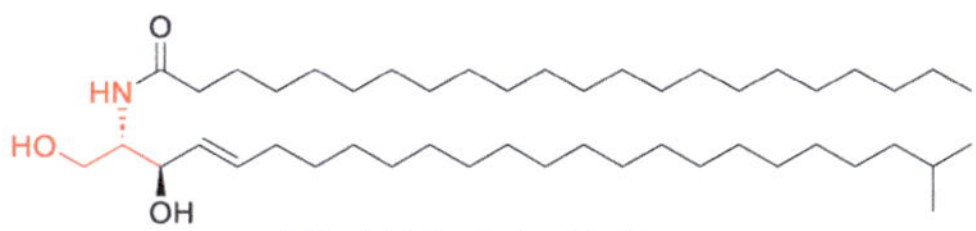

600. (–)-iotrochotamide II

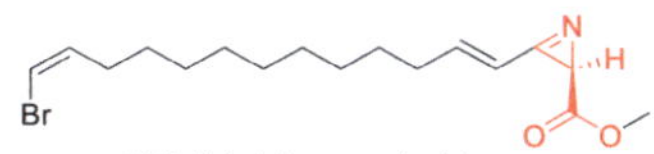

601. (–)-debromoantazirine

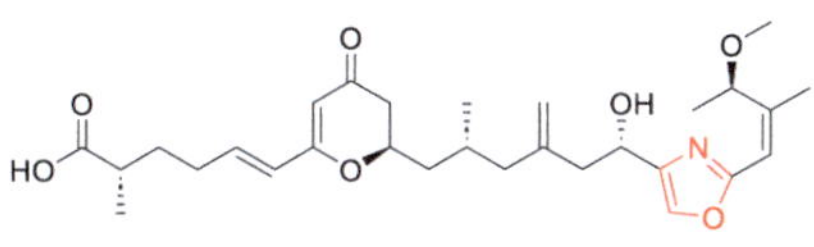

602. enigmazole D

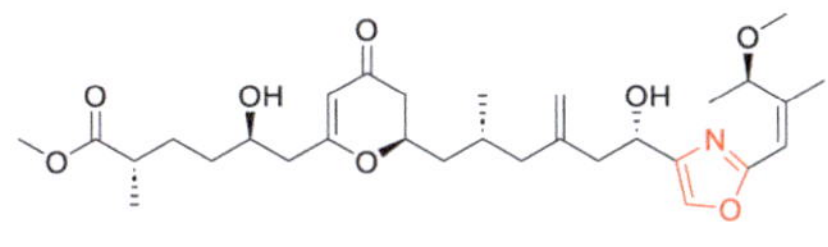

603. enigmazole E

Fig. 17 (continued)

glycine and polyketide [327, 328]. Finally, purine base (**625**, **626**) or related purine base **627–629** could be derived from glycine amino acid, as one of its precursors [329, 330].

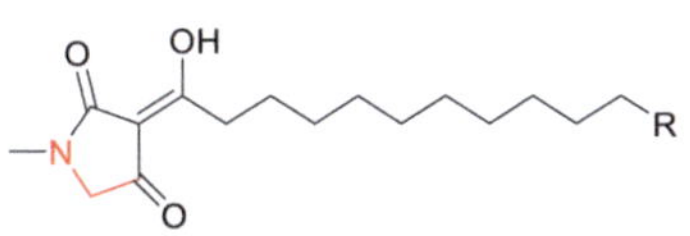

604/605. melophlin A (R = $(CH_2)_4CH_3$)
606/607. melophlin D (R = $(CH_2)_3CH_3$)
608/609. melophlin E (R = $(CH_2)_2CH(CH_3)_2$)
610/611/612/613. (–)-melophlin F (R = $CH_2CHCH_3CH_2CH_3$)
614/615. melophlin G (R = $CH_2CH_2CH_3$)
616/617. melophlin H (R = $CH_2CH(CH_3)_2$)
618/619/620/621. (–)-melophlin I (R = $CHCH_3CH_2CH_3$)
622/623. melophlin K (R = $CH(CH_3)_2$

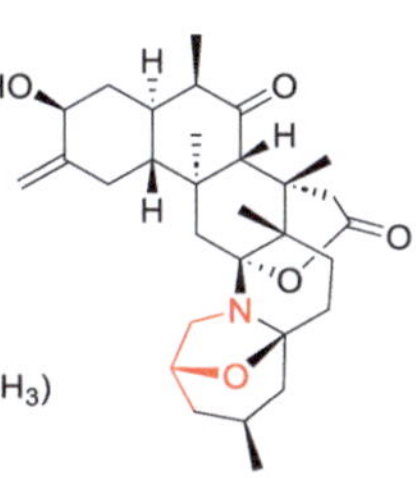

624. (+)-lobozoanthamine

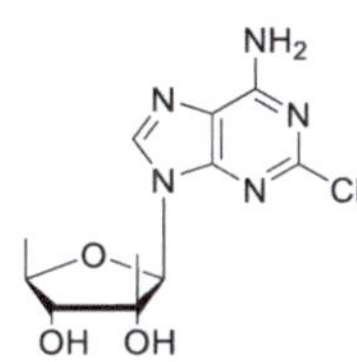

625. (+)-kumusine

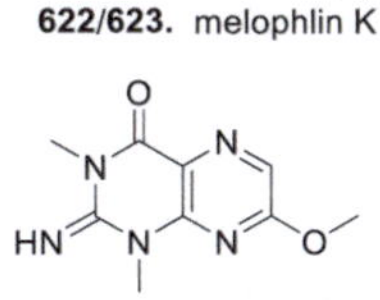

626. 1,3,O^7-trimethyl isoxanthopterin

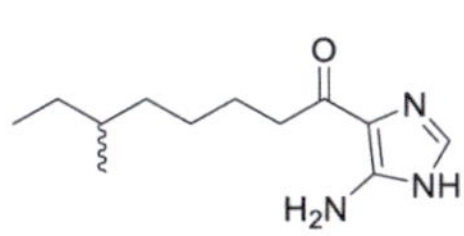

627/628. nocarimidazole C

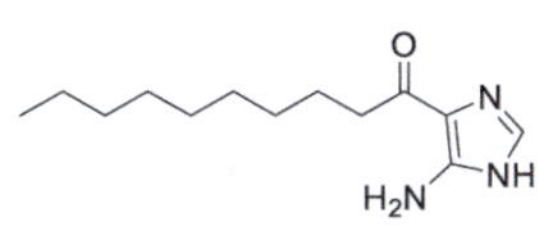

629. nocarimidazole D

Fig. 18 Structures of marine alkaloid derived glycine (**604–629**) from Indonesian waters. The marine alkaloid derived glycine building blocks were proposed with red color

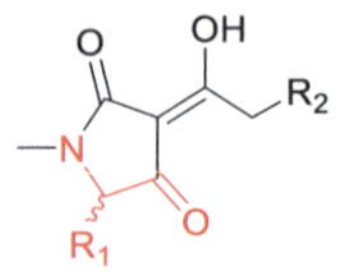

630–633. (–)-melophlin B
$R_1 = \alpha\text{–}CH_3$, $R_2 = CH_2CHCH_3(CH_2)_7CH_3$
634–641. (–)-melophlin C
$R_1 = CH_3$, $R_2 = (CH_2)_2CHCH_3(CH_2)_6CH_3$
642–643. (–)-melophlin J
$R_1 = CH_3$, $R_2 = (CH_2)_{11}CH_3$
644–645. (–)-melophlin L
$R_1 = CH_3$, $R_2 = (CH_2)_{10}CH_3$
646–647. (–)-melophlin M
$R_1 = CH_3$, $R_2 = (CH_2)_9CH_3$
648–649. melophlin N
$R_1 = CH_3$, $R_2 = (CH_2)_8CH(CH_3)_2$
650–653. melophlin O
$R_1 = CH_3$, $R_2 = (CH_2)_7CHCH_3CH_2CH_3$

Fig. 19 Structures of marine alkaloid derived alanine (**630–653**) from Indonesian waters. Marine alkaloid derived alanine building block was colored with red

3.7 *Alkaloid Derived Alanine*

Alanine amino acid was able to identify as a building block of (−)-melophlin B (**630/631/632/633**) [320], (−)-melophlin C (**634/635/636/637/638/639/640/641**), (−)-melophlin J (**642/643**), (−)-melophlin L (**644/645**), (−)-melophlin M (**646/647**), melophlin N (**648/649**), and melophlin O (**650/651/652/653**) [321]. The source of melophlin natural products was the marine sponge *Melophlus sarasinorum.* All compounds were obtained as a mixture of tautomers and/or diastereomers. The AC of (−)-melophlin B (**630/631/632/633**) was determined by chemical correlation, Marfey and ECD analyses. The structures of marine alkaloid derived alanine are shown in Fig. 19.

3.8 *Miscellaneous Alkaloid Derived Amino Acids*

Six amino acid building blocks can be identified from this subclass including L-aspartic acid, L-histidine, L-phenylalanine, L-valine, L-asparagine, and L-cysteine. Three marine alkaloids cinachyrazoles A–C (**654–656**) [331] possessed L-aspartic acid building block were isolated from MLU collection of a marine sponge *Cinachyrella* sp. Pyrrazole metabolites (**654–656**) could be derived from L-aspartic acid and L-glutamic acid to form 1,3-diaminopropane as an intermediate [332]. In addition to pyrazole metabolites, pyrimidine skeleton was also expected derived from L-aspartic acid as in **657** [333]. The metabolite was characterized from *Kaliapsis* sp. (BLI). (−)-Reticulatins A and B (**658**, **659**) possessing L-histidine building block [190] were isolated from the marine sponge *Hyrtios reticulatus.* The ACs of (−)-**658** and **659** were analyzed by comparison between experimental and calculated ECD data. The marine sponge *Clathria basilana* (UEP and MLU) was the source of clathyrimine A (**660**) [334] and (−)-(*E*)-2-amino-3-methyl-*N*-styrylbutanamide (**661**) [335]. The former was likely derived from L-phenylalanine, while the latter was biosynthesized from L-valine. The AC of (−)-**661** was confirmed by Marfey analysis. Compound **660** contained quinolizine moiety. (−)-Aurantoside J (**662/663**) obtained as a mixture of tautomers was isolated from the marine sponge *Theonella swinhoei* [336] possessing L-asparagine building block. The AC of (−)-**662/663** was determined by a comparison of experimental and calculated NMR and optical rotations. (+)-Biakamides A (**664**), B (**665**), (−)-biakamides C (**666**), D (**667**), the hybrid PKS-NRPS metabolites [337] were isolated from PUA collection of a marine sponge *Petrosaspongia* sp. (Fig. 20). The ACs of the (+)-**664**, **665** and (−)-**666**, **667** were established by ECD analysis and total synthesis [337]. The metabolite contained thiazole moiety which was possibly derived from L-cysteine amino acid.

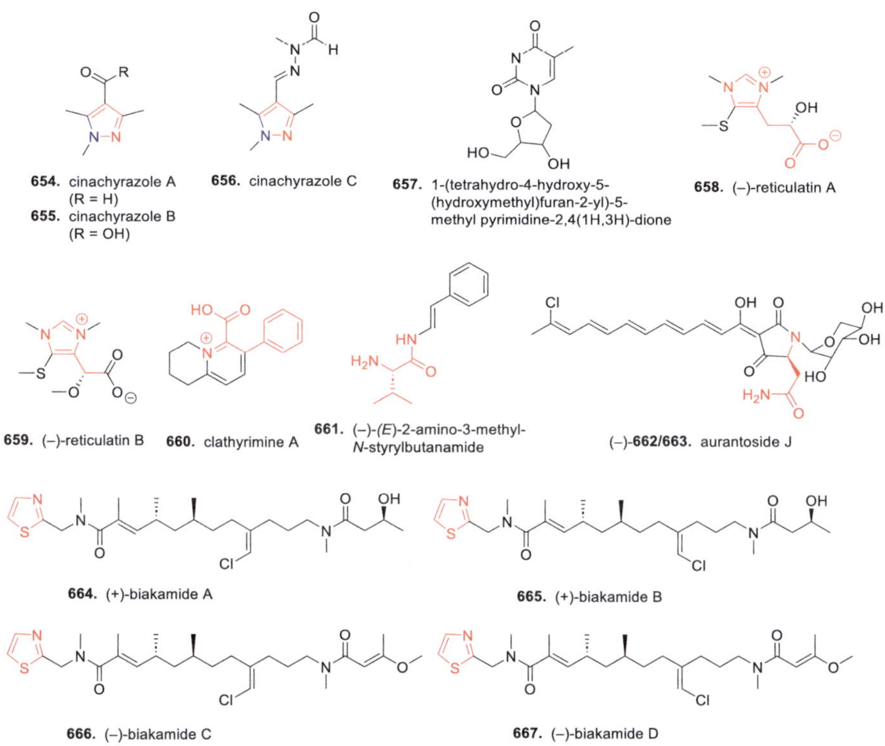

Fig. 20 Structures of marine miscellaneous alkaloid derived amino acid (**654–667**). The marine miscellaneous alkaloid building blocks were proposed with red color

4 Peptide

Biosynthetically, marine peptides can be classified into nonribosomal peptides (NRPs) and ribosomally synthesized and posttranslationally modified peptides (RiPPs). The characteristic of NRPS metabolites is recognized by the presence of nonproteinogenic amino acids (>200), while RiPPs metabolites were limited by a set of 20 proteinogenic amino acids (22 including selenocysteine and pyrrolysine) [338]. As cyclization plays important role in biological activities, the Indonesian marine peptides are classified into two subclasses: linear peptide and cyclic peptide. The latter can be further divided into two families: cyclopeptide and cyclodepsipeptide. The building blocks of Indonesian marine peptides can be described as in Fig. 21 consisted of proteinogenic amino acids, nonproteinogenic amino acids, and their modifications.

4.1 Linear Peptide

The class of Indonesian marine linear peptides can further be divided into at five subclasses including dipeptides [8 molecules: (−)-**668**, **674**, **675**, (+)-**669–673**), tripeptides [1 molecule: (+)-**676**], pentapeptide [1 molecule: (−)-**677**], heptapeptide [1 molecule: (+)-**678**], tridecapeptide [1 molecule: (−)-**679**]. Moreover, some features marine linear peptides are identified possessing D-amino acid [7 molecules: (+)-**669–673**, (+)-jasplakinolide Z5 (revised), (+)-**676**], lipopeptide [6 molecules: (−)-**668**, (+)-**669–673**], heterocyclization [3 molecules: (−)-**668**, **677**, **679**], β-amino acid [1 molecule: (+)-**676**], and *N*-methylamide [6 molecules: (−)-**668**, **674**, (+)-jasplakinolide Z5 (revised), (+)-**676**, (−)-**677**, (−)-**679**], *N*-acylated tetramic acid [5 molecules: (+)-**669–673**], and disulfide bond [1 molecule: (−)-**679**]. The structures of Indonesian marine linear peptides are shown in Fig. 22.

Eight marine linear dipeptides were distributed into two types of sponges: *Dysidea* sp. [(−)-dysithiazolamide (**668**, NMU) [339] and (+)-sintokamides A–E (**669–673**, CJV) [340] and *Callyspongia* sp. [(−)-callyspongiamides A, B (**674**, **675**)] [341]. The RC of (−)-**668** was assigned by JBCA and NMR calculation [339], while its AC was confirmed by total synthesis [342]. The ACs of (+)-**669** was revealed by X-ray analysis [340] and total synthesis featured with a concise, unified strategy involving no protecting-group strategy [343], while the ACs of (+)-**669**, (+)-**670**, and (+)-**673** were disclosed by total synthesis [343]. (+)-Jasplakinolide Z5 was proposed as a revised structure [344] after reisolation work from the marine sponge *Jaspis splendens* (EKM) and reinvestigation of its 1D and 2D NMR spectra. The original structure of (+)-jasplakinolide Z5 was isolated from the Fijian sponge *Jaspis splendens* in 2011 [345]. In addition to this work, a new linear tripeptide, (+)-jasplakinolide Z6 (**676**) was isolated from EKM waters [344]. One linear pentapeptide, (−)-seribunamide A (**677**) was obtained from a JSCR marine sponge of *Ircinia* sp. [346]. The AC of (−)-**677** was assigned by application of Marfey's method. The

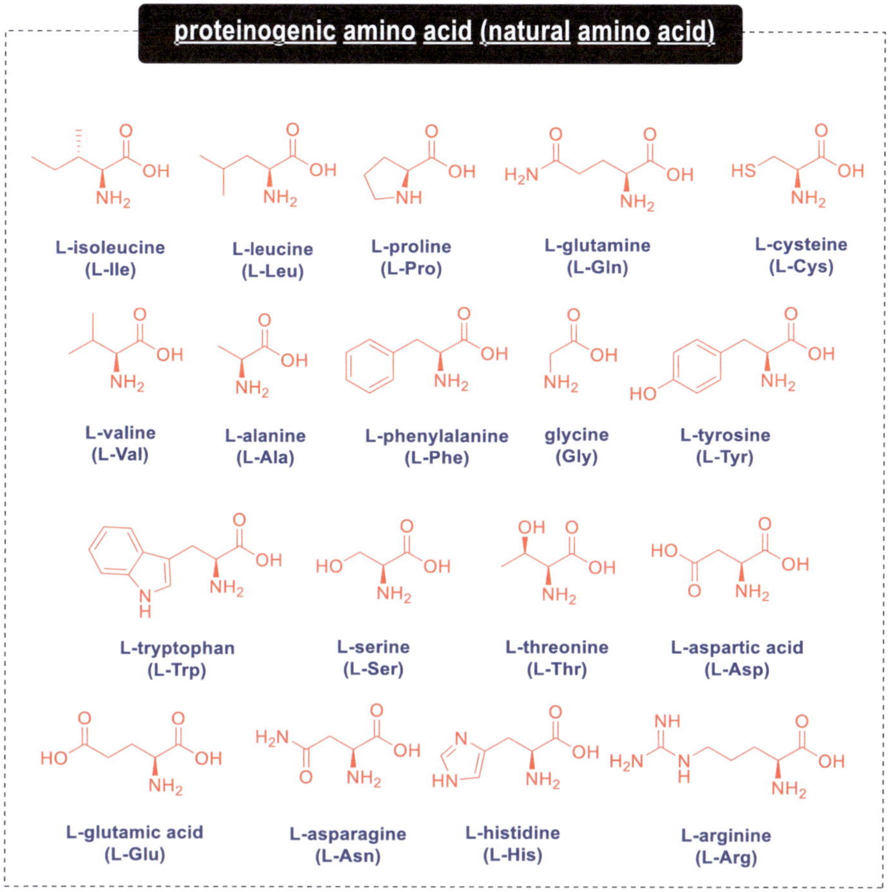

Fig. 21 Marine peptide building blocks found from Indonesian marine organisms consisted of proteinogenic and nonproteinogenic amino acids as well as their modifications. Building blocks drawn in green were new and found in marine organisms, those in red were common in terrestrial and/or marine organisms

marine sponge *Callyspongia subarmigera* collected from SSW waters biosynthesized a new linear heptapeptide, (+)-subarmigeride A (**678**) [347]. Its AC was determined using Marfey's analysis. (−)-Kendarimide (**679**), a linear tridecapeptide was characterized from a SES collection of a marine sponge *Haliclona* sp. [348]. Its AC was confirmed through Marfey's method [348] and synthesis on four diastereomers of MeCys-MeCys [349].

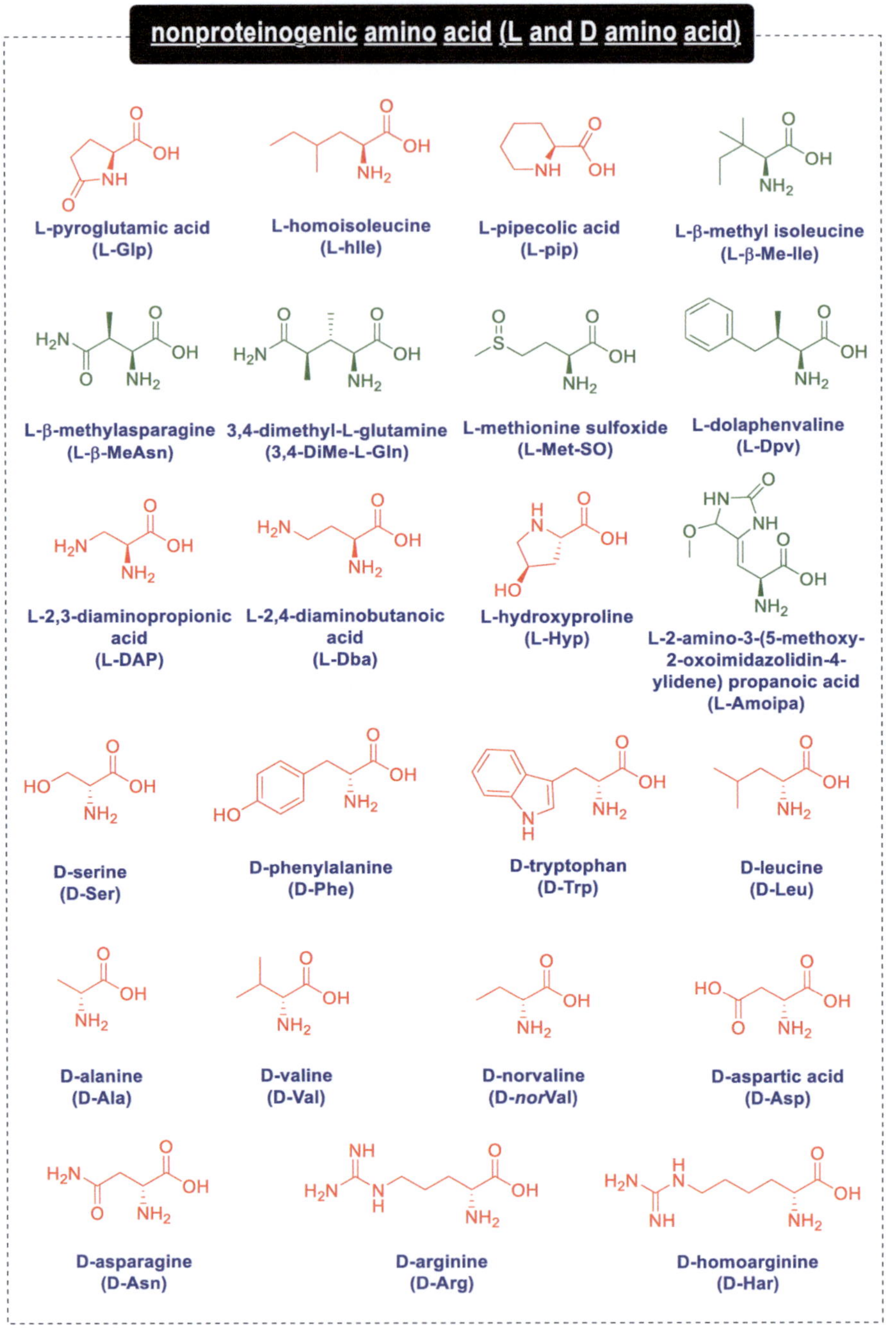

Fig. 21 (continued)

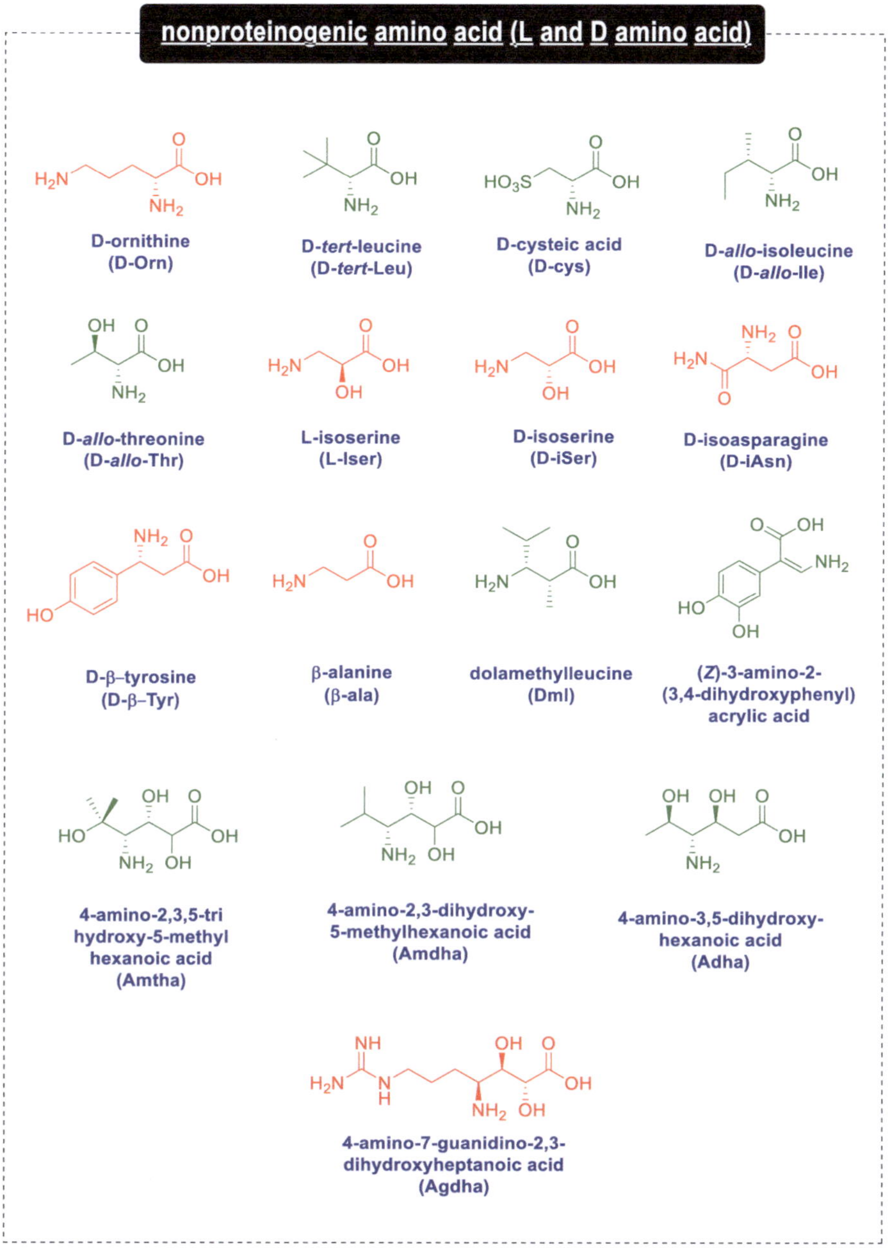

Fig. 21 (continued)

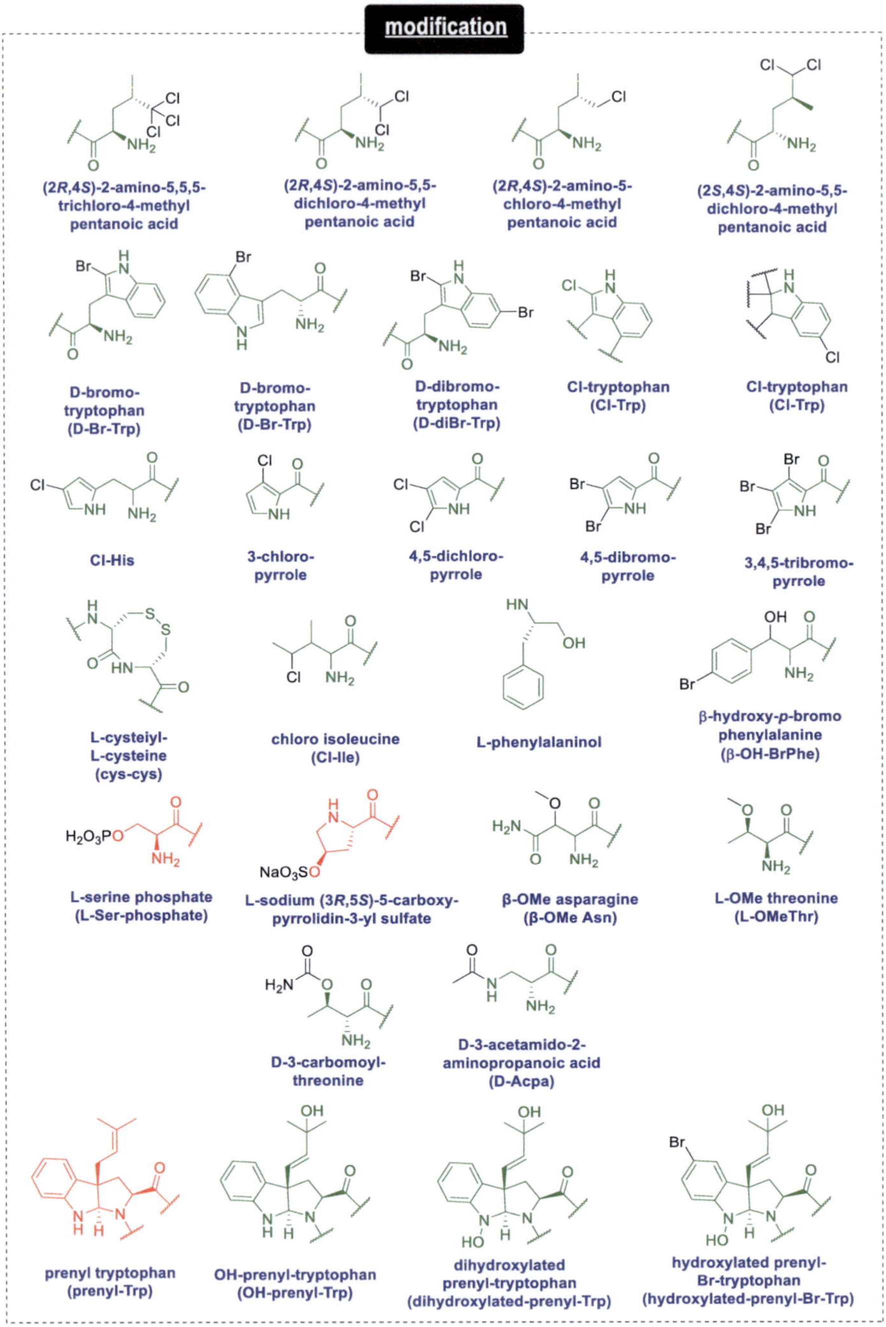

Fig. 21 (continued)

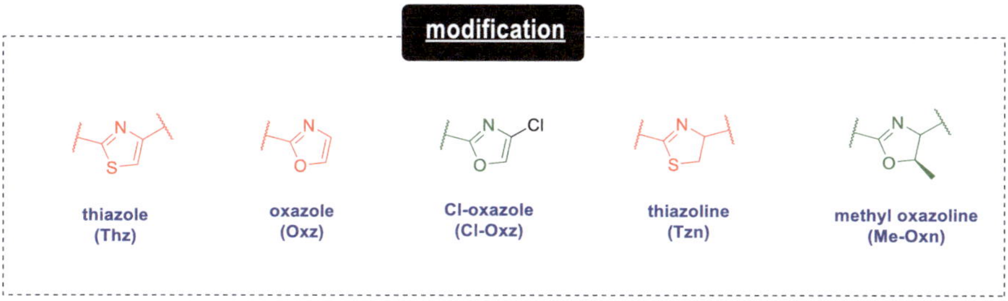

Fig. 21 (continued)

4.2 Cyclopeptide

The cyclization of peptides plays an important role in exerting biological functions. The Indonesian marine cyclopeptides (Fig. 23) can be subclassified on the basis of their cyclization: head-to-side chain (1 molecule: (−)-**680**), head-to-tail (26 molecules: (−)-**681**, **683**, **685–693a**, **696–702**, **703b**, **704–706** and (+)-**682** as well as **684**, **694**, **695**), tail-to-side chain (15 molecules: (−)-**707–721**), side chain-to-side chain (5 molecules: (−)-**722–724**, (+)-**725**, **726**). Some of them possessed D-amino acid (14 molecules: (−)-**681**, **683**, **687**, **691**, **693b**, **707**, **708** and (+)-**682**, **703b**, **704–706** as well as **694**, **695**), heterocyclization (16 molecules: (−)-**681**, **683**, **685–693a** and (+)-**682** as well as **684**, **694**, **695**), β-amino acid (5 molecules: (−)-**681**, (+)-**703b**, **704–706**), *N*-methylamide (5 molecules: (−)-**681**, (+)-**703b**, **704–706**), disulfide bond (5 molecules: (−)-**722–726**), proline residue (35 molecules: (+)-**682** and (−)-**683**, **685**, **692**, **693b**, **696–702**, **709–722**, **725**, **726** as well as **684**, **694**, **695**), isoprene unit (7 molecules: (+)-**682** and (−)-**683**, **699–702** as well as **684**). Meanwhile, the Indonesian marine macrocyclic peptides can also be subclassified on the basis of their biosynthetic origin: nonribosomal (NRPS) and ribosomally synthesized and posttranslationally modified peptides (RiPPs). In general, regular amino acids are characteristic of RiPPs, whereas nonproteinogenic amino acids are constituent of NRPS. Other characteristics of NRPS metabolites are structure modifications via epimerization, hydroxylation, methylation, heterocyclization, and/or oxidative cross linking. On the other hand, RiPPs cyclic peptides are typically produced by cyanobacteria including *Prochloron* sp. and *Anabaena* sp. featuring with thiazoline, methyloxazoline, and oxazoline, or their oxidized counterparts. At least one proline residue is always present when the RiPPs do not possess heterocyclized amino acid. Prenylation of serine, threonine or tyrosine residues is another typical reaction of RiPPs cyclic peptides [350]. Under this category, Indonesian NRPS macrocyclic peptides can putatively be counted as in (−)-**681**, **688–693a**, **693b**, **703a**, **707–721** and (+)-**703b–706**, while Indonesian RiPPs cyclic peptides can be proposed as in (−)-**681**, **683**, **685–687**, **696–702**, **722–724** and (+)-**682**, **725**, **726** as well as **684**, **694**, **695**.

One head-to-side chain cylcotripeptide, (−)-neopetromin (**680**) containing two L-Tyr and L-Trp residues, was isolated from a marine sponge *Neopetrosia* sp.

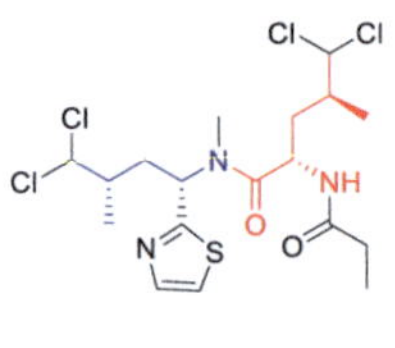
668. (–)-dysithiazolamide
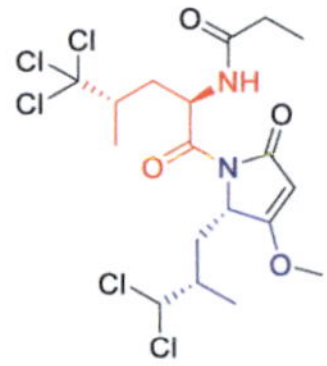
669. (+)-sintokamide A
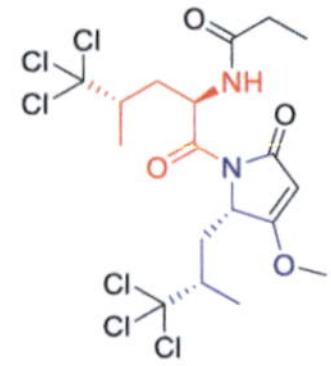
670. (+)-sintokamide B
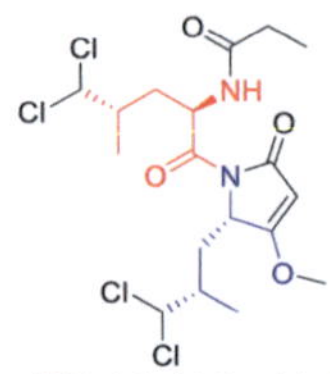
671. (+)-sintokamide C
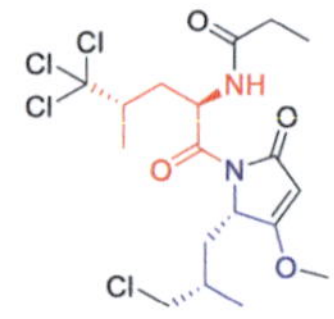
672. (+)-sintokamide D

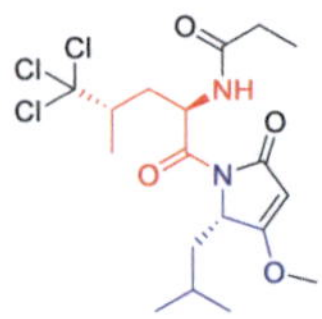
673. (+)-sintokamide E
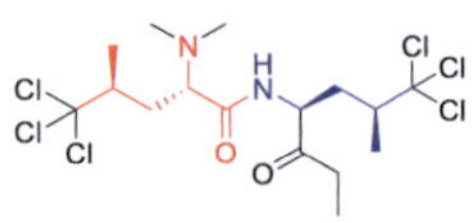
674. (–)-callyspongiamide A
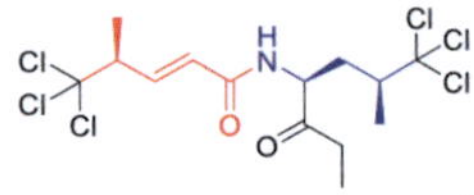
675. (–)-callyspongiamide B
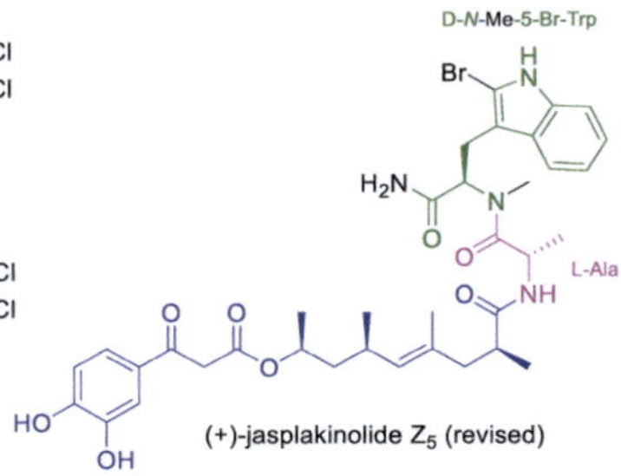

(+)-jasplakinolide Z_5 (revised)

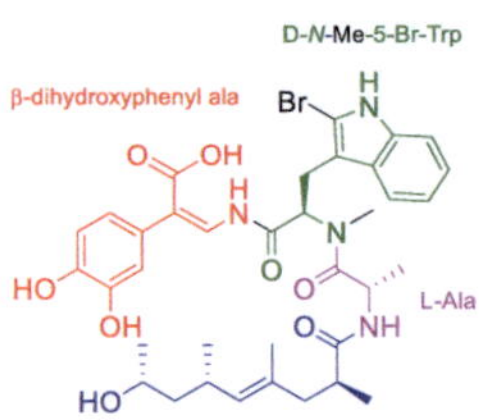

676. (+)-jasplakinolide Z_6
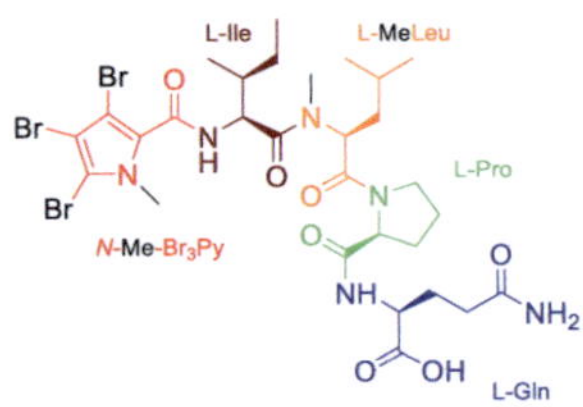

677. (–)-seribunamide A
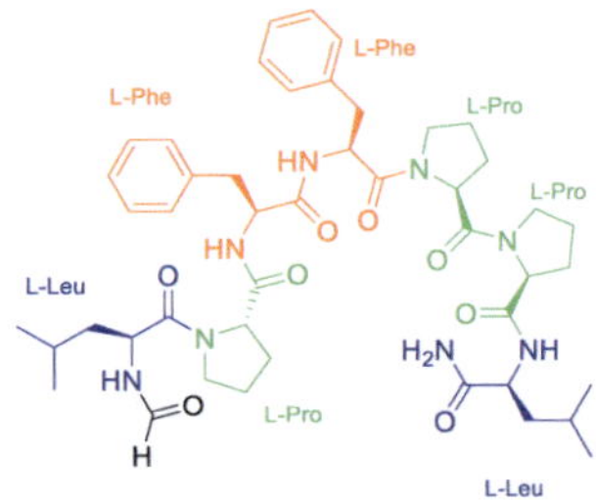

678. (+)-subarmigeride A

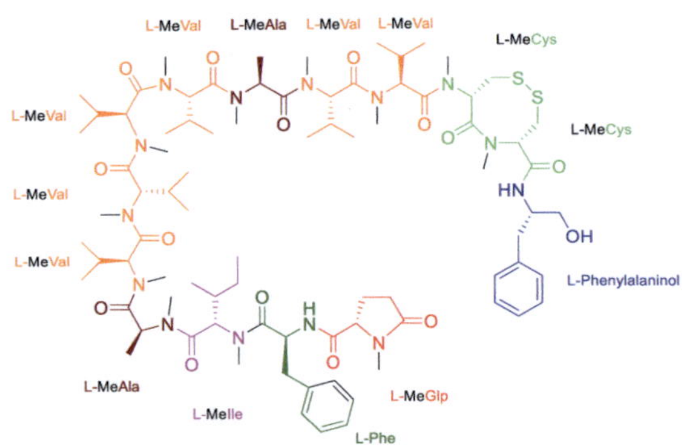

679. (–)-kendarimide

Fig. 22 Structures of marine linear peptide (**668–679**) from Indonesian waters. Amino acid building blocks and their modification were drawn with different color

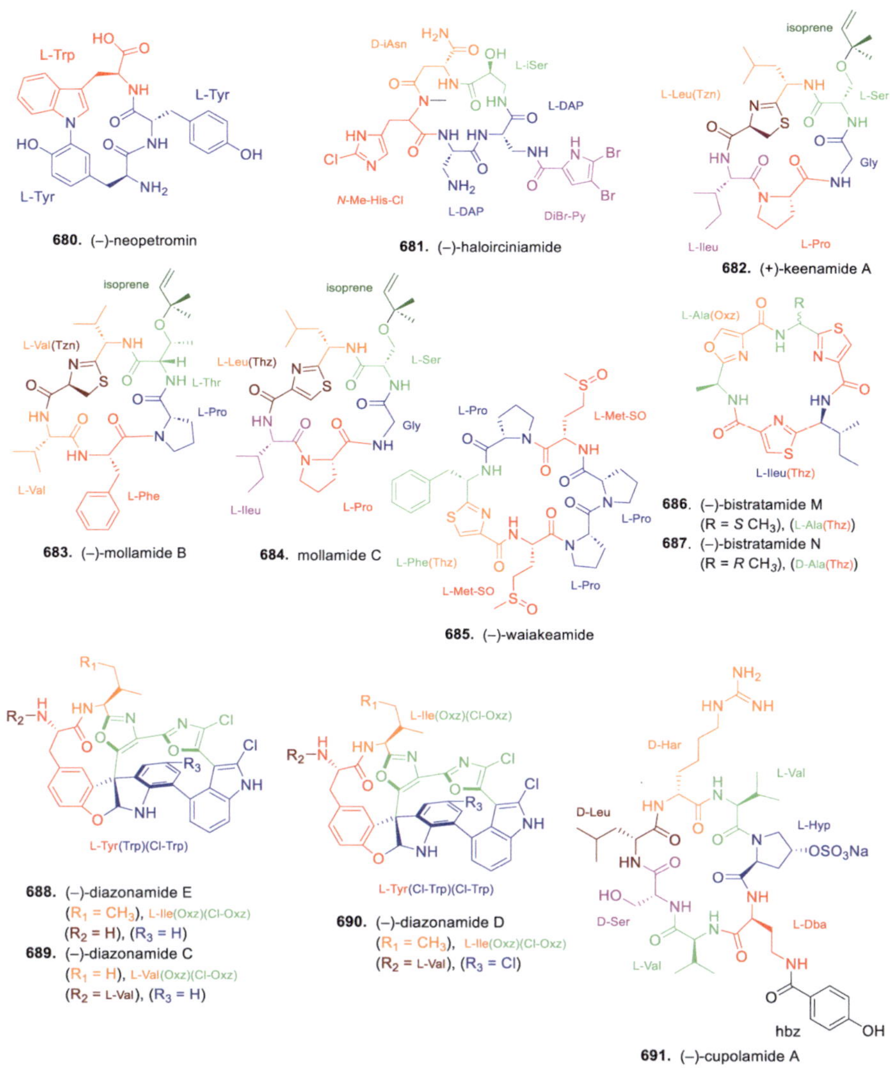

Fig. 23 Structures of marine cyclopeptide (**680–726**) from Indonesian waters. Amino acid building blocks and their modification were drawn with different color

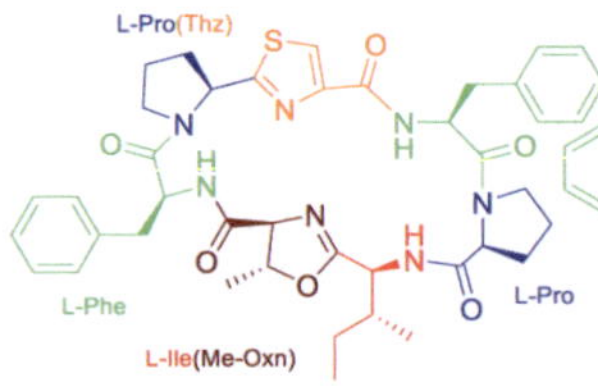

692. (–)-*cis,cis*-ceratospongamide

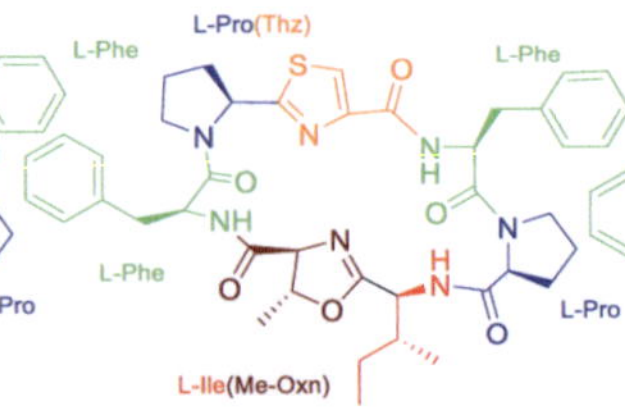

693a. (–)-*trans,trans*-ceratospongamide

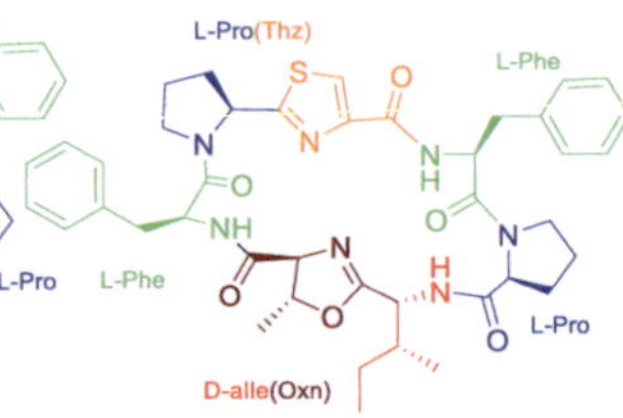

693b. (–)-*trans, trans*-ceratospongamide

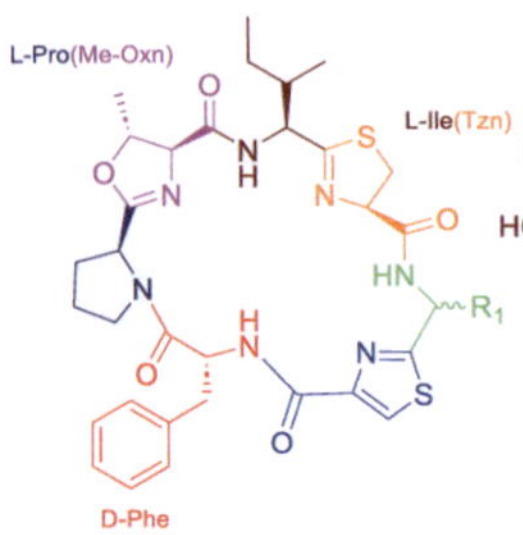

694. lissoclinamide 9
(R_1 = *R* $CHCH_3CH_2CH_3$)
D-Val(Thz)
695. lissoclinamide 10
(R_1 = *S* $CHCH_3CH_2CH_3$)
L-Ile(Thz)

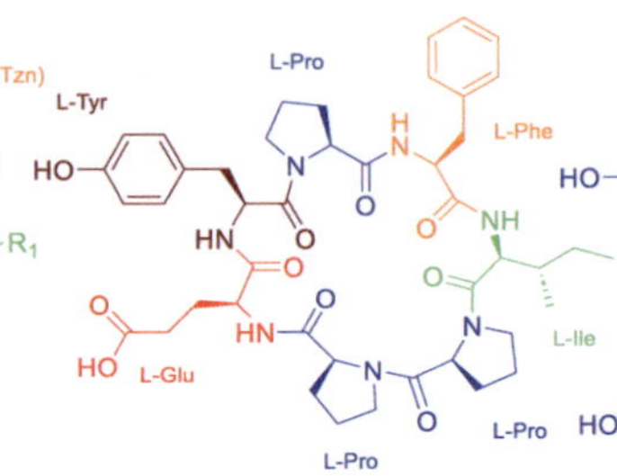

696. (–)-carteritin A

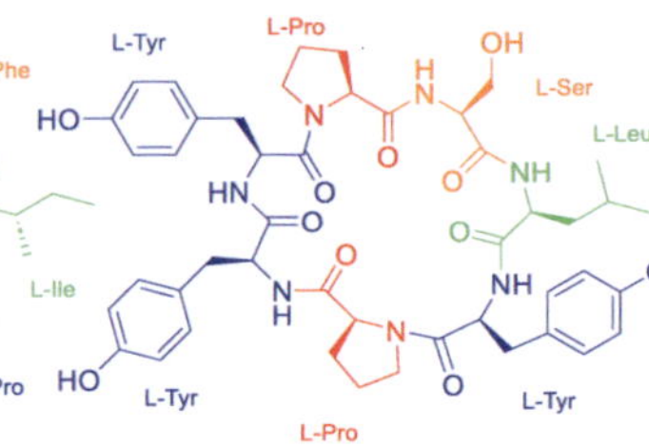

697. (–)-carteritin B

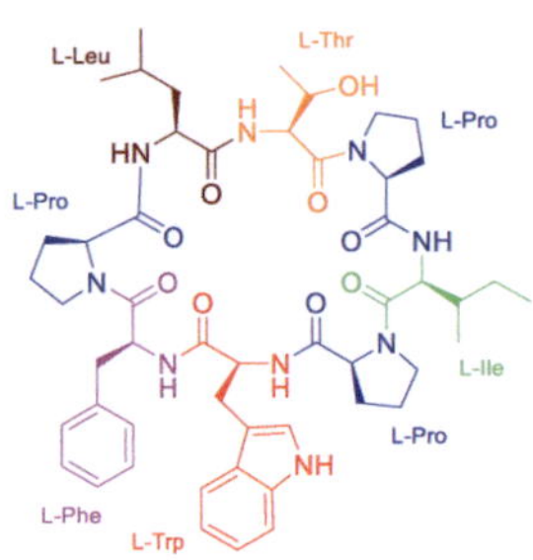

698. (–)-stylissamide X

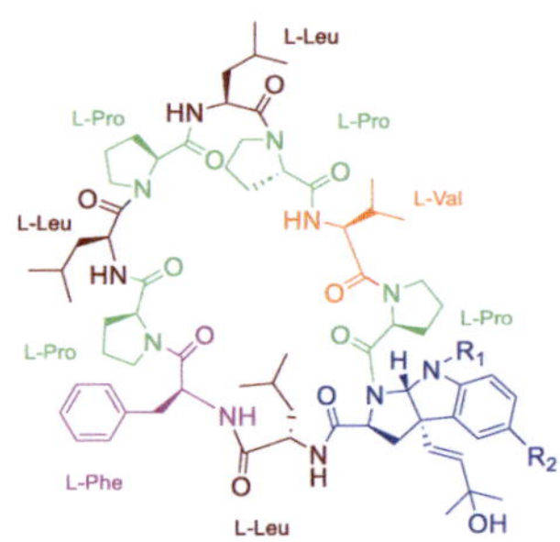

700. (–)-trikoramide B
($R_1 = R_2$ = H)
OH-Prenyl-Trp
701. (–)-trikoramide C
(R_1 = OH, R_2 = H)
dihydroxylated prenyl-Trp
702. (–)-trikoramide D
(R_1 = H, R_2 = Br)
hydroxylated prenyl-BrTrp

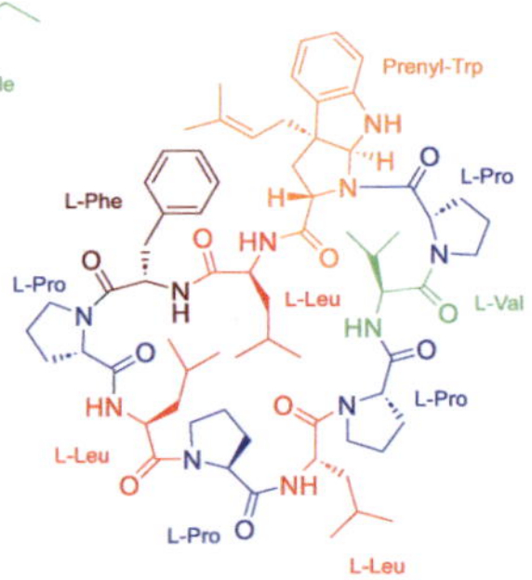

699. (–)-trikoramide A

Fig. 23 (continued)

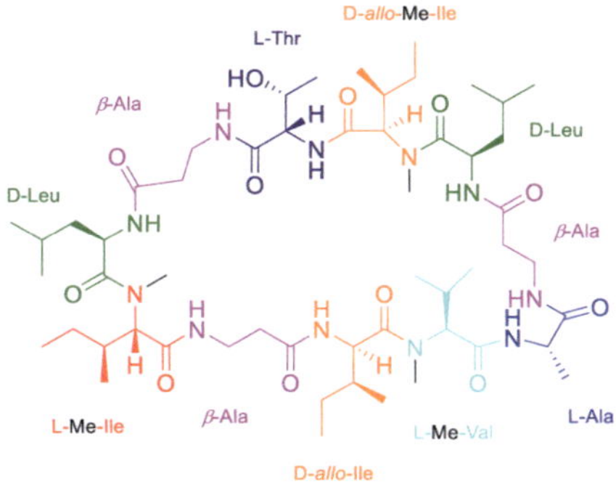

703a. (–)-barangamide A
703b. (+)-barangamide A

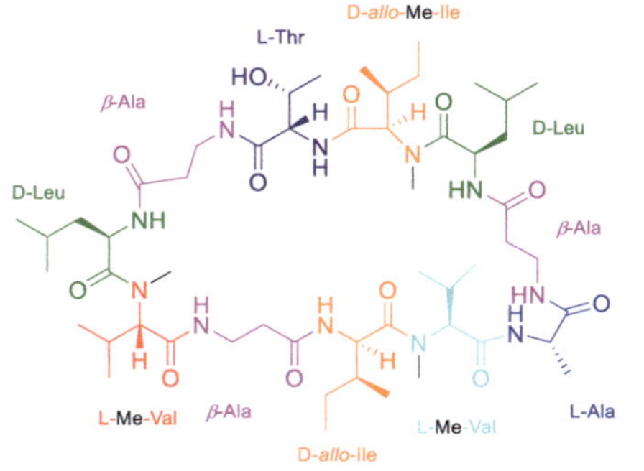

704. (+)-barangamide B

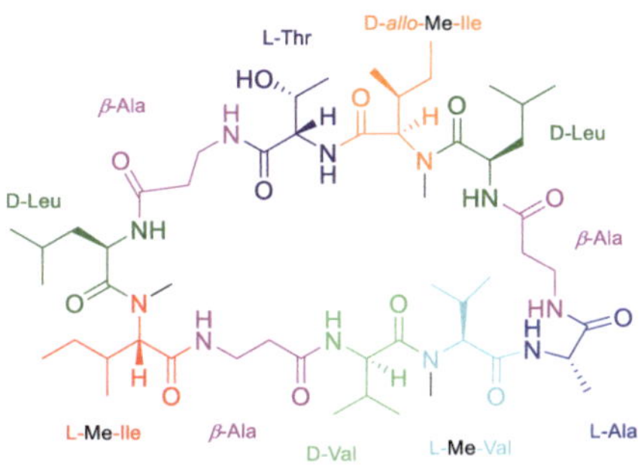

705. (+)-barangamide C

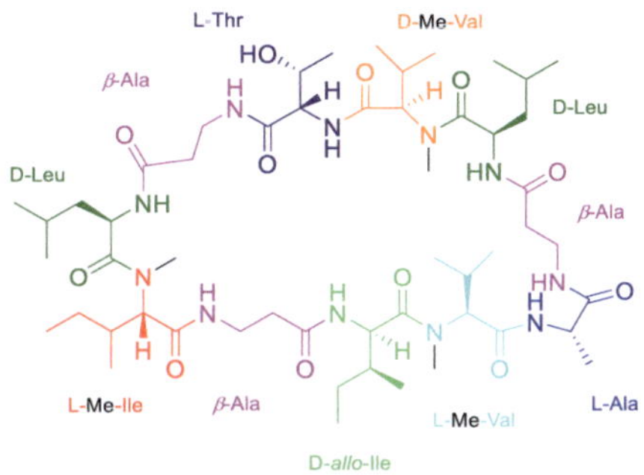

706. (+)-barangamide D

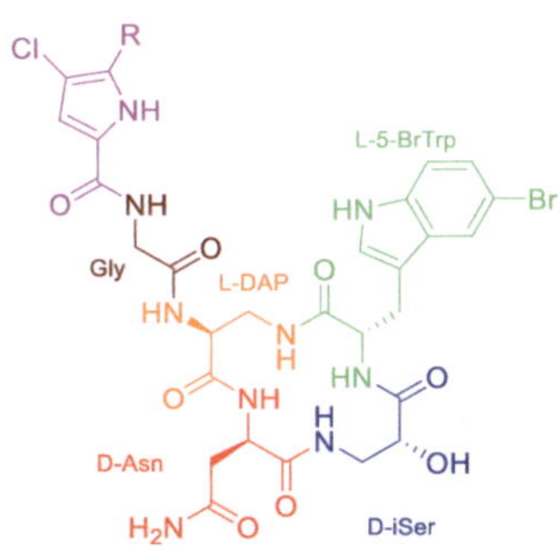

707. (–)-cyclopsammocinamide A
(R = H) 4-Cl-Py
708. (–)-cyclopsammocinamide B
(R = Cl) 4,5-DiCl-Py

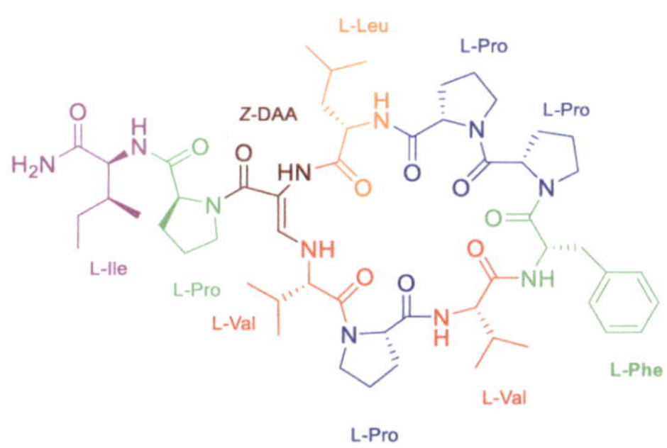

709. (–)-callyaerin H

Fig. 23 (continued)

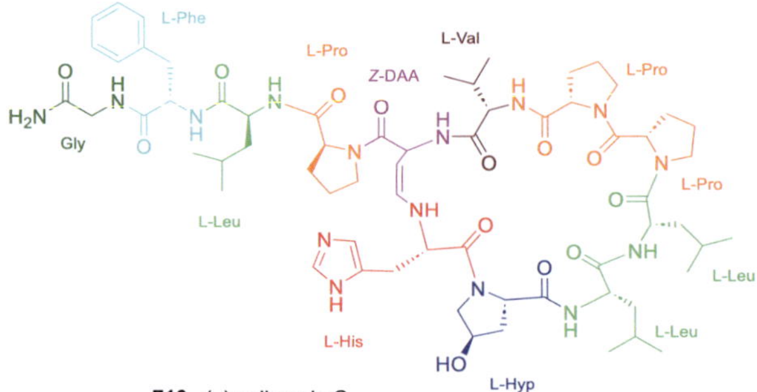

710. (–)-callyaerin C

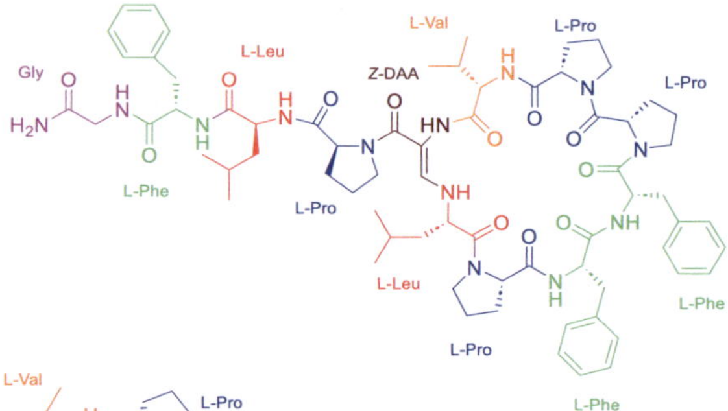

711. (–)-callyaerin I

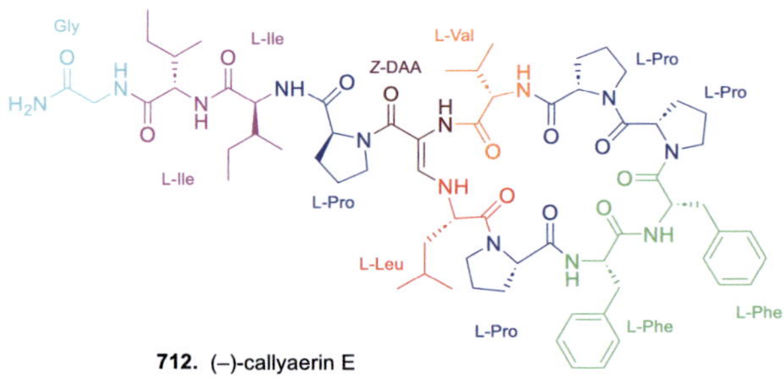

712. (–)-callyaerin E

Fig. 23 (continued)

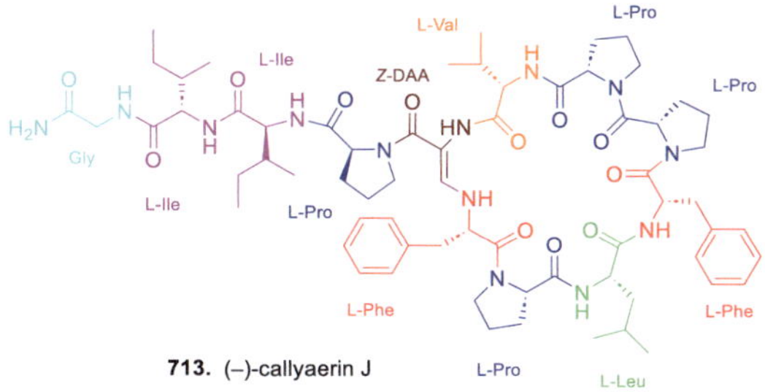

713. (–)-callyaerin J

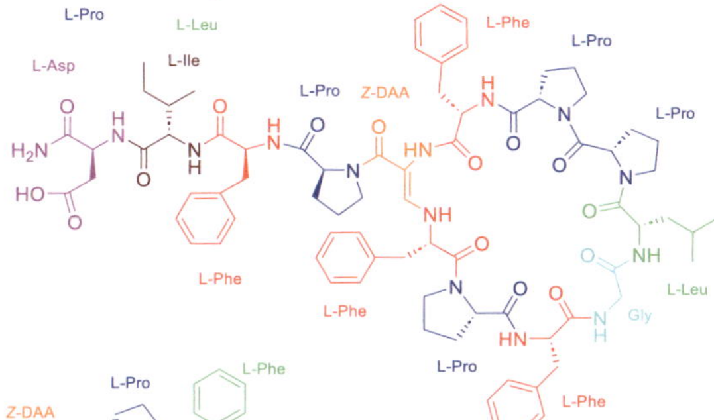

714. (–)-callyaerin K

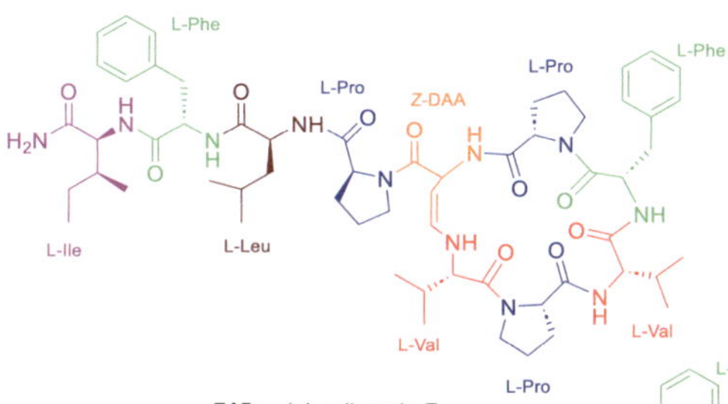

715a. (–)-callyaerin F

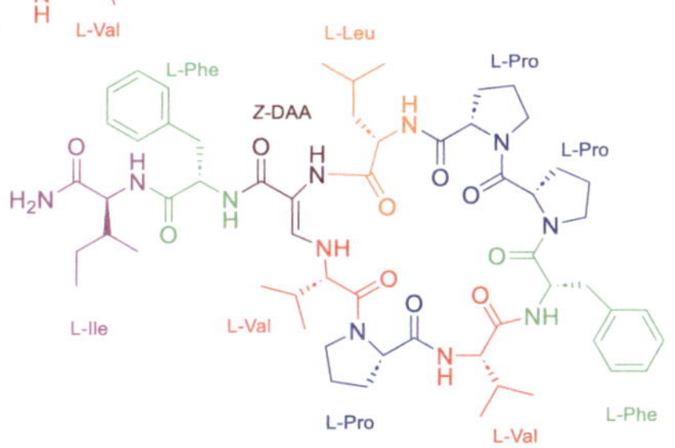

715b. (–)-callyaerin F

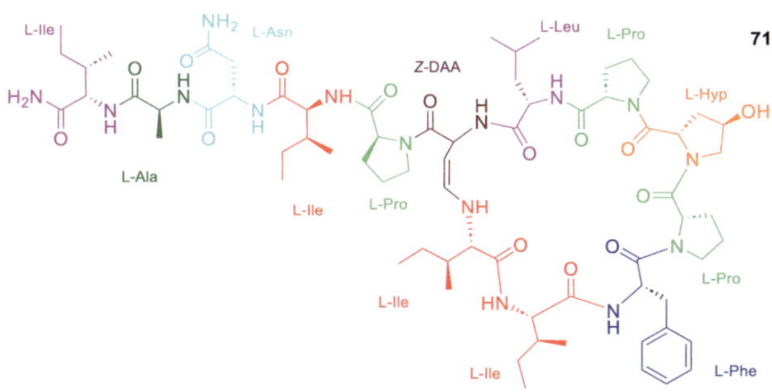

716a. (–)-callyaerin D

Fig. 23 (continued)

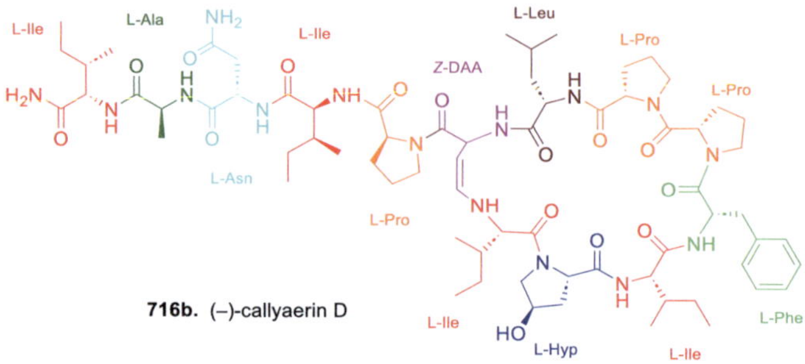

716b. (–)-callyaerin D

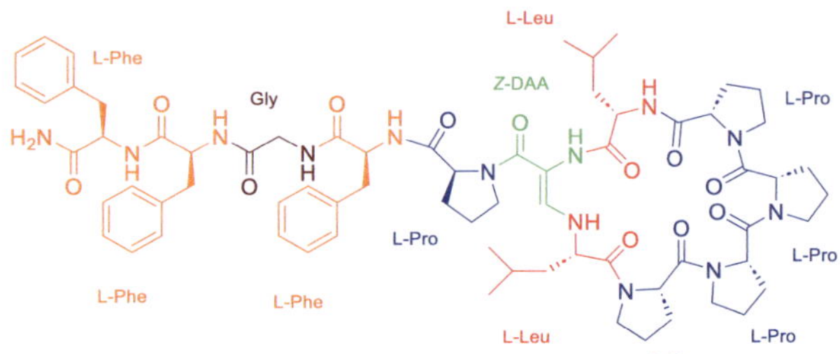

717a. (–)-callyaerin G

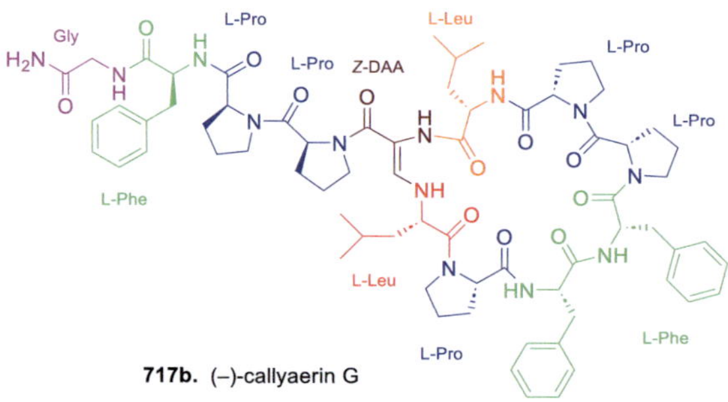

717b. (–)-callyaerin G

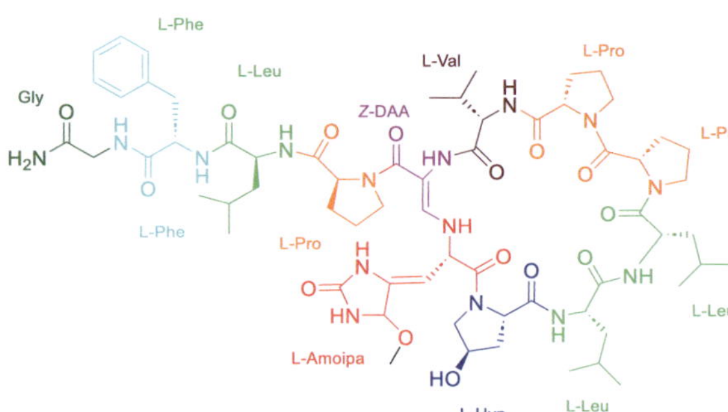

718. (–)-callyaerin M

Fig. 23 (continued)

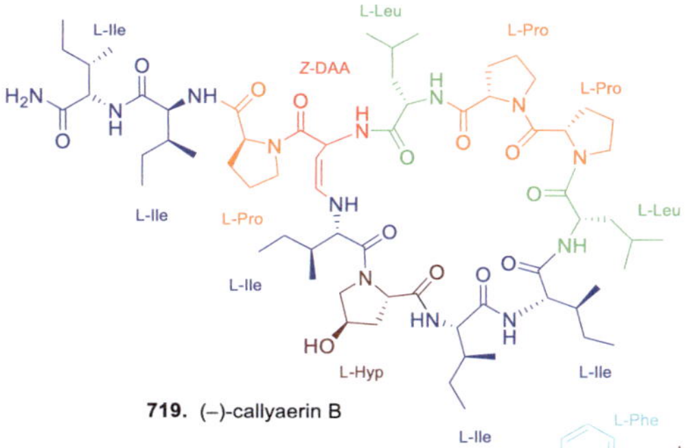

719. (–)-callyaerin B

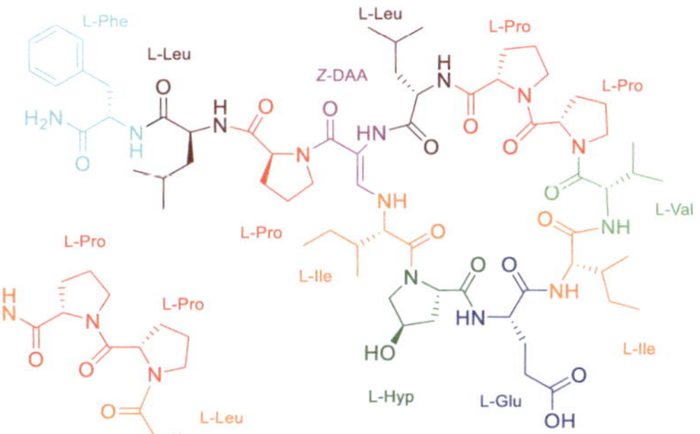

720. (–)-callyaerin L

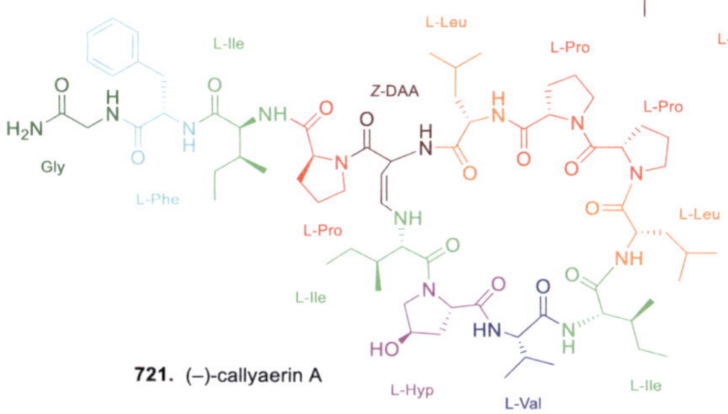

721. (–)-callyaerin A

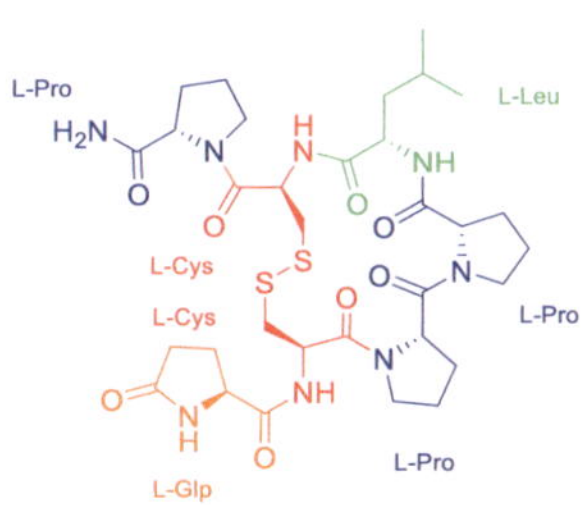

722. (–)-gombamide D

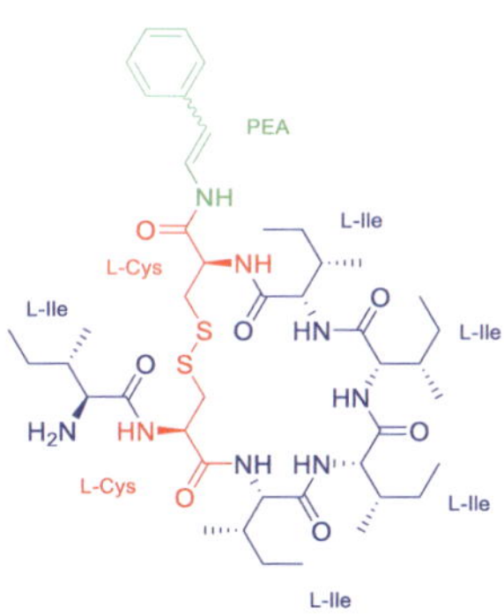

723. (–)-microcionamide C
E-PEA
724. (–)-microcionamide D
Z-PEA

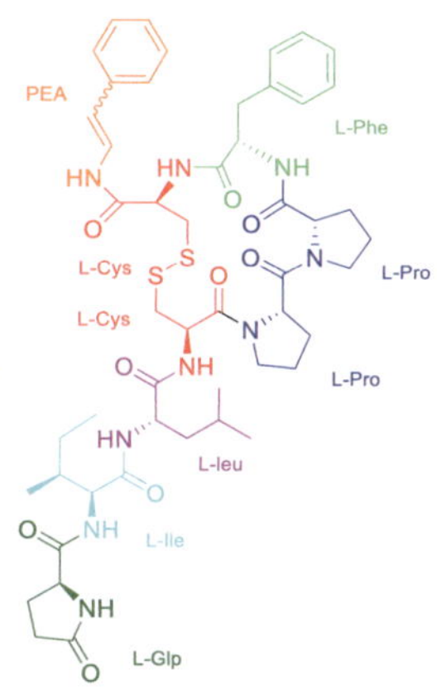

725. (+)-gombamide B
E-PEA
726. (+)-gombamide C
Z-PEA

Fig. 23 (continued)

(NSW) [351]. Its AC was secured by Marfey's method and chemical transformation including ozonolysis. Twenty-six head-to-tail peptides were isolated from various specimens including sponges: (−)-haloirciniamide (**681**), *Ircinia* sp, JSCR [346]; (−)-waiakeamide (**685**), *Ircinia dendroides*, NSW [352]; (−)-cupolamide A (**691**), *Theonella cupola*, NSW [353]; (−)-carteritins A, B (**696**, **697**, respectively), *Stylissa carteri*, NSW [354]; (−)-stylissamide X (**698**), *Stylissa* sp., PUA [355]; (+)-barangamides A–D (**703b–706**), *Theonella swinhoei*, SSW [356, 357], molluscs: (+)-keenamide A (**682**), *Pleurobranchus forskalii*, NSW [358], tunicates: (−)-mollamide B (**683**) and mollamide C (**684**), *Didemnum molle*, NSW [359]; (−)-bistratamides M, N (**686**, **687**, respectively), *Lissoclinum bistratum*, SWP [360]; (−)-diazonamides E, C, D (**688**, **689**, **690**, respectively), *Diazona* sp., PUA [361]; lissoclinamides 9, 10 (**694**, **695**, respectively), *Lissoclinum patella*, NMU [362], red algae: (−)-*cis,cis*-ceratospongamide (**692**), (−)-*trans,trans*-ceratospongamide (**693a**), *Ceratodicyton spongiosum*, NSW [363], cyanobacteria: (−)-trikoramides A–D (**699–702**), *Symploca hydnoides*, RUI [364, 365]. Molecular modelling was used to assign the RCs of (−)-**692**, **693a** [363], and **693b** [366]. The ACs of (−)-**681**, (−)-**685–693a**, (−)-**696–700**, (+)-**703b–706** were mainly elucidated by Marfey's method. In addition to this strategy, chemical derivatization was used to confirmed the ACs of (−)-**683**, **684**. In some cases, total synthesis established the ACs of (−)-**692** [367–369], **693b** [366] leading to a revision of **693a**, and (−)-**698** [370]. X-ray analysis also confirmed the ACs of (−)-**692** [366]. Hydrolysis and chiral TLC were used to investigate the ACs of **694**, **695** [362]. The ACs of (−)-trikoramides B (**700**) and D (**702**) were also assigned by ECD analysis [365].

Fifteen cyclopeptides were subclassified by tail-to-side chain cyclization represented by (−)-cyclopsammocinamides A, B (**707**, **708**, respectively), (−)-callyaerins H (**709**), C (**710**), I (**711**), E (**712**), J (**713**), K (**714**), F (**715a**, **715b**), D (**716a**, **716b**), G (**717a**, **717b**), M (**718**), B (**719**), L (**720**), and A (**721**). Two cyclotetrapeptides with dipeptide side chain, (−)-**707**, **708** were isolated from a marine sponge *Psammocinia* sp. (NSW) [371]. Their ACs were determined by Marfey's method. Thirteen callyaerin cyclopeptides (**709–721**) consisted of one cycloheptapeptide with dipeptide side chain: (−)-callyaerin H (**709**), MLU [372], five cycloheptapeptides with tetrapeptide side chain: (−)-callyaerins C, I, E, J, K (**710–714**), MLU, SSW [372, 373], one cyclopentapeptide with tetrapeptide side chain: (−)-callyaerin F (**715a**), MLU [372], one cycloheptapeptide with dipeptide side chain: revised (−)-callyaerin F (**715b**), SSW [373], three cycloheptapeptides with pentapeptide side chain: (−)-callyaerin D (**716a**), its revision (−)-**716b**, (−)-callyaerin G (**717a**), MLU, SSW [372–374], one cyclooctapeptide with tetrapeptide side chain: revised (−)-callyaerin G (**717b**), SSW [373], one cycloheptapeptide with tetrapeptide side chain: (−)-callyaerin M (**718**), SSW [373], two cyclooctapeptides with tripeptide side chain: (−)-callyaerins B, L (**719**, **720**, respectively), MLU, SSW [372, 373], one cyclooctapeptide with tetrapeptide side chain: (−)-callyaerin A (**721**), MLU [372] were isolated from the marine sponge *Callyspongia aerizusa*. The ACs of callyaerin cyclopeptides were mainly determined by Marfey's method. In addition to Marfey's method, only (−)-callyaerin A (**721**) were confirmed its AC by total synthesis [375].

Three callyaerins D, F, G (**716a**, **715a**, **717a**) were revised by reinvestigation of 2D NMR and MS data.

An investigation of the marine sponge *Clathria basilana* collected from MLU yielded the isolation of five new side chain-to-side chain cyclopeptides, (−)-gombamide D (**722**), (−)-microcionamides C (**723**), D (**724**), (+)-gombamides B (**725**) and C (**726**) [335]. The ACs of the five cyclothioheptapeptides were determined using Marfey's analysis.

4.3 Cyclodepsipeptide

About twenty-six marine cyclodepsipeptides (Fig. 24) have been reported from Indonesian waters consisted of thirteen molecules possessing head-to-tail and twelve molecules having tail-to-side chain cyclizations. Thirteen molecules were (−)-jaspamide Q (**727**, EKM), (−)-jaspamide R (**728a**, EKM) [376] and its revision (−)-jaspamide R_1 (**728b**) [344], (−)-trikoveramides A–C (**729–731**, RUI) [377], (−)-celebesides A–C (**732–734**, UEP) [378], (−)-triproamide A (**735**, RUI), (−)-pemukaianalides A, B (**736**, **737**, respectively, RUI) [379], (+)-streptodepsipeptide SV21 (**738**, LPG) [380], and (−)-PM170453 (**739**, MLU) [381]. The RCs of (−)-**732–734** were assigned by JBCA and quantum mechanical calculations. As usual, the ACs of cyclodepsipeptides were also determined by Marfey's method. In addition, chiral HPLC was also used to determine the ACs of (−)-**732–734**, while total synthesis was employed to secure the ACs of (−)-**739**. The source of those cyclodepsipeptides was the marine sponge *Jaspis splendens* for (−)-jaspamides Q (**727**), R (**728a**) and its revision R1 (**727b**), *Symploca hydnoides* for (−)-trikoveramides A–C (**729–731**), (−)-triproamide A (**735**), and (−)-pemukaianalides A, B (**736**, **737**), the marine sponge *Siliquariaspongia mirabilis* for (−)-celebesides A–C (**732–734**), the marine bacteria *Streptomyces cavourensis* derived from *Stichopus vastus* for (+)-streptodepsipeptide SV21 (**738**), and a marine cyanobacterium *Lyngbya* sp. for (−)-PM170453 (**739**). Structure revision of (−)-jaspamide R_1 was proposed after reisolation work and NMR reanalysis of Fijan *Jaspis splendens* [345].

About thirteen cyclodepsipeptides possessing tail-to-side chain cyclization were isolated from four different species of marine sponges *Daedalopelta* sp., (+)-daedophamide (**740**), ENT [382]; *Siliquariaspongia mirabilis*, (+)-theopapuamides B–D (**741–743**), UEP [378]; *Sidonops microspinosa* (+)-microspinosamide (**744**), MLU [383]; and *Theonella swinhoei*, (−)-theonellapeptolides If (**745**, NSW) [384], IIa (**746**, SSW), IIb (**747**, SSW), IIc (**748**, SSW) [385], IIe (**749**, SSW) [356], (−)-sulfinyltheonellapeptolide (**750**, NSW) [384], (−)-cyclolithistide A (**751a**, NSW) [386], and (−)-gunungamide A (**752**, MLU) [387]. The RCs of (+)-**741–743** were initially assigned by JBCA and quantum mechanical calculations. Marfey's method was used to assign the ACs of (+)-**740–744** and (−)-**745–750**. In case of (−)-**752**, JBCA was mainly applied instead of Marfey's amino acid derivatization to determine its RC due to epimerization of threonine residue during hydrolysis. In 2014, the AC of

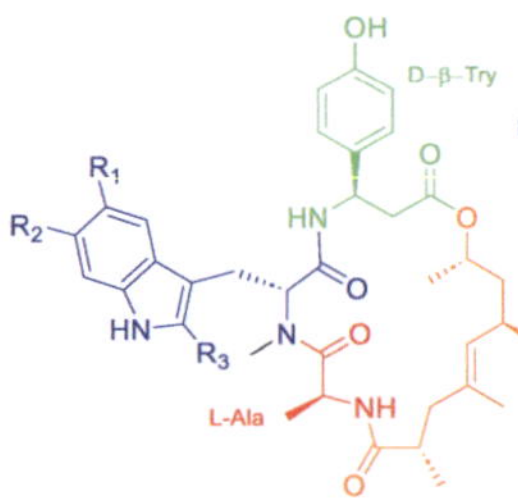

727. (−)-jaspamide Q
$R_1 = R_2 = R_3 = H$
D-*N*-Me-Trp

728a. (−)-jaspamide R
$R_1 = R_3 = Br, R_2 = H$
D-*N*-Me-BrTrp

728b. (−)-jaspamide R_1
$R_2 = R_3 = Br, R_1 = H$
D-*N*-Me-BrTrp

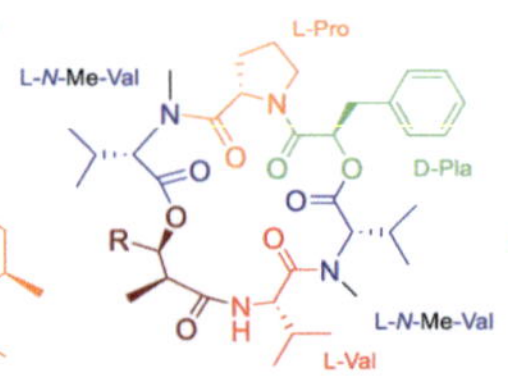

729. (−)-trikoveramide A
$R = CH_2CH_2CH_2C{\equiv}CH$
Hmoya

730. (−)-trikoveramide B
$R = CH_2CH_2CH_2CH{=}CH_2$
Hmoea

731. (−)-trikoveramide C
$R = (CH_2)_4CH_3$
Hmoaa

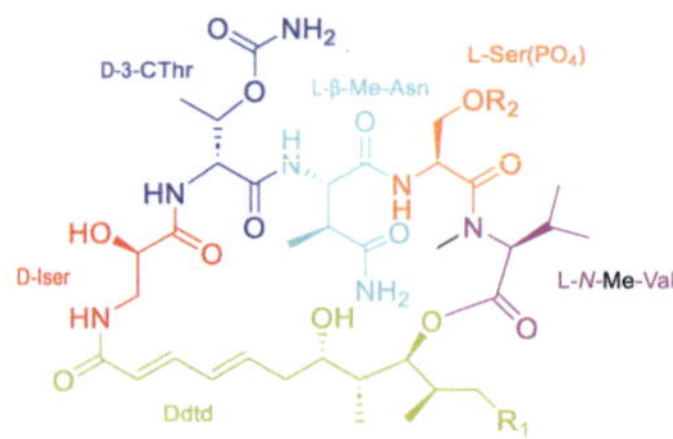

732. (−)-celebeside A
$R_1 = CH_2CH_3$ Ddtd
$R_2 = PO_3H_2$ L-Ser(PO4)

733. (−)-celebeside B
$R_1 = CH_3$ Ddtd
$R_2 = PO_3H_2$ L-Ser(PO4)

734. (−)-celebeside C
$R_1 = CH_2CH_3$ Ddtd
$R_2 = H$ L-Ser

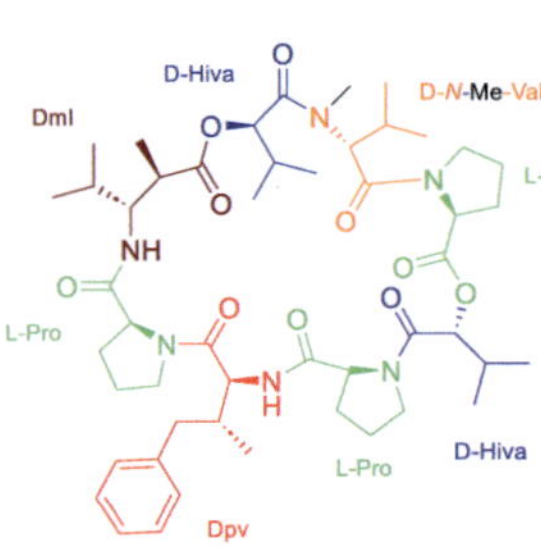

735. (−)-triproamide A

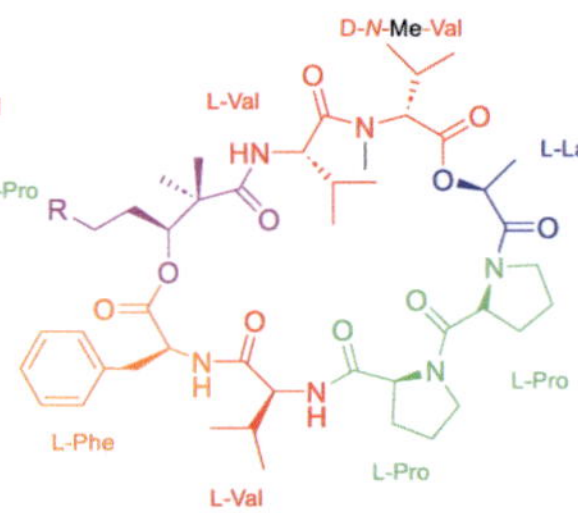

736. (−)-pemukaianalide A
$R = CH_2CH{=}CH_2$
Dhoea

737. (−)-pemukaianalide B
$R = CH_2CH_2CH_3$
Dhoaa

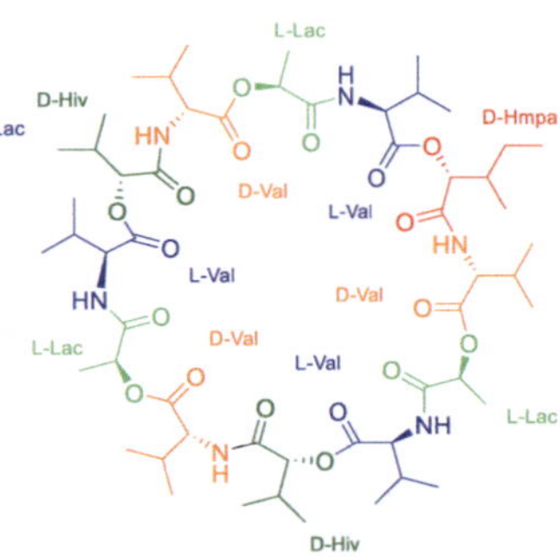

738. (+)-streptodepsipeptide SV21

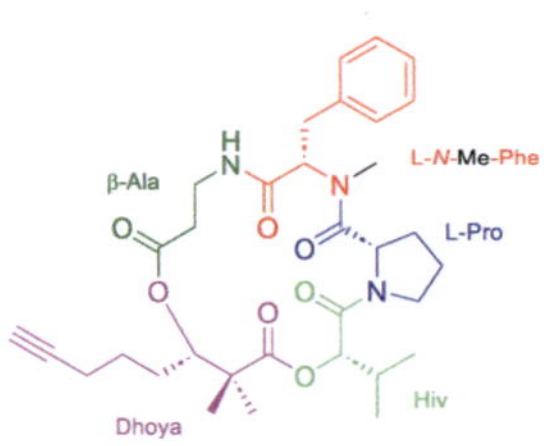

739. (−)-PM170453

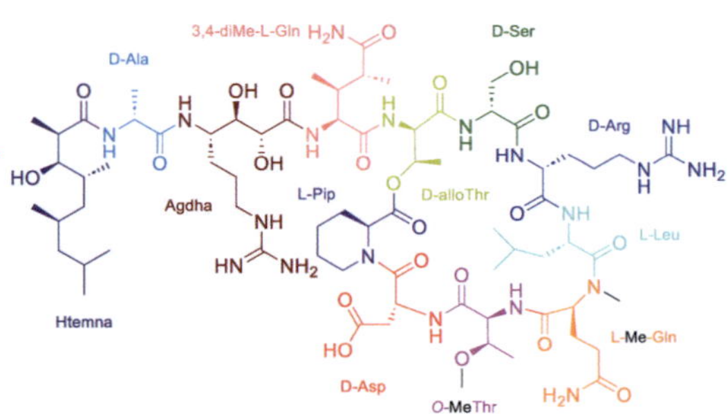

740. (+)-daedophamide

Fig. 24 Structures of marine cyclodepsipeptides (**727–752**) from Indonesian waters. Amino acid building blocks, and their modification as well as other parts were drawn with different color

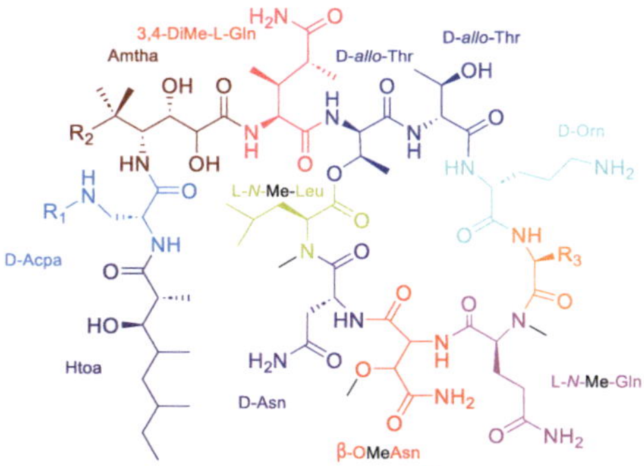

741. (+)-theopapuamide B
R_1 = Ac D-Acpa
R_2 = OH Amtha
R_3 = $CH_2CH(CH_3)_2$ L-Leu

742. (+)-theopapuamide C
R_1 = Ac D-Acpa
R_2 = H Amdha
R_3 = $CH_2CH(CH_3)_2$ L-Leu

743. (+)-theopapuamide D
R_1 = Ac D-Acpa
R_2 = H Amdha
R_3 = $CH_2CHCH_3CH_2CH_3$ L-hIle

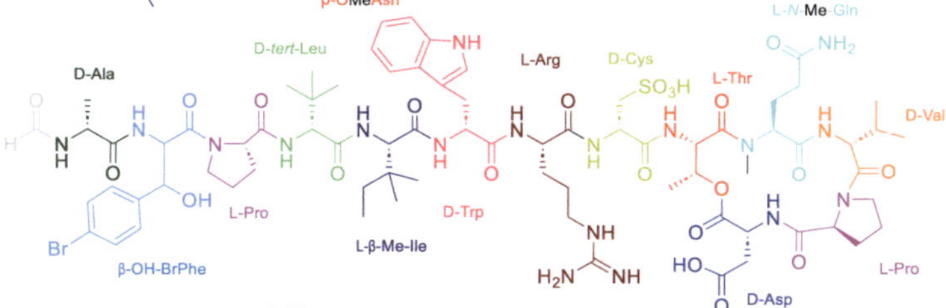

744. (+)-microspinosamide

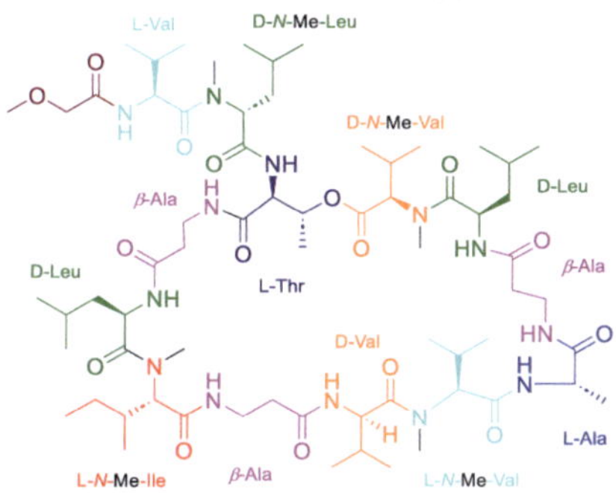

745. (–)-theonellapeptolide If

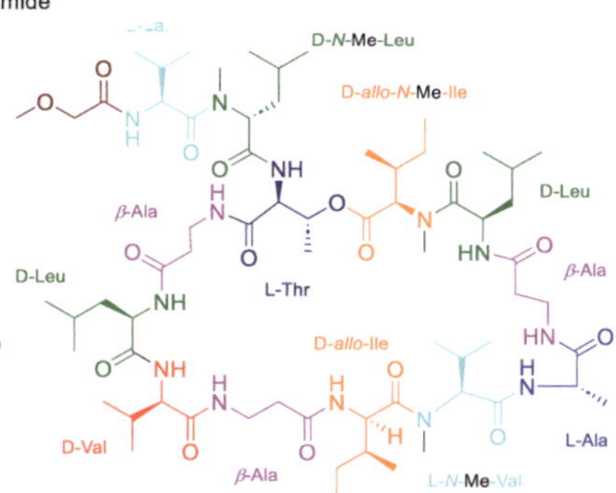

746. (–)-theonellapeptolide IIa

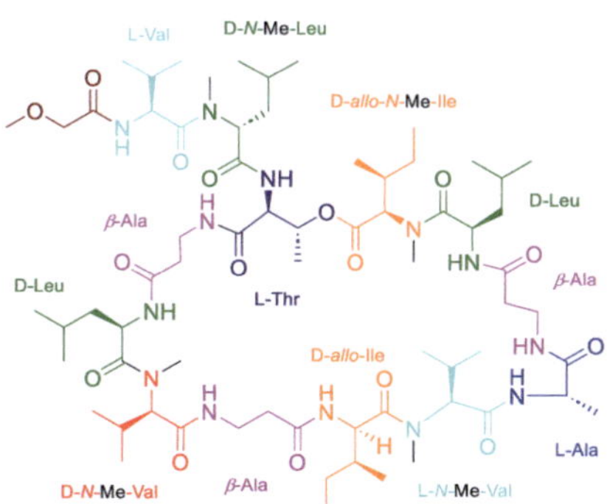

747. (–)-theonellapeptolide IIb

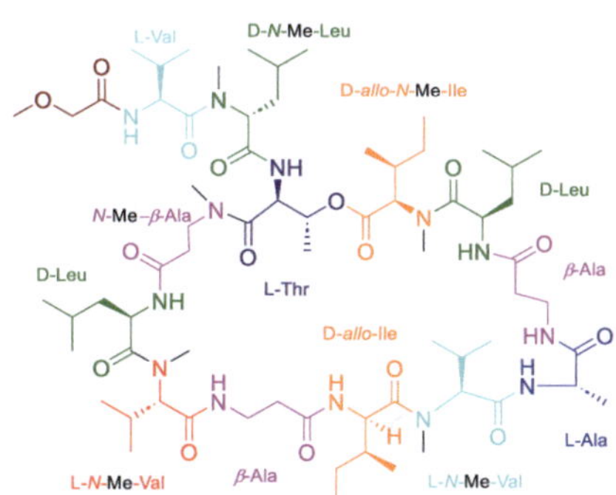

748. (–)-theonellapeptolide IIc

Fig. 24 (continued)

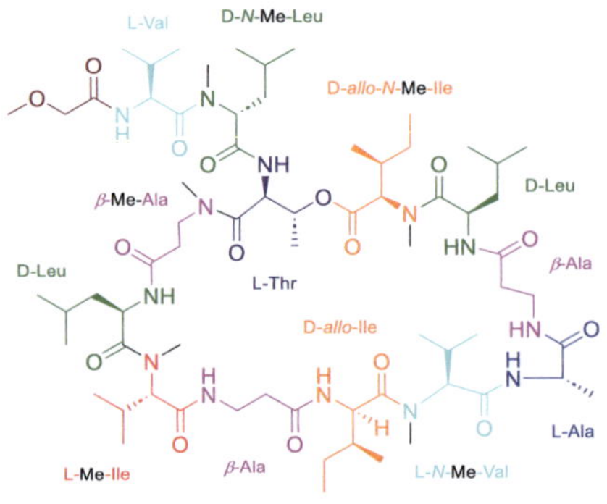

749. (–)-theonellapeptolide IIe

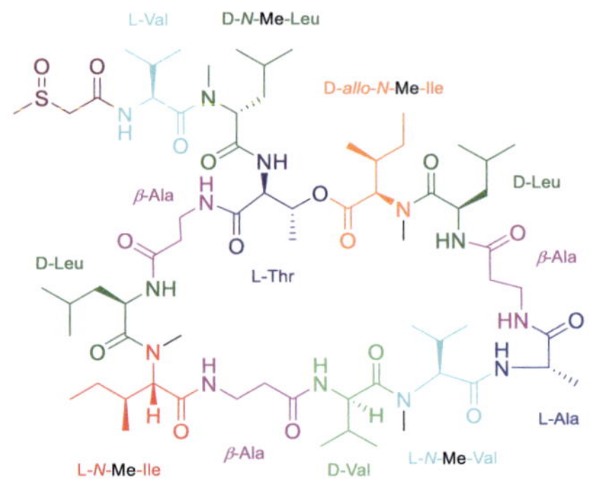

750. (–)-sulfinyltheonellapeptolide

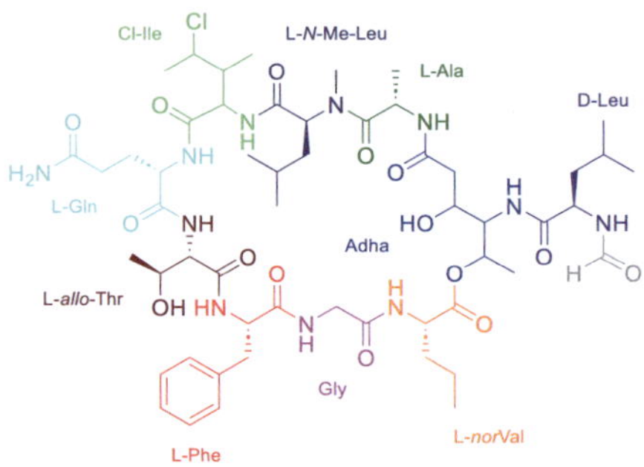

751a. (–)-cyclolithistide A

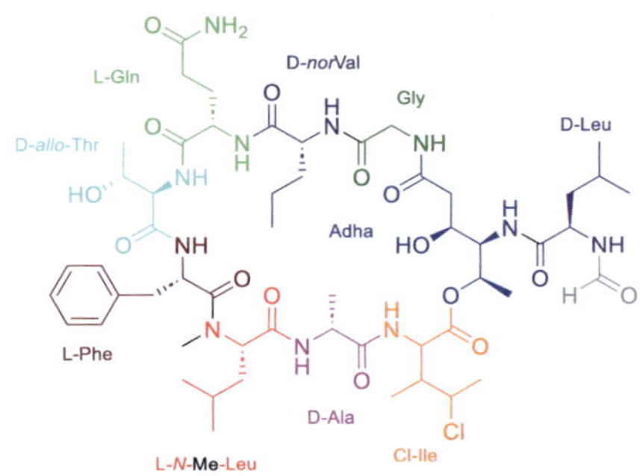

751b. (–)-cyclolithistide A

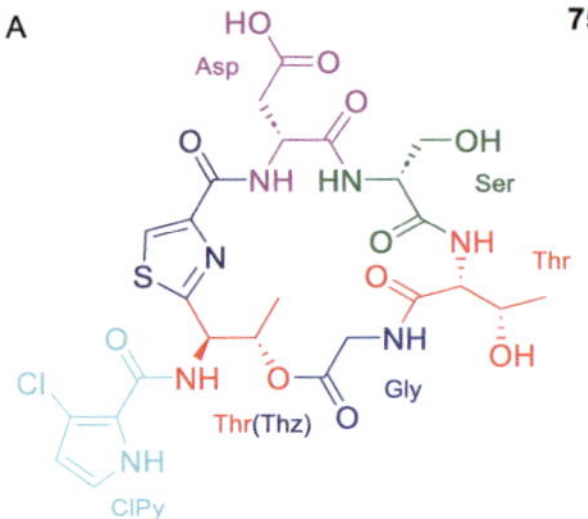

752. (–)-gunungamide A

Fig. 24 (continued)

(−)-cyclolithistide (**751a**) initially assigned by chiral TLC and chemical transformation was revised by reisolation work and 2D NMR reanalysis of (−)-cyclolithistide A (**751b**) isolated from the Japanese marine sponge *Discodermia japonica* [388]. Its AC was assigned by Marfey's and Mosher's methods.

5 Fatty Acid

A fatty acid building block is derived from starter and extender units. Acetyl-CoA (C_2) building block is likely a common starter unit found in Indonesian marine fatty acid as in (−)-**753–755** and (+)-**756–760**, while unusual starter units such as C_3, C_6C_1 are able to identify from metabolites (−)-**761**, **762** and (+)-**763**, respectively. The common extender units found in the class are likely malonyl-CoA (C_2), whereas a mixed of C_2 and C_3 is found in (−)-**761**. Indonesian fatty acid marine natural products (**753–763**) (Fig. 25) are subclassified as unsaturated fatty acid (+)-**756–760** and (−)-**761**, a mixed biosynthesis of fatty acid-C_1 building block (−)-**753–755**, fatty acid-polyketide (−)-**762**, and fatty acid-polyketide-alkaloid (+)-**763**.

An investigation of a marine sponge *Cinachyrella* sp. collected from NMU waters yielded (−)-cinachylenic acids B–D (**753–755**) possessing $1 \times C_2 + 8 \times C_2 + 1 \times C_1$ building blocks [331]. The ACs of (−)-**753–755** were determined by ECD, optical rotation, and biogenetic consideration. (+)-(3*S*,18*S*,4*E*,16*E*)-Eicosa-1,19-diyne-3,18-diol-4,16-diene (**756**) was isolated from *Callyspongia pseudoreticulata* (SSW) [389]. Its AC was determined by Mosher chemistry. Compound **757** identified as polyunsaturated brominated fatty acid possessing acetylenic bonds collected ENT waters [390]. Both compounds (+)-**756** and **757** are consisted of $1 \times C_2 + 9 \times C_2$ building blocks. Three fatty acid derivatives assigned as (+)-lembehynes A–C (**758–760**) were isolated from *Haliclona* sp. (NSW) [391, 392]. The ACs of (+)-**758–760** was confirmed by Mosher analysis [392] and total synthesis [393–395]. Compounds (+)-**758** and **759** share the same building block $1 \times C_2 + 17 \times C_2$, while compound (+)-**760** contains $1 \times C_2 + 18 \times C_2$ units. A new polyacetylenic alcohol (−)-**761** was isolated from a marine sponge *Callyspongia* sp. [396]. Its AC was determined by Mosher chemistry. A marine sponge *Plakinastrella* sp. was the source of (−)-elenic acid (**762**) [397]. The AC of (−)-**762** was determined by *R*- and *S*-phenylglycinyl methyl ester (PGME) reagent. Moreover, the AC of (−)-**762** was confirmed by total synthesis [398]. (+)-Lanesoic acid (**763**) was obtained from *Theonella* sp. (CSW) [399]. Its RC was determined by JBCA and NMR calculation. The last three metabolites (−)-**761**, **762** and (+)-**763** are likely consisted of $1 \times C_3 + 14 \times C_2$, $1 \times C_6C_1 + 8 \times C_2 + 2 \times C_3$, and $1 \times C_6C_1 + 4 \times C_2$, respectively.

6 Polyketide

Around 133 molecules were classified as polyketides consisted of cyclic peroxide (23 molecules), aromatic polyketide (71 molecules), lactone (35 molecules), and miscellaneous polyketide (4 molecules). The building block of polyketides is mainly derived from acetate (C_2) unit based on the isotope labelling experiment [400].

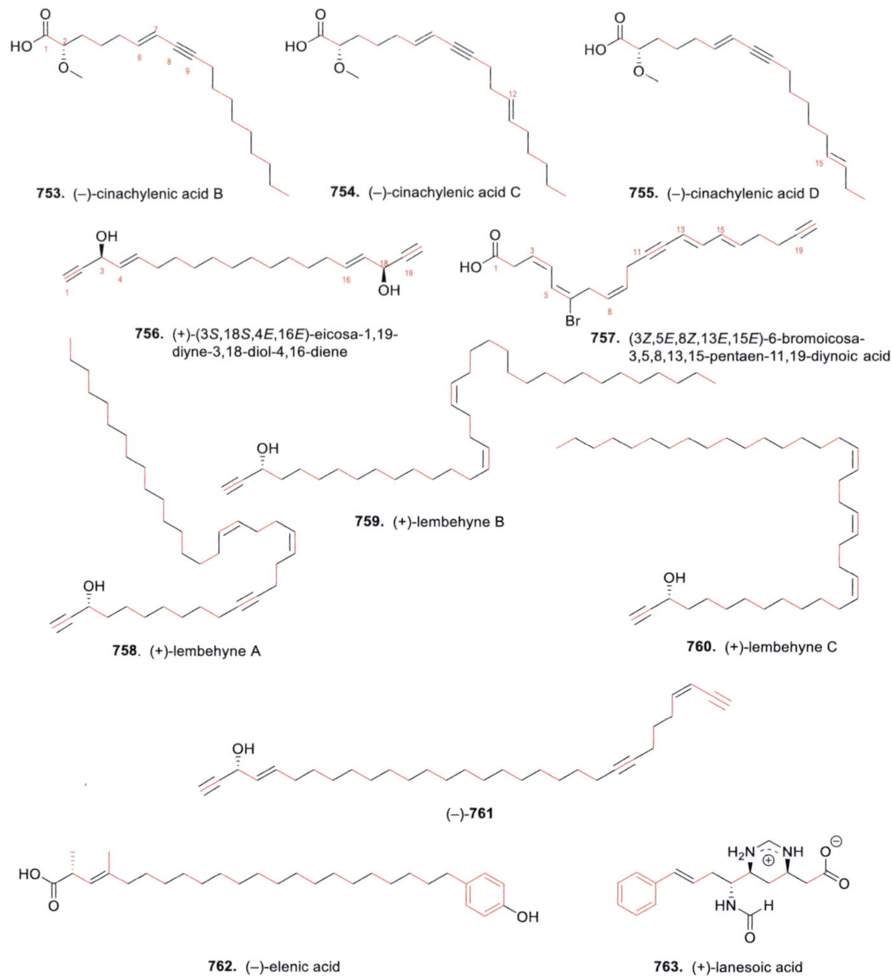

Fig. 25 Structures of marine fatty acid (**753–763**) from Indonesian waters. C_2 building blocks were proposed with red color

6.1 Cyclic Peroxide

(+)-Manadic acids A (**764**) and B (**765**) [401] were discovered from a marine sponge *Plakortis* sp. (NSW). The AC of (+)-**764** was determined by chemical transformation and Mosher's method. The marine sponge *Plakortis* cfr. *simplex* was the source of (−)-manadoperoxides A–K (**766–772, 774, 782/783, 784/785, 786**) [402, 403]. The AC of (−)-**766** was determined by chemical transformation and Mosher's method. In addition to this determination, the ACs of (−)-**768, 769** were determined by chemical transformation, Riguera and Mosher's method. Manadoperoxides I (**782/783**) and J (**784/785**) were obtained as a mixture of stereoisomers. (−)-Peroxyplakoric ester C

(**773**), a new isomer of (−)-manadoperoxide C (**770**) [403], was isolated from NSW collection of *Plakortis* cfr. *lita*. Two new (−)-(9*S*, 10*S*) or (9*R*, 10*R*)-plakorstatin 1 and (−)-(9*R*, 10*R*) or (9*S*, 10*S*)-plakorstatin 2 (**775**, **776**) [404] were obtained from *Plakortis nigra* (NSW). Another species of *Plakortis*, *Plakortis bergquistae* (NSW) was the source of five new cyclic peroxides (−)-manadodioxans A–C (**777–779**) and (+)-manadodioxans D, E (**780**, **781**) [405]. The structures of Indonesian cyclic peroxides are presented in Fig. 26.

6.2 Aromatic

About 71 MNPs were classified as aromatic polyketides (Fig. 27) which were distributed into Ascomycota (32 molecules), Actinobacteria (2 molecules), Porifera (36 molecules), Plantae (1 molecule). Furthermore, 32 MNPs derived from Ascomycota can further be distributed into 12 genera including *Cladosporium* (2 molecules), *Cosmophora* (3 molecules), *Penicillium* (3 molecules), *Drechslera* (2 molecules), *Trichoderma* (1 molecule), *Aspergillus* (13 molecules), *Curvularia* (1 molecule), *Fusarium* (2 molecules), *Daldinia* (2 molecules), *Phomopsis* (1 molecule), *Mycelium* (1 molecule), unidentified fungi (2 molecules).

The genus of *Cladosporium* was the source of acetyl sumiki's acid (**787**) [406] and (−)-cladosporamide A (**788**) [407]. The former was collected at BLI associated with a sponge *Callyspongia aerizusa*, while the latter was in NSW waters. Three aromatic polyketides cosmochlorins A–C (**789–791**) [408] were isolated from the marine fungus *Cosmopora vilior* associated with the mangrove *Sonneratia alba* (WJV). Three aromatic polyketides, (+)-peniprenylphenol A (**792**) [409], a fungal diphenyl ether **793** [410], and (+)-penicitrinone G (**794**) [411] were obtained from the marine fungus *Penicillium chrysogenum* (SSW), *Penicillium albobiverticillium* (NSW), and *Penicillium citrinum* (SES). The ACs of (+)-**792** were determined by comparison of experimental and calculated ECD and OR data. The marine fungus *Drechslera hawaiiensis* associated with *Callyspongia aerizusa* (BLI) was the source of two new aromatic polyketides, (−)-butoxyl spiciferin (**795**) and (+)-spiciferol A (**796**) [412]. One aromatic hexaketide assigned as (+)-epoxysorbicillinol (**797**) [413] was obtained from the marine fungus *Tricoderma longibrachiatum* associated with *Haliclona* sp. (UEP). Its AC was assigned by ECD [413] and confirmed by total synthesis [414].

Thirteen MNPs **798a–803a**, BLI) [415] and their revision **798b1** [416], **798b2** [417, 418], **799b1** [416], **799b2** [418, 419], **800b**, **801b**, **802b1** [416], **802b2** [420], **803b** [420]; **804–808** [421, 422]; **809a** [415, 416] and its revision **809b** [423] were classified as aromatic polyketides obtained from the *Aspergillus* marine fungi. Among them, seven molecules (**798a–803a**) were revised and established their ACs by total synthesis for **798b2**, **799b2**, **802b2**, and **809b**. Moreover, the AC of **799b2** was also determined by X-ray analysis. An anthraquinoid polyketide, lunatin (**810**) was isolated from the marine fungus *Curvularia lunata* associated with *Niphates olemda* (BLI) [424]. Two other aromatic polyketides as in karimunones A and B

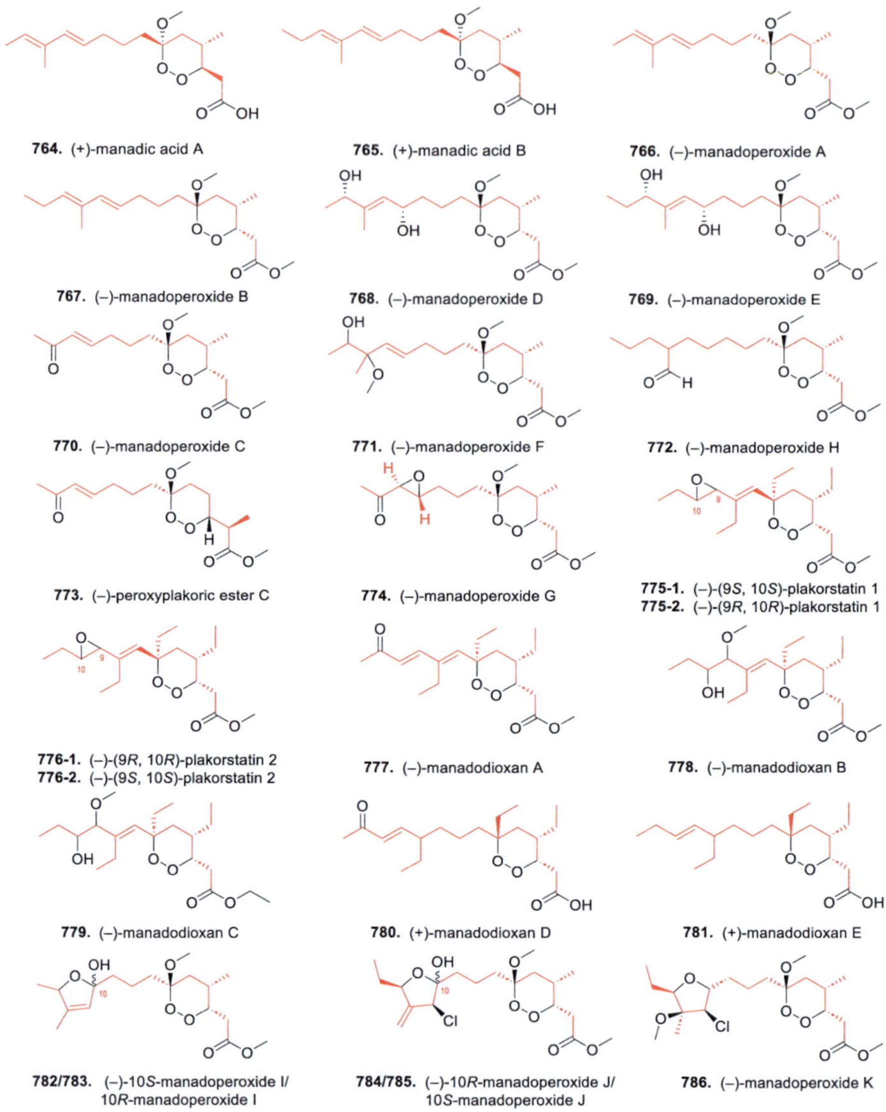

Fig. 26 Structures of marine cyclic peroxide (**764–786**) from Indonesian waters. Polyketide skeleton containing C_2 building blocks was proposed with red color

(**811** and **812**, respectively) [425] were obtained from a marine fungus *Fusarium* sp. associated with *Xestospongia* sp. (CJV). In addition to NMR and MS measurement, the planar structure of **811** was confirmed by DFT-based NMR calculation. The marine fungus *Daldinia eschscholtzii* associated with *Xestospongia* sp. (CJV) gave two inseparable molecules (+)-(*R*)-karimanone/(*S*)-karimanone (**813/814**) [426].

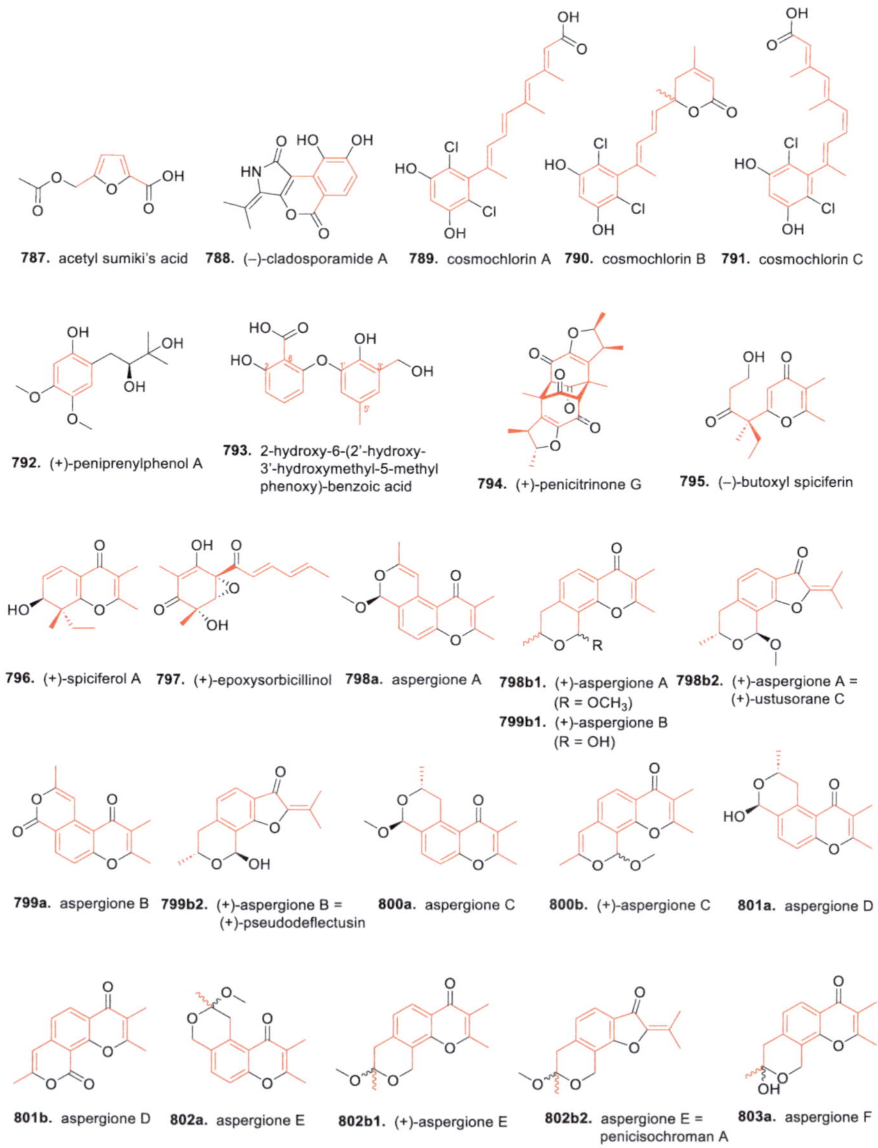

Fig. 27 Structures of marine aromatic polyketides (**787–857**) from Indonesian waters. Polyketide skeleton containing C_2 building blocks was proposed with red color

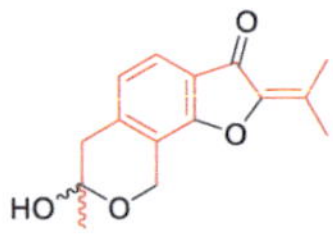

803b. (−)-aspergione F = pergillin

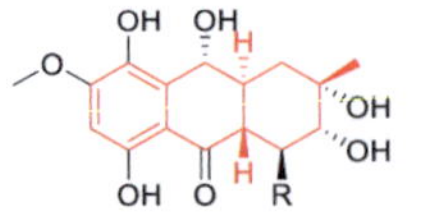

804. (−)-tetrahydrobostrycin (R = OH)
805. (−)-1-deoxytetrahydrobostrycin (R = H)

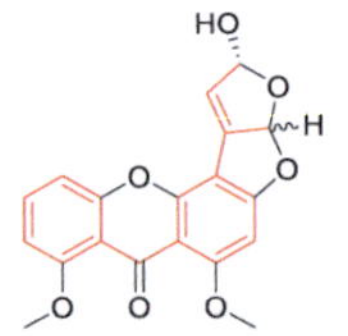

806. (−)-oxisterigmatocystin J

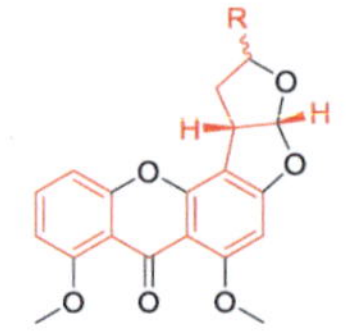

807. (−)-oxisterigmatocystin K (R = α–CH_3)
808. (−)-oxisterigmatocystin L (R = β–CH_3)

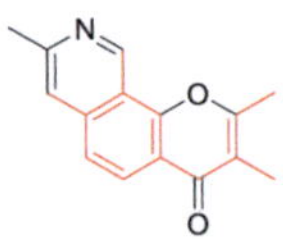

809a. aspergillitine

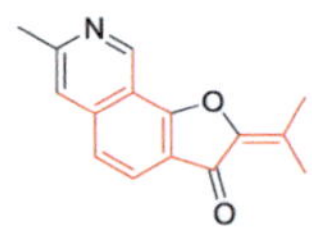

809b. TMC-120B

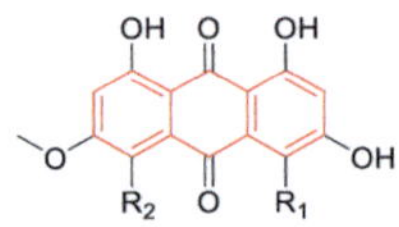

810. lunatin (R_1 = R_2 = H)
811. karimunone A (R_1 = $COCH_3$; R_2 = OH)

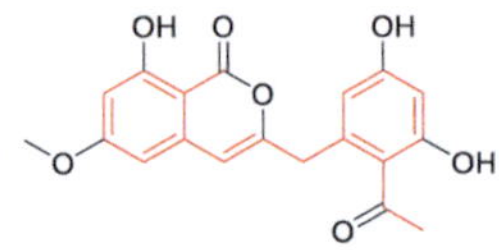

812. karimunone B

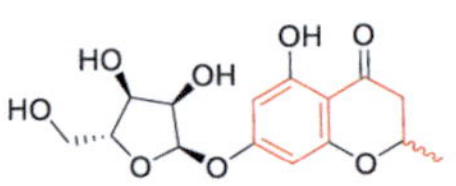

813/814. (+)-(*R*)-karimanone/(*S*)-karimanone

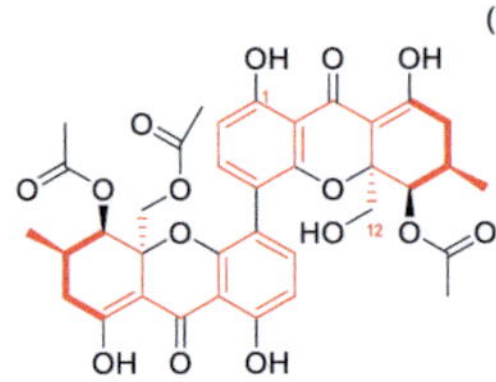

815. (+)-12-*O*-deacetylphomoxanthone A

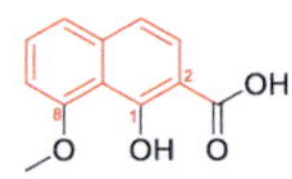

816. 2-carboxy-8-methoxynaphthalene-1-ol

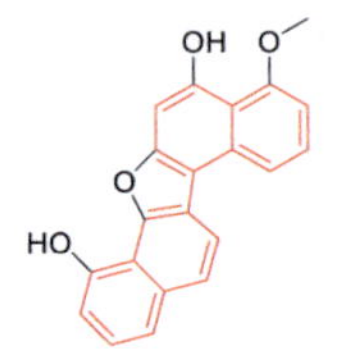

817. xylarianaphthol-1

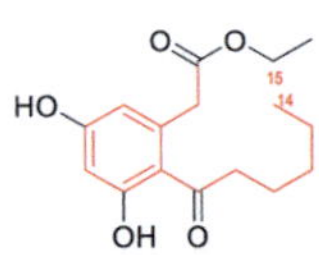

818. 14,15-seco-curvularin

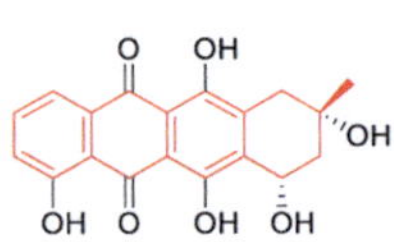

819. (+)-komodoquinone B

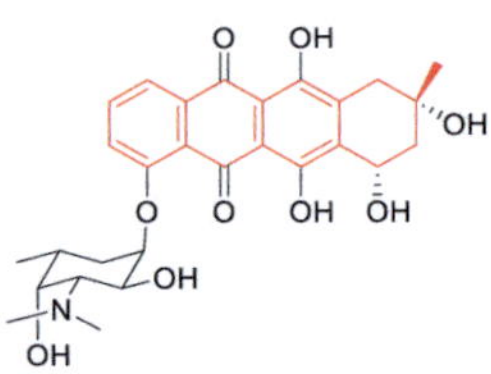

820. (+)-komodoquinone A

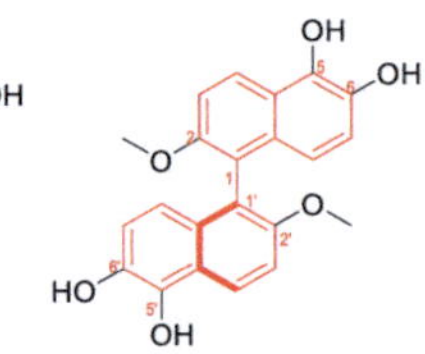

821. (+)-(*S*)-2,2'-dimethoxy-1,1'-binaphthyl-5,5',6,6'-tetraol

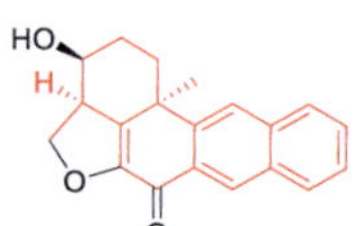

822. (−)-xestosaprol L

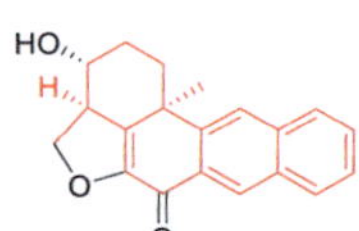

823. (−)-xestosaprol I

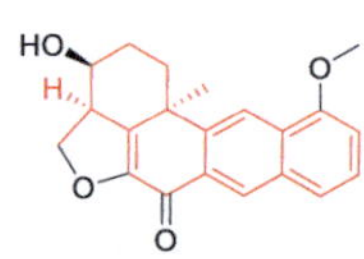

824. (−)-xestosaprol J

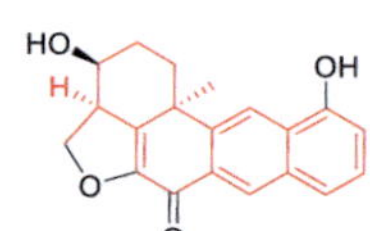

825. (−)-xestosaprol K

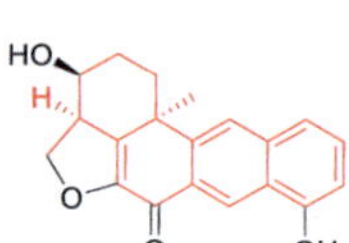

826. (−)-xestosaprol G

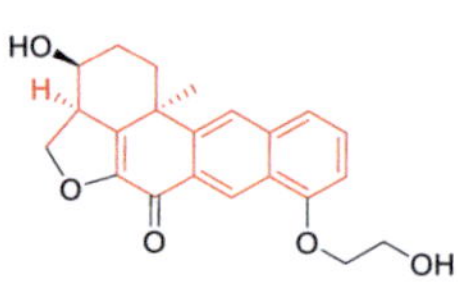

827. (+)-xestosaprol F

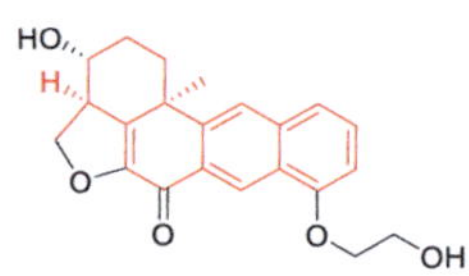

828. (−)-xestosaprol H

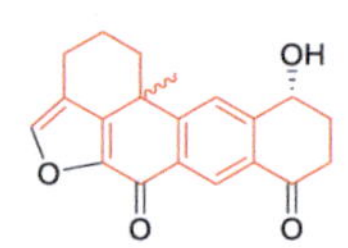

829. (+)-xestosaprol D

Fig. 27 (continued)

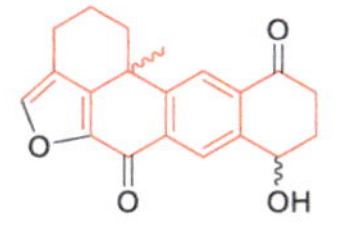
830. (–)-xestosaprol E

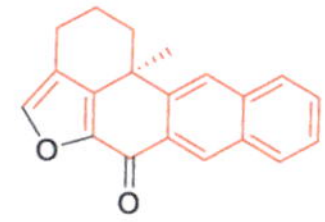
831. (+)-xestosaprol M

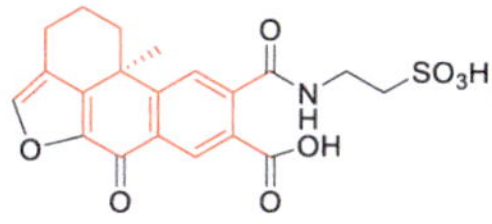
832/833. xestoadociaquinone A

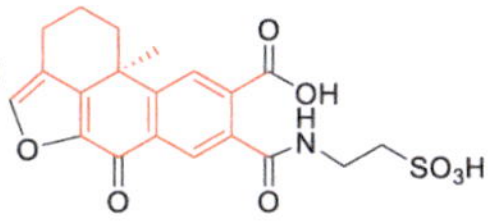
832/833. xestoadociaquinone B

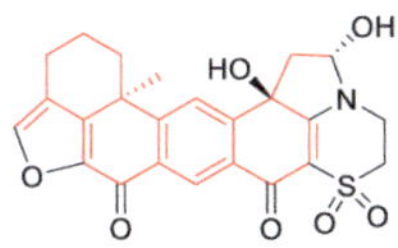
834. (–)-xestoadociaminal A/ petroquinone I

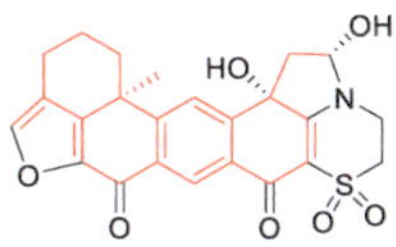
835. xestoadociaminal B

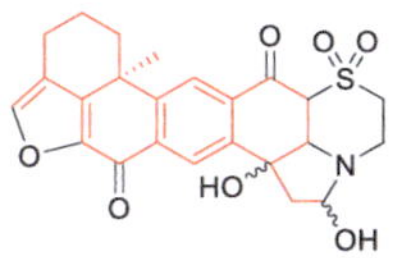
836/837. xestoadociaminals C/D

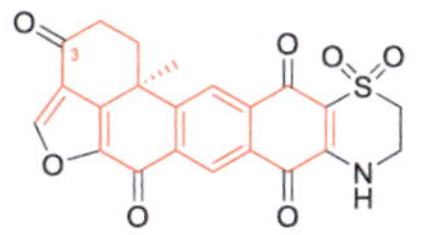
838. (+)-3-ketoadociquinone B

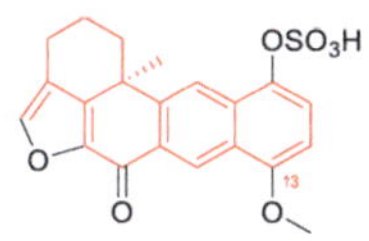
839. (+)-13-*O*-methyl-xestoquinol sulfate

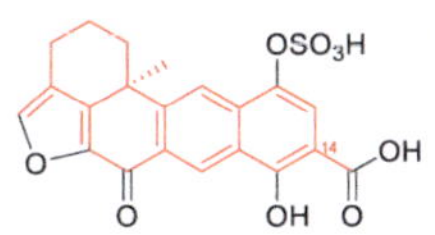
840. (–)-14-carboxyxestoquinol sulfate

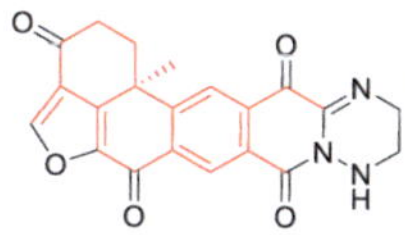
841. (+)-noelaquinone

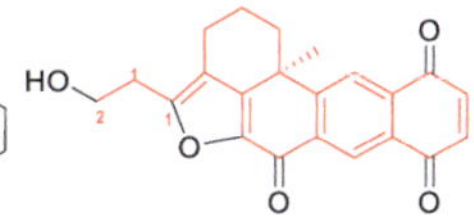
842. (–)-1-(2-hydroxyethyl) xestoquinone

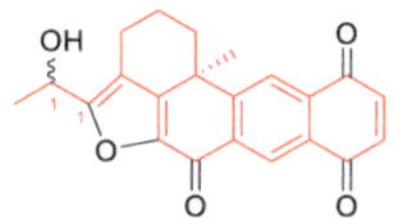
843/844. 1-(1-hydroxyethyl) xestoquinone

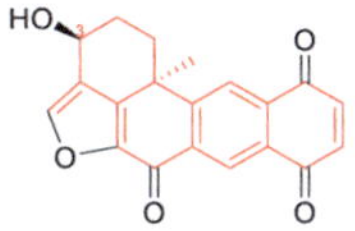
845. (–)-3*S*-3-hydroxy xestoquinone

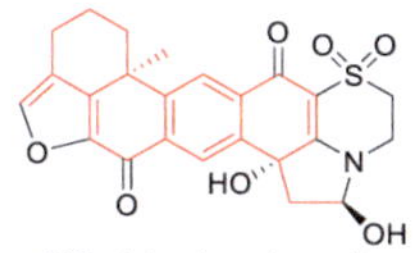
846. (+)-petroquinone K

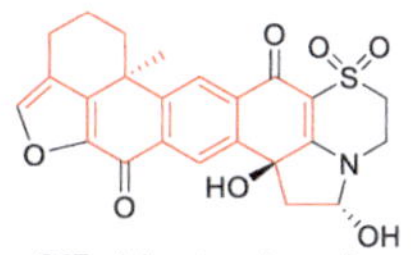
847. (+)-petroquinone L

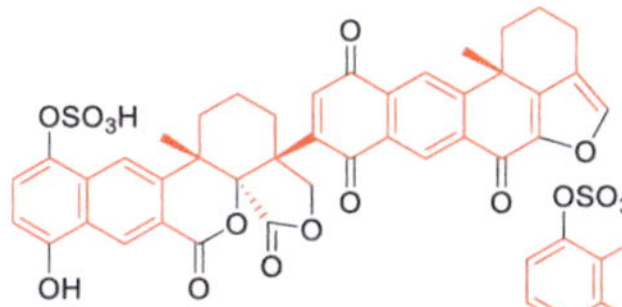
848. (–)-petroquinone C

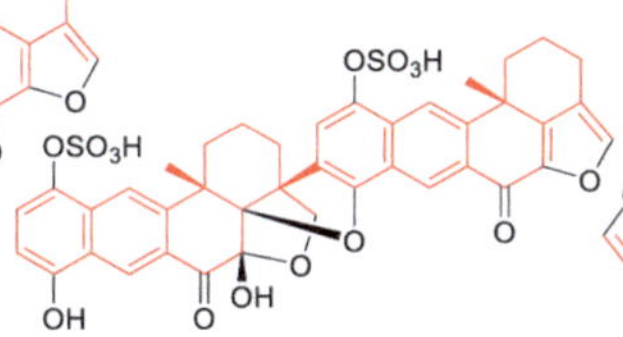
849. (+)-petroquinone D

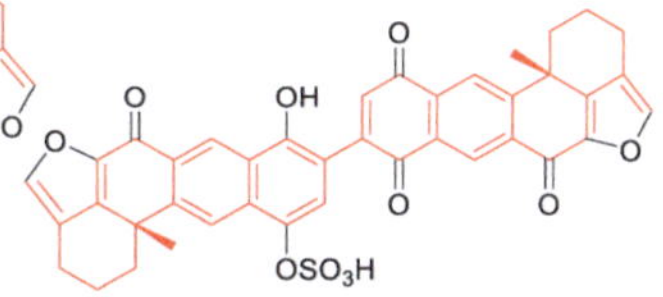
850. (+)-petroquinone E

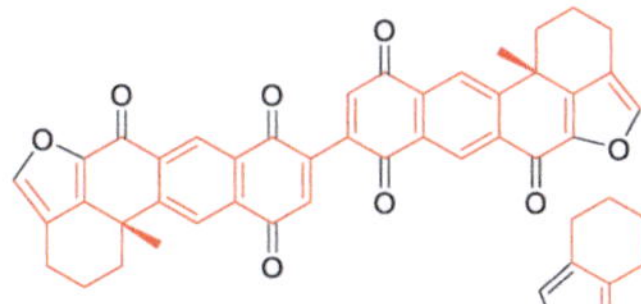
851. (+)-petroquinone F

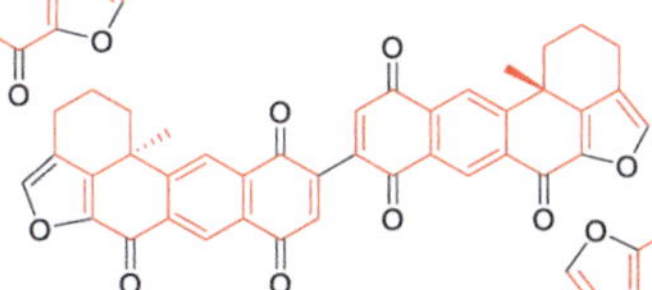
852. (+)-petroquinone G

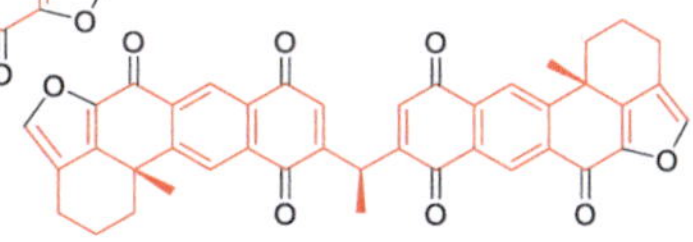
853. (+)-petroquinone H

Fig. 27 (continued)

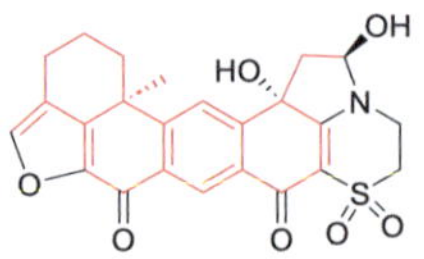

854. (+)-petroquinone J

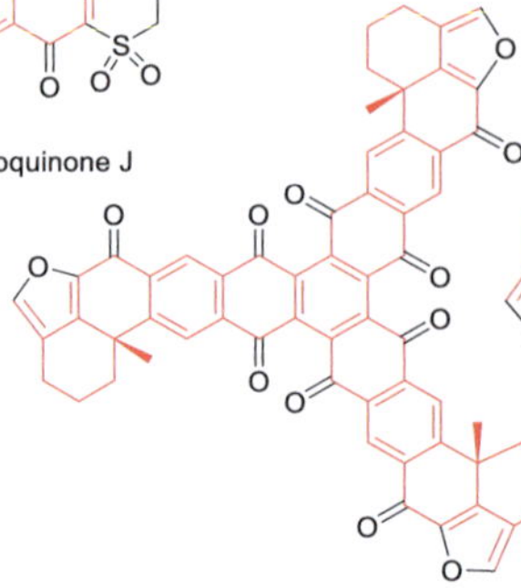

855. (+)-petroquinone A

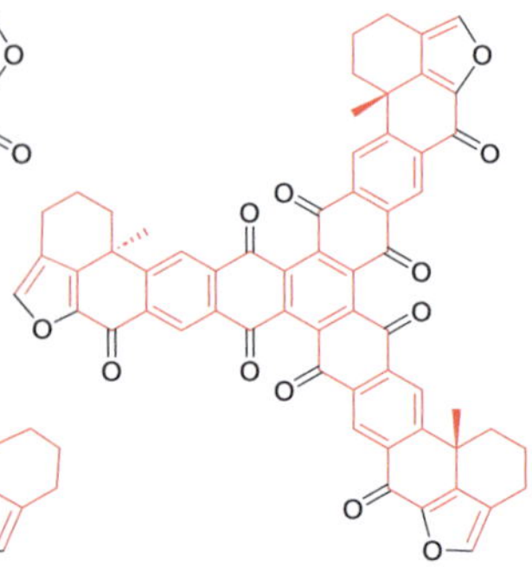

856. (–)-petroquinone B

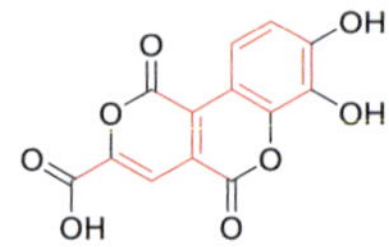

857. lumnitzeralactone

Fig. 27 (continued)

(+)-12-*O*-deacetylphomoxanthone A (**815**) [427] was identified from the rice culture of *Phomopsis* sp. isolated from the mangrove plant *Rhizopora mucronata* (JSCR). 2-Carboxy-8-methoxynaphthalene-1-ol (**816**) was obtained from *Mycelium sterilium* associated with algae (UEP) [428]. In 2014, xylarianaphthol-1 (**817**) was isolated from a fungus (Xylariales) associated with a sponge [429]. The structure of **817** was elucidated as the dimer of **816**. Another aromatic polyketide, 14,15-seco-curvularin (**818**) obtained from unidentified fungus isolated from the marine sponge *Spirastrella vagabunda* (CSW) [430]. Only two aromatic polyketides were reported from marine *Streptomyces* sp.: (+)-komodoquinone B (**819**) and (+)-komodoquinone A (**820**) [431]. The ACs of (+)-**819**, **820** were determined by Mosher's chemistry and chemical transformation [431, 432].

Marine sponges' genera *Lendenfeldia* (1 molecule), *Xestospongia* (18 molecules), *Petrosia* (15 molecules), and *Xestospongia/Petrosia* (2 molecules) were contributors of 36 MNP aromatic polyketides. (+)-(*S*)-2,2′-Dimethoxy-1,1′-binaphthyl-5,5′,6,6′-tetraol (**821**) were isolated from a marine sponge *Lendenfeldia* sp. (UEP) [96]. Eighteen aromatic polyketides, (−)-**822–826**, **828**, **830** (EKM) [433, 434], **840** (NSW) [435] and (+)-**827**, **829**, **831** (EKM) [433, 434], **838**, **839** (NSW) [436], **841** (EKM) [437] as well as **832/833**, **836/837** (NSW) [435] were isolated from *Xestospongia* sponge. Xestoadociaquinones A/B (**832/833**) and xestoadociaminal C/D (**836/837**) were identified as a mixture. The genus *Petrosia* gave fifteen aromatic polyketides including (−)-**842**, **845**, **848**, **856** and (+)-**846**, **847**, **849–855** as well as **843/844** (NSW) [438]. Compounds **843/844** were a mixture of stereoisomers with ratio *R*/*S* 1:1. Two aromatic polyketides, (−)-xestoadociaminal A (**834**) and xestoadociaminal B (**835**) [435] was isolated from *Xestospongia* sponge (NSW), while the same molecule assigned as petroquinone I (**834**) [438] was obtained from *Petrosia* sponge (NSW). The ACs of petroquinones family were assigned by comparison of calculated and experimental ECD. Finally, one new aromatic polyketide assigned as lumnitzeralactone (**857**) was discovered from mangrove plant *Lumnitzera racemosa* (NMU) [439]. The planar structure of **857** was confirmed by NMR calculation.

6.3 Lactone

About 35 marine lactones (**858–892**) were discovered from Indonesian waters with a variety size of rings. Furthermore, marine lactones can be classified into three subclasses including γ-(**858–861**), δ-(**862–872**), and macrolactones (**873–892**). Four γ-lactones as in (−)-plakofuranolactone (**858**) [440], (−)-herbaric acid (**859**) [424], (+)-biaketide (**860**) [318], (−)-*iso*-cladospolide B (**861**) [441] were isolated from the marine sponge *Plakortis* cf. *lita* (NSW), the marine fungus *Cladosporium herbarum* associated with the marine sponge *Aplysina aerophoba* (BLI), a marine sponge *Dysidea* sp. (PUA), and a fungus associated with a sponge (SSW), respectively. The AC of (−)-**858** was assigned by experimental and calculated optical rotation data. Total synthesis was used to establish the AC of (−)-**859** [442] and (−)-**861** [443]. Moreover, the AC of (−)-**861** was secured by X-ray analysis [444]. Among the eleven δ-lactones, herbarin B (**862**) and herbarin A (**864**) [424], were isolated from the marine fungi *Cladosporium herbarum* associated with *Calyspongia aerizusa* (BLI), compound **863** was obtained from *Fusarium* sp. associated with *Rhizophora mucronata* [445], and (+)-helicascolide C (**865**) was purified from the extract of *Daldinia eschscholzii* associated with *Gracilaria* sp. [446]. The planar structures of herbarin B (**862**) and herbarin A (**864**) were confirmed by total synthesis [447]. The AC of (+)-**865** was established by X-ray and CD analysis [446] and confirmed by total synthesis [448]. The marine sponge *Theonella* cf. *swinhoei* (NSW) was the source of (+)-bitungolides A–F (**866**, **867a**, **868**, **869**, **870a**, **871a**, respectively) [449]. The AC of (+)-**866** was determined by X-ray analysis. Total synthesis was used to revise the structures of (+)-**867a** to (−)-**867b** [450], (+)-**870a** to **870b** [450, 451], and **871a** to **871b** [452]. The marine sponge *Discodermia du Bocage* obtained from MLU yielded a nonribosomal peptide-polyketide, (+)-PM742 (**872**) [453]. Its AC was determined using Marfey's analysis.

Twenty macrolactones including (+)-xestodecalactones A–C (**873–875**, *Penicillium* cf. *montanense* associated with *Xestospongia exigua*, BLI) [454], (−)-pandangolides 1, 2 (**876**, **877**, a marine fungus isolated from a marine sponge, BLI) [441], (−)-pandangolides 3, 4 (**878–879**, *Cladosporium herbarum* isolated from *Callyspongia aerizusa,* SSW) [406], (−)-callyspongiolide (**880a**, *Callyspongia* sp., MLU) [455], (+)-phormidolide (**881a**, *Phormidium* sp. UEP) [456], a mixture of candidaspongiolide-like compounds (**882/883/884/885/886**) and (+)-**887** (*Candidaspongia* sp., ENT) [457], (−)-laulimalide or (−)-fijianolide B (**888**), (−)-isolaulimalide or (−)-fijianolide A (**889**) (*Hyattella* sp. and *Chromodoris lochi*, NSW) [458, 459], (−)-isoswinholide B (**890**), (−)-swinholide K (**891**) (*Theonella swinhoei*, NSW) [460], and (+)-enimagzole C (**892**, *Homophymia* sp., GTO) [319]. The ACs of (+)-**874–875** were confirmed by total synthesis [461, 462]. Initial attempt to determine the AC of (−)-**876** using MPA ester with addition of $Ba(ClO_4)_2$ and

CD from another (−)-**876** isolated from the Red sea sponge *Niphates rowi* [463]. Although the NMR and optical rotation data of synthetic **876** were not consistent with the natural product, the structure of pandangolide 1 (**876**) was secured by the single crystal X-ray analysis [464]. During stereochemical determination via total synthesis [465], (−)-callyspongiolide (**880a**) was revised to (−)-callyspongiolide (**880b**). Several strategies were applied to determine correct stereochemistry of (+)-phormidolide (**881a**) for example using Mosher chemistry [456], synthesis and chelation control as in (**881b1**) [466], biosynthetic gene cluster analysis or genomic configurational prediction (**881b2**) [467], analysis of quantum chemical calculation and anisotropic NMR parameters (**881b3**) [468]. The AC of (+)-phormidolide (**881b3**) proposed by quantum chemical calculation and anisotropic parameters was finally confirmed by synthesis [469]. The AC of (−)-laulimalide (**888**) was secured by X-ray diffraction analysis [470] and confirmed by total synthesis [471]. In addition to (−)-laulimalide (**888**), the AC of (−)-isolaulimalide/(−)-fijianolide A (**889**) was established by total synthesis [472]. Mosher's analysis was used to determine the stereochemistry of (+)-enimagzole C (**892**) [319]. The structures of Indonesian marine lactones are presented in Fig. 28.

6.4 Miscellaneous Polyketides

Two polyol marine natural products (Fig. 29) assigned as (+)-karatungiols A and B (**893**, **894**) were isolated from *Amphidinium* sp. associated with acoel flatworm [473]. (−)-seco-Patulolide C (**895**) was isolated from a fungus associated with a sponge [441]. Its AC was determined by Mosher's chemistry [474]. The marine sponge *Theonella swinhoei* was the source of aurantoic acid (**896**) [176]. The building block of (+)-**893** and **894** is proposed on the basis of the amphidinol A isotopic labelling [328, 400].

7 Miscellaneous

Based on biosynthetic gene cluster analysis, a polybrominated diphenyl ether subclass can be derived from chorismic acid [475] involving flavin-dependent decarboxylase-brominase and cytochrome P450 (CYP450) enzymes [475, 476]. About eight Indonesian MNPs (**894–901a**, Fig. 30) are categorized as polybrominated diphenyl ethers found in *Dysidea* (*Lamellodysidea*) *herbacea* collected from WST (**897–900**) [477] and BTN (**901–904a**) [478] waters. Compounds **898/899** were found as a mixture. To confirm the structures, the compounds **901–904a** were correlated using methylation and debromination reaction. Compound **904a** was revised to be **904b** based on the X-ray crystallography data [479]. Two carbohydrate metabolites were identified as (+)-sinularioside (**905**) and (−)-kakelokelose (**906**). A mixed biosynthesis of fatty acid and carbohydrate yielded a triacetylated

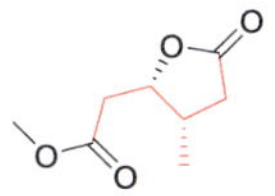

858. (–)-plakofuranolactone

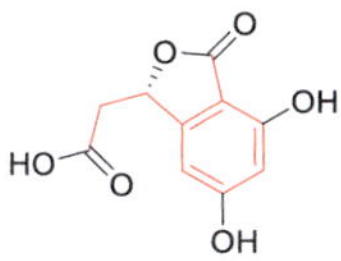

859. (–)-herbaric acid

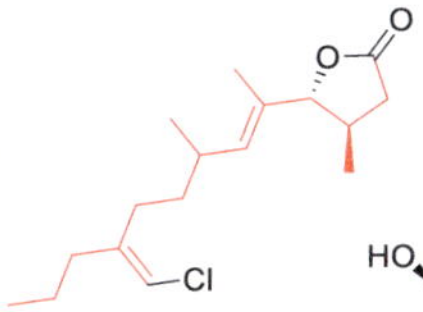

860. (+)-biaketide

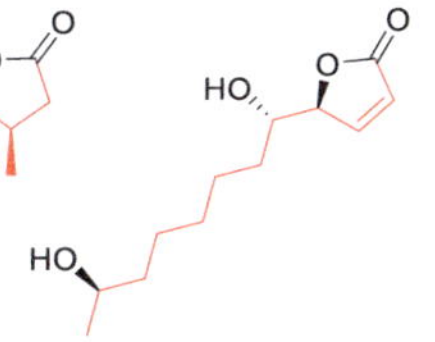

861. (–)-*iso*-cladospolide B

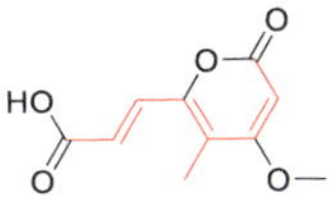

862. herbarin B

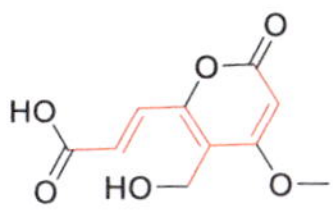

863

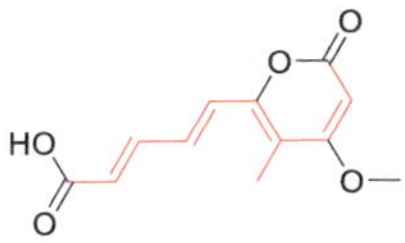

864. herbarin A

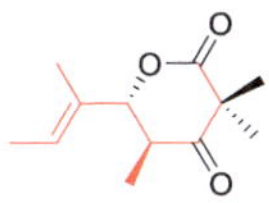

865. (+)-helicascolide C

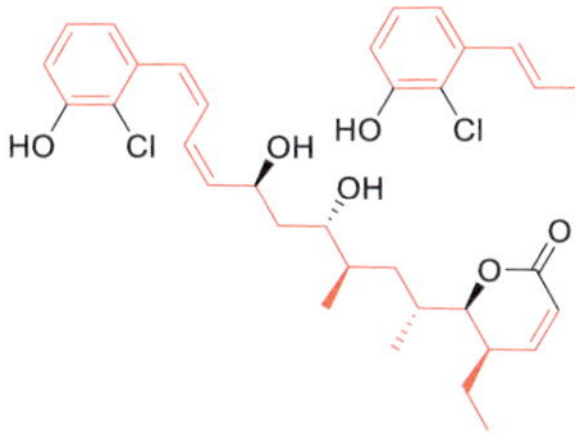

866. (+)-bitungolide A

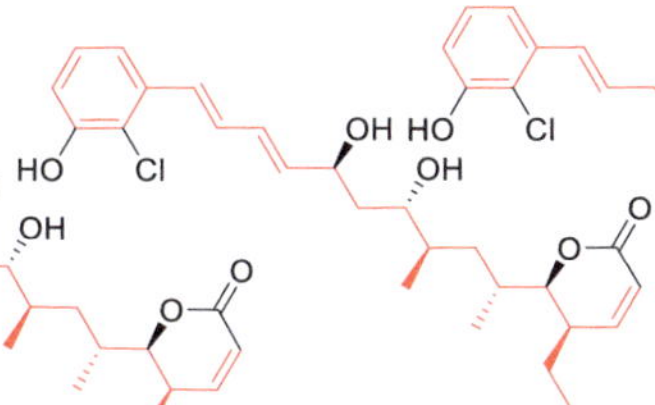

867a. (+)-bitungolide B

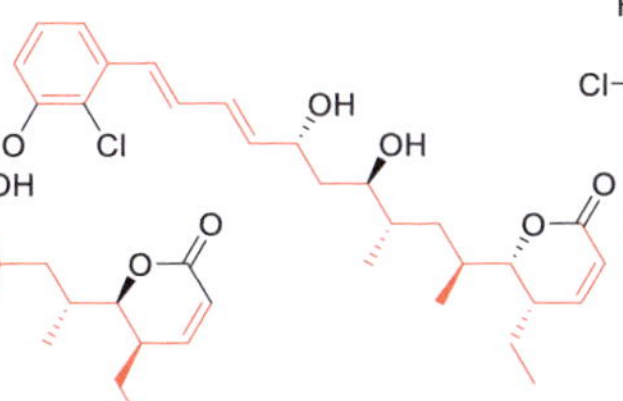

867b. (–)-bitungolide B

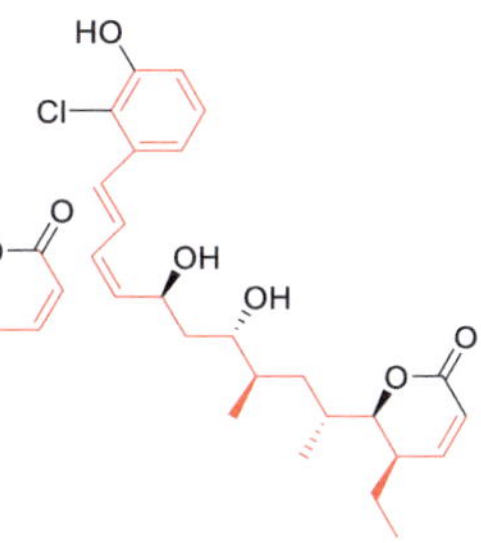

868. (+)-bitungolide C

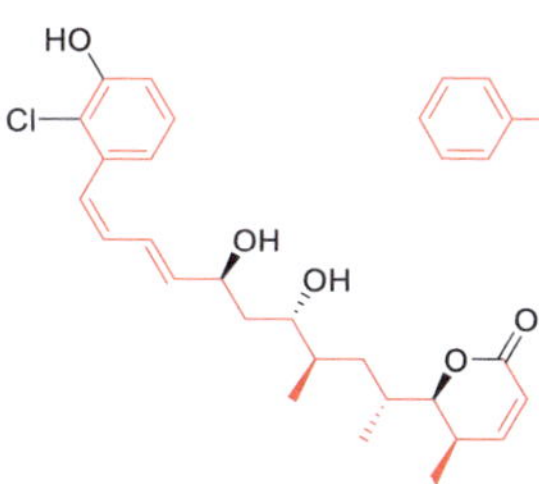

869. (+)-bitungolide D

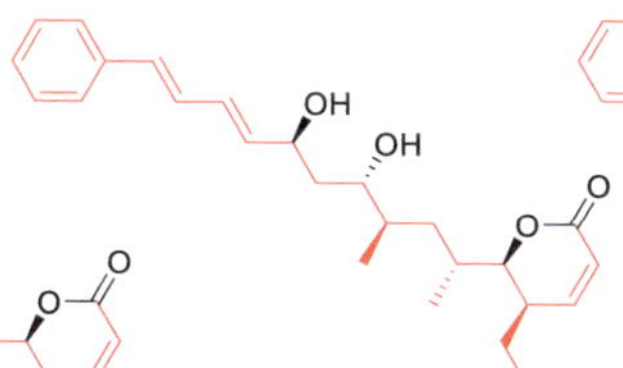

870a. (+)-bitungolide E

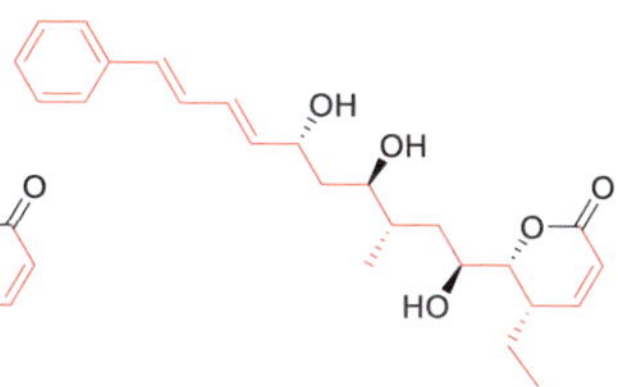

870b. (–)-bitungolide E

Fig. 28 Structures of marine lactones (**858–892**) from Indonesian waters. Polyketide skeleton containing C_2 building blocks was proposed with red color

glycolipid, (+)-sinularioside **905** isolated from the Indonesian soft coral *Sinularia* sp. [480]. The AC of (+)-**905** was determined using comparison of CD spectrum of methyl α-arabinopyranoside perbenzoate from (+)-**905** and an authentic sample of methyl α-D-arabinopyranoside perbenzoate. (−)-Kakelokelose (**906**), an unusual sulfated mannose homopolysaccharide was obtained from the NSW collection of *Didemnum molle* [481]. The monomer of (−)-**906** was determined to be sequence of 2,3-disulfated mannose units joined through β-(1,6) glycosidic linkages. Five Indonesian MNPs could be derived from tyrosine amino acid, but the molecules did not contain nitrogen atom as in **907–911**. Ethyl 2-(4-methoxyphenyl)-2-oxoacetate

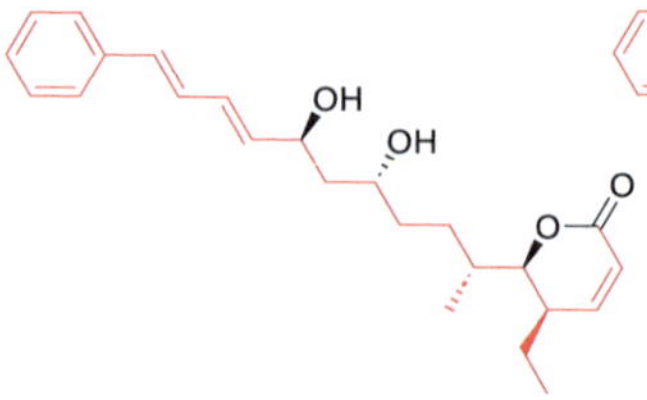

871a. (+)-bitungolide F

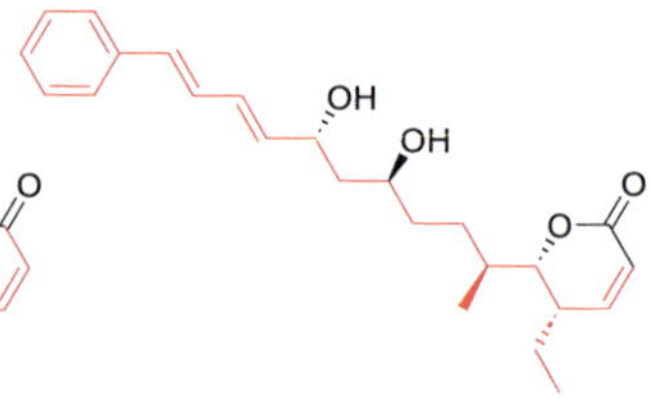

871b. (–)-bitungolide F

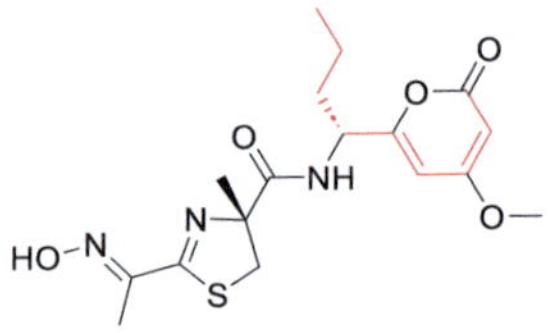

872. (+)-PM742

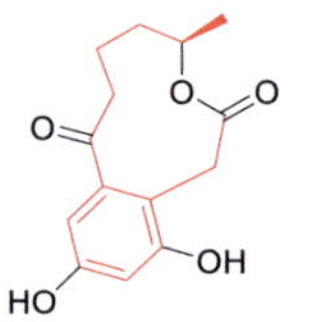

873. (+)-xestodecalactone A

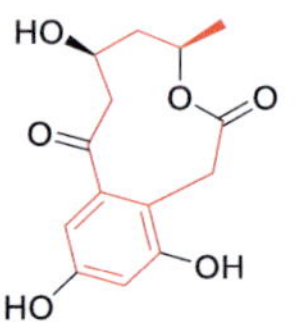

874. (+)-xestodecalactone B

875. (+)-xestodecalactone C

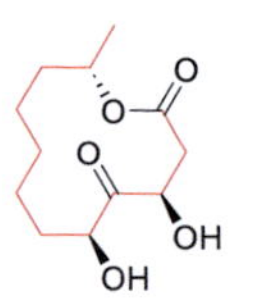

876. (–)-pandangolide 1

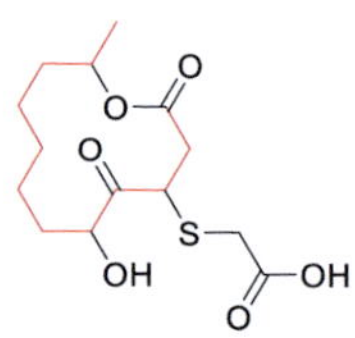

877. (–)-pandangolide 2

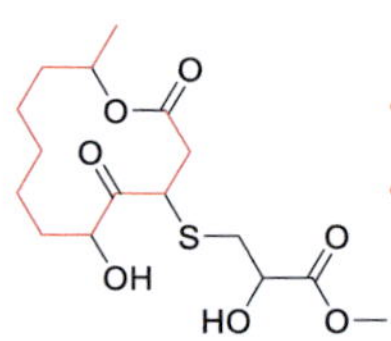

878. (–)-pandangolide 3

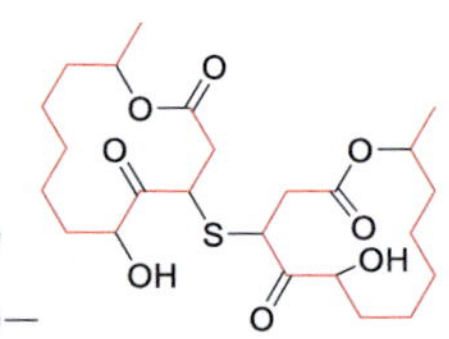

879. (–)-pandangolide 4

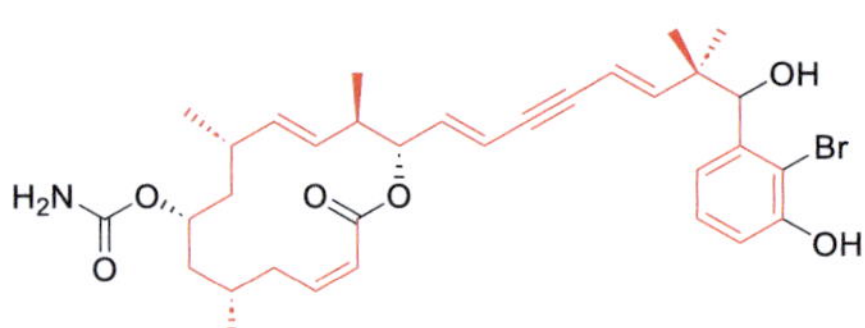

880a. (–)-callyspongiolide

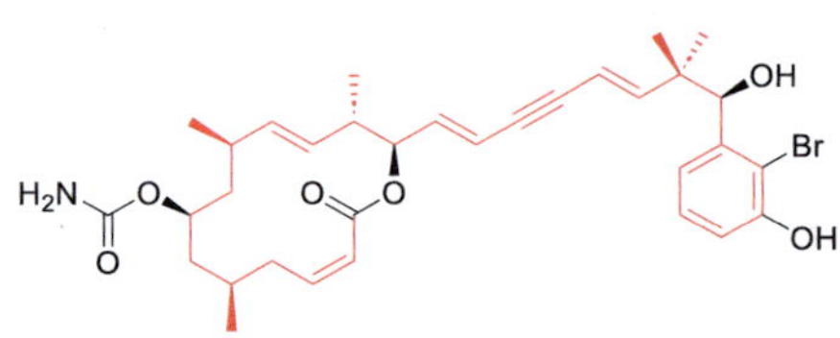

880b. (–)-callyspongiolide

Fig. 28 (continued)

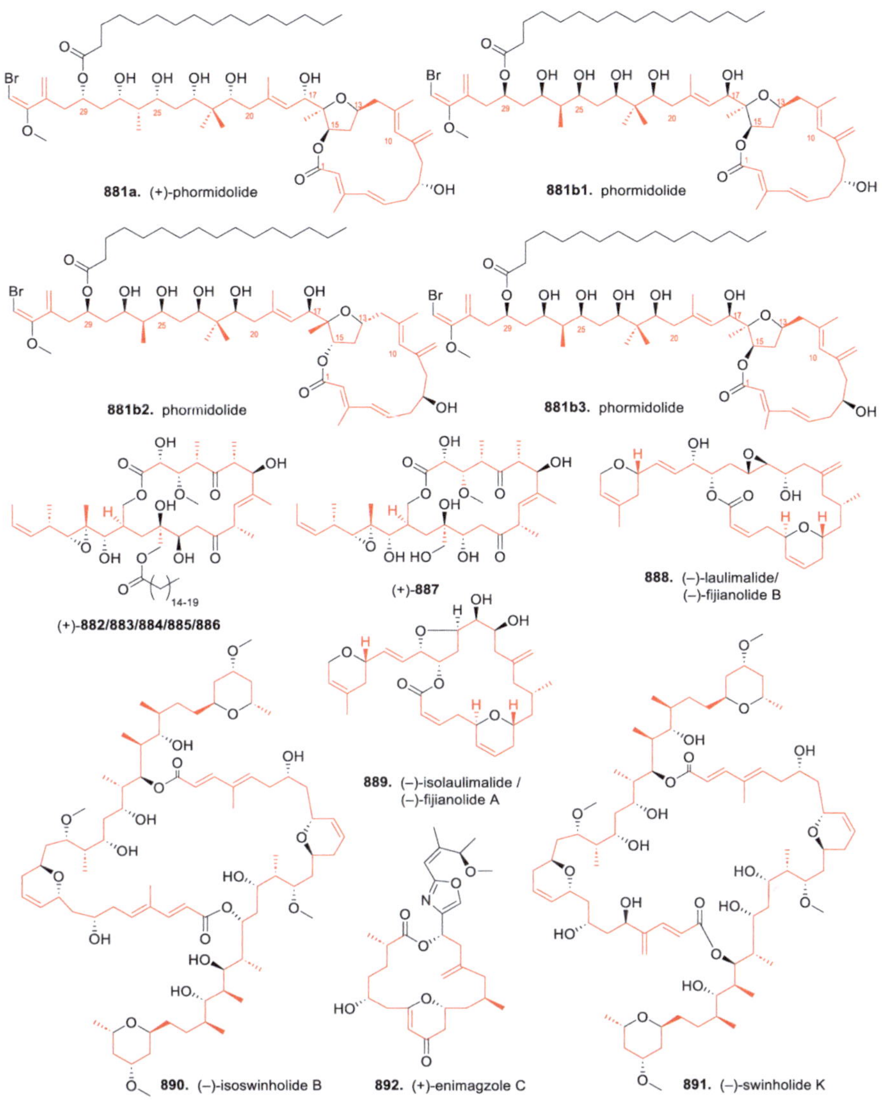

Fig. 28 (continued)

(**907**) and methyl 2-(4-hydroxyphenyl)-2-oxoacetate (**908**) [307] were isolated from the ascidian *Polycarpa aurata* collected from NSW waters. An ascidian *Botryllus* sp. (NSW) was the source of cadiolides A and B (**909, 910**) [482]. (−)-Vidalenolone (**911**) was obtained from an algae *Vidalia* sp. [483]. The AC of (−)-**911** was determined by CD, DFT-based optical rotation calculation, and NMR calculation [484].

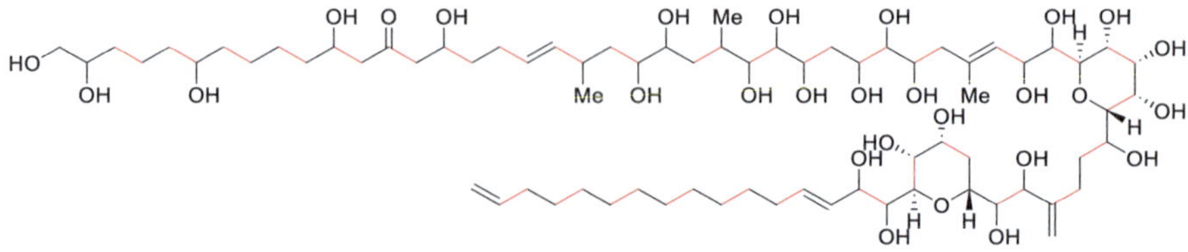

893. (+)-karatungiol A

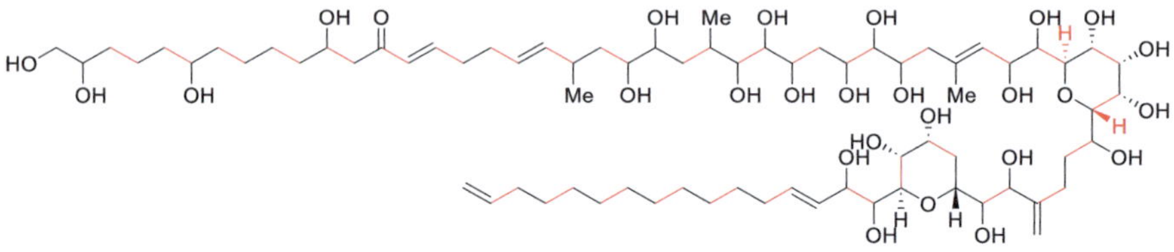

894. (+)-karatungiol B

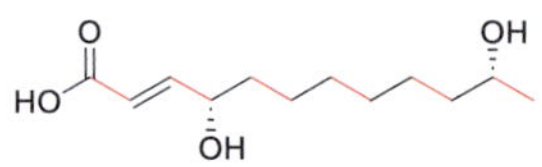

895. (–)-*seco*-patulolide C

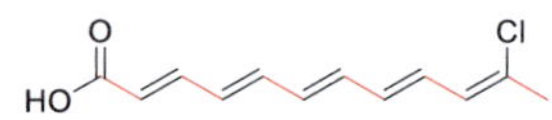

896. aurantoic acid

Fig. 29 Structures of marine miscellaneous polyketide (**893–896**) from Indonesian waters. C_2 building blocks were proposed with red color

8 Significant Biological Activities

It is common to express biological activities of MNPs using statistical parameters at the 50% of the inhibitory concentration (IC_{50}), effective concentration (EC_{50}), growth inhibition (GI_{50}), effective dose (ED_{50}), lethal concentration (LC_{50}) or cytotoxic concentration (CC_{50}). Potent cytotoxic, antiproliferative, anticancer, antiparasite, and immunomodulator activities of Indonesian MNPs are evaluated based on the values of IC_{50}, EC_{50}, or $GI_{50} < 0.1$ μM [16]. For antibacterial activity, the result is frequently expressed as minimum inhibitory concentration (MIC). The potent antibacterial activity is considered to have MIC < 1.0 μg/mL [16]. Within this category, 37 potent Indonesian MNPs are able to identify and approximately distributed into six biological activities: cytotoxic (7 molecules), antiproliferative (9 molecules), anticancer (1 molecule), antibacterial (18 molecules), antiparasite (1 molecule), and immunomodulator (1 molecule) activities. The significant biological activities of Indonesian MNPs are compiled as in Tables 1, 2, 3, 4, 5 and 6.

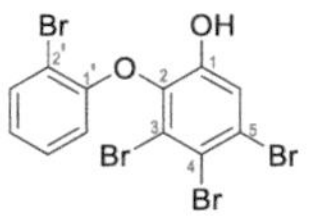

897. 3,4,5-tribromo-2-(2'-bromophenoxy)phenol

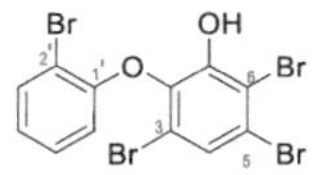

898. 3,5,6-tribromo-2-(2'-bromophenoxy)phenol

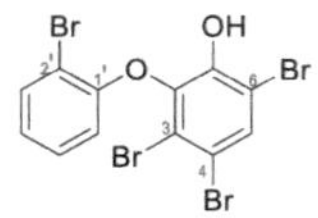

899. 3,4,6-tribromo-2-(2'-bromophenoxy)phenol

900. 3,4,6-tribromo-2-(2'-bromophenoxy)anisole

901. 2,3,5-tribromo-6-(3′,5′-dibromo-2′-methoxyphenoxy)phenol

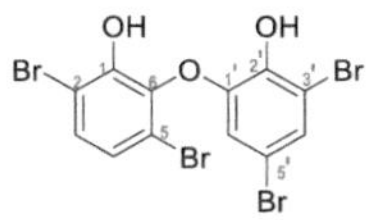

902. 2,5-dibromo-6-(3′,5′-dibromo-2′-hydroxyphenoxy)phenol

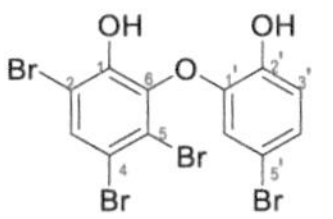

903. 2,4,5-tribromo-6-(5′-bromo-2′-hydroxyphenoxy)phenol

904a. 2,4,5-tribromo-6-(3′,5′-dibromo-2′-hydroxyphenoxy)anisole

904b. 2,4,5-tribromo-6-(3′,5′-dibromo-2′-hydroxyphenoxy)anisole

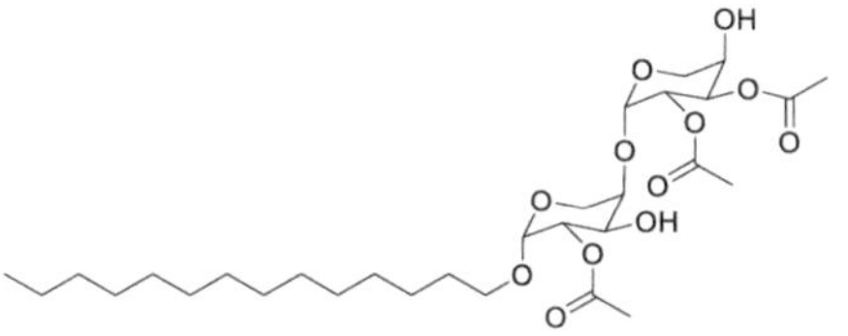

905. (+)-sinularioside

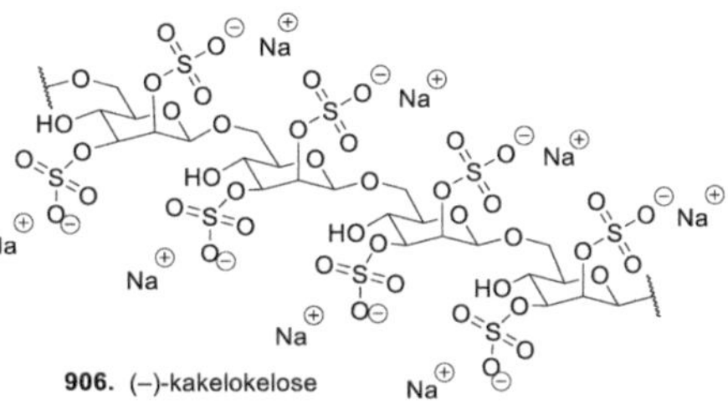

906. (–)-kakelokelose

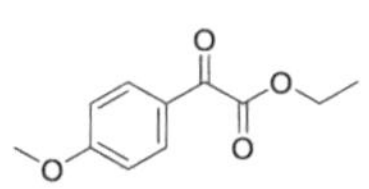

907. ethyl 2-(4-methoxyphenyl)-2-oxoacetate

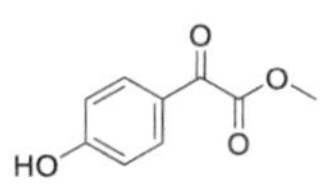

908. methyl 2-(4-hydroxyphenyl)-2-oxoacetate

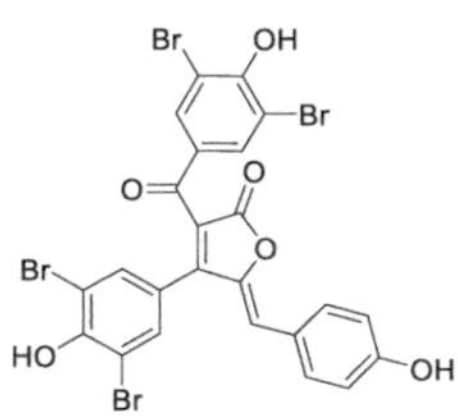

909. cadiolide A

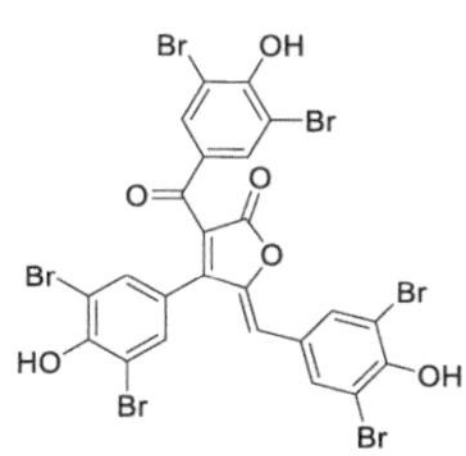

910. cadiolide B

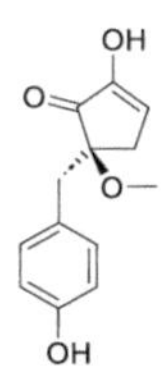

911. (–)-vidalenolone

Fig. 30 Structures of marine miscellaneous (**897–911**) from Indonesian waters

Table 1 Significant cytotoxic activity of Indonesian MNPs

Origin	Type	Cell lines	Cytotoxic activity
Human	Cervical	HeLa	**368** $IC_{50} = 0.09$ μM [206]
Human	Lymphoma	Jurkat J16 T	(−)-**880a** $IC_{50} = 0.07$ μM [455]
Human	Lymphoma	Jurkat	(−)-**880b** $IC_{50} = 0.0111$ μM [465]
Human	Lymphoma	Ramos B lymphocyte	(−)-**880a** $IC_{50} = 0.06$ μM [455]
Human	Epidermoid	KB	(−)-**888** $IC_{50} = 0.029$ μM [458]
Human	Ovarian	SK-OV-3	(−)-**888** $IC_{50} = 0.01153 \pm 0.00053$ μM [485]
Human	Ovarian	1A9	(−)-**888** $IC_{50} = 0.0039 \pm 0.0004$ μM [486]
Human	Ovarian	PTX10	(−)-**888** $IC_{50} = 0.006 \pm 0.001$ μM [486]
Human	Ovarian	PTX22	(−)-**888** $IC_{50} = 0.0063 \pm 0.001$ μM [486]
Human	Ovarian	A8	(−)-**888** $IC_{50} = 0.0092 \pm 0.002$ μM [486]
Human	Ovarian	B10	(−)-**888** $IC_{50} = 0.015 \pm 0.0002$ μM [486]
Human	Ovarian	A2780/AD10	(−)-**888** $IC_{50} = 0.031 \pm 0.0006$ μM [486]
Human	Ovarian	A-2780	(+)-**872** $GI_{50} = 0.025$ μM [453]
Human	Ovarian	IGROV-1	(+)-**872** $GI_{50} = 0.028$ μM [453]
Human	Breast	MDA-MB-435	(−)-**888** $IC_{50} = 0.00574 \pm 0.00058$ μM [485] (−)-**888** $IC_{50} = 0.0023 \pm 0.0002$ μM [487]
Human	Breast	MDA-MB-231	(+)-**872** $GI_{50} = 0.0397$ μM [453]
Human	Breast	MCF-7	(−)-**888** $IC_{50} = 0.0038$ μM [488] (−)-**888** $IC_{50} = 0.0116–0.0005$ μM [489]
Human	Breast	NCI/ADR	(−)-**888** $IC_{50} = 0.036$ μM [488]
Human	Breast	MaTu	(−)-**888** $IC_{50} = 0.0038$ μM [488]
Human	Breast	MaTu/ADR	(−)-**888** $IC_{50} = 0.006$ μM [488]
Human	Prostate	PC-3 M	(−)-**888** $IC_{50} = 0.0059 \pm 0.0003$ μM [489]
Human	Colon	HT-29	(−)-**888** $IC_{50} = 0.0069$ μM [487]
Human	Colon	HCT-116	(−)-**888** $IC_{50} = 0.0078 \pm 0.0008$ μM [489]
Human	Colon	HT-29	(+)-**872** $GI_{50} = 0.0303$ μM [453] (−)-**888** $IC_{50} = 0.019–0.097$ μM [490]
Human	Colon	LoVo	(+)-**872** $GI_{50} = 0.0160$ μM [453]
Human	Lung	A-549	(+)-**872** $GI_{50} = 0.0303$ μM [453] (−)-**888** $IC_{50} = 0.019–0.097$ μM [490]
Human	Lung	NCI-H460	(+)-**872** $GI_{50} = 0.0382$ μM [453]
Human	Pancreatic	PSN-1	(+)-**872** $GI_{50} = 0.0442$ μM [453]
Human	Skin	MEL-28	(−)-**888** $IC_{50} = 0.019–0.097$ μM [490]
Murine	Lymphoma	P388	(−)-**888** $IC_{50} = 0.019–0.097$ μM [490]
Murine	Lymphoma	L5178Y	(+)-jasplakinolide Z_5 $IC_{50} < 100$ nM [344]

(continued)

Table 1 (continued)

Origin	Type	Cell lines	Cytotoxic activity
Murine	Bladder	NBT-T2	(+)-**882/883/884/885/886** $IC_{50} = 0.005\ \mu M$ [457]
Hamster	Lung	V79	**413** $IC_{50} = 0.08\ \mu M$ [491]

Table 2 Significant antiproliferative activity of Indonesian MNPs

Origin	Type	Cell lines	Antiproliferative activity
Human	Umbilical cord	HUVEC	(+)-**266** $IC_{50} = 0.064\ \mu M$ [154] (−)-**273** $IC_{50} = 0.009\ \mu M$ [154] (−)-**278** $IC_{50} = 0.06\ \mu M$ [154] (+)-**328** $IC_{50} = 0.002 \pm 0.001\ \mu M$ [184, 187] (−)-**330** $IC_{50} = 0.019\ \mu M$ [184] (−)-**332** $IC_{50} = 0.008\ \mu M$ [186, 492] (−)-**333** $IC_{50} = 0.04\ \mu M$ [186, 492] (−)-**334** $IC_{50} = 0.023\ \mu M$ [186]
Human	Breast	MDA-MB-231	(−)-**888** $GI_{50} = 0.0052 \pm 0.0009\ \mu M$ [493]
Human	Breast	HCC1806	(−)-**888** $GI_{50} = 0.0007 \pm 0.0002\ \mu M$ [493]
Human	Breast	BT-549	(−)-**888** $GI_{50} = 0.004 \pm 0.0016\ \mu M$ [493]

Table 3 Significant anticancer activity of Indonesian MNPs

Origin	Type	System	Anticancer activity
Human	Ubiquitin–proteasome system (UPS) inhibitor	Ubc13-Uev1A	(+)-**94** $IC_{50} = 0.09\ \mu M$ [91]

9 Supply Challenge of Talented MNPs

MNPs drug discovery research program usually faces major challenges, including identification of novel structures, supply issues, defining biological targets, and mechanisms of action. A variety of structure determination studies has been described in the previous section. The next task is how to provide a sufficient quantity of talented MNPs with a valid chemistry for thorough biological investigation including total synthesis (large-scale synthesis), semi-synthesis, heterologous expression, mariculture/aquaculture, and fermentation [496]. Under this category, a few significant bioactive MNPs are briefly discussed including (+)-cortistatin A (**328**), (−)-laulimalide or fijianolide B (**888**), (+)-manzamine A (**912**) and its analogues, (−)-$\Delta^{9(12)}$-capnellene (**4**), and polybrominated diphenyl ethers.

Although the remarkable total and semi-syntheses of (+)-cortistatin A (**328**) have been achieved [497], the supply of (+)-cortistatin A (**328**) is still hampered by its low yield indicating that newer strategies to construct the molecule are required. The low yield issue is also faced by the total synthesis of the polyketide-macrolide, (−)-laulimalide or fijianolide B (**888**) [498]. It was found that (−)-laulimalide or

Table 4 Significant antibacterial activity of Indonesian MNPs

Microorganisms	Antibacterial activity
Staphylococcus epidermis	(−)-**167** MIC < 0.037 μg/mL [126]
Staphylococcus aureus ATCC 6538p	(+)-**531** MIC = 0.025 μg/mL [292] (+)-**542** MIC = 0.013 μg/mL [292] (+)-**551** MIC = 0.002 μg/mL [292] (+)-**552** MIC = 0.032 μg/mL [292] (+)-**553** MIC = 0.032 μg/mL [292] (+)-**557** MIC = 0.008 μg/mL [292]
Staphylococcus aureus	(+)-**538** IC_{50} = 0.9 μg/mL [494]
Bacillus subtilis ATCC 6633	(+)-**531** MIC = 0.0063 μg/mL [292] (+)-**541** IC_{50} = 0.1 μg/mL [292] (+)-**542** MIC = 0.0063 μg/mL [292] (+)-**551** MIC = 0.004 μg/mL [292] (+)-**553** MIC = 0.064 μg/mL [292] (+)-**557** MIC = 0.032 μg/mL [292]
Kocuria rhizophila NBRC 12708	(+)-**531** MIC = 0.0063 μg/mL [292] (+)-**541** MIC = 0.1 μg/mL [292] (+)-**542** MIC = 0.0031 μg/mL [292] (+)-**551** MIC = 0.004 μg/mL [292] (+)-**553** MIC = 0.064 μg/mL [292] (+)-**557** MIC = 0.016 μg/mL [292]
Salmonella enterica ATCC14028	(+)-**531** MIC = 0.0016 μg/mL [292] (+)-**541** MIC = 0.5 μg/mL [292] (+)-**542** MIC = 0.0031 μg/mL [292] (+)-**551** MIC = 0.008 μg/mL [292] (+)-**552** MIC = 0.032 μg/mL [292] (+)-**557** MIC = 0.016 μg/mL [292]
Proteus hauseri NBRC 3851	(+)-**531** MIC = 0.013 μg/mL [292] (+)-**542** MIC = 0.025 μg/mL [292] (+)-**551** MIC = 0.002 μg/mL [292] (+)-**553** MIC = 0.032 μg/mL [292] (+)-**557** MIC = 0.008 μg/mL [292]
Escherichia coli ATCC 35270	(+)-**551** MIC = 0.016 μg/mL [292] (+)-**557** MIC = 0.064 μg/mL [292]
Mycobacterium tuberculosis H37Rv	(+)-**538** MIC = 0.9 μg/mL [494] (+)-**543** MIC = 0.4 μg/mL [297]
Mycobacterium smegmatis, propionate	**614/615** MIC = 0.4 μg/mL [495] **616/617** MIC = 0.4 μg/mL [495] **618/619/620/621** MIC = 0.4 μg/mL [495]

Table 5 Significant antiprotozoal activity of Indonesian MNPs

Microorganisms	Antiprotozoal activity
Trypanosoma brucei rhodesiense	(−)-**767** IC_{50} = 0.0088 μM [403]

Table 6 Significant immunomodulator activity of Indonesian MNPs

	Immunomodulator Activity
MLR	(+)-**764** $IC_{50} = 0.05\ \mu M$ [401]

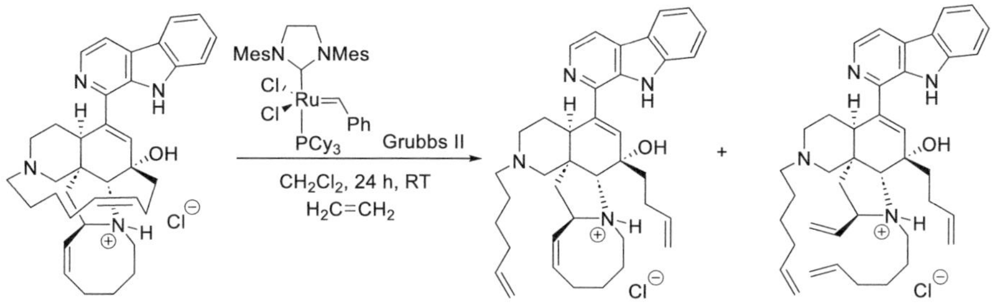

Fig. 31 Cleavage of (+)-manzamine A (**912**) to the corresponding structures with Grubbs II

fijianolide B (**888**) exhibits side effect, such as toxicity and mortality with minimal tumor growth inhibition (in vivo).

In another case, (+)-manzamine A (**912**) was originally discovered from two Okinawan sponges [499, 500] and later from the Indonesian marine sponge *Acanthostrongylophora ingens*. The access of manzamine analogues (+)-**532**, (+)-**541–543**, (+)-**551**, (+)-**553**, (+)-**557** could be achieved by the fermentation of the true producer of manzamine A, *Micromonospora* sp. [501]. Moreover, semi-synthesis of (+)-manzamine A (**912**) [502] via 2nd-generation Grubbs catalyst was able to give (+)-**913** and (−)-**914** (Fig. 31). Compound (+)-**913** possessed potent antibacterial against *Mycobacterium intracellulare* with IC_{50} 0.0001 μM and antiprotozoal activities against *Leishmania donovani* with IC_{50} 0.0021 μM. Additionally, compound (+)-**913** is about 1.8-fold more potent than that of ciprofloxacin (IC_{50} 0.0018 μM), a drug that is frequently used against antibiotic-resistant infections.

An example of heterologous expression strategy includes capnellane synthase (CiTC) gene isolated from the soft coral *Capnella imbricata* genome and applied in *Escherichia coli* to give (−)-$\Delta^{9(12)}$-capnellene (**4**) [503]. Although scalability is not necessarily straightforward, this discovery has overturned the long-held belief that the coral-derived compound was the product of symbiotic or associated marine microorganisms. In another case, the cyanobacterial symbionts *Hormoscilla spongeliae* (formerly *Oscillatoria spongeliae*) has been identified as the true producer of polybrominated diphenyl ethers [475].

10 Summary and Future Prospects

The earliest of MNP from Indonesian waters was (−)-25-hydroxy-24ξ-methylcholesterol (**292**) reported by Engelbrecht et al. in 1972 from a marine soft coral *Nepthea* sp. (NST) [169]. Fifty-two years later, about 911 molecules including 879 original MNPs are approximately discovered from 108 species belonging to 116 genera, 76 families, 37 orders, 17 classes, and 12 marine phyla including Actinobacteria (6 molecules), Cyanobacteria (12 molecules), Ascomycota (55 molecules), Dinoflagellate (2 molecules), Ciliophora (2 molecules), Rhodophyta (3 molecules), Tracheophyte (1 molecule), Porifera (622 molecules), Cnidaria (127 molecules), Chordata (66 molecules), Mollusca (14 molecules) and Echinodermata (1 molecule) from 22 provinces (NST, RUI, WST, LPG, BTN, JSCR, WJV, CJV, BLI, WNT, ENT, EKM, NSW, GTO, CSW, SSW, SES, NMU, MLU, SWP, CPA, PUA) and unidentified exact province (UEP). The molecules can be grouped into terpenoid (338 molecules), alkaloid (329 molecules), peptide (85 molecules), fatty acid (11 molecules), polyketide (133 molecules), and miscellaneous (15 molecules). The organic molecules were mainly discovered from central (651 molecules) followed by eastern (177 molecules) and western (83 molecules) parts of Indonesia. Of 911 molecules, only 37 MNPs are categorized to have potent biological activities.

From a structural point of view, unique functional groups and mixed biosynthetic-pathway Indonesian marine metabolites are frequently found as a reflection of enzymes with unique specificities and reactivities. In a certain case, both enantiomers formed in a single species such as *Acanthostrongylophora ingens* or its symbiont are remarkable. Moreover, skeletons, building blocks, or scaffolds are conserved or evolved within marine organisms. Among the 92 skeletal types encountered in the terpenoid metabolites, approximately 27 are unprecedented skeletons. Meanwhile, alkaloid and peptides are derived from 20 proteinogenic, 39 nonproteinogenic, and 33 modifications of amino acids. In case of fatty acid and polyketide metabolites, the MNPs are mainly derived from C_2 building block. To uncover the true structure of bioactive MNPs, orthogonal strategies in structure determination have been applied such as NMR, HRMS, IR, UV, specific optical rotation, X-ray, ECD, chemical transformations (derivatization or selective degradation, total synthesis), molecular modelling, and quantum chemical calculations. In some cases, the true structure of elusive, complex, or reactive MNPs can be anticipated by the use of LC-HR-MS/MS (molecular networking), analysis of microbial genomes, and their combination with those of above strategies. The true structures of MNPs reflect their true function in nature which can be utilized as important materials in many fields.

Indonesian shallow waters and mesophotic areas are still under explored implying new strategies should be developed based on the current situation. New technologies in sample collection, modern isolation work based on reactivity of MNPs, orthogonal strategies in structure determination, integrated chemical and genomic analyses, and the use of databases such as MIBiG [504], NP atlas [505] including the use of artificial intelligence [506] will accelerate the discovery of important marine drugs. Moreover, supply challenge of talented MNPs will be focused on the total synthesis (large-scale synthesis), semi-synthesis, heterologous expression, mariculture/aquaculture, and fermentation [507]. This comprehensive review provides an understanding why and how nature generates molecules, what is the relationship between those molecules, and their application in other scientific fields mainly in medicine by using classic and modern approaches in natural product chemistry. This review also underscores the collaborative contributions of Indonesian and international research teams in the advancement of marine natural products [13, 39, 40, 73, 74, 79, 390, 478, 504, 508–512].

References

1. Myers N, Mittermeier RA, Mittermeier CG et al (2000) Biodiversity hotspots for conservation priorities. Nature 403:853–858
2. Cabral R, Cruz-Trinidad A, Geronimo R et al (2012) Opportunities and challenges in the coral triangle. Environ Sci Technol 46:7930–7931
3. Provinces of Indonesia Available online: https://en.wikipedia.org/wiki/ProvincesofIndonesia. Accessed on 09 June 2024
4. Glaser M, Baitoningsih W, Ferse SCA et al (2010) Whose sustainability? Top-down participation and emergent rule in marine protected area management in Indonesia. Mar Policy 34:1215–1225
5. Blunt JW, Copp BR, Keyzers RA et al (2016) Marine natural products. Nat Prod Rep 33:382–431
6. Blunt JW, Carroll AR, Copp BR et al (2018) Marine natural products. Nat Prod Rep 35:8–53
7. Carroll AR, Copp BR, Davis RA et al (2019) Marine natural products. Nat Prod Rep 36:122–173
8. Carroll AR, Copp BR, Davis RA et al (2020) Marine natural products. Nat Prod Rep 37:175–223
9. Carroll AR, Copp BR, Davis RA et al (2021) Marine natural products. Nat Prod Rep 38:362–413
10. Carroll AR, Copp BR, Davis RA et al (2022) Marine natural products. Nat Prod Rep 39:1122–1171
11. Carroll AR, Copp BR, Davis RA et al (2023) Marine natural products. Nat Prod Rep 40:275–325
12. Carroll AR, Copp BR, Grkovic T et al (2024) Marine natural products. Nat Prod Rep 40:275–325
13. Hanif N, Murni A, Tanaka C et al (2019) Marine natural products from Indonesian waters. Mar Drugs 17:364
14. Goyer E, Lavaud C, Massiot G (2023) Meroterpenoids? A historical and critical review of the biogenetic determinant. Nat Prod Rep 40:1071–1077
15. Han J, Jiang L, Zhang L et al (2021) Peculiarities of meroterpenoids and their bioproduction. Appl Microbiol Biotechnol 105:3987–4003
16. Holland DC, Carroll AR (2023) Marine indole alkaloid diversity and bioactivity. What do we know and what are we missing? Nat Prod Rep 40:1595–1607
17. Daloze D, Braekman JC, Georget P et al (1977) Two novel sesquiterpenes from soft corals of the genera *Lemnalia* and *Paralemnalia* (Coelenterata, Octocorallia, Alcyonacea). Bull Soc Chim Belg 86:47–54
18. Tursch B, Colin M, Daldze D et al (1975) Chemical studies of marine invertebrates. XII. Lemnacarnol, a novel sesquiterpene from the soft coral *Lemnalia carnosa*. Bull Soc Chim Belg 84:81–82

19. Karlsson R, Losman D (1976) Structure and absolute configuration of the sesquiterpene lemnacarnol. Acta Crystallogr Sect B Struct Sci Cryst Eng Mater B32:1614–1616
20. Losman D, Karlsson R, Shieldrick GM (1977) 2-Deoxy-12-oxolemnacarnol: a case of twinning. Acta Crystallogr Sect B Struct Sci Cryst Eng Mater B33:1959–1962
21. Ayanoglu E, Gebreyesus T, Beechan CM et al (1978) Terpenoids LXXV. $\Delta^{9(12)}$-capnellene, a new sesquiterpene hydrocarbon from the soft coral *Capnella imbricata*. Tetrahedron Lett 19:1671–1674
22. Morris LA, Jaspars M, Adamson K et al (1998) The capnellenes revisited: new structures and new biological activity. Tetrahedron 54:12953–12958
23. Sheikh YM, Singy G, Kaisin M et al (1976) Terpenoids—LXXI: chemical studies of marine invertebrates—XIV. Four representatives of a novel sesquiterpene class—the capnellane skeleton Tetrahedron 32:1171–1178
24. Kaisin M, Sheikh YM, Durham LJ et al (1974) Capnellane—a new tricyclic sesquiterpene skeleton from the soft coral *Capnella imbricata*. Tetrahedron Lett 15:2239–2242
25. Sheikh YM, Djerassi C, Braekman JC et al (1977) Terpenoids—LXXII[1]: chemical studies of marine invertebrates—XXVI[2] $\Delta^{9(12)}$-capnellene-3β,8β,10α,14-tetrol, a novel polyoxygenated sesquiterpene from the alcyonarian *Capnella imbricata*. Tetrahedron 33:2115–2117
26. Wu IT, Fan YC, Lin GZ et al (2023) A new capnellene skeleton from the octocoral *Capnella imbricata*. J Mol Struct 1271:133995
27. Karlsson R (1977) The structure and absolute configuration of $\Delta^{9(12)}$-capnellene-3β,8β,10α-triol. Acta Crystallogr Sect B Struct Sci Cryst Eng Mater B33:1143–1147
28. Ohshima T, Kagechika K, Adachi M et al (1996) Asymmetric Heck reaction−carbanion capture process. Catalytic asymmetric total synthesis of (–)-$\Delta^{9(12)}$-capnellene. J Am Chem Soc 118:7108–7116
29. Wang GYS, Abrell LM, Avelar A et al (1998) New hirsutane based sesquiterpenes from salt water cultures of a marine sponge-derived fungus and the terrestrial fungus *Coriolus consors*. Tetrahedron 54:7335–7342
30. Li HJ, Lan WJ, Lam CK et al (2011) Hirsutane sesquiterpenoids from the marine-derived fungus *Chondrostereum* sp. Chem Biodiversity 8:317–324
31. Li HJ, Xie YL, Xie ZL et al (2012) Chondrosterins A-E, triquinane-type sesquiterpenoids from soft coral-associated fungus *Chondrostereum* sp. Mar Drugs 10:627–638
32. Kutateladze AG, Kuznetsov DM (2017) Triquinanes and related sesquiterpenes revisited computationally: structure corrections of hirsutanols B and D, hirsutenol E, cucumin B, antrodins C-E, chondroterpenes A and H, chondrosterins C and E, dichrocephone A, and pethybrene. J Org Chem 82:10795–10802
33. Asai R, Mitsuhashi S, Shigetomi K et al (2011) Absolute configurations of (–)-hirsutanol A and (–)-hirsutanol C produced by *Gloeostereum incarnatum*. J Antibiot 64:693–696
34. Tursch B, Braekman JC, Daloze D et al (1974) Chemical studies of marine invertebrates. VIII africanol, an unusual sesquiterpene from *Lemnalia africana* (Coelenterata, Octocorallia, Alcyonacea). Tetrahedron Lett 15:747–750
35. Karlsson R (1976) Measurement of Bijvoet differences. Structure and absolute configuration of africanol, a sesquiterpene. Acta Crystallogr Sect B Struct Sci Cryst Eng Mater B32:2609–2614
36. Beechan CM, Djerassi C, Eggert H (1978) Terpenoids-LXXIV[1]: the sesquiterpenes from the soft coral *Sinularia mayi*. Tetrahedron 34:2503–2508
37. Carreras J, Livendahl M, McGonigal PR et al (2014) Gold(I) as an artificial cyclase: short stereodivergent syntheses of (–)-epiglobulol and (–)-4β,7α- and (–)-4α,7α-aromadendrane-diols. Angew Chem Int Ed 53:4896–4899
38. Handayani D, Edrada RA, Proksch P et al (1997) New oxygenated sesquiterpenes from the Indonesian soft coral *Nephthea chabrolii*. J Nat Prod 60:716–718
39. Hanif N, Murni A, Yamauchi M et al (2015) A new trinor-guaiane sesquiterpene from an Indonesian soft coral *Anthelia* sp. Nat Prod Commun 10:1907–1910
40. Hanif N, Dinelsa FF, Yanti HD et al (2023) Stereochemical determination of NMR chemical shifts in marine terpenoids, antheliol and sangiangol B, using DFT calculations. Nat Prod Res 37:3170–3176

41. Ayanoglu E, Gebreyesus T, Beechan CM et al (1979) Terpenoids—LXXVI[1]: precapnella-diene, a possible biosynthetic precursor of the capnellane skeleton. Tetrahedron 35:1035–1039
42. Grote D, Hänel F, Dahse H-M et al (2008) Capnellenes from the Soft Coral *Dendronephthya rubeola*. Chem Biodiversity 5:1683–1693
43. Takenaka Y, Ito H, Iguchi K (2007) Enantioselective formal synthesis of (+)-precapnelladiene by chiral copper-catalyzed asymmetric [2+2]-cycloaddition reaction. Tetrahedron 63:510–513
44. Harrison B, Crews P (1997) The structure and probable biogenesis of helianane, a heterocyclic sesquiterpene, from the Indo-Pacific sponge *Haliclona fascigera*. J Org Chem 62:2646–2648
45. Green JC, Jimenez-Alonso S, Brown ER et al (2011) Total synthesis and repudiation of the helianane family. Org Lett 13:5500–5503
46. Martorano LH, Valverde AL, Ribeiro CMR et al (2020) Unraveling the helianane family: a complementary quantum mechanical study. New J Chem 44:8055–8060
47. Beechan CM, Djerassi C, Finer JS et al (1977) Terpenoids LXXIII. Sinularene, a sesquiterpene hydrocarbon based on a novel skeleton from the soft coral, *Sinularia mayi*. Tetrahedron Lett 18:2395–2398
48. Torii M, Kato H, Hitora Y et al (2017) Lamellodysidines A and B, sesquiterpenes isolated from the marine sponge *Lamellodysidea herbacea*. J Nat Prod 80:2536–2541
49. Kamo S, Kurosawa H, Matsuzawa A et al (2022) Total synthesis of (–)-lamellodysidine A via an intramolecular Diels-Alder reaction. Org Lett 24:921–923
50. Kapojos MM, Abdjul DB, Yamazaki H et al (2018) Protein tyrosine phosphatase 1B inhibitory polybromobiphenyl ethers and monocyclofarnesol-type sesquiterpenes from the Indonesian marine sponge *Lamellodysidea* cf. *herbacea*. Phytochem Lett 24:10–14
51. Hérin M, Tursch B (1976) Chemical studies of marine invertebrates. XXV. Flexibilene, an unprecedented fifteen-membered ring diterpene hydrocarbon from the soft coral *Sinularia flexibilis* (Coelenterata, Octocorallia). Bull Soc Chim Belg 85:801–803
52. McMurry J, Matz JR, Kees KL et al (1982) Synthesis of flexibilene, a naturally occurring 15-membered-ring diterpene. Tetrahedron Lett 23:1777–1780
53. Shaker KH, Müller M, Ghani MA et al (2010) Terpenes from the soft corals *Litophyton arboreum* and *Sarcophyton ehrenbergi*. Chem Biodiversity 7:2007–2015
54. Januar HI, Chasanah E, Motti CA et al (2010) Cytotoxic cembranes from Indonesian specimens of the soft coral *Nephthea* sp. Mar Drugs 8:2142–2152
55. Hérin M, Tursch B (1976) Chemical studies of marine invertebrates. XXIV. Minor cembrane diterpenes from the soft coral *Sinularia flexibilis* (Coelenterata, Octocorallia). Bull Soc Chim Belg 85:707–719
56. Tursch B, Braekman JC, Daloze D (1975) Chemical studies of marine invertebrates-XIII. 2-hydroxynepthenol, a novel cembrane diterpene from the soft coral *Lithophyton viridis* (Coelenterata, Octocorallia, Alconacea). Bull Soc Chim Belg 84:767–774
57. Suzuki M, Shimada A, Kato T (1978) Cyclization of polyenes xxx[1] synthesis of 2-hydroxy-and 1,2-dehydro-7,8-oxido-nephthenols and cembrenol. Chem Lett 7:759–762
58. Tursch B, Braekman JC, Daloze D et al (1974) Chemical studies of marine invertebrates. X. Lobophytholide, a new cembranolide diterpene from the soft coral *Lobophytum cristagalli* (Coelenterata, Octocorallia, Alyconacea). Tetrahedron Lett 15:3769–3772
59. Karlsson R (1977) Lobophytolide, a cembrane diterpene. Acta Crystallogr Sect B Struct Sci Cryst Eng Mater B33:2032–2034
60. Januar HI, Zamani NP, Soedharma D et al (2017) New cytotoxic cembranoid from Indonesian soft coral *Sarcophyton* sp. Pharmacogn Res 9:65–68
61. Fattorusso E, Romano A, Taglialatela-Scafati O et al (2009) Oxygenated cembranoids of the decaryiol type from the Indonesian soft coral *Lobophytum* sp. Tetrahedron 65:2898–2904
62. Tursch B, Braekman JC, Daloze D et al (1975) Chemical studies of marine invertebrates–XI: sinulariolide, a new cembranolide diterpene from the soft coral *Sinularia flexibilis* (Coelenterata, Octocorallia, Alcyonacea). Tetrahedron 31:129–133
63. Karlsson R (1977) The structure and absolute configuration of sinulariolide, a cembranolide diterpene. Acta Crystallogr Sect B Struct Sci Cryst Eng Mater B33:2027–2031

64. Tursch B (1976) Some recent developments in the chemistry of Alcyonaceans. Pure Appl Chem 48:1–6
65. Lin YS, Chen CH, Liaw CC et al (2009) Cembrane diterpenoids from the Taiwanese soft coral *Sinularia flexibilis*. Tetrahedron 65:9157–9164
66. Hu LC, Yen WH, Su JH et al (2013) Cembrane derivatives from the soft corals, *Sinularia gaweli* and *Sinularia flexibilis*. Mar Drugs 11:2154–2167
67. Fattorusso E, Luciano P, Putra MY et al (2011) Chloroscabrolides, chlorinated norcembranoids from the Indonesian soft coral *Sinularia* sp. Tetrahedron 67:7983–7988
68. Anta C, González N, Santafé G et al (2002) New xenia diterpenoids from the Indonesian soft coral *Xenia* sp. J Nat Prod 65:766–768
69. Chen YH, Tai CY, Su YD et al (2011) Discovery of new eunicellins from an Indonesian octocoral *Cladiella* sp. Mar Drugs 9:934–943
70. Chen YH, Hwang TL, Su YD et al (2012) New 6-hydroxyeunicellins from a soft coral *Cladiella* sp. Chem Pharm Bull 60:160–163
71. Tai CY, Chen YH, Hwang TL et al (2011) Cladielloides C and D: novel eunicellin-based diterpenoids from an Indonesian octocoral *Cladiella* sp. Bull Chem Soc Jpn 84:531–536
72. Xu B, Tantillo DJ, Rudolf JD (2021) Mechanistic insights into the formation of the 6,10-bicyclic eunicellane skeleton by the bacterial diterpene synthase Bnd4. Angew Chem Int Ed 60:23159–23163
73. Hanif N, Murni A, Tanaka J (2020) Sangiangols A and B, two new dolabellanes from an Indonesian marine soft coral, *Anthelia* sp. Molecules 25:3803
74. Murni A, Hanif N, Tanaka J (2013) A new cytotoxic dolabellane from the Indonesian soft coral *Anthelia* sp. Indones J Chem 13:216–220
75. El-Desoky AH, Kato H, Tsukamoto S (2017) Ceylonins G-I: spongian diterpenes from the marine sponge *Spongia ceylonensis*. J Nat Med 71:765–769
76. Parrish SM, Yoshida WY, Kondratyuk TP et al (2014) Spongiapyridine and related spongians isolated from an Indonesian *Spongia* sp. J Nat Prod 77:1644–1649
77. Li SW, Mudianta W, Cuadrado C et al (2021) Litosetoenins A−E, diterpenoids from the soft coral *Litophyton setoensis*, backbone rearranged through divergent cyclization achieved by epoxide reactivity inversion. J Org Chem 86:1171–11781
78. Kato H, Nehira T, Matsuo K et al (2015) Niphateolide A: isolation from the marine sponge *Niphates olemda* and determination of its absolute configuration by an ECD analysis. Tetrahedron 71:6956–6960
79. Hanif N, Dinelsa FF, Murni A et al (2024) Lamellodysideol, a new flexible marine norditerpenoid from the Indonesian marine sponge *Lamellodysidea herbacea*. Nat Prod Res 1−6, in press
80. Hehre W, Klunzinger P, Deppmeier B et al (2019) Efficient protocol for accurately calculating ^{13}C chemical shifts of conformationally flexible natural products: scope, assessment, and limitations. J Nat Prod 82:2299–2306
81. Craig KS, Williams DE, Hollander I et al (2002) Novel sesterterpenoid and norsesterterpenoid RCE-protease inhibitors isolated from the marine sponge *Hippospongia* sp. Tetrahedron Lett 43:4801–4804
82. Elkhayat E, Edrada RA, Ebel R et al (2004) New luffariellolide derivatives from the Indonesian sponge *Acanthodendrilla* sp. J Nat Prod 67:1809–1817
83. Afifi AH, Kagiyama I, El-Desoky AH et al (2017) Sulawesins A-C, furanosesterterpene tetronic acids that inhibit USP7, from a *Psammocinia* sp. marine sponge. J Nat Prod 80:2045–2050
84. Ibrahim SRM, Ebel R, Wray V et al (2008) Diacarperoxides, norterpene cyclic peroxides from the sponge *Diacarnus megaspinorhabdosa*. J Nat Prod 71:1358–1364
85. Salmoun M, Braekman JC, Dewelle J et al (2007) New terpenoids from two Indonesian marine sponges. Nat Prod Res 21:149–155
86. Williams DE, Tahir A, Andersen RJ (1999) A new acyclic diketotriterpenoid isolated from the Indonesian marine sponge *Hyrtios erectus*. J Nat Prod 62:653–654

87. Enders D, Schüßeler T (2002) First highly efficient asymmetric synthesis of the *Hyrtios erectus* diketotriterpenoid. Synthesis 15:2280–2288
88. Fouad M, Edrada RA, Ebel R et al (2006) (2006) Cytotoxic isomalabaricane triterpenes from the marine sponge *Rhabdastrella globostellata*. J Nat Prod 69:211–218
89. Wang W, Lee JS, Nakazawa T et al (2009) (25*S*)-Cholesten-26-oic acid derivatives from an Indonesian soft coral *Minabea* sp. Steroids 74:758–760
90. Aoki S, Naka Y, Itoh T et al (2002) Lembehsterols A and B, novel sulfated sterols inhibiting thymidine phosphorylase, from the marine sponge *Petrosia strongylata*. Chem Pharm Bull 50:827–830
91. Ushiyama S, Umaoka H, Kato H et al (2012) Manadosterols A and B, sulfonated sterol dimers inhibiting the Ubc13–Uev1A interaction, isolated from the marine sponge *Lissodendryx fibrosa*. J Nat Prod 75:1495–1499
92. Whitson EL, Bugni TS, Chockalingam PS et al (2009) Fibrosterol sulfates from the Philippine sponge *Lissodendoryx (Acanthodoryx) fibrosa*: sterol dimers that inhibit PKC ζ. J Org Chem 74:5902–5908
93. El-Desoky AH, Eguchi K, Kagiyama I et al (2023) Aaptocarbamates A−G, chlorinated terpene carbamates with antiosteoclastogenic activities from the marine sponge *Aaptos* sp. Phytochemistry 216:113872
94. Yamazaki H, Nakayama W, Takahashi O et al (2015) Verruculides A and B, two new protein tyrosine phosphatase 1B inhibitors from an Indonesian ascidian-derived *Penicillium verruculosum*. Bioorg Med Chem Lett 25:3087–3090
95. Bogdanov A, Papu A, Kehraus S et al (2020) Metabolome of the *Phyllidiella pustulosa* species complex (Nudibranchia, Heterobranchia, Gastropoda) reveals rare dichloroimidic sesquiterpene derivatives from a phylogenetically distinct and undescribed clade. J Nat Prod 83:2785–2796
96. Dai J, Liu Y, Zhou Y-D et al (2007) Cytotoxic metabolites from an Indonesian sponge *Lendenfeldia* sp. J Nat Prod 70:1824–1826
97. Issa HH, Tanaka J, Rachmat R et al (2003) Floresolides, new metacyclophane hydroquinone lactones from an ascidian, *Aplidium* sp. Tetrahedron Lett 44:1243–1245
98. Nicolau KC, Xu H (2006) Total synthesis of floresolide B and $\Delta^{6,7}$-*Z*-floresolide B. Chem Commun 6:600–602
99. Chen Y, Harmata M (2011) Benzothiazines in organic synthesis. Tetrahedron Lett 52:4069–4071
100. Kotoku N, Matsumoto H, Hisa T et al (2021) Fasciospyrinadinone and fasciospyrinadinol, novel 3-alkylpyridine sesquiterpenoid from an Indonesian marine sponge, as selective growth inhibitors of the cancer cells under nutrient starvation. Heterocycles 103:827–838
101. Capon RJ (2020) Extracting value: mechanistic insight into the formation of natural product artifacts-case studies in marine natural products. Nat Prod Rep 37:55–79
102. Liu M, Li S (2024) Nitrile biosynthesis in nature: how and why. Nat Prod Rep 41:649–671
103. Hitora Y, Ogura K, El-Desoky AHH et al (2021) Halichonic acid B, a rearranged nitrogenous bisabolene-type sesquiterpene from a marine sponge *Axinyssa* sp. Chem Pharm Bull 69:802–805
104. Raiju K, Hitora Y, Kato H et al (2019) Halichonic acid, a new rearranged bisabolene-type sesquiterpene from a marine sponge *Halichondria* sp. Tetrahedron Lett 60:1079–1081
105. Reber KP, Niner EL (2022) Synthesis of (–)-halichonic acid and (–)-halichonic acid B. Beilstein J Org Chem 18:1629–1635
106. Ebel R, Rusman Y, Brauers G et al (2006) Novel oxygenated meroterpenoids from the sponge-derived fungus *Penicillium citreonigrum*. Planta Med 72:961–1089
107. Xu J, Takasaki A, Kobayashi H et al (2006) Four new macrocyclic trichothecenes from two strains of marine-derived fungi of the genus *Myrothecium*.. J Antibiot 59:451–455
108. Williams DE, Patrick BO, Tahir A et al (2004) Boneratamides A-C, new sesquiterpenoids isolated from the marine sponge *Axinyssa aplysinoides*. J Nat Prod 67:1752–1754
109. Ooka K, Nakanishi K, Udagawa Y et al (2022) Bioinspired total synthesis of boneratamides A-C. Org Biomol Chem 20:8236–8242

110. Kapojos M, Mangindaan REP, Nakazawa T et al (2008) Three new nardosinane type sesquiterpenes from an Indonesian soft coral *Nephthea* sp. Chem Pharm Bull 56:332–334
111. Piña IC, Sanders ML, Crews P (2003) Puupehenone congeners from an Indo-Pacific *Hyrtios* sponge. J Nat Prod 66:2–6
112. Wahab HA, Pham NB, Muhammad TST et al (2017) Merosesquiterpene congeners from the Australian sponge *Hyrtios digitatus* as potential drug leads for atherosclerosis disease. Mar Drugs 15:6
113. Li F, Renata H (2021) A chiral-pool-based strategy to access *trans-syn*-fused drimane meroterpenoids: chemoenzymatic total syntheses of polysin, *N*-acetyl-polyveoline and the chrodrimanins. J Am Chem Soc 143:18280–18286
114. Ebada SS, de Voogd N, Kalscheuer R et al (2017) Cytotoxic drimane meroterpenoids from the Indonesian marine sponge *Dactylospongia elegans*. Phytochem Lett 22:154–158
115. Aoki S, Kong D, Matsui K et al (2004) Sesquiterpene aminoquinones, from a marine sponge, induce erythroid differentiation in human chronic myelogenous leukemia, K562 cells. Chem Pharm Bull 52:935–937
116. Daletos G, de Voogd NJ, Muller WEG et al (2014) Cytotoxic and protein kinase inhibiting nakijiquinones and nakijiquinols from the sponge *Dactylospongia metachromia*. J Nat Prod 77:218–226
117. Balansa W, Mettal U, Wuisan ZG et al (2019) A new sesquiterpenoid aminoquinone from an Indonesian marine sponge. Mar Drugs 17:158
118. Suna H, Arai M, Tsubotani Y et al (2009) Dysideamine, a new sesquiterpene aminoquinone, protects hippocampal neuronal cells against iodoacetic acid-induced cell death. Bioorg Med Chem 17:3968–3972
119. Ohte S, Yamazaki H, Takahashi et al (2021) Inhibitory effects of sesquiterpene lactones from the Indonesian marine sponge *Lamellodysidea* cf. *herbacea* on bone morphogenetic protein-induced osteoblastic differentiation. Bioorg Med Chem Lett 35:127783
120. Yasman ERA, Wray V et al (2003) New 9-thiocyanatopupukeanane sesquiterpenes from the nudibranch *Phyllidia varicosa* and its sponge-prey *Axinyssa aculeata*. J Nat Prod 66:1512–1514
121. Putra MY, Ianaro A, Panza E et al (2012) Sinulasulfoxide and sinulasulfone, sulfur-containing alkaloids from the Indonesian soft coral *Sinularia* sp. Tetrahedron Lett 53:3937–3939
122. Trianto A, Hermawan I, De Voogd NJ et al (2011) Halioxepine, a new meroditerpene from an Indonesian Sponge *Haliclona* sp. Chem Pharm Bull 59:1311–1313
123. Tarazona G, Benedit G, Fernández R et al (2018) Can stereocluster separated by two methylene groups be related by DFT studies? The case of cytotoxic meroditerpenes halioxepines. J Nat Prod 81:343–348
124. Poock C, Kalesse M (2021) Total synthesis and structure revision of halioxepine. Chem Eur J 27:1615–1619
125. Williams DE, Telliez JB, Liu J et al (2004) Meroterpenoid MAPKAP (MK2) inhibitors isolated from the Indonesian marine sponge *Acanthodendrilla* sp. J Nat Prod 67:2127–2129
126. Hertiani T, Edrada-Ebel R, Ortlepp S et al (2010) From anti-fouling to biofilm inhibition: new cytotoxic secondary metabolites from two Indonesian *Agelas* sponges. Bioorg Med Chem 18:1297–1311
127. Basabe P, Martín M, Bodero O et al (2010) Synthesis of (+)-makassaric acid, a protein kinase MK2 inhibitor. Tetrahedron 66:6008–6012
128. Fattorusso E, Romano A, Taglialatela-Scafati O et al (2006) Coelodiol and coeloic acid, *ent*-isocopalane diterpenes from the Indonesian sponge *Coelocarteria* cfr. *singaporensis*. Tetrahedron Lett 47:2197–2200
129. El-Desoky AH, Kato H, Angkouw ED et al (2016) Ceylonamides A−F, nitrogenous spongian diterpenes that inhibit RANKL-induced osteoclastogenesis, from the marine sponge *Spongia ceylonensis*. J Nat Prod 79:1922–1928
130. El-Desoky AH, Kato H, Kagiyama I et al (2017) Ceylonins A−F, spongian diterpene derivatives that inhibit RANKL-induced formation of multinuclear osteoclasts, from the marine sponge *Spongia ceylonensis*. J Nat Prod 80:90–95

131. Jomori T, Setiawan A, Sasaoka M et al (2019) Cytotoxicity of new diterpene alkaloids, ceylonamides G-I, isolated from Indonesian marine sponge of *Spongia* sp. Nat Prod Commun 14:1–7
132. Tursch B, Braekman JC, Daloze D et al (1978) Chemical studies of marine invertebrates. XXXI. Crassolide, a highly oxygenated diterpene from the soft coral *Lobophytum crassum* (Coelenterata, Octocorallia, Alcynoacea). Bull Soc Chim Belg 87:75–81
133. Grote D, Dahse HM, Seifert K (2008) Furanocembranoids from the soft corals *Sinularia asterolobata* and *Litophyton arboretum*. Chem Biodiversity 5:2449–2456
134. Hsieh PW, Chang FR, McPhail AT et al (2003) New cembranolide analogues from the Formosan soft coral *Sinularia flexibilis* and their cytotoxicity. Nat Prod Res 17:409–418
135. Kapojos MM, Lee J-S, Oda T et al (2010) Two unprecedented cembrene-type terpenes from an Indonesian soft coral *Sarcophyton* sp. Tetrahedron 66:641–645
136. Fattorusso E, Romano A, Taglialatela-Scafati O et al (2008) Xenimanadins A-D, a family of xenicane diterpenoids from the Indonesian soft coral *Xenia* sp. Tetrahedron 64:3141–3146
137. Fu X, Schmitz FJ, Williams GC (1999) Malayenolides A–D, novel diterpenes from the Indonesian sea pen *Veretillum malayense*. J Nat Prod 62:584–586
138. Aoki S, Okano M, Matsui K et al (2001) Brianthein A, a novel briarane-type diterpene reversing multidrug resistance in human carcinoma cell line, from the gorgonian *Briareum excavatum*. Tetrahedron 57:8951–8957
139. Rodríguez J, Nieto RM, Jiménez C (1998) New briarane stecholide diterpenes from the Indonesian Gorgonian *Briareum* sp. J Nat Prod 61:313–317
140. Tanaka C, Yamamoto Y, Otsuka M et al (2004) Briarane diterpenes from two species of octocorals, *Ellisella* sp. and *Pteroeides* sp. J Nat Prod 67:1368–1373
141. García M, Rodríguez J, Jiménez C (1999) Absolute structures of new briarane diterpenoids from *Junceella fragilis*. J Nat Prod 62:257–260
142. Chen YH, Tai CY, Hwang TL et al (2010) Cladielloides A and B: new eunicellin-type diterpenoids from an Indonesian octocoral *Cladiella* sp. Mar Drugs 8:2936–2945
143. Ibrahim SRM (2012) Diacarperoxide S, new norterpene cyclic peroxide from the sponge *Diacarnus megaspinorhabdosa*. Nat Prod Commun 7:9–12
144. Ibrahim SRM, Al Haidari RA, Mohamed GA (2014) Megaspinoxide A: new norterpene cyclic peroxide from the sponge *Diacarnus megaspinorhabdosa*. Nat Prod J 4:38–42
145. Mudianta IW, White AM, Garson MJ (2015) Oxygenated terpenes from Indo-Pacific nudibranchs: scalarane sesterterpenes from *Glossodoris hikuerensis* and 12-acetoxy dendrillolide A from *Goniobranchus albonares*. Nat Prod Commun 10:865–868
146. Hertzer C, Kehraus S, Böhringer N et al (2020) Antibacterial scalarane from *Doriprismatica stellata* nudibranchs (Gastropoda, Nudibranchia), egg ribbons, and their dietary sponge *Spongia* cf. *agaricina* (Demospongiae, Dictyoceratida). Beilstein J Org Chem 16:1596–1605
147. Roy MC, Tanaka J, de Voogd N et al (2022) New scalarane class sesterterpenes from an Indonesian sponge, *Phyllospongia* sp. J Nat Prod 65:1838–1842
148. Jiménez JI, Yoshida WY, Scheuer PJ et al (2000) Scalarane-based sesterterpenes from an Indonesian sponge *Strepsichordaia aliena*. J Nat Prod 63:1388–1392
149. Jiménez JI, Yoshida WY, Scheuer PJ et al (2000) Honulactones: new bishomoscalarane sesterterpenes from the Indonesian sponge *Strepsichordaia aliena*. J Org Chem 65:6837–6840
150. Zhang H, Crews P, Tenney K et al (2017) Cytotoxic phyllactone analogs from the Marine Sponge *Phyllospongia papyrecea*. Med Chem 13:295–300
151. Williams DE, Hollander I, Feldberg L et al (2009) Scalarane-based sesterterpenoid RCE-protease inhibitors isolated from the Indonesian marine sponge *Carteriospongia foliascens*. J Nat Prod 72:1106–1109
152. Abdjul DB, Yamazaki H, Takahashi O et al (2015) Two new protein tyrosine phosphatase 1B inhibitors, hyattellactones A and B, from the Indonesian marine sponge *Hyattella* sp. Bioorg Med Chem Lett 25:904–907
153. Crews P, Harrison B (2000) New triterpene-ketides (Merotriterpenes), haliclotriols A and B, from an Indo-Pacific *Haliclona* Sponge. Tetrahedron 56:9039–9046

154. Aoki S, Sanagawa M, Watanabe Y et al (2007) Novel isomarabarican triterpenes, exhibiting selective anti-proliferative activity against vascular endothelial cells, from marine sponge *Rhabdastrella globostellata*. Bioorg Med Chem 15:4818–4828
155. Costantino V, Sala GD, Mangoni A et al (2010) Blurring the boundary between bio- and geohopanoids: plakohopanoid, a C_{32} biohopanoid ester from *Plakortis* cf. *lita*. Eur J Org Chem 27:5171–5176
156. Guella G, Dini F, Pietra F (1999) Metabolites with a novel C_{30} Backbone from marine ciliates. Angew Chem Int Ed 38:1134–1136
157. Nicolaou KC, Zhang H, Ortiz A (2008) The true structures of the vannusals, part 1: initial forays into suspected structures and intelligence gathering. Angew Chem Int Ed 48:5642–5647
158. Nicolaou KC, Ortiz A, Zhang H et al (2010) Total synthesis and structural revision of vannusals A and B: synthesis of the originally assigned structure of vannusal B. J Am Chem Soc 132:7138–7152
159. Nicolaou KC, Ortiz A, Zhang H et al (2010) Total synthesis and structural revision of vannusals A and B: synthesis of the true structures of vannusals A and B. J Am Chem Soc 132:7153–7176
160. Saielli G, Nicolaou KC, Ortiz A et al (2011) Addressing the stereochemistry of complex organic molecules by density functional theory-NMR: vannusal B in retrospective. J Am Chem Soc 133:6072–6077
161. Nicolaou KC, Zhang H, Ortiz A et al (2008) Total synthesis of the originally assigned structure of vannusal B. Angew Chem Int Ed 47:8605–8610
162. Nicolaou KC, Ortiz A, Zhang H (2009) The true structures of the vannusals, part 2: total synthesis and revised structure of vannusal B. Angew Chem Int Ed 48:5648–5652
163. Rothberg I, Tursch BM, Djerassi C (1973) Terpenoids. LXVIII. 23ξ-acetoxy-17-deoxy-7,8-dihydroholothurinogenin, a new triterpenoid sapogenin from a sea cucumber. J Org Chem 38:209–214
164. Dai HF, Edrada RA, Ebel R et al (2005) Norlanostane triterpenoidal saponins from the marine sponge *Melophlus sarassinorum*. J Nat Prod 68:1231–1237
165. Maarisit W, Yamazaki H, Kanno SI et al (2017) A tetramic acid derivative with protein tyrosine phosphatase 1B inhibitory activity and a new nortriterpene glycoside from the Indonesian marine sponge *Petrosia* sp. Bioorg Med Chem Lett 27:999–1002
166. Mak JYW, Pouwer RH, Williams CM (2014) Natural products with anti-Bredt and bridgehead double bonds. Angew Chem Int Ed 53:13664–13688
167. Sadahiro Y, Hitora Y, Fukumoto A et al (2020) Melophluosides A and B, new triterpene galactosides from the marine sponge *Melophlus sarasinorum*. Tetrahedron Lett 61:151852
168. Ksebati MB, Schmitz FJ, Gunasekera SP (1988) Pouosides A−E, novel triterpene galactosides from a marine sponge, *Asteropus* sp. J Org Chem 53:3917–3921
169. Engelbrecht JP, Tursch B, Djerassi C (1972) A new sterol from an alcyonarian. Steroids 20:121–126
170. Bortolotto M, Braekman JC, Daloze D et al (1976) Chemical studies of marine invertebrates. XVIII. Four novel polyhydroxylated steroids from *Sinularia dissecta* (Coelenterata, Octocorallia, Alyconacea). Bull Soc Chim Belg 85:27–34
171. Putra MY, Bavestrello G, Cerrano C et al (2012) Polyhydroxylated sterols from the Indonesian soft coral *Sinularia* sp. and their effect on farnesoid X-activated receptor. Steroids 77:433–440
172. González N, Barral MA, Rodríguez J et al (2001) New cytotoxic steroids from the gorgonian *Isis hippuris*. Structure-activity studies. Tetrahedron 57:3487–3497
173. Chianese G, Fattorusso E, Taglialatela-Scafati O et al (2011) Desulfohaplosamate, a new phosphate-containing steroid from *Dasychalina* sp., is a selective cannabinoid CB 2 receptor ligand. Steroids 76:998–1002
174. Bortolotto M, Braekman JC, Daloze D et al (1976) Chemical studies of marine invertebrates. XXIII. A novel polyhydroxylated sterol from the soft coral *Litophyton viridis* (coelenterata, octocorallia, alcyonacea). Steroids 28:461–466
175. Bortolotto M, Braekman JC, Daloze D et al (1977) Chemical studies of marine invertebrates. XXIX. 4α-methyl-3β,8β-dihydroxy-5α-ergost-24(28)-en–23-one, a novel polyoxygenated sterol from the soft coral *Litophyton viridis*. (Coelenterata, octocorallia, alcyonacea). Steroids 30:159–164

176. Angawi RF, Calcinai B, Cerrano C et al (2009) Dehydroconicasterol and aurantoic acid, a chlorinated polyene derivative, from the Indonesian sponge *Theonella swinhoei*. J Nat Prod 72:2195–2198
177. Losman D, Karlsson R (1978) 25-Methylenecholest-5-ene-3β,7β,19-triol. A case of pseudotranslation. Calculation of structure invariants from partial structure information. Acta Crystallogr Sect B Struct Sci Cryst Eng Mater B34:2586–2589
178. Dai J, Sorribas A, Yoshida WY et al (2010) Topsentinols, 24-isopropyl steroids from the marine sponge *Topsentia* sp. J Nat Prod 73:1597–1600
179. Jurek J, Scheuer PJ, Borges K (1994) Two steroidal alkaloids from a sponge *Corticium* sp. J Nat Prod 57:1004–1007
180. Murtihapsari M, Salam S, Kurnia D et al (2019) A new antiplasmodial sterol from Indonesian marine sponge, *Xestospongia* sp. Nat Prod Res 35:937–944
181. Sahidin I, Sabandar CW, Wahyuni et al (2018) A-nor steroids from the marine sponge, *Clathria* species. Malays J Anal Sci 22:375–382
182. Anta C, Gonzáles N, Rodríguez J et al (2002) A new secosterol from the Indonesian Octocoral *Pachyclavularia violacea*. J Nat Prod 65:1357–1359
183. Morris LA, Christie EM, Jaspars M et al (1998) A bioactive secosterol with an unusual A and B-ring oxygenation pattern isolated from an Indonesian soft coral *Lobophytum* sp. J Nat Prod 61:538–541
184. Aoki S, Watanabe Y, Sanagawa M et al (2006) Cortistatins A, B, C, and D, anti-angiogenic steroidal alkaloids, from the marine sponge, *Corticium simplex*. J Am Chem Soc 128:3148–3149
185. Watanabe Y, Aoki S, Tanabe D et al (2007) Cortistatins E, F, G, and H, four novel steroidal alkaloids from marine sponge *Corticium simplex*. Tetrahedron 63:4074–4079
186. Aoki S, Watanabe Y, Tanabe D et al (2007) Cortistatins J, K, L, novel *abeo*-9(10–19)-androstane-type steroidal alkaloids with isoquinoline unit, from marine sponge *Corticium simplex*. Tetrahedron Lett 48:4485–4488
187. Nicolaou KC, Peng X-S, Sun Y-P et al (2009) Total synthesis and biological evaluation of cortistatins A and J and analogues thereof. J Am Chem Soc 131:10587–10597
188. Shaker KH, Göhl M, Müller T et al (2015) Indole alkaloids from the sea anemone *Heteractis aurora* and homarine from *Octopus cyanea*. Chem Biodiversity 12:1746–1755
189. Yamanokuchi R, Imada K, Miyazaki M et al (2012) Hyrtioreticulins A−E, indole alkaloids inhibiting the ubiquitin-activating enzyme, from the marine sponge *Hyrtios reticulatus*. Bioorg Med Chem 20:4437–4442
190. Imada K, Sakai E, Kato H et al (2013) Reticulatins A and B and hyrtioreticulin F from the marine sponge *Hyrtios reticulatus*. Tetrahedron 69:7051–7055
191. Sakai E, Kato H, Rotinsulu H et al (2014) Variabines A and B: new β-carboline alkaloids from the marine sponge *Luffariella variabilis*. J Nat Med 68:215–219
192. Doro-Goldsmith E, Song Q, Li X-L et al (2024) Absolute configuration of 12*S*-deoxynortryptoquivaline from ascidian-derived fungus *Aspergillus clavatus* determined by anisotropic NMR and chiroptical spectroscopy. J Nat Prod 87:381–387
193. Abe T, Yamada K (2017) Concise syntheses of hyrtioreticulins C and D via a C-4 Pictet-Spengler reaction: revised signs of specific rotations. J Nat Prod 80:241–245
194. Lorig-Roach N, Hamkins-Indik F, Johnson TA et al (2018) The potential of achiral sponge-derived and synthetic bromoindoles as selective cytotoxins against PANC-1 tumor cells. Tetrahedron 74:217–223
195. Carney JR, Scheuer PJ, Kelly-Borges M (1993) Makaluvamine G a cytotoxic pigment from an Indonesian Sponge *Histodermella* sp. Tetrahedron 49:8483–8486
196. Salmoun M, Devijver C, Daloze D et al (2002) 5-Hydroxytryptamine-derived alkaloids from two marine sponges of the genus *Hyrtios*. J Nat Prod 65:1173–1176
197. Rao KV, Santarsiero BD, Mesecar AD et al (2003) New manzamine alkaloids with activity against infectious and tropical parasitic diseases from an Indonesian sponge. J Nat Prod 66:823–828

198. Ibrahim S, Mohamed G, Haidari RA et al (2018) Ingenine F: a new cytotoxic tetrahydrocarboline alkaloid from the Indonesian marine sponge *Acanthostrongylophora ingens*. Phcog Mag 14:231–234
199. Ibrahim SRM, Ebel RA, Ebel R et al (2008) Acanthomine A, a new pyrimidine-β-carboline from the sponge *Acanthostrongylophora ingens*. Nat Prod Commun 3:175–178
200. Ibrahim SRM, Homaned GA, Zayed MF et al (2015) Ingenines A and B, two new alkaloids from the Indonesian sponge *Acanthostrongylophora ingens*. Drug Res 65:361–365
201. Ibrahim SRM, Mohamed GA (2016) Ingenines C and D, new cytotoxic pyrimidine−β-carboline alkaloids from the Indonesian sponge *Acanthostrongylophora ingens*. Phytochem Lett 18:168–171
202. Ibrahim SRM, Mohamed GA (2017) Ingenine E, a new cytotoxic β-carboline alkaloid from the Indonesian sponge *Acanthostrongylophora ingens*. J Asian Nat Prod Res 19:504–509
203. Van Wagoner RM, Jompa J, Tahir A et al (1999) Trypargine alkaloids from a previously undescribed *Eudistoma* sp. ascidian. J Nat Prod 62:794–797
204. Zhang PY, Wan SB, Ren SM et al (2010) First total synthesis of marine alkaloid hyrtiosulawesine. Chin Chem Lett 21:1307–1309
205. Szabó T, Hazai V, Volk B et al (2019) First total synthesis of the β-carboline alkaloids trigonostemine A, trigonostemine B and a new synthesis of pityriacitrin and hyrtiosulawesine. Tetrahedron Lett 60:1471–1475
206. Segraves NL, Robinson SJ, Garcia D et al (2004) Comparison of fascaplysin and related alkaloids: a study of structures, cytotoxicities, and sources. J Nat Prod 67:783–792
207. Segraves NL, Lopez S, Johnson TA et al (2003) Structures and cytotoxicities of fascaplysin and related alkaloids from two marine phyla—*Fascaplysinopsis sponges* and *Didemnum tunicates*. Tetrahedron Lett 44:3471–3475
208. Zhidkov ME, Smirnova PA, Tryapkin OA et al (2019) Total syntheses and preliminary biological evaluation of brominated fascaplysin and reticulatine alkaloids and their analogues. Mar Drugs 17:496
209. Zhidkov ME, Baranova OV, Balaneva NN et al (2007) The first syntheses of 3-bromofascaplysin, 10-bromofascaplysin and 3,10-dibromofascaplysin—marine alkaloids from *Fascaplysinopsis reticulata* and *Didemnum* sp. by application of a simple and effective approach to the pyrido[1,2-a:3,4-b′]diindole system. Tetrahedron Lett 48:7998–8000
210. Zhidkov ME, Sidorova MA, Lyakhova IA (2018) One-step transformation of the marine alkaloid fascaplysin into homofascaplysins B and B-1. The first syntheses of 3-bromohomofascaplysin B and 3–bromohomofascaplysin B-1. Tetrahedron Lett 59:1417–1420
211. Chepyshef SV, Lujan-Montelongo A, Chao A et al (2017) Alkenyl isocyanide conjugate additions: a rapid route to γ-carbolines. Angew Chem Int Ed 56:4310–4313
212. Copp BR, Jompa J, Tahir A et al (1988) Styelsamines A−D: new tetracyclic pyridoacridine alkaloids from the Indonesian ascidian *Eusynstyela latericius*. J Org Chem 63:8024–8026
213. Skyler D, Heathcock CH (2001) A simple biomimetic synthesis of styelsamine B. Org Lett 3:4323–4324
214. Nakahara S, Kubo A (2003) Total synthesis of styelsamine C, a cytotoxic fused tetracyclic aromatic alkaloid. Heterocycles 60:2017–2018
215. Skyler D, Heathcock CH (2002) The pyridoacridine family tree: a useful scheme for designing synthesis and predicting undiscovered natural products. J Nat Prod 65:1573–1581
216. Aoki S, Ho W, Matsui K et al (2003) Pyridoacridine alkaloids inducing neuronal differentiation in a neuroblastoma cell line, from marine sponge *Biemna fortis*. Bioorg Med Chem 11:1969–1973
217. Moran DAP, Takada K, Ise Y et al (2015) Two cell differentiation inducing pyridoacridines from a marine sponge *Biemna* sp. and their chemical conversions. Tetrahedron 71:5013–5018
218. Barnes EC, Said NABM, Williams ED et al (2010) Ecionines A and B, two new cytotoxic pyridoacridine alkaloids from the Australian marine sponge, *Ecionemia geodides*. Tetrahedron 66:283–287
219. Ibrahim SRM, Mohamed GA, Elkhayat ES et al (2013) Sagitol C, a new cytotoxic pyridoacridine alkaloid from the sponge *Oceanapia* sp. Bull Fac Pharm 51:229–232

220. Thale Z, Johnson T, Tenney K et al (2002) Structures and cytotoxic properties of sponge-derived bisannulated acridines. J Org Chem 67:9384–9391
221. Smith CJ, Venables DA, Hopmann C et al (1997) Plakinidine D, a new pyrroloacridine alkaloid from two ascidians of the genus *Didemnum*. J Nat Prod 60:1048–1050
222. Kudo Y, Kato H, Rotinsulu H et al (2014) Aaptoline A, a new quinoline alkaloid from the marine sponge *Aaptos suberitoides*. Heterocycles 88:591–594
223. Kim S, Yang W, Cha DS et al (2021) Synthesis of proposed structure of aaptoline A, a marine sponge-derived 7,8-dihydroxyquinoline, and its neuroprotective properties in *C. elegans*. Molecules 26:5964
224. Mokhlesi A, Hartman R, Achten E et al (2016) Lissodendrins A and B: 2-aminoimidazole alkaloids from the marine sponge *Lissodendroryx (Acanthodoryx) fibrosa*. Eur J Org Chem 639–643
225. Wang W, Oda T, Fujita A et al (2007) Three new sulfur-containing alkaloids, polycarpaurines A, B, and C, from an Indonesian ascidian *Polycarpa aurata*. Tetrahedron 63:409–412
226. Hassan WHB, Al-Taweel A, Proksch P (2009) Two new imidazole alkaloids from *Leucetta chagosensis* sponge. Saudi Pharm J 17:295–298
227. Wei X, Hu X, Yu R et al (2020) Efficient total synthesis of lissodendrin B, 2-aminoimidazole marine alkaloids from *Lissodendroryx (Acanthodoryx) fibrosa*. Mar Drugs 18:36
228. Guo P, Wang Z, Li G et al (2016) First discovery of polycarpine, polycarpaurines A and C, and their derivatives as novel antiviral and antiphytopathogenic fungus agents. J Agric Food Chem 64:4264–4272
229. Bennington F, Morin RD, Clark Jr et al (1958) Psychopharmacological activity of ring- and side chain-substituted β-phenethylamines. J Org Chem 23:1979–1984
230. Calcul L, Longeon A, Al Mourabit A et al (2003) Novel alkaloids of the aaptamine class from an Indonesian marine sponge of the genus *Xestospongia*. Tetrahedron 59:6539–6544
231. Pham CD, Hartmann R, Müller WEG et al (2013) Aaptamine derivatives from the Indonesian sponge *Aaptos suberitoides*. J Nat Prod 76:103–106
232. Herlt A, Mander L, Rombang W et al (2004) Alkaloids from marine organisms. Part 8: isolation of bisdemethylaaptamine and bisdemethylaaptamine-9-*O*-sulfate from an Indonesian *Aaptos* sp. marine sponge. Tetrahedron 60:6101–6104
233. Arai M, Han C, Yamano Y et al (2014) Aaptamines, marine spongean alkaloids, as anti-dormant mycobacterial substances. J Nat Med 68:372–376
234. Fukumoto A, Hitora Y, Kai A et al (2018) Isolation of aaptic acid from the marine sponge *Aaptos lobata* and inhibitory effect of aaptamines on RANKL-induced formation of multinuclear osteoclasts. Heterocycles 97:1219–1225
235. Liu H, Fujiawara T, Nishikawa T et al (2005) Lissoclibadins 1–3, three new polysulfur alkaloids, from the ascidian *Lissoclinum* cf. *badium*. Tetrahedron 61:8611–8615
236. Nakazawa T, Xu J, Nishikawa T et al (2007) Lissoclibadins 4–7, polysulfur aromatic alkaloids from the indonesian Ascidian *Lissoclinum* cf. *badium*. J Nat Prod 70:439–442
237. Wang W, Takahashi O, Oda T et al (2009) Lissoclibadins 8–14, polysulfur dopamine-derived alkaloids from the colonial ascidian *Lissoclinum* cf. *badium*. Tetrahedron 65:9598–9603
238. Nagasawa Y, Kato H, Rotinsulu H et al (2011) Spironaamidine, a new spiroquinone-containing alkaloid from the marine sponge *Leucetta microraphis*. Tetrahedron Lett 52:5342–5344
239. Hassan W, Edrada RA, Ebel R et al (2004) New imidazole alkaloids from the Indonesian sponge *Leucetta chagosensis*. J Nat Prod 67:817–822
240. Tsukamoto S, Kawabata T, Kato H et al (2007) Naamidines H and I, cytotoxic imidazole alkaloids from the Indonesian marine sponge *Leucetta chaogosensis*. J Nat Prod 70:1658–1660
241. Guo P, Li G, Liu Y et al (2018) Naamines and naamidines as novel agents against a plant virus and phytopathogenic fungi. Mar Drugs 16:311
242. Koswatta PB, Lovely CJ (2010) Concise total synthesis of naamine G and naamidine H. Chem Commun 46:2148–2150
243. Das J, Koswatta PB, Jones JD et al (2012) Total syntheses of kealiinines A–C. Org Lett 14:6210–6213

244. Li G, Guo J, Wang Z et al (2018) Marine natural products for drug discovery: first discovery of kealiinines A-C and their derivatives as novel antiviral and antiphytopathogenic fungus agents. J Agric Food Chem 66:7310–7318
245. Gibbons JB, Gligorich KM, Welm BE et al (2012) Synthesis of the reported structures for kealiinines B and C. Org Lett 14:4734–4737
246. McCauley EP, Lam H, Lorig-Roach N et al (2017) Investigation of the physical and bioactive properties of bromo-and iodo-containing sponge-derived compounds possessing an oxyphenylethanime core. J Nat Prod 80:3225–3266
247. Tarazona G, Santamaria G, Cruz PG et al (2017) Cytotoxic anomoian B and aplyzanzine B, new bromotyrosine alkaloids from Indonesian sponges. ACS Omega 2:3494–3501
248. Dai J, Parrish SM, Yoshida WY et al (2016) Bromotyrosine-derived metabolites from an Indonesian marine sponge in the family Aplysinellidae (order Verongiida). Bioorg Med Chem Lett 26:499–504
249. Mudianta IW, Skinner-Adams T, Andrews KT et al (2012) Psammaplysin derivatives from the Balinese marine sponge *Aplysinella strongylata*. J Nat Prod 75:2132–2143
250. Shashi S, Hussain MA, Khan FA (2019) Total synthesis of enisorine D and its analogues. Synthesis 51:4601–4610
251. Park SK, Jurek J, Carney JR et al (1994) Two more bastadins, 16 and 17, from an Indonesian sponge *Ianthella basta*. J Nat Prod 57:407–410
252. Couladourous EA, Pitsinos EN, Moutsos VI et al (2004) A general method for the synthesis of bastranes and isobastranes: first total synthesis of bastadins 5, 10, 12, 16, 20, and 21. Chem Eur J 11:406–421
253. Niemann H, Lin W, Müller WEG et al (2013) Trimeric hemibastadin congener from the marine sponge *Ianthella basta*. J Nat Prod 76:121–125
254. Mohanty I, Moore SG, Yi D et al (2020) Precursor-guided mining of marine sponge metabolomes lends insight into biosynthesis of pyrrole-imidazole alkaloids. ACS Chem Biol 15:2185–2194
255. Gamal AE, Agarwal V, Diethelm S et al (2016) Biosynthesis of coral settlement cue tetrabromopyrrole in marine bacteria by a uniquely adapted brominase–thioesterase enzyme pair. Proc Natl Acad Sci USA 113:3797–3802
256. Ebada SS, Edrada-Ebel R, de Voogd NJ et al (2009) Dibromopyrrole alkaloids from the marine sponge *Acanthostylotella* sp. Nat Prod Commun 4:47–52
257. Ebada SS, Linh MH, Longeon A et al (2015) Dispacamide E and other bioactive bromopyrrole allkaloids from two Indonesian marine sponges of the genus *Stylissa*. Nat Prod Res 29:231–238
258. Eder C, Proksch P, Wray V et al (1999) New alkaloids from the Indopacific sponge *Stylissa carteri*. J Nat Prod 62:184–187
259. Fouad M, Debbab A, Wray V et al (2012) New bioactive alkaloids from the marine sponge *Stylissa* sp. Tetrahedron 68:10176–10179
260. Eder C, Proksch P, Wray V et al (1999) New bromopyrrole alkaloids from the Indopacific sponge *Agelas nakamurai*. J Nat Prod 62:1295–1297
261. Abjul DB, Yamazaki H, Kanno S et al (2017) An anti-mycobacterial bisfunctionalized spingolipid and new bromopyrrole alkaloid from the Indonesian marine sponge *Agelas* sp. J Nat Med 71:531–536
262. Katsuki A, Kato H, Ise Y et al (2019) Agesamines A and B, new dibromopyrrole alkaloids from the sponge *Agelas* sp. Heterocycles 98:558–561
263. Supriyono A, Schwarz B, Wray V et al (1995) Bioactive alkaloids from the tropical marine sponge *Axinella carteri*. Z Naturforsch C 50:669–674
264. O'Malley DP, Li K, Maue M et al (2007) Total synthesis of dimeric pyrolle-imidazole alkaloids; sceptrin, ageliferin, nagelamide E, oxysceptrin, nakamuric acid, and the axinellamine carbon skeleton. J Am Chem Soc 129:4762–4775
265. Sun Y-T, Lin B, Li S-G et al (2017) New bromopyrrole alkaloids from the marine sponge *Agelas* sp. Tetrahedron 73:2786–2792
266. Singh RP, Bhandari MR, Torres FM et al (2020) Total synthesis of (±)-2-debromohymenin via gold-catalyzed intramolecular alkyne hydroarylation. Org Lett 22:3412–3417

267. Hirozumi R, Kudo Y, Cho Y et al (2025) Total synthesis and structural revision of (±)-mauritamide B. J Nat Prod 88:806–814
268. Linnington RG, Williams DE, Tahir A et al (2003) Latonduines A and B, new alkaloids isolated from the marine sponge *Stylissa carteri*: structure elucidation, synthesis, and biogenetic implications. Org Lett 5:2735–2738
269. Issa HH, Tanaka J, Rachmat R et al (2005) Polycitorols A and B, new tricyclic alkaloids from an ascidian. Mar Drugs 3:78–83
270. Kaga A, Chiba S (2018) Synthesis of tricyclic marine alkaloids, cylindricines, lepadiformines, fasicularin, and polycitorols: a recent update. Synthesis 50:685–699
271. Xi H, Nie X, Gao F et al (2023) A bacterial spermidine biosynthetic pathway via carboxyaminopropylagmatine. Sci Adv 9:eadj9075
272. Long Q, Zhou W, Zhou H et al (2024) Polyamine-containing natural products: structure, bioactivity, and biosynthesis. Nat Prod Rep 41:525–564
273. Piwko AT, Miller BG, Smith JM (2023) Revisiting the manzamine biosynthetic hypothesis. Nat Prod Rep 40:964–971
274. Silva SBL, Oberhänsli F, Tribalat MA et al (2019) Insights into the biosynthesis of cyclic guanidine alkaloids from Crambeidae marine sponges. Angew Chem Int Ed 58:520–525
275. Anstis DG, Davison EK, Sperry J (2024) 1,2,4-thiadiazole alkaloids−isolation, biological activity and synthesis. Tetrahedron 150:133767
276. Zhang H, Loveridge ST, Tenney K et al (2016) A new 3-alkylpyridine alkaloid from the marine sponge *Haliclona* sp. and its cytotoxic activity. Nat Prod Res 30:1262–1265
277. Maarisit W, Abdjul D, Yamazaki H et al (2017) Anti-mycobacterial alkaloids, cyclic 3-alkyl pyridinium dimers, from the Indonesian marine sponge *Haliclona* sp. Bioorg Med Chem Lett 27:3503–3506
278. Arai M, Kamiya K, Shin D et al (2016) N-Methylniphatyne A, a new 3-alkylpyridine alkaloid as an inhibitor of the cancer cells adapted to nutrient starvation, from an Indonesian marine sponge of *Xestospongia* sp. Chem Pharm Bull 64:766–771
279. Jiménez JI, Goetz G, Mau CMS et al (2000) ‘Upenamide: an unprecedented macrocyclic alkaloid from the Indonesian sponge *Echinochalina* sp. J Org Chem 65:8465–8469
280. Hitora Y, Maeda R, Honda K et al (2021) Neopetrosidines A-D, pyridine alkaloids isolated from the marine sponge *Neopetrosia chaliniformis* and their cell cycle elongation activity. Bioorg Med Chem 50:116461
281. Unsworth WP, Gallagher KA, Jean M et al (2013) Direct imine acylation: synthesis of the proposed structures of upenamide. Org Lett 15:262–265
282. Unsworth WP, Taylor RJK (2013) ‘Upenamide: trials and tribulations. Org Biomol Chem 11:7250–7261
283. Jaspars M, Pasupathy V, Crews P (1994) A tetracyclic diamine alkaloid, halicyclamine A, from the marine sponge *Haliclona* sp. J Org Chem 59:3253–3255
284. Mudianta IW, Katavic PL, Lambert LK et al (2010) Structure and absolute configuration of 3-alkylpiperidine alkaloids from an Indonesian sponge of the genus *Halichondria*. Tetrahedron 66:2752–2760
285. Dewi AS, Hadi TA, Fajarningsih ND et al (2014) Acanthocyclamine A from the Indonesian marine sponge *Acanthostrongylophora ingens*. Aust J Chem 67:1205–1210
286. Esposito G, Bourguet-Kondracki M-L, Mai LH et al (2016) Chloromethylhalicyclamine B, a marine-derived protein kinase CK1δ/ ε Inhibitor. J Nat Prod 79:2953–2960
287. Kato H, El-Desoky AH, Takeishi Y et al (2019) Tetradehydrohalicyclamine B, a new proteasome inhibitor from the marine sponge *Acanthostrongylophora ingens*. Bioorg Med Chem Lett 29:8–10
288. Esposito G, Mai LH, Longeon A et al (2019) A collection of bioactive nitrogen-containing molecules from the marine sponge *Acanthostrongylophora ingens*. Mar Drugs 17:472
289. Arai M, Ishida S, Setiawan A et al (2009) Haliclonacyclamines, tetracyclic alkylpiperidine alkaloids, as anti-dormant mycobacterial substances from a marine sponge of *Haliclona* sp. Chem Pharm Bull 57:1136–1138

290. Dalling AG, Spät G, Fürstner A (2022) Total synthesis of the tetracyclic pyridinium alkaloid *epi*-tetradehydrohalicyclamine B. Angew Chem Int Ed 61:e202209651
291. Liang Z, Sulzmaier FJ, Yoshida WY et al (2015) Neopetrocyclamines A and B, polycyclic diamine alkaloids from the sponge *Neopetrosia* cf. *exigua*. J Nat Prod 78:543–547
292. Kim C-K, Riswanto R, Won T-H et al (2017) Manzamine alkaloids from an *Acanthostrongylophora* sp. sponge. J Nat Prod 80:1575–1583
293. Yousaf M, El Sayed K, Rao KV et al (2002) 12,34-Oxamanzamines, novel biocatalytic and natural products from manzamine producing Indo-Pacific sponges. Tetrahedron 58:7397–7402
294. Yousaf M, Hammond NL, Peng J et al (2004) New manzamine alkaloids from an Indo-Pacific sponge. Pharmacokinetics, oral availability, and the significant activity of several manzamines against HIV-I, AIDS opportunistic infections, and inflammatory disease. J Med Chem 47:3512–3517
295. Ichiba T, Corgiat JM, Scheuer PJ et al (1994) 8-Hydroxymanzamine A, a β-carboline alkaloid from a sponge, *Pachypellina* sp. J Nat Prod 57:168–170
296. Rao KV, Donia MS, Peng J et al (2006) Manzamine B and E and ircinal A related alkaloids from an Indonesian *Acanthostrongylophora* sponge and their activity against infectious, tropical parasitic, and Alzheimer's diseases. J Nat Prod 69:1034–1040
297. Rao KV, Kasanah N, Wahyuono S et al (2004) Three new manzamine alkaloids from a common Indonesian sponge and their activity against infectious and tropical parasitic diseases. J Nat Prod 67:1314–1318
298. Wahba AE, Fromentin Y, Zou Y et al (2012) Acantholactone, a new manzamine related alkaloid with an unprecedented δ-lactone and ε-lactam ring system. Tetrahedron Lett 53:6329–6331
299. El-desoky A, Kato H, Eguchi K et al (2014) Acantholactam and pre-*neo*-kauluamine, manzamine-related alkaloids from the Indonesia *Acanthostrongylophora ingens*. J Nat Prod 77:1536–1540
300. Furusato A, Kato H, Nehira T et al (2014) Acanthomanzamines A−E with new manzamine frameworks from the marine sponge *Acanthostrongylophora ingens*. Org Lett 16:3888–3891
301. AlTarabeen M, Daletos G, Ebrahim W et al (2015) Ircinal E, a new manzamine derivative from the Indonesian marine sponge *Acanthostrongylophora ingens*. Nat Prod Commun 10:1951–1953
302. El Sayed KA, Kelly M, Kara UAK et al (2001) New manzamine alkaloids with potent activity against infectious diseases. J Am Chem Soc 123:1804–1808
303. Peng J, Hu J-F, Kazi AB et al (2003) Manadomanzamines A and B: a novel alkaloid ring system with potent activity against *Mycobacteria* and HIV-1. J Am Chem Soc 125:13382–13386
304. Ohtani II, Ichiba T, Isobe M et al (1995) Kauluamine: an unprecedented manzamine dimer from an Indonesian marine sponge, *Prianos* sp. J Am Chem Soc 117:10743–10744
305. Kasmiati K, Yoshioka Y, Okamoto Y et al (2018) New crambescidin-type alkaloids from the Indonesian marine sponge *Clathria bulbotoxa*. Mar Drugs 16:84
306. Pham CD, Weber H, Hartmann R et al (2013) New Cytotoxic 1,2,4-thiadiazole alkaloids from the ascidian *Polycarpa aurata*. Org Lett 15:2230–2233
307. Casertano M, Imperatore C, Luciano P et al (2019) Chemical investigation of the Indonesian tunicate *Polycarpa aurata* and evaluation of the effects against *Schistosoma mansoni* of the novel alkaloids polyaurines A and B. Mar Drugs 17:278
308. González N, Rodríguez J, Jiménez C (1999) Didemniserinolipids A−C, unprecedented serinolipids from the tunicate *Didemnum* sp. J Org Chem 64:5705–5707
309. Mansour MP, Volkman JK, Holdsworth DG et al (1999) Very-long-chain (C_{28}) unsaturated fatty acids in marine dinoflagellates. Phytochemistry 50:541–548
310. Castro LFC, Tocher DR, Monroig O (2016) Long-chain polyunsaturated fatty acid biosynthesis in chordates: insights into the evolution of Fads and Elovl gene repertoire. Prog Lipid Res 62:25–40
311. Kiyota H, Dixon DJ, Luscombe CK et al (2002) Synthesis, structure revision, and absolute configuration of (+)-didemniserinolipid B, a serinol marine natural product from a tunicate *Didemnum* sp. Org Lett 4:3223–3226

312. Ren J, Tong R (2014) Asymmetric total synthesis of (+)-didemniserinolipid B via Achmatowicz rearrangement/bicycloketalization. J Org Chem 79:6987–6995
313. Nakamura Y, Kato H, Nishikawa T et al (2013) Siladenoserinols A–L: new sulfonated serinol derivatives from a tunicate as inhibitors of p53-Hdm2 interaction. Org Lett 15:322–325
314. Torii M, Hitora Y, Kato H et al (2018) Siladenoserinols M-P, sulfonated serinol derivatives from a tunicate. Tetrahedron 74:7516–7521
315. Yoshida M, Saito K, Kato H et al (2018) Total synthesis and biological evaluation of siladenoserinol A and its analogues. Angew Chem Int Ed 57:5147–5150
316. Ibrahim SRM, Mohamed GA, El-Khayat ES et al (2008) Strepsiamides A–C, new ceramides from the marine sponge *Strepsichordaia lendenfeldi*. Nat Prod Commun 3:205–209
317. Ibrahim SRM, Mohamed GA, Fouad MA et al (2009) Iotrochotamides I and II: new ceramides from the Indonesian sponge *Iotrochota purpurea*. Nat Prod Res 23:86–92
318. Trianto A, de Voogd NJ, Tanaka J (2014) Two new compounds from an Indonesian sponge *Dysidea* sp. J Asian Nat Prod Res 16:163–168
319. Tarazona G, Fernández R, Pérez M et al (2022) Enigmazole C: a cytotoxic macrocyclic lactone and its ring-opened derivatives from a new species of *Homophymia* sponge. J Nat Prod 85:1059–1066
320. Aoki S, Higuchi K, Ye Y et al (2000) Melophlins A and B, novel tetramic acids reversing the phenotype of ras-transformed cells, from the marine sponge, *Melophlus sarassinorum*. Tetrahedron 56:1833–1836
321. Wang C-Y, Wang B-G, Wiryowidagdo S et al (2003) Melophlins C–O, thirteen novel tetramic acids from the marine sponge *Melophlus sarassinorum*. J Nat Prod 66:51–56
322. Fattorusso E, Romano A, Taglialatela-Scafati O et al (2008) Lobozoanthamine, a new zoanthamine-type alkaloid from the Indonesian soft coral *Lobophytum* sp. Tetrahedron Lett 49:2189–2192
323. Ichiba T, Nakao Y, Scheuer PJ (1995) Kumusine, a chloroadenine riboside from a sponge, *Theonella* sp. Tetrahedron Lett 36:3977–3980
324. Van Wagoner RM, Jompa J, Tahir A et al (2001) A novel modified pterin from a *Eudistoma* species ascidian. J Nat Prod 64:1100–1101
325. Karim MdRUI, Harunari E, Sharma AR et al (2020) Nocarimidazoles C and D, antimicrobial alkanoylimidazoles from a coral-derived actinomycete *Kocuria* sp.: application of $^{1}J_{C,H}$ coupling constants for the unequivocal determination of substituted imidazoles and stereochemical diversity of anteisoalkyl chains in microbial metabolites. Beilstein J Org Chem 16:2719–2727
326. Mohanty I, Podell S, Biggs JS et al (2020) Multi-omic profiling of *Melophlus* sponges reveals diverse metabolomic and microbiome architectures that are non-overlapping with ecological neighbors. Mar Drugs 18:124
327. Behena DC, Stockdill JL, Stoltz BM (2008) The biology and chemistry of the zoanthamine alkaloids. Angew Chem Int Ed 47:2365–2386
328. Wagoner RMV, Satake M, Wright JLC (2014) Polyketide biosynthesis in dinoflagellates: what makes it different? Nat Prod Rep 31:1101
329. Zhang Y, Morar M, Ealick SE (2008) Structural biology of the purine biosynthetic pathway. Cell Mol Life Sci 65:3699–3724
330. Keyzers RA, Gray CA, Schleyer MH et al (2006) Malonganenones A-C, novel tetraprenylated alkaloids from the Mozambique gorgonian *Leptogorgia gilchristi*. Tetrahedron 62:2200–2206
331. Mokhlesi A, Hartmann R, Kurtán T et al (2017) New 2-methoxy acetylenic acids and pyrazole alkaloids from the marine sponge *Cinachyrella* sp. Mar Drugs 15:356–366
332. Waldman AJ, Ng TL, Wang P et al (2017) Heteroatom–heteroatom bond formation in natural product biosynthesis. Chem Rev 117:5784–5863
333. Setyowati EP, Jenie UA, Sudarsono et al (2018) Identification of cytotoxic constituent of Indonesian sponge *Kaliapsis* sp. Pak J Biol Sci 11:2560–2566
334. Sperry S, Crews PA (1996) A novel alkaloid from the Indo-Pacific sponge *Clathria basilana*. Tetrahedron Lett 37:2389–2390

335. Mokhlesi A, Stuhldreier F, Wex KW et al (2017) Cyclic cysteine-bridged peptides from the marine sponge *Clathria basilana* induce apoptosis in tumor cells and depolarize the bacterial cytoplasmic membrane. J Nat Prod 80:2941–2952
336. Angawi RF, Bevastrello G, Calcinai B et al (2011) Aurantoside J: new tetramic acid glycoside from *Theonella swinhoei*. Insight into the antifungal potential of aurantosides. Mar Drugs 9:2809–2817
337. Kotoku N, Ishida R, Matsumoto H et al (2017) Biakamides A–D, unique polyketides from a marine sponge, act as selective growth inhibitors of tumor cells adapted to nutrient starvation. J Org Chem 82:1705–1718
338. Mordhost S, Ruijne F, Vagstad AL et al (2023) Emulating nonribosomal peptides with ribosomal biosynthetic strategies. RSC Chem Biol 4:7–36
339. Ardá A, Rodríguez J, Nieto RM et al (2005) NMR *J*-based analysis of nitrogen-containing moieties and application to dysithiazolamide, a new polychlorinated dipeptide from *Dysidea* sp. Tetrahedron 61:10093–10098
340. Sadar MD, Williams DE, Mawji NR et al (2008) Sintokamides A to E, chlorinated peptides from the sponge *Dysidea* sp. that inhibit transactivation of the *N*-terminus of the androgen receptor in prostate cancer cells. Org Lett 10:4947–4950
341. Kapojos MM, Abdjul DB, Yamazaki H et al (2018) Callyspongiamides A and B, sterol *O*-acyltransferase inhibitors, from the Indonesian marine sponge *Callyspongia* sp. Bioorg Med Chem Lett 28:1911–1914
342. Ardá A, Soengas RG, Nieto MI et al (2008) Total synthesis of (–)-dysithiazolamide. Org Lett 10:2175–2178
343. Gu Z, Zakarian A (2010) Concise total synthesis of sintokamides A, B, and E by a unified, protecting-group-free strategy. Angew Chem Int Ed 49:9702–9705
344. Ebada SS, Müller WEG, Lin W et al (2019) New acyclic cytotoxic jasplakinolide derivative from the marine sponge *Jaspis splendens*. Mar Drugs 17:100
345. Watts KR, Morinaka BI, Amagata T et al (2011) Biostructural features of additional jasplakinolide (jaspamide) analogues. J Nat Prod 74:341–351
346. Fernández R, Bayu A, Hadi TA et al (2020) Unique polyhalogenated peptides from the marine sponge *Ircinia* sp. Mar Drugs 18:396
347. Castaldi A, Teta R, Esposito G et al (2022) Computational metabolomics tools reveal subarmigerides, unprecedented linear peptides from the marine sponge holobiont *Callyspongia subarmigera*. Mar Drugs 20:673
348. Aoki S, Cao L, Matsui K et al (2004) Kendarimide A, a novel peptide reversing P-glycoprotein-mediated multidrug resistance in tumor cells, from a marine sponge of *Haliclona* sp. Tetrahedron 60:7053–7059
349. Kotoku N, Cao L, Aoki S et al (2005) Absolute stereostructure of kendarimide A, a novel MDR modulator, from a marine sponge. Heterocycles 65:563–578
350. Wakimoto T, Tan K, Tajima H et al (2015). Cytotoxic cyclic peptides from the marine sponges. In: Kim SK (eds) Handbook of anticancer drugs from marine origin. Springer, Cham
351. Hitora Y, El-Desoky AH, Sadahiro Y et al (2024) Neopetromin, a cyclic tripeptide with a C-N cross-link, from the marine sponge *Neopetrosia* sp., that causes vacuole fragmentation in tobacco BY-2 cells. J Nat Prod 87:1197–1202
352. Mau CMS, Nakao Y, Yoshida WY et al (1996) Waiakeamide, a cyclic hexapeptide from the sponge *Ircinia dendroides*. J Org Chem 61:6302–6304
353. Bonnington LS, Tanaka J, Higa T et al (1997) Cupolamide A: a cytotoxic cyclic heptapeptide from two samples of the sponge *Theonella cupola*. J Org Chem 62:7765–7767
354. Afifi AH, El-Desoky A, Kato H et al (2016) Carteritins A and B, cyclic heptapeptides from the marine sponge *Stylissa carteri*. Tetrahedron Lett 57:1285–1288
355. Arai M, Yamano Y, Fujita M et al (2012) Stylissamide X, a new proline-rich cyclic octapeptide as an inhibitor of cell migration, from an Indonesian marine sponge of *Stylissa* sp. Bioorg Med Chem Lett 22:1818–1821
356. Roy MC, Ohtani I, Tanaka J et al (1999) Barangamide A, a new cyclic peptide from the Indonesian sponge *Theonella swinhoei*. Tetrahedron Lett 40:5373–5376

357. Roy MC, Ohtani II, Ichiba T et al (2000) New cyclic peptides from the Indonesian sponge *Theonella swinhoei*. Tetrahedron 56:9079–9092
358. Weeson KJ, Hamann MT (1996) Keenamide A, a bioactive cyclic peptide from the marine mollusk *Pleurobranchus forskalii*. J Nat Prod 59:629–631
359. Donia MS, Wang B, Dunbar DC et al (2008) Mollamides B and C, cyclic hexapeptides from the Indonesian tunicate *Didemnum molle*. J Nat Prod 71:941–945
360. Urda C, Fernández R, Rodríguez J et al (2017) Bistratamides M and N, oxazole-thiazole containing cyclic hexapeptides isolated from *Lissoclinum bistratum* interaction of zinc (II) with bistratamide K. Mar Drugs 15:209–220
361. Fernández R, Martín MJ, Rodríguez-Acebes R et al (2008) Diazonamides C–E, new cytotoxic metabolites from the ascidian *Diazona* sp. Tetrahedron Lett 49:2283–2285
362. Morris LA, van den Bosch JJK, Versluis K et al (2000) Structure determination and MS^n analysis of two new lissoclinamides isolated from the Indo-Pacific ascidian *Lissoclinum patella*: NOE restrained molecular dynamics confirms the absolute stereochemistry derived by degradative methods. Tetrahedron 56:8345–8353
363. Tan LT, Williamson RT, Gerwick WH et al (2000) *Cis, cis*-and *trans, trans*-ceratospongamide, new bioactive cyclic heptapeptides from the Indonesian red alga *Ceratodicyton spongiosum* and symbiotic sponge *Sigmadocia symbiotica*. J Org Chem 65:419–425
364. Phyo MY, Ding CYG, Goh HC et al (2019) Trikoramide A, a prenylated cyanobactin from the marine cyanobacterium *Symploca hydnoides*. J Nat Prod 82:3482–3488
365. Phyo MY, Goh TMB, Goh JX et al (2021) Trikoramides B-D, bioactive cyanobactins from the marine cyanobacterium *Symploca hydnoides*. Mar Drugs 19:548
366. Yokokawa F, Sameshima H, In Y et al (2002) Total synthesis and conformational studies of ceratospongamide, a bioactive cyclic heptapeptide from marine origin. Tetrahedron 58:8127–8143
367. Yokokawa F, Sameshima H, Shiori T (2001) Total synthesis of *cis,cis*-ceratospongamide, a bioactive thiazole-containing cyclic peptide from marine origin. Synlett SI:968–988
368. Deng S, Taunton J (2002) Kinetic control of proline amide rotamers: total synthesis of *trans, trans*-and *cis, cis*-ceratospongamide. J Am Chem Soc 124:916–917
369. Kutsumura N, Sata NU, Nishiyama S (2002) Synthetic studies on ceratospongamides, cyclic heptapeptides containing thiazole and oxazoline units: total synthesis of *cis, cis*-ceratospongamide. Bull Chem Soc Jpn 75:847–850
370. Huang T, Zou Y, Wu MC et al (2015) Total synthesis of proline-rich cyclic octapeptide stylissamide X. Chem Nat Compd 51:523–526
371. El-Desoky AHH, Hitora Y, Onodera K et al (2022) Cyclopsammocinamides A and B, enantiomeric cyclic peptides of cyclocinamide A, from the marine sponge *Psammocinia* sp. Chem Pharm Bull 70:818–822
372. Ibrahim SR, Min CC, Teuscher F et al (2010) Callyaerins A-F and H, new cytotoxic cyclic peptides from the Indonesian marine sponge *Callyspongia aerizusa*. Bioorg Med Chem 18:4947–4956
373. Daletos G, Kalscheuer R, Koliwer-Brandl H et al (2015) Callyaerins from the marine sponge *Callyspongia aerizusa*: cyclic peptides with antitubercular activity. J Nat Prod 78:1910–1925
374. Ibrahim SRM, Edrada-Ebel RA, Mohamed GA et al (2008) Callyaerin G, a new cytotoxic cyclic peptide from the marine sponge *Callyspongia aerizusa*. *ARKIVOC* xii:164–171
375. Zhang S, Rodriguez LMDL, Leung IKH et al (2018) Total synthesis and conformational study of callyaerin A: anti-tubercular cyclic peptide bearing a rare rigidifying (Z)-2,3-diaminoacrylamide moiety. Angew Chem Int Ed 57:3631–3635
376. Ebada SS, Wray V, de Voogd NJ et al (2009) Two new jaspamide derivatives from the marine sponge *Jaspis splendens*. Mar Drugs 7:435–444
377. Phyo MY, Katermeran NP, Goh JX et al (2021) Trikoveramides A–C, cyclic depsipeptides from the marine cyanobacterium *Symploca hydnoides*. Phytochemistry 190:112879
378. Plaza A, Bifulco G, Keffer JL et al (2009) Celebesides A–C and theopapuamides B–D, depsipeptides from an Indonesian sponge that inhibit HIV-1 entry. J Org Chem 74:504–512

379. Phyo MY, Goh JX, Tan LT (2022) Triproamide and pemukainalides, cyclic depsipeptides from the marine cyanobacterium *Symploca hydnoides*. J Nat Prod 85:485–492
380. Wibowo JT, Kellerman MY, Köck M et al (2021) Anti-infective and antiviral activity of valinomycin and its analogues from a sea cucumber-associated bacterium, *Streptomyces* sp. SV 21. Mar Drugs 19:81
381. Fernández R, Pérez M, Losada A et al (2024) Isolation and total synthesis of PM170453, a new cyclic depsipeptides isolated from *Lyngbya* sp. Mar Drugs 22:303
382. Urda C, Fenández R, Rodríguez J et al (2017) Daedophamide, a cytotoxic cyclodepsipeptide from a *Daedalopelta* sp. collected in Indonesia. J Nat Prod 80:3054–3059
383. Rashid MA, Gustafson KR, Cartner LK et al (2001) Microspinosamide, a new HIV-inhibitory cyclic depsipeptide from the marine sponge *Sidonops microspinosa*. J Nat Prod 64:117–121
384. Sinisi A, Calcinai B, Cerrano C et al (2013) New tridecapeptides of the theonellapeptolide family from the Indonesian sponge *Theonella swinhoei*. Beilstein J Org Chem 9:1643–1651
385. Haedar JR, Uria AR, Lallo S et al (2022) New theonellapeptolides from Indonesian marine sponge *Theonella swinhoei* as anti-austerity agents. Mar Drugs 20:661
386. Clark DP, Carrol J, Naylor S et al (1998) An antifungal cyclodepsipeptide, cyclolithistide A, from the sponge *Theonella swinhoei*. J Org Chem 63:8757–8764
387. Tarazona G, Fernández R, Cruz PG et al (2019) Combining JBCA and Marfey's methodology to determine the absolute configuration of threonines: the case of gunungamide A, a new cyclic depsipeptide containing chloropyrrole from the sponge *Discodermia* sp. Org Chem Front 6:15–21
388. Tajima H, Wakimoto T, Takada K et al (2014) Revised structure of cylclolithistide A, a cyclic depsipeptide from the marine sponge *Discodermia japonica*. J Nat Prod 77:154–158
389. Braekman JC, Daloze D, Devijver C et al (2003) A new C-20 polyacetylene from the sponge *Callyspongia pseudoreticulata*. J Nat Prod 66:871–872
390. Aratake S, Trianto A, Hanif N et al (2009) A new polyunsaturated brominated fatty acid from a *Haliclona* sponge. Mar Drugs 7:523–527
391. Aoki S, Matsui K, Tanaka K et al (2000) Lembehyne A, a novel neuritogenic polyacetylene, from a Marine Sponge of *Haliclona* sp. Tetrahedron 56:9945–9948
392. Aoki S, Matsui K, Wei H et al (2002) Structure–activity relationship of neuritogenic spongean acetylene alcohols, lembehynes. Tetrahedron 58:5417–5422
393. Murakami N, Nakajima T, Kobayashi M (2001) Total synthesis of lembehyne A, a neuritogenic spongean polyacetylene. Tetrahedron Lett 42:1941–1943
394. Dzhemileva LU, D'yakonov VA, Makarov AA et al (2017) The first total synthesis of the marine acetylenic alcohol, lembehyne B-a selective inducer of early apoptosis in leukemia cancer cells. Org Biomol Chem 15:470−476
395. Dzhemileva LU, D'yakonov VA, Makarov AA et al (2020) Total synthesis of natural lembehyne C and investigation of its cytotoxic properties. J Nat Prod 83:2399−2409
396. Balansa W, Trianto A, de Voogd NJ et al (2017) A new cytotoxic polyacetylenic alcohol from a sponge *Callyspongia* sp. Nat Prod Commun 12:1909–1911
397. Juagdan EG, Kalidindi RS, Scheuer PJ et al (1995) Elenic acid, an inhibitor of topoisomerase II, from a sponge, *Plakinastrella* sp. Tetrahedron Lett 36:2905–2908
398. Takahanashi S, Takagi M, Takikawa H et al (1998) Synthesis of elenic acid, an inhibitor of topoisomerase II from the sponge *Plakinastrella* sp. J Chem Soc Perkin Trans 1:1603–1606
399. Rodríguez J, Jimenez C, Blanco M et al (2016) Lanesoic acid: a cytotoxic zwitterion from *Theonella* sp. Org Lett 18:5832–5835
400. Hou A, Dickshat JS (2023) Labelling studies in the biosynthesis of polyketides and non-ribosomal peptides. Nat Prod Rep 40:470–499
401. Ichiba T, Scheuer PJ, Kelly-Borges M (1995) Two cytotoxic 3,6-epidioxy fatty acids from an Indonesian sponge, *Plakortis* sp. Tetrahedron 51:12195–12202
402. Fattorusso C, Persico M, Calcinai B et al (2010) Manadoperoxides A−D from the Indonesian sponge *Plakortis* cfr. *simplex*. Further insights on the structure-activity relationships of simple 1,2-dioxane antimalarials. J Nat Prod 73:1138–1145

403. Chianese G, Fattorusso E, Scala F et al (2012) Manadoperoxides, a new class of potent antitrypanosomal agents of marine origin. Org Biomol Chem 10:7197–7207
404. Pettit GR, Nogawa T, Knight JC et al (2004) Antineoplastic agents. 535. Isolation and structure of plakorstatins 1 and 2 from the Indo-Pacific sponge *Plakortis nigra.* J Nat Prod 67:1611–1613
405. Gushiken M, Kagiyama I, Kato H et al (2015) Manadodioxans A−E: polyketide endoperoxides from the marine sponge *Plakortis bergquistae.* J Nat Med 69:595–600
406. Jadulco R, Proksch P, Wray V et al (2001) New macrolides and furan carboxylic acid derivative from the sponge-derived fungus *Cladosporium herbarum.* J Nat Prod 64:527–530
407. Rotinsulu H, Yamazaki H, Sugai S et al (2018) Cladosporamide A, a new protein tyrosine phosphatase 1B inhibitor, produced by an Indonesian marine sponge-derived *Cladosporium* sp. J Nat Med 72:779–783
408. Shiono Y, Miyazaki N, Murayama T et al (2016) GSK-3β inhibitory activities of novel dichroloresorcinol derivatives from *Cosmospora vilior* isolated from a mangrove plant. Phytochem Lett 18:122–127
409. Newaz AW, Yong K, Yi W et al (2023) Antimicrobial metabolites from the Indonesian mangrove sediment-derived fungus *Penicillium chrysogenum* sp. ZZ1151. Nat Prod Res 37:1702–1708
410. Sumilat DA, Yamazaki H, Endo K et al (2017) A new biphenyl ether derivative produced by Indonesian ascidian-derived *Penicillium albobiverticillium.* J Nat Med 71:776–779
411. Sabdaningsih A, Liu Y, Mettal U et al (2020) A new citrinin derivative from the Indonesian marine sponge-associated fungus *Penicillium citrinum.* Mar Drugs 18:227
412. Edrada RA, Wray V, Berg A et al (2000) Novel spiciferone derivatives from the fungus *Drechslera hawaiiensis* isolated from the marine sponge *Callyspongia aerizusa.* Z Naturforsch 55c:218−221
413. Spery S, Samuels GJ, Crews P (1998) Vertinoid polyketides from the salt culture of the fungus *Tricoderma longibrachiatum* separated from a *Haliclona* sponge. J Org Chem 63:10011–10014
414. Sib A, Gulder TAM (2018) Chemo-enzymatic total synthesis of oxosorbicillinol, sorrentanone, rezishanones B and C, sorbicatechol A, bisvertinolone, and (+)-epoxysorbicillinol. Angew Chem Int Ed 57:14650–14653
415. Lin WH, Fu HZ, Li J et al (2001) Novel chromone derivatives from marine fungus *Aspergillus versicolor* isolated from the sponge *Xestospongia exigua.* Chin Chem Lett 12:235–238
416. Lin W, Brauers G, Ebel R et al (2003) Novel chromone derivatives from the fungus *Aspergillus versicolor* isolated from the marine sponge *Xestospongia exigua.* J Nat Prod 66:57–61
417. Lu Z, Wang Y, Miao C et al (2009) Sesquiterpenoids and benzofuranoids from the marine-derived fungus *Aspergillus ustus* 094102. J Nat Prod 72:1761–1767
418. Kuramochi K, Saito F, Nakazaki A et al (2010) Synthesis of pseudodeflectusin and ustusorane C: structural revision of aspergiones A and B. Biosci Biotechnol Biochem 74:1635–1640
419. Saito F, Kuramochi K, Nakazaki A et al (2006) Synthesis and absolute configuration of (+)-pseudodeflectusin: structural revision of aspergione B. Euro J Org Chem 4796−4799
420. Kuramochi K, Tsubaki K (2015) Synthesis and structural characterization of natural benzofuranoids. J Nat Prod 78:1056–1066
421. Xu J, Nakazawa T, Ukai K et al (2008) Tetrahydrobostrycin and 1-deoxytetrahydrobostrycin, two new hexahydroanthrone derivatives, from a marine-derived fungus *Aspergillus* sp. J Antibiot 61:415–419
422. Artasasta MA, Yanwirasti Y, Taher M et al (2021) Apoptotic activity of new oxisterigmatocystin derivatives from the marine-derived fungus *Aspergillus nomius* NC06. Mar Drugs 19:631
423. Simonetti SO, Larghi EL, Bracca ABJ et al (2012) Synthesis of the unique angular tricyclic chromone structure proposed for aspergillitine, and its relationship with alkaloid TMC-120B. Org Biomol Chem 10:4124–4134
424. Jadulco R, Brauers G, Edrada RA et al (2002) New metabolites from sponge-derived fungi *Curvularia lunata* and *Cladosporium herbarum.* J Nat Prod 65:730–733

425. Sibero MT, Zhou T, Fukaya K et al (2019) Two new aromatic polyketides from a sponge-derived *Fusarium*. Beilstein J Org Chem 15:2941–2947
426. Sibero MT, Zhou T, Igarashi Y et al (2020) Chromanone-type compounds from marine sponge-derived *Daldinia eschscholtzii* KJMT FP 4.1. J Appl Pharm Sci 10:001–007
427. Shiono Y, Sasaki T, Shibuya F et al (2013) Isolation of a phomoxanthone A derivative, a new metabolite of tetrahydroxanthone, from a *Phomopsis* sp. isolated from mangrove, *Rhizhopora mucronata*. Nat Prod Commun 8:1735–1737
428. Tarman K, Lindequist U, Wende K et al (2011) Isolation of a new natural product and cytotoxic and antimicrobial activities of extracts from fungi of Indonesian marine habitats. Mar Drugs 9:294–306
429. Kotoku N, Higashimoto K, Kurioka M et al (2014) Xylarianaphthol-1, a novel dinaphthofuran derivative, activates p21 promoter in a p53-independent manner. Bioorg Med Chem Lett 24:3389–3391
430. Abrell LM, Bargeson B, Crews P (1996) A new polyketide, secocurvularin, from the salt water culture of a sponge derived fungus. Tetrahedron Lett 37:8983–8984
431. Itoh T, Kinoshita M, Aoki S et al (2003) Komodoquinone A, a novel neuritogenic anthracycline, from marine *Streptomyces* sp. KS3. J Nat Prod 66:1373–1377
432. Itoh T, Kinoshita M, Wei H et al (2003) Stereostructure of komodoquinone A, a neuritogenic anthracyline, from marine *Streptomyces* sp. KS3. Chem Pharm Bull 51:1402–1404
433. Dai J, Sorribas A, Yoshida WY et al (2010) Xestosaprols from the Indonesian Marine Sponge *Xestospongia* sp. J Nat Prod 73:1188–1191
434. Millán-Aguiñaga N, Soria-Mercado IE, Williams P (2010) Xestosaprols D and E from the Indonesian marine sponge *Xestospongia* sp. Tetrahedron Lett 51:751–753
435. He F, Mai LH, Longeon A et al (2015) Novel adociaquinone derivatives from the Indonesian sponge *Xestospongia* sp. Mar Drugs 13:2617–2628
436. Cao S, Foster C, Brisson M, Lazo JS, Kinston DGI (2005) Halenaquinone and xestoquinone derivatives, inhibitors of Cdc25B phosphatase from a *Xestospongia* sp. Bioorg Med Chem 13:999–1003
437. Zhu Y, Yoshida WY, Kelly-Borges M et al (1998) Noelaquinone, a new hexacyclic triazine quinone from an Indonesian *Xestospongia* sp. Heterocycles 49:355–360
438. Tanokashira N, Kukita S, Kato H et al (2016) Petroquinones: trimeric xestoquinone derivatives isolated from the marine sponge *Petrosia alfiani*. Tetrahedron 72:5530–5540
439. Kappen J, Manurung J, Fuchs T et al (2023) Challenging structure elucidation of lumnitzeralactone, an ellagic acid derivative from the mangrove *Lumnitzera racemosa*. Mar Drugs 21:242
440. Costantino V, Sala GD, Saurav K et al (2017) Plakofuranolactone as a quorum quenching agent from the Indonesian sponge *Plakortis* cf. *lita*. Mar Drugs 15:59–71
441. Smith CJ, Abbanat D, Bernan VS et al (2000) Novel polyketide metabolites from a species of marine fungi. J Nat Prod 63:142–145
442. Choi PJ, Sperry J, Brimble MA (2010) Heteroatom-directed reverse Wacker oxidations. Synthesis of the reported structure of (–)-herbaric acid. J Org Chem 75:7388–7392
443. Trost BM, Aponick A (2006) Palladium-catalyzed asymmetric allylic alkylation of meso- and dl-1,2-divinylethylene carbonate. J Am Chem Soc 128:3931–3933
444. Zhang F-Z, Li X-L, Li X et al (2019) Polyketides from the mangrove-derived endophytic fungus *Cladosporium cladosporioides*. Mar Drugs 17:296
445. Shiono Y, Shibuya F, Koseki T et al (2014) A new α-pyrone metabolite from a mangrove plant endophytic fungus *Fusarium* sp. J Asian Nat Prod Res 17:403–408
446. Tarman K, Palm GJ, Porzel A et al (2012) Helicascolide C, a new lactone from an Indonesian marine algicolous strain of *Daldinia eschscholzii* (Xylariaceae, Ascomycota). Phytochem Lett 5:83–86
447. Heimberger J, Cade HC, Padgett J et al (2015) Total synthesis of herbarin A and B, determination of their antioxidant properties and toxicity in zebrafish embryo model. Bioorg Med Chem Lett 25:1192–1195

448. Haydl AM, Berthold D, Spreider PA et al (2016) Stereodivergent and protecting-group-free synthesis of the helicascolide family: a rhodium-catalyzed atom-economical lactonization strategy. Angew Chem Int Ed 55:5765–5769
449. Sirirath S, Tanaka J, Ohtani II et al (2002) Bitungolide A–F, new polyketides from the Indonesian sponge *Theonella* cf. *swinhoei*. J Nat Prod 65:1820–1823
450. Reddy KM, Shashidar J, Ghosh SA (2014) Concise approach for the synthesis of bitungolides: total syntheses of (–)-bitungolides B & E. Org Biomol Chem 12:4002–4012
451. Shashidar J, Reddy KM, Ghosh S (2011) The first total synthesis of (–)-bitungolide E. Tetrahedron Lett 52:3106–3109
452. Ghosh S, Kumar SU, Shashidhar J (2008) Total synthesis of (–)-bitungolide F and determination of its absolute stereochemistry. J Org Chem 73:1582–1585
453. Cruz PG, Fernández R, Rodríguez-Acebes R et al (2024) From sea sponge to clinical trials: starting the journey of the novel compound PM742. Mar Drugs 22:339
454. Edrada RA, Heubes M, Brauers G et al (2002) Online analysis of xestodecalactones A-C, novel bioactive metabolites from the fungus *Penicillium* cf. *montanense* and their subsequent isolation from the sponge *Xestospongia exigua*. J Nat Prod 65:1598–1604
455. Pham CD, Hartmann R, Böhler P et al (2014) Callyspongiolide, a cytotoxic macrolide from the marine sponge *Callyspongia* sp. Org Lett 16:266–269
456. Willamson RT, Boulanger A, Vulpanovici A et al (2002) Structure and absolute stereochemistry of phormidolide, a new toxic metabolite from the marine cyanobacterium *Phormidium* sp. J Org Chem 67:7927–7936
457. Trianto A, Hermawan I, Suzuka T et al (2011) Two new cytotoxic candidaspongiolides from an Indonesian sponge. ISRN Pharm 1–6
458. Corley DG, Herb R, Moore RE et al (1988) Laulimalides: new potent cytotoxic macrolides from a marine sponge and a nudibranch predator. J Org Chem 53:3644–3646
459. Quiñoà E, Kakou Y, Crews P (1988) Fijianolide, polyketide heterocycles from a marine sponge. J Org Chem 53:3642–3644
460. Sinisi A, Calcinai B, Cerrano C et al (2013) Isoswinholide B and swinholide K, potently cytotoxic dimeric macrolides from *Theonella swinhoei*. Bioorg Med Chem Lett 21:5332–5338
461. Bringmann G, Lang G, Michel M et al (2004) Stereochemical assignment of the fungal metabolite xestodecalactone A by total synthesis. Tetrahedron Lett 45:2829–2831
462. Liang Q, Zhang J, Quan W et al (2007) The first asymmetric total syntheses and determination of absolute configurations of xestodecalactones B and C. J Org Chem 72:2694–2697
463. Gesner S, Cohen N, Ilan M, Yarden O, Carmeli S (2005) Pandangolide 1a, a metabolite of the sponge-associated fungus *Cladosporium* sp., and the absolute stereochemistry of pandangolide 1 and *iso*-cladospolide B. J Nat Prod 68:1350–1353
464. Show K, Gonnade RG, Kumar P (2018) First total synthesis of the proposed structure of pandangolide 1. Eur J Org Chem 25:3352–3364
465. Zhou J, Gao B, Xu Z et al (2016) Total synthesis and stereochemical assignment of callyspongiolide. J Am Chem Soc 138:6948–6951
466. Lam NYS, Muir G, Challa VR et al (2019) A counterintuitive stereochemical outcome from a chelation-controlled vinylmetal aldehyde addition leads to the configurational reassignment of phormidolide A. Chem Commun 55:9717–9720
467. Helfrich EJN, Ueoka R, Dolev A et al (2019) Automated structure prediction of *trans*-acyltransferase polyketide synthase products. Nat Chem Biol 15:813–821
468. Ndukwe IE, Wang X, Lam NYS et al (2020) Synergism of anisotropic and computational NMR methods reveals the likely configuration of phormidolide A. Chem Commun 56:7565–7568
469. Muir G, Paterson I, Lam NYS et al (2024) Synthesis of the phormidolide A macrocycle supports the proposed configurational reassignment. Org Lett 26:8548–8553
470. Jefford CW, Bernardinelli G, Tanaka J et al (1996) Structures and absolute configurations of the marine toxins, latrunculin A and laulimalide. Tetrahedron Lett 37:159–162
471. Ghosh AK, Wang Y (2000) Total synthesis of (–)-laulimalide. J Am Chem Soc 122:11027–11028

472. Gollner A, Mulzer J (2008) Total synthesis of neolaulimalide and isolaulimalide. Org Lett 10:4701–4704
473. Washida K, Koyama T, Yamada K et al (2006) Karatungiols A and B, two novel antimicrobial polyol compounds, from the symbiotic marine dinoflagellate *Amphidinium* sp. Tetrahedron Lett 47:2521–2525
474. Zhu M, Gao H, Wu C et al (2015) Lipid-lowering polyketides from a soft coral-derived fungus *Cladosporium* sp. TZP29. Bioorg Med Chem Lett 25:3606–3609
475. Agarwal V, Blanton J, Podell S et al (2017) Metagenomic discovery of polybrominated diphenyl ether biosynthesis by marine sponges. Nat Chem Biol 13:537–543
476. Agarwal V, El Gamal A, Yamanaka K (2014) Biosynthesis of polybrominated aromatic organic compounds by marine bacteria. Nat Chem Biol 10:640–647
477. Handayani D, Edrada RA, Proksch P. Wray V et al (1997) Four new bioactive polybrominated diphenyl ethers of the sponge *Dysidea herbacea* from West Sumatra, Indonesia. J Nat Prod 60:1313−1316
478. Hanif N, Tanaka J, Setiawan A et al (2007) Polybrominated diphenyl ethers from the Indonesian sponge *Lamellodysidea herbacea*. J Nat Prod 70:432–435
479. Calcul L, Chow R, Oliver AG et al (2009) NMR strategy for unravelling structures of bioactive sponge-derived oxy-polyhalogenated diphenyl ethers. J Nat Prod 72:443–449
480. Putra MY, Ianaro A, Panza E et al (2012) Sinularioside, a triacetylated glycolipid from the Indonesian soft coral *Sinularia* sp., is an inhibitor of NO release. Bioorg Med Chem Lett 22:2723–2725
481. Riccio R, Kinnel RB, Bifulco G et al (1996) Kakelokelose, a sulfated mannose polysaccharide with anti-HIV activity from the Pacific tunicate *Didemnum molle*. Tetrahedron Lett 37:1979–1982
482. Smith CJ, Jompa J, Tahir A et al (1998) Cadiolides A and B, new metabolites from an ascidian of the genus *Botryllus*. J Org Chem 63:4147–4150
483. Yoo H-D, Ketchum SO, France D et al (2002) Vidalenolone, a novel phenolic metabolite from the tropical red alga *Vidalia* sp. J Nat Prod 65:51–53
484. Lan S, Sheng-Ding Z, Hua-Jie Z (2011) Absolute configuration determination for natural products (1)-absolute configuration determination for ketone, lactone, and alcohol by comparing computed optical rotations and ^{13}C NMR with the experimental results. Chem J Chin Univ 32:2568–2573
485. Moobery SL, Tien G, Hernandez AH et al (1999) Laulimalide and isolaulimalide, new-paclitaxel-like microtubule-stabilizing agents. Cancer Res 59:653–660
486. Pryor DE, O'Brate A, Blicer G et al (2002) The microtubule stabilizing agent laulimalide does not bind in the taxoid site, kills cells resistant to paclitaxel and epothilones, and may not require its epoxide moiety for activity. Biochemistry 41:9109–9115
487. Gallagher BM Jr, Fang FG, Johannes CW et al (2004) Synthesis and biological evaluation of (–)-laulimalide analogues. Bioorg Med Chem Lett 14:575–579
488. Ahmed A, Hoegenauer EK, Enev VS et al (2003) Total synthesis of the microtubule stabilizing antitumor agent laulimalide and some nonnatural analogues: the power of Sharpless asymmetric epoxidation. J Org Chem 68:3026–3042
489. Gollner A, Altmann K−H, Gertsch J et al (2009) The laulimalide family: total synthesis and biological evaluation of neolaulimalide, isolaulimalide, laulimalide and a nonnatural analogue. Chem Eur J 15:5979−5997
490. Tanaka J, Higa T, Bernardinelli G et al (1996) New cytotoxic macrolides from the sponge *Fasciospongia rimosa*. Chem Lett 25:255–256
491. Oda T, Kamoshita K, Maruyama S et al (2007) Cytotoxicity of lissoclibadins and lissoclinotoxins, isolated from a tropical ascidian *Lisscoclinum* cf. *badium*, against human solid-tumor-derived cell lines. Biol Pharm Bull 30:385–387
492. Aoki S, Watanabe Y, Tanabe D et al (2007) Structure–activity relationship and biological property of cortistatins, anti-angiogenic spongean steroidal alkaloids. Bioorg Med Chem Lett 15:6758–6762

493. Morris JD, Takahashi-Ruiz L, Persi LN et al (2022) Re-evaluation of the fijianolide/laulimalide chemotype suggests an alternate mechanism of action for C-15/C-20 analogs. ACS Omega 7:8824–8832
494. Kubota T, Kurimoto S, Kobayashi J (2020). The manzamine alkaloids. In: Knölker HJ (ed) The alkaloids: chemistry and biology. Elsevier
495. Arai M, Yamano Y, Kamiya K et al (2016) Anti-dormant mycobacterial activity and target molecule of melophlins, tetramic acid derivatives isolated from a marine sponge of *Melophlus* sp. J Nat Med 70:467–475
496. Liang X, Luo D, Luesch H (2019) Advances in exploring the therapeutic potential of marine natural products. Pharmocol Res 147:104373
497. Indu S, Kaliappan KP (2020) Synthetic approaches towards cortistatins: evolution and progress through its ages. Org Biomol Chem 18:3965–3995
498. Cao YN, Zheng LL, Wang D et al (2018) Recent advances in microtubule-stabilizing agents. Eur J Med Chem 143:806–828
499. Sakai R, Higa T, Jefford CW et al (1986) Manzamine A, a novel antitumor alkaloid from a sponge. J Am Chem Soc 108:6404–6405
500. Ichiba T, Sakai R, Kohmoto S et al (1988) New manzamine alkaloids from a sponge of the genus *Xestospongia.* Tetrahedron Lett 29:3083–3086
501. Waters AL, Peraud O, Kasanah N et al (2014) An analysis of the sponge *Acanthostrongylophora ingens*' microbiome yields an actinomycete that produce natural product manzamine A. Front Mar Sci 1:1–15
502. Winkler JD, Londregan AT, Hamann MT (2007) Synthetic modification of manzamine A via Grubbs metahathesis. Novel structures with enhanced antibacterial and antiprotozoal properties. Org Lett 9:4467–4469
503. Burkhardt I, de Rond T, Chen PYT et al (2022) Ancient plant-like terpene biosynthesis in corals. Nat Chem Biol 18:664–669
504. Zdouc M, Blin K, Louwen NLL et al (2025) MIBiG 4.0: advancing biosynthetic gene cluster curation through global collaboration. Nucleic Acids Res 53:D678–D690
505. Poynton EF, van Santen JA, Pin M et al (2025) The natural products atlas 3.0: extending the database of microbially derived natural products. Nucleic Acids Res 53:D691–D699
506. Gangwal A, Lavecchia A (2025) Artificial intelligence in natural product drug discovery: current applications and future perspectives. J Med Chem 68:3948–3969
507. Martínez H, Santos M, Pedraza L et al (2025) Advanced technologies for large scale supply of marine drugs. Mar Drugs 23:69
508. Hanif N, Tyas TA, Hidayati L et al (2021) Oxy-polybrominated diphenyl ethers from the Indonesian marine sponge, *Lamellodysidea herbacea*: X-ray, SAR, and computational studies. Molecules 26:6328
509. Hanif N, Miftah JA, Yanti HD et al (2025) Integrated biological and chemical investigation of Indonesian marine organisms targeting anti-quorum-sensing, anti-biofilm, anti-biofouling, and anti-biocorrosion activities. Molecules 30:1202
510. Hanif N, Ardan MS, Tohir D et al (2019) Polybrominated diphenyl ethers with broad spectrum antibacterial activity from the Indonesian marine sponge *Lamellodysidea herbacea.* J Appl Pharm Sci 9:001–006
511. Hanif N, Ardianti R, Ahmadi P et al (2018) Ichthyotoxic principles against zebrafish embryos from the Indonesian marine sponge *Neopetrosia chaliniformis.* J Appl Pharm Sci 8:044–048
512. Hanif N, Nuraulia BS, Ohlinger WS et al (2024) Stereochemical assignment of varicosenone, a merosteroid flexible side chain, from the Indonesian sea slug *Phyllidia varicosa* using NMR and DFT-based NMR calculations. J Appl Pharm Sci 14:240–244

Chapter 3
Marine Natural Products from Malaysia

Abujunaid Habib Khan and Chin-Soon Phan

1 Introduction

Malaysia is a Southeast Asia country with a land area of 330,803 km^2. It is separated into Peninsular Malaysia and Borneo's East Malaysia by the South China Sea. In term of biodiversity, Malaysia is ranked 12th according to the National Biodiversity Index and is one of the world's megadiverse countries. Based on the Convention on Biological Diversity (www.cbd.int), Malaysia contained about 15,000 species of vascular plants, 306 species of mammals, 742 species of birds, 242 species of amphibians, 567 species of reptiles, over 400 species of freshwater fish, over 500 species of marine fish and more than 150,000 species of invertebrates.

In 1994, the Asian Symposium on Medicinal Plants and Spices (ASOMPS VII) was held in Melaka, Malaysia. Malaysian Natural Products Society (MNPS, https://mymnps.org/) was formed as a spin off activity from ASOMPS VII and formally registered in 1995. One of the aims of MNPS is to (1) promote natural products research and (2) regularly organizes a natural product conference in Malaysia, also known as International Conference on Natural Products (ICNP). Many early natural products studies in Malaysia were focused on terrestrial natural products, also known as phytochemistry, a research area that remains active even today.

To the best of our knowledge, marine natural products have been isolated from Malaysian soft corals and red algae since 2000s, and this represent one of the earliest studies of marine natural products in Malaysia. Most of these marine organisms were collected near Sabah waters and numerous marine natural products have been isolated. Sabah is a state in Malaysia, occupying the northern part of Borneo Island and supports more than 75% of Malaysia's coral reefs. Sabah's northern and eastern

A. H. Khan · C.-S. Phan (✉)
Latvian Institute of Organic Synthesis, Aizkraukles Street 21, LV-1006 Riga, Latvia
e-mail: chinsoon@osi.lv

A. H. Khan
e-mail: abujunaid.habib.khan@osi.lv

L. T. Tan and N. Hanif (eds.), *Marine Natural Products Research in South-East Asia*,
https://doi.org/10.1007/978-3-032-01098-8_3

coasts are part of the Coral Triangle Region, hence the marine biodiversity found within the coastal waters (covering area of 5,436,000 ha) is among the richest in the world, according to the Sabah Biodiversity Strategy 2012–2022 (https://sabc.sabah.gov.my/).

2 Marine Natural Products Chemistry

This section will present all marine natural products isolated from soft corals, red algae, sponges and cyanobacteria collected in Malaysia, with Malaysian soft corals and red alga being the most studied marine organisms. The marine natural products described here include terpenes, polyketides, peptides and alkaloids, with terpenes being the most abundant and mainly found in soft corals and red algae. Terpenes are the largest class of natural products with more than 80,000 known members [1]. In this section, we captured 58 research articles from natural products or closely related journals and revealed 228 different marine natural products and bioactivities from Malaysian soft corals, red algae, sponges and cyanobacteria.

2.1 Soft Corals

The first natural products from Malaysian soft corals were reported in 2007. Yao et al. discovered four new decalin-type bicyclic diterpenoids, lemnalosides A-D (**S1**-**S4**) and a known compound, lemnabourside (**S5**) from soft coral *Lemnalia* sp. collected in the South China Sea, Malaysia (Fig. 1) [2]. Lemnalosides A (**S1**) and B (**S2**) displayed inhibition against hyphae formation of *Streptomyces* 85E at a concentration of 20.0 and 2.5 μg/disc, respectively.

The next chemical investigation on soft coral *Lemnalia* sp. (Mengalum Island, Sabah) was carried out by Phan et al., which characterized two new nardosinane-type sesquiterpenoids, paralemnolins V (**S6**) and W (**S7**), and five known compounds, 4,10-diacetoxy-2,8-neolemnadien-5-one (**S8**), 10-*O*-methyl alismoxide (**S9**), germacrene D (**S10**), 1β-hydroxy-4(15),5*E*,10(14)-germacratriene (**S11**), and paralemnolin L (**S12**) (Fig. 1) [3]. All compounds **S6**-**S12** showed negligible inhibition with MIC > 250 μg/mL against human pathogenic bacteria, *Staphylococcus aureus* (ATCC 25923), *Escherichia coli* (ATCC 11775), *Bacillus cereus* (ATCC 11778), *Yersinia enterocolitica* (ATCC 23715), *Listeria monocytogenes* (ATCC 12932), *Salmonella typhimurium* (ATCC 13311), *Salmonella enteritidis* (ATCC 13076) and *Pseudomonas aeruginosa* (ATCC 10145). Only paralemnolins V (**S6**) and W (**S7**) displayed fungistatic effects against hyphal growth of fungus *Lagenidium thermophilum* IPMB 1401 at MIC 25 μg/mL, and compound **S9** showed fungistatic effects against hyphal growth of fungus *Exophiala* sp. NJM 1551 at MIC 25 μg/mL.

In 2022, Tani et al. isolated one new nardosinane-type sesquiterpenoid, parathyrsoidin K (**S13**) and seven known compounds, including parathyrsoidin J

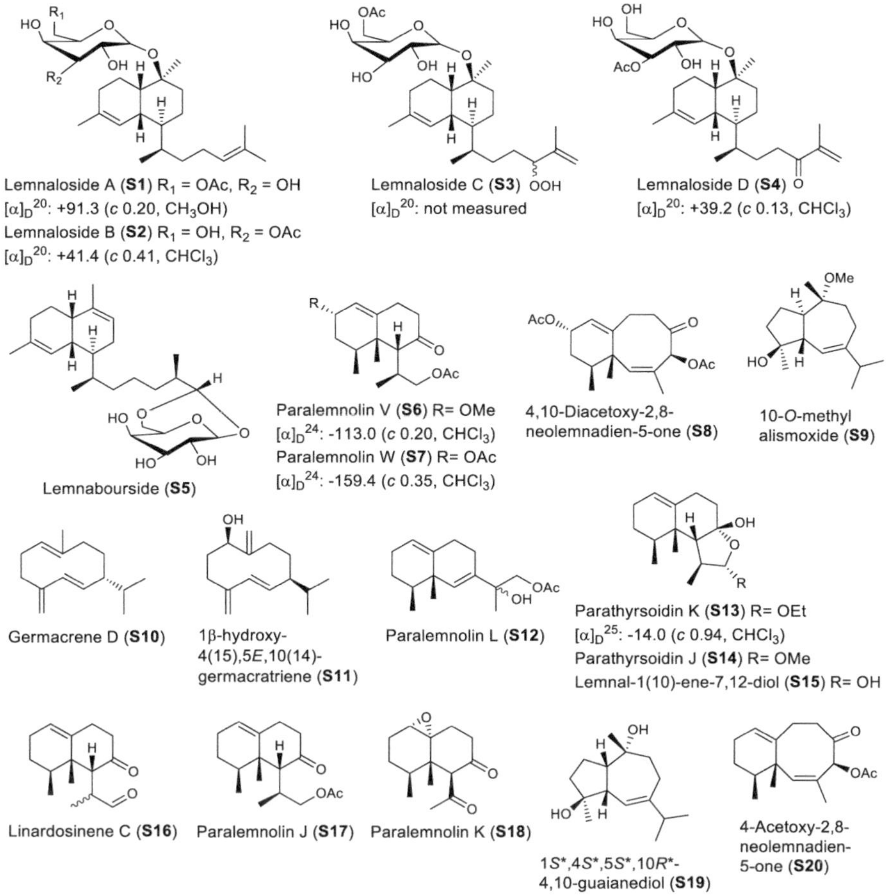

Fig. 1 Structures of marine natural products **S1-S20** isolated from soft corals *Lemnalia* species

(**S14**), lemnal-1(10)-ene-7,12-diol (**S15**), linardosinene C (**S16**), paralemnolins J (**S17**) and K (**S18**), 1*S**,4*S**,5*S**,10*R**-4,10-guaianediol (**S19**) and 4-acetoxy-2,8-neolemnadien-5-one (**S20**) from soft coral *Lemnalia* sp. (Ghuan Bay, North Sabah) (Fig. 1) [4]. In the anti-inflammatory assay, lemnal-1(10)-ene-7,12-diol (**S15**) and linardosinene C (**S16**) inhibited LPS-induced nitric oxide production in RAW 264.7 macrophages by more than 70% at 12.5 μg/mL.

In 2009, Ishii et al. investigated two populations of soft corals *Nephthea* sp. collected at Malaysia. A new steroid, 4α-methyl-ergosta-6,8(14),22*E*-triene-3β-ol (**S21**) and one known compound, cholesterol (**S22**) was discovered from a population of soft coral collected at Sepanggar Island, Sabah (Fig. 2) [5]. While another population of soft coral collected at Sibuan Island, Sabah, contained one new germacrene-type norsesquiterpenoid, 1-acetoxy-germacra-5*E*,10(14)-diene-4-one

(**S23**) and three known compounds, germacra-4(15),5*E*,10(14)-trien-1-ol (**S24**), 1-acetoxy-germacra-4(15),5*E*,10(14)-triene (**S25**) and 24-methylenecholesterol (**S26**) (Fig. 2) [6].

A year later, Ishii et al. reported three known compounds 24-methylenecholesterol (**S26**), 5β,8β-epidioxy-11-hydroxy-6-eudesmene (**S27**), 5β,8β-epidioxy-11-hydroperoxy-6-eudesmene (**S28**) and 24-methyl-cholesta-5,24(28)-diene-3ß,19-diol-7ß-monoacetate (**S29**) from three populations of soft corals *Nephthea* sp.

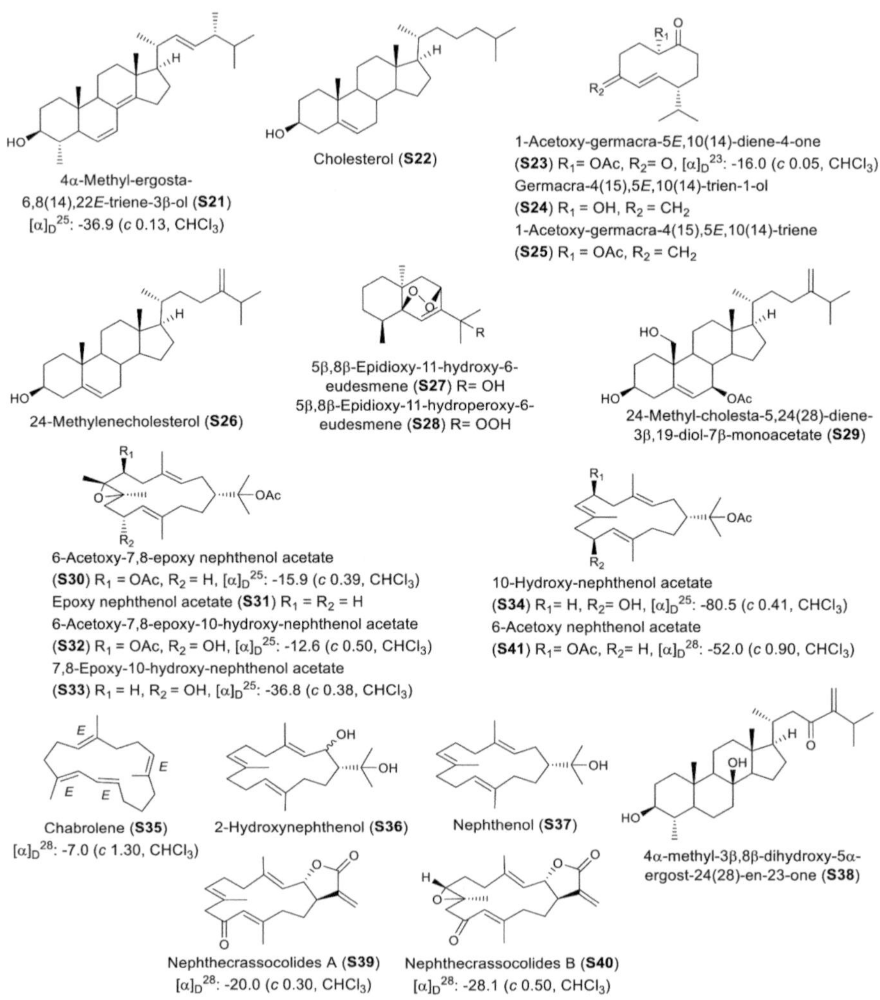

Fig. 2 Structures of marine natural products **S21**-**S41** isolated from soft corals *Nephthea* species

(Dinawan island, Lankayan Island and Sebangkat, Sabah) (Fig. 2) [7]. 5β,8β-epidioxy-11-hydroperoxy-6-eudesmene (**S28**) displayed antimicrobial activity against human pathogenic bacteria *Pseudomonas aurelis*, *Proteus mirabilis*, *Escherichia coli*, *Salmonella enteridis*, *Salmonella thyphymunium*, *Staphylococcus aereus* and *Vibrio cholerae* at 30 μg/disc, but no inhibition zone was found against *Salmonella* sp., *Salmonella typii* and *Listeria monocytogenes*. In addition, compound **S28** also showed strong cytotoxicity against P-388 murine leukemia cells at IC_{50} 0.36 mg/mL. Within the same year, Ishii et al. isolated a new cembrane-type diterpenoid, 6-acetoxy-7,8-epoxynephthenol acetate (**S30**) and a known compound, epoxynephthenol acetate (**S31**) from soft coral *Nephthea* sp. (Layangan Island, Sabah) (Fig. 2) [8].

In 2016, Ishii et al. characterized three new cembrane-type diterpenoids, 6-acetoxy-7,8-epoxy-10-hydroxy-nephthenol acetate (**S32**), 7,8-epoxy-10-hydroxy-nephthenol acetate (**S33**) and 10-hydroxy-nephthenol acetate (**S34**), along with one known compound, 6-acetoxy-7,8-epoxy-nephthenol acetate (**S31**) from soft coral *Nephthea* sp. (Layangan Island, Sabah) (Fig. 2) [9]. Compounds **S33** and **S34** exhibited weak antibacterial activity against *S. aureus* and *E. coli*, and weak cytotoxic activity against human cervical cancer (HeLa) and human breast cancer (MCF-7) cells.

Two years later, Ishii et al. identified a new norditerpene, chabrolene (**S35**) along with three known compounds, 2-hydroxynephthenol (**S36**), nephthenol (**S37**) and 4α-methyl-3β,8β-dihydroxy-5α-ergost-24(28)-en-23-one (**S38**) from soft coral *Nephthea* sp. (Mantanani Island, Sabah) (Fig. 2) [10]. Compound **S35** showed repellent activity against maize weevil *Sitophilus zeamais* at a concentration of 25 μg/cm^2.

In 2019, Tani et al. isolated two new cembrane-type diterpenoids, nephthecrassocolides A (**S39**) and B (**S40**) together with three known compounds, 6-acetoxy nephthenol acetate (**S41**), nephthenol (**S37**), epoxy nephthenol acetate (**S30**), and 6-acetoxy-7,8-epoxy nephthenol acetate (**S31**) from soft coral *Nephthea* sp. (Sepanggar Island, Sabah) (Fig. 2) [11]. Compounds **S39** and **S37** displayed fungistatic effects against hyphal growth of fungus *Lagenidium thermophilum* IPMB 1401 with MIC 12.5 μg/mL.

The first chemical investigation on soft coral *Sinularia* (Semporna Island, Sabah) was conducted in 2012. Vairappan et al. isolated two known diterpenes, (3*S*,4*S*,11*S*,12*S*,1*E*,7*E*)-3,4-11,12-bisepoxycembra-1,7-diene (**S42**) and (1*E*,3*E*,7*E*)-11,12-epoxycembra-1,3,7-triene (**S43**) from *Sinularia flexibilis* (Fig. 3) [12]. Compounds **S42** and **S43** exhibited anti-bacterial activity against *S. aureus* (MIC 20 mg/disc) and *E. coli* (MIC 30 mg/disc).

Two years later, Palaniveloo et al. investigated another *Sinularia flexibilis* (Mantanani Island, Sabah) and reported ten known cembrane-type diterpenoids, cembrene A (**S44**), diepoxycembrene A (**S45**), (1*R*,3*S*,4*S*,7*E*,11*E*)-3,4-epoxycembra-7,11,15-triene (**S46**), not matched compound (**S47**), sinulariolide (**S48**), 11-epoxysinulariolide (**S49**), 5-dehydrosinulariolide (**S50**), sinulariolide acetate (**S51**) dihydrosinuflexolide (**S52**) and flexilarin B (**S53**) (Fig. 3) [13]. **S47** was identified as flexibilide/sinularin, however we found the proposed structure **S47**

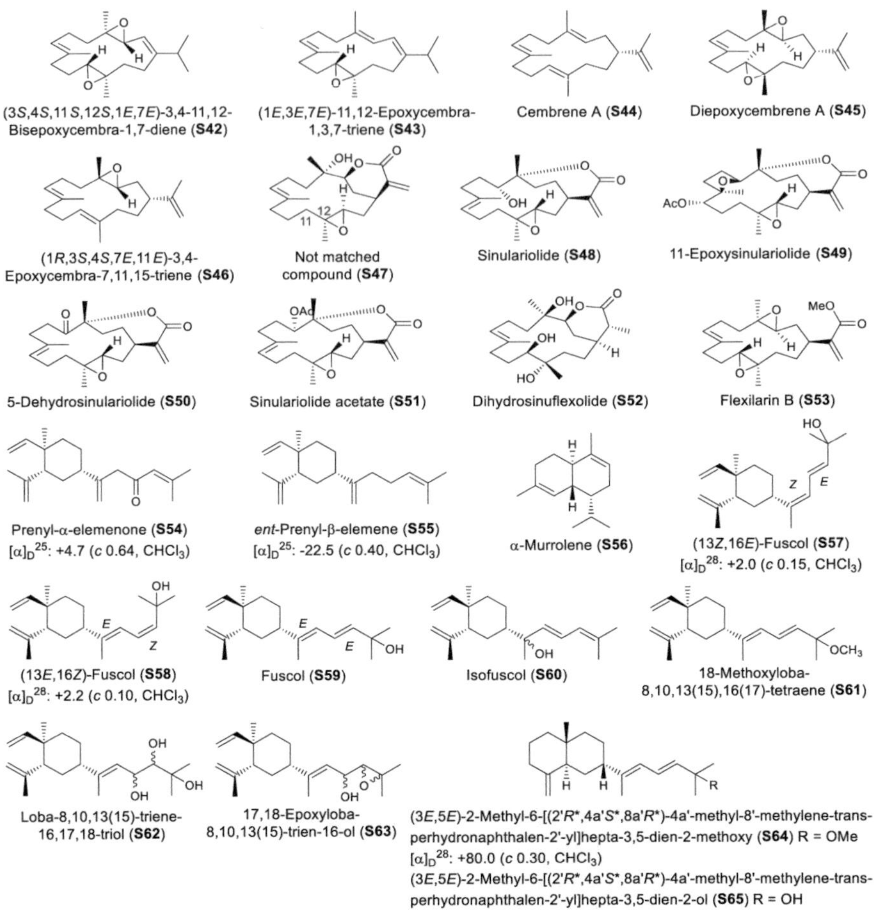

Fig. 3 Structures of marine natural products **S42-S65** isolated from soft corals *Sinularia* species

does not match to flexibilide/sinularin, and we suspect that the correct structure of **S47** should be **S68**, in which the epoxide ring should be located at C-11/C-12 instead of C-12/C-13.

In 2016, Phan and Vairappan isolated two new lobane-type diterpenoids, prenyl-α-elemenone (**S54**) and *ent*-prenyl-β-elemene (**S55**) together with a known compound, α-murrolene (**S56**) from soft coral *Sinularia* sp. (Mantanani Island, Sabah) (Fig. 3) [14]. In the same year, Kamada et al. investigated two populations of soft corals *Sinularia flexibilis* collected at Malaysia [15]. Two new lobane-type diterpenoids, (13*Z*,16*E*)-fuscol (**S57**) and (13*E*,16*Z*)-fuscol (**S58**), and five known compounds, fuscol (**S59**), isofuscol (**S60**), 18-methoxyloba-8,10,13(15),16(17)-tetraene (**S61**), loba-8,10,13(15)-triene-16,17,18-triol (**S62**), and 17,18-epoxyloba-8,10,13(15)-trien-l6-ol (**S63**) were discovered from a population collected at Dinawan Island, Sabah (Fig. 3). While another population collected at Banggi Island, Sabah contained

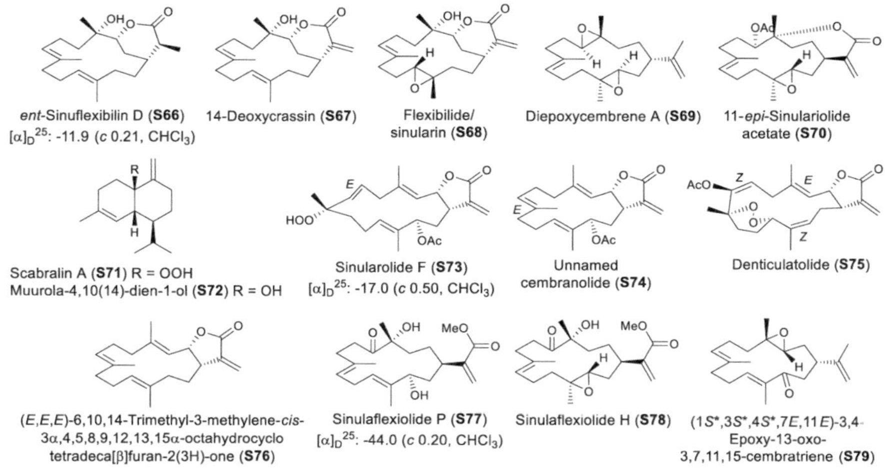

Fig. 4 Structures of marine natural products **S66-S79** isolated from soft corals *Sinularia* species

one new lobane-type diterpenoid, (3*E*,5*E*)-2-methyl-6-[(2′*R**,4a′*S**,8a′*R**)-4a′-methyl-8′-methylene-trans-perhydronaphthalen-2′-yl]hepta-3,5-dien-2-methoxy (**S64**) and a known compound (3*E*,5*E*)-2-methyl-6-[(2′*R**,4a′*S**,8a′*R**)-4a′-methyl-8′-methylene-*trans*-perhydronaphthalen-2′-yl]hepta-3,5-dien-2-ol (**S65**) (Fig. 3). Compounds **S57**, **S58**, **S59** and **S60** exhibited antibacterial activities with MIC 10.0–12.5 μg/disc against *Escherichia coli* HP0808, *Salmonella thypi* HP0809, *Staphylococcus aureus* HP0810 and *Vibrio cholera* HP0811.

In 2018, Kamada et al. investigated two populations of soft corals *Sinularia* collected at Malaysia. One new cembrane-type diterpenoid, *ent*-sinuflexibilin D (**S66**) and seven known compounds, 14-deoxycrassin (**S67**), sinularin (**S68**), diepoxycembrene A (**S69**), 11-*epi*-sinulariolide acetate (**S70**), 5-dehydrosinulariolide (**S50**), scabralin A (**S71**) and muurola-4,10(14)-dien-1-ol (**S72**) were isolated from a population of soft coral collected at Mengalum Island, Sabah (Fig. 4) [16]. While another population of soft coral collected at Mantanani Island, Sabah, contained one new cembrane-type diterpenoid, sinularolide F (**S73**), and three known compounds, unnamed cembranolide (**S74**), denticulatolide (**S75**) and (*E*,*E*,*E*)-6,10,14-trimethyl-3-methylene-*cis*-3α,4,5,8,9,12,13,15α-octahydrocyclo tetradeca[β]furan-2(3H)-one (**S76**) (Fig. 4) [17]. Compounds **S66**, **S67**, **S68**, **S70** and **S50** displayed cytotoxic activity against adult T-cell leukemia (S1T) cells with IC_{50} values ranging from 3.15 to 7.78 μg/mL. Compounds **S67** and **S68** exhibited fungistatic effect with MIC 12.5 μg/mL against hyphal growth of fungus *Exophiala* sp. NJM 1551 and *Lagenidium thermophilum* IPMB 1401, respectively. In the anti-inflammatory assay, compounds **S73** and **S75** inhibited LPS-induced nitric oxide, interleukin-1β and interleukin-6 production in RAW 264.7 macrophages by more than 90% at 25.0 μg/mL. Western blot result showed that

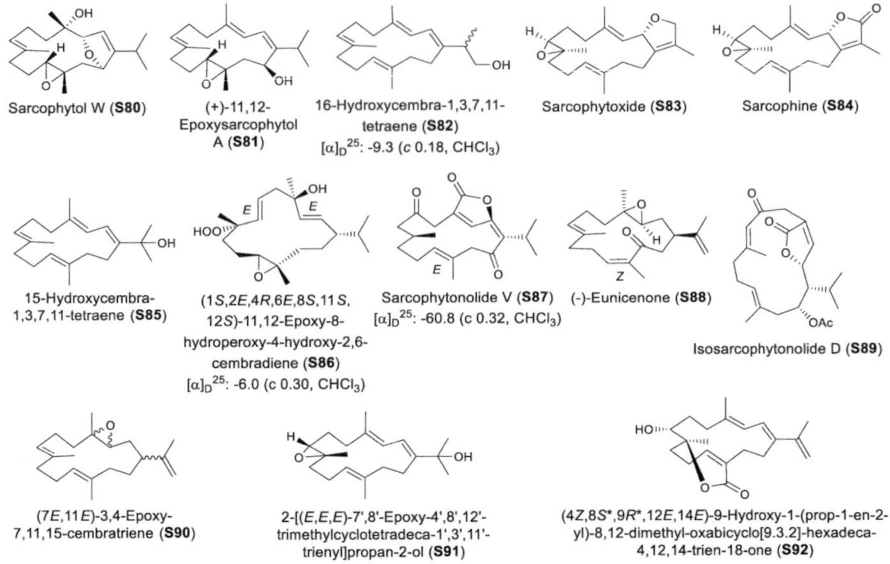

Fig. 5 Structures of marine natural products **S80-S92** isolated from soft corals *Sarcophyton* species

the anti-inflammatory activity of compounds **S73** and **S75** was achieved by down-regulating inducible nitric oxide synthase (iNOS) expression. In addition, compounds **S73** and **S75** triggered the up-regulation of pro-apoptotic protein (BAX), down regulation of anti-apoptotic protein (B-cell lymphoma-extra large, Bcl-xL), and apoptosis activation (caspase-3) in the human leukemia (HL-60) cells.

One year later, Phan et al. isolated one new cembrane-type diterpenoid, sinulaflexiolide P (**S77**) and three known compounds sinulaflexiolide H (**S78**), 11-*epi*-sinulariolide acetate (**S70**), and (1*S**,3*S**,4*S**,7*E*,11*E*)-3,4-epoxy-13-oxo-3,7,11,15-cembratriene (**S79**) from *Sinularia flexibilis* (Mantanani Island, Sabah) (Fig. 4) [18].

The first chemical investigation on soft coral *Sarcophyton* was conducted in 2015. Phan and Vairappan identified two known compounds sarcophytol W (**S80**) and (+)-11,12-epoxysarcophytol A (**S81**) from soft coral *Sarcophyton* sp. (Mantanani Island, Sabah) (Fig. 5) [19]. A year later, Kamada et al. isolated a new cembrane-type diterpenoid, 16-hydroxycembra-1,3,7,11-tetraene (**S82**) along with three known compounds, sarcophytoxide (**S83**), sarcophine (**S84**) and 15-hydroxycembra-1,3,7,11-tetraene (**S85**) from soft coral *Sarcophyton* sp. (Karah Island, Terengganu) (Fig. 5) [20]. In 2018, Kamada et al. investigated another population of soft coral *Sarcophyton* sp. (Sempona Island, Sabah) and discovered one new cembrane-type diterpenoid, (1*S*,2*E*,4*R*,6*E*,8*S*,11*S*,12*S*)-11,12-epoxy-8-hydroperoxy-4-hydroxy-2,6-cembradiene (**S86**) and a known compound, cholesterol (**S22**) (Fig. 5) [21]. A year later, Tani et al. reported a new cembrane-type diterpenoid, sarcophytonolide V (**S87**) and six known compounds, (−)-eunicenone

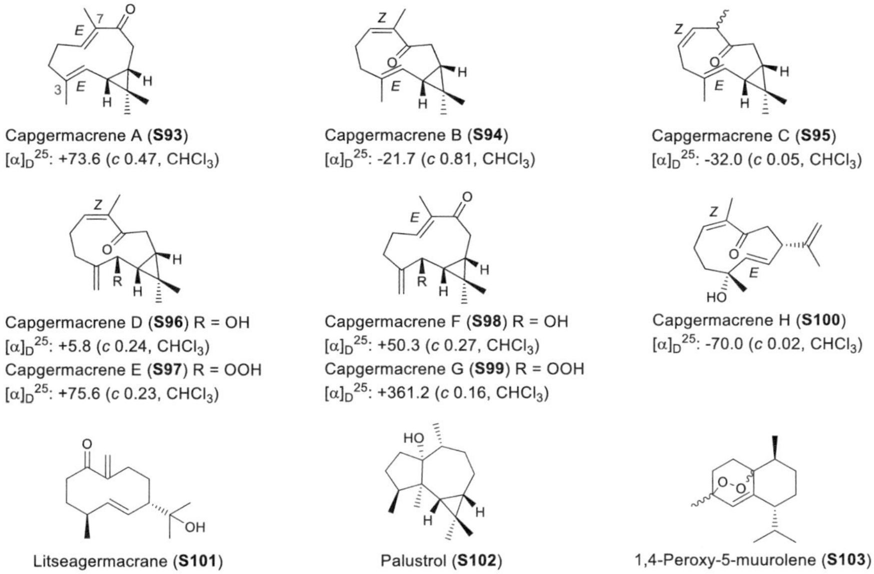

Fig. 6 Structures of marine natural products **S93-S103** isolated from soft corals *Capnella* species

(**S88**), isosarcophytonolide D (**S89**), (1*S**,3*S**,4*S**,7*E*,11*E*)-3,4-epoxy-13-oxo-7,11,15-cembratriene (**S79**), (7*E*,11*E*)-3,4-epoxy-7,11,15-cembratriene (**S90**), 2-[(*E*,*E*,*E*)-7′,8′-epoxy-4′,8′,12′-trimethylcyclotetradeca-1′,3′,11′-trienyl]propan-2-ol (**S91**), and (4*Z*,8*S**,9*R**,12*E*,14*E*)-9-hydroxy-1-(prop-1-en-2-yl)-8,12-dimethyl-oxabicyclo[9.3.2]-hexadeca-4,12,14-trien-18-one (**S92**) from soft coral *Sarcophyton* sp. (Sepanggar Island, Sabah) (Fig. 5) [22]. Sarcophytonolide V (**S87**) exhibited fungistatic effect against hyphal growth of fungus *Haliphthoros sabahensis* IPMB 1402 at MIC 6.25 μg/mL.

From 2015 to 2019, Phan et al. isolated eight new germacrane-type sesquiterpenoids, capgermacrenes A (**S93**), B (**S94**), C (**S95**), D (**S96**), E (**S97**), F (**S98**), G (**S99**) and H (**S100**), together with three known compounds, litseagermacrane (**S101**), palustrol (**S102**) and 1,4-peroxy-5 muurolene (**S103**) from soft coral *Capnella* sp. (Mantanani Island, Sabah) (Fig. 6) [23–26]. In the anti-inflammatory assay, capgermacrene A (**S93**) inhibited LPS-induced nitric oxide, prostaglandin E_2 and interleukin-1β production in RAW 264.7 macrophages by more than 60% at 20.0 μg/mL, while capgermacrene B (**S94**) was inactive. Western blot result showed that the anti-inflammatory activity of capgermacrene A (**S93**) was achieved by down-regulating inducible nitric oxide synthase (iNOS) expression. In addition to anti-inflammatory property, capgermacrenes A (**S93**), B (**S94**), D (**S96**), E (**S97**), F (**S98**) and G (**S99**) displayed cytotoxic activity against adult T-cell leukemia (S1T) cells with IC_{50} of 0.79, 7.79, 6.19, 4.75, 2.39 and 3.97 μg/mL, respectively. Despite having the same planar structure and different only in one olefin configuration, capgermacrene A (**S93**) is more potent than capgermacrene B (**S94**), indicating that both trisubstituted olefins at C-2/C-3 and C-6/C-7 with *trans* (*E*) configuration are important for the activity.

From 2018 to 2019, Phan et al. investigated three populations of soft corals *Xenia* sp. collected at Malaysia. Population A (Mengalum Island, Sabah) contained one new guaiane-type sesquiterpenoid, 10β-*O*-methyl-1αH,5αH-guaia-6-en-4β-ol (**S104**) and two known compounds alismoxide (**S105**) and 10-*O*-methyl alismoxide (**S106**) [27], population B (Mantanani Island, Sabah) contained one new xenicane-type diterpenoid, 15-deoxy-isoxeniolide-A (**S107**) and four known compounds, xeniolide-A (**S108**), isoxeniolide-A (**S109**), 9-deoxy-isoxeniolide-A (**S110**) and coraxeniolide-B (**S111**) [28], and population C (Mengalum Island, Sabah) contained one new xenicane-type diterpenoid, 12-*epi*-9-deacetoxyxenicin (**S112**) along with a hydroperoxide product, 12-*epi*-9-deacetoxy-8-hydroperoxyxenicin (**S113**) and two known compounds, germacrene C (**S114**) and alismol (**S115**) (Fig. 7) [29]. Compounds **S111, S112** and **S113** displayed cytotoxic activity against adult T-cell leukemia (S1T) cells with IC_{50} of 4.12, 5.89 and 6.45 μg/mL, respectively.

In 2014, Bahl et al. characterized a briarane-type diterpenoid, 2-acetoxyverecynarmin C (**S116**) from soft coral *Pennatula aculeata* (Dinawan Island, Sabah) (Fig. 8) [30]. In 2022, Tani and Vairappan isolated two known

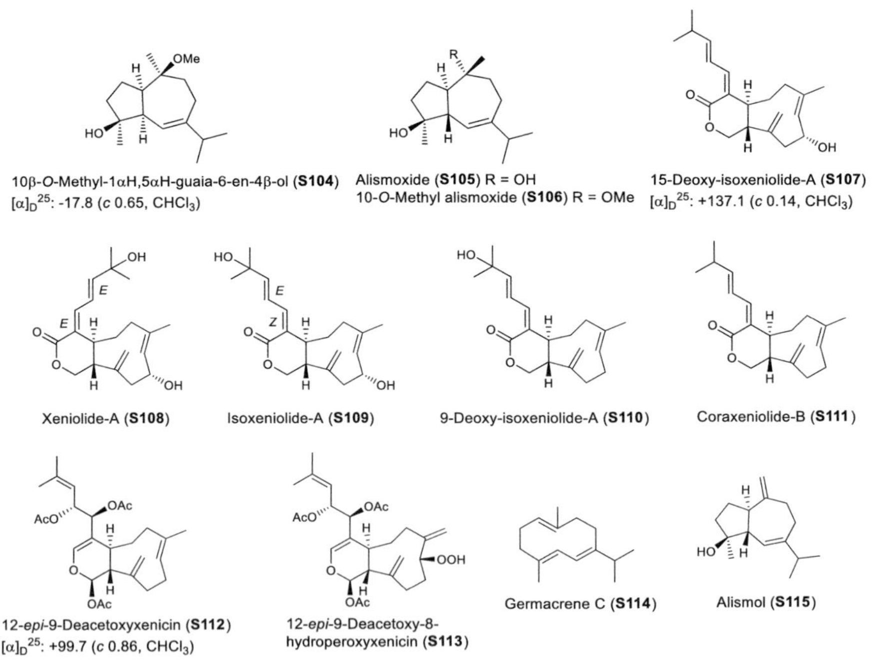

Fig. 7 Structures of marine natural products **S104-S115** isolated from soft corals *Xenia* species

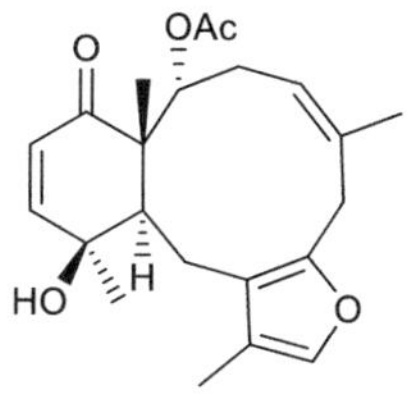

2-Acetoxyverecynarmin C (**S116**)
$[\alpha]_D^{20}$: -7.5 (*c* 1.00, EtOH)

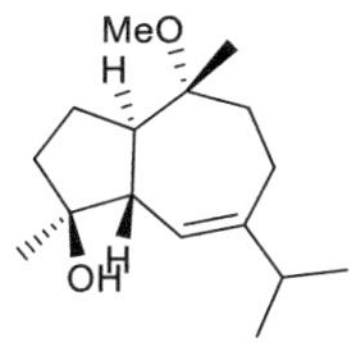

10α-Methoxy-4β-hydroxy guaian-6-ene (**S117**)

Fig. 8 Structures of **S116** and **S117** isolated from soft corals *Pennatula* and *Litophyton*, respectively

compounds, alismol (**S115**) and 10α-methoxy-4β-hydroxy guaian-6-ene (**S117**) (Fig. 8) from soft coral *Litophyton arboreum* (Sepanggar Island, Sabah) [31]. Compound **S117** displayed fungistatic effect against hyphal growth of fungus *Haliphthoros milfordensis* IPMB 1603 at MIC 12.5 μg/mL.

2.2 Red Algae

The first natural products isolated from Malaysian red algae were reported in 2001. Suzuki et al. discovered a new C_{15}-acetogenin, (+)-3*Z*-chlorofucin (**R1**) and two new chamigrane-type sesquiterpenoids, pannosane (**R2**) and pannosanol (**R3**) from red alga *Laurencia pannosa* (Talang-Talang Kecil Island, Sarawak) (Fig. 9) [32]. All isolated compounds were tested against marine bacteria. Pannosanol (**R3**) displayed antibacterial activity MIC 60 μg/disc against *Proteus mirabilis* and MIC 100 μg/disc against *Chromobacterium violaceum* and *Vibrio cholerae*. (3Z)-chlorofucin (**R1**) and pannosane (**R2**) exhibited antibacterial activity against *Chromobacterium violaceum* with MIC 100 and 60 μg/disc, respectively. In the same year, Suzuki group reported another two new C_{15}-acetogenins, lembyne-A (**R4**) and lembyne-B (**R5**) from new species of red alga *Laurencia* (Lembu Island, Kedah). While two known compounds, elatol (**R6**) and *iso*-obtusol (**R7**) were isolated from five populations of red algae *Laurencia majuscula* (Satang Besar Island, Sarawak; Redang Island, Terengganu; Tikus Island, Nunuyan Laut Island and Bankawan Island, Sabah) (Fig. 9) [33]. All isolated compounds were tested against marine bacteria. Elatol (**R6**) showed antibacterial activity against *P. mirabilis*, *Clostridium cellobioparum* and *Flavobacterium helmiphilum* with MIC 5, 5 and 10 μg/disc, respectively. *Iso*-obtusol (**R7**) displayed activity against *C. cellobioparum* and *P. mirabilis* with MIC 10 and 15 μg/disc, respectively.

Two years later, Vairappan isolated two known chamigrane-type sesquiterpenoids, elatol (**R6**) and *iso*-obtusol (**R7**) from red algae *L. majuscula* (Tikus Island and Bankawan Island, Sabah). Elatol (**R6**) displayed antibacterial activity with MIC < 10 mg/mL against pathogen *Klebsiella pneumoniae* (BP984), *Salmonella* sp.

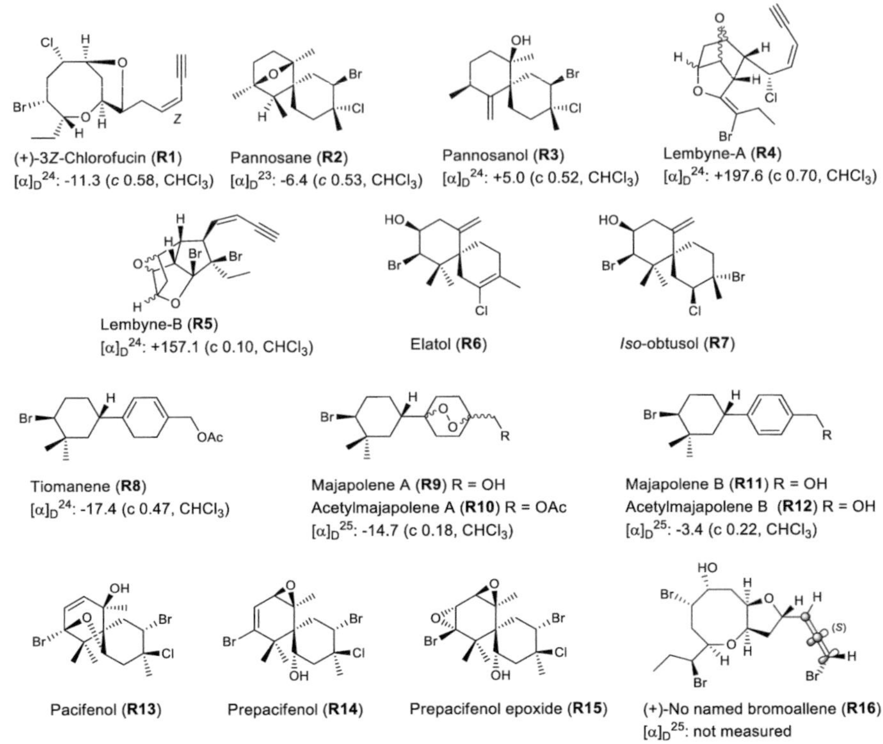

Fig. 9 Structures of **R1**-**R16** isolated from red algae *Laurencia*

(BP986), *Staphylococcus aureus* (BP987), *Staphylococcus epidermis* (BP988) and *Pseudomonas* sp. (BP985) [34].

In 2008, Vairappan et al. investigated three populations of red algae *Laurencia* sp. Population A collected at Tioman Island, Pahang contained three new compounds, tiomanene (**R8**), acetylmajapolenes A (**R10**) and B (**R12**) along with two known compounds, majapolenes A (**R9**) and B (**R11**) (Fig. 9) [35]. Population B (Karah Island, Terengganu) contained three known chamigrane-type sesquiterpenoids, pacifenol (**R13**), prepacifenol (**R14**) and prepacifenol epoxide (**R15**). Population C (Nvireh Island, Terengganu) found two known C_{15}-acetogenins, (+)-3*Z*-chlorofucin (**R1**) and an unnamed (+)-bromoallene-related molecule (**R16**). Compounds **R8**-**R11** displayed antibacterial activity with MIC < 40 μg/disc against marine bacteria *Proteus mirabilis*, *Proteus vulgaris*, *Chromobacterium violaceum*, *Erwinia* sp., *Vibrio parahaemolyticus* and *Vibrio alginolyticus*.

In 2009, Vairappan and Lee isolated a known C_{15}-acetogenin, (+)-3*Z*-dihydrorhodophytin (**R17**) from red alga *Laurencia nangii* Masuda (Banggi Island, Sabah) (Fig. 10) [36]. (+)-3*Z*-dihydrorhodophytin (**R17**) exhibited antibacterial activity at MIC 1.25 μg/disc against *Salmonella enteritidis*, and 2.25 μg/disc against *Staphalococcus aureus* and *Vibrio Cholerae.* The inhibition of *E. coli*,

Salmonella thphymunium and *Salmonella typhii* were at MIC 7.25 μg/disc. In 2010, Vairappan et al. reported a new compound, 10-acetoxyangasiol (**R18**) and four known compounds, an unnamed chamigrane (**R19**), 1-methyl-2,3,5-tribromoindole (**R20**), cupalaurenol (**R21**) and aplysidiol (**R22**) from four populations of red algae *Laurencia* sp. (Lohok Butun, Selakan Island, Layangan Island and Gaya Island, Sabah) (Fig. 10) [37]. All isolated compounds showed antibacterial activity against *Staphyloccocus* sp. at 30 mg/disc.

In 2011, Vairappan group isolated five known synderane-type sesquiterpenes, palisadin A (**R23**), palisadin B (**R24**), 5β-acetoxypalisadin B (**R25**), aplysistatin (**R26**), and 5β-hydroxypalisadin B (**R27**) and two known chamigrane-type sesquiterpenes, 4-bromo-β-chamigren-8-one (**R28**) and (2*Z*,6*S**,9*R**)-chamigr-2,5(14)-dien-8-one (**R29**) from three populations of red algae *L. snackeyi* (Bak-Bak Beach, Carrington Reef and Lankayan Island, Sabah) (Fig. 10) [38]. The researcher proposed the first chemotaxonomical markers for *L. snackeyi* in Malaysia. One year later, Kamada and Vairappan characterized one new compound, (+)-dihydroitomanallene B (**R30**), and two known compounds, (+)-itomanallene B (**R31**) and (+)-pannosallene (**R32**) from six populations of red algae *L. nangii* (Tun Sakaran Marine Park, Sabah) (Fig. 10) [39].

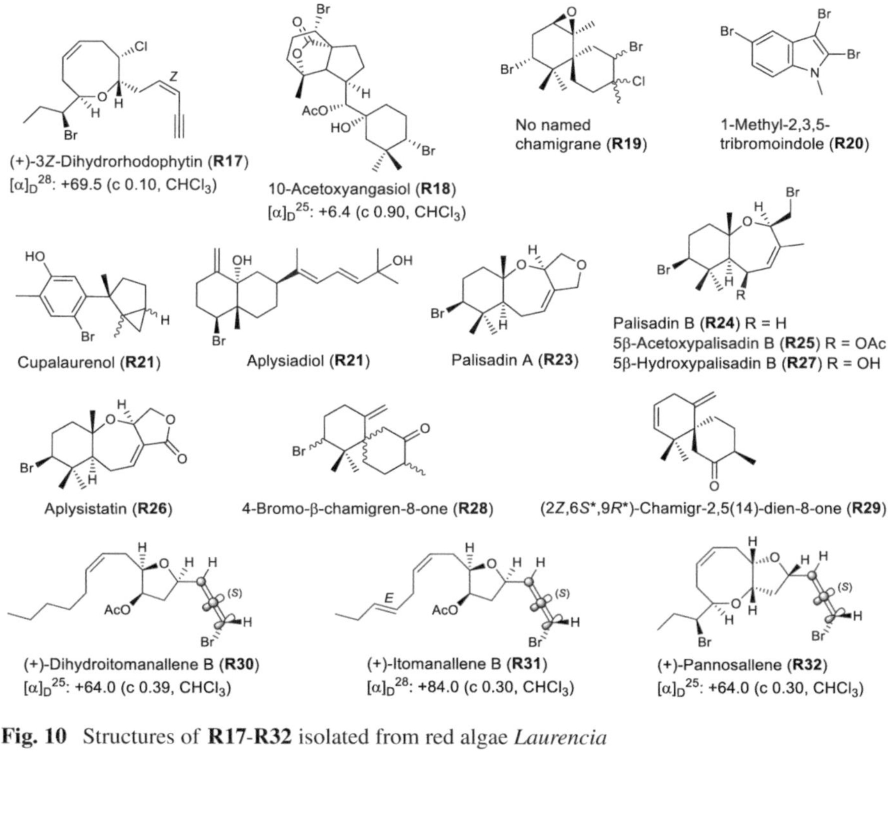

Fig. 10 Structures of **R17-R32** isolated from red algae *Laurencia*

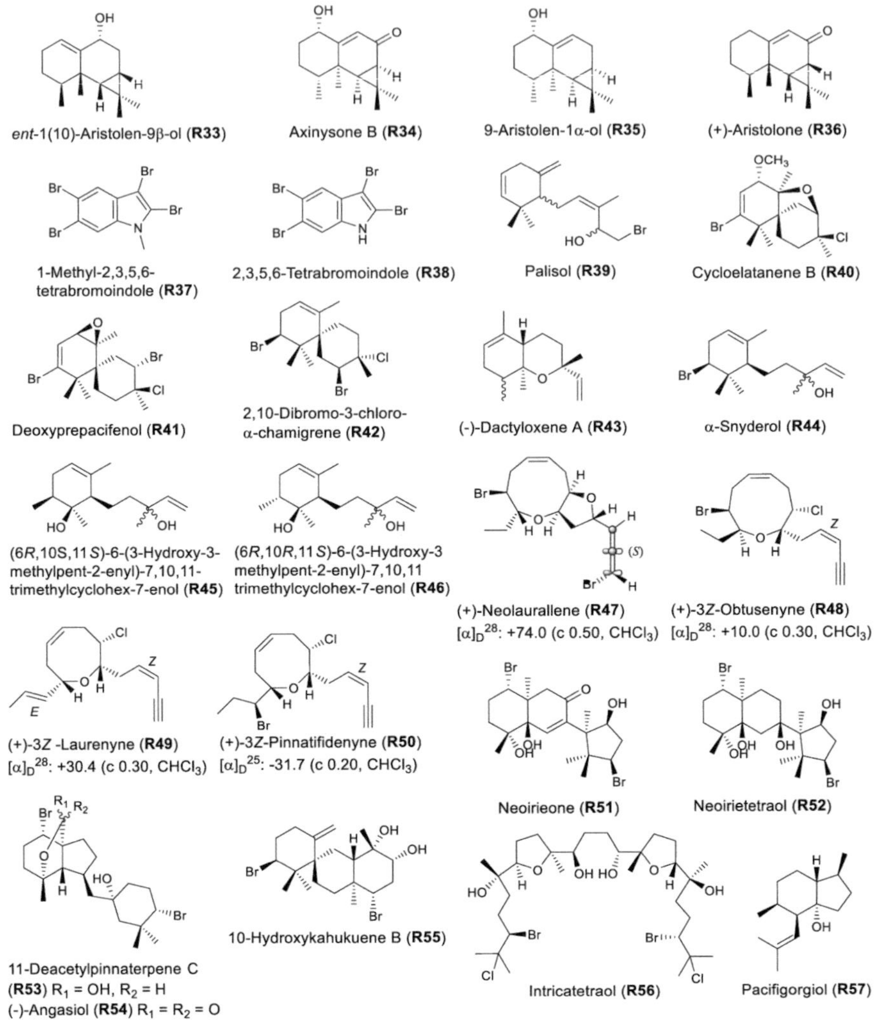

Fig. 11 Structures of **R33-R57** isolated from red algae *Laurencia*

In 2013, Kamada and Vairappan reported one new aristolane-type sesquiterpenoid, *ent*-1(10)-aristolen-9β-ol (**R33**) along with six known compounds axinysone B (**R34**), 9-aristolen-1α-ol (**R35**), (+)-aristolone (**R36**), 1-methyl-2,3,5-tribromoindole (**R19**), 1-methyl-2,3,5,6-tetrabromoindole (**R37**) and 2,3,5,6-tetrabromoindole (**R38**) from red alga *L. similis* (Sepanggar Island, Sabah) (Fig. 11) [40]. 2,3,5,6-Tetrabromoindole (**R38**) displayed antibacterial activity with MIC < 7.5 μg/disc against *Staphylococcus aureus* (HP0808), *Staphylococcus* sp. (HP1008), *Streptococcus pyrogenes* (HP1208) and *Salmonella enteritidis* (HP0608), and cytotoxic activity with IC_{50} < 8.0 μg/mL against human cervical cancer (HeLa), human breast cancer (MCF-7) and adult T-cell leukemia (S1T) cells. Within the same year, Vairappan et al. isolated four known compounds, palisol (**R39**), palisadin A (**R23**), 5β-acetoxypalisadin B (**R25**), and aplysistatin (**R26**) from red alga *L. snackeyi* (Sepanggar Island, Sabah) (Fig. 11) [41]. In the anti-inflammatory assay, aplysistatin (**R26**) inhibited LPS-induced nitric oxide production in RAW 264.7 macrophages by more than 95% at 20.0 μg/mL. In addition, prostaglandin E_2 (PGE_2), tumor necrosis factor alpha (TNF-α) and interleukins-1β and -6 were reduced by over 60% at 20.0 μg/mL. Western blot result showed that the anti-inflammatory activity of compound **R26** was achieved by down-regulating both inducible nitric oxide synthase (iNOS) and cyclooxygenase-2 (COX-2) expression by more than 95% and 60%, respectively.

In 2014, Vairappan et al. investigated seven populations of red algae *L. nangii* (North Borneo Island) and isolated 20 known compounds, cycloelatanene B (**R40**), deoxyprepacifenol (**R41**), 2,10-dibromo-3-chloro-α-chamigrene (**R42**), dactyloxene A (**R43**), α-snyderol (**R44**), (6*R*,10*S*,11*S*)-6-(3-hydroxy-3-methyl-pent-2-enyl)-7,10,11-trimethylcyclo-hex-7-enol (**R45**), (6*R*,10*R*,11*S*)-6-(3-hydroxy-3-methyl-pent-2-enyl)-7,10,11-trimethylcyclo-hex-7-enol (**R46**), (+)-3*Z*-dihydrorhodophytin (**R17**), (+)-itomanallene B (**R31**), (+)-pannosallene (**R32**), (+)-neolaurallene (**R47**), (+)-3*Z*-obtusenyne (**R48**), (+)-3*Z*-laurenyne (**R49**), (+)-3*Z*-pinnatifidenyne (**R50**), neoirieone (**R51**), neoirietetrol (**R52**), 11-deacetylpinnaterpene C (**R53**), (−)-angasiol (**R54**), 10-hydroxykahukuene B (**R55**) and intricatetraol (**R56**) (Fig. 11) [42]. This large-scale sampling of seven populations of red algae *L. nangii* covered four coastal areas of Sabah (Tunku Abdul Rahman Marine Park, Dinawan Island, Tun Mustapha Marine Park and Tun Sakaran Marine Park) was able to show concrete evidence of chemotaxonomical markers existing in *L. nangii*. The first chemotaxonomy in red algae *Laurencia* was proposed by Prof. William Fenical (Scripps Institution of Oceanography, University of California, San Diego) [43]. He showed that physical environmental factors on red algae *Laurencia* had little effect in their natural products synthesis (chemotaxonomical marker), he also showed that morphologically identical but genetically distinct sympatric populations of *L. pacifica* can produce different types of natural products, suggesting that the unique natural product types among *Laurencia* populations reflect their genetic differences. Subsequently, the structure of cycloelatanene B (**R40**) was revised by total synthesis [44], the configuration of intricatetraol (**R56**) was confirmed by total synthesis [45], and the experimental optical rotation of (+)-3*Z*-pinnatifidenyne (**R50**) was opposite to the reported ones.

In the same year of 2014, Wijesinghe et al. isolated four known compounds, pacifigorgiol (**R57**), palisadin B (**R24**), 5β-hydroxypalisadin B (**R27**) and palisol (**R39**) from red alga *L. snackeyi* (Sulug Island, Sabah) (Fig. 11). In the anti-inflammatory assay, 5β-hydroxypalisadin B (**R27**) inhibited LPS-induced nitric oxide production in RAW 264.7 macrophages by more than 90% at 50.0 μM and only showed only weak inhibition for prostaglandin E_2 (PGE_2), tumor necrosis factor alpha (TNF-α) and interleukins-1β and -6 [46]. Further investigation using zebrafish embryo in vivo model demonstrated a protective effect against the formation of stress-induced reactive oxygen species (ROS) [47].

In 2015, Kanmada and Vairappan reported one new laurene-type sesquiterpenoid, neolaurene (**R58**) and five known compounds, cycloelatanene B (**R40**), deoxyprepacifenol (**R41**), 2,10-dibromo-3-chloro-α-chamigrene (**R42**), (+)-neolaurallene (**R47**) and intricatetraol (**R56**) from red alga *L. nangii* (Lohok Batun, Sabah) (Fig. 12) [48]. Neolaurene (**R58**) displayed antibacterial activity with MIC 7.5–12.5 μg/disc against *E. coli*, *Vibrio cholerae*, *Staphylococcus aureus* and *Salmonella thypi*.

Two years later, Kanmada and Vairappan characterized two new synderane-type sesquiterpenoids, snakediol (**R59**) and snakeol (**R60**) along with nine known compounds, α-bromocuparane (**R61**), debromolaurinterol (**R62**), palisadin D (**R63**), (*R*,*Z*)-3,3-dimethyl-5-methylene-4-(3-methylpenta-2,4-dien-1-yl)cyclohex-1-ene (**R64**), palisadin A (**R23**), palisadin B (**R24**), 5β-acetoxypalisadin B (**R25**),

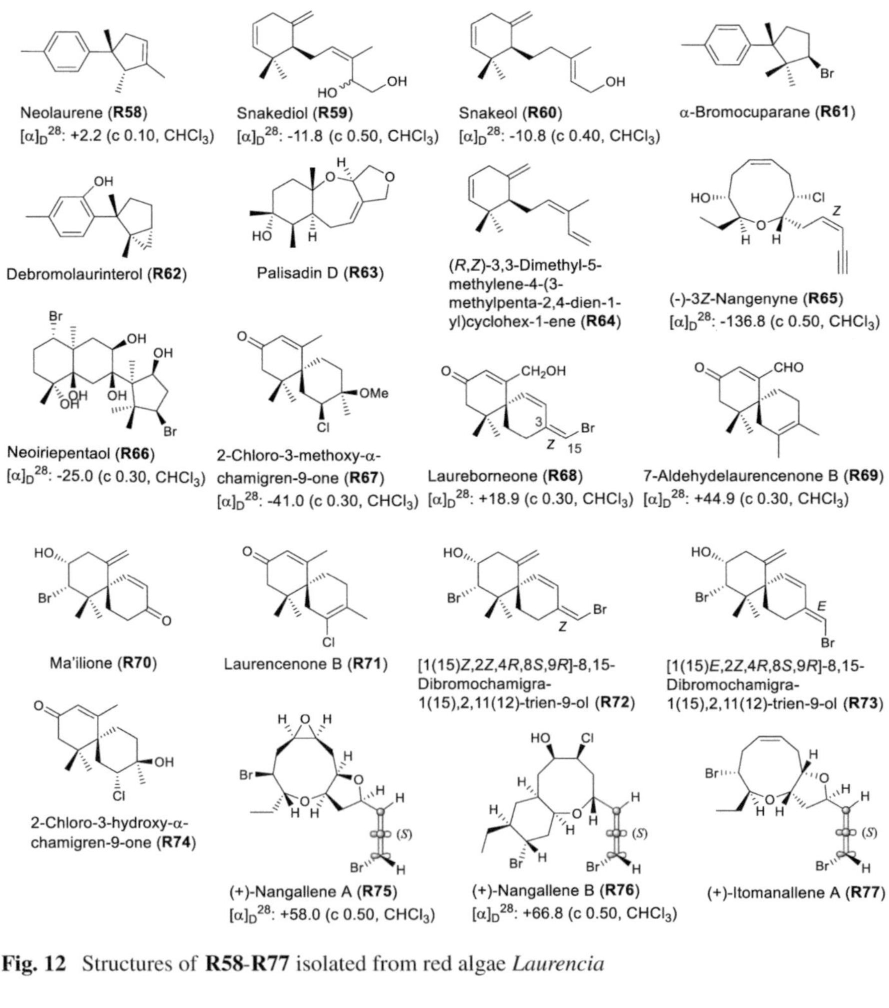

Fig. 12 Structures of **R58**-**R77** isolated from red algae *Laurencia*

palisol (**R39**) and pacifigorgiol (**R57**) from the red algae *L. snackeyi* (Bum Bum Island, Sabah) (Fig. 12) [49]. Snakediol (**R59**), snakeol (**R60**), α-bromocuparane (**R61**) and debromolaurinterol (**R62**) showed antibacterial activity with MIC 0.36–0.58 mM against *E. coli* and *Salmonella thypi*.

In 2018, Kamada et al. discovered two new compounds, nangenyne (**R65**) and neoiriepentaol (**R66**) along with two known compounds, neoirietetraol (**R52**) and (−)-dactyloxene A (**R43**) from red alga *L. nangii* (Mantanani Island, Sabah) (Fig. 12) [50]. Neoiriepentaol (**R66**) exhibited fungistatic effects against hyphal growth of fungus *Lagenidium thermophilum* IPMB 1401 at MIC 12.5 μg/mL and fungicidal effects against *L. thermophilum* IPMB 1401 at 50 μg/mL after 24 h treatment.

In the same year of 2018, Kamada et al. investigated another population of red alga *L. majuscula* collected at Dinawan Island, Sabah and found three new chamigrane-type sesquiterpenoids, 2-chloro-3-methoxy-α-chamigren-9-one (**R67**), laureborneone (**R68**) and 7-aldehydelaurencenone B (**R69**) together with five known compounds, ma'ilione (**R70**), laurencenone B (**R71**), [1(15)*E*,2*Z*,4*R*,8*S*,9*R*]-8,15-dibromochamigra-1(15),2,11(12)-trien-9-ol (**R72**), [1(15)*Z*,2*Z*,4*R*,8*S*,9*R*]-8,15-dibromochamigra-1(15),2,11(12)-trien-9-ol (**R73**) and 2-chloro-3-hydroxy-α-chamigren-9-one (**R74**) from red alga *L. majuscula* (Dinawan Island, Sabah) (Fig. 12) [51]. All isolated compounds **R67**-**R74** were evaluated for their cytotoxic activity against human cervical cancer (HeLa), human breast cancer (MCF-7) and mouse murine leukemia (P388) cells. Only compounds **R68**, **R72** and **R73** showed cytotoxic activity with IC_{50} values of 2.26–2.47 μg/mL, while compounds **R67**, **R70**, **R71** and **R74** were inactive, suggesting bromo-exomethylene at C-3/C-15 might be important for the activity.

In 2019, Kamada et al. investigated three populations of red algae *Laurencia* sp. Population A *L. nangii* (Semporna Island, Sabah) discovered two new C_{15}-acetogenins, (+)-nangallenes A (**R75**) and B (**R76**), along with two known compounds, (+)-itomanallene A (**R77**) and 2,10-dibromo-3-chloro-α-chamigrene (**R42**) (Fig. 12) [52]. Population B *L. majuscula* (Selalong Island, Sabah) contained three new halogenated tricyclic sesquiterpenoids, omphalaurediol (**R78**) and rhodolaurenones B (**R79**) and C (**R80**), along with nine known compounds, rhodolaurenone A (**R81**), elatol (**R6**), laurencenone B (**R71**), 2-chloro-3-hydroxy-α-chamigren-9-one (**R74**), rhodolaureol (**R82**), isorhodolaureol (**R83**), cartilagineol (**R84**), (+)-deschloroelatol (**R85**) and laurencenone D (**R86**) (Fig. 13) [53]. Population C *L. majuscula* (Mantanani Island, Sabah) contained two new chamigrane-type sesquiterpenoids, lauremantanones A (**R87**) and B (**R88**), together with seven known compounds, elatol (**R6**), laurencenone B (**R71**), 2-chloro-3-hydroxy-α-chamigren-9-one (**R74**), cartilagineol (**R84**), obtusol (**R89**), dendroidiol (**R90**) and puertitol A (**R91**) (Fig. 13) [54]. Subsequently, the structure of (+)-itomanallene A (**R77**) was revised by total synthesis [55].

In 2020, Phan et al. discovered one new C_{15}-acetogenin, (+)-4-*epi*-isolaurallene A (**R92**) and a first time isolated as natural product, (+)-4-*epi*-itomanallene A (**R93**) from red alga *L. nangii* (Carrington Reef, Sabah) (Fig. 13) [56]. Subsequently, the structure of (+)-4-*epi*-isolaurallene A (**R92**) was revised by total synthesis of

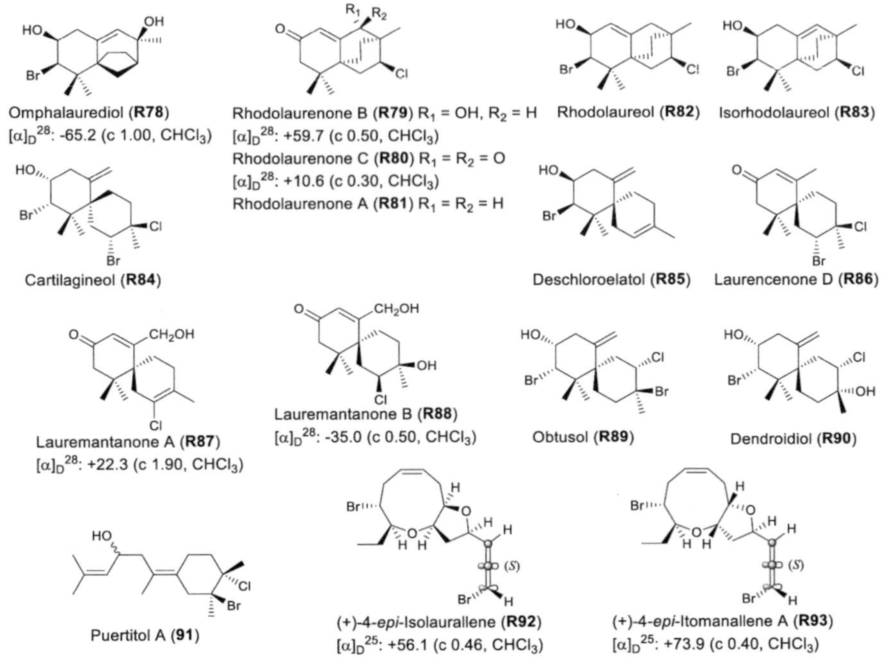

Fig. 13 Structures of **R78-R93** isolated from red algae *Laurencia*

its enantiomer, (−)-4-*epi*-isolaurallene A (Fig. 21) [57], and the structure of (+)-4-*epi*-itomanallene A (**R93**) was first synthesized in 2010 and isolated as natural product for the first time in 2020 [55]. In 2023, Palaniveloo et al. reported three known compounds, palisadin A (**R23**), 5β-acetoxypalisadin B (**R25**) and aplysistatin (**R26**) from red alga *L. snackeyi* (Terumbu Island, Negeri Sembilan) [58].

2.3 Sponges

The first natural products isolated from Malaysian sponges were reported in 2015. Hamada group isolated one new manoalide-type sesterterpenoid, (4*E*,6*E*)-dehydro-25-*O*-methylmanoalide (**P1**) and a known compound, (4*E*,6*E*)-dehydromanoalide (**P2**) from sponge *Luffariella variabilis* (Sepanggar Island, Sabah) (Fig. 14) [59]. Compounds **P1** and **P2** displayed cytotoxic activity against adult T-cell leukemia (S1T) cells with IC_{50} of 1.70 and 1.90 μM, respectively. Three years later, Phan et al. reported four known scalarane-type sesterterpenoids, scalarolide (**P3**), scalarolide acetate (**P4**), methyl 18-hydroxy-19-norscalar-16-en-20-carboxylate (**P5**) and 12-*O*-deacetyl-12-*epi*-19-*O*-methylscalarin (**P6**) from sponge *Spongia* sp. (Mengalum Island, Sabah) (Fig. 14) [60]. Compounds **P3**, **P4** and **P6** exhibited cytotoxic activity against adult T-cell leukemia (S1T) cells with IC_{50} values of 3.93, 5.16 and 2.31 μg/

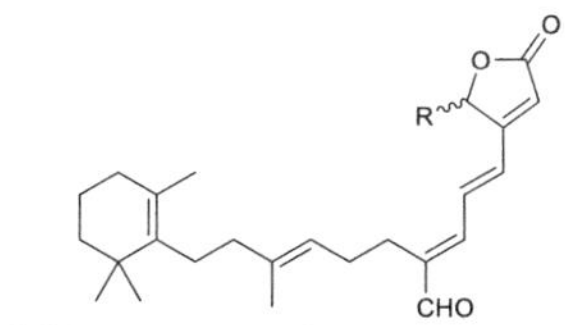

(4*E*,6*E*)-Dehydro-25-*O*-methylmanoalide (**P1**) R= OMe
$[\alpha]_D^{25}$: +2.0 (*c* 0.05, CH_3OH)
(4*E*,6*E*)-Dehydromanoalide (**P2**) R= OH

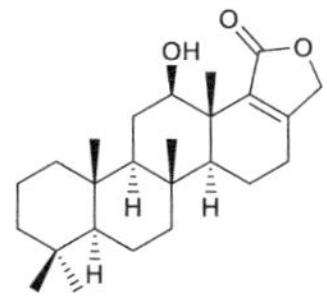

Scalarolide (**P3**)

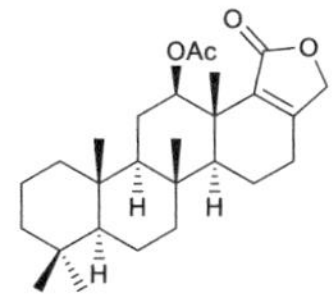

Scalarolide acetate (**P4**)

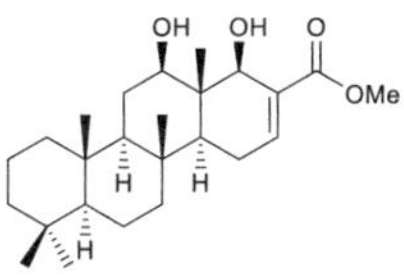

Methyl 18-hydroxy-19-norscalar-16-en-20-carboxylate (**P5**)

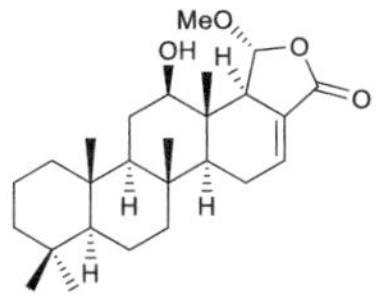

12-*O*-Deacetyl-12-*epi*-19-*O*-methylscalarin (**P6**)

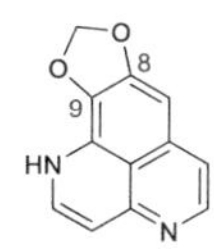

1,3-Dioxolo [4,5-d] benzo [de]-1,6-naphthyridine (**P7**)
$[\alpha]_D^{25}$: not measured

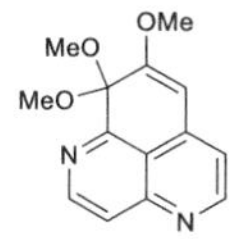

8,9,9-trime-thoxy-9H-benzo [de]-1,6-naphthyridine (**P8**)

Fig. 14 Structures of marine natural products **P1-P8** isolated from marine sponges

mL, respectively. In 2019, Hamada group characterized a new aaptamine-type alkaloid, 1,3-dioxolo [4,5-d] benzo [de]-1,6-naphthyridine also known as methylenedioxyaaptamine (**P7**) and one known compound, 8,9,9-trimethoxy-9H-benzo [de]-1,6-naphthyridine (**P8**) from sponge *Aaptos aaptos* (Sepanggar Island, Sabah) (Fig. 14) [61]. Methylenedioxyaaptamine (**P7**) possessed cytotoxic activity against adult T-cell leukemia (S1T) cells with IC_{50} of 0.29 μM, while compound (**P8**) was inactive, indicated dioxolane moiety at C-8/C-9 is important for the activity.

2.4 Marine Cyanobacteria

The first natural products isolated from Malaysian marine cyanobacteria were reported in 2017. Okino group characterized two new chlorinated fatty acid amides, columbamides D (**C1**) and E (**C2**), and LC–MS detected three known compounds, apratoxins A (**C6**) and C (**C7**) and wewakazole (**C8**), from the cyanobacterium *Moorea bouillonii* (Mantanani Island, Sabah) (Fig. 15) [62]. The absolute configurations of **C1** (10*R*,20*R*) were determined using three approaches, (1) Marfey's analysis of **C1** with *R*- and *S*-synthetic standards, (2) comparison of NMR spectra of **C1** with all four synthetic isomers (10*R*,20*R*, 10*S*,20*R*, 10*R*,20*S* and 10*S*,20*S*) and (3) chiral-phase HPLC analysis of **C1** with all four synthetic isomers. Three years later, further study carried out by Okino group found another three new chlorinated fatty acid amides, columbamides F (**C3**), G (**C4**), and H (**C5**) from cyanobacterium *Moorea bouillonii* collected at Mantanani Island, Sabah [63], and two known compounds lyngbyabellin A (**C9**) and homohydroxydolabellin (**C10**) from *Moorea bouillonii* collected unknown island, Sabah (Fig. 15) [64]. The absolute configurations of **C3** (10*R*,20*R*) were determined using two approaches, (1) Marfey's analysis

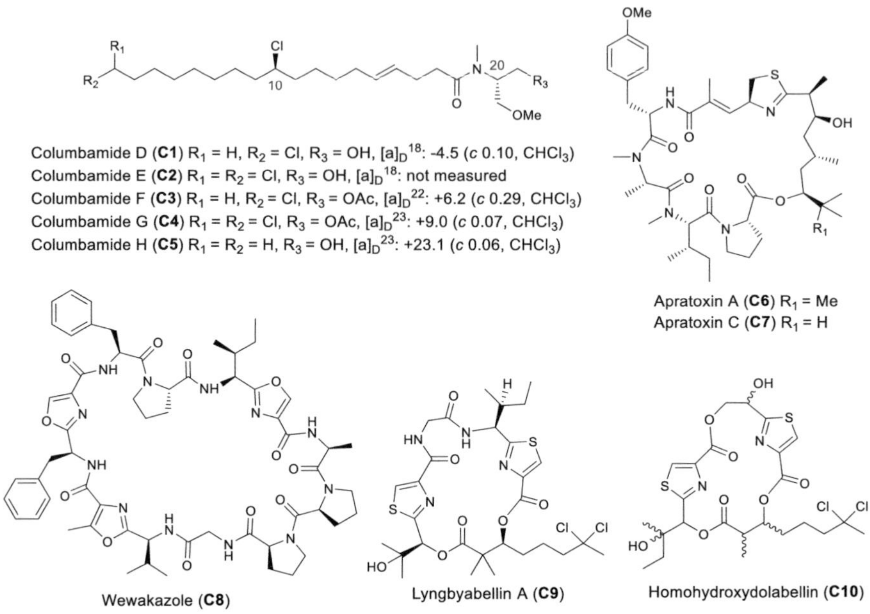

Fig. 15 Structures of marine natural products **C1-C10** isolated from marine cyanobacteria

of **C3** with *R*- and *S*-synthetic standards, and (2) chiral-phase HPLC analysis after conversion and esterification with a chiral labeling reagent Ohrui's acid, (1*S*,2*S*)-2-(anthracene-2,3-dicarboximido)cyclohexanecarboxylic acid. Columbamide D (**C1**) showed biosurfactant activity with critical micelle concentrations of about 0.34 mM. In addition, lyngbyabellin A (**C9**) and homohydroxydolabellin (**C10**) displayed inhibitory activity against human pathogenic parasite *Plasmodium falciparum* with IC_{50} values of 6.4 and 0.3 μM, respectively.

3 Geographic Sampling, Isolation Protocols, and NMR Analysis of Marine-Derived Compounds

This section will summarize information on: (1) sampling location of marine soft corals, red algae, sponges and cyanobacteria for natural products studies, (2) natural products isolation workflow for marine soft corals, red algae, sponges and cyanobacteria, and (3) ^{1}H and ^{13}C NMR spectroscopy for structure characterization of terpenes isolated from marine soft corals and red algae.

3.1 Sampling Location

The samples of all marine soft corals, red algae, sponges and cyanobacteria were collected in Malaysia. We collected all the Global Positioning System (GPS) coordinates from 58 research articles as described in Sect. 2 and generated a map to visualize the sampling locations (Fig. 16). Most of the sampling locations were determined by Prof. Charles S. Vairappan at Universiti Malaysia Sabah, Sabah, Malaysia. Therefore, most of these samples were collected from the waters surrounding Sabah, such as Mantanani, Mengalum, Sepanggar, Banggi and Semporna Islands. Typically, a sample collection expedition consists of a group of people, with divers collecting underwater samples, while others on the boat to document their GPS coordinates and process the samples for voucher specimen (in 70% ethanol) and chemical extraction (in MeOH).

The exact GPS coordinates extracted from the publications described in this section are as follows:

- Soft corals *Lemnalia* (6° 21′ 180″ N, 115° 44′ 000″ E; 7° 33′ 8711″ N, 117° 35′ 017″ E), *Nephthea* (6° 04′ 017″ N, 116° 04′ 836″ E; 4° 39′ 089″ N, 118° 39′ 579″ E; 5° 50′ 750″ N, 115° 59′ 585″ E; 5° 19′ 588″ N, 115° 12′ 020″ E; 5° 46′ 629″ N, 115° 53′ 393″ E; 6° 70′ 245″ N, 116° 31′ 930″ E; 6° 4.683′ N, 116° 4.710′ E), *Sinularia* (4° 06′ 53.35″ N, 118° 37′ 41.60″ E; 6° 43′ 15.89″ N, 116° 20′ 36.38″ E; 6° 42.313″ N, 116° 19.335″ E; 5° 50.750′ N, 115° 59.585′ E; 7° 21.711′ N, 117° 15.223′ E; 6° 10′ 7.44″ N, 115° 34′ 32.94″ E; 6° 44.015′ N, 116° 19.202′ E; 6° 42′ 4.19″ N, 116° 19′ 58.43″ E), *Sarcophyton* (6° 42.775″ N, 116° 19.758″ E; 5° 35′ 57.9″ N, 103° 03′ 56.5″ E; 4° 06′ 53.4″ N, 118° 37′ 41.6″ E; 6° 4.683′ N, 116° 4.710′ E), *Capnella* (6° 43.059″ N, 116° 20.189″ E), *Xenia* (6° 12′ 007″ N, 115° 35′ 011″ E; 6° 43′ 09.31″ N, 116° 20′ 26.23″ E; 6° 10.124′ N, 115° 34.549′ E) and *Litophyton* (6° 4.683′ N, 116° 4.710′ E).

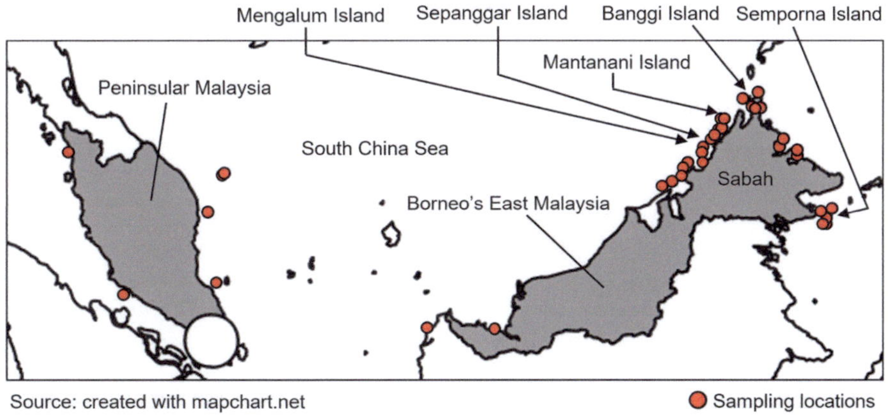

Fig. 16 Sampling location of Malaysian soft corals, red algae, sponges and cyanobacteria

- Red algae *Laurencia* sp. (6° 4′ 20″ N, 100° 03′ 21″ E; 2° 46′ 52″ N, 104° 12′ 35″ E; 5° 35′ 43″ N, 103° 03′ 50″ E; 4° 50′ 46″ N, 103° 39′ 55″ E; 4° 27′ 23″ N, 118° 41′ 12″ E; 4° 35′ 00″ N, 118° 42′ 04″ E; 5° 46′ 63″ N, 115° 53′ 39″ E; 6° 00′ 47″ N, 116° 03′ 14″ E), *L. pannosa* (1° 53′ 50″ N, 109° 45′ 59″ E;), *L. majuscula* (1° 47′ 20″ N, 110° 09′ 53″ E; 5° 45′ 17″ N, 103° 01′ 16″ E; 6° 04′ 06″ N, 117° 57′ 48″ E; 5° 56′ 00″ N, 118° 06′ 36″ E; 6° 04′ 56″ N, 117° 59′ 50″ E; 6° 04′ 06″ N, 117° 57′ 48″ E; 6° 04′ 56″ N, 117° 59′ 50″ E; 05° 84.330″ N, 115° 99.082″ E; 04° 55.931″ N, 118° 67.328″ E; 06° 59.728″ N, 116° 34.830″ E), *L. nangii* (7° 21.711′ N, 117° 15.223″ E; 5° 19′ 58″ N, 115° 12′ 02″ E; 04° 04′ 933″ N, 118° 59′ 833″ E; 04° 57′ 509″ N, 118° 70′ 352″ E; 07° 21′ 711″ N, 117° 15′ 223″ E; 07° 13′ 321″ N, 117° 25′ 002″ E; 05° 58′ 100″ N, 116° 00′ 000″ E; 05° 95′ 952″ N, 115° 99′ 562″ E; 05° 84′ 330″ N, 115° 99′ 082″ E; 04° 27.233″ N, 118° 41.133″ E; 6° 43′ 08.52″ N, 116° 20′ 25.99″ E; 4° 34′ 38.4″ N 118° 41′ 50.6″ E; 7° 07.60.4″ N, 117° 15.020″ E), *L. snackeyi* (6° 56.483″ N, 116° 50.267″ E; 7° 08.547″ N, 117° 14.797″ E; 06° 30.301″ N, 117° 55.022″ E; 04° 38.218″ N, 118° 59.535″ E; 2°463.456″ N, 101° 845.939″ E) and *L. similis* (04° 06′ 53.35″ N, 118° 37′ 41.60″ E).
- Sponges (6° 03′ N, 116° 04′ E; 6° 12′ 073″ N, 115° 36′ 062″ E; 6° 03′ N, 116° 04′ E).
- Cyanobacteria (6° 43′ 07.6″ N, 116° 20′ 20.8″ E; 06° 42′ 16″ N, 116° 19′ 23″ E).

3.2 Compound Isolation Workflow

We reviewed all compound isolation workflows from 58 research articles as described in Sect. 2 and summarized natural product discovery workflow from chemical extraction to compounds isolation. The natural product discovery workflows described here is generally guided by the presence of (1) major compounds or (2) biological activity and LC–MS analysis.

At the Charles S. Vairappan laboratory, the downstream workflow for all soft corals, red algae and sponges is processed in a similar manner (Fig. 17a). Samples collected from the ocean were cut into smaller pieces and soaked in methanol (MeOH) for 3–7 days before chemical extraction. The resulted MeOH crude extract was solvent–solvent partitioned between water (H_2O) and ethyl acetate (EtOAc) to yield H_2O and EtOAc fractions. Only EtOAc fraction was fractionated into six fractions by silica gel column chromatography with a gradient of hexane–EtOAc solvent system (9:1, 8:2, 7:3, 6:4, 1:1 and 1:0 ratio). The H_2O faction was not further investigated. The resulted fractions 1–6 were profiled for the presence of major compounds by silica gel thin layer chromatography (TLC) and ^{1}H NMR measurement in $CDCl_3$. The measured ^{1}H NMR spectra were manually analyzed against in-house database to dereplicate known compounds. Once a sample was deemed to contain new compounds, all compounds were purified by preparative silica gel TLC only or further purified by preparative reversed-phase high performance liquid chromatography (HPLC). In preparative silica gel TLC, fractions were applied on a

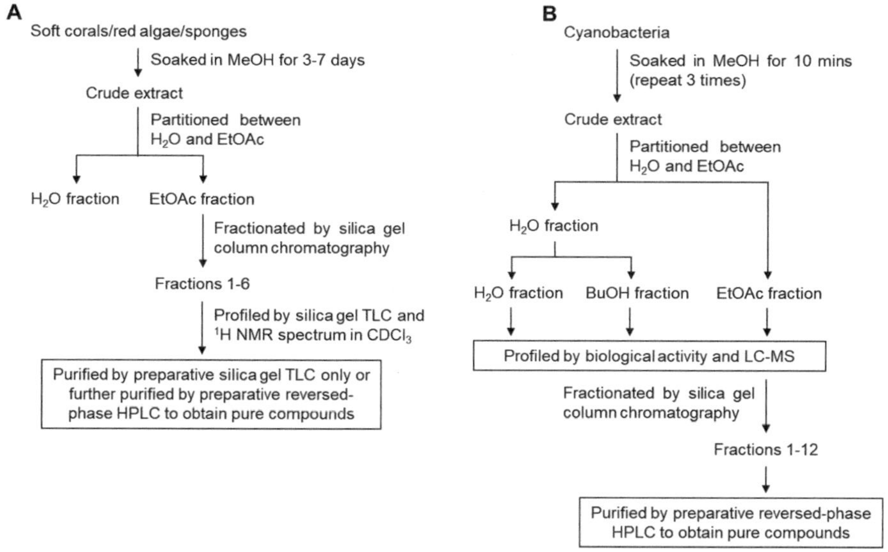

Fig. 17 Natural product discovery workflow from chemical extraction to pure compounds. **a** Compound isolation workflow for soft corals, red algae and sponges. **b** Compound isolation workflow for marine cyanobacteria

horizontal line 1 cm above the bottom of TLC plate (20 × 20 cm) and developed in a closed chamber using toluene, hexane, EtOAc, chloroform ($CHCl_3$) or a combination thereof as the developing solvent system. If the compound required further purification by preparative reversed-phase HPLC, the compound was typically separated with a C_{18} (octadecylsilane) column (10 × 250 mm, 5 μm) using gradient elution of H_2O-acetonitrile with or without 0.1% formic acid.

Marine cyanobacterial compounds were also isolated in a similar manner, except that all fractions (H_2O, BuOH and EtOAc; where BuOH is butanol) were considered for downstream process (Fig. 17b). The bioactive fractions were analyzed by liquid chromatography-mass spectrometry (LC–MS) and dereplicated using MarinLit database by searching for mass (*m/z*) and predicted molecular weight. Then, the fractions were purified by several rounds of preparative reversed-phase HPLC.

For the dereplication based on ^1H NMR spectra ($CDCl_3$, 600 MHz NMR JEOL), we focused on the downfield (deshielded) region of ^1H chemical shifts > 4.0 ppm for aromatic, vinylic and oxygenated methine/methylene protons, and upfield (shielded) region of ^1H chemical shifts 1.0–3.5 ppm for methoxy, acetoxy and vinylic methyls. Before starting compound isolation, we manually compared the ^1H NMR signals of fractions 1–6 (Fig. 17a) in the downfield (> 4.0 ppm) and upfield regions (1.0–3.5 ppm) with our in-house database (isolated compounds and personal experience). The acquired pure compounds were measured by one- and two-dimensional (1D and 2D) NMR for structure characterization. Typical NMR measurement of a compound with 2–4 mg using 600 MHz of NMR spectroscopy in a 5 mm NMR tube requires 5 min of ^{1}H NMR, 12 h of ^{13}C NMR, 4 h of HSQC, 8–12 h of HMBC, 2 h of COSY and 12 h of NOESY.

3.3 ^{1}H and ^{13}C NMR Structure Characterization

Although sensitivity and resolution have greatly improved since the invention of nuclear magnetic resonance (NMR) spectroscopy in the 1940s, many minor compounds cannot be fully characterized by NMR due to limited amount. Here, we describe chemical shifts and coupling constants extracted from 1D NMR that can be used to guide configuration assignments for some marine natural products presented in Sect. 2.

Many marine terpenes have trisubstituted and disubstituted olefins (double bond). The ^{13}C NMR data and coupling constant values can provide valuable information about olefin configuration (Fig. 18). The ^{13}C NMR data of vinylic methyl at *trans*-trisubstituted olefin was appeared at δ_C 13.3 ppm in capgermacrene A (**S94**) and δ_C 15.0 ppm in (13*E*,16*Z*)-fuscol (**S58**), while *Z*-configuration has a larger chemical shifts δ_C 21.9 ppm in capgermacrene B (**S95**) and δ_C 20.7 ppm in (13*Z*,16*E*)-fuscol (**S57**) [14, 23]. The proton-proton vicinal coupling constant ($^{3}J_{HH}$) for the *cis*-disubstituted olefin was $^{3}J_{HH} = 11.7$ Hz in (13*E*,16*Z*)-fuscol (**S58**), and a larger value $^{3}J_{HH} = 15.1$ Hz was observed for the *trans*-disubstituted olefin in (13*Z*,16*E*)-fuscol (**S57**) [14, 23]. Note that chemical shift values in ppm, ^{13}C NMR indicated as δ_C, proton-proton vicinal coupling constant indicated as $^{3}J_{HH}$, coupling constant values in Hz, and *trans* and *cis* indicated as *E* and *Z* configurations, respectively.

A number of xenicane-type diterpenoids from soft corals have acetoxy (OAc) group at C-12 and C-13 [65]. We summarized all such xenicane-type diterpenoids

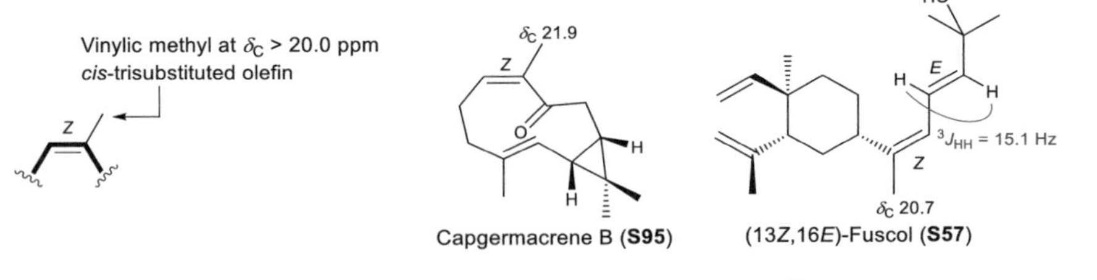

Fig. 18 Configuration assignment of di/trisubstituted olefin based on ^{13}C NMR data and coupling constant values

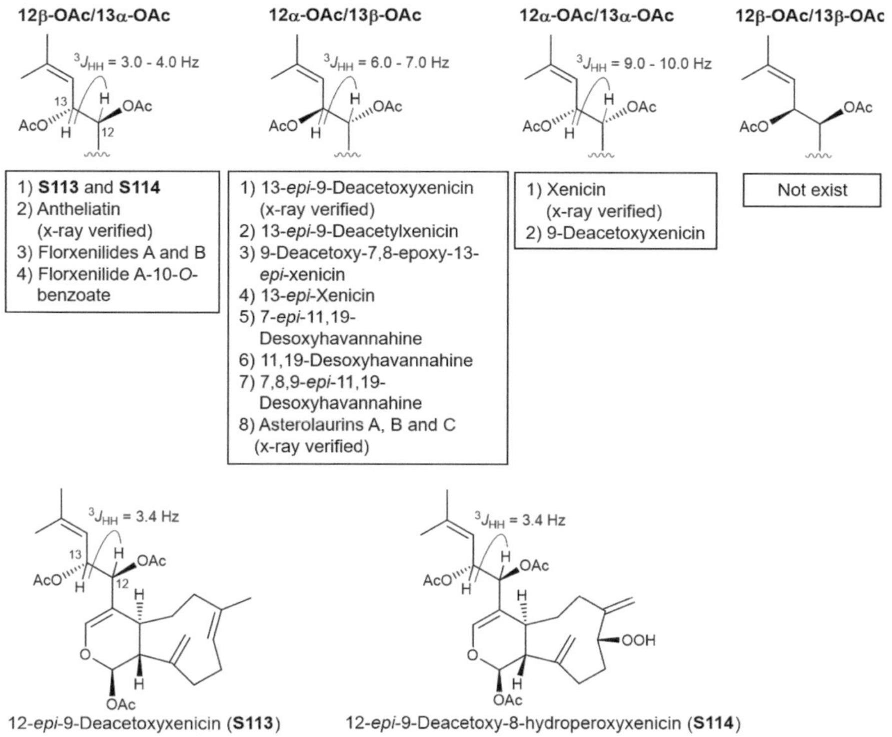

Fig. 19 Stereochemical assignment of 12-OAc/13-OAc in xenicane-type diterpenoids based on coupling constant values between H-12 and H-13

and found that the coupling constant values between H-12 and H-13 can guide the stereochemical assignment of 12-OAc/13-OAc (Fig. 19). In the 12β-OAc/13α-OAc configuration, the proton-proton vicinal coupling constant ($^3J_{HH}$) values between H-12 and H-13 was $^3J_{HH}$ = 3.0–4.0 Hz, as observed in 12-*epi*-9-deacetoxyxenicin (**S113**, $^3J_{HH}$ = 3.4 Hz), 12-*epi*-9-deacetoxy-8-hydroperoxyxenicin (**S114**, $^3J_{HH}$ = 3.4 Hz), antheliatin ($^3J_{HH}$ = 4.2 Hz), florxenilides A ($^3J_{HH}$ = 3.6 Hz) and B ($^3J_{HH}$ = 3.6 Hz), and florxenilide A-10-*O*-benzoate ($^3J_{HH}$ = 3.6 Hz) [29, 66, 67]. The coupling constant values in 12α-OAc/13β-OAc was $^3J_{HH}$ = 6.0–7.0 Hz, as observed in 9-deacetoxy-7,8-epoxy-13-*epi*-xenicin ($^3J_{HH}$ = 6.3 Hz), 13-*epi*-xenicin ($^3J_{HH}$ = 7.0 Hz), 13-*epi*-9-deacetoxyxenicin ($^3J_{HH}$ = 6.6 Hz), 13-*epi*-9-deacetylxenicin ($^3J_{HH}$ = 6.8 Hz), havannahine ($^3J_{HH}$ = 6.0 Hz), 11,19-desoxyhavannahine ($^3J_{HH}$ = 6.0 Hz), 7,8,9-*epi*-11,19-desoxyhavannahine ($^3J_{HH}$ = 6.0 Hz), and asterolaurins A ($^3J_{HH}$ = 6.3 Hz), B ($^3J_{HH}$ = 5.7 Hz) and C ($^3J_{HH}$ = 7.2 Hz) [68–72]. While 12α-OAc/13α-OAc configuration had coupling constant values of $^3J_{HH}$ = 9.0–10.0 Hz, as observed in xenicin ($^3J_{HH}$ = 9.0 Hz) and 9-deacetoxyxenicin ($^3J_{HH}$ = 10.0 Hz) [68, 73]. To date, 12β-OAc/13β-OAc xenicane-type diterpenoids have not been isolated from soft corals. Note that attempt to search for 12β-OAc/13β-OAc analogs at the SciFinder platform were unsuccessful.

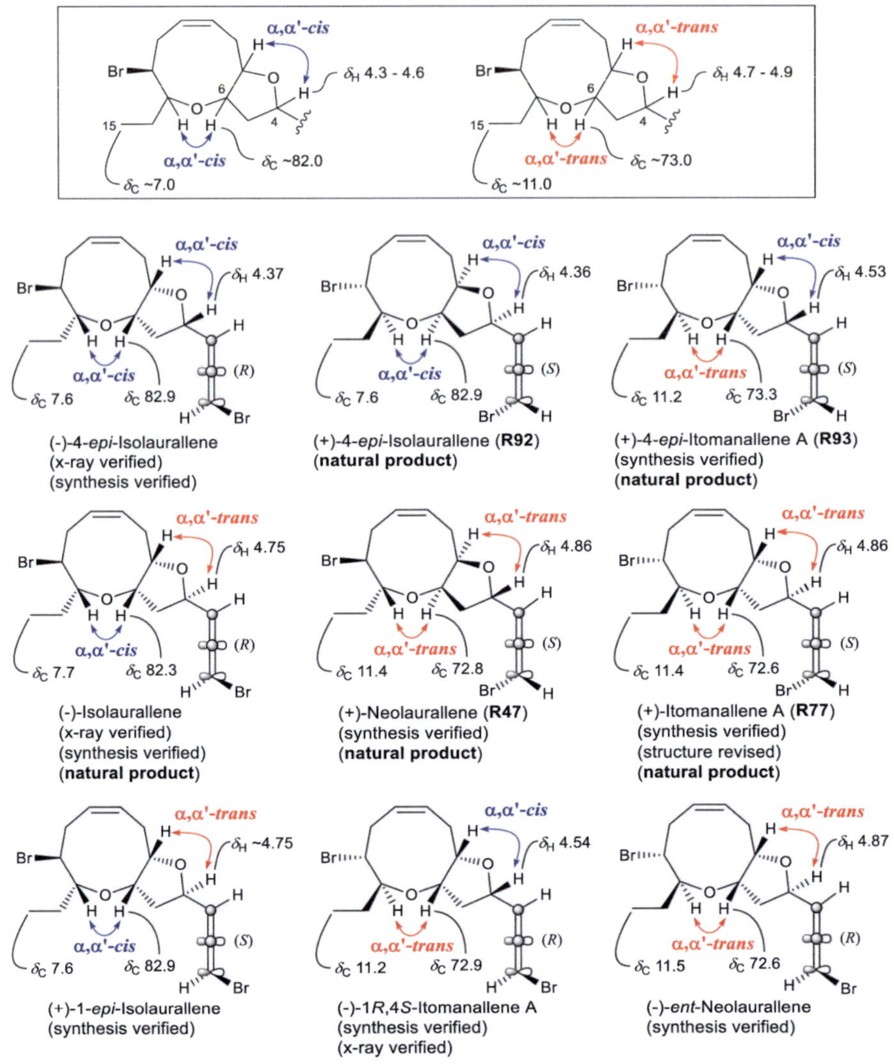

Fig. 20 Structural assignment of α,α′-*cis* or α,α′-*trans* in C_{15}-acetogenins based on ^{1}H NMR at H-4 and ^{13}C NMR at C-6/C-15

To date, only five diastereomers of dioxabicyclic bromoallene marine natural products, (+)-4-*epi*-isolaurallene (**R92**), (+)-4-*epi*-itomanallene A (**R93**), (−)-isolaurallene, (+)-neolaurallene (**R47**) and (+)-itomanallene A (**R77**) have been reported. Their complex fused bicyclic structures possessed multiple stereocenters have attracted a lot of attention from synthetic chemists. This led to a structural revision in itomanallene A through total synthesis [55]. We summarized all such C_{15}-acetogenins both natural and synthetic, and found that the ^{1}H NMR of H-4 and ^{13}C

NMR of C-6/C-15 can guide the structural assignment of α,α′-*cis* or -*trans* (Fig. 20). In the α,α'-*cis* configuration at H-4/H-7, the ^{1}H NMR chemical shift at H-4 was appeared at δ_H 4.3–4.6 ppm, as observed in synthetic (−)-4-*epi*-isolaurallene (δ_H 4.36), natural (+)-4-*epi*-isolaurallene (**R92**, δ_H 4.36), natural (+)-4-*epi*-itomanallene A (**R93**, δ_H 4.53) and synthetic (−)-1*R*,4*S*-itomanallene A (δ_H 4.54) [55–57]. In contrast, H-4 chemical shift was appeared at δ_H 4.7–4.9 ppm for the α,α′-*trans* configuration at H-4/H-7, as observed in natural (+)-itomanallene A (**R77**, δ_H 4.86), natural (−)-isolaurallene (δ_H 4.75), natural (+)-neolaurallene (**R47**, δ_H 4.86), synthetic (−)-*ent*-neolaurallene (δ_H 4.87) and synthetic (+)-*epi*-isolaurallene (δ_H 4.75) [55, 74–76]. As for the α,α′-*cis* configuration at H-6/H-13, ^{13}C NMR chemical shift at C-6 and C-15 was appeared at δ_H ~ 82 and ~ 7.0 ppm, respectively, as observed in synthetic (−)-4-*epi*-isolaurallene (δ_C 82.9 and 7.6), natural (+)-4-*epi*-isolaurallene (**R92**, δ_C 82.9 and 7.7), natural (−)-isolaurallene (δ_C 82.3 and 7.7) and synthetic (+)-*epi*-isolaurallene (δ_C 82.9 and 7.7) [56, 57, 74, 76]. In contrast, C-6 and C-15 chemical shifts for the α,α′-*trans* configuration at H-6/H-13 were appeared at ~ 73 and ~ 11.0 ppm, respectively, as observed in natural (+)-4-*epi*-itomanallene A (**R93**, δ_C 73.3 and 11.2), natural (+)-itomanallene A (**R77**, δ_C 72.6 and 11.4), natural (+)-neolaurallene (**R47**, δ_C 71.8 and 11.4), synthetic (−)-*ent*-neolaurallene (δ_C 72.6 and 11.5) and synthetic (−)-1*R*,4*S*-itomanallene A (δ_C 72.9 and 11.2) [55, 56, 75]. Note that chemical shift values in ppm, ^{13}C and ^{1}H NMR indicated as δ_C and δ_H, respectively, α,α′-*cis* configuration indicated two protons orientated in the same face, α,α′-*trans* configuration indicated two protons orientated in opposite direction, and absolute configuration of bromoallene moiety indicated as (*S*) or (*R*) that assigned based on Lowe's rule, optical rotation (+) for (*S*) configuration and optical rotation (−) for (*R*) configuration.

In addition to C_{15}-acetogenins, another group of natural products that often encountered in red algae is chamigrane-type sesquiterpenoids. Many chamigrane-type sesquiterpenoids have a *E*- or *Z*-configuration of bromo-exomethylene at C-3/C-15, we summarized such chamigrenes and found that the ^{1}H NMR of H-15 can guide the configuration assignment of bromo-exomethylene (Fig. 21). In the *cis*-configuration at C-3/C-15, the ^{1}H NMR chemical shift at H-15 was appeared at δ_H 5.9–6.0 ppm, as observed in laureborneone (**R68**, δ_H 6.03), [1(15)*Z*,2*Z*,4*R*,8*S*,9*R*]-8,15-dibromochamigra-1(15),2,11(12)-trien-9-ol (**R72**, δ_H 5.90) and no named chamigrenes (δ_H 5.96, 5.97 and 6.00) [51, 77, 78]. In contrast, H-15 chemical shifts for the *trans*-configuration at C-3/C-15 were appeared at δ_H 6.1–6.2 ppm, as observed in [1(15)*E*,2*Z*,4*R*,8*S*,9*R*]-8,15-dibromochamigra-1(15),2,11(12)-trien-9-ol (**R73**, δ_H 6.11) and unnamed chamigrenes (δ_H 6.23 and 6.26) [51, 77, 78]. Note that chemical shift values in ppm, and ^{1}H NMR indicated as δ_H.

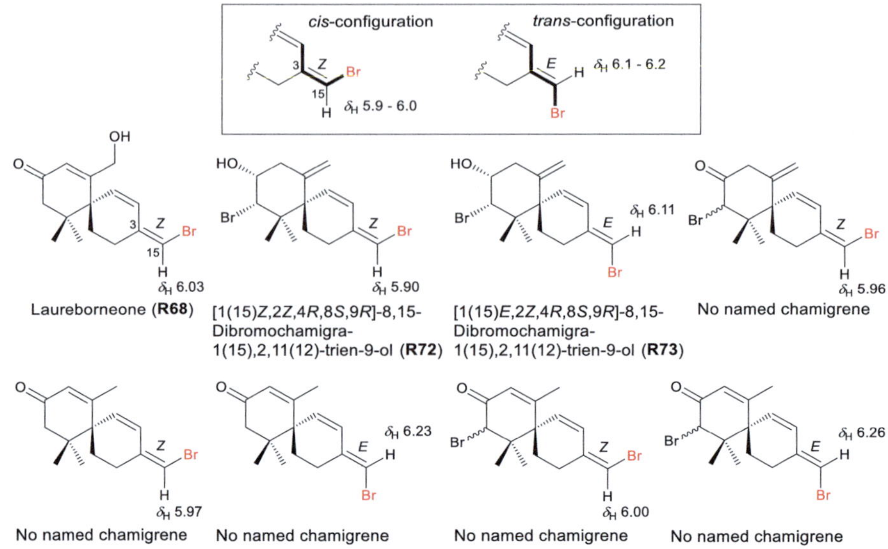

Fig. 21 Configuration assignment of *E*- or *Z*-bromo-exomethylene in chamigrenes based on ^{1}H NMR at H-15

4 Marine Natural Products Biosynthesis

To date, there are no studies on natural products biosynthesis in Malaysian marine organisms. This is partially due to the challenges of acquiring the genomes of these marine organisms, which (1) have a complex mixture present in their cells and (2) are heavily contaminated by other organisms. Natural products from sessile marine animals are thought to be produced by symbiotic microbes rather than the host. This section will present some prominent examples of marine natural products biosynthesis and explain the origin of marine natural products.

4.1 Origin of Sponge-Derived Natural Products

The true producers of marine natural products have been controversial. A well-known example is the polytheonamides A and B (Fig. 22) isolated from marine sponge *Theonella swinhoei* at Japan in 1994 by the Nobuhiro Fusetani and Shigeki Matsunaga group [79, 80]. The absolute configurations and structures of polytheonamides A and B were so complex that it took a decade to solve [81, 82]. In early studies on natural products biosynthesis, peptide natural product that composed of non-proteinogenic and D-amino acids were assumed to be non-ribosomal peptide synthetase (NRPS). However, it was discovered later that these polytheonamides were actually produced by the sponge-associated bacteria Entotheonella by Jörn Piel group [83]. They used a degenerate PCR/pool-dilution strategy of a ca. 1 million clone metagenomic DNA library of *T. swinhoei* and found the biosynthetic gene cluster for polytheonamides [84]. This also showed that polytheonamides were not NRPS products, but belong to the ribosomally synthesized and post-translationally modified peptides (RiPPs).

In the polytheonamide biosynthesis, precursor peptide PoyA was heavily modified by six post-translationally modifying proteins PoyB, PoyC, PoyD, PoyD, PoyF and

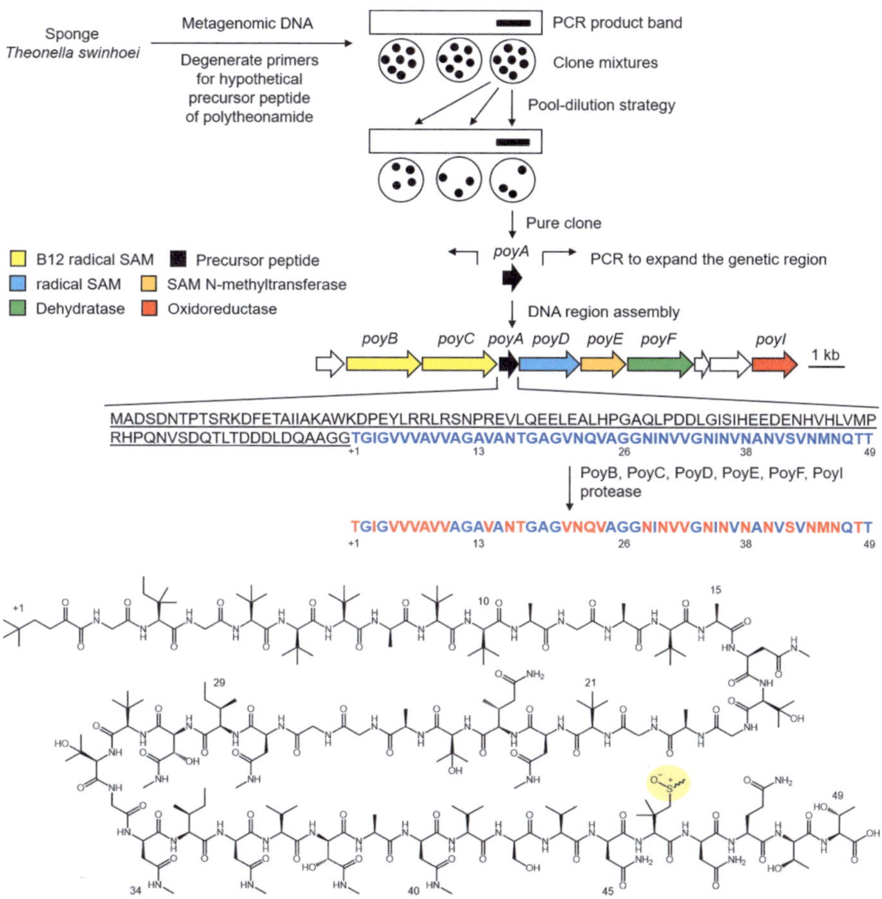

Fig. 22 The degenerate PCR/pool-dilution strategy, biosynthetic gene cluster and structure of polytheonamides. The structures of polytheonamides A and B with a different configuration at the methionine sulfoxide are highlighted in yellow circle. Biosynthetic gene clusters and the genes are colored based on annotation given to translated proteins. The leader peptide sequences are underlined. The core peptide sequences are shown as blue colored bold letters and numbered '+ 1' from the start of the core sequence. The modified amino acid sequences are shown as red colored bold letters

PoyI to achieve 17 C-methylations, 8 N-methylations, 4 hydroxylations, 1 dehydration, and 18 α-carbon epimerizations. Functional study to verify enzymatic activity was investigated using heterologous expression in *Escherichia coli* [83]. The order for post-translational modification and protease involved in leader peptide cleavage remain to be studied. The detailed about polytheonamide biosynthesis has been reviewed elsewhere [85].

4.2 Origin of Coral-Derived Natural Products

Similar to sponges, the origin of soft coral-derived natural products has been debated, and is thought to come from bacterial symbionts [86–88]. In 2022, two research teams, Bradley S. Moore's group and Eric W. Schmidt's group, independently investigated this research topic and revealed that terpenes are produced by corals [89, 90]. Bradley S. Moore's group used hidden Markov model (HMM) to search for coral-encoded terpene cyclases from the public octocoral genomes and transcriptomes, and biochemically characterized 11 coral-encoded terpene cyclases (HcTC-3, PbTC-1, BaTC-2, TmTC-1, BaTC-1, ErTC-2, XsTC-1, RmTC-1, HcTc-2, DgTC-1, and TmTC-2) to their main products, as shown in Fig. 23a [89]. The group also metagenome sequenced of the soft coral *Capnella imbricata*, a known producer of the bioactive capnellane sesquiterpenoids, which led to the identification and enzymatic activity verification of one coral-encoded terpene cyclase, CiTC-1 (Fig. 23b). Eric W. Schmidt's group used HMM to search for coral-encoded terpene cyclases from the private metatranscriptome of *Erythropodium caribaeorum* that they sequenced, and biochemically characterized two coral-encoded terpene cyclases (EcTPS1, and EcTPS6) to their main products, as shown in Fig. 23c [90]. The group carried out a followed-up study on coral-encoded terpene cyclase HcTC-3 aiming to identify cytochrome P450 enzyme involved in the biosynthesis of pseudopterosin A, a potent subcutaneous analgesic in mice with ED_{50} of 3.12 mg/kg. The characterized coral-encoded terpene cyclase HcTC-3 was found to colocalize with the P450 enzyme HcCYP on a single contig of 263 kb in length sequenced from the soft coral *Heliopora coerulea*, separated by 20 kb [91]. The function of HcCYP was characterized by coexpression in yeast *Saccharomyces cerevisiae* and found that HcCYP alone was sufficient to catalyze the conversion of elisabethatriene to 7,8-dihydroxyerogorgiaene, as shown in Fig. 23d.

The discovery from Bradley S. Moore's and Eric W. Schmidt's groups raises the question of whether other marine sessile animals also harbor natural products encoding biosynthetic genes. Many marine soft coral natural products possess drug-like scaffolds that are important for pharmaceutical industry. In this regard, Eric W. Schmidt's group has developed a semisynthetic strategy to produce eunicellane scaffold at gram scale in yeast [92]. This highlights the potential biotechnological application in drug discovery.

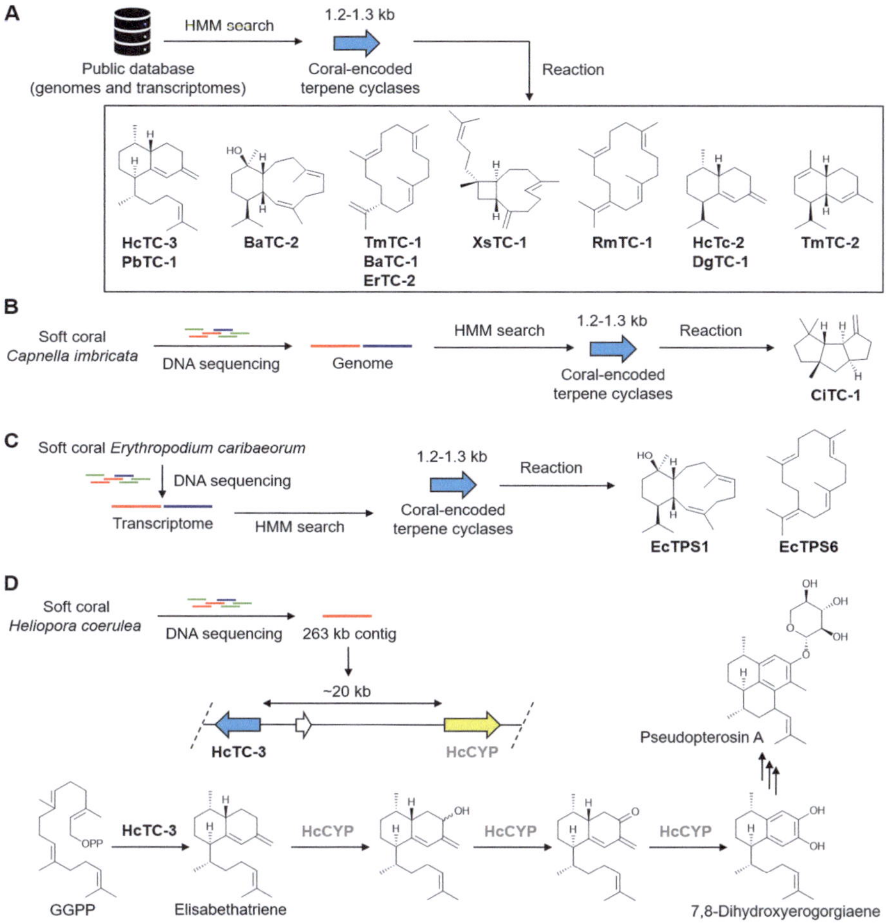

Fig. 23 **a** The coral-encoded terpene cyclases were discovered from public database and their catalyzed reactions. **b, c** The coral-encoded terpene cyclases were discovered from DNA sequencing of *C. imbricata* and *E. caribaeorum*, and their catalyzed reactions. **d** The coral-encoded P450 enzyme was discovered from DNA sequencing of *H. coerulea*, and their catalyzed reactions. The coral-encoded terpene cyclases are shown as black colored bold letters. The coral-encoded P450 enzymes are shown as grey colored bold letters

4.3 Laurencia Sesquiterpene Synthase

Many of the terpenes reviewed in this chapter have been isolated from red algae *Laurencia*, therefore we focus on the biosynthesis of terpenes found in *Laurencia*. In 2017, Makoto Fujita and Jing-Ke Weng group jointly studied the first terpene synthase from red alga *Laurencia pacifica* [93]. The research group performed DNA sequencing and assembly of the transcriptomes of *L. pacifica*, and identified three red alga-encoded terpene synthases (LphTPS-A to C), but only LphTPS-A was selected for biochemical characterization through heterologous expression in yeast *Saccharomyces cerevisiae* BY4743 strain, as shown in Fig. 24a. The study demonstrated an emerging protein family of terpene synthases in red algae *Laurencia.* However, the characterized product by LphTPS-A was not related to halogenated terpenoids, which are the signature natural products of red algae. In 2024, Bradley S. Moore group reported 13 red alga-encoded terpene synthases, six of which were found in *Laurencia subopposita* (LsTS-1 to 6), and three were previously identified in *L. pacifica* (LphTPS-A to C) [94]. The research group performed DNA sequencing and assembly of the transcriptomes of *L. pacifica,* and identified three red alga-encoded terpene synthases (LphTPS-A to C), but only LphTPS-A was selected for biochemical characterization through heterologous expression in yeast *Saccharomyces cerevisiae* BY4743 strain, as shown in Fig. 24b. The characterized red alga-encoded terpene synthase LsTS-2 was found to colocalize with the vanadium-dependent haloperoxidase LsVHPO-1 on a single contig sequenced from the red alga *L. subopposita,* separated by 25 kb (Fig. 24c). The research group suspected that colocalized vanadium-dependent haloperoxidase LsVHPO-1 may be involved in the biosynthesis of oppositol, a halogenated terpenoid isolated from *L. subopposita.*

To date, most vanadium-dependent haloperoxidase identified from red algae were not characterized. In 2022, Tatsufumi Okino group biochemically characterized vanadium-dependent haloperoxidases from red alga *Laurencia okamurae* and proposed an asymmetric bromination that triggers cyclization, in which the cationic intermediate attacks an electron-rich olefinic bond and results in ring closure (Fig. 24d) [95]. In the near future, undoubtedly these research groups will reveal concrete evidence for the biosynthesis of halogenated terpenoids from red algae.

4.4 Biosynthesis Cyanobacterial-Derived Natural Products

Unlike sponges, soft corals and red algae, much work has been done on the biosynthesis of natural products from cyanobacteria. Patellamides possess unique structural features and potent bioactivities. However, the true producer of patellamides was remained a mystery until the discovery of the patellamide biosynthetic gene cluster. In 1981, patellamides were isolated from the marine tunicate *Lissoclinum patella* [96]. In the early cell-separation studies, it has been proposed that patellamides isolated from the tunicate *L. patella* are synthesized by the symbiontic cyanobacterium *Prochloron didemni* [97, 98]. But a decade later, patellamides were found distributed in both tunicates and cyanobacteria [99]. In 2005, Eric W. Schmidt's group identified a gene cluster encoding patellamide in the genome of cyanobacterium *P. didemnid,* confirming that the true producer of patellamide is the cyanobacterium *P. didemnid* [100]. This conclusion was also supported by another two genomic studies [101, 102].

In the first discovery of patellamide gene cluster, Eric W. Schmidt's group used tblastn to search for hypothetical precursor peptide gene from the *P. didemnid* genome and heterologous expressed the whole pathway (PatA to G) to produce patellamide

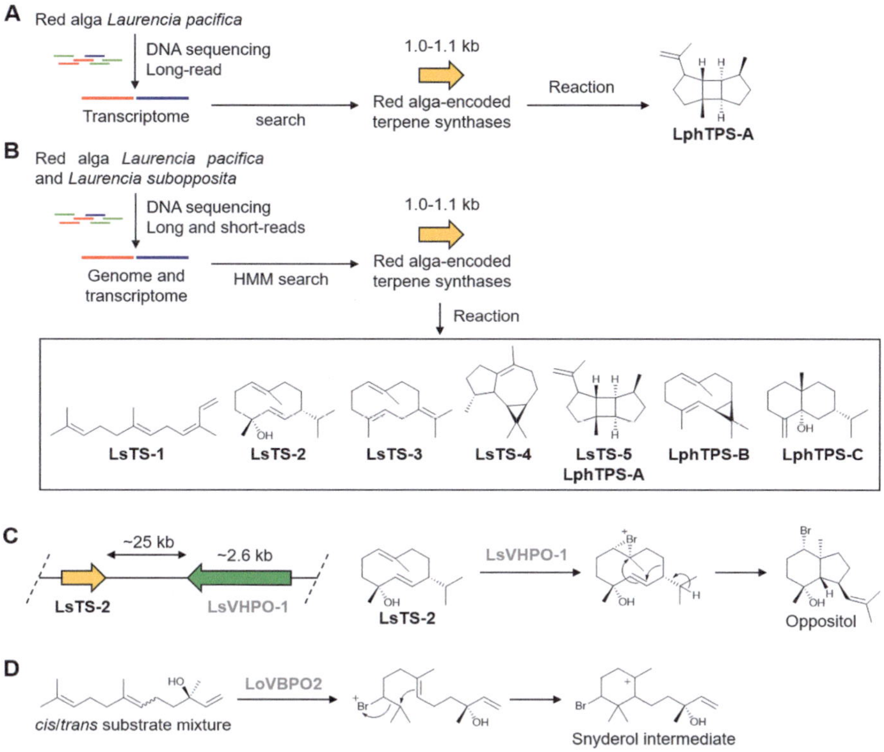

Fig. 24 **a**, **b** The red alga-encoded terpene synthases were discovered from DNA sequencing of *L. pacifica* and *L. subopposita*, and their catalyzed reactions. **c** The red alga-encoded terpene synthase and vanadium-dependent haloperoxidase were found colocalize on a single contig sequenced from *L. subopposite*, and proposed biosynthesis pathway of oppositol. **d** The proposed reactions of red alga-encoded vanadium-dependent haloperoxidase from *L. okamurae*. The red alga-encoded terpene synthases are shown as black colored bold letters. The red alga-encoded vanadium-dependent haloperoxidases are shown as grey colored bold letters

in *E. coli* (Fig. 25a) [100]. Since then, many such cyclic peptides have been isolated from different free-living and symbiotic cyanobacteria, and the family of these cyclic peptides are collectively known as cyanobactin [103]. The precursor peptide is designated as “E” (PatE in patellamide biosynthesis), serine proteases “A” and “G” (PatA and PatG), YcaO cyclodehydratase “D” (PatD) and prenyltransferase “F” (PatF). The detailed about cyanobactin biosynthesis has been reviewed elsewhere [103–105]. In Jan 2025, SciFinder was searched with the cyanobactin as query and found more than 100 publications, showing that cyanobactins has been extensively studied. In addition to cyanobactins isolation, many of these studies focused on the discovery of cyanobactin prenyltransferases “F”. LynF is the first characterized representative of the cyanobactin prenyltransferase protein family (Fig. 25b) [105]. To date, 13 active cyanobactin prenyltransferases have been biochemically characterized, LynF [105,

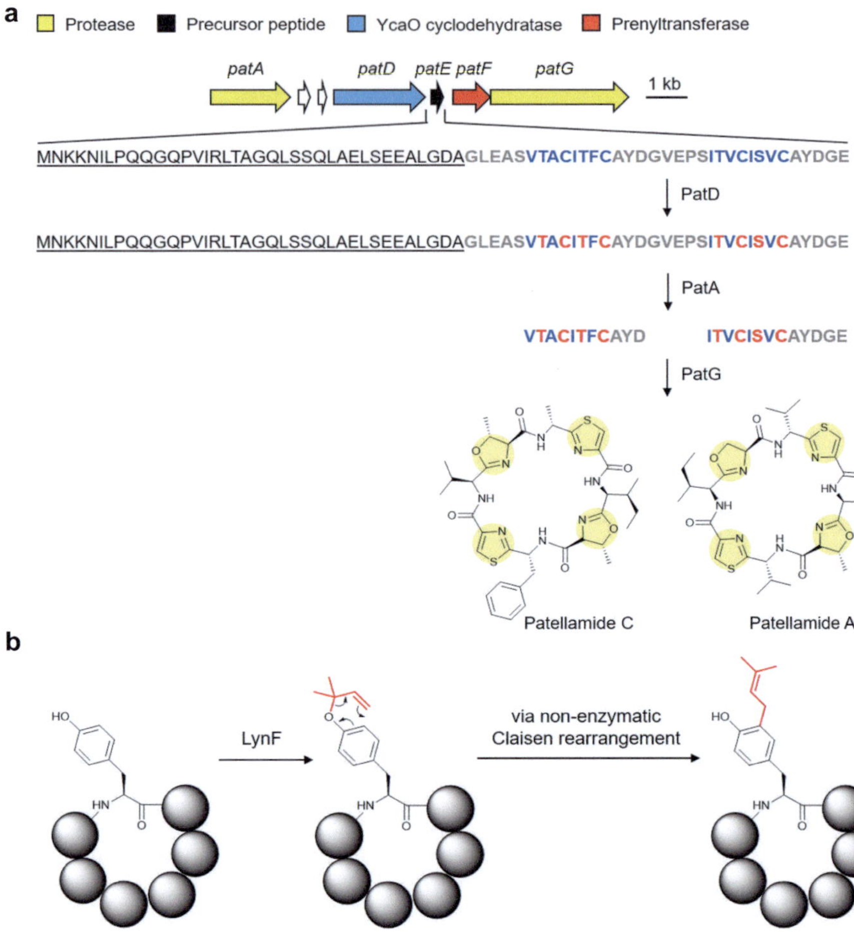

Fig. 25 **a** The biosynthetic gene cluster, biosynthetic pathway and structure of cyanobactin. **b** The reaction of cyanobactin prenyltransferase LynF. The structures of azolines and azoles installed by the YcaO cyclodehydratase PatD are highlighted in yellow circles. Biosynthetic gene clusters and the genes are colored based on annotation given to translated proteins. The leader peptide sequences are underlined. The core peptide sequences are shown as blue colored bold letters. The modified amino acid sequences are shown as red colored bold letters. The recognition amino acids sequences for PatA/PatG are shown as grey colored bold letters

106], PagF [107], PirF [108, 109], KgpF [110], AcyF [111], TruF1 [112], TolF [113], AgcF [114], AutF [115], LimF [116], AgeMTPT [117], MusF1 and MusF2 [118].

5 Perspective

D. John Faulkner, a Professor at Scripps Institution of Oceanography, University of California, San Diego (UCSD), was a prominent figure in the field of marine natural products chemistry [119, 120]. Although he passed away in 2002, his work continues to inspire many young researchers, including researchers from Southeast Asia. This chapter highlights marine natural products isolated from Malaysian soft corals, red algae, sponges and cyanobacteria based on:

- Publications authored by Chin-Soon Phan, the corresponding author of this chapter; or
- Compounds encountered by him during his research training in the following laboratories: Charles S. Vairappan laboratory (2014–2018) produced 29 publications on Malaysian soft corals and 21 publications on Malaysian red algae; Charles S. Vairappan/Toshiyuki Hamada laboratory (2016) contributed 3 publications on Malaysian sponges; and Tatsufumi Okino laboratory (2019–2021) published 3 research articles on Malaysian marine cyanobacteria.

Furthermore, Chin-Soon Phan's contributions to the field have been recognized with the D. John Faulkner Award from the American Society of Pharmacognosy (2024) and Best PhD Thesis Award from the Malaysian Natural Products Society (2018), further underscoring his expertise and qualification to author this chapter.

The most active period of marine natural products in Malaysia was during 2016–2020, accounting for 31/58 papers published during that time (Fig. 26). However, since 2020, active research on marine natural products in Malaysia has significantly decreased by 90%, with no new marine natural products reported after 2022. We hope that this chapter will offer valuable insights and reignite the enthusiasm for studying natural products from marine organisms in Malaysia.

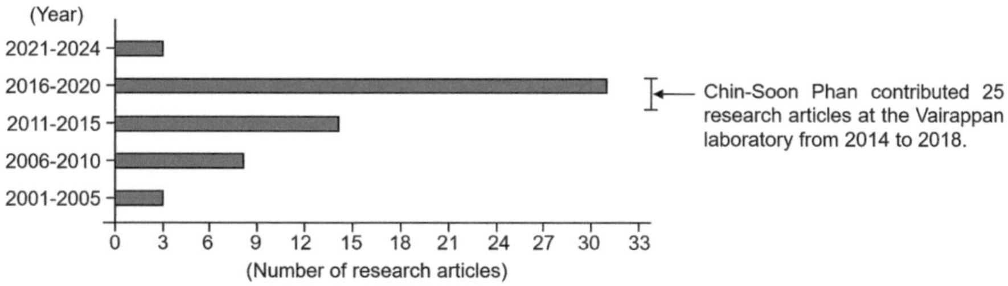

Fig. 26 A total of 58 research articles on marine natural products isolated from soft corals, red algae, sponges and cyanobacteria collected in Malaysia between 2001 to 2024

6 Conclusion

In 1984, the first review on marine natural products was published by D. John Faulkner, one of the pioneers in marine natural products chemistry [121]. After his death in 2002, Natural Products Reports journal continues to regularly publish this series of useful reviews on marine natural products. The first series was published by John W. Blunt and his colleagues in 2003 [122] and taken over by Carroll et al. [123], which continues today. Marine natural products have stimulated numerous new investigations in other disciplines, which may not otherwise have been possible. Nowadays, marine natural products have successfully attracted attention from, (1) synthetic chemists to challenge the total synthesis of these complex structures, (2) ecologists and biologists to reveal the ecological or physiological functions of these molecules, and (3) natural product chemists and synthetic biologists to unveil the biosynthesis of these molecules, access the cryptic natural products, and harness the biosynthetic enzymes for biocatalyst development.

To date, 58 research articles (from natural products or closely related journals) reported the isolation of 228 different marine natural products from soft corals, red algae, sponges and cyanobacteria collected in Malaysia. Of these 228 compounds, 73 were identified as new marine natural products, which also means that the chance of discovering new molecules from Malaysian marine organisms is as high as 32%. We believe there are still many unexplored marine organisms in Malaysia that may contain unprecedented chemical scaffold with medicinal property. Recent advances in analytical tools and molecular biology have become possible to study marine natural products in Malaysia at a previously unseen scale, in three aspects: (1) detection, isolation and characterization of minor natural products, (2) genome sequencing to study biosynthetic enzymes, and (3) connecting chemistry back to their genes, as described in the Sect. 4 "marine natural products biosynthesis".

Acknowledgements The authors would like to acknowledge the funding from the EU project No. 101087181 (Natural Products Research at Latvian Institute of Organic Synthesis as a Driver for Excellence in Innovation) supported Dr. Abujunaid Habib Khan and Dr. Chin-Soon Phan. Next, the authors would like to acknowledge Prof. Charles S. Vairappan (Universiti Malaysia Sabah), Prof. Tatsufumi Okino (Hokkaido University), Prof. Toshiyuki Hamada (Kagoshima University) and their group members and other dedicated researchers (as listed in the references) for their contributions in the discovery of marine natural products from Malaysian soft corals, red algae, sponges and cyanobacteria. Finally, the author would like to acknowledge Prof. You-Jin Jeon (Jeju National University) for his efforts in determining the anti-inflammatory activity of marine natural products from Malaysia [17, 23, 41, 46, 47], Prof. Deukjoon Kim (Seoul National University) and Prof. Hyoungsu Kim (Ajou University) for their efforts in the synthesis of oxacyclic halogenated C_{15}-acetogenins and structure revision of (+)-itomanallene A (**R77**) and (+)-4-*epi*-isolaurallene A (**R92**) [55, 57, 76], Prof. Makoto Fujita (The University of Tokyo) for his efforts in the synthesis of C_{16}-chamigrenes and structure revision of cycloelatanene B (**R40**) [44], and Prof. Yoshiki Morimoto (Osaka City University) for his efforts in the synthesis of oxasqualenoid and stereochemistry assignment of intricatetraol (**R56**) [45].

References

1. Christianson DW (2017) Structural and chemical biology of terpenoid cyclases. Chem Rev 117:11570–11648
2. Yao G, Vidor NB, Foss AP, Chang LC (2007) Lemnalosides A-D, decalin-type bicyclic diterpene glycosides from the marine soft coral *Lemnalia* sp. J Nat Prod 70:901–905
3. Phan CS, Kamada T, Hatai K, Vairappan CS (2018) Paralemnolins V and W, new nardosinane-type sesquiterpenoids from a Bornean soft coral, *Lemnalia* sp. Chem Nat Compd 54:903–906
4. Tani K, Kamada T, Phan CS, Vairappan CS (2022) New bioactive sesquiterpenoid from Malaysian soft coral genus *Lemnalia*. Heterocycles 104:1469–1476
5. Ishii T, Matsuura H, Zhaoqi Z, Vairappan CS (2009) A new 4α-methylated sterol from a *Nephthea* sp. (Nephtheidae) Bornean soft coral. Molecules 14:3360–3366
6. Ishii T, Matsuura H, Zhaoqi Z, Vairappan CS (2009) A new norsesquiterpenoid from a Bornean soft coral genus *Nephthea*. Molecules 14:4591–4596
7. Ishii T, Ueoka R, Matsunaga S, Vairappan CS (2010) Bioactive secondary metabolites from the Bornean soft corals of the genus *Nephthea*. Malays J Sci 29:262–268
8. Ishii T, Zhaoqi Z, Vairappan CS (2010) A new cembrane diterpene from the Bornean soft coral *Nephthea* sp. Molecules 15:3857–3862
9. Ishii T, Kamada T, Vairappan CS (2016) Three new cembranoids from the Bornean soft coral *Nephthea* sp. J Asian Nat Prod Res 18:415–422
10. Ishii T, Kamada T, Phan CS, Vairappan (2018) Chabrolene, a novel norditerpene from the Bornean soft coral *Nephthea* sp. 47:319–322
11. Tani K, Kamada T, Phan CS, Vairappan CS (2019) New cembrane-type diterpenoids from Bornean soft coral Nephthea sp. with antifungal activity against *Lagenidium thermophilum*. Nat Prod Res 33:3343–3349
12. Vairappan CS, Zanil II, Kamada T, Palaniappan M (2012) Isopropyl(ene)-type cembrene diterpene an important secondary metabolite in soft coral *Sinularia flexibilis* of tun Sakaran Marine Park, Malaysia. J Trop Biol Conserv 9:200–205
13. Palaniveloo K, Ramachandram RT, Vairappan CS (2014) Cembrane diterpenes as chemotax-onomical markers for *Sinularia flexibilis*. J Trop Biol Conserv 11:103–116
14. Phan CS, Ng SY, Kamada T, Vairappan CS (2016) Two new lobane diterpenes from a Bornean soft coral *Sinularia* sp. Nat Prod Commun 11:899–900
15. Kamada T, Wong WT, Phan CS, Vairappan CS (2016) Anti-bacterial diterpenes from Bornean soft corals *Sinularia flexibilis*. Int J Res Appl Nat Soc Sci 4:117–124
16. Kamada T, Phan CS, Hamada T, Hatai K, Vairappan CS (2018) Cytotoxic and antifungal terpenoids from Bornean soft coral, *Sinularia flexibilis*. 13:17–19
17. Kamada T, Kang MC, Phan CS, Zanil II, Jeon YJ, Vairappan CS (2018) Bioactive cembranoids from the soft coral genus *Sinularia* sp. in Borneo. Mar Drugs 16:99
18. Phan CS, Yee CS, Vairappan CS, Ishii T, Kamada T (2019) Sinulaflexiolide p, a cembrane-type diterpenoid from Bornean soft coral *Sinularia flexibilis*. Chem Nat Compd 55:285–288
19. Phan CS, Vairappan CS (2015) Isopropyl(ene)-type cembrane diterpene an important chemotaxonomical marker in Bornean soft coral genus *Sarcophyton*. J Trop Biol Conserv 12:137–143
20. Kamada T, Phan CS, Tin HS, Vairappan CS, Muhammad TST (2016) 16-Hydroxycembra-1,3,7,11-tetraene, a new cembrane diterpene from Malaysian soft coral genus *Sarcophyton*. Nat Prod Commun 11:1077–1078
21. Kamada T, Zanil II, Phan CS, Vairappan CS (2018) A new cembrane, from soft coral genus *Sarcophyton* in Borneo. Nat Prod Commun 13:123–124
22. Tani K, Kamada T, Phan CS, Vairappan CS (2018) A new bioactive cembranolide sarco-phytonolide V from Bornean soft coral genus *Sarcophyton*. Nat Prod Commun 14.https://doi.org/10.1177/1934578X19868375
23. Phan CS, Ng SY, Kim EA, Jeon YJ, Palaniveloo K, Vairappan CS (2015) Capgermacrenes A and B, bioactive secondary metabolites from a Bornean soft coral, *Capnella* sp. Mar Drugs 13:3103–3115

24. Ishii T, Phan CS, Kamada T, Vairappan CS (2016) Capgermacrene C, a new sesquiterpenoid from a Bornean soft coral, *Capnella* sp. Nat Prod Commun 11:1065–1066
25. Phan CS, Vairappan CS (2017) Capgermacrenes D-G, new sesquiterpenoids from a Bornean soft coral, *Capnella imbricata.* Nat Prod Res 31:742–748
26. Phan CS, Kamada T, Ishii T, Hamada T, Vairappan CS (2019) Cytotoxic sesquiterpenoids from soft coral *Capnella imbricata.* Nat Prod Commun 14.https://doi.org/10.1177/1934578X19857494
27. Phan CS, Kamada T, Ishii T, Hamada T, Vairappan CS (2018) A new guaiane-type sesquiterpenoid from a Bornean soft coral, *Xenia stellifera.* Nat Prod Commun 13:15–16
28. Phan CS, Kamada T, Kobayashi K, Hamada T, Vairappan CS (2018) 15-deoxy-isoxeniolide-A, new diterpenoid from a Bornean soft coral, *Xenia* sp. Nat Prod Res 32:202–207
29. Phan CS, Kamada T, Ishii T, Hamada T, Vairappan CS (2019) 12-*Epi*-9-deacetoxyxenicin, new cytotoxic diterpenoid from a Bornean soft coral, *Xenia* sp. Nat Prod Res 33:808–813
30. Bahl A, Jachak SM, Palaniveloo K, Ramachandram T, Vairappan CS, Chopra HK (2014) 2-Acetoxyverecynarmin C, a new briarane COX inhibitory diterpenoid from *Pennatula aculeata.* Nat Prod Commun 9:1139–1141
31. Tani K, Vairappan CS (2022) The chemotaxonomic identification using structure types of secondary metabolites and their bioactivities of Bornean *Litophyton arboretum.* J Trop Biol Conserv 19:67–76
32. Suzuki M, Daitoh M, Vairappan CS, Abe T, Masuda M (2001) Novel halogenated metabolites from the Malaysian *Laurencia pannosa.* J Nat Prod 64:597–602
33. Vairappan CS, Daitoh M, Suzuki M, Abe T, Masuda M (2001) Antibacterial halogenated metabolites from the Malaysian *Laurencia* species. Phytochemistry 58:291–297
34. Vairappan CS (2003) Potent antibacterial activity of halogenated metabolites from Malaysian red algae, *Laurencia majuscula* (Rhodomelaceae, Ceramiales). Biomol Eng 20:255–259
35. Vairappan CS, Suzuki M, Ishii T, Okino T, Abe T, Masuda M (2008) Antibacterial activity of halogenated sesquiterpenes from Malaysian *Laurencia* spp. Phytochemistry 69:2490–2494
36. Vairappan CS, Lee TK (2009) C-15 halogenated acetogenin with antibacterial activity against food pathogens. Malays J Sci 28:263–268
37. Vairappan CS, Ishii T, Lee TK, Suzuki M, Zhaoqi Z (2010) Antibacterial activities of a new brominated diterpene from Borneon *Laurencia* spp. Mar Drugs 8:1743–1749
38. Tan KL, Matsunaga S, Vairappan CS (2011) Halogenated chamigranes of red alga *Laurencia snackeyi* (Weber-van Bosse) Masuda from Sulu-Sulawesi Sea. Biochem Syst Ecol 39:213–215
39. Kamada T, Vairappan CS (2012) A new bromoallene-producing chemical type of the red alga *Laurencia nangii* masuda. Molecules 17:2119–2125
40. Kamada T, Vairappan CS (2013) New bioactive secondary metabolites from Bornean red alga, *Laurencia similis* (Ceramiales). Nat Prod Commun 8:287–288
41. Vairappan CS, Kamada T, Lee WW, Jeon YJ (2013) Anti-inflammatory activity of halogenated secondary metabolites of *Laurencia snackeyi* (Weber-van Bosse) Masuda in LPS-stimulated RAW 264.7 macrophages. J Appl Phycol 25:1805–1813
42. Vairappan CS, Zanil II, Kamada T (2014) Structural diversity and geographical distribution of halogenated secondary metabolites in red algae, *Laurencia nangii* Masuda (Rhodomelaceae, Ceramiales), in the coastal waters of North Borneo Island. J Appl Phycol 26:1189–1198
43. Howard BM, Nonomura AM, Fenical W (1980) Chemotaxonomy in marine algae: secondary metabolite synthesis by *Laurencia* in unialgal culture. Biochem Syst Ecol 8:329–336
44. Lee S, Hoshino M, Fujita M, Urban S (2017) Cycloelatanene A and B: absolute configuration determination and structural revision by the crystalline sponge method. Chem Sci 8:1547–1550
45. Morimoto Y, Okita T, Takaishi M, Tanaka T (2007) Total synthesis and determination of the absolute configuration of (+)-intricatetraol. Angew Chem Int Ed Engl 46:1132–1135
46. Wijesinghe WAJP, Kang MC, Lee WW, Lee HS, Kamada T, Vairappan CS, Jeon YJ (2014) 5β-Hydroxypalisadin B isolated from red alga *Laurencia snackeyi* attenuates inflammatory response in lipopolysaccharide-stimulated RAW 264.7 macrophages. Algae 29:333–341

47. Wijesinghe WAJP, Kim EA, Kang MC, Lee WW, Lee HS, Vairappan CS, Jeon YJ (2014) Assessment of anti-inflammatory effect of 5β-hydroxypalisadin B isolated from red seaweed *Laurencia snackeyi* in zebrafish embryo in vivo model. Environ Toxicol Pharmacol 37:110–117
48. Kamada T, Vairappan CS (2015) New laurene-type sesquiterpene from Bornean *Laurencia nangii*. Nat Prod Commun 10:843–844
49. Kamada T, Vairappan CS (2017) Non-halogenated new sesquiterpenes from Bornean *Laurencia snackeyi*. Nat Prod Res 31:333–340
50. Kamada T, Phan CS, Vairappan CS (2018) Neoiriepentaol and nangenyne, halogenated diterpenoid and C_{15}-acetogenin from red alga *Laurencia nangii* Masuda collected in Borneo. J Appl Phycol 30:3379–3386
51. Kamada T, Phan CS, Sien VST, Vairappan CS (2018) Halogenated chamigrane sesquiterpenes from Bornean *Laurencia majuscula*. J Appl Phycol 30:3373–3378
52. Kamada T, Phan CS, Vairappan CS (2019) New anti-bacterial halogenated tricyclic sesquiterpenes from Bornean *Laurencia majuscula* (Harvey) Lucas. Nat Prod Res 33:464–471
53. Kamada T, Phan CS, Vairappan CS (2019) Nangallenes A and B, halogenated nonterpenoid C_{15}-acetogenins from the Bornean red alga *Laurencia nangii*. J Asian Nat Prod Res 21:241–247
54. Kamada T, Phan CS, Okino T, Vairappan CS (2019) Cytotoxicity and antibacterial potential of halogenated chamigrenes from Malaysian red alga, *Laurencia majuscula*. Planta Med Int Open 6:e36–e40
55. Jeong W, Kim MJ, Kim H, Kim S, Kim D, Shin KJ (2010) Substrate-controlled asymmetric total synthesis and structure revision of (+)-itomanallene A. Angew Chem Int Ed Engl 49:752–756
56. Phan CS, Kamada T, Vairappan CS (2020) Two new epimers of C_{15}-acetogenin, 4-*epi*-isolaurallene and 4-*epi*-itomanallene A as diastereomeric model. Nat Prod Res 34:1008–1013
57. Park Y, Jang H, Kwak SY, Lim S, Lee D, Kim H, Kim D (2025) Formal total synthesis of (−)-isolaurallene and structure determination of (+)-4-*epi*-isolaurallene via stereodivergent intramolecular amide enolate alkylation. Org Lett 27:3199–3203
58. Palaniveloo K, Ong KH, Satriawan H, Abdul Razak S, Suciati S, Hung HY, Hirayama S, Rizman-Idid M, Tan JK, Yong YS, Phang SM (2023) In vitro and in silico cholinesterase inhibitory potential of metabolites from *Laurencia snackeyi* (Weber-van Bosse) M. Masuda. 3 Biotech 13:337
59. Hamada T, Harada D, Hirata M, Yamashita K, Palaniveloo K, Okamura H, Iwagawa T, Arima N, Iriguchi T, de Voogd NJ, Vairappan CS (2015) Manoalide-related sesterterpene from the marine sponge *Luffariella variabilis*. Nat Prod Commun 10:863–864
60. Phan CS, Kamada T, Hamada T, Vairappan CS (2018) Cytotoxic sesterterpenoids from Bornean sponge *Spongia* sp. Rec Nat Prod 12:643–647
61. Hamada T, Matsumoto Y, Phan CS, Kamada T, Onitsuka S, Okamura H, Iwagawa T, Arima N, Tani F, Vairappan CS (2019) Aaptamine-related alkaloid from the marine sponge *Aaptos aaptos*. Nat Prod Commun 14:1–3
62. Lopez JAV, Petitbois JG, Vairappan CS, Umezawa T, Matsuda F, Okino T (2017) Columbamides D and E: chlorinated fatty acid amides from the marine cyanobacterium *Moorea bouillonii* collected in Malaysia. Org Lett 19:4231–4234
63. Mehjabin JJ, Wei L, Petitbois JG, Umezawa T, Matsuda F, Vairappan CS, Morikawa M, Okino T (2020) Biosurfactants from marine cyanobacteria collected in Sabah, Malaysia. J Nat Prod 83:1925–1930
64. Fathoni I, Petitbois JG, Alarif WM, Abdel-Lateff A, Al-Lihaibi SS, Yoshimura E, Nogata Y, Vairappan CS, Sholikhah EN, Okino T (2020) Bioactivities of lyngbyabellins from cyanobacteria of *Moorea* and *Okeania* genera. Molecules 25:3986
65. Ng SY, Phan CS, Ishii T, Kamada T, Hamada T, Vairappan CS (2020) Terpenoids from marine soft coral of the genus *Xenia* in 1977 to 2019. Molecules 25:5386
66. Rudi A, Ketzinel S, Goldberg I, Stein Z, Kashman Y, Benayahu Y, Schleyer M (1995) Anthe-liatin and zahavins A and B, three new cytotoxic xenicane diterpenes from two soft corals. J Nat Prod 58:1581–1586

67. Cheng YB, Jang JY, Khalil AT, Kuo YH, Shen YC (2006) Xenicane-type diterpeneswith cytotoxicity from *Xenia florida*. J Nat Prod 69:675–678
68. Braekman JC, Daloze D, Tursch B, Declercq JP, Germain G, van Meersscche M (1979) Chemical studies of marine invertebrates. XXXIX. Three novel diterpenoids from the soft coral *Xenia novae-britanniae*. Bull Soc Chim Belg 88:71–77
69. Almourabit A, Gillet B, Ahond A, Beloeil JC, Poupat C, Potier P (1989) Invertébrés marins du lagon néo-calédonien, XI. Les desoxyhavannahines, nouveaux metabolites de *Xenia membranacea*. J Nat Prod 52:1080–1087
70. Bowden BF, Cusack BJ, Dangel A (2003) 13-*Epi*-9-deacetoxyxenicin, a cytotoxic diterpene from the soft coral *Asterospicularia laurae* (Alcyonacea). Mar Drugs 1:18–26
71. El-Gamal AAH, Chiang CY, Huang SH, Wang SK, Duh CY (2005) Xenia diterpenoids from the Formosan soft coral *Xenia blumi*. J Nat Prod 68:1336–1340
72. Lin YC, Abd El-Razek MH, Hwang TL, Chiang MY, Kuo YH, Dai CF, Shen YC (2009) Asterolaurins A-F, xenicane diterpenoids from the Taiwanese soft coral *Asterospicularia laurae*. J Nat Prod 72:1911–1916
73. Bowden BF, Coll JC, Ditzel E, Mitchell SJ, Robinson WT (1982) Studies of Australian soft coral. XXVIII the structure determination of two new diterpenes from the genus *Xenia* (Alcyonacea). Aust J Chem 35:997–1002
74. Kurata K, Furusaki A, Suehiro K, Katayama C, Suzuki T (1982) Isolaurallene, a new nonterpenoid C15-bromoallene, from the red alga *Laurencia nipponica* Yamada. Chem Lett 11:1031–1034
75. Suzuki M, Kurosawa E, Furusaki A, Katsuragi S, Matsumoto T (1984) Neolaurallene, a new halogenated C-15 nonterpenoid from the red alga *Laurencia okamurai* Yamada. Chem Lett 13:1033–1034
76. Kim MJ, Sohn T, Kim D, Paton RS (2012) Concise substrate-controlled asymmetric total syntheses of dioxabicyclic marine natural products with 2,10-dioxabicyclo[7.3.0]dodecene and 2,9-dioxabicyclo[6.3.0]undecene skeletons. J Am Chem Soc 134:20178–20188
77. Davyt D, Fernandez R, Suescun L, Mombrú AW, Saldaña J, Domínguez L, Coll J, Fujii MT, Manta E (2001) New sesquiterpene derivatives from the red alga *Laurencia scoparia*. Isolation, structure determination, and anthelmintic activity. J Nat Prod 64:1552–1555
78. Suzuki M, Kurosawa E (1978) Two new halogenated sesquiterpenes from the red alga *Laurencia majuscula* Harvey. Tetrahedron Lett 48:4805–4808
79. Hamada T, Sugawara T, Matsunaga S, Fusetani N (1994) Polytheonamides, unprecedented highly cytotoxic polypeptides from the marine sponge *Theonella swinhoei* 1. Isolation and component amino acids. Tetrahedron Lett 35:719–720
80. Hamada T, Sugawara T, Matsunaga S, Fusetani N (1994) Polytheonamides, unprecedented highly cytotoxic polypeptides from the marine sponge *Theonella*swinhoei 2. Structure elucidation. Tetrahedron Lett 35:609–612
81. Hamada T, Matsunaga S, Yano G, Fusetani N (2005) Polytheonamides A and B, highly cytotoxic, linear polypeptides with unprecedented structural features, from the marine sponge, *Theonella swinhoei*. J Am Chem Soc 127:110–118
82. Hamada T, Matsunaga S, Fujiwara M, Fujita K, Hirota H, Schmucki R, Güntert P, Fusetani N (2010) Solution structure of polytheonamide B, a highly cytotoxic nonribosomal polypeptide from marine sponge. J Am Chem Soc 132:12941–12945
83. Freeman MF, Gurgui C, Helf MJ, Morinaka BI, Uria AR, Oldham NJ, Sahl H-G, Matsunaga S, Piel J (2012) Metagenome mining reveals polytheonamides as posttranslationally modified ribosomal peptides. Science 338:387–390
84. Gurgui C, Piel J (2010) Metagenomic approaches to identify and isolate bioactive natural products from microbiota of marine sponges. In: Streit WR, Daniel R (eds) Metagenomics: methods and protocols, methods in molecular biology. Humana Press, pp 247–264
85. Freeman MF, Vagstad AL, Piel J (2016) Polytheonamide biosynthesis showcasing the metabolic potential of sponge-associated uncultivated 'Entotheonella' bacteria. Curr Opin Chem Biol 31:8–14

86. McCauley EP, Piña IC, Thompson AD, Bashir K, Weinberg M, Kurz SL, Crews P (2020) Highlights of marine natural products having parallel scaffolds found from marine-derived bacteria, sponges, and tunicates. J Antibiot 73:504–525
87. Mydlarz LD, Jacobs RS, Boehnlein J, Kerr RG (2003) Pseudopterosin biosynthesis in *Symbiodinium* sp., the dinoflagellate symbiont of *Pseudopterogorgia elisabethae*. Chem Biol 10:1051–1056
88. Frenz-Ross JL, Enticknap JJ, Kerr RG (2008) The effect of bleaching on the terpene chemistry of *Plexaurella fusifera*: evidence that zooxanthellae are not responsible for sesquiterpene production. Mar Biotechnol 10:572–578
89. Burkhardt I, de Rond T, Chen PYT, Moore BS (2022) Ancient plant-like terpene biosynthesis in corals. Nat Chem Biol 18:664–669
90. Scesa PD, Lin Z, Schmidt EW (2022) Ancient defensive terpene biosynthetic gene clusters in the soft corals. Nat Chem Biol 18:659–663
91. Scesa PD, Schmidt EW (2025) Pseudopterosin biosynthesis: unravelling a decades old problem in animal specialized metabolism. J Am Chem Soc 147:3072–3079
92. Scesa PD, Schmidt EW (2023) Biomimetic approach to diverse coral diterpenes from a biosynthetic scaffold. Angew Chem Int Ed Engl 62:e202311406
93. Kersten RD, Lee S, Fujita D, Pluskal T, Kram S, Smith JE, Iwai T, Noel JP, Fujita M, Weng JK (2017) A red algal bourbonane sesquiterpene synthase defined by microgram-scale NMR-coupled crystalline sponge X-ray diffraction analysis. J Am Chem soc 139:16838–16844
94. Steele TS, Burkhardt I, Moore ML, de Rond T, Bone HK, Barry K, Bunting VM, Grimwood J, Handley LH, Rajasekar S, Talag J, Michael TP, Moore BS (2024) Biosynthesis of haloterpenoids in red algae via microbial-like type I terpene synthases. ACS Chem Biol 19:185–192
95. Ishikawa T, Washio K, Kaneko K, Tang XR, Morikawa M, Okino T (2022) Characterization of vanadium-dependent bromoperoxidases involved in the production of brominated sesquiterpenes by the red alga *Laurencia okamurae*. Appl Phycol 3:120–131
96. Ireland CM, Durso AR Jr, Newman RA, Hacker MP (1982) Antineoplastic cyclic peptides from the marine tunicate *Lissoclinum patella*. J Org Chem 47:1807–1811
97. Degnan BM, Hawkins CJ, Lavin MF, McCaffrey EJ, Parry DL, vanden Brenk AL, Watters DJ (1989) New cyclic peptides with cytotoxic activity from the ascidian *Lissoclinum patella*. J Med Chem 32:1349–1354
98. Biard JF, Grivois C, Verbist JF, Debitus C, Carre JB (1990) Origin of bistramide a identified in *Lissoclinum bistratum* (Urochordata): possible involvement of symbiotic Prochlorophyta. J Mar Biol Assoc UK 70:741–746
99. Salomon CE, Faulkner DJ (2002) Localization studies of bioactive cyclic peptides in the ascidian *Lissoclinum patella*. J Nat Prod 65:689–692
100. Schmidt EW, Nelson JT, Rasko DA, Sudek S, Eisen JA, Haygood MG, Ravel J (2005) Patellamide A and C biosynthesis by a microcin-like pathway in *Prochloron didemni*, the cyanobacterial symbiont of *Lissoclinum patella*. Proc Natl Acad Sci USA 102:7315–7320
101. Long PF, Dunlap WC, Battershill CN, Jaspars M (2005) Shotgun cloning and heterologous expression of the patellamide gene cluster as a strategy to achieving sustained metabolite production. ChemBioChem 6:1760–1765
102. Behrendt L, Raina JB, Lutz A, Kot W, Albertsen M, Halkjær-Nielsen P, Sørensen SJ, Larkum AW, Kühl M (2018) In situ metabolomic- and transcriptomic-profiling of the host-associated cyanobacteria *Prochloron* and *Acaryochloris marina*. ISME J 12:556–567
103. Arnison PG, Bibb MJ, Bierbaum G, Bowers AA, Bugni TS, Bulaj G, Camarero JA, Campopiano DJ, Challis GL, Clardy J, Cotter PD, Craik DJ, Dawson M, Dittmann E, Donadio S, Dorrestein PC, Entian KD, Fischbach MA, Garavelli JS, Göransson U, Gruber CW, Haft DH, Hemscheidt TK, Hertweck C, Hill C, Horswill AR, Jaspars M, Kelly WL, Klinman JP, Kuipers OP, Link AJ, Liu W, Marahiel MA, Mitchell DA, Moll GN, Moore BS, Müller R, Nair SK, Nes IF, Norris GE, Olivera BM, Onaka H, Patchett ML, Piel J, Reaney MJ, Rebuffat S, Ross RP, Sahl HG, Schmidt EW, Selsted ME, Severinov K, Shen B, Sivonen K, Smith L, Stein T, Süssmuth RD, Tagg JR, Tang GL, Truman AW, Vederas JC, Walsh CT,

Walton JD, Wenzel SC, Willey JM, van der Donk WA (2013) Ribosomally synthesized and post-translationally modified peptide natural products: overview and recommendations for a universal nomenclature. Nat Prod Rep 30:108–160
104. Montalbán-López M, Scott TA, Ramesh S, Rahman IR, van Heel AJ, Viel JH, Bandarian V, Dittmann E, Genilloud O, Goto Y, Grande Burgos MJ, Hill C, Kim S, Koehnke J, Latham JA, Link AJ, Martínez B, Nair SK, Nicolet Y, Rebuffat S, Sahl HG, Sareen D, Schmidt EW, Schmitt L, Severinov K, Süssmuth RD, Truman AW, Wang H, Weng JK, van Wezel GP, Zhang Q, Zhong J, Piel J, Mitchell DA, Kuipers OP, van der Donk WA (2021) New developments in RiPP discovery, enzymology and engineering. Nat Prod Rep 38:130–239
105. McIntosh JA, Donia MS, Nair SK, Schmidt EW (2011) Enzymatic basis of ribosomal peptide prenylation in cyanobacteria. J Am Chem Soc 133:13698–13705
106. McIntosh JA, Lin Z, Tianero MDB, Schmidt EW (2013) Aestuaramides, a natural library of cyanobactin cyclic peptides resulting from isoprene-derived Claisen rearrangements. ACS Chem Biol 8:877–883
107. Hao Y, Pierce E, Roe D, Morita M, McIntosh JA, Agarwal V, Cheatham TE, Schmidt EW, Nair SK (2016) Molecular basis for the broad substrate selectivity of a peptide prenyltransferase. Proc Natl Acad Sci USA 113:14037–14042
108. Estrada P, Morita M, Hao Y, Schmidt EW, Nair SK (2018) A single amino acid switch alters the isoprene donor specificity in ribosomally synthesized and post-translationally modified peptide prenyltransferases. J Am Chem Soc 140:8124–8127
109. Morita M, Hao Y, Jokela JK, Sardar D, Lin Z, Sivonen K, Nair SK, Schmidt EW (2018) Post-translational tyrosine geranylation in cyanobactin biosynthesis. J Am Chem Soc 140:6044–6048
110. Parajuli A, Kwak DH, Dalponte L, Leikoski N, Galica T, Umeobika U, Trembleau L, Bent A, Sivonen K, Wahlsten M, Wang H, Rizzi E, De Bellis G, Naismith J, Jaspars M, Liu X, Houssen W, Fewer DP (2016) A unique tryptophan C-prenyltransferase from the kawaguchipeptin biosynthetic pathway. Angew Chem Int Ed Engl 55:3596–3599
111. Dalponte L, Parajuli A, Younger E, Mattila A, Jokela J, Wahlsten M, Leikoski N, Sivonen K, Jarmusch SA, Houssen WE, Fewer DP (2018) N-Prenylation of tryptophan by an aromatic prenyltransferase from the cyanobactin biosynthetic pathway. Biochemistry 57:6860–6867
112. Tianero MD, Pierce E, Raghuraman S, Sardar D, McIntosh JA, Heemstra JR, Schonrock Z, Covington BC, Maschek JA, Cox JE (2016) Metabolic model for diversity-generating biosynthesis. Proc Natl Acad Sci USA 113:1772–1777
113. Purushothaman M, Sarkar S, Morita M, Gugger M, Schmidt EW, Morinaka BI (2021) Genome mining based discovery of the cyclic peptide tolypamide and TolF, a Ser/Thr forward O-prenyltransferase. Angew Chem Int Ed 60:8460–8465
114. Phan CS, Matsuda K, Balloo N, Fujita K, Wakimoto T, Okino T (2021) Argicyclamides A-C unveil enzymatic basis for guanidine bis-prenylation. J Am Chem Soc 143:10083–10087
115. Clemente C, Johnson N, Ouyang X, Popin RV, Dall'Angelo S, Wahlsten M, Jokela J, Colombano A, Nardone B, Fewer DP, Houssen WE (2022) Biochemical characterization of a cyanobactin arginine-N-prenylase from the autumnalamide biosynthetic pathway. Chem Commun 58:12054–12057
116. Zhang Y, Hamada K, Nguyen DT, Inoue S, Satake M, Kobayashi S, Okada C, Ogata K, Okada M, Sengoku T, Goto Y, Suga H (2022) LimF is a versatile prenyltransferase for histidine-C-geranylation on diverse non-natural substrates. Nat Catal 5:682–693
117. Sardar D, Hao Y, Lin Z, Morita M, Nair SK, Schmidt EW (2017) Enzymatic N- and C-protection in cyanobactin RiPP natural products. J Am Chem Soc 139:2884–2887
118. Mattila A, Andsten RM, Jumppanen M, Assante M, Jokela J, Wahlsten M, Mikula KM, Sigindere C, Kwak DH, Gugger M (2019) Biosynthesis of the bis-prenylated alkaloids muscoride A and B. ACS Chem Biol 14:2683–2690
119. Andersen RJ, Ireland CM, Molinski TF, Bewley CA (2004) Research achievement award–D. John Faulkner. J Nat Prod 67:1239–1251
120. Fenical W (2003) D. John Faulkner (1942–2002): marine natural products chemistry and marine chemical ecology. Angew Chem Int Ed Engl 42:1438–1439

121. Faulkner DJ (1984) Marine natural products: metabolites of marine algae and herbivorous marine molluscs. Nat Prod Rep 1:251–280
122. Blunt JW, Copp BR, Munro MHG, Northcote PT, Prinsep MR (2003) Marine natural products. Nat Prod Rep 20:1–48
123. Carroll AR, Copp BR, Davis RA, Keyzers RA, Prinsep MR (2020) Marine natural products. Nat Prod Rep 37:175–223

Chapter 4
Charting the Philippine Seas—Progress in Marine Natural Products Research in the Philippines in the Last 25 Years

Lilibeth A. Salvador-Reyes, Shalice R. Susana-Guevarra, and Ernest Guiller S. Pineda

1 Introduction

The Philippines, recognized as one of the 18 megadiverse countries in the world, is home to over 52,000 plants and animals (https://www.cbd.int/countries/profile?country=ph). Situated at the Coral Triangle, a region characterized by significant marine biodiversity, this archipelagic nation harbors an extensive array of marine life. Among documented marine organisms in the Philippines are 1755 species of marine fishes, 820 species of algae, 648 species of mollusks, 468 species of scleractinian corals, and 19 species of seagrass [1, 2]. This exceptional biodiversity and the intricate relationships among organisms result in a substantial chemical and biomolecular diversity.

The high biodiversity in the Philippines has spurred significant interest in marine natural products by local and foreign researchers. One of the earliest works on the discovery of bioactive and chemically diverse natural products from Philippine marine organisms is was conducted by National Scientist Lourdes J. Cruz and Professor Baldomero M. Olivera on cone snails during the 1970s. Their pioneering study sought to elucidate the mechanism by which the gastropod *Conus geographus* proves fatal to fishermen who encounter its sting [3]. The country's extensive marine biodiversity, particularly cone snails, and limited research on this species propelled the comprehensive investigation on these predatory gastropods. Separation and biological activity assay demonstrated that the venom of cone snails contains

Present Address:
L. A. Salvador-Reyes (✉) · S. R. Susana-Guevarra · E. G. S. Pineda
Marine Science Institute, College of Science, University of the Philippines Diliman, Quezon City, Philippines
e-mail: lsreyes@msi.upd.edu.ph

S. R. Susana-Guevarra
e-mail: shalicesusana@msi.upd.edu.ph

L. T. Tan and N. Hanif (eds.), *Marine Natural Products Research in South-East Asia*,
https://doi.org/10.1007/978-3-032-01098-8_4

a mixture of complex and potent peptide-based molecules that elicited a variety of symptomology when injected intracranially [4]. This basic research on *Conus* venom has transformed the field of natural products and neuropharmacology. It has yielded pharmacological probes to interrogate ion channels and receptors and an alternative drug to morphine, Prialt [5, 6]. Parallel to the discovery of bioactive conopeptides was the interest in understanding the taxonomy, distribution, and biology of cone snails and other gastropods, demonstrating that drug discovery and basic research in organismal biology can complement one another.

Collaborations and support for research in marine natural products in the Philippines over the last 25 years have strengthened the field through human resource development, the establishment of critical infrastructure for natural products analysis, capacity building, and research exchange. Government support from the Philippines Department of Science and Technology, through the PharmaSeas Drug Discovery Program, the Discovery and Development of Health Products, and Tuklas Lunas, has been instrumental in establishing research centers focused on marine and terrestrial natural products. Collaborations with the National Cooperative Drug Discovery Group [7] and the Philippine Mollusk Symbiont—International Cooperative Biodiversity Group have forged linkages and partnerships with foreign marine natural products research groups.

The collection of samples from the Philippines is guided by RA 9147, or the Wildlife Resources Conservation and Protection Act. This outlines the guidelines for collecting and utilizing marine organisms and their by-products. The collection of marine organisms is overseen by the Department of Agriculture-Bureau of Fisheries and Aquatic Resources, (DA-BFAR) (https://www.officialgazette.gov.ph/2001/07/30/republic-act-no-9417/). A Gratuitous Permit is executed with DA-BFAR and requires prior informed consent from local communities in the collection sites. Foreign entities are mandated to have a local institution actively involved in the collection, research, and development of marine organisms. Access and benefit sharing in the Philippines is guided by the Convention on Biological Diversity and, more recently, the Nagoya Protocol on Access and Benefit Sharing. In 2023, the Philippines also signed the treaty under the UN Convention on the Law of the Sea (UNCLOS) on the Conservation and Sustainable Use of Marine Biological Diversity of Areas beyond National Jurisdiction (BBNJ) agreement (https://pcaf.da.gov.ph/index.php/2023/10/09/philippines-signs-bbnj-treaty-protecting-marine-biodiversity/), highlighting the country's shared commitment to protect the high seas for the future of humanity.

2 Small Molecule Compounds from Marine Macroorganisms

Marine macroorganisms, particularly invertebrates, have been widely studied for their natural products. Benthic organisms have evolved natural products because of the intense competition for space, the need for chemical defenses to evade predators, and adaptation to the dynamic nature of the oceans [8]. Anticancer activity has been the major theme among natural products from marine macroorganisms. To date, ten marine natural products or their derivatives have obtained clinical approval as anticancer agents (https://www.marinepharmacology.org/approved). The first marine-derived anticancer agent, cytarabine, was developed based on the pyrimidine nucleoside isolated from the marine sponge *Cryptotethya crypta* [9]. Cytarabine is an FDA-approved drug for treating malignant acute myeloid, lymphoid, and myelogenous leukemia, and non-Hodgkin's lymphoma. Eribulin mesylate was derived from the macrolide halichondrin B from *Halichondria okadai* and was approved in 2010 by the FDA as a treatment for metastatic breast cancer [9]. Trabectedin (ET-743) was discovered in the tunicate *Ecteinascidia turbinata*. At the same time, antibody drug conjugates of monomethyl auristatin E, an analogue of the cyanobacterium compound dolastatin 10, are now used in multiple cancer therapies.

2.1 Sponges

Marine sponges have been a prolific source of structurally diverse and biomedically relevant natural products. They are sessile filter feeders and lack the physical structure for protection; hence, they rely on bioactive secondary metabolites as chemical defense against competitors, predators, and pathogens. Sponges also utilize these chemicals to adapt to their environment and to form close associations with marine microorganisms, including bacteria, fungi, and viruses. These relationships, as well as the conspicuous morphology of sponges and ease of collection and handling, enable researchers to easily initiate the investigation of sponge chemistry among other marine invertebrates.

Earlier studies conducted on the sponges from the reefs of the Caribbean, Pacific, Mediterranean, Red Sea, and Antarctic established that secondary metabolites are used as chemical defense by sponges against predation and competition [10]. For example, various sponge species, including *Clathria flabellata*, *Dendrilla membranosa*, and *Haliclona rudis* in the western Antarctic peninsula, demonstrated unpalatability and chemical defense as feeding deterrence against major predators such as sea stars [11]. Similar observations for *Melophlus sarasinorum* collected in Guam were reported, wherein the sponge had minimal to no contact with fish and sea stars [12]. Feeding deterrence against the spongivores pufferfish *Canthigaster solandri* and sea urchin *Diadema savignyi* was also observed in *M. sarasinorum* and *Stylissa massa*, both collected in Guam's reefs [10]. In addition, sponges in reefs around Zanzibar in the western Indian Ocean are still abundant despite exposure to various anthropogenic factors such as untreated sewage and runoff from the city [13]. Crude extracts of the ten most dominant sponge species in the region exhibited potent antimicrobial properties. Seven of these sponges displayed cytotoxicity and allelopathic properties against the coral *Porites* sp. as their competitor [13]. The physical structures and production or accumulation of secondary metabolites may have contributed to the defense of sponges against predators and competitors.

Sponge samples (Fig. 1) collected in the Philippines were studied for their chemistry and/or biological activities. From 2000 to 2025, 205 compounds were purified from 32 different genera of marine sponges. Most new compounds were purified from *Dysidea* sp. (n = 25), followed by *Myriastra* sp. and *Spongia* sp., each with 12 compounds reported. Novel compounds were also purified from *Xestospongia* sp. and *Theonella* sp., yielding eight compounds from each species. *Acanthella* sp. yielded nine known compounds.

Anti-infective activity screening of sponge compounds from the Philippines is more infrequent than antiproliferative activity assessment. A breadth of pathogens is used, including pathogenic Gram (+) and Gram (–) microorganisms and protozoal parasites. While viral activity evaluation is challenging, several sponge compounds were tested in relevant enzymes in viral diseases, such as HIV-AIDS. At the same time, screening against relevant proteins was also carried out for several compounds.

Fig. 1 Different sponge species from the Philippines investigated for their chemistry and bioactivity. **a** Xestoquinone-producing *Neopetrosia compacta*, **b** Renieramycin-producing *Xestospongia* sp., **c** Luffariellolide producer, **d** unidentified sponge from Cebu, Philippines

In 2008, West and Faulkner reported acanthosulfate (**1**) (Fig. 2) as the first merosesterterpene with a *trans–antitrans–syn–anti* configuration in the scalarane-type structural backbone based on spectroscopic data [14]. The novel disulfated sesterterpenoid was isolated from the methanolic extract of *Acanthodendrilla* sp. (order Dendroceratida, family Dictyodendrillidae) collected in Aklan, Philippines. Acanthosulfate inhibited the proteosome function, with $IC_{50} = 11.5$ μM. However, it was not pursued for in vivo studies based on the bioassay results using BMS Oncology Diverse Cell Panel (ODCA), revealing that the compound is non-selective and has low potency [14]. Semi-synthesis of the triacetate derivative was also achieved via acid hydrolysis and acetylation of acanthosulfate.

Purification of *Suberea* sp. collected in Palawan, Philippines, identified the chemical components with substantial antibacterial activities, which include sesterterpenes luffariellolide (**2**) and the oxygenated analogues 18-hydroxy-luffariellolide (**3**), acantholide A (**4**), and acantholide C (**5**) (Fig. 2) [15]. Luffariellolide moderately inhibited the growth of *Salmonella enterica.* At the same time, anti-*Klebsiella pneumoniae* activity was evident with the treatment of 18-hydroxy-luffariellolide and acantholide C [15]. Acantholide A was not tested due to insufficient material. Since luffariellolide was collected in high yield, it was chemically modified and assessed for structure-antimicrobial activity dependency. Increased potency against *K. pneumoniae* was observed with modifications in the γ-hydroxybutenolide moiety [15]. Moreover, the mechanism of anti-*K. pneumoniae* may be independent of the proposed covalent bond formation between γ-hydroxybutenolide and a functional group in the binding site of the target protein [15].

Luffariellolide was subsequently purified from another sponge genus, *Cacospongia* sp. (order Dictyoceratida, family Thorectidae) collected in Davao, Philippines, along with the other natural products belonging to diterpenes and mixed biogenesis metabolites [16]. They were identified as cacospongins A–D (**6**–**9**) (Fig. 2), which were determined to be new chemical entities, and the two compounds named 15,16-epoxy-8(17),13(16),14-labdatriene (**10**) and ambliofuran (**11**) (Fig. 2). Only cacospongins B and D treatments resulted in a 12 mm zone of inhibition against *Staphylococcus aureus* and methicillin-resistant *S. aureus* (MRSA). Notably, this was the first report on the purification of furanolabdane, furanoditerpenes, and diterpene-benzenoids, and in particular, 15,16-epoxy-8(17),13(16),14-labdatriene sourced from a marine organism [16].

Antimicrobial scalarane-type sesterterpenes were purified from *Hyrtios* sp. (order Dictyoceratida, family Thorectidae) collected in Bohol, Philippines [17], including a 12-*O*-desacetylfuroscalarol epimer (**12**) (Fig. 3) regarded as a new chemical entity consisting of an α-axial orientation at H-12. Known compounds: sesterstatin 4 (**13**), heteronemin (**14**), 12-deacetyl-12-*epi*-scalarobutenolide (**15**), 16-hydroxyscalarolide (**16**), 16-*O*-deacetyl-16-*epi*-scalarobutenolide (**17**), 12-*O*-deacetyl-19-deoxyscalarin (**18**), hyrtiosin A (**19**), and hyrtiosal (**20**) (Fig. 3) were also purified from *Hyrtios* sp. These sesterterpenes were tested against Gram (+) and Gram (−) bacteria. The most potent compound, heteronemin, displayed an MIC = 2 μg/mL and MIC = 4 μg/mL against *Bacillus subtilis* and *Micrococcus luteus*, respectively. The moderately active

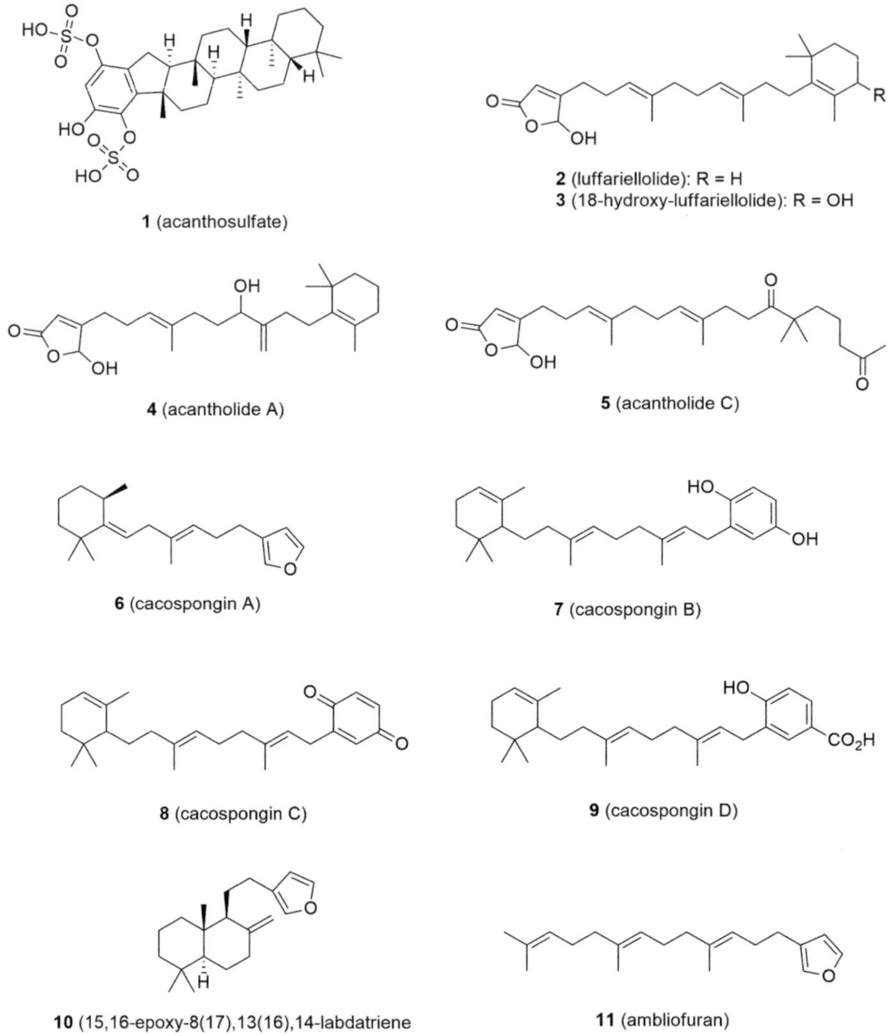

Fig. 2 Chemical structures of compounds **1–11**

12-*O*-deacetyl-19-deoxyscalarin and hyrtiosal have MIC = 16 μg/mL for *B. subtilis*. The other compounds showed weak to no activity against Gram (+) bacteria [17].

Two different species of the genus *Acanthella* were studied for their chemistry and bioactivity. Diterpenes were isolated through a bioactivity-guided purification of *Acanthella cavernosa* collected at two different sites. The sponge sample collected in Aurora, Philippines, afforded the known kalihinols J (**21**), X (**22**), and Y (**23**) (Fig. 4) [18]. A sample of the same sponge collected in Davao Gulf, Philippines yielded the previously described kalihinols F (**24**) and G (**25**) and kalihinene (**26**) as

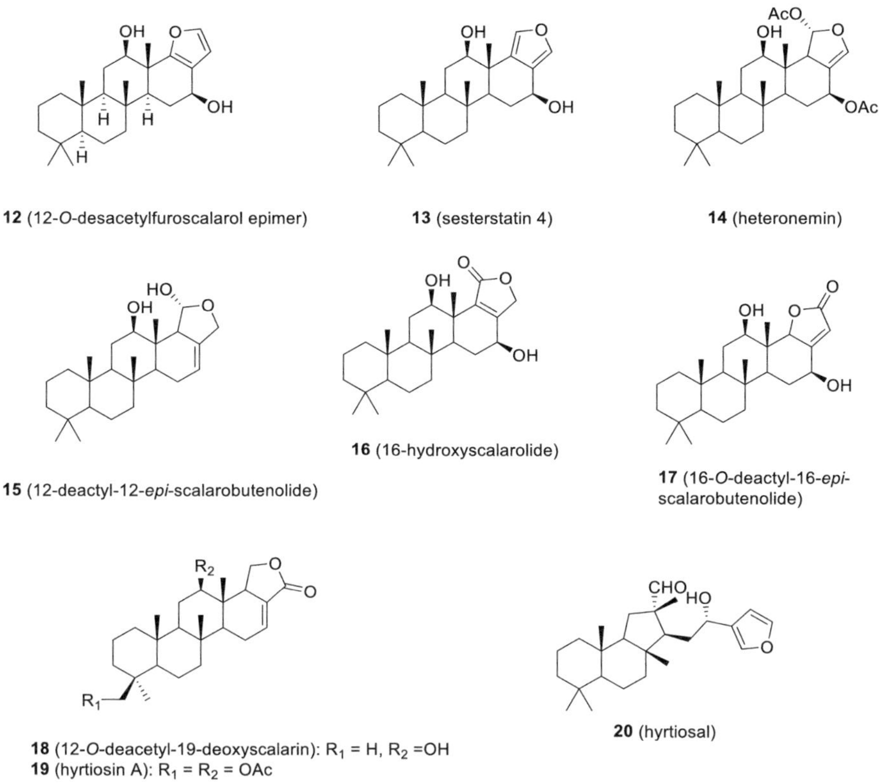

Fig. 3 Chemical structures of compounds **12–20**

well as the two new metabolites, 10-formamido-kalihinol F (**27**) and 15-formamido-kalihinol F (**28**) (Fig. 4) [18]. Interestingly, these two new compounds are the first reported derivatives of kalihinol F bearing a formamido moiety. These diterpenes were assessed for their inhibitory effects in bacterial folate biosynthesis using β-galactosidase induction and rescue assays. Kalihinols F, G, X, and Y, kalihinene, and 15-formamido-kalihinol F induced folate biosynthesis at 3.1–25.0 induction ratios at 8 μg/mL. Significant differential activity in the rescue assay was observed with the treatment of kalihinol F, G, and kalihinene [18]. Moreover, kalihinols X and Y exhibited potent antibacterial activity with MIC = 1.56 μg/mL, while the new diterpenes displayed weak activity. Substitutions at C-10 and the presence of a furanyl moiety influence the potency and specificity of activity in folate biosynthesis. The weak antimicrobial activity suggests the negative effect of the formamido moiety on cellular uptake or the specificity of the compounds [18].

Scalimides A–L (**29–40**) (Fig. 5) were purified from *Spongia* sp. (order Dicytoceratida) collected from Bohol, Philippines [19]. These compounds were tested on Gram (+) and Gram (−) bacteria and MCF-7 breast cancer cells. Scalimides B, E,

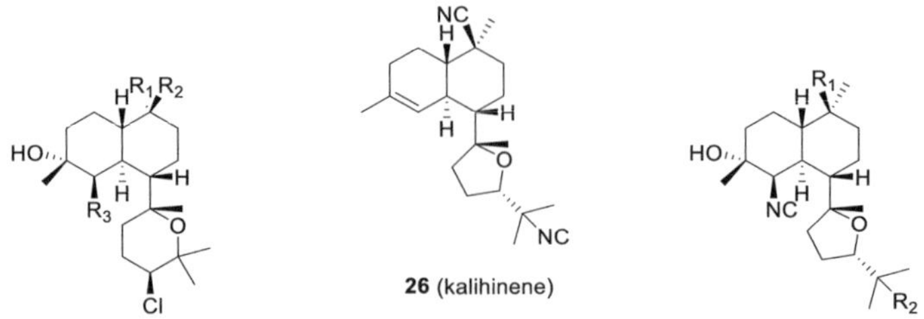

26 (kalihinene)

21 (kalihinol J): R_1 = NCS, R_2 = CH_3, R_3 = NHCHO
22 (kalihinol X): R_1 = NCS, R_2 = CH_3, R_3 = NC
23 (kalihinol Y): R_1 = R_2 = CH_3, R_3 = NC

24 (kalihinol F): R_1 = R_2 = NC
25 (kalihinol G): R_1 = NC, R_2 = NCS
27 (10-formamido-kalihinol F): R_1 = NHCHO, R_2 = NC
28 (15-formamido-kalihinol F): R_1 = NC, R_2 = NHCHO

Fig. 4 Chemical structures of compounds **21–28**

I, and L were inactive against all the bacteria tested at 128 μg/mL, in contrast to the moderate activity observed for MCF-7 cells (EC_{50} = 20.7–73.9 μM). Moderate anticancer effects have been documented on the remaining compounds. At the same time, increased potency towards Gram (+) bacteria has been shown in scalimides A, C, D, F, G, H, and K [19]. Notably, scalimide J exhibited broad-spectrum inhibitory activity against *Micrococcus luteus* and *Bacillus subtilis* with MIC = 4–8 μg/mL. The potency of scalimides is influenced by the functional groups at the C-3', C-16, and C-25 positions [19]. The carboxylic moiety at C-3' in scalimides D and K, and the methyl ether at C-16 in scalimides C and D are important for increased antibacterial and/or cytotoxic activity. Potency was enhanced by modifying the carbonyl group to a lactam in the imide moiety at C-25. The opposite trend was observed for the reduction of the carbonyl group to a hydroxy or methylene at C-24 [19].

Pyrrole alkaloids are commonly produced by sponges from different genera, including *Axinella*. The extraction and purification of *Axinella carteri* (Dendy 1889) (Demospongiae, Halichondrida, Axinellidae) from Palawan, Philippines, afforded the novel metabolite ugibohlin (**41**) and its known analogue dibromoisophakellin (**42**) (Fig. 6) [20]. High resolution NMR spectroscopy led to the reassignment of the chemical shifts for dibromoisophakellin. Bromopyrrole-imidazole alkaloids spongiacidin D (**43**), 3-bromohymenialdisine (**44**), and dihydrospongiacidine (**45**) (Fig. 6) were purified from *Acanthella carteri* Dendy (Axinellidae) collected from Quezon, Philippines [21]. Bioactivity-guided purification of *Mycophora* sp. from Tawi-Tawi, Philippines, identified crambescidin 800 (**46**) (Fig. 6) as an antimalarial compound against the chloroquine-sensitive 3D7 (IC_{50} = 160 nM) and chloroquine-resistant FCR3 (IC_{50} = 240 nM) *Plasmodium falciparum* strains [22].

Instead of antiviral assays, several sponge compounds were tested in relevant enzymes involved in viral replication in HIV-AIDS. HIV-1 integrase is an essential enzyme that facilitates and catalyzes the insertion and incorporation of the viral DNA into the host cell genome. It is critical for the life cycle of HIV, particularly in replication, maturation, and persistence within the host [23]. On the other hand, protein kinase C (PKC) facilitates the phosphorylation of several proteins involved

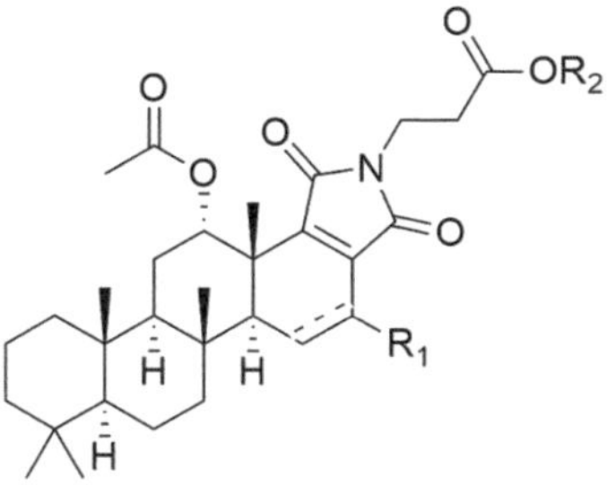

29 (scalimide A): R_1 = β-OH, R_2 = H
30 (scalimide B): R_1 = α-OH, R_2 = H
31 (scalimide C): R_1 = β-OCH_3, R_2 = H
32 (scalimide D): R_1 = α-OCH_3, R_2 = H
33 (scalimide E): R_1 = α-OCH_3, R_2 = CH_3
34 (scalimide F): R_1 = R_2 = H, Δ^{15}

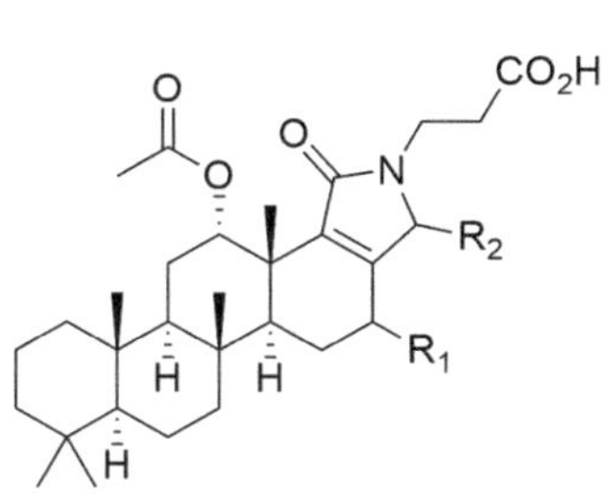

35 (scalimide G): R_1 = α-OCH_3, R_2 = β-OH
36 (scalimide H): R_1 = α-OCH_3, R_2 = β-OCH_3
37 (scalimide I): R_1 = OH, R_2 = H

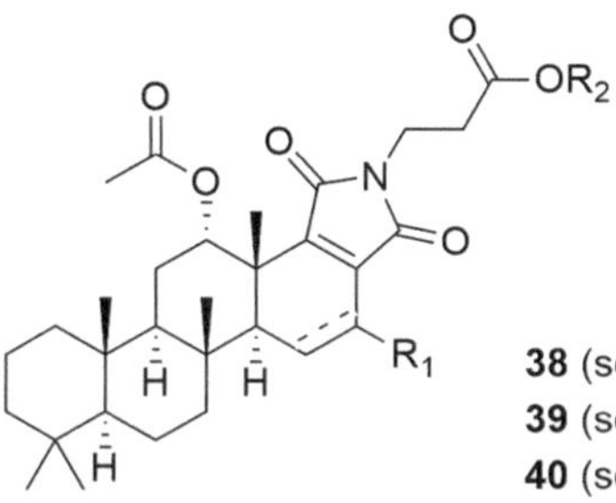

38 (scalimide J): R_1 = β-OH, R_2 = H
39 (scalimide K): R_1 = R_2 = H, Δ^{15}
40 (scalimide L): R_1 = H, R_2 = CH_3, Δ^{15}

Fig. 5 Chemical structures of compounds **29**–**40**

in cellular signal transduction pathways and physiological responses [24]. PKC-induced activation of transcription factors such as NF-κB, NF-AT, and AP-1 increases the transcription of HIV and consequently reactivates the latent HIV. Activating latent HIV will eventually lead to the immune system recognizing and eliminating the reactivated virus [24].

Diplynes A–E (**47**–**51**) (Fig. 7) are new sponge-derived brominated polyacetylenes extracted and purified from *Diplastrella* sp. (order Hadromerida, family Spirastrellidae) collected at Boracay Island, Aklan, Philippines [25]. The known sulfated compounds, diplyne A 1-sulfate (**52**), diplyne C 1-sulfate (**53**), and 2-deoxydiplyne D sulfate (**54**) (Fig. 7), were also identified. HIV-1 integrase was not inhibited by the brominated fatty acids (IC_{50} > 50 μg/mL), indicating they are not the bioactive principles contributing to the activity of the sponge extract. On the other hand, the sulfated analogues exhibited integrase inhibitory activity at IC_{50} ranging from ~ 30 to 90 μg/mL. The mild activity can be attributed to the partial degradation of the compound during the assay.

Protein kinase C (PKC), specifically the atypical isoform ζ (PKCζ) and the transcription factor NF-κB, were the targets of the sulfated sterols from *Spheciospongia*

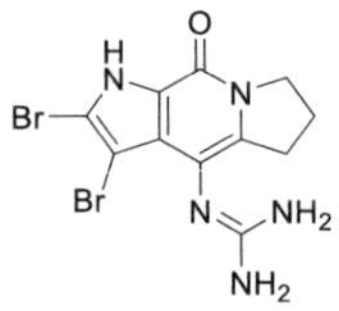

41 (ugibohlin)

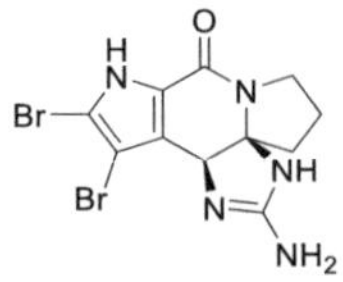

42 (dibromoisophakellin)

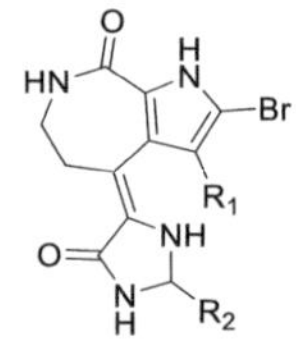

43 (spongiacidin D): R_1 = H, R_2 = O
44 (3-bromohymenialdisine): R_1 = Br, R_2 = NH_2

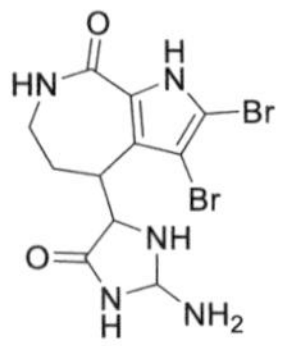

45 (dihydrospongiacidine)

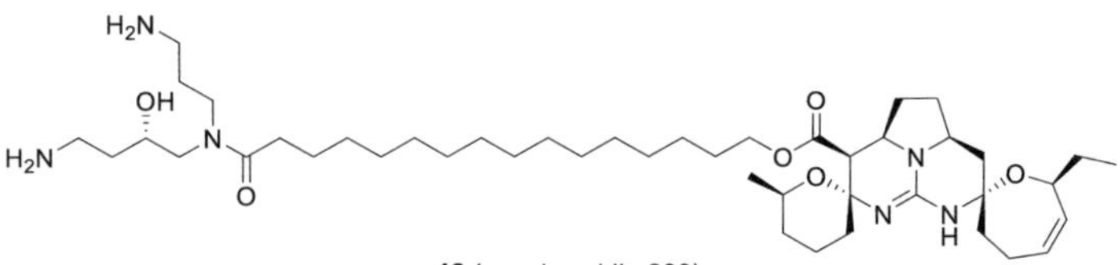

46 (crambescidin 800)

Fig. 6 Chemical structures of compounds **41–46**

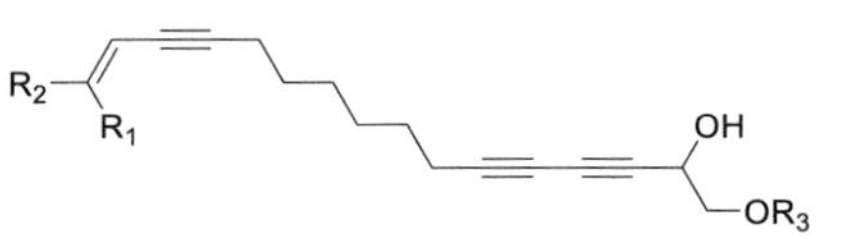

47 (diplyne A): R_1 = R_3 = H, R_2 = Br
48 (diplyne B): R_1 = Br, R_2 = R_3 = H
52 (diplyne A 1-sulfate): R_1 = H, R_2 = Br, R_3 = SO_3H

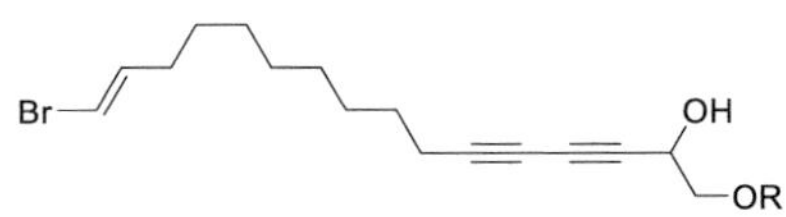

49 (diplyne C): R = H
53 (diplyne C 1-sulfate): R = SO_3H

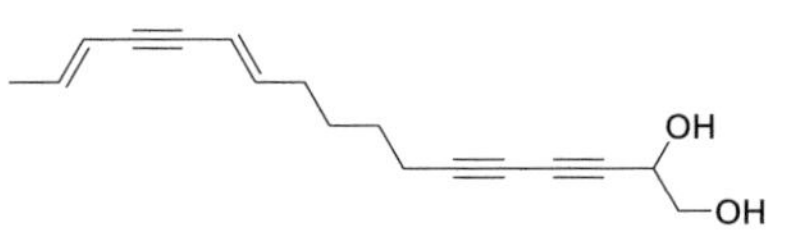

51 (diplyne E)

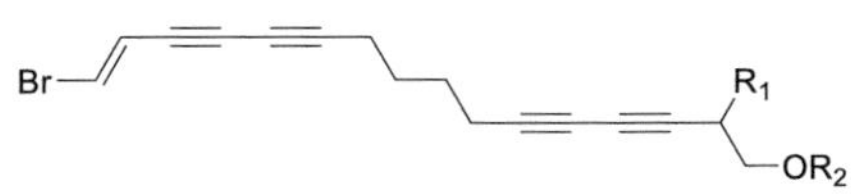

50 (diplyne D): R_1 = OH, R_2 = H
54 (2-deoxydiplyne D sulfate): R_1 = H, R_2 = SO_3H

Fig. 7 Chemical structures of compounds **47–54**

sp. (Hadromerida, Clionaidae) from Cagayan de Oro, Philippines [26]. The novel sterols spheciosterol sulfates A–C (**55–57**) (Fig. 8) and the known topsentiasterol sulfate E (**58**) (Fig. 8) inhibited PKCζ at $IC_{50} = 0.11–0.59$ μM. Moderate activation of NF-κB was also observed with $EC_{50} = 12–64$ μM. Based on the SAR analysis, the longer side chain in the sterol moiety increases PKCζ inhibition [26]. Side chains present in these novel sterols were formed through a series of methylations in the triterpene precursor parkeol, facilitated by *S*-adenosylmethionine (SAM), based on the proposed biosynthetic pathway.

New PKCζ inhibitors belonging to the same class of compounds were purified from a *Lissodendoryx* (*Acanthodoryx*) *fibrosa* collected in Coron, Palawan, Philippines [27]. These inhibitors were identified as fibrosterol sulfates A (**59**) (IC_{50} = 16.4 μM) and B (**60**) (IC_{50} = 5.6 μM), while fibrosterol sulfate C (**61**) (Fig. 9)

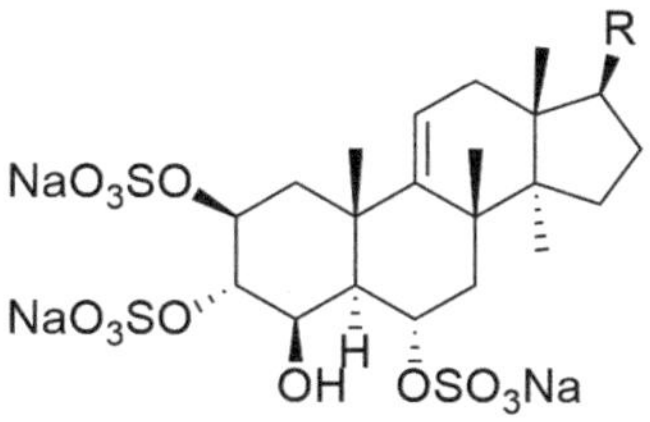

55 (spheciosterol sulfate A): R = 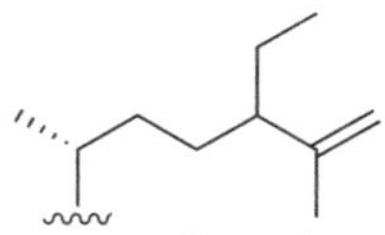

56 (spheciosterol sulfate B): R = 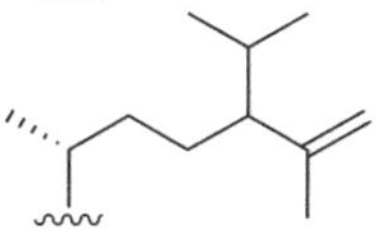

57 (spheciosterol sulfate C): R = 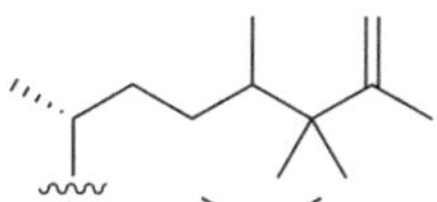

58 (topsentiasterol sulfate E): R = 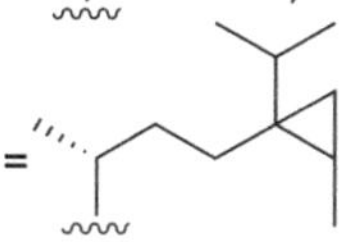

Fig. 8 Chemical structures of compounds **55**–**58**

was not tested due to limited yield. The bioactivity suggested the influence of the number of sulfate groups on PKCζ inhibition. Notably, the oxygenation pattern in the steroid rings in the spheciosterol sulfates and the fibrosterol sulfates influenced the bioactivity, indicating the relevance in inhibiting PKCζ [27].

In line with the global trend, most sponge compounds from the Philippines were also tested for cytotoxic activity against cancer cell lines or in reporter assays targeted towards cancer-relevant proteins and pathways. Potential applications for cancer treatment, as monotherapy and a combinatorial regimen, were also explored. Some compounds were screened on isogenic cell lines and purified enzymes to evaluate potential anticancer utility. This is in addition to phenotypic screening for changes in cancer cell proliferation, as assessed by changes in protein synthesis, mitochondrial respiration, or membrane integrity. The transcription factor p53 involved in tumor suppression, is commonly mutated in cancer cells and hence, targeted in bioactivity testing. p53 induces the expression of several genes, including p21, in response to DNA damage [28, 29]. p21, on the other hand, is a protein that facilitates the arrest of the cell cycle by inhibiting cyclin-dependent kinases (CDKs) when over expressed, which in turn, allows cellular repair prior to the continuation of the cell cycle. Expression of p21 can be dependent or independent of p53 regulation. Mutations in either p53 or p21 genes may lead to loss of cellular checkpoints as well as genetic instability

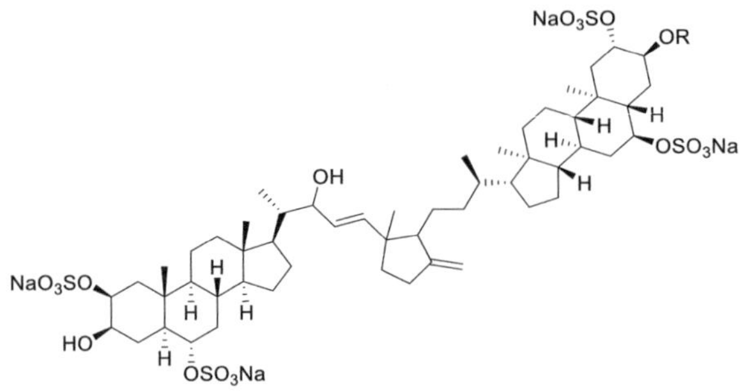

59 (fibrosterol sulfate A): R = H
60 (fibrosterol sulfate B): R = SO_3Na

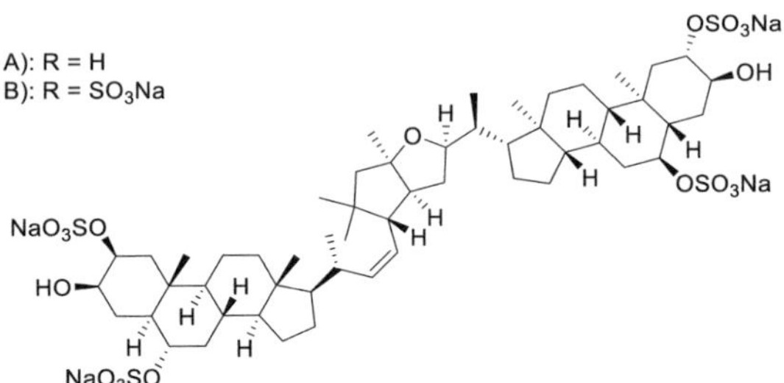

61 (fibrosterol sulfate C)

Fig. 9 Chemical structures of compounds **59–61**

and consequently, tumor progression [28, 29]. As such, compounds that target either the p53 or p21 present an attractive starting point for pharmacological modulation.

Bioactivity-guided purification of the methanolic extracts of *Didiscus* sp. collected in Batanes, Philippines, yielded two new bisabolane-type sesquiterpenes, (+)-curcuphenol (**62**) and (+)-curcudiol (**63**) (Fig. 10) [29]. The known compounds, β-sitosterol (**64**) and phenethylamine (**65**) (Fig. 10), were also identified, and notably, this was the first report on their isolation from the sponge genus *Didiscus*. Tasdemir et al. [28] purified stelletins A–D (**66–69**), (−)-stelletin E (**70**), rhabdastrellic acid-A (**71**) (Fig. 10), and the newly described stelletins H (**72**) and I (**73**) (Fig. 10) from a *Rhabdastrella globostella* collected in Zamboanga del Sur, Philippines. The isomalabaricane triterpenes were proposed as chemotaxonomic markers for the marine sponge *R. globostellata* [28].

Compounds from *Didiscus* sp. and *R. globostella* were tested on wild-type, p21-deficient, and p53-deficient HCT116 cell lines, using a tetrazolium-based assay to provide insights into their possible mode of anticancer activity. A p53-independent mechanism of HCT116 cytotoxic activity of (+)-curcuphenol was evident, which was not observed in (+)-curcudiol treatment [29]. The majority of the isomalabaricanes exhibited weak to no activity. Potent and selective cytotoxicity towards $p21^{-/-}$ cells was observed with the treatment of stelletin B ($IC_{50} = 0.034$ μM) and (−)-stelletin E ($IC_{50} = 0.039$ μM). Possible mechanisms of action for these bioactive metabolites include p21-dependent cytotoxicity and cell sensitivity.

Aldisine (**74**), 2-bromoaldisine (**75**), 10*Z*-debromoaldisine (**76**), 10*E*-hymenialdisine (**77**), 10*Z*-hymenialdisine (**78**), hymenin (**79**), oroidin (**80**),

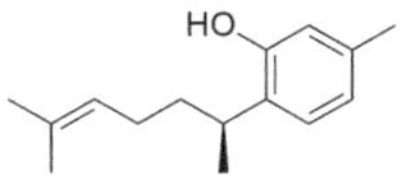

62 ((+)-curcuphenol)

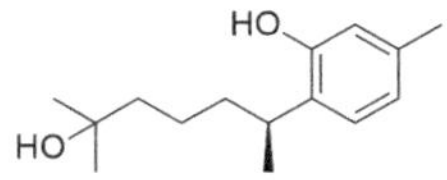

63 ((+)-curcudiol)

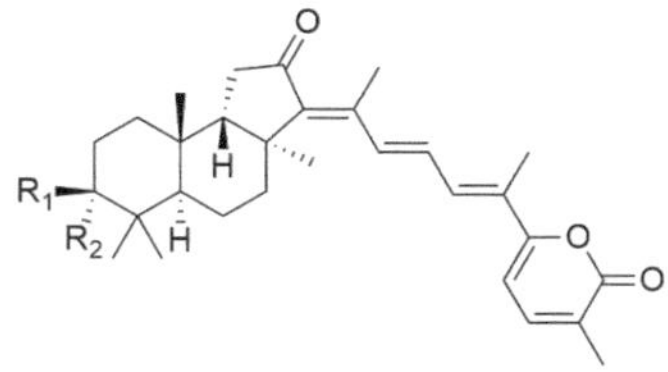

66 (stelletin A): $R_1 = R_2 = O$
68 (stelletin C): $R_1 = OCOCH_3$, $R_2 = H$

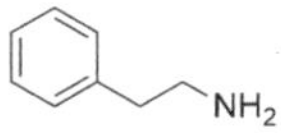

65 (phenethylamine)

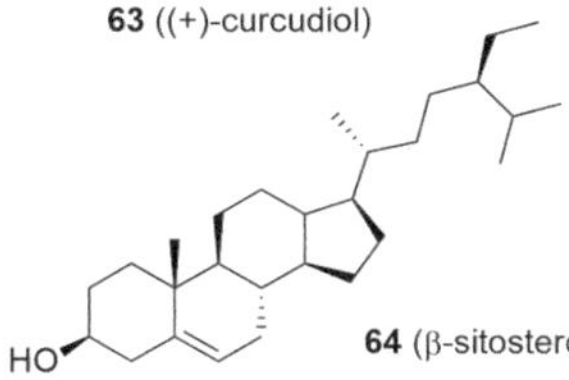

64 (β-sitosterol)

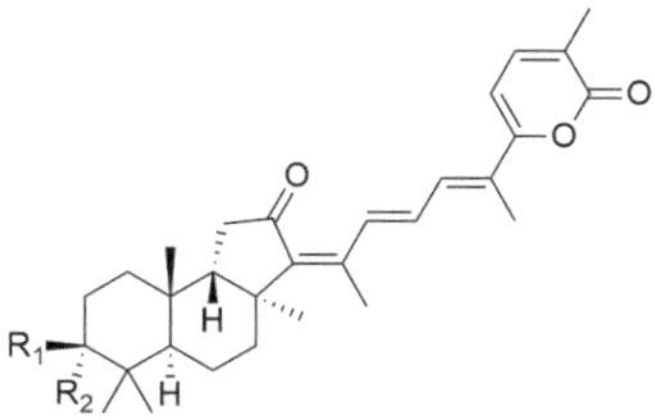

67 (stelletin B): $R_1 = R_2 = O$
69 (stelletin D): $R_1 = OCOCH_3$, $R_2 = H$

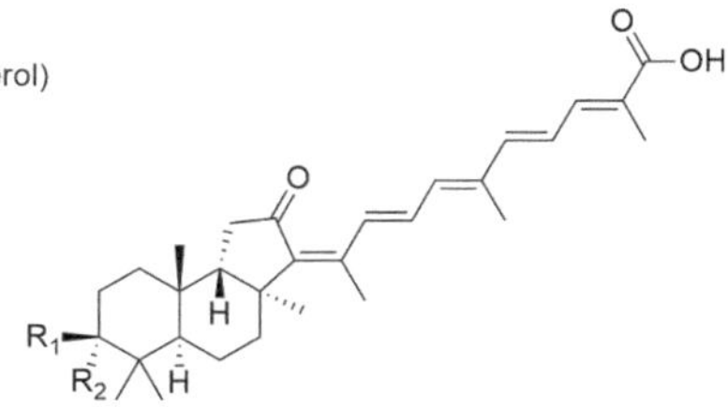

70 ((-)stelletin E): $R_1 = R_2 = O$
73 (stelletin I): $R_1 = OCOCH_3$, $R_2 = H$

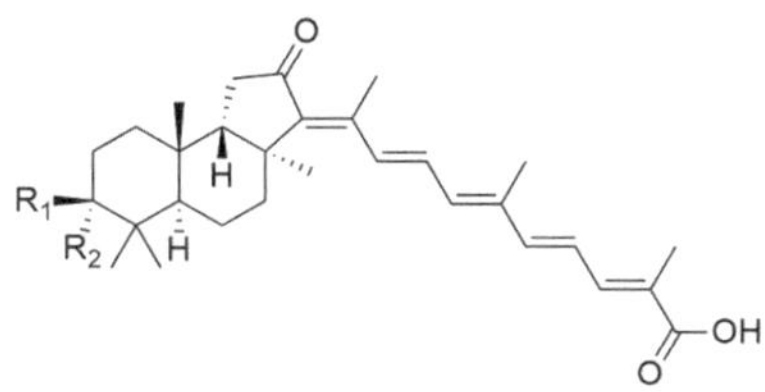

71 (rhabdastrellic acid-A): $R_1 = R_2 = O$
72 (stelletin H): $R_1 = OCOCH_3$, $R_2 = H$

Fig. 10 Chemical structures of compounds **62–73**

and 4,5-dibromopyrrole-2-carobinamide (**81**) (Fig. 11), were extracted and purified from *Stylissa massa* collected in Surigao, Philippines [30]. Compounds **74–81** were tested for mitogen-activated protein kinase (MAPK) kinase-1 (MEK-1) inhibitory activity. MEK-1 phosphorylates and activates MAPKs that are translocated into the nucleus to facilitate cell proliferation and differentiation. Unregulated activation of MEK-1 and MAPKs may lead to cancer progression. IC_{50}s at nanomolar levels, 3 nM for *E* isomer and 6 nM for *Z* isomer, were observed for the two stereoisomers of 10-hymenialdisine [30]. The authors proposed that the *E* isomer is unstable and irreversibly converts to the *Z* configuration. Moreover, it has been suggested that the aminoimidazolidinone ring and bromine moiety are important functionalities for the potent enzyme inhibitory activity, as in the case of 10*Z*-hymenialdisine [30]. Since the Raf/MEK-1/MAPK signaling cascade is a relevant target for cancer treatment, the alkaloids were tested for their anti-tumor activity. Consistent with initial findings, 10-hymenialdisine isomers were the most potent among the compounds against LoVo and Caco-2 cells, with the Z configuration being less sensitive [30].

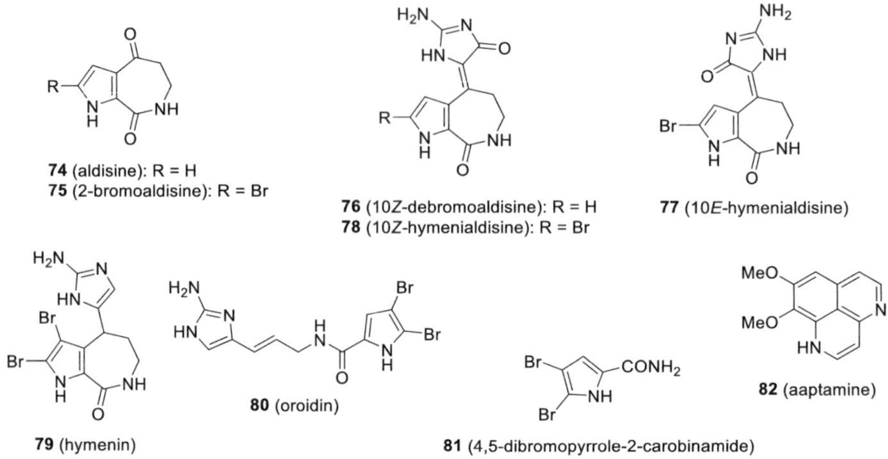

Fig. 11 Chemical structures of compounds **74–82**

Another report by Luyao et al. [31] and Mazo et al. [32] on *S. massa* (Surigao, Philippines) and *Desmacella* sp. (Agusan del Norte, Philippines), respectively, yielded the known alkaloid aaptamine (**82**) (Fig. 11). Profiling of aaptamine to determine its possible target receptors was done using PathHunter β-arrestin functional assay on a panel of 168 agonist and antagonist G-protein coupled receptors (GPCR) at 20 μM [31]. For agonist determination, aaptamine was incubated with PathHunter cell lines to induce a response. For antagonist determination, PathHunter cells were preincubated with the compound then an agonist at EC_{50} concentration was added. To validate the hit, aaptamine was subjected to an agonist, antagonist, and positive allosteric modulation (PAM) secondary screening that utilizes 13 GPCR biosensor assays (Arrestin) against ADRA2A, ADRA2C, ADRB2, CCR1, CCR3, CXCR3, CXCR7, DRD2L, DRD2S, DRD4, HTR1E, OPRK1, and SSTR1 as targets [31]. Aaptamine targets chemokines, adrenoreceptors, and dopamine receptors based on the GPCR profiling. These receptors are drug targets for treating noncommunicable diseases (NCDs), including cancer, cardiovascular diseases, and diabetes [31]. These findings corroborate the anti-inflammatory and antidiabetic properties of aaptamine [32].

The anti-inflammatory activity of aaptamine was assessed using a microplate-based membrane stabilization test of erythrocytes and albumin denaturation assay. Aaptamine inhibited the heat-induced hemolysis of erythrocytes at $IC_{50} = 0.47 \pm 1.26$ μg/mL and the albumin denaturation at IC_{50} 15.12 ± 1.39 μg/mL [32]. The antidiabetic activity of aaptamine was determined based on the α-amylase enzyme inhibition using a microplate-based starch-iodine assay. The enzyme activity was inhibited by aaptamine at $IC_{50} = 9.49$ μg/mL [32]. Furthermore, specific activities of aaptamine were also validated, such as antagonism of the α- (ADRA2C, IC_{50} 11.9 μM) and β-adrenoreceptors (ADRB2, IC_{50} 0.20 μM) and dopamine receptor D4 (DRD4, IC_{50} 6.9 μM), agonist on selected chemokine receptors, by itself (CXCR7,

EC_{50} 6.2 μM; CCR1, EC_{50} 11.8 μM) or as a potentiator of agonist activity (CXCR3, EC_{50} 31.8 μM; CCR3, EC_{50} 16.2 μM) [31].

Xestolactone A (**83**) and xestosaprol O (**84**) (Fig. 12) are compounds related to adociaquinones and xestoquinones [33]. These compounds were purified from *Xestospongia vansoesti* from the Philippines, though the exact collection site was not specified. Xestolactone A and xestosaprol O were evaluated for their ability to inhibit the enzyme indoleamine-2,3-dioxygenase (IDO) in search of anticancer agents with unique cellular targets. IDO is a monomeric heme enzyme that exhibits immunosuppressive effects via tryptophan (Trp) catabolism in kynurenine pathways, allowing cancer cells to evade the immune system [33]. In addition, immunological rejection and immune escape of the cancer cells are attributed to the inhibition of T-cell activation and proliferation after localizing the IDO-mediated Trp degradation and to kynurenine-binding to aryl hydrocarbon receptor (AhR), which induces a tumor microenvironment pathogenic inflammation [33]. Xestosaprol O presents potent inhibitory activity with $IC_{50} = 4$ μM, while xestolactone A displays moderate inhibition at $IC_{50} = 81$ μM. Synthesis of analogues using xestosaprol O as the structural basis was also achieved. The more potent synthetic analogues **85** ($IC_{50} = 0.11$ μM) and **86** ($IC_{50} = 1.4$ μM) (Fig. 12) were developed and validated the importance of the orientation of the dioxydihydrothiazine ring fusion to C-14/C-15. Notably, **85** was synthesized for the first time using Sato photochemical coupling reaction in combination with highly effective oxidation steps.

Another set of novel meroterpenoids with IDO inhibitory activity was isolated from a different sponge genus, *Haliclona* sp. (order Haplosclerida, family Chalinidae) collected at Culasian Point, Leyte, Philippines [34]. Halicloic acids A (**87**) and B (**88**) (Fig. 12) exhibited moderate inhibition of IDO with $IC_{50} = 10$ μM and 11 μM, respectively. Their respective aldehydic decomposition products exhibited decreased

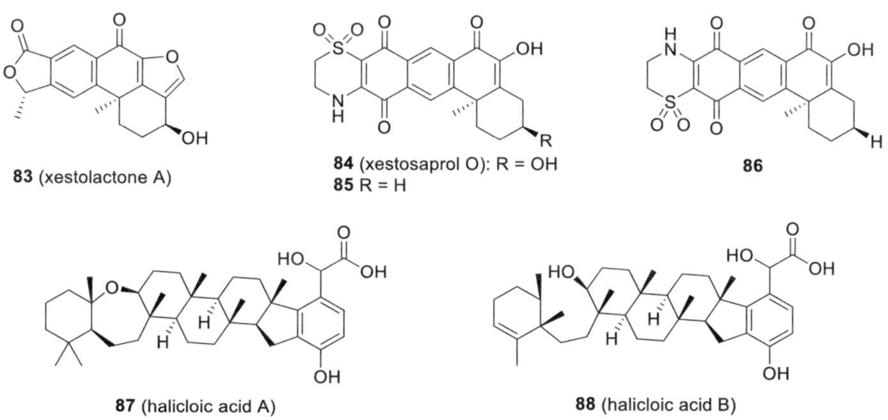

83 (xestolactone A)

84 (xestosaprol O): R = OH
85 R = H

86

87 (halicloic acid A)

88 (halicloic acid B)

Fig. 12 Chemical structures of compounds **83–88**

potency compared to their natural precursors (IC_{50} = 29–53 μM). The proposed late-stage biogenesis of halicloic acids A and B suggested the complete relative configuration of halicloic acid B as 2*S**,3*S**,6*S**,7*R**,10*R**,11*S**,14*S**15*S**,18*S**,23*R** with the assumption that the same carbocation intermediate leads to halicloic acids A and B and that suprafacial migrations of the Wagner-Meerwein 1,2-proton and methyl are required for the conversion of the intermediate to halicloic acid B [34]. On the other hand, the proposed mechanism of oxidative decarboxylation of halicloic acids A and B in the presence of DMSO/TFA suggested that TFA can react with the phenol functional group in these two acids, forming the TFA ester analogue. Then, the ester moiety reacts with DMSO, resulting in a Moffat-Swern intermediate, which reacts with the phenol moiety of halicloic acids A and B, followed by decarboxylation and tautomerization to generate the aldehydic decomposition products [34].

Extraction and purification of the cytotoxic aqueous extract of *Haliclona* sp. (order Haplosclerida, family Haliclonidae), from Jessie Beazley Reef, Sulu, Philippines afforded the novel isoquinoline alkaloid, 1-hydroxymethyl-7-methoxyisoquinolin-6-ol (**89**) and mimosamycin (**90**) (Fig. 13) [35]. 1-hydroxymethyl-7-methoxyisoquinolin-6-ol was inactive against melanoma LOX and ovarian OVCAR-3 cells at 50 μg/mL, while mimosamycin was identified as the bioactive principle with IC_{50} = 10 μg/mL. The sponge *Haliclona cymaeformis* (Esper) (Haplosclerida) was found to be in a symbiotic relationship with the red algae *Ceratodictyon spongiosum* Zanardini (Rhodymentiales) in Mahatao, Batanes, Philippines [36]. Two pyruvic acid derivatives, *p*-sulfooxyphenylpyruvic acid (**91**) and *p*-hydroxyphenylpyruvic acid (**92**) (Fig. 13), were purified and assessed for tyrosine kinase inhibitory activity. However, none of the pyruvic acids inhibited tyrosine kinase.

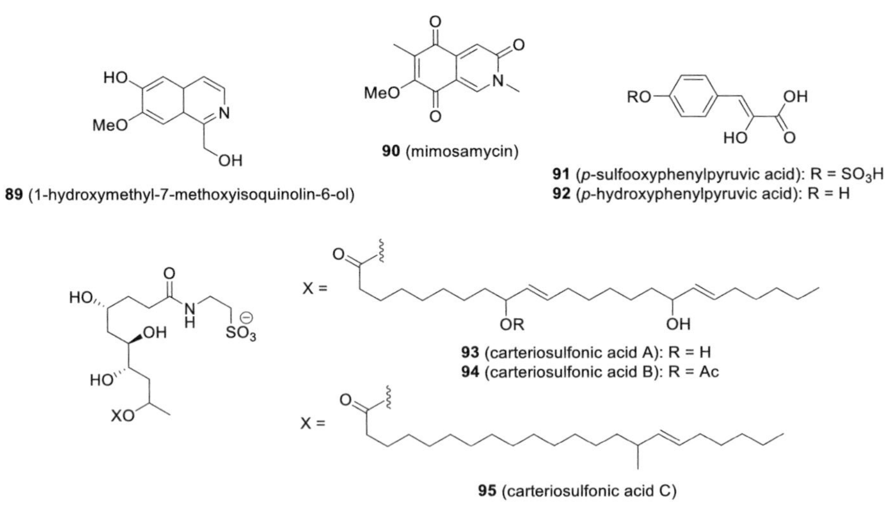

Fig. 13 Chemical structures of compounds **89–95**

Glycogen synthase kinase (GSK)3β and Wnt/β-catenin pathway have been targeted for treating various diseases since they regulate many fundamental cellular biological processes, including cell survival and proliferation [37]. GSK-3β regulates Wnt pathway through phosphorylation of β-catenin, leading to its degradation. Dysregulation and activation of Wnt signaling pathway may lead to the development of cancer and tumor suppression, respectively [37]. Screening of Wnt signaling modulators and subsequently GSK-3β inhibitors prioritized *Carteriospongia* sp. for purification [37]. Carteriosulfonic acids A–C (**93–95**) (Fig. 13) showed modest inhibition of GSK-3β at $IC_{50} = 6.8–12.5$ μM [37]. The importance of the long chain fatty acid in inhibiting the kinase activity was evident with the comparison of the sulfonic acids and the prepared hydrolysis product desacyl-carteriosulfonic acid.

Corticium niger (Homosclerophorida, Plakinidae) was the source of the cytotoxic steroidal alkaloid plakinamines. *C. niger* from Aklan, Philippines afforded four new plakinamines I–K (**96–98**) and dihydroplakinamine K (**99**) (Fig. 14) [38]. These metabolites exhibited moderate anti-HCT116 activity ($IC_{50} = 1.4–10.6$ μM), with plakinamine K and dihydroplakinamine K as the most potent. Except for dihydroplakinamine K, the compounds were also tested on an 11-cancer cell line panel. The results corroborated with the activity against HCT116 cells. Notably, the cell line panel testing showed that plakinamine I is highly selective [38].

Plakinamines I and J were re-isolated from a different collection of *C. niger* Pulitzer-Finali 1996 from Batangas, Philippines, along with the newly described plakinamines N (**100**) and O (**101**) (Fig. 14) [39]. Consistent with the previous study, these plakinamines possess growth inhibitory activities. In the National Cancer Institute (NCI) 60 human tumor cell line anticancer drug assay, the mean GI_{50}s of plakinamine N, O, and J were 11.5, 2.4, and 1.4 μM against different human colon cancer cell lines, respectively. These results support the possible selectivity of plakinamines toward colon cancer cells. Plakinamine I exhibited weak to moderate

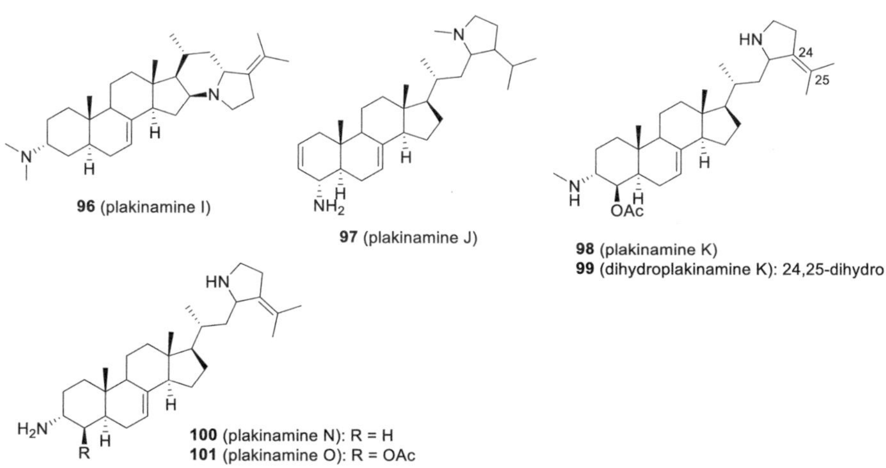

Fig. 14 Chemical structures of compounds **96–101**

activity in the preliminary screening and was not pursued for further testing in the NCI-60 cell line panel [39]. The modest activity may be attributed to its structure having an additional fused piperidine ring system.

A novel 4-methylene sterol, 7α-hydroxytheonellasterol (**102**) (Fig. 15), was purified from *Theonella swinhoe* Gray, 1868 (Lithistida) collected in Bohol, Philippines. 7α-hydroxytheonellasterol exhibited moderate cytotoxicity against an eight cell-line panel with mean $IC_{50} = 29.5$ μM [40]. Three known compounds named theonellasterol (**103**), its diol derivative **104**, and swinhosterol C (**105**) (Fig. 15) were purified from the same sponge. Only theonellasterol was tested for its cytotoxic activity due to yield limitations, with weak to no activity at 100 μM against the panel of cell lines. The toxicity exhibited by 7α-hydroxytheonellasterol suggested that the 7α-hydroxy group is critical in enhancing the bioactivity [40].

A more recent publication on *T. swinhoe* from the same collection site reported the purification of new oxygenated 4-*exo*-methylene sterols identified as theonellasterol-5,8-oxide (**106**), 8β-hydroxytheonellasterol (**107**), 15β-hydroxytheonellasterol (**108**), 8α-hydroxytheonellasterol (**109**), 8α-hydroxy-14,15-β-epoxy-theonellasterol (**110**), 8β-hydroxy-7α-formyl-B-northeonellasterol (**111**), 28-homoswinhoeisterol (**112**) (Fig. 15) and 7α-hydroxytheonellasterol [41]. The stereochemical configuration at C-7 of 7α-hydroxytheonellasterol was reassigned. Related analogues, which can be artefacts 15α-hydroxytheonellasterol (**113**) and theonellasterol-3,7-dimethyl ether (**114**) (Fig. 15), were also extracted from this sponge [41].

Novel sesquiterpenoids furodysin lactone (**115**) and pyrodysionic acid (**116**) (Fig. 16) were purified from the methanolic extracts of *Dysidea* sp. (NCI 815) from Siquijor, Philippines [42]. The known metabolite 3,6,11-trihydroxy-9,11-secoergostane-7,24(28)-dien-9-one (**117**) (Fig. 16) was also reported for the first time from *Dysidea* sp. The same genus of sponge collected in Bohol, Philippines, afforded 15 new scalarane-type sesterterpenoids (**118–132**) (Fig. 16) and the previously isolated inseparable mixture of 12-*epi*-phyllactones D and E (**133**) (Fig. 16) [43]. Compounds **127**, **128**, and **130** were named 16α-hydroxyhonulactone C, 24α-hydroxyhonulactone C, and 24α-hydroxyphyllofolactone H, respectively. 12-*epi*-phyllactones D and E were stereochemically reassigned due to discrepancies in the experimental and published NMR data. The cytotoxic activity against the human breast cancer MDA-MB-231 cell line was assessed for the sesterterpenoids. The highest activity was observed for **124** with $GI_{50} = 4.2$ μM, while the remaining compounds were either moderately cytotoxic or inactive. Structure–activity relationship (SAR) studies showed decreased cytotoxicity with the presence of the $\Delta^{15,16}$-olefin moiety in compounds **119**, **120**, **130**, and **131** and with the oxidation at C-20 as observed in compounds **119**, **126**, **127**, **128**, **129**, and **130** [43].

Other compounds purified from *Dysidea* sp. include the new polychlorinated compounds dysideaprolines A–F (**134–139**) (Fig. 17) as well as barbaleucamides A (**140**) and B (**141**) (Fig. 17) from the same sponge collected in Palawan, Philippines, along with the known compound dysidenin (**142**) (Fig. 17) [44]. Dysideaprolines D and E were interesting due to the presence of chlorine in one of the fatty acid or peptide-derived units [44]. The structures of barbaleucamides A and B also showed

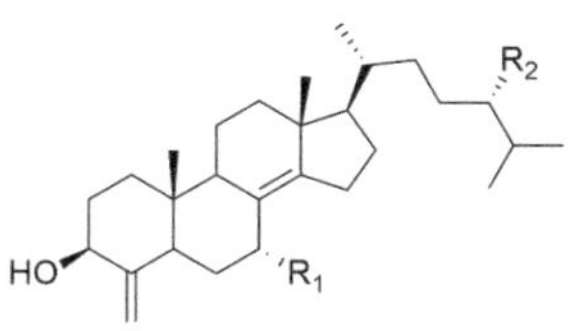

102 (7α-hydroxytheonellasterol): R_1 = OH, R_2 = CH_2CH_3
103 (theonellasterol): R_1 = H, R_2 = CH_2CH_3
104 R_1 = OH, R_2 = CH_3
105 (swinhosterol C): R_1 = OCH_3, R_2 = CH_2CH_3

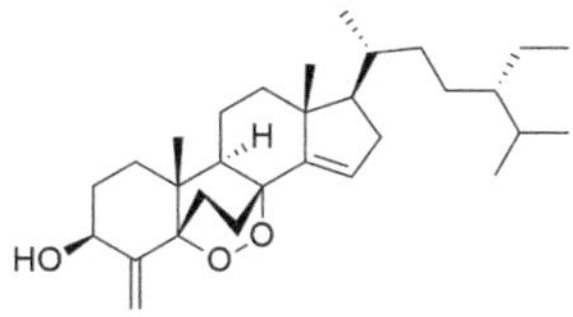

106 (theonellasterol-5,8-oxide)

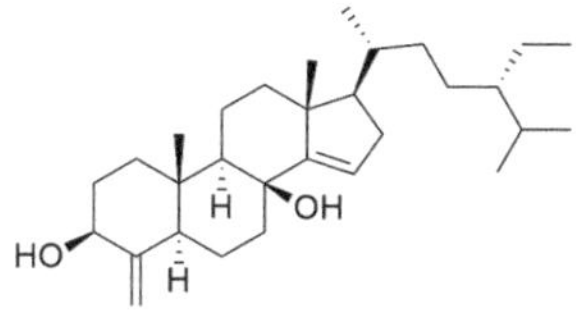

107 (8β-hydroxytheonellasterol)

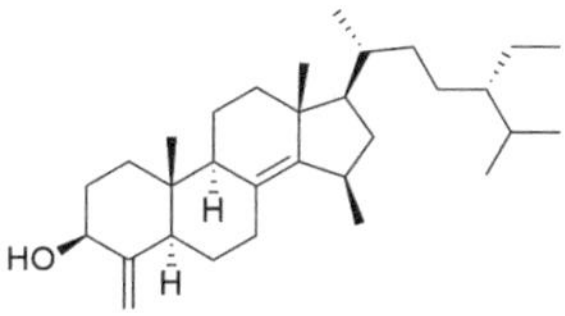

108 (15β-hydroxytheonellasterol)

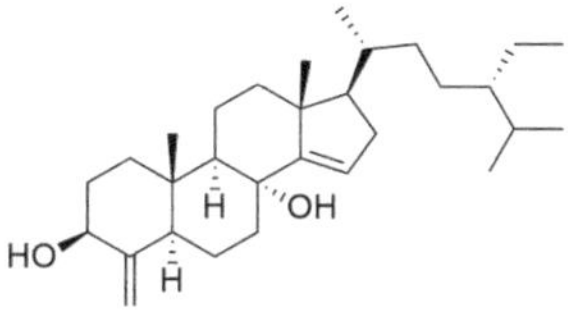

109 (8α-hydroxytheonellasterol)

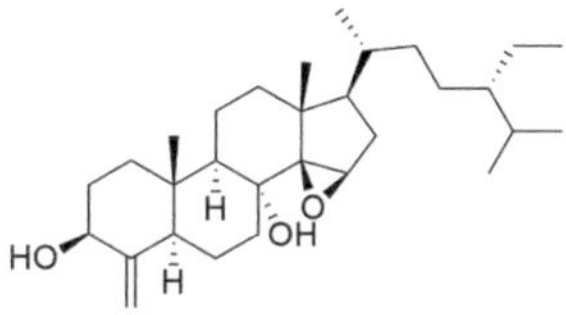

110 (8α-hydroxy-14,15-β-epoxy-theonellasterol)

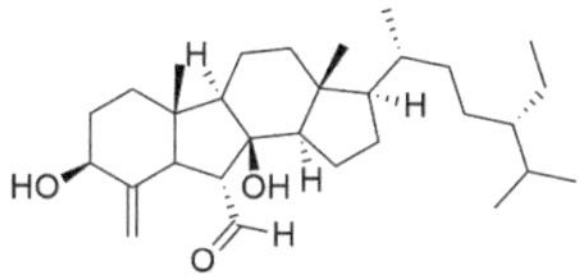

111 (8β-hydroxy-7α-formyl-B-northeonellasterol)

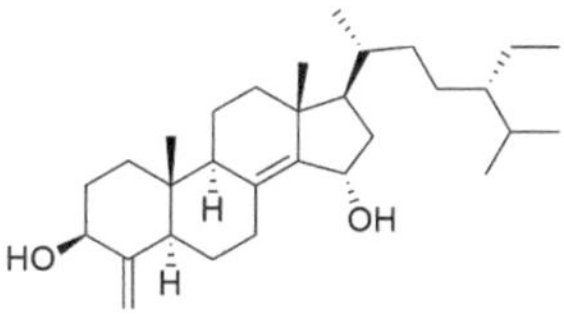

113 (15α-hydroxytheonellasterol)

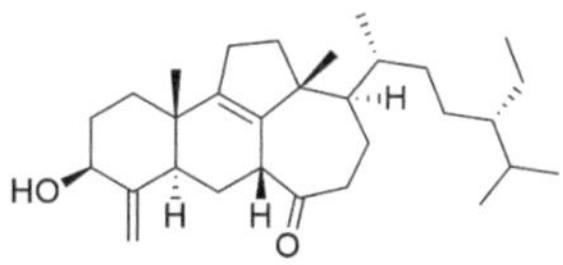

112 (28-homoswinhoeisterol)

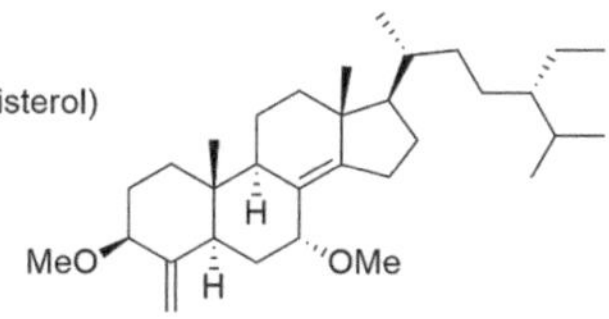

114 (theonellasterol-3,7-dimethyl ether)

Fig. 15 Chemical structures of compounds **102–114**

similarity with the cyanobacterial compound barbamide. These observations led to the hypothesis that the associated cyanobacteria biosynthesize the new compounds.

To date, the most potent cytotoxic compounds from sponges from the Philippines are members of the renieramycin family of compounds. These compounds are frequently purified from the blue sponge *Xestospongia* sp. (Order Haplosclerida, Family Petrosiidae) collected from Puerto Galera, Oriental Mindoro, Philippines

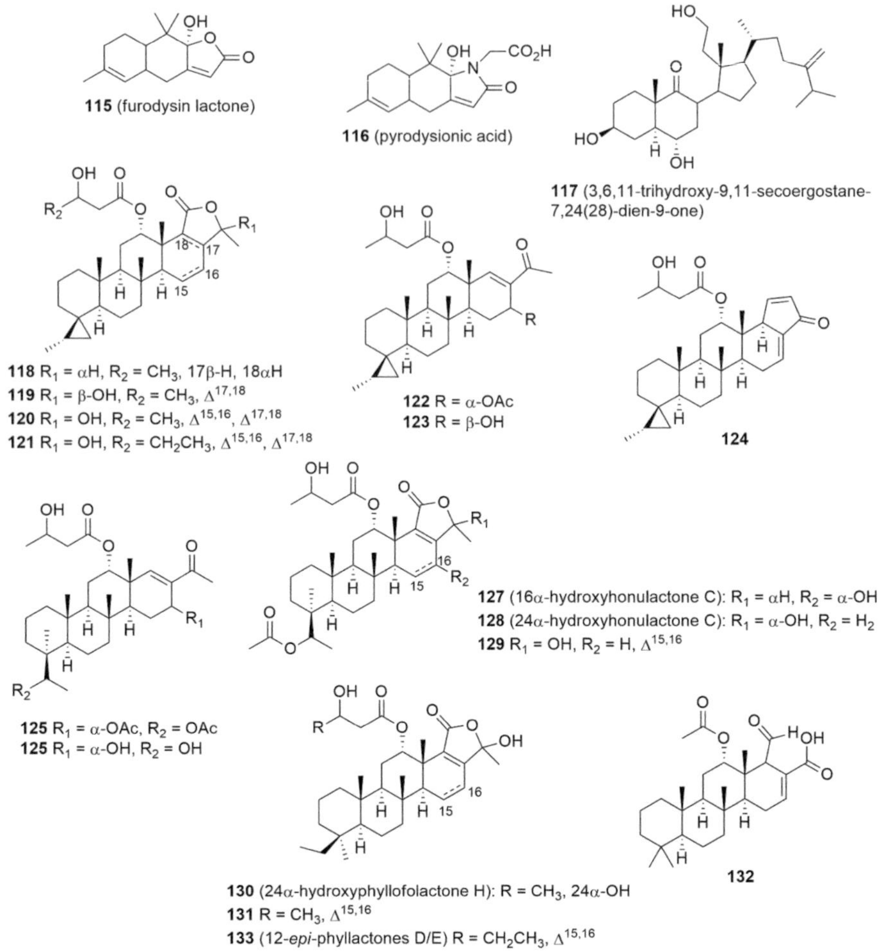

Fig. 16 Chemical structures of compounds **115–133**

(Fig. 1b). New derivatives of renieramycin were extracted and purified by Tatsukawa et al. [45] and Saito et al. [46]. These new compounds were identified as renieramycins W (**143**), X (**144**), and Y (**145**) , and 7-demethylrenieramycin O (**146**) (Fig. 18) [45], with 7-demethylrenieramycin O displaying weak activity against HCT116 and DU145 cells [46]. Known analogues, renieramycins M (**147**) and T (**148**) (Fig. 18), were also isolated [45].

Renieramycin M is a stabilized derivative from KCN treatment during extraction. While renieramycin M has been extensively studied as a single treatment in various types of cancer [47–49], the work of Concepcion et al. in the Philippines focused on the combinatorial regimen of renieramycin M (RM) and the clinically approved drug doxorubicin (DOX) against breast cancer in vitro [50] and in vivo [51]. Monotherapy using renieramycin M displayed highly potent cytotoxic effects with $IC_{50} = 6.0 \pm$

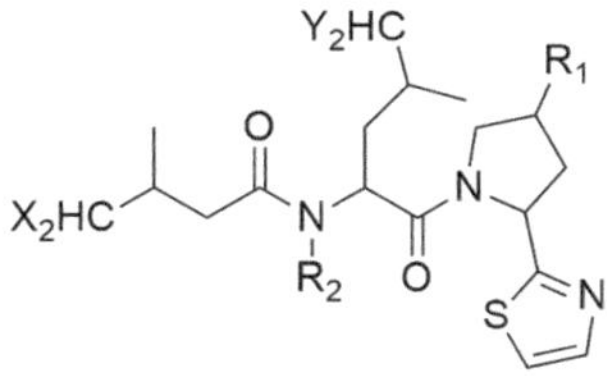

134 (dysideaproline A): R_1 = H, R_2 = CH_3, X = Y = Cl
135 (dysideaproline B): R_1 = R_2 = CH_3, X = Y = Cl
136 (dysideaproline C): R_1 = R_2 = H, X = Y = Cl
137 (dysideaproline D): R_1 = H, R_2 = CH_3, X = H, Y = Cl
138 (dysideaproline E): R_1 = H, R_2 = CH_3, X = Cl, Y = H
139 (dysideaproline F): R_1 = H, R_2 = CH_3, X = Cl, Y = ClH

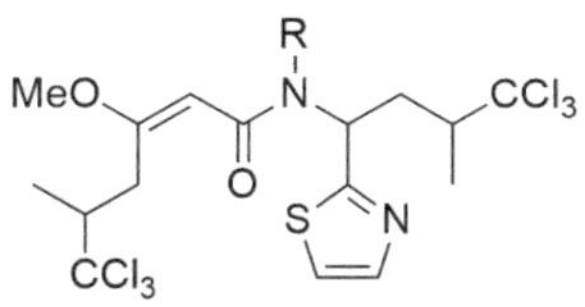

140 (barbaleucamide A): R = H
141 (barbaleucamide B): R = CH_3

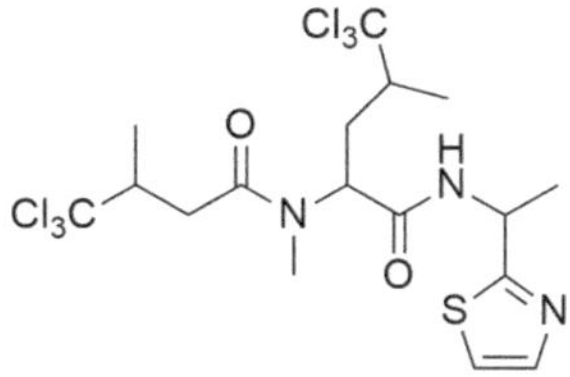

142 (dysidenin)

Fig. 17 Chemical structures of compounds **134**–**142**

0.5 nM, compared to doxorubicin ($IC_{50} = 356 \pm 25$ nM) after a 72-h treatment. The concentration-dependent cytotoxicity of renieramycin M was also reflected in the real-time cell analysis (RTCA), wherein time-dependent and faster cytotoxic effects compared to doxorubicin were also apparent [50]. Moreover, the RTCA profile and morphological changes observed with renieramycin M indicated similarity in the mechanism of action with DNA-damaging agents. The effects of renieramycin M on the cell cycle, specifically at S and G2/M phases, were only evident at high concentrations (25 and 100 nM). Transcriptome analysis further supported these observations, wherein genes involved in DNA replication, DNA repair, cell cycle, transcription, and apoptosis were differentially expressed [50]. Renieramycin M induced the downregulation of genes, including *CCNA2*, *CDC45*, *DNA2*, *GINS2*, *MCM6/7/10*, *POLA1*, *RFC4*, and *TOPIIA* for DNA replication, *BRCA1* for double-strand-break repair, *ATM* and *CHEK1* as DNA damage mediators, and ErbB/PI3K-Akt pathway gene *ERBB4*. Expression of several genes involved in proteosomal degradation (*PSMB8/9/10*), tumor suppression (*GADD45A*), and ErbB/PI3K-Akt pathway (*ERBB2/3* and *HBEGF*) increased in renieramycin M-treated cells [50].

Simultaneous administration of renieramycin and doxorubicin led to synergism for inhibiting the proliferation of MCF-7 cells. The concentration to inhibit cell growth by 95% was reduced with combinatorial treatment compared to single drug treatment. At 1:50 RM:DOX combination ratio, synergistic effects were significantly evident, and cell-based assays indicated the mechanism of cytotoxic activity to be affecting the DNA replication and cell cycle, as well as the induction of apoptosis. Like monotherapy, accelerated cytotoxicity and RTCA profile for DNA-damaging agents were also documented for RM + DOX treatment. However, there was no observed transient increase in the cell index, suggesting increased DNA damage with no recovery and decreased number of senescent cells [50]. Gene ontology

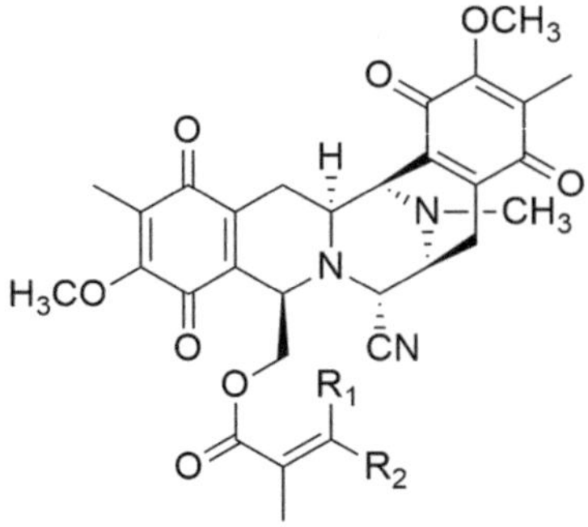

143 (renieramycin W): R_1 = H, R_2 = CH_3
147 (renieramycin M): R_1 = CH_3, R_2 = H

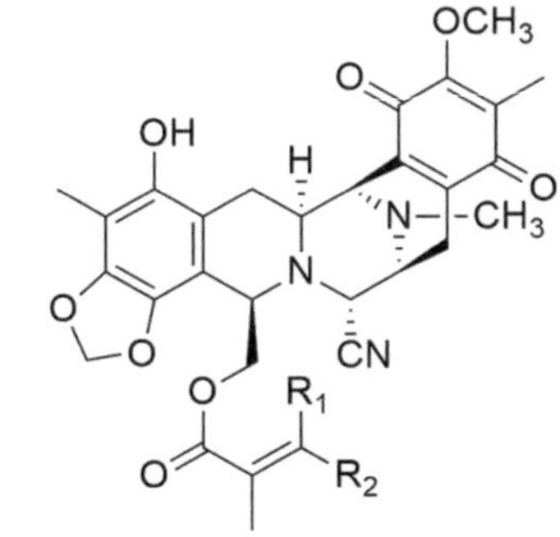

144 (renieramycin X): R_1 = H, R_2 = CH_3
148 (renieramycin T): R_1 = CH_3, R_2 = H

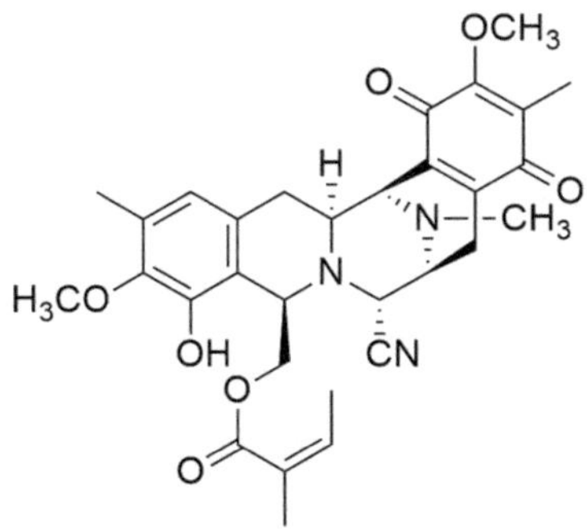

145 (renieramycin Y)

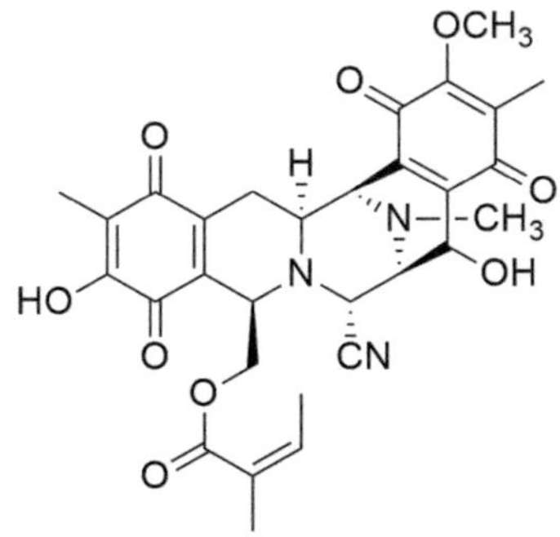

146 (7-demethylrenieramycin O)

Fig. 18 Chemical structures of compounds **143–148**

statistical overrepresentation test revealed several biological processes and cellular pathways to be regulated only in combinatorial treatment aside from those activated or repressed by monotherapy of RM and/or DOX. These include G-protein coupled receptor (GPCR) signaling pathway, oxidative stress response, mitochondrial cytochrome C release, interferon signaling pathway, apoptosis, endosomal/vacuolar pathway, post-translational protein modifications, RNA translocation, and receptor tyrosine kinase signaling pathway [50]. Notably, genes involved in DNA replication (*ANAPC10*, *CDC27*, *ORC5*, *RPA2*, *UBE2D1*), G2 checkpoint kinase-mediated DNA repair (*WEE1*), and release of mitochondrial cytochrome C (*IL6*, *JUN*, *MLLT11*, *OPA1*, *SOS1*, and *TRIAP1*) were regulated uniquely with combinatorial therapy. Results suggest possible mechanisms of the synergistic antiproliferative properties of RM + DOX combination therapy in relation to overlapping and unique cellular targets of the two cytotoxic compounds. Renieramycin M and doxorubicin, in combination, may have repressed, enhanced, or complemented related and/or different pathways, resulting in amplification of DNA injury and repression of DNA repair among other biological processes involved in cell survival [50].

Using the 4T1 breast cancer murine model, renieramycin M and doxorubicin combinatorial regimen were tested in vivo for their antitumor and antimetastatic activities [51]. The study highlighted the reduced mortality with the combinatorial treatment relative to the single-drug treatment. Notably, attenuated tumor volume and decreased metastasis in the liver were observed in 5 + 1 mg/kg DOX + RM combination [51]. These findings support the potential of the sponge-derived renieramycin M as a therapeutically-relevant molecule in combating cancer, especially breast cancer.

Cognizant of the supply problem in developing marine natural products as therapeutics, an open sea mariculture of the *Xestospongia* sp. blue sponge (Fig. 19) was undertaken [52]. Explants were collected from wild populations of the blue sponge in a coastal lagoon in Puerto Galera, Oriental Mindoro, Philippines. The open sea mariculture was placed adjacent to the wild explants. The survival and growth of the sponge, renieramycin M production, and bioactivity of the extracts were quantified to determine the effects of harvesting regime, culture period, temperature, sponge translocation, and farming methods. Renieramycin M production and antiproliferative activity were independent of the farming method and frequency of harvesting [52]. In contrast, the culture period and location significantly influenced these post-harvest parameters. Renieramycin M was lowest in sponges collected in May, as evidenced by the lack of bioactivity of the extracts. Subsequent increase in the renieramycin M content from maricultured *Xestospongia* sp. fragments was observed during the latter half of the year. Interestingly, translocation of sponges to Bolinao, Pangasinan validated the spatial variation in renieramycin M content [52]. The two sites are > 100 km apart and while explants from Puerto Galera were transferred to Bolinao, sponge explants eventually ceased producing renieramycin M.

Fig. 19 Addressing the supply issue using sponge mariculture. **a** Open sea mariculture of the renieramycin-producing *Xestospongia* sp. in Puerto Galera, Oriental Mindoro, Philippines. **b** Individual sponge fragments reared using the tube method

More recently, Lalas et al. [53] provided more information on the distribution, ecology, and improvement of the open-sea mariculture set up initiated by Santiago et al. [52]. The distribution and abundance of blue sponge colonies varied inside and outside Puerto Galera Bay based on various biotic and abiotic factors. High distribution and abundance of sponge colonies were observed in the embayed area characterized by weak tidal currents, flow, and turbulence, and high nutrient levels resulting in more juvenile input, healthier, and undisturbed colonies. Predation and disease-like mortality also contribute to these variations. These factors including the number of oscula during transplantation and effective farming method, were considered in developing the mariculture protocol [53]. Herein, the use of PVC pipes as an artificial substrate for sponge transplantation is highly recommended along with proper site selection and regular monitoring and maintenance of the set up. In selecting a mariculture site, natural abundance of the selected species, minimal exposure to abiotic factors such as wind and waves (depending on the preference of the species), and depth should be considered to ensure that the environmental conditions are ideal for their survival. In addition, these parameters should also consider the maintenance work that will be done and the effects on the benthic communities [53].

Researchers from Thailand also devised an aquaria-based setup for the blue sponge [54]. In this setup, renieramycin M production was stimulated by changes in the Ca/Mg levels, with 430/1230 ppm being most favorable to the compound biosynthesis [54]. The biosynthetic origin of renieramycins from the blue/purple *Haliclona* sponges was traced to its intracellular symbiont *Candidatus* Endohaliclona renieramycinifaciens [55]. Sponge samples were collected from various geographical locations in the Pacific Ocean, specifically in Palau, Papua New Guinea, and Bali, Indonesia. Renieramycins, particularly renieramycin E and the degradation products renierone and *N*-formyl dehydrorenierone, were confirmed to be present in all sponge samples. Metagenomics identified the biosynthetic gene cluster *ren* encoding for all the enzymes involved in the pentacyclic core formation and modifications and the bacterial producer [55]. The reduced genome of *Ca.* E. renieramycinifaciens and the specificity of its symbiotic relationship with one clade of *Haliclona* sponges indicated that *Ca.* E. renieramycinifaciens is an obligate symbiont to the host sponge, which provides the substrates and cofactors necessary for renieramycin production [55].

The degradation products of renieramycin, renierone (**149**), and its analogue *O*-demethylrenierone (**150**) (Fig. 20) were isolated from *Petrosia* sp. (order Haplosclerida, family Petrosiidae) collected at Murcielagos Island, Zamboanga, Philippines [56]. Novel cribrostatin 7 (**151**) (Fig. 20) was purified from the same sponge genus collected at Coron, Palawan, Philippines [56]. All three compounds were cytotoxic against HCT116 with $IC_{50} = 24–45$ μg/mL, wherein renierone is the most potent. On the other hand, *Petrosia nigricans* Lindgren 1897 (Strongylophora) from Lipata, Surigao, Philippines yielded two new sarasinosides, 5,8-epoxysarasinoside (**152**) and 8,9-epoxysarasinoside (**153**) (Fig. 20) [57]. These metabolites exhibited weak activities against HCT116 and A549 cell lines.

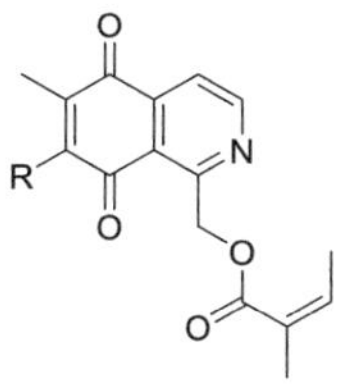

149 (renierone): R = OCH_3
150 (*O*-demethylrenierone): R = OH
151 (cribrostatin 7): R = $NHCH_2CH_2SO_3H$

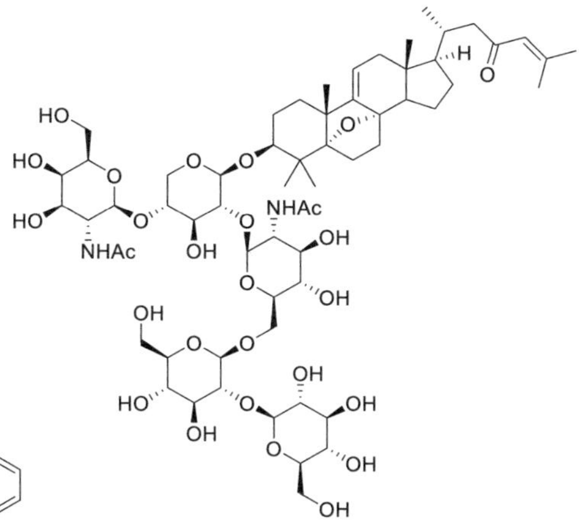

152 (5,8-epoxysarasinoside)

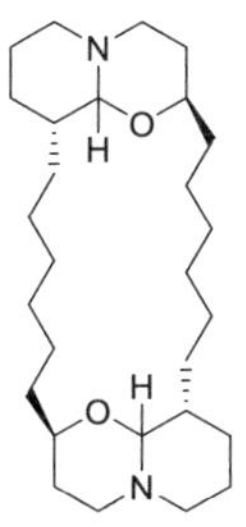

154

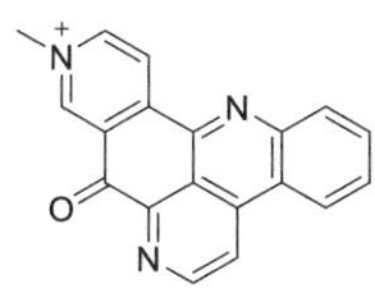

155 (deoxyamphimedine)

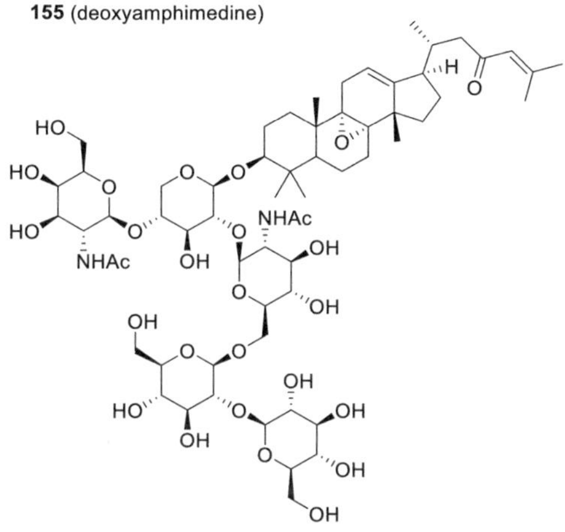

153 (8,9-epoxysarasinoside)

Fig. 20 Chemical structures of compounds **149**–**155**

Secondary metabolites belonging to other classes of compounds were also purified from a yellow *Xestospongia* sp., unrelated to the renieramycin-producing blue sponge. Xestospongin-type molecule (**154**) (Fig. 20) inhibited the collagen-induced and the epinephrine-induced platelet aggregation at 2 and 20 μg/mL, respectively [58]. This molecule was identified as a new natural product from *Xestospongia* sp. collected at Quezon, Philippines. The same sponge genus collected from Surigao, Philippines, was the source of deoxyamphimedine (**155**) (Fig. 20) [59]. It is a new pyroacridine alkaloid that presents cytotoxic effects via DNA damage through the induction of DNA cleavage and topoisomerase-independent generation of reactive oxygen species (ROS).

Zyzza fuliginosa (Carter, 1879) (phylum Porifera, order Poecilosclerida) from Batanes, Philippines was the source of a variety of molecules reported by Tasdemir et al. [60]. This set of compounds includes a new purine 3,7-dimethylguanine (**156**) (Fig. 21), with no cytotoxic effects against CCRF CEM leukemia and HCT116 colon cancer cells at 100 and 10 μg/mL, respectively. Known metabolites, including 4-hydroxyphenylacetic acid (**157**) and its methyl ester analogue (**158**), 4-hydroxyphenethyl alcohol (**159**), and the amino acids L-phenylalanine (**160**) and L-tryptophan (**161**) (Fig. 21) were also isolated from the same sponge. *Z. fuliginosa* also yielded the alkaloids makaluvamines A (**162**), C (**163**), and K (**164**) (Fig. 21) [60].

Interestingly, the analogue makaluvamine O (**165**) was purified from a *Smenospongia* sp., collected in the same site [28]. This sponge proved to be a rich source of indole alkaloids and terpenes. Other known compounds from this sponge include 5,6-dibromotryptamine (**166**), aureol (**167**), and furospinulosin 1 (**168**) (Fig. 21) while four new compounds 5-bromo-L-tryptophan (**169**),

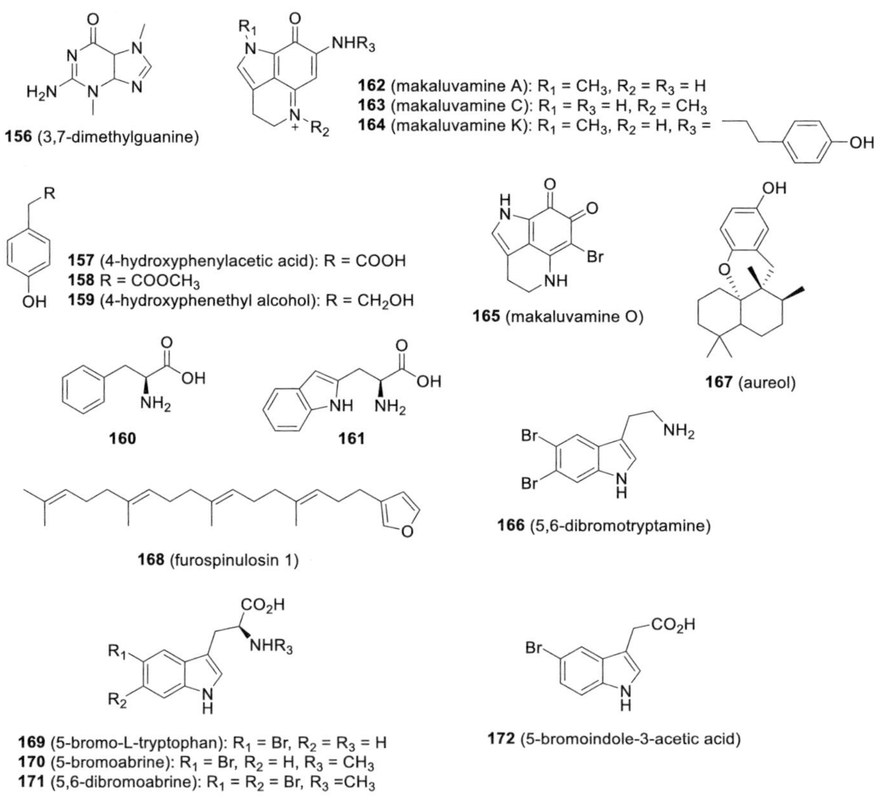

Fig. 21 Chemical structures of compounds **156–172**

5-bromoabrine (**170**), 5,6-dibromoabrine (**171**), and 5-bromoindole-3-acetic acid (**172**) (Fig. 21) were also obtained. Although previously synthesized, the study demonstrated the natural occurrence of 5-bromo-L-tryptophan and 5-bromoindole-3-acetic acid, with the aforementioned bromoamino acid possibly biosynthesized from indole and serine catalyzed by a hydrocarbon-utilizing bacterium, *Achromobacter petrophilum*, as a dietary metabolite [28]. The secondary metabolites were hypothesized to be produced by the sponge-associated bacteria or synthesized as a diet-derived metabolite. All new compounds were inactive against $p53^{+/+}$ and $p53^{-/-}$ HCT116 cells, while p53-dependent mechanism of cytotoxicity was observed for 5,6-dibromotryptamine, aureol, and furospinulosin 1. Makaluvamine O displayed increased potency against $p21^{-/-}$ cells relative to the parental cells. This was the first report of *N*-methylated 5-bromo-L-tryptophan from a marine sponge and furospinulosin 1 from a *Smenospongia* sp. *Aplysina* sp. from Davao was also a source of 5,6-dibromotryptamine [58]. The natural product presented 51% inhibition of epinephrine-induced platelet aggregation at 20 μg/mL, which can be attributed to its possible stimulatory activity of adrenergic receptors.

A new polyketide-derived macrolide lactam, chondropsin C (**173**) (Fig. 22), was purified from *Ircinia* sp. (collection site not specified), and its cytotoxicity against melanoma LOX and leukemia MOLT-4 cells was assessed [61]. The IC_{50}s generated for LOX and MOLT-4 cells were 0.8 and 0.2 ng/mL, respectively. The isolation of the new psammaplysin derivatives irciniaplysins A–D (**174–177**) (Fig. 22) from *Ircinia* sp. (Luga-it, Misamis Oriental, Philippines) was reported in 2024 [62]. Unlike other psammaplysins, irciniaplysins showed no significant antiproliferative effects on HCT116 cells, even at the highest test concentration (60 μM). Known ceratinamide B (**178**), psammaplysin U (**179**), and 19-hydroxypsammaplysin U (**180**) (Fig. 22) were also extracted from the same sponge sample. From the same collection site, the sponge *Aaptos suberitoides* (Bronsted, 1934) was collected for chemical analysis [63]. The sponge extracts exhibited cytotoxic effects in the brine shrimp lethality assay and antimicrobial activity against *S. aureus* and *Escherichia coli*. Pyrimidines thymine (**181**) and uracil (**182**) (Fig. 23) were purified from the bioactive sponge extracts.

Two novel dimeric macrolides named clavosolides A (**183**) and B (**184**) (Fig. 23) have been identified from the purification of the cytotoxic *Myriastra clavosa* (Order Astrophorida, Family Ancorinidae) from two different collections, at Boracay Island [64] and at Pamilican Island in Bohol, Philippines [65]. In contrast to the cytotoxic sponge extracts, Rao and Faulkner [64] reported that the purified compounds were non-cytotoxic. It was hypothesized that the true producers of these compounds are cyanobacteria associated with the sponge, based on their observed presence upon microscopic examination. The extraction of *M. clavosa* also afforded two new glycosides, clavosolides C (**185**) and D (**186**) (Fig. 23) [65]. Inactivity at 10 μg/mL against ten different human tumor cell lines was observed for clavosolide C. This observation was consistent with the activity of clavosolides A and B. Clavosolide D was not tested due to insufficient material.

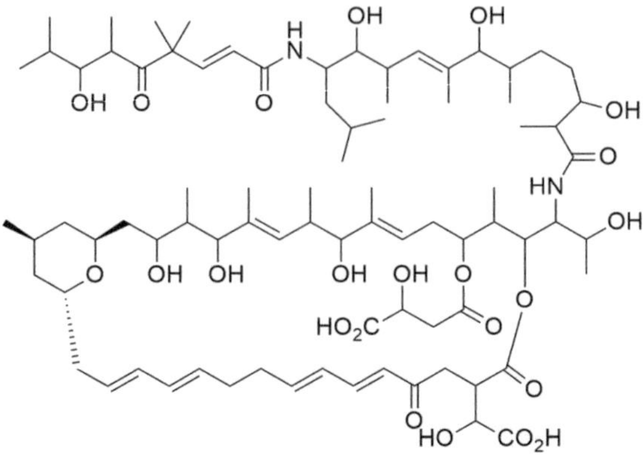

173 (chondropsin)

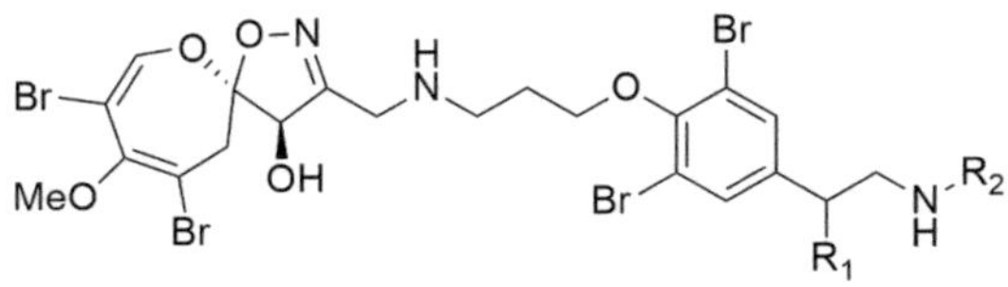

174 (irciniaplysin A): R_1 = H, R_2 = 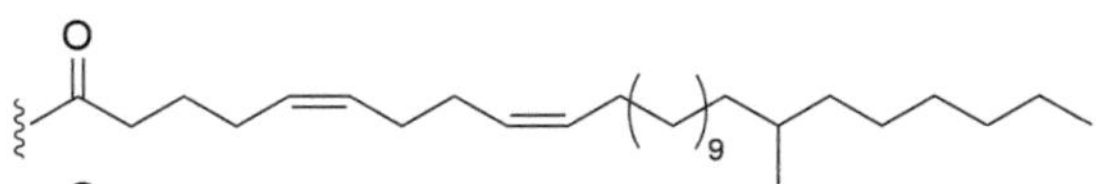

175 (irciniaplysin B): R_1 = OH, R_2 = 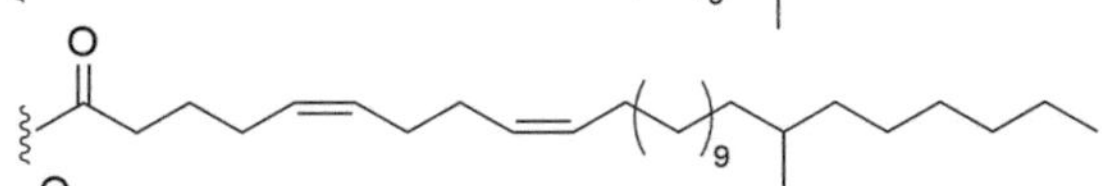

176 (irciniaplysin C): R_1 = H, R_2 = 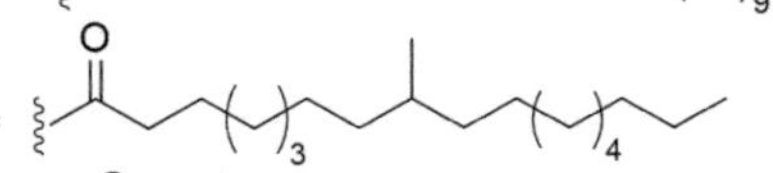

177 (irciniaplysin D): R_1 = OH, R_2 = 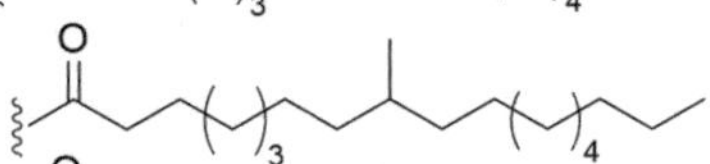

178 (ceratinamide B): R_1 = H, R_2 = 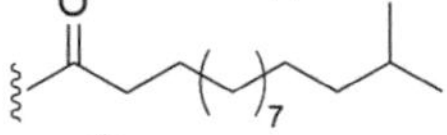

179 (psammaplysin U): R_1 = H, R_2 = 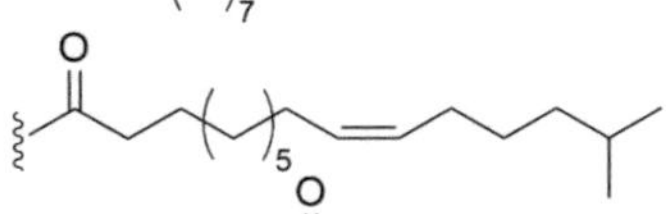

180 (19-hydroxypsammaplysin U): R_1 = OH, R_2 = 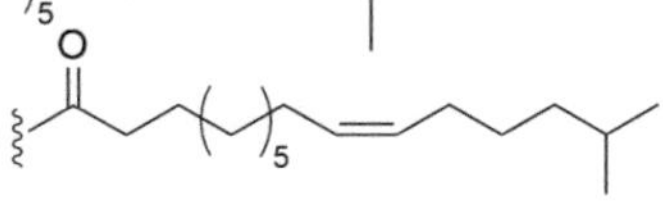

Fig. 22 Chemical structures of compounds **173–180**

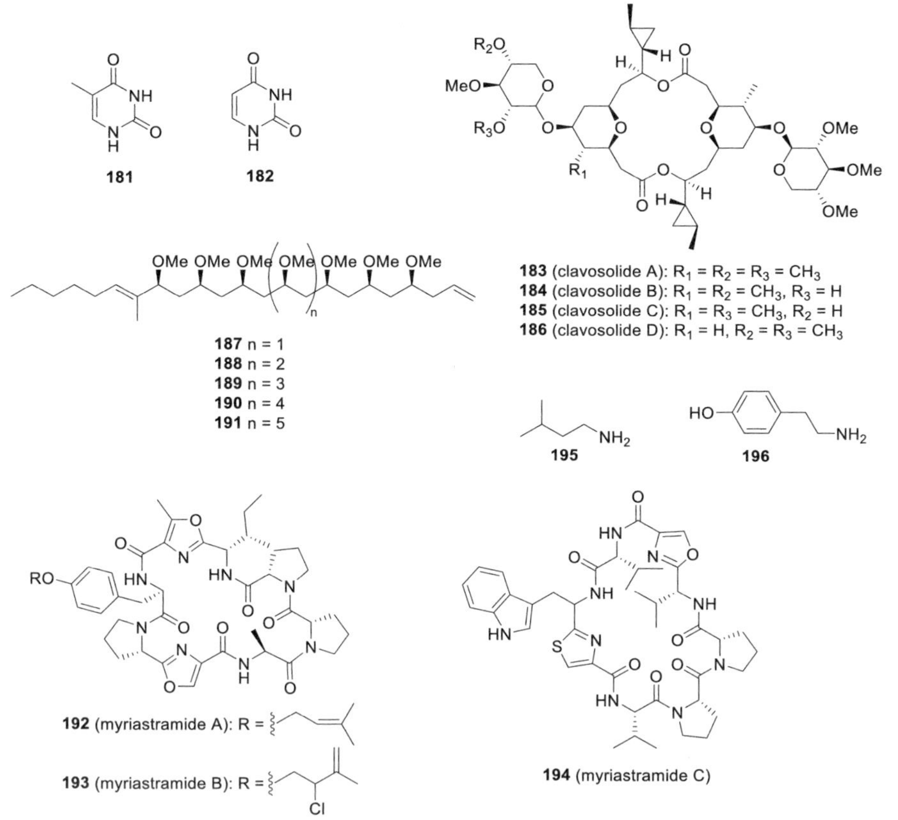

Fig. 23 Chemical structures of compounds **181–196**

Five new isotactic polymethoxydienes from *M. clavosa* named 4S,6S,8S,10S,12S,14S,16S,17E)-4,6,8,10,12,14,16-heptamethoxy-17-methyltricosa-1,17-diene (**187**), (4S,6S,8S,10S,12S,14S,16S,18S,19E)-4,6,8,10,12,14,16,18-octamethoxy-19-methylpentacosa-1,19-diene (**188**), (4S,6S,8S,10S,12S,14S,16S,18S,20S,21E)-4,6,8,10,12,14,16,18,20-nanomethoxy-21-methylheptacosa-1,21-diene (**189**), (4S,6S,8S,10S,12S,14S,16S,18S,20S,22S,23E)-4,6,8,10,12,14,16,18,20,22-decamethoxy-23-methylnonacosa-1,23-diene (**190**), and (4S,6S,8S,10S,12S,14S,16S,18S,20S,22S,24S,25E)-4,6,8,10,12,14,16,18,20,22,24-undecamethoxy-25-methyluntriacont-1,25-diene (**191**) (Fig. 23) exhibited moderate cytotoxicity (IC_{50} = 75–119 μM) on human colon cancer HTC116 cells [64].

Myriastra clavosa proved to be a good source of chemically-diverse molecules, including amino acid-based metabolites. Modified cyclic octapeptides myriastramides A–C (**192–194**) (Fig. 23) were isolated from this sponge collected in Pamilican Island, Bohol, Philippines [66]. Myriastramide A was inactive against ten human tumor cell lines at 10 μg/mL. So far, the compounds identified from *M. clavosa* collected in the Philippines were non-cytotoxic. Insufficiency of the yield has been a common issue for this sponge; hence, testing compounds such as myriastramides B and C was not possible. Isopentylamine (**195**) and tyramine (**196**) (Fig. 23) were purified from *Desmacella* sp. sponge collected off the coast of Agusan del Norte, Philippines [32]. These two compounds displayed antimicrobial activity against *Pseudomonas aeruginosa,* with higher potency observed in isopentylamine. Anti-inflammatory activities of these natural products were also evident based on the red blood cell hemolysis and albumin denaturation assays. Significant α-amylase inhibitory effect was observed with the treatment of tyramine ($IC_{50} = 0.17$ μg/mL).

Bioactivity-guided purification yielded two cyclic peptides, microcionamides A (**197**) and B (**198**) (Fig. 24), from *Clathria* (*Thalysias*) *abietina* from Tigtabon Island, Zamboanga, Philippines [67]. Potent antiproliferative effects on MCF-7 and SKBR-3 breast tumor cell lines were observed for both peptides, with $IC_{50} = 98$–177 nM. Morphological changes in the cells were also evident and indicated the induction of apoptosis as the mechanism of cytotoxic activity. These changes include cytoplasmic membrane blebbing, cytoplasmic condensation and shrinkage, cell-to-cell contact loss, membrane-bound vesicle formation, and extensive DNA fragmentation [67]. Microcionamides A and B also inhibited the growth of *Mycobacterium tuberculosis* H37Ra at MIC = 5.7 μM.

Analogues of these peptides, which only differ in the terminal amino acid residue, were also purified from an Indonesian *Clathria basilana* [68]. Microcionamides C and D bear an isoleucine (Ile) in the *N*-terminal instead of valine (Val) as in microcionamides A and B. These compounds also displayed antiproliferative activity against lymphoma, leukemia, and ovarian cancer cells and inhibited the growth of *Staphylococcus aureus* and *Enterococcus faecium.* These bioactivities were correlated to their possible mechanism of antiproliferative activity through induction of apoptosis, inhibition of starvation-induced autophagy, while for the Gram (+)

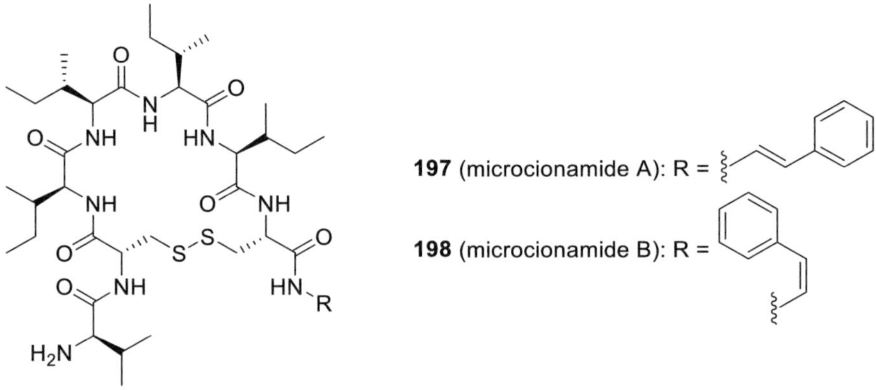

Fig. 24 Chemical structures of compounds **197** and **198**

bacteria, microcionamides A and C inhibited the growth via depolarization of the bacterial cytoplasmic membrane [68].

In 2022, Inocentes et al. [69] fully synthesized modified peptides inspired by microcionamides using a standard Fmoc-based solid phase peptide synthesis (SPSS) strategy followed by an on-resin cyclization step. To simplify the overall synthesis of microcionamides, the 2-phenylethylenamine (2-PEA) moiety was replaced with a tryptophan (Trp) residue and then modified to synthesize its alanine (Ala) analogues. Antiproliferative activity against MCF-7 and HK-2 cell lines and antibacterial effects on *S. aureus*, *P. aeruginosa*, and *K. pneumoniae* of the compounds were evaluated. The Trp and Ala analogues were tested to determine the SAR based on the *N*-terminal Val, the hydrophobicity, steric-hindrance in the isoleucine (Ile) core moiety, and the cyclic scaffold [69]. The Trp analogue exhibited comparable anti-*S. aureus* (MIC = 9.1 μM) and anti-*P. aeruginosa* (MIC = 15.0 μM) with the positive controls oxacillin and nalidixic acid, respectively. The Ala analogues were inactive. Based on the SAR analysis, the *N*-terminal Val residue, the hydrophobic property, steric-hindrance or bulkiness, and the cyclic structure are essential for the antibacterial activity. None of the compounds inhibited the proliferation of MCF-7 and HK-2 cells, which suggests that the 2-PEA functional group is necessary for the antiproliferative activity against mammalian cell lines [69].

Lipopolysaccharide-induced inflammation in macrophage cells serves as the primary screening platform for discovering anti-inflammatory compounds from marine sponges. Hits are subsequently tested for impacts on pro-inflammatory markers. A screening of 231 sponge extracts from the Philippines afforded 30 extracts that decreased NO production to < 30% without significant effects on cell viability. Quinone compounds (Fig. 25), including xestoquinone (**199**) and its monosubstituted derivatives 14-hydroxymethylxestoquinone (**200**), 15-hydroxymethylxestquinone (**201**), and the inseparable mixture of 14-methoxyxestoquinone and 15-methoxyxestoquinone (**202**) at 2:1 ratio, were identified as the anti-inflammatory components from the semipolar extract of *Neopetrosia compacta* (Fig. 1). Adociaquinone B (**203**), along with its structural isomer adociaquinone A (**204**) (Fig. 25) were also purified from this sponge [70]. This sponge was collected in Puerto Galera, Oriental Mindoro and Bolinao, Pangasinan, Philippines.

Potent cytotoxicity of adociaquinones A and B was evident in LPS-stimulated RAW264.7 macrophage cells and corroborates the observation of published studies [71, 72]. The adociaquinones were previously reported as cytotoxic to HCT116 with IC_{50} ~ 20 μM [73]. Despite the flat and planar structure, these compounds did not induce DNA intercalation [73]. Dose-dependent anti-inflammatory activity was observed based on the reduced NO production in cells treated with xestoquinone and monosubstituted derivatives [70]. The monosubstituted analogues present a wider therapeutic window and increased selectivity index compared to xestoquinone. Based on SAR analysis, substitution at the para-quinone moiety affects the cytotoxic activity of these quinoid compounds. Furthermore, the presence of α,β-unsaturated carbonyl groups in the quinone moiety suggests their anti-inflammatory activity may partly involve the Nrf2-ARE signaling pathway. Activation of Nrf2 was apparent after treatment with xestoquinone, hydroxymethylxestoquinones, and methoxyxestoquinone

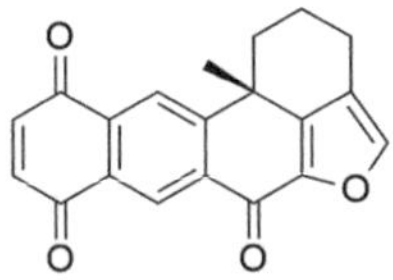

199 (xestoquinone)

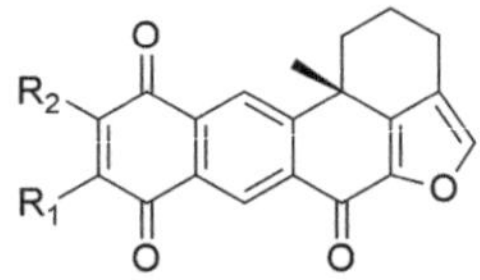

200 (14-hydroxymethylxestoquinone): $R_1 = CH_2OH$, $R_2 = H$
201 (15-hydroxymethylxestoquinone): $R_1 = H$, $R_2 = CH_2OH$

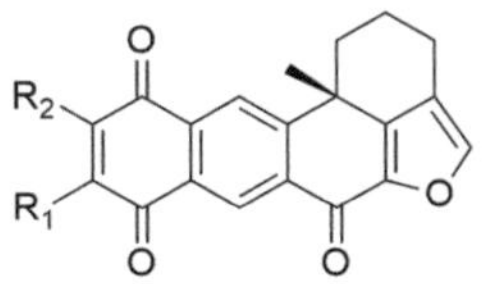

202a (14-methoxyxestoquinone): $R_1 = OCH_3$, $R_2 = H$
202b (15-methoxyxestoquinone): $R_1 = H$, $R_2 = OCH_3$

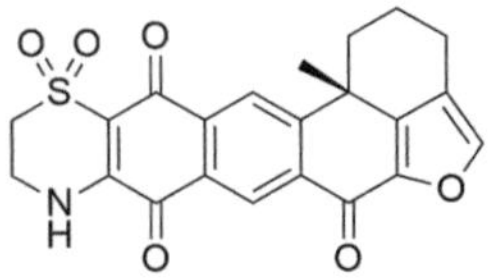

203 (adociaquinone B)

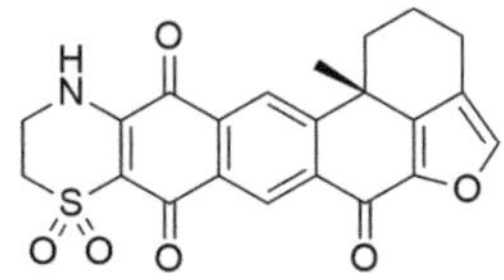

204 (adociaquinone A)

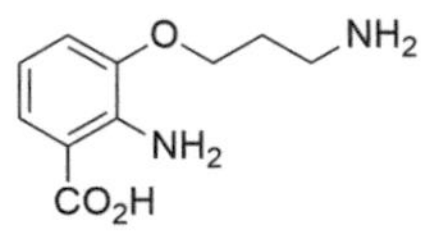

205 (oscarellin)

Fig. 25 Chemical structures of compounds **199–205**

mixture using Nrf2-luciferase reporter MCF-7 cells [70]. Despite the reduced potency in stimulating Nrf2 compared to xestoquinone, the monosubstituted derivatives offer more favorable effects due to attenuated cytotoxicity. At the transcript level, treatment with xestoquinone and monosubstituted derivatives decreased the expression of pro-inflammatory genes *Nos2* and *Il1b*. Methoxyxestoquinone mixture presents upregulation of the cytoprotective *Nqo1* genes [70]. The differential gene expression and Nrf2 activation may be implicated in the attenuated NO production.

The reduction of NO levels after treatment with xestoquinones may be related to the HIF-1 (Hypoxia-Inducible Factor) inhibitory activity of xestoquinones, which reduces the proliferation of tumor cells. Xestoquinone, adociaquinones A and B, and the hydroxymethylated xestoquinone analogues were obtained from a *Petrosia alfiani* collected in Malaysia. The compounds were tested for their ability to inhibit the activation of HIF-1 selectively [74]. Adociaquinones A and B exhibited equipotent ($IC_{50} = 0.2$ μM) and selective inhibition of iron chelator-induced HIF-1 activation in T47D cells. On the other hand, xestoquinone and the hydroxymethylated compounds inhibited the hypoxia-induced and iron chelator-induced HIF-1 activation at lower IC_{50} values = 1.2–12.3 μM. The study highlighted the importance of the 1,4-quinone moieties in HIF-1 inhibition [74].

Anti-inflammatory compounds were also purified from *Oscarella stillans*. The natural product oscarellin (**205**) (Fig. 25) is a new anthranillic acid derivative obtained from *O. stillans* (Bergquist and Kelly) collected in Honda Bay, Palawan, Philippines [75]. Cell-based assay indicated oscarellin as noncytotoxic on RAW264.7 cells, and the production of the inflammatory mediators TNF-α and IL-6 was reduced with pretreatment of oscarellin at 0.1–10 μM. The mechanism of immunomodulatory activity in LPS-stimulated macrophage RAW264.7 cells of oscarellin involved activating the transcription factors NF-κB and AP-1 [75]. An opposite trend was observed for ATF-3. Furthermore, results indicated the relevance of inactivating JNK and ERK1/2 pathways to facilitate the anti-inflammatory activity of oscarellin [75].

While sponges contribute significantly to shaping coral reefs through their interaction with other marine organisms, chemical ecological studies remain a minor theme in the Philippines. In 2018, the study on the chemistry and bioactivities of *Callyspongia samarensis* sponge extracts was conducted by Azcuna et al. (Bolinao, Pangasinan) [76] and Resuello et al. (collection site not specified) [77]. Although no compounds were purified, it is important to include their findings as these are the only published studies on the Philippines *C. samarensis* currently available. Fractionation of the bioactive methanol sponge extracts afforded semi-pure fractions from column chromatography (OCF) and HPLC-purification (HPLC). The fraction with the highest yield, OCF-G, was tested using hard coral *Porites cylindrica* fragments. Based on the gel-coral assay, this induced coral bleaching in *P. cylindrica* [76]. Moreover, HPLC fractions from OCF-G were tested for their antiproliferative activity against MCF-7, HCT116, and MDCK cells. HPLC 8 displayed potency among the fractions at 50 μg/mL and was tested for bleaching-inducing activity. Fused aromatic rings and/or conjugated double bonds may be present in the bioactive principle in HPLC 8 based on the UV profile [76]. Non-competitive inhibition of β-secretase 1 and AMPK activation was observed with the treatment of *C. samarensis* methanol extract. This activity shows the potential of the sponge as a source of drug leads for the treatment of Alzheimer's disease. A total of 15 putative compounds in the bioactive methanol sponge extract were identified based on the mass fragmentation pattern using HPLC/MS–MS analysis; nine displayed similarity with the compounds in the literature and databases [77]. The remaining compounds were tagged as unknown, with one compound hypothesized to cross the blood–brain barrier (BBB) based on the BBB permeability assay of the extract. The extract was non-cytotoxic (LD_{50} > 2000 mg/kg) in the acute oral toxicity test [77].

2.2 *Other Marine Macroorganisms*

Unlike sponges, other marine invertebrates and marine plants from the Philippines have fewer reported purification and bioactivity studies. Only select genera of soft corals, echinoderms, ascidians, and seaweeds from the Philippines have described compounds. Interestingly, seaweeds have an extensive documented ethnomedicinal use in the Philippines and are highly biodiverse. More recently, a seminal analysis of heavily defended gastropods, like the cone snails, demonstrated that these organisms also produce a suite of small molecule compounds. Venom-producing gastropods were thought to rely on their peptide-based molecules and shells for protection. Surprisingly, small molecules from gastropods have diverse chemoecological roles.

2.2.1 Soft Corals

Bioactivity analysis of extracts from soft corals indicated potential cytotoxicity. Methanolic extracts from *Sarcophyton crassocaule, Isis hippuris, Cespitularia stolonifera, Sinularia* sp., *Si. flexibilis, Si. polydactyla, Sarcophyton ehrenbergi, S. glaucum, Lobophytum crassum, L. durum,* and *Dendronephthya hemprichii* , collected from a fringing reef from Sipaca Point, Talisayan, Misamis Oriental, Northern Mindanao, Philippines, were tested in a brine shrimp lethality assay [78]. Highly toxic species were *S. crassocaule* (LC_{50} = 0.40 ppm), *L. durum* (LC_{50} = 3.45 ppm), *I. hippuris* (LC_{50} = 3.67 ppm), *L. crassum* (LC_{50} = 4.25 ppm), *S. ehrenbergi* (LC_{50} = 4.25 ppm), *C. stolonifera* (LC_{50} = 26.92 ppm), and *D. hemprichii* (LC_{50} = 28.18 ppm). The observed toxicity was independent of the wave exposure in the collection site and suggests the innate chemical defenses present in these soft coral species [78]. However, the identity of the bioactive compounds from these soft corals was not reported.

Three cytotoxic compounds were purified from the soft coral *Heteroxenia* sp. (Family Xeniidae) collected in Mindoro Island, Philippines [79]. The sesquiterpenes (+)-α-muurolene (**206**), a novel derivative (+)-6-hydroxy-α-muurolene (**207**), and the polyhydroxysterol, sarcoaldosterol A (**208**) (Fig. 26) were identified as the bioactive components in this extract. (+)-6-hydroxy-α-muurolene showed the most potent activity against the Gram (+) bacterium *Bacillus subtilis* and showed cytotoxicity in brine shrimp, with LC_{50} = 6.49 μg/mL [79]. A species of *Lobophytum* sp. showed anti-HIV property, primarily in the cembranoid-enriched fraction of the aqueous extract [80]. The diterpenes lobohedleolide (**209**), (7*Z*)-lobohedleolide (**210**), and a new compound, 17-dimethylaminolobohedleolide (**211**) (Fig. 26), were purified as the anti-HIV components, causing a 55–70% protection. The compounds showed moderate activity using an XTT-based in vitro anti-HIV assay, with EC_{50} = 3.6 μg/mL for lobohedleolide, EC_{50} = 4.6 μg/mL for (7*Z*)-lobohedleolide, and EC_{50} = 3.3 μg/mL for 17-dimethylaminolobohedleolide [80].

2.2.2 Echinoderms

Echinoderms belonging to the Family Holothuriidae are the most extensively investigated in the Philippines. Mariculture of *Holothuria scabra* and other members of this family is well-documented in Pangasinan, Philippines [81]. An initial bioactivity screening and mass spectrometric analysis of three species indicated that *H. scabra* Jaeger 1833, *Holothuria fuscocinerea* Jaeger 1833, and *Holothuria impatiens* Forskål 1775 contained bioactive triterpene glycosides [82]. The semi-pure butanolic extract from *H. fuscocinerea* body wall exhibited the highest membranolytic activity (approx. 0.46 mg eq. plant saponins/mL body wall), followed by *H. scabra* (approx. 0.16 mg eq./mL), and *H. impatiens* (approx. 0.02 mg eq./mL). HPLC fractions containing mixtures of saponins were also assayed, showing varying levels of hemolytic activity [82]. The sulfated tetraosides tend to have higher membranolytic activity and hydrophilicity than non-sulfated hexaosides. The structural diversity of

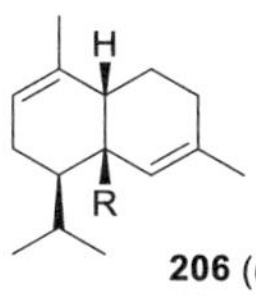

206 ((+)-α-muurolene): R = H
207 ((+)-6-hydroxy-α-muurolene): R = OH

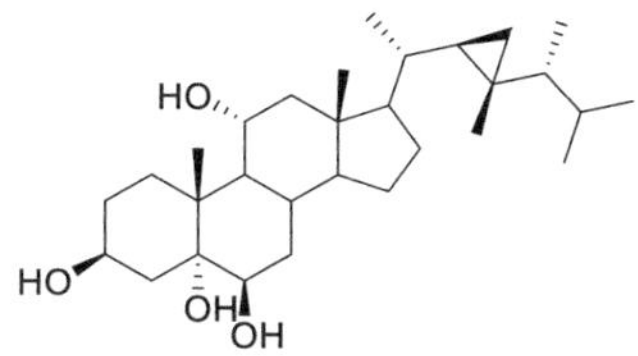

208 (sarcoaldosterol A)

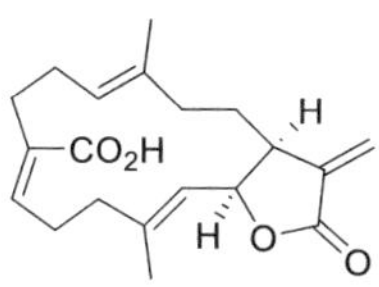

209 (lobohedleolide)

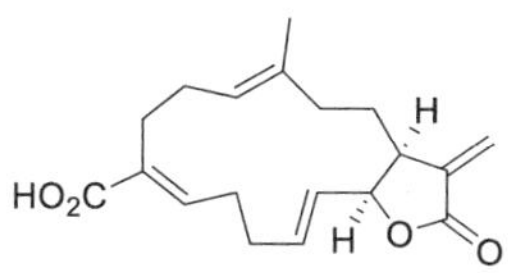

210 ((7*Z*)-lobohedleolide)

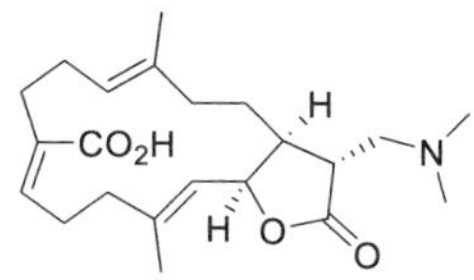

211 (17-dimethylaminolobohedleolide)

Fig. 26 Chemical structures of compounds **206**–**211**

saponins arises from variations in both the triterpene aglycone (holostane type) and the carbohydrate chain (number/type of sugars like xylose, glucose, quinovose, 3-*O*-methylglucose, and presence/position of sulfate groups) [82]. Phylogenetic mapping suggests an evolutionary trend within Holothuriidae from non-sulfated (primitive, e.g., in *H. forskali*) to sulfated (derived) saponins, potentially driven by selection for increased hydrophilicity and/or membranolytic activity for chemical defense [82].

The presence of holothurinosides in all three species suggested that these compounds are possibly a primitive, shared character in the genus. Conversion of desholothurin A (**212**) and holothurinoside C (**213**) to holothurinosides G (**214**) and H (**215**) (found in *H. forskali*) involves the addition of a disaccharide unit, thus increasing hydrophilicity (Fig. 27). A follow-up study in *H. scabra* indicated that triterpene glycosides are localized in the body wall [83]. New saponins were identified based on mass spectrometric analysis using molecular networking. Phospholipids, including phosphatidyl choline (PC), phosphatidyl ethanolamine (PE), phosphatidyl inositol (PI), phosphatidyl serine (PS), and plasmalogens, were detected as abundant compounds in *H. scabra* [83]. These compounds are involved in structure, metabolism, signaling, digestion, nutrient transport, membrane fluidity, autophagy, and stress response in *H. scabra*. A defense mechanism of *H. scabra* is evisceration, mediated partly by sphingolipids (cerebrosides, ceramides) in the body wall. Sulfated steroids such as cholestanol sulfate were also identified and suggested to protect the sea cucumber against its own saponin [83].

A collection of the feather star, *Comanthus* sp. (Order Comatulida, Family Comasteridae) from Mindoro, Philippines, yielded ten compounds consisting of naphthopyrones and anthraquinones [84]. Two angular naphthopyrones, comaparvin (**216**) and 6-methoxycomaparvin (**217**) (Fig. 28), significantly inhibited TNF-α-induced NF-κB activation in H4IIE rat hepatoma cells and MCF-7 human breast cancer cells using reporter gene assays [84]. Comaparvin also inhibited NF-κB activation

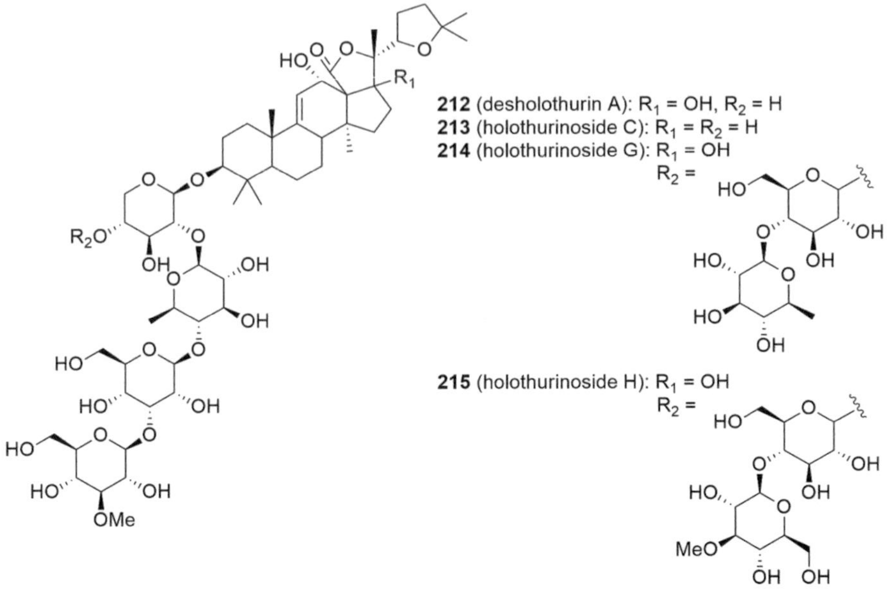

Fig. 27 Chemical structures of compounds **212–215**

by interleukin-1β and phorbol 12-myristate 13-acetate. 6-methoxycomaparvin-5-methyl ether (**218**) (Fig. 28) was inactive at tested concentrations (up to 125 μM). Structure–activity relationship highlighted the importance of the C-5 position for activity, while modification at C-6 does not affect the induction of NF-κB activation [84]. Comaparvin inhibited TNF-α-induced nuclear translocation and phosphorylation (Ser536) of the p65 NF-κB subunit, DNA binding of NF-κB (EMSA), and TNF-α-induced phosphorylation of the inhibitor IκBα. This compound also downregulated TNF-α-induced expression of NF-κB target genes involved in anti-apoptosis (XIAP, A20) and antioxidant defense (MnSOD), but not the pro-apoptotic Bax. Modulation of NF-κB activity was unrelated to the antioxidant activity, with the naphthopyrones having negligible radical scavenging activity based on the DPPH assay [84].

2.2.3 Ascidians

Two ascidian species from the Philippines yielded modified cyclic hexapeptides and a novel alkaloid. The hexacyclic and halogenated alkaloid, perophoramidine (**219**) (Fig. 29), was purified from the ascidian *Perophora namei* from Zamboanga del Norte, Philippines [85]. Perophoramidine exhibited cytotoxicity against the HCT116 human colon carcinoma cell line with an $IC_{50} = 60$ μM and induced apoptosis in HCT116 cells. A collection of *Lissoclinum bistratum* Sluiter (Phylum Urochordata, Family Didemnidae) from Tablas Strait, Romblon, Philippines yielded six modified cyclic hexapeptides, bistratamides E–J (**220–225**) (Fig. 29), along with the known

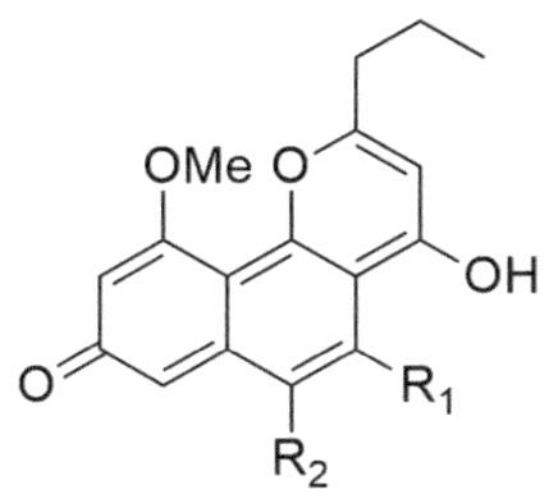

216 (comaparvin): R_1 = OH, R_2 = H
217 (6-methoxycomaparvin): R_1 = OH, R_2 = OCH_3
218 (6-methoxycomaparvin-5-methyl ether): R_1 = R_2 = OCH_3

Fig. 28 Chemical structures of compounds **216–218**

bistratamides C (**226**) and D (**227**) (Fig. 29) [86]. These peptides incorporate modified amino acid residues like thiazole, oxazole, oxazoline, or methylated oxazole rings. Bistratamides E–J exhibited weak to moderate cytotoxicity against the HCT116 cell line [86]. Compounds containing two thiazole rings (bistratamides H and J) showed higher activity compared to those with mixed thiazole/oxazole rings (bistratamides G and I) or oxazoline rings (bistratamides E and F) [86].

2.2.4 Mollusks

Beyond its peptide components, the venom of *Conus imperialis* also contains a variety of non-peptidic bioactive molecules. The distal venom gland of *C. imperialis* and *Conus genuanus* showed a red pigmentation, indicative of other components in the venom of these snails [87]. Serotonin (**228**) (Fig. 30), a well-known neurotransmitter, has been identified in the venom of *C. imperialis*, suggesting a role in modulating the physiological responses of prey during envenomation [88]. Genuanine (**229**) (Fig. 30), a novel guanine derivative, was purified from *C. imperialis* and *C. genuanus* [87]. This compound caused paralysis in mice at a low concentration of 40 nmol, indicating its contribution to the venom's overall toxicity [87]. Furthermore, *C. imperialis* venom contains conazoliums A (**230**) and B (**231**), which are small molecules that mimic the mating pheromones of polychaete worms [89]. These compounds can induce mating behaviors in the worms, suggesting a strategy of aggressive mimicry employed by the snail to facilitate prey capture [89]. Notably, conazolium A has been shown to act as a competitive antagonist of the human α7 nicotinic acetylcholine receptor [89]. Other small molecules, such as guanosine (**232**), adenosine (**233**), and homarine (**234**) (Fig. 30), were also detected in the venom, although their specific functions remain to be fully elucidated [89].

Significant genetic and chemical distinctions were observed between populations of *C. imperialis* inhabiting deep-water versus shallow-water environments in the Philippines [89]. Deep-water specimens possess conazoliums A and B as prominent

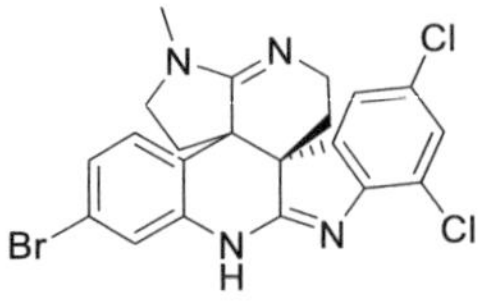

219 (perophoramidine)

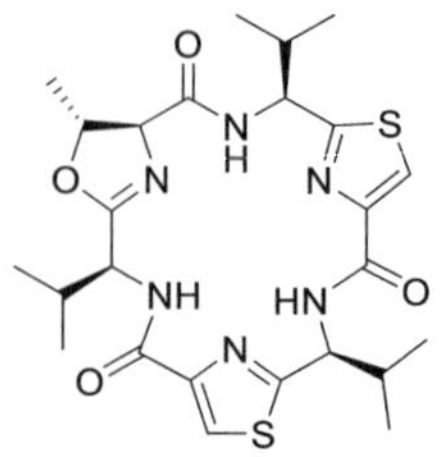

220 (bistratamide E)

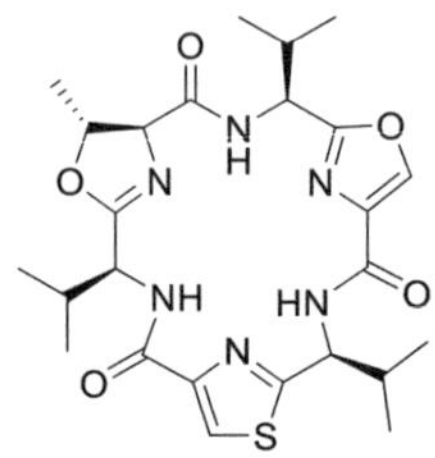

221 (bistratamide F)

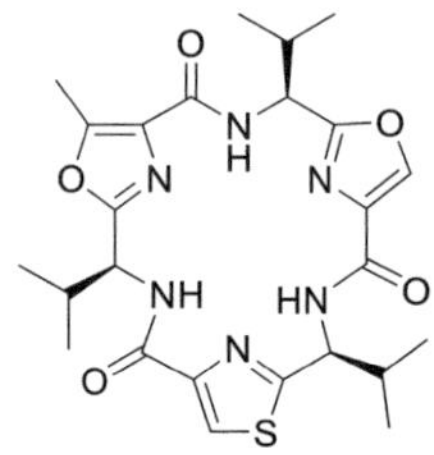

222 (bistratamide G)

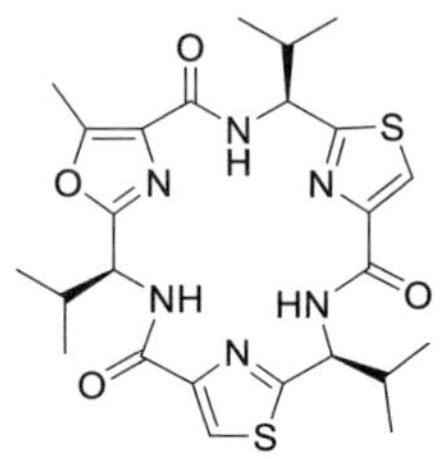

223 (bistratamide H)

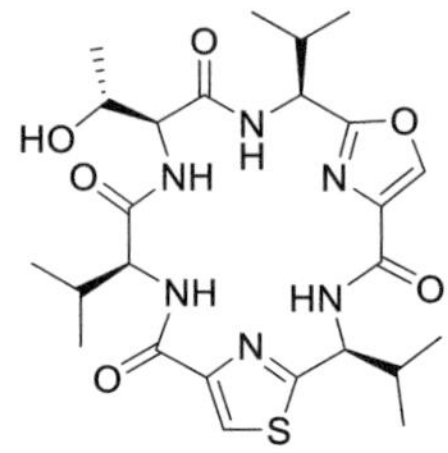

224 (bistratamide I)

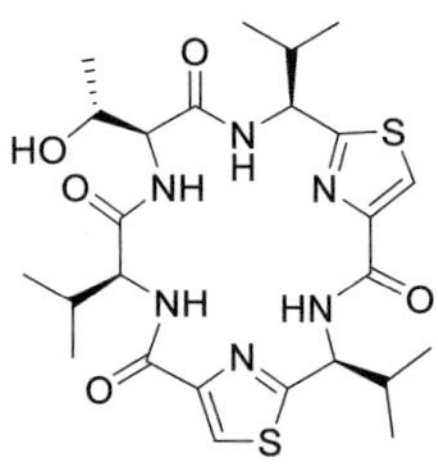

225 (bistratamide J)

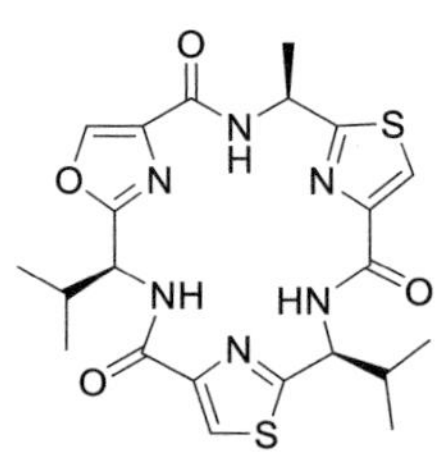

226 (bistratamide C)

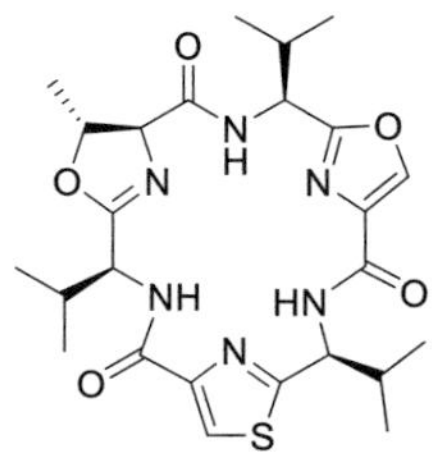

227 (bistratamide D)

Fig. 29 Chemical structures of compounds **219–227**

venom components [89]. In contrast, shallow-water specimens lack these compounds but exhibit an abundance of neurotransmitters such as serotonin, glutamate, and other related molecules [89]. Interestingly, genuanine is present in the venom of both deep-water and shallow-water *C. imperialis* populations [89]. The observed divergence in venom composition between deep-water and shallow-water *C. imperialis* populations suggests the possibility of cryptic speciation within this species. This adaptation likely reflects the different prey availability or hunting strategies required in these distinct marine habitats within the Philippine archipelago. The presence of non-peptide molecules like serotonin, genuanine, and conazoliums, which can directly influence prey physiology and behavior through neurotransmitter action and pheromone mimicry, highlights the remarkable evolutionary adaptations in this marine predator.

While conazoliums are small molecules used by *C. imperialis* as a predation strategy, a set of six new steroids as potential biomedical agents were purified from

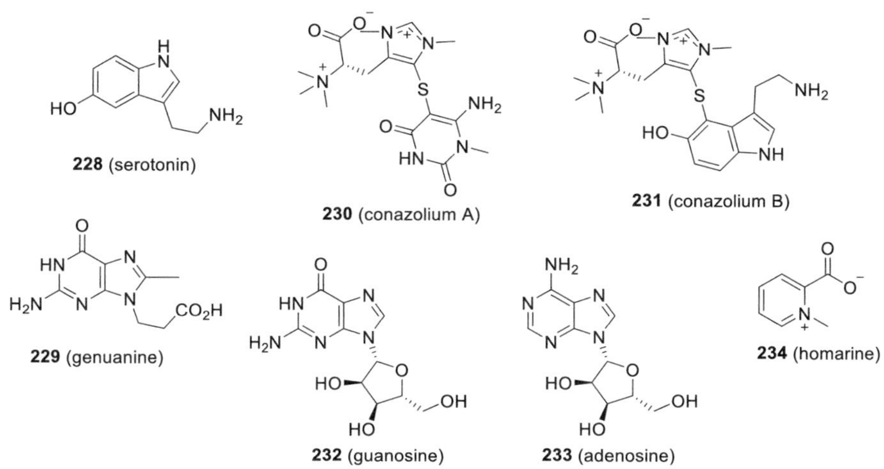

Fig. 30 Chemical structures of compounds **228–234**

the hypobranchial gland of *Conus geographus* from Cebu, Philippines. The methanol extract of the hypobranchial gland was screened for neuroactivity using dorsal root ganglion neurons. The crude extract attenuated the response of DRG neurons to ATP and KCl. Bioactivity-guided purification yielded five steroidal compounds named conosteroids A–E (**235–239**) (Fig. 31) as the neuroactive components [90]. The conosteroids are among the growing cholesterol analogues that have been previously identified from other cone snail species [91, 92]. Conosteroid A was identified as 24(*S*),3β,25-trihydroxycholest-5-en-7-one, while conosteroid B shares the same planar structure but bears opposite configurations at C-3 and C-24 [90]. Conosteroid C is the dehydroanalogue of conosteroid A, lacking the hydroxy group at C-3 and bearing an additional double bond at this position. Conosteroid F is structurally-related to conosteroid C, missing only the hydroxy group at C-25. Conosteroids D and E are the keto analogues of conosteroids C and D, respectively. Conosteroids D and E are oxidized at C-24, bearing a C=O group instead of an alcohol. Conosteroids were determined to constitute a significant portion of the hypobranchial gland extract, with 0.025% yield. Given the small size of the hypobranchial gland of *C. geographus*, the yield of the conosteroids was insufficient to pursue in-depth biological activity analysis [90]. Instead, chemical synthesis of a conosteroid analogue, cholesta-3,5-dien-7-one (**240**) (Fig. 31), was prepared for further biological analysis.

The conosteroids showed varied effects on DRG neurons stimulated with ATP or KCl [90]. Similar to the crude extract, conosteroid A caused a block effect on Ca^{2+} influx in KCl-stimulated DRG neurons. In contrast, conosteroids B–E caused the opposite effect. A direct effect on DRG neurons was also observed for conosteroids. Conosteroids B and C amplified the Ca^{2+} influx in DRG cells treated with ATP. The conosteroid analogue cholesta-3,5-dien-7-one showed the same effects as conosteroid A in DRG neurons [90]. Single cell transcriptome analysis of responsive

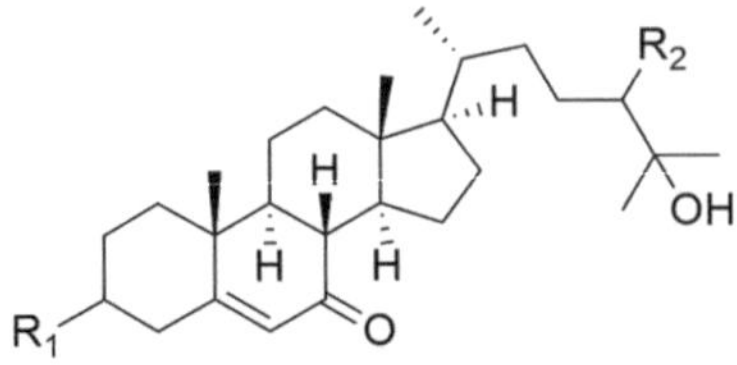

235 (conosteroid A): R_1 = ▬OH , R_2 = ····OH
236 (conosteroid B): R_1 = ····OH , R_2 = ▬OH

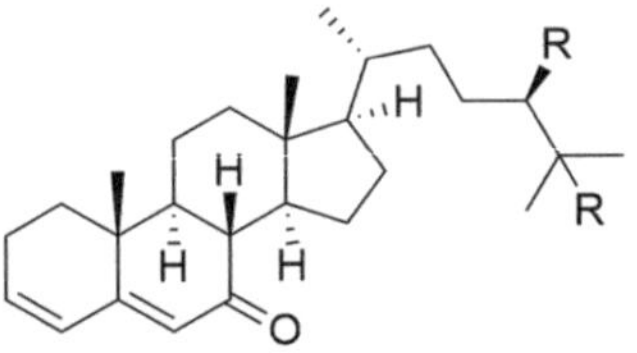

237 (conosteroid C): R = OH
240 (cholesta-3,5-dien-7-one): R = H

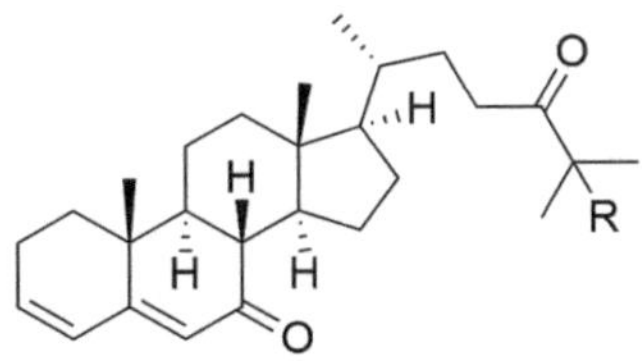

238 (conosteroid D): R = OH
239 (conosteroid E): R = H

Fig. 31 Chemical structures of compounds **235–240**

and unresponsive neurons to cholesta-3,5-dien-7-one indicated that the neuroactivity of this compound may be related to $GABA_AR$ subtypes [90]. Patch clamp assay indicated that cholesta-3,5-dien-7-one is a negative allosteric modulator of $GABA_AR$ $\alpha_1\beta_1\gamma_2$ and $\alpha_4\beta_3\gamma_2$, with $IC_{50} = 1.5$ μM and 1.0 μM, respectively [90]. This conosteroid analogue also showed in vivo analgesic activity using the hot plate method [90.

From the shell-less mollusk, *Placobranchus ocellatus*, collected from Buyong, Cebu, Philippines, six new propionate-derived compounds were purified [93]. The major compound isolated was the (+)-enantiomer of 9,10-deoxytridachione (**241**) (Fig. 32). Tridachiapyrones G–J (**242–245**) (Fig. 32) are structurally related to 9,10-deoxytridachione, bearing modifications principally at the alkenyl chain. Tridachiapyrones G and H have an exomethylene in comparison to 9,10-deoxytridachione. Tridachiapyrone I has an additional OH group at C-13, relative to tridachiapyrones G and H. Tridachiapyrone J is the C-14 keto analog of 9,10-deoxytridachiapyrone. The isomers tridachiahydropyrones B (**246**) and C (**247**) (Fig. 32) are the peroxide analogues of tridachiahydropyrone, bearing a peroxide bridge between C-7 and C-16 [93]. Tridachiahydropyrone B is the *E*-isomer, while tridachiahydropyrone C bears the opposite configuration at the C10-C11 double bond. No bioactivity was presented for these compounds.

2.2.5 Seaweeds

Among the most diverse organisms in the Philippines are the eukaryotic algae or seaweeds. As of 2020, there are 1065 recorded seaweed species in the Philippines,

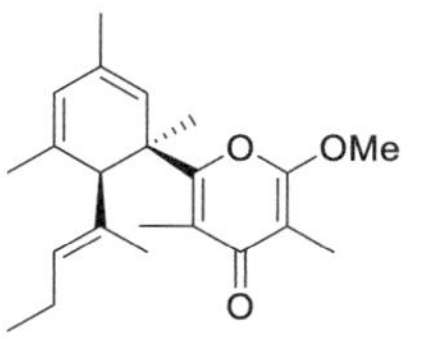

241 (9,10-deoxytridachione)

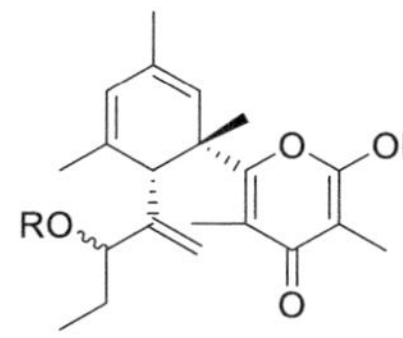

242 (tridachiapyrone G): R = H, 13*R* or *S*
243 (tridachiapyrone H): R = H, 13*S* or *R*
245 (tridachiapyrone J): R = OH, 13*R* or *S*

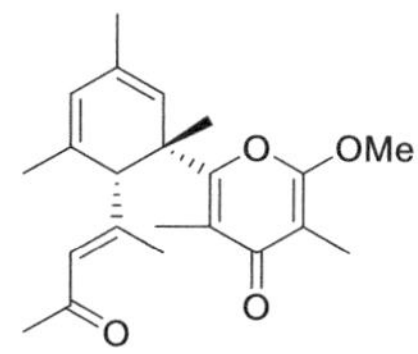

244 (tridachiapyrone I)

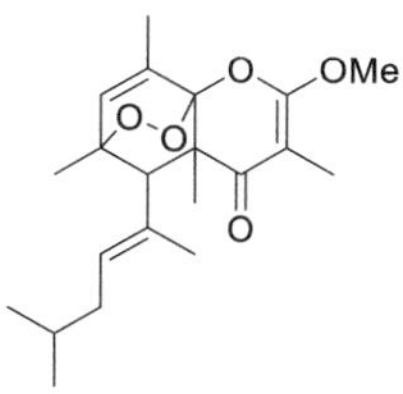

246 (tridachiahydropyrone B)

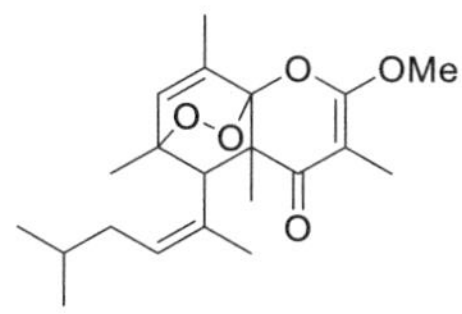

247 (tridachiahydropyrone C)

Fig. 32 Chemical structures of compounds **241**–**247**

600 are red seaweeds, 272 are green seaweeds, and 193 are brown seaweeds [94]. The Philippines is the second top seaweed producer, with a yield of ~ 1.5 million tonnes [95]. Macroalgae are also among the top marine organisms used for marine biotechnology in pharmaceutical and food applications [96]. Seaweeds are traditionally gathered for food, used as feeds, and as a traditional medicine. Ethnobotanical studies on seaweeds from the Philippines documented the traditional use of seaweeds from Ilocos Norte Province [97], Eastern Sorsogon [98], and Eastern Camarines Sur Province [99]. Ilocano communities use a total of 34 distinct seaweed species for the treatment of indigestion, diarrhea, and stomach aches, pulmonary disorders such as cough and asthma, and glandular issues, specifically goiter [97]. The most used are 16 species of Rhodophyta (red algae), seven species of Ochrophyta (brown algae), and 11 species of Chlorophyta (green algae), belonging to 12 families and 20 genera [97]. There was a significant overlap between medicinal and culinary uses, with nearly 80% of the medicinally utilized species being consumed as food by the Ilocanos [97]. Sea grapes or *Caulerpa* were highly utilized by Filipinos residing in Eastern Camarines Sur, Philippines [99]. Locals from Eastern Sorsogon, Philippines, utilized 12 seaweed species, with 10 species primarily used for culinary purposes and medicinal applications [98]. *Chondria armata* and *Digenea simplex* were reported as anti-helminthics and align with known scientific information [98]. *C. armata* contains domoic acid (**248**) and *D. simplex* contains kainic acid (**249**) (Fig. 33), both of which are potent neurotoxins known to have anti-helminthic properties [98, 100].

Treatment of goiter is remarkably consistent among different seaweed genera, particularly *Codium* spp. (*arabicum, intricatum, cf. tenue*) and *Halimeda* spp. (*incrassata, macroloba, opuntia*) in Ilocos Norte [97], *Caulerpa chemnitzia* in Eastern Sorsogon [98], and both *Caulerpa lentillifera* and *C. racemosa* in Eastern Camarines

Fig. 33 Chemical structures of compounds **248** and **249**

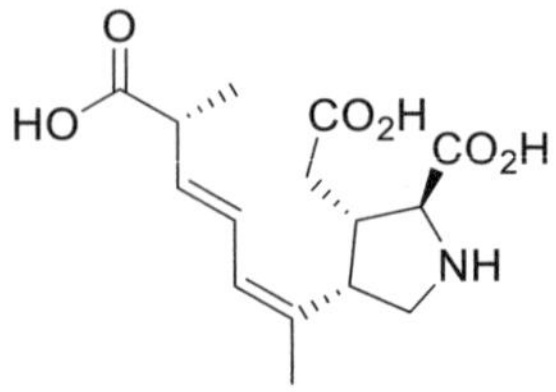

248 (domoic acid)

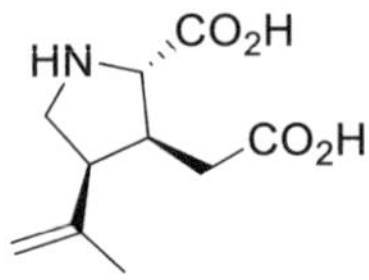

249 (kainic acid)

Sur [99]. This ethnomedicinal use of seaweeds against goiter is linked to the capacity of many seaweeds to accumulate iodine from seawater [101]. A wide variety of seaweed species are used for digestive problems, such as *Caulerpa* spp. [99], *Padina* spp. (*australis, minor*) [97], *Chondrophycus undulatus* [97], *Laurencia okamurae, Palisada perforata* [97], numerous *Gracilaria* species [97], and *Halymenia durvillei* [97]. The frequent application of seaweeds for these common issues may reflect the high content of dietary fibers known to influence gut motility and health, or the presence of specific compounds with soothing, antimicrobial, or anti-inflammatory effects on the digestive tract [102]. Among the most utilized seaweeds based on ethnobotanical studies are *Caulerpa, Gracilaria, Codium, Gelidiella*, and *Digenea* genera, principally via an oral route.

The ethnomedicinal use of seaweeds from the Philippines can be traced to the many bioactive compounds produced by these organisms [103]. Among these are algal polysaccharides, such as carrageenans, agar, fucoidans, and ulvans [104]. Phenolic compounds such as phlorotannins contribute to the free radical scavenging activity of seaweeds [105]. Pigments from seaweeds, including fucoxanthin (**250**) and β-carotene (**251**) (Fig. 34), are documented to have anti-inflammatory and anti-cancer effects [106]. Polyunsaturated fatty acids and modified fatty acids in seaweeds were shown to activate the ARE-Nrf2 pathway [107, 108].

Phytochemical screening and bioactivity analysis of the hexane, dichloromethane, and methanol extracts of the red seaweed *Gracilaria coronopifolia* J. G. Agardh were done via a GC–MS analysis and chorioallantoic membrane (CAM) angiogenesis assay [109]. GC–MS identified compounds like caryophyllene (**252**), 11-octadecenoic acid methyl ester (**253**), hexadecanoic acid methyl ester (**254**), and n-hexadecanoic acid (**255**) (Fig. 35) in the extracts. TLC screening indicated the presence of flavonoids, phenols, anthraquinones, steroids, and essential oils. The dichloromethane extract was the most potent in the CAM and the AMPK activation assays, with $IC_{50} = 1.21$ μg/mL and $EC_{50} = 70.2$ μg/mL, respectively. Angiogenesis inhibition correlated with decreased levels of Fe, Zn, and Cu in the digested CAM, measured by flame atomic absorption spectroscopy (FAAS) and specific colorimetric assays (thiocyanate for Fe; 4-(2-pyridylazo) resorcinol [PAR] for Zn and Cu). Significant lowering of metal levels was observed, especially at 1000 μg/mL extract concentration. The anti-angiogenic effect is proposed to be mediated via AMPK activation. AMPK activation can inhibit mTOR signaling (possibly via TSC1/TSC2), reducing protein synthesis, cell proliferation, and survival [109].

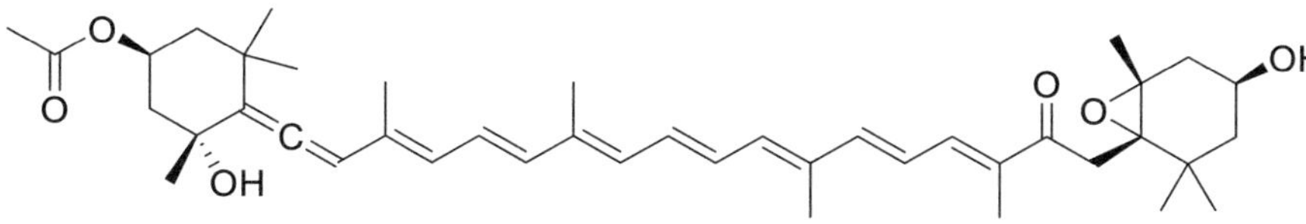

250 (fucoxanthin)

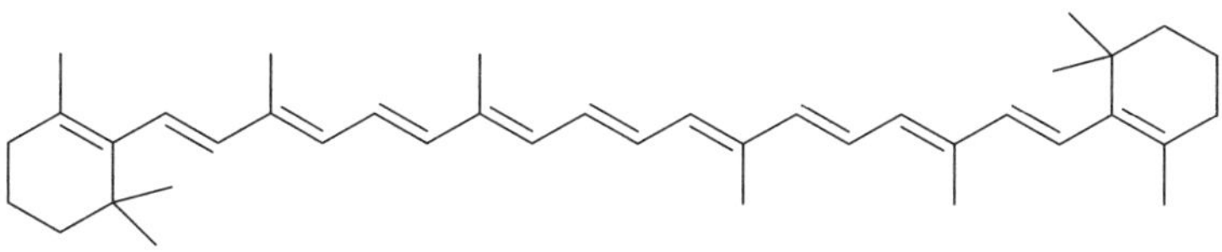

251 (β-carotene)

Fig. 34 Chemical structures of compounds **250** and **251**

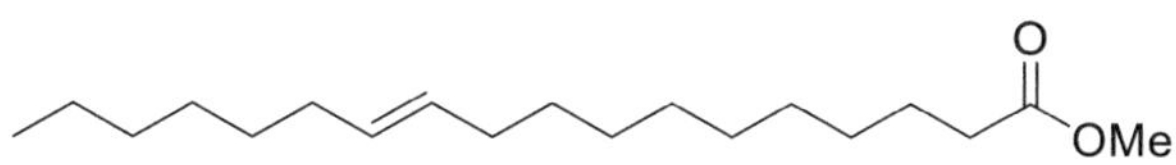

253

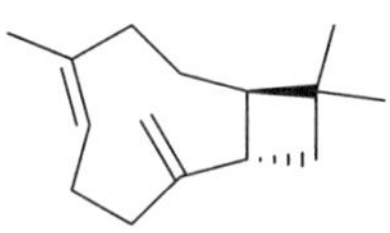

252 (caryophyllene)

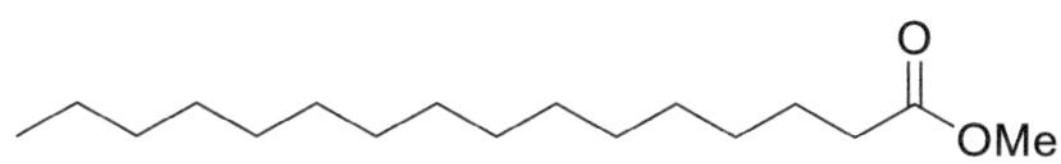

254

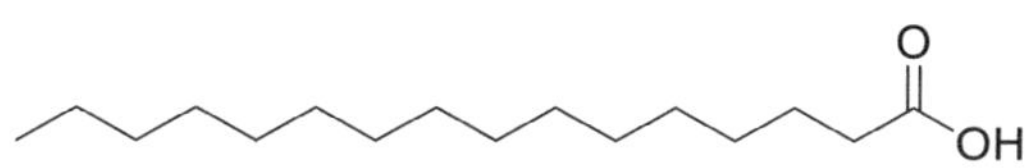

255

Fig. 35 Chemical structures of compounds **252–255**

The acidified methanolic extract of *Sargassum aquifolium* (Turner) C. Agardh underwent proximate analysis, phytochemical screening, antioxidant, antimicrobial, and enzyme assays [110]. *Sargassum aquifolium* had high carbohydrate (32.29 ± 0.17%), ash (30.19 ± 0.14%), protein (16.89 ± 0.09%), crude fiber (10.03 ± 0.15%), crude fat (3.86 ± 0.07%), and moisture (6.74 ± 0.10%) content. The Total Phenolic Content (TPC) was calculated at 5.74 ± 0.04 mg GAE/g (Folin-Ciocalteu method), while the total Flavonoid Content was 539.07 ± 6.36 mg RE/g ($AlCl_3$ colorimetric method, rutin standard) [110]. The extract of *S. aquifolium* showed potent antioxidant activity, comparable to the positive control ascorbic acid. *Sargassum aquifolium* extract gave an IC_{50} = 107 μg/mL and IC_{50} = 21.01 μg/mL in ABTS + radical scavenging and Copper Reducing Antioxidant Capacity (CUPRAC), respectively [110]. Potent enzyme inhibitory activity for *S. aquifolium* extract was also observed against tyrosinase, elastase, and α-glocosidase. Antimicrobial activity was observed in Gram (+) microorganisms: *B. cereus* (MIC = 125, MBC = 250 μg/mL), MRSA, *S. epidermidis*, *S. saprophyticus* (MIC = 250, MBC = 500 μg/mL each); and Gram (−): *P. aeruginosa* (MIC = 500, MBC = 1000 μg/mL). While *S. aquifolium* was inactive (> 1000 μg/mL) against *L. monocytogenes*, *E. coli*, *S. marcescens*, *A. hydrophila*, and *P. fluorescens* [110].

A similar study on *Ulva pertusa* Kjellman detected alkaloids, flavonoids, phenols, tannins, saponins, and terpenoids using phytochemical tests [111]. Total Phenolic Content (TPC) was calculated at 20.54 ± 2.08 mg GAE/g extract (Folin-Ciocalteu method, gallic acid standard), while the total Flavonoid Content was determined to be 539.07 ± 6.36 mg RE/g extract ($AlCl_3$ colorimetric method, rutin standard) [111]. The extract of *U. pertusa* showed a dose-dependent antioxidant activity in a DPPH antioxidant assay, with IC_{50} = 1.53 ± 0.012 mg/mL. Strongest inhibition of bacteria was observed, based on the bacterial clearing zone, at 100 mg/mL assay concentration for *Staphylococcus aureus* (30.67 mm zone), *Aeromonas hydrophila* (~ 19–20 mm), *Streptococcus agalactiae* (~ 18 mm), *A. sobria* (~ 18 mm), *Escherichia coli* (~ 15 mm), and *Vibrio harveyi* (9 mm) [111].

While several studies have documented the bioactivity and phytochemicals in seaweeds from the Philippines, few have dealt with the purification of bioactive components. Two different collections of the red algae *Laurencia* sp. yielded 12 identified compounds. A collection of *Laurencia* sp. from Palawan, Philippines afforded two vinyl bromides (**256** and **257**), cartilagineol (**258**), ma'ilione (**259**), ma'iliohydrin (**260**), and the enone analogue (**261**) (Fig. 36) [112]. Ma'iliohydrin, a new cytotoxic tribrominated chamigrene sesquiterpenoid, bears a dibromohydrin moiety ($CHBr_2$-CH(OH)). These compounds have cytotoxic activity against cancer cell lines. The highest activity of ma'iliohydrin was observed against NCI/ADR-RES (multidrug-resistant breast cancer cell line) with $GI_{50} = 10^{-8}$ M. The proposed biosynthetic pathway involves the addition of BrOH (hypobromous acid, likely from enzymatic bromination) across the double bond of the vinyl bromide side chain in **256** or **257** (Fig. 36) to generate the dibromohydrin moiety of ma'iliohydrin [112]. Subsequent (hypothetical) loss of dibromomethane (CH_2Br_2) from ma'iliohydrin could lead to the norchamigrene ma'ilione.

Six cyclic ether acetogenins, named laurefurenynes A–F (**262**–**267**) (Fig. 37), were isolated from a *Laurencia* sp. from an unidentified collection site in the Philippines [113]. These compounds were purified as three diastereomeric pairs differing in double bond geometry (*cis*/*trans* enyne): laurefurenynes A (*cis*) and B (*trans*) [2,2′-bifuran core]; laurefurenynes C (*cis*) and D (*trans*) [2,8-dioxa-bicyclo[5.2.1]decane core]; laurefurenynes E (*cis*, 12-Br) and F (*trans*, 12-Br) [2,8-dioxa-bicyclo[5.2.1]decane core] [113]. The proposed biosynthetic pathway of laurefurenynes A–F starts from a highly unsaturated C15 linear precursor, potentially forming intermediate laurediols [113]. These intermediates undergo the proposed epoxide- or bromonium ion (Br^+)-mediated cyclizations involving hydroxyl groups attacking double bonds to form the different tetrahydrofuran-containing bicyclic systems (2,2′-bifuran or 2,8-dioxa-bicyclo[5.2.1]decane) found in laurefurenynes A–F [113]. Laurefurenynes A–F were evaluated for differential cytotoxicity in soft agar

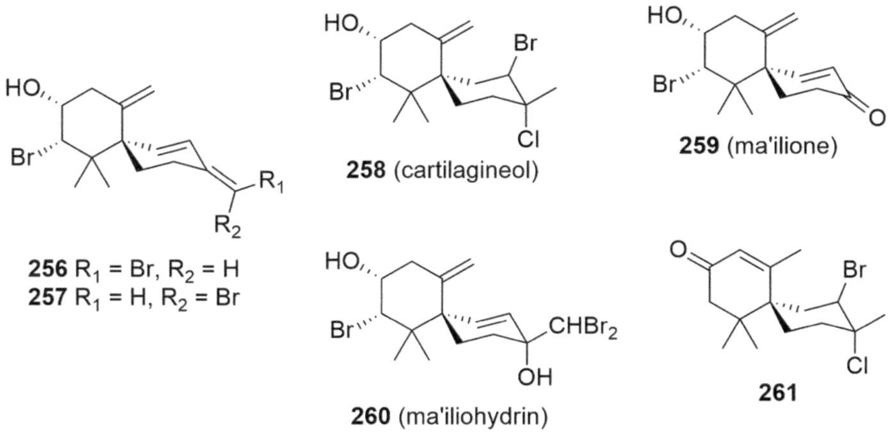

256 R_1 = Br, R_2 = H
257 R_1 = H, R_2 = Br
258 (cartilagineol)
259 (ma'ilione)
260 (ma'iliohydrin)
261

Fig. 36 Chemical structures of compounds **256–261**

assay against murine leukemia (L1210), murine solid tumor (colon 38), human solid tumors (colon H116, lung H125), and normal murine/human hematopoietic progenitor cells (CFU-GM) [113]. Laurefurenynes A and B were inactive, while laurefurenynes D and E showed weak activity against colon 38 cells. Laurefurenynes C and F both showed moderate non-selective cytotoxicity. Laurefurenynes C inhibited the growth of L1210, colon 38, and H-125 cells, while laurefurenynes F had cytotoxic activity in these three cell lines and H-116, and human CFU-GM as well [113].

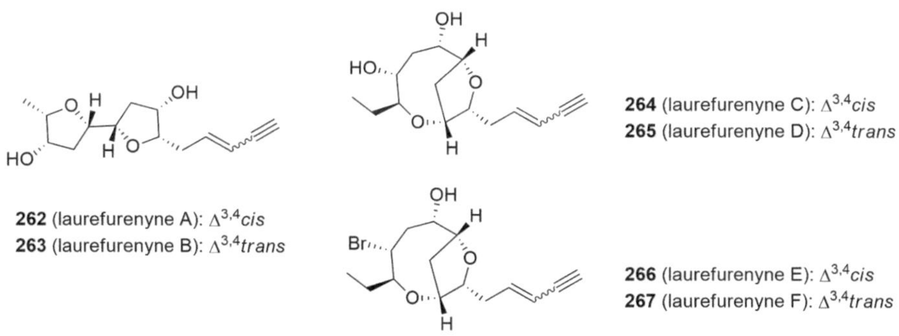

262 (laurefurenyne A): $\Delta^{3,4}$*cis*
263 (laurefurenyne B): $\Delta^{3,4}$*trans*
264 (laurefurenyne C): $\Delta^{3,4}$*cis*
265 (laurefurenyne D): $\Delta^{3,4}$*trans*
266 (laurefurenyne E): $\Delta^{3,4}$*cis*
267 (laurefurenyne F): $\Delta^{3,4}$*trans*

Fig. 37 Chemical structures of compounds **262–267**

3 Small Molecule Compounds from Marine Microorganisms

By the early 2000s, and with the advent of next-generation sequencing platforms, it became evident that the true producers of biomedically important molecules from marine organisms are their associated microorganisms [114]. Microorganisms closely associated with marine macroorganisms may be classified as obligate, facultative, or casual. Obligate microbial symbionts have a significantly reduced genome, indicating the lack of capability to live independently from the host [115]. Facultative microorganisms do not completely rely on the host, but there is a demonstrated role in microbial interactions with the host. Casual microorganisms have no direct association with the host and may result from the host's interaction with microorganisms from the environment. Free-living microorganisms are also isolable from sediments and the water column.

Marine sponges were among the first macroorganisms to be investigated for their associated microorganisms (Fig. 38). The filter-feeding nature of sponges, together with structural similarity of sponge compounds with microbial compounds, indicated the impact of microbial symbiosis on the chemical defenses in sponges [116]. Analysis of 81 sponge species revealed that 39 microbial phyla, including Proteobacteria, Actinobacteria, and Cyanobacteria, compose the majority of the sponge microbiome [117]. The most widely studied sponge-microbiome interaction for natural products biosynthesis is between the sponge *Theonella swinhoei* and *Candidatus* Entotheonella. This unculturable bacterium *Ca.* Entotheonella is the recognized producer of the bioactive sponge compounds polytheonamide and onnamide A (**268**) (Fig. 39) [118]. The success of terrestrial microorganisms as a source of antibiotics also significantly contributed to the interest in discovering bioactive compounds from marine microorganisms.

3.1 Sponge-Associated Microorganisms

Among the early reported microbial-derived compounds from the Philippines is the red pigment heptylprodigiosin (**269**) (Fig. 40) or 16-methyl-15-heptyl-prodiginine from *Pseudovibrio denitrificans* Z143-1 [119]. *P. denitrificans* Z143-1 was isolated from an unknown ascidian collected from Zamboanga del Norte, Philippines. This microorganism is characterized by circular, nonluminescent, and translucent colonies and turns into opaque red colonies over time with incubation in marine agar [119]. A two-day culture of *P. denitrificans* Z143-1 in marine broth yielded heptylprodigiosin after solvent partitioning, silica, and ion exchange chromatography [119]. Heptylprodigiosin and related prodigiosins have multiple bioactivities, including antimicrobial, cytotoxic, and antiprotozoal activities [120]. Bioactivity screening on heptylprodigiosin showed antimicrobial activity against methicillin-resistant and methicillin-sensitive *Staphylococcus aureus* [119]. Antimalarial activity assay of

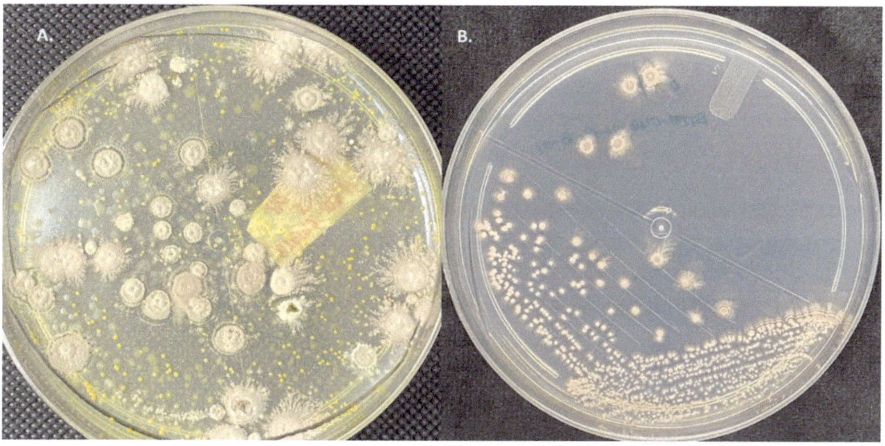

Fig. 38 Culturable sponge-associated marine microorganisms as starting points for drug discovery. **a** Distinct colonies of microorganisms are observed in the mother plates of sponge extracts. **b** Repeated plate streaking of morphologically-distinct colonies yields purified microbial colonies

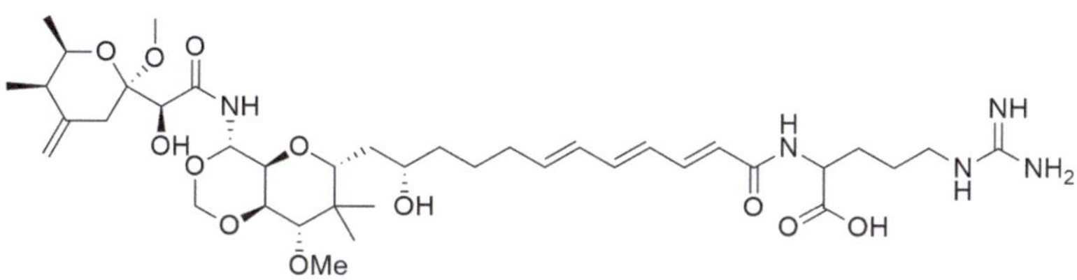

268 (onnamide A)

Fig. 39 Chemical structures of onnamide A

heptylprodigiosin in chloroquine-sensitive *Plasmodium falciparum* 3D7 showed similar potency with the antimalarial drug quinine [121]. Lazaro et al. tested the in vivo antimalarial activity of heptylprodigiosin in mice infected with *Plasmodium berghei* ANKA strain [121]. Oral administration of heptylprodigiosin did not yield any protective effect from *P. berghei.* Subcutaneous injections at 5 and 20 mg/kg showed an increased median survival of the infected animals [121]. This highlights that the activity of heptylprodigiosin may be affected by its stability during administration. While heptylprodigiosin did not show mutagenic effects, sclerotic lesions were observed during subcutaneous injections [121].

The potential utility of heptylprodigiosin as an anticancer agent was also investigated as a single agent and in combination with the marine compound adociaquinone B. Heptylprodigiosin was tested in MCF-7 breast cancer cells and CD95-sensitive (S-Jurkat) and CD95-resistant ($Jurkat^R$) Jurkat T cells [122]. Heptylprodigiosin gave an IC_{50} ~ 2.0 μM in S-Jurkat and induced apoptosis within 16 h of treatment. Initiation of apoptosis was accompanied by an increase in subdiploid cells, changes in cell

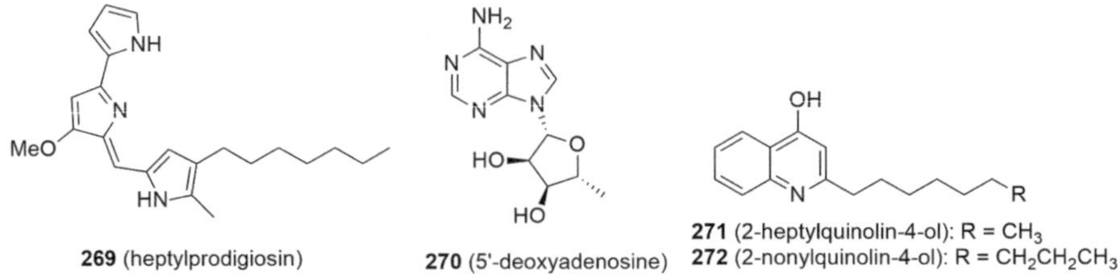

Fig. 40 Chemical structures of compounds **269–272**

morphology, and a decrease in G0/G1 cells [122]. The apoptotic effect of heptylprodigiosin was mediated by caspases and occurs via the extrinsic pathway [122]. Further, the action of heptylprodigiosin was mediated by CD95 receptor, FADD, and caspase 8. Increased mitochondrial potential was observed with the treatment of heptylprodigiosin. Necrosis was, however, observed at high concentrations of heptylprodigiosin [122]. To limit the toxicity from HPDG, a combinatorial regimen with adociaquinone B was designed. Three combinations of heptyprodigiosin:adociaquinone B (1:1, 3:1, 1:3) were tested in MCF-7 breast cancer cells, alongside the individual compounds [123]. A significant shift in IC_{50} for the compounds in combination, compared to the individual compounds, was observed. Isobologram analysis and fractional inhibitory concentration indicated a synergistic effect between adociaquinone B and heptylprodigiosin, with the 3:1 adociaquinone B:heptylprodigiosin giving the highest synergistic effect.

From the marine sponge *Haliclona* sp. Nov. from Pangasinan, Philippines, was isolated the β-proteobacterium *Achromobacter xylosoxidans* strain ISP2-142-O-2-A [124]. Using a liquid culture of *A. xylosoxidans* in ISP2 medium, quinolones and a nucleoside derivative were purified by a series of column fractionations. The nucleoside 5′-deoxyadenosine (**270**) caused a significant cytoprotective effect in an HIV cytoprotection assay, causing 68% and 45% survival at 0.5 μg/mL and 0.05 μg/mL, respectively [124]. This compound did not reactivate HIV in a latency assay. Further, 5’-deoxyadenosine was non-cytotoxic to normal mammalian MDCK cells, with no antimicrobial activity [124].

The antimicrobial activity of the extract of *A. xylosoxidans* ISP2-142-O-2-A can be attributed to the quinolone compounds produced by this microorganism. 2-Heptylquinolin-4-ol (**271**) (Fig. 40) showed antimicrobial activity against *Staphylococcus aureus* ATCC 12600 (MIC = 1.5 μg/mL), with moderate cytotoxicity to MDCK cells (IC_{50} = 7.5 μg/mL) [124]. 2-nonylquinolin-4-ol (**272**) (Fig. 40) showed more potent antimicrobial activity against *S. aureus* ATCC 12600, with MIC = 0.1 μg/mL [124]. Cytotoxicity to MDCK cells was observed with IC_{50} = 3.0 μg/mL. These quinolone compounds were inactive against methicillin-resistant *S. aureus* and the Gram (−) pathogens *Klebsiella pneumoniae* and *Pseudomonas aeruginosa* [124].

Whole genome sequencing followed by biosynthetic gene cluster (BGC) analysis of *A. xylosoxidans* ISP2-142-O-2-A revealed ten putative BGCs in total, indicating the potential to produce siderophores, arylpolyenes, terpenes, type-1 polyketides, resorcinol, ectoine, phosphonates, and betalactones [124]. Biosynthesis of the quinolines in *A. xylosoxidans* was inferred by homology to the known 4-hydroxy-2-alkylquinoline (HAQ) pathway in *P. aeruginosa*, involving the *pqsABCDE* operon. Homologous genes were found in the *A. xylosoxidans* ISP2-142-O-2-A draft genome, including anthranilate synthase components (*trpE, trpG*) for the precursor anthranilic acid, homologs of *pqsA* (anthranilate-CoA ligase: *abmG* (2-aminobenzoate-CoA ligase) and benzoate-CoA ligase, homologs of *pqsB/pqsC* (β-ketoacyl-ACP synthases): *fabH* and 3-oxoacyl-[ACP] synthase III, and homologs of *pqsD* (condensing enzyme/transacetylase) were also *fabH* and 3-oxoacyl-[ACP] synthase III [124].

The actinobacteria, *Streptomyces cacaoi* R2A-843A from an unidentified sponge from Cebu, Philippines, showed superior biosynthetic capability for peptide-derived compounds based on genomics and metabolomics profiling [125]. *S. cacaoi* R2A-843A dedicates 12% of its genome (by length) to secondary metabolite biosynthesis [125]. Biosynthetic gene clusters for nonribosomal peptides were the most represented, including those for cacaoidin and the cyclic peptide BE-18257A (**273**) [125]. The genome analysis was corroborated by GNPS molecular networking, which identified BE-18257A, indole-3-carbinol, and 5-methylbenzotriazole with high confidence. Manual annotation of the molecular network also identified pentaminomycins A–D and a new analogue bearing a D-Leu and homocysteine at positions 1 and 3, respectively [125]. The glycosylated lanthipeptide cacaoidin was also identified from the metabolome analysis as a minor component. Purification of the fermentation broth of *S. cacaoi* R2A-843A afforded the peptides BE-18257A (**273**) and BE-18257B (**274**) (Fig. 41) as the major isolable peptides [125]. Time course monitoring of these major peptides showed appreciable production starting at 48 h, with BE-18257B being the more abundant peptide [125].

Microorganisms associated with macroorganisms from unique ecological niches also showed superior biosynthetic potential. An unidentified Chalinidae family of sponge from the mesophotic zone (62 m) was collected from Negros Oriental, Philippines [126]. From the tissue of this sponge, a fungus *Pleosporales* sp. NBUF144 was purified. The extract from the broth and mycelia of *Pleosporales* sp. NBUF144

Fig. 41 Chemical structures of compounds **273** and **274**

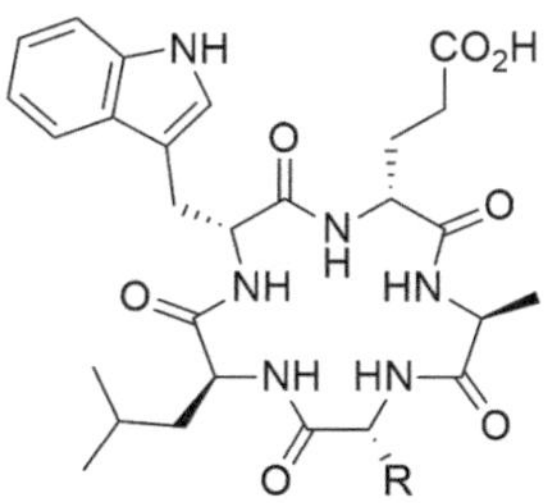

273 (BE-18257A): R = D-valine
274 (BE-18257B): R = D-leucine

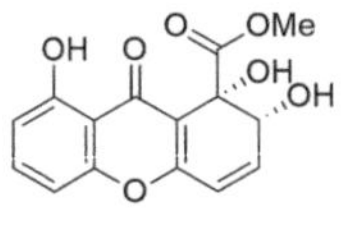

275 (globosuxanthone F)

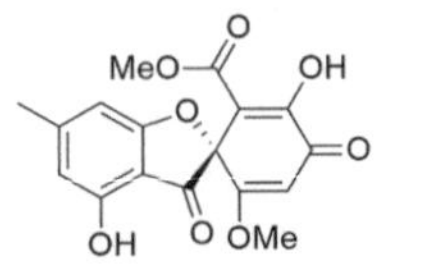

276 (2'-hydroxy bisdechlorogeodin)

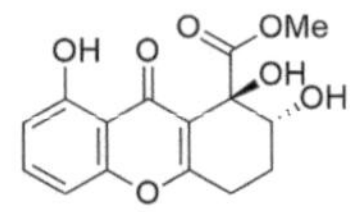

277 (3,4-dihydroglobosuxanthone A)

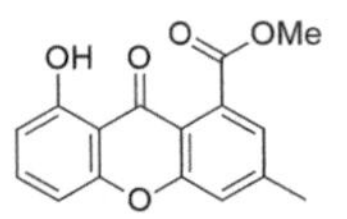

278 (8-hydroxy-3-methylxanthone-1-carboxylate)

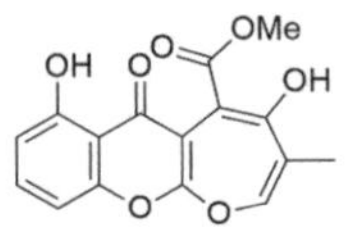

279 (crosphaeropsone C)

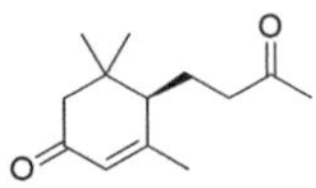

280 (4-megastigmen-3,9-dione)

Fig. 42 Chemical structures of compounds **275–280**

cultured in potato dextrose broth medium was purified by a combination of size exclusion and reversed phase chromatography [126]. This afforded the new compounds globosuxanthone F (**275**) and 2′-hydroxy bisdechlorogeodin (**276**), and the known compounds 3,4-dihydroglobosuxanthone A (**277**), 8-hydroxy-3-methylxanthone-1-carboxylate (**278**), crosphaeropsone C (**279**), and 4-megastigmen-3,9-dione (**280**) (Fig. 42) [126]. Globosuxanthone F is structurally-related to globosuxanthone A, and 2′-hydroxy bisdechlorogeodin is related to bisdechlorogeodin, both known fungal polyketides, suggesting typical fungal PKS pathways are involved in the biosynthesis. Cytotoxicity assessment on the compounds was done in CCRF-CEM human acute lymphatic leukemia cells. Globosuxanthone F showed the most potent cytotoxic activity, with $IC_{50} = 0.46$ μM. 2′-hydroxy bisdechlorogeodin showed moderate activity, with an estimated IC_{50} of 2 μM [126]. Other compounds were inactive with $IC_{50} > 20$ μM. Structure–activity analysis suggests the $\Delta^{3,4}$ double bond and the C-1/C-2 hydroxy groups are crucial for the cytotoxicity of this xanthone scaffold [126].

An *Aaptos* sponge from a mesophotic reef (103 m) in Negros Oriental, Philippines afforded the fungus *Cymostachys* sp. NBUF082 [127]. Untargeted metabolomics (LC–MS/MS molecular networking) and the OSMAC approach allowed the prioritization of this fungus for chemical investigation. The first reported natural products from this fungus consist of the polyphenols bearing dihydroisobenzofuram moieties named cymopolyphenols A–F (**281–286**) (Fig. 43), with cymopolyphenols D–F being dimers or trimers of cymopolyphenol A [127]. The known compounds F1839-1 (**287**) and stachybotrylactone (**288**) (Fig. 43) were also purified from the extract of this fungus [127]. The basic dihydroisobenzofuran monomer in cymopolyphenol A is likely derived from a polyketide pathway. The polymeric analogues are proposed to be biosynthesized de novo from the monomer, possibly via radical coupling or enzyme-catalyzed reactions (e.g., quinone methide pathway), rather than being extraction artifacts [127]. Cymopolyphenols A–F were tested for antimicrobial activity, cytotoxicity, iron chelation properties, and ability to induce plant defense [127]. Except for cymopolyphenol F, other compounds purified showed weak to no antibiotic activity against aquatic pathogens *Pseudoalteromonas carrageenovora*, *Vibrio shilanii*, *V.*

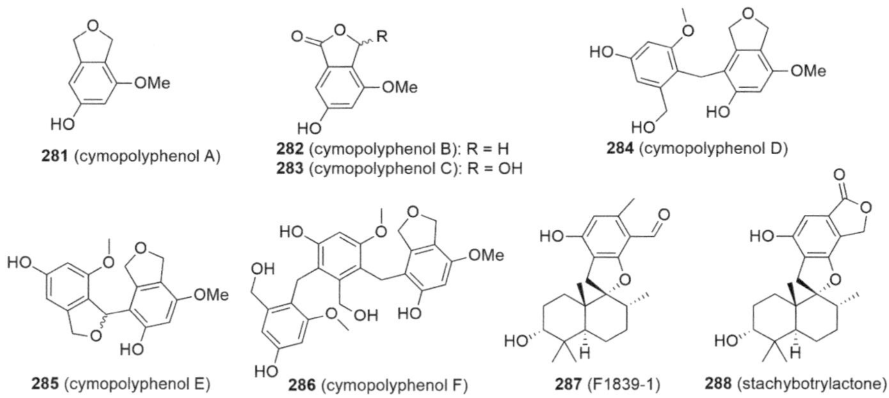

Fig. 43 Chemical structures of compounds **281–288**

scophthalmi, *V. alginolyticus*, and human pathogens *Salmonella typhi*, *Pseudomonas aeruginosa*, *Staphylococcus aureus*, and *Bacillus pumilus*. Cymopolyphenol F was most active against *V. scophthalmi* (MIC 16 μg/mL). The compounds were also noncytotoxic to U87 human glioblastoma and CCRF-CEM human T lymphoblast cells. Cymopolyphenol E had comparable Fe^{2+} chelation activity with EDTA at 10 μM. Stachybotrylactone induced expression of the pathogenesis-related protein 1 (PR1) marker in PR1:GUS transgenic *Arabidopsis thaliana* plants at 10 μM, indicating activation of plant disease resistance pathways [127].

3.2 Mollusk-Associated Microorganisms

Other marine animals, such as bryozoans, shipworms, and ascidians, are also known to harbor microbial symbionts. Heavily defended invertebrates such as cone snails were initially perceived to lack microbial symbionts and small molecule compounds for their defense because of their highly effective peptide-based toxins. A preliminary analysis by Peraud et al. [128] on *Conus pulicarius*, *C. tribblei*, and *C. rolani*, indicated the presence of culturable Actinobacteria. The extracts from these microorganisms were demonstrated to be neuroactive in the dorsal root ganglion assay and highlighted the potential impacts of these microorganisms on the chemical weapons of cone snails [128]. Polyketides and peptide-based molecules have been discovered, to date, from bacteria associated with cone snails. Several of these compounds were shown to target various receptors related to pain and neurological processes.

A follow-up study on *C. rolani* and *C. tribblei* provided an actinomycete *Nocardiopsis alba* strain CR167 from the venom ducts of these cone snails collected from Mactan Island, Cebu, Philippines [129]. The crude extract of *N. alba* strain CR167 showed neuroactivity in a calcium imaging assay. Bioactivity-guided purification yielded the new γ-pyrones named nocapyrones H–Q (**289–298**) (Fig. 44) together with the known compounds nocapyrones A–C (**299–301**) (Fig. 44). Apart from nocapyrone J, other analogues showed neuroactivity in the calcium imaging assay using mouse dorsal root ganglion (DRG) neurons. The effects of nocapyrones were cell-type dependent, with inhibition and amplification of Ca^{2+} influx observed. Nocapyrones B and H caused a decrease in Ca^{2+} influx in large-diameter capsaicin-resistant DRG cells. An increased Ca^{2+} was seen in small-diameter DRG neurons that are responsive to capsaicin [129]. These compounds showed modest to weak cytotoxicity in MCF-7 breast adenocarcinoma cells and normal CHO-AA8 cells. Nocapyrones B and H were tested in cell lines overexpressing transient receptor potential (TRP) channels (TRPV1, TRPA1, TRPM8). Following the DRG assay, inhibition from treatment with nocapyrones B and H was observed in cells overexpressing the menthol-sensitive TRPM8 [129].

The biosynthesis of nocapyrones was elucidated to confirm the presence of these compounds in the gastropod host. The biosynthetic gene cluster of nocapyrones was mapped to the ~ 6 kb *ncp* cluster. This contained four genes hypothesized to be essential for biosynthesis. NcpD, a putative acyltransferase, initiates the biosynthesis by loading the fatty acyl-CoA esters as the starting unit [129]. Thewo units of methylmalonyl-CoA extend the fatty acid via the NcpC PKS domain (Type I PKS). The pyrone ring formation occurs during cleavage, and the core structure is then modified by NcpB methyltransferase to yield the O-methylated product. Further oxidation is undertaken by the oxidoreductase NcpA. PCR confirmed the presence

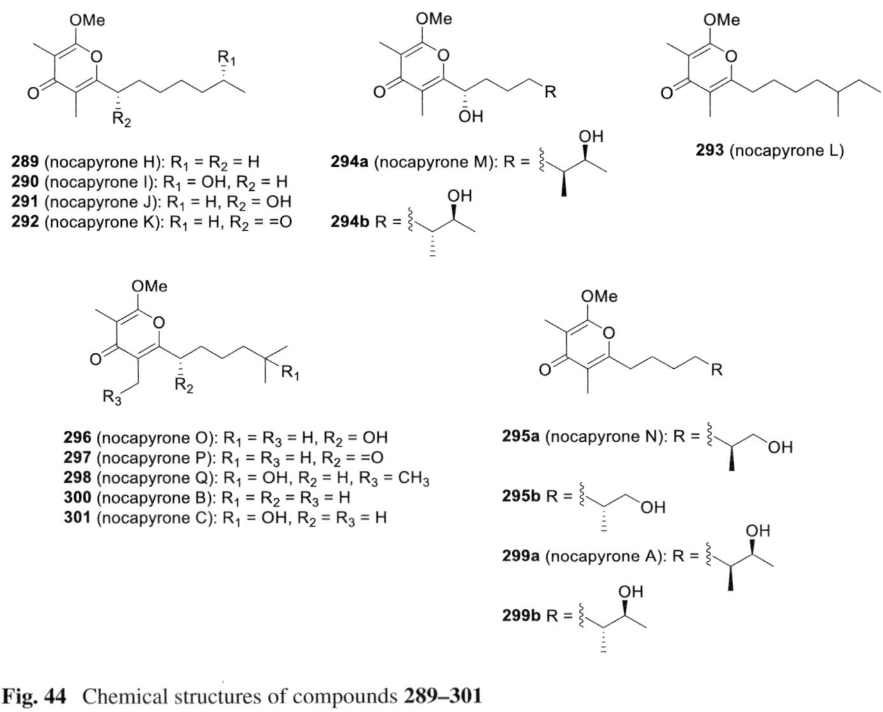

Fig. 44 Chemical structures of compounds **289–301**

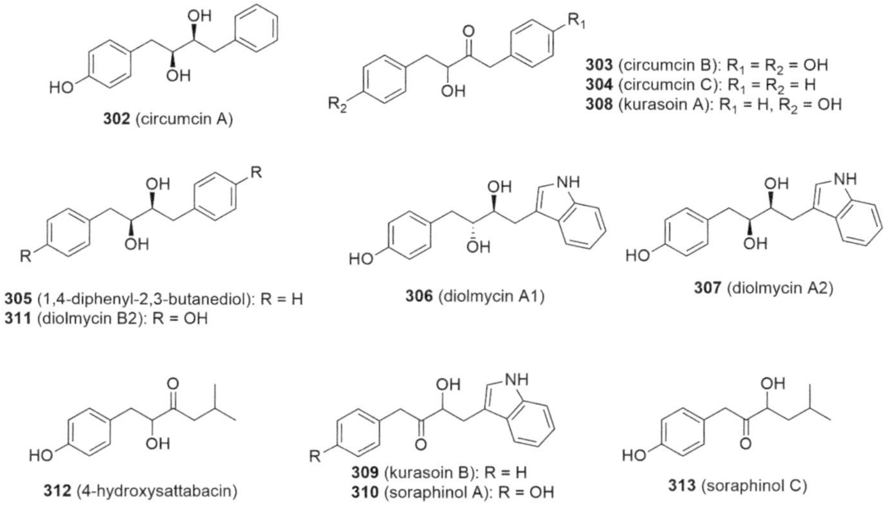

Fig. 45 Chemical structures of compounds **302–313**

of *ncpB*, *ncpC*, and the *N. alba* 16S rRNA gene in whole snail tissue of *C. rolani* and *C. tribblei*. At the same time, nocapyrones were detected in the snail body, venom duct, and mucus [129].

The actinobacteria *Gordonia* spp. have been observed to be associated with multiple species of cone snails. The crude extract of *Gordonia* spp. 647W.R1a.05 associated with *Conus circumcisus* collected from Cebu, Philippines, showed neuroactivity in the DRG assay, causing an increased Ca^{2+} influx [130]. Bioactivity- and chemistry-guided purification yielded a series of butanediol derivative, circumcins A–C (**302–304**) (Fig. 45), 1,4-diphenyl-2,3-butanediol (**305**), diolmycins A1 (**306**) and A2 (**307**), kurasoins A (**308**) and B (**309**), soraphinol A (**310**), diolmycin B2 (**311**), 4-hydroxysattabacin (**312**), and soraphinol C (**313**) (Fig. 45). Circumcin B and soraphinol A elicited the same phenotype as the crude extract, suggesting these as the bioactive components responsible for the observed DRG phenotype in the extract. Screening of circumcin B against a panel of 44 human receptors, transporters, and channels in the National Institute of Mental Health's Psychoactive Drug Screening Program (PDSP), however, failed to identify a target. Kurasoin B and soraphinol A showed potent binding to human norepinephrine transporter (NET) with K_i of 2757 nM and 867 nM, respectively [130].

Another set of aromatic compounds, griseorhodins D–F (**314–316**) (Fig. 46) were purified from *Streptomyces* CN48+ [131]. This actinomycete was isolated from the venom duct tissue of *Chicoreus nobilis* collected near Mactan Island, Cebu, Philippines. Griseorhodins D–F amplified the KCl-induced Ca^{2+} influx in mouse DRG cells. Screening in the PDSP assay showed only griseorhodin E having significant binding to human dopamine active transporter, with $K_i = 557$ nM. A different target

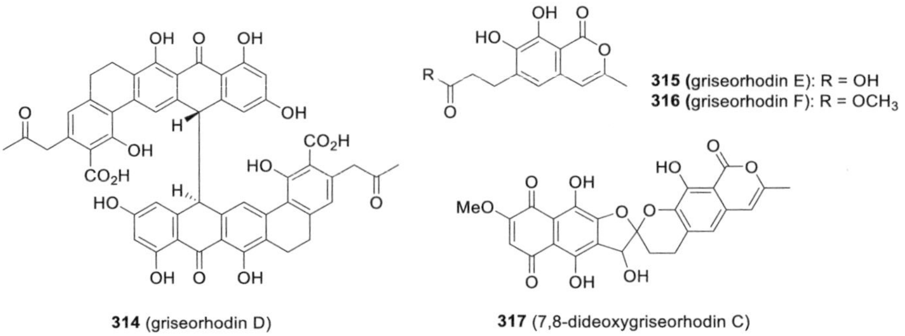

Fig. 46 Chemical structures of compounds **314–317**

receptor may be responsible for the neuroactivity of griseorhodins. While griseorhodin E showed significant binding with dopamine active transporter, this receptor is, however, absent in DRG neurons [131].

Another griseorhodin analogue, 7,8-dideoxygriseorhodin C (**317**) (Fig. 46), was purified from the *Streptomyces* sp. strain 1425S.R.1a.1, isolated from the body tissue homogenate of the snail *Truncatella guerinii* from Cebu, Philippines [132]. No indication of neuroactivity was given for 7,8-dideoxygriseorhodin C. It was instead tested for antibiotic properties in single and combination treatment. 7,8-dideoxygriseorhodin C inhibited the growth of Gram (+) bacteria, including *Staphylococcus aureus*, methicillin-resistant *S. aureus* (MRSA), *Enterococcus faecium*, and *Bacillus subtilis*. The compound showed selectivity for Gram (+) bacteria, with no significant antibiotic activity against Gram (−) bacteria nor cytotoxicity against normal MDCK and CHO-AA8 cells [132]. Because of the potent and selective activity of 7,8-dideoxygriseorhodin against MRSA, a combination treatment with oxacillin was devised. Oxacillin gave a minimum inhibitory concentration (MIC) of 1.59–6.24 μg/mL against MRSA, while 7,8-dideoxygriseorhodin gave an MIC of 0.08–0.12 μg/mL. Combination of 7,8-dideoxygriseorhodin and oxacillin at 1 × MIC ratio indicated a synergistic activity. The combination therapy caused a 100-fold decrease in microbial density after 24 h [132].

A series of spirotetronate polyketides was purified from *Streptomyces* sp. strain 1053U.I.1a.3b, associated with the hepatopancreas of *Lienardia totopotens* collected from Cebu, Philippines [133]. New compounds lobophorins H (**318**) and I (**319**) were purified using a chemistry-guided purification, together with lobophorins B (**320**), C (**321**), and F (**322**) (Fig. 47). Besides lobophorin H, the other analogues showed inhibitory activity against *M. tuberculosis* with MIC = 1.3–7.8 μg/mL. The lack of activity of lobophorin H highlights the importance of the sugar moiety for the antibiotic activity of this class of compounds. Further structure modification can be undertaken on the lobophorin scaffold to improve its selectivity index. The bioactive lobophorin showed a narrow selectivity window, with IC_{50} = 0.3–8.6 μg/mL [133].

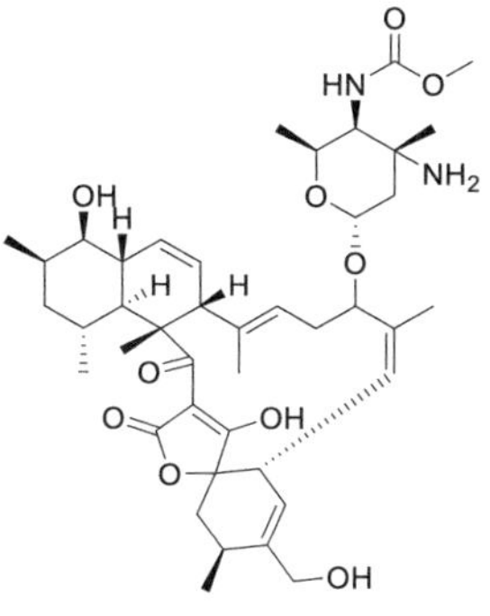

318 (lobophorin H)

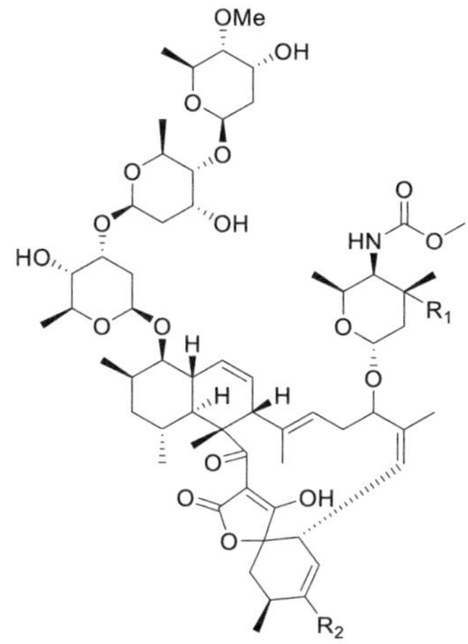

319 (lobophorin I): $R_1 = NO_2$, $R_2 = CHO$
320 (lobophorin B): $R_1 = NO_2$, $R_2 = CH_2OH$
321 (lobophorin C): $R_1 = NH_2$, $R_2 = CH_2OH$

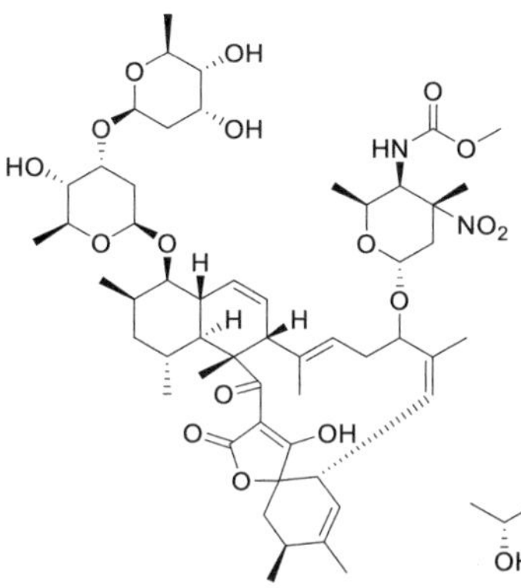

322 (lobophorin F)

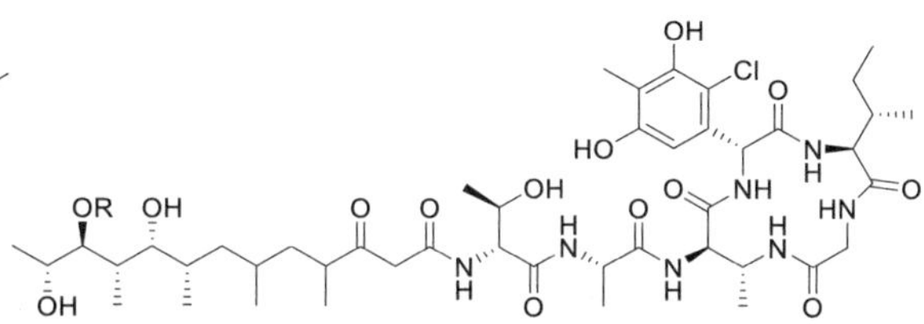

323 (totopotensamide A): R =

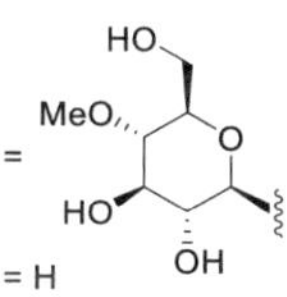

324 (totopotensamide B): R = H

Fig. 47 Chemical structures of compounds **318–324**

Two new polyketide-peptide hybrids were also purified from *Streptomyces* sp. strain 1053U.I.1a.3b [134]. Totopotensamide A (**323**) is a glycosylated macrolactam, while totopotensamide B (**324**) (Fig. 47) is the aglycone analogue. Both compounds bear novel amino acid and polyketide moieties, 4-chloro-5,7-dihydroxy-6-methylphenylglycine, ClPhg, and 2*R*,3*R*-diaminobutyric acid, respectively. The bioactivity and potential functions of totopotensamides A and B are yet to be determined.

Bioactive peptide-based natural products were also represented secondary metabolites from mollusk-associated microorganisms. The modified peptides nobilamides A–H (**325–332**) (Fig. 48) were purified from *Streptomyces* sp. strains CN 48 and CT3a from the mollusk hosts *Chicoreus nobilis* and *Conus tribblei*, respectively [135]. Nobilamides A–C are linear heptapeptides bearing a conserved D-Leu,

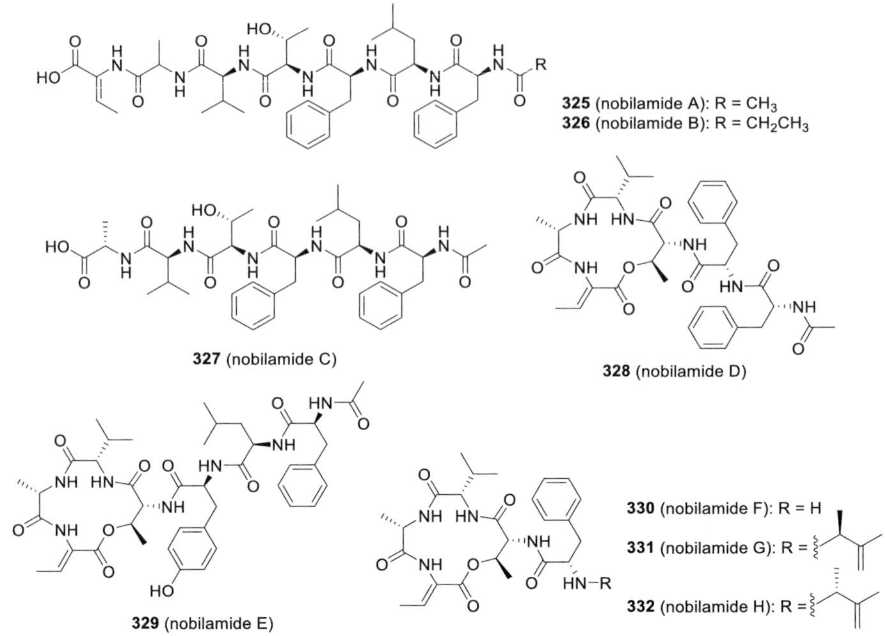

Fig. 48 Chemical structures of compounds **325–332**

D-Phe, and D-*allo*-Thr. Nobilamides A and B differ in the N-terminal modifying unit. Nobilamide C lacks the characteristic Z-dehydrobutyrine unit present in nobilamides A and B. Nobilamides D–H are cyclic depsipeptide analogues of nobilamides A–C. The carboxylic acid group in the C-terminal amino acid forms a depsipeptide linkage with the OH group of the Thr moiety. Nobilamides D–H differ in the pendant side chain, while the core macrocycle is identical in these analogues [135].

The nobilamides were identified as the bioactive components in the crude extract of *Streptomyces* sp. CN48. The neuroactivity of nobilamides was traced to their effect on capsaicin-sensitive cells and, consequently, direct inhibition of the capsaicin target receptor TRPV1. Overexpression of TRPV1 in HEK293 cells allowed for a competition binding assay. The nobilamides showed antagonistic activity, which was abolished in the absence of the *Z*-dehydrobutyrine moiety, indicating the importance of this moiety for bioactivity [135]. The natural combinatorial library showed other important moieties such as the D-amino acids Leu and Phe, for TRPV1 antagonism. The central role of the *Z*-dehydrobutyrine activity in the activity of TRPV1 further suggested a covalent interaction between this moiety and a nucleophilic group of an amino acid in TRPV1. Six TRPV1 mutants: C621A, C578A, N604A, E649A, E600V, and F660A were transfected in HEK-293 cells. Nobilamide B and the related compound A-3302 B inhibited the N604A, E649A, and E600V-bearing TRPV1 mutants. On the other hand, mutants bearing C621A, C578A, or F660A were unaffected by nobilamides. These suggest that the nobilamides are covalent inhibitors of TRPV1, binding at the pore-loop segment [135].

Chemical synthesis of nobilamide B via solid phase synthesis, with the *Z*-dehydrobutyrine moiety introduced via Staudinger ligation [136] or EDC/CuCl induced elimination [137]. An Ala scan was undertaken to further probe the impacts of individual amino acid changes on the bioactivity of nobilamide B [138]. Nobilamide B and six Ala-substituted analogues were chemically synthesized. L-Ala systematically replaced the *Z*-dehydrobutyrine, L-Val-3, D-*allo*-Thr-4, L-Phe-5, D-Leu-6, D-Phe-7. TRPV1 antagonism was abrogated in nobilamide analogues with substituted D-Phe-7, D-Leu-6, and D-*allo*-Thr-4. The replacement of the *Z*-dehydrobutryine moiety caused a decrease in activity, although was less pronounced. Other analogues have comparable neuroactivity to the synthetic nobilamide B.

A screening of 254 extracts of cone snail-associated microorganisms in the DRG assay afforded the extract of *Streptomyces* sp. strain CP32, isolated from the cone snail *Conus pulicarius* from Cebu, Philippines, as neuroactive [139]. This extract amplified Ca^{2+} release in response to KCl stimulation. Using a bioactivity-guided purification, aerugine-related compounds bearing a thiazoline moiety were purified. The new compounds named pulicatins A–E (**333–337**) (Fig. 49) and five known compounds were traced as the neuroactive components [139]. Pulicatin A is the 4′-methylated analogue of aerugine, while pulicatin B is a diastereomer of pulicatin A, differing in the absolute configuration at the 5″ position. Pulicatin C is the 4′,5′-dehydro derivative of pulicatin A, featuring a vinylic methyl group. Pulicatin D contains a C-7′ aldehyde group conjugated to the C4′-C5′ double bond. Pulicatin E is the C-7′ carboxamide analogue of pulicatin D. Pulicatins F (**338**) and G (**339**) and watasemycins A (**340**) and B (**341**) (Fig. 49) were also identified from this extract [139]. Although the pulicatins, aeruginaldehyde, and watasemycins are structurally-related, their effects on Ca^{2+} influx in DRG cells stimulated by KCl showed differences. Pulicatin A caused a block effect on Ca^{2+} influx despite only differing by a single methyl group from aeruginaldehyde. Aeruginaldehyde and pulicatin G, in contrast, caused an amplification in Ca^{2+} influx. Aeruginaldehyde, watasemycins A and B, pulicatins A, B, D, F, and G were tested against a panel of 46 targets in the PDSP screening platform. Pulicatin A showed the most potent activity in the human serotonin 5-HT_{2B} receptor, with $K_i = 505$ nM [139]. Other compounds showed higher K_i in 5-HT_{2B} receptor, compared to pulicatin A. Additionally, pulicatin G inhibited the K-opiod receptor, while pulicatin F also showed binding to the histamine H1 receptor. The pulicatins were selective for 5-HT_{2B} receptors over other 5-HT serotonin receptors [139].

The potent and selective activity of pulicatin A against serotonin 5-HT_{2B} receptor prompted its chemical synthesis and 30 other natural and synthetic analogues [140]. The design of the analogues focused on modifications to the thiazoline side chain or the aromatic ring to identify the critical and tunable elements for 5-HT_{2B} binding. The pulicatins showed antagonist activity in 5-HT_{2B}, with the modification of the thiazoline side chain to an imidazole causing improvement in the potency, but either decreasing or shifting the selectivity of activity [140]. Introducing changes to the aromatic ring showed detrimental effects on the activity and also changes to the

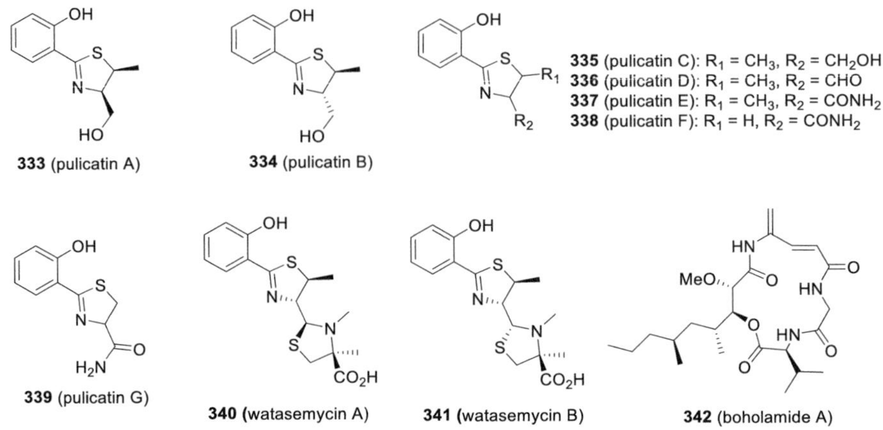

Fig. 49 Chemical structures of compounds **333–342**

selectivity profile. In vivo assay of pulicatin A showed modest protection in the 6 Hz seizure test (up to 100 mg/kg) without motor toxicity (rotarod), and significant effect in the hot plate test (100 mg/kg) but not the tail flick, suggesting potential antinociceptive activity [140].

A screen of Ca^{2+} modulation in glioblastoma (U87MG cells) yielded the crude extracts of *Nocardiopsis* sp. strain 3158H.R.1a.03 as a bioactive component [141]. This microorganism was obtained from the tissue of *Truncatella* sp. (PMS-3143 T) collected in Bohol, Philippines. Bioactivity-guided purification afforded a new member of the 4-amido-2,4-pentadieneoate (APD)-class peptide named boholamide A (**342**) (Fig. 49). Boholamide A also bears aglycine, valine, and a 3-hydroxy-4,6-dimethyl-2-methoxy-nonanoic acid (HDMN) moiety [141]. Boholamide A is structurally related to rakicidin A, vinylamycin A, and microtermolide. Using a tetrazolium-based cytotoxicity assay, boholamide A was detremined to be cytotoxic against a panel of cancer lines: U87MG (human glioblastoma, $IC_{50} = 410$ nM), CH157 (human meningioma, $IC_{50} = 98$ nM), HUH7 (human liver carcinoma, $IC_{50} = 190$ nM), A549 (human lung adenocarcinoma, $IC_{50} = 390$ nM) and to the normal CHO-AA8 ($IC_{50} = 390$ nM) [141]. Similar to rakicidin, boholamide A also caused cytotoxicity in U87MG cells under hypoxic conditions, with a 3.6-fold increase in potency compared to normoxic conditions, as evidenced by a decrease in IC_{50} to 120 nM [141]. Further validation using image-based analysis using DRG neurons indicated that boholamide A caused a direct Ca^{2+} amplification in DRG neurons, with thermosensory N15 neurons being largely affected. Vesiculation and eventual cell lysis were observed in DRG neurons and mouse cortex cells [141].

3.3 Shipworm-Associated Microorganisms

Aside from the high biodiversity of cone snails, the Philippines also boasts several unique species of shipworms. Shipworms are teredinid bivalves typically found in estuarine or marine environments. Shipworms are distributed in tropical regions, but are absent at higher latitudes [142]. Canonical shipworms feed primarily on woody material [143–145]. Shipworms are regarded as pests and also as important marine organisms. Because of their xylophagous diet, shipworms have wreaked havoc in wooden structures immersed in water. Nonetheless, shipworms play an important role in the marine carbon cycle, particularly in liberating recalcitrant carbon sequestered in lignocellulosic material [145]. Currently, there are 18 accepted genera of teredinids composed of around 70 species throughout the world's oceans [144–146]. Aside from the wood-boring shipworms (Fig. 50), the mud-dwelling *Kuphus polythalamius* [146, 147], and the rock-boring *Lithoredo abatanica* [145] have been reported from the Philippines.

Similar to other wood-consuming terrestrial organisms, shipworms rely on their microbial symbionts (Fig. 50) to digest wood. Cellulolytic activity for carbohydrate and lignin degradation is mainly carried out by the shipworm symbiont *Teredinibacter turnerae*. Five cellulolytic shipworm endosymbionts have been successfully cultured and isolated in the laboratory. Four additional *Teredinibacter* species have been isolated from the wood-boring shipworm *Bankia setacea*, which include *Teredinibacter waterburyi, Teredinibacter haidensis, Teredinibacter purpureus*, and *Teredinibacter franksiae* [148, 149]. Among these bacterial species, only *T. haidensis* was confirmed to be capable of fixing atmospheric dinitrogen and respiring nitrate under anaerobic conditions [149]. These findings suggest that diazotrophy is not a common feature among all *Teredinibacter* species.

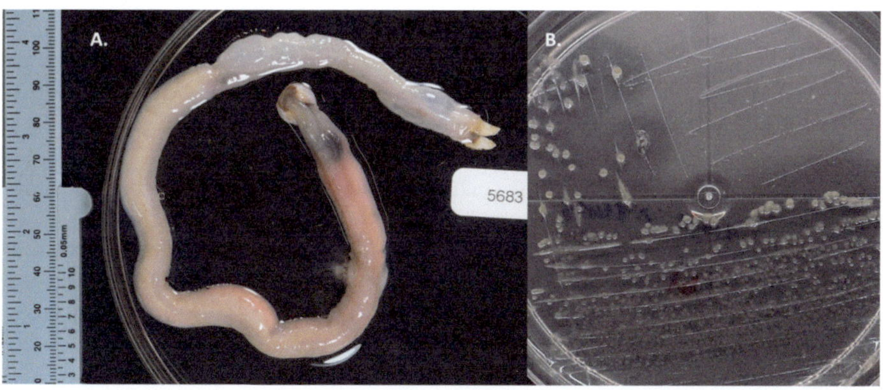

Fig. 50 Teredinids and their associated microorganism. **a** Shipworm from Pangasinan, Philippines after extraction from wood. **b** Associated microorganisms with shipworms grown as axenic cultures. Clearing zone around the colonies in an agar plate indicates cellulolytic activity

A recent comprehensive analysis of gill metagenomes from varying shipworm species and their cultivated symbiont genomes indicated ~ 200 uncharacterized biosynthethic gene clusters (BGCs) believed to express important families of secondary metabolites [150]. Of these, only four BGCs are identified to be ubiquitous across all wood-eating shipworms, regardless of species and sample location. Meanwhile, seven BGCs are common in all *T. turnerae* genomes and gill metagenomes of shipworms with *T. turnerae* as their major symbiont. These findings highlight that *T. turnerae* is one of the richest sources of biosynthetic gene clusters among shipworm symbionts with sequenced genomes [150].

Teredinibacter turnerae T7901 is a cellulolytic and diazotrophic endosymbiont found in the bacteriocytes of the wood-boring shipworm *Bankia gouldi* [151]. Similar strains of *T. turnerae* have also been harvested from the gill tissues of 23 species of teredinid bivalves, which include strains T7902, T7903, T8402, T8602, and T8201 isolated from the wood-boring shipworms *Lyrodus pedicellatus, Teredo navalis, Teredora malleolus, Dicyathifer manni*, and *Psiloteredo healdi*, respectively [152, 153]. Sequence analysis of protein-coding genes rpoB, gyrB, recA, sseA, and celAB in 25 *T. turnerae* isolates from 13 host species across a wide geographical range alludes to the emergence of two distinct lineages in this ubiquitous shipworm symbiont species [152]. Genetic variance for these protein-coding loci was observed among *T. turnerae* strains from the same host species but collected at different sampling locations [152]. Despite belonging to the same bacterial species, different strains of *T. turnerae* have varying biosynthetic gene clusters [150].

Genomic studies of *T. turnerae* T7901 revealed that the symbiont is a prolific source of enzymes that confer nutritional advantage to the host organism. It has acquired an unusually high number of multifunctional carbohydrate-active enzymes and protein domains involved in complex polysaccharide degradation, which include glycoside hydrolases, carbohydrate esterases, pectin lyases, and carbohydrate-binding molecules [151]. Several secretion systems that play a role in translocating gene products across the cell membrane were detected. These secretion systems may assist in releasing carbohydrate-degrading enzymes to the extracellular milieu to interact with the target substrate (e.g. cellulose, hemicellulose, pectin). Most of the carbohydrate-active enzymes synthesized by *T. turnerae* T7901 are substrates for the Type 2 secretion system, most of which are predicted to have an N-terminal signal peptide for secretion [151]. A small fraction of these genes is also predicted to express C–N hydrolases. Likewise, it also possesses a complete set of nitrogen fixation genes and about 40 genes predicted to express enzymes for nitrogen assimilation (most of which are dedicated to urea metabolism and transport) to compensate for its nitrogen-deficient diet of wood [151]. Genes predicted to encode quorum-sensing receptors, such as GacS and GacA orthologs, were also detected.

A mass spectrometric-based metabolomics analysis of four different strains of *T. turnerae* was done to assess the chemical diversity of small molecule compounds from this microorganism and compare the biosynthetic potential of the different strains [154]. Using GNPS molecular networking, 93 clusters were made from the detected metabolites. Only nine clusters were identified, belonging to fatty acid amides, diketopiperazines, rhamnolipids, phosphatidylethanolamines, tartrolons, turnercyclamycins, teredinibactins, lyso-ornithine lipids (LOLs), and macrolides [154]. As with the metagenomics analysis, the metabolomics result suggested yet to be identified chemistry from the extract of *T. turnerae*. Further, differences in metabolomic potential of the four *T. turnerae* strains were observed. Several compounds, such as the turnercyclamycins, were produced by *T. turnerae* T7901 and *T. turnerae* 1022X.S.1B.7A [154]. Oxylipins were more abundant in *T. turnerae* 991H.S.0A.06B. Putative analogues of the known compounds tartrolon D and teredinibactin were also observed [154].

The first compounds purified from *T. turnerae* T7901 are two macrodiolide polyketides belonging to the tartrolon class, tartrolon D (**343**) and its boronated derivative tartrolon E (**344**) (Fig. 51) [155]. Tartrolons are known pharmacological macrodiolides created by dimers or pseudodimers consisting of two polyketide chains linked as diesters. Gene disruption studies confirmed that the proposed GCF_11 (region 2 in strain T7901 genome) is the tartrolon biosynthetic gene cluster, *trt*, which includes ten core biosynthetic genes (*trtA-trtJ*) [150, 151, 155]. It is proposed that *trtDEF* are three multimodular PKS open reading frames (ORFs) that encode *trans* acyltransferase (AT) type I PKSs. The two putative oxygenases *trtG* and *trtI* located downstream of *trtDEF* may play a role in the decoration of an acidic hydroxy substituent. Meanwhile, *trtAB* are hypothetical acyltransferases and may act as proofreaders. The putative polyketide cyclase *trtJ* may be involved in the cyclization of tartrolons. A standalone type II thioesterase, *trtH*, was also identified and may cause regeneration of misprimed thiolation domains.

Tartrolon E suppressed the growth of the marine pathogen *Vibrio anguilarum* and inhibited *P. aeruginosa, S. aureus,* and methicillin resistant *S. aureus,* with MIC values of 0.31 μM, 0.08 μM, and 1.25 μM, respectively [155]. Cytotoxicity was also observed, with an IC_{50} of 2 μM against the MCF-7 breast cancer cell line [155]. Most remarkably, tartrolon E has shown nanomolar to picomolar broad anti-apicomplexan

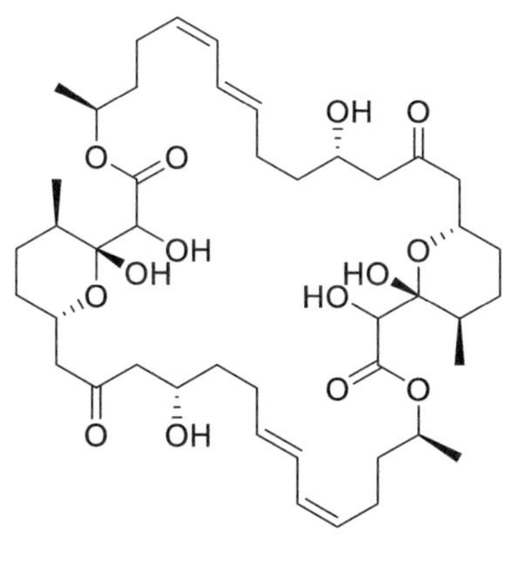

343 (tartrolon D)

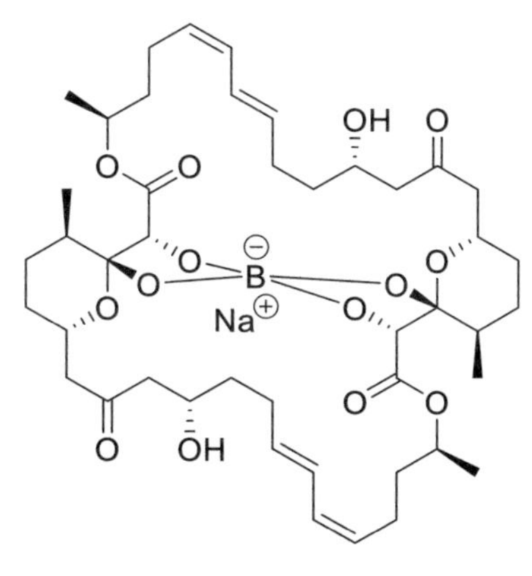

344 (tartrolon E)

Fig. 51 Chemical structures of compounds **343** and **344**

activity against *Babesia*, *Theileria*, *Plasmodium*, *Cryptosporidium*, and *Sarcocystis* [156]. Interestingly, tartrolon E could suppress the growth of the shipworm symbiont strain Bs02, which was later identified as *Teredinibacter waterburyi* [148]. These findings support the hypothesis that bioactive metabolites produced by shipworm symbionts may play a role in shaping the host microbiota by contributing to bacterial suppression in the cecum. It is also suggested that tartrolons may be involved in detoxification by pumping out excess boron due to the high concentration of boron in the marine environment [155]. Additionally, its pan anti-apicomplexan activity indicates that tartrolon E is likely to be protective against gregarines, a ubiquitous apicomplexan parasite in marine and terrestrial invertebrates [156].

A linear trimeric ester of 2,3-dihydroxybenzoate (DHB)-L-Orn-L-Ser triscatecholate siderophore named turnerbactin (**345**) (Fig. 52), along with its monomer (**346**), dimer (**347**), dehydrated dimer (**348**), and dehydrated trimer (**349**) (Fig. 52), were isolated from *T. turnerae* T7901 [157] via a chrome azurol sulfonate-guided activity. Turnerbactin acts as a siderophore, acquiring iron from the environment. Genome analysis of strain *T. turnerae* T7901 identified region 3 (GCF_8) as a putative NRPS gene cluster for siderophore biosynthesis, *tnb*, which includes six core biosynthetic genes (*tnbA-tnbF*) [150, 151, 157]. Interestingly, genes *tnbCEBA* share a high degree of homology to the enterobactin biosynthetic genes *entCEBA*. Turnerbactin and enterobactin are highly similar, with the latter assuming a cyclic trimer of DHB-L-serine. Biosynthesis was attributed to the incorporation of 2,3-dihydroxybenzoate to an NRPS pathway. It is suggested that TnbC isomerizes chorismic acid into isochorismic acid, which is subsequently hydrolyzed into 2,3-dihydro-DHB by TnbB prior to oxidization by TnbA to form DHB. Afterwards, TnbE activates and transfers DHB to the aryl carrier C-terminal portion of EntB before loading the substrate into the NRPS TnbF. This NRPS comprises two modules where L-ornithine and L-serine are loaded and sequentially attached into DHB. In addition to these biosynthetic genes, two putative Fur box sequences have been identified, which play a role in regulating siderophore production in response to iron availability, as confirmed by the increased Fe^{3+} binding activity of strain *T. turnerae* T7901 under iron-limited conditions [157].

Teredinibactins are another group of metal binding compounds purified from *T. turnerae* T7901 [158]. A phenolate-thiazoline moiety characterizes these compounds. Carbon labeling analysis indicated that the thiazoline moiety is derived from glutathione or its metabolites via a noncanonical pathway [159]. Teredinibactin A (**350**) (Fig. 52) binds copper in a 1:1 molar ratio [158]. Siderophores are important in acquiring iron, which is essential for multiple metabolic processes, including oxygen transport, DNA synthesis, and electron transport. On the other hand, it can also deter bacterial growth by depleting other microorganisms of this important cofactor. Combined with the antiproliferative properties of tartrolons, both molecules may inhibit the growth of microorganisms that could compete with carbohydrate uptake from digested wood in the shipworm cecum.

Recently, a new family of cyclic lipodepsipeptide, turnercyclamycins, was isolated from *T. turnerae* T7901 [160]. Turnercyclamycins A–D (**351**–**354**) (Fig. 53) have a common peptide macrocycle and differ in the lipophilic fatty acid conjugated to the *N*-terminus of the peptide chain. For the major compounds, turnercyclamycins A and

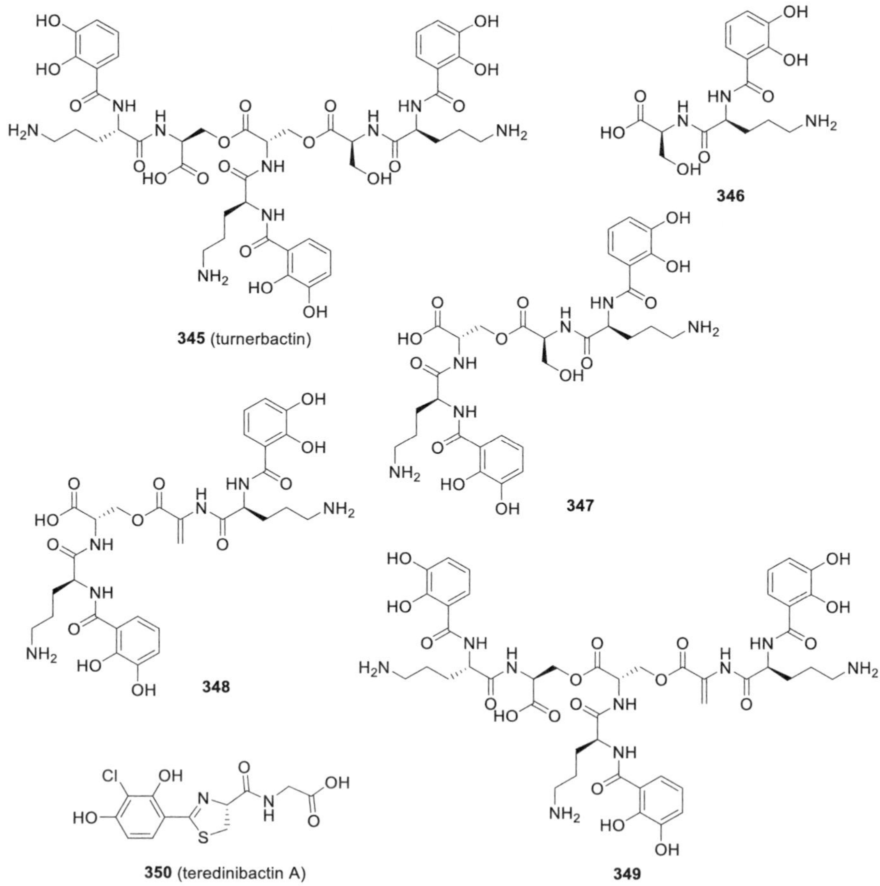

Fig. 52 Chemical structures of compounds **345–350**

B, the fatty acid substituents are myristic acid (C14:0) and *cis*-palmitoleic acid (C16:1 Δ^7), respectively [160]. Genome analysis provided further information on the gene cluster responsible for the biosynthesis of turnercyclamycins [160]. The proposed GCF_1 (regions 3 and 5 in strain *T. turnerae* T7901 genome) is the turnercyclamycin biosynthetic gene cluster, *tur*, which contains 12core biosynthetic genes (*turA-turL*) [150, 151, 160]. It is proposed that genes *turABCDE* include NRPS megasynthase modules, each containing condensation, adenylation, and thiolation domains with predicted substrate identities that are in close agreement with the natural amino acids found in the sequence of the peptide moiety. The prediction for the modules incorporating β-OH-Ile and β-OH-Asp corresponded to isoleucine and aspartic acid, respectively [160]. Meanwhile, modules that are predicted to incorporate 2,4-diamino-3-hydroxybutanoic acid (DAHB) and homoserine have not been assigned. It is also

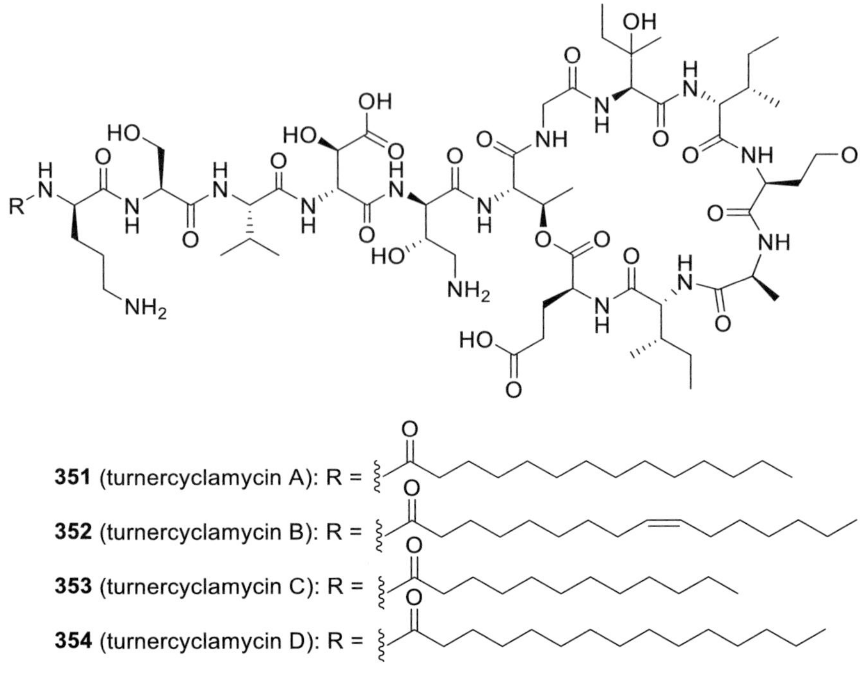

Fig. 53 Chemical structures of compounds **351– 354**

likely that the *turfFGH* gene products are involved in β-hydroxylation found to occur in DAHB, β-OH-Ile, and β-OH-Asp [160].

These lipopeptides belong to a family of antibiotics that are known to inhibit several types of Gram (+) and Gram (−) bacteria. Both turnercyclamycins A and B displayed identical and significant activity against *E. coli* (MIC = 1 μg/mL), *K. pneumoniae* (MIC = 2 μg/mL), and *Acinetobacter baumanii* (MIC = 8 μg/mL), while noncytotoxic against HEK-293 cells and murine erythrocytes [160].

Chemically-modified lipids known as oxylipins were isolated from *T. turnerae* strain 991H.S.0a.06, including (*E*)-9-oxohexadec-10-enoic acid (**355**), (*E*)-11-oxooctadec-12-enoic acid (**356**), and a novel C18 oxylipin with an epoxide group at C-12 and C-13, referred to as turneroic acid (**357**) (Fig. 54) [161]. Of these three isolated compounds, only turneroic acid inhibited biofilm formation in *S. epidermidis*, with no antiproliferative activity against mammalian cells [161]. Interestingly, turneroic acid displayed reduced activity against planktonic *S. epidermidis*, as indicated by its four-fold reduction of its MIC against planktonic *S. epidermidis* [161]. These results suggest that turneroic acid is selective for biofilm over planktonic bacterial cells. The existence of various antiproliferative compounds may constitute a chemical repertoire as part of a complex and multi-pronged strategy of keeping the cecum nearly sterile, in which turneroic acid inhibits accretion of opportunistic microbes in biofilms, tartrolons and turnercyclamycins limit the growth of planktonic microorganisms, while turnerbactin deprives them of an important cofactor [155, 156, 160, 161].

An epoxide ring in C18 oxylipins improves selectivity against biofilm inhibition and attenuates cytotoxicity towards mammalian cells [161]. Further analysis on the structure-activity relationship and mechanism of action of modified lipids was undertaken using readily available synthetic analogues of turneroic acid. The presence of a trans-epoxide moiety is important for the biofilm inhibiting activity, but the position of this moiety in the aliphatic was tolerated [162]. The related *trans*-12,13-epoxyoctadecanoic acid was unable to disperse mature biofilms of *S. epidermidis*, but decreased the metabolic activity, extracellular polymeric substances, and adherent microcolonies of *S. epidermidis* [162]. *Trans*-12,13-epoxyoctadecanoic acid also differentially regulated the expression of quorum-sensing-relevant genes, *icaR* and *icaB* [162].

Two strains of *T. turnerae* (strains 4401H.S.1B.02-A and 4409 W.S.0A.02-C) from shipworms collected from Batangas, Philippines yielded the lyso-ornithine lipid *N*-3-hydroxyoleoyl ornithine (**358**) (Fig. 54) [163]. This compound showed membrane targeting activity, with preferential binding to mammalian cells. This leads to differential inhibition of cancer growth and normal cells, as well as Gram (+) microorganisms. Further, the production of *N*-3-hydroxyoleoyl ornithine was regulated by the phosphate concentration in the culture medium [163].

As of this writing, no compounds have yet been isolated from the chemoautotrophic microorganism *Thiosocius teredinicola* associated with *K. polythalamius.* Despite the relatively lower biosynthetic capacity of its thioautotrophic symbiont to express secondary metabolites, a class of pyoluteorin-based antimicrobial compounds has been successfully purified from an associated microorganism of the giant shipworm *K. polythalamius* [164]. *Pseudomonas aeruginosa* strain 1682U.R.0a.27 is not a significant component of the *K. polythalamius* metagenome and was obtained from only a single specimen [150, 164]. This microorganism is likely a casual associate rather than a consistent *K. polythalamius* gill microbial community member.

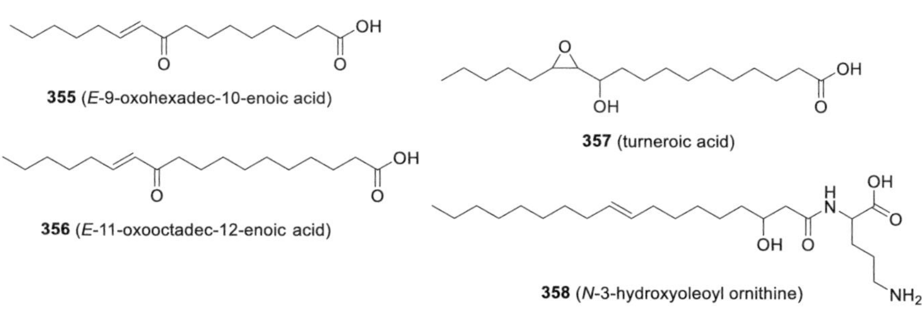

Fig. 54 Chemical structures of compounds **355–358**

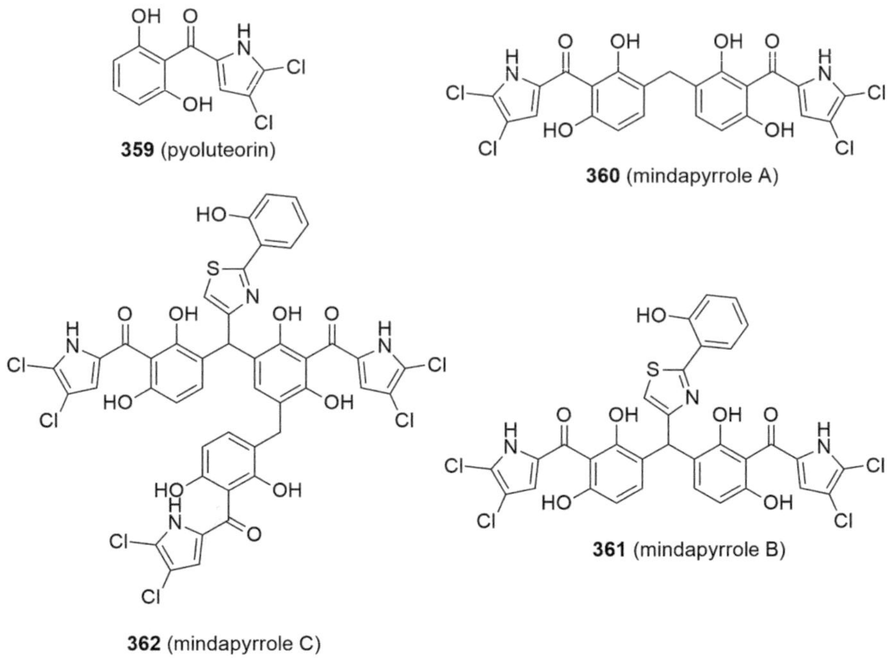

Fig. 55 Chemical structures of compounds **359** –**362**

The extract of *P. aeruginosa* strain 1682U.R.0a.27 was determined to contain unique natural products based on the Antibiotic Mode of Action Profiling (BioMAP) results. The extract showed activity against Gram (+) pathogens (*S. aureus, B. subtilis, S. epidermidis,* and methicillin resistant *S. aureus* (MRSA)) and forms a BioMAP cluster with monensin, a known ionophore [164]. Bioactivity-guided purification yielded pyoluteorin (**359**) and its analogues, mindapyrroles A–C (**360**–**362**) (Fig. 55). Mindapyrrole B displayed the highest antimicrobial activity and selectivity among these compounds over mammalian cells. It is suggested that the attachment of a hydroxylphenylthiazole moiety to the methylene bridge of the pyoluteorin dimer is postulated to increase the potency of mindapyrrole B against these pathogens and reduced mammalian cytotoxicity [164].

3.4 *Other Hosts-Associated Microorganisms*

A limited number of compounds have been described from microorganisms associated with other macroorganisms from the Philippines. An undescribed marine fungus, *Aspergillus* sp. (strain CNC-688) was isolated from the surface of the brown alga *Sargassum* sp. collected in Mactan Island, Cebu, Philippines [165]. A static culture fermentation in seawater-based YPG medium was done, and the

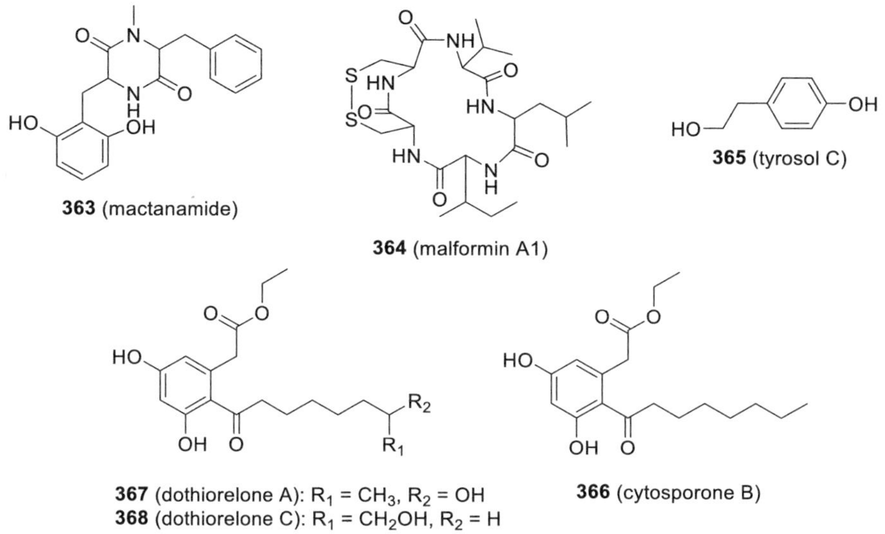

Fig. 56 Chemical structures of compounds **363–368**

mycelia were extracted with ethyl acetate. The lipophilic extract was purified by repeated column chromatography and a recrystallization step to afford a diketopiperazine named mactanamide (**363**) (Fig. 56). Mactanamide contains an *N*-methyl-D-phenylalanine and the uncommon amino acid D-2,6-dihydroxyphenylalanine. Mactanamide displayed fungistatic action against *Candida albicans* at 2.5 mg/mL in liquid culture, with no cytotoxicity observed at this concentration [165].

From the seagrass *Enhalus acoroides* collected in Negros Oriental, Philippines, the fungi *Aspergillus tubingensis* IFM 63452 was isolated [166]. *A. tubingensis* IFM 63452 showed adaptation to marine conditions and grew better with salt supplementation. *A. tubingensis* IFM 63452 was grown in a stationary condition using malt extract broth supplemented with marine salt for four weeks. The lipophilic extract from the broth was chromatographed to afford the cyclic pentapeptide malformin A1 (**364**) (Fig. 56). Malformin A1 was tested for trypanocidal, cytotoxic, and histone-modifying activities. Malformin A1 showed activity against *Trypanosoma congolense* IL 3000 (bloodstream form) with $IC_{50} = 15.08$ ng/mL (~ 28.4 nM) [166]. Cytotoxicity was observed against HeLa cervical cancer and P388 murine leukemia cells, with $IC_{50} = 50.15$ ng/mL (~ 94.6 nM) and $IC_{50} = 70.38$ ng/mL (~ 132.7 nM), respectively [166]. A 3.33–4.67-fold selectivity was observed against *T. congolense*. The mechanism of action of malformin A1 was investigated by monitoring the effects on epigenetic changes in histone proteins. Malformin A1 induced epigenetic changes at 10 and 100 ng/mL. Treatment of HeLa cells with malformin A1 caused a significant increase in levels of repressive marks H3K9me2 and H3K9me3, and activating mark H4K16ac [166]. At 10 ng/mL, there was a significant decrease in repressive

H3K27me3. At 100 ng/mL, activating marks were downregulated in H3K27ac and H4K5ac [166].

A potentially new species of the fungus *Phyllosticta* was isolated as an endophyte from a mangrove in the Philippines [167]. Chromatographic separation of the ethyl acetate extract of *Phyllosticta* yielded the compounds tyrosol C (**365**), cytosporone B (**366**), dothiorelone A (**367**), and dothiorelone C (**368**) (Fig. 56) [167]. These compounds were previously reported to have a variety of bioactivities, such as antibiotic, cytotoxic, and antioxidant [167].

3.5 Sediment-Derived Microorganisms

Marine sediment provides a unique and favorable habitat for marine microorganisms, which allows the microorganisms, in turn, to synthesize secondary metabolites with structural diversity and ecological and medical importance. Marine-derived sediments were collected via SCUBA diving at 15–30 m deep using a 110-cm sediment core sampler from 11 geographically distinct sites across west-central Philippines including Occidental Mindoro, Antique, Iloilo, Negros Occidental, Negros Oriental, Central and South Central Visayas, and Tubbataha Reefs, Palawan, Philippines [168]. A total of 92 strains out of the 2212 marine-derived actinomycetes isolated were confirmed to be antibiotic-producing *Streptomyces* species. PCR-based screening of biosynthetic genes detected the presence of NRPS genes in all 92 active strains. In contrast, detection of PKS type-I and -II genes across different species and sometimes within strains of the same species was variable [168]. Phylogenetic analysis using 16S rRNA and *rpoB* gene sequences identified 19 distinct *Streptomyces* species, with *S. parvulus* being the most dominant. Moreover, three species coded DSD742, DSD1006, and DSD3025 were described as potentially novel [168]. Results of the antibacterial screening of crude extracts against three Gram (-) bacteria (*E. coli* ATCC 25922, *P. aeruginosa* ATCC 27853 and *E. aerogenes* ATCC 13048) and a multidrug-resistant strain *S. aureus* ATCC BAA-44 and anticancer activity on ovarian A2780 (ECACC 93112519) cells demonstrated the significant variation in the strength and spectrum of the bioactivity among different strains of *Streptomyces* species, or simply, intraspecies heterogeneity [168].

Screening of marine sediments collected 200–500 m offshore near Alad and Lugbong Islands in Romblon, Philippines, at 20–30 m depth led to the identification of the bioactive *Streptomyces griseorubens* DSD069 [169]. This actinobacteria inhibited the growth of multidrug-resistant *Staphylococcus aureus* (MDRSA) ATCC BAA-44 by 96.4%, the highest antibacterial activity among the 35 isolates tested. Two anthracycline shunt metabolites were isolated, named bisanhydroaklavinone (**369**) and 1-hydroxybisanhydroaklavinone (**370**) (Fig. 57), and exhibited antibacterial activity at MIC = 6.25 μg/mL and 50 μg/mL, respectively. Cell-based assay results suggested the disruption of bacterial cell membrane integrity as the mechanism of antibacterial activity based on the observed leakage of intracellular

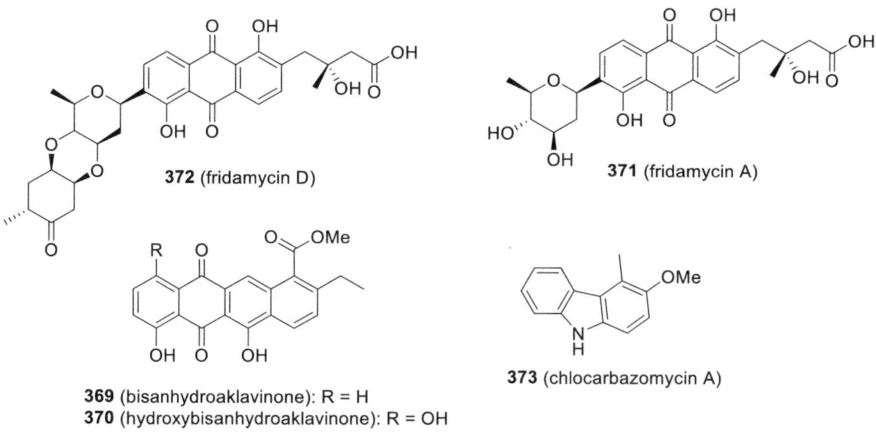

Fig. 57 Chemical structures of compounds **369–373**

contents such as DNA and proteins, cell shrinkage, and increased membrane permeability [169]. Moreover, an anthracycline biosynthesis pathway, which typically produces anticancer drugs including doxorubicin and daunorubicin, was proposed for the production of bisanhydroaklavinone and 1-hydroxybisanhydroaklavinone. Bisanhydroaklavinone was formed via PKS pathway through enzymatic methylation and cyclization of propionyl-CoA and malonyl-CoA precursors to form aklavinone, which subsequently undergoes spontaneous double dehydration. Then, hydroxylation of bisanhydroaklavinone at the C-1 position, possibly catalyzed by an enzyme similar to SnoaW/SnoaL2 or AclR, consequently produces 1-hydroxybisanhydroaklavinone [169]. The study highlighted marine sediments as the source for antibacterial drug leads.

Fifteen culturable actinomycetes identified as *Streptomyces* species based on 16S rRNA gene analysis were isolated from the marine sediments collected via SCUBA at a 20–30 m depth off the coast of the Visayan Sea, specifically in Islas de Gigantes, Carles, Iloilo, Philippines [170]. Four strains tolerated up to 12% NaCl (DSD040, DSD01, DSD042, and DSD043) and were identified to be closely related to *Streptomyces albogriseolus*. LCMS-based metabolic profiling revealed differences between the closely related *S. albogriseolus* strains, and known compounds microeunicellol A and albogrisin C were detected in some strains. A total of ten extracts showed anti-ESKAPE activity with MIC values 0.15 to > 20 mg/mL, and the same number of isolates displayed > 50% antiproliferative activity against ovarian A2780 cells. *S. albogriseolus* DSD041 and DSD042 displayed activity against Gram (-) pathogens but were inactive on *E. faecium, S. aureus,* and A2780 cells [170]. *S. albogriseolus* DSD04 demonstrated selectivity towards inhibition of A2780 cells. Furthermore, *S. albogriseolus* DSD043 was only active against *A. baumannii* and A2780 cell line. These observations showed significant variation in terms of bioactivity among the 15 isolates, particularly among the four closely related to *S. albogriseolus*. PCR amplification and detection of BGCs confirmed the presence of Type I PKS, Type II PKS, and NRPS gene domains in most isolates, including the four *S. albogriseolus* strains. Variations in the amplification patterns were also observed [170].

Screening of actinomycete isolates DSD011, DSD017, and DSD042 derived from marine sediments (Islas de Gigantes, Carles, Iloilo, Philippines) for their antiproliferative activity against the model organism *Saccharomyces cerevisiae* identified the crude extract of DSD011 as an inhibitor (86.1% inhibition at 2500 μg/mL), while the other two were inactive [171]. This isolate was pursued for further investigations to provide insights into the possible mode of action [171]. Treatment with DSD011 extracts induced cell death via necrosis or membrane damage, which was confirmed using flow cytometry, and budding yeast analysis suggested G1 cell cycle arrest in nearly 50% of the yeast population. These observations indicated that the bioactive principle(s) in the extract might interfere with proteins or signaling pathways relevant to G1/S cell cycle transition, which is similar to proximicin C, a known actinomycete-derived anticancer compound.

The isolate DSD011 afforded theangucycline glycosides fridamycins A (**371**) and D (**372**) (Fig. 57) with activity against MDRSA ATCC BAA-44 at MIC = 500 μg/mL and 62.5 μg/mL, respectively [172]. The structures and identity of these polycyclic aromatic polyketides were determined using Liquid Chromatography Mass Spectrometry-Time of Flight (LCMS-TOF), tandem mass spectrometry (MS/MS) fragmentation, and Global Natural Product Social molecular networking (GNPS) database. 16S rRNA gene sequencing revealed a new species of *Streptomyces*, closely related to two terrestrial *Streptomyces* strains, NBRC 13027 and NRRL B-2889, based on the multi-locus sequence analysis (MLSA) of the housekeeping genes *atp*D, *recA*, *rpoB*, and *trpB*. Targeted screening of PCR amplified genes indicated that the *Streptomyces* sp. strain DSD011 produces the compounds based on the presence of type II polyketide synthase (PKS II) and non-ribosomal peptide synthase (NRPS) in its genome [172]. The angucycline core originates from a decaketide backbone followed by cyclization reactions facilitated by PKS II. The mechanism of antibacterial activity of fridamycins A and D may be correlated to the anticancer activity of related angucyclins, which induce DNA cleavage.

More recently, the potentially new species of actinobacteria *Streptomyces tubbatahanensis* sp. nov. strain DSD3025^T was further investigated. The strain was isolated from the marine sediment collected at 26–50 cm depth subsurface layer in Tubbataha Reefs Natural Park, Cagayancillo, Palawan, Sulu Sea, Philippines using a core sampler [173]. The bacterium was isolated using a dry stamp method on trehalose-containing marine agar. It was identified based on polyphasic taxonomy, which includes whole-genome sequencing, 16S rRNA, MLSA (using *atpD*, *gyrB*, *rpoB*, *recA*, *trpB*), DNA-DNA hybridization (dDDH), and orthologous average nucleotide identity (orthoANI) as well as phenotypic and chemotaxonomic characterizations.

Streptomyces tubbatahanensis sp. nov. strain DSD3025^T was cultivated on MM1 agar, and the biomass was extracted using ethyl acetate [173]. Further purification using solid-phase extraction and mass-directed HPLC yielded chlorinated alkaloid chlocarbazomycin A (**373**) (Fig. 57). Metabolite profiling using UPLC-ESI-QTOF-MS also detected related compounds, including chlocarbazomycin B/C, chlocarbazomycin E, brocarbazomycin A, thiocarbazomycins A and B, carbazomycin B, and several potentially new chlorinated compounds that were not fully characterized [173]. *Streptomyces tubbatahanensis* sp. nov. strain DSD3025^{T} crude extract displayed bactericidal activity against multidrug-resistant *Staphylococcus aureus* (MDRSA) ATCC BAA-44 via membrane damage at 5 mg/mL, based on flow cytometry assay by calcein-PI dual staining assay. Dose-dependent antiproliferative activity against breast MCF-7, colon HCT116, ovarian A2780, kidney HK-2, and liver HepG2 cells were also observed, with HepG2 as the least susceptible upon crude extract treatment [173].

Chlocarbazomycin A exhibited moderate to potent antibacterial and antiproliferative activities against ESKAPE pathogens and various mammalian cell lines [173]. The alkaloid inhibited the growth of *Staphylococcus aureus* ATCC BAA-44 (MIC_{90} = 128 μg/mL) and *Streptococcus pyogenes* (MIC_{90} = 64 μg/mL) by 98% and 95%, respectively. Proliferation of HCT116 and A2780 cells was also inhibited by 55.4% and 50.2%, respectively, at 125 μg/mL test concentration. growth inhibition against colon HCT116 and ovarian A2780 cells, respectively. Varying degrees of cytotoxicity from no to high cytotoxic effects were also observed on liver (HepG2), kidney (HK-2), and cardiac (H9c2) cells after treatment with chlocarbazomycin A. The putative biosynthetic pathway for chlocarbazomycin A was achieved using comprehensive genome mining and bioinformatics. The pathway involved tryptophan, pyruvate, and acetate as precursors with a candidate biosynthetic gene cluster *chlCz* containing enzymes such as aminotransferase (chlCz1), ThDP-dependent enzyme (chlCz2), KASIII/ACAT (chlCz3), ACP (chlCz4), carbazole synthase (chlCz5), oxidoreductase (chlCz6a/b, chlCz7a/b), and *O*-methyltransferase (chlCz8) [173]. Chlorination at C-4 is proposed to occur after carbazole ring formation, catalyzed by a putative flavin-dependent halogenase (chlCz9) outside the main BGC cluster. A tryptophan 6-halogenase in BGC region 1 was deemed unlikely to be involved in the chlorination of chlocarbazomycin A at C-4.

4 Medium-Sized Peptide-Based Molecules from Marine Mollusks

The superfamily Conoidea (formerly known as Toxoglossa) is one of the most biodiverse marine organisms, consisting of over 10,000 species [174]. Among the members of this superfamily are cone snails (Conidae, ~ 700 species), auger snails (Terebridae, ~ 300 species), and turrid snails (Turridae, > 10,000 species) (Figs. 58, 59 and 60). These mollusks are known carnivores, preying on fish (piscivore), worms (vermivore), or other mollusks (molluscivore) [174, 175]. The feeding behavior of Conoideans influences their prey hunting strategies that generally employ an arsenal of peptide-based molecules. Species relying on these chemical weapons possess a venom duct, a specialized tubular esophageal gland, where toxin synthesis occurs [176, 177]. A venom bulb pumps out the toxins on the distal end of the venom gland.

Fig. 58 Different members of the Superfamily Conoidea. Left to right: Fish-hunting *Conus geographus*, auger snail *Terebra sulcata*, and turrids *Unedogemmula bisaya*, and *Crassispira cerithina*

The toxins are delivered through an extendable proboscis and a radula, which functions similarly to a barbed hypodermic needle [176, 177]. This specialized radula functions like a harpoon, enabling the snail to secure its prey and inject the venom [178]. A flexible membrane-like structure with chitinous plates is observed in some Conoideans without a radula. These structures function as a grazing organ like those found in non-predatory snails.

The most widely studied among these venom peptides are the conopeptides and conotoxins of cone snails. A single snail is estimated to produce 50–200 distinct conopeptides and represents the largest library of potential therapeutics [179]. The diversity of the conopeptides is due to the different ecological niches of Conoideans. The ConoServer database (ConoServer.org) describes the largest and most comprehensive repository for information on conopeptides, described either at the nucleic acid and protein levels [180]. To date, over 8500 protein sequences of conopeptides from 123 species are documented in the ConoServer database.

Conopeptides are produced via a ribosomal pathway and post-translationally modified. Precursor peptides have variable lengths that are cleaved at specific sites via proteolysis to yield the mature toxin [181]. The *N*-terminus contains a conserved signal sequence (~ 20 amino acids long) and a propeptide region that is thought to play a role in toxin maturation [181]. The C-terminus contains a hypervariable region, the mature toxin region, ranging in size from 12 to 30 amino acids. The most

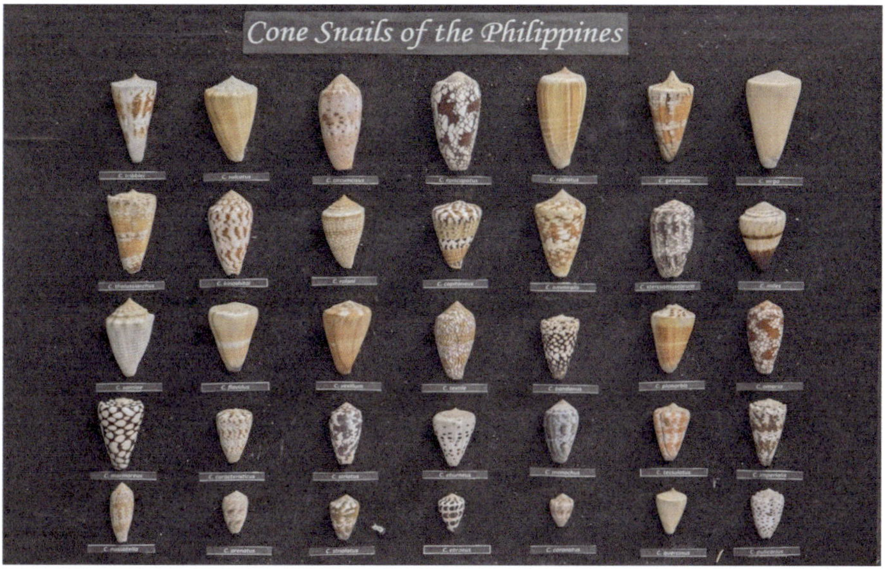

Fig. 59 Cone snails are well-loved by collectors because of their different shell patterns. Apart from this, venom components of cone snails are highly diverse, dictated in part by the unique ecological niches of members of the *Conus* species

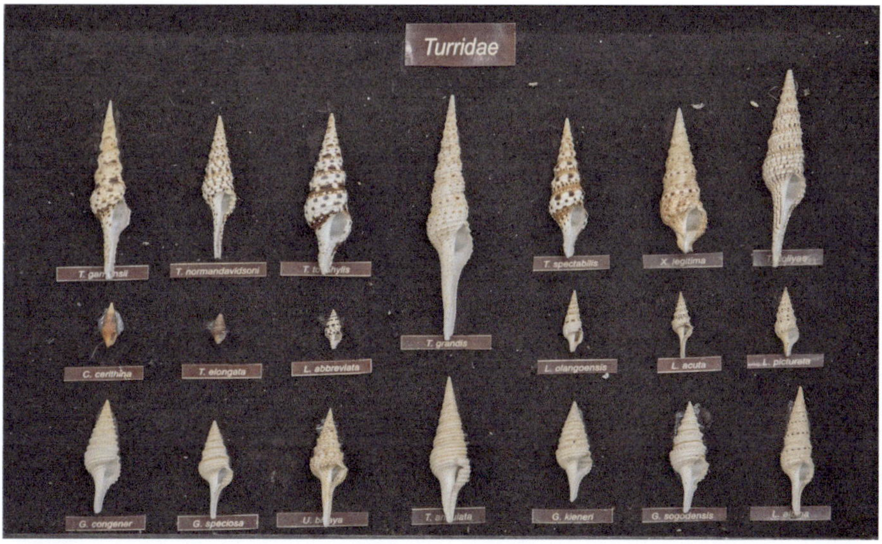

Fig. 60 Turrids from the Philippines. The small size of some species has been extra challenging in discovering turripeptides

common posttranslational modifications in the mature toxin include Cys oxidation, hydroxylation, γ-carboxylation, epimerization, and bromination [182]. Conopeptide gene superfamilies are classified based on the signal sequence similarity. The mature toxin sequence shows the highest sequence divergence among conopeptides.

The cone snail venom repertoire consists of conopeptides classified according to the number of Cys residues. Those with less than two Cys are referred to as conopeptides, while those possessing more Cys residues that form two or more disulfide bonds are called conotoxins. Conopeptides are classified as contulakins, conantokins, conorfamides, conolysins, conopressins, contryphans, conophans, conomarphins, conomaps, cono-CAPs, cono-NPYs and conoGAYs [183]. Among the heavily disulfide-bonded conotoxins, these are classified according to their cysteine framework, depending on the number of cysteine residues, loop size, and distinct disulfide bond connectivity [182]. There are 28 cysteine frameworks identified so far.

Conopeptides can also be grouped according to their pharmacological targets. Pharmacological families and nomenclature of conopeptides with known mechanisms of action are labeled by Greek letters [180]. As such, in the nomenclature of conopeptides, the Greek letter is indicated for those with known receptor targets and omitted if currently undetermined [182]. This is followed by one or two Arabic letters (i.e., G, M, Im, eb) to indicate the species source of the conopeptide. The Roman numeral indicates the Cys framework. As an example, for the conopeptide ω-MVIIA, which yielded the anti-pain drug Prialt, this conotoxin targets voltage-gated calcium channels (ω), was sourced from *Conus magus* (M), and possesses a Cys framework VII (C–C–CC–C–C). The last letter, A, denotes the conotoxin variant.

4.1 Cone Snails

Conus geographus is widely regarded as the most dangerous cone snail, responsible for most documented human fatalities attributed to *Conus* envenomations. Historical and clinical reports indicate fatality rates as high as 65–70% in untreated cases [184]. Envenomations typically occur accidentally when humans handle the snails, often affecting the palms or fingers. A human lethal dose of 0.038–0.029 mg/kg is estimated [185]. *Conus geographus* employs a unique predatory strategy known as "net-hunting," often targeting schools of small fish hiding in reef crevices at night [186]. The snail approaches with its false mouth (rostrum) highly distended, using it like a net to engulf multiple fish simultaneously before injecting venom into each captured individual. This strategy necessitates a particularly rapid and overwhelming venom effect. *Conus geographus* utilizes a specific combination of toxins, termed the "nirvana cabal," which likely involves inducing sensory disorientation and sedation, followed by neuromuscular paralysis [179]. Furthermore, the composition of *Conus* venom is not static; it can vary depending on the context, such as predation versus defense [185]. Studies on *C. geographus* defense-evoked venom revealed a consistent core of highly abundant paralytic conotoxins, including α-GI, α-GII, μ-GIIIA, ω-GVIA, and ω-GVIIA, which aligns with the severe neuromuscular symptoms observed in human envenomations [185].

Transcriptomic and proteomic analyses of the *C. geographus* venom duct revealed a remarkable functional compartmentalization, with distinct regional variations in gene expression and protein abundance along the duct's length [187]. The M, T, and P superfamilies of conopeptides decline distally. At the same time, other venom proteins like conoporins, conodipines, con-ikot-ikots, and astacin proteases exhibit increased concentration and specific expression clusters in different venom duct segments [187]. This spatial organization strongly suggests a sophisticated mechanism for tailoring venom composition, potentially allowing for the sequential deployment or differential mixing of toxins optimized for specific situations.

Adding another layer to its envenomation strategy, *C. geographus* employs metabolic disruption alongside neurotoxicity. The discovery of a specialized, fish-like insulin, Con-Ins G1, as a highly expressed component of the venom represents a paradigm shift [188]. This "weaponized" insulin induces rapid hypoglycemic shock in fish prey, complementing the paralytic effects of neurotoxins [189]. The subsequent identification of a venom somatostatin analogue, Consomatin G1, that selectively blocks glucagon release further underscores this multi-pronged attack on prey glucose homeostasis, likely accelerating incapacitation [190]. This metabolic warfare dimension highlights the diverse and sophisticated biochemical adaptations underlying the predatory success of *C. geographus*.

The A-superfamily is one of the most extensively studied groups of conotoxins, belonging to the α-conotoxin pharmacological family, which targets nicotinic acetylcholine receptors (nAChRs) [191]. α-Conotoxins are < 20 amino acids in length, with two disulfide linkages. Amino acid residues in the intercysteine loops determine the selectivity of the α-conotoxins. α-Conotoxins in the venom of *C. geographus* contribute to neuromuscular blockade and paralysis [187]. To date, there are five identified α-conotoxins from *C. geographus* that have been extensively analyzed for pharmacological activity (Table 1).

The 13-amino acid α-conotoxin GI (Table 1) is the first characterized toxin from the venom of a cone snail [192, 193]. This conotoxin was purified from *C. geographus* collected from Cebu and Marinduque, Philippines. This 13-amino acid peptide causes paralysis in vertebrates [193] and causes death in mice at a dose of 40 μg/kg [194]. α-GI (Table 1) features Cysteine Framework I and an α3/5 loop structure, with the native, most stable globular isomer having disulfide bonds Cys2-Cys7 and Cys3-Cys13 [195]. The three-dimensional structure of α-Conotoxin GI, determined by NMR and X-ray crystallography, reveals a compact fold stabilized by the disulfide bonds and intramolecular hydrogen bonds [195].

α-GI is a potent and selective competitive antagonist of the muscle-type nAChR (subtypes containing α1, β1, δ, and either γ [fetal] or ε [adult] subunits), with IC_{50} = 1.3–995 nM [194, 196–199]. α-GI acts postsynaptically at the neuromuscular junction by physically blocking the binding of acetylcholine to the nAChR [193, 200]. This inhibition prevents channel opening and subsequent muscle depolarization, leading to flaccid paralysis. α-GI reduced the amplitude of miniature endplate potentials (MEPPs) and currents (MEPCs) without significantly affecting quantal

Table 1 Selected conopeptides from different cone snails from the Philippines

Conopeptide	Sequence	Source	Reference
α-GI	ECCNPACGRHYSC-NH2	*C. geographus*	[193]
α-GIA	ECCNPACGRHYSCGK-NH2	*C. geographus*	[193]
α-GII	ECCHPACGKHFSC-NH2	*C. geographus*	[193]
α-GIC	GCCSHPACAGNNQHIC-NH2	*C. geographus*	[202]
α-GID	IRDγCCSNPACRVNNOHVC	*C. geographus*	[203]
μ-GIIIA	RDCCTOOKKCKDRQCKOQRCCA-NH2	*C. geographus*	[204]
μ-GIIIB	RDCCTOORKCKDRRCKOMKCCA-NH2	*C. geographus*	[204]
μ-GIIIC	RDCCTOOKKCKDRRCKOLKCCA-NH2	*C. geographus*	[204]
μO§-GVIIJ	G(Btr)CGDOGATCGKLRLYCCSGFCD(Scc)YTKTCKDKSSA	*C. geographus*	[205]
ω-GVIA	CKSOGSSCSOTSYNCCRSCNOYTKRCY(NH2)	*C. geographus*	[206]
ω-GVIB	CKSOGSSCSOTSYNCCRSCNOYTKRCYG	*C. geographus*	[207]
ω-GVIC	CKSOGSSCSOTSYNCCRSCNOYTKRC	*C. geographus*	[207]
Contulakin-G	ZSEEGGGSNATKKPYIL (Z = pyroglutamate)	*C. geographus*	[208]
ω-MVIIA	CKGKGAKCSRLMYDCCTGSCRSGKC(NH_2)	*C. magus*	[207, 209]
ω-MVIIB	CKGKGASCHRTSYDCCTGSCNRGKC(NH_2)	*C. magus*	[209]
ω-MVIIC	CKGKGAPCRKTMYDCCSGSCGRRGKC(NH_2)	*C. magus*	[210]
ω-MVIID	CQGRGASCRKTMYNCCSGSCNRGRC(NH_2)	*C. magus*	[211]
ImI	GCCSDPRCAWRC-NH_2	*C. imperialis*	[212]
Conomarphin Eb1	GWVYHAHPE**P***NSFWT	*C. eburneus*	[213]
Conomarphin Eb2	GWVYHANP**E***ANSWWT	*C. eburneus*	[213]

content, confirming its postsynaptic site of action [201]. Alanine-scanning mutagenesis identified the amino acid residues critical for activity, including Glu1, Pro5, Gly8, Arg9, and Tyr11, providing insights into the toxin-receptor interface [198].

α-GIA and α-GII were purified at the same time as α-GI (Table 1). α-GIA is a longer analogue of α-GI, with additional Gly and Lys-NH_2 at the N-terminal [193], while α-GII differs from α-GI by an Arg to Lys substitution at position 9 [193]. GIC (Table 1) was identified from the DNA of *C. geographus* hepatopancreas and was subsequently prepared by chemical synthesis [202]. The amino acid sequence at the N-terminus shows more divergence with GI [202]. Interestingly, a bioassay-guided purification of *C. geographus* from the Great Barrier Reef yielded the α-GID (Table 1), with 19 amino acid residues, the extension resulting from the additional four amino acids in the *N*-terminus [203]. α-GID also features additional posttranslational modifications such as ɣ-carboxylation in Glu and hydroxylation in Pro [203].

The structurally-related α-GI, α-GIA, and α-GII showed similar physiological effects [193]. α-GI and α-GII are also present in high concentration in the defense-evoked venom of *C. geographus* [185]. Structurally-divergent α-GIC and α-GID show different selectivity against acetylcholine receptors compared to α-GI. Unlike α-GIA, α-GIC did not cause paralysis in fish and mice [202]. α-GIC also did not inhibit the muscle nAChRs but showed potent activity in human α3β2 nAChR. The binding of α-GIC in nAChR can be reversed [202]. α-GID potently inhibited $\alpha7$ and $\alpha3\beta2$ nAChRs with $IC_{50} = 3.0–5.0$ nM [203]. Deleting the four amino acid residues on the N-terminus of GID tunes the potency and selectivity of α-GID binding to nAChRs. The Arg residue at position 12 also influences the activity of GID.

μ-Conotoxins, primarily found within the M-superfamily, are potent inhibitors of voltage-gated sodium channels (Na_V), with a particular preference for the muscle subtype ($Na_V1.4$) [214]. They are key components of the paralytic arsenal of piscivorous cone snails like *C. geographus* [215]. M-superfamily peptides typically exhibit Cysteine Framework III (CC–C–C–CC) and are rich in basic residues and often contain hydroxyproline (Hyp) PTMs [214].

The M-superfamily toxins (GIIIA, GIIIB, and GIIIC, Table 1) represent another set of pharmacological agents from *C. geographus.* These μ-conotoxins are potent and selective blockers of skeletal muscle sodium channel, $Na_V1.4$ [204]. μ-GIIIA is present in large amounts in the defensive venom of *C. geographus* [185]. Seven μ-conotoxins have been identified from *C. geographus*, with four variants being posttranslational modification intermediates, with the hydroxyproline being replaced by proline in these variants [204]. Radioligand binding assay using ^{125}I-labeled GIIIA indicated that μ-GIIIA binding to sodium channel was abolished with tetrodotoxin [204]. μ-GIIIA acts as a pore blocker, physically occluding the outer vestibule of the Na_V channel with a 1:1 stoichiometry [216]. The positively charged Arg13 residue is critical and thought to interact directly with the channel's ion selectivity filter within the pore [217]. Other basic residues (Lys11, Arg19) also contribute significantly to binding and toxicity [217].

μO§-GVIIJ (Table 1) is a unique conopeptide with its μ-conotoxin-like activity (Na_V block) combined with a reported O1/Framework VI/VII structure and a distinct covalent binding mechanism [205]. GVIIJ bears bromotryptophan, hydroxyproline, and S-cysteinylated cysteine. GVIIJ blocks a broad range of Na_V subtypes with high affinity: $Na_V1.1$, $Na_V1.2$, $Na_V1.3$, $Na_V1.4$, and $Na_V1.7$ ($K_d = 4.7$–41 nM), with lower affinity for $Na_V1.6$ ($K_d = 360$ nM) [205]. This conotoxin binds to a novel site on Na_V channels, designated as Site 8, with a covalent disulfide bond between a cysteine residue on μO§-GVIIJ and a specific cysteine residue on the sodium channel's extracellular surface [205]. This mechanism is distinct from the pore-blocking action of GIIIA/B/C.

The O-superfamily is another major group of conotoxins found in *C. geographus* [186]. This superfamily exhibits significant functional diversity. It includes the well-known ω-conotoxins, which primarily target voltage-gated calcium channels (Ca_V) [218], and other peptides like Contulakin-G (Table 1) that target G-protein coupled receptors (GPCRs) [208]. ω-conotoxins typically possess Cysteine Framework VI/VII (C–C–CC–C–C) and adopt an inhibitor cystine knot (ICK) or knottin fold [182]. Common posttranslational modifications (PTMs) include hydroxyproline and C-terminal amidation [206, 207].

The purification of the 27-amino acid peptide ω-GVIA (Table 1) was guided by the shaking phenotype observed in venom extracts of *C. geographus* [207]. ω-GVIA is a key component of the defensive venom [185]. The shaking phenotype from ω-GVIA was linked to the block in Ca^{2+} entry into the nerve terminal during the presynaptic action potential [219]. ω-GVIA is an extremely potent and highly selective blocker of N-type voltage-gated calcium channels [220]. It binds with picomolar affinity, with $IC_{50} = 3.7$–38 pM in displacement assays and $ED_{50} = 68$ pM in functional assays [221, 222]. It shows very low affinity for P/Q-type ($Ca_V2.1$) channels [222]. By blocking presynaptic $Ca_V2.2$ channels, GVIA inhibits neurotransmitter release, leading to paralysis [207]. GVIA exhibits potent analgesic effects in animal models, significantly exceeding morphine's potency [223]. Key residues Lys2 [224] and Tyr13 [225] are critical for high-affinity binding.

The 16-amino acid contulakin-G (CGX-1160) (Table 1) features two distinctive PTMs: N-terminal pyroglutamate and an O-linked disaccharide (β-D-Gal-(1 → 3)-α-D-GalNAc) attached to Thr10 [208]. The C-terminal sequence of contulakin-G shows similarity with endogenous neurotensin [208], suggesting the potential target of this conopeptide to be neurotensin receptors (NTSRs). Contulakin-G can bind to human NTSR1, rat NTSR1/NTSR2, and mouse NTSR3/SORT1 [208]. Compared to neurotensin, contulakin-G has lower affinity and agonist potency, with the most potent agonist activity observed in hNTS1. Contulakin-G has an $EC_{50} \sim 0.96$ μM for hNTS1 activation, compared to neurotensin with $EC_{50} = 1.4$ nM [208]. Contulakin-G, however, induces significantly less receptor desensitization and internalization [226]. The deglycosylated contulakin G showed a more potent binding affinity to hNTR1, with $EC_{50} = 23$ nM [208]. On the other hand, glycosylation in contulakin-G improves the metabolic stability of this peptide, increasing the half-life [226]. At the same time, reduced desensitization of cell surface receptors was also observed in contulakin-G, in part due to its glycosylation [226]. Contulakin-G exhibits potent analgesic activity in various preclinical pain models (acute, inflammatory, neuropathic). It received orphan drug designation and progressed to Phase II clinical trials for intractable pain under the code CGX-1160.

Conantokins represent a distinct class of *Conus* peptides, they are conopeptides, Cys-poor components of the venom peptide, lacking a disulfide linkage. Conantokins are also characterized by multiple residues of γ-carboxyglutamate (Gla) and often lack disulfide bonds, adopting helical structures in the presence of divalent cations. Conantokins also have a distinct pharmacological activity, as antagonists of *N*-methyl-D-aspartate (NMDA) subtype of glutamate receptors. Conantokin-G (Con-G, Sleeper peptide, CGX-1007, Conotoxin GV) is a 17-amino acid linear peptide [227, 228], which in the presence of divalent cations like Ca^{2+} or Mg^{2+} can adopt a stable α-helical conformation [229].

Conantokin was initially named the "sleeper peptide" because it induced a sleep-like state when injected into young mice [228]. In contrast, adult mice injected intracranially with conantokin-G became hyperactive [230]. Conantokin-G is a selective NMDA antagonist. It abrogates cGMP elevation in response to NMDA but not kainic acid [230]. In various preclinical studies, it demonstrated potent anticonvulsant potential, analgesic activity in neuropathic pain models, and neuroprotective effects. Two allelic variants of Conantokin-G are among the highly expressed transcripts in *C. geographus*, indicating the importance of conantokin in the envenomation strategy of this snail [186]. The conantokins are part of the 'nirvana cabal' in the cone snail arsenal, which causes sensory disorientation of its prey. The presence of conantokins in *C. geographus* is validated by the high expression in the transcriptome [186].

Another piscivorous cone snail is *Conus magus*, commonly known as the magician cone snail, produces pharmacologically-related conotoxins as *C. geographus*. A total of 278 conopeptides have been identified from nucleic acid and peptide sequences of *C. magus* (ConoServer.org, retrieved March 2025). Among these, 11 have been purified from the venom peptide. Just like *C. geographus*, *C. magus* venom also contains the shaker peptides, including ω-MVIIA, ω-MVIIB, ω-MVIIC, and ω-MVIID (Table 1) [207, 209]. Unlike the ω-conotoxins of *C. geographus*, those from *C. magus* do not have hydroxyproline units; instead, these are replaced with positively charged amino acids [209]. While the ω-conotoxin of *C. geographus* and *C. magus* have the same Cys framework, the size of the intercysteine loops between the conserved Cys residues is different. Like ω-GVIIA, ω-MVIIA also targeted N-type voltage-gated calcium channels, albeit with narrower selectivity [209, 231]. The direct effect of ω-MVIIA and the less toxic effects on mammalian cells allowed this conotoxin to be developed as a non-opioid analgesic [232]. ω-MVIIA showed anti-nociceptive effects in animal models of morphine resistance [232]. Further, ω-MVIIA did not cause tolerance with prolonged use. These advantages of ω-MVIIA allowed for its development as an alternative treatment to pain in patients with morphine tolerance. Prialt (Ziconotide) was approved as an alternative to morphine treatment in 2004 (https://www.accessdata.fda.gov/drugsatfda_docs/nda/2004/21-060_Prialt.cfm).

The imperial cone (*Conus imperialis*) is a vermivore, with worms constituting its main prey. Little is known about the lifestyle and hunting strategy of vermivorous cone snails compared to their piscivorous counterparts. A collection of *C. imperialis* from Cape Bolinao, Philippines yielded the α-conotoxin ImI (Table 1). While α-conotoxins are predominantly found in piscivorous species, the work by McIntosh et al. [212] demonstrated the presence of divergent α-conotoxins from a vermivorous species. ImI was purified by a combination of size exclusion and reversed phase HPLC, with a molecular weight cutoff of 3000 Da, based on the criteria for α-conotoxins [212]. Intracranial injection of 3–5 nmol of ImI caused tremors and seizures in mice [212]. Bioactivity analysis of ImI in an electrophysiological assay indicated that Im1 is 100-fold more potent in an amphibian system than in fish, with IC_{50} of 0.25–0.5 μM and 25–50 μM, respectively.

Individual variations in the venom composition of *C. imperialis* were also observed, based on transcriptomics and proteomics analysis using 454 pyrosequencing and mass spectrometry, respectively [233]. Abundant transcripts in *C. imperialis* were common in individual specimens, while low level transcripts showed more variability [233]. This suggests that highly expressed conopeptides are essential for prey capture, while minor venom components may be diversified. At the same time, the translation of the highly expressed conotoxin gene superfamily correlated well with the transcript levels. Production of novel conopeptides was observed in individual *C. imperialis* and highlights the importance of analyzing single specimens, instead of pooled samples, to access new chemical diversity.

Other conopeptides characterized from *C. imperialis* include im23a and im23b, and Im10A [234]. *N*-terminal Edman sequencing and subsequent cDNA cloning using 3′-and 5′-RACE methods established the full amino acid sequences of the mature toxins [234]. Im10A formed a type I β-turn facilitated by the stable disulfide connectivity [235]. Using a partial sciatic nerve ligation, Im10A was demonstrated to have a partial analgesic activity. However, Im10A did not show activity in neuronal nicotinic acetylcholine receptor, DRG voltage-dependent sodium, potassium and calcium ion channels and opioid receptor [235]. The disulfide bond connectivity of im23a was meticulously determined using chemical mapping involving partial reduction and differential alkylation, combined with NMR structural analysis. Both approaches converged on a I-II, III-IV, V-VI pattern, linking Cys4–Cys11, Cys15–Cys25, and Cys26–Cys41 [234]. This connectivity was further validated through the total chemical synthesis of im23a using orthogonal cysteine protection strategies. Solution structure determination of recombinant ^{15}N-labeled im23a using multidimensional NMR spectroscopy (PDB ID: 2LMZ) revealed a novel helical hairpin fold. This structure consists of two α-helical segments (approximately residues 8–17 and 22–31) connected by a turn or loop region.

Analysis of the precursor cDNA sequences of im23a and im23b revealed a highly conserved *N*-terminal signal peptide region distinct from those defining previously established conotoxin gene superfamilies. This unique signal sequence led to the definition of a new K-superfamily of conotoxins [234]. Furthermore, the arrangement of cysteine residues within the mature peptide sequences (C–C–C–CC–C) defined a novel cysteine framework, designated as Framework XXIII. The biological effects of native and recombinant im23a and im23b were assessed through in vivo assays in mice [234]. Intracranial injection into young mice (P14, postnatal day 14) elicited distinct excitatory symptoms affecting the central nervous system (CNS). A dose of 20 μg/mouse caused squinting, pronounced hypersensitivity to touch (tactile allodynia), and distress vocalizations upon touching, with the effects persisting for approximately 5 min. A higher dose of 40 μg/mouse resulted in more severe neurological symptoms, including uncontrolled motor activity (circular or random running), seizure-like activity, or, conversely, periods of lethargy. Interestingly, older mice (postnatal day 42) exhibited significantly milder symptoms when administered at the same dose. In contrast to the potent CNS effects, intraperitoneal (IP) or intravenous (IV) administration of 40 μg of im23a into P14 mice failed to produce any observable symptoms.

Conus eburneus is a less-studied vermivorous snail, with < 40 conopeptides referenced in ConoServer (ConoServer.org). Transcriptomic analysis of the venom duct of *C. eburneus* mapped 149 unique conopeptide transcripts to 28 different gene superfamilies [213]. The highest expression was observed in the M-, T-, O1-, and O2 superfamilies. The transcriptome expression was corroborated by the global proteome analysis, with the M and T superfamilies being the most highly expressed [236]. The conomarphins, cysteine-free peptides, typically 15 amino acids long, belonging to the M-superfamily are characterized by conserved motifs (A/T)(H/N)P and NS(W/F)W(T/S) [213]. Multiple variants of conomarphins were observed in *C. eburneus*, differing in amino acid sequence (positions 7–10) and post-translational modifications (PTMs). Identified variants include Eb1 (Table 1), Eb2 (Table 1), and Eb3 (homologous to Bt3), with PTMs like proline hydroxylation (Hyp/O), γ-carboxylation of glutamic acid (Gla/γE), and D-amino acid epimerization (D-Phe) in Eb1.

Because of the abundance of conomarphins in the venom of *C. eburneus*, biological activity analysis was carried out in mice, fish, and freshwater snails. Native conomarphins Eb1/Eb2 caused paralysis-like symptoms (sluggishness, retraction of siphon, foot, tentacles) in the freshwater mollusk *Pomacea padulosa* [213]. There was no observed activity in fish (*Carassius auratus*) or mice (intracranial injection) even at a high concentration of 10 nmol. Structure–activity studies in *P. padulosa* using synthetic analogues showed that D-Phe is critical for mollusk activity, while modifications at Hyp (removal of hydroxylation or substitution with Ala) are tolerated.

The α-conopeptide Eu1.6, a 16-amino acid peptide with the sequence GCCSNPACMLKNPNLC-NH_2, was also purified from *C. eburneus* [237]. Eu1.6 showed potent analgesic activity in rat neuropathic pain models. Eu1.6 caused a 62% increase in pain threshold after 2 h of administration of 24.9 μg/kg (15 nmol/kg). Eu1.6 directly inhibits high voltage-activated (HVA) N-type calcium channels ($Ca_v2.2$) in a concentration-dependent, reversible, voltage-dependent, and state-dependent manner [237]. This inhibition is independent of $GABA_B$ receptor activation. It slows $Ca_v2.2$ activation and deactivation kinetics and shifts steady-state inactivation to more hyperpolarized potentials.

4.2 Turrids

A seminal paper by Watkins et al. [238] indicated that turrids, like cone snails, also possess diverse peptide-based toxins. cDNA cloning of *Lophiotoma olangoensis* identified 16 different toxin-encoding gene families, with highly divergent sequences from conopeptides. After almost 20 years, the venoms of turrids remain understudied compared to conotoxins, partly due to challenges in accessing specimens, which often inhabit deeper waters, and their typically small size, which limits the amount of venom obtainable for traditional biochemical analysis [239]. Unique local collection methods, such as tangle nets used in deep water and the specialized *lumun-lumun* technique, using fine mesh nets deployed on the seabed to collect micromolluscs [240], have been instrumental in providing access to these often-elusive species for toxinology research. The application of molecular techniques, particularly transcriptome sequencing and cDNA cloning, has been crucial in overcoming the limitation of small venom quantities, allowing for the discovery of numerous putative turritoxin gene families and providing a foundation for subsequent characterization.

To date, only five genera of turrids from the Philippines have undergone an extensive chemical and pharmacological investigation (Fig. 60). Gonzales and Saloma [241] analyzed the transcriptome of three turrid species: *Unedogemmula bisaya*, *Crassispira cerithina*, and *Gemmula speciosa. Unedogemmula bisaya, and Gemmula* sp. were collected from Cavite, Batangas, and Manila Bay, Philippines, while *C. cerithina* was from Olango Island, Cebu, Philippines. Total mRNA was extracted from the venom ducts of these three species and subjected to sequencing using HiSeq 2000. The sequences were analyzed against the ConoServer toxin database and a custom NCBI database to identify toxin-related sequences. A total of 74 putative turrid toxin genes were identified from *G. speciosa*, including sequences with precursor structures similar to conotoxin superfamilies D, I2, L, M, O1, O2, and P. A non-disulfide-rich putative toxin (B2 signal sequence) and novel cysteine frameworks, including one with 14 Cys residues were also identified [241]. Combination of cDNA cloning and high performance liquid chromatography afforded the purification of two turripeptides from *G. speciosa*, gsp9a and gsp9b (Table 2) [239]. These peptides belong to the Pg-gene superfamily, characterized by a conserved signal sequence and a Cys framework IX (C–C–C–C–C–C) in the mature region. Gsp9a and gsp9b are extensively posttranslationally modified, with all Pro and Glu residues bearing hydroxylation at position 4 and γ-cayboxylation, respectively.

Transcriptome analysis of *Unedogemmula bisaya* identified 41 putative turrid toxin genes. Similar precursor types to *G. speciosa* were found in *U. bisaya*, including conotoxin-like superfamilies D, I2, L, M, O1, O2, P; non-disulfide rich B2 signal; novel frameworks including 14-Cys [241]. Bioactivity-guided purification of the venom duct extract from *Unedogemmula bisaya* afforded a 17-amino acid turripeptide named Ubi3a (Table 2) [242]. This peptide belongs to the M-superfamily of conopeptides and has a Cys framework III pattern (CC–C–C–CC). However, solution-based NMR analysis highlighted that the Cys connectivity in Ubi3a differs from other conopeptides in Framework III in the length of the intercysteine regions [242]. Intracranial mouse bioassay indicated that Ubi3a caused prolonged tremors, starting at 1.4 nmol. The peptide inhibited the activity of the $\alpha9\alpha10$ nicotinic acetylcholine receptor with $IC_{50} = 10.2$ μM. Initial Constellation Pharmacology screening of Ubi3a (1.0 μM) revealed an excitatory activity, with amplified responses

Table 2 Turripeptides from five turrid species from the Philippines

Species	Peptide	Sequence/Framework	Reference
Gemmula speciosa	gsp9a	IDOORYCNHIICYγDSγCSQWCTAGCNSITSKCDT	[239]
Gemmula speciosa	gsp9b	GDOORFCRDKLCSGDGDCSVWCTAGCNHDMGKCDTL	[239]
Crassispira cerithina	cce9a	GSCGLPCHENRRCGWACYCDDGICKPLRV	[243]
Crassispira cerithina	cce9b	GSCGLPCHEN-RRCGWACYCDDGICKPLRV	[244]
Unedogemmula bisaya	ubi3a	DCCOCOAGAVRCRFACC-NH_2	[242]
Clavus spp.	cdg14a	YDCIGDADCSGGTYCVGGYCVVG-NH_2	[248]

in medium-sized isolatecttin B^{4+} and small-diameter capsaicin-sensitive CGRP expressing mouse dorsal root ganglion neurons.

Transcriptome analysis *of Crassispira cerithina* identified 22 putative turrid toxin genes [241], with precursors resembling conotoxin superfamilies D, I2, L, M, O1, O2, and P. Bioactivity-guided purification of the venom extract from *C. cerithina* afforded the 29-amino acid long peptide cce9a (Table 2) [243]. This peptide lacks any posttranslational modification, apart from a Cys disulfide. The Cys connectivity is classified under Framework IX (C–C–C–C–C–C). Intracranial injection of cce9a in 12–16 days old mice showed age-dependent effects of this peptide. Younger mice (12–14 days) showed lethargy and unresponsiveness with intracranial administration of (5–40 nmol) cce9a. However, older mice (16 days) showed hyperactivity with cce9a at the same peptide concentration. The age-dependent effects of cce9a were similar to the observed effects of conantokins such as conantokin G [207]. This similarity in the biological effects of cce9a and conantokin G promoted the screening of cce9a using the constellation pharmacology.

Further characterization of related peptides in *C. cerithina* and *Inquisitor intertincta* by cDNA cloning further identified cce9b and iqi9a (Table 2) [244]. These peptides showed a conserved arrangement of cysteine residue, with high similarity to the P-conotoxin family from *Conus* [245, 246], where the cysteine framework of these peptides is designated as Framework IX. cce9b (Table 2) differs from cce9a in the *N*-terminal region [244]. Iqi9a, cce9a, cce9b, and iqi9b (DVCSGSCYYHYQCSRSCYCHYSHCRDKYEK) are now classified as the P-like crassipeptide [244]. These peptides were chemically synthesized using a resin-based method. The synthetic derivatives were tested in the intracranial mouse bioassay at 5–10 nmol, and their effects were compared. All peptides caused lethargy upon intracranial injection, while other symptomatology was surprisingly different in 12–14 days old mice, suggesting that cce9a and cce9b had different molecular targets. Constellation pharmacology was used to identify the subset of dorsal root ganglion neurons affected by P-like crassipeptide and provide insights into the potential cellular target. Ca^{2+} influx in Fura-2 AM-labeled DRG neurons was observed, and the effects were compared to the conopeptide pl14a. More than 89% of the DRG neurons were unaffected by the P-crassipeptides and pl14a. DRG neurons affected by pl14a were significantly more affected compared to the P-crassipeptides. Cce9a only affected small-diameter, capsaicin sensitive DRG neurons affected by pl14a [244], indicating that the P-crassipeptide may have identical targets, which overlap with those of pl14a. The DRG effects suggested that the P-crassipeptides may be modulators of potassium channels instead of NMDA receptors. Electrophysiological analysis in *Drosophila* shaker and human Kv1.1 channel was done. While cce9a and iqi9a did not affect potassium channels, cce9b showed antagonistic activity toward human $K_v1.1$ and shaker-K427D channels, with IC_{50} of 1.1 μM and 2.9 μM, respectively. cce9b did not affect other hK_v1 channels [244].

Another genus of turrids, *Clavus*, was collected from Olango Island, Cebu, Philippines and subjected to transcriptome analysis. Comparative transcriptome analysis of two specimens of *Clavus davidgilmorii* and *Clavus canalicularis* identified 1176 putative venom peptides, termed "drillipeptides" [247]. These were classified into 158 gene families based on signal sequence and precursor similarity. Bioactivity-guided purification of the venom duct extracts of *C. davidgilmorii* identified a 23-amino acid peptide named Cdg14a (Table 2) [248]. Sequence analysis from mass spectrometry and Edman degradation indicated that Cdg14a has four Cys residues, with Cys connectivity similar to Framework XIV of conotoxins (–C–C–C–C–) and defined as the J-type drillipeptide superfamily. Transcriptome analysis indicated that several other peptides belong to this superfamily, bearing an identical Cys framework, and the amino acids between the Cys show divergence [248]. Chemical synthesis identified that the native Cdg14a had a globular fold. Intracranial bioassay showed that Cdg14a caused a characteristic paw thumping phenotype in the intracranial bioassay. Thumping of the hind paw was observed starting at 2.5 nmol of Cdg14a, with higher concentrations causing seizure and/or death. Cdg14a caused an amplification of Ca^{2+} influx in KCl depolarized DRG neurons. Constellation pharmacology of Cdg14a indicated that this peptide affects small diameter neurons strongly labeled by isolectin and responsive to allyl isothiocyanate [248].

4.2.1 Auger Snails

Among the members of the Superfamily Conoidea, auger snails have been the least studied in the Philippines. Three peptides were purified from *Terebra subulata* collected from Marinduque, Philippines [249]. The purification was targeted towards small and hydrophilic peptides. This afforded the peptides Agx-s6a, Agx-s7a, and Agx-11a. These peptides are 40–41 amino acids long and Cys-rich, forming 3–4 disulfide linkages [249]. While the mature peptides showed a similar Cys framework to conopeptides, the precursor sequences were unlike those of the conopeptides. The bioactive component from *T. subulata*, which elicits twisting in *Caenorhabditis elegans*, was traced to Agx-s7a. The purified teretoxins did not elicit any activity in intracranial injection in mice [249].

5 Future Prospects

Over 400 compounds identified from the Philippines were collected from 30 provinces. Most of the collections were done in Palawan, which is dubbed the last frontier. Other highly sampled sites were linked to previous success in sampling and documented high biodiversity. For example, most cone snail and turrid collections were done in Cebu, Bohol, and Marinduque. The Visayan Sea has been hailed as the epicenter of Conoidean biodiversity [250]. The Verde Island Passage, which spans the provinces of Batangas, Oriental and Occidental Mindoro, Marinduque, and Romblon, was also a highly represented sampling site. Less represented in the sampling sites were provinces in the Cagayan and Eastern Visayas regions. The majority of the collections were also done in coastal provinces, with areas in the Exclusive Economic Zone of the Philippines having been unexplored. Samples were mainly collected from the photic zone readily accessible by SCUBA. Only two studies reported collection from the mesophotic zone. Hence, the deeper parts of the Philippine Seas have potentials that remain largely undiscovered. Exploring these would hinge on available technologies for deep sea collections, such as ROVs or submersibles.

Sponges, Conoidean snails, and microorganisms are the most represented source organisms for marine natural products discovery in the Philippines. While macroorganisms have been preferentially the source organisms since the 1970's and 1980's, marine microorganisms have been gaining ground as a popular starting point for compound discovery. Philippine researchers have adopted the paradigm shift in marine natural products, moving from an invertebrate-centric discovery to now looking into simple microorganisms for their biosynthetic potential. Works on marine microorganisms have focused on culturable microorganisms that can express their secondary metabolites in solid or liquid media. Cryptic metabolites continue to be observed in genomes of marine microorganisms. Expression of these silent biosynthetic gene clusters may be facilitated by using multiple culture conditions, elicitor compounds, or cocultivation [251]. More recently, metagenome analysis has become a significant trend in the Philippines. Paired with recombinant expression, this can lead to novel discoveries on unculturable microorganisms and cryptic metabolites from the Philippines.

Bioactivity screening among marine natural products from the Philippines focuses on antiproliferative and antibiotic activity. This is partly due to the ease of endpoint assays utilizing only one or two reagents and relying on spectrophotometric readouts to assess changes in growth. Simple, fast, and relatively economical assays are attractive for researchers from developing countries like the Philippines. The intracranial mouse bioassay has promoted the discovery of neuroactive compounds in the Philippines and allowed for extensive testing of conopeptides. Again, this assay is simple and can be readily performed by local researchers. More recently, research exchanges and capacity building between the Philippines and US-based researchers via the PMS-ICBG project funded by the US NIH-Fogarty International, allowed the establishment of the dorsal root ganglion assay in the Philippines. This high content and medium throughput screen allow for testing more samples than the intracranial mouse bioassay. The DRG also provides initial information on targeted neuronal populations and potential mechanisms of neuroactivity of bioactive compounds [252]. Drug discovery, particularly for neglected diseases that remain a significant concern in the Philippines, like malaria, schistosomiasis, and tuberculosis, may be relevant for in-country drug discovery in the Philippines. The existing marine natural products chemical library can serve as a starting point. Collaborating with relevant agencies like the Research Institute for Tropical Medicine and the Virology Institute of the Philippines could hasten this effort.

Deciphering the mechanism of action and cellular targets among marine natural products from the Philippines has been a minor theme. Only select compounds have so far been analyzed for in-depth mechanistic studies. This would be critical in moving bioactive compounds to the next level of the development pipeline. While enzyme assays provide information on the direct target of marine natural products, certain drawbacks, such as a lack of activity in phenotypic assays due to problems with stability and biotransformation, or insufficient recapitulation of the disease model in enzyme assays [253]. The more widely used cell viability assay measures the changes in cell proliferation in cancer cell lines, but does not provide information on the potential cellular target of bioactive compounds [253]. High content screening and reporter-based assay provide information on the targeted pathways and proteins of bioactive compounds, while still gaining insights into metabolic and cellular stability. Chemical synthesis of marine natural products or their analogues has also been a minor theme. Developing this field would address the supply issue and enable the design of new generations of analogues with better potency and selectivity. Few compounds have also been tested in animal models due to low potency or limited yield.

With the impending changes of the world's oceans from climate change, ocean acidification, and eutrophication, the potential impacts on marine biodiversity are more widely documented than the effects on chemical diversity [254]. The Philippines is among the countries projected to be highly impacted by climate change [255]. Changes in biodiversity risk the loss of chemical diversity and potentially important compounds. For example, a model on the distribution of *C. geographus* and *C. textile* under different scenarios for 2050 and 2090, with different impacts on the two organisms but both with inherent consequences. The prediction suggested that future conditions in Southeast Asia and Micronesia may become more favorable for *C. geographus*, risking more human encounters and possible fatalities [256]. In contrast, conditions will negatively affect *C. textile,* and consequently, other bioactive molecules that remain to be discovered [256]. The effects of climate change on cellular processes and the associated microbiome also impact the biosynthetic potential of marine natural products. However, the organismal response to increased temperature and ocean acidification is greatly variable. The sponge *Haliclona tubifera* from the Philippines showed different gene responses following prolonged or short-term temperature stress [257]. Transcripts related to heat shock proteins, antioxidants, signal transduction, and innate immunity were differentially regulated under short-term exposure [257]. In contrast, sustained temperature stress caused a significant change in transcripts relating to cell damage repair, apoptosis, signaling, and transcription [257]. Microbiome shift also varies among sponges. The sponge *Stylissa flabelliforme* showed reduced γ-proteobacteria, while α-proteobacteria were increased in ocean warming scenario. The response of the microbiome in *S. flabelliforme* under lower pH showed shifts in prioritized metabolic pathways [258]. Potentially opportunistic and pathogenic microorganisms in *Spheciospongia vesparium* were observed to increase under high temperature [259]. In contrast, the Philippine sponge *Neopetrosia compacta* showed more resilient microbial populations that can cope with ocean warming and acidification [260]. Increased allelopathic potency in certain seaweed and seagrass species may point to more resilient traits for these organisms under stressful oceanic conditions. Extracts of *Canistocarpus* (=*Dictyota*) *cervicornis* showed more heightened effects against the coral *Acropora intermedia* under elevated CO_2 [261]. The eelgrass *Zostera marina* in a simulated heatwave scenario also had a higher potential to deter microbial colonizers [262]. In the face of climate change, there will be winners and losers in the marine environment. Given this, interrogating the impacts of climate change on biomedically important marine organisms from the Philippines is warranted. This will allow for better management of these resources.

Overall, the Philippines has significantly contributed to efforts to discover marine natural products. Research in the country remains primarily focused on biomedical discoveries, utilizing the secondary metabolites of marine organisms. Developing high value products from marine resources forms part of the efforts in harnessing the country's Blue Economy. The field continues to be relevant and productive by aligning local research with global trends in marine natural products and implementing continuous capacity building and human resource development. The Philippine Seas continue to abound with potential high value products and economically important organisms. This, therefore, warrants the protection and conservation of this fragile ecosystem.

Acknowledgements We are grateful for the funding support from the Philippine Council for Health Research and Development through the Discovery and Development of Health Products—Marine Component (DDHP-MC). We thank the DDHP-MC project for the specimen photos. We dedicate this paper to National Scientist Lourdes J. Cruz, Professor Baldomero M. Olivera, and Professor Gisela P. Concepcion for their tireless efforts in promoting marine natural products research in the Philippines and inspiring Filipino researchers in the field of natural products. This is MSI Contribution No. 1618.

References

1. Carpenter KE, Springer VG (2005) The center of the center of marine shore fish biodiversity: the Philippine Islands. Environ Biol Fishes 72:467–480
2. Roleda MY, Benayahu Y (2023) The Philippine seas: biodiversity and ecological impacts of natural and anthropogenic stressors in tropical reef systems, 2nd ed. Frontiers Media SA

3. Tecson-Mendoza EM (2010) National Scientist Lourdes Jansuy Cruz: From Rice to Conus Research—the Journey. Philipp Sci Lett 1:15–20
4. Olivera BM, Cruz LJ, Yoshikami D (1999) Effects of *Conus* peptides on the behavior of mice. Curr Opin Neurobiol 9:772–777
5. Dutertre S, Lewis RJ (2010) Use of venom peptides to probe ion channel structure and function. J Biol Chem 285:13315–13320
6. Safavi-Hemami H, Brogan SE, Olivera BM (2019) Pain therapeutics from cone snail venoms: from Ziconotide to novel non-opioid pathways. J Proteomics 190:12–20
7. Concepcion GP, Anas ARJ, Azcuna MAA (2014) Anticancer compounds from Philippine marine organisms act on major pathways in cancer. Philipp Sci Lett 7:207–227
8. Banday AH, Azha N, Farooq R, Sheikh SA, Ganie MA, Parray MN, Mushtaq H, Hameed I, Lone MA (2024) Exploring the potential of marine natural products in drug development: A comprehensive review. Phytochem Lett 59:124–135
9. Saeed AFUH, Su J, Ouyang S (2021) Marine-derived drugs: recent advances in cancer therapy and immune signaling. Biomed Pharmacother 134:111091
10. Rohde S, Nietzer S, Schupp PJ (2015) Prevalence and mechanisms of dynamic chemical defenses in tropical sponges. PLoS ONE 10:e0132236
11. Peters KJ, Amsler CD, McClintock JB, van Soest RWM, Baker BJ (2009) Palatability and chemical defenses of sponges from the western Antartic Peninsula. Mar Ecol Prog Ser 385:77–85
12. Rohde S, Schupp PJ (2011) Allocation of chemical and structural defenses in the sponge *Melophlus sarasinorum*. J Exp Mar Biol Ecol 399:76–83
13. Helber SB, Hoeijmakers DJJ, Muhando CA, Rohde S, Schupp PJ (2018) Sponge chemical defenses are a possible mechanism for increasing sponge abundance on reefs in Zanzibar. PLoS ONE 13:e0197617
14. West LM, Faulkner DJ (2008) Acanthosulfate, a sulfated hydroxyhydroquinone sesterterpenoid from the sponge *Acanthodendrilla* sp. J Nat Prod 71:269–271
15. Lee J, Shin A-Y, Lee H-S (2017) Isolation and synthesis of luffariellolide derivatives and evaluation of antibacterial activities against gram-negative bacteria. Bull Korean Chem Soc 38:804–807
16. Tasdemir D, Concepción GP, Mangalindan GC, Harper MK, Hajdu E, Ireland CM (2000) New Terpenoids from a *Cacospongia* sp. from the Philippines. Tetrahedron 56:9025–9030
17. Choi J-H, Lee H-S, Campos WL (2020) Scalarane-type sesterterpenes from the Philippines sponge *Hyrtios* sp. Ocean Polar Res 42:15–20
18. Bugni TS, Singh MP, Chen L, Arias DA, Harper MK, Greenstein M, Maiese WM, Concepción GP, Mangalindan GC, Ireland CM (2004) Kalihinols from two *Acanthella cavernosa* sponges: inhibitors of bacterial folate biosynthesis. Tetrahedron 60:6981–6988
19. Shin A-Y, Lee H-S, Lee J (2022) Isolation of scalimides A-L: β-alanine-bearing scalarane analogs from the marine Sponge *Spongia* sp. Mar Drugs 20:726
20. Goetz GH, Harrigan GG, Likos J (2001) Ugibohlin: a new dibromo-seco-isophakellin from *Axinella carteri*. J Nat Prod 64:1581–1582
21. Macabeo APG, Guce F (2014) Bromopyrrole-imidazole alkaloids from *Acanthella carteri* Dendy (Axinellidae). Res J Pharm Biol Chem Sci 5:720–723
22. Lazaro JEH, Nitcheu J, Mahmoudi N, Ibana JA, Mangalindan GC, Black GP, Howard-Jones AG, Moore CG, Thomas DA, Mazier D, Ireland CM, Concepcion GP, Murphy PJ, Diquet B (2006) Antimalarial activity of crambescidin 800 and synthetic analogues against liver and blood stage of *Plasmodium* sp. J Antibiot 59:583–590
23. Elliott JL, Kutluay SB (2020) Going beyond integration: The emerging role of HIV-1 integrase in virion morphogenesis. Viruses 12:1005
24. Irrinki AM, Kaur J, Randhawa B, McFadden R, Snyder C, Truong H, Soohoo D, Hu E, Yu H, Murray BP, Lu B, Kornyeyev D, Irwan ID, Nguyen L, Yang Y-S, Belzile J-P, Schmitz U, Appleby TC, Schultz B, Lalezari J, Deeks S, Cihlar T, Murry JP (2025) Activating PKC-ε induces HIV expression with improved tolerability. PLoS Pathog 21:e1012874

25. Lerch ML, Harper MK, Faulkner DJ (2003) Brominated polyacetylenes from the Philippines Sponge *Diplastrella* sp. J Nat Prod 66:667–670
26. Whitson EL, Bugni TS, Chockalingam PS, Concepcion GP, Harper MK, He M, Hooper JNA, Mangalindan GC, Ritacco F, Ireland CM (2008) Spheciosterol sulfates, PKCζ inhibitors from a Philippine Sponge *Spheciospongia* sp. J Nat Prod 71:1213–1217
27. Whitson EL, Bugni TS, Chockalingam PS, Concepcion GP, Feng X, Jin G, Harper MK, Mangalindan GC, McDonald LA, Ireland CM (2009) Fibrosterol sulfates from the Philippine Sponge *Lissodendoryx* (*Acanthodoryx*) *fibrosa*: sterol dimers that inhibit PKCζ. J Org Chem 74:5902–5908
28. Tasdemir D, Mangalindan GC, Concepción GP, Verbitski SM, Rabindran S, Miranda M, Greenstein M, Hooper JNA, Harper MK, Ireland CM (2002) Bioactive isomalabaricane triterpenes from the marine sponge *Rhabdastrella globostellata*. J Nat Prod 65:210–214
29. Tasdemir D, Bugni TS, Mangalindan GC, Concepcion GP, Harper MK, Ireland CM (2003) Bisabolane type sesquiterpenes from a marine *Didiscus* sponge. Turk J Chem 27:273–279
30. Tasdemir D, Mallon R, Greenstein M, Feldberg LR, Kim SC, Collins K, Wojciechowicz D, Mangalindan GC, Concepción GP, Harper MK, Ireland CM (2002) Aldisine alkaloids from the Philippine Sponge *Stylissa massa* are potent inhibitors of mitogen-activated protein Kinase Kinase-1 (MEK-1). J Med Chem 45:529–532
31. Luyao H, Luesch H, Uy M (2021) GPCR pharmacological profiling of aaptamine from the Philippine Sponge *Stylissa* sp. extends its therapeutic potential for noncommunicable diseases. Molecules 26:5618
32. Mazo RAMC, Gelani CD, Lavilla CA, Uy MM, Tabugo SRM, Ohta E, Ohta S (2023) Antibacterial, anti-inflammatory, and antidiabetic studies of the amines isolated from the Philippine marine Sponge *Desmacella* sp. Curr Bioact Compd 19:52–61
33. Centko RM, Steinø A, Rosell FI, Patrick BO, de Voogd N, Mauk AG, Andersen RJ (2014) Indoleamine 2, 3-dioxygenase inhibitors isolated from the sponge *Xestospongia vansoesti*: Structure elucidation, analogue synthesis, and biological activity. Org Lett 16:6480–6483
34. Williams DE, Steinø A, de Voogd NJ, Mauk AG, Andersen RJ (2012) Halicloic acids A and B isolated from the marine sponge *Haliclona* sp. collected in the Philippines inhibit indoleamine 2,3-dioxygenase. J Nat Prod 75:1451–1458
35. Rashid MA, Gustafson KR, Boyd MR (2001) A new isoquinoline alkaloid from the marine Sponge *Haliclona* species. J Nat Prod 64:1249–1250
36. Bugni TS, Concepción GP, Mangalindan GC, Harper MK, James RD, Ireland CM (2002) *P*-sulfooxyphenylpyruvic acid from the red macro alga *Ceratodictyon spongiosum* and its sponge symbiont *Haliclona cymaeformis*. Phytochemistry 60:361–363
37. McCulloch MWB, Bugni TS, Concepcion GP, Coombs GS, Harper MK, Kaur S, Mangalindan GC, Mutizwa MM, Veltri CA, Virshup DM, Ireland CM (2009) Carteriosulfonic acids A−C, GSK-3β inhibitors from a *Carteriospongia* sp. J Nat Prod 72:1651–1656
38. Ridley CP, Faulkner DJ (2003) New cytotoxic steroidal alkaloids from the Philippine sponge *Corticium niger*. J Nat Prod 66:1536–1539
39. Sunassee SN, Ransom T, Henrich CJ, Beutler JA, Covell DG, McMahon JB, Gustafson KR (2014) Steroidal alkaloids from the marine sponge *Corticium niger* that inhibit growth of human colon carcinoma cells. J Nat Prod 77:2475–2480
40. Qureshi A, Faulkner DJ (2000) 7α-hydroxytheonellasterol, a cytotoxic 4-methylene sterol from the Philippines sponge *Theonella swinhoei*. J Nat Prod 63:841–842
41. Shin A-Y, Lee H-S, Lee Y-J, Lee JS, Son A, Choi C, Lee J (2020) Oxygenated theonellastrols: Interpretation of unusual chemical behaviors using quantum mechanical calculations and stereochemical reassignment of 7α-hydroxytheonellasterol. Mar Drugs 18:607
42. Goetz GH, Harrigan GG, Likos J (2001) Furodysin lactone and pyrodysinoic acid, new sesquiterpenes from a Philippines *Dysidea* species. J Nat Prod 64:1486–1488
43. Shin A-Y, Son A, Choi C, Lee J (2021) Isolation of scalarane-type sesterterpenoids from the marine sponge *Dysidea* sp. and stereochemical reassignment of 12-*epi*-phyllactone D/E. Marine Drugs 19:627

44. Harrigan GG, Goetz GH, Luesch H, Yang S, Likos J (2001) Dysideaprolines A–F and barbaleucamides A–B, novel polychlorinated compounds from a *Dysidea* species. J Nat Prod 64:1133–1138
45. Tatsukawa M, Punzalan LLC, Magpantay HDS, Villaseñor IM, Concepcion GP, Suwanborirux K, Yokoya M, Saito N (2012) Chemistry of renieramycins. Part 13: isolation and structure of stabilized renieramycin type derivatives, renieramycins W-Y, from Philippine blue sponge *Xestospongia* sp., pretreated with potassium cyanide. Tetrahedron 68:7422–7428
46. Saito N, Hiramatsu A, Hirade H, Kubota M, Toyoshima R, Fujino A, Sirimangkalakitti N, Suwanborirux K, Concepcion G (2017) Chemistry of renieramycins. 16. Structure of 7-demethylrenieramycin O (= 14α-hydroxyrenieramycin S) from blue sponge, *Xestospongia* sp. Heterocycles 95:748–752
47. Charupant K, Daikuhara N, Saito E, Amnuoypol S, Suwanborirux K, Owa T, Saito N (2009) Chemistry of renieramycins. Part 8: Synthesis and cytotoxicity evaluation of renieramycin M–jorunnamycin A analogues. Bioorg Med Chem 17:4548–4558
48. Halim H, Chunhacha P, Suwanborirux K, Chanvorachote P (2011) Anticancer and antimetastatic activities of renieramycin M, a marine tetrahydroisoquinoline alkaloid, in human non-small cell lung cancer cells. Anticancer Res 31:193–201
49. Buaban K, Innets B, Petsri K, Sinsook S, Chanvorachote P, Chansriniyom C, Suwanborirux K, Yokoya M, Saito N, Chamni S (2023) Semisynthesis of 5-*O*-ester derivatives of renieramycin T and their cytotoxicity against non-small-cell lung cancer cell lines. Sci Rep 13:21485
50. Tun JO, Salvador-Reyes LA, Velarde MC, Saito N, Suwanborirux K, Concepcion GP (2019) Synergistic cytotoxicity of renieramycin M and doxorubicin in MCF-7 breast cancer cells. Mar Drugs 17:536
51. Acyatan ZG, Susana-Guevarra SR, Picart MRD, Belen EL, Salvador-Reyes LA, Concepcion GP (2024) *In vivo* anti-tumor and anti-metastatic activity of renieramycin M and doxorubicin combination treatments on 4T1 breast cancer murine model. SciEnggJ 17:379–390
52. Santiago VS, Manzano GG, Yu CC, Aliño PM, Salvador-Reyes LA (2019) Mariculture potential of renieramycin-producing Philippine blue sponge *Xestospongia* sp. (Porifera: Haplosclerida). Aquaculture 502:356–364
53. Lalas JAA, Manzano GG, Desabelle LAB, Salvador-Reyes LA, Aliño PM, Baria-Rodriguez MV (2023) Notes on the distribution, ecology and open sea mariculture trials of the Philippine blue sponge, *Xestospongia* sp. (Porifera). Aquac Fish Fish 3:132–148
54. Yokseng K, Darumas U, Pedpradab P (2021) Culture of a blue marine sponge, *Xestospongia* sp., in semi-circulate close system stressed with calcium and magnesium concentration. J Aquac Mar Biol 10:230–235
55. Tianero MD, Balaich JN, Donia MS (2019) Localized production of defence chemicals by intracellular symbionts of *Haliclona* sponges. Nat Microbiol 4:1149–1159
56. Sandoval IT, A. DR, S. BT, P. CG, Kay HM, and Ireland CM (2004) Cytotoxic isoquinoline quinones from sponges of the genus *Petrosia*. Nat Prod Res 18:89–93
57. Mama RL, Gelani CD, Daluz JMT, Uy MM, Ohta E, Ohta S (2024) Two new sarasinosides from marine sponge *Petrosia nigricans*. Nat Prod Res 38:2395–2403
58. Pimentel SMV, Bojo ZP, Roberto AVD, Lazaro JEH, Mangalindan GC, Florentino LM, Lim-Navarro P, Tasdemir D, Ireland CM, Concepcion GP (2003) Platelet aggregation inhibitors from Philippine marine invertebrate samples screened in a new microplate assay. Mar Biotechnol 5:395–400
59. Marshall KM, Andjelic CD, Tasdemir D, Concepción GP, Ireland CM, Barrows LR (2009) Deoxyamphimedine, a pyridoacridine alkaloid, damages DNA via the production of reactive oxygen species. Mar Drugs 7:196–209
60. Tasdemir D, Mangalindan GC, Concepci GP, Harper MK, Ireland CM (2001) 3,7-dimethylguanine, a new purine from a Philippine sponge *Zyzzya fuliginosa*. Chem Pharm Bull 49:1628–1630
61. Rashid MA, Gustafson KR, Boyd MR (2001) New chondropsin macrolide lactams from marine sponges in the genus *Ircinia*. Tetrahedron Lett 42:1623–1626

62. Ang AMG, Uy MM, Ohta E, Ômura H, Ohta S (2024) Irciniaplysins A–D: new psammaplysin derivatives from Philippine marine sponge *Ircinia* sp. Chem Biodivers 21:e202400962
63. Quiao MA, Uy M (2013) Pyrimidines from the Philippine marine sponge *Aaptos suberitoides*. Int J Eng Res 4:656–659
64. Rao MR, Faulkner DJ (2002) Clavosolides A and B, dimeric macrolides from the Philippines sponge *Myriastra clavosa*. J Nat Prod 65:386–388
65. Erickson KL, Gustafson KR, Pannell LK, Beutler JA, Boyd MR (2002) New dimeric macrolide glycosides from the marine sponge *Myriastra clavosa*. J Nat Prod 65:1303–1306
66. Erickson KL, Gustafson KR, Milanowski DJ, Pannell LK, Klose JR, Boyd MR (2003) Myriastramides A-C, new modified cyclic peptides from the Philippines marine sponge *Myriastra clavosa*. Tetrahedron 59:10231–10238
67. Davis RA, Mangalindan GC, Bojo ZP, Antemano RR, Rodriguez NO, Concepcion GP, Samson SC, de Guzman D, Cruz LJ, Tasdemir D, Harper MK, Xidong F, Carter GT, Ireland CM (2004) Microcionamides A and B, bioactive peptides from the Philippine sponge *Clathria* (*Thalysias*) *abietina*. J Org Chem 69:4170–4176
68. Mokhlesi A, Stuhldreier F, Wex KW, Berscheid A, Hartmann R, Rehberg N, Sureechatchaiyan P, Chaidir C, Kassack MU, Kalscheuer R, Brötz-Oesterhelt H, Wesselborg S, Stork B, Daletos G, Proksch P (2017) Cyclic cystine-bridged peptides from the marine sponge *Clathria basilana* induce apoptosis in tumor cells and depolarize the bacterial cytoplasmic membrane. J Nat Prod 80:2941–2952
69. Inocentes CRV, Salvador-Reyes LA, Villaraza AJL (2023) Total synthesis and bioactivity evaluation of hydrophobic microcionamide-inspired peptides. Chem Biodivers 20:e202200832
70. Susana SR, Salvador-Reyes LA (2022) Anti-inflammatory activity of monosubstituted xestoquinone analogues from the marine sponge *Neopetrosia compacta*. Antioxidants 11:607
71. Cao S, Murphy BT, Foster C, Lazo JS, Kingston DG (2009) Bioactivities of simplified adociaquinone B and naphthoquinone derivatives against Cdc25B, MKP-1, and MKP-3 phosphatases. Bioorg Med Chem 17:2276–2281
72. He F, Mai LH, Longeon A, Copp BR, Loaëc N, Bescond A, Meijer L, Bourguet-Kondracki M-L (2015) Novel adociaquinone derivatives from the indonesian sponge *Xestospongia* sp. Mar Drugs 13:2617–2628
73. Concepcion GP, Foderaro TA, Eldredge GS, Lobkovsky E, Clardy J, Barrows LR, Ireland CM (1995) Topoisomerase II-mediated DNA cleavage by adocia- and xestoquinones from the Philippine sponge *Xestospongia* sp. J Med Chem 38:4503–4507
74. Du L, Mahdi F, Datta S, Jekabsons MB, Zhou Y-D, Nagle DG (2012) Structures and mechanisms of antitumor agents: xestoquinones uncouple cellular respiration and disrupt HIF signaling in human breast tumor cells. J Nat Prod 75:1553–1559
75. Kwon I-S, Kwak JH, Pyo S, Lee H-W, Kim A, Schmitz FJ (2017) Oscarellin, an anthranilic acid derivative from a Philippine sponge, *Oscarella stillans*, as an inhibitor of inflammatory cytokines in macrophages. J Nat Prod 80:149–155
76. Azcuna M, Tun JO, Yap HT, Concepcion GP (2018) *Callyspongia samarensis* (Porifera) extracts exhibit anticancer activity and induce bleaching in *Porites cylindrica* (Scleractinia). Chem Ecol 34:397–411
77. Resuello DL, Lirio SB, Porto AE, Macabeo APG, Huang H-Y, Corpuz MJ-AT, Villaflores OB (2020) β-secretase 1 inhibitory activity and AMP-activated protein kinase activation of *Callyspongia samarensis* extracts. Nat Prod Res 34:525–529
78. Ayllon M, Earl H, Metillo E, Uy M (2019) Toxicity of crude extracts from soft corals (Anthozoa, Alcyonacea) collected at varying wave exposure sites in Talisayan, Northern Mindanao, Philippines. Int J Biosci 14:116–129
79. Edrada RA, Wray V, Witte L, Lv O, Proksch P (2000) Bioactive terpenes from the soft coral *Heteroxenia* sp. from Mindoro, Philippines. Z Naturforsch C 55:82–86
80. Rashid MA, Gustafson KR, Boyd MR (2000) HIV-inhibitory cembrane derivatives from a Philippines collection of the soft coral *Lobophytum* species. J Nat Prod 63:531–533
81. Dumalan RJP, Bondoc KGV, Juinio-Meñez MA (2019) Grow-out culture trial of sandfish *Holothuria scabra* in pens near a mariculture-impacted area. Aquaculture 507:481–492

82. Bondoc KGV, Lee H, Cruz LJ, Lebrilla CB, Juinio-Meñez MA (2013) Chemical fingerprinting and phylogenetic mapping of saponin congeners from three tropical holothurian sea cucumbers. Comp Biochem Physiol B: Biochem Mol Biol 166:182–193
83. Torreno VPM, Molino RJEJ, Junio HA, Yu ET (2023) Comprehensive metabolomics of Philippine *Stichopus* cf. *horrens* reveals diverse classes of valuable small molecules for biomedical applications. PLoS ONE 18:e0294535
84. Chovolou Y, Ebada SS, Wätjen W, Proksch P (2011) Identification of angular naphthopyrones from the Philippine echinoderm *Comanthus* species as inhibitors of the NF-κB signaling pathway. Eur J Pharmacol 657:26–34
85. Verbitski SM, Mayne CL, Davis RA, Concepcion GP, Ireland CM (2002) Isolation, structure determination, and biological activity of a novel alkaloid, perophoramidine, from the Philippine ascidian *Perophora namei*. J Org Chem 67:7124–7126
86. Perez LJ, Faulkner DJ (2003) Bistratamides E−J, modified cyclic hexapeptides from the Philippines ascidian *Lissoclinum bistratum*. J Nat Prod 66:247–250
87. Neves JLB, Lin Z, Imperial JS, Antunes A, Vasconcelos V, Olivera BM, Schmidt EW (2015) Small molecules in the cone snail arsenal. Org Lett 17:4933–4935
88. McIntosh JM, Foderaro TA, Li W, Ireland CM, Olivera BM (1993) Presence of serotonin in the venom of *Conus imperialis*. Toxicon 31:1561–1566
89. Torres JP, Lin Z, Watkins M, Salcedo PF, Baskin RP, Elhabian S, Safavi-Hemami H, Taylor D, Tun J, Concepcion GP, Saguil N, Yanagihara AA, Fang Y, McArthur JR, Tae H-S, Finol-Urdaneta RK, Özpolat BD, Olivera BM, Schmidt EW (2021) Small-molecule mimicry hunting strategy in the imperial cone snail, *Conus imperialis*. Sci Adv 7:eabf2704
90. Niu C, Leavitt LS, Lin Z, Paguigan ND, Sun L, Zhang J, Torres JP, Raghuraman S, Chase K, Cadeddu R, Karthikeyan M, Bortolato M, Reilly CA, Hughen RW, Light AR, Olivera BM, Schmidt EW (2021) Neuroactive type-A γ-aminobutyric acid receptor allosteric modulator steroids from the hypobranchial gland of marine mollusk, *Conus geographus*. J Med Chem 64:7033–7043
91. Aknin M, Viracaoundin I, Faure R, Gaydou EM (1998) 5α,8α-epidioxycholest-6-en-3-β-ol from three cone snails of the Indian ocean. J Am Oil Chem Soc 75:1679–1681
92. Lee Y-J, Han S, Kim SH, Lee H-S, Shin HJ, Lee JS, Lee J (2017) Three new cytotoxic steroidal glycosides isolated from *Conus pulicarius* collected in Kosrae, Micronesia. Mar Drugs 15:379
93. Fu X, Hong EP, Schmitz FJ (2000) New polypropionate pyrones from the Philippine sacoglossan mollusc *Placobranchus ocellatus*. Tetrahedron 56:8989–8993
94. Lastimoso JM, Santiañez WJ (2021) Updated checklist of the benthic marine macroalgae of the Philippines. Philipp J Sci 150:29–92
95. FAO (2022) The state of world fisheries and aquaculture 2022. Towards blue transformation. FAO, Rome
96. Daniotti S, Re I (2021) Marine biotechnology: challenges and development market trends for the enhancement of biotic resources in industrial pharmaceutical and food applications. A statistical analysis of scientific literature and business models. Mar Drugs 19:61
97. Dumilag RV, Javier RF (2022) Ethnobotany of medicinal seaweeds of Ilocos Norte, Philippines. Philippine J Sci 151:1135–1156
98. Dumilag RV, Belgica THR, Mendoza LC, Hibay JM, Arevalo AE Jr, Malto MAD, Orgela EG, Longavela MR, Corral LEH, Olipany RD (2022) Seaweed ethnobotany of eastern Sorsogon, Philippines. Algae 37:227–237
99. Sindo EL, Clores MA, Bigcas EJJ, Dumilag RV (2025) Seaweed ethnobotany of Eastern Camarines Sur, the Philippines. Philipp J Sci 154:27–35
100. Higa T, Kuniyoshi M (2000) Toxins associated with medicinal and edible seaweeds. J Toxicol: Toxin Rev 19:119–137
101. Yeh TS, Hung NH, Lin TC (2014) Analysis of iodine content in seaweed by GC-ECD and estimation of iodine intake. J Food Drug Anal 22:189–196
102. Jiménez-Escrig A, Sánchez-Muniz FJ (2000) Dietary fibre from edible seaweeds: Chemical structure, physicochemical properties and effects on cholesterol metabolism. Nutr Res 20:585–598

103. Peñalver R, Lorenzo JM, Ros G, Amarowicz R, Pateiro M, Nieto G (2020) Seaweeds as a functional ingredient for a healthy diet. Mar Drugs 18:301
104. De Jesus Raposo MF, De Morais AMB, De Morais RMSC (2015) Marine polysaccharides from algae with potential biomedical applications. Mar Drugs 13:2967–3028
105. Khan F, Jeong G-J, Khan MSA, Tabassum N, Kim Y-M (2022) Seaweed-derived phlorotannins: A review of multiple biological roles and action mechanisms. Mar Drugs 20:384
106. Chuyen HV, Eun J-B (2017) Marine carotenoids: bioactivities and potential benefits to human health. Crit Rev Food Sci Nutr 57:2600–2610
107. Ratnayake R, Liu Y, Paul VJ, Luesch H (2013) Cultivated sea lettuce is a multiorgan protector from oxidative and inflammatory stress by enhancing the endogenous antioxidant defense system. Cancer Prev Res 6:989–999
108. Wang R, Paul VJ, Luesch H (2013) Seaweed extracts and unsaturated fatty acid constituents from the green alga *Ulva lactuca* as activators of the cytoprotective Nrf2–ARE pathway. Free Radical Biol Med 57:141–153
109. Villaflores OB, Ortega KMM, Empaynado-Porto A, Lirio S, Yak H-K, Albano DR, Corpuz MJ-AT (2019) Anti-angiogenic activity of *Gracilaria coronopifolia* J.G. Agardh extract by lowering the levels of trace metals (iron, zinc and copper) in duck chorioallantoic membrane and in vitro activation of AMP-kinase. Mol Biol Rep 46:4151–4160
110. Arguelles E, Sapin A (2022) Chemical composition and bioactive properties of *Sargassum aquifolium* (Turner) C. Agardh and its potential for pharmaceutical application. Philipp J Sci 151:9–24
111. Pakingking RVJ, Usero R, de Jesus-Ayson EGT, Logronio DJ, Caipang MC (2022) Phytochemical composition antioxidant and antibacterial activity of the Philippine marine green alga (*Ulva pertusa*). Int Aquat Res 14:51–62
112. Francisco MEY, Erickson KL (2001) Ma'iliohydrin, a cytotoxic chamigrene dibromohydrin from a Philippine *Laurencia* species. J Nat Prod 64:790–791
113. Abdel-Mageed WM, Ebel R, Valeriote FA, Jaspars M (2010) Laurefurenynes A-F, new cyclic ether acetogenins from a marine red alga, *Laurencia* sp. Tetrahedron 66:2855–2862
114. Gerwick William H, Moore Bradley S (2012) Lessons from the past and charting the future of marine natural products drug discovery and chemical biology. Chem Biol 19:85–98
115. McCutcheon JP, Moran NA (2012) Extreme genome reduction in symbiotic bacteria. Nat Rev Microbiol 10:13–26
116. Brinkmann CM, Marker A, Kurtböke DI (2017) An overview on marine sponge-symbiotic bacteria as unexhausted sources for natural product discovery. Diversity 9:40
117. Pita L, Rix L, Slaby BM, Franke A, Hentschel U (2018) The sponge holobiont in a changing ocean: from microbes to ecosystems. Microbiome 6:46
118. Wilson MC, Mori T, Rückert C, Uria AR, Helf MJ, Takada K, Gernert C, Steffens UAE, Heycke N, Schmitt S, Rinke C, Helfrich EJN, Brachmann AO, Gurgui C, Wakimoto T, Kracht M, Crüsemann M, Hentschel U, Abe I, Matsunaga S, Kalinowski J, Takeyama H, Piel J (2014) An environmental bacterial taxon with a large and distinct metabolic repertoire. Nature 506:58–62
119. Sertan-de Guzman AA, Predicala RZ, Bernardo EB, Neilan BA, Elardo SP, Mangalindan GC, Tasdemir D, Ireland CM, Barraquio WL, Concepcion GP (2007) *Pseudovibrio denitrificans* strain Z143–1, a heptylprodigiosin-producing bacterium isolated from a Philippine tunicate. FEMS Microbiol Lett 277:188–196
120. Paul T, Bandyopadhyay TK, Mondal A, Tiwari ON, Muthuraj M, Bhunia B (2022) A comprehensive review on recent trends in production, purification, and applications of prodigiosin. Biomass Conv Bioref 12:1409–1431
121. Lazaro JEH, Nitcheu J, Predicala RZ, Mangalindan GC, Nesslany F, Marzin D, Concepcion GP, Diquet B (2002) Heptyl prodigiosin, a bacterial metabolite, is antimalarial in vivo and non-mutagenic in vitro. J Nat Toxins 11:367–377
122. Ranches GD, Rudy A, Vollmar A, Tun JO, Concepcion GP (2013) Heptylprodigiosin induces apoptosis in Jurkat leukemia T cells via CD95 death receptor and evades antiapoptotic Bcl-2 and Bcl-xL proteins. Philippine Sci Lett 6:153–167

123. Bojo ZP, Deaño CD, Jacinto SD, Concepcion GP (2010) Synergistic *in vitro* cytotoxicity of adociaquinone B and heptyl prodigiosin against MCF-7 breast cancer cell line. Philipp Sci Lett 3:48–58
124. Azcuna M, Salvador-Reyes L, Tun JO, Lluisma A, Uy ID, Cunanan L, Siringan MA, Concepcion GP (2019) Characterization of the biosynthetic potential of the β-proteobacterium *Achromobacter xylosoxidans* strain ISP2-142-O-2-A using microbiological, chemical, and genomic approaches. Philipp Sci Lett 148:199–209
125. Malto ZBL, Reyes JM, Lo BLD, Davis KBS, Concepcion GP, Salvador-Reyes LA (2023) Genomics and metabolomics-based assessment of the biosynthetic potential of the sponge-associated microorganism *Streptomyces cacaoi* strain R2A–843A from the Philippines. SciEnggJ 16:67–89
126. Zhou J, Zhang H, Ye J, Wu X, Wang W, Lin H, Yan X, Lazaro JEH, Wang T, Naman CB, He S (2021) Cytotoxic polyketide metabolites from a marine mesophotic zone Chalinidae sponge-associated fungus *Pleosporales* sp. NBUF144. Mar Drugs 19:186
127. Wang T, Zhou J, Zou J, Shi Y, Zhou W, Shao P, Yu T, Cui W, Li X, Wu X, Ye J, Yan X, Naman CB, Lazaro JEH, He S (2021) Discovery of cymopolyphenols A–F from a marine mesophotic zone Aaptos sponge-associated fungus *Cymostachys* sp. NBUF082. Front Microbiol 12
128. Peraud O, Biggs JS, Hughen RW, Light AR, Concepcion GP, Olivera BM, Schmidt EW (2009) Microhabitats within venomous cone snails contain diverse actinobacteria. Appl Environ Microbiol 75:6820–6826
129. Lin Z, Torres JP, Ammon MA, Marett L, Teichert RW, Reilly CA, Kwan JC, Hughen RW, Flores M, Tianero MD, Peraud O, Cox JE, Light AR, Villaraza AJL, Haygood MG, Concepcion GP, Olivera BM, Schmidt EW (2013) A bacterial source for mollusk pyrone polyketides. Chem Biol 20:73–81
130. Lin Z, Marett L, Hughen RW, Flores M, Forteza I, Ammon MA, Concepcion GP, Espino S, Olivera BM, Rosenberg G, Haygood MG, Light AR, Schmidt EW (2013) Neuroactive diol and acyloin metabolites from cone snail-associated bacteria. Bioorg Med Chem Lett 23:4867–4869
131. Lin Z, Zachariah MM, Marett L, Hughen RW, Teichert RW, Concepcion GP, Haygood MG, Olivera BM, Light AR, Schmidt EW (2014) Griseorhodins D-F, neuroactive intermediates and end products of post-PKS tailoring modification in griseorhodin biosynthesis. J Nat Prod 77:1224–1230
132. Miller BW, Torres JP, Tun JO, Flores MS, Forteza I, Rosenberg G, Haygood MG, Schmidt EW, Concepcion GP (2020) Synergistic anti-methicillin-resistant *Staphylococcus aureus* (MRSA) activity and absolute stereochemistry of 7,8-dideoxygriseorhodin C. J Antibiot 73:290–298
133. Lin Z, Koch M, Pond CD, Mabeza G, Seronay RA, Concepcion GP, Barrows LR, Olivera BM, Schmidt EW (2014) Structure and activity of lobophorins from a turrid mollusk-associated *Streptomyces* sp. J Antibiot 67:121–126
134. Lin Z, Flores M, Forteza I, Henriksen NM, Concepcion GP, Rosenberg G, Haygood MG, Olivera BM, Light AR, Cheatham TE III, Schmidt EW (2012) Totopotensamides, polyketide-cyclic peptide hybrids from a mollusk-associated bacterium *Streptomyces* sp. J Nat Prod 75:644–649
135. Lin Z, Reilly CA, Antemano R, Hughen RW, Marett L, Concepcion GP, Haygood MG, Olivera BM, Light A, Schmidt EW (2011) Nobilamides A-H, long-acting transient receptor potential vanilloid-1 (TRPV1) antagonists from mollusk-associated bacteria. J Med Chem 54:3746–3755
136. Yamashita T, Matoba H, Kuranaga T, Inoue M (2014) Total syntheses of nobilamides B and D: application of traceless Staudinger ligation. Tetrahedron 70:7746–7752
137. Jacinto MPV, Flores MS, Lin Z, Concepcion GP, Schmidt EW, Faulkner S, Villaraza AJL (2014) Synthesis and bioactivity of nobilamide B. RSC Adv 4:37609–37612
138. Belleza OJV, Tun JO, Concepcion GP, Villaraza AJL (2018) On the inhibition of capsaicin response in dorsal root ganglion neurons by nobilamide B and analogues: A structure–activity relationship study. MedChemComm 9:1673–1678

139. Lin Z, Antemano RR, Hughen RW, Tianero MDB, Peraud O, Haygood MG, Concepcion GP, Olivera BM, Light A, Schmidt EW (2010) Pulicatins A–E, neuroactive thiazoline metabolites from cone snail-associated bacteria. J Nat Prod 73:1922–1926
140. Lin Z, Smith MD, Concepcion GP, Haygood MG, Olivera BM, Light A, Schmidt EW (2017) Modulating the serotonin receptor spectrum of pulicatin natural products. J Nat Prod 80:2360–2370
141. Torres JP, Lin Z, Fenton DS, Leavitt LU, Niu C, Lam P-Y, Robes JM, Peterson RT, Concepcion GP, Haygood MG, Olivera BM, Schmidt EW (2020) Boholamide A, an APD-class, hypoxia-selective cyclodepsipeptide. J Nat Prod 83:1249–1257
142. Voight JR (2015) Xylotrophic bivalves: aspects of their biology and the impacts of humans. J Molluscan Stud 81:175–186
143. Shipway JR, Altamia MA, Rosenberg G, Concepcion GP, Haygood MG, Distel DL (2019) *Tamilokus mabinia*, a new, anatomically divergent genus and species of wood-boring bivalve from the Philippines. PeerJ 7:e6256
144. Shipway JR, Altamia MA, Rosenberg G, Concepcion GP, Haygood MG, Distel DL (2019) A rock-boring and rock-ingesting freshwater bivalve (shipworm) from the Philippines. Proc R Soc B: Biol Sci 286:20190434
145. Shipway JR, Rosenberg G, Concepcion GP, Haygood MG, Savrda C, Distel DL (2019) Shipworm bioerosion of lithic substrates in a freshwater setting, Abatan River, Philippines: Ichnologic, paleoenvironmental and biogeomorphical implications. PLoS ONE 14:e0224551
146. Shipway JR, Altamia MA, Haga T, Velásquez M, Albano J, Dechavez R, Concepcion GP, Haygood MG, Distel DL (2018) Observations on the life history and geographic range of the giant chemosymbiotic shipworm *Kuphus polythalamius* (Bivalvia: Teredinidae). Biol Bull 235:167–177
147. Distel DL, Altamia MA, Lin Z, Shipway JR, Han A, Forteza I, Antemano R, Limbaco MGJP, Tebo AG, Dechavez R (2017) Discovery of chemoautotrophic symbiosis in the giant shipworm *Kuphus polythalamia* (Bivalvia: Teredinidae) extends wooden-steps theory. Proc Natl Acad Sci 114:E3652–E3658
148. Altamia MA, Shipway JR, Stein D, Betcher MA, Fung JM, Jospin G, Eisen J, Haygood MG, Distel DL (2020) *Teredinibacter waterburyi*sp. nov., a marine, cellulolytic endosymbiotic bacterium isolated from the gills of the wood-boring mollusc *Bankia setacea* (Bivalvia: Teredinidae) and emended description of the genus *Teredinibacter*. Int J Syst Evol Microbiol 70:2388–2394
149. Altamia MA, Shipway JR, Stein D, Betcher MA, Fung JM, Jospin G, Eisen J, Haygood MG, Distel DL (2021) *Teredinibacter haidensis* sp. nov., *Teredinibacter purpureus* sp. nov. and *Teredinibacter franksiae* sp. nov., marine, cellulolytic endosymbiotic bacteria isolated from the gills of the wood-boring mollusc *Bankia setacea* (Bivalvia: Teredinidae) and emended description of the genus Teredinibacter. Int J System Evol Microbiol 71
150. Altamia MA, Lin Z, Trindade-Silva AE, Uy ID, Shipway JR, Wilke DV, Concepcion GP, Distel DL, Schmidt EW, Haygood MG (2020) Secondary metabolism in the gill microbiota of shipworms (Teredinidae) as revealed by comparison of metagenomes and nearly complete symbiont genomes. mSystems 5:e00261-20
151. Yang JC, Madupu R, Durkin AS, Ekborg NA, Pedamallu CS, Hostetler JB, Radune D, Toms BS, Henrissat B, Coutinho PM, Schwarz S, Field L, Trindade-Silva AE, Soares CAG, Elshahawi S, Hanora A, Schmidt EW, Haygood MG, Posfai J, Benner J, Madinger C, Nove J, Anton B, Chaudhary K, Foster J, Holman A, Kumar S, Lessard PA, Luyten YA, Slatko B, Wood N, Wu B, Teplitski M, Mougous JD, Ward N, Eisen JA, Badger JH, Distel DL (2009) The complete genome of *Teredinibacter turnerae* T7901: an intracellular endosymbiont of marine wood-boring bivalves (shipworms). PLoS ONE 4:e6085
152. Altamia MA, Wood N, Fung JM, Dedrick S, Linton EW, Concepcion GP, Haygood MG, Distel DL (2014) Genetic differentiation among isolates of *Teredinibacter turnerae*, a widely occurring intracellular endosymbiont of shipworms. Mol Ecol 23:1418–1432

153. Distel DL, Morrill W, MacLaren-Toussaint N, Franks D, Waterbury J (2002) *Teredinibacter turnerae* gen. nov., sp. nov., a dinitrogen-fixing, cellulolytic, endosymbiotic gammaproteobacterium isolated from the gills of wood-boring molluscs (Bivalvia: Teredinidae). Int J Syst Evol Microbiol 52:2261–2269
154. Villacorta JB, Rodriguez CV, Peran JE, Batucan JD, Concepcion GP, Salvador-Reyes LA, Junio HA (2022) Mining small molecules from *Teredinibacter turnerae* strains isolated from Philippine teredinidae. Metabolites 12:1152
155. Elshahawi SI, Trindade-Silva AE, Hanora A, Han AW, Flores MS, Vizzoni V, Schrago CG, Soares CA, Concepcion GP, Distel DL, Schmidt EW, Haygood MG (2013) Boronated tartrolon antibiotic produced by symbiotic cellulose-degrading bacteria in shipworm gills. Proc Natl Acad Sci 110:E295–E304
156. O'Connor RM, Nepveux VFJ, Abenoja J, Bowden G, Reis P, Beaushaw J, Bone Relat RM, Driskell I, Gimenez F, Riggs MW, Schaefer DA, Schmidt EW, Lin Z, Distel DL, Clardy J, Ramadhar TR, Allred DR, Fritz HM, Rathod P, Chery L, White J (2020) A symbiotic bacterium of shipworms produces a compound with broad spectrum anti-apicomplexan activity. PLoS Pathog 16:e1008600
157. Han AW, Sandy M, Fishman B, Trindade-Silva AE, Soares CAG, Distel DL, Butler A, Haygood MG (2013) Turnerbactin, a novel triscatecholate siderophore from the shipworm endosymbiont *Teredinibacter turnerae* T7901. PLoS ONE 8:e76151
158. Miller BW, Schmidt EW, Concepcion GP, Haygood MG (2022) Halogenated metal-binding compounds from shipworm symbionts. J Nat Prod 85:479–484
159. Sung Y, Miller BW, Schmidt EW (2025) Glutathione in the biosynthesis of glutazoline teredinibactins. Org Lett 27:1774–1778
160. Miller BW, Lim AL, Lin Z, Bailey J, Aoyagi KL, Fisher MA, Barrows LR, Manoil C, Schmidt EW, Haygood MG (2021) Shipworm symbiosis ecology-guided discovery of an antibiotic that kills colistin-resistant *Acinetobacter*. Cell Chem Biol 28:1628-1637.e1624
161. Lacerna NM, Ramones CMV, Robes JMD, Picart MRD, Tun JO, Miller BW, Haygood MG, Schmidt EW, Salvador-Reyes LA, Concepcion GP (2020) Inhibition of biofilm formation by modified oxylipins from the shipworm symbiont *Teredinibacter turnerae*. Mar Drugs 18:656
162. Peran JE, Salvador-Reyes LA (2024) Modified oxylipins as inhibitors of biofilm formation in *Staphylococcus epidermidis*. Front Pharmacol 15:1379643
163. Pineda EGS, Peran JE, Batucan JD, Acyatan ZG, Susana-Guevarra SR, Salvador-Reyes LA (2025) Characterization of a new lyso-ornithine lipid from the shipworm symbiont *Teredinibacter turnerae*. Phytochem Lett 67:102960
164. Lacerna NM II, Miller BW, Lim AL, Tun JO, Robes JMD, Cleofas MJB, Lin Z, Salvador-Reyes LA, Haygood MG, Schmidt EW, Concepcion GP (2019) Mindapyrroles A-C, pyoluteorin analogues from a shipworm-associated bacterium. J Nat Prod 82:1024–1028
165. Lorenz P, Jensen PR, Fenical W (1998) Mactanamide, a new fungistatic diketopiperazine produced by a marine *Aspergillus* sp. Nat Prod Lett 12:55–60
166. Notarte KI, Nakao Y, Yaguchi T, Bungihan M, Suganuma K, dela Cruz TE (2017) Trypanocidal activity, cytotoxicity and histone modifications induced by malformin A1 isolated from the marine-derived fungus *Aspergillus tubingensis* IFM 63452. Mycosphere 8:111–120
167. Tan MA, dela Cruz TEE, Apurillo CCS, Proksch P (2015) Chemical constituents from a Philippine mangrove endophytic fungi *Phyllosticta* sp. Der Pharma Chem 7:43–45
168. Tenebro CP, Trono DJVL, Vicera CVB, Sabido EM, Ysulat JA, Macaspac AJM, Tampus KA, Fabrigar TAP, Saludes JP, Dalisay DS (2021) Multiple strain analysis of *Streptomyces* species from Philippine marine sediments reveals intraspecies heterogeneity in antibiotic activities. Sci Rep 11:17544
169. Paderog MJV, Suarez AFL, Sabido EM, Low ZJ, Saludes JP, Dalisay DS (2020) Anthracycline shunt metabolites from Philippine marine sediment-derived *Streptomyces* destroy cell membrane integrity of multidrug-resistant *Staphylococcus aureus*. Front Microbiol 11
170. Sabido EM, Tenebro CP, Trono DJVL, Vicera CVB, Leonida SFL, Maybay JJWB, Reyes-Salarda R, Amago DS, Aguadera AMV, Octaviano MC, Saludes JP, Dalisay DS (2021) Insights into the variation in bioactivities of closely related *Streptomyces* strains from marine sediments of the Visayan Sea against ESKAPE and ovarian cancer. Mar Drugs 19:441

171. Maglonzo JRM, Sabido EM, Salibay CC, Dalisay DS, Saludes JP (2022) Antiproliferative activity of Philippine marine sediment-derived actinomycetes. Pharm Sci Res 9:139–148
172. Sabido EM, Tenebro CP, Suarez AFL, Ong SDC, Trono DJVL, Amago DS, Evangelista JE, Reynoso AMQ, Villalobos IGM, Alit LDD, Surigao CF, Villanueva CA, Saludes JP, Dalisay DS (2020) Marine sediment-derived *Streptomyces* strain produces angucycline antibiotics against multidrug-resistant *Staphylococcus aureus* harboring SCCmec type 1 gene. J Mar Sci Eng 8:734
173. Tenebro CP, Trono DJVL, Balida LAP, Bayog LKA, Bruna JR, Sabido EM, Caspe DPC, Santos ELCdL, Saludes JP, Dalisay DS (2023) Synergy between genome mining, metabolomics, and bioinformatics uncovers antibacterial chlorinated carbazole alkaloids and their biosynthetic gene cluster from *Streptomyces tubbatahanensis* sp. nov., a novel actinomycete isolated from Sulu Sea, Philippines. Microbiol Spectr 11:e03661–e03622
174. Puillandre N, Bouchet P, Duda TF, Kauferstein S, Kohn AJ, Olivera BM, Watkins M, Meyer C (2014) Molecular phylogeny and evolution of the cone snails (Gastropoda, Conoidea). Mol Phylogenet Evol 78:290–303
175. Olivera BM, Seger J, Horvath MP, Fedosov AE (2015) Prey-capture strategies of fish-hunting cone snails: Behavior, neurobiology and evolution. Brain Behav Evol 86:58–74
176. Favreau P, Stöcklin R (2009) Marine snail venoms: Use and trends in receptor and channel neuropharmacology. Curr Opin Pharmacol 9:594–601
177. Olivera BM, Rivier J, Clark C, Ramilo CA, Corpuz GP, Abogadie FC, Mena EE, Woodward SR, Hillyard DR, Cruz LJ (1990) Diversity of *Conus* neuropeptides. Science 249:257–263
178. Kohn AJ, Nishi M, Pernet B (1999) Snail spears and scimitars: a character analysis of Conus radular teeth. J Molluscan Stud 65:461–481
179. Olivera BM (1997) *Conus* venom peptides, receptor and ion channel targets, and drug design: 50 million years of neuropharmacology. Mol Biol Cell 8:2101–2109
180. Kaas Q, Yu R, Jin A-H, Dutertre S, Craik DJ (2011) ConoServer: updated content, knowledge, and discovery tools in the conopeptide database. Nucleic Acids Res 40:D325–D330
181. Robinson SD, Norton RS (2014) Conotoxin gene superfamilies. Mar Drugs 12:6058–6101
182. Jin A-H, Muttenthaler M, Dutertre S, Himaya SWA, Kaas Q, Craik DJ, Lewis RJ, Alewood PF (2019) Conotoxins: Chemistry and biology. Chem Rev 119:11510–11549
183. Lebbe EK, Tytgat J (2016) In the picture: Disulfide-poor conopeptides, a class of pharmacologically interesting compounds. J Venom Anim Toxins Incl Trop Dis 22:30
184. Cruz LJ, White J (2017) Clinical toxicology of Conus snail stings. In: White J, Meier J (eds) Handbook of clinical toxicology of animal venoms and poisons. CRC Press, Boca Raton, FL, pp 117–127
185. Dutertre S, Jin A-H, Alewood PF, Lewis RJ (2014) Intraspecific variations in *Conus geographus* defence-evoked venom and estimation of the human lethal dose. Toxicon 91:135–144
186. Hu H, Bandyopadhyay PK, Olivera BM, Yandell M (2012) Elucidation of the molecular envenomation strategy of the cone snail *Conus geographus* through transcriptome sequencing of its venom duct. BMC Genomics 13:284
187. Safavi-Hemami H, Hu H, Gorasia DG, Bandyopadhyay PK, Veith PD, Young ND, Reynolds EC, Yandell M, Olivera BM, Purcell AW (2014) Combined proteomic and transcriptomic interrogation of the venom gland of *Conus geographus* uncovers novel components and functional compartmentalization. Mol Cell Proteomics 13:938–953
188. Menting JG, Gajewiak J, MacRaild CA, Chou DH-C, Disotuar MM, Smith NA, Miller C, Erchegyi J, Rivier JE, Olivera BM, Forbes BE, Smith BJ, Norton RS, Safavi-Hemami H, Lawrence MC (2016) A minimized human insulin-receptor-binding motif revealed in a *Conus geographus* venom insulin. Nat Struct Mol Biol 23:916–920
189. Dutt M, Dutertre S, Jin A-H, Lavergne V, Alewood PF, Lewis RJ (2019) Venomics reveals venom complexity of the piscivorous cone snail, *Conus tulipa*. Mar Drugs 17:71
190. Ramiro IBL, Bjørn-Yoshimoto WE, Imperial JS, Gajewiak J, Salcedo PF, Watkins M, Taylor D, Resager W, Ueberheide B, Bräuner-Osborne H, Whitby FG, Hill CP, Martin LF, Patwardhan A, Concepcion GP, Olivera BM, Safavi-Hemami H (2022) Somatostatin venom analogs

evolved by fish-hunting cone snails: from prey capture behavior to identifying drug leads. Sci Adv 8:eabk1410

191. Santos AD, McIntosh JM, Hillyard DR, Cruz LJ, Olivera BM (2004) The A-superfamily of conotoxins: Structural and functional divergence. J Biol Chem 279:17596–17606
192. Cruz LJ, Gray WR, Olivera BM (1978) Purification and properties of a myotoxin from *Conus geographus* venom. Arch Biochem Biophys 190:539–548
193. Gray WR, Luque A, Olivera BM, Barrett J, Cruz LJ (1981) Peptide toxins from *Conus geographus* venom. J Biol Chem 256:4734–4740
194. Ma X, Huang Q, Yu S, Xu S, Huang Y, Zhao Z, Xiao X, Dai Q (2021) The 3/4- and 3/6-subfamily variants of α-conotoxins GI and MI exhibit potent inhibitory activity against muscular nicotinic acetylcholine receptors. Mar Drugs 19:705
195. Guddat LW, Martin JA, Shan L, Edmundson AB, Gray WR (1996) Three-dimensional structure of the α-conotoxin GI at 1.2 Å resolution. Biochemistry 35:11329–11335
196. Johnson DS, Martinez J, Elgoyhen AB, Heinemann SF, McIntosh JM (1995) Alpha-Conotoxin ImI exhibits subtype-specific nicotinic acetylcholine receptor blockade: Preferential inhibition of homomeric alpha 7 and alpha 9 receptors. Mol Pharmacol 48:194–199
197. Groebe DR, Gray WR, Abramson SN (1997) Determinants involved in the affinity of α-conotoxins GI and SI for the muscle subtype of nicotinic acetylcholine receptors. Biochemistry 36:6469–6474
198. Ning J, Li R, Ren J, Zhangsun D, Zhu X, Wu Y, Luo S (2018) Alanine-scanning mutagenesis of α-Conotoxin GI reveals the residues crucial for activity at the muscle acetylcholine receptor. Mar Drugs 16:507
199. Wilhelm P, Luna-Ramirez K, Chin YKY, Dekan Z, Abraham N, Tae H-S, Chow CY, Eagles DA, King GF, Lewis RJ, Adams DJ, Alewood PF (2022) Cysteine-rich α-Conotoxin SII displays novel interactions at the muscle nicotinic acetylcholine receptor. ACS Chem Neurosci 13:1245–1250
200. Blount K, Johnson A, Prior C, Marshall IG (1992) α-Conotoxin GI produces tetanic fade at the rat neuromuscular junction. Toxicon 30:835–842
201. McManus OB, Musick JR, Gonzalez C (1981) Peptides isolated from the venom of *Conus geographus* block neuromuscular transmission. Neurosci Lett 25:57–62
202. McIntosh JM, Dowell C, Watkins M, Garrett JE, Yoshikami D, Olivera BM (2002) Alpha-Conotoxin GIC from *Conus geographus*, a novel peptide antagonist of nicotinic acetylcholine receptors. J Biol Chem 277:33610–33615
203. Nicke A, Loughnan ML, Millard EL, Alewood PF, Adams DJ, Daly NL, Craik DJ, Lewis RJ (2003) Isolation, structure, and activity of GID, a novel Alpha 4/7-Conotoxin with an extended N-terminal sequence. J Biol Chem 278:3137–3144
204. Cruz LJ, Gray WR, Olivera BM, Zeikus R, Kerr L, Yoshikami D, Moczydlowski E (1985) *Conus geographus* toxins that discriminate between neuronal and muscle sodium channels. J Biol Chem 260:9280–9288
205. Gajewiak J, Azam L, Imperial J, Walewska A, Green BR, Bandyopadhyay PK, Raghuraman S, Ueberheide B, Bern M, Zhou HM, Minassian NA, Hagan RH, Flinspach M, Liu Y, Bulaj G, Wickenden AD, Olivera BM, Yoshikami D, Zhang M-M (2014) A disulfide tether stabilizes the block of sodium channels by the conotoxin μO§-GVIIJ. Proc Natl Acad Sci 111:2758–2763
206. Olivera BM, McIntosh JM, Curz LJ, Luque FA, Gray WR (1984) Purification and sequence of a presynaptic peptide toxin from *Conus geographus* venom. Biochemistry 23:5087–5090
207. Olivera BM, Gray WR, Zeikus R, McIntosh JM, Varga J, Rivier J, de Santos V, Cruz LJ (1985) Peptide neurotoxins from fish-hunting Cone snails. Science 230:1338–1343
208. Raig AG, Norberg T, Griffin D, Hoeger C, Akhtar M, Schmidt K, Low W, Dykert J, Richelson E, Navarro V (1999) Contulakin-G, an O-glycosylated invertebrate neurotensin. J Biol Chem 274:13752–13759
209. Olivera BM, Cruz LJ, De Santos V, LeCheminant G, Griffin D, Zeikus R, McIntosh JM, Galyean R, Varga J (1987) Neuronal calcium channel antagonists. Discrimination between calcium channel subtypes using.omega.-conotoxin from *Conus magus* venom. Biochemistry 26:2086–2090

210. Hillyard DR, Monje VD, Mintz IM, Bean BP, Nadasdi L, Ramachandran J, Miljanich G, Azimi-Zoonooz A, McIntosh JM, Cruz LJ, Imperial JS, Olivera BM (1992) A new conus peptide ligand for mammalian presynaptic Ca^{2+} channels. Neuron 9:69–77
211. Monje VD, Haack JA, Naisbitt SR, Miljanich G, Ramachandran J, Nasdasdi L, Olivera BM, Hillyard DR, Gray WR (1993) A new *Conus* peptide ligand for Ca channel subtypes. Neuropharmacology 32:1141–1149
212. McIntosh JM, Yoshikami D, Mahe E, Nielsen DB, Rivier JE, Gray WR, Olivera BM (1994) A nicotinic acetylcholine receptor ligand of unique specificity, alpha-conotoxin ImI. J Biol Chem 269:16733–16739
213. Mendoza CB, Masacupan DJM, Batoctoy DCR, Yu ET, Lluisma AO, Salvador-Reyes LA (2019) Conomarphins cause paralysis in mollusk: Critical and tunable structural elements for bioactivity. J Pept Sci 25:e3179
214. Jacob RB, McDougal OM (2010) The M-superfamily of conotoxins: A review. Cell Mol Life Sci 67:17–27
215. Cruz LJ, Kupryszewski G, LeCheminant GW, Gray WR, Olivera BM, Rivier J (1989) Mu-Conotoxin GIIIA, a peptide ligand for muscle sodium channels: Chemical synthesis, radiolabeling and receptor characterization. Biochemistry 28:3437–3442
216. Li RA, Ennis IL, French RJ, Dudley SC Jr, Tomaselli GF, Marbán E (2001) Clockwise domain arrangement of the sodium channel revealed by (mu)-conotoxin (GIIIA) docking orientation. J Biol Chem 276:11072–11077
217. Sato K, Ishida Y, Wakamatsu K, Kato R, Honda H, Ohizumi Y, Nakamura H, Ohya M, Lancelin JM, Kohda D (1991) Active site of mu-conotoxin GIIIA, a peptide blocker of muscle sodium channels. J Biol Chem 266:16989–16991
218. Schroeder CI, Lewis RJ (2006) ω-conotoxins GVIA, MVIIA and CVID: SAR and clinical potential. Mar Drugs 4:193–214
219. Kerr LM, Yoshikami D (1984) A venom peptide with a novel presynaptic blocking action. Nature 308:282–284
220. Vink S, Alewood PF (2012) Targeting voltage-gated calcium channels: developments in peptide and small-molecule inhibitors for the treatment of neuropathic pain. Br J Pharmacol 167:970–989
221. Favreau P, Gilles N, Lamthanh H, Bournaud R, Shimahara T, Bouet F, Laboute P, Letourneux Y, Ménez A, Molgó J, Le Gall F (2001) A new ω-conotoxin that targets N-type voltage-sensitive calcium channels with unusual specificity. Biochemistry 40:14567–14575
222. Lewis RJ, Nielsen KJ, Craik DJ, Loughnan ML, Adams DA, Sharpe IA, Luchian T, Adams DJ, Bond T, Thomas L, Jones A, Matheson J-L, Drinkwater R, Andrews PR, Alewood PF (2000) Novel w-conotoxins from *Conus catus* discriminate among neuronal calcium channel subtypes. J Biol Chem 275:35335–35344
223. Rajendra W, Armugam A, Jeyaseelan K (2004) Toxins in anti-nociception and anti-inflammation. Toxicon 44:1–17
224. Sato K, Park NG, Kohno T, Maeda T, Kim JI, Kato R, Takahashi M (1993) Role of basic residues for the binding of ω-conotoxin GVIA to N-type calcium channels. Biochem Biophys Res Commun 194:1292–1296
225. Kim JI, Takahashi M, Ogura A, Kohno T, Kudo Y, Sato K (1994) Hydroxyl group of Tyr13 is essential for the activity of omega-conotoxin GVIA, a peptide toxin for N-type calcium channel. J Biol Chem 269:23876–23878
226. Lee H-K, Zhang L, Smith MD, Walewska A, Vellore NA, Baron R, McIntosh JM, White HS, Olivera BM, Bulaj G (2015) A marine analgesic peptide, Contulakin-G, and neurotensin are distinct agonists for neurotensin receptors: uncovering structural determinants of desensitization properties. Front Pharmacol 6
227. McIntosh JM, Olivera BM, Cruz LJ, Gray WR (1984) Gamma-carboxyglutamate in a neuroactive toxin. J Biol Chem 259:14343–14346
228. Olivera BM, McIntosh JM, Clark C, Middlemas D, Gray WR, Cruz LJ (1985) A sleep-inducing peptide from *Conus geographus* venom. Toxicon 23:277–282

229. Rigby AC, Baleja JD, Li L, Pedersen LG, Furie BC, Furie B (1997) Role of γ-carboxyglutamic acid in the calcium-induced structural transition of conantokin G, a conotoxin from the marine snail *Conus geographus*. Biochemistry 36:15677–15684
230. Mena EE, Gullak MF, Pagnozzi MJ, Richter KE, Rivier J, Cruz LJ, Olivera BM (1990) Conantokin-G: A novel peptide antagonist to the *N*-methyl-D-aspartic acid (NMDA) receptor. Neurosci Lett 118:241–244
231. Yoshikami D, Bagabaldo Z, Olivera BM (1989) The inhibitory effects of omega-conotoxins on Ca channels and synapses. Ann N Y Acad Sci 560:230–248
232. Miljanich GP (2004) Ziconotide: neuronal calcium channel blocker for treating severe chronic pain. Curr Med Chem 11:3029–3040
233. Jin A-H, Dutertre S, Dutt M, Lavergne V, Jones A, Lewis RJ, Alewood PF (2019) Transcriptomic-proteomic correlation in the predation-evoked venom of the Cone snail, *Conus imperialis*. Mar Drugs 17:177
234. Ye M, Khoo KK, Xu S, Zhou M, Boonyalai N, Perugini MA, Shao X, Chi C, Galea CA, Wang C, Norton RS (2012) A Helical conotoxin from *Conus imperialis* has a novel cysteine framework and defines a new superfamily. J Biol Chem 287:14973–14983
235. Yu S, Du T, Liu Z, Wu Q, Feng G, Dong M, Zhou X, Jiang L, Dai Q (2016) Im10A, a short conopeptide isolated from *Conus imperialis* and possesses two highly concentrated disulfide bridges and analgesic activity. Peptides 81:15–20
236. Itang CEMM, Gaza JT, Masacupan DJM, Batoctoy DCR, Chen Y-J, Nellas RB, Yu ET (2020) Identification of conomarphin variants in the *Conus eburneus* venom and the effect of sequence and PTM variations on conomarphin conformations. Mar Drugs 18:503
237. Liu Z, Bartels P, Sadeghi M, Du T, Dai Q, Zhu C, Yu S, Wang S, Dong M, Sun T, Guo J, Peng S, Jiang L, Adams DJ, Dai Q (2018) A novel α-conopeptide Eu1.6 inhibits N-type (CaV2.2) calcium channels and exhibits potent analgesic activity. Sci Rep 8:1004
238. Watkins M, Hillyard DR, Olivera BM (2006) Genes expressed in a turrid venom duct: Divergence and similarity to conotoxins. J Mol Evol 62:247–256
239. Heralde FM, Imperial J, Bandyopadhyay PK, Olivera BM, Concepcion GP, Santos AD (2008) A rapidly diverging superfamily of peptide toxins in venomous *Gemmula* species. Toxicon 51:890–897
240. Seronay RA, Fedosov AE, Astilla MAQ, Watkins M, Saguil N, Heralde FM, Tagaro S, Poppe GT, Aliño PM, Oliverio M, Kantor YI, Concepcion GP, Olivera BM (2010) Accessing novel conoidean venoms: biodiverse lumun-lumun marine communities, an untapped biological and toxinological resource. Toxicon 56:1257–1266
241. Gonzales DTT, Saloma CP (2014) A bioinformatics survey for conotoxin-like sequences in three turrid snail venom duct transcriptomes. Toxicon 92:66–74
242. Omaga CA, Carpio LD, Imperial JS, Daly NL, Gajewiak J, Flores MS, Espino SS, Christensen S, Filchakova OM, López-Vera E, Raghuraman S, Olivera BM, Concepcion GP (2017) Structure and biological activity of a turripeptide from *Unedogemmula bisaya* Venom. Biochemistry 56:6051–6060
243. Cabang AB, Imperial JS, Gajewiak J, Watkins M, Corneli PS, Olivera BM, Concepcion GP (2011) Characterization of a venom peptide from a crassispirid gastropod. Toxicon 58:672–680
244. Imperial JS, Cabang AB, Song J, Raghuraman S, Gajewiak J, Watkins M, Showers-Corneli P, Fedosov A, Concepcion GP, Terlau H, Teichert RW, Olivera BM (2014) A family of excitatory peptide toxins from venomous crassispirine snails: Using constellation pharmacology to assess bioactivity. Toxicon 89:45–54
245. Lirazan MB, Hooper D, Corpuz GP, Ramilo CA, Bandyopadhyay P, Cruz LJ, Olivera BM (2000) The spasmodic peptide defines a new conotoxin superfamily. Biochemistry 39:1583–1588
246. Miles LA, Dy CY, Nielsen J, Barnham KJ, Hinds MG, Olivera BM, Bulaj G, Norton RS (2002) Structure of a novel P-superfamily spasmodic conotoxin reveals an inhibitory cystine Knot Motif. J Biol Chem 277:43033–43040

247. Lu A, Watkins M, Li Q, Robinson SD, Concepcion GP, Yandell M, Weng Z, Olivera BM, Safavi-Hemami H, Fedosov AE (2020) Transcriptomic profiling reveals extraordinary diversity of venom peptides in unexplored predatory gastropods of the genus *Clavus*. Genome Biol Evol 12:684–700
248. Chua VM, Gajewiak J, Watkins M, Espino SS, Ramiro IBL, Omaga CA, Imperial JS, Carpio LPD, Fedosov A, Safavi-Hemami H, Salvador-Reyes LA, Olivera, BM, Concepcion GP (2020) Purification and characterization of the pink-Floyd Drillipeptide, a bioactive venom peptide from *Clavus davidgilmouri* (Gastropoda: Conoidea: Drilliidae). Toxins 12:508
249. Imperial JS, Watkins M, Chen P, Hillyard DR, Cruz LJ, Olivera BM (2003) The augertoxins: biochemical characterization of venom components from the toxoglossate gastropod *Terebra subulata*. Toxicon 42:391–398
250. Jenkins CN, Van Houtan KS (2016) Global and regional priorities for marine biodiversity protection. Biol Cons 204:333–339
251. Abdelmohsen UR, Grkovic T, Balasubramanian S, Kamel MS, Quinn RJ, Hentschel U (2015) Elicitation of secondary metabolism in actinomycetes. Biotechnol Adv 33:798–811
252. Teichert RW, Schmidt EW, Olivera BM (2015) Constellation pharmacology: A new paradigm for drug discovery. Annu Rev Pharmacol Toxicol 55:573–589
253. Al-Awadhi F, Salvador L, Luesch H (2016) Screening strategies for drug discovery and target identification. In: Baker BJ (ed) Marine biomedicine: from beach to bedside. CRC Press, Boca Raton, FL, pp 135–166
254. Wudu K, Abegaz A, Ayele L, Ybabe M (2023) The impacts of climate change on biodiversity loss and its remedial measures using nature based conservation approach: A global perspective. Biodivers Conserv 32:3681–3701
255. Birkmann J, Jamshed A, McMillan JM, Feldmeyer D, Totin E, Solecki W, Ibrahim ZZ, Roberts D, Kerr RB, Poertner H-O, Pelling M, Djalante R, Garschagen M, Leal Filho W, Guha-Sapir D, Alegría A (2022) Understanding human vulnerability to climate change: a global perspective on index validation for adaptation planning. Sci Total Environ 803:150065
256. Siqueira-Silva T, Martinez PA (2023) Impacts of climate change on the distribution of venomous *Conus* (Gastropoda: Conidae) species in the Indo-Pacific region. Mar Environ Res 192:106237
257. Guzman C, Conaco C (2016) Gene expression dynamics accompanying the sponge thermal stress response. PLoS ONE 11:e0165368
258. Botté ES, Nielsen S, Abdul Wahab MA, Webster J, Robbins S, Thomas T, Webster NS (2019) Changes in the metabolic potential of the sponge microbiome under ocean acidification. Nat Commun 10:4134
259. Chai G, Li J, Li Z (2024) The interactive effects of ocean acidification and warming on bioeroding sponge *Spheciospongia vesparium* microbiome indicated by metatranscriptomics. Microbiol Res 278:127542
260. Posadas N, Baquiran JIP, Nada MAL, Kelly M, Conaco C (2021) Microbiome diversity and host immune functions influence survivorship of sponge holobionts under future ocean conditions. ISME J 16:58–67
261. Del Monaco C, Hay ME, Gartrell P, Mumby PJ, Diaz-Pulido G (2017) Effects of ocean acidification on the potency of macroalgal allelopathy to a common coral. Sci Rep 7:41053
262. Guan C, Saha M, Weinberger F (2020) Simulated heatwaves lead to upregulated chemical defense of a marine foundation macrophyte against microbial colonizers. Front Mar Sci 7:463

Chapter 5
Marine Biodiversity and Marine Natural Products Research in Singapore

Lik Tong Tan, Beverly Pi Lee Goh, Loke Ming Chou, and Wei Liang Nicholas Yap

1 Introduction

Singapore, despite its small size and highly urbanized environment, hosts a remarkable diversity of marine life. The island nation's marine ecosystems include coral reefs, mangroves and seagrass meadows, which support a wide range of marine organisms. However, extensive coastal development, land reclamation and climate change have significantly altered the seascape, leading to habitat loss and biodiversity shifts. Over the decades, Singapore's marine biodiversity has faced challenges such as sedimentation, pollution and overexploitation, yet resilience in species richness has been observed. This chapter provides an overview of marine biodiversity in Singapore and highlights the importance of ongoing conservation efforts and research initiatives.

Alongside marine biodiversity studies, marine natural products research in Singapore has gained increasing attention. Marine organisms, including corals, sponges, tunicates and cyanobacteria, are rich sources of bioactive compounds with pharmaceutical potential. Investigations into these natural products have led to the discovery of antimicrobial, anticancer and cytotoxic agents, offering promising applications in drug development. With advancements in biotechnology and molecular techniques, researchers in Singapore continue to explore the chemical diversity of marine organisms, uncovering novel compounds that may contribute to medicine, biotechnology and environmental management.

L. T. Tan (✉) · B. P. L. Goh
Natural Sciences and Science Education, National Institute of Education, Nanyang Technological University, Singapore, Singapore
e-mail: liktong.tan@nie.edu.sg

L. M. Chou
Department of Biological Sciences, National University of Singapore, Singapore, Singapore

W. L. N. Yap
Tropical Marine Science Institute, St John's Island National Marine Laboratory, National University of Singapore, Singapore, Singapore

L. T. Tan and N. Hanif (eds.), *Marine Natural Products Research in South-East Asia*,
https://doi.org/10.1007/978-3-032-01098-8_5

Given Singapore's urban coastal environment, understanding the interplay between biodiversity conservation and marine natural product research is critical. In this chapter, the impacts of urbanization on marine life and the role of marine bioprospecting in sustainable development is explored. The chapter includes sections on the status of marine biodiversity in Singapore, such as sponge, anemone and soft coral diversity, marine natural products and marine chemical ecology research as well as perspectives on marine natural products research in Singapore. By integrating conservation science with drug discovery, Singapore aims to balance ecological protection with scientific innovation, ensuring the long-term viability of its marine resources.

2 Current State of Marine Biodiversity in Singapore

Coastal urbanisation since the 1960s has vastly altered Singapore's seascape. Land reclamation realigned shores while increasing the country's land area by 25% [1] at the expense of its much-limited sea space. More than two-thirds of the new coastline are now lined by seawalls [2]. The restricted sea space itself is heavily used by various sectors including shipping, industry, military, aquaculture and recreation. Ships dot the harbour while port operations proceed round the clock in one of the world's busiest ports.

Physical shrinkage of the marine environment was concomitant with loss of intertidal and marine habitats. Over 60% of coral reefs have since been buried by reclaimed land [3]. Mangrove forests covering 13% of land area in the past now occupy about 1% [4], while seagrass meadows were confined to an estimated 201 hectares [5]. Between 1953 and 1993, intertidal sand and mud flats were reduced from 33 to 8 km^2 [3]. Biodiversity flourishing on intertidal flats were severely constrained by conditions on seawalls. The restricted space of a steep slope with rapid tidal fluctuations is completely different to a natural gentle sloping shore. Besides, the hard structure of seawalls does not support burrowing animals or have crevices for species to seek shelter [6].

Along the urban shore, species richness and abundance differed widely between locations [7] and were influenced by habitat conditions [8]. The more stable conditions in the subtidal zone of seawalls favoured natural recruitment and development of corals [9]. This was also observed for jetty piles [10], while in modified habitats such as marinas, piles and other structures such as pontoons provided settlement opportunities for various species [11] and favourably enhanced marine biodiversity.

Population decline of many species was inevitable from habitat loss and urbanisation impacts. Over-harvesting of targeted species for food (like sea cucumbers) or the live aquarium trade (like anemones and anemone fishes) in the past almost decimated them from Singapore reefs [12]. However, coral species richness remained resilient as community composition changed and adapted to these impacts [1, 13].

Land reclamation and dredging of navigational channels in past decades proceeded without impact assessments, contributing to massive sediment loading of the waters and reducing average underwater visibility from 10 m in the 1960s to 2 m today [1]. Impact assessments and effective management from the late 1990s reduced sediment loading but constant water motion generated by numerous vessels did not permit fast settlement of sediment and resuspended bottom sediment. The National Biodiversity Centre under the National Parks Board was established in 2006 and for the first time, there was a specific agency to manage and protect marine biodiversity. Development projects were now required to minimise impacts to the coastal and marine environment and implement necessary mitigation measures. Improved management effectiveness supported by robust science and greater public awareness set the stage for marine biodiversity enhancement in the urbanising seascape [14]. A 5-year project to systematically survey marine habitats concluded in 2015 with the discovery of over 40 species new to science and over 120 species previously unrecorded from Singapore waters [15]. Ecological restoration efforts have been stepped up [16] and natural recovery has been documented [17]. These are all most relevant to maintaining ecosystem services as urbanisation continues.

Species richness has remained healthy despite physical transformation of the coastal and marine environment. Strong management measures and continuing research will contribute to ensuring the enhancement of marine biodiversity. Investigations of bioactive compounds from marine species began in the 1990s with work on the scleractinian coral, *Galaxea fascicularis.* Cytotoxic and anti-cancer agents were isolated from its mucus and characterised [18, 19]. Apoptotic activity against a multiple drug resistant leukaemia cell line was also demonstrated [20]. Bioactivity of gorgonians was also studied [21] for antimicrobial properties and toxicity to brine shrimp and the rodent malaria parasite, *Plasmodium berhei* [22]. Bioactive compounds conferring chemical defence to gorgonians [23] or have antifungal properties [24] have also been investigated.

3 Marine Natural Products and Marine Chemical Ecology Research in Singapore

3.1 Bioactivity Screening of Benthic Marine Invertebrates

3.1.1 Octocorals

The study on soft corals in Singapore's Southern Islands identified seven genera: *Carijoa, Cladiella, Sinularia, Lobophytum, Sarcophyton, Stereonephthya* and *Nephthea*, with an additional unidentified genus found at Kusu Island [25]. *Cladiella* sp. was the most abundant, while *Sarcophyton* sp. exhibited the largest colony coverage. Preliminary bioactivity tests using the brine shrimp toxicity assay revealed high toxicity levels in extracts from *Sarcophyton* and *Cladiella* sp., suggesting their potential as sources of bioactive compounds for drug discovery.

3.1.2 Gorgonians

The biodiversity of gorgonians in Singapore is notable given the challenging environmental conditions of the region. A checklist of 31 species from 12 genera and six families (including Anthothelidae, Subergorgiidae, Melithaeidae, Acanthogorgiidae, Plexauridae and Ellisellidae) has been compiled, highlighting the diversity of these octocorals despite high sedimentation and poor water transparency [26]. Singapore's gorgonian species are primarily found at depths of 12–18 m, in contrast to the deeper occurrence of gorgonians in other reefs. In addition, the highest number of gorgonians, in terms of abundance and species richness, was found on the lower slopes and bottom of reefs in Singapore [27]. Many species documented in Singapore tolerate sedimentation levels that are typically unsuitable for gorgonians, indicating their resilience. In fact, the growth of several species of gorognians were monitored and found to have relatively high rate of growth and regeneration [28]. The research by Goh and Chou also revealed 14 new or potentially new records, emphasizing the need for further taxonomic studies [26].

The bioactivity screening of Singapore gorgonians revealed diverse biological properties, including antimicrobial activity, ichthyotoxicity and toxicity to *Artemia salina* [22, 29]. For instance, nine out of ten gorgonian extracts displayed antimicrobial activity, with stronger inhibition against Gram-positive bacteria (*Bacillus subtilis*) than Gram-negative (*Escherichia coli*). Yeast inhibition (*Saccharomyces cerevisiae*) was specific to the family Ellisellidae, while no antifungal activity was detected against *Aspergillus* sp. [22]. Toxicity assays with *A. salina* showed variation across extracts, with chronic exposure (24 h) being more lethal than acute exposure (6 h). Their LD_{50} values ranged from 3.2 to 125.9 mg/mL, indicating differential toxicity levels. Of the ten gorgonian extracts, the extracts that exhibited the highest microbial activity were *Junceella* (*Dichotella*) cf. *gemmacea* and a species of *Echinogorgia*, while *Echinogorgia* sp. E and *Ctenocella* (*Umbracella*) sp. A were the most toxic in the brine shrimp toxicity assay. However, none of the gorgonian extracts demonstrated significant efficacy against *Plasmodium berghei* in malaria-infected mice, as survival rates and parasitemia levels remained comparable to untreated controls. These findings suggest that while Singapore gorgonians contain bioactive compounds with antimicrobial and toxic properties, their potential for anti-malarial applications remains uncertain. In spite of significant biological activity, none of the gorgonian species have been investigated for their natural products chemistry. Several chemical ecological studies have been conducted on gorgonians from Singapore and results are summarized in Sect. 3.2.1.

3.1.3 Sponges

The marine sponge diversity in Singapore comprises 82 species from 43 genera and 27 families, with common species including *Cinachyrella australiensis*, *Neopetrosia chaliniformis*, *Oceanapia sagittaria* and *Xestospongia testudinaria* [30, 31]. Sponge assemblages are highly variable across sites, with species richness ranging from 12 to 38 species per site. Western sites, such as Pulau Semakau and Terumbu Pempang Tengah, exhibit higher diversity, while eastern and southern sites, including St John's Island and Raffles Lighthouse, show more disturbance- adapted assemblages [32]. Chronic sedimentation from land reclamation has likely shaped sponge communities, favoring sediment- and low-light-resistant species while reducing overall biodiversity. The findings suggest that sponge communities in Singapore, though impacted, remain diverse but are subject to ongoing environmental pressures.

As part of a pilot scale project on drug discovery from local marine organisms, 19 intertidal marine organisms from Singapore waters were collected and screened for the presence of biologically-active natural products [33]. The organic extracts of these organisms were prepared and screened in the brine shrimp lethality, the cytotoxicity (MCF-7 and MOLT-4 cell lines) and the quorum sensing inhibition (*Chromobacterium violaceum* CV026) assays. A majority of the marine-derived extracts (about 60% or 12 species), with the exception of *Ceratodictyon spongiosum*, *Dactylospongia* sp.3, *Spheciospongia* sp.3 and *Haliclona* sp.1, exhibited significant activity (> 80% mortality) at the highest concentration of 1000 ppm. Furthermore, extracts of the marine sponge, *Dysidea* sp., exhibited 100% mortality rates for all test concentrations at 10 ppm and 1000 ppm.

3.2 *Marine Chemical Ecology Studies*

3.2.1 Chemical Ecology of Gorgonians from Singapore

Gorgonians found on Singapore's reefs are constantly subjected to environmental stressors, including heavy sedimentation, which may compromise the host's immune system. Several studies on local gorgonians have been carried out to investigate the ecological roles of gorgonians' natural products in conferring resilience of these invertebrates.

The first such chemical ecology studies involved assessing the toxicity and palatability of gorgonians' extracts [34]. Eight of ten gorgonian extracts were toxic to goldfish, with toxicity increasing over time and concentration, while *Euplexaura* cf. *pinnata*, a non-toxic extract, caused nervous behavior. Goldfish rejected pellets containing extracts at natural concentrations at rates ranging from 10% (*Subergorgia suberosa*, *Echinogorgia* sp. C) to 90% (*Ctenocella* cf. *umbraculum*), but extracts from *Echinogorgia* sp. A and *Euplexaura* cf. *pinnata* were not rejected. Higher extract concentrations led to greater rejection, with the lowest rate at 70% for *Junceella* cf. *gemmacea* at 5 times the concentration. The palatability of gorgonian extracts in Singapore is similar to that observed in Caribbean gorgonians.

Another study by Koh et al. focused on the chemical and physical defenses of Singapore gorgonians through secondary metabolites and structural defenses [23]. Bioassays using fish demonstrated that extracts from five species in the Ellisellidae family and three from the Plexauridae family exhibited deterrent properties. The deterrent effect was attributed to both chemical compounds and the presence of calcium carbonate-based sclerites, which physically deterred predation in some species. The study found that chemical defenses were more widespread than physical defenses, with some species showing a synergistic effect between different chemical compounds. However, only two species from the Ellisellidae family exhibited significant deterrence due to sclerites. These findings highlight the complex interplay between chemical and structural defenses in gorgonians, contributing to their survival in predator-rich coral reef environments.

Lastly, a study by Koh et al. examined whether gorgonians from Singaporean reefs could resist fungal settlement and invasion [24]. Extracts from 10 gorgonian species across three families were tested against nine gorgonian-derived deuteromycetous fungal species using Minimum Inhibitory Concentration (MIC) experiments. Several species, including *Euplexaura* cf. *pinnata* and *Echinogorgia* sp. C, exhibited antifungal activity, with MIC values ranging from 1.5 to 18.0 mg/mL. However, most gorgonian extracts required concentrations higher than their natural levels to inhibit fungal growth. Only extracts from *Echinogorgia* sp. C, *Ctenocella* cf. *umbraculum* and *Subergorgia suberosa* showed inhibitory effects at concentrations lower than their natural levels.

3.2.2 Prevalence of Chemical Defense in Marine Sponges at Pulau Hantu

A study by Ho and co-workers investigated the chemical defences of seven common intertidal marine sponge species, namely *Pseudoceratina purpurea*, *Spheciospongia* cf. *vagabunda*, *Halichondria cartilaginea*, *Haliclona* sp., *Stelletta clavosa*, *Clathria reinwardti* and *Hippospongia* sp., from Pulau Hantu, Singapore [35]. The chemical defence potential of these sponges were assessed using the brine shrimp (*Artemia salina*) toxicity assay. Such an assay has been shown to provide reasonable indication of the potential toxicity of marine natural products to fish [36]. The findings revealed that the relatively polar fractions of sponge extracts exhibited high toxicity, with at least one polar fraction from each species showing more than 50% toxicity at 100 ppm. Notably, *Hippospongia* sp. fractions displayed significant cytotoxicity against MOLT-4 leukemic cancer cell lines, suggesting a promising potential for anticancer agents. The study also utilized ^{1}H-NMR spectroscopy and MS-based molecular networking techniques to characterize the chemical nature of bioactive fractions, identifying molecular families, such as fatty acid amides, ceramides and bromotyrosine alkaloids, as well as novel metabolites. The main conclusion of the study was that chemical defences, evidenced by brine shrimp toxicity, were prevalent among the common intertidal sponge species at Pulau Hantu.

3.2.3 Chemical Variations in *Speciospongia*cf. *Inconstans*

A recent study (unpublished) was conducted to determine if three morphotypes of the sponge *Speciospongia* cf. *inconstans* collected from Pulau (P.) Hantu, Singapore

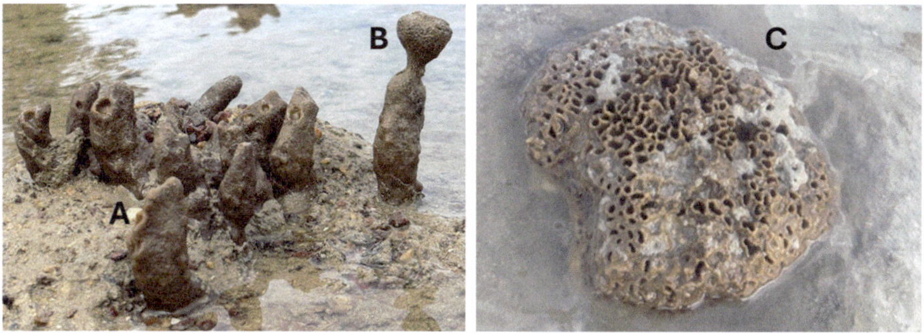

Fig. 1 Three morphotypes (A to C) of *S.* cf. *inconstans* collected from P. Hantu. Note globular projection on the top of morphotype B

were chemically diverse and distinct (Fig. 1). The clionaid endopsammic sponge, *Spheciospongia* cf. *inconstans*, is a common benthic invertebrate found at P. Hantu and has a highly variable morphology ranging from large cushion shapes to masses buried within the sand, with thick, tapering fistules protruding from the substrate [37]. Sponge samples from three distinct morphotypes (Fig. 1) were collected and extracted with MeOH and CH_2Cl_2 for chemical analysis through ^{1}H NMR spectroscopy. Quorum Sensing Inhibition (QSI) assay was also used to determine if inhibitory activity would correlate with differences in chemical variability. Results of ^{1}H NMR spectroscopy were analyzed and no statistically significant chemical variability was evident among the different morphotypes at the crude extract level. However, promising results of chemical dissimilarity was observed between positions of morphotype 'B' at the extract level from the QSI assay with extracts from the 'top' of this morphotype ranking high in inhibitory activity compared to other morphotypes (Fig. 2). Interestingly, the top of morphotype 'B' has a globular attachment. These globular projections break off relatively easily and may form new sponge colonies. As such, it would be interesting to investigate the chemical constituents of the globular lumps if it is found to be more chemically defended. Results obtained on the VLC-derived fractions of morphotype 'A' also showed dissimilarities, with fractions 2, 3 and 4 indicating the presence of unique compounds likely related to the actinosporin structural class [38].

3.3 Marine-Derived Antifoulants

3.3.1 Antifouling Activity of Sponge and Soft Coral Extracts

A few studies, involving field-based and lab-based experiments, have been conducted on antifouling activity of extracts derived from several benthic marine macroorganisms, including sponges, soft corals and algae. One study reported that crude extracts, particularly from the soft coral, *Dendronephthya* sp., significantly reduced

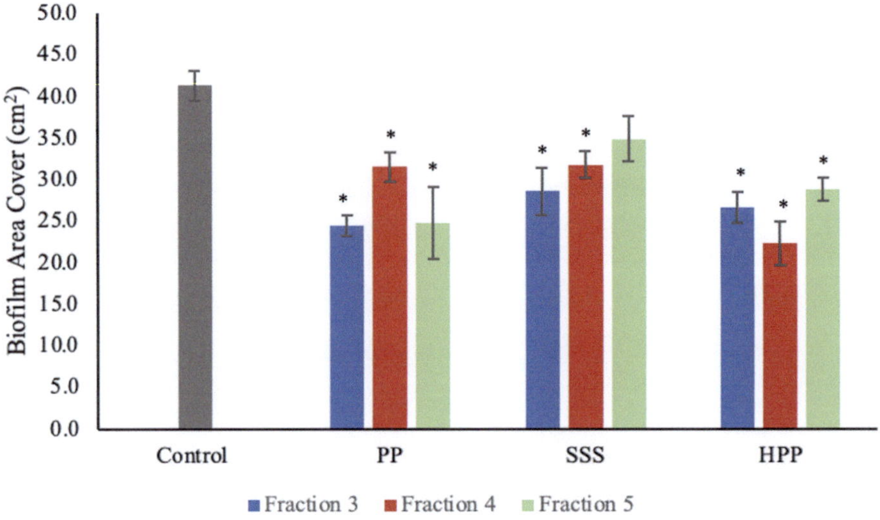

Fig. 2 Mean area cover of biofilm on Phytagels after one week of exposure to three polar fractions of sponges tested at 500 ppm concentration in field assay. (All bars are mean values of five replicates, vertical bars indicate standard error). Treatments include: Control = negative control Phytagel, PP = *Pseudoceratina purpurea*, SSS = *Speciospongia* sp., HPP = *Hippospongia* sp. Means with asterisks (*) indicate significantly lower biofilm cover compared to Controls ($p < 0.05$; n = 9). Treatments of all polar fractions except Fraction 5 of *Speciospongia* sp. resulted in significantly lower biofilm cover compared to the control from Dunnett's post-hoc test

biofouling, including biofilm formation and invertebrate larval settlement, while extracts from the calcareous green alga, *Halimeda opuntia*, did not show significant differences from controls in 28-day field testing [39]. Unpublished research on fractions from extracts of various sponge species found that, during the first week of a 28-day field study, treatments containing polar fractions from *Pseudoceratina purpurea*, *Sphecìospongia* sp. and *Hippospongia* sp. resulted in significantly lower biofilm cover compared to controls. (Fig. 2). Additionally, the study (unpublished) showed that fractions from extracts of five sponge species, namely, *Pesudoceratina purpurea, Speciospongia* sp., *Hippospongia* sp., *Haliclona* sp. and *Stellata clavosa*, elicited lower production of byssal threads in the green mussel, *Perna viridis* in 12 h laboratory-based assays, compared with controls (Fig. 3). Antifouling activity in the sponge genus *Hippospongia* has also been reported by others. For instance, extracts from *Hippospongia*, have been linked to inhibition of barnacle settlement [40, 41].

3.3.2 Antifoulants from Filamentous Marine Cyanobacteria

In 2010, Tan and co-workers showed that dolastatin 16 (**1**) (Fig. 4), isolated from *Lyngbya majuscula* from P. Hantu, exhibited significant activity against the larval settlement and metamorphosis of the barnacle, *A. amphitrite*, with EC_{50} and LC_{50}

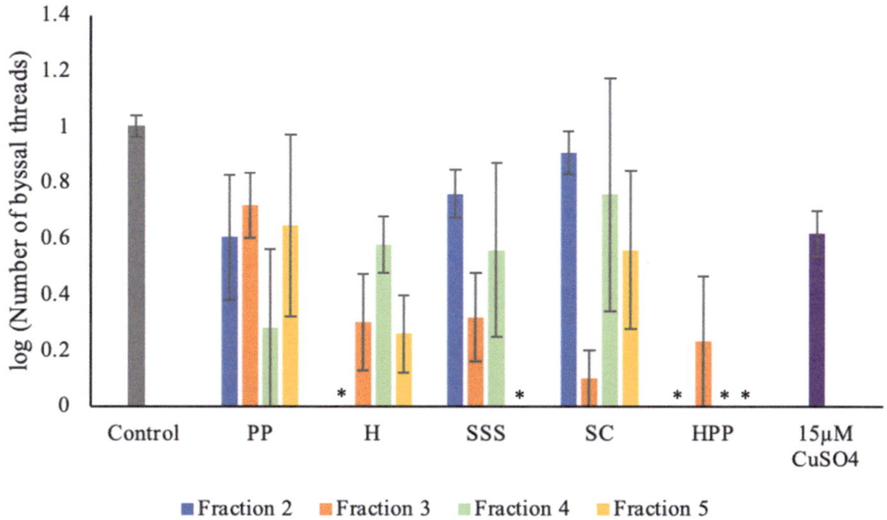

Fig. 3 Logarithmic transformed mean number of byssal threads produced by mussels after 12 h exposure to four fractions of sponges tested at 50 ppm in byssal thread assay (bars have mean values of three replicates, vertical bars indicate standard error). Treatments include: Control = negative control with EtOAc, PP = *Pseudoceratina purpurea*, H = *Haliclona* sp., SSS = *Speciospongia* sp., SC = *Stelleta clavosa*, HPP = *Hippospongia* sp. Byssal thread production exposed to four fractions that were significantly lower than that in the controls were Fraction 2 of *Haliclona* sp., Fraction 5 of *Speciospongia* sp., and Fractions 2, 4 and 5 of *Hippospongia* sp., from Dunnett's post-hoc test ($p < 0.05$; n = 3)

values of 0.003 and 20 µg/mL, respectively [42]. This finding highlighted dolastatin 16 potential as a lead compound for the development of novel antifouling agents. Dolastatin 16 is a natural cyclic depsipeptide previously isolated from the sea hare, *Dolabella auricularia* [43]. The molecule features two rare amino acids, dolaphenvaline and dolamethylleuin, and was first discovered by Pettit and colleagues in 1997 [43]. Due to the exquisite biological activity of dolastatin 16, including anticancer activity, there are at least three reports on its total synthesis and derivatives [44–46].

Subsequent chemical studies in 2017 revealed that synthetic dolastatin 16 and its two intermediates, namely the northern carboxylic acid fragment **2** and southern amine fragment **3** (Fig. 4), displayed varying levels of antifouling activity [44]. The total synthesis of dolastatin 16 was achieved using a convergent method, involving the scalable and concise preparation of the southern and northern segments **2** and **5** (Fig. 4), and efficient assembly of the two segments to construct the macrocyclic skeleton of dolastatin 16. While synthetic dolastatin 16 exhibited highly potent activity, fragments **2** and **3** showed moderate to low activity (EC_{50} > 10 and 1.17 µg/mL, respectively) against the settlement of *A. amphitrite*.

In 2022 a team of researchers in Japan synthesized and evaluated derivatives of dolaphenvaline, dolamethylleuine and dolastatin 16 to explore their potential as eco-friendly antifouling materials [46]. All tested compounds exhibited antifouling

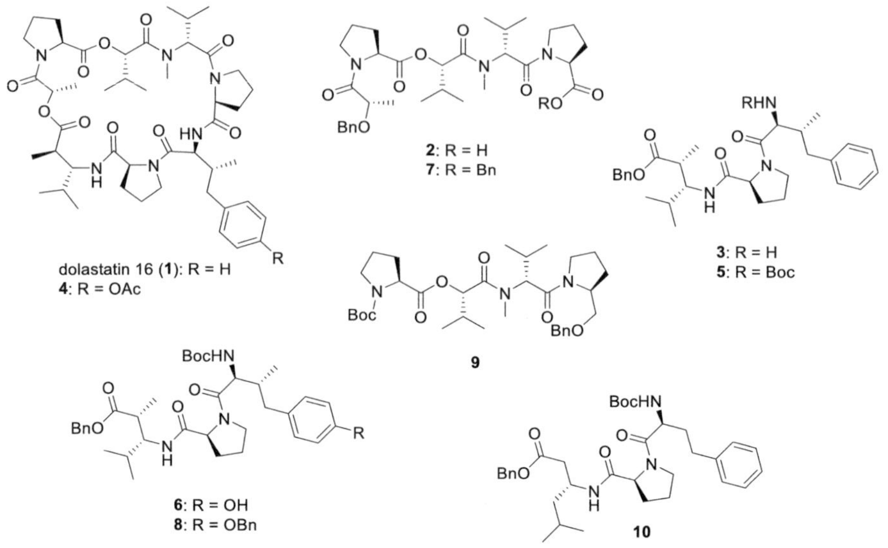

Fig. 4 Dolastatin 16 and synthetic derivatives as antifouling agents

properties with low toxicity toward cyprid larvae of *A. amphitrite*. Structure–activity relationship (SAR) analysis revealed that modifying the aromatic ring of dolastatin 16 by introducing a functional group, i.e., **4** (Fig. 4), reduced its antifouling activity to a moderate level with EC_{50} of 1.74 μg/mL. In contrast, compounds **5**, **6** and **7** (Fig. 4) demonstrated higher activity, with EC_{50} values below 1 μg/mL. In addition, protecting the southern fragment with a Boc group improved activity (**5**, EC_{50} = 0.79 μg/mL), likely due to increased hydrophobicity compared to the unprotected fragment **3** (EC_{50} = 1.17 μg/mL). The antifouling activity of the southern fragment was also influenced by the steric bulkiness of functional groups at the para-position of the aromatic ring. A hydroxy group (**6**, EC_{50} = 0.60 μg/mL) slightly enhanced activity compared to **5**, while a benzyloxy group (**8**, EC_{50} = 4.62 μg/mL) significantly reduced activity. For the northern fragment, a benzyl ester (**7**, EC_{50} = 0.90 μg/mL) showed much greater potency than the more polar fragment **2** (EC_{50} > 10 μg/mL), suggesting that higher hydrophobicity enhances antifouling activity. Conversely, a benzyl ether (**9**, EC_{50} = 3.27 μg/mL) decreased activity, indicating that the lactate moiety or carbonyl group in the northern fragment plays a critical role in maintaining antifouling efficacy [46].

Inspired by the structure of dolastatin 16, a recent paper details the efficient syntheses of tripeptides containing β-amino acids and their settlement inhibition activities against *A. amphitrite* and the blue mussel, *Mytilus galloprovincialis* [47]. For instance, the EC_{50} value of the most active compound, **10** (Fig. 4), was found to be more potent than the original tripeptide, **5**, which consisted of dolamethylluecine and dolaphenvaline. Additionally, the EC_{50} value of **10** was similar to that of $CuSO_4$ (EC_{50} = 0.29 mg/mL). Furthermore, these oligopeptides could be easily synthesized in just a few steps from commercially available amino acids, providing a cost-effective route for large-scale production.

3.4 *Toxins from Marine Invertebrates*

3.4.1 Sea Anemone

Sea anemones are a successful group of cnidarians that can be found in a wide range of habitats, from the tropical intertidal shores to the deep sea [48, 49]. Many species are engaged in vital ecological roles, e.g., symbiosis with clownfish, and are important in both the hobbyist-aquarium trade, and in heritage cuisines [48]. These marine invertebrates are also well-known for their ability to deliver a painful sting [50]. Some species, such as *Dolfeinia armata* Wassilieff, 1908, been identified be a threat to public health, due to the potent venom for which they harbour in their stings [50]. Like all cnidarians, sea anemones secrete an intra-cellular capsule known as 'nematocyst' from the Golgi Apparatus of a specialized cell [51]. Nematocysts are used by the animal for prey capture and defense, for each contains a barb that and venom that when a stimulus applied, explosively discharges and pierces into the victim, for which the venom is dispensed.

In Singapore, over 40 species of sea anemones have been documented, although much of this inventory work took place over the past four recent decades [50–53]. For the longest time, despite being abundant and commonly encountered in many marine habitats of Singapore, studies pertaining to sea anemones in Singapore and nearby waters were largely neglected [51]. Past neglect of sea anemones research in the region stems from the fact that these animals are difficulty to identify to species, and taxonomic uncertainties for many common taxa encountered in Singapore persist [51, 54]. In recent years, name-bearing type specimens for many Indo-Pacific species have been clarified, aiding in efforts in documenting these invertebrates from Singapore [49, 54]. Likewise, the integrative use of molecular data has helped clarify the identities of sea anemones in Singapore [50].

Similar to the limited studies relating on the biology and ecology of sea anemones in Singapore, research on their natural products remains scarce. Only *Radianthus magnifica* (previously known as *Heteractis magnifica*) have been researched locally, of which the sea anemone's toxins were characterized. As such, species of sea anemones that are recognizable and are known to harbour a potent sting should be investigated. Among the diversity of sea anemones that occur in Singapore, three potential species that are known to deliver potent stings, occur in the intertidal and should be investigated for novel toxins, including *Dofleinia armata*, *Actinodendron arboreum* and *Actinostephenus haeckeli*. Collectively, these three species are known among citizen scientists and nature enthusiasts as the "Hell's fire anemones" so named because these can deliver sting that gives an intense, burning sensation. While primary characterization of the toxins of both *D. armata* and *A. haeckeli* have been conducted by researchers elsewhere [55]; the venom from *A. arboreum* has yet to be characterized *Heteractis magnifica*, commonly known as the magnificent sea anemone, is a large and colorful species found in tropical Indo-Pacific waters. It forms symbiotic relationships with clownfish, providing them protection while benefiting from their waste nutrients. This anemone produces toxins, primarily composed of peptides and proteins, which are delivered through specialized stinging cells called nematocysts. These toxins help in capturing prey and deterring predators by causing paralysis or irritation. While not usually dangerous to humans, contact with its tentacles can result in painful stings, redness or allergic reactions in sensitive individuals.

The research on toxins from *Heteractis magnifica* collected from Singapore has yielded significant insights into their structure, biochemical nature and biological activity. These toxins can be broadly categorized into neurotoxins and cytolysins, both of which have unique properties and mechanisms of action. A key discovery is HmK, a potassium channel toxin from *H. magnifica* [56]. HmK is a 35-amino acid peptide with a molecular weight of 4055 Da, structurally stabilized by three disulfide bonds. It inhibits the binding of α-dendrotoxin to rat brain synaptosomal membranes, with a high binding affinity (Ki $\approx$ 1 nM), and effectively blocks voltage-gated K^+ currents in Kv1.2 channels. Additionally, HmK enhances acetylcholine release at the neuromuscular junction, suggesting a role in modulating synaptic activity. Notably, the toxin's gene has been cloned, expressed in *E. coli*, and purified as a recombinant protein, showing identical properties to the native form. Structural comparisons with other sea anemone K^+ channel toxins, such as ShK from *Stichodactyla helianthus* and BgK from *Bunodosoma granulifera*, indicate that specific conserved residues play critical roles in binding to K^+ channels [57].

Another major group of toxins from *H. magnifica* are the cytolysins, collectively referred to as magnificalysins (HMgs). These are 19.5 kDa proteins belonging to a multigene family [58–60]. The diversity in cytolysin isoforms suggests a possible evolutionary strategy for the anemone to enhance its defensive and predatory capabilities. These cytolysins exhibit haemolytic activity, with recombinant versions displaying lower potency compared to the native forms. Studies have shown that the N-terminal region is crucial for membrane insertion and pore formation, and modifications in this domain significantly alter haemolytic activity. The research also highlights that cytolysins from *H. magnifica* share similarities with other actinoporins, such as equinatoxins from *Actinia equina* [61]. These toxins function by binding to lipid membranes and forming pores, which lead to cell lysis. Structural studies suggest that specific hydrophobic and aromatic residues in the toxin molecule facilitate membrane interaction and insertion.

The findings on *H. magnifica* toxins by local researchers have important implications in neuroscience and pharmacology. HmK's ability to selectively block Kv1.2 channels and modulate neurotransmitter release makes it a potential candidate for studying ion channel functions and developing neuroactive drugs. Meanwhile, the cytolysins, with their potent membrane-disrupting activity, could have applications in biomedical research, such as targeted cancer therapy, where pore-forming toxins are explored for their ability to selectively lyse cancer cells. As such, research on

Heteractis magnifica toxins has provided a detailed biochemical and functional characterization of both potassium channel toxins and cytolysins. These studies offer a deeper understanding of venomous mechanisms in marine organisms and open avenues for biotechnological applications [62].

3.4.2 Hard Corals

Galaxea fascicularis, commonly known as the galaxy coral or starburst coral, is a stony coral species found in tropical reef environments across the Indo-Pacific region. It is characterized by its striking, star-like polyps with elongated, tapered tentacles that often display vibrant colors, such as green, brown or white. These corals form large colonies with a branching or encrusting growth pattern, contributing to reef-building processes. *Galaxea fascicularis* is a photosynthetic coral, housing symbiotic zooxanthellae that provide energy through photosynthesis. In addition, two different types of tentacles are present in *G. fascilcularis*, including catch tentacles, used for catching prey and deterring predators, and sweeper tentacles, specialized tentacles used for territorial aggression. Studies by Yosef and co-workers found that catch tentacles possess higher neurotoxicity to blowfly larvae and hemolytic activity compared to the sweeper tentacles [63]. This is consistent with a role in prey capture for catch tentacles. In contrast, sweeper tentacles have higher phospholipase A2 activity, which has been suggested they may have a role in inducing tissue damage during territorial aggression. Furthermore, the expression of genes encoding cytolytic toxins and phospholipases differs between the tentacle types.

Early studies on the bioactive components derived from *G. fascilcularis* were conducted in the mid-1990s, primarily by Ding and associates at the National University of Singapore. The mucus of this hard coral species has been demonstrated to exhibit cytotoxic activity against normal liver cells (BL8L), with a lethal dose (LD_{50}) of 7.49 μg protein/mL [18]. However, after five repeated inoculations over a 10-day period, the BL8L cells developed resistance to the coral mucus. When the cells were allowed to grow for four weeks in the absence of MS, subsequent exposure to the mucus failed to induce cell death. In comparison, MS exhibited twice the cytotoxicity against transformed rat hepatoma cells (JB1), with an LD_{50} of 3.49 μg protein/mL. These cancerous hepatoma cells became completely non-viable within 96 h of treatment.

Chromatographic separation of the crude mucus using CM Sepharose revealed 15 distinct peaks, nine of which exhibited cytotoxic activity against rat liver cells [19]. Further purification of the most cytotoxic peak through HPLC yielded a compound designated as S2. This compound, with a molecular weight of 742 Da, was identified as an *N*-linked sialylated glycan containing the NeuNacα(2–6)Galβ(1–4)GlcNAc moiety. S2 demonstrated selective cytotoxicity, being most effective against a multidrug-resistant murine leukemia cell line (P388/VCR), with an LD_{50} of 0.3 μg/mL. The compound's cytotoxic mechanism involved anti-topoisomerase activity [64]. Specifically, S2 inhibited the DNA relaxation function of both topoisomerase I and II. Since elevated topoisomerase II expression is associated with higher-grade tumors, cells expressing high levels of this enzyme were more susceptible to S2's effects. This specificity explains the selective cytotoxicity of S2 toward the P388/VCR cell line, while sparing normal fetal human lung cells [19]. Additionally, S2 stabilized the topoisomerase I-DNA cleavable complex, disrupting DNA replication and inducing cell death, underscoring its potential as an anticancer therapeutic candidate.

Another bioactive component isolated from *G. fascicularis* mucus, designated P6, demonstrated a distinct mechanism of action. P6 specifically induced in situ DNA digestion in P388/VCR cells within 30 min of treatment [20]. After three days of incubation, DNA degradation coincided with complete cell death. Fluorescent labeling of fragmented DNA revealed that P6 triggered apoptosis as early as 1.5 h after treatment. By day three, all treated leukemia cells were apoptotic. In contrast, P6 exhibited no DNA digestion or apoptosis-inducing effects on normal BL8L liver cells. These findings highlight the potential of P6 as an anticancer agent, warranting further studies to isolate and characterize the apoptotic factor within P6 and its role in cancer therapy development.

3.4.3 *Lophozozymus pictor* (Thunder Crab)

Lophozozymus pictor, commonly known as the mosaic reef crab, is a highly toxic species of marine crab found in coral reefs and rocky shorelines across the Indo-Pacific region. This crab is renowned for its striking, colorful shell patterns and its potent toxins, including saxitoxin and tetrodotoxin [65, 66]. Even in small amounts, these toxins, such as saxitoxin, can be lethal, making the crab unfit for human consumption. As such, food poisoning incidents caused by the ingestion of this crab have been reported [67]. Despite its toxicity, *L. pictor* plays an important ecological role within its habitat, contributing to the reef's biodiversity and acting as both predator and prey in the marine food web. Its vivid appearance and unique adaptations continue to attract scientific interest, particularly in studies of marine toxins and ecological interactions.

Lophozozymus pictor has been extensively studied for its potent toxins. Research spanning several decades has explored the crab's toxicity, toxin composition, mode of action, bioactivity and implications for human health. Early investigations demonstrated that aqueous extracts from *L. pictor* were highly lethal to mice when administered intraperitoneally or orally [68]. Toxicity studies revealed a dose-dependent relationship, with symptoms including limb paralysis, labored respiration, convulsions and ultimately, death. The toxin was found to be a potent respiratory depressant, leading to a progressive decline in blood pressure without affecting neuromuscular transmission. Further pharmacological analyses identified that the toxin affects sodium and potassium ion transport, similar to the mechanism of palytoxin (PTX), one of the most potent marine biotoxins known [69]. Comparative studies showed that *L. pictor* toxin (LPTX) differs from other marine neurotoxins such as tetrodotoxin and saxitoxin in its prolonged toxicity effects, as evidenced by extended survival times in lower doses.

Chemical analyses have provided insights into the molecular nature of LPTX. Mass spectrometry and chromatography studies revealed that LPTX shares similarities with PTX but differs in its spectral properties, displaying a unique fluorescence profile [70]. The toxin's molecular weight ranges between 1000 and 5000 Da, suggesting a complex polypeptide or large non-ribosomal peptide structure. Antibody-based studies indicated that LPTX cross-reacts with anti-PTX antibodies, reinforcing the structural relatedness of these toxins [71]. However, its distinct bioactivity profile and chromatographic behavior suggest the presence of novel functional groups or modifications that differentiate it from known marine toxins.

Beyond its systemic toxicity, LPTX exhibits strong cytotoxicity against mammalian cell lines, particularly HeLa cells [72]. Studies demonstrated that the toxin disrupts cell adhesion and induces morphological changes characteristic of apoptosis and necrosis. The minimum cytotoxic concentration was found to be 0.006 mouse units (MU)/mL, making it significantly more potent than conventional chemotherapeutic agents at equivalent doses. Neuropharmacological investigations revealed that LPTX is a potent inhibitor of synaptosomal GABA uptake in rat brain tissue [73, 74]. This suggests that the toxin could modulate neurotransmission by interfering with inhibitory synaptic function, potentially contributing to its lethal effects through sustained excitotoxicity.

Early reports suggested a uniform distribution of the toxin throughout the crab's body. However, later research using bioassays and chemical extraction techniques demonstrated that toxicity was concentrated in the hepatopancreas and gut, while the muscle and carapace exhibited significantly lower toxicity [75]. This localization suggests that the toxin is acquired through diet and subsequently accumulated in the digestive organs. Interestingly, *L. pictor* specimens maintained in captivity for extended periods (24 days) exhibited a marked decline in toxicity, with almost complete toxin depletion observed [75]. This finding strongly supports the hypothesis that *L. pictor* does not synthesize its own toxins but rather bioaccumulates them from dietary sources, possibly toxic dinoflagellates or cyanobacteria in coral reef ecosystems.

The consumption of *L. pictor* has been linked to several cases of severe poisoning and fatalities in Singapore and neighboring regions [67]. Symptoms of poisoning closely resemble those induced by PTX, including cardiovascular collapse and neuromuscular paralysis. Given the variation in individual crab toxicity, accidental ingestion poses a severe public health risk. Additionally, the presence of LPTX in reef-associated food chains raises concerns about bioaccumulation in other marine species, potentially impacting seafood safety. Future studies are needed to explore the ecological role of *L. pictor* toxins and their persistence in marine environments.

The research on *Lophozozymus pictor* in Singapore has significantly expanded our understanding of marine natural products and biotoxins. The discovery of LPTX as a potent cytotoxin and neurotoxin, structurally related to palytoxin, underscores the ecological and biomedical significance of reef-associated organisms. The findings also highlight the importance of monitoring marine food sources to prevent toxin-related health risks. Future studies should focus on elucidating the precise biosynthetic origins of LPTX and its broader applications in pharmacology and marine toxicology.

3.5 Isolation of Bioactive Marine Natural Products

3.5.1 Tunicates

Reports on the isolation and structural determination of natural products from marine organisms found in Singapore water are limited to several taxonomic groups, including tunicates, sponges and marine cyanobacteria. To date, about four compounds have been reported from four marine invertebrate species. One notable discovery is patellamide E (**11**) (Fig. 5), a cyclic peptide isolated from the ascidian, *Lissoclinum patella*, collected at Pulau Salu, Singapore [76]. Patellamide E was characterized and purified using extensive NMR spectroscopy and HPLC analysis, which confirmed its unique structural features, including thiazole and oxazoline rings. The compound exhibited mild cytotoxicity against human colon tumor cells. Previous studies revealed that the patellamides are highly bioactive compounds that, along with other cyanobactins, are produced in the symbiotic relationship between didemnid ascidians and the cyanobacterium *Prochloron* [77]. The biosynthetic pathway of patellamide synthesis is well characterized, with the associated operons identified within the *Prochloron* genome [78].

From a collection of an unidentified ascidian sample, collected from a reef slope at 10 m off Pulau Subar Laut, Singapore, three new pyridoacridine alkaloids, including kuanoniamines E (**15**) and F (**16**) and subarine (**17**) (Fig. 5), were obtained [79]. In addition, known pyridoacridine alkaloids, such as ascididemin (**18**) and kuanoniamines A (**12**), C (**13**) and D (**14**) (Fig. 5), were also identified [80, 81]. Pyridoacridine alkaloids are unique marine nitrogen-containing molecules that represent a large family of alkaloids [82]. They have been reported from diverse marine organisms, such as sponges, ascidians, anemones, prosobranch mollusk and tunicates. These alkaloids are known for their diverse biological activities, including anti-HIV, topoisomerase II inhibition, DNA intercalation and insecticidal properties. Subarine, a ring-opened pyridoacridine alkaloid, was found to potentially represent a biosynthetic intermediate in the formation of other pentacyclic pyridoacridines.

3.5.2 Sponges

A new bioactive monogalactosyl monoglyceride, spongilipid (**19**) (Fig. 5), was isolated from a collection of the sponge, *Spongia* cf. *hispida*, obtained from Terumbu Pempang Laut [83]. The structure and absolute configuration of spongilipid was subsequently established by X-ray crystallography. Preliminary biological assays

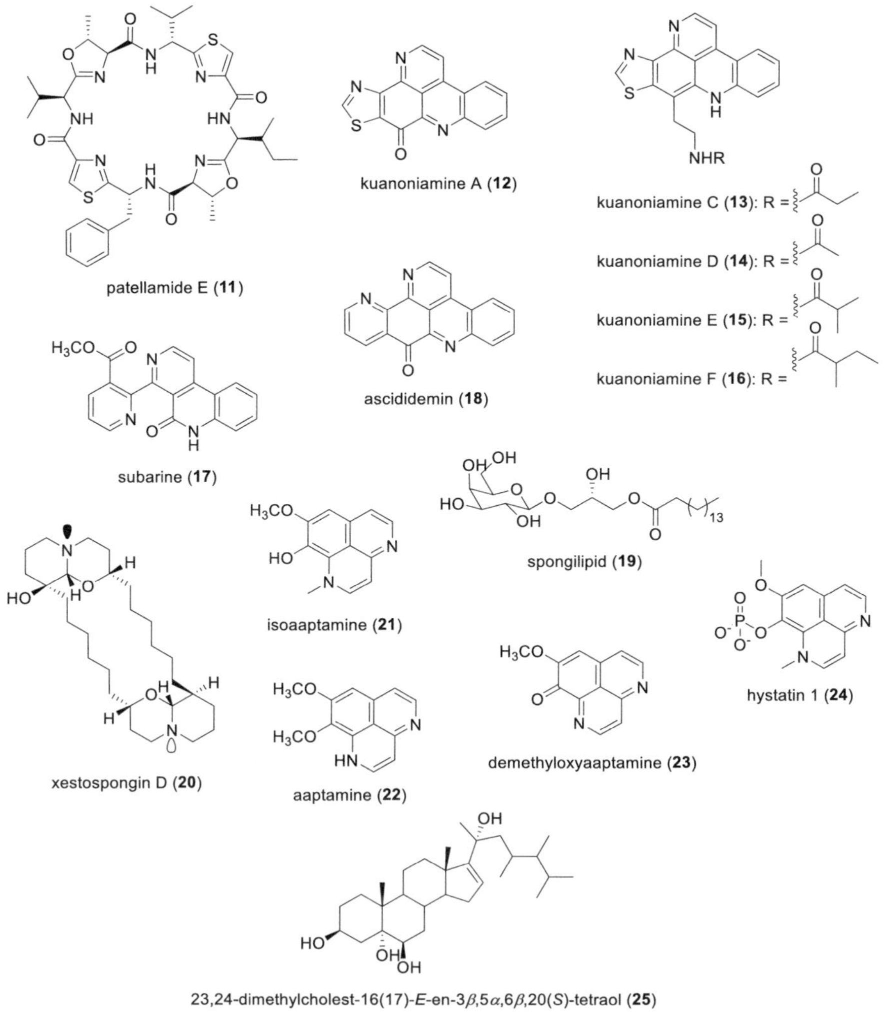

Fig. 5 Chemical structures of natural products isolated from Singapore-derived marine invertebrates

suggested that spongilipid exhibits potential bioactivity, highlighting its relevance for pharmaceutical or ecological applications. Spongilipid was found to specifically inhibit growth of the fecal opportunist *Enterococcus faecalis.*

From the marine sponge, *Niphates* sp., researchers isolated (±)-xestospongin D (**20**) (Fig. 5), a macrocyclic alkaloid with a 1-oxa-quinolizidine skeleton [84]. The sponge samples were collected using SCUBA at Beting Bemban Beset reef, Singapore. Unlike previous reports of xestospongin D from *Xestospongia* and *Haliclona* species, the compound obtained from Singapore sponge was racemic, suggesting potential environmental or biosynthetic influences on its chirality. Biological assays revealed that (±)-xestospongin D inhibited the growth of leukemia and breast cancer cell lines, as well as the bacterial strain *Micrococcus luteus*.

Another significant contribution to marine natural products research in Singapore was the isolation of isoaaptamine (**21**) (Fig. 5) from the sponge *Hymeniacidon* sp. obtained from north side of Terumbu Pemalang Besar Reef, Singapore [85]. The structural determination of isoaaptamine was achieved using X-ray crystallography, which confirmed its isoquinoline framework and provided insights into its bioactivity. This compound, alongside aaptamine (**22**) and demethyloxyaaptamine (**23**) (Fig. 5), exhibited potent cancer cell growth inhibition, particularly against leukemia cell lines. However, due to its inherent instability, isoaaptamine was converted into a more stable phosphate prodrug, hystatin 1 (**24**) (Fig. 5), which retained its biological activity.

3.5.3 Soft Coral

A new cytotoxic polyhydroxysterol, identified as 23,24-dimethylcholest-16(17)-*E*-en-3*β*,5*α*,6*β*,20(*S*)-tetraol (**25**) (Fig. 5) was isolated from the soft coral, *Sarcophyton trocheliophorum*, collected from Pulau Hantu at a depth of 10 m [86]. Additionally, nine known compounds were isolated, including gorosten-5(*E*)-3*β*-ol, 24-methylcholestane-3*β*,5*a*,6*β*,25-tetraol 25-monoacetate, 2,3-dihydroxypropyloctadecyl ether, tetradecyl octadecenoate, tetradecyl-9-*Z*-ocatadecenoate, isoneocembrene A, 7,8-epoxy-1(*E*),3(*E*),11(*E*)-cembratriene, 7,8-epoxy-1(*E*),3(*E*),11(*E*)-cembratrien-15-ol and sarcophine. Compound **25** demonstrated strong growth inhibitory effects against human HL60 leukemia, M14 melanoma and MCF7 breast carcinoma cell lines, with EC_{50} values of 2.8, 4.3, and 4.9 μg/mL, respectively, while showing minimal cytotoxicity toward normal human peripheral blood lymphocytes.

3.5.4 Marine Cyanobacteria

Marine cyanobacteria in Singapore exhibit significant diversity, particularly along the Southern Islands, where studies have recorded new species previously undocumented in local waters [87, 88]. Research has identified cyanobacteria belonging to the orders Oscillatoriales and Nostocales, with genera such as *Leptolyngbya*, *Lyngbya*, *Moorea* and *Nostoc* being prevalent. Many species showed wide morphological variability and were phylogenetically distinct, suggesting the potential presence of novel species [87]. Some species, such as *Moorea producens* and *Lyngbya majuscula*, are known for producing biologically active compounds with ecological and health implications. Given the role of cyanobacteria as primary producers and nitrogen fixers, their presence enhances ecological microdiversity in Singapore's coastal marine environment. However, the detection of toxin-producing strains underscores the need for continuous monitoring to mitigate potential harmful algal blooms (HABs) and their associated risks.

Aromatic Polyketides

A series of aromatic polyketides, including aplysiatoxin (**26**), debromoaplysiatoxin (**27**) and anhydrodebromoaplysiatoxin (**28**) (Fig. 6), along with two newly identified analogues, 3-methoxyaplysiatoxin (**29**) and 3-methoxydebromoaplysiatoxin (**30**) (Fig. 6), were reported for the first time from the marine cyanobacterium, *Trichodesmium erythraeum*, found at Pulau Seringat Kias, Singapore in 2012 [89]. The cyanobacterial strain responsible for the bloom was identified through phylogenetic analysis of its 16S rRNA sequences. Aplysiatoxins are structurally characterized by a complex polyether-lactone framework and are known for their potent biological activities, including tumor-promoting and cytotoxic effects [90]. Additionally, the antiviral potential of these marine toxins was evaluated against Chikungunya virus (CHIKV) in infected cells. Post-treatment assays with the debrominated analogues, debromoaplysiatoxin, anhydrodebromoaplysiatoxin and 3-methoxydebromoaplysiatoxin, demonstrated a dose-dependent inhibition of CHIKV at concentrations ranging from 0.1 to 10.0 μM. Notably, debromoaplysiatoxin (**27**) and 3-methoxydebromoaplysiatoxin (**30**) (Fig. 6) exhibited significant anti-CHIKV activity, with EC_{50} values of 1.3 μM and 2.7 μM, respectively, and selectivity indices of 10.9 and 9.2, respectively.

Since the discovery of aplysiatoxins from *Trichodesmium erythraeum* in 2012, subsequent marine cyanobacterial strains found at coastal waters were found to contain such cyanotoxins. Due to the prevalence of these cyanotoxins, a pilot study

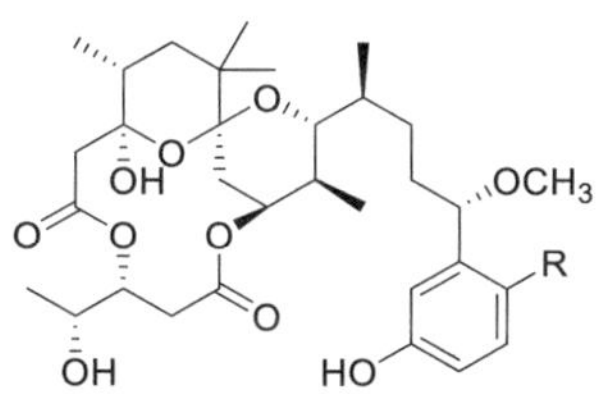

aplysiatoxin (**26**): R = Br
debromoaplysiatoxin (**27**): R = H

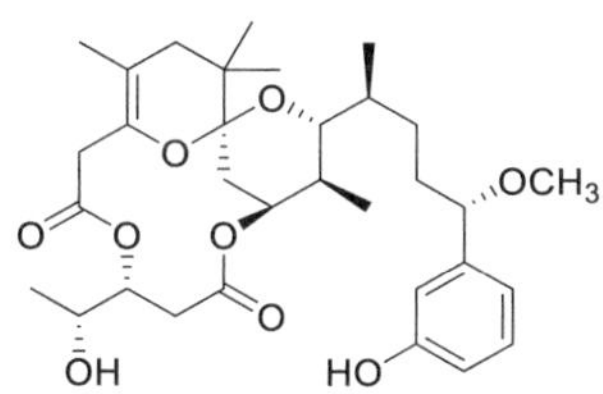

anhydrodebromoaplysiatoxin (**28**)

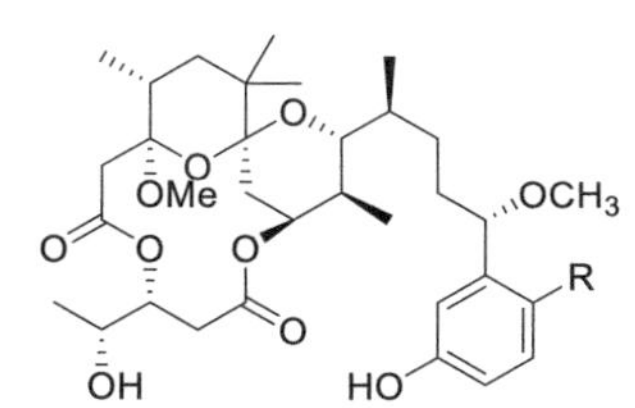

3-methoxyaplysiatoxin (**29**): R = Br
3-methoxydebromoaplysiatoxin (**30**): R = H

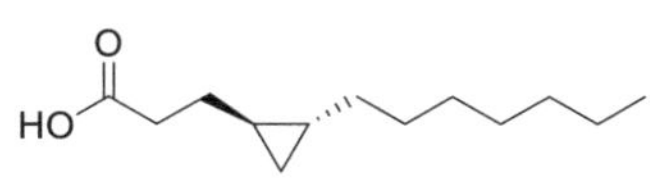

lyngbyoic acid (**31**)

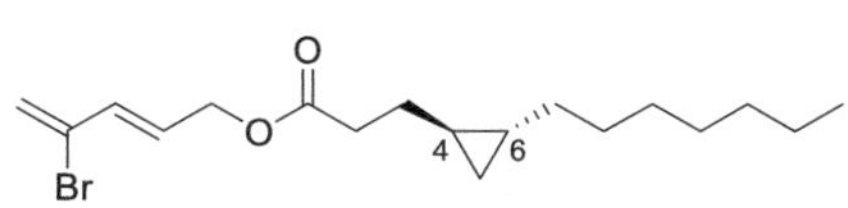

rel-(4*R*,6*R*)-benderadiene (**32**)

Fig. 6 Marine cyanobacterial polyketides and modified fatty acid derivatives

based on LC–MS/MS and molecular networking was employed as a rapid analytical approach to detect aplysiatoxins in four environmental marine cyanobacterial samples collected from intertidal areas in Singapore [91]. Phylogenetic analysis based on 16S-ITS rRNA gene sequences identified these filamentous cyanobacterial samples from Pulau Hantu as *Trichodesmium erythraeum*, *Oscillatoria* sp. PAB-2 and *Okeania* sp. PNG05-4.

Organic extracts were prepared and analyzed using LC-HRMS/MS, followed by Global Natural Product Social Molecular Networking (GNPS) to screen for aplysiatoxin-related molecules. The molecular networking analysis revealed the presence of six known compounds, debromoaplysiatoxin, anhydrodebromoaplysiatoxin, 3-methoxydebromoaplysiatoxin, aplysiatoxin, oscillatoxin A and 31-noroscillatoxin B, along with potential new analogues. Additionally, variations in molecular networking clusters associated with the aplysiatoxin molecular family were observed between *Trichodesmium erythraeum* samples collected from two different locations, as well as among different cyanobacterial species found at Pulau Hantu. The unprecedented nature of this pilot study underscores the importance of the MS/MS-based molecular networking approach as a rapid biomonitoring analytical technique to detect cyanotoxins, such as the aplysiatoxins, from environmental samples.

Modified Fatty Acid Derivatives

Marine cyanobacteria are well known for producing structurally diverse bioactive specialized metabolites, particularly during bloom events [92, 93]. These ecologically active allelochemicals serve as chemical defenses, protecting the microalgae from competing microbes and herbivores. From a collection of the marine cyanobacterium cf. *Lyngbya* sp., a small quantity of a new cyclopropane-containing molecule, benderadiene (**32**) (Fig. 6), along with known lyngbyoic acid (**31**) (Fig. 6), was isolated and characterized using spectroscopic techniques [94]. Quorum-sensing (QS) inhibitory activity of both compounds was evaluated using live reporter assays with *Pseudomonas aeruginosa* PAO1 *lasB-gfp* and *rhlA-gfp* strains. Both metabolites demonstrated dose-dependent inhibition of QS-regulated gene expression. Notably, lyngbyoic acid exhibited stronger activity against the PAO1 *lasB-gfp* biosensor strain, with an IC_{50} of 20.4 μM, and also displayed anti-biofilm activity when tested against wild-type *P. aeruginosa*. The quorum-sensing inhibitory effect of lyngbyoic acid is consistent with the findings of Kwan and co-workers, where the free acid was found to strongly affect LasR when screened against a panel of reporters based on four different AHL receptors, including LuxR, AhyR, TraR and LasR [95]. The relatively high abundance of lyngbyoic acid suggests its ecological role as a defensive allelochemical, interfering with microbial QS systems while also serving as a precursor for related analogs. Preliminary molecular docking studies further support this mode of action, indicating potential interactions between these cyanobacterial free acids and the ligand-binding site of the *P. aeruginosa* LasR transcriptional regulator.

Lipopeptides

Several bioactive lipopeptides have been identified from marine cyanobacteria in Singapore. Besarhanamides A (**33**) and B (**34**) (Fig. 7) are fatty acid amides isolated from the marine cyanobacterium *Lyngbya majuscula*, collected from Pulau Hantu, Singapore [96]. The structural elucidation of these secondary metabolites was accomplished through extensive 2D NMR spectroscopic analysis, complemented by chemical derivatization, including Marfey's method. Additionally, besarhanamide A exhibited moderate toxicity in the brine shrimp bioassay, with an LD_{50} value of 13 μM.

Cyclic Depsipeptides

Marine cyanobacterial cyclic depsipeptides are characterised by their distinct *N*-methylation, α- and β-hydroxy acids, β-amino acids and fatty acyl units. In addition, their biosynthesis involves a hybrid pathway of non-ribosomal peptide synthetases (NRPS) and polyketide synthases (PKS). The kulolide superfamily of cyclic depsipeptides is one such hybrid polyketide-peptide structural class. Since the first discovery of kulolide by Scheuer and co-workers [97] from the cephalaspidean mollusk, *Philinopsis speciosa*, a variety of kulolide-related molecules have been reported from filamentous marine cyanobacteria, including *Lyngbya majuscula* (*Moorea producens*), *Moorea* sp., *Symploca* cf. *hydnoides*, *Oscillatoria* sp. and *Rivularia* sp. [98–101]. Members of the kulolide superfamily are characterized by the presence of either a 2,2-dimethyl-3-hydroxy-7-octynoic acid (Dhoya) or a 3-hydroxy-2-methyl-7-octynoic acid (Hmoya) moiety included in their macrocyclic backbone. Many of the reported compounds showed moderate cytotoxic and antiparasitic activities. The significant structural similarities within this large metabolite class, along with their diverse reported metabolic sources, suggest that the kulolide superfamily may have originated in cyanobacteria through an ancient evolutionary process. Due to the prevalence of such unique compounds, it is not surprising that kulolide related molecules were also uncovered in our chemical analysis of marine cyanobacterial samples.

Hantupeptins

The marine cyanobacterium *Lyngbya majuscula* (*Moorea producens*) from Pulau Hantu Besar, Singapore, produced a series of cyclodepsipeptides known as hantupeptins A (**35**)–C (**37**) (Fig. 7) [102, 103]. All three variants exhibited 100% brine shrimp mortality at concentrations of 100 and 10 ppm, demonstrating significantly higher toxicity than their closest analog, trungapeptin A, which was only mildly toxic to brine shrimp [104]. In vitro cytotoxicity assays against the leukemia cell line MOLT-4 revealed IC_{50} values of 32 nM for hantupeptin A, 0.2 μM for hantupeptin B and 3.0 μM for hantupeptin C. Against the breast cancer cell line MCF-7, the IC_{50} values were 4.0 μM, 0.5 μM, and 1.0 μM, respectively. In contrast, trungapeptin A was reported to be inactive when tested at 10 μg/mL against KB or LoVo cells [104]. Notably, the hantupeptins possess 2*R*,3*S* absolute configurations, whereas the corresponding stereocenters in trungapeptin are 2*S*,3*R*, which may contribute to their differing bioactivities [105]. There are at least two publications on partial stereoselective synthesis of fragments in hantupeptins [106, 107].

Benderamide A

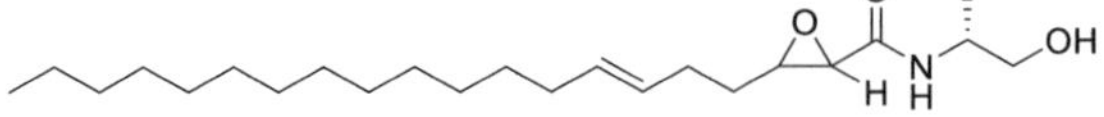

besarhanamide A (**33**)

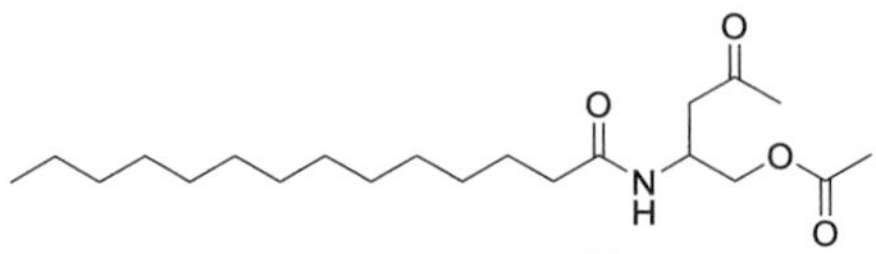

besarhanamide B (**34**)

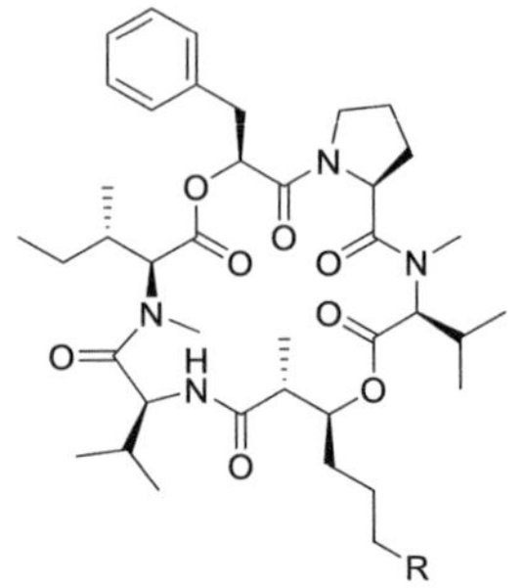

hantupeptin A (**35**): R =

hantupeptin B (**36**): R =

hantupeptin C (**37**): R =

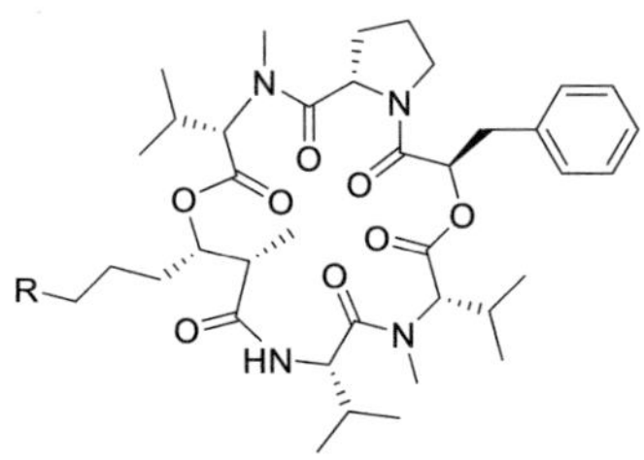

trikoveramide A (**40**): R =

trikoveramide B (**41**): R =

trikoveramide C (**42**): R =

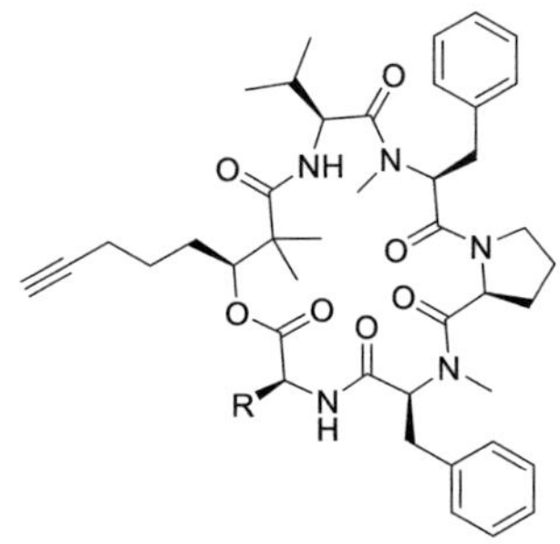

benderamide A (**38**): R = CH_3
cocosamide B (**39**): R = H

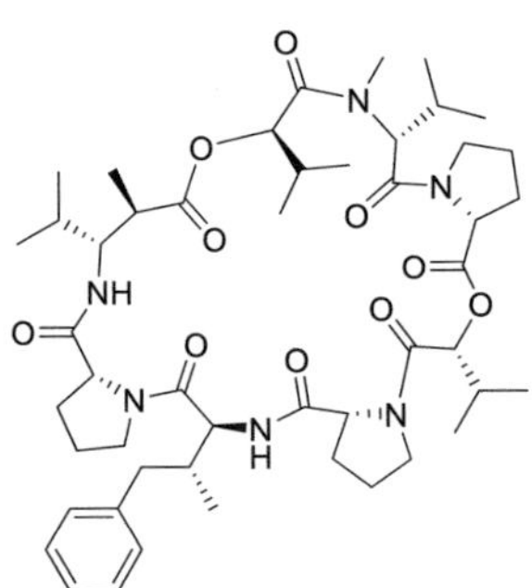

triproamide (**43**)

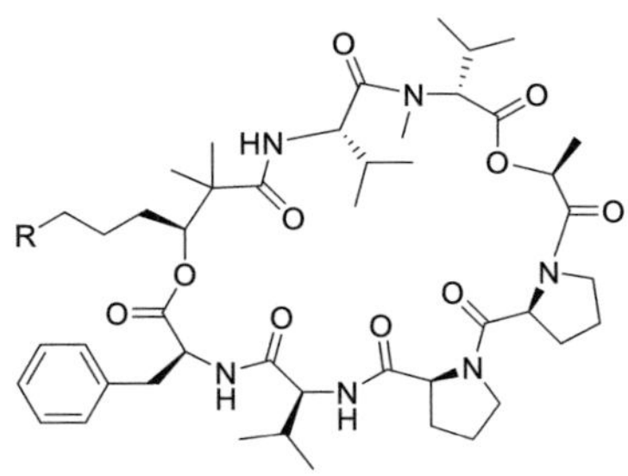

kulokainalide-1 (**44**): R =

pemukainalide A (**45**): R =

pemukainalide B (**46**): R =

Fig. 7 Marine cyanobacterial linear lipopeptides and cyclic depsipeptides

Benderamide A (**38**) (Fig. 7), a cyclic depsipeptide containing (*S*)-2,2-dimethyl-3-hydroxy-7-octynoic acid, was isolated from a Singaporean collection of cf. *Lyngbya* sp. using a bioassay-guided approach [108]. Its planar structure was determined through a combination of 1D and 2D NMR experiments and MS analysis, while its absolute configuration was established using Marfey's analysis, nuclear Overhauser effect spectroscopy (NOESY) and one-dimensional nuclear Overhauser effect (1D-NOE) studies. Although benderamide A showed no antiproliferative activity against MCF7 breast cancer cells, the substitution of Ala in place of Gly may explain its reduced cytotoxicity compared to the closely related cocosamide B (**39**) (Fig. 7) [109].

Trikoveramides

Trikoveramides A (**40**)–C (**42**) (Fig. 7), members of the kulolide superfamily of cyclic depsipeptides, were isolated from the marine cyanobacterium *Symploca hydnoides*, collected from Bintan Island, Indonesia [110]. This collection of *S. hydnoides* is particularly rich in chemistry as other structural classes, such as the cytotoxic cyanobactins (e.g., trikoramides) and a new dolastatin 16 analogue (e.g., triproamide), have been isolated. The trikoveramides planar structures were determined using a combination of NMR spectroscopy and high-resolution mass spectrometry. Additionally, the relative stereochemistry of the 3-hydroxy-2-methyl-7-octynoic acid (Hmoya) unit in trikoveramide A was established using *J*-based configuration analysis and NOE correlations. The cytotoxic activity of the trikoveramides was evaluated against MOLT-4 human leukemia cells, yielding IC_{50} values of 9.3 μM for trikoveramide B, 35.6 μM for trikoveramide C and 48.8 μM for trikoveramide A. Furthermore, all three compounds exhibited weak to moderate inhibition in a quorum sensing assay using *Pseudomonas aeruginosa lasB-gfp* and *rhlA-gfp* bioreporter strains.

Triproamide and Permukainalides

A new cyclic depsipeptide, triproamide (**43**) (Fig. 7), was isolated from the polar VLC-derived fraction of extracts from the marine cyanobacterium, *Symploca hydnoides* [111]. Triproamide contains the rare 4-phenylvaline (dolaphenvaline, Dpv) and a β-amino acid, dolamethylleucine (Dml), which was originally identified in dolastatin 16. Alongside triproamide, the known compound kulokainalide-1 (**44**) (Fig. 7) [112] and two new analogues, pemukainalides A (**45**) and B (**46**) (Fig. 7), were also isolated. Their planar structures were determined through extensive NMR and mass spectrometric analysis, while their absolute and relative configurations were established using a combination of Marfey's method, *J*-based configuration analysis and chiral-phase HPLC. Kulokainalide-1 and pemukainalide A exhibited cytotoxic activity against the MOLT-4 leukemia cell line, with IC_{50} values of 5.9 μM and 5.6 μM, respectively.

Lagunamides

Three cytotoxic cyclic depsipeptides, lagunamides A (**47**), B (**48**) and C (**49**) (Fig. 8), were isolated from *Lyngbya majuscula* (*Moorea producens*) collected at Pulau Hantu, Singapore [113, 114]. Additional analog, lagunamide D (**50**) (Fig. 8), was subsequently isolated from a collection of marine cyanobacteria from Loggerhead Key in the Dry Tortugas, Florida [115]. These compounds share a planar macrocyclic scaffold consisting of both peptide and polyketide substructures, with structural differences primarily in the polyketide portion. Collectively, these cyclic depsipeptides belong to aurilides, a class of naturally occurring cyclic depsipeptides known for their potent cytotoxic properties, often isolated from marine cyanobacteria [116]. Lagunamides A and B are 26-membered macrocycles, whereas the original structure of lagunamide C was reported to contain an additional methylene carbon in its polyketide segment. A recent total synthesis of lagunamide C led to a structural revision, identifying it as the related analog odoamide [117]. All three lagunamides exhibited potent activity against the *Plasmodium falciparum* NF54 strain, with IC_{50} values of 0.19 μM, 0.91 μM and 0.29 μM for lagunamides A, B and C, respectively. The reduced activity of lagunamide B may be attributed to the presence of a double bond in its side chain.

Lagunamide A exhibits cytotoxicity against various cancer cell lines, including HCT8 (human colorectal cancer), P388 (murine leukemia), PC3 (human prostate cancer), A549 (human lung cancer) and SK-OV3 (human ovarian cancer), with IC_{50} values ranging from 1.6 to 6.4 nM [118]. The molecule induces an antiproliferative effect through mitochondria-mediated apoptosis in HCT8 and MCF7 (human breast cancer) cell lines. Lagunamide B was found to be relatively less active as it demonstrated cytotoxicity against P388 murine leukemia cells, with IC_{50} value of 20.5 nM [65]. In another study, Lagunamide C displayed potent cytotoxicity against A549, PC3, P388, HCT8 and SK-OV3 cell lines, with IC_{50} values ranging from 2.1 to 24.4 nM [114]. The antiproliferative activity of lagunamide D was evaluated in A549 human lung adenocarcinoma cells using MTT assays, and the determined IC_{50} was at 7.1 nM, which was comparable with lagunamide A at 6.7 nM [115].

Due to the potent anticancer activities of the lagunamides, their partial and total syntheses, including derivatives, have been reported by several synthetic groups [117, 119–125]. Being the most potent in the series, the first report on the total synthesis of lagunamide A in 2012 by Dai and co-workers led to its stereochemical revision of the original proposed structure [119]. Subsequent synthetic efforts by Huang and co-workers led to the elaboration of diverse asymmetrical total synthesis of lagunamide A and its five derivatives (e.g., **51–55**) (Fig. 8) starting from commercially available (*S*)-2-methylbutan-1-ol. The key feature in this diverse synthesis includes the elaboration of four consecutive chiral centers at C37–40 and the final macrocyclization.

Previous syntheses primarily used aldol reactions to establish the stereochemically complex polyketide moiety in lagunamide A. In contrast, Gorges and Kazmaier strategically employed sequential Matteson homologations to establish the polyketide fragment in lagunamide A [123]. The Matteson homologation is a powerful method for synthesizing complex structures by using a chiral boronic ester to sequentially generate multiple stereogenic centers along a growing carbon chain [126]. This approach typically achieves exceptional stereoselectivity, often yielding diastereomeric ratios as high as 1000:1. Starting with methylboronic acid ester (**56**), which was subjected to Matteson homologation with ethyl Grignard reagent, the

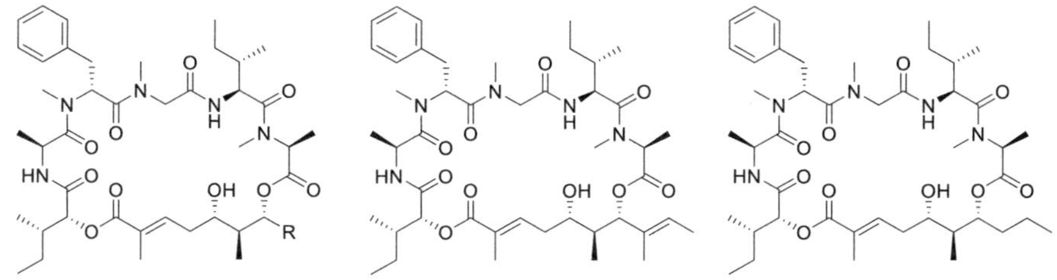

lagunamide A (**47**): R =

lagunamide C (**49**): R =

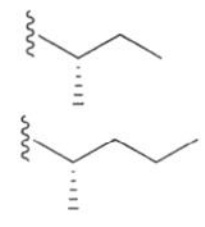

lagunamide B (**48**)

lagunamide D (**50**)

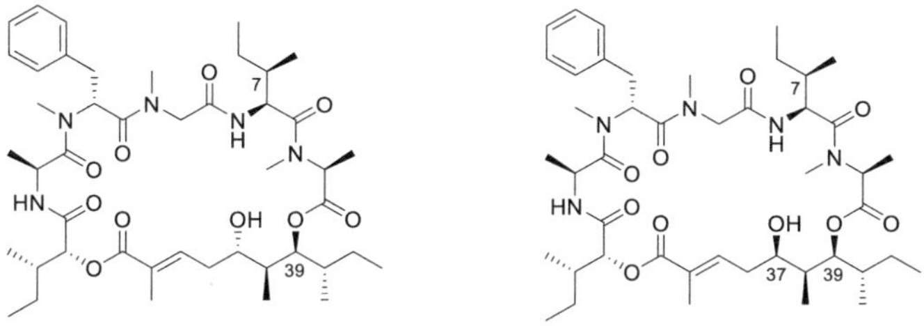

7,39-*epi*-lagunamide A (**51**)

7,37,39-*epi*-lagunamide A (**52**)

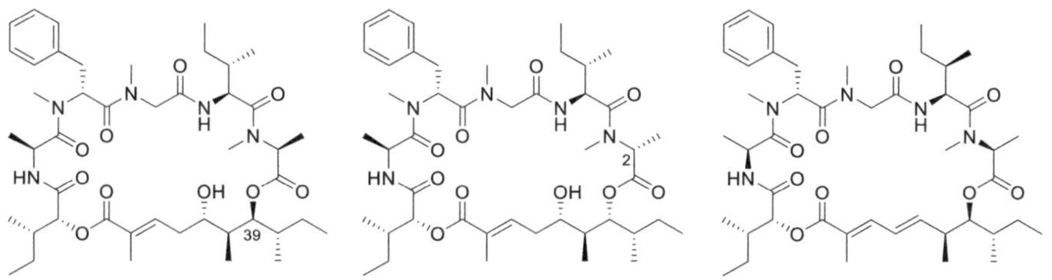

39-*epi*-lagunamide A (**53**)

2-*epi*-lagunamide A (**54**)

55

Fig. 8 Lagunamides and synthetic derivatives

boronic acid ester (**57**) was formed diastereomerically pure (Scheme 1). This was followed by sequential Matteson homologations as well as functional group interconversions to obtain aldehyde **58** with high diastereoselectivity [123]. This was followed by Horner–Wadsworth–Emmons olefination using phosphonate **59** and subsequent peptide couplings to produce cyclization precursor **60**. The synthesis of (−)-lagunamide A was completed through HATU-mediated macrolactamization (Scheme 1).

The structure–activity relationships (SARs) of lagunamide A and its analogues against a diverse panel of cancer cell lines, including A549 (lung), HeLa (cervical), U2OS (osteosarcoma), HepG2 and BEL-7404 (liver), BGC-823 (gastric), HCT116

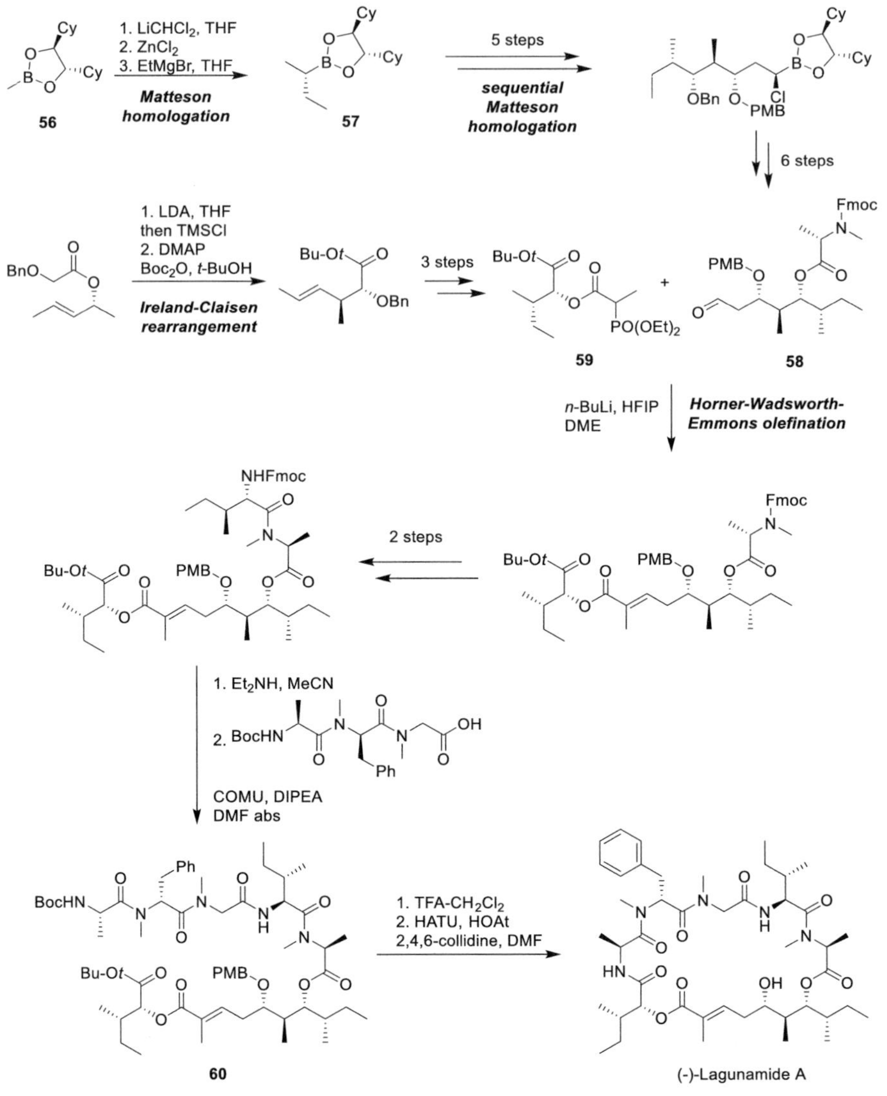

Scheme 1 Total synthesis of (−)-lagunamide A by Gorges and Kazmaier [123]

(colorectal), MCF-7 (breast), HL-60 (leukemia) and A375 (melanoma), were investigated by Huang and co-workers [127]. Their findings identified the *R* configuration at the C-39 position as a key structural determinant of lagunamide A's anticancer activity. Further mechanistic studies in A549 cells revealed that lagunamide A induces caspase-mediated mitochondrial apoptosis. This process involves mitochondrial dysfunction, characterized by the dissipation of mitochondrial membrane potential and excessive reactive oxygen species production, ultimately leading to cell death. Additionally, both anti-apoptotic and pro-apoptotic members of the B-cell lymphoma 2 (Bcl-2) family play a role in this apoptosis, with myeloid cell leukemia-1 (Mcl-1) being particularly significant. These findings suggested that lagunamide A exerts its anticancer effects through mitochondrial apoptosis and provided valuable insights for the design of novel anticancer drugs based on lagunamide derivatives.

In another study by Hu and co-workers, lagunamide A exhibited significant antiproliferative and pro-apoptotic effects by modulating DNA damage repair [128]. Thermal proteome profiling identified EYA3, a key player in the DNA damage response, as a stabilized protein upon lagunamide A treatment. In addition, functional studies confirmed that lagunamide A inhibits DNA damage repair by interfering with γH2AX foci formation, enhancing the sensitivity of tumor cells to doxorubicin. The formation of γH2AX foci is a key indicator of DNA damage, and the number and size of these foci can be used to assess the extent and type of DNA damage [129]. As such, the modulation of lagunamide A on DNA damage repair shifts the balance toward apoptosis rather than repair, suggesting a potential therapeutic approach for cancer treatment.

Cyanobactins

Cyanobactins are a diverse class of ribosomally synthesized and post-translationally modified peptides (RiPPs) produced by cyanobacteria [130, 131]. These secondary metabolites exhibit a wide range of biological activities, including antimicrobial, anticancer and enzyme-inhibitory properties. Cyanobactins are characterized by their cyclic or linear peptide structures, often containing unusual amino acids and extensive modifications such as prenylation, heterocyclization and thiazoline or oxazoline ring formation. They are biosynthesized through a highly conserved pathway involving precursor peptides and tailoring enzymes encoded within cyanobactin gene clusters [132]. Notable examples include patellamides, trunkamides and aesturamide, some of which have shown potential as drug leads. The structural diversity and bioactivity of cyanobactins make them an important focus for natural product research and biotechnological applications.

Phyo et al. [133] reported the isolation of a new cyclic decapeptide, trikoramide A (**61**) (Fig. 9), from the marine cyanobacterium *Symploca hydnoides*, collected from the intertidal shores of Trikora Beach, Bintan Island, Indonesia. Trikoramide A is a *C*-prenylated cyclotryptophan-containing cyanobactin, and its absolute configuration was determined using Marfey's method and NOESY correlation analysis. The compound exhibited cytotoxic activity against the MOLT-4 and AML2 cancer cell lines, with IC_{50} values of 4.8 μM and 8.2 μM, respectively. Subsequently, three additional cyanobactins, namely trikoramides B (**62**)–D (**64**) (Fig. 9), were isolated from the same cyanobacterial samples [134]. These compounds are structural analogues of trikoramide A, with variations in the *C*-prenylated cyclotryptophan unit. Their absolute configurations were determined using a combination of Marfey's method, electronic circular dichroism (ECD) spectroscopy, and NOESY correlations. Among them, trikoramides B and D demonstrated cytotoxicity against the MOLT-4 cell line, with IC_{50} values of 5.2 μM and 4.7 μM, respectively. Interestingly, the kulolide-related compounds, trikoveramides and pemukainalides, as well as triproamide were

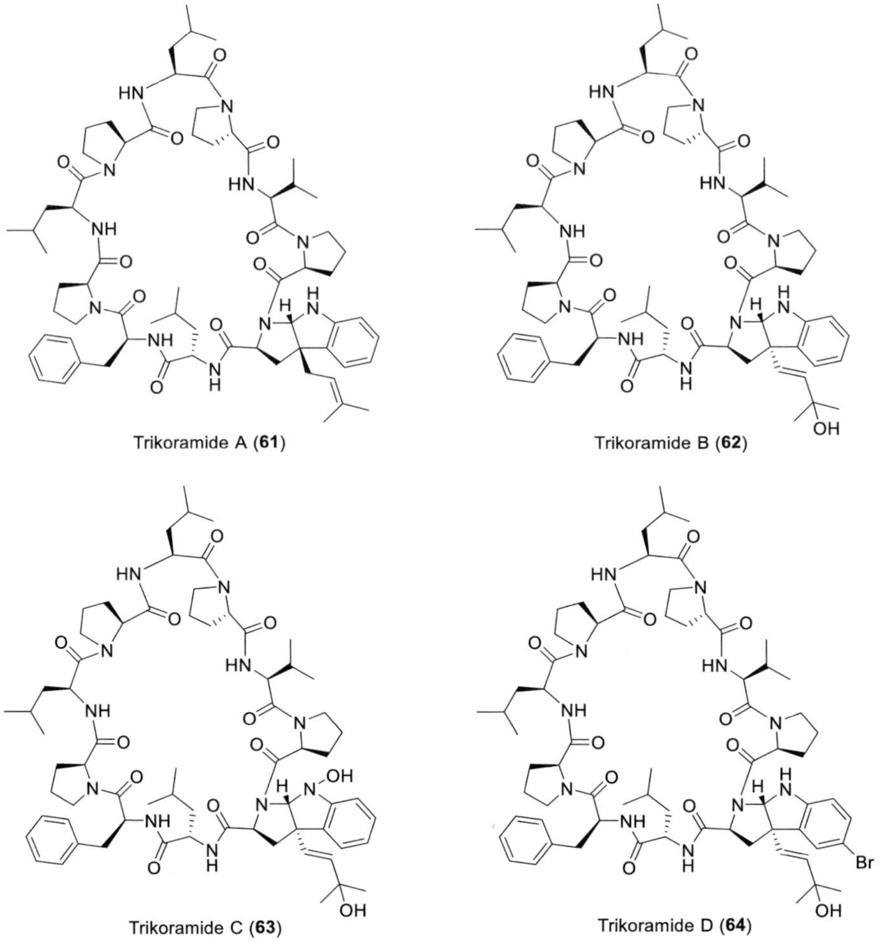

Fig. 9 Trikoramides from *Symploca hydnoides*

also obtained from the trikoramide-containing *S. hydnoides* samples. The presence of diverse structural classes found within the same cyanobacterial samples highlights *Symploca* as a chemically endowed cyanobacterial genus.

3.5.5 Marine Actinobacteria

A genome-guided investigation led to the discovery of sungeidines A (**65**)–H (**72**) (Fig. 10), a novel class of microbial secondary metabolites with distinctive structural characteristics, isolated from *Micromonospora* sp. MD118, obtained from mangrove

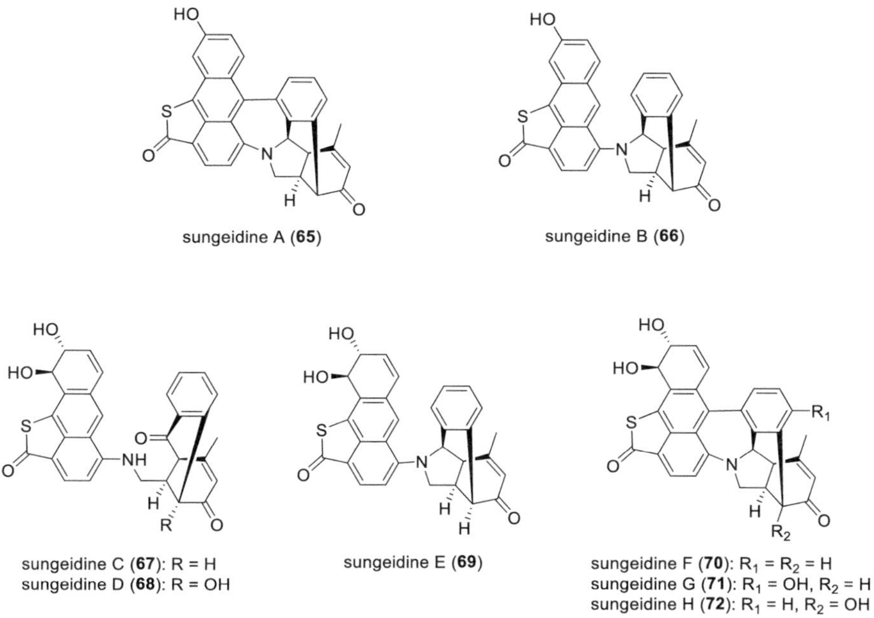

Fig. 10 Sungeidines from *Micromonospora* sp. MD118

sediment at the Sungei Buloh Wetland Reserve, Singapore [135]. Although sungeidines share an evolutionary lineage with dynemicin-type enediynes, they are synthesized by a biosynthetic gene cluster that displays notable deviations from known enediyne BGCs. The findings suggested that sungeidines are biosynthesized from two octaketide chains, which undergo a different processing mechanism compared to those in dynemicin-type enediynes. The biosynthetic pathway uniquely involves an activating sulfotransferase that facilitates a dehydration reaction. The absence of several genes, including a putative epoxidase gene, likely contributes to the divergence of the sungeidine biosynthetic pathway from canonical enediyne systems.

3.6 *Marine Invertebrate-Derived Microbes as Sources of Bioactive Compounds*

The marine microbiome represents a vast and largely untapped reservoir of microbial diversity, harboring an extraordinary array of microorganisms that thrive in diverse and often extreme environments [136]. These microbial communities, which include bacteria, archaea, fungi and viruses, have evolved unique biochemical pathways to adapt to their specific niches, such as deep-sea hydrothermal vents, polar ice and coral reefs. This metabolic versatility makes the marine microbiome a rich source of natural products with potential applications in medicine, agriculture and industry

[137]. Over the past few decades, advances in genomic and metagenomic technologies have enabled researchers to explore this hidden world, revealing a treasure trove of bioactive compounds, including antibiotics, anticancer agents and enzymes with industrial relevance [138]. The marine microbiome, therefore, holds immense promise for the discovery of novel natural products that can address pressing global challenges, such as antibiotic resistance and the need for sustainable biotechnological solutions.

Marine invertebrates, such as sponges, corals and tunicates, are known to host diverse microbial communities that play a crucial role in their survival and ecological interactions [139]. The close association between marine invertebrates and their microbial symbionts often leads to the production of secondary metabolites with applications in drug discovery, including antibiotics, antivirals, and anticancer agents [140]. For instance, the anticancer compound ET-743, derived from a tunicate-associated microbe, has been approved for clinical use, highlighting the therapeutic potential of these microbial resources [141]. Advances in metagenomics, metabolomics and culture-independent techniques have enabled researchers to uncover the biosynthetic potential of these microbes, even when they are unculturable in the laboratory [142]. As a result, marine invertebrate-derived microbes are increasingly recognized as a promising source of novel bioactive compounds, offering new opportunities for addressing global health challenges and advancing biotechnological innovation [143].

Our research group has investigated symbiotic microbes associated with various common benthic marine invertebrates, such as sponges and soft corals, found in Singapore's waters, focusing on their potential to produce bioactive natural products. By employing an integrated approach that combines culture-dependent methods, MS-based molecular networking, bioinformatics and bioassay screening, we have identified several marine microbial strains capable of producing bioactive compounds. Additionally, these strains harbor unique biosynthetic gene clusters, underscoring their potential for novel natural product discovery. The following sections provide an overview of key findings, including unpublished data, from our laboratory as well as others, highlighting the promising bioactive capabilities of these marine invertebrate-associated microbes.

3.6.1 Marine Bacteria Associated with Common Intertidal Benthic Invertebrates from Singapore

In view of the importance of marine microbiome associated with benthic macroorganisms as sources of bioactive compounds, a pilot scale study was initiated to investigate the biomedical potential of these microbes isolated from common benthic invertebrates found in Singapore. A total of six common marine invertebrates were selected, including three sponges, *Gelliodes fibulata*, *Clathria reinwardti* and *Xestospongia testudinaria*, as well as three soft corals, *Sacrophyton tenuispiculatum*, *Sinularia* sp. and *Lobophytum crassum* (unpublished). The culturable marine bacteria were isolated in three different marine media, such as M1, M2 and M3 marine media.

Small scale fermentations of these isolated bacterial strains were performed and their organic extracts prepared for biological assessment in a panel of assays, including the brine shrimp (*Artemia salina*) toxicity, cytotoxicity (MOLT-4, human leukaemic cell line) and antibacterial assays. Antibacterial activity was tested against three gram-positive, *Bacillus cereus*, Methicillin-resistant *Staphylococcus aureus* (MRSA) and *Enterococcus faecalis* V583 as well as five gram-negative bacteria, namely *Escherichia coli, Escherichia coli* EC 598, *Pseudomonas aeruginosa, Burkholderia cenocepacia* and *Acinetobacter baumannii.*

In total, 57 bacterial isolates were isolated from these six benthic marine invertebrates based on three different marine media. After 3 weeks of incubation, 32, 17 and 8 bacterial isolates were isolated in M3, M2 and M1 medium, respectively. The most growth was observed in general medium M2, while M3 medium yielded the highest diversity based on morphology of marine bacterial isolates. Bacteria showing growth on M2 medium use starch, yeast extract and peptone as a primary source of energy as it does not contain any glucose, free amino acids and vitamins. The nitrogen sources for these microorganisms were proteins and peptones. M3 medium contains glucose along with proteins and peptones. M1 medium has low concentrations of vitamins and free amino acids when compared with M3 medium. Of a total of 31 bacterial isolates obtained from three sponges, 5, 10 and 16 bacterial isolates were obtained from *Geolloides fibulata, Clathria reinwardti* and *Xestospongia testudinaria*, respectively. Twenty-six bacterial isolates were isolated from soft corals, with 8, 6 and 12 isolated obtained from *Sacrophyton tenuispiculatum, Sinularia* sp. and *Lobophytum crassum*, respectively.

Small scale fermentation (1L) was performed on all 57 marine bacterial isolates using marine medium from which they are originally isolated. Fermentation flasks were then incubated at 28 °C on a rotary shaker at 150 rpm for a period of 21 days. Organic solvents, including ethyl acetate and methanol, were used for extraction of natural products for further testing in the brine shrimp (*Artemia salina*) toxicity, antibacterial and cytotoxic (MOLT-4) assays. Twelve (21%) of the crude bacterial extracts displayed activity (> 25%) in the brine shrimp toxicity assay when tested at 1000 ppm. These bacterial isolates were isolated from sponges *Gelliodes fibulata* (S1 and S4), *Clathria reinwardti* (S6) and *Xestospongia testudinaria* (S16) as well as from soft corals *Sacrophyton tenuispiculatum* (C3, C4, C5 and C8) and *Sinularia* sp., (C9, C10, C11 and C13). In addition, eight bacterial extracts (S1, S6, C3, C4, C8, C9, C11 and C13) exhibited significant toxicity of 100% at 1000 ppm.

In vitro cytotoxicity assay based on MOLT-4 cancer cells line was used to investigate anticancer potential of these bacterial isolates. Of the 57 crude bacterial extracts tested, 26 and 6 of the bacterial extracts displayed cytotoxicity of > 25% when tested at 1000 ppm and 100 ppm, respectively. Ten (S6, S10, S22, S24, C6, C8, C9, C10 and C11) and two (S6 and C8) bacterial extracts exhibited 100% activity at 1000 ppm and 100 ppm, respectively. Of the 26 (~ 46%) bacterial extracts which exhibited activity (> 25%) at the 1000 ppm, 14 and 12 bacterial extracts came from bacteria isolated from sponges and soft corals, respectively.

Hospital-acquired infection (HAI) is one of the major threatens to public health and is a substantial cause of morbidity and mortality [144]. The majority of HAIs are caused by a few groups of bacterial species, including *Escherichia coli, Pseudomonas aeruginosa* and *Acinetobacter baumannii,* which are commonly found in hospital environments [145]. With the increasing antibiotic usage, as well as misapply of these drugs, there is significant increase in antibiotics resistant bacterial strains for all of the dominant species causing HAIs, which further complicates the therapeutic strategies. Thus, the development of new drugs or control strategies as alternative therapies is clearly a matter of urgency. The vast majority of currently available antibiotics have been isolated from terrestrial sources, and also recent research strongly suggests that the marine microbes represents an untapped source of new antibacterial compounds [146].

Crude organic extracts of 57 marine bacterial isolates from sponge and soft corals were subjected for primary and secondary screening against bacteria, using well diffusion Mueller Hinton agar plates for zone of inhibition (ZOI). In the primary screening of antibacterial activity of the 57 bacterial extracts, 14 (24%) bacterial extracts displayed ZOI $\geq$ 5 mm against *Escherichia coli* and/or *Bacillus cereus,* when tested at 1000 ppm. This included eight (S4, S7, S8, S14, S18, S19, S20 and S21) and six (C3, C14, C19, C20, C22 and C26) bacterial extracts obtained from sponges and soft corals, respectively. At 1000 ppm, 12 and four bacterial extracts displayed activity against *E. coli* (ZOI ranging from 9.5 mm to 29 mm) and *B. cereus* (ZOI ranging from 8 to 16 mm), respectively. At a lower concentration of 100 ppm, nine bacterial extracts displayed activity against *E. coli* (ZOI ranging from 5 mm to 18.5 mm). However, no bacterial extracts displayed activity against *B. cereus* at 100 ppm. Results of primary screening encouraged us to further test marine bacterial extracts in secondary assay using pathogenic bacterial strains.

Fourteen marine bacterial extracts that displayed activity in the primary screening at 1000 ppm were selected for secondary screening against six antibiotic resistant bacteria, including Methicillin-Resistant *Staphylococcus aureus* (MRSA), *Enterococcus faecalis* V583, *Escherichia coli* EC 598, *Pseudomonas aeruginosa, Burkholderia cenocepacia* and *Acinetobacter baumannii.* Nine bacterial extracts displayed activity (ZOI $\approx$ 17 mm to 19 mm) against resistant pathogens, *B. cenocepacia, A. baumannii* and/or *P. aeruginosa* at 1000 ppm that is either comparable or greater than positive control standard antibiotics, kanamycin sulfate, tetracycline, chloramphenicol and ampicillin (ZOI $\approx$ 11 mm to 22 mm). At a lower concentration of 100 ppm, none of these bacterial extracts displayed any activity. However, bacterial extracts of S14 (ZOI $\approx$ 18 mm) and C22 (ZOI $\approx$ 19 mm) displayed activity only against *B. cenocepacia* and *P. aeruginosa,* respectively, while S8, S18. S20, C14, C19, C20 and C26 (ZOI $\approx$ 17 mm to 19 mm) displayed activity only against *A. baumannii* at 1000 ppm.

Of the 57 bacterial isolates, only eight bacterial isolates had successful pure DNA extractions, using liquid nitrogen method for DNA extractions, with absorbance ratio of 260/280 greater than 1.8. Phylogenetic tree based on 16S rDNA gene sequences analysis of these bacterial isolates revealed that S21 and S16 representing 88% homology between each other whereas bacterial isolate S4 related closely with *Arthrobacter* genus with 55% homology. The genus *Arthrobacter* is identified as a member of the high G + C contents branch of Gram-positive bacteria 'actinomycete', and phylogenetically related to other coryneform genera, such as *Aureobacterium, Cellulomonas, Curtobacterium* and *Microbacterium*. Phylogenetic analysis of five bacterial isolates, C14, C16, C20, C22 and C26, revealed that soft coral samples *Sinularia* sp. and *Lobophytum crassum* have bacterial communities related with *Staphylococcus* genus of Firmicutes phylum.

This unpublished study presented a pilot scale analysis of the biomedical potential of marine bacteria isolated from marine invertebrates (sponges and soft corals) from reefs in Singapore. About 65% (37 of 57) of bacterial isolates displayed activities in which approximately 61% (19 of 31) and 69% (18 of 26) of bacterial extracts were obtained from sponges and soft coral respectively. Significantly, 21%, 46% and 24% of the bacterial extracts displayed activity in brine shrimp toxicity (> 25%), cytotoxicity (anticancer) (> 25%) and in antibacterial assay (ZOI $\geq$ 5 mm), respectively. Interestingly, one (C3) of the bacterial extract obtained from soft coral *Sarcophyton tenuispiculatum* displayed activities in all three bioassays. These results underscore the potential of marine bacteria isolated from marine invertebrates collected from Singapore for drug discovery, as they are still widely unexplored and might provide a valuable resource for new pharmaceutical lead compounds. The combination of different bioactivity assays with chemical analysis is a useful strategy to prioritize bacterial isolates for further bioassay-guided isolation work.

3.6.2 Integrated Approach to Drug Discovery from Microbes Associated with Benthic Marine Organisms

To facilitate discovery efforts for anti-quorum sensing compounds from microbes associated with marine samples, an integrated-omic approach, including metabolomics and genomics, was employed [147]. Using a culture-dependent method, marine bacteria were isolated from 13 subtidal samples, including sponges and sediments, collected using a rectangular dredge in the Singapore Strait. Initial organic crude extracts prepared via small scale culturing of isolated marine bacterial colonies were screened using the *Pseudomonas aeruginosa* reporter strain for anti-quorum sensing activity. Bacterial extracts with quorum sensing inhibitory (QSI) activity were then analyzed using a mass-spectrometry based metabolomics Global Natural Products Social Molecular Networking platform (GNPS; https://gnps.ucsd.edu/) for compound dereplication [148]. In addition, marine bacterial strains that showed biological activity were subjected to whole genome sequencing for annotation of biosynthetic gene clusters using an antiSMASH bioinformatics tool [149]. The integration of both metabolomics and genomic techniques employed in the study is an effective and informed decision-making approach for the selection of marine bacterial strains with high probability of discovering novel bioactive compounds.

Homogenates from the 13 marine samples were prepared and plated on eight different marine media selected based on previous publications on similar isolation work [150–153]. Colonies displaying interesting morphology, such as bright colors, matte textures or unique colony shapes, were identified as colonies of interest. This resulted in a total of 102 bacterial colonies of interest obtained over a period of two months incubation. In summary, sponge samples *Geodia* sp. yielded the highest number of bacterial isolates (20), followed by *Stelletta* sp. (13), *Coelocarteria singaporensis* (11), *Haliclona* sp. (9), *Dysidea* sp. (8), *Xestospongia testudinaria* (7), *Rhabdastrella globostellata*, cf. *Leiodermatium* sp. (2) and *Ircinia* sp. (1).

With 102 colonies of interest isolated from 13 marine samples, their organic extracts were prepared via small scale culturing and screened against the *Pseudomonas aeruginosa* reporter strain, PAO1 *lasB-gfp*, [154] for quorum sensing inhibitory activity at 50 mg/mL. Any extract that showed a reduction in the green fluorescent protein (GFP) expression in relation to the controls were considered as a positive result. The induction of the *lasB* gene, encoding elastase, is a good indicator for LasR activity and any decrease in GFP would indicate the presence of an antagonist of the 3-oxo-C12-HSL, which inhibits the expression of *lasB*. It is well known that many bacterial types, including pathogenic strains, use quorum-sensing system to coordinate virulence and biofilm development [155]. The quorum sensing (QS) system in the reporter strain, *Pseudomonas aeruginosa*, comprises two distinct circuits that utilize acyl-homoserine lactone signals, along with a third system that employs quinolone signals [156]. These three interconnected QS systems collectively control the expression of hundreds of genes, many of which encode virulence factors. *P. aeruginosa* has become a model microorganism for investigating the molecular mechanisms of QS and exploring the ecology and evolution of group behaviors in bacteria [157].

From the screening efforts, only five marine bacterial extracts, namely #24, #27, #33, #34 and #52, were found to inhibit the GFP expression in *P. aeruginosa* PAO1 *lasB-gfp*. They were further tested to confirm the reduction of the GFP expression in a dose dependent manner without affecting the growth rates of the tested bioreporter strain. Of the five extracts, four bacterial strains, namely #24, #27, #33 and #34, were isolated from the sponge, *Geodia* sp. and one extract was prepared from bacterial strain #52 isolated from sponge, *Coelocarteria singaporensis*.

The chemical nature of five bacterial extracts that exhibited anti-quorum sensing activity were further analysed using the MS-based molecular networking platform (Wang et al. 2016). Unfortunately, two of these bacterial extracts, #27 and #33, did not achieve optimal ionization during the tandem mass spectrometry based on their respective observed total ion chromatogram (TIC) profile. The analysis of the remaining extracts, namely #24, #34 and #52, for compound dereplication and comparative analyses. This dereplication process would serve as a rapid analytical tool to shortlist the diverse crude extracts prepared from the culturing of the isolated bacterial strains. From the analysis of the dereplication results of the three marine bacterial extracts that exhibited anti-quorum sensing activity, the extracts did not show any matched known compounds within the GNPS database. In addition, each of the two bacterial strains, #24 and #52, had produced unique clusters of ionizable compounds. The unique cluster from strain #24 showed it contained relatively higher molecular weight compounds than that from strain #52.

Isolation and purification of genomic material from selected bacterial strains were also performed based on their positive quorum sensing inhibitory activity and molecular networking data. Several marine bacterial strains isolated from sponge, *Geodia* sp., that did not show positive bioassay activity were also included for the phylogenetic analysis as there was a high number of bacterial strains isolated from this sponge species. Only 29 strains were eventually identified based on their 16S rRNA gene sequencing. It was revealed that these 29 strains were affiliated to 14 different bacterial genera, of which one falls under the uncultured category, after comparing against sequences in the NCBI GenBank database. Two out of the three priority strains that are of particular interest to us, based on positive bioassay activity and molecular networking analysis, exhibited more than 98% confirmation to uncultured [158] bacterial clones. The findings suggested that the two bacterial strains probably belong to a putatively novel bacterial genus class or that the marine bacteria strains in question were not previously studied in detail.

The highest number of colonies of interest isolated were found to be affiliated with the genus *Kocuria* (nine colonies), followed by *Micrococcus* and uncultured (three colonies each), *Micromonospora*, *Gordonia* and *Bacillus* (two colonies each). The majority of the colonies of interest are classified under the Actinomycetales phylum (*Kocuria*, *Micrococcus*, *Micromonospora*, *Streptomonospora*, *Gordonia*) and the rest are Proteobacteria phylum (*Kiloniella*, *Pseudomonas*, *Alcanivorax*, *Vibrio*) and Firmicutes phylum (*Bacillus*, *Staphylococcus*, *Paenibacillus*). The phylogenetic analysis revealed that the uncultured bacterial clones, #24, #25 and #52, form a distinct clade possibly belonging to a novel genus that is closely related with *Kiloniella* and *Bacillus* due to its closest relative proximity.

Whole genome sequencing of two of the priority strains, namely #24 and #52, were carried out for annotation of the natural product biosynthetic gene clusters (BGC) using antiSMASH 4.0.2 [159]. The preliminary analysis conducted will form the basis for future studies into these two putative novel marine bacteria that were shown to contain anti-quorum sensing compounds. The genome for #24 and #52 were found to be 3.8 Mb and 5.6 Mb in length, with a GC content of 41.3% and 32.5%, respectively. The biosynthetic potential of the bacteria were assessed using antiSMASH 4.0.2. Out of six putative gene clusters identified that are housed in the #24 bacterium genome, there is one siderophore cluster (50% similar to the carotenoid BGC), one bacteriocin cluster, one terpene cluster, one type III polyketide/saccharide hybrid cluster, one non-ribosomal peptide synthetase gene cluster (85% similar to the lichenysin BGC), and one others cluster that is 85% similar to the bacilysin BGC. There are 13 putative gene clusters identified that are housed in the #52 bacterium genome. Of these 13 clusters, there is one siderophore cluster (100% similar to the petrobactin BGC), four non-ribosomal peptide synthetase gene clusters, one terpene cluster, four bacteriocin clusters, one sactipeptide cluster (100% similar to the thurincin H BGC), one arylpolyene cluster, and one others cluster.

In summary, this study is the first in Singapore to use an integrated biological, genomic and metabolomic approach for the discovery of anti-quorum sensing molecules from subtidal marine samples. From the findings, five of the isolated marine bacteria produce compounds that have anti-quorum sensing activity based on their bacterial organic extracts. The compounds produced by three of these marine bacteria, obtained from the sponges, *Geodia* sp. and *Coelocarteria singaporensis*, are possibly novel or new analogues as they do not have any compound match based on the GNPS library search. The results highlighted the importance of a streamlined and targeted approach for bioprospecting novel bioactive compounds. Molecular networking-based dereplication is a valuable tool for prioritizing microbial isolates for downstream processing and chemical investigation [160]. Such an approach has provided at least three marine bacterial isolates for further chemical analysis, including compound isolation and purification via large scale culturing.

Microbes Associated with the Marine Neptune's Cup Sponge, Cliona patera

This study employed both culture-dependent and culture-independent approaches to explore the diversity and biomedical potential of marine bacteria associated with the Neptune's Cup sponge, *Cliona patera*, from Singapore [161]. *C. patera*, one of the most iconic sponge species globally, was first described as *Spongia patera* from Singapore [162, 163]. Typically shaped like a wine glass, the sponge can grow to over a meter in height and diameter. *C. patera* was abundant in Singapore waters in the early nineteenth century, but due to its impressive size, it became highly sought after by natural history museums and private collectors. By the early twentieth century, it had not been recorded in the region and was presumed extinct [164]. However, in early 2011 the sponge species was rediscovered off Singapore's southern islands during a routine survey dive [164].

The rediscovery of *C. patera* in Singapore offers a unique opportunity to study the sponge in its natural habitat, including its associated microbial community. This is achieved by assessing the diversity of the prokaryotic community associated with the top oval concave disk and stalk of the gamma-stage (massive, free-living) *C. patera* using 16S rRNA amplicon sequencing. Additionally, bacteria cultured from two sponge colonies were screened for quorum sensing inhibitory activity using extracts tested on *Pseudomonas aeruginosa* PAO1 *lasB-gfp* biosensor strains. The disruption of quorum sensing systems in pathogenic bacteria via small molecules offers a novel therapeutic strategy against infectious diseases [165]. A mass spectrometry-based metabolomics approach, using the Global Natural Product Social Molecular Networking platform, was also employed to annotate the metabolomes of selected QSI-active marine bacterial strains [166].

The study found that *Cliona patera* harbors a complex and diverse prokaryotic microbiome, with more than twice the number of distinct OTUs compared to the surrounding seawater. Consistent with previous studies on marine sponge microbiomes, particularly within the Clionaidae family, the microbial communities of *C. patera* were predominantly composed of Proteobacteria, accounting for 70% of the relative abundance. Specifically, Gammaproteobacteria and Alphaproteobacteria were the most prevalent groups [167–169]. This high prevalence of Proteobacteria supports the classification of *C. patera* as a low microbial abundance (LMA) sponge, as predicted by Moitinho-Silva and co-workers [170]. Other major taxa included Bacteroidetes (6.05%), Actinobacteria (1.82%), Acidobacteria (1.66%), Archaea (0.77%), Planctomycetes (0.22%), Verrucomicrobia (0.2%), Nitrospirae (0.2%), and Marinimicrobia (0.02%), with 10.5% of the bacterial community remaining unclassified. Unlike many other marine sponges, *C. patera* did not have a dominant presence of Chloroflexi [171]. Additionally, the commonly reported sponge-specific phylum Poribacteria was virtually absent in *C. patera*. The genus *Ruegeria*, from the phylum Proteobacteria, was present in all *C. patera* sponge samples but absent in the seawater group. *Ruegeria* (class Alphaproteobacteria) is primarily composed of marine bacteria and has been consistently isolated and cultured from various sponge species [172]. This genus is part of the Roseobacter clade, an ecologically significant marine group known for its wide range of metabolic activities, including the production of novel bioactive compounds.

Using culture-dependent methods, 110 bacterial colonies were isolated from *Cliona patera* sponge samples. The diversity of media used in the study played a crucial role, as no single medium accounted for the majority of bacterial isolates. Approximately 30.9% of bacterial extracts tested at 100 µg/mL showed moderate (60–80%) to significant (> 80%) reductions in fluorescence in the *Pseudomonas aeruginosa* PAO1 *lasB-gfp* biosensor strain. This relatively high percentage of quorum sensing inhibitory (QSI) active bacterial strains was found in both the cup and stem regions of the sponge. This is the first study to report the presence of marine bacterial strains with QS inhibitory activity isolated from different regions of a sponge host. Notably, the QSI-active bacterial strains belonged to either the phylum Firmicutes (class Bacilli) or Proteobacteria (class Alphaproteobacteria), both of which are known for producing diverse bioactive compounds.

MS-based molecular networking of organic extracts from selected quorum sensing inhibitory (QSI) active marine bacterial strains, belonging to the classes Bacilli and Alphaproteobacteria, revealed a diverse chemical composition. A total of 591 precursor ions were identified from the initial analysis of five QSI active bacterial extracts, each showing over 50% QS inhibitory activity. These ions corresponded to a variety of natural product molecular family clusters, including siderophores (e.g., dehydroxynocardamine), pyrrolidine derivatives (e.g., anisomycin), biosurfactants (e.g., surfactin D), diketopiperazines [e.g., cyclo(L-Val-L-Pro)], and indole alkaloids (e.g., harmaline). Additionally, other structural classes were detected, such as aromatic polyketides (e.g., 2,6-dihydroxyanthraquinone) and pyrone derivatives (e.g., pyrenocine B). Further analysis of organic extracts from 11 QSI active strains revealed additional surfactin derivatives, expanding the chemical space.

In summary, this study sheds light on the unique microbiome of the Neptune's Cup sponge, *Cliona patera*. The sponge hosted 5222 distinct OTUs, predominantly from the phylum Proteobacteria, particularly the classes Gammaproteobacteria and Alphaproteobacteria. Additionally, several marine bacteria associated with *C. patera*, specifically from the classes Bacilli and Alphaproteobacteria, emerged as promising candidates for novel quorum sensing inhibitory molecules, as indicated by their MS-based molecular networking profiles.

Biomedical Potential of Terrestrial and Marine Fungi from St John's Island, Singapore

A recent study by Munusamy and co-workers investigated the biosynthetic potential and antimicrobial properties of 30 Ascomycete fungi isolated from St. John's Island, Singapore [173]. All isolates belonged to the phylum Ascomycota, spanning six orders and 18 genera, with Hypocreales being the most represented (37%).

Fungal extracts were tested for antimicrobial activity against *Staphylococcus aureus*, methicillin-resistant *S. aureus* (MRSA) and *Candida albicans*. Bioassay-guided fractionation of two promising isolates led to the isolation of seven compounds, including penilumamides A, D and E from terrestrial fungal strain F4335 (*Aspergillus flavipes*), and xanthomegnin, viomellein, pretrichodermamide C and vioxanthin from marine fungal strain F7180 (*Penicillium janthinellum*). Vioxanthin showed the strongest antibacterial activity, with IC_{50} values of 3.0 μM against *S. aureus* and 1.6 μM against MRSA. Viomellein exhibited weak antiproliferative activity against A549 cells, with an IC_{50} of 42 μM. These findings provide valuable insights into the fungal diversity and biosynthetic potential of this unique habitat, forming a foundation for further exploration of selected strains to discover novel natural products.

4 Challenges and Future Marine Natural Products Research in Singapore

Marine natural products research in Singapore holds immense potential due to the country's strategic location in one of the most biodiverse marine regions in the world, coupled with its advanced infrastructure and interdisciplinary expertise. However, the field faces several challenges that need to be addressed to fully realize its potential. One of the primary challenges is the limited natural marine biodiversity resulting from Singapore's small territorial waters and extensive coastal urbanization [174]. Land reclamation, pollution and habitat destruction have significantly altered marine ecosystems, reducing the availability of marine organisms for research. Sustainability is another critical concern where overharvesting of marine resources for research and commercial purposes can lead to ecological imbalances and biodiversity loss.

Additionally, the high costs associated with the collection, isolation, and characterization of marine natural products, particularly those from deep-sea environments, pose a significant barrier. The need for specialized equipment, bioinformatics tools and skilled personnel further compounds these challenges.

Despite these challenges, the future of marine natural products research in Singapore is promising, particularly with the integration of innovative technologies and interdisciplinary approaches. Advances in metabolomics [175, 176], genomics [177, 178] and synthetic biology [179] are revolutionizing the field, enabling researchers to identify and characterize bioactive compounds more efficiently. Metabolomics, for instance, allows for the comprehensive analysis of metabolites produced by marine organisms, providing insights into their chemical diversity and potential pharmaceutical applications [180]. Genomics, on the other hand, facilitates the identification of biosynthetic gene clusters responsible for the production of bioactive compounds, even in unculturable marine microbes [181]. Synthetic biology offers the possibility of replicating and modifying these compounds in the laboratory, reducing the need for large-scale extraction from marine environments and addressing sustainability concerns [182].

Another promising area is the use of artificial intelligence (AI) and machine learning in drug discovery [183]. AI-driven platforms can analyze vast datasets to predict the bioactivity of marine natural products, identify potential drug candidates, and optimize their chemical structures. This approach can significantly accelerate the drug discovery process, reducing the time and cost associated with traditional methods. Furthermore, the development of marine biotechnology parks or specialized research hubs could foster innovation and commercialization in marine natural products research. These hubs could serve as centers for collaborative research, bringing together experts from academia, industry and government to translate scientific discoveries into marketable products.

The exploration of marine microbiomes represents another frontier in marine natural products research [184]. Marine invertebrates, such as sponges, corals and tunicates, host diverse microbial communities that produce a wide range of bioactive compounds. Advances in metagenomics and culture-independent techniques have enabled researchers to tap into this hidden reservoir of microbial diversity, uncovering novel compounds with potential applications in medicine, agriculture and industry [185]. For example, the discovery of ET-743, an anticancer compound derived from a tunicate-associated microbe, highlights the therapeutic potential of marine invertebrate-derived microbes [186]. In addition to technological innovations, there is a need for interdisciplinary education and training programs to address skill gaps in marine biology, chemistry, pharmacology, and bioinformatics. By nurturing a new generation of researchers with expertise in these areas, Singapore can strengthen its position as a leader in marine natural products research. Collaborative networks with biodiverse countries in Southeast Asia, such as Indonesia and Malaysia, can also enhance access to marine biodiversity and facilitate knowledge exchange.

In conclusion, while marine natural products research in Singapore faces significant challenges, the integration of innovative technologies, interdisciplinary approaches and sustainable practices offers a pathway to unlocking its full potential. By focusing on areas such as metabolomics, genomics, synthetic biology and marine microbiomes, Singapore can continue to make significant contributions to drug discovery, biotechnology and environmental conservation.

References

1. Chou L, Huang D, Tan K, Toh T, Goh B, Tun K (2019) Singapore. In: Hamel J-F (ed) World seas: an environmental evaluation. Volume II: The Indian Ocean to the Pacific. Elsevier, London, pp 539–558
2. Lai S, Loke LH, Hilton MJ, Bouma TJ, Todd PA (2015) The effects of urbanisation on coastal habitats and the potential for ecological engineering: a Singapore case study. Ocean Coast Manag 103:78–85
3. Hilton MJ, Manning SS (1995) Conversion of coastal habitats in Singapore: indications of unsustainable development. Environ Conserv 22:307–322
4. Yee A, Ang W, Teo S, Liew S, Tan H (2010) The present extent of mangrove forests in Singapore. Nat Singap 3:139–145

5. Yaakub SM, McKenzie LJ, Erftemeijer PL, Bouma T, Todd PA (2014) Courage under fire: seagrass persistence adjacent to a highly urbanised city–state. Mar Pollut Bull 83:417–424
6. Chou LM (2021) "Living" coastal protection and nearshore biodiversity reclamation. In: Ting DSK, Carriveau R (eds) Sustaining tomorrow via innovative engineering. World Scientific, Singapore, pp 291–310
7. Lee AC, Tan KS, Sin TM (2009) Intertidal assemblages on coastal defence structures in Singapore I: a faunal study. Raffles Bull Zool Suppl 22:237–254
8. Lim LJW, Loh JBY, Lim AJS, Tan BYX, Ip YCA, Neo ML, Tan R, Huang D (2020) Diversity and distribution of intertidal marine species in Singapore. Raffles Bull Zool 68:396–403
9. Ng CSL, Chen D, Chou LM (2012) Hard coral assemblages on seawalls in Singapore. In: Tan K (ed) Contributions to marine science. National University of Singapore, Singapore, Tropical Marine Science Institute, pp 75–79
10. Chou LM (2021) Enhancing marine biodiversity with artificial structures. In: Ting DSK, Stagner JA (eds) Sustainable engineering technologies and architectures. AIP Publishing, Melville, New York, pp 5–1 to 5–22
11. Chou LM, Ng CSL, Toh KB, Tun K, Cheo PR, Ng JY (2023) Marinas can co-function as biodiversity sanctuaries. In: Ting DSK, Stagner JA (eds) Nourishing tomorrow: clean engineering and nature-friendly living. World Scientific, Singapore, pp 179–212
12. Chou LM (2024) Coral reefs in an urban seascape. In: Koh T, Lye LH, Lum S (eds) Peace with nature. World Scientific, Singapore, pp 39–45
13. Guest JR, Tun K, Low J, Vergés A, Marzinelli EM, Campbell AH, Bauman AG, Feary DA, Chou LM, Steinberg PD (2016) 27 years of benthic and coral community dynamics on turbid, highly urbanised reefs off Singapore. Sci Rep 6:36260
14. Chou LM (2023) Chapter 10—Sustaining marine biodiversity in Singapore's heavily urbanized coast. In: Ting D, O'Brien P (eds) Progress in sustainable development. Elsevier, Amsterdam, Netherlands, pp 265–282
15. Siang TK, Siong KK, Goh L (2015) Taking stock of Singapore's marine life: the comprehensive marine biodiversity survey Johor Straits international workshop 2012. Raffles Bull Zool Suppl 31:1–6
16. Todd PA, Tanzil JTI, Yap NWL, Martin P (2021) Advancing marine science in Singapore: the marine science research & development programme (2015–21). 2021, Singapore: Marine Science Research & Development Programme, National Research Foundation, Prime Minister's Office, Singapore, p 205
17. Tan YJ, Tham JK, Paul A, Rana U, Ang HP, Nguyen NT, Yee AT, Leong BP, Drummond S, Tun KP (2023) Remote sensing mapping of the regeneration of coastal natural habitats in Singapore: implications for marine conservation in tropical cities. Singap J Trop Geogr 44:130–148
18. Ding JL, Fung FMY, Chou LM (1994) Cytotoxic effects of mucus from coral *Galaxea fascicularis*. J Mar Biotechnol 2:27–33
19. Fung FMY, Tachibana S, Chou LM, Ding JL (1997) Cytotoxic and anticancer agents in mucus of *Galaxea fascicularis*: purification and characterization. J Mar Biotechnol 5:50–57
20. Ding JL, Fung FMY, Ng GWS, Chou LM (1999) Novel bioactivities from a Coral, *Galaxea fascicularis*: DNase-like activity and apoptotic activity against a multiple-drug-resistant Leukemia Cell Line. Mar Biotechnol 1:328–336
21. Goh NKC, Sim TS, Chou LM (1995) Bioactivity of gorgonians (subclass Octocorallia) in Singapore: preliminary studies. Mar Res 4:33–46
22. Goh NKC, Chou LM (1998) Bioactivity screening of Singapore gorgonians: antimicrobial activity, toxicity to *Artemia salina* and efficacy against *Plasmodium berghei*. Zoolog Sci 15:805–812
23. Koh LL, Goh NKC, Chou LM, Tan YW (2000) Chemical and physical defenses of Singapore gorgonians (Octocorallia: Gorgonacea). J Exp Mar Biol Ecol 251:103–115
24. Koh LL, Tan TK, Chou LM, Goh NKC (2002) Antifungal properties of Singapore gorgonians: a preliminary study. J Exp Mar Biol Ecol 273:121–130

25. Goh BPL, Tan GE, Tan LT (2009) Diversity, distribution and biological activity of soft corals (Octocorallia, Alcyonacea) in Singapore. J Coast Develop 12:89–98
26. Goh NKC, Chou LM (1996) An annotated checklist of the gorgonians (Anthozoa: Octocorallia) of Singapore, with a discussion of gorgonian diversity in the Indo-West Pacific. Raffles Bull Zool 44:435–459
27. Goh NKC, Chou LM (1994) Distribution and biodiversity of Singapore gorgonians (subclass Octocorallia)—a preliminary survey. In: Sasekumar A, Marshall N, Macintosh DJ (eds) Ecology and conservation of Southeast Asian Marine and freshwater environments including wetlands. Springer, Netherlands, Dordrecht, pp 101–109
28. Goh NKC, Chou LM (1995) Growth of five species of gorgonians (Sub-Class Octocorallia) in the sedimented waters of Singapore. Mar Ecol 16:337–346
29. Goh NKC, Sim TS, Chou LM (1995) Bioactivity of gorgonians (Sub-class Octocorallia) in Singapore: preliminary studies. Mar Res 4:33–46
30. de Voogd NJ, Cleary DFR (2009) Variation in sponge composition among Singapore reefs. Raffles Bull Zool Suppl 22:59–67
31. Lim S-C, de Voogd NJ, Tan K-S (2012) Biodiversity of shallow-water sponges (Porifera) in Singapore and description of a new species of *Forcepia* (Poecilosclerida: Coelosphaeridae). Contrib Zool 81:55–71
32. Lim SF, Tan LT, Goh BPL (2023) Distribution and abundance of intertidal sponge (Porifera) communities in coral reefs of Singapore. Ocean Sci J 58:30
33. Seng AWP, Ong YZ, Lee WY, Lee PPF, Lim SC, Tan KS, Tan LT (2007) Biomedical potentials of intertidal marine organisms from Singapore. J Coast Develop 11:1–13
34. Goh NK, Chou L (1998) Toxicity and palatability of gorgonian (Octocorallia: Gorgonacea) extracts to goldfish *Carassius auratus* (Linnaeus, 1758). Asian Marine Biol 15:199–210
35. Ho YY, Salleh NF, Goh JX, Oluwabusola ET, Tan LT (2023) Chemical nature of specialised defensive metabolites from seven common intertidal marine sponges found at Pulau Hantu, Singapore. Nat Singapore 16:e2023108
36. Chan W, Shaughnessy AEP, van den Berg CP, Garson MJ, Cheney KL (2021) The validity of brine shrimp (*Artemia* sp.) toxicity assays to assess the ecological function of marine natural products. J Chem Ecol 47:834–846
37. Schönberg CHL, Lim S-C (2019) Psammobiosis and bioerosion: examiningecological strategies in sponges using the case example *Coelocarteria singaporensis*. Facies 65:14
38. Abdelmohsen UR, Cheng C, Viegelmann C, Zhang T, Grkovic T, Ahmed S, Quinn RJ, Hentschel U, Edrada-Ebel R (2014) Dereplication strategies for targeted isolation of new Antitrypanosomal Actinosporins A and B from a marine sponge associated-*Actinokineospora* sp. EG49. Mar Drugs 12:1220–1244
39. Chua CH, Tan LT, Goh BPL (2023) Field testing of potential antifouling compounds from four marine organisms found in Singapore. Nat Singap 16:e2023077
40. Hellio C, Tsoukatou M, Maréchal J-P, Aldred N, Beaupoil C, Clare AS, Vagias C, Roussis V (2005) Inhibitory effects of Mediterranean sponge extracts and metabolites on larval settlement of the barnacle *Balanus amphitrite*. Mar Biotechnol 7:297–305
41. Longo C, Trani R, Marzano CN, Mercurio M, Lazic T, Cotugno P, Santobianchi E, Gravina MF (2021) Anti-fouling activity and toxicity of the Mediterranean alien sponge *Paraleucilla magna* Klautau, Monteiro & Borojevic, 2004 (Porifera, Calcarea). PeerJ 9:e12279
42. Tan LT, Goh BPL, Tripathi A, Lim MG, Dickinson GH, Lee SSC, Teo SLM (2010) Natural antifoulants from the marine cyanobacterium *Lyngbya majuscula*. Biofouling 26:685–695
43. Pettit GR, Xu J-p, Hogan F, Williams MD, Doubek DL, Schmidt JM, Cerny RL, Boyd MR (1997) Isolation and structure of the human cancer cell growth inhibitory cyclodepsipeptide dolastatin 16,1. J Nat Prod 60:752–754
44. Casalme LO, Yamauchi A, Sato A, Petitbois JG, Nogata Y, Yoshimura E, Okino T, Umezawa T, Matsuda F (2017) Total synthesis and biological activity of dolastatin 16. Org Biomol Chem 15:1140–1150
45. Pettit GR, Smith TH, Arce PM, Flahive EJ, Anderson CR, Chapuis J-C, Xu J-P, Groy TL, Belcher PE, Macdonald CB (2015) Antineoplastic agents. 599. Total synthesis of dolastatin 16. J Nat Prod 78:476–485

46. Casalme LO, Katayama K, Hayakawa Y, Nakamura K, Yamauchi A, Nogata Y, Yoshimura E, Matsuda F, Umezawa T (2022) Effective synthesis and antifouling activity of dolastatin 16 derivatives. Mar Drugs 20:124
47. Umezawa T, Sari IN, Yoshimura E, Nogata Y (2025) Novel settlement inhibition oligopeptides containing β-amino acids. RSC Sustain 3:540–545
48. Rodríguez E, Barbeitos MS, Brugler MR, Crowley LM, Grajales A, Gusmão L, Häussermann V, Reft A, Daly M (2014) Hidden among sea anemones: the first comprehensive phylogenetic reconstruction of the Order Actiniaria (Cnidaria, Anthozoa, Hexacorallia) reveals a novel group of hexacorals. PLoS ONE 9:e96998
49. Fautin DG (2016) Catalog to families, genera, and species of orders Actiniaria and Corallimorpharia (Cnidaria: Anthozoa). Zootaxa 4145:1–449
50. Yap NWL, Mitchell ML, Quek ZBR, Tan R, Tan KS, Huang D (2023) Taxonomy and molecular phylogeny of the Sea Anemone *Macrodactyla* (Haddon, 1898) (Cnidaria, Actiniaria), with a description of a new species from Singapore. Zool Stud 62:e29
51. Fautin DG (2009) Structural diversity, systematics, and evolution of cnidae. Toxicon 54:1054–1064
52. England KW (1987) Certain *Actiniaria* (Cnidaria, Anthozoa) from the Red Sea and tropical Indo-Pacific Ocean 53:205–292
53. Fautin DG, Tan R, Liang W, Yap N, Hee TS, Crowther A, Goodwill R, Sanpanich K, Chieh TY (2015) Sea anemones (Cnidaria: Actiniaria) of Singapore: shallow-water species known also from the Indian subcontinent. Raffles Bull Zool
54. Yap NWL, Tan R, Yong CLX, Tan KS, Huang D (2019) Sea anemones (Cnidaria, Actiniaria) of Singapore: redescription and taxonomy of *Phymanthus pinnulatus* Martens in Klunzinger, 1877. Zookeys 840:1–20
55. Honma T, Iso T, Ishida M, Nagashima Y, Shiomi K (2003) Occurrence of type 3 sodium channel peptide toxins in two species of sea anemones (*Dofleinia armata* and *Entacmaea ramsayi*). Toxicon 41:637–639
56. Gendeh GS, Young LC, de Medeiros CLC, Jeyaseelan K, Harvey AL, Chung MCM (1997) A new potassium channel toxin from the Sea Anemone *Heteractis magnifica*: isolation, cDNA cloning, and functional expression. Biochemistry 36:11461–11471
57. Gendeh GS, Chung MCM, Jeyaseelan K (1997) Genomic structure of a potassium channel toxin from *Heteractis magnifica*. FEBS Lett 418:183–188
58. Wang Y, Yap LL, Chua KL, Khoo HE (2008) A multigene family of Heteractis magnificalysins (HMgs). Toxicon 51:1374–1382
59. Wang Y, Chua KL, Khoo HE (2000) A new cytolysin from the sea anemone, *Heteractis magnifica*: isolation, cDNA cloning and functional expression. Biochim et Biophys Acta Protein Struct Mol Enzymol 1478:9–18
60. Khoo KS, Kam WK, Khoo HE, Gopalakrishnakone P, Chung MCM (1993) Purification and partial characterization of two cytolysins from a tropical sea anemone, *Heteractis magnifica*. Toxicon 31:1567–1579
61. Khoo HE, Lim JPC, Tan CH (1995) Effects of sea anemone (*Heteractis magnifica* and *Actinia equina*) cytolysins on synaptosomal uptake of gaba and choline. Toxicon 33:1365–1371
62. Menezes C, Thakur NL (2022) Sea anemone venom: ecological interactions and bioactive potential. Toxicon 208:31–46
63. Yosef O, Popovits Y, Malik A, Ofek-Lalzer M, Mass T, Sher D (2020) A tentacle for every occasion: comparing the hunting tentacles and sweeper tentacles, used for territorial competition, in the coral *Galaxea fascicularis*. BMC Genomics 21:548
64. Fung FMY, Ding JL (1998) A novel antitumour compound from the mucus of a coral, *Galaxea fascicularis*, inhibits topoisomerase I and II. Toxicon 36:1053–1058
65. Tsai Y-H, Hwang D-F, Chai T-J, Jeng S-S (1995) Occurrence of tetrodotoxin and paralytic shellfish poison in the Taiwanese crab *Lophozozymus pictor*. Toxicon 33:1669–1673
66. Llewellyn LE, Endean R (1989) Toxicity and paralytic shellfish toxin profiles of the xanthid crabs, *Lophozozymus pictor* and *Zosimus aeneus*, collected from some Australian coral reefs. Toxicon 27:596–600

67. Tan CTT, Lee EJD (1988) A fatal case of crab toxin (*Lophozozymus pictor*) poisoning. Asia Pac J Pharmacol 3:7–9
68. Teh YF, Gardiner JE (1970) Toxin from the coral reef crab, *Lophozozymus pictor*. Pharmacol Res Commun 2:251–256
69. Lau CO, Tan CH, Li QT, Ng FH, Yuen R, Khoo HE (1995) Bioactivity and mechanism of action of *Lophozozymus pictor* toxin. Toxicon 33:901–908
70. Lau CO, Khoo HE, Yuen R, Wan M, Tan CH (1993) Isolation of a novel fluorescent toxin from the coral reef crab, *Lophozozymus pictor*. Toxicon 31:1341–1345
71. Lau CO, Tan CH, Khoo HE, Li QT, Yuen R (1995) Ethanolic extraction, purification, and partial characterization of a fluorescent toxin from the coral reef crab, *Lophozozymus pictor*. Nat Toxins 3:87–90
72. Tan CH, Teh YF (1972) The effect of *Lophozozymus pictor* toxin on Hela cells. Experientia 28:46–46
73. Tan CH, Abkowitz SJ (1979) *Lophozozymus pictor* toxin: a potent inhibitor of synaptosomal GABA uptake. Neurosci Lett 14:339–342
74. Lau CO, Ng FH, Khoo HE, Yuen R, Tan CH (1996) Inhibition of sodium-dependent uptake processes in purified rat brain synaptosomes by *Lophozozymus pictor* toxin and palytoxin. Neurochem Int 28:385–390
75. Chia DGB, Lau CO, Ng PKL, Tan CH (1993) Localization of toxins in the poisonous mosaic crab, *Lophozozymus pictor* (Fabricius, 1798) (Brachyura, Xanthidae). Toxicon 31:901–904
76. McDonald LA, Ireland CM (1992) Patellamide E: a new cyclic peptide from the Ascidian *Lissoclinum patella*. J Nat Prod 55:376–379
77. Baur P, Kühl M, Comba P, Behrendt L (2022) Possible functional roles of patellamides in the ascidian-*Prochloron* symbiosis. Mar Drugs 20:119
78. Koehnke J, Bent AF, Houssen WE, Mann G, Jaspars M, Naismith JH (2014) The structural biology of patellamide biosynthesis. Curr Opin Struct Biol 29:112–121
79. Nilar SPJ, Carté BK, Butler MS (2002) Three new pyridoacridine type alkaloids from a Singaporean Ascidian. J Nat Prod 65:1198–1200
80. Ji K, Cheng J-f, Nakamura H, Ohizumi Y, Hirata Y, Sasaki T, Ohta T, Nozoe S (1988) Ascididemin, a novel pentacyclic aromatic alkaloid with potent antileukemic activity from the okinawan tunicate *Didemnum* sp. Tetrahedron Lett 29:1177–1180
81. Carroll AR, Scheuer PJ (1990) Kuanoniamines A, B, C, and D: Pentacyclic alkaloids from a tunicate and its prosobranch mollusk predator *Chelynotus semperi*. J Org Chem 55:4426–4431
82. Ibrahim SR, Mohamed GA (2016) Marine pyridoacridine alkaloids: biosynthesis and biological activities. Chem Biodivers 13:37–47
83. Pettit GR, Bond TJ, Herald DL, Penny M, Doubek DL, Williams MD, Pettit RK, Hooper JNA (1997) Isolation and structure of spongilipid from the Republic of Singapore marine porifera *Spongia* cf. *hispida*. Can J Chem 75:920–925
84. Pettit GR, Orr B, Herald DL, Doubek DL, Tackett L, Schmidt JM, Boyd MR, Pettit RK, Hooper JNA (1996) Isolation and X-ray crystal structure of racemic Xestospongin D from the Singapore marine sponge *Niphates* sp1. Bioorg Med Chem Lett 6:1313–1318
85. Pettit GR, Hoffmann H, McNulty J, Higgs KC, Murphy A, Molloy DJ, Herald DL, Williams MD, Pettit RK, Doubek DL, Hooper JNA, Albright L, Schmidt JM, Chapuis J-C, Tackett LP (2004) Antineoplastic agents. 380. Isolation and X-ray crystal structure determination of isoaaptamine from the Republic of Singapore *Hymeniacidon* sp. and conversion to the phosphate prodrug hystatin 1. J Nat Prod 67:506–509
86. Dong H, Gou Y-L, Kini RM, Xu H-X, Chen S-X, Teo SLM, But PP-H (2000) A new cytotoxic polyhydroxysterol from soft coral *Sarcophyton trocheliophorum*. Chem Pharm Bull 48:1087–1089
87. Curren E, Leong SCY (2018) New records of benthic coastal cyanobacteria observed from the Southern Islands of Singapore. Algol Stud 154:93–112
88. Curren E, Leong SCY (2018) *Lyngbya regalis* sp. nov. (Oscillatoriales, Cyanophyceae), a new tropical marine cyanobacteri*um*. Phytotaxa 367:120–132

89. Gupta DK, Kaur P, Leong ST, Tan LT, Prinsep MR, Chu JJH (2014) Anti-chikungunya viral activities of aplysiatoxin-related compounds from the marine cyanobacterium *Trichodesmium erythraeum*. Mar Drugs 12:115–127
90. Zanchett G, Oliveira-Filho EC (2013) Cyanobacteria and cyanotoxins: from impacts on aquatic ecosystems and human health to anticarcinogenic effects. Toxins 5:1896–1917
91. Ding CYG, Pang LM, Liang Z-X, Goh KKK, Glukhov E, Gerwick WH, Tan LT (2018) MS/MS-based molecular networking approach for the detection of aplysiatoxin-related compounds in environmental marine cyanobacteria. Mar Drugs 16:505
92. Paul DGN, JV (1999) Production of secondary metabolites by filamentous tropical marine cyanobacteria: ecological functions of the compounds. J Phycol 35:1412–1421
93. Sneed JM, Meickle T, Engene N, Reed S, Gunasekera S, Paul VJ (2017) Bloom dynamics and chemical defenses of benthic cyanobacteria in the Indian River Lagoon, Florida. Harmful Algae 69:75–82
94. Salleh NF, Wang J, Kundukad B, Oluwabusola ET, Goh DXY, Phyo MY, Tong JJL, Kjelleberg S, Tan LT (2023) Cyclopropane-containing specialized metabolites from the marine cyanobacterium cf. *Lyngbya* sp. Molecules 28:3965
95. Kwan JC, Meickle T, Ladwa D, Teplitski M, Paul V, Luesch H (2011) Lyngbyoic acid, a “tagged” fatty acid from a marine cyanobacterium, disrupts quorum sensing in *Pseudomonas aeruginosa*. Mol BioSyst 7:1205–1216
96. Tan LT, Chang YY, Ashootosh T (2008) Besarhanamides A and B from the marine cyanobacterium *Lyngbya majuscula*. Phytochemistry 69:2067–2069
97. Reese MT, Gulavita NK, Nakao Y, Hamann MT, Yoshida WY, Coval SJ, Scheuer PJ (1996) Kulolide: a cytotoxic depsipeptide from a cephalaspidean mollusk, *Philinopsis speciosa*. J Am Chem Soc 118:11081–11084
98. Iwasaki A, Shiota I, Sumimoto S, Matsubara T, Sato T, Suenaga K (2017) Kohamamides A, B, and C, cyclic depsipeptides from an *Okeania* sp. marine cyanobacterium. J Nat Prod 80:1948–1952
99. Almaliti J, Malloy KL, Glukhov E, Spadafora C, Gutiérrez M, Gerwick WH (2017) Dudawalamides A-D, antiparasitic cyclic depsipeptides from the marine cyanobacterium *Moorea producens*. J Nat Prod 80:1827–1836
100. Boudreau PD, Byrum T, Liu W-T, Dorrestein PC, Gerwick WH (2012) Viequeamide A, a cytotoxic member of the kulolide superfamily of cyclic depsipeptides from a marine button cyanobacterium. J Nat Prod 75:1560–1570
101. Levert A, Alvariño R, Bornancin L, Abou Mansour E, Burja AM, Genevière A-M, Bonnard I, Alonso E, Botana L, Banaigs B (2018) Structures and activities of tiahuramides A-C, cyclic depsipeptides from a Tahitian collection of the marine cyanobacterium *Lyngbya majuscula*. J Nat Prod 81:1301–1310
102. Tripathi A, Puddick J, Prinsep MR, Lee PPF, Tan LT (2009) Hantupeptin A, a cytotoxic cyclic depsipeptide from a Singapore collection of *Lyngbya majuscula*. J Nat Prod 72:29–32
103. Tripathi A, Puddick J, Prinsep MR, Lee PPF, Tan LT (2010) Hantupeptins B and C, cytotoxic cyclodepsipeptides from the marine cyanobacterium *Lyngbya majuscula*. Phytochemistry 71:307–311
104. Bunyajetpong S, Yoshida WY, Sitachitta N, Kaya K (2006) Trungapeptins A−C, cyclodepsipeptides from the marine cyanobacterium *Lyngbya majuscula*. J Nat Prod 69:1539–1542
105. Gupta DK, Ding GCY, Teo YC, Tan LT (2016) Absolute stereochemistry of the β-hydroxy acid unit in hantupeptins and trungapeptins. Nat Prod Commun 11:1934578X1601100120
106. Avula SN, Sunitha M, Chakunta RG (2016) Stereoselective synthesis of southern fragment of hantupeptin-A. Acta Chim Slov 63:344–350
107. Srinivas A, Sunitha M, Rao CG (2016) Stereoselective synthesis of hantupeptins A, B and C common fragment. Indian J Chem 55B:1239–1242
108. Ding CYG, Ong JFM, Goh HC, Coffill CR, Tan LT (2018) Benderamide A, a Cyclic depsipeptide from a Singapore collection of marine cyanobacterium cf *Lyngbya* sp. Mar Drugs 16:409

109. Gunasekera SP, Owle CS, Montaser R, Luesch H, Paul VJ (2011) Malyngamide 3 and cocosamides A and B from the marine cyanobacterium *Lyngbya majuscula* from Cocos Lagoon, Guam. J Nat Prod 74:871–876
110. Phyo MY, Katermeran NP, Goh JX, Tan LT (2021) Trikoveramides A-C, cyclic depsipeptides from the marine cyanobacterium *Symploca hydnoides*. Phytochemistry 190:112879
111. Phyo MY, Goh JX, Tan LT (2022) Triproamide and pemukainalides, cyclic depsipeptides from the marine cyanobacterium *Symploca hydnoides*. J Nat Prod 85:485–492
112. Nakao Y, Yoshida WY, Szabo CM, Baker BJ, Scheuer PJ (1998) More peptides and other diverse constituents of the marine mollusk *Philinopsis speciosa*. J Org Chem 63:3272–3280
113. Tripathi A, Puddick J, Prinsep MR, Rottmann M, Tan LT (2010) Lagunamides A and B: cytotoxic and antimalarial cyclodepsipeptides from the marine cyanobacterium *Lyngbya majuscula*. J Nat Prod 73:1810–1814
114. Tripathi A, Puddick J, Prinsep MR, Rottmann M, Chan KP, Chen DY-K, Tan LT (2011) Lagunamide C, a cytotoxic cyclodepsipeptide from the marine cyanobacterium *Lyngbya majuscula*. Phytochemistry 72:2369–2375
115. Luo D, Putra MY, Ye T, Paul VJ, Luesch H (2019) Isolation, structure elucidation and biological evaluation of lagunamide D: a new cytotoxic macrocyclic depsipeptide from marine cyanobacteria. Mar Drugs 17:83
116. Michon S, Cavelier F, Salom-Roig XJ (2021) Synthesis and biological activities of cyclodepsipeptides of aurilide family from marine origin. Mar Drugs 19:55
117. Hagimoto K, Tojo S, Teruya T, Yoshida M, Kigoshi H (2024) Structure revision of lagunamide C to odoamide by total synthesis and biological evaluation. Tetrahedron 154:133871
118. Tripathi A, Fang W, Leong DT, Tan LT (2012) Biochemical studies of the lagunamides, potent cytotoxic cyclic depsipeptides from the marine cyanobacterium *Lyngbya majuscula*. Mar Drugs 10:1126–1137
119. Dai L, Chen B, Lei H, Wang Z, Liu Y, Xu Z, Ye T (2012) Total synthesis and stereochemical revision of lagunamide A. Chem Commun 48:8697–8699
120. Huang W, Ren R-G, Dong H-Q, Wei B-G, Lin G-Q (2013) Diverse synthesis of marine cyclic depsipeptide Lagunamide A and its analogues. J Org Chem 78:10747–10762
121. Liu H-M, Chang C-Y, Lai Y-C, Yang M-D, Chang C-Y (2014) An efficient synthesis of the C27–C45 fragment of lagunamide A, a cyclodepsipeptide with potent cytotoxic and antimalarial properties. Tetrahedron: Asymmetry 25:187–192
122. Banasik BA, Wang L, Kanner A, Bergdahl BM (2016) Further insight into the asymmetric vinylogous Mukaiyama aldol reaction (VMAR); application to the synthesis of the C27–C45 segment of lagunamide A. Tetrahedron 72:2481–2490
123. Gorges J, Kazmaier U (2018) Matteson homologation-based total synthesis of lagunamide A. Org Lett 20:2033–2036
124. Pal S, Chakraborty TK (2014) Toward the total synthesis of a lagunamide B analogue. Tetrahedron Lett 55:3469–3472
125. Fatino A, Weese C, Valdez S, Jiménez-Somarribas A, Rafferty RJ (2018) Synthetic studies towards lagunamide C: polyketide assembly investigations. Tetrahedron Lett 59:624–627
126. Matteson DS (2013) Boronic esters in asymmetric synthesis. J Org Chem 78:10009–10023
127. Huang X, Huang W, Li L, Sun X, Song S, Xu Q, Zhang L, Wei B-G, Deng X (2016) Structure determinants of lagunamide A for anticancer activity and its molecular mechanism of mitochondrial apoptosis. Mol Pharm 13:3756–3763
128. Hu Y, Mostert D, Orgler C, Andler O, Zischka H, Kazmaier U, Vollmar AM, Braig S, Sieber SA, Zahler S (2024) Thermal proteome profiling reveals insight to antiproliferative and pro-apoptotic effects of lagunamide A in the modulation of DNA damage repair. ChemBioChem 25:e202400024
129. Zhou C, Li Z, Diao H, Yu Y, Zhu W, Dai Y, Chen FF, Yang J (2006) DNA damage evaluated by γH2AX foci formation by a selective group of chemical/physical stressors. Mutat Res/Genet Toxicol Environ Mutagen 604:8–18
130. Sivonen K, Leikoski N, Fewer DP, Jokela J (2010) Cyanobactins—ribosomal cyclic peptides produced by cyanobacteria. Appl Microbiol Biotechnol 86:1213–1225

131. Martins J, Vasconcelos V (2015) Cyanobactins from cyanobacteria: current genetic and chemical state of knowledge. Mar Drugs 13:6910–6946
132. Czekster CM, Ge Y, Naismith JH (2016) Mechanisms of cyanobactin biosynthesis. Curr Opin Chem Biol 35:80–88
133. Phyo MY, Ding CYG, Goh HC, Goh JX, Ong JFM, Chan SH, Yung PYM, Candra H, Tan LT (2019) Trikoramide A, a prenylated cyanobactin from the marine cyanobacterium *Symploca hydnoides*. J Nat Prod 82:3482–3488
134. Phyo MY, Goh TMB, Goh JX, Tan LT (2021) Trikoramides B-D, bioactive cyanobactins from the marine cyanobacterium *Symploca hydnoides*. Mar Drugs 19:548
135. Low ZJ, Ma G-L, Tran HT, Zou Y, Xiong J, Pang L, Nuryyeva S, Ye H, Hu J-F, Houk KN, Liang Z-X (2020) Sungeidines from a non-canonical Enediyne biosynthetic pathway. J Am Chem Soc 142:1673–1679
136. Alanzi AR (2024) Exploring microbial dark matter for the discovery of novel natural products: characteristics, abundance challenges and methods. J Microbiol Biotechnol 35:e2407064
137. Paul SI, Majumdar BC, Ehsan R, Hasan M, Baidya A, Bakky MAH (2021) Bioprospecting potential of marine microbial natural bioactive compounds. J Appl Biotechnol Rep 8:96–108
138. Trindade M, van Zyl LJ, Navarro-Fernández J, Abd Elrazak A (2015) Targeted metagenomics as a tool to tap into marine natural product diversity for the discovery and production of drug candidates. Front Microbiol 6
139. O'Brien PA, Webster NS, Miller DJ, Bourne DG (2019) Host-microbe coevolution: applying evidence from model systems to complex marine invertebrate holobionts. mBio 10. https://doi.org/10.1128/mbio.02241-02218
140. Liu L, Zheng Y-Y, Shao C-L, Wang C-Y (2019) Metabolites from marine invertebrates and their symbiotic microorganisms: molecular diversity discovery, mining, and application. Marine Life Sci Technol 1:60–94
141. Ramesh C, Tulasi BR, Raju M, Thakur N, Dufossé L (2021) Marine natural products from tunicates and their associated microbes. Mar Drugs 19:308
142. Engelberts JP, Robbins SJ, Damjanovic K, Webster NS (2021) Integrating novel tools to elucidate the metabolic basis of microbial symbiosis in reef holobionts. Mar Biol 168:175
143. Rizzo C, Lo Giudice A (2018) Marine invertebrates: underexplored sources of bacteria producing biologically active molecules. Diversity 10:52
144. Sunarti LS (2024) Hospital acquired infection revisited: route of transmission. Int J Pathogen Res 13:24–35
145. Ventola CL (2015) The antibiotic resistance crisis: part 1: causes and threats. Pharm Ther 40:277
146. Monciardini P, Iorio M, Maffioli S, Sosio M, Donadio S (2014) Discovering new bioactive molecules from microbial sources. Microb Biotechnol 7:209–220
147. Ong JFM, Goh HC, Lim SC, Pang LM, Chin JSF, Tan KS, Liang Z-X, Yang L, Glukhov E, Gerwick WH, Tan LT (2019) Integrated genomic and metabolomic approach to the discovery of potential anti-quorum sensing natural products from microbes associated with marine samples from Singapore. Mar Drugs 17:72
148. Allard P-M, Péresse T, Bisson J, Gindro K, Marcourt L, Pham VC, Roussi F, Litaudon M, Wolfender J-L (2016) Integration of molecular networking and in-silico MS/MS fragmentation for natural products dereplication. Anal Chem 88:3317–3323
149. Blin K, Kim HU, Medema MH, Weber T (2019) Recent development of antiSMASH and other computational approaches to mine secondary metabolite biosynthetic gene clusters. Brief Bioinform 20:1103–1113
150. Magarvey NA, Keller JM, Bernan V, Dworkin M, Sherman DH (2004) Isolation and characterization of novel marine-derived actinomycete taxa rich in bioactive metabolites. Appl Environ Microbiol 70:7520–7529
151. Choi EJ, Nam S-J, Paul L, Beatty D, Kauffman CA, Jensen PR, Fenical W (2015) Previously uncultured marine bacteria linked to novel alkaloid production. Chem Biol 22:1270–1279
152. Duncan KR, Crüsemann M, Lechner A, Sarkar A, Li J, Ziemert N, Wang M, Bandeira N, Moore BS, Dorrestein PC (2015) Molecular networking and pattern-based genome mining

improves discovery of biosynthetic gene clusters and their products from Salinispora species. Chem Biol 22:460–471

153. Maansson M, Vynne N, Klitgaard A, Nybo J, Melchiorsen J, Nguyen D, Sanchez L, Ziemert N, Dorrestein P, Andersen M (2016) An integrated metabolomic and genomic mining workflow to uncover the biosynthetic potential of bacteria. mSystems 1:e00028–15
154. Rasmussen TB, Bjarnsholt T, Skindersoe ME, Hentzer M, Kristoffersen P, Kote M, Nielsen J, Eberl L, Givskov M (2005) Screening for quorum-sensing inhibitors (QSI) by use of a novel genetic system, the QSI selector. J Bacteriol 187:1799–1814
155. Haque S, Ahmad F, Dar SA, Jawed A, Mandal RK, Wahid M, Lohani M, Khan S, Singh V, Akhter N (2018) Developments in strategies for quorum sensing virulence factor inhibition to combat bacterial drug resistance. Microb Pathog 121:293–302
156. Soukarieh F, Williams P, Stocks MJ, Cámara M (2018) *Pseudomonas aeruginosa* quorum sensing systems as drug discovery targets: current position and future perspectives. J Med Chem 61:10385–10402
157. Grandy S, Scur M, Dolan K, Nickerson R, Cheng Z (2023) Using model systems to unravel host–*Pseudomonas aeruginosa* interactions. Environ Microbiol 25:1765–1784
158. Garza DR, Dutilh BE (2015) From cultured to uncultured genome sequences: metagenomics and modeling microbial ecosystems. Cell Mol Life Sci 72:4287–4308
159. Blin K, Wolf T, Chevrette MG, Lu X, Schwalen CJ, Kautsar SA, Suarez Duran HG, De Los Santos EL, Kim HU, Nave M (2017) AntiSMASH 4.0—improvements in chemistry prediction and gene cluster boundary identification. Nucleic Acids Res 45:W36–W41
160. Purves K, Macintyre L, Brennan D, Hreggviðsson GÓ, Kuttner E, Ásgeirsdóttir ME, Young LC, Green DH, Edrada-Ebel R, Duncan KR (2016) Using molecular networking for microbial secondary metabolite bioprospecting. Metabolites 6:2
161. Ho XY, Katermeran NP, Deignan LK, Phyo MY, Ong JFM, Goh JX, Ng JY, Tun K, Tan LT (2021) Assessing the diversity and biomedical potential of microbes associated with the Neptune's cup sponge, *Cliona patera*. Front Microbiol 12:631445
162. Hardwicke T (1822) Description of a zoophyte, commonly found about the coasts of Singapore Island, with a plate. Asiatic Res 14:180–181
163. Low ME (2012) The date of publication of *Cliona patera* (Hardwicke), the 'sponge plant from the shores of Singapore' (Porifera: Hadromerida: Clionaidae). Nat Singapore 5:223–227
164. Lim S, Tun K, Goh E (2012) Rediscovery of the Neptune's cup sponge in Singapore: *Cliona* or *Poterion*. Contrib Mar Sci 2012:49–56
165. Saurav K, Bar-Shalom R, Haber M, Burgsdorf I, Oliviero G, Costantino V, Morgenstern D, Steindler L (2016) In search of alternative antibiotic drugs: Quorum-quenching activity in sponges and their bacterial isolates. Front Microbiol 7:416
166. Nguyen DD, Wu C-H, Moree WJ, Lamsa A, Medema MH, Zhao X, Gavilan RG, Aparicio M, Atencio L, Jackson C (2013) MS/MS networking guided analysis of molecule and gene cluster families. Proc Natl Acad Sci 110:E2611–E2620
167. Webster NS, Thomas T (2016) The sponge hologenome. MBio. https://doi.org/10.1128/mbio.00135-00116
168. Pita L, Rix L, Slaby BM, Franke A, Hentschel U (2018) The sponge holobiont in a changing ocean: from microbes to ecosystems. Microbiome 6:46
169. Sacristán-Soriano O, Turon X, Hill M (2020) Microbiome structure of ecologically important bioeroding sponges (family Clionaidae): the role of host phylogeny and environmental plasticity. Coral Reefs 39:1285–1298
170. Moitinho-Silva L, Steinert G, Nielsen S, Hardoim CC, Wu Y-C, McCormack GP, López-Legentil S, Marchant R, Webster N, Thomas T (2017) Predicting the HMA-LMA status in marine sponges by machine learning. Front Microbiol 8:752
171. Thomas T, Moitinho-Silva L, Lurgi M, Björk JR, Easson C, Astudillo-García C, Olson JB, Erwin PM, López-Legentil S, Luter H (2016) Diversity, structure and convergent evolution of the global sponge microbiome. Nat Commun 7:11870
172. Karimi E, Keller-Costa T, Slaby BM, Cox CJ, da Rocha UN, Hentschel U, Costa R (2019) Genomic blueprints of sponge-prokaryote symbiosis are shared by low abundant and cultivatable Alphaproteobacteria. Sci Rep 9:1999

173. Munusamy M, Tan K, Nge CE, Gakuubi MM, Crasta S, Kanagasundaram Y, Ng SB (2023) Diversity and biosynthetic potential of fungi isolated from St. John's Island, Singapore. Int J Mol Sci 24:1033
174. Chou LM (2006) Marine habitats in one of the World's Busiest harbours. In: Wolanski E (ed) The environment in Asia Pacific harbours. Springer, Dordrecht, Netherlands, pp 377–391
175. Stuart KA, Keira W, Clare WM, Edrada-Ebel R (2020) Metabolomic tools used in marine natural product drug discovery. Expert Opin Drug Discov 15:499–522
176. Bayona LM, de Voogd NJ, Choi YH (2022) Metabolomics on the study of marine organisms. Metabolomics 18:17
177. Chu L, Huang J, Muhammad M, Deng Z, Gao J (2020) Genome mining as a biotechnological tool for the discovery of novel marine natural products. Crit Rev Biotechnol 40:571–589
178. Berde CV, Berde VB, Reddy PN, Veera Bramhachari P (2023) Bioprospection of marine sponge microbiome for bioactive metabolites employing advanced metagenomics tools. In: Veera Bramhachari P, Berde CV (eds) Marine bioactive molecules for biomedical and pharmacotherapeutic applications. Springer Nature, Singapore, pp 15–38
179. Reen FJ, Gutiérrez-Barranquero JA, Dobson ADW, Adams C, Gara F (2015) Emerging concepts promising new horizons for marine biodiscovery and synthetic biology. Mar Drugs 13:2924–2954
180. Reverter M, Rohde S, Parchemin C, Tapissier-Bontemps N, Schupp PJ (2020) Metabolomics and marine biotechnology: coupling metabolite profiling and organism biology for the discovery of new compounds. Front Marine Sci 7
181. Kaari M, Manikkam R, Baskaran A (2022) Exploring newer biosynthetic gene clusters in marine microbial prospecting. Mar Biotechnol 24:448–467
182. Costantini M (2020) Genome mining and synthetic biology in marine natural product discovery. Mar Drugs 18:615
183. Almaliti J, Gerwick WH (2023) Methods in marine natural product drug discovery: what's new? Expert Opin Drug Discov 18:687–691
184. Tan LT (2023) Impact of marine chemical ecology research on the discovery and development of new pharmaceuticals. Mar Drugs 21:174
185. Leal MC, Sheridan C, Osinga R, Dionísio G, Rocha RJM, Silva B, Rosa R, Calado R (2014) Marine microorganism-invertebrate assemblages: perspectives to solve the "supply problem" in the initial steps of drug discovery. Mar Drugs 12:3929–3952
186. Uria A, Piel J (2009) Cultivation-independent approaches to investigate the chemistry of marine symbiotic bacteria. Phytochem Rev 8:401–414

Chapter 6
Natural Products from Phytoplankton in Singapore's Coastal Waters

Emily Curren, Jerome Wai Kit Kok, Jun Yi Loh, and Sandric Chee Yew Leong

1 Introduction

Phytoplankton are crucial primary producers in many waters around the world and support entire aquatic ecosystems. They are single-celled organisms that can rapidly proliferate and adapt to changing conditions in the waters. Phytoplankton therefore exhibits a wide range of physiologies, which makes them an ideal organism for bioprospecting [1]. Several thousand compounds which have been identified from marine phytoplankton have tended to come from temperate strains [2]. However, tropical marine waters are known to have much higher phytoplankton biodiversity [3] and should therefore also harbour a very high potential for bioprospecting [4]. This chapter provides a review of findings related to tropical phytoplankton and examines the organisms that can provide useful bioactive substances of commercial value.

Today, phytoplankton bioprospecting has yielded significant commercial value in the sectors of human nutrition and animal feed. Studies have observed several medicinal properties of carotenoids, such as β-carotene, astaxanthin, fucoxanthin, and violaxanthin [1]. These are pigments produced by phytoplankton in response to varying environmental conditions [5]. Components of sunblock and other skincare/cosmetic products have also been extracted from phytoplankton, especially cyanobacteria [6]. These mycosporine-like amino acids (MAA) that are used in sunblock products have been observed to have increased production in the presence of ultraviolet (UV) radiation [7], which tends to be more persistently higher in tropical climates throughout the year. Increasingly, techniques are also being developed

E. Curren (✉) · J. W. K. Kok · J. Y. Loh · S. C. Y. Leong
St. John Island National Marine Laboratory, Tropical Marine Science Institute (TMSI), National University of Singapore, Singapore, Singapore
e-mail: e0013223@u.nus.edu

J. W. K. Kok
National University of Singapore College, Singapore, Singapore

L. T. Tan and N. Hanif (eds.), *Marine Natural Products Research in South-East Asia*,
https://doi.org/10.1007/978-3-032-01098-8_6

for phytoplankton to be an important source of alternative protein that consists of the complete nine essential amino acids needed for human nutrition [8].

Tropical coastal waters are additionally known to be highly productive. Places in Southeast Asia have very high density of urbanised coastlines, which give rise to typically eutrophic waters [9]. Of the different countries, Singapore is especially unique because the bloom-causative species tend to differ from the surrounding waters [10, 11]. Causative species have typically tended to involve members of the Kareniaceae family, which produce a range of toxins that are known to be ichthyotoxic [12]. However, the chemical function of these substances remains poorly understood. In another study carried out on freshwater pico-cyanobacteria, tropical *Synechoccus* sp. from Singapore waters was reported for the first time to be able to produce the cylindrospermopsin toxin [13]. The propensity for tropical waters to yield unique phytoplankton physiologies needs to be closely examined. The diversity and variation observed across tropical species might correspond with genetic adaptations that yield previously unrecognised bioactive products.

This chapter sets out to explore some of the bioactive products that have been characterised for phytoplankton species observed in tropical waters. It will also examine studies that test tropical physiologies to assess whether strains in tropical Singapore may harbour novel, valuable bioactive products. We focus specifically on non-toxic metabolites as well as toxic substances that are typically implicated in harmful algal bloom occurrences. Additionally, we also look at carbohydrate and protein production which are more recent trends that respectively support the biofuel and alternative protein industries.

2 Metabolites

Marine phytoplankton are a source of novel natural products which have high biological activity. These natural products are classified into primary and secondary metabolites, which have different functions in the organism. Primary metabolites are compounds which are directly involved in the growth, development and reproduction of the organism [14]. These metabolites can result from both anabolic and catabolic metabolism. Products from anabolic metabolism include nucleotides and amino acids that are used by the cell as building blocks for macromolecules. Some might also be converted to coenzymes such as vitamins [15]. Catabolic metabolism produces primary metabolites such as lactic acid, ATP and pyruvate, which is used for energy production and other metabolic pathways [16]. In phytoplankton, these metabolites are essential for basic metabolic processes, such as photosynthesis, respiration and nucleotide metabolism. These metabolites are usually non-toxic as they are fundamentally required for the growth and development of the organism [17]. For instance, chlorophyll and other carotenoid pigments are integral for photosynthesis and have protective functions, which we will explore more in later sections of this chapter.

In contrast, secondary metabolites are described as compounds that are not directly required for primary metabolism in that organism [17]. These compounds were once thought to have minor significance to the organism and are usually expressed in greatest quantities from the active to stationary phase. These metabolites have specific functions and are produced in response to various physical or biological stresses in the surrounding environment, such as nutrient limitation or drought stress [17]. Secondary metabolites also function as toxins to deter predators or inhibit the growth of competing species [18]. There is a big degree of structural variation across different metabolite classes. These secondary metabolites can be broadly categorized into the following main groups: terpenes, alkaloids and phenolics. Terpenoids are a diverse and extensive class of secondary metabolites. They are a mix of linear to polycyclic hydrocarbons which are constructed through the condensation of isoprene (C_5H_8) units. Terpenes are generally classified based on the number of monoterpene (C_{10}), sesquiterpene (C_{15}) and diterpene (C_{20}) units. A commonly studied terpenoid is fucoxanthin, which is a carotenoid that is responsible for the brown and orange colour found in living organisms. This pigment is also a powerful antioxidant that reduces oxidative stress by neutralizing free radicals. Diatoms, such as *Thalassiosira* sp. [19] and *Chaetoceros*, are found to contain this pigment [20].

Many compounds, such as limonene and dimethyl sulfides (DMS), are commonly produced in phytoplankton, including the cyanobacterium *Synechocystis* sp. [21] and dinoflagellate *Symbiodinium* sp. [22]. Both compounds have allelopathic and chemical defense functions. Limonene protects cyanobacteria from predation due to its distinct scent [23]. DMS is a trace gas that has protective effects in organisms in relation to oxidative stress and can function as grazing defense mechanisms [22].

Alkaloids are a diverse group of secondary metabolites consisting of more than 12,000 compounds. These alkaloids are basic compounds and bioactive. A well-studied alkaloid is saxitoxin (STX), which is a neurotoxin that is commonly associated with harmful algal blooms and fish kill events around the world [24]. A widespread producer of STX is the dinoflagellate *Alexandrium* sp., which results in fatal paralytic shellfish poisoning. STX binds voltage-gated calcium and sodium ion channels, resulting in neural pore blockage in nerves and muscles, resulting in death [25]. Anatoxin-a is also a toxic alkaloid which is found in cyanobacteria genera, such as *Anabaena* [26] and *Planktothrix* [27].

Some alkaloids are also peptides, which include non-ribosomal peptides (NRPs). Non-ribosomal peptide synthases (NRPS) are enzymatic modules which produce these NRPs, by integrating specific amino acids into peptide chains. Lyngbyatoxin-a is a metabolite produced by *Moorena producens* (previously *Lyngbya majuscula*) via the NRPS pathway. This toxin binds and activates protein kinase C (pkC) [28], leading to the production of inflammatory cytokines and tumor promotion [29]. Polyketides are another large class of secondary metabolites, which are synthesized from acetyl-CoA using polyketide synthases (PKS). A group of toxic polyketides are the brevetoxins, which is produced by the dinoflagellate *Karenia brevis*. Brevetoxins are cyclic polyether compounds which bind to sodium ion channels and neurotoxic shellfish poisoning (NSP) [30]. These toxins will be further elaborated in a later section.

A commonly known group of compounds are the mycosporine-like amino acids (MAAs), which are found in dinoflagellates, diatoms and cyanobacteria. These MAAs, consisting of aminocyclohexenone or aminocycloheximine rings, are low molecular weight UV photoprotectors, which have cosmeceutical applications [31]. In contrast, phenolics are characterized by the presence of one or more hydroxyl groups attached to an aromatic ring. They can range from simple (C_6), to long chain polymers, such as lignins $(C_6\text{-}C_3)_n$ and tannins $(C_6\text{-}C_3\text{-}C_6)_n$ [32]. A well-known phenolic compound is scytonemin, which is an indole pigment produced in the sheaths of many cyanobacteria genera, such as *Nostoc* and *Calothrix* [33].

In this section, we will explore the various primary and secondary metabolites produced by key species of phytoplankton found in Singapore's coastal waters. It is important to understand the production and functions of these products to comprehend their ecological roles and impact on marine organisms and the surrounding ecosystem. This knowledge will enhance our understanding of harmful algal blooms (HABs) and their effects on human health and marine life.

2.1 *Pigments*

The marked decline in the stratospheric ozone has led to an increase in ultraviolet radiation flow to the earth [34]. UV radiation has widespread effects on organisms and has affected growth and development by inducing DNA damage, resulting in cytotoxicity [35]. However, many organisms are still able to survive under large amounts of UV exposure due to the presence of UV-absorbing compounds. These compounds are found across various phytoplankton groups, such as the cyanobacteria, diatoms and dinoflagellates.

2.1.1 Cyanobacteria

All phytoplankton are known to contain pigment compounds. These primary metabolites mainly function in photosynthesis but also protect against harmful UV radiation. Marine *Spirulina* sp. is known to exist among cyanobacterial mats in the central regions of the Johor Straits. This genus of cyanobacteria contains phycocyanobilin, which gives *Spirulina* its blue-green colour [36]. The chief biliprotein in this genus is C-phycocyanin. This pigment is protective and is also known to be an antioxidant, as it can quench different derivatives of reactive oxygen species [36]. Other pigments, such as β-carotene and xanthophylls (e.g., diadinoxanthin and diatoxanthin) have demonstrated high scavenging activity with the DPPH radical [37]. Astaxanthin (**1**) (Fig. 1) is also another pigment produced in *Spirulina platensis* that has anti-inflammatory and anti-apoptotic activities [38]. Oral ingestion of astaxanthin supplements was linked to lower levels of inflammation and risk of cardiovascular disease in healthy human individuals. This supplement also decreased the levels of total cholesterol, triglycerides, high-density lipoprotein (HDL) and low-density

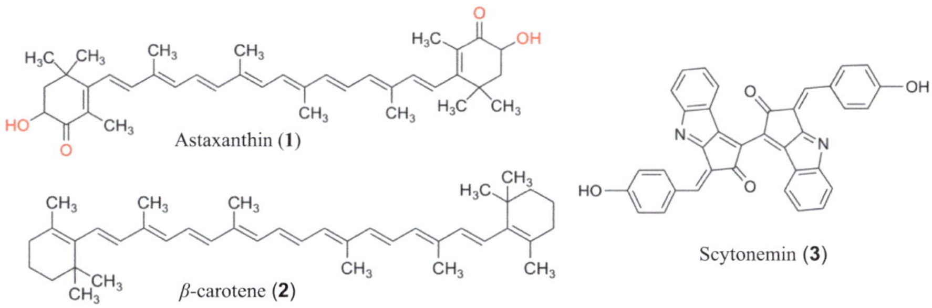

Fig. 1 Chemical structures of cyanobacterial pigments

lipoprotein (LDL) levels in human volunteers. Astaxanthin has also shown efficacy in the treatment of ocular diseases, such as cataract and macular degeneration [39].

As a dietary nutrient, *β*-carotene supplementation has shown positive effects on multiple diseases. *β*-carotene **(2)** (Fig. 1) is a compound that humans are unable to synthesise. This compound after ingestion is converted to vitamin A and then retinoic acid (RA) [40]. An isomer of RA, all-trans RA (ATRA) is shown to block the cell cycle, increase apoptosis and DNA fragmentation in gastric cancer cell lines MKN-45 and MKN-28 [41]. ATRA induces cell cycle arrest at the G0/G1 phase, downregulates c-myc and upregulates the expression of p53 [42]. Furthermore, a *β*-carotene rich diet was found to be preventative for cardiac-related issues. This compound is primarily transported in LDL and its binding will prevent oxidation and reduce incidence of atherogenesis [43].

Other pigment-rich cyanobacteria have also been found in the intertidal regions of Singapore [44]. Three heterocystous genera, *Calothrix, Rivularia* and *Nostoc*, were previously identified from the Southern Islands of Singapore. These cyanobacteria contained the sheath pigment scytonemin **(3)** (Fig. 1), which blocks up to 90% of UV-A radiation from entering the cells [45]. This lipid-soluble compound is produced significantly in extremophilic genera, and these cyanobacteria can survive in thermal springs, intertidal zones and subaerial surfaces. Under field conditions, this pigment is golden to dark brown and shows strong absorbance with peaks in UV-A (384 nm) and UV-C (250 nm) [46]. This is useful for the cyanobacteria that is exposed to high solar irradiance in tropical regions, especially when they are present on intertidal rocks and experience direct sun exposure during the day.

Scytonemin is an indole-alkaloid and is present in more than 300 cyanobacteria species. It can form up to 5% of the dry weight of the organism. The synthesis of scytonemin is primarily triggered by UV-A radiation, where its presence will upregulate all 18 genes in the pathway of scytonemin synthesis [47]. Other stress factors, such as increased salinity and temperatures, desiccation, nitrogen limitation and osmotic stress, can also induce synthesis of this compound. In recent reports, scytonemin demonstrated anti-inflammatory effects both in vitro and in vivo. Topical application of scytonemin suppressed the expressions of inducible nitric oxide synthase (iNOS) and tumor necrosis factor-*α* (TNF-*α*) in mice [48]. Both

TNF-α and iNOS are well-studied mediators of inflammation, and its expression is increased in inflamed cells. Hence, a down-regulation of these inflammatory molecules after scytonemin application will protect against a variety of skin diseases, such as eczema and dermatitis. Reduced scytonemin that was isolated from the heterocystous cyanobacterium *Nostoc commune* suppressed nitric oxide production by lipopolysaccharides in murine macrophage RAW264 cells [49]. Furthermore, reduced scytonemin also decreased the activity of nuclear factor-κβ (NF-κβ) and mitogen-activated protein kinases (MAPKs) [49]. Hence, besides UV protection, reduced scytonemin also has anti-inflammatory property.

2.1.2 Diatoms

The marine diatom genera *Skeletonema* and *Thalassiosira* are often observed in surface water samples of the Singapore Strait. Photoprotective pigments exist in *Skeletonema* sp., where light intensity regulated the concentration of phenolic compounds, such as flavonoids, and ascorbic acid. In this diatom, phenols accounted for scavenging activity with gradual light intensity changes, and flavonoids primarily responded in times of great light variations [50].

The marine diatom *Thalassiosira weissflogii* demonstrated long-term acclimation to UV radiation [51]. Cells were exposed to 14 h of PAR and UV-A daily, with exposure to UV-B for 4 h a day [51]. Initially, the growth rate of *T. weissflogii* declined over the first two weeks. However, recovered after two weeks till the end of the 40-day treatment. It is interesting to note that major photosynthetic pigments, such as chlorophyll *α*, fucoxanthin and *β*-carotene, did not change significantly with time [51]. *T. weissflogii* has previously been investigated as a potential target for metabolic engineering and production of the pigment fucoxanthin [19]. Fucoxanthin (**4**) (Fig. 2) is a major carotenoid among marine microalgae. Together with several proteins, this pigment binds to chlorophyll *a* and forms fucoxanthin-chlorophyll *a*-complexes to harvest light energy in the thylakoids [52]. Fucoxanthin has anti-inflammatory effects, and this is observed by inhibiting inflammatory cytokines in RAW 264.7 macrophages [53]. Additionally, fucoxanthin was able to inhibit the expression of cyclooxygenase 2 protein and inducible nitric oxide synthase. A recent study evaluated the health benefits of fucoxanthin in the prevention of chronic illnesses in animal studies [54]. Fucoxanthin has a hepatoprotective effect, where it inhibits hepatic lipid accumulation in mice fed with a high-fat diet. A supplementation of fucoxanthin at a 0.2% level by weight lowered plasma triglyceride levels [55]. Other photoprotective carotenoids, such as the xanthophyll pigments diadinoxanthin (**5**) and diatoxanthin (**6**) (Fig. 2), were also identified in *Thalassiosira gravida*, each being produced in concentrations up to 80 mol per 100 mol of Chlorophyll a content [56].

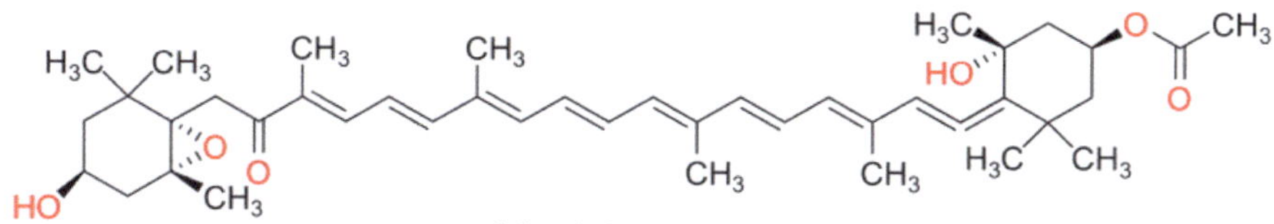

Fucoxanthin (**4**)

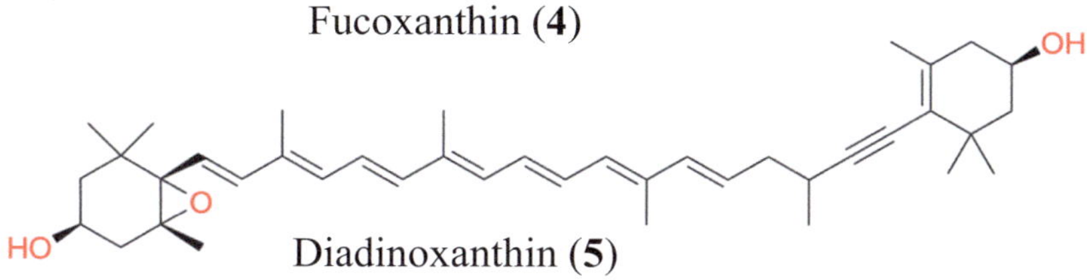

Diadinoxanthin (**5**)

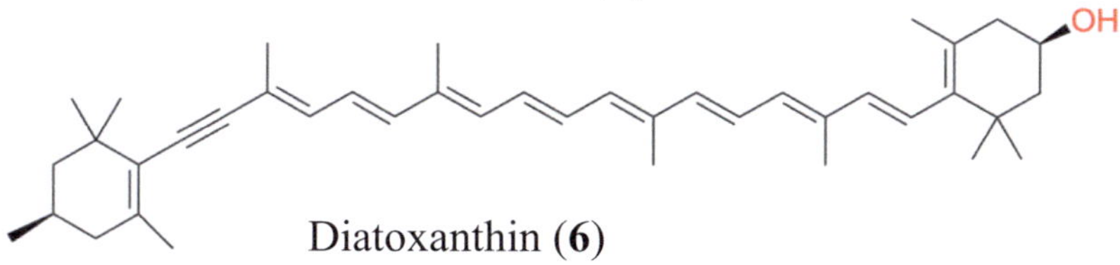

Diatoxanthin (**6**)

Fig. 2 Chemical structures of diatom-derived pigments

2.1.3 Dinoflagellates

Dinoflagellates also produce UV-absorbing compounds named mycosporine-like amino acids (MAAs). MAAs are derivatives of mycosporines (**7**) (Fig. 3) which is a family of more than 20 small, water-soluble compounds that absorb UV strongly between 310 and 362 nm [57]. They are also photostable in marine environments. MAAs typically consist of a cyclohexanone or cycloheximine ring that is conjugated to other amino acids or amino alcohols. These compounds dissipate absorbed radiation as heat and do not produce reactive oxygen species [58]. MAAs are commonly studied in dinoflagellates, such as *Scrippsiella* and *Alexandrium.*

Members of the genus *Scrippsiella* have been observed from the coastal waters of the Johor Straits. In the dinoflagellate *Scrippsiella sweeneyae*, five types of MAAs were observed: palythene (**8**), mycosporine-glycine (**9**), porphyra-334 (**10**), shinorine (**11**), and palythene (**12**) (Fig. 3), [59]. In another study, *Scrippsiella trochoidea* had the highest MAA concentration among 33 species of microalgae, across 13 classes, including the Bacillariophyceae, Chlorophyceae and Dinophyceae, at > 6000 fg/cell [60]. In another dinoflagellate *Alexandrium tamarense,* 11 MAAs were identified in response to the increase in photosynthetically active radiation (PAR). The top two compounds which exhibited the greatest increases in the first 2 h were porphyra-334 and palythene [61]. Porphyra-334 (**10**) has been investigated in antiphotoaging experiments and found to suppress ROS production and increase production of extracellular matrix components, such as type I collagen and elastin [62]. A less well-known function of MAAs is in wound healing, where mycosporine-glycine (**9**) and

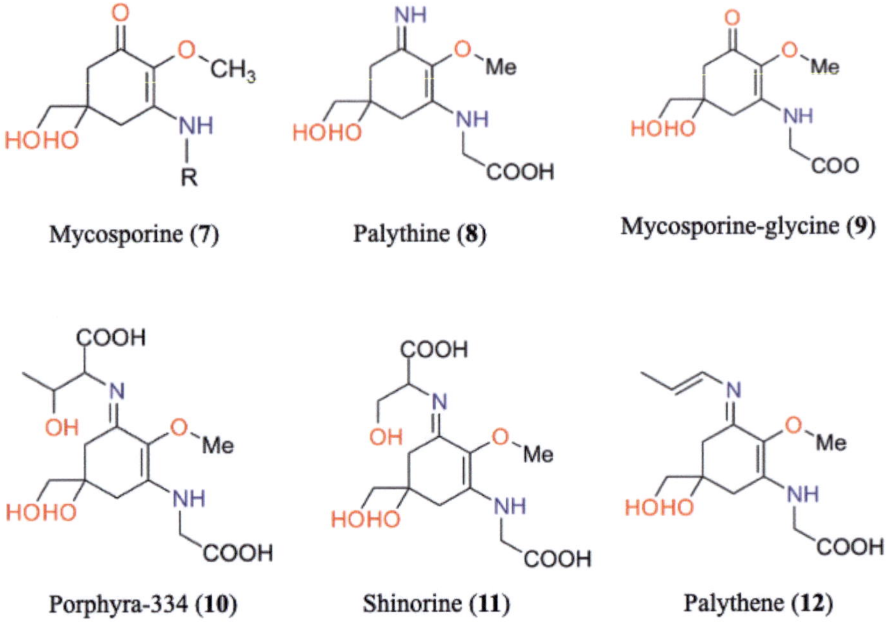

Fig. 3 Chemical structures of UV-absorbing mycosporine-like amino acids

shinorine (**11**), in concentrations of 0.1 mg/mL and 0.05 mg/mL accelerated wound repair on human keratinocyte HaCaT cells [63].

Dinoflagellates from the family Kareniaceae have been observed in Singapore waters. Athecate dinoflagellates from the family Kareniaceae contain chl-*a*, fucoxanthin, chl-c_2, diatoxanthin, diadinoxanthin and β-carotene [64, 65]. They have fucoxanthin as their major secondary pigment, as opposed to the frequently reported peridinin in dinoflagellates [65, 66]. The dinoflagellate genus *Takayama* has also been recorded from Singapore waters since 2004 [67]. In a study involving exposure of *T. acrotrocha* to various nitrogen forms (NO_3^-, NH_4^+ and urea) and concentrations, *T. acrotrocha* was found to vary its pigment profile as a response to metabolic needs. When urea was supplied as the sole form of nitrogen, an increase in light-harvesting pigments, including chl-*a*, fucoxanthin and chl-c_2 were found [68], potentially to fulfil higher energetic needs associated with the additional metabolic steps in urea utilisation [69]. Under exposure to a high-NH_4^+ regime, photoprotection xanthophylls and β-carotene were observed to increase in proportions [68]. This was a potential stress-response triggered by known intracellular toxication of high-NH_4^+ exposure [70]. For example, accumulations of reactive oxygen species (ROS) in chloroplasts, are known to cause photo-oxidative damage [71]. Xanthophylls and β-carotene prevent the formation of ROS by dissipating excess energy in the photosystem [72, 73]. Similar protective functions were also observed to reduce oxidative stress under phosphate deprivation in the Kareniaceae species, *Karlodinium veneficum* [74]. In essence, pigment profiles can serve to elucidate metabolic needs of phytoplankton.

2.2 *Harmful Metabolites*

Phytoplankton can produce harmful metabolites that have a widespread impact on the marine ecosystem. In this section, we will highlight the harmful products that are produced by key species of phytoplankton from various classes such as the Cyanophyceae, Dinophyceae and Bacillariophyceae which are found in the coastal waters of Singapore.

2.2.1 Cyanobacteria

The genus *Neolyngbya* is a group of filamentous cyanobacteria is widely distributed along Singapore's shores and known to produce harmful natural products. This genus forms mats on both the Singapore and Johor Strait and has been discovered as early as 2018 from the West Singapore Strait [44]. To date, a total of five records of *Neolyngbya* mats exist. These mats exist on various substrates, such as concrete seawalls and breakwater sandbags in the intertidal zone. The dominant species of *Neolyngbya* mats in Singapore is *Neolyngbya regalis,* which has straight to wavy, yellow–brown filaments (Fig. 4a–c). Under field conditions, *N. regalis* dominant mats can vary from dark green, brown to black, depending on the site of isolation. Cultured *N. regalis* was dark yellow-green and grew in a clump (Fig. 4D).

Neolyngbya strains that were isolated from the South China Sea is known to produce cytotoxic metabolites wenchangamides A (**13**) and B (**14**) (Fig. 5) [75]. Wenchangamide A induced apoptosis in HCT116 human colon cancer cells in vitro with IC_{50} of 38 μM. Cell cycles were arrested as the cells observed to halt at the G2/M phase then accumulated in the sub-G1 phase, which suggests apoptotic cell death [75]. The structure of wenchangamide A is related to minnamide A (**15**) (Fig. 5), which is a natural compound isolated from the filamentous cyanobacterium *Okeania hirsuta* from Minna Island, Okinawa, Japan. The molecular weight of wenchangamide A is

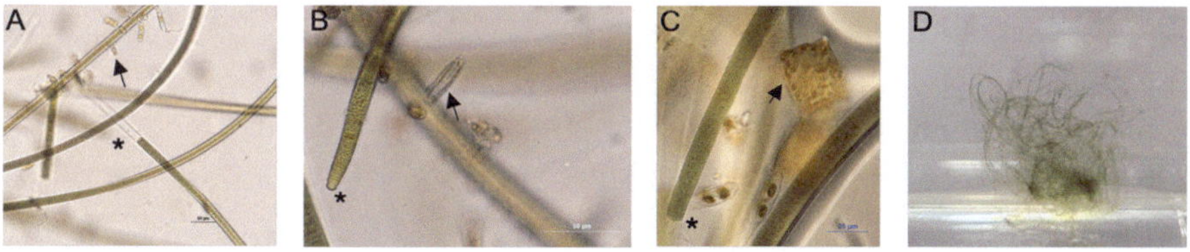

Fig. 4 Field samples of *Neolyngbya regalis* isolated from the West Singapore Strait with attached diatoms. The asterisks (*) shows individual filaments (A-C) and the arrows denote the diatoms. Cultured *N. regalis* in f/2 media (D)

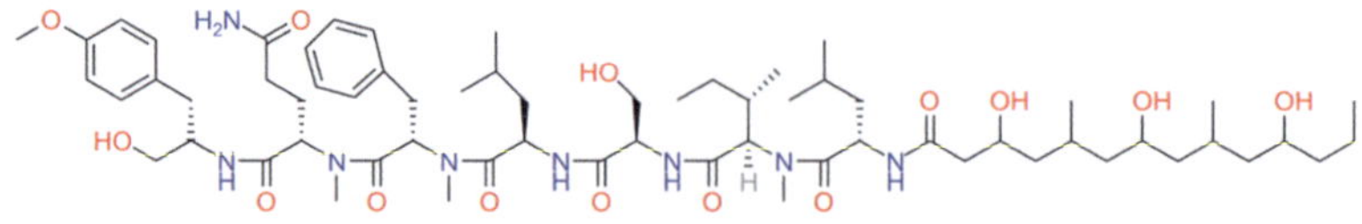

Wenchangamide A (**13**)

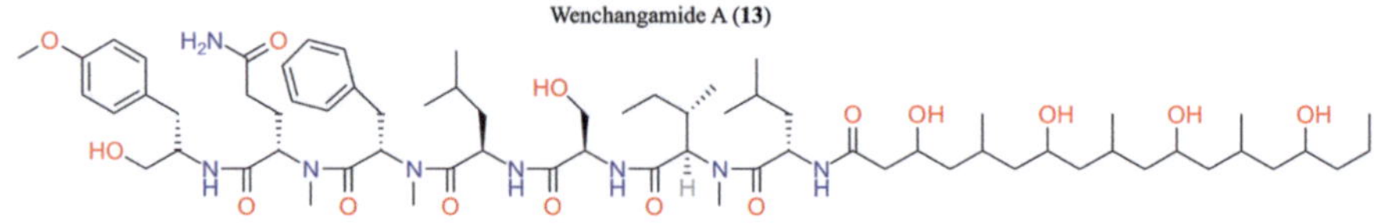

Wenchangamide B (**14**)

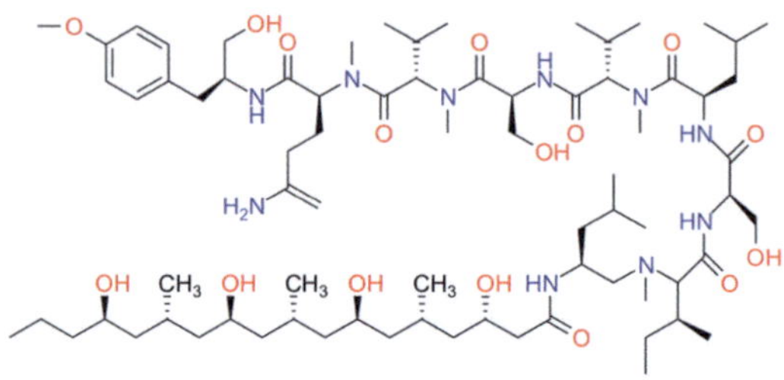

Fig. 5 Chemical structures of marine cyanobacterial wenchangamides and minnamide A

238 Da lower than that of minnamide A and has other structural differences [76]. The core scaffold of polypeptides is different between the two compounds, namely minnamide A has an *N*-Me-Val-Ser-*N*-Me-Val moiety in place of the *N*-Me-Phe group in wenchangamide A.

Another toxic filamentous cyanobacterium, *Moorena producens*, also occurs in Singapore's coastal waters. *M. producens* (previously named *Lyngbya majuscula*) is recorded from three different location: one from the north (Pulau Ubin) and two from southern islands (Sentosa; [77]) and St. John's Island [78]. In 2008, recurrent blooms of *M. producens* occurred from July to December in a residential waterway on southern island Sentosa, with thick floating mats on the water surface [77]. This cyanobacterium often exists as dark patchy mats on concrete breakwater seawalls along the coasts of Singapore. *M. producens* is cosmopolitan with worldwide distribution.

The genus *Moorena* is known to be one of the most prolific producer of bioactive compounds among the cyanobacteria, with 111 natural products being isolated from species in this genus. In 2005, an isolation of *M. producens* from the benthic sediments of an island from the Singapore Straits at Pulau Hantu Besar, led to the identification of a cyclodepsipeptide, Hantupeptin A (**16**) (Fig. 6) [79]. This compound exhibited cytotoxicity against leukemia cell line MOLT-4 and breast cancer cell lines MCF-7 with IC_{50} values of 32 nM and 4.0 μM, respectively [79].

Blooms of *M. producens* are known to have numerous effects upon contact or ingestion. This cyanobacterium has been termed "fireweed" due to the burning sensation experienced when swimmers contact blooms of this organism. Aerosolised

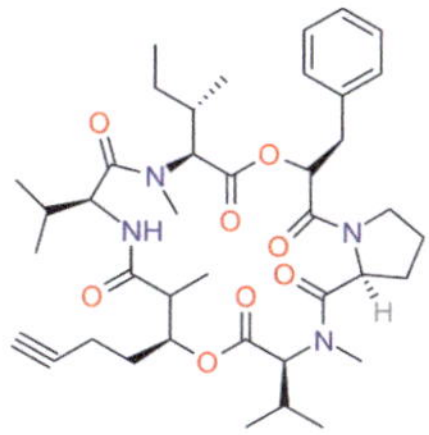

Hantupeptin A (**16**)

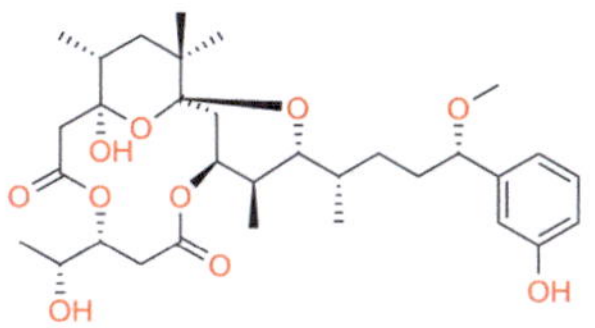

Debromoaplysiatoxin (**17**)

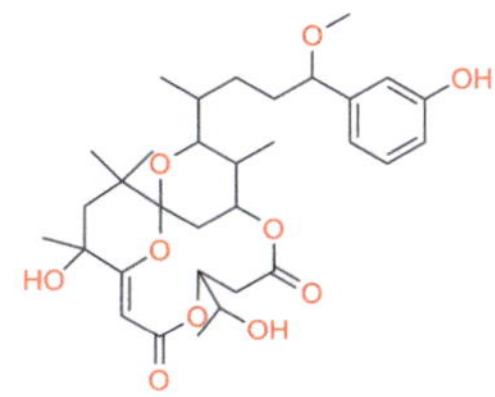

Oscillatoxin B2 (**18**)

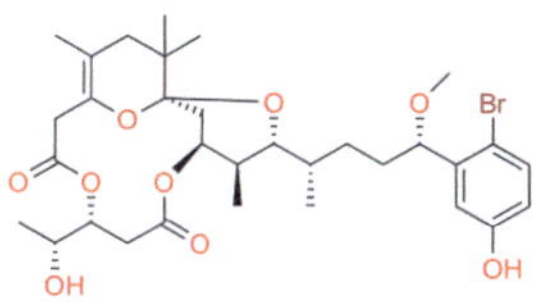

Anhydroaplysiatoxin (**19**)

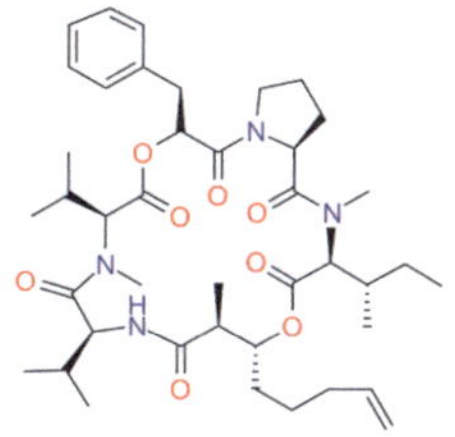

Tiahuramide B (**20**)

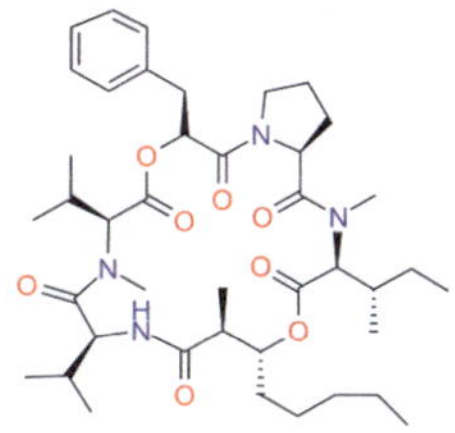

Tiahuramide C (**21**)

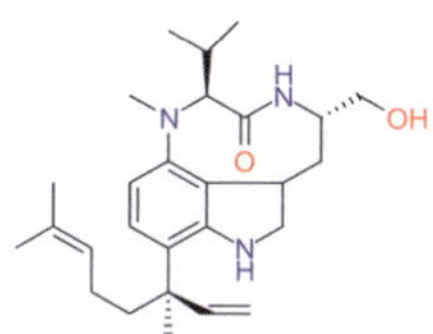

Lyngbyatoxin A (**22**)

Fig. 6 Chemical structure of marine cyanobacterial natural products

metabolites from *M. producens* have resulted in beachgoers developing facial rashes, eye inflammation and conjunctivitis in Gushikawa Beach, Okinawa [80]. Furthermore, filaments of this cyanobacterium were found in air and window scrapings during bloom seasons [81]. In another report, *M. producens* that was isolated from a bloom in Kuba Beach, Okinawa, Japan, contained nine cytotoxins including debromoaplysiatoxin (**17**), oscillatoxin B2 (**18**) and anhydroaplysiatoxin (**19**) (Fig. 6), which were cytotoxic to L1210 mouse lymphoma cells in concentrations of 10–90% at 10 μg/mL [82]. Other compounds of *Moorena* strains also showed toxicity against various cancer cell lines. Majusculamides are lipopeptides which were found to be cytotoxic to CCRF-CEM cells at 0.2 μg/mL as well as PANC-1 and U251N cancer cell lines [83]. *M. producens* from Tiahura Atoll, Moorea Island in the French Polynesia produced the hexadepsipeptides tiahuramides B (**20**), and C (**21**) (Fig. 6) [84]. These compounds were moderately toxic to SH-SY5Y human neuroblastoma cells at an IC_{50} value of 14 and 6 μM, respectively [84].

In 2010, there was a report of a mass human poisoning event of chelonitoxism, following the consumption of a poisoned hawksbill turtle *Eretmochelys imbricata* in the Federated States of Micronesia [85]. This turtle feast resulted in the death of three

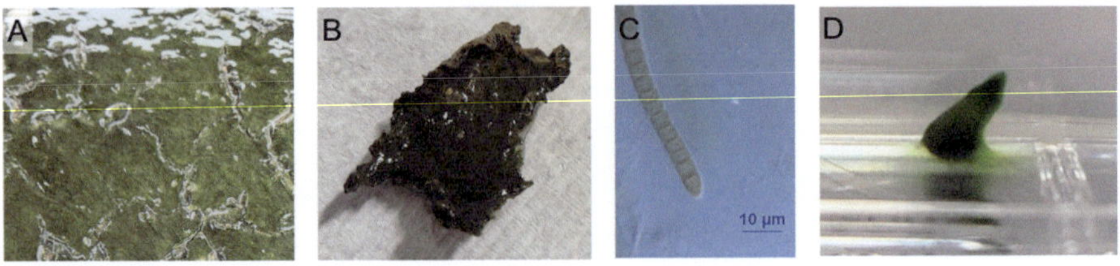

Fig. 7 Field samples of *Symploca* sp. isolated from the breakwater sandbags at East Coast Park along the Singapore Strait. **a** Mat of *Symploca* sp. **b** Section of field sample measuring 0.5 mm in thickness **c** Individual filament of *Symploca* sp. **d** Culture of *Symploca* sp. in f/2 media

children and the poisoning of 20 adult individuals and was found to have consumed large amounts of toxic cyanobacterium *Moorena producens*. However, the toxin which caused the poisoning event was not identified. However, from another fatally poisoned sea turtle *Chelonia mydas*, lyngbyatoxin A (**22**) (Fig. 6) was previously isolated [86]. Lyngbyatoxins are ichthyotoxic cyclic depsipeptides and act as potent tumor promoters [87]. Hence, the accumulation of lyngbyatoxin A in marine turtles is evident and the consumption of these contaminated seafood can severely threaten human health.

Another bioactive genus of cyanobacteria that is found on our shores is *Symploca*. To date, there are only two records of *Symploca,* and both are isolated from the Singapore Strait. The first were colonies of *Symploca hydnoides* growing on benthic sediments near Lazarus Island [88]. The second record of *Symploca* sp. was isolated from the synthetic surfaces of breakwater sandbags in the intertidal zone of East Coast Park along the Singapore Strait (Fig. 7a). Field samples were isolated (Fig. 7b) and cultured in the laboratory. Individual filaments measured about 5 μm in diameter (Fig. 7c) and the thallus had the morphology of a right-leaning cone (Fig. 7d).

Samples of *S. hydnoides* from Lazarus Island were collected and a new molecule structurally like santacruzamate A, was identified [88]. Santacruzamate A (**23**) (Fig. 8) is a potent histone acetylase (HDAC) inhibitor, which was cytotoxic to HCT-116 human colon carcinoma, dermal fibroblast (hDF) and huT-78 cutaneous T-cell lymphoma cells, with GI_{50} values of 29.4, > 100 and 1.4 μM, respectively [89].

In another study, symplocamide A (**24**) (Fig. 8) from *Symploca* sp. of Papua New Guinea was cytotoxic to various cancer cell lines [90]. Symplocamide A is a depsipeptide and a potent cancer cell cytotoxin which inhibits serine proteases with a 200-fold greater inhibition of chymotrypsin over trypsin [90]. In neuro-2a neuroblastoma and H-460 lung cancer cells, the IC_{50} was 29 nM and 40 nM, respectively [90].

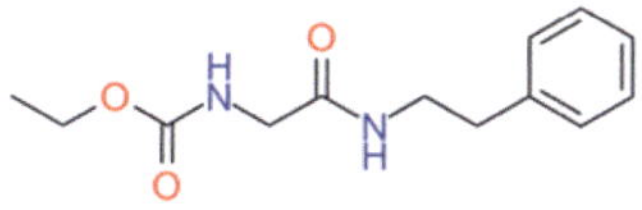

Santacruzamate A (23)

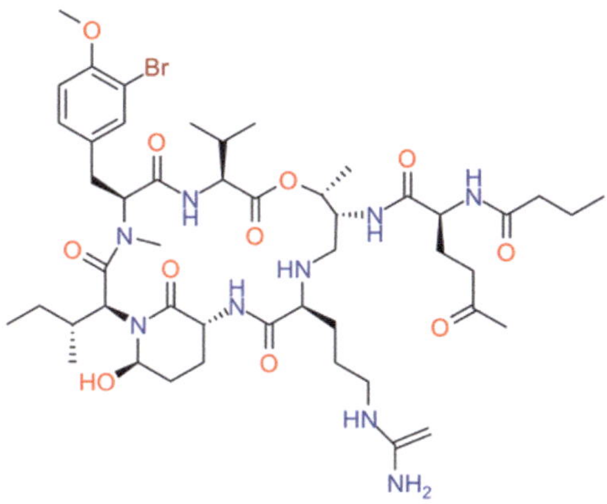

Symplocamide A (24)

Fig. 8 Chemical structures of *Symploca*-derived natural products

2.2.2 Diatoms

Toxic diatoms have been identified from Singapore's coastal waters. Five species of *Pseudo-nitzschia* were recorded since 2016 [91]. Here we focus on the pennate diatom *Pseudo-nitzschia pungens*, which produces domoic acid (DA). This compound is a potent neurotoxin that is responsible for amnesic shellfish poisoning (ASP). Domoic acid (**25**) (Fig. 9) binds to glutamate receptors in the brain, overstimulating the neurons, resulting in neuronal damage [92]. ASP in humans can result in disorientation, seizures and even death [93]. Memory loss suffered from this poisoning event can last up to five years in individuals [94]. In coastal areas, outbreaks of *Pseudo-nitzschia* blooms are a concern for public health as DA accumulates in various seafood, such as king scallop *Pecten maximus* [95] and commercially important fish, including striped anchovy and white mullet [96]. Contaminated seafood exceeding 20 $\mu g\ g^{-1}$ of DA is not allowed to be harvested. Furthermore, blooms of this genus are increasingly reported in Southeast Asian waters in recent years, resulting in shellfish contamination. In 2018, blooms of *Pseudo-nitzschia cuspidata* in Borneo, Malaysia resulted in the accumulation of DA of up to 8 $\mu g\ g^{-1}$ tissue of Pacific oyster *Magallana* sp. [97].

The centric diatom *Thalassiosira* is also observed in both the Johor and Singapore Straits. A recent report showed that *Thalassiosira rotula* produced cytotoxic glycerolipids, that recorded antiproliferative activity. Adele et al. [98] showed that 24-methylene cholesterol (**26**) (Fig. 9) had cytotoxic activity, with apoptosis and cell cycle arrest on the human breast cancer MCF7 and lung cancer A549 cell lines.

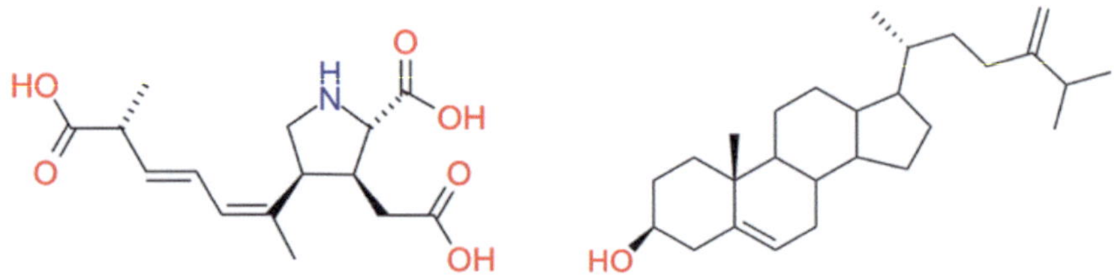

Domoic acid (25) **24-methylene cholesterol (26)**

Fig. 9 Chemical structures of natural products from diatoms

2.2.3 Dinoflagellates

In the past, there have been blooms of toxic dinoflagellates in Singapore, resulting in massive fish kills. The unarmored dinoflagellates *Karlodinium australe* and *Karenia mikimotoi* are known to be present in the Johor Strait. This species was responsible for the algal blooms associated with fish kill events in 2009 and is known to be a toxic species [10]. This fish kill event in the East Johor Strait resulted in substantial economic losses [10]. Another HAB from January to February 2014 and February 2015 resulted in the loss of 160 million tons of fish [10]. These were due to multi-species blooms, where *K. australe* and *K. mikimotoi* were the dominant species.

From *K. veneficum*, two families of karlotoxins namely the KmTx1 and KmTx2 groups, have been described since 2002 [99]. Karlotoxin analogs differ from each other in terms of toxicity levels, geographic distribution and UV absorbance maxima. The effects of karlotoxins are numerous. They are reported to induce cellular lysis, resulting in a range of cytotoxic and hemolytic effects. Other effects recorded have been fish gill damage and prey immobilization [100]. In times of an algal bloom, karlotoxins produced from *K. veneficum* have been found to immobilize competing species, allowing the dinoflagellate to obtain inorganic nutrients under conditions of nutrient limitation, resulting in increased growth rates [100].

From the *Karenia* genus, the species *K. mikimotoi* were part of multi-species blooms in the Johor Strait in 2014 and 2015 [11], leading to massive economic losses due to fish death in aquaculture farms. In one report, extracts of *K. mikimotoi* had harmful effects on four mammalian cell lines, including human hepatocellular cancer cell line Bel-7402 and mouse neuroblastoma cell line Neuro-2a [101]. *K. mikimotoi* are known to produce gymnocins, which are cytotoxic polyether compounds and are ichthyotoxic [102]. Their blooms have caused fish death in many instances, resulting in oxidative stress mediated muscle damage [103]. Gymnocins displayed moderate toxicity against mouse lymphoid P388 cells by inducing lipid peroxidation and altering cellular morphology [104]. Hence, blooms of *K. mikimotoi* are a potential threat to food security.

From the same genus, another species *Karenia brevis* is a known toxic dinoflagellate that produces brevetoxins. Brevetoxins (PbTx) are polyethers which have two natural parent molecules which are brevetoxins A (PbTx-1) and B (PbTx-2) [105]. These compounds cause neurotoxic shellfish poisoning (NSP) and breathing difficulties in humans upon consumption and contact. After exposure to this toxin, symptoms, such as diarrhea, nausea and vomiting, are observed. In 1992, following a major *K. brevis* bloom in New Zealand, a total of 180 people experienced NSP after consumption of contaminated shellfish, such as oysters and mussels [30]. *K. brevis* has been blooming at the western shelf of the Florida peninsula since 1971 [106], and aquaculture farms are mandated to close if *K. brevis* cell abundances exceed 5000 cells L^{-1} [107]. Previous blooms of *K. brevis* have resulted in numerous hospitalizations of individuals consuming contaminated shellfish, such as clams [107].

Armoured dinoflagellates are known to be toxic in Singapore waters. These dinoflagellates have theca covering their cells which forms protective layer, unlike athecate dinoflagellates. The armoured dinoflagellate genus *Alexandrium* and *Gymnodinium* is known to be bioactive, producing a diversity of toxin compounds of different classes. One well known toxin group is the hydrophilic paralytic shellfish toxins (PSTs) which include saxitoxin (STX). Isolates of *Gymnodinium catenatum* from east Johor Strait and west Singapore Strait showed the presence of gonyautoxins (GTX1-4), neosaxitoxin (neoSTX) and STX [108].

PSTs neurotoxins, which are well studied because they have great toxicity when contaminated shellfish are consumed, resulting in paralytic shellfish poisoning (PSP) and serious effects on human health [109]. PSTs have a basic tetrahydropurine ring that have substitutions at various positions. Analogs of PST are categorized according to their R4 side chain [110]. To date, there are more than 50 PSTs with different structures and toxicities, and these groups include lipophilic cycloimines, such as gymnodimines and spirolides [111]. The common saxitoxins produced by *Alexandrium* species include STX (**27**), neoSTX (**28**) and GTX1-4 (**29–32**) (Fig. 10). During blooms of *Alexandrium*, seafood harvesting bans are implemented when the STX toxin exceeds safe levels of 80 μg of STX per 100 g of tissue [112]. In surrounding regions, HAB episodes of *Alexandrium tamiyavanichii* in Kuantan Port of Pahang, Malaysia, resulted in the hospitalization of 10 individuals with PSP symptoms [113]. Toxicity levels during one event reached a peak of 3500 μg of STX of 100 g of shellfish [113].

PSTs act on voltage-gated ion channels, resulting in the blockage of ion transport and a loss of action potential in excitable membranes in vertebrates [114]. The onset of PSP typically occurs within 30 min to a few hours following the ingestion of contaminated shellfish and includes symptoms, such as nausea, vomiting and abdominal pain. In more severe instances, PSP can lead to paralysis, breathing difficulties and death, as soon as 3-4 h after consumption [115].

The dinoflagellate *Alexandrium* is also known to produce tetrodotoxin (TTX), which also contributes to PSTs. Blooms of *A. tamarense* have led to the accumulation of TTX (**33**) (Fig. 10) in the scallop *Patinopecten yessoensis* in various countries [116, 117. More recently, TTX had been discovered in bivalve mollusks from the Adriatic Sea [118]. This toxin binds to a different site of voltage-gated sodium channels

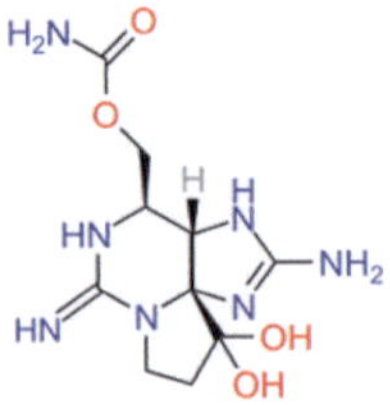

Saxitoxin (**27**)

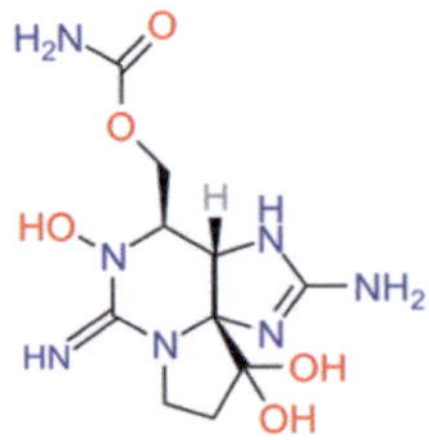

Neo-saxitoxin (**28**)

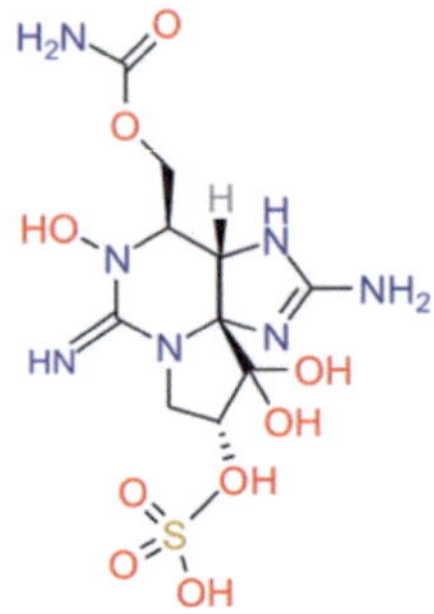

GTX-1 (**29**)

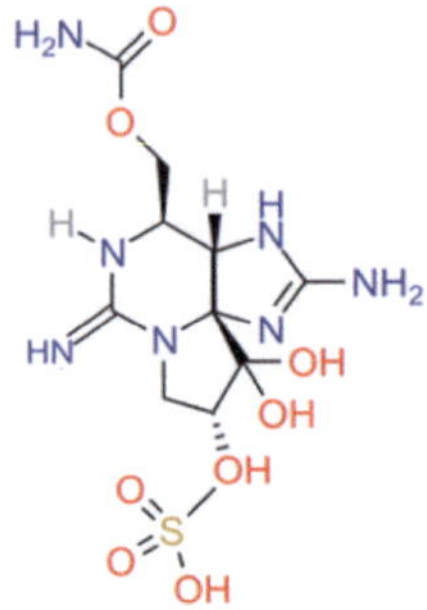

GTX-2 (**30**)

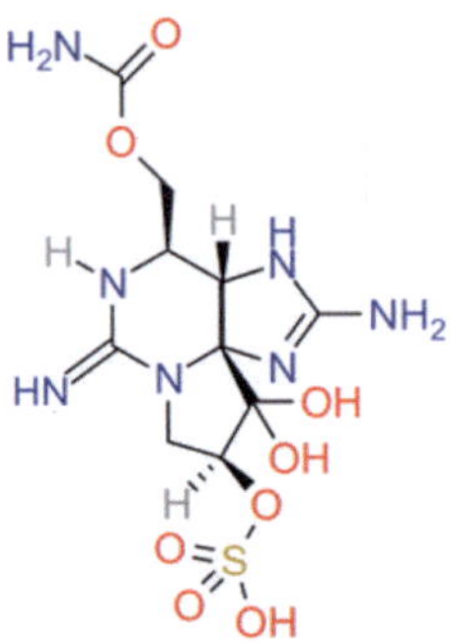

GTX-3 (**31**)

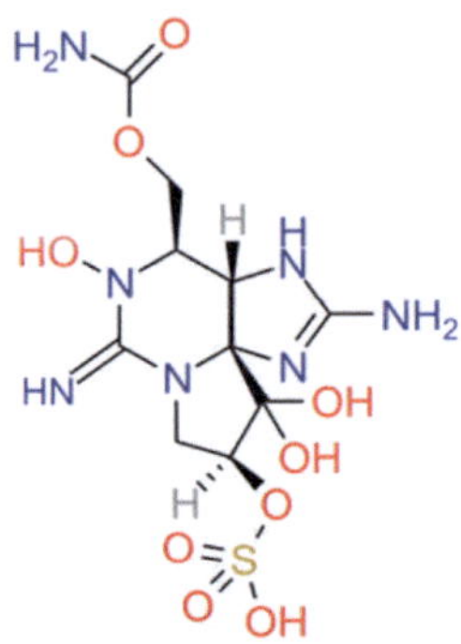

GTX-4 (**32**)

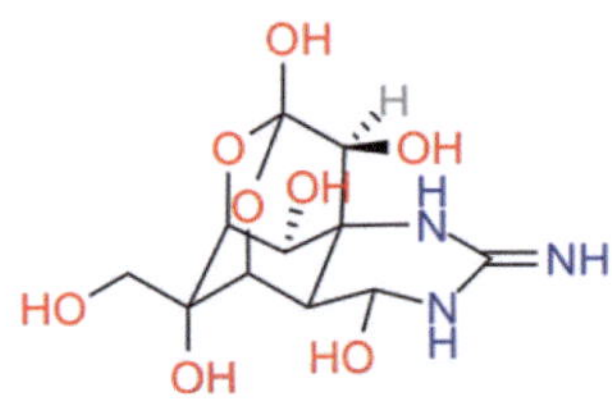

Tetrodotoxin (**33**)

Fig. 10 Chemical structures of dinoflagellate-derived toxins

compared to STX in nerve cells and have caused muscle weakness and respiratory paralysis in poisoned individuals, even resulting in death [119].

2.3 *Carbohydrate and Protein Pools*

As the dominant primary producer of aquatic systems, phytoplankton is a crucial source of energy for the entire ecosystem [120]. Through photosynthesis, carbon is captured and forms carbohydrates, lipids, or protein biomolecules. The general composition of these compounds is 40–60% protein; 17–40% carbohydrates; and 16–26% lipids [121]. However, under nitrogen (N)-depleted conditions, lipid content could increase up to around 50% [122]. This phytoplankton chemistry drives two major industries in the form of biofuels and as a source of dietary protein. One major benefit is that phytoplankton do not require arable land-space and plenty of water

[123] which is a typical requirement for biofuel crops. In addition, they also tend to have shorter harvesting cycles (around 1–2 weeks depending on the desired density) compared to biofuel crops which require several months up to a year.

Arthrospira platensis and *Synechococcus* are common cyanobacteria chosen for biofuel production. Other cyanobacteria genera have also been considered for biofuel cultivation. One such genus is *Leptolyngbya,* which is a cosmopolitan, widespread genus with global distribution. An alkalihalophilic strain of *Leptolyngbya* sp. had carbohydrate content ranging from 56 to 65% per biomass concentrations [124]. In another study, marine *Leptolyngbya valderiana* contained 27% of fermentable sugar that was suitable for bioethanol production. Furthermore, at pH 5.5, a 45% conversion to bioethanol was yielded, at 0.16 g/g. Apart from cyanobacteria, lipid production in diatoms have also received a lot of attention [125]. Some studies have suggested that cyanobacteria in nature would tend to store energy as carbohydrates, whereas other eukaryotic phytoplankton would tend toward lipid stores [126]. Although carbohydrates and lipids can both be used as biofuel feedstock, lipids have around twice the energy potential of carbohydrates, like starch [127]. The differences therefore have implications for the biofuel density that can be achieved across the different phytoplankton taxa.

Carbohydrate-based biofuels, such as biohydrogen, bioethanol and biobutanol, are thought to be sustainable alternatives to petroleum-based fuel, being a third-generation biofuel. Phytoplankton can take up carbon dioxide through photosynthesis, which typically produces starch as the primary energy source and lipids as the secondary source [128]. There is generally a trade-off between lipid accumulation and growth rates [129], although several exceptions have been documented especially with genetic engineering methods. Targets for these processes involve down-regulating lipid metabolism and altering fatty acid synthesis pathways [130]. In a study of the freshwater green alga *Chlamydomonas* sp., Kato et al. [131] demonstrated that mutation of the *ISA1* gene disrupted the starch debranching enzyme (DBE) and enhanced lipid composition by about twice that of the natural parent strain. In contrast, growth rates of the mutant strain were lowered by only about 20%. Genetic engineering therefore presents unique potential to circumvent the constraints imposed by a trade-off in nature between lipid accumulation and cellular growth rates.

Across the phytoplankton taxa, there has generally been no consistent indicator of how varying environmental conditions influence lipid production. In a comparison between the *Phaeodactylum tricornutum* diatom and the *Prorocentrum minimum* dinoflagellate, increasing temperature up to 24 °C triggered a sharp increase of around 150% in the dinoflagellate, but no changes were observed for the diatom [132]. In a study of four diatom strains from the Red Sea [133], short-term increase in temperature to 30 °C decreased lipid production. However, prolonged incubation at these temperatures promoted a recovery of the production levels closer to the original rates. These results contrasted results from Rousch et al. [134], where diatom cultures transferred from 24 °C to a range of temperatures up to 40 °C yielded an increase across different fatty acid types. Both studies examined *Chaetoceros* sp., which is commonly also found in tropical waters including Singapore. The range in lipid responses to similar treatment conditions indicate that there is high strain plasticity, and efforts to culture and harvest these phytoplankton cells should therefore include careful examination of local strains.

One of the major challenges of biofuels as a viable energy source continues to be production cost. Advances have been made to manage some of these challenges. The main two directions are to repurpose wastewater nutrients, and to co-produce other valuable bioproducts, such as pigments, which have been described earlier [135]. The coupling of cell culturing bioreactors using nutrients from wastewater presents a particular boon for Singapore because there is existing infrastructure to monitor and manage the nutrients that enters the urban watershed [136]. The rapid growth of phytoplankton presents a dual benefit of having a nutrient source while remediating and removing nutrients from wastewater.

Apart from biofuel, phytoplankton is increasingly becoming an important source of protein in the human diet. History records the use microalgae in food since the pre-Columbian era, during the 1400s [137]. Microalgae as a food source gained a resurgence in prominence during the 1970s when the National Aeronautics and Space Administration (NASA) recognised *Arthrospira* sp. (or *Spirulina*) as viable sources of nutrients for survival in space [138]. Subsequent studies have found *Arthrospira* sp. to possess valuable anti-inflammatory, antioxidant, and immunomodulating properties that support cardiovascular and brain health [139, 140—amongst others.

At present, the global demand for protein is 202 million tons per year, but this level includes large portions of the undernourished population who have less access to such nutrition [141]. Some estimates examined by this study propose a 50% increase in the demand for protein by 2050. This trend is challenging because traditional rates of meat and protein crop production are unlikely to match the increased demand. In Singapore, there has been wide-ranging moves in developing capacity relating to alternative proteins. Some key milestones include being the first country in the world to legalise lab-grown meats in 2020 for commercial purposes; a five-year research programmed launched in 2022 called Proteins4Singapore, which included growing microalgae as an alternative protein source; the opening of biofermentation labs in collaboration with the University of Illinois Urbana Champaign (UIUC) in 2024; and the launch of the world's largest methionine industry manufacturer through Evonik's, a German company in 2024. Heavy investment in this sector highlights the unique requirements in Singapore that makes it important to assess the potential of phytoplankton as a viable alternative source of protein.

One of the key benefits of phytoplankton is that they can have very high protein composition which consists of all the essential amino acids which humans cannot produce themselves [142]. In one study, it was found that up to 60% of the biomass of the green algae *Chlorella* sp. consisted of these essential amino acids [143], highlighting their importance as a non-meat alternative that supplies the full dietary requirement for proteins. Other benefits include the small land footprint which allows better control of growth conditions (e.g., using bioreactors or indoor environments) and therefore culturing at higher stocking densities. At present, there is only a small handful of phytoplankton taxa which have been approved for consumption. They mainly include green algae (e.g., *Chlamydomonas* sp., *Dunaliella* sp.) and cyanobacteria (e.g., *Arthrospira* sp., *Aphanizomenon* sp., *Nostoc* sp.) [144–146].

In practice, industrial cultivation of *Chlorella* sp. typically involves heterotrophic cultures to promote higher stocking densities especially within an enclosed bioreactor [147, 148]. Maximum protein content of up to 74% were reported for heterotrophic *Chlorella vulgaris* cells grown under nitrate-replete conditions of 1.5 g-N L^{-1}, pH 7, and 28 °C temperature [148]. In contrast, *Tetraselmis* sp. grown under light conditions in a photobioreactor using urban wastewater yielded about 30% protein, 25% fibre, 18% carbohydrates, and 7% lipids [149]. Different cultures therefore yield a range of substances that can be optimised under a range of environmental conditions. This pattern presents very high potential for different cost-profit models, but careful study of each individual strain is necessary. The role of nitrate-replete conditions in driving protein accumulations also presents a dilemma for lipid production processes, which typically require nutrient-depleted environments.

Of the species which have been used for nutritional protein, some taxa (i.e., *Aphanizomenon* sp.) have been known to produce cyanotoxins, such as cylindrospermopsin [150]. Specifically, the continued discovery of novel toxins and toxicity dynamics [151] makes it difficult for assurance about food safety. As a result, getting safety approvals for phytoplankton as a human food source has been challenging. Today, it tends to be much more widely used as the protein source in aquaculture feed. This approach also alleviates the need to valorise and functionalise protein compounds in the phytoplankton cells before being added to feed because aquaculture livestock is naturally able to digest the cells [152]. Apart from just proteins, phytoplankton also possesses other nutritional benefits, such as pigments as well as other vitamins and minerals [153].

Presently, increasing urbanisation around the world is rapidly elevating the demand for energy. Coupled with increasingly unpredictable weather patterns that threaten the reliability of biofuel crops, phytoplankton presents a viable alternative source of biofuel. This outcome is supported by ongoing research to improve lipid production and accumulation in phytoplankton cells, which increases efficiency and can help to reduce costs. In general, regardless of whether for biofuel or as an alternative protein, phytoplankton presents a lot of potential because they can be cultured to high densities given the appropriate environment. One major challenge would be the vast number of biomolecules associated with phytoplankton. Extraction and purification of the target substance requires proper methods and techniques. Ongoing research and development can help to develop and optimise the required technologies. Furthermore, this outcome presents the potential for multiple income streams to be established. Some of the key groups of biomolecules have been discussed in this chapter. The value of phytoplankton is further enhanced if methods are developed to allow nutrients to be taken up from wastewater, which in turn contributes toward ecosystem health and sustainability goals.

3 Discussion and Conclusion

Primary and secondary metabolites from phytoplankton have potential pharmaceutical applications. Solubility is a factor affecting the effectiveness of pigments being used as commercial sunscreen compounds. Although scytonemin is a useful compound to combat high irradiation, it has low solubility in water [47]. However, fat-soluble scytonemin can blend with oil-based stabilisers to increase the efficacy of sunscreen. One of the greatest challenges of using pigments produced by phytoplankton, such as scytonemin, mycosporines or carotenoids, is the upscaling and to increase biomass production. The presence of biological and physical contaminants increases the difficulty of mass culture for compound extraction [47]. Compared to open-reactor systems, biomass productivity in closed photobioreactor systems is higher as they provide a controlled environment with lower contamination risks. However, scaling up in closed systems is more expensive and requires higher maintenance. Besides harvesting scytonemin from cultured cyanobacteria, this compound is also considered for commercial production through genetic engineering. Scytonemin needs to first be isolated and produced in non-photosynthetic organisms, such as *Escherichia coli*, before upscaling can begin [154]. However, only the monomer and not the dimer of Scytonemin was isolated thus far, which exhibits the required spectrum of protective properties [154].

It is also important to consider other organisms that can produce greater concentrations of specific pigments. Currently, fucoxanthin cannot be artificially synthesized, so the majority of fucoxanthin production lies with brown algae [155]. However, researchers are exploring alternative organisms which have a high production of this compound. Diatoms can produce fucoxanthin from 0.2 to 2% of dry weight. This is observed in *Phaeodactylum tricomutum*, which produces fucoxanthin 100× per dry weight compared to that of brown seaweed, which is the present commercial source of this compound [52]. This diatom is relatively easy to culture given that they require basic nutrients including phosphorous and silicate. However, the main challenge of upscaling fucoxanthin for production is the conditions of the photobioreactor. Sufficient light is crucial for maximizing fucoxanthin production, yet it is necessary to minimize photoinhibition, which reduces productivity [156].

One major challenge associated with phytoplankton protein has been the difficulty in extracting digestible proteins that can be readily incorporated into human diets. Soto-Sierra et al. [157] described the various ways that phytoplankton proteins can be presented. The simplest process is for phytoplankton to be directly dried and prepared as whole-cell protein. However, studies have emerged to suggest that these cells may not be easily broken down within the human digestive system, and the protein may therefore not be bioavailable in our diets. Specifically, *Chlorella* sp. is known to have more complex cell walls compared to *Arthrospira* sp., making the former a less bioavailable source of protein in its whole-cell form. In general, full functionality of cellular protein would involve mechanical treatment, such a bead miller as well as several steps of chemical extraction and physical separation processes [158, 159]. Cell morphologies and life stages can also affect the yield, with the more motile cells being more fragile than vegetative ones in the green alga *Tetraselmis chui* [160]. The process can separate and purify lipids as well as other medicinal substances, creating additional income streams and helps to increase profits of the process.

While the production of beneficial pigments and proteins are a focus for commercial applications, it is also important to examine harmful metabolites, which can have negative consequences on human health and food security. The production of harmful metabolites is a complex process in different phytoplankton. Since it is well-known that these dinoflagellates produce toxins, it is necessary to know which conditions induce a greater expression of these harmful metabolites. PSTs are known to be produced at different growth phases [161]. Extracellular PSTs were secreted mostly during the exponential phase, where GTX1 and 4 were mainly produced. However, in the stationary phase GTX5 was the dominant toxin produced. It was also noted that PST production was regulated by bacterial presence, and this was observed in two *Alexandrium* species, *Alexandrium lusitanicum* NEPCC 253 and *Alexandrium tamarense* NEPCC 407 [162]. Additionally, bacteria-free *Alexandrium* still can produce PSTs. Axenic *A. lusitanicum* produced a greater concentration of each toxin, especially GTX 3 and 4, across the different growth phases [162]. However, in *A. tamarense* the opposite was true, where the non-axenic culture produced more toxins compared to the axenic culture, with more saxitoxin equivalents produced.

In conditions that deviated from the optimum, there was an increased levels of toxin production in the *Alexandrium* dinoflagellate. In *A. pacificum,* temperature was a crucial factor affecting the levels and production of STX genes [163]. For instance, maximum growth of *A. pacificum* was recorded in waters of 16 °C, where STX cellular contents (*sxtA* and *sxtG*) were increased after exposure to cold stress [163]. Cells in this optimal condition were also larger in size, due to their increased toxin accumulation. In temperatures of 20 °C, *A. pacificum* produces neoSTX and GTX4 in decreased concentrations of 0.76 and 18.29 fmol/cell, respectively [163]. This could mean that periods of temperature drop due to climate change in the coastal waters in Singapore could affect the toxin expression of this harmful dinoflagellate. However, temperature regulation of toxin production is observed differently in other HAB species. In the toxic cyanobacterium *Aphanizomenon gracile,* warmer temperatures of 30 °C resulted in the upregulation of the *sxtA* gene by fourfold [164]. The growth optimum of this cyanobacterium was lower at 23 °C [164].

Nutrient conditions will also influence the toxin levels in dinoflagellate cells, such as in *Alexandrium* species [165]. It is generally accepted that various sources of nitrogen will influence toxin production in dinoflagellates. Ammonium induced the highest concentration of intracellular toxins, such as GTX1, GTX3 and GTX4, followed by urea then nitrate [166]. In another study of *A. tamarense*, levels of GTX5 increased with increasing concentrations of N [167]. In *A. minutum,* transcriptional amounts of genes *sxtA* and *sxtG* decreased with deficiencies in N and P [168]. Another environmental parameter that influences STX production would be salinity. In one report, *A. pacificum* had an optimal growth at 30 PSU and had the greatest expression of STX in the exponential phase at this salinity [169]. This suggests that this dinoflagellate expresses the greatest levels of STX at optimal salinity. However, another species *A. minutum* produces more toxin in lower salinity levels [170]. Changes in cell size in response to different environmental conditions will also result in different levels of toxin expression [171].

Usually in algal blooms, the major and classical toxins are reported and identified. However, there are many uncharacterized bioactive extracellular compounds that are not well studied toxins yet affect the surrounding marine organisms. These compounds have ecological importance due to their potential harmful effects on the surrounding marine organisms and are also known as allelopathic chemicals. Furthermore, these allelopathic chemicals are not regulated by other existing groups of dinoflagellate toxins. The effects of these chemicals are characterized by their lytic activity against other organisms, from standard haemolytic assays or other target prey. Anti-pathogen, anti-protist and anti-grazing activities were observed from allelopathic activities of *Alexandrium* with other organisms.

Some dinoflagellates produce toxins and yet they are able to survive. There are a few reasons explaining this phenomenon. Studies have shown that organisms that contain 4-α-methyl sterols do not bind karlotoxins, resulting in an insensitivity to this toxin [172]. The predominant sterol in *K. veneficium* is gymnodinosterol, which does not bind karlotoxins. Other sterols, such as brassicasterol, cholesterol and ergosterol, interact with karlotoxins [172]. As a result, organisms primarily consisting of these sterols will then experience the harmful effects of karlotoxins. In the case of armoured dinoflagellates, they can protect against toxins or allelochemicals by forming temporary cysts and shedding their theca [173]. In conditions where the concentration of allelochemicals decreases, these dinoflagellates germinate from their pellicle cysts [174]. However, this is not to say that these toxins have no effect on the organism. Dinoflagellates possess antioxidant defense systems which help mitigate oxidative damage. *Alexandrium* is known to contain both enzymatic and non-enzymatic antioxidants which have mitigated oxidative stress under changing environmental conditions. Enzymatic antioxidants are enzymes, such as superoxide dismutase (SOD) and catalase. Non-enzymatic antioxidants are compounds, such as ascorbic acid and glutathione [175]. Hence, it is possible that the exposure of dinoflagellates to their own toxins during a bloom induces oxidative stress and its related response pathways but are able to recover from this damage and maintain cellular homeostasis through these protective mechanisms. This enables the dinoflagellate to be resilient in times of an algal bloom.

In the past few years, there have been more reports of *Alexandrium* species in the coastal waters of Singapore, along both the Johor and Singapore Straits. In the West Johor Strait, incubation studies of six different plastic polymers including polypropylene, polyethylene and polyvinyl chloride were carried out. Identification of phytoplankton via the 28S rRNA gene region revealed the presence of *Alexandrium* on these plastic types [176]. This dinoflagellate was also identified to be present on beach microplastics isolated from the Southern Strait of Singapore from 2022 to 2023 [177]. The increased presence of this toxic dinoflagellate highlights the potential of future blooms, threatening food security and human health.

While we have discussed the negative implications surrounding toxic metabolites from phytoplankton, it is also important to highlight the clinical applications of these bioactive compounds [178]. For instance, neo-STX was previously used in severe cases of achalasia, to lower esophageal sphincter relaxation and aid swallowing. This administration was without side effects and inflammation. This compound was also used as a local anesthetic in subcutaneous injections and was shown to be effective and safe without adverse effects after 2 weeks [179]. Neo-STX showed promise as a long-acting local anesthetic, with pain scores lower after treatment with the neurotoxin compared to bupivacaine, which is commonly used for epidurals [179]. TTX has previously been explored in various treatments for cancer. TTX is known to reduce the proliferation of human SW620 colorectal cancer cells [180]. This compound has also been considered as a strategy to treat visceral pain in functional gastrointestinal disorders. The application of twice daily injections of TTX for 4 days demonstrated a strong analgesic effect against visceral pain in cancer patients [181] As of now, TTX is produced in various organisms, such as marine pufferfish [182] and xanthid crabs [183]. Harvesting TTX from phytoplankton is safer and would offer multiple advantages over pufferfish and crabs as these larger organisms require careful handling and have external surfaces that are highly toxic.

Given the potency of certain metabolites, it would be useful to enhance their existing activity through structural modifications. Analogues can be designed to create more robust and potent compounds that have greater potential for clinical development. For instance, structural modifications of santacruzamate A resulted in a potent class of class I HDAC hydrazide inhibitors [184]. These compounds resulted in apoptosis in acute myeloid leukemia (AML) cells alone but resulted in more pronounced anti-AML effects when used in combination with venetoclax, which is an existing drug used to treat blood cancers [184].

In the field of natural product research, it is essential to identify useful or harmful metabolites that enhance our scientific understanding to protect public health. However, this is only possible if there is a comprehensive scientific database that facilitates product identification. These databases enable researchers to explore the biological activities, chemical structures and biosynthetic pathways of natural products. Hence, the availability of publicly accessible databases is crucial. Commercial, databases such as Antibase and MarinLit, provides information of secondary metabolites only through subscription access. Existing open access databases such as the NORINE database and ALGALTOX list are useful to a limited extent as they do not contain the full list of metabolites [185]. Certain essential structural information regarding the natural product is also absent. From cyanobacteria, over 2600 metabolites have been isolated. One useful database for the identification of cyanobacterial compounds is CyanoMetDB, which details the structural information about secondary metabolites [186]. This aids the identification of novel compounds by providing detailed chemical structures and metadata that is associated with secondary metabolites in cyanobacteria. CyanoMetDB is also able to identify BGCs linked to these metabolites, facilitating studies on enzymatic and genetic pathways linked to their production. The toxin diversity of dinoflagellates can be studied using the Global Natural Products Social Molecular Networking (GNPS), an open-access database for raw to annotated compounds, which is useful for identifying novel secondary metabolites [187] New okadaic acid analogues from dinoflagellate *Dinophysis* was discovered with the help of GNPS [188]. In another study, novel toxin families that were analogues of palytoxin (PLTX) were identified from strains of *Ostreopsis* cf. *ovata* [189].

It is useful to characterize metabolites that are produced. However, there are many products that are not produced under a given environmental condition in an organism. A useful approach to identifying metabolites that are produced is through genome mining, and this is a powerful technique to uncover natural products from various phytoplankton groups. Genome analysis can uncover unexpressed natural products by identifying silent or cryptic gene clusters that are not actively expressed under standard laboratory or field conditions. For instance, the novel filamentous cyanobacterium *Sphaerothrix gracilis* was isolated from foam microplastics off the southern coast in Singapore [190]. Although it has not yet been tested for expressed natural products the draft genome analysis of this species revealed the presence of multiple PKS genes from 7 putative biosynthetic gene clusters (BGCs) including puwainaphycin D from *Cylindrospermum alatosporum* CCALA 988 [190]. The use of bioinformatics in genome mining enables the identification of novel biosynthetic proteins which have not yet been associated with existing natural product gene clusters. Putative secondary metabolite gene clusters are usually identified using antiSMASH [191], which is a leading tool in this domain. This software allows comprehensive analysis of microbial genomes, predicting and annotating BGCs. This tool can be used in combination with the genomic database MIBiG (Minimum Information about a Biosynthetic Gene cluster) to enhance the discovery of BGCs and their products [192]. The output from antiSMASH is cross-referenced with the MIBiG database for validation [192]. Even though certain natural products are not expressed and reported, the identification of putative BGCs and their secondary metabolites through such pipelines is important, for the discovery of strategies to activate their production. This approach will enhance our understanding of metabolites in phytoplankton for biotechnological and pharmaceutical applications.

References

1. Abida H, Ruchaud S, Rios L et al (2013) Bioprospecting marine plankton. Marine Drugs 11(11):4594–4611. https://doi.org/10.3390/MD11114594
2. Blunt JW, Copp BR, Munro MHG, Northcote PT, Prinsep MR (2011) Marine natural products. Nat Prod Rep 28(2):196–268. https://doi.org/10.1039/C005001F
3. Righetti D, Vogt M, Gruber N, Psomas A, Zimmermann NE (2019) Global pattern of phytoplankton diversity driven by temperature and environmental variability. Sci Adv 5(5):6253–6268. https://doi.org/10.1126/SCIADV.AAU6253/SUPPL_FILE/AAU6253_SM.PDF
4. Katayama T, Rahman NA, Khatoon H et al (2022) Bioprospecting of tropical microalgae for high-value products: n-3 polyunsaturated fatty acids and carotenoids. Aquac Rep 27:101406. https://doi.org/10.1016/J.AQREP.2022.101406
5. Kok JWK, Yeo DCJ, Leong SCY (2019) Growth, pigment, and chromophoric dissolved organic matter responses of tropical *Chattonella subsalsa* (Raphidophyceae) to nitrogen enrichment. Phycological Res 67(2):134–144. https://doi.org/10.1111/PRE.12360
6. Rastogi RP, Shree A, Patel HM, Chaudhry S, Madamwar D (2023) Characterization, UV-induction, antioxidant function and role in photo-protection of mycosporine-like amino acids

(MAAs) in a unicellular cyanobacterium, *Euhalothece* sp.WR7. Algal Res 70:103030. https://doi.org/10.1016/J.ALGAL.2023.103030
7. Llewellyn CA, Greig C, Silkina A, Kultschar B, Hitchings MD, Farnham G (2020) Mycosporine-like amino acid and aromatic amino acid transcriptome response to UV and far-red light in the cyanobacterium *Chlorogloeopsis fritschii* PCC 6912. Scient Reports 10(1):1–13. https://doi.org/10.1038/s41598-020-77402-6
8. Espinosa-Ramírez J, Mondragón-Portocarrero AC, Rodríguez JA, Lorenzo JM, Santos EM (2023) Algae as a potential source of protein meat alternatives. Front Nutr 10:1254300. https://doi.org/10.3389/FNUT.2023.1254300/BIBTEX
9. Yñiguez AT, Lim PT, Leaw CP et al (2021) Over 30 years of HABs in the Philippines and Malaysia: what have we learned? Harmful Algae 102:101776. https://doi.org/10.1016/J.HAL.2020.101776
10. Leong SCY, Lim LP, Chew SM, Kok JWK Teo SLM (2015) Three new records of dinoflagellates in Singapore's coastal waters, with observations on environmental conditions associated with microalgal growth in the Johor Straits. Accessed August 14, 2024. http://zoobank.org/urn:lsid:zoobank.org:pub:8039D2A1-A032-4E5F-882D-D69BA18711E6
11. Kok JWK, Leong SCY (2019) Nutrient conditions and the occurrence of a *Karenia mikimotoi* (Kareniaceae) bloom within East Johor Straits, Singapore. Reg Stud Mar Sci 27:100514. https://doi.org/10.1016/J.RSMA.2019.100514
12. Li X, Yan T, Yu R, Zhou M (2019) A review of *Karenia mikimotoi*: bloom events, physiology, toxicity and toxic mechanism. Harmful Algae 90:101702. https://doi.org/10.1016/J.HAL.2019.101702
13. Gin KYH, Sim ZY, Goh KC et al (2021) Novel cyanotoxin-producing *Synechococcus* in tropical lakes. Water Res 192:116828. https://doi.org/10.1016/J.WATRES.2021.116828
14. Sanchez S, Demain AL (2008) Metabolic regulation and overproduction of primary metabolites. https://doi.org/10.1111/j.1751-7915.2007.00015.x
15. Drew SW, Demain AL (1977) Effect of primary metabolites on secondary metabolism. Annu Rev Microbiol 31:343–356. https://doi.org/10.1146/ANNUREV.MI.31.100177.002015/CITE/REFWORKS
16. Fernández M, Zúñiga M (2006) Amino acid catabolic pathways of lactic acid bacteria. Crit Rev Microbiol 32(3):155–183. https://doi.org/10.1080/10408410600880643
17. Böttger A, Vothknecht U, Bolle C, Wolf A (2018) Plant secondary metabolites and their general function in plants, pp 3–17. https://doi.org/10.1007/978-3-319-99546-5_1
18. Harborne JB (1990) Role of secondary metabolites in chemical defence mechanisms in plants. Ciba Found Symp 154:126–139. https://doi.org/10.1002/9780470514009.CH10
19. Marella TK, Tiwari A (2020) Marine diatom *Thalassiosira weissflogii* based biorefinery for co-production of eicosapentaenoic acid and fucoxanthin. Bioresour Technol 307:123245. https://doi.org/10.1016/J.BIORTECH.2020.123245
20. Tachihana S, Nagao N, Katayama T et al (2020) High productivity of eicosapentaenoic acid and fucoxanthin by a marine diatom *Chaetoceros gracilis* in a semi-continuous culture. Front Bioeng Biotechnol. 8:602721. https://doi.org/10.3389/FBIOE.2020.602721/BIBTEX
21. Vincent M, Blanc-Garin V, Chenebault C et al (2024) Impact of carbon fixation, distribution and storage on the production of farnesene and limonene in *Synechocystis* PCC 6803 and *Synechococcus* PCC 7002. Int J Mol Sci 25(7):3827. https://doi.org/10.3390/IJMS25073827/S1
22. Steinke M, Brading P, Kerrison P, Warner ME, Suggett DJ (2011) Concentrations of Dimethylsufoniopropionate and Dimethyl sulfide are strain-specific in symbiotic dinoflagellates (*Symbiodinium* sp., Dinophyceae). J Phycol 47(4):775–783. https://doi.org/10.1111/J.1529-8817.2011.01011.X
23. Koksharova OA (2020) Cyanobacterial VOCs as allelopathic tools. In: Bacterial volatile compounds as mediators of airborne interactions, pp 257–280. https://doi.org/10.1007/978-981-15-7293-7_11
24. Orr RJS, Stüken A, Murray SA, Jakobsen KS (2013) Evolution and distribution of saxitoxin biosynthesis in dinoflagellates. Marine Drugs 11(8):2814–2828. https://doi.org/10.3390/MD11082814

25. Hille B (1975) The receptor for tetrodotoxin and saxitoxin. A structural hypothesis. Biophys J 15(6):615. https://doi.org/10.1016/S0006-3495(75)85842-5
26. Osswald J, Rellán S, Gago A, Vasconcelos V (2007) Toxicology and detection methods of the alkaloid neurotoxin produced by cyanobacteria, anatoxin-a. Environ Int 33(8):1070–1089. https://doi.org/10.1016/J.ENVINT.2007.06.003
27. Viaggiu E, Melchiorre S, Volpi F et al (2004) Anatoxin-a toxin in the cyanobacterium Planktothrix rubescens from a fishing pond in northern Italy. Environ Toxicol 19(3):191–197. https://doi.org/10.1002/TOX.20011
28. Basu A, Lazo JS, Kozikowski AP (1992) Structural requirements of Lyngbyatoxin A for activation and downregulation of protein kinase C. Biochemistry 31(15):3824–3830. https://doi.org/10.1021/BI00130A013/ASSET/BI00130A013.FP.PNG_V03
29. Pooja SP, Niveshika N (2022) Insight into the potential cyanobacterial metabolites and their screening strategies. Biosci Biotechnol Res Asia 19(1):255–279. https://doi.org/10.13005/BBRA/2983
30. Baden DG (1989) Brevetoxins: unique polyether dinoflagellate toxins. FASEB J 3(7):1807–1817. https://doi.org/10.1096/FASEBJ.3.7.2565840
31. Vega J, Schneider G, Moreira BR, Herrera C, Bonomi-Barufi J, Figueroa FL (2021) Mycosporine-like amino acids from Red Macroalgae: UV-photoprotectors with potential cosmeceutical applications. Appl Sci 11(11):5112. https://doi.org/10.3390/APP11115112
32. Geissman TA (1963) Flavonoid compounds, tannins, lignins and related compounds. Comprehen Biochem 9:213–250. https://doi.org/10.1016/B978-1-4831-9718-0.50018-7
33. Mushir S, Deep S, Fatma T et al (2007) Screening of cyanobacterial strains for UV screening compound Scytonemin-environmental perspectives. Int J Innov Res Sci Eng Technol 3297(2):2319–8753. Accessed August 1, 2024. www.ijirset.com
34. McKenzie RL, Aucamp PJ, Bais AF, Björn LO, Ilyas M, Madronich S (2011) Ozone depletion and climate change: impacts on UV radiation. Photochem Photobiol Sci 10(2):182–198. https://doi.org/10.1039/C0PP90034F
35. Holm-Hansen O, Lubin D, Helbling EW (1993) Ultraviolet radiation and its effects on organisms in aquatic environments. In: Environmental UV photobiology, pp 379–425. https://doi.org/10.1007/978-1-4899-2406-3_13
36. Li Y (2022) The bioactivities of phycocyanobilin from spirulina. J Immunol Res 2022(1):4008991. https://doi.org/10.1155/2022/4008991
37. Sommella E, Conte GM, Salviati E et al (2018) Fast profiling of natural pigments in different *Spirulina* (*Arthrospira platensis*) dietary supplements by DI-FT-ICR and evaluation of their antioxidant potential by pre-column DPPH-UHPLC assay. Molecules 23(5):1152. https://doi.org/10.3390/MOLECULES23051152
38. An J, Gao F, Ma Q, Xiang Y, Ren D, Lu J (2017) Screening for enhanced astaxanthin accumulation among *Spirulina platensis* mutants generated by atmospheric and room temperature plasmas. Algal Res 25:464–472. https://doi.org/10.1016/J.ALGAL.2017.06.006
39. Giannaccare G, Pellegrini M, Senni C, Bernabei F, Scorcia V, Cicero AFG (2020) Clinical applications of astaxanthin in the treatment of ocular diseases: emerging insights. Marine Drugs 18(5):239. https://doi.org/10.3390/MD18050239
40. Nagao A (2004) Oxidative conversion of carotenoids to retinoids and other products. J Nutr 134(1):237S-240S. https://doi.org/10.1093/JN/134.1.237S
41. Fang JY, Xiao SD (1998) Effect of trans-retinoic acid and folic acid on apoptosis in human gastric cancer cell lines MKN-45 and MKN-28. J Gastroenterol 33(5):656–661. https://doi.org/10.1007/S005350050152/METRICS
42. Dimberg A, Bahram F, Karlberg I, Larsson LG, Nilsson K, Öberg F (2002) Retinoic acid–induced cell cycle arrest of human myeloid cell lines is associated with sequential down-regulation of c-Myc and cyclin E and posttranscriptional up-regulation of p27Kip1. Blood 99(6):2199–2206. https://doi.org/10.1182/BLOOD.V99.6.2199
43. Dagenais GR, Marchioli R, Tognoni G, Yusuf S (2000) Beta-carotene, vitamin C, and vitamin e and cardiovascular diseases. Curr Cardiol Rep 2(4):293–299. https://doi.org/10.1007/S11886-000-0084-4/METRICS

44. Curren E, Leong SCY. New records of benthic coastal cyanobacteria observed from the Southern Islands of Singapore. https://doi.org/10.1127/algol_stud/2018/0302
45. Dillon JG, Castenholz RW (2003) The synthesis of the UV-screening pigment, scytonemin, and photosynthetic performance in isolates from closely related natural populations of cyanobacteria (Calothrix sp.). Environ Microbiol 5(6):484–491. https://doi.org/10.1046/J.1462-2920.2003.00436.X
46. Madrahi GS, Naeimpoor F (2022) UV induced biosynthesis of cyano-sunscreen "Scytonemin" by *Leptolyngbya mycodia* and its effectual antioxidant activity: scytonemin production in cyanobacterium *Leptolyngbya mycodia*. Iranian J Pharm Sci 18(1):19–33. https://doi.org/10.22037/IJPS.V18.41360
47. Sen S, Mallick N (2022) Scytonemin: unravelling major progress and prospects. Algal Res 64:2211–9264. https://doi.org/10.1016/j.algal.2022.102678
48. Kang MR, Jo SA, Lee H et al (2020) Inhibition of skin inflammation by scytonemin, an ultraviolet sunscreen pigment. Marine Drugs 18(6):300. https://doi.org/10.3390/MD18060300
49. Itoh T, Koketsu M, Yokota N, Touho S, Ando M, Tsukamasa Y (2014) Reduced scytonemin isolated from Nostoc commune suppresses LPS/IFNγ-induced NO production in murine macrophage RAW264 cells by inducing hemeoxygenase-1 expression via the Nrf2/ARE pathway. Food Chem Toxicol 69:330–338. https://doi.org/10.1016/J.FCT.2014.04.019
50. Smerilli A, Balzano S, Maselli M et al (2019) Antioxidant and photoprotection networking in the coastal diatom Skeletonema marinoi. Antioxidants 8(6):154. https://doi.org/10.3390/ANTIOX8060154
51. Zudaire L, Roy S (2001) Photoprotection and long-term acclimation to UV radiation in the marine diatom *Thalassiosira weissflogii*. J Photochem Photobiol B 62(1–2):26–34. https://doi.org/10.1016/S1011-1344(01)00150-6
52. Kim SM, Jung YJ, Kwon ON et al (2012) A potential commercial source of fucoxanthin extracted from the microalga *Phaeodactylum tricornutum*. Appl Biochem Biotechnol 166(7):1843–1855. https://doi.org/10.1007/S12010-012-9602-2/TABLES/2
53. Heo SJ, Ko SC, Kang SM et al (2008) Cytoprotective effect of fucoxanthin isolated from brown algae *Sargassum siliquastrum* against H_2O_2-induced cell damage. Eur Food Res Technol 228(1):145–151. https://doi.org/10.1007/S00217-008-0918-7/METRICS
54. Bae M, Kim MB, Park YK, Lee JY (2020) Health benefits of fucoxanthin in the prevention of chronic diseases. Biochimica et Biophysica Acta (BBA)—Molecular and Cell Biology of Lipids 1865(11):158618. https://doi.org/10.1016/J.BBALIP.2020.158618
55. Woo MN, Jeon SM, Kim HJ et al (2010) Fucoxanthin supplementation improves plasma and hepatic lipid metabolism and blood glucose concentration in high-fat fed C57BL/6N mice. Chem Biol Interact 186(3):316–322. https://doi.org/10.1016/J.CBI.2010.05.006
56. Sansone C, Del Mondo A, Pistelli L, Smerilli A, Saggiomo M, Brunet C (2024) Opportunities and challenges of diatom cell factory for human health. Diatom Photosynthesis. Published online July 22, 2024:477–527. https://doi.org/10.1002/9781119842156.CH16
57. Carreto JI, Carignan MO (2011) Mycosporine-like amino acids: relevant secondary metabolites. chemical and ecological aspects. Marine Drugs 9(3):387–446. https://doi.org/10.3390/MD9030387
58. Geraldes V, Pinto E (2021) Mycosporine-like amino acids (MAAs): biology, chemistry and identification features. Pharmaceuticals 14(1):63. https://doi.org/10.3390/PH14010063
59. Taira H, Taguchi S (2017) Cellular Mycosporine-like amino acids protect photosystem II of the Dinoflagellate *Scrippsiella sweeneyae* from ultraviolet radiation damage. J Photochem Photobiol B 174:27–34. https://doi.org/10.1016/J.JPHOTOBIOL.2017.07.015
60. Llewellyn CA, Airs RL (2010) Distribution and abundance of MAAs in 33 species of microalgae across 13 classes. Marine Drugs 8(4):1273–1291. https://doi.org/10.3390/MD8041273
61. Callone AI, Carignan M, Montoya NG, Carreto JI (2006) Biotransformation of mycosporine like amino acids (MAAs) in the toxic dinoflagellate *Alexandrium tamarense*. J Photochem Photobiol B 84(3):204–212. https://doi.org/10.1016/J.JPHOTOBIOL.2006.03.001

62. Ryu J, Park SJ, Kim IH, Choi YH, Nam TJ (2014) Protective effect of porphyra-334 on UVA-induced photoaging in human skin fibroblasts. Int J Mol Med 34(3):796–803. https://doi.org/10.3892/IJMM.2014.1815
63. Chrapusta E, Kaminski A, Duchnik K, Bober B, Adamski M, Bialczyk J (2017) Mycosporine-like amino acids: potential health and beauty ingredients.Marine Drugs 15(10):326. https://doi.org/10.3390/MD15100326
64. De Salas MF, Bolch CJS, Botes L, Nash G, Wright SW, Hallegraeff GM (2003) *Takayama* gen. nov. (Gymnodiniales, Dinophyceae), a new genus of unarmored dinoflagellates with sigmoid apical grooves, including the description of two new species. J Phycol 39(6):1233–1246. https://doi.org/10.1111/J.0022-3646.2003.03-019.X
65. Mooney BD, Nichols PD, De Salas MF, Hallegraeff GM (2007) Lipid, fatty acid, and sterol composition of eight species of Kareniaceae (dinophyta): Chemotaxonomy and putative lipid phycotoxins. J Phycol 43(1):101–111. https://doi.org/10.1111/J.1529-8817.2006.00312.X
66. Yoon HS, Hackett JD, Van Dolah FM, Nosenko T, Lidie KL, Bhattacharya D (2005) Tertiary endosymbiosis driven genome evolution in dinoflagellate algae. Mol Biol Evol 22(5):1299–1308. https://doi.org/10.1093/molbev/msi118
67. Tang YZ, Kong L, Morse RE, Gobler CJ, Holmes MJ (2012) Report of a bloom-forming dinoflagellate Takayama acrotrochoa from tropical coastal waters of Singapore. J Phycol 48(2):455–466. https://doi.org/10.1111/J.1529-8817.2012.01135.X
68. Loh JY (2022) Nitrogen-induced responses of tropical athecate dinoflagellates in harmful algal blooms
69. Antia NJ, Harrison PJ, Oliveira L (1991) The role of dissolved organic nitrogen in phytoplankton nutrition, cell biology and ecology. Phycologia 30(1):1–89. https://doi.org/10.2216/I0031-8884-30-1-1.1
70. Collos Y, Harrison PJ (2014) Acclimation and toxicity of high ammonium concentrations to unicellular algae. Marine Poll 80(1–2):8–23. https://doi.org/10.1016/j.marpolbul.2014.01.006
71. Shi D, Zhuang K, Chen Y, Xu F, Hu Z, Shen Z (2020) Effects of excess ammoniacal nitrogen (NH4+-N) on pigments, photosynthetic rates, chloroplast ultrastructure, proteomics, formation of reactive oxygen species and enzymatic activity in submerged plant Hydrilla verticillata (L.f.) Royle. Aq Toxicol 226. https://doi.org/10.1016/J.AQUATOX.2020.105585
72. Dubinsky Z, Stambler N (2009) Photoacclimation processes in phytoplankton: mechanisms, consequences, and applications. Aq Microbial Ecol 56(2–3):163–176. https://doi.org/10.3354/ame01345
73. Brunet C, Johnsen G, Lavaud J, Roy S (2011) Pigments and photoacclimation processes. In: Phytoplankton pigments: characterization, chemotaxonomy and applications in oceanography. Accessed September 16, 2024. https://hal.science/hal-01101814/
74. Cui Y, Zhang H, Lin S (2017) Enhancement of non-photochemical quenching as an adaptive strategy under phosphorus deprivation in the dinoflagellate *Karlodinium veneficum*. Front Microbiol 8(Mar). https://doi.org/10.3389/FMICB.2017.00404/FULL
75. Ding L, Bar-Shalom R, Aharonovich D et al (2021) Metabolomic characterization of a cf. Neolyngbya cyanobacterium from the south china sea reveals wenchangamide a, a lipopeptide with in vitro apoptotic potential in colon cancer cells. Mar Drugs 19(7):397. https://doi.org/10.3390/MD19070397/S1
76. Sumimoto S, Kobayashi M, Sato R et al (2019) Minnamide A, a linear Lipopeptide from the marine cyanobacterium *Okeania hirsuta*. Org Lett 21(4):1187–1190. https://doi.org/10.1021/ACS.ORGLETT.9B00135/SUPPL_FILE/OL9B00135_SI_002.PDF
77. Ng LCS, Ong JL, Chou LM (2024) *Lyngbya majuscula* blooms in an enclosed marine environment. Published online 2008. Accessed July 25, 2024. www.tshe.org/EA
78. Curren E, Leong SCY (2019) Global phylogeography of toxic cyanobacteria Moorea producens reveals distinct genetic partitioning influenced by Proterozoic glacial cycles. Harmful Algae 86:10–19. https://doi.org/10.1016/J.HAL.2019.04.010
79. Tripathi A, Puddick J, Prinsep MR, Lee PPF, Lik TT (2009) Hantupeptin A, a cytotoxic cyclic depsipeptide from a Singapore collection of *Lyngbya majuscula*. J Nat Prod 72(1):29–32. https://doi.org/10.1021/NP800448T/SUPPL_FILE/NP800448T_SI_001.PDF

80. Moore RE (1981) Toxins from marine blue-green algae. The Water Environ:15–23. https://doi.org/10.1007/978-1-4613-3267-1_2
81. Osborne NJT, Webb PM, Shaw GR (2024) The toxins of *Lyngbya majuscula* and their human and ecological health effects. Accessed August 14, 2024. www.elsevier.com/locate/envint
82. Nagai H, Watanabe M, Sato S et al (2019) New aplysiatoxin derivatives from the Okinawan cyanobacterium Moorea produces. Published online 2019. https://doi.org/10.1016/j.tet.2019.03.020
83. Moore RE, Entzeroth M (1988) Majusculamide D and Deoxymajusculamide D, two cytotoxins from *Lyngbya majuscula*. Phytochemistry 27(10):3101–3103
84. Bonnard I, Bornancin L, Dalle K et al (2020) Assessment of the chemical diversity and potential toxicity of benthic cyanobacterial blooms in the lagoon of Moorea Island (French Polynesia). J Mar Sci Eng 8(6):406. https://doi.org/10.3390/JMSE8060406
85. Pavlin BI, Musto J, Pretrick M et al (2015) Mass poisoning after consumption of a hawksbill turtle, Federated States of Micronesia, 2010. Western Pac Surveill Response J 6(1):25. https://doi.org/10.2471/WPSAR.2014.5.3.006
86. Yasumoto T (1998) Fish poisoning due to toxins of microalgal origins in the Pacific. Toxicon 36(11):1515–1518. https://doi.org/10.1016/S0041-0101(98)00142-1
87. Fujiki H, Mori M, Nakayasu M, Terada M, Sugimura T, Moore RE (1981) Indole alkaloids: Dihydroteleocidin B, teleocidin, and lyngbyatoxin A as members of a new class of tumor promoters. Proc Natl Acad Sci USA 78(6 I):3872–3876. https://doi.org/10.1073/PNAS.78.6.3872
88. Tan LT, Parveen Katermeran N, Yadanar Phyo M (2023) Biodiversity record: new Singapore record of the marine cyanobacterium, *Symploca hydnoides*. Nature Singapore 16. https://doi.org/10.26107/NIS-2023-0063
89. Pavlik CM, Wong CYB, Ononye S et al (2013) Santacruzamate A, a potent and selective histone deacetylase inhibitor from the panamanian marine cyanobacterium cf. *Symploca* sp. J Nat Prod 76(11):2026–2033. https://doi.org/10.1021/NP400198R/SUPPL_FILE/NP400198R_SI_001.PDF
90. Linington RG, Edwards DJ, Shuman CF, McPhail KL, Matainaho T, Gerwick WH (2008) Symplocamide A, a potent cytotoxin and chymotrypsin inhibitor from the marine cyanobacterium *Symploca* sp. J Nat Prod 71(1):22–27. https://doi.org/10.1021/NP070280X/SUPPL_FILE/NP070280XSI20070612_060408.PDF
91. Tan TH, Leaw CP, Leong SCY et al (2024) Marine micro-phytoplankton of Singapore, with a review of harmful microalgae in the region. Accessed September 18, 2024. http://zoobank.org/urn:lsid:zoobank.org:pub:866484F0-5681-4E09-9789-CD5739335FF0
92. Pérez-Gómez A, Tasker RA (2022) Domoic acid as a neurotoxin. In: Handbook of neurotoxicity, Second Edition, 2, pp 873–897. https://doi.org/10.1007/978-3-031-15080-7_87
93. Lefebvre KA, Robertson A (2010) Domoic acid and human exposure risks: a review. Toxicon 56(2):218–230. https://doi.org/10.1016/J.TOXICON.2009.05.034
94. Todd ECD (1993) Domoic acid and amnesic shellfish poisoning—a review. J Food Prot 56(1):69–83. https://doi.org/10.4315/0362-028X-56.1.69
95. Berengere H, Tania HF, Romain LG, Mathilde S, Annie C. Two decades of *Pseudo-nitzschia* spp. blooms and king scallop (*Pecten maximus*) contamination by domoic acid along the French Atlantic and English Channel. 51:26–39. https://doi.org/10.1016/j.hal.2015.10.017
96. Liefer JD, Robertson A, MacIntyre HL, Smith WL, Dorsey CP (2013) Characterization of a toxic *Pseudo-nitzschia* spp. bloom in the Northern Gulf of Mexico associated with domoic acid accumulation in fish. Harmful Algae 26:20–32. https://doi.org/10.1016/J.HAL.2013.03.002
97. Teng ST, Abdullah N, Hanifah AH et al (2021) Toxic bloom of *Pseudo-nitzschia cuspidata* (Bacillariophyceae) and domoic acid contamination of bivalve molluscs in Malaysia Borneo. Toxicon 202:132–141. https://doi.org/10.1016/J.TOXICON.2021.09.018
98. Cutignano A, Conte M, Tirino V et al (2022) Cytotoxic potential of the marine diatom *Thalassiosira rotula*: insights into bioactivity of 24-methylene cholesterol. Mar Drugs 20(10):595. https://doi.org/10.3390/MD20100595/S1

99. Deeds JR, Hoesch RE, Place AR, Kao JPY (2015) The cytotoxic mechanism of karlotoxin 2 (KmTx 2) from *Karlodinium veneficum* (Dinophyceae). Aquat Toxicol 159:148–155. https://doi.org/10.1016/J.AQUATOX.2014.11.028
100. Hallegraeff GM, Anderson DM, Davidson K et al (2023) Fish-killing marine algal blooms: causative organisms, ichthyotoxic mechanisms, impacts and mitigation. UNESCO IOC Manuals and Guides, 93. https://doi.org/10.25607/OBP-1964
101. Chen Y, Yan T, Yu R, Zhou M (2011) Toxic effects of *Karenia mikimotoi* extracts on mammalian cells. Chin J Oceanol Limnol 29(4):860–868. https://doi.org/10.1007/S00343-011-0514-8/METRICS
102. Hort V, Abadie E, Arnich N, Dechraoui Bottein MY, Amzil Z (2021) Chemodiversity of brevetoxins and other potentially toxic metabolites produced by *Karenia* spp. and their metabolic products in marine organisms. Marine Drugs 19(12):656. https://doi.org/10.3390/MD19120656
103. Kwok CSN, Lai KKY, Lam W, Xu SJL, Lam SW, Lee FWF (2021) Proteome analysis of whole-body responses in medaka experimentally exposed to fish-killing dinoflagellate *Karenia mikimotoi*. Int J Mol Sci 22(21):11625. https://doi.org/10.3390/IJMS222111625/S1
104. Tanaka Y, Satake M, Yotsu-Yamashita M, Oshima Y (2013) Gymnocin-A carboxylic acid and Gymnocin-A2, cytotoxic polyethers from the red tide dinoflagellate *Karenia mikimotoi*. Heterocycles 87(10):2037–2046. https://doi.org/10.3987/COM-12-12814
105. Poli MA, Musser SM, Dickey RW, Eilers PP, Hall S (2000) Neurotoxic shellfish poisoning and brevetoxin metabolites: a case study from Florida. Toxicon 38(7):981–993. https://doi.org/10.1016/S0041-0101(99)00191-9
106. Tullis-Joyce P, Sen RS (2021) Occurrence of *Karenia brevis* near Southwest Florida coast 1971 to 2017: a geospatial analysis. J Coast Conserv 25(6):1–14. https://doi.org/10.1007/S11852-021-00844-1/METRICS
107. Reich A, Lazensky R, Faris J et al (2015) Assessing the impact of shellfish harvesting area closures on neurotoxic shellfish poisoning (NSP) incidence during red tide (*Karenia brevis*) blooms. Harmful Algae 43:13–19. https://doi.org/10.1016/J.HAL.2014.12.003
108. Holmes MJ, Bolch CJS, Green DH, Cembella AD, Teo SLM (2002) Singapore isolates of the dinoflagellate *Gymnodinium catenatum* (Dinophyceae) produce a unique profile of paralytic shellfish poisoning toxins. J Phycol 38(1):96–106. https://doi.org/10.1046/J.1529-8817.2002.01153.X
109. Raposo MIC, Gomes MTSR, Botelho MJ, Rudnitskaya A (2020) Paralytic shellfish toxins (PST)-transforming enzymes: a review. Toxins 12(5):344. https://doi.org/10.3390/TOXINS12050344
110. Laabir M, Collos Y, Masseret E et al (2013) Influence of environmental factors on the paralytic shellfish toxin content and profile of Alexandrium catenella (Dinophyceae) isolated from the Mediterranean Sea. Mar Drugs 11(5):1583–1601. https://doi.org/10.3390/MD11051583
111. Martens H, Tillmann U, Harju K, Dell'aversano C, Tartaglione L, Krock B (2017) Toxin variability estimations of 68 *Alexandrium ostenfeldii* (Dinophyceae) strains from The Netherlands reveal a novel abundant gymnodimine. Microorganisms 5(2):29. https://doi.org/10.3390/MICROORGANISMS5020029
112. Meng D, Shi J, Li M et al (2023) Identification of monitoring organ in bivalves for early warning of paralytic shellfish toxins accumulation. J Ocean Univ China 22(1):251–257. https://doi.org/10.1007/S11802-023-5402-2/METRICS
113. Mohammad-Noor N, Adam A, Lim PT et al (2018) First report of paralytic shellfish poisoning (PSP) caused by *Alexandrium tamiyavanichii* in Kuantan Port, Pahang, East Coast of Malaysia. Phycological Res 66(1):37–44. https://doi.org/10.1111/PRE.12205
114. Algal toxins in seafood and drinking water
115. García C, Bravo MDC, Lagos M, Lagos N (2004) Paralytic shellfish poisoning: post-mortem analysis of tissue and body fluid samples from human victims in the Patagonia fjords. Toxicon 43(2):149–158. https://doi.org/10.1016/J.TOXICON.2003.11.018
116. Kodama M, Sato S, Sakamoto S, Ogata T (1996) Occurrence of tetrodotoxin in Alexandrium tamarense, a causative dinoflagellate of paralytic shellfish poisoning. Toxicon 34(10):1101–1105. https://doi.org/10.1016/0041-0101(96)00117-1

117. Numano S, Kudo Y, Cho Y, Konoki K, Yotsu-Yamashita M (2019) Temporal variation of the profile and concentrations of paralytic shellfish toxins and tetrodotoxin in the scallop, *Patinopecten yessoensis*, cultured in a Bay of East Japan. Marine Drugs 17(12):653. https://doi.org/10.3390/MD17120653
118. Bordin P, Dall'Ara S, Tartaglione L et al (2021) First occurrence of tetrodotoxins in bivalve mollusks from Northern Adriatic Sea (Italy). Food Control 120:107510. https://doi.org/10.1016/J.FOODCONT.2020.107510
119. How CK, Chern CH, Huang YC, Wang LM, Lee CH (2003) Tetrodotoxin poisoning. Am J Emerg Med 21(1):51–54. https://doi.org/10.1053/AJEM.2003.50008
120. Roy S (2018) Distributions of phytoplankton carbohydrate, protein and lipid in the world oceans from satellite ocean colour. ISME J 12(6):1457–1472. https://doi.org/10.1038/S41396-018-0054-8
121. Jónasdóttir SH (2019) Fatty acid profiles and production in marine phytoplankton. Mar Drugs 17(3):151. https://doi.org/10.3390/MD17030151
122. Morales M, Aflalo C, Bernard O (2021) Microalgal lipids: A review of lipids potential and quantification for 95 phytoplankton species. Biomass Bioenergy 150. https://doi.org/10.1016/J.BIOMBIOE.2021.106108
123. Chen CY, Zhao XQ, Yen HW et al (2013) Microalgae-based carbohydrates for biofuel production. Biochem Eng J 78:1–10. https://doi.org/10.1016/J.BEJ.2013.03.006
124. Zhu C, Che J, Zhai X, Wang J, Kong F, Chi Z (2020) Cost-effective and efficient production of carbohydrates from an Alkalihalophilic Leptolyngbya sp. in a photobioreactor with periodical mixing. ACS Sustain Chem Eng 8(40):15310–15316. https://doi.org/10.1021/ACSSUSCHEMENG.0C05499/SUPPL_FILE/SC0C05499_SI_001.PDF
125. Yi Z, Xu M, Di X, Brynjolfsson S, Fu W (2047) Exploring valuable lipids in diatoms. Front Mar Sci 4(Jan). https://doi.org/10.3389/FMARS.2017.00017/FULL
126. Becker KW, Collins JR, Durham BP et al (2018) Daily changes in phytoplankton lipidomes reveal mechanisms of energy storage in the open ocean. Nat Commun 9(1):5179. https://doi.org/10.1038/S41467-018-07346-Z)
127. Khoo KS, Ahmad I, Chew KW, Iwamoto K, Bhatnagar A, Show PL (2023) Enhanced microalgal lipid production for biofuel using different strategies including genetic modification of microalgae: a review. Prog Energy Combust Sci 96. https://doi.org/10.1016/J.PECS.2023.101071
128. Vitova M, Bisova K, Kawano S, Zachleder V (2014) Accumulation of energy reserves in algae: From cell cycles to biotechnological applications. Biotechnol Adv 33(6):1204–1218. https://doi.org/10.1016/J.BIOTECHADV.2015.04.012
129. Mulgund A (2022) Increasing lipid accumulation in microalgae through environmental manipulation, metabolic and genetic engineering: a review in the energy NEXUS framework. Energy Nexus 5. https://doi.org/10.1016/J.NEXUS.2022.100054
130. Muñoz CF, Südfeld C, Naduthodi MIS, Weusthuis RA, Barbosa MJ, Wijffels RH, D'Adamo S (2021) Genetic engineering of microalgae for enhanced lipid production. Biotechnol Adv 52. https://doi.org/10.1016/j.biotechadv.2021.107836
131. Kato Y, Oyama T, Inokuma K, et al. Enhancing carbohydrate repartitioning into lipid and carotenoid by disruption of microalgae starch debranching enzyme. Commun Biol 4(1). https://doi.org/10.1038/s42003-021-01976-8
132. Cao Z, Bi R, Zhang C et al (2023) Quantification of multiple environmental controls on lipid biomarkers in common marine diatoms and dinoflagellates.Mar Biol 170(11). https://doi.org/10.1007/S00227-023-04284-3
133. Jin P, Gonzàlez G et al (2020) Long-term exposure to increasing temperature can offset predicted losses in marine food quality (fatty acids) caused by ocean warming. Evol Appl 13(9):2497–2506. https://doi.org/10.1111/eva.13059
134. Rousch JM, Bingham SE, Sommerfeld MR (2003) Changes in fatty acid profiles of thermo-intolerant and thermo-tolerant marine diatoms during temperature stress. J Exp Mar Biol Ecol 295(2):145–156. https://doi.org/10.1016/S0022-0981(03)00293-4

135. Rafa N, Ahmed SF, Badruddin IA, Mofijur M, Kamangar S (2021) Strategies to produce cost-effective third-generation biofuel from microalgae. Front Energy Res 9. https://doi.org/10.3389/FENRG.2021.749968/FULL
136. Chua LHC, Lo EYM, Shuy EB, Tan SBK (2009) Nutrients and suspended solids in dry weather and storm flows from a tropical catchment with various proportions of rural and urban land use. J Environ Manage 90(11):3635–3642. https://doi.org/10.1016/J.JENVMAN.2009.07.001
137. Aaronson S (1986) A role for algae as human food in antiquity. Food Foodways 1(3):311–315. https://doi.org/10.1080/07409710.1986.9961891
138. Karkos PD, Leong SCY, Karkos CD, Sivaji N, Assimakopoulos DA (2011) *Spirulina* in clinical practice: evidence-based human applications. Evidence-Based Complemen Alt Med 2011. https://doi.org/10.1093/ecam/nen058
139. Trotta T, Porro C, Cianciulli A, Nutrients MP (2022). Beneficial effects of spirulina consumption on brain health. Nutrients 14(3). https://doi.org/10.3390/nu14030676
140. Prete V, Carmelita Abate A, Di Pietro P, De Lucia M, Vecchione C, Carrizzo A (2024) Beneficial effects of spirulina supplementation in the management of cardiovascular diseases. Nutrients 16(5). https://doi.org/10.3390/nu16050642
141. Henchion M, Hayes M, Mullen A, Fenelon M, Tiwari B (2017) Future protein supply and demand: strategies and factors influencing a sustainable equilibrium. Foods 6(7):1–21. https://doi.org/10.3390/foods6070053
142. Diaz CJ, Douglas KJ, Kang K et al (2023) Developing algae as a sustainable food source. Front Nutr 9. https://doi.org/10.3389/FNUT.2022.1029841/FULL
143. Katiyar R, Banerjee S, Arora A (2021) Recent advances in the integrated biorefinery concept for the valorization of algal biomass through sustainable routes. Biofuels, Bioprod Biorefin 15(3):879–898. https://doi.org/10.1002/BBB.2187
144. Torres-Tiji Y, Fields FJ, Mayfield SP (2020) Microalgae as a future food source. Biotechnol Adv 41. https://doi.org/10.1016/J.BIOTECHADV.2020.107536
145. Wu JY, Tso R, Teo HS, Haldar S (2023) The utility of algae as sources of high value nutritional ingredients, particularly for alternative/complementary proteins to improve human health. Front Nutr 10. https://doi.org/10.3389/FNUT.2023.1277343/FULL
146. Naik B, Mishra R, Kumar V et al (2024) Micro-algae: Revolutionizing food production for a healthy and sustainable future. J Agric Food Res 15. https://doi.org/10.1016/J.JAFR.2023.100939
147. Xie T, Xia Y, Zeng Y, Li X (2017) Nitrate concentration-shift cultivation to enhance protein content of heterotrophic microalga *Chlorella vulgaris*: over-compensation strategy. Bioresour Technol. Accessed September 16, 2024. https://www.sciencedirect.com/science/article/pii/S0960852417302286
148. Canelli G, Tarnutzer C, Carpine R et al (2020) Biochemical and nutritional evaluation of *Chlorella* and *Auxenochlorella* biomasses relevant for food application. Front Nutr 7. https://doi.org/10.3389/FNUT.2020.565996/FULL
149. Pereira H, Silva J, Santos T et al (2019) Nutritional potential and toxicological evaluation of *Tetraselmis* sp. CTP4 microalgal biomass produced in industrial photobioreactors. Molecules 24(17). https://doi.org/10.3390/molecules24173192
150. Preußel K, Stüken A, Wiedner C, Chorus I, Fastner J (2006) First report on cylindrospermopsin producing *Aphanizomenon flos-aquae* (Cyanobacteria) isolated from two German lakes. Toxicon 47(2):156–162. https://doi.org/10.1016/J.TOXICON.2005.10.013)
151. Krock B, Busch JA, Tillmann U et al (2017) LC-MS/MS detection of karlotoxins reveals new variants in strains of the marine dinoflagellate *Karlodinium veneficum* from the Ebro Delta (NW Mediterranean). Marine Drugs 15(12):391. https://doi.org/10.3390/md15120391
152. Ahmad A, Hassan SW, Bioengineered FB (2022) An overview of microalgae biomass as a sustainable aquaculture feed ingredient: food security and circular economy. Bioengineered 13(4):9521–9547. https://doi.org/10.1080/21655979.2022.2061148
153. Nagappan S, Das P, AbdulQuadir M et al (2021) Potential of microalgae as a sustainable feed ingredient for aquaculture. J Biotechnol 341:1–20. https://doi.org/10.1016/J.JBIOTEC.2021.09.003

154. Malla S, Sommer MOA (2014) A sustainable route to produce the scytonemin precursor using Escherichia coli. Green Chem 16(6):3255–3265. https://doi.org/10.1039/C4GC00118D
155. Pajot A, Huynh GH, Picot L, Marchal L, Nicolau E (2022) Fucoxanthin from algae to human, an extraordinary bioresource: insights and advances in up and downstream processes. Marine Drugs 20(4):222. https://doi.org/10.3390/MD20040222
156. Zarekarizi A, Hoffmann L, Burritt D (2019) Approaches for the sustainable production of fucoxanthin, a xanthophyll with potential health benefits. J Appl Phycol 31(1):281–299. https://doi.org/10.1007/S10811-018-1558-3/METRICS
157. Soto-Sierra L, Stoykova P, Nikolov ZL (2018) Extraction and fractionation of microalgae-based protein products. Algal Res 36:175–192. https://doi.org/10.1016/J.ALGAL.2018.10.023
158. Lu K, Zhao X, Ho S et al (2019) Biorefining and the functional properties of proteins from lipid and pigment extract residue of Chlorella pyrenoidosa. ChenMarine Drugs 17(8). https://doi.org/10.3390/md17080454
159. Anjos L, Estêvão J, Infante C, Mantecón L, Power DM (2022) Extracting protein from microalgae (Tetraselmis chuii) for proteome analysis. MethodsX 9. https://doi.org/10.1016/J.MEX.2022.101637
160. Mear H, Gillon P, Gifuni I, Lavenant L, Poidevin A, Couallier E (2023) Extraction of soluble proteins by bead milling from *Tetraselmis chui* in two different physiological states. Algal Res 74. https://doi.org/10.1016/J.ALGAL.2023.103180
161. Song W, Song X, Shen H, Ding Y, Cheng R, Yu Z (2023) Degradation of paralytic shellfish toxins during flocculation of *Alexandrium pacificum* by an oxidized modified clay: a laboratory experiment. Ecotoxicol Environ Saf 253:114667. https://doi.org/10.1016/J.ECOENV.2023.114667
162. Hold GL, Smith EA, Harry Birkbeck T, Gallacher S (2001) Comparison of paralytic shellfish toxin (PST) production by the dinoflagellates *Alexandrium lusitanicum* NEPCC 253 and Alexandrium tamarense NEPCC 407 in the presence and absence of bacteria. FEMS Microbiol Ecol 36(2–3):223–234. https://doi.org/10.1111/J.1574-6941.2001.TB00843.X
163. Wang H, Kim H, Park H, Ki JS (2022) Temperature influences the content and biosynthesis gene expression of saxitoxins (STXs) in the toxigenic dinoflagellate *Alexandrium pacificum*. Sci Total Environ 802:149801. https://doi.org/10.1016/J.SCITOTENV.2021.149801
164. Cirés S, Delgado A, González-Pleiter M, Quesada A (2017) Temperature influences the production and transport of saxitoxin and the expression of sxt genes in the cyanobacterium *Aphanizomenon gracile*. Toxins 9(10):322. https://doi.org/10.3390/TOXINS9100322
165. Leong SCY, Maekawa M, Taguchi S (2010) Carbon and nitrogen acquisition by the toxic dinoflagellate Alexandrium tamarense in response to different nitrogen sources and supply modes. Harmful Algae 9(1):48–58. https://doi.org/10.1016/J.HAL.2009.07.003
166. Leong SCY, Murata A, Nagashima Y, Taguchi S (2004) Variability in toxicity of the dinoflagellate *Alexandrium tamarense* in response to different nitrogen sources and concentrations. Toxicon 43(4):407–415. https://doi.org/10.1016/J.TOXICON.2004.01.015
167. Chen T, Zhang H, Dong W, Bu K, Chen X (2024) Toxin production and transcriptomic response to nitrate concentrations in the toxic dinoflagellate *Alexandrium tamarense*. Mar Environ Res 198:106550. https://doi.org/10.1016/J.MARENVRES.2024.106550
168. Meng FQ, Song JT, Zhou J, Cai ZH (2019) Transcriptomic profile and sexual reproduction-relevant genes of *Alexandrium minutum* in response to nutritional deficiency. Front Microbiol 10:465340. https://doi.org/10.3389/FMICB.2019.02629/BIBTEX
169. Bui QTN, Kim H, Park H, Ki JS (2021) Salinity affects saxitoxins (STXs) toxicity in the dinoflagellate *Alexandrium pacificum*, with low transcription of SXT-biosynthesis genes sxtA4 and sxtG. Toxins 13(10):733. https://doi.org/10.3390/TOXINS13100733
170. Lim PT, Ogata T (2005) Salinity effect on growth and toxin production of four tropical *Alexandrium* species (Dinophyceae). Toxicon 45(6):699–710. https://doi.org/10.1016/J.TOXICON.2005.01.007
171. Abadie E, Muguet A, Berteaux T et al. Toxin and growth responses of the neurotoxic dinoflagellate *Vulcanodinium rugosum* to varying temperature and salinity. https://doi.org/10.3390/toxins8050136

172. Deeds JR, Place AR (2006) Sterol-specific membrane interactions with the toxins from *Karlodinium micrum* (Dinophyceae)—a strategy for self-protection? Afr J Mar Sci 28(2):421–425. https://doi.org/10.2989/18142320609504190
173. Smayda TJ (2010) Adaptations and selection of harmful and other dinoflagellate species in upwelling systems 1 Morphology and adaptive polymorphism. Prog Oceanogr 85(1–2):53–70. https://doi.org/10.1016/J.POCEAN.2010.02.004
174. Bravo I, Figueroa RI (2014) Towards an ecological understanding of dinoflagellate cyst functions. Microorganisms 2(1):11–32. https://doi.org/10.3390/MICROORGANISMS2010011
175. Li Y, Zhu H, Guan C et al (2014) Towards molecular, physiological, and biochemical understanding of photosynthetic inhibition and oxidative stress in the toxic Alexandrium tamarense induced by a marine bacterium. Appl Microbiol Biotechnol 98(10):4637–4652. https://doi.org/10.1007/S00253-014-5578-X/FIGURES/9
176. Lee AE, Chan TJR, Gan AJH et al (2024) Colonization of plastics by microorganisms. Unpublished data. Published online 2024
177. Yu DCY, Chan TJR, Chu QE et al (2024) Profiling beach plastics and their hitchhiking phytoplankton. Unpublished data. Published online 2024
178. Lagos N (2014) Clinical applications of paralytic shellfish poisoning toxins. In: Toxins and biologically active compounds from microalgae: Volume 2: biological effects and risk management, 2, pp 309–329. https://doi.org/10.1201/B16806-16/CLINICAL-APPLICATIONS-PARALYTIC-SHELLFI-SH-POISONING-TOXINS-N
179. Rodríguez-Navarro AJ, Berde CB, Wiedmaier G et al (2011) Comparison of neosaxitoxin versus bupivacaine via port infiltration for postoperative analgesia following laparoscopic cholecystectomy: a randomized double-blind trial. Reg Anesth Pain Med 36(2):103–109. https://doi.org/10.1097/AAP.0B013E3182030662
180. Bae YH, Kim H, Lee SJ (2022) Effect of tetrodotoxin on the proliferation and gene expression of human SW620 colorectal cancer cells. Biomed Sci Lett 28(1):42–49. https://doi.org/10.15616/BSL.2022.28.1.42
181. Campos-Ríos A, Rueda-Ruzafa L, Herrera-Pérez S, Rivas-Ramírez P, Lamas JA (2021) Tetrodotoxin: a new strategy to treat visceral pain? Toxins 13(7):496. https://doi.org/10.3390/TOXINS13070496
182. Saoudi M, Abdelmouleh A, El Feki A (2010) Tetrodotoxin: a potent marine toxin. Toxin Rev 29(2):60–70. https://doi.org/10.3109/15569543.2010.487631
183. Tsai YH, Ho PH, Hwang CC, Hwang PA, Cheng CA, Hwang DF (2006) Tetrodotoxin in several species of xanthid crabs in southern Taiwan. Food Chem 95(2):205–212. https://doi.org/10.1016/J.FOODCHEM.2004.12.032
184. Hao W, Wang L, Xu T et al (2024) Marine cytotoxin santacruzamate A derivatives as potent HDAC1-3 inhibitors and their synergistic anti-leukemia effects with venetoclax. Mar Drugs 22(6):250. https://doi.org/10.3390/MD22060250/S1
185. Van Santen JA, Kautsar SA, Medema MH, Linington RG (2021) Microbial natural product databases: moving forward in the multi-omics era. Nat Prod Rep 38(1):264–278. https://doi.org/10.1039/D0NP00053A
186. Jones MR, Pinto E, Torres MA et al (2021) CyanoMetDB, a comprehensive public database of secondary metabolites from cyanobacteria. Water Res 196:117017. https://doi.org/10.1016/J.WATRES.2021.117017
187. Global Natural Products Social Molecular Networking (GNPS)—Google Scholar. Accessed August 12, 2024. https://scholar.google.com/scholar?hl=en&as_sdt=0%2C5&q=Global+Natural+Products+Social+Molecular+Networking+%28GNPS%29+&btnG
188. Sibat M, Réveillon D, Antoine C et al (2021) Molecular networking as a novel approach to unravel toxin diversity of four strains of the dominant *Dinophysis* species from French coastal waters. Harmful Algae 103:102026. https://doi.org/10.1016/J.HAL.2021.102026
189. Ternon E, Glukhov E, Trytten E, Lemée R, Gerwick WH (2022) On the hunt for new toxin families produced by a mediterranean strain of the benthic dinoflagellate *Ostreopsis* cf. *ovata*. Toxins (Basel) 14(4):234. https://doi.org/10.3390/TOXINS14040234/S1

190. Curren E, Kuwahara VS, Yoshida T, Leong SCY (2024) Draft genome of novel cyanobacteria *Sphaerothrix gracilis* isolated from coastal microplastics reveal insights to chemical ecology, bloom and plastic-utilization potential. Funct Integr Genomics 24(2):1–4. https://doi.org/10.1007/S10142-024-01328-9/METRICS
191. Blin K, Shaw S, Kloosterman AM et al (2021) antiSMASH 6.0: improving cluster detection and comparison capabilities. Nucleic Acids Res 49(W1):W29–W35. https://doi.org/10.1093/NAR/GKAB335
192. Terlouw BR, Blin K, Navarro-Muñoz JC et al (2023) MIBiG 3.0: a community-driven effort to annotate experimentally validated biosynthetic gene clusters. Nucleic Acids Res 51(D1):D603–D610. https://doi.org/10.1093/NAR/GKAC1049

Chapter 7
Discovery and Development of Potential Therapeutic Agents from Sponge-Derived Marine Natural Products in Thailand

Lik Tong Tan and Sumaitt Putchakarn

1 Introduction

Thailand has a central position in Southeast Asia and consists of more than 3100-km coastline which is uniquely surrounded by the vast Andaman Sea to the west and Gulf of Thailand to the east. The marine ecosystems of Thailand, situated at the crossroads of the Andaman Sea and the Gulf of Thailand, represent a vital component of the global Indo-Pacific biodiversity hotspot [1]. As part of Indian Ocean, the Andaman Sea is bordered by the Andaman and Nicobar Islands to the west and the Malay Peninsula to the east. The Andaman coast is 1093 km long, with about 545 offshore islands, and dominated by pocket beaches, cliffs and extensive tidal flats. The Gulf of Thailand, on the other hand, is a semi-enclosed inlet of the South China Sea, with a southern boundary line defined by Cape Bai Bung in south Vietnam to Kota Bharu on the coast of Malaysia. The Gulf of Thailand is relatively shallow, having an average depth of 45 m, and characterized by extensive and wide mainland beaches with lagoons and bays. Both the Andaman and Gulf of Thailand coasts are also influenced by the Asian tropical wet monsoonal conditions and support diverse tropical biotopes, including mangroves, seagrass beds and coral reefs with tremendous diversity of marine flora and fauna. Moreover, Thailand's marine diversity is shaped by a complex interplay of environmental factors, anthropogenic influences and oceanographic conditions [2].

L. T. Tan (✉)
Natural Sciences and Science Education, National Institute of Education, Nanyang Technological University, Singapore, Singapore
e-mail: liktong.tan@nie.edu.sg

S. Putchakarn
Institute of Marine Science, Burapha University, Chonburi, Thailand
e-mail: sumaitt@bims.buu.ac.th

L. T. Tan and N. Hanif (eds.), *Marine Natural Products Research in South-East Asia*,
https://doi.org/10.1007/978-3-032-01098-8_7

The coral reef coverage in Thai waters is extensive and varies from 153 to 238 cm^2 [3]. In addition, their reef structure and distribution are influenced by several factors, including oceanographic conditions, exposure to the southwest monsoon, geographical distance from the mainland, currents, substrata as well as human activities. Diverse marine species can be found in these coral reef systems. For instance, more than 270 stony coral species have been reported, together with countless fishes and benthic invertebrate animals [4]. Within the Gulf of Thailand, reef areas are mainly consisting of fringing reefs surrounding 250 islands located in the eastern, western and upper Gulf region. The reef areas in the Andaman Sea are estimated to be about 117 km^2, with the best developed reefs located at the Similan and Surin Archipelago where water clarity is the highest. Without a doubt, Thailand has enjoyed the luxury of marine biodiversity and these bioresources represent rich sources for drug discovery and development efforts.

One such group of marine benthic organisms that have attracted the attention of natural products chemists are the marine sponges (Porifera) [5]. Sponges are a key component of many coral reef ecosystems, and in some biogeographic regions they are considered the dominant benthic fauna. Several sponge community studies conducted in Thailand revealed high diversity. For instance, within the Gulf of Thailand, more than 90 sponge species have been documented [6]. Moreover, certain sponge species, such as *Aaptos suberitoides*, *Acanthella cavernosa*, *Biemna fortis*, *Cinachyrella australiensis*, *Clathria* (*Thalysias*) *reinwardti*, *Coelocarteria singaporensis*, *Echinodictyum asperum*, *Hyrtios erectus*, *Haliclona* (*Gellius*) *cymaeformis*, *Iotrochota baculifera*, *I. purpurea*, *Mycale* (*Zygomycale*) *parishii*, *Neopetrosia exigua*, *Oceanapia sagittaria*, *Spheciospongia vagabunda*, *Xestospongia testudinaria*, are widespread. Several research groups in Thailand have been actively involved in the natural products research of Thai sponges as well as other benthic organisms. These researchers include Suwanborirux's group at Chulalongkorn University, Hunsa Prawat's group at Chulabhorn Research Institute and Chulabhorn Graduate Institute and Anuchit Plubrukarn's group at Prince of Songkla University. Together with their international collaborators, such as Drs. Werner Herz (Florida State University, USA), Late Tatsuo Higa and Junichi Tanaka (University of the Ryukyus, Okinawa), Naoki Saito (Meiji Pharmaceutical University) and Anake Kijjoa (University of Porto), they have uncovered the diverse natural products chemistry and their biological activities of many Thai sponge species.

In view of their importance as sources of potential pharmaceutical agents, this chapter provides a comprehensive coverage on the chemistry and biology of secondary metabolites reported from Thai sponges, based on publications in international journals that have appeared between 1988 and Dec 2023. In addition, the medicinal chemistry of selected molecules as well as their semi-synthetic derivatives, having potent activities, will be highlighted to showcase the impact of these compounds in drug development efforts, particularly in the search of drug leads as potential anti-infective and anticancer agents.

2 Overview on the Biodiversity of Marine Sponges in Thailand

The biodiversity study of marine sponges from Thailand began in 1925 when Dr. E. Topsent described a new species of sponge from the Gulf of Siam as *Prostylissa siamensis* Topsent, 1925, which has now been revised to *Amorphenopsis siamensis* (Topsent 1925) [7]. The study of marine sponges in Thailand began in 1962 with the work of Kannika Bunyathithi. Bunyathithi examined sponges collected during field trips conducted by the Biology Department of the Faculty of Science, Chulalongkorn University, as well as from surveys of the South China Sea carried out under the NAGA Project (1959–1960). The study reported the presence of two classes, seven orders and ten families of marine sponges. Later in 1990–1991, there was a joint survey and collection of marine invertebrate samples by the Australian Institute of Marine Science and the Institute of Marine Science, Burapha University in the coastal areas of the Gulf of Thailand and Andaman Sea. About 30 species of sea sponges from the Gulf of Thailand were found, of which four species were reported in the areas of Koh Samui and Koh Tao, Surat Thani Province [8]. In 2000, Hooper et al. catalogued marine sponges in the South China Sea, of which the Gulf of Thailand is a subregion, and found over 1500 species of sponges [9]. They reported 22 species of marine sponges found in the Gulf of Thailand, of which 12 were found in the Koh Samui area. A year later, Kritsanapuntu et al. reported 126 species of sponges living on coral reefs in the eastern Gulf of Thailand [10]. In addition to these early key reports, other numerous diversity, ecological and taxonomic studies have been conducted on Thai sponges and the results are summarized in Table 1. Examples of selected sponge species with biomedical importance are highlighted in Fig. 1.

3 Bioactive Natural Products from Thai Sponges

The discussion of natural products derived from Thai sponges in this section will be based on structural classes, including terpenoids, steroids, polyketides, and alkaloids. These molecules and sponge species are also summarized in Table 2.

3.1 Terpenoids

Early studies on natural products from Thai sponges by Prof. Phillip Crews's research group at University of California, Santa Cruz, yielded a series of unique terpenoids, including sesquiterpenes and diterpenes (Fig. 2). The sponge collections stemmed from an expedition to areas around Phuket with funding support from the National Oceanic and Atmospheric Administration (NOAA), National Sea Grant College Program, Department of Commerce through the California Sea Grant

Table 1 Summary list of taxonomic and diversity studies conducted on Thai sponges

Authors (Year)	Results	Refs.
Hooper et al. (1993)	*Oceanapia sagittaria* (Sollas) is redescribed from shallow water, soft benthic communities surrounding the limestone islands of Ko Samui, southwest Gulf of Thailand, with additional new records from southern Papua New Guinea	[11]
Asa et al. (2000)	*Petrosia* sp., from coral communities in the Gulf of Thailand, is a viviparous hermaphrodite with asynchronous gamete development, whose reproductive activity and larval production are significantly affected by environmental events like coral bleaching	[12]
Putchakarn et al. (2006)	Study reported 44 species of marine sponges in the class Demospongiae from 31 genera, 26 families and 11 orders from coral reef surveys around Koh Kram and adjacent islands, Chonburi province	[13]
Putchakarn et al. (2004)	Reported 52 species of Demospongiae and two species of Calcarea from marine invertebrate species diversity survey in coral reefs in Chonburi Province	[14]
Putchakarn et al. (2004)	Description of a new sponge species, *Cladocroce burapha*, found in Bangsaen Beach, Chonburi Province	[15]
Putchakarn (2006)	PhD thesis reported 56 marine sponges found in the Gulf of Thailand	[16]
Putchakarn (2007)	Study reported 45 species of marine sponges from Surat Thani and Nakhon Si Thammarat Provinces	[17]
Hongphatthakhiri et al. (2008)	Reported 13 marine sponges found in the coral reefs around Koh Ka, Chumphon Province	[18]
Putchakarn et al. (2008)	Reported 72 marine sponges around the eastern coast of the Gulf of Thailand divided into Chon Buri and Rayong Provinces found 62 marine sponges from 11 orders, 33 families and 39 genera, of these seven species were first recorded in Thai waters; Chanthaburi to Trat province found 72 species from 11 orders, 37 families and 50 genera, of which three species were new recorded in Thai Waters	[19]

(continued)

Table 1 (continued)

Authors (Year)	Results	Refs.
Pumbua et al. (2009)	Reported 67 species of Demospongiae, from 11 orders, 32 families and 40 genera along the Nangrong Beach, Ko Jorakhe and Mu Ko Chuang, Amphoe Sattahip, Chon Buri province. Four species were new recorded in Thai Waters, namely *Cliona utriculara*, *Timea aurantiaca*, *Diplastrella* sp. and *Clathria* (*Thalysias*) *tingens*. Widely distribution species throughout the study area were *Xestospongia testudinaria*, *Iotrochota baculifera*, *Monanchora unguiculata*, *Oceanapia sagittaria* and *Neopetrosia* sp. These species were also commonly found in Gulf of Thailand and the Indo-West Pacific region	[20]
Chobram et al. (2009)	Reported 72 marine sponge species of Demospongiae from 13 orders, 29 families and 41 genera from Mu Ko Tarutao and Mu Ko Adang-Rawi, Satun province, Andaman Sea. 60 species were found from Tarutao Islands and 22 species from Adang–Rawi Islands. Out of these, nine species were the new records in Thai Waters, namely *Holoxea valida*, *Cliona albimarginata*, *Desmanthus rhabdophorus*, *Echinodictyum clathratum*, *Stylissa conulosa*, *Axinyssa mertoni*, *Axinyssa oinops*, *Agelas ceylonica* and *Gelliodes fibulata*	[21]
Hongpatarakiri and Putchakarn (2010)	Reported 10 species of demosponges around Koh Losin, Pattani province, Gulf of Thailand	[22]
Putchakarn (2011)	Reported 50 species of 10 orders, 25 families, 35 genera of marine sponges found in the Koh Samui area, Surat Thani Province, in the research project on "The Diversity of marine sponges along the coast of the Western Gulf of Thailand" between 28 January and 5 February 2010, including 7 survey points. Of these, two species, *Corticium niger* and *Higginsia massalis* were the first recorded in Thailand. Order Haplosclerida was the most abundant group of sponges (17 species), followed by the Order Poecilosclerida with nine species	[23, 24]

(continued)

Table 1 (continued)

Authors (Year)	Results	Refs.
Putchakarn (2012)	Study reported 48 species of Demospongiae from 12 orders, 24 families and 36 genera from Phi Phi Islands, Krabi province Andaman Sea. Out of these, four species, *Erylus carteri*, *Aciculites* sp., *Merlia tenuis* and *Petrosia* (*Strongylophora*) sp. were new records in Thai waters	[25]
Putchakarn (2013)	Reported 11 orders, 24 families, 35 genera and 50 species of marine sponges in the Chumphon archipelago, Chumphon Province, with a total of 8 survey points. Order Haplosclerida is the most abundant group of sponges (17 species) followed by Order Poecilosclerida with 13 species. *Desmapsamma* sp. was first recorded in Thailand while the most prominent and widespread marine sponges were *Peneres nux*, *Neopetrosia* sp. "blue" and *Gelliodes* sp. "purple"	[26]
Putchakarn (2013)	Reported 12 orders, 24 families, 35 genera and 51 species of marine sponges in the area of Koh Tao, Surat Thani Province at 11 survey points. Order Haplosclerida is the most abundant group of sponges (18 species), followed by Order Poecilosclerida (12 species). *Lissodendoryx* (*Ectyodoryx*) sp. and *Cladocroce* sp. "blue" were first recorded. The most predominant and widespread distribution is *Xestospongia* sp. "purple", followed by *Neopetrosia* sp. "blue" and *Cacospongia* sp. "black"	[26]
Hongpatarakiri and Putchakarn (2013)	Reported 47 marine sponge species from Mu Ko Surin, Phangnga Province	[27]
Lim et al. (2016)	Complied and updated the inventory checklist of the sponge fauna from the Singapore Strait to the Taiwan Strait along the western coastline of the South China Sea. 421 sponge species belonging to 27 orders, 84 families and 172 genera were listed, and 95 marine sponge species in the Gulf of Thailand were updated	[6]

College Program as well as NIH. Collected in winter of 1988, the sponge, *Topsentia* sp., was found to contain spirocyclic sesquiterpene, axisonitrile 3 (**1**), while *Acanthella cavernosa* provided a series of known diterpenes, kalihinols X (**2**) and Y (**3**), and new kalihinols J (**4**) and I (**5**) (Fig. 2) [28]. Axisonitrile 3 was also reported from the Thai sponge, *Leucophloeus fenestrata*, from Phuket [70]. These terpenoids

Fig. 1 Selected Thai sponge species, having biomedical importance. **a** *Myrmekioderma* sp.; **b** *Dysidea arenaria*; **c** *Halichondria* sp.; **d** *Hyrtios erectus*; **e** *Neopetrosia* sp. "blue"; **f** *Neopetrosia* sp. (*Xestospongia* sp.) and *Jorunna funebris*; **g** *Corticium* sp.; **h** *Petrosia (Strongylophora)* sp.; **i** *Stylissa massa*; **j** *Acanthella cavernosa*; **k** *Axinyssa* sp.; **l** *Tetilla japonica*; **m** *Haliclona baeri*; **n** *Pseudoceratina purpurea*; **o** *Acanthodendrilla* sp.; **p** *Pachastrissa nux*; **q** *Mycale (Zygomycale) parishii*; **r** *Mycale (Mycale) grandis*; **s** *Suberea* aff. *praetensa*; **t** *Aciculites* sp.; **u** *Erylus carteri*; **v** *Xestospongia* sp. "purple"; **w** *Monanchora unguiculata*; **x** *Ircinia mutans*; **y** *Xestospongia testudinaria*; **z** *Oceanapia sagittaria*

were evaluated for antiparasitic activity against *Nippostrongylus brasiliensis* and only axisonitrile 3, kalihinols J, I and Y, were found to be active [28]. In addition, axisonitrile 3 and kalihinol X exhibited significant antimicrobial activity against *Staphylococcus aureus*, *Candida albicans* and *Trichophyton mentagrophytes.*

From another collection of the bright blue sponge, *Dysidea arenaria*, two sesquiterpene ethers, arenarans A (**6**) and B (**7**) were reported (Fig. 2) [29]. The chemical structures of these molecules were confirmed via spectroscopic measurements as well as chemical transformations. Arenarans A (**6**) and B (**7**) were assessed for their cytotoxicity properties in several whole-cell assays and only the former molecule was cytotoxic against several cell lines, including A549, human lung carcinoma, HT-29, human colon adenocarcinoma, HCT-29, human colon adenocarcinoma and P-388 murine leukemia with IC_{50} values of 9.51, 9.11, 5.28 and 3.17 μg/mL, respectively. It was only in 2016, more than 20 years after their first report, that the total synthesis of the arenarans were accomplished by Torres and co-workers using commercially available (–)-sclareol [71].

Table 2 List of new marine natural products reported from Thai sponges for the period of 1988 to 2023

Class/ Sub-class	*Sponge* compound	Location	Refs.
Terpenoids			
Early works by Crews's research group			
	Acanthella cavernosa (O. Bubarida) kalihinols J (**4**) and I (**5**)	All sponge collections were collected based on an expedition to areas around Phuket	[28]
	Dysidea arenaria (O. Dictyoceratida) arenarans A (**6**) and B (**7**)		[29]
	Dactylospongia elegans (O. Dictyoceratida) (–)-dactylosponol (**8**) and (–)-dactylospontriol (**9**)		[30]
Sesquiterpenes			
Aromatic Sesquiterpenes			
	Myrmekioderma sp. (O. Axinellida) 3-oxoabolene (**15**), 1-oxocurcuphenol (**16**)	Phi Phi Islands National Park, Krabi Province	[31]
	Myrmekioderma sp. (O. Axinellida) (*R*)- and (*S*)-4,6″-biscurcudiols (**19** and **20**), myrmekioperoxides A (**21**), B (**22**), myrmekiodermaral (**23**)		[32]
Nitrogenous Sesquiterpenes			
	Axinyssa sp. (O. Suberitida) (1Z,4Z)-7α*H*-11-aminogermacra-1(10),4-diene (**24**) and *N*,*N*′-11-bis[(1Z,4Z)-7α*H*-germacra-1(10),4-dienyl]urea (**25**)	Andaman Sea, Trang Province	[33]
	Axinyssa sp. (O. Suberitida) axinyssines A (**26**)–L (**37**)	Andaman Sea	[34]
	Halichondria sp. (CRI328) (O. Suberitida) (–)-axisonitrile-3 (**38**), (+)-axamide-3 (**39**) and (–)-halichamine (**42**)	Phi Phi Island, Andaman Sea, Krabi Province	[35]
	Halichondria sp. (O. Suberitida) halichons A (**43**), B (**44**), C (**45**), D (**47**), E (**48**), F (**49**), G (**50**), H (**51**) and 4-*epi*-halichon C (**46**)	Phi Phi Island, Andaman Sea, Krabi Province	[36]
Merosesquiterpenes			
	Smenospongia sp. (O. Dictyoceratida) 6′-iodoaureol (**52**), 6′-aureoxyaureol (**53**)	Phi Phi Island, Andaman Sea, Krabi Province	[37]

(continued)

Table 2 (continued)

Class/ Sub-class	*Sponge* compound	Location	Refs.
	Verongula cf. *rigida* (O. Verongiida) quintaquinone (**54**), 5-*epi*-nakijiquinone L (**55**), 3-farnesyl-2-hydroxy-5-methoxyquinone (**56**)	nearby Ko Ha Islets, Krabi Province	[38]
Diterpenes	*Ciocalapata* sp. (O. Suberitida) 8-isocyanoamphilecta-11(20),15-diene (**61**)	Koh-Tao, Surat-Thani Province	[39]
	Stylissa cf. *massa* (O. Scopalinida) 8-isocyanato-15-formamidoamphilect-11(20)-ene (**64**) and 8-isothiocyanato-15-formamidoamphilect-11(20)-ene (**65**)		[40]
Sesterterpenes	*Mycale* sp. (O. Poecilosclerida) mycaperoxides A (**68**) and B (**69**)	Kang Ta Sin of Sichang Island	[41]
	Mycale sp. (O. Poecilosclerida) mycaperoxide H (**70**)	Sichang Island	[42]
	Xestospongia sp. (O. Haplosclerida) Compound **71**	Sikao Sea, Trang Province	[43]
	Brachiaster sp. (O. Tetractinellida) 12-deacetoxyscalarin 19-acetate (**74**), (*E*)- and (*Z*)-neomanoalide 24,25-diacetates (**75** and **76**)	Koh-Tao, Surat-Thani Province	[44]
	Hyrtios gumminae (O. Dictyoceratida) similan A (**81**), 12*β*,20-dihydroxy-16β-acetoxy-17-scalaren-19,20-olide (**82**), 12*β*-acetoxy-20-hydroxy-17-scalaren-19,20-olide (**83**) and 12*β*,16*α*,20-trihydroxy-17-scalaren-19,20-olide (**84**)	Similan Island, Andaman Sea, Phang-Nga Province	[45]
	H. erectus (O. Dictyoceratida) erectusolide A (**89**), erectuseneols A (**90**)–F (**95**)	Similan Island, Phang-Nga Province	[46]
	H. erectus (O. Dictyoceratida) erectusolides B (**96**)–D (**99**), seco-manoalide-25-methyl ether (**98**), erectusfuranones A (**100**) and B (**101**)		[47]
Triterpenes	*Haliclona* sp. (O. Haplosclerida) phuklona sulfate (**111**)	South Andaman Sea in the vicinity of Phuket Island	[48]

(continued)

Table 2 (continued)

Class/ Sub-class	*Sponge* compound	Location	Refs.
Sterols	*Corticium* sp. (O. Homosclerophorida) 4-acetoxy-plakinamine B (**112**)	Koh-Tao, Surat-Thani Province	[49]
	Petrosia (Strongylophora) sp. (O. Haplosclerida) steroids **113–120**	Similan Islands, Andaman Sea, Phang Nga	[50]
Hydroxypyran-2-ones			
	Tetilla japonica (O. Tetractinellida) tetillapyrone (**122**) and nortetillapyrone (**123**)	Captain Yuth Beach, Chonburi Province	[51]
Alkaloids			
Maleimide-5-oximes			
	Haliclona baeri (O. Haplosclerida) maleimide-5-oxime (**124**)	Coastal area of Chonburi Province and from Koh Chang, Trad Province	[52]
	Pseudoceratina purpurea (O. Podocopida) 3-methylmaleimide-5-oxime (**125**)		[53, 54]
Halogenated Tyrosine-Containing Alkaloids			
	Suberea aff. *praetensa* (O. Verongiida) 11,17-dideoxyagelorins A (**130**) and B (**131**), subereatensin (**136**)	seashore of Ban Phae Village, Gulf of Thailand, Rayong Province	[55–57]
	Pseudoceratina purpurea (O. Podocopida) purpuroceratic acids A (**141**) and B (**142**)	Kho Chang Island, Trat Province	[53]
	Acanthodendrilla sp. (O. Dendroceratida) 13-oxosubereamolline D (**163**)	Koh-Ha Islets, Krabi	[58]
	Acanthodendrilla sp. (O. Dendroceratida) acanthodendrilline (**176**)		[59]
Pyridoacridine Alkaloids			
	Petrosia sp. (O. Haplosclerida) 2-bromoamphimedine (**181**)	intertidal region of Phuket Island	[60]

(continued)

Table 2 (continued)

Class/ Sub-class	*Sponge* compound	Location	Refs.
Indole-Containing Alkaloids			
	Smenospongia sp. (O. Dictyoceratida) four brominated indole alkaloids (**182–185**)	Phi Phi Island in the Andaman Sea, Krabi Province	[37]
	Damiria sp. (O. Poecilosclerida) damirines A (**196**) and B (**197**)	vicinity of Phuket Island	[61]
Trisoxazole Macrolides			
	Pachastrissa nux (O. Tetractinellida) kabiramides F (**205**)–L (**215**)	Sichang Island; vicinity of Koh-Tao, Surat-Thani Province; Chumphon Islands National Park, Chumphon Province	[62–64]
Renieramycins			
	Xestospongia sp. (O. Haplosclerida) renieramycins J (**225**)–O (**230**), 7-desmethylrenieramycin O (**231**) and Q (**232**)–V (**239**)	vicinity of Sichang Island,	[65–68]
	Jorunna funebris (O. Nudibranchia) jorunnamycins A (**240**)–C (**242**); compounds are probably sequestered from sponge diet	vicinity of Ko Sichang	[69]

The cytotoxic activity of crude extracts prepared from the sponge *Dactylospongia elegans*, collected from three different Indo Pacific areas, including Phuket, Thailand, led to the isolation of 17 cytotoxic sesquiterpenes. From two sponge collections from Thailand, two new sesquiterpenes, (–)-dactylosponol (**8**) and (–)-dactylospontriol (**9**), along with five known sesquiterpenes, including (–)-smenodiol (**10**) well as compounds **11** to **14** (Fig. 2) were reported [30].

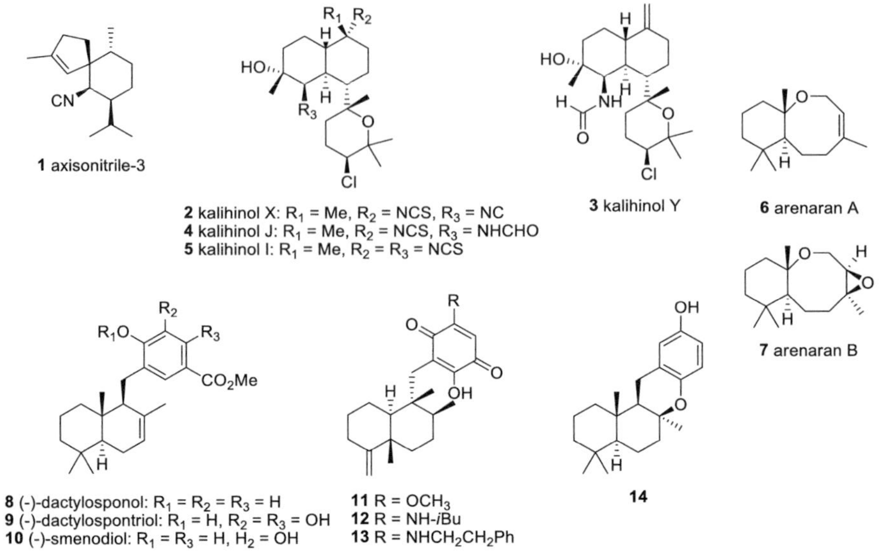

Fig. 2 Chemical structures of terpenoids from Thai marine sponges collected by Crews's research group at University of California, Santa Cruz

3.1.1 Sesquiterpenes

Aromatic Sesquiterpenes

A sponge specimen of *Myrmekioderma* sp. obtained from Phi Phi Islands National Park, Krabi Province, was initially screened for biological activity and found cytotoxicity against a panel of cancer cell lines [31]. Subsequent chemical investigations led to the report of two new aromatic bisabolanes, including 3-oxoabolene (**15**) and 1-oxocurcuphenol (**16**), along with two known compounds, (+)-curcuphenol (**17**) and curcudiol (**18**) (Fig. 3). In addition, curcudiol, 3-oxoabolene and 1-oxocurcuphenol exhibited potent cytotoxicity against HT-29 cancer cells with IC_{50} values of 0.01, 0.67 and 0.017 μM, respectively. From another collection of the sponge, *Myrmekioderma* sp. made in early 2009 from Phi Phi Islands at the Andaman Sea, Krabi Province, additional unique new bisabolanes, such as myrmekioperoxides A (**21**) and B (**22**), and their aromatic dimers, including (*R*)- and (*S*)-4,6″-biscurcudiols (**19** and **20**) and myrmekiodermaral (**23**), were discovered, along with seven known bisabolane sesquiterpenes (Fig. 3) [32]. The absolute stereochemistry of (*R*)- and (*S*)-4,6″-biscurcudiols (**19** and **20**) were confirmed based on experimental and calculated ECD spectra as well as comparison of their ^{1}H NMR spectral data with those of the known dicurcuphenols B and C. When tested in a panel of six cancer cell lines, the new compounds **19**, **20** and **23** showed weakly to moderate cytotoxicity. For instance, (*R*)-4,6″-biscurcudiol and myrmekiodermaral had IC_{50} values of 31.0 and 23.50 μM in MOLT-3, respectively, while the IC_{50}'s of **20** was 28.47 μM in HeLa cell lines.

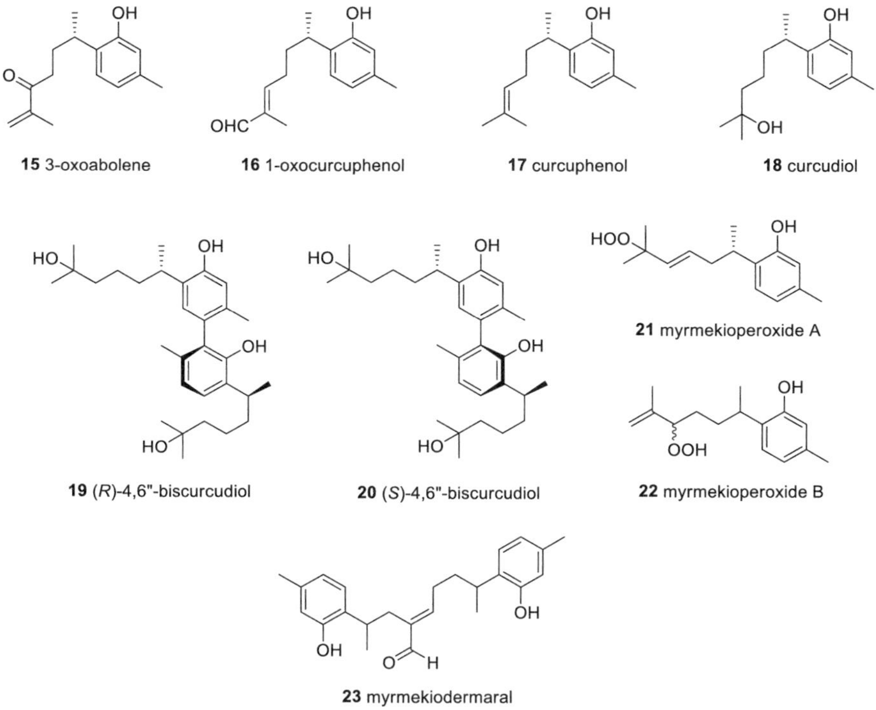

Fig. 3 Chemical structures of aromatic sesquiterpenes

Nitrogenous Sesquiterpenes

Nitrogenous sesquiterpenoids from natural sources are relatively rare and they possess diverse biological activities [72]. At least two Thai sponge species have been investigated for the production of these unique nitrogenous sesquiterpenoids. From an antimicrobial active EtOAc extracts prepare from samples of the sponge, *Axinyssa* sp., two new nitrogenous germacrane molecules, (1*Z*,4*Z*)-7*αH*-11-aminogermacra-1(10),4-diene (**24**) and *N*,*N*-11-bis[(1*Z*,4*Z*)-7*αH*-germacra-1(10),4-dienyl]urea (**25**), were isolated (Fig. 4) [33]. Sponge samples were collected from the Andaman Sea, Trang Province, Thailand. Compound **24** exhibited significant antifungal activity against *C. albicans*, having an inhibition zone of 27 mm as well as against *S. aureus* and *B. subtilis* with inhibition zones of 23 and 22 mm, respectively. However, compound **25** did not exhibit any antimicrobial activity. A series of new formamidobisbolene-based compounds, including axinyssines A (**26**)–L (**37**), were reported from another collection of sponge sample, *Axinyssa* sp., obtained from the Andaman Sea, Thailand (Fig. 4) [34]. Nitrogen-containing bisabolene sesquiterpenes are unique natural products found from marine organisms, and they are characterized by the presence of isothiocyanate, formamide, isonitrile, thiocyanate or amine groups found either at C-3 or C-7 of the bisabolene backbone. The structures of the axinyssines were determined based on extensive NMR, mass spectroscopy, ICD spectral data as well as Mosher reaction and chemical conversion for the assignment of absolute configurations. Axinyssines C, F–H and L showed growth inhibitory activity (inhibitory rate of 95% in a dose of 125 mg/mL) against the gram-positive bacteria, including *Bacillus subtilis* and *Staphylococcus haemolyticus*, and gram-negative bacteria, such as *Pseudomonas lachrymans*, *Agrobacterium tumefaciens* and *Xanthomonas vesicatoria*.

Initial screening on the methanolic extract of *Halichondria* sp. (CRI328), collected off Phi Phi Island in the Andaman Sea, showed moderate cytotoxic activity against a panel of cancer cell lines. Subsequent chemical workup on the EtOAc-soluble fraction this extract resulted in the isolation of five nitrogenous sesquiterpenes (–)-axisonitrile-3 (**38**), (+)-axamide-3 (**39**), axamide-2 (**40**), (3*S**,5*R**,6*R**,9*R**)-3-formamido-1(10)-cadinene (**41**) and (–)-halichamine (**42**), of which **39** and **42** were new while **38** was isolated from natural source for the first time, as well as two steroids, ergosterol and ergosterol peroxide (Fig. 4) [35]. The enantiomers of (–)-axisonitrile-3 and (+)-axamide-3 had been isolated from sponges previously [73]. When tested for cytotoxicity, only (–)-axisonitrile-3 (**1**) exhibited strong activity against the HepG2 cell line with an IC_{50} value of 1.3 μM.

Additional nine new nitrogenous sesquiterpenes, including three isothiocyanates, halichons A (**43**), B (**44**) and H (**51**), two isocyanides, halichon C (**45**) and 4-*epi*-halichon C (**46**), two thiocyanates, halichons D (**47**) and E (**48**) and two formamides, halichons F (**49**) and G (**50**), were isolated together with 11 known compounds, from the Thai marine sponge, *Halichondria* sp., collected from Phi Phi Island in the Andaman Sea, Krabi Province (Fig. 4) [36]. Sixteen of the isolated compounds were assessed for cytotoxicity against a panel of cancer cell lines and only halichon C, 4-*epi*-halichon C, and known molecule, (–)-10-isocyano-4-cadinene, showed moderate cytotoxicity against MOLT-3 cell lines with IC_{50} values of 20.9, 29.0, and 9.1 μM, respectively. In addition, halichon C exhibited moderate cytotoxicity against HepG2 and MDA-MB-231 cell lines with IC_{50} values of 24.3 and 19.3 μM, respectively.

Merosesquiterpenes

The term "meroterpenes" is used to define terpenoid compounds that contain a terpene moiety coupled to a nonterpene unit [74]. Marine derived meroterpenes have been reported mainly from diverse organisms, including brown algae, sponges, tunicates and microorganisms. Sponge-derived merosesquiterpenes are a subgroup of meroterpenes, many of which contain either a drimane or a 4,9-friedodrimane terpenoid skeleton connected to a polyketide-derived quinone or hydroquinone. Merosesquiterpenes have been reported from at least two Thai sponges, *Smenospongia* sp. and *Verongula* cf. *rigida*.

Two new merosesquiterpenes, 6′-iodoaureol (**52**), an iodo-sesquiterpene hydroquinone and 6′-aureoxyaureol (**53**), a bissesquiterpene hydroquinone, were isolated with four new brominated indole alkaloids (see Sect. 3.4) from the Thai sponge *Smenospongia* sp. at Phi Phi Island in the Andaman Sea, Krabi Province (Fig. 5)

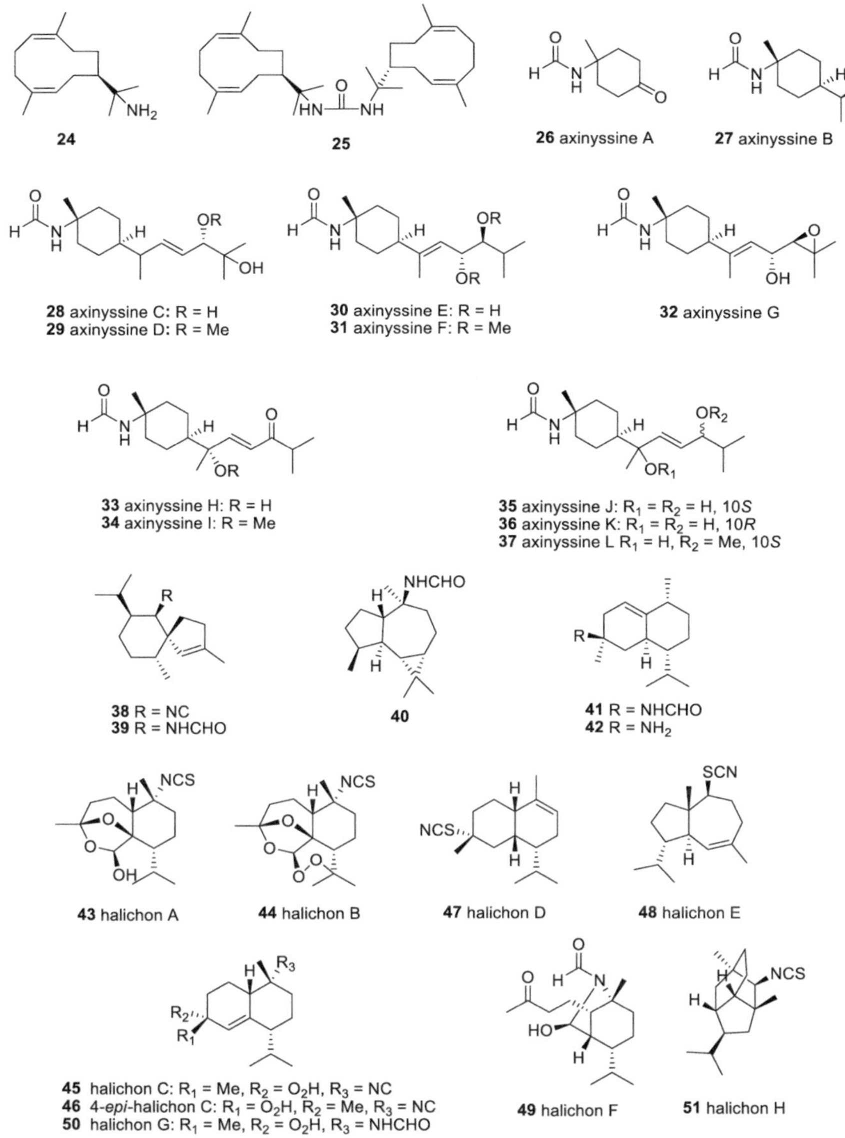

Fig. 4 Chemical structures of nitrogenous sesquiterpenes

[37]. In addition, fourteen known compounds, including four sesquiterpene hydroquinones and ten brominated indole alkaloids, were found in the sponge sample. 6′-Iodoaureol was the first reported iodoesesquiterpene hydroquinone. However, only the known compound, aureole, exhibited moderate cytotoxicity against LH-60 cell

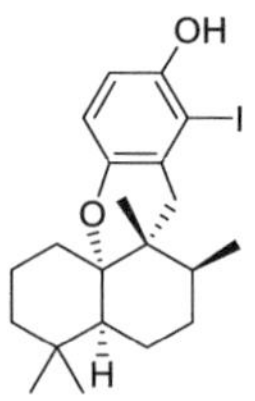

52 6'-iodoaureol

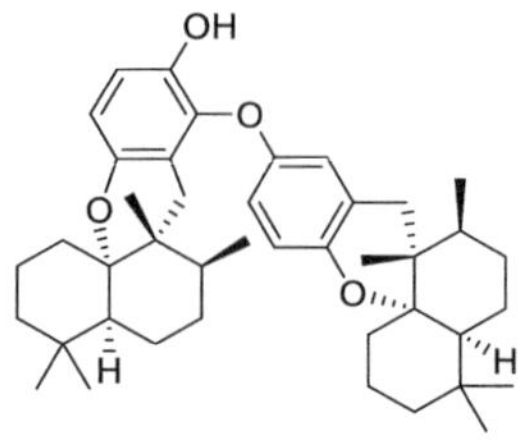

53 6'-aureoxyaureol

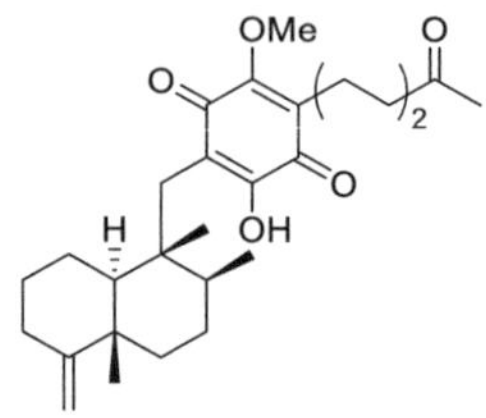

54 quintaquinone

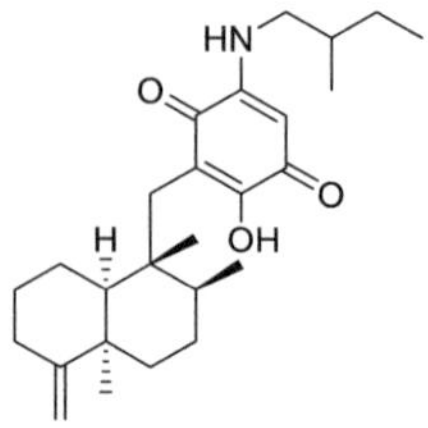

55 5-*epi*-nakijiquinone L

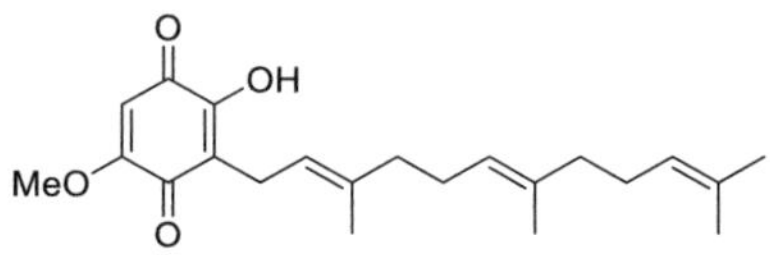

56 3-farnesyl-2-hydroxy-5-methoxyquinone

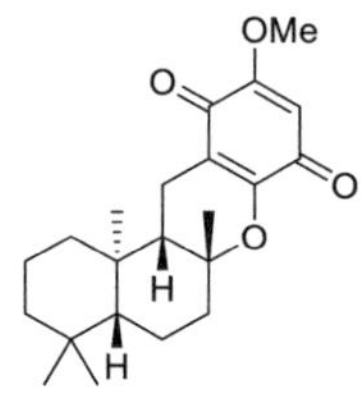

57 cyclospongiaquinone 1

Fig. 5 Chemical structures of merosesquiterpenes

with the IC_{50} value 14.6 μM and weak activity in A549 cell with the IC_{50} value 76.4 μM.

From a collection of the sponge *Verongula* cf. *rigida*, obtained from nearby Ko Ha Islets, Krabi Province, thirteen merosesquiterpenes, including three new compounds, quintaquinone (**54**), 5-*epi*-nakijiquinone L (**55**), and 3-farnesyl-2-hydroxy-5-methoxyquinone (**56**), were reported (Fig. 5) [38]. Only the known merosesquiterpene, cyclospongiaquinone 1 (**57**) (Fig. 5), exhibited strong steroid 5*α*-reductase inhibitory activity in the same range as the standard, finasteride.

3.1.2 Diterpenes

At least three new and five known diterpenes were isolated from two different Thai sponge species, *Ciocalapata* sp. and *Stylissa* cf. *massa.* Several of these molecules exhibited antimalarial properties. From a collection of the sponge, *Ciocalapata* sp., obtained from the vicinity of Koh-Tao, Surat-Thani Province, a new isonitrile diterpene of the amphilectane family, namely 8-isocyanoamphilecta-11(20),15-diene (**61**), together with three known isonitriles, 8,15-diisocyano-11(20)-amphilectene (**58**), 7-isocyanoamphilecta-11(20),15-diene (**59**) and 8-isocyanoamphilecta-11(20),14-diene (**60**), and two known steroidal peroxides, including ergosterol peroxide (**62**) and 5α,9α-epidioxy-8α,14α-epoxy-(22*E*)-ergosta-6,22-dien-3β-ol (**63**), were obtained (Fig. 6) [39]. All four isonitriles were found to exhibit potent activity against *Plasmodium falciparum* K1 with IC_{50}

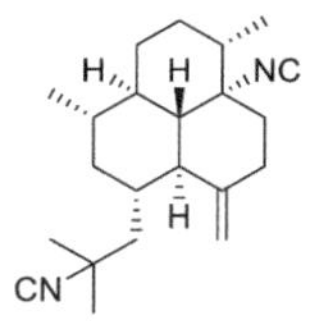

58 8,15-diisocyano-11(20)-amphilectene

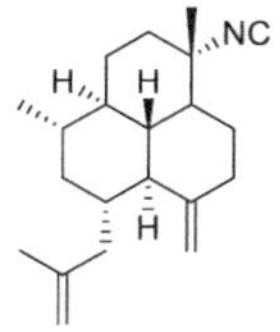

59 7-isocyanoamphilecta-11(20),15-diene

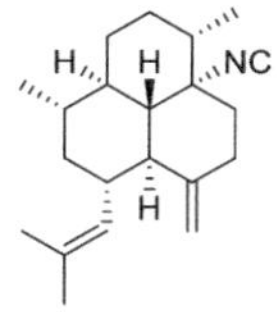

60 8-isocyanoamphilecta-11(20),14-diene

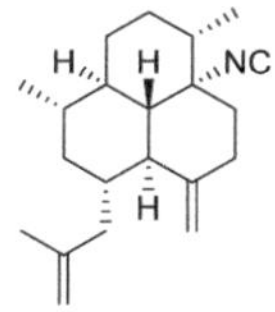

61 8-isocyanoamphilecta-11(20),15-diene

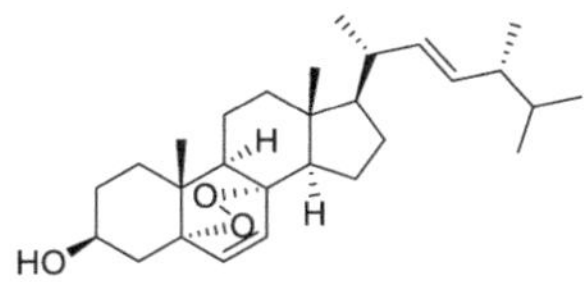

62 ergosterol peroxide

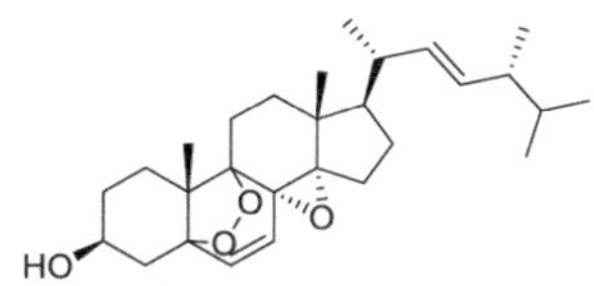

63 5α,9α-epidioxy-8α,14α-epoxy-(22*E*)-ergosta-6,22-dien-3β-ol

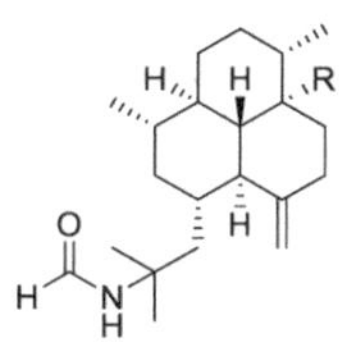

64 8-isocyanato-15-formamidoamphilect-11(20)-ene: R = NCO
65 8-isothiocyanato-15-formamidoamphilect-11(20)-ene: R = NCS
66 8-isocyano-15-formamidoamphilect-11(20)-ene: R = NC

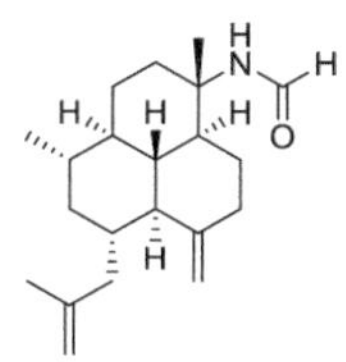

67 7-formamidoamphilect-11(20),15-diene

Fig. 6 Chemical structures of diterpenes

values ranging from 0.09–1.07 μmol/L. However, the steroidal peroxides **62** and **63** showed reduced antimalarial activity but exhibited potent cytotoxicity against MCF-7 cancer cell line at 0.025 and 0.003 μmol/L, respectively. 8,15-Diisocyano-11(20)-amphilectene (**58**) showed moderate cytotoxicity against MCF-7 and fibroblast cell lines.

Investigation on the chemistry of the sponge *Stylissa* cf. *massa* led to the discovery of two new amphilectane-type diterpenes, 8-isocyanato-15-formamidoamphilect-11(20)-ene (**64**) and 8-isothiocyanato-15-formamidoamphilect-11(20)-ene (**65**), along with two known derivatives, 8-isocyano-15-formamidoamphilect-11(20)-ene (**66**) and 7-formamidoamphilect-11(20),15-diene (**67**) (Fig. 6) [40]. This sponge sample was collected by SCUBA from the vicinity of Koh-Tao, Surat-Thani Province, Thailand. In the antimalarial assay based on *Plasmodium falciparum* K1, 8-isocyano-15-formamidoamphilect-11(20)-ene (**66**) was found to be 10 times more active than 8-isocyanato-15-formamidoamphilect-11(20)-ene (**64**) and 8-isothiocyanato-15-formamidoamphilect-11(20)-ene (**65**). It is proposed that the binding of the isonitrile unit, present in the diterpenes, to Fe(III) in heme resulted in their antimalarial activity. The resulting complex formed between isonitrile terpenoids and Fe(III) could prevent heme detoxification in the *Plasmodium* parasites. This is supported by previous studies by Wright and co-workers on the complex formation between isonitrile terpenoids and Fe(III) which inhibited heme detoxification in *Plasmodium* parasites based on the modeling studies using diterpene isonitriles [75].

3.1.3 Sesterterpenes

More than 15 molecules belonging to the sesterterpene class of terpenoids have been described from at least five different sponge species from Thailand. Early studies on a collection of *Mycale* sp., collected from Kang Ta Sin of Sichang Island, Thailand, yielded two new bioactive norsesterterpene 1,2-dioxanes, mycaperoxides A (**68**) and B (**69**) (Fig. 7) [41]. Mycaperoxides are natural bioactive endoperoxides and they have been reported from *Mycale* species in different geographical regions. The discovery of mycaperoxides A and B was conducted through collaborative efforts of three research groups, mainly by Drs Higa (Okinawa), Suwanborirux (Thailand) and Jefford (Switzerland). Their complete structures were determined by spectroscopic (e.g., X-ray crystallography) and chemical methods, such as the modified Mosher's method. Mycaperoxides A (**68**) and B (**69**) exhibited significant cytotoxicity against three cancer cell lines of P-388, A-549 and HT-29, with IC_{50} values ranging from 0.5–1.0 μg/mL). In addition, mycaperoxides A and B showed antiviral activity (IC_{50} 0.25–1.0 μg/mL) against vesicular stomatitis virus and herpes simplex virus type-1 as well as inhibited the growth of the gram-positive bacteria *Bacillus subtilis* and *Staphylococcus aureus*. Further chemical investigation of the *Mycale* sp. collected at Sichang Island in mid-2000, led to the isolation of another analogue, mycaperoxide H (**70**) (Fig. 7) [42]. As mycaperoxide B was also obtained in relatively higher quantity, it was used to confirm the absolute stereochemistry of mycaperoxide H via conversion of **69** to the methyl ester derivative of mycaperoxide H. Mycaperoxide H was found to be cytotoxic against HeLa cancer cells with an IC_{50} value of 0.8 μg/mL.

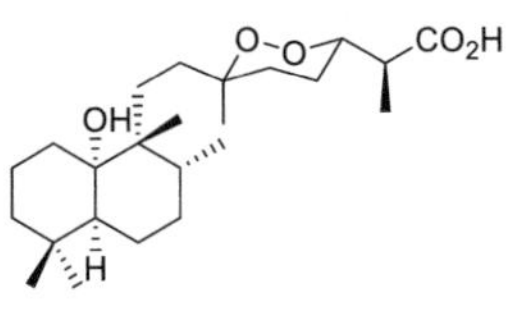

68 mycaperoxide A

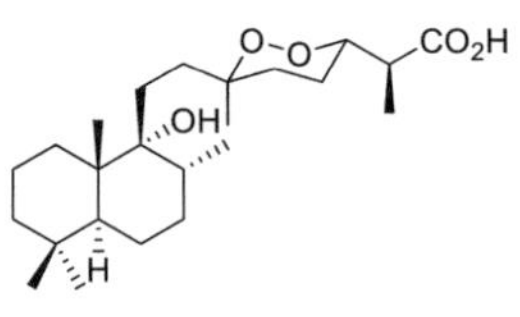

69 mycaperoxide B

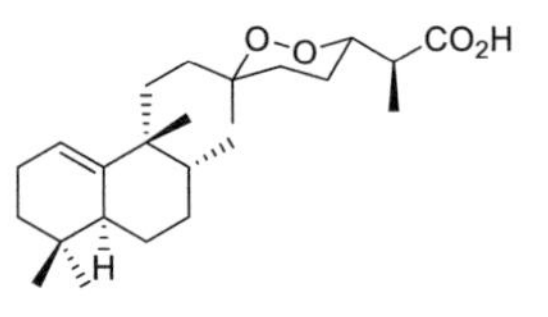

70 mycaperoxide H

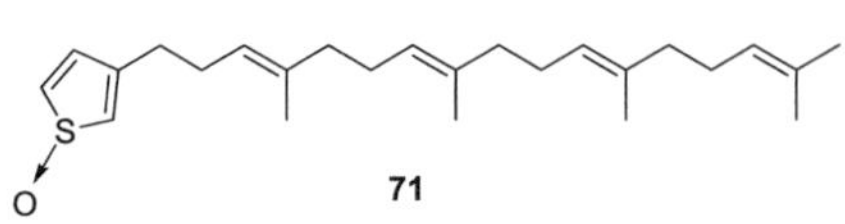

71

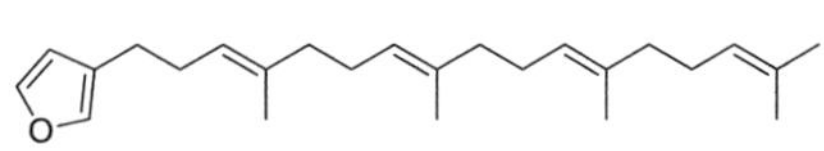

72 furospinosulin-1

Fig. 7 Chemical structures of sesterterpenes

The first discovery of an acyclic sesterterpene consisting of a thiophene-*S*-oxide unit (**71**) was purified from a collection of *Xestospongia* sp. at Sikao Sea, Trang Province (Fig. 7) [43]. This discovery is unusual as *Xestospongia*-derived molecules are usually alkaloids, polycyclic quinones, polyacetylenic derivatives, amino alcohols and sterols. The presence of a thiophene-*S*-oxide moiety contrasted with the furan ring found in the well-known furospinosulin-1 (**72**) (Fig. 7). While furospinosulin-1 arrests cancer cells by hypoxia-selective growth inhibition [76], **71** showed weak cytotoxicity against Vero cells with an IC_{50} value of 31 μM.

Chemical investigation on a sponge sample, *Brachiaster* sp., provided a new scalarane-type sesterterpene, 12-deacetoxyscalarin 19-acetate (**74**), as well as two new derivatives of manoalide-type sesterterpenes, (*E*)- and (*Z*)-neomanoalide 24,25-diacetates (**75** and **76**) (Fig. 7) [44]. In addition, five known sesterterpenes, including heteronemin (**73**), heteronemin acetate (**77**), 12-*epi*-19-deoxyscalarin (**78**), 12-deacetyl-12-*epi*-19-deoxyscalarin (**79**) and manoalide 25-acetate (**80**), were also reisolated from the sponge samples (Fig. 8). The sponge samples, appearing lumpy and irregular shaped with a tannish-gray color externally underwater, were collected from Koh-Tao, Surat-Thani, over two separate expeditions made in April 2001 and April 2002. Within the scalarane-type metabolites, compounds **73**, **74** and **77** exhibited strong antitubercular activity with the MICs in a range of 10°–10^{2} μM. This is consistent with the observations made by Crews and Bescansa where the oxygenation pattern at C-19 of the scalarane-type sesterterpenes could assert strong influence on their biological activities [77]. However, in the cytotoxicity assay, only compounds **73** and **77** showed potent cytotoxicity. Among the manoalide series of compounds, only **80** showed significant antitubercular and cytotoxic activities.

Marine sponges of the genus *Hyrtios* have been shown to be prolific producers of diverse bioactive natural products, including alkaloids, sesterterpenes and sesquiterpenes [78]. At least two *Hyrtios* species from Thailand, including *H. gumminae* and *H. erectus*, were investigated for their secondary metabolites. Additional natural scalarane-type metabolites, including four new sesterterpenoids similan A (**81**), 12*β*,20-dihydroxy-16-acetoxy-17-scalaren-19,20-olide (**82**), 12*β* -acetoxy-20-hydroxy-17-scalaren-19,20-olide (**83**) and 12*β*,16*α*,20-trihydroxy-17-scalaren-19,20-olide (**84**), were reported from the Thai sponge, *Hyrtios gumminae*, obtained from Similan Island in the Andaman Sea (Fig. 8) [45]. Seven known molecules, such as hyrtiosal (**85**), scalarafuran (**86**), scalarolide, 16-acetoxyscalarolide or sesterstatin 7, 12-*epi*-*O*-deacetyl-19-deoxyscalarin (**87**), hyrtiolide (**88**) and cholesterol, were also isolated (Fig. 8). Due to the inseparable mixture of 12*β*,20*α*-dihydroxy-16*β*-acetoxy-17-scalaren-19,20-olide and 12*β*,20*β*-dihydroxy-16*β*-acetoxy-17-scalaren-19,20-olide, their structures were determined by conversion, using Ac_2O and pyridine, into their diacetate derivatives **82a** and **82b**, which were then easily separated by preparative TLC. A similar method was used to elucidate the structures of **83** and **84** by conversion into their separable diacetate derivatives **83a** and **83b** as well as **84a** and **84b**, respectively (Fig. 8). The biosynthesis of the unique similan A (**81**) was also proposed by the authors to derive from **86** via **85**. When tested in a panel

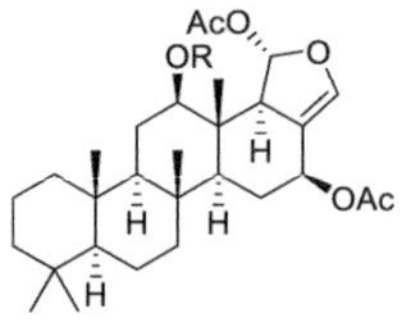

73 heteronemin: R = H
77 heteronemin acetate: R = Ac

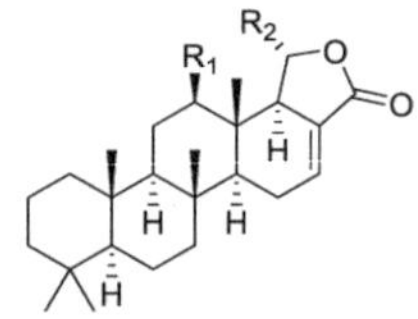

74 12-deacetoxyscalarin 19-acetate: R_1 = H, R_2 = OAc
78 12-*epi*-19-deoxyscalarin: R_1 = OAc, R_2 = H
79 12-deacetyl-12-*epi*-19-deoxyscalarin: R_1 = OH, R_2 = H

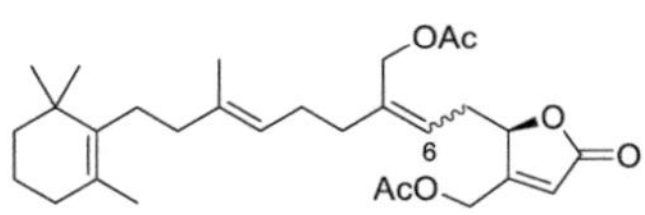

75 (*E*)-neomanoalide 24,25-diacetates: 6*E*
76 (*Z*)-neomanoalide 24,25-diacetates: 6*Z*

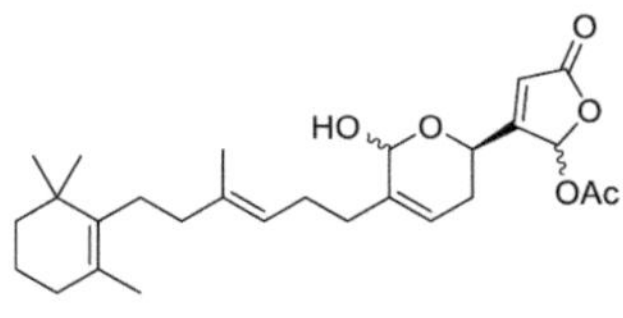

80 manoalide 25-acetate

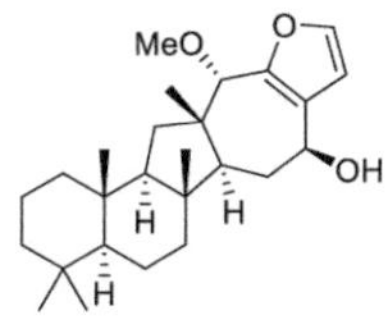

81 similan A

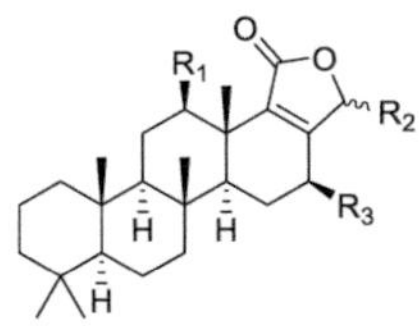

82 12β,20-dihydroxy-16β-acetoxy-17-scalaren-19,20-olide: R_1 = OH, R_2 = OH, R_3 = OAc
82a R_1 = OH, R_2 = α-OAc, R_3 = OAc
82b R_1 = OH, R_2 = β-OAc, R_3 = OAc
83 12β-acetoxy-20-hydroxy-17-scalaren-19,20-olide: R_1 = OAc, R_2 = OH, R_3 = H
83a R_1 = OAc, R_2 = α-OAc, R_3 = H
83b R_1 = OAc, R_2 = β-OAc, R_3 = H

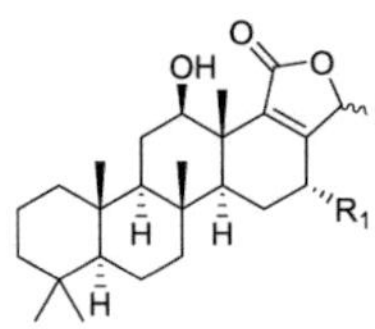

84 12β,16α,20-trihydroxy-17-scalaren-19,20-olide: R_1 = OH, R_2 = OH
84a R_1 = OAc, R_2 = α-OAc
84b R_1 = OAc, R_2 = β-OAc

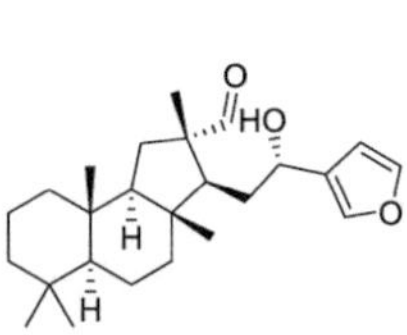

85 hyrtiosal

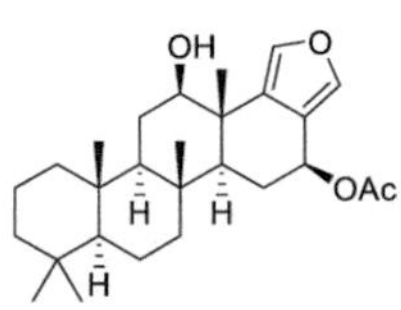

86 scalarafuran

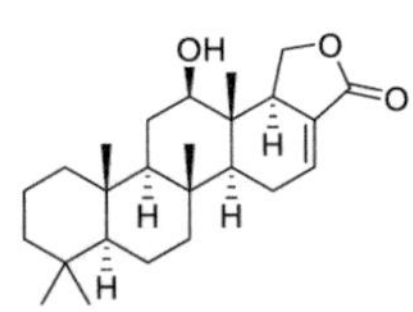

87 12-epi-*O*-deacetyl-19-deoxyscalarin

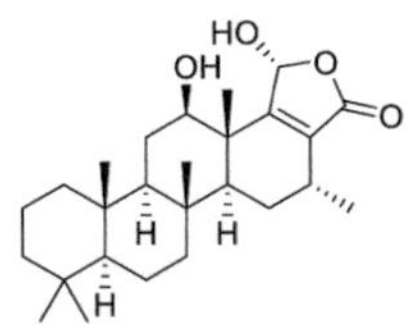

88 hyrtiolide

Fig. 8 Chemical structures of sesterterpenes and derivatives

of six cancer cell lines, only the known compounds hyrtiosal and 12-*epi*-*O*-deacetyl-19-deoxyscalarin showed moderate cytotoxic activity with IC_{50} values ranging from 5.2 to 57 μM [45].

In addition to *H. gumminae*, the natural products of another species, *H. erectus*, were extensively investigated by Dr. Prawat research group at the Chulabhorn

Research Institute. From a collection of *H. erectus* from Similan Island, Phang-Nga Province, seven new molecules, including the sesterterpene, erectusolide A (**89**) and six phenolic alkenes, erectuseneols A (**90**)–F (**95**), along with nine known compounds were reported from the EtOAc-derived extracts (Fig. 9) [46]. These known compounds included luffalactone, luffariolide E, (6*E*)- and (6*Z*)-neomanoalide 24,25-diacetates, 6,6-dimethylundecane-2,5,10-trione, *threo*- and *erythro*-cavernosines, (4*E*,6*E*)-dehydromanoalide and echinoclerodane A. The 1D NMR spectra of erectusolide A (**89**) were similar to the known luffariolide E and was described as a mixture of 3*S**,4*R**- and 3*R**,4*R**-compound 1 in 1:0.3 ratio. Cytotoxicity of several molecules were evaluated against a panel of six cancer cell lines, including MOLT-3 (acute lymphoblastic leukemia), HepG2 (hepatocarcinoma), HeLa (human cervical carcinoma), HuCCA-1 (human cholangiocarcinoma), A549 (human lung cancer), and MDA-MB-231 (hormone independent breast cancer). However, only the known terpenoids, (4*E*,6*E*)-dehydromanoalide and echinoclerodane A, exhibited good cytotoxic activity against MOLT-3 cell line, with IC_{50} values of 3.79 and 5.82 μM, respectively. The isolation of the six rare new phenolic alkenes has been suggested to have chemotaxonomic relevance for this species.

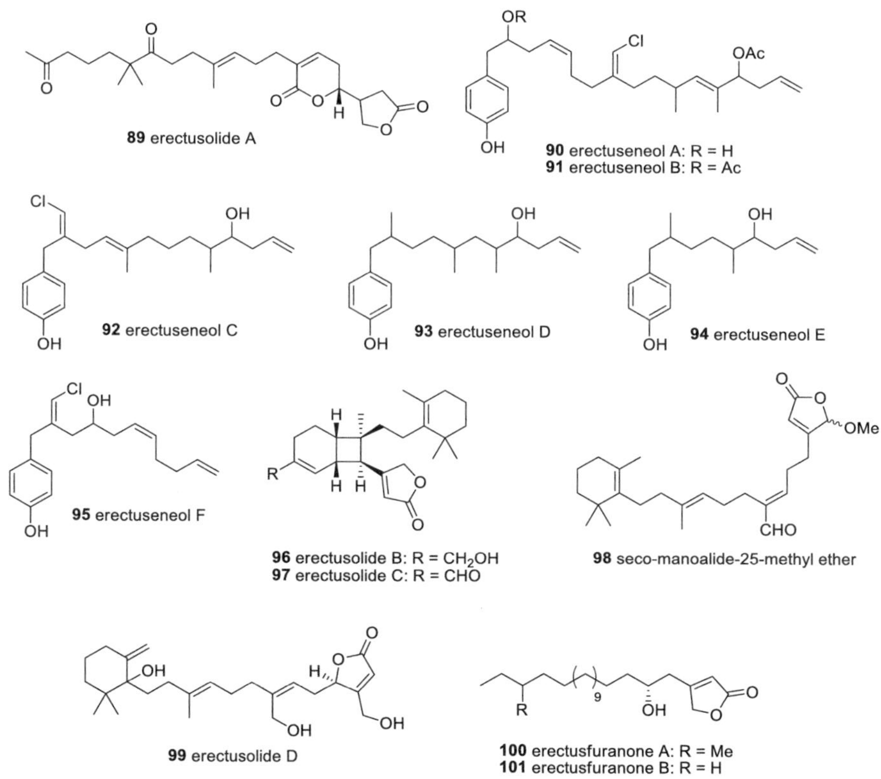

Fig. 9 Chemical structures of natural products from Thai sponge, *H. erectus*

Further investigation of the EtOAc soluble extract led to isolation of four new sesterterpenes, erectusolides B (**96**), C (**97**), D (**99**) and seco-manoalide-25-methyl ether (**98**), as well as two new 2-furanone derivatives, erectusfuranones A (**100**) and B (**101**) (Fig. 9) [47]. These new compounds were accompanied by seventeen known compounds, including thirteen known sesterterpenes [namely, (6*Z*)-neomanoalide-24-acetate, two diastereomers of 24-*O*-methylmanoalide, luffariolide B, manoalide, (6*E*)- and (6*Z*)-neomanoalide, seco-manoalide, scalarafuran, 12-acetylscalarolide, 12-*epi*-*O*-deacetyl-19-deoxyscalarin, 12-*epi*-scalarin and 12-*O*-deacetyl-12-*epi*-scalarin], three indole alkaloids (e.g. 5-hydroxy-1H-indole-3-carbaldehyde, hyrtiosine A and variabine B) and one norterpene (cavernosine). When evaluated for cytototoxicity, only the two diastereomers of 24-*O*-methylmanoalide, seco-manoalide and 12-*O*-deacetyl-12-*epi*-scalarin showed significant cytotoxic activity against MOLT-3 cell line with IC_{50} values of 1.77, 1.30, 6.68, and 7.64 μM, respectively. Moreover, the two diastereomers of 24-*O*-methylmanoalide exhibited cytotoxic activity against HeLa, HuCCA-1 and A549 cell lines with IC_{50} values of 1.51–16.77 μM [47].

Chemistry and Biology of Natural and Semi-synthetic Heteronemin Derivatives

Investigations on the antitubercular and cytotoxic properties of fourteen heteronemin-related sesterterpenes were carried out by Jaisamut and co-worker in 2009 [79]. These heteronemin-related compounds included six naturally occurring sesterterpenes, e.g., heteronemin (**73**), heteronemin acetate (**77**), 12-deacetyl-12-*epi*-19-deoxyscalarin (**79**), 12-*epi*-19-deoxyscalarin (**78**), 12-deacetoxyscalarin acetate (**74**) and 12-epideacetyl-19*α*-acetoxy-20*α*-methoxyscalaran (**102**) (Fig. 10). The latter molecule, 12-epideacetyl-19*α*-acetoxy-20*α*-methoxyscalaran (**102**), was isolated from the Thai sponge *Hyrtios* sp. collected from Koh-Tao, Surat-Thani Province, as a minor scalarane derivative [79]. In addition, the isolation of heteronemin (**73**) as a major component with very good yield of more than 1.5 g, obtained from 101 g of freeze-dried sponge sample, provided an opportunity to use heteronemin as starting material to synthesize new synthetic derivatives. These synthetic heteronemin derivatives were: 12-oxoheteronemin (**103**), 12-deacetyl-12-*epi*-20-deoxoscalarin (**104**), scalarafuran (**105**), 16-deacetoxy-15,16-dehydroscalarafuran (**106**), 12-deacetyl-12-episcalaradial (**107**), 12-deacetyl-12,18-diepiscalaradial (**108**), 12-deacetyl-12-oxoscalaradial (**109**) and 12-episcalaradial (**110**) (Fig. 10) (refer to Scheme 1 in reference 57 for summary of synthesis).

Using a nonvirulent strain of *M. tuberculosis* H37Ra, compound **103** exhibited the most potent antitubercular activity, with MIC value of 0.23 μM. Likewise, in the cytotoxic assay, **103** was the most potent when tested against human oral fibroblast cell lines with IC_{50} value of 0.91 μM. Further analysis of the biological data led to the conclusion that two main regions in **103** were implicated in the antitubercular activity, including the 19*α*-acetoxy group and 16*β*-OAc group. Subsequent cytotoxic evaluation of heteronemin and 12-oxoheteronemin against six cancer cell lines

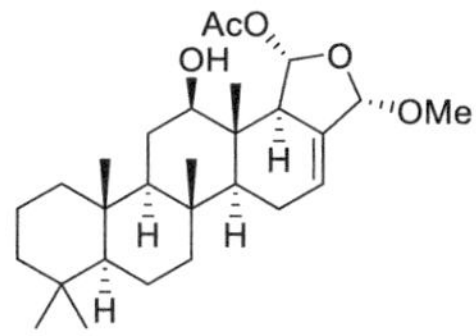

102 12-epideacetyl-19α-acetoxy-20α-methoxyscalaran

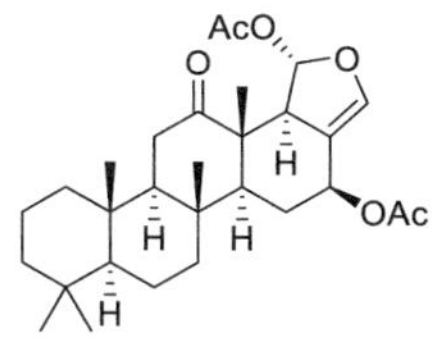

103 12-oxoheteronemin

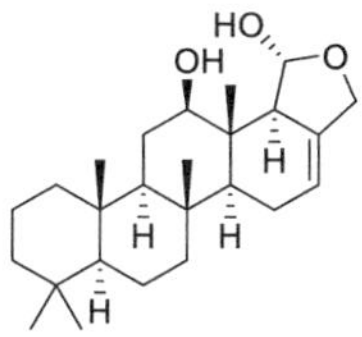

104

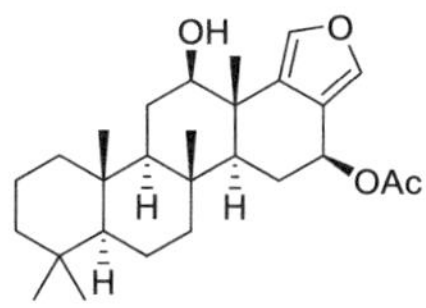

105

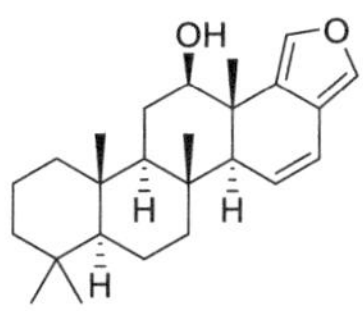

106

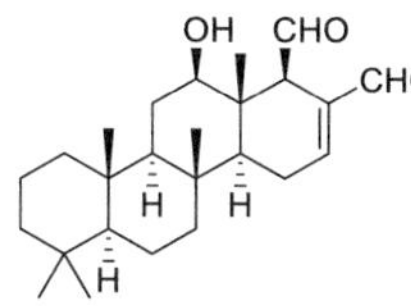

107

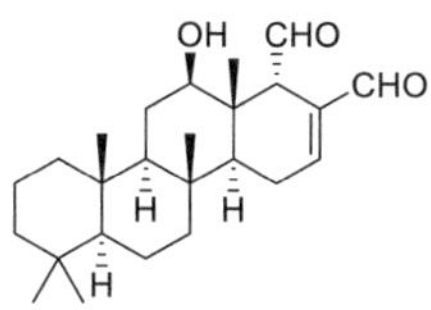

108

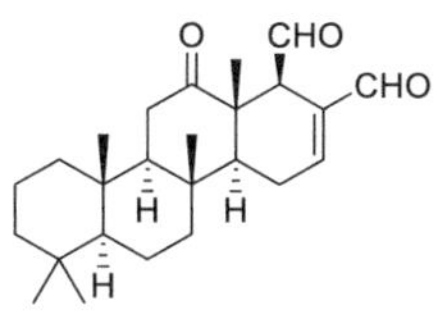

109

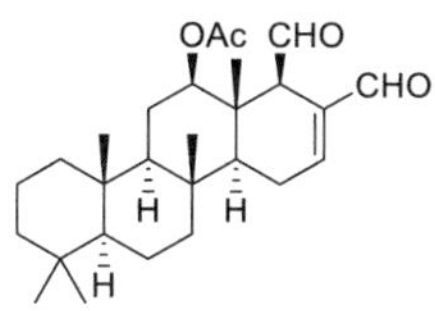

110

Fig. 10 Chemical structures of heteronemin derivatives

revealed their IC_{50} values ranging from 0.66 to 1.35 μM [80]. When tested against two estrogenic receptor-positive cell lines, namely MCF-7 and T47D, heteronemin was found to be relatively more potent. Furthermore, these two molecules showed no overt effects on cell cycle distribution in HeLa cells, but rapidly initiated apoptosis as revealed by increased sub-G1 populations of cells and caspase-dependent PARP cleavage. Due to the unique antiproliferative activity of heteronemin, this molecule has been receiving extensive attention on its development as an anticancer agent [81].

3.1.4 Triterpenes

A sample of the sponge, *Haliclona* sp., collected in the South Andaman Sea in the vicinity of Phuket Island, yielded an unusual pentacyclic sulfated hydroquinone, phuklona sulfate (**111**) (Fig. 11) [48]. The sponge samples were collected under contract for the National Cancer Institute with funding support from federal funds from the National Cancer Institute, National Institutes of Health. The structure of this unusual molecule was deduced solely on 1D and 2D NMR spectroscopy. Phuklona sulfate exhibited modest cytoprotection against HIV in the NCI assay with an EC_{50} of 30 μg/mL and an IC_{50} of 117 μg/mL.

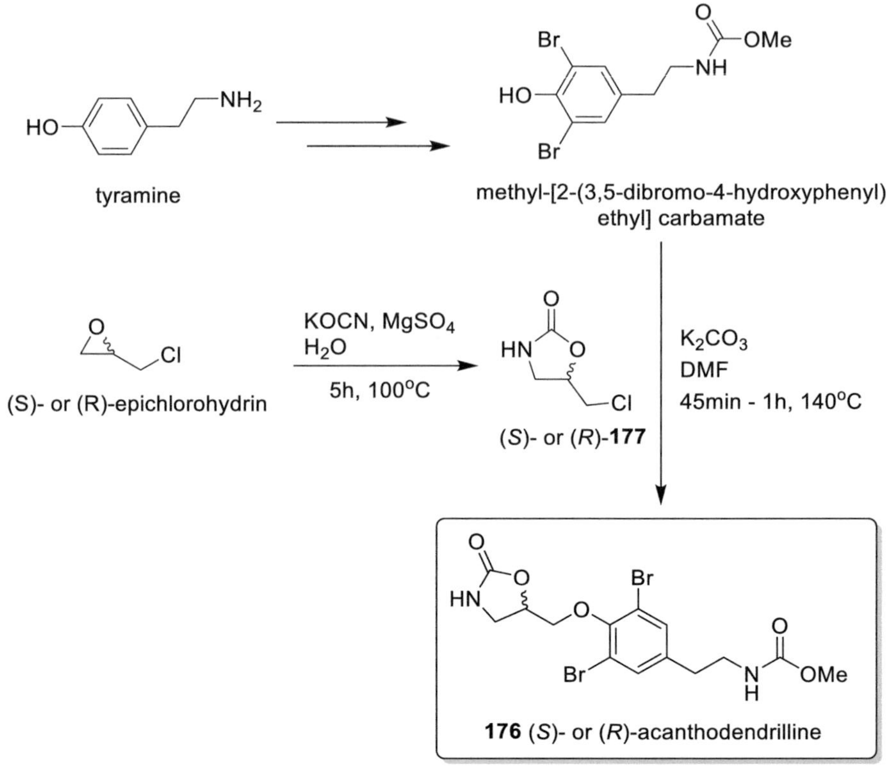

Scheme 1 Concise synthesis of **176**

Fig. 11 Chemical structure of phuklona sulfate

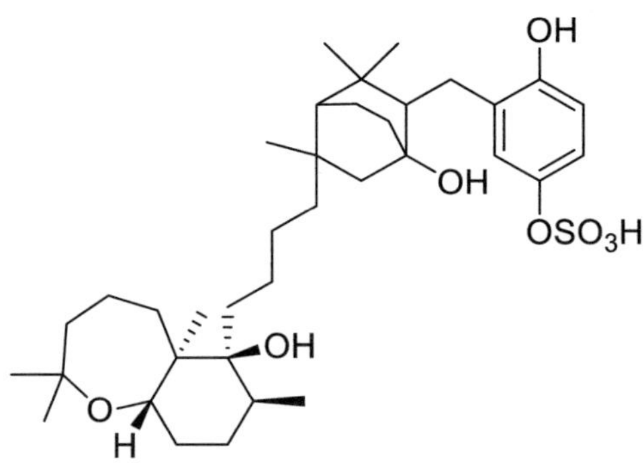

111 phuklona sulfate

3.2 Sterols

New steroids have been reported mainly from two sponge species, *Corticium* sp. and *Petrosia* (*Strongylophora*) sp. from two research groups in Thailand. A new stigmastane-type steroidal alkaloid, 4-acetoxy-plakinamine B (**112**), was purified from the Thai sponge *Corticium* sp. collected from Koh-Tao, Surat-Thani Province (Fig. 12) [49]. This sponge was reported to appear small and flat, with colony size ranging from 3 to 15 cm wide and 0.2 to 0.4 cm thick. Sponge texture was leathery, with a dark brownish grey color on the outside. Compound **112** was evaluated for acetylcholinesterase inhibitory activity and found to exhibit high inhibitory property with IC_{50} of 3.75 μM. This result is important as compounds with acetylcholinesterase inhibitory activity are potential drugs for the treatment of Alzheimer's disease dementia phase. Study on the kinetics of enzyme inhibition suggested an unusual mixed-competitive mode of inhibition. When it was reported in 2007, 4-acetoxy-plakinamine B is the first steroidal alkaloid having a stigmastane skeleton to exhibit good potency in the acetylcholinesterase inhibition bioassay.

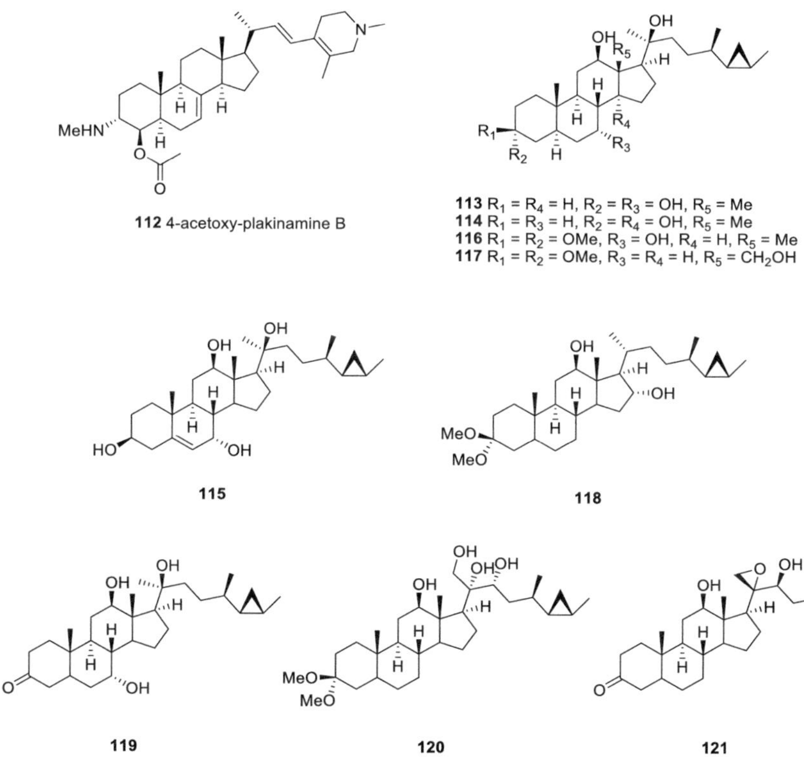

Fig. 12 Chemical structures of sponge-derived sterols **112–121**

From another collection of sponge species belonging to the genus *Petrosia (Strongylophora)*, a series of eight new steroids **113–120**, along with eight known steroids, were discovered (Fig. 12) [50]. The sponge samples were obtained from the Similan Islands in the Andaman Sea, Phang Nga, Thailand. Several of these steroids were evaluated for their cytotoxicity against a panel of human cancer cell lines, such as MOLT-3 (acute lymphoblastic leukemia), HepG2 (hepatocarcinoma), A549 (human lung cancer), HuCCA-1 (human cholangiocarcinoma), HeLa (cervical carcinoma) and MDA-MB-231 (hormone-independent breast cancer) as well as a normal cell line, MRC-5 (normal human embryonic lung fibroblasts). However, all molecules except for the known steroid **121** (Fig. 12) exhibited weak to moderate cytotoxicity, with the IC_{50} values ranging from 11.2 to 103.5 μM. Compound **121** showed cytotoxicity with IC_{50} values of 7.10 and 6.11 μM against HepG-2 and HeLa cell lines, respectively [50].

3.3 Hydroxypyran-2-Ones

From a collection of *Tetilla japonica*, obtained from Captain Yuth Beach, Chonburi, Thailand, two unusual tetrahydrofurylhydroxypyran-2-ones, namely tetillapyrone (**122**) and nortetillapyrone (**123**), were isolated (Fig. 13) [51]. Their structures were deduced based on NMR spectrometry and an X-ray analysis of tetillapyrone. These two molecules can be considered as α-substituted glutaconic anhydrides, which similar to synthetic β-substituted glutaconic anhydrides, would be in equilibrium with the enol or 6-hydroxypyran-2-one form. Since the report of these metabolites, they were also found in other Thai sponge species as well as other places [82–84].

Tetillapyrone and/or nortetillapyrone were also found in samples of *Haliclona baeri* and *Haliclona cymaeformis* collected from the Gulf of Thailand [52]. Using a panel of seven yeast, such as *Candida albicans, C. glabata, C. dubliniensis, C. tropicalis, C. krusei, C. parapsilosis* and *Cryptococcus neoformans*, as well as eight filamentous fungi, including three non- dermatophyte filamentous fungi (*Aspergillus fumigatus, A. flavus, A. niger*) and five dermatophyte filamentous fungi

122 tetillapyrone **123** nortetillapyrone

Fig. 13 Chemical structures of hydroxypyran-2-ones, tetillapyrone (**122**) and nortetillapyrone (**123**)

(*Microsporum gypseum*, *M. canis*, *Trichophyton mentagrophytes*, *T. rubrum* and *Epidermophyton floccosum*), nortetillapyrone showed antifungal activity, particularly against dermatophytic filamentous fungi [52]. For instance, nortetillapyrone showed good activity against *C. glabrata* (62.5–31.25 μg/mL), *C. dubliniensis* (62.5 μg/mL) and *Cryptococcus neoformans* (31.25 μg/mL) and significant antifungal activity against dermatophytes with the MIC values ranging from 31.25 to 125 μg/mL.

3.4 Alkaloids

3.4.1 Maleimide-5-Oximes

Compounds containing maleimide scaffold have been shown to possess various biological activity, ranging from antibacterial to anticancer [85]. Maleimide-5-oxime alkaloids have been reported from several species of Thai sponges, including *Pseudoceratina purpurea*, *Haliclona baeri* and *Cliona patera* [52–54]. For instance, an investigation on bioactive natural products from the marine sponges, including *Haliclona baeri* and *Haliclona cymaeformis*, obtained from the Gulf of Thailand at the coastal area of Chonburi Province and from Koh Chang, Trad Province, led to a series of molecules [52]. From *Haliclona baeri*, maleimide-5-oxime (**124**) (Fig. 14) and 3,4-dihydroxybenzoic acid were obtained while *Haliclona cymaeformis* provided *p*-hydroxybenzaldehyde, phenylacetic acid and nortetillapyrone (**123**) (Fig. 13). In addition, tetillapyrone (**122**) was purified from both *Haliclona* species. Tetillapyrone (**122**) and nortetillapyrone (**123**) were previously isolated as new compounds from the Thai sponge, *Tetilla japonica* [51]. When tested against three human tumor cell lines, including the estrogen-dependent ER(+) MCF-7, the estrogen-independent ER(–) MDA-MB-231 and NCI-H460, maleimide-5-oxime, tetillapyrone and nortetillapyrone were found to be inactive.

A methyl analog of maleimide-5-oxime, 3-methylmaleimide-5-oxime (**125**) (Fig. 14), was also reported from sponges, *Pseudoceratina purpurea* and *Cliona patera*, collected from Gulf of Thailand near Kho Chang Island, Trad Province and Koh Lan, Chonburi Province, respectively [53, 54]. Maleimide-5-oxime has also

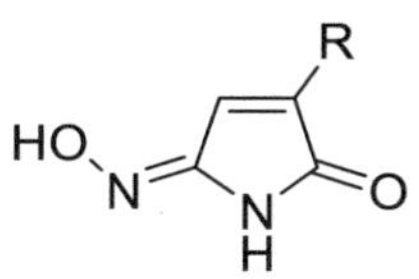

124 maleimide-5-oxime: R = H
125 3-methylmaleimide-5-oxime: R = CH_3

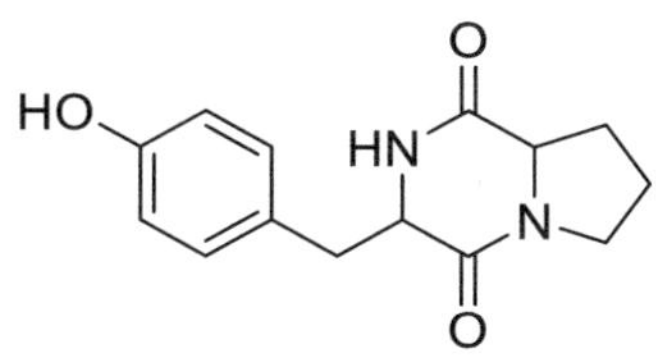

126

Fig. 14 Chemical structures of maleimide-5-oximes and diketopiperazine

been isolated from *Cliona patera* [54]. The fact that the simple maleimide-5-oxime alkaloids were isolated along with other classes of compounds, such as tetillapyrone (**122**), nortetillapyrone (**123**) and diketopiperazine (**126**) (Fig. 14), from various sponge species may suggest their microbial sources.

3.4.2 Halogenated Tyrosine-Containing Alkaloids

Chemical investigations on the sponge, *Suberea* aff. *praetensa*, obtained over three separate collections using trawl net on the seashore of Ban Phae Village at the Gulf of Thailand, Rayong Province, yielded a series of tyrosine-derived alkaloids. The first sponge collection made in 1998 provided known compounds, clionasterol, fistularin-3 (**127**), agelorins A (**128**) and B (**129**), along with two new molecules, 11,17-dideoxyagelorins A (**130**) and B (**131**) (Fig. 15) [55]. A second sponge collection made in late 1999 gave additional known compounds as 5-chloro- (**132**) and 5-bromocavernicolin (**133**) while a third sample collection in early 2001 provided known compounds **134**, **135** and **137–139** as well as the unusual new alkaloid, subereatensin (**136**) (Fig. 15) [56, 57]. The biosynthesis of these alkaloids is derived from tyrosine, but they differ in the degree of halogenation and posterior cyclization. It is proposed that the biosynthesis of subereatensin (**136**) is derived from rearrangement of tyrosine skeleton, via an arene oxide intermediate, and deviates from the formation of halotyrosine catalyzed by haloperoxidase [57]. An acetate derivative of **132**, compound **140**, was revealed to be cytotoxic against MCF-7, SF-268 and UACC-62 cancer cell lines, with GI_{50} values below 10 μM (Fig. 15) [56]. It is unclear on the origin of **140** in the publication by Kijjoa and co-workers [56] but is assumed that the molecule was formed via acetylation reaction of **132**.

Numerous bromotyrosine alkaloids, both known and new, have been identified from the marine sponge, *Pseudoceratina purpurea*, collected from Thailand. From a collection of *Pseudoceratina purpurea*, collected by SCUBA near Kho Chang Island, Trat Province, Gulf of Thailand, two new bromotyrosine alkaloids, purpuroceratic acids A (**141**) and B (**142**) (Fig. 16), along with known compounds, including aplysamine-2 (**143**) (Fig. 16), maleimide 5-oxime, clionasterol and 1-tetradecene, were isolated [53]. When tested in a panel of cancer cell lines, only aplysamine-2 (**143**) exhibited, after incubation for 48 h, moderate cytotoxicity against three cell lines, such as MCF-7 (breast), NCI-H460 (lung) and SF-268 (CNS), with IC_{50} values ranging from 25.7 to 40.9 μM.

From another sponge collection of *Pseudoceratina* cf. *purpurea*, from Koh-Ha Islets, Thailand, fifteen know bromotyrosine alkaloids, such as aeroplysinin (**144**), araplysillins I (**145**) and II (**146**), purealidins B (**147**), J (**148**), Q (**149**) and R (**150**), fistularin 3 (**151**), hemifistularin 3 (**152**), isoanomoian A, (**153**), aplyzanzine A (**154**), aplysamine 2 (**155**), purpureamine J (**156**) and aerophobins 1 (**157**) and 2 (**158**), were purified and their structures confirmed based on spectroscopic analysis as well as comparison with literature (Fig. 16) [86]. These molecules were isolated due to the initial discovery of strong acetylcholinesterase (AChE)-inhibiting activity of the sponge extracts, which showed greater than 90% inhibition at 0.1 mg/mL. All

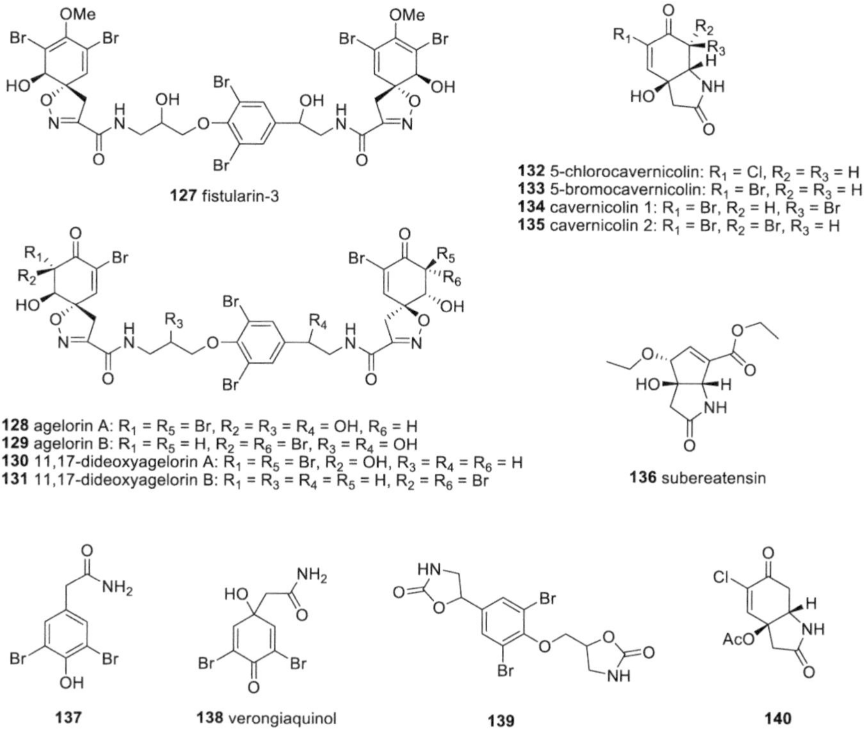

Fig. 15 Chemical structures of alkaloids isolated from the Thai sponge, *Suberea* aff. *praetensa*

the isolated compounds were eventually assessed for their AChE inhibitory properties. Compounds **149** and **155** as well as **153** and **154** showed strong and moderate AChE inhibitory activity, respectively. In addition, compounds **149** and **153–155** were subjected to enzyme inhibiting kinetics study and their V_{max} measurements on AChE indicated that they inhibited the enzyme in a non-competitive way. These assay results revealed interesting SAR data. For instance, the pre-existing quaternary ammonium moiety of **149** and **155** seem to be important for activity as compared to the tertiary amino group of **153** and **154**. Also, the electronic change of the amino group to *N*-oxide in compound **156** significantly reduced the activity. Furthermore, the presence of oxime in **155** or an isoxazole in **149** could result in enhanced activity over the amino group in **153** and **154**. The authors recommended that further studies, especially on enzyme-binding simulation, could facilitate the identification of pharmacophores as well as development of new AChE-inhibitors based on the structures of the bromotyrosines [86]. When tested against MCF-7 and human fibroblast cells, only **157** exhibited cytotoxicity with IC_{50} values of 0.8 and 4.2 μM, respectively.

In a study by Sirimangkalakitti and co-workers, twenty bromotyrosine alkaloids, including a new compound 13-oxosubereamolline D (**163**) (Fig. 17), were isolated from the MeOH crude extract of the marine sponge, *Acanthodendrilla*

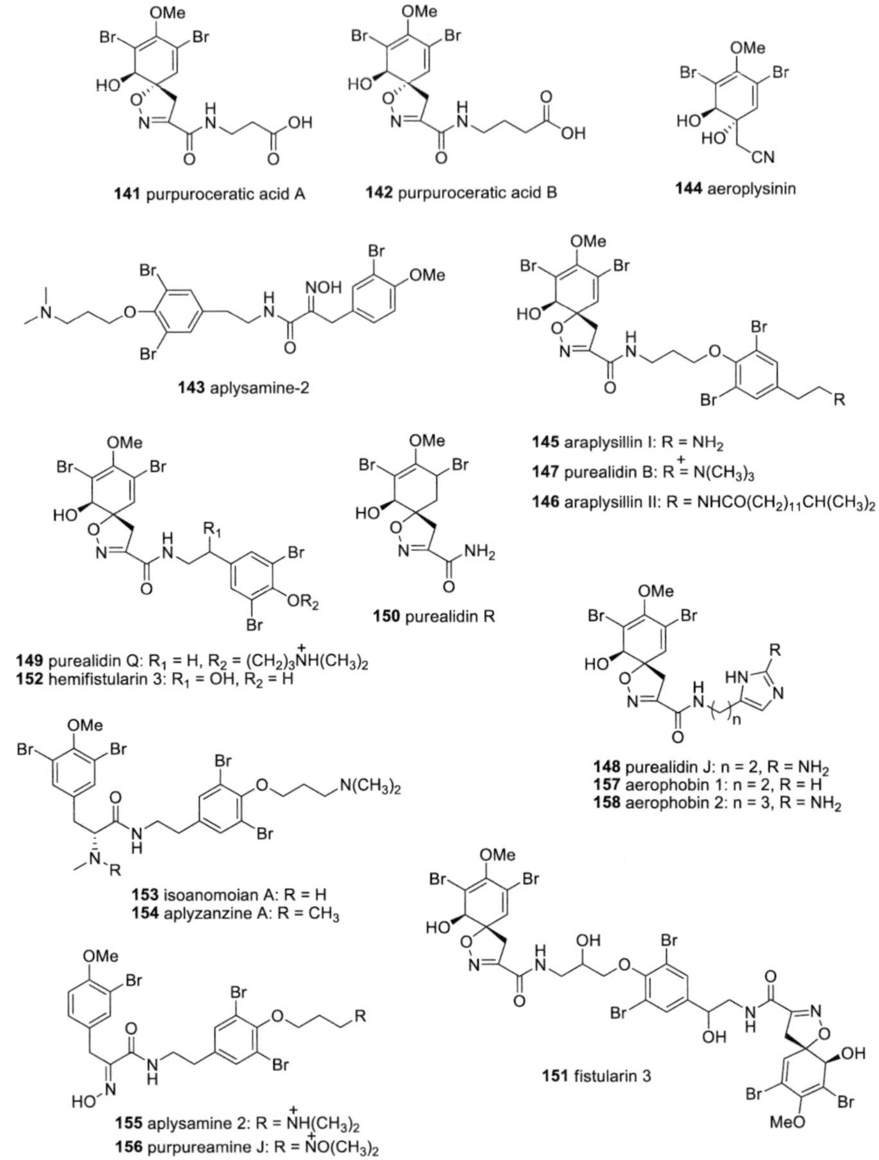

Fig. 16 Chemical structures of bromotyrosine-derived alkaloids isolated from the Thai sponge, *Pseudoceratina purpurea*

sp. collected from Koh-Ha Islets, Krabi, Thailand [58]. As bromotyrosine-derived alkaloids are usually reported from sponges of the order Verongida [87], their study is the first report of such alkaloids from a sponge species belonging to a different order Dendroceratida. Based on spectroscopic analysis, these twenty alkaloids were identified as subereamolline C (**159**), subereamolline D (**160**), **161**, **162**, 13-oxosubereamolline D (**163**), aerothionin (**164**), homoaerothionin (**165**), 11-oxoaerothionin (**166**), oxohomoaerothionin (**167**), fistularin 1 (**168**), 11,19-dideoxyfistularin 3 (**169**), 19-deoxyfistularin 3 (**170**), 3,5-dibromo-1-hydroxy-4,4-dimethoxy-2,5-cyclohexadiene-1-acetamide (**171**), verongiaquinol (**138**), cavernicolin-1 (**134**), cavernicolin-2 (**135**), 7*β*-bromo-5-chlorocavernicolin (**172**), 7*α*-bromo-5-chlorocavernicolin (**173**), 5,7*β*-dichloro-cavernicolin (**174**) and 5,7*α*-dichloro-cavernicolin (**175**) (Fig. 17). The most potent alkaloid against AChE was revealed as homoaerothionin (**165**), having IC_{50}s of 2.9, 4.5 and 6.2 μM against EeAChE, hrAChE and BChE, respectively. In comparison with other alkaloids, it is likely that the 1-hydroxy-2,4-dibromo-3-methoxy-8-carbamoylspirocyclohexadienylisoxazole (SHI) moiety and specific length of the alkyl diamine chain in **165** are important for the inhibitory activity.

Subsequent investigation of the crude EtOAc extract from *Acanthodendrilla* sp. yielded another new bromotyrosine alkaloid, acanthodendrilline (**176**), based on spectroscopic analysis as well as synthesis (Fig. 17) [59]. This compound was the first example of a bromotyrosine alkaloid that consisted of dibromotyramine, methylcarbamate and 2-oxazolidone units in the structure. In addition, the concise synthesis of enantiomers of acanthodendrilline were prepared using commercially available starting materials, such as tyramine and (*S*)- and (*R*)-epichlorohydrins. The synthetic molecules were used to determine the absolute configuration of C-11 in acanthodendrilline as the *S* configuration. However, both enantiomers exhibited weak AChE inhibitory activity, but (*S*)-**176** was approximately threefold more cytotoxic against the human non-small cell lung cancer (H292) cell line than (*R*)-**176**.

In the initial synthesis of (*S*)- and (*R*)-**176**, commercially available tyramine was used to generate methyl-[2-(3,5-dibromo-4-hydroxyphenyl) ethyl] carbamate (56% yield) in two steps [59]. Separately, (*S*)- or (*R*)-epichlorohydrin was transformed in one synthetic step into the respective enantiomers, 5-chloromethyl-2-oxazolidinone [(*S*)-**177** or (*R*)-**177**, each about 40% yield]. Potassium cyanate was used as the nucleophile to obtain the oxazolidinone motif from the oxirane ring in (*S*)- or (*R*)-epichlorohydrin. Subsequent coupling of (*S*)-**177** or (*R*)-**177** with methyl-[2-(3,5-dibromo-4-hydroxyphenyl) ethyl] carbamate, via base mediated nucleophilic substitution, provided (*S*)-**176** (48% yield) or (*R*)-**176** (37% yield), respectively (Scheme 1).

3.4.3 Pyridoacridine Alkaloids

Marine-derived pyridoacridine alkaloids are unique colored nitrogenous metabolites that represent a large family of alkaloids. They have been reported from a wide range of benthic marine invertebrates, such as sponges, ascidians, anemones, prosobranch

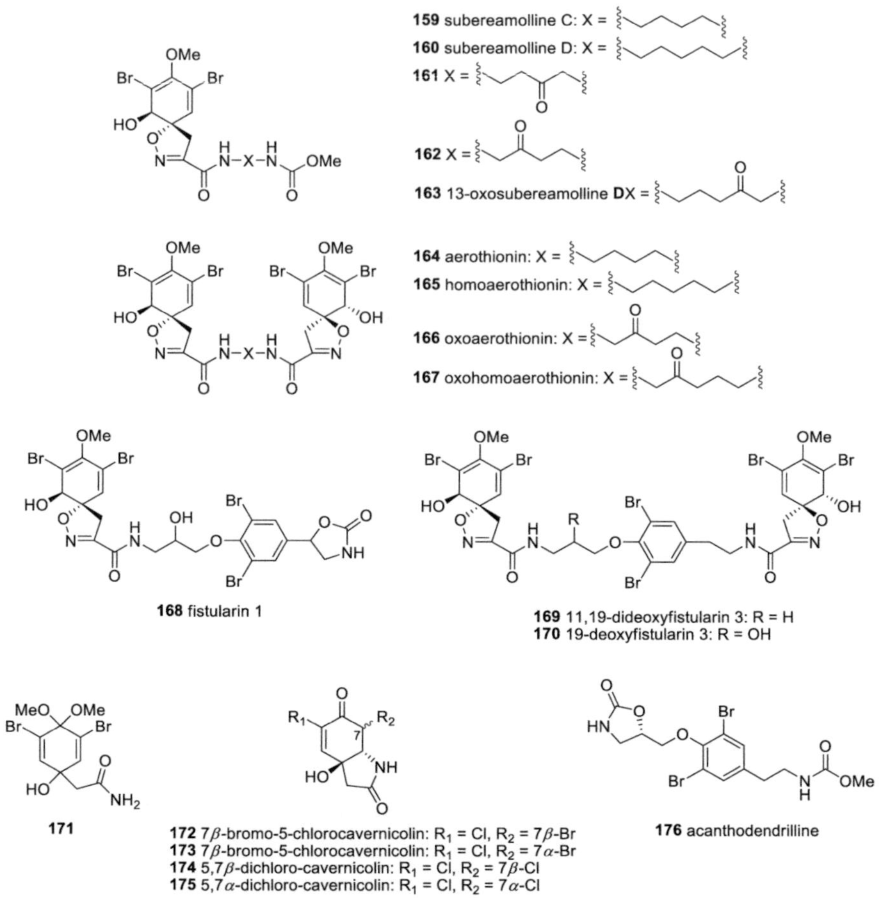

Fig. 17 Chemical structures of halogenated tyrosine-derived alkaloids isolated from the Thai sponge, *Acanthodendrilla* sp

mollusk and tunicates [88, 89]. In addition, attention to pyridoacridines has risen due to their significant biological activities, including antiviral, antibacterial, anticancer and anti-parasitic properties [89–91]. It has been proposed that the pyridoacridine alkaloids are biosynthesized via a pathway that utilizes tryptophan and DOPA as precursors.

Kuanoniamines A and C

In an investigation on bioactive natural products from marine sponges in the Gulf of Thailand, several known compounds, including 24α-methylcholestanol, *p*-hydroxybenzaldehyde, phydroxybenzoic acid, phenylacetic acid, 3-formylindole,

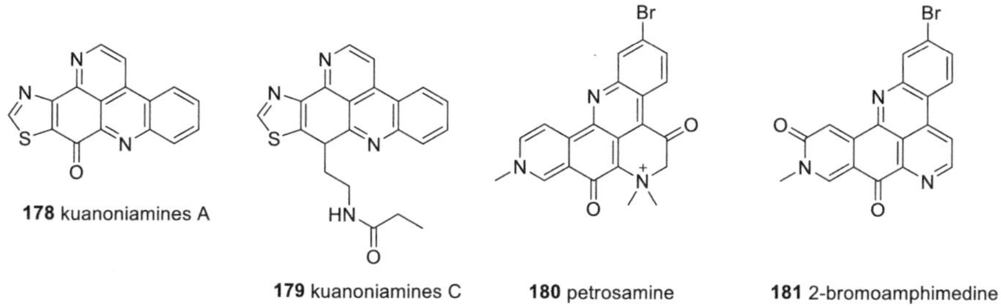

Fig. 18 Chemical structures of pyridoacridine alkaloids

kuanoniamines A (**178**) and C (**179**), were purified from the EtOAc extract of *Oceanapia sagittaria*, collected from the Gulf of Thailand near Koh Samed (Fig. 18) [92]. Members belonging to the genus *Oceanapia* are relatively abundant and widely distributed. To date, more than 100 metabolites, mainly alkaloids, lipids and steroids, have been reported from various *Oceanapia* species [93]. The alkaloids, kuanoniamines A and C, were previously reported from an unidentified purple colonial tunicate and its prosobranch mollusc predator, *Chelynotus semprei*, and possessed anti proliferative activity against human pharyngeal cancer cell lines [94]. The sponge-derived kuanoniamines A and C were evaluated for their effects on five human cancer cell lines, including MCF-7 (breast carcinoma, estrogen–dependent ER+), MDA-MB-231 (breast carcinoma, estrogen–independent ER−), SF-268 (glioma), NCI-H460 (non-small cell lung cancer), UACC-62 (melanoma), as well as a non-tumour human cell line MRC-5 (diploid embryonic lung fibroblast) and peripheral human lymphocytes. Between these alkaloids, kuanoniamine A was found to be more cytotoxic against these cancer cell lines, with IC_{50} values ranging from 0.12 to 1.83 μM. In addition, kuanoniamine A was found to be a more potent inhibitor of DNA synthesis when compared to kuanoniamine C as well as causing significant reduction in MCF-7 cells in the G2/M phase and increase in apoptotic cells. Although less cytotoxic, kuanoniamine C was revealed to have high selectivity toward the estrogen dependent (ER+) breast cancer cell line.

At least two studies were conducted on the molecular mechanisms of kuanoniamine C (**179**). A series of pharmacological studies by Machihara and Namba revealed that co-treatment of kuanoniamine C and an approvied anticancer agent, bortezomib, could represent novel therapeutic approach for the treatment of osteosarcoma [95]. Osteosarcoma is a type of bone cancer that mostly develops in the osteoblast cells, which are involved in bone formation. This disease usually happens in children and young adults. It was found that co-treatment of kuanoniamine C and bortezomib led to suppression of glucose-regulated protein 78 (GRP78) expression, which is important for stimulation of cell death induced by the later anticancer agent [95]. In addition, kuanoniamine C suppressed GRP78 expression through GRP78 mRNA degradation in an endoplasmic reticulum stress response-independent manner and the downregulation of GRP78 expression was regulated by p53 signaling. In a recent study by Oki and co-workers, new molecular target for kuanoniamine C on the inhibition of mitochondrial respiratory complex II was uncovered [96]. By modulating mitochondrial function, kuanoniamine C suppressed the differentiation of pre-adipose cells into white adipocytes as well as inhibited white adipose tissue expansion in the early phase of high-fat-diet-induced obesity model.

Petrosamine

Additional pyridoacridine alkaloids, including a known petrosamine (**180**) and a new 2-bromoamphimedine (**181**), were isolated from a Thai greenish black sponge *Petrosia* sp. obtained from the intertidal region of Phuket Island, Thailand (Fig. 18) [60]. Petrosamine was previously isolated by Molinski and Faulkner in 1988 from *Petrosia* sp. collected at Carrie Bow Cay, Belize but no bioactivities were reported [97]. The Thai research group found petrosamine to exhibit significant anticholinesterase activity, about six times higher than that of the standard inhibitor, galanthamine. Acetylcholinesterase inhibitors prevent the breakdown of the neurotransmitter, acetylcholine into choline and acetate, and thus increasing the concentration and duration of the neurotransmitter in the central nervous system, autonomic ganglia and neuromuscular junctions. Such inhibitors can be used for the symptomatic treatment of neurodegenerative diseases, such as dementia of Alzheimer's disease [98]. Subsequent evaluation of petrosamine, using computational docking study with the acetylcholinesterase from the electric eel *Torpedo californica* (*Tc*AChE), revealed that the ammonium group of the alkaloid is important for the petrosamine-*Tc*AChE interaction.

3.4.4 Indole-Containing Alkaloids

The sponge genus *Smenospongia* consists of about 19 described species and known to be prolific producers of diverse classes of secondary metabolites, including phenyl alkenes, indole alkaloids, terpenoids, chromenes, γ-pyrones, naphthoquinones and polyketides [99]. From a collection of *Smenospongia* sp. collected from Phi Phi Island in the Andaman Sea, Krabi Province, six new molecules, including a sesquiterpene, a bissesquiterpene and four brominated indole alkaloids (**182**–**185**), were described (Fig. 19) [37]. Along with the new compounds, 16 known compounds, such as four sesquiterpene hydroquinones, ten brominated indole alkaloids [compounds **186** to **195** (Fig. 19)], ergosterol and furospinosulin-1, were also isolated. Of the fifteen compounds evaluated for cytotoxicity against a panel of cancer cell lines, including MOLT-3, HepG2, A549, HuCCA-1, HeLa, HL-60 and MDA-MB-231, only the known compound **194** displayed cytotoxicity against MOLT-3 and HeLa cells with the IC_{50} values 5.4 and 9.4 μM, respectively.

Two new isomeric forms of the indolocarbazole molecules, damirines A (**196**) and B (**197**), were obtained from an understudied sponge, *Damiria* sp., collected in the vicinity of Phuket Island, Thailand (Fig. 19) [61]. These new alkaloids represent a new hexacyclic carbon–nitrogen framework consisting of an indolo[3,2-*a*]carbazole fused

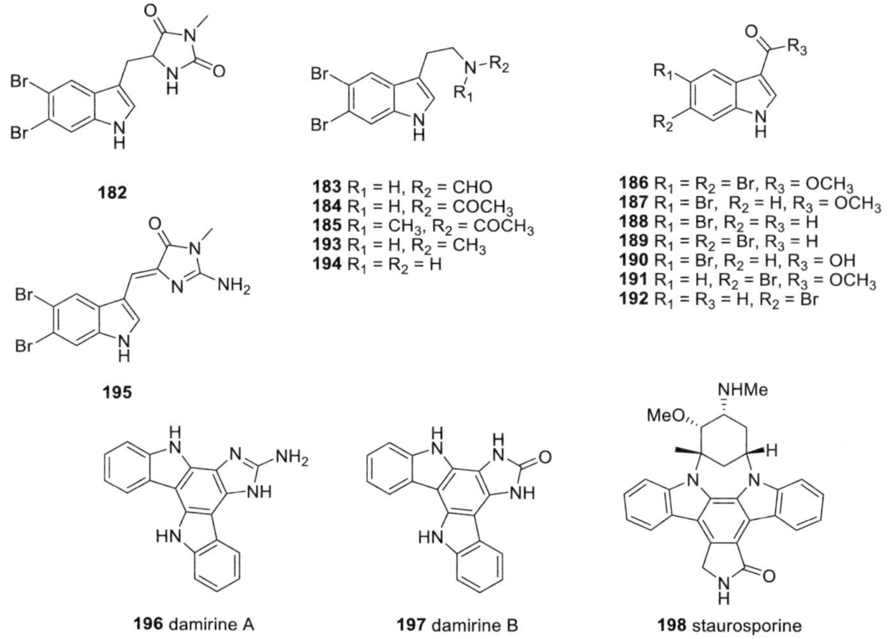

Fig. 19 Chemical structures of indole-containing alkaloids

with either an aminoimidazole or a imidazolone ring. Interestingly, the indolo[3,2-*a*]carbazole scaffold has been generated in previous synthetic studies, but rarely found in a natural product. The sponge collection was made under contract with the Coral Reef Research Foundation for the Natural Products Branch, National Cancer Institute (NCI). The indolocarbazole family of bisindole alkaloids is represented by the natural product staurosporine (**198**) (Fig. 19), a protein kinase C inhibitor that belongs to the indolo[2,3-*a*]carbazole structural class. However, isomeric forms of indolocarbazole alkaloids have rarely been reported from nature. Due to the lack of protons in the damirines, the use of the HMBC (optimized for 3 Hz couplings) and LR-HSQMBC pulse sequence, a long-range heteronuclear correlation experiment, had been particularly useful for their structural elucidation [100]. Upon biological evaluation in the NCI 60 cancer cell lines, damirine A showed selective cytotoxic activity against six different cancer cell lines, including one melanoma (MALME-3 M, $GI_{50} = 1.9$ μM), one breast (MDA-MB-468, $GI_{50} = 2.0$ μM), two colon (SW-620, $GI_{50} = 3.3$ μM; HCC-2998, $GI_{50} = 2.3$ μM) and two leukemia (MOLT-4, $GI_{50} = 1.9$ μM; K-562, $GI_{50} = 2.2$ μM) cell lines.

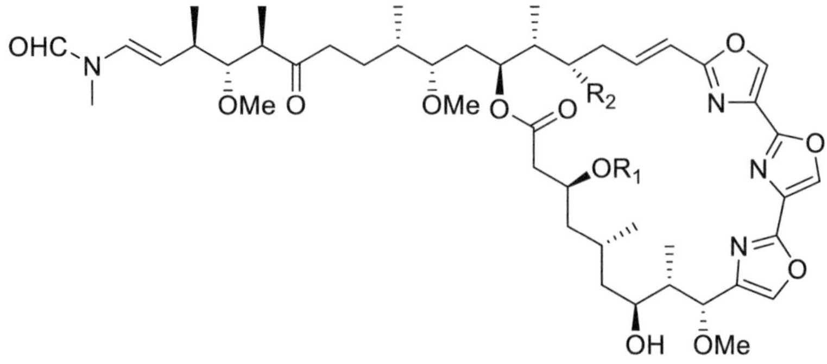

199 kabiramide B: R_1 = $CONH_2$, R_2 = OH
200 kabiramide C: R_1 = $CONH_2$, R_2 = OMe
201 kabiramide D: R_1 = H, R_2 = OMe

Fig. 20 Chemical structures of kabiramides B (**199**)–D (**201**)

3.4.5 Trisoxazole Macrolides

Kabiramides

The kabiramides are marine-derived trisoxazole macrolides known to possess cytotoxic and antifungal properties. Structurally, they consist of a characteristic macrolide moiety, where three contiguous oxazole units are connected, as well as a side-chain with an *N*-methyl-vinylformamide terminus. Members of this class of compounds, namely kabiramides A to E, were initially reported from the egg-ribbons of the *Hexabranchus* nudibranch obtained from Ishigaki Island, Okinawa, and later found in other marine invertebrates, such as marine sponges [101, 102]. The major antifungal compound, kabiramide C (**200**) (Fig. 20), was isolated in high amounts, about 0.25% wet weight of the egg-ribbons and their planar structures were initially reported in 1986 for kabiramide C and 1989 for minor metabolites, kabiramides A, B (**199**), D (**201**) and E. Kabiramides B (**199**) and C (**200**) were also isolated from a sponge, *Halichondria* sp., suggesting the sponge origin of kabiramides (Fig. 20) [103]. It was only in 2003, almost two decades after the report of kabiramide C, that its absolute stereochemistry was determined by X-ray crystallography via immersing a solution of the molecule in preformed crystals of actin [104]. Incidentally, kabiramide C used in the X-ray crystallography study was obtained from the black sponge, *Ircinia* sp. from the Gulf of Thailand.

Kabiramide C (**200**), isolated from *Ircinia* sp. in Thailand, was subsequently used to characterize the interactions between the molecule with G-actin as well as its underlying mechanism [105]. Tanaka and co-workers found that the G-actin-kabiramide C complex is stable and is formed via a two-step binding reaction, including initial fast binding of the macrocyclic portion, followed by slow step binding of the side chain of the compound [105]. Interestingly, the X-ray crystallography of a complex between actin and kabiramide C (**200**) as well as related compounds revealed that the side chain unit of these macrolides was found inserted into the hydrophobic cavity, which associates with the hydrophobic stretch of another actin protein to form F-actin [104]. Kabiramide C and related molecules can be viewed as biomolecular mimics of well-studied proteins, such as gelsolin and CapG, known to bind to the barbed end of the actin filaments. In subsequent studies, functional fluorescent derivatives of kabiramide C, including compounds **203** and **204**, were synthesized from synthetic compound **202** as starting material, and used as molecular probes for imaging changes in the distribution of barbed ends in cells as well as for quantitative analysis of actin in vitro (Fig. 21) [106, 107].

From a collection of the Thai black sponge, *Pachastrissa nux*, obtained from several expeditions, including Sichang Island in 1997, vicinity of Koh-Tao, Surat-Thani Province in 2004 and 2006, and Chumphon Islands National Park, Chumphon Province in 2008, a series of ten kabiramides, namely kabiramides B–D and F (**205**)–L (**215**) (Fig. 22), were isolated [62–64]. The known kabiramides B–D were previously reported by Matsunaga and co-workers [102]. It has been observed that colonies of *P. nux* exist in two different growth forms: the protruding gorgonian-like shaped capitum sprouting from the base and the substratum-attached irregular-shaped base. Both growth forms are soft and tender with a brown khaki inner coloration. In the course of a project aimed at isolating kabiramide C from *P. nux*, collected from Sichang Island, four new related metabolites, kabiramides F (**205**)–I (**208**) (Fig. 22), along with known kabiramides B and D, were obtained. Kabiramide C was isolated as the major component of the sponge samples derived from two collections, yielding a total of about 1.69 g. The structures of these new molecules were deduced solely by extensive spectroscopic methods [62].

The significant amount of kabiramide C from *Pachastrissa nux* facilitated the semisynthesis of four kabiramide C derivatives, including 19,20-dihydrokabiramide C (**209**), 19,20,34,35-tetrahydrokabiramide C (**210**), Δ^6-kabiramide C (**211**) and 7-amino-19,20,34,35-tetrahydrokabiramide C (**212**), to be evaluated for cytotoxicity (Fig. 23). When tested against a panel of five cancer cell lines, kabiramides B–D, F (**205**), G (**206**) and kabiramide derivatives **209–211** exhibited similar range of cytotoxicity, suggesting that the oxazole rings within the intact macrocyclic ring is important for cytotoxic properties. Based on previous crystallographic studies on kabiramide C with actin where the macrocyclic ring was found posited in a hydrophobic region, the polar functional groups present in kabiramides H (**207**) and I (**208**) resulted in less G-actin binding and hence showed weaker cytotoxicity [62].

Further investigation of the methanolic extracts of sponge samples collected from other locations in the Gulf of Thailand furnished additional new kabiramide analogues, namely kabiramides J (**213**)–L (**215**) (Fig. 24) [63, 64]. In addition to their cytotoxic, the antimalarial properties of these compounds were investigated and revealed that they showed strong antimalarial activity against *Plasmodium falciparum* K1. In particular, kabiramides J and K exhibited significant antimalarial activity with IC_{50} values of 0.31 and 0.39 μM, respectively [63].

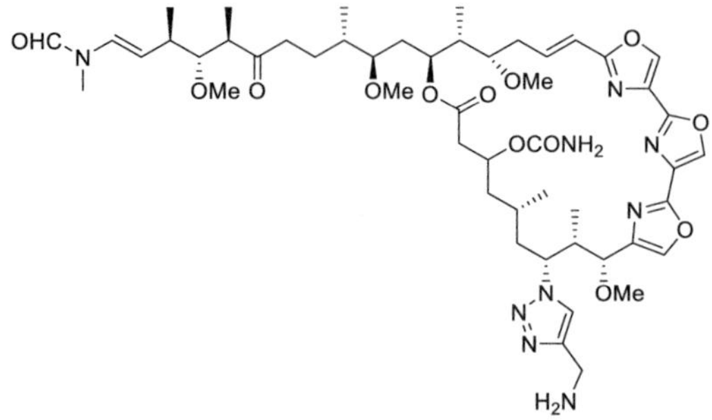

202 7-(4-aminomethyl-1*H*-1,2,3-triazol-1-yl)kabiramide C

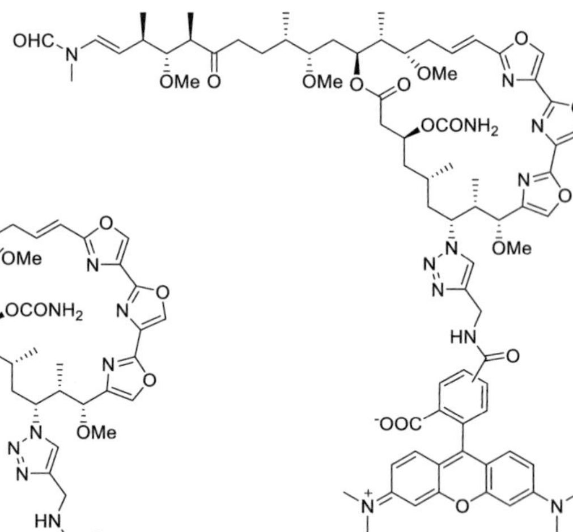

203 TMR-KabC

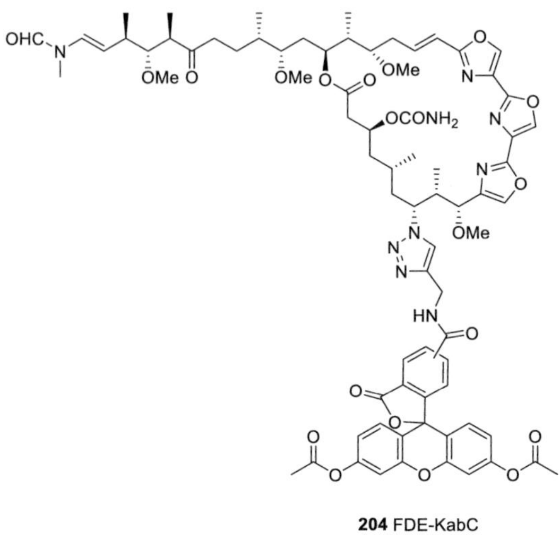

204 FDE-KabC

Fig. 21 Chemical structures of florescent derivatives of kabiramide C

Chemical Ecology of the Kabiramides

It was observed by Sirirak and co-workers that the extract derived from capitum of *P. nux* showed consistently strong biological activities, such as cytotoxicity, antimalarial and antitubercular activities, as compared to extract obtained from the sponge base, which was barely active. Based on this observation, they investigated various parts of the sponge to see whether there was any trade-off in the physical structural composition and the chemical distribution [108]. Using kabiramides C and G as chemical markers, they found that these alkaloids accumulated in the capitum than the

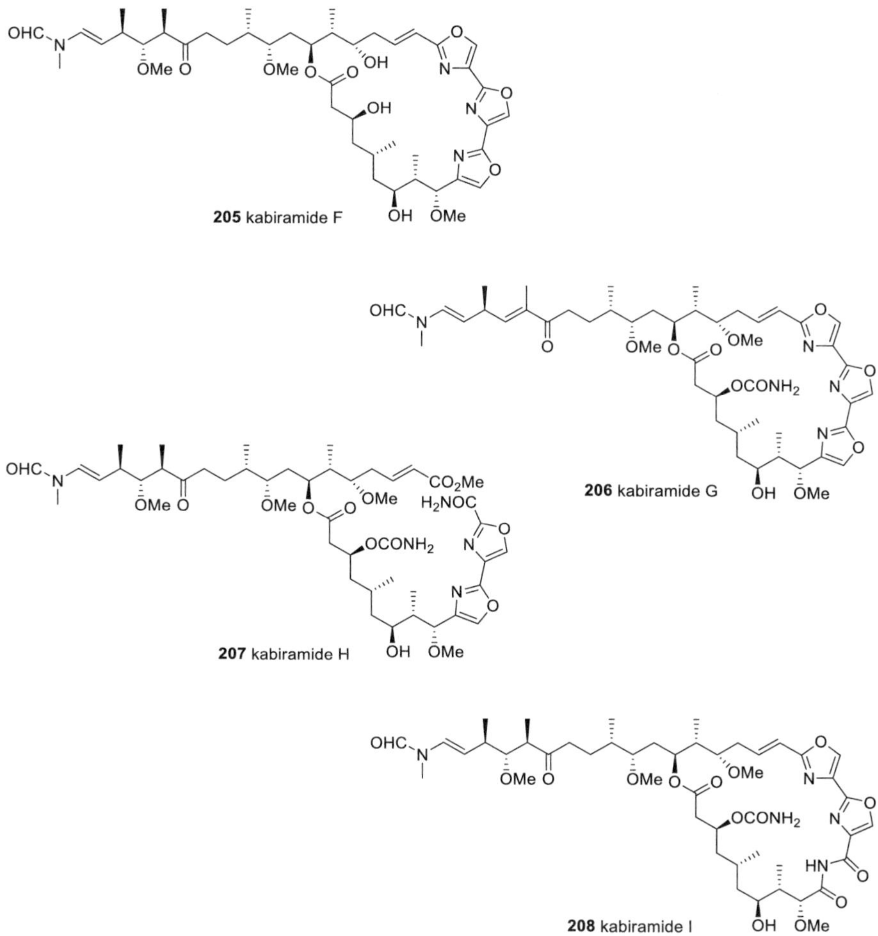

Fig. 22 Chemical structures of kabiramides F (**205**)–I (**208**)

base. They hypothesized that the allocation of kabiramides was a trade-off with the structural materials used to reinforce the strength of the sponge animal.

NMR-based metabolomics methods were used to track the connection between the chemical allocation and the bud differentiation in *Penares* cf. *nux* collected from two locations, including Koh-Tao, Surat-Thani province and Saiburi, Pattani province [109]. It was found that geographic locations had no significant effects on the sponge's chemical profiles. Based on Principal Component Analysis (PCA) of the $CHCl_3$-derived extracts, differences in chemical allocations, especially trisoxazole macrolides and the sterols, were discovered in the sponge capitums and bases, suggesting that the sponge was able to allocate the secondary metabolites in the former region. However, PCA analysis of the aqueous MeOH-derived extracts, containing mainly the water-soluble primary metabolites, was found scattered

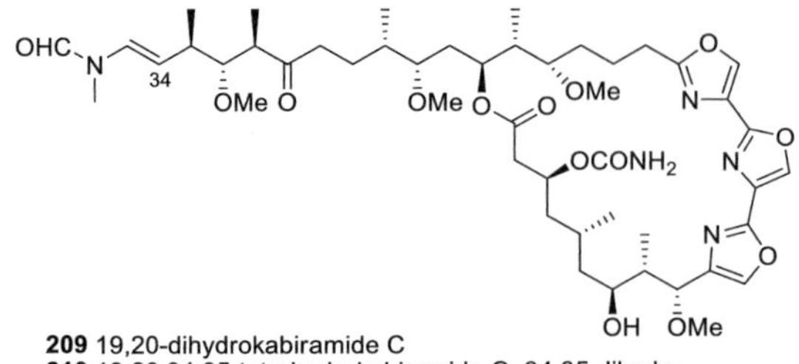

209 19,20-dihydrokabiramide C
210 19,20,34,35-tetrahydrokabiramide C: 34,35 dihydro

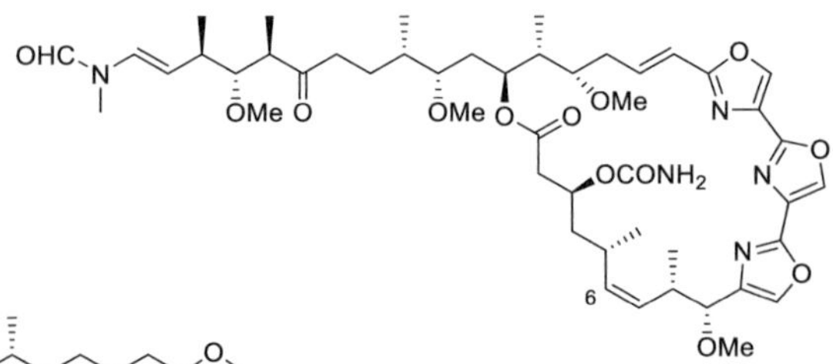

211 Δ^6- kabiramide C

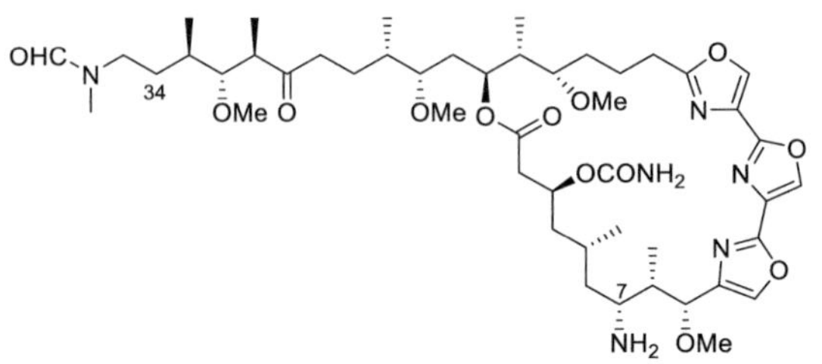

212 7-amino-19,20,34,35-tetrahydrokabiramide C

Fig. 23 Chemical structures of kabiramide C derivatives **209–212**

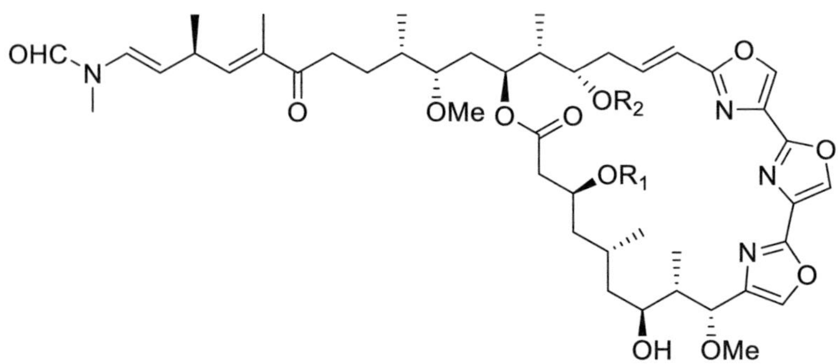

213 kabiramide J: R_1 = $CONH_2$, R_2 = H
214 kabiramide K: R_1 = H, R_2 = CH_3
215 kabiramide L: R_1 = R_2 = H

Fig. 24 Chemical structures of kabiramides J (**213**)–L (**215**)

throughout the score plot. This suggested that there were no chemical differences between the primary metabolisms in the capitums and bases.

A recent study by Butsuri et al. used matrix-assisted laser desorption ionization-time of flight imaging mass spectrometry (MALDI-TOF IMS) to study the distribution of trisoxazole macrolides in the sponge *P. nux* [110]. The major compounds, kabiramides, were found primarily in the sponge's outer tissues, including the pinacoderm, mesohyl, ectosome and outer choanosome. Additionally, the presence of individual macrolides varied among specimens, with no mutual influence or tissue/cell-type specificity. These findings highlight a targeted distribution of toxic metabolites, supporting the optimal defense theory, suggesting that *P. nux* evolved to concentrate its chemical defenses in areas most vulnerable to predation or damage.

In another study, researchers revealed the impact of the alkaloids, including kabiramides B, C and G, allocation on the sponge, *P. nux*, surface-attached bacteria [111]. The bacterial population density found on the surface of both the capitum and the appendage extending from the core base was significantly less than that on the surface of the sponge base. This could suggest the repelling property of the kabiramides against the fouling bacteria, particularly on the capitums and appendages [112]. In addition, almost two dozen bacterial strains were analyzed but none of the strains were producers of the alkaloids. The sponge's specific name in this study is *Penares* cf. *nux*, probably due to realization of a transfer of *Pachastrissa nux* to the *Penares* genus [113].

3.4.6 Isoquinolines

Isoquinoline Quinones

Three known single-unit isoquinoline quinones, *N*-formyl-1,2-dihydrorenierone (**216**), *O*-demethylrenierone (**217**) and mimosamycin (**218**), were isolated from the Thai purplish-blue sponge, *Cribrochalina* sp., obtained in the vicinity of Koh-Tao, Surat-Thani Province, Thailand (Fig. 25) [114]. Their chemical structures were confirmed by comparison of NMR spectral data with previously published data by Frincke and Faulkner (1982) for **216** and **217** and Fukumi et al. (1977) for **218** [115, 116]. Interestingly, the optical activity of the isolated *N*-formyl-1,2-dihydrorenierone was measured as +230° as compared to that reported previously by Frincke and Faulkner at –227°, suggesting their enantiomeric relationship. Compounds **216** to **218** were evaluated for cytotoxicity, using sulforhodamine B (SRB) method, against four solid tumor cell lines, including KB, HT-29, MCF-7 and HeLa. All three alkaloids showed significant selective cytotoxicity, particularly against MCF-7 and HeLa cell lines. Compound **218** was less discriminative when compared to the other two alkaloids. This could be due to the presence of hydrophobic and electrophilic nature of the angelate side chain in **216** and **217**.

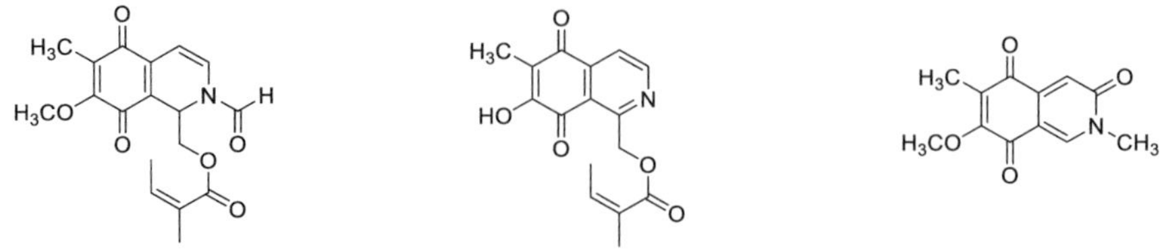

216 *N*-formyl-1,2-dihydrorenierone **217** *O*-demethylrenierone **218** mimosamycin

Fig. 25 Chemical structures of isoquinoline quinones

Renieramycins

The tetrahydroisoquinoline class of alkaloids, including saframycins, ecteinascidins, renieramycins and safracins, have attracted chemical and biological interests due to their diverse structures and potent biological activities, ranging from antitumor to antimicrobial properties. The main structural differences between ecteinascidins, saframycins and renieramycins are the groups at the C-22 position. For instance, C-22 positions in the saframycins and renieramycins are connected to the C-N and C-O bond, respectively, while a more complicated extension structure of the sulfur-containing lactone bridge ring is present in the ecteinascidins [117]. Ecteinascidin-743 (ET-743) (**219**) (Fig. 26), initially isolated from the Caribbean tunicate, *Ecteinascidia turbinata*, is the first marine-derived antitumor drug approved by the European Commission in 2007 and the U.S. Food and Drug Administration (FDA) in 2015 for treatment of soft tissue sarcoma and ovarian carcinoma [118]. In addition, a recently approved synthetic analog of natural ecteinascidin-736, lurbinectedin (**220**) (Fig. 26), is used as second line treatment of metastatic small cell lung cancer [119]. At least one other synthetic molecule, ecubectedin (PM14), which was developed from the medicinal chemistry program of lurbinectedin at PharmaMar, is in Phase II clinical testing for treatment of solid tumors [120].

Renieramycin alkaloids are usually isolated from several sponge species, including *Reniera* sp., *Haliclona cribricutis*, *Cribrochalina* sp., *Neopetrosia* sp. and *Xestospongia* sp., and they possess a pentacyclic bis-tetrahydroisoquinolinequinone moiety as the core scaffold [117, 121]. The first series of this class of alkaloids was discovered in 1982 by Frincke and Faulkner where four new natural products, renieramycins A (**221**)–D (**224**), were purified from the sponge *Reniera* sp. (Fig. 27) [115]. To date, more than twenty renieramycins have been reported [121]. The core structure of renieramycin is similar to that of the saframycin group of tetrahydroisoquinoline quinones, which have been reported from disparate bacterial strains, including free living terrestrial Myxobacteria and Actinobacteria (the saframycins) [122–124], Proteobacteria (the safracins) [125, 126] and obligate symbionts of marine ascidians (the ecteinascidins) [127, 128]. Subsequent investigations by Donia's research group led to the discovery of renieramycin producing symbiont *Candidatus* Endohaliclona renieramycinifaciens from the renieramycin-containing sponge *Haliclona* sp. collected from different locations in the Pacific Ocean, including Papua New

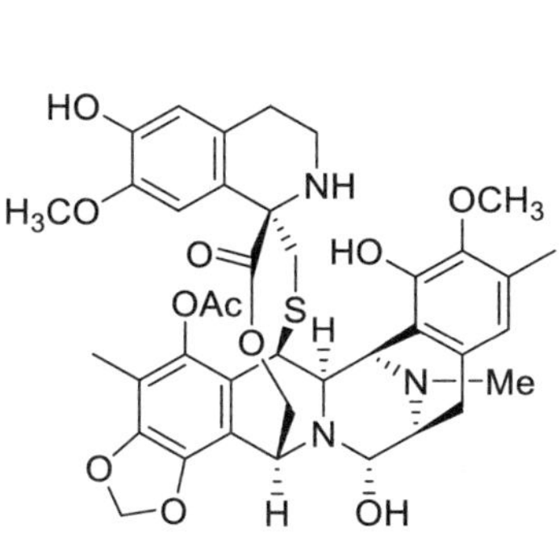

219 ecteinascidin 743

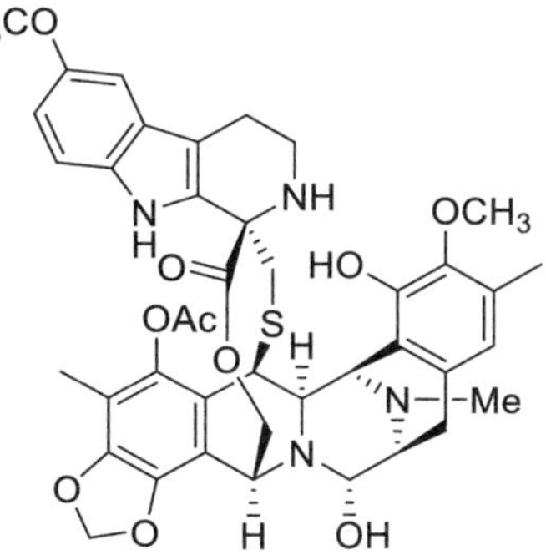

220 lurbinectedin

Fig. 26 Chemical structures of approved anticancer drugs, ecteinascidin-743 (**219**) and lurbinectedin (**220**)

Guinea and Palau [129]. It was discovered that microbial symbionts are found localized within unique sponge cellular reservoirs, known as chemobacteriocytes, where they can obtain nutrients from the host as well as avoiding competition from other bacteria.

A majority of renieramycins were reported as artifacts arising from attempts to improve the stability of renieramycin alkaloids [130]. For instance, several potent renieramycins have been isolated from the bright Thai blue sponge, *Xestospongia* sp. and reported mainly by Drs. Saito and Suwanborirux research groups beginning in the early 2000s [65]. Their collaborative efforts were initiated by an international cooperative research project "Medicinal Chemistry on Alkaloids from Plants in the Malaya Area" supported by the Japan Society for the Promotion of Science (JSPS). The structures of renieramycins J (**225**)–O (**230**), 7-desmethylrenieramycin O (**231**) and renieramycins Q (**232**)–S (**234**), isolated from the Thai sponge *Xestospongia* sp. (collected by scuba divers in the vicinity of Sichang Island) using pretreatment of extracts with potassium cyanide (KCN), were reported in 2003, 2004 and 2017 (Fig. 27) [66–68, 131].

In 2002, a similar approach of using KCN to stabilize natural products was reported by the same Thai research group on the tunicate extracts of *Ecteinascidia thurstoni* to obtain alkaloids, such as ecteinascidins 770 (**235**) and 786 (**236**) (Fig. 27) [132]. Incidentally, the discovery of ecteinascidin-related compounds from *E. thurstoni* was the first example of the Asian tunicate, containing this class of molecules, found growing around Phuket Island.

This additional step of adding KCN in phosphate buffer solution (pH 7.0) improved the isolated yields of renieramycins by approximately 100-fold as well as facilitated the subsequent discovery of minor components of the alkaloids in the sponge, such as renieramycins T (**237**) and U (**238**), which contain an unique ecteinascidin-renieramycin hybrid structure, as well as renieramycin V (**239**), which has a sterol unit attached to the C14 position (Fig. 28) [133, 134]. Also, the presence

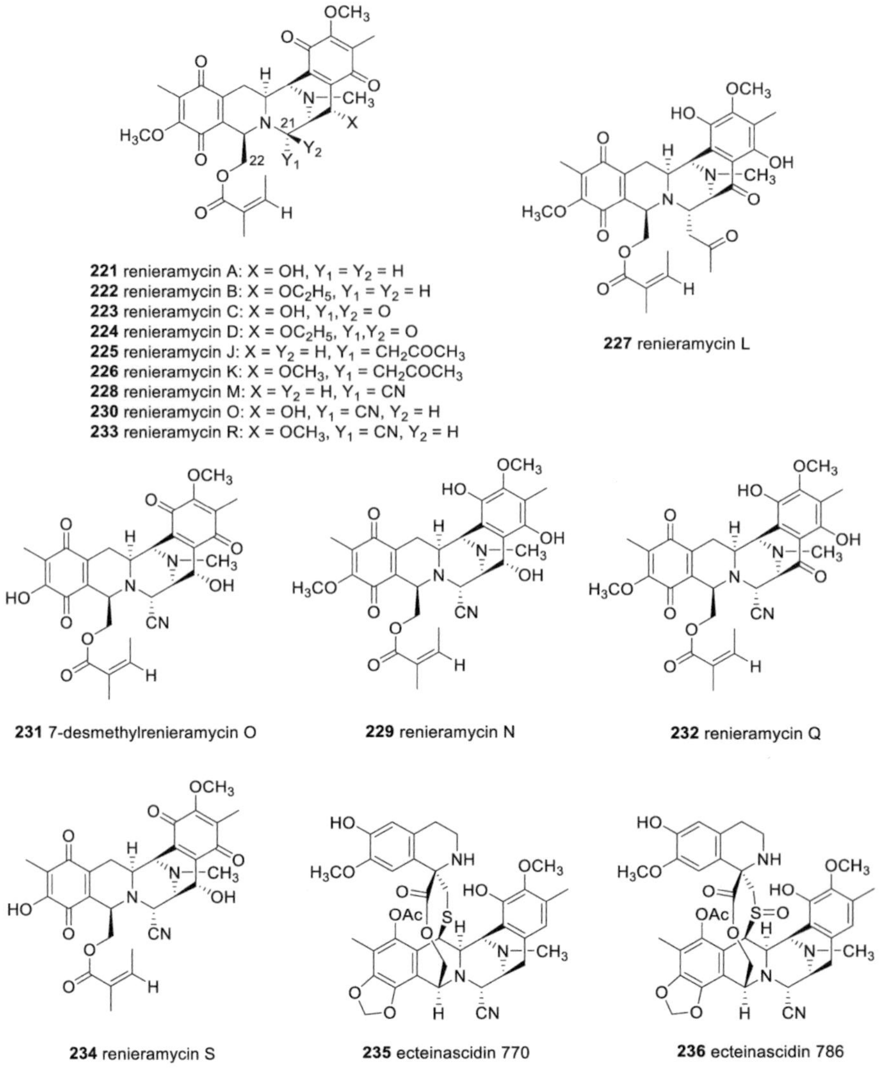

Fig. 27 Chemical structures of renieramycins **221**–**234**, ecteinascidins 770 (**235**) and 786 (**236**)

of high amounts of these alkaloids with DNA-damaging activity prevented the extraction of high quality and quantity of sponge DNA for molecular studies. As such, an efficient method on the removal of renieramycins, especially renieramycin M, from the sponge samples using simple methanolic extraction before DNA extraction was developed to dramatically increase the yield and purity of sponge DNA [69].

Related renieramycin analogues, jorunnamycins A (**240**)–C (**242**) were isolated from the aqueous KCN-pretreated nudibranch, *Jorunna funebris*, obtained in the vicinity of Ko Si Chang, Thailand (Fig. 28) [135]. As *Jorunna* nudibranchs are

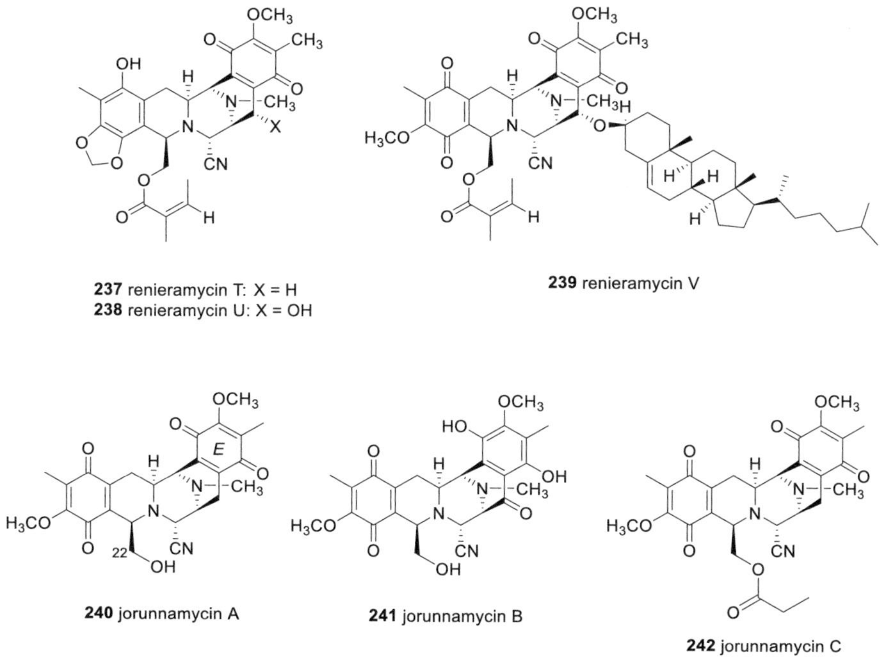

Fig. 28 Chemical structures of renieramycins T (**237**)–V (**239**) and jorunnamycins A (**240**)–C (**242**)

carnivorous and feed exclusively on sponges, including *Xestospongia* sp., *Haliclona* sp., *Euplacella* cf. *australis* and *Oceanapia* sp., it is possible that the precursors of jorunnamycins are sequestered by *J. funebris* from *Xestospongia* sp. and chemically modified for chemical defense [136]. The presence of these renieramycin-related molecules was investigated in the nudibranch's mantle, visceral organs (including digestive glands and gonads) as well as their egg ribbons separately. All three parts were homogenized with phosphate buffer (pH 7.0), followed by adding of aqueous KCN solution to each suspension. Each mixture was further macerated with methanol and the extract filtered before further purification using flash column chromatography on silica gel to afford renieramycin type compounds. From 700 mg of nudibranch's mantle, several molecules, including jorunnamycin A (7.5 mg), renieramycins M (25.3 mg), N (0.5 mg) and O (6.4 mg) were isolated. Visceral organs (980 mg) contained jorunnamycin B (23.6 mg), renieramycins N (9.2 mg), O (2.0 mg) and Q (1.3 mg) while only jorunnamycins A (3.8 mg), C (21.7 mg) and renieramycin M (62.1 mg) were detected in the egg ribbons (270 mg). Interestingly, esterase activity in converting jorunnamycin A from renieramycin M was detected by Waropastrakul and co-workers in the crude enzyme prepared from the visceral portion of *J. funebris* [137]. This was achieved via the catalytic hydrolysis of the angeloyl unit in the later compound and detected using RP-HPLC. In addition, partial purification of the esterase enzyme was carried out using 55–75% ammonium sulfate after initial

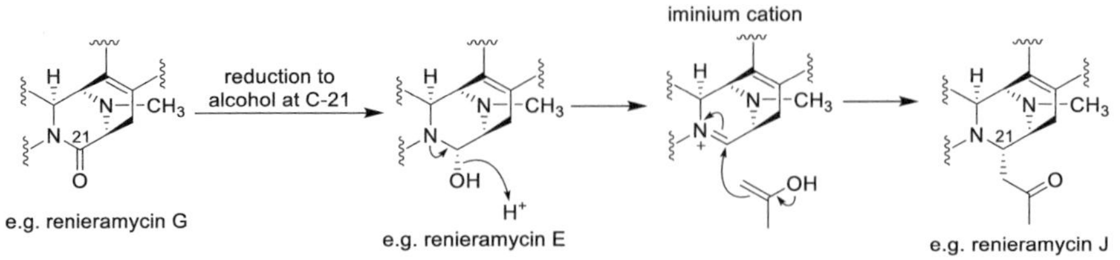

Scheme 2 Proposed pathway to the generation of propan-2-one unit at C-21 in renieramycins as artifacts of the isolation process

precipitation of the visceral proteins with 55% ammonium sulfate. The discovery of a specific enzyme activity could potentially be used to replace harmful reagents for such specific chemical reactions.

It is postulated that renieramycins J (**225**)–O (**230**) (Fig. 27), having a propan-2-one unit at C-21, are artifacts of the isolation process [67]. They could be generated from partial reduction of the lactam carbonyl (such as renieramycin G) to provide carbinolamine (e.g., renieramycin E). An enol nucleophilic propan-2-one can then be added at C-21 to give the Mannich adduct (e.g., renieramycin J) via the iminium cation intermediate stereoselectively (Scheme 2). Based on subsequent work using model compounds for such conversion [138] as well as preliminary earlier results by Arai research group on the conversion of saframycin S to saframycin A using NaCN [139], Saito and co-workers attempted successfully in stabilizing renieramycin E to form renieramycin M (**228**) by pretreatment of the homogenized sponge samples with KCN. For instance, in 1980 Arai and co-workers isolated a labile minor compound, saframycin S with a hydroxyl group at C-21. Saframycin S was successfully transformed into saframycin A using NaCN in phosphate buffer with good yield of 60%. The reverse reaction was also demonstrated using 0.1 N H_2SO_4 at 120 °C for about 40 min with a yield of 41%.

From the KCN-treated crude fraction (16.2 g), about 1.822 g of renieramycin M (**228**) was obtained together with renieramycin N (**229**) (247 mg) [67]. They found that the addition of KCN to the homogenized sponge samples stabilized the labile amino alcohol moiety by conversion to a more stable amino nitrile group. This method of adding KCN allowed gram-scale preparation of renieramycins for further drug development. In particular, the gram isolation of renieramycin M (**228**) has been instrumental in providing sufficient quantity for the generation of many semi-synthetic analogs for SAR studies as well as biological testing.

4 Medicinal Chemistry of Renieramycins and Their Semi-synthetic Analogues

4.1 Renieramycins M–O, T, U and Jorunnamycin A

Several renieramycins, isolated from the Thai blue sponge, *Xestospongia* sp., and their synthetic derivatives display potent anticancer property with nanomolar IC_{50}s against several cancer cell lines. For example, renieramycins M (**228**), N (**229**) and O (**230**) (Fig. 27) possess potent cytotoxicity against colon (HCT116) and lung carcinoma (QG56) [66, 67]. In addition, renieramycin M has been reported to show strong cytotoxicity against various human cancer cell lines, including colon (DLD1), lung (NCI-H460), pancreatic adenocarcinoma (AsPC1) and ductal breast epithelial (T47D) cells [66, 140].

Using oligonucleotide microarray analysis to profile effects on cellular transcription, renieramycin M (**228**) was found to down regulate PTPRK gene, which can be a useful biomarker to study the transcriptional effects of this class of cytotoxic alkaloids [141]. A study by Halim and co-workers found that renieramycin M induced lung cancer cells apoptosis via p53-dependent pathway and the molecule could inhibit progression and metastasis of lung cancer cells [140]. Renieramycin M was also found to induce apoptosis of U373MG human glioblastoma cells and in the process of apoptosis, downregulated the expression of 274 genes, many of which were associated with the ErbB (EGFR) signaling pathway [142]. Renieramycin M can sensitize resistance to anoikis via decreasing cellular levels of survival and apoptotic proteins (including phospho-Akt, phospho-ERK, Bcl-2 and Mcl-1) and suppress cancer stem cell-like phenotypes in H460 lung cancer cells [143, 144]. When renieramycin M was used in combination with another cytotoxic compound, doxorubicin, synergistic cytotoxicity was observed in MCF-7 breast cancer cells [145].

The renieramycin–ecteinascidin hybrid alkaloids, renieramycins T (**237**) and U (**238**) (Fig. 28), also exhibit potent cytotoxic activity against non-small-cell lung cancer (NSCLC) cells [133]. The cytotoxicity mechanisms of the renieramycin–ecteinascidin hybrid alkaloids against NSCLC cells have been investigated. Renieramycin T (**237**) was found to enhance apoptosis induction via the degradation of the myeloid cell leukemia 1 (Mcl-1) protein in NSCLC [146, 147]. Furthermore, it suppressed mouse melanoma (B16F10) cell metastasis and migration through the downregulation of NF-E2-related factor 2 [148].

Cytotoxicity of jorunnamycins A (**240**)–C (**242**) (Fig. 28) against DU145, HCT116 and QG56 cacancer cell lines were also evaluated, revealing jorunnamycin C to display anti-proliferative activity against these cancer cell lines with IC_{50} values of 0.32, 1.5 and 2.8 nM, respectively [135]. These results indicated that the substituents on C-22 and E ring contributed to the improvement of cytotoxicity. In addition, jorunnamycin A was found to inhibit epithelial to mesenchymal transition, suppressing anchorage-independent survival as well as sensitizing anoikis in lung cancer cells via p53-dependent up-regulation of Bax and down-regulation of Mcl-1 and Bcl-2 [149].

Moreover, the suppressive effect of jorunnamycin A on cancer spheroid initiation and self-renewal in the cancer stem cells (CSCs) of human lung cancer cells has been revealed in a recent study [150].

4.2 Semi-synthetic Renieramycin M Derivatives

Several structure–activity relationship studies of potent renieramycin M (**228**) (Fig. 27) and its semi-synthetic derivatives having various linear and aromatic ester groups at C-5 and C-22 have been conducted and tested against non-small-cell lung cancer cell lines, listed as one of the world's leading causes of death. In spite of the promising anticancer effects of renieramycin M, its further development as a potential drug agent is hampered by its unwanted toxicity. Specifically, the presence of two quinone moieties in renieramycin M was shown to induce accidental necrosis as well as increase reactive oxygen species concentrations [151]. To overcome this side effect, one quinone moiety of renieramycin M was modified to provide the 5-*O*-acetylated hydroquinone derivative (**243**) (Fig. 29). This simple transformation significantly reduced the accidental necrosis inducing effect while maintaining the apoptotic property of the renieramycin M when tested on lung cancer H23 cells [151]. Subsequent transformation of the quinone systems in renieramycin M to bishydroquinone derivative (**244**) (Fig. 29) further enhanced its apoptotic effects on H292 human lung cancer cells. Protein analysis showed that the bishydroquinone renieramycin M-mediated apoptosis was associated with an increase of pro-apoptotic B-cell lymphoma 2 associated X (BAX) protein and the decrease of anti-apoptosis myeloid cell leukemia 1 (MCL1) and B-cell lymphoma 2 (BCL2) proteins [152].

Based on the cytotoxic profile on 5-*O*-acetyl hydroquinone renieramycin M (**243**), it is hypothesized that the hydroquinone moiety on ring A is important for its cytotoxicity. In fact, this moiety resembles the partial structure of the approved anticancer drug, trabectedin (ecteinascidin 743) (**219**) as well as the more stable ecteinascidin 770 (**235**). Thus, using 5-*O*-acetyl hydroquinone renieramycin M (**243**) as a lead molecule, development of a new generation of the renieramycins possessing the hydroquinone as highly potent and apoptotic inducing anti-NSCLC (non-small-cell lung cancer cells) drug agents was carried out.

The first series of twenty hydroquinone 5-*O*-monoester analogues of renieramycin M with different substituents was semisynthesized by Chamni and co-workers in 2017 [153]. These twenty hydroquinone 5-*O*-monoester analogues of renieramycin M were semisynthesized from renieramycin M by initial Pd-catalyzed hydrogenation to provide bishydroquinonerenieramycin M (**244**), followed by regioselective esterification (Scheme 3). Of these series of synthetic analogues, compound **245** (Fig. 29) exhibited tenfold increased cytotoxicity against the H292 NSCLC cell line (IC_{50} of 2.3 nM) as compared to renieramycin M (IC_{50} 24 nM) [153]. Another analog, 5-*O*-cinnamoyl ester hydroquinone renieramycin M (**246**) (Fig. 29), was revealed to significantly induced apoptosis by increasing expression of apoptosis-inducing

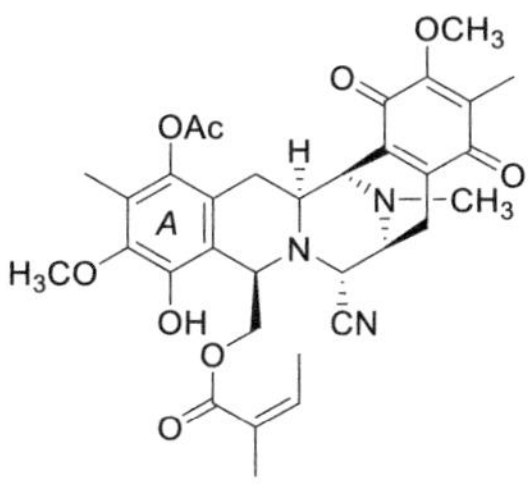

243 5-*O*-acetylated hydroquinone derivative of renieramycin M

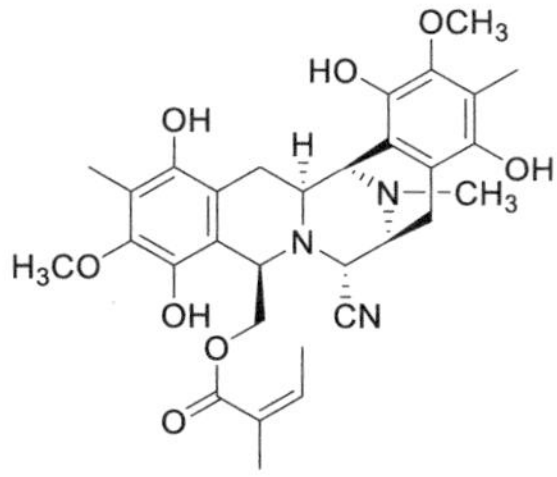

244 bishydroquinone renieramycin M

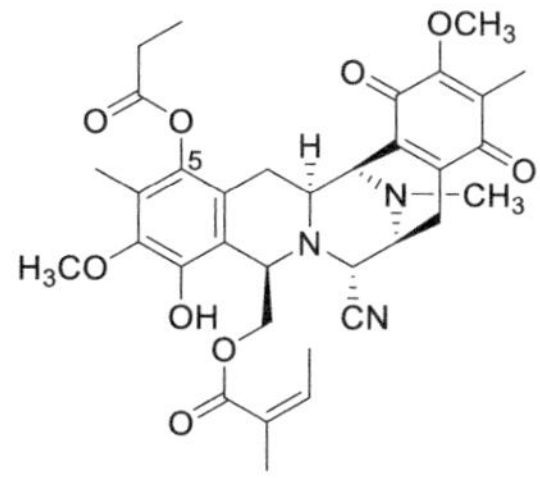

245

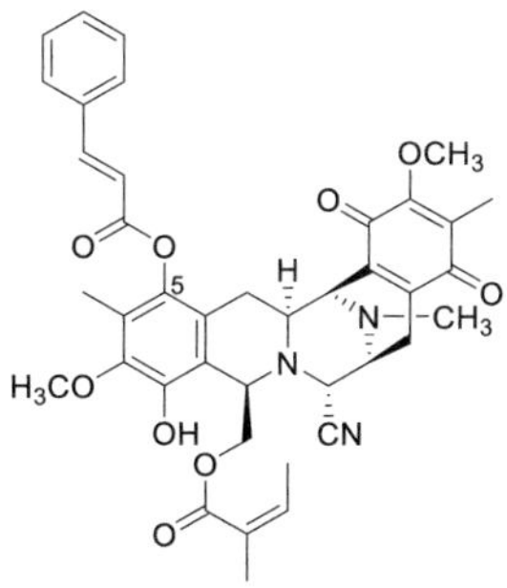

246 hydroquinone 5-*O*-cinnamoyl ester derivative of renieramycin M

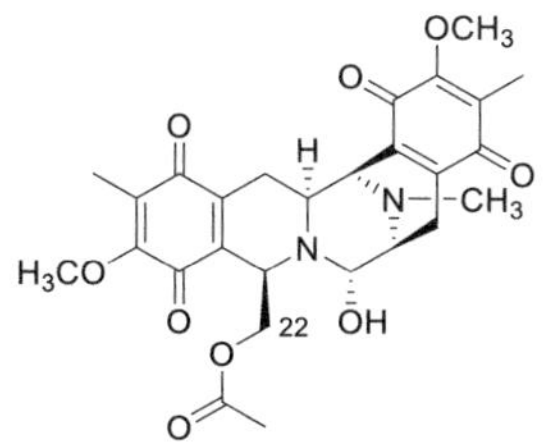

247 Jorumycin

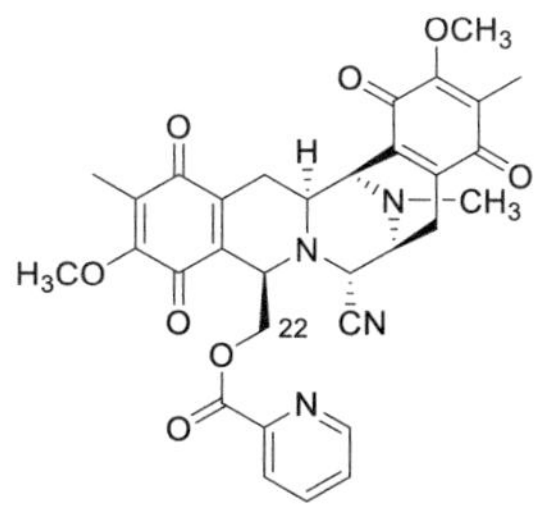

248

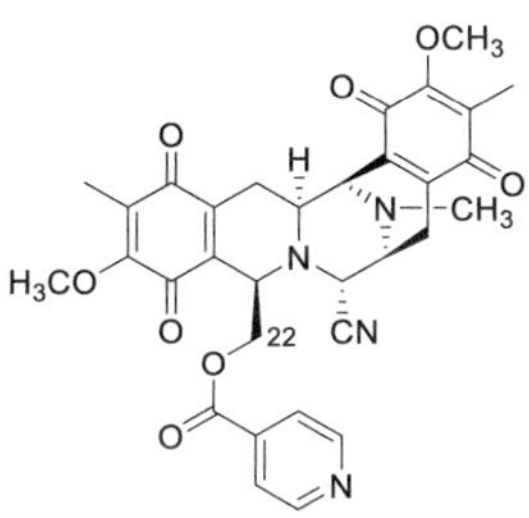

249

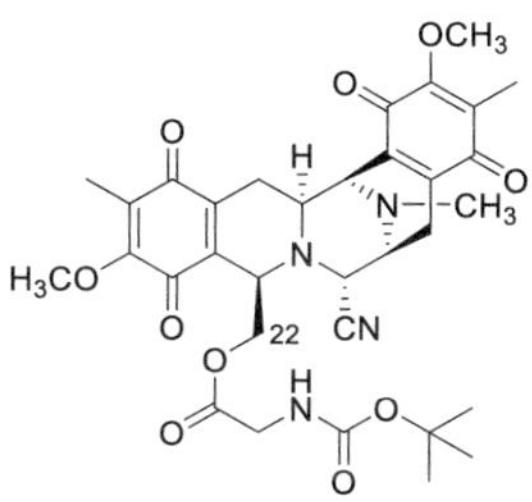

250 22-*O*-(*N*-Boc-L-glycine) ester of renieramycin M

Fig. 29 Chemical structures of renieramycin M derivatives **243–250** and jorumycin

factors in H292 lung cancer calls [154]. 5-*O*-Cinnamoyl ester hydroquinone renieramycin M (**246**) was synthesized easily by a two-step chemical modification of renieramycin M via Pd catalyzed hydrogenation and Steglich esterification [154]. Interestingly, this analog was also shown to suppress lung cancer stem cells by targeting Akt and destabilizing c-Myc [155]. Studies have shown that the abnormal overexpression/activation of Akt has been observed in many cancers, such as ovarian, lung and pancreatic cancers, and is also associated with increased cancer cell proliferation and survival. As such, targeting Akt is potentially an important approach for cancer prevention and therapy [156]. In addition, c-Myc is a transcription factor that

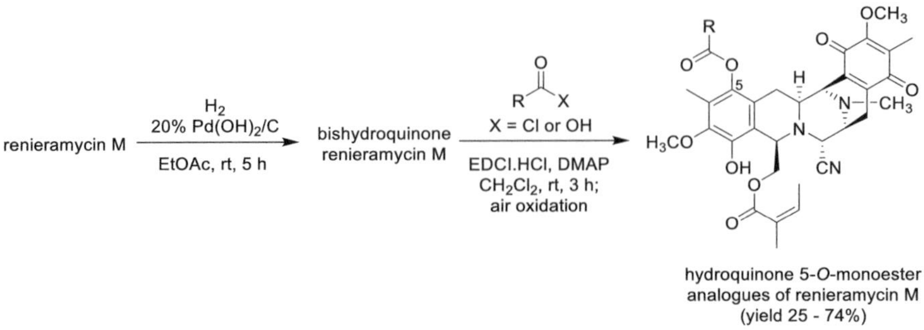

Scheme 3 Preparation of hydroquinone 5-*O*-monoester analogues of renieramycin M

is constitutively expressed in more than 70% of human cancers, and its inhibition has been shown to induce rapid tumor regression in animal model [157]. These series of semisynthetic works demonstrated that the hydroquinone 5-*O*-monoester analogues are a new generation of renieramycins that can be further developed as potential drug candidates for lung cancer treatment.

Based on observations that the different side chains at C-22 of renieramycin M (**228**), jorunnamycins A (**240**) and C (**242**) can affect cytotoxicity, the preparation and cytotoxic assessment of 22-*O*-ester derivatives of jorunnamycin A ensued. The semisynthesis of jorunnamycin A can be prepared by the well-known three step procedure, including hydrogenation, hydride reduction and air oxidation, from renieramycin M (Scheme 4) [158, 159]. Once jorunnamycin A is formed, it can be further transformed to antitumor jorumycin (**247**), which was previously isolated in minute amounts from the mantle and mucus of the Pacific nudibranch, *Jorunna funebris* [160]. Further biological assessment of jorumycin and jorunnamycins A and C showed growth inhibition at micromolar concentrations when tested against two human carcinoma cell lines, HCT-116 and QG 56. These series of preliminary biological evaluations provided the basis for initiating investigations on the biological effects of various ester side chains at C-22.

The first study by Charupant and co-workers focused on the cytotoxic evaluation of twenty-four 22-*O*-ester derivatives prepared from renieramycin M via jorunnamycin A [159]. They found that introducing nitrogen-heterocyclic esters, such as pyridinecarbonyl and quinolinecarbonyl units, resulted in either similar or increased cytotoxicity (e.g., compound **248**) (Fig. 28) compared to renieramycin M against colon and breast cancer cell lines. An additional series of eighteen 22-*O*-ester derivatives of jorunnamycin A, having benzoyl and trans-cinnamoyl moieties with electronegative and electron withdrawing functional groups together with nitrogen-heterocyclic acryloyl moieties, was prepared from renieramycin M [161]. Many of these synthetic analogues were also prepared from starting material, renieramycin M, via jorunnamycin A. Acylation of jorunnamycin A with a corresponding acid chloride in the presence of pyridine as a solvent and 4-dimethylaminopyridine (DMAP) as a catalyst at −17 °C for 1 h gave an acyclic ester as well as aromatic ester analogues

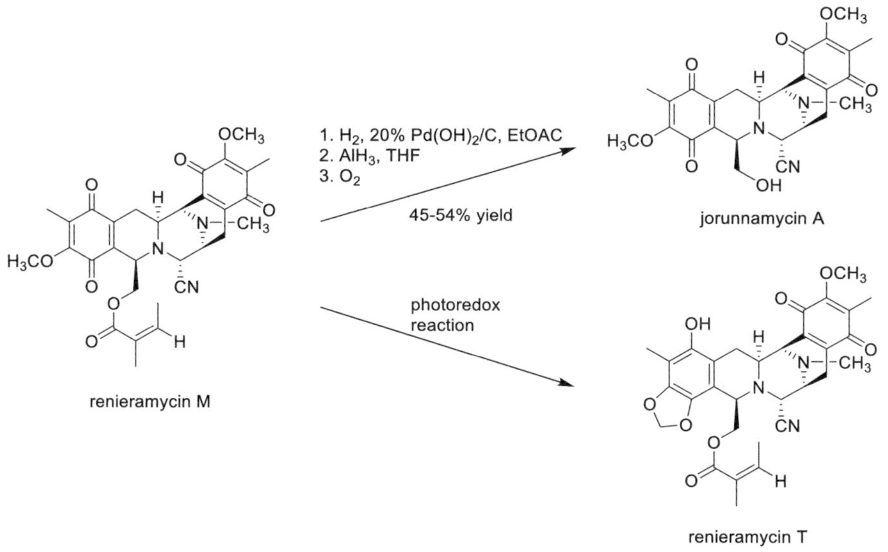

Scheme 4 Semisynthesis of jourunnamycin A and renieramycin T from renieramycin M

in 55–82% yields (Scheme 5). It was revealed that the 22-*O*-(4-pyridinecarbonyl) ester **249** (Fig. 28) is the most potent molecule with IC_{50} values of 1.1 and 1.6 nM against the H292 and H460 NSCLC cell lines, respectively. In addition, its in vitro anticancer activity showed strong cytotoxic and apoptosis-inducing effects in H460, H292 and A549 NSCLC cells. Furthermore, immunoblotting confirmed that **249** induced apoptotic cell death in an ERK/MEK/Bcl-2-dependent manner [162]. In a recent study, **249** inhibited tumor cell invasion and tube formation in human umbilical vein endothelial cells (HUVECs) as well as suppression of Akt/mTOR/p70S6K signaling which led to the downregulation of matrix metalloproteinases (MMP-2 and MMP-9), hypoxia-inducible factor-1α (HIF-1α) and vascular endothelial growth factor (VEGF) [140]. More importantly, this analog suppressed in vivo metastasis by decreasing the number of cancer colonies in the lung, showing that this metabolite could be useful for further clinical applications [163].

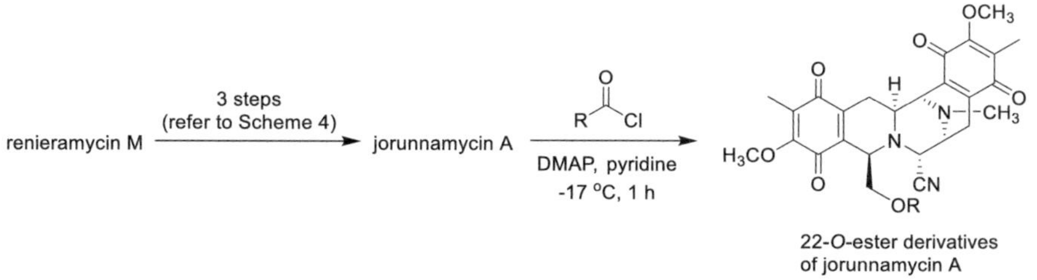

Scheme 5 Preparation of 22-*O*-ester derivatives of jorunnamycin A from renieramycin M

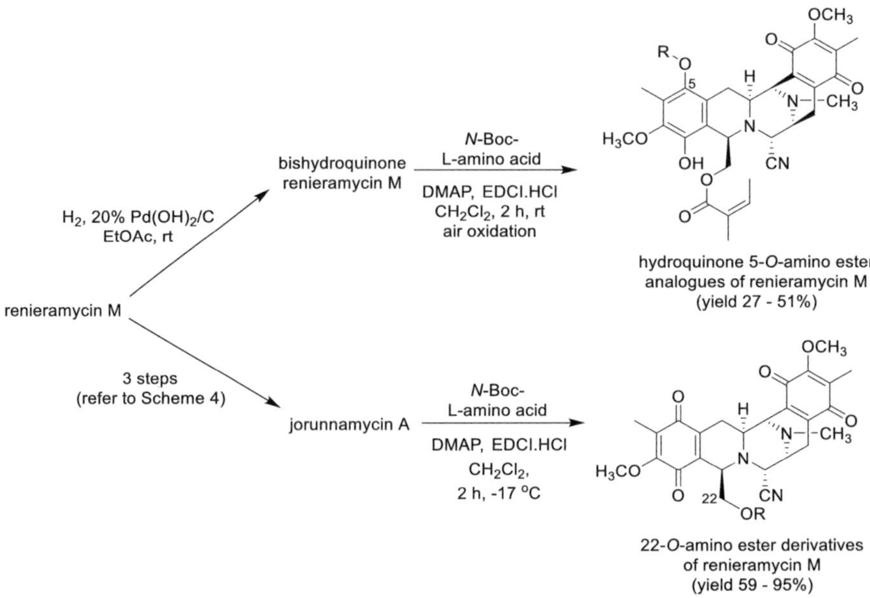

Scheme 6 Semi-syntheses of hydroquinone 5-*O*-amino ester and 22-*O*-amino ester derivatives from renieramycin M

Taking an inspiration from the structure of ecteinascidin 743 (**219**), where the presence of the aminoacyl-containing ester bridge on ring B of the tetrahydroisoquinoline core contributes to interactions with macromolecules, two new series of the 22-*O*-amino ester and hydroquinone 5-*O*-amino ester derivatives of renieramycin M were designed (Scheme 6) [164]. From these new series, the most cytotoxic derivative, 22-*O*-(*N*-Boc-L-glycine) ester of renieramycin M (**250**), was revealed to be seven-fold higher potency than renieramycin M (IC_{50} 24.56 nM) and 61-fold more than jorunnamycin A (IC_{50} 217.43 nM) against H292 cells. This analog was also shown to be a promising drug candidate for anti-metastasis treatment in lung cancer via inhibition of migratory features associated with suppression on epithelial-to-mesenchymal transition (EMT) in human lung cancer cells [165].

4.3 Renieramycin T Derivatives and Simplified Analogues

Having spending considerable ongoing productive efforts in generating semisynthetic analogues of renieramycin M/jorunnamycin A, the Thai and Japanese research groups have recently turned their attention to another potent cytotoxic analogue, renieramycin T (**237**) (Fig. 28). Both renieramycins M and T were isolated from the Thai blue sponge, *Xestospongia* sp., with the later molecule obtained as a minor component [133]. While renieramycin M (**228**) is classified as a bistetrahydroiso-quinolinequinone alkaloid, renieramycin T (**237**) is a renieramycin–ecteinascidin hybrid molecule, containing a 1,3-dioxole moiety on ring A. This moiety is similar to the known chemotherapeutic drugs trabectedin and lurbinectedin. Due to its limited quantity, few biological studies have been conducted on renieramycin T and its semisynthetic analogs.

Renieramycins M and T have shown potent nanomolar range cytotoxicity against a panel of cancer cell lines, including colon, lung, pancreatic and breast cacinomas, as well as induce apoptosis-mediated cell death [133, 146]. These findings are compelling evidence that modifications of the core structure with a solitary quinone moiety with a 1,3-dioxole unit on ring A of renieramycin T has the potential to generate anti-lung cancer therapeutic agents. In addition, it is known that renier-amycin–ecteinascidin hybrid alkaloids exhibits potent cytotoxicity, with unique cyto-toxic mechanism that is devoid of unwanted toxicity and amenable to mild synthetic approach. Further semisynthetic studies on renieramycin T are facilitated by its trans-formation from renieramycin M through ambient-induced intramolecular cyclization to renieramycin T in excellent yield (Scheme 4) [166, 167]. It is envisioned that the transformation involves an initial 1,5-H abstraction to generate a biradical species. Subsequent electron transfer provides a zwitterion and intramolecular cyclization to give a dioxole ring (Scheme 7).

Initial synthetic studies involved the use of naturally derived renieramycin T, via esterification method, generated only two derivatives, namely 5-*O*-acetyl renier-amycin T (**251**) and 5-*O*-(*N*-Boc-L-alanine)-renieramycin T (**252**) (Fig. 30) [168, 169]. 5-*O*-Acetyl renieramycin T (**251**) was found to induce cell death in H292 lung cancer cells, via p53-dependent apoptosis, by suppression of the antiapoptotic B-cell lymphoma-2 protein as well as reduction of the proapoptotic Bax protein [168]. The second analog, 5-*O*-(*N*-Boc-L-alanine)-renieramycin T (**252**), induced spheroid formation and apoptosis in lung cancer cells, while inhibiting CSC signals by suppressing the Akt protein [169].

Scheme 7 Proposed mechanism of photoredox process to generate the dioxole ring in renieramycin T

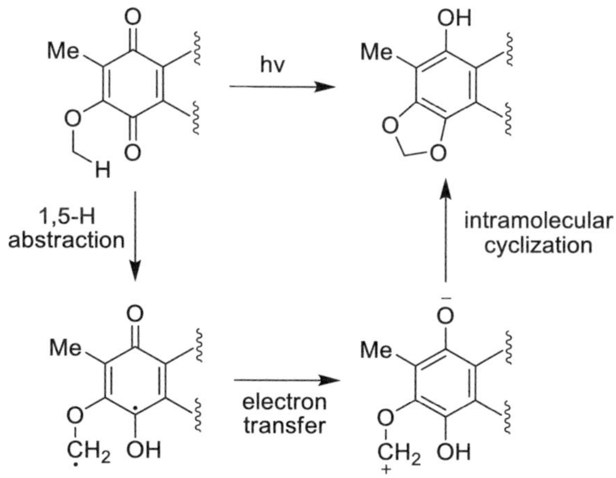

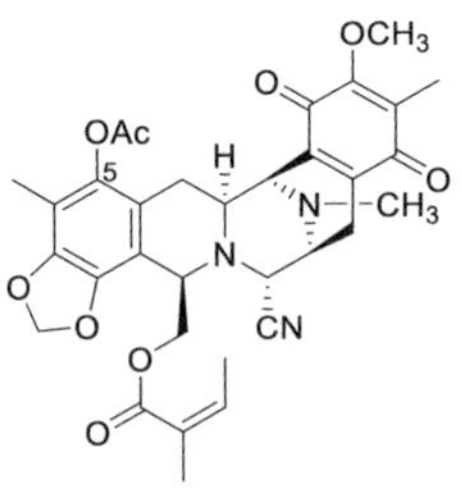

251 5-*O*-acetyl-renieramycin T

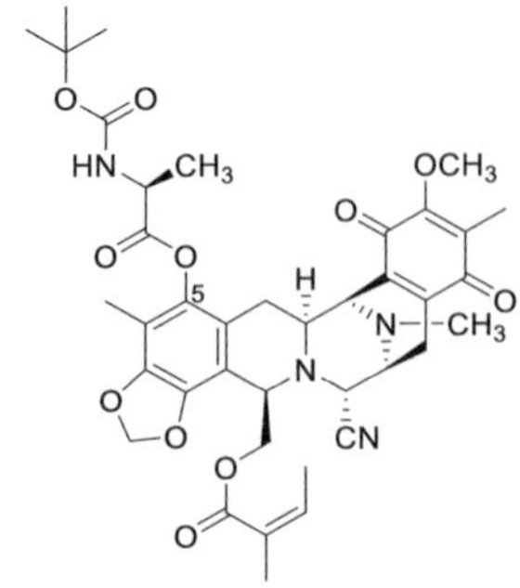

252 5-*O*-(*N*-Boc-L-alanine)-renieramycin T

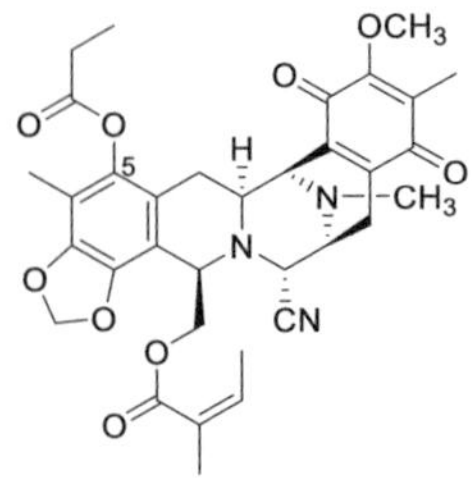

253 5-*O*-(3-propanoyl) ester of renieramycin T

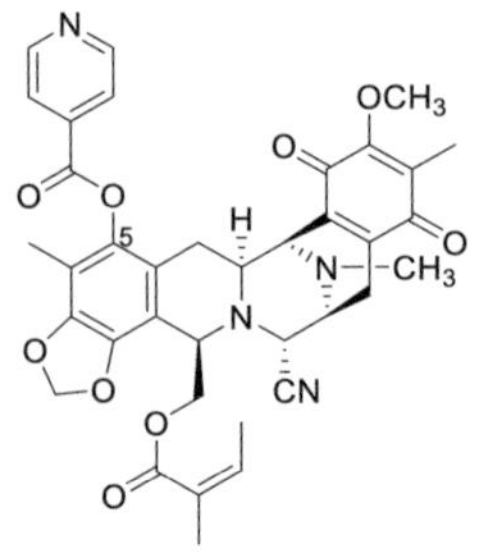

254 5-*O*-(4'-pyridinecarbonyl) renieramycin T

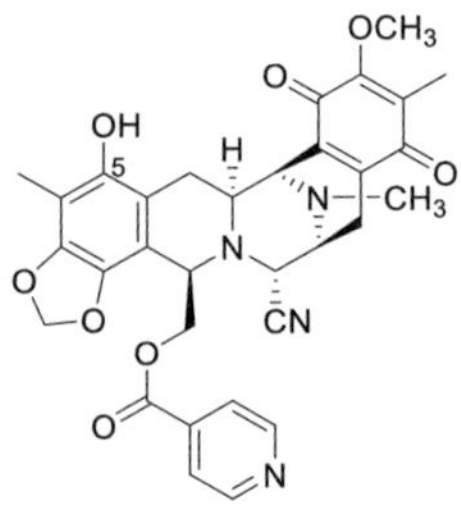

255 22-*O*-(4'-pyridinecarbonyl) renieramycin T

Fig. 30 Chemical structures of renieramycin T derivatives **251–255**

Using intramolecular photoredox reaction on renieramycin M to form renieramycin T, a series of 5-*O*-ester and 22-*O*-ester derivatives of renieramycin T were then generated [170, 171]. Initial optimization reaction conditions using 18-W fluorescent lamp (white light) in CH_2Cl_2 for 24 h transformed renieramycin M to renieramycin T in 64% yield. Subsequent improvement to the yield of 81% for renieramycin T was achieved by exposing the reaction mixture to irradiation by a 4-W light-emitting diode lamp (blue light) in CH_2Cl_2 (Scheme 8) [171]. Of the tested analogues, 5-*O*-(3-propanoyl) ester of renieramycin T (**253**) (Fig. 30) exhibited potent cytotoxic activity with IC_{50} values at 33.4 and 33.8 nM against H292 and H460 cell lines, respectively as well as induced apoptosis with minimal necrosis [170]. Another renieramycin T analog, 5-*O*-(4′-pyridinecarbonyl) renieramycin T (**254**) (Fig. 30), showed similar cytotoxicity with IC_{50} values of 35.27 and 34.77 nM against H290 and H460 cells, respectively [171]. Compound **254** was synthesized from initial irradiation of renieramycin M to form renieramycin T under the 4-W LED light in CH_2Cl_2. Once formed, renieramycin T underwent Steglich esterification using isonicotinoyl chloride as acylating agent, 1-ethyl-3-(3-dimethylaminopropyl) carbodiimide (EDCI) as a coupling agent and 4-dimethylaminopyridine (DMAP) as a nucleophilic base catalyst to form compound **254** in 20% yield (Scheme 8). Compounds **253** and **254** were two-fold more potent than that of renieramycin T but similar in potency to

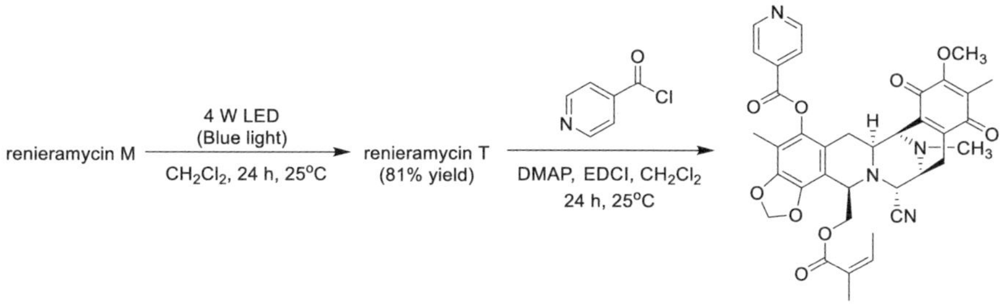

5-*O*-(4'-pyridinecarbonyl)
renieramycin T

Scheme 8 Semisynthesis of 5-*O*-(4 -pyridinecarbonyl) renieramycin T (**254**)

that of renieramycin M. Surprisingly, other renieramycin T derivatives [e.g. 22-*O*-(4′-pyridinecarbonyl) renieramycin T (**255**) (Fig. 30)] with a hydroxyl group and 4′-pyridinecarbonyl at C-5 and C-22, respectively, exhibited weaker cytotoxicity.

Recently, simplified right-half and left-half derivatives of renieramycin T were also generated as potential anticancer agents for SAR studies and treatment of lung cancers. These simplified analogues were created due to the need to understand the SARs of several chemical moieties present in the complex structure of renieramycin T. These synthetic efforts concluded that the right-half derivatives of renieramycin T exhibited higher levels of anticancer activity than the left-half [172, 173].

Such studies were initiated by developing a general protocol for the asymmetrical synthesis of 3-*N*-arylmethylated right-half model molecules of renieramycins [172]. This initial work led to the identification of compound **257** (Fig. 31) to exhibit potent cytotoxicity against human prostate cancer DU145 and colorectal cancer HCT116 cell lines with IC_{50} values of 11.9 and 12.5 nM, respectively. Further work by Petsri and co-workers found that of the five simplified synthetic molecules (namely **256–260**), two analogues, namely TM-(–)-18 (**256**) and TM-(–)-4a (**257**) (Fig. 31) as well as renieramycin T exerted anti-cancer activities with the loss of myeloid cell leukemia 1 (Mcl-1) and partly reduced B-cell lymphoma 2 (Bcl-2) [147]. In particular, the cyanide and benzene ring units in renieramycin T structure were identified to be essential for its Mcl-1-targeting activity. It is proposed that the interaction of renieramyin T or simplified analogs with Mcl-1 protein will lead to ubiquitination and proteasomal degradation of Mcl-1 protein. It is known that Mcl-1 has important function to inhibit apoptosis by interacting and sequestering the pro-apoptotic Bak (Bcl-2 antagonist killer 1) protein [174]. The reduced Mcl-1 protein, due to interaction with renieramycin T/simplified analogs, could then facilitate Bak dimerization, mitochondrial membrane destabilization, cytochrome c release and eventual apoptosis. The syntheses of compounds **256–260** (Fig. 31) were achieved through a key molecule, **261** (Fig. 31), which is easily synthesized from L-Tyr (Scheme 9) [173, 175].

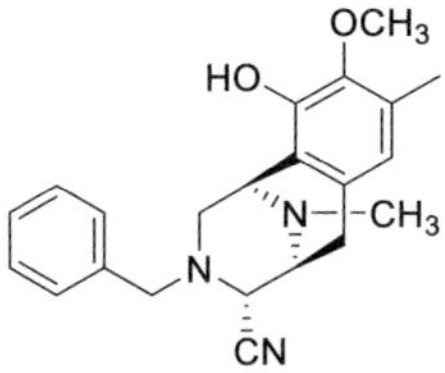
256 TM-(-)-18

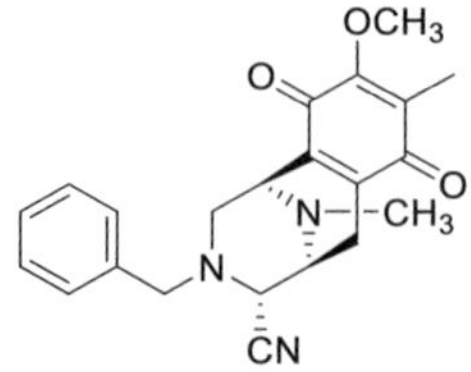
257 TM-(-)-4a

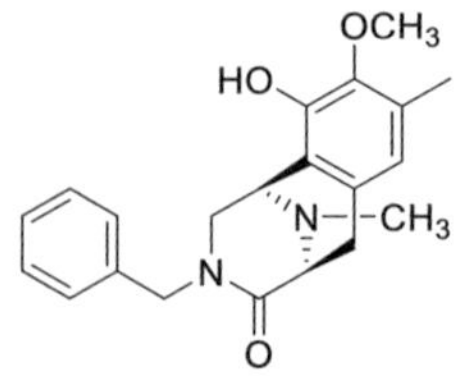
258 TM-(-)-45

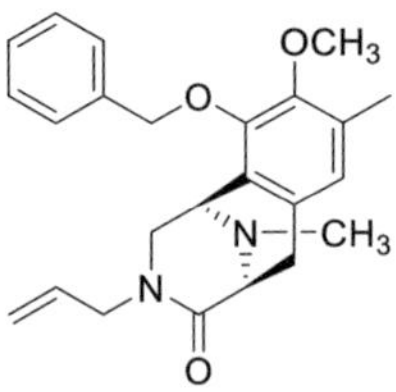
259 TM-(-)-52

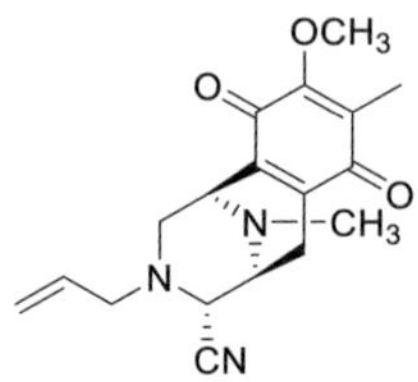
260 TM-(-)-55

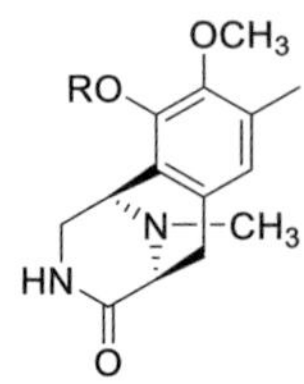
261 R = H
262 R = Bn

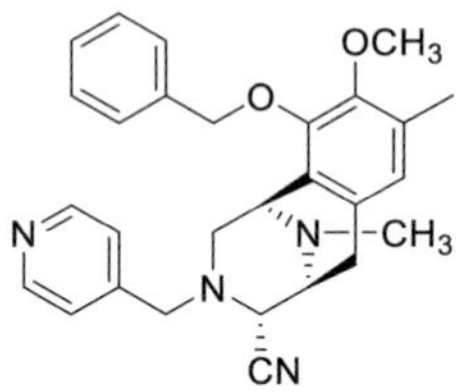
263 DH_22

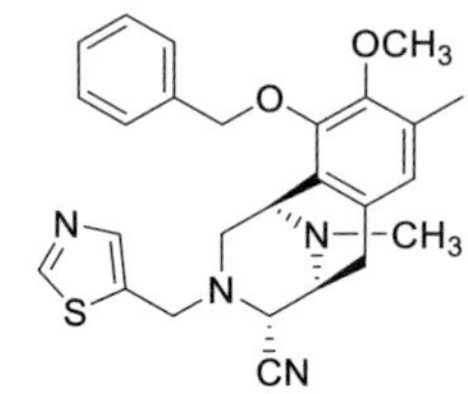
264 DH_25

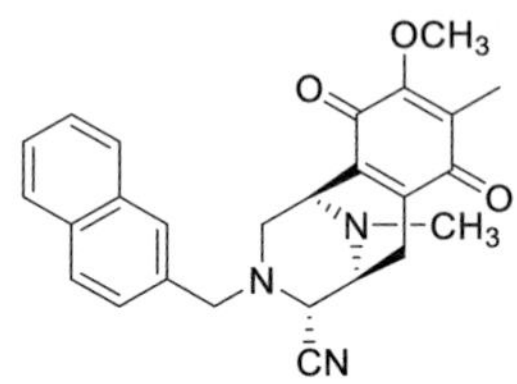
265 DH_32

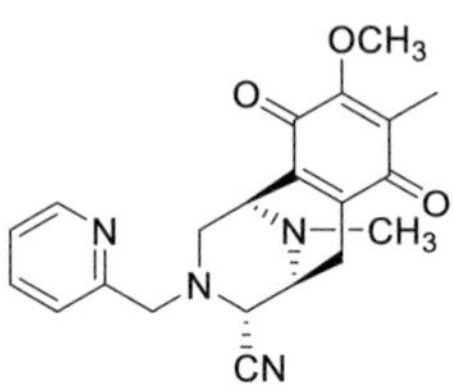
266 DH_18

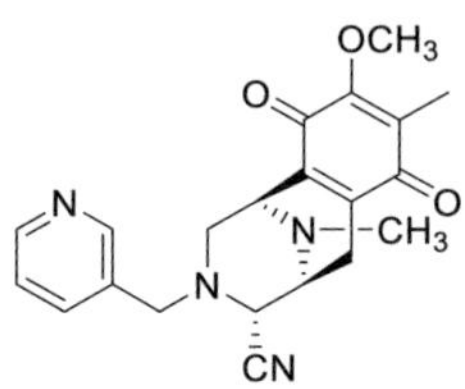
267 DH_21

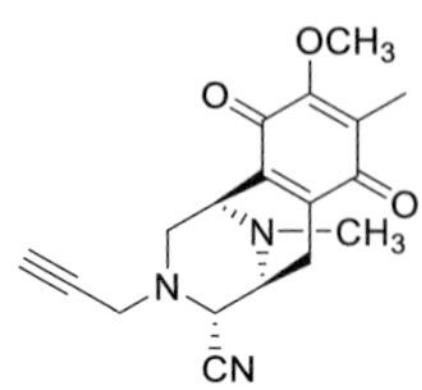
268 DH_35

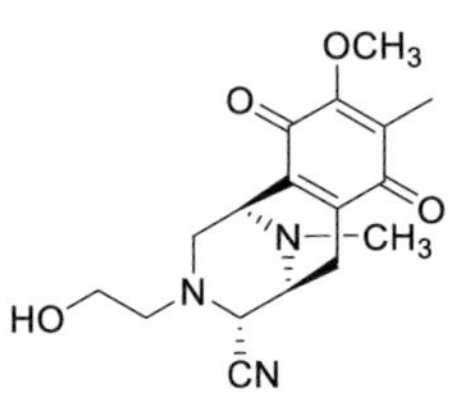
269 DH_38

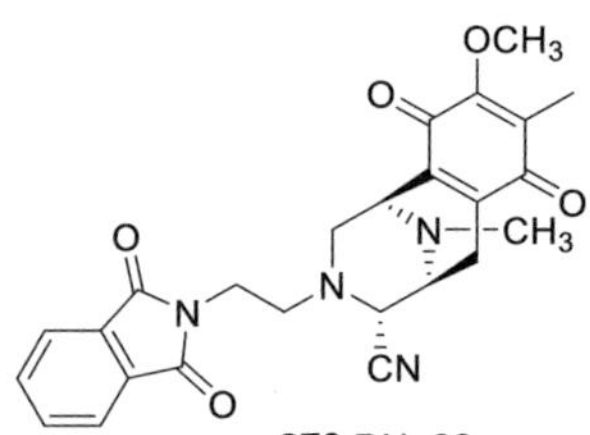
270 DH_39

Fig. 31 Chemical structures of simplified renieramycin T derivatives **256–270**

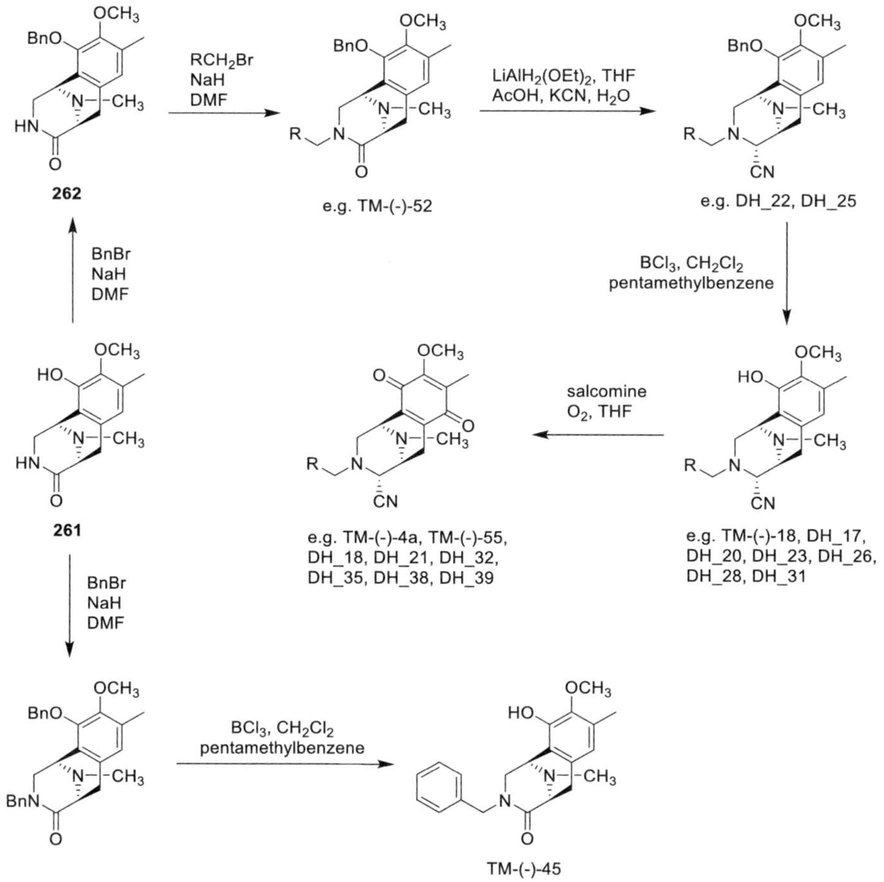

Scheme 9 Asymmetric synthesis of derivatives of renieramycin T right-half analogs

Three additional simplified analogues, such as DH_22 (**263**), DH_25 (**264**) and DH_32 (**265**) (Fig. 31), were also identified to possess potent anti-lung cancer activities by targeting various pathways/proteins, including degradation of β-catenin through ubiquitin–proteasomal, p53-dependent apoptotic as well as the Akt signaling pathways [176–178]. Similar to **256–260**, simplified compounds **263–270** were synthesized from **262** (Fig. 31), which in turn can be synthesized from L-Tyr (Scheme 9). The formation of DH_22, DH_25 and related derivatives was carried out in two steps, including activation of the lactam nitrogen with sodium hydride, adding of in situ prepared bromide as well as reduction with $LiAlH_2(OEt)_2$ in THF, from **262** (Scheme 9) [177]. Additional steps, such as debenzylation using BCl_3

in the presence of pentamethyl-benzene as well as oxidation with O_2 in the presence of salcomine, were needed to afford *p*-quinone present in DH_32 and related compounds (Scheme 9) [178].

DH_22 exhibited cytotoxicity with an IC_{50} value of 13.27 μM when tested against the NSCLC cell line A549. Based on a series of biochemical analyses, the compound contributed to the activation of p53-dependent apoptosis and decreased the cellular level of c-Myc [176]. Compound DH_25, having a thiazole ring, was also revealed to have the ability to induce apoptosis in lung cancer cells, which was accompanied by an increase in PARP cleavage, a decrease in Bcl-2 as well as Mcl-1 [177]. In addition, virtual molecular docking showed a high-affinity interaction between DH_25 and an Akt at the allosteric binding site, which could result in Akt inhibition. Akt is a known key regulatory protein of cancer stem cells and is responsible for cancer aggressiveness and metastasis. As such, Akt is a validated target for the development of anticancer drugs [179]. Lastly, from a series of simplified *p*-quinone-containing analogues (**265**–**270**) (Fig. 31), DH_32 (**265**) was found to exhibit the highest efficacy against various lung cancer cell lines, including A549, H23 and H292 cells, with IC_{50} values ranging from 1.52 to 5.76 μM [178]. DH_32 was also found to significantly suppress the formation of tumor spheroids as compared to renieramycin T. It also inhibited CSCs by promoting the degradation of β-catenin via ubiquitin-proteasomal pathways. Since aberrant Wnt/β-catenin signaling pathway facilitates cancer stem cell renewal, cell proliferation and differentiation, developing therapeutic agents targeting this pathway has potential in cancer therapy [180]. A high affinity binding of DH_32 to β-catenin was revealed using molecular docking analysis.

In a recent synthetic work by Phookphan and co-workers, they investigated the effect of reneiramycin T right-half analogs on STAT3 in lung cancer cells [181]. These new simplified analogues, including DH_17 (**271**), DH_20 (**272**), DH_23 (**273**), DH_26 (**274**), DH_28 (**275**), DH_30 (**276**) and DH_31 (**277**) (Fig. 32), were synthesized by converting the benzyl ether to phenol in the E ring and incorporating the pyridyl, thiazolyl or naphthalyl group in the ring C of the reneiramycin T right-half analog. This led to the discovery of a simplified analogue, DH_31 (**277**), that showed potent activity against H292 and H460 cells, with IC_{50} values of 5.54 μM and 2.9 μM, respectively [181]. In addition, the anticancer properties of DH_31 in non-small cell lung cancer (NSCLC) include its ability to inhibit cell migration and anoikis resistance by downregulating the protein expression of STAT3 and its downstream targets [181]. Moreover, DH_31 promotes STAT3 depletion through enhanced ubiquitin-proteasomal degradation. These findings suggest that DH_31 holds promise as a therapeutic candidate for metastatic cancer treatment. Consequently, DH_31 should be explored further as a lead compound for the development of STAT3-targeted anticancer drugs. Taken together, results from these series of simplified analogues showed their great potential for further development as new anticancer drugs and should be further evaluated in in vivo testing.

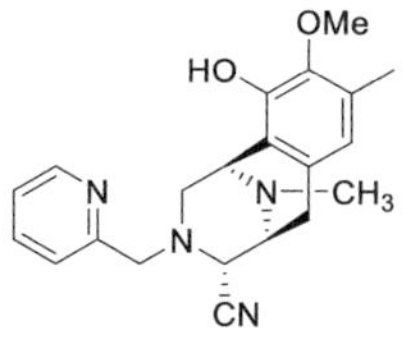

271 DH_17

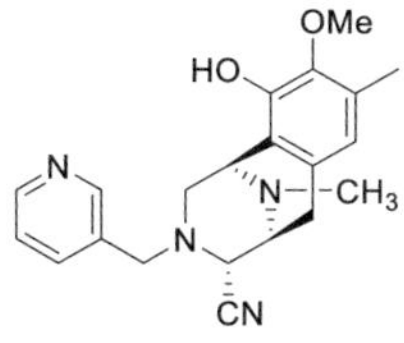

272 DH_20

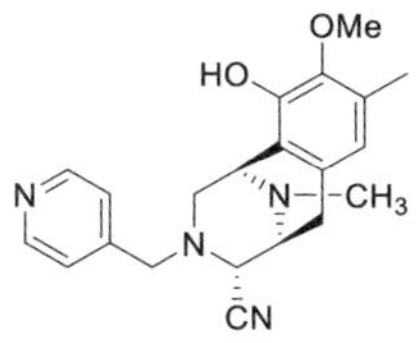

273 DH_23

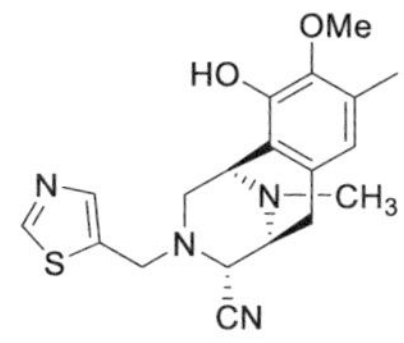

274 DH_26

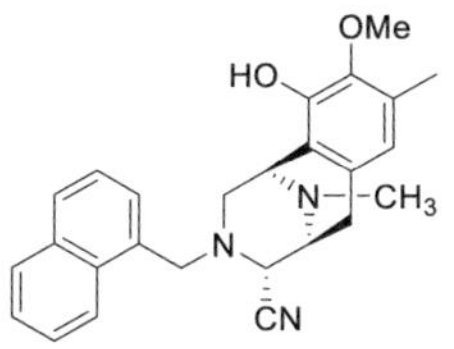

275 DH_28

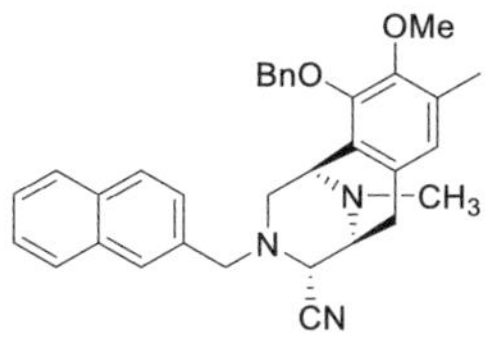

276 DH_30

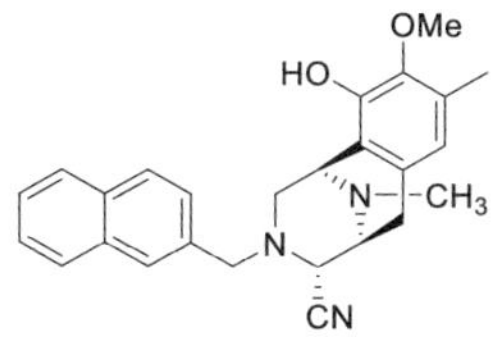

277 DH_31

Fig. 32 Chemical structures of simplified renieramycin T derivatives **271–277**

5 Ecology and Aquaculture of Renieramycin-Containing Marine Sponges

Upon the discovery of the Thai blue sponge, *Xestospongia* sp., to contain renieramycin metabolites, a study by Daumas and co-workers was initiated to investigate the distribution patterns of this sponge species and associated reef organisms from the Gulf of Thailand. Samples of this sponge species, obtained from the coral reefs of the Gulf of Thailand, were found associated with other reef organisms, including corals (e.g. *Porites lutea*) and zoanthid (e.g. *Palythoa caesia*), as well as attached to rock beds and dead coral rubble [182]. Differences in renieramycin M concentrations were also found at different collection sites and were dependent on partnership with coexisting marine species. For instance, there were direct correlations between the highest maximum compound concentration and maximum average frequency of marine organisms found associated with *Xestospongia* sp. This was observed at Samui, Hin Kob and Sumpayu where partnership between the sponge and algae, *P. lutea* and *P. caesia*, respectively, recorded the maximum renieramycin M concentration [182]. This suggested that *Xestospongia* sp. could produce renieramycin M for spatial competition possibly via the tissue-contact mode. In another recent study by Darumas and co-workers investigated the variation of carbon–nitrogen contents and renieramycin M concentrations within different regions, including edge, inner and outer areas, of Xestospongia sp. coexisting with *Porites lutea* and *Palythoa caesia* in laboratory setting [183]. The study revealed that the carbon–nitrogen contents as well as renieramycin M concentrations of *Xestospongia* sp. coexisting with different organisms were not significantly different, confirming the concentration of renieramycin M in the sponge to be uniform. In addition, renieramycin M did not show any allelopathic effect on *P. lutea* and *P. caesia*. However, the compound inhibited settlement of acorn barnacle (*Semibalanus balanoides*) but did not prevent/inhibit the settlement of pelecypods and growth of aerobic bacteria in any treatments [183].

Due to the exquisite cytotoxic nature of renieramycin M as well as using it as starting material to generate synthetic analogs, the culturing of the Thai blue sponge, *Xestospongia* sp., to produce the alkaloid in clinical investigations was explored in several studies. As the sponge has brilliant blue color, its cultivation in hatchery, using water circulation close system with vertical long line rope method, as an ornamental species for local market in Thailand was initially explored [184]. Subsequent maricultivation in natural environment for production of the alkaloid and derivatives was carried out over a period of 10 months and results were promising [185]. The blue sponge was cultured in the Trang coastline using the vertical long line method on three different substrates, including cement, plastic rope and PVC pipe. Growth of the sponge animal varies according to season with no observable trends. The sponge growth fluctuated within a short range in summer and beginning of the monsoon period (February to July). This was followed by a sudden decrease in growth in the middle of the monsoon season from August to November, suggesting that the water quality, such as salinity, DO and dissolved silicate, could affect the animal growth. The use of cement as substrate was preferred as the sponge growth showed plump shape with brilliant blue color comparable to sponges in found in nature. More importantly, the maximum concentration of renieramycin M production was found during monsoon season in October coinciding with a decrease in sponge growth [185]. The authors suggested that metabolite production could be induced by stress and energy budget processes. Similar open-sea mariculture approach was also conducted for the renieramycin-producing Philippine blue sponge, *Xestospongia* sp. [186].

In addition to open-sea maricultivation of *Xestospongia* sp., at least two studies, using semi-circulate close system, were conducted to investigate abiotic and biotic factors involved in stimulation of renieramycin M production in sponges. For instance, a study by Kieattisak and co-workers was able to demonstrate the production of alkaloid by exposing the sponge samples to different concentrations of calcium/magnesium (Ca/Mg) in natural seawater [187]. Their results revealed that a Ca/Mg concentration at 430/1230 ppm stimulated maximum alkaloid accumulation in the sponge tissue (1.74 mg/1500 mg tissue), while sponges supplemented with Ca/Mg at a level of 410/1170 ppm showed the most growth. In another recent study, two phytoplankton species, *Chaetocerose gracilis* and *Nannochloropsis* sp., were used as live feed to determine the growth and production of renieramycin M in *Xestospongia* sp. [188]. Sponges fed by *C. gracilis* weighed significantly higher than sponges fed by *Nannochloropsis* sp. In spite of relatively higher maximum alkaloid accumulation found in sponges fed by *Nannochloropsis* sp. showed maximum RM accumulation (0.32 mg/1500 mg tissue), the metabolite accumulation was not significantly different among treatment groups. The study also found the presence of *Ca. E. renieramycinifaciens* in sponge, suggesting that renieramycin M is most likely biosynthesized by the bacterial symbiont.

6 Conclusions

This comprehensive review on sponge-derived natural products reported by Thai researchers and their collaborators revealed many structurally unique bioactive molecules. A majority of these compounds belong to the terpenoid and alkaloid structural classes, isolated from sponges belonging to several taxonomic orders, including order Dictyoceratida, Suberitida, Haplosclerida, Poecilosclerida, Tetractinellida, Axinellida, Verongiida, Podocopida, Dendroceratida, Scopalinida, Homosclerophorida and Clionaida. Several of these sponge orders are also known to be prolific producers of bioactive natural products. For instance, order Dictyoceratida is the highest producer among all the sponge orders, contributing over 20% of new secondary metabolites [189]. Not surprisingly, these Thai sponges targeted by Thai researchers were collected from marine areas with high sponge diversity, namely from Krabi (Phi Phi, Ko Ha), Trang (Sikao), Phang Nga (Similan) Provinces at the Andaman Sea as well as Surat Thani (Koh-Tao), Chonburi (Ko Sichang), Rayong (Koh Semed), Trat and Chumphon (Chumphon Islands National Park) Provinces within the Gulf of Thailand.

Amongst the many bioactive compounds isolated from Thai sponges, the most noteworthy are the kabiramides and renieramycins reported from the Thai black sponge, *Pachastrissa nux*, and the blue sponge, *Xestospongia* sp., respectively. In addition to strong antimalarial activity of kabiramides, e.g., kabiramides J and K, a variety of fluorescent probes based on kabiramide C were developed as molecular probes for staining the leading edge of actin polymerization and could be useful for staining motile cells. The development of these probes was facilitated by the reisolation of kabiramide C from the Thai sponge, *P. nux*. Significant efforts have been accomplished by Thai researchers and international collaborators on the medicinal chemistry of renieramycin M and their semi-synthetic derivatives as potential anticancer drugs, especially for the treatment of lung cancers. The isolation of renieramycin M in high amounts allowed its chemical transformation into more than 90 semisynthetic derivatives. Moreover, simplified synthetic renieramycin T right-half analogues, such as DH_22 (**263**), DH_25 (**264**) and DH_32 (**265**), were also generated as potential anticancer agents for SAR studies and treatment of lung cancers. Several simplified analogues have also shown their potent interactions with the Akt pathway and downregulation of anti-apoptotic proteins, such as Mcl-1 and Bcl-2 proteins. It is without doubt that continued medicinal chemistry research by Thai researchers and their collaborators on local sponges-derived compounds and synthetic derivatives will lead to the uncovering and development of potent pharmaceutical agents.

References

1. Myers N, Mittermeier RA, Mittermeier CG, da Fonseca GAB, Kent J (2000) Biodiversity hotspots for conservation priorities. Nature 403:853–858

2. Brown B, Clarke K, Warwick R (2002) Serial patterns of biodiversity change in corals across shallow reef flats in Ko Phuket, Thailand, due to the effects of local (sedimentation) and regional (climatic) perturbations. Mar Biol 141:21–29
3. Phongsuwan N, Chankong A, Yamarunpatthana C, Chansang H, Boonprakob R, Petchkumnerd P, Thongtham N, Paokantha S, Chanmethakul T, Panchaiyapoom P (2013) Status and changing patterns on coral reefs in Thailand during the last two decades. Deep Sea Res Part II 96:19–24
4. Yeemin T, Sutthacheep M, Pettongma R (2006) Coral reef restoration projects in Thailand. Ocean Coast Manag 49:562–575
5. Varijakzhan D, Loh J-Y, Yap W-S, Yusoff K, Seboussi R, Lim S-HE, Lai K-S, Chong C-M (2021) Bioactive compounds from marine sponges: fundamentals and applications. Mar Drugs 19:246
6. Lim S-C, Putchakarn S, Thai M-Q, Wang D, Huang YM (2016) Inventory of sponge fauna from the Singapore Strait to Taiwan Strait along the western coastline of the South China Sea. Raffles Bull Zool Suppl 34:104–129
7. Erpenbeck D, van Soest RWM (2002) Family halichondriidae gray, 1867. In: Hooper JNA, van Soest RWM, Willenz P (eds) Systema Porifera: a guide to the classification of sponges. Springer, Boston, MA, pp 787–815
8. McCauley RD, Riddle MJ, Sorokin SJ, Murphy PT, Goldsworthy PM, McKenna AJ, Baker JT, Kelley RA (1993) AIMS bioactivity unit marine invertebrate collection. I: Western Australia, in AIMS Report Number 8. Australian Institute of Marine Science, Perth, Western Australia, p 76
9. Hooper JNA, Kennedy JA, van Soest RWM (2000) Annotated checklist of sponges (Porifera) of the South China Sea region. Raffles Bull Zool Suppl 8:125–207
10. Kritsanapuntu S, Chaitanawisuti N, Yeemin T, Putchakan S (2001) First investigation on biodiversity of marine sponges associated with reef coral habitats in the eastern Gulf of Thailand. Asian Mar Biol 18:105–115
11. Hooper JNA, Kelly-Borges MM, Riddle MMJ (1993) *Oceanapia saggitaria* from the Gulf of Thailand. Memoirs of the Queensland Museum 33:61–72
12. Asa S, Yeemin T, Chaitanawisuti N, Kritsanapuntu S (2000) Sexual reproduction of a marine sponge, *Petrosia* sp. from coral communities in the Gulf of Thailand. In: Proceedings 9th international coral reef symposium. Bali, Indonesia
13. Putchakarn S, Sonchaeng P, van Soest RWM (2006) The demosponges dwelling in the coral reefs from Khram Islands, the eastern coast of the Gulf of Thailand. In: Custodio MR, Lobo-Hajdu G, Hajdu E, Muricy G (eds) 7th International sponge symposium. biodiversity, innovation, sustainability. Armacao de Buzios, 7–13 May, 2006, Museu Nacional Rio de Janeiro Serie Livros
14. Putchakarn S, Monkongsomboon S, Noiraksa T, Sonchaeng P (2004) Biodiversity of marine animals in coral reefs along the eastern coast of Thailand (Chonburi Province). Burapha University Research Grant, Thailand
15. Putchakarn S, Weerdt WH, Soncheang P, van Soest R (2004) A new species of *Cladocroce* Topsent, 1892 (Porifera, Haplosclerida) from the Gulf of Thailand. Beaufortia 54:113–116
16. Putchakarn S (2006) Biodiversity of sponges (Demospongiae, Porifera) in the gulf of Thailand. Burapha University, p 200
17. Putchakarn S (2007) Species diversity of marine sponges dwelling in coral reefs in Had Khanom—Mo Ko Thale Tai National Park, Nakhon Si Thammarat Province, Thailand. J Mar Biol Assoc UK 87:1635–1642
18. Hongpadharakiree K, Seehabutr V, Putchakarn S, (2008) Preliminary study on species diversity of sponges in coral reef area around Koh Ka, Chumphon province, Gulf of Thailand, Thailand. In: Proceedings of 46th Kasetsart University Annual Conference: Fisheries. Kasetsart University, Bangkok
19. Putchakarn S, Monkongsomboon S, Sanpanich K, Desakulwattana C (2008) Biodiversity of sponges (Demospongiae, Porifera) from the eastern coast of Thailand: Chon Buri and Rayong Provinces. Burapha University Research Grant, Thailand

20. Pumbua W, Chobram W, Putchakarn S, Matchacheep S, Sanpanich K, Manthachitra V (2009) Species diversity of marine sponge along Nangrong Beach, Koh Jorake, and Mu Koh Chuang, Amphoe Sattahip, Chon Buri Province. In: Chuthamas P (ed) Proceeding of the 4th RSPG researchers club conference "Thai Resources: turn to the new way in Thai Way (Thai Innovation)". Khao Kheow Open Zoo, Si Racha, Chon Buri Province, Thailand
21. Chobram W, Pumbua W, Putchakarn S, Matchacheep S, Sanpanich K, Manthachitra V (2009) Species diversity of marine sponges (Phylum Porifera) from Mu Ko Tarutao and Mu Ko Adang-Rawi, Satun Province. In: Chuthamas P (ed) Proceeding of the 4th RSPG researchers club conference "Thai Resources: turn to the new way in Thai Way (Thai Innovation)". Chon Buri Province, Thailand
22. Hongpatarakiri K, Putchakarn S (2010) Species diversity of demosponges (Demospongiae, Porifera) around Koh Losin, Pattani province, Gulf of Thailand, Thailand. In: Proceedings of 48th Kasetsart University Annual Conference: Plants. Kasetsart University, Bangkok
23. Putchakarn S (2011) Species diversity of marine sponges along Chanthaburi and Trat Provinces, the eastern coast of the Gulf of Thailand. Publications of the Seto Marine Biological Laboratory 41:17–23
24. Putchakarn S (2011b) Species diversity of marine sponges (Demospongiae, Porifera) along the western coasts of the Gulf of Thailand, in Burapha University Research Grant. Burapha University, p 67
25. Putchakarn S (2012) Species diversity of marine sponges along the Phi Phi Islands., in The Thailand Research Fund, Fiscal Year 2009. Contract No. TRG5280006: Thailand, p 75
26. Putchakarn S (2013) Species diversity of marine sponges (Demospongiae, Porifera) along the western coasts of the Gulf of Thailand. Burapha University Research Grant, Thailand
27. Hongpatarakiri K, Putchakarn S (2013) Species diversity of marine sponges around Mu Ko Surin, Phangnga province, Thailand. In: Proceedings of 51st Kasetsart University Annual Conference: Veterinary Medicine, Fisheries. Kasetsart University, Bangkok
28. Alvi KA, Tenenbaum L, Crews P (1991) Anthelmintic polyfunctional nitrogen-containing terpenoids from marine sponges. J Nat Prod 54:71–78
29. Horton PA, Crews P (1995) The arenarans, sesquiterpene ethers from the marine sponge *Dysidea arenaria.* J Nat Prod 58:44–50
30. Rodriguez J, Quiñoá E, Riguera R, Peters BM, Abrell LM, Crews P (1992) The structures and stereochemistry of cytotoxic sesquiterpene quinones from *Dactylospongia elegans.* Tetrahedron 48:6667–6680
31. Yegdaneh A, Putchakarn S, Yuenyongsawad S, Ghannadi A, Plubrukarn A (2013) 3-oxoabolene and 1-oxocurcuphenol, aromatic bisabolanes from the sponge *Myrmekioderma* sp. Nat Product Commun 8, 1934578X1300801002
32. Wongbundit S, Mahidol C, Kaweetripob W, Prachyawarakorn V, Eurtivong C, Sahakitpichan P, Tuntiwachwuttikul P, Ruchirawat S, Prawat H (2020) Biscurcudiols, myrmekioperoxides, and myrmekiodermaral from the Thai marine sponge *Myrmekioderma* sp. Tetrahedron 76:131162
33. Satitpatipan V, Suwanborirux K (2004) New nitrogenous germacranes from a Thai marine sponge, *Axinyssa* n. sp. J Nat Prod 67:503–505
34. Cheng W, Liu D, Zhang F, Zhang Q, Pedpradab P, Proksch P, Liang H, Lin W (2014) Formamidobisabolene-based derivatives from a sponge *Axinyssa* sp. Tetrahedron 70:3576–3583
35. Prawat H, Mahidol C, Wittayalai S, Intachote P, Kanchanapoom T, Ruchirawat S (2011) Nitrogenous sesquiterpenes from the Thai marine sponge *Halichondria* sp. Tetrahedron 67:5651–5655
36. Prawat H, Mahidol C, Kaweetripob W, Prachyawarakorn V, Tuntiwachwuttikul P, Ruchirawat S (2016) Sesquiterpene isocyanides, isothiocyanates, thiocyanates, and formamides from the Thai sponge *Halichondria* sp. Tetrahedron 72:4222–4229
37. Prawat H, Mahidol C, Kaweetripob W, Wittayalai S, Ruchirawat S (2012) Iodo–sesquiterpene hydroquinone and brominated indole alkaloids from the Thai sponge *Smenospongia* sp. Tetrahedron 68:6881–6886

38. Jiso A, Kittiwisut S, Chantakul R, Yuenyongsawad S, Putchakarn S, Schäberle TF, Temkitthaworn P, Ingkaninan K, Chaithirayanon K, Plubrukarn A (2020) Quintaquinone, a merosesquiterpene from the yellow sponge *Verongula* cf. *rigida* Esper. J Nat Prod 83:532–536
39. Wattanapiromsakul C, Chanthathamrongsiri N, Bussarawit S, Yuenyongsawad S, Plubrukarn A, Suwanborirux K (2009) 8-Isocyanoamphilecta-11(20),15-diene, a new antimalarial isonitrile diterpene from the sponge *Ciocalapata* sp. Can J Chem 87:612–618
40. Chanthathamrongsiri N, Yuenyongsawad S, Wattanapiromsakul C, Plubrukarn A (2012) Bifunctionalized amphilectane diterpenes from the sponge *Stylissa* cf. *massa*. J Nat Prod 75:789–792
41. Tanaka J, Higa T, Suwanborirux K, Kokpol U, Bernardinelli G, Jefford CW (1993) Bioactive norsesterterpene 1,2-dioxanes from a Thai sponge, *Mycale* sp. J Org Chem 58:2999–3002
42. Phuwapraisirisan P, Matsunaga S, Fusetani N, Chaitanawisuti N, Kritsanapuntu S, Menasveta P (2003) Mycaperoxide H, a new cytotoxic norsesterterpene peroxide from a Thai marine sponge *Mycale* sp. J Nat Prod 66:289–291
43. Pedpradab P, Suwanborirux K (2011) A new acyclic thiophene sesterterpene from the Sikao Bay sponge, *Xestospongia* sp. J Asian Nat Prod Res 13:879–883
44. Wonganuchitmeta S-n, Yuenyongsawad S, Keawpradub N, Plubrukarn A (2004) Antitubercular sesterterpenes from the Thai sponge *Brachiaster* sp. J Nat Prod 67:1767–1770
45. Mahidol C, Prawat H, Sangpetsiripan S, Ruchirawat S (2009) Bioactive scalaranes from the Thai sponge *Hyrtios gumminae*. J Nat Prod 72:1870–1874
46. Kaweetripob W, Mahidol C, Wongbundit S, Tuntiwachwuttikul P, Ruchirawat S, Prawat H (2018) Sesterterpenes and phenolic alkenes from the Thai sponge *Hyrtios erectus*. Tetrahedron 74:316–323
47. Kaweetripob W, Mahidol C, Tuntiwachwuttikul P, Ruchirawat S, Prawat H (2018) Cytotoxic sesterterpenes from Thai marine sponge *Hyrtios erectus*. Mar Drugs 16:474
48. Bokesch HR, Stull AC, Pannell LK, McKee TC, Boyd MR (2002) A new pentacyclic sulfated hydroquinone from the marine sponge *Haliclona* sp. Tetrahedron Lett 43:3079–3081
49. Langjae R, Bussarawit S, Yuenyongsawad S, Ingkaninan K, Plubrukarn A (2007) Acetylcholinesterase-inhibiting steroidal alkaloid from the sponge *Corticium* sp. Steroids 72:682–685
50. Pailee P, Mahidol C, Ruchirawat S, Prachyawarakorn V (2017) Sterols from Thai marine sponge *Petrosia* (*Strongylophora*) sp. and their cytotoxicity. Mar Drugs 15:54
51. Watanadilok R, Sonchaeng P, Kijjoa A, Damas AM, Gales L, Silva AM, Herz W (2001) Tetillapyrone and nortetillapyrone, two unusual hydroxypyran-2-ones from the marine sponge *Tetilla japonica*. J Nat Prod 64:1056–1058
52. Wattanadilok R, Sawangwong P, Rodrigues C, Cidade H, Pinto M, Pinto E, Silva A, Kijjoa A (2007) Antifungal activity evaluation of the constituents of *Haliclona baeri* and *Haliclona cymaeformis*, collected from the Gulf of Thailand. Mar Drugs 5:40–51
53. Kijjoa A, Bessa J, Wattanadilok R, Sawangwong P, Nascimento MSJ, Pedro M, Silva AMS, Eaton G, Rv S, Herz W (2005) Dibromotyrosine derivatives, a maleimide, aplysamine-2 and other constituents of the Marine sponge *Pseudoceratina purpurea*. Zeitschrift für Naturforschung B 60:904–908
54. Sawangwong P, Wattanadilok R, Kijjoa A, Silva AMS, Eaton G, Herz W (2008) Secondary metabolites from a marine sponge *Cliona patera*. Biochem Syst Ecol 36:493–496
55. Kijjoa A, Watanadilok R, Sonchaeng P, Silva AMS, Eaton G, Herz W (2001) 11,17-Dideoxyagelorin A and B, new bromotyrosine derivatives and analogs from the marine sponge *Suberea* aff. *praetensa*. Zeitschrift für Naturforschung C 56:1116–1119
56. Kijjoa A, Watanadilok R, Sonchaeng P, Sawangwong P, Pedro M, Nascimento MSJ, Silva AMS, Eaton G, Herz W (2002) Further halotyrosine derivatives from the marine sponge *Suberea* aff. *praetensa*. Zeitschrift für Naturforschung C 57:732–738
57. Kijjoa A, Sonchaeng P, Puchakarn S, Sawangwong P, Herz W (2004) Bromotyrosine derivatives from the marine sponge *Suberea* aff. *praetensa*. Sponge Science in the Millennium 68:391–397

58. Sirimangkalakitti N, Olatunji OJ, Changwichit K, Saesong T, Chamni S, Chanvorachote P, Ingkaninan K, Plubrukarn A, Suwanborirux K (2015) Bromotyrosine alkaloids with acetylcholinesterase inhibitory activity from the Thai sponge *Acanthodendrilla* sp. Nat Prod Commun 10, 1934578X1501001135
59. Sirimangkalakitti N, Yokoya M, Chamni S, Chanvorachote P, Plubrukrn A, Saito N, Suwanborirux K (2016) Synthesis and absolute configuration of acanthodendrilline, a new cytotoxic bromotyrosine alkaloid from the Thai marine sponge *Acanthodendrilla* sp. Chem Pharm Bull 64:258–262
60. Nukoolkarn VS, Saen-oon S, Rungrotmongkol T, Hannongbua S, Ingkaninan K, Suwanborirux K (2008) Petrosamine, a potent anticholinesterase pyridoacridine alkaloid from a Thai marine sponge *Petrosia* n. sp. Bioorg Med Chem 16:6560–6567
61. Tran TD, Cartner LK, Bokesch HR, Henrich CJ, Wang XW, Mahidol C, Ruchirawat S, Kittakoop P, O'Keefe BR, Gustafson KR (2021) NMR characterization of rearranged staurosporine aglycone analogues from the marine sponge *Damiria* sp. Magn Reson Chem 59:534–539
62. Petchprayoon C, Asato Y, Higa T, Garcia-Fernandez L, Pedpradab S, Marriott G, Suwanborirux K, Tanaka J (2007) Four new Kabiramides from the Thai Sponge, *Pachastrissa nux*. Heterocycles 69:447–456
63. Sirirak T, Kittiwisut S, Janma C, Yuenyongsawad S, Suwanborirux K, Plubrukarn A (2011) Kabiramides J and K, trisoxazole macrolides from the sponge *Pachastrissa nux*. J Nat Prod 74:1288–1292
64. Sirirak T, Brecker L, Plubrukarn A (2013) Kabiramide L, a new antiplasmodial trisoxazole macrolide from the sponge *Pachastrissa nux*. Nat Prod Res 27:1213–1219
65. Saito N (2021) Chemical research on antitumor isoquinoline marine natural products and related compounds. Chem Pharm Bull 69:155–177
66. Suwanborirux K, Amnuoypol S, Plubrukarn A, Pummangura S, Kubo A, Tanaka C, Saito N (2003) Chemistry of renieramycins. Part 3. Isolation and structure of stabilized renieramycin type derivatives possessing antitumor activity from Thai sponge *Xestospongia* Species, pretreated with potassium cyanide. J Nat Prod 66:1441–1446
67. Amnuoypol S, Suwanborirux K, Pummangura S, Kubo A, Tanaka C, Saito N (2004) Chemistry OF Renieramycins. Part 5. Structure elucidation of renieramycin-type derivatives O, Q, R, and S from Thai marine sponge *Xestospongia* species pretreated with potassium cyanide. J Nat Prod 67:1023–1028
68. Oku N, Matsunaga S, van Soest RWM, Fusetani N (2003) Renieramycin J, a highly cytotoxic tetrahydroisoquinoline alkaloid, from a marine sponge *Neopetrosia* sp. J Nat Prod 66:1136–1139
69. Cheun-Arom T, Chuanasa T (2021) An efficient DNA extraction for a blue *Xestospongia* sp. Sponge and its associated microorganisms containing cytotoxic substances. Mar Biotechnol 23:917–927
70. Omar S, Tenenbaum L, Manes LV, Crews P (1988) Novel marine sponge derived amino acids 7. The fenestins. Tetrahedron Letters 29:5489–5492
71. Torres A, Gutierrez P, Alvarez-Manzaneda R, Chahboun R, Alvarez-Manzaneda E (2016) Preparation of oxocene terpenes. the first enantiospecific synthesis of cytotoxic arenaran A. Org Biomol Chem 14:9836–9845
72. Chen D-L, Wang B-W, Sun Z-C, Yang J-S, Xu X-D, Ma G-X (2020) Natural nitrogenous sesquiterpenoids and their bioactivity: a review. Molecules 25:2485
73. Di Blasio B, Fattorusso E, Magno S, Mayol L, Pedone C, Santacroce C, Sica D (1976) Axisonitrile-3, axisothiocyanate-3 and axamide-3. Sesquiterpenes with a novel spiro [4, 5] decane skeleton from the sponge *Axinella cannabina*. Tetrahedron 32:473–478
74. Menna M, Imperatore C, D'Aniello F, Aiello A (2013) Meroterpenes from marine invertebrates: structures, occurrence, and ecological implications. Mar Drugs 11:1602–1643
75. Wright AD, Wang H, Gurrath M, König GM, Kocak G, Neumann G, Loria P, Foley M, Tilley L (2001) Inhibition of heme detoxification processes underlies the antimalarial activity of terpene isonitrile compounds from marine sponges. J Med Chem 44:873–885

76. Arai M, Kawachi T, Setiawan A, Kobayashi M (2010) Hypoxia-selective growth inhibition of cancer cells by furospinosulin-1, a furanosesterterpene isolated from an Indonesian marine sponge. ChemMedChem 5:1919–1926
77. Crews P, Bescansa P (1986) Sesterterpenes from a common marine sponge, *Hyrtios erecta*. J Nat Prod 49:1041–1052
78. Shady NH, El-Hossary EM, Fouad MA, Gulder TA, Kamel MS, Abdelmohsen UR (2017) Bioactive natural products of marine sponges from the genus *Hyrtios*. Molecules 22:781
79. Jaisamut S, Thengyai S, Yuenyongsawad S, Karalai C, Plubrukarn A, Suwanborirux K (2009) Structure-activity relationships of antitubercular scalaranes: heteronemin revisited. Pure Appl Chem 81:1019–1026
80. Kittiwisut S, Rohena CC, Yuenyongsawad S, Mooberry SL, Plubrukarn A (2014) Antiproliferative effects of 12-oxoheteronemin vs heteronemin. Nat Prod Commun 9, 1934578X1400900318
81. Wang K, Chen Y-F, Yang Y-CS, Huang H-M, Lee S-Y, Shih Y-J, Li Z-L, Whang-Peng J, Lin H-Y, Davis PJ (2022) The power of heteronemin in cancers. J Biomed Sci 29:41
82. El-Gamal AA, Al-Massarani SM, Shaala LA, Alahdald AM, Al-Said MS, Ashour AE, Kumar A, Abdel-Kader MS, Abdel-Mageed WM, Youssef DT (2016) Cytotoxic compounds from the Saudi Red Sea sponge *Xestospongia testudinaria*. Mar Drugs 14:82
83. Al-Massarani SM, El-Gamal AA, Al-Said MS, Abdel-Kader MS, Ashour AE, Kumar A, Abdel-Mageed WM, Al-Rehaily AJ, Ghabbour HA, Fun H-K (2016) Studies on the red sea sponge *Haliclona* sp. for its chemical and cytotoxic properties. Pharmacognosy Magazine 12:114
84. Lakshmi V, Raghubir R, Gupta P (2008) New ceramides from the sponge *Cinachyra cavernosa*. J Asian Nat Prod Res 10:747–751
85. Ma Z, Qiu S, Chen H-C, Zhang D, Lu Y-L, Chen X-L (2022) Maleimide structure: a promising scaffold for the development of antimicrobial agents. J Asian Nat Prod Res 24:1–14
86. Olatunji OJ, Ogundajo AL, Oladosu IA, Changwichit K, Ingkaninan K, Yuenyongsawad S, Plubrukarn A (2014) Non-competitive inhibition of acetylcholinesterase by bromotyrosine alkaloids. Nat Prod Commun 9, 1934578X1400901107
87. Ferreira Montenegro P, Pham GN, Abdoul-Latif FM, Taffin-de-Givenchy E, Mehiri M (2024) Marine bromotyrosine derivatives in spotlight: bringing discoveries and biological significance. Mar Drugs 22:132
88. Ibrahim SRM, Mohamed GA (2016) Marine pyridoacridine alkaloids: biosynthesis and biological activities. Chem Biodivers 13:37–47
89. Sandjo LP, Kuete V, Biavatti MW (2015) Pyridinoacridine alkaloids of marine origin: NMR and MS spectral data, synthesis, biosynthesis and biological activity. Beilstein J Org Chem 11:1667–1699
90. Ibrahim SRM, Mohamed GA (2016) Pyridoacridine alkaloids from deep-water marine organisms: structural elucidation. Bulletin of Faculty of Pharmacy, Cairo University 54:107–135
91. Cadelis MM, Copp BR (2023) Chapter Two—Marine pyridoacridine, pyridoacridone and pyrroloacridine alkaloids. In: Knölker H-J (ed) The alkaloids: chemistry and biology. Academic Press, Londong, UK, pp 97–157
92. Kijjoa A, Wattanadilok R, Herz W, Campos N, Nascimento MSJ, Pinto M (2007) Anticancer activity evaluation of kuanoniamines A and C isolated from the marine sponge *Oceanapia sagittaria*, collected from the Gulf of Thailand. Mar Drugs 5:6–22
93. Xu M-J, Zhong L-J, Chen J-K, Bu Q, Liang L-F (2022) Secondary metabolites from marine sponges of the genus *Oceanapia*: chemistry and biological activities. Mar Drugs 20:144
94. Carroll AR, Scheuer PJ (1990) Kuanoniamines A, B, C, and D: pentacyclic alkaloids from a tunicate and its prosobranch mollusk predator *Chelynotus semperi*. J Org Chem 55:4426–4431
95. Machihara K, Namba T (2020) Kuanoniamine C stimulates bortezomib-induced cell death via suppression of glucose-regulated protein 78 in osteosarcoma. Biochem Biophys Res Commun 527:289–296
96. Oki S, Kageyama S, Machihara K, Namba T (2023) Kuanoniamine C suppresses adipogenesis and white adipose tissue expansion by modulating mitochondrial function. Biol Pharm Bull 46:1787–1796

97. Molinski TF, Fahy E, Faulkner DJ, Van Duyne GD, Clardy J (1988) Petrosamine, a novel pigment from the marine sponge *Petrosia* sp. J Org Chem 53:1340–1341
98. Walczak-Nowicka ŁJ, Herbet M (2021) Acetylcholinesterase inhibitors in the treatment of neurodegenerative diseases and the role of acetylcholinesterase in their pathogenesis. Int J Mol Sci 22:9290
99. Ibrahim SRM, Fadil SA, Fadil HA, Hareeri RH, Abdallah HM, Mohamed GA (2022) Genus *Smenospongia*: untapped treasure of biometabolites—biosynthesis, synthesis, and bioactivities. Molecules 27:5969
100. Williamson RT, Buevich AV, Martin GE, Parella T (2014) LR-HSQMBC: a sensitive NMR technique to probe very long-range heteronuclear coupling pathways. J Org Chem 79:3887–3894
101. Matsunaga S, Fusetani N, Hashimoto K, Koseki K, Noma M (1986) Bioactive marine metabolites. Part 13. Kabiramide C, a novel antifungal macrolide from nudibranch eggmasses. J Am Chem Soc 108:847–849
102. Matsunaga S, Fusetani N, Hashimoto K, Koseki K, Noma M, Noguchi H, Sankawa U (1989) Bioactive marine metabolites. 25. Further kabiramides and halichondramides, cytotoxic macrolides embracing trisoxazole, from the Hexabranchus egg masses. J Org Chem 54:1360–1363
103. Kernan MR, Molinski TF, Faulkner DJ (1988) Macrocyclic antifungal metabolites from the Spanish dancer nudibranch *Hexabranchus sanguineus* and sponges of the genus *Halichondria.* J Org Chem 53:5014–5020
104. Klenchin VA, Allingham JS, King R, Tanaka J, Marriott G, Rayment I (2003) Trisoxazole macrolide toxins mimic the binding of actin-capping proteins to actin. Nat Struct Mol Biol 10:1058–1063
105. Tanaka J, Yan Y, Choi J, Bai J, Klenchin VA, Rayment I, Marriott G (2003) Biomolecular mimicry in the actin cytoskeleton: mechanisms underlying the cytotoxicity of kabiramide C and related macrolides. Proc Natl Acad Sci 100:13851–13856
106. Petchprayoon C, Suwanborirux K, Miller R, Sakata T, Marriott G (2005) Synthesis and characterization of the 7-(4-aminomethyl-1H-1,2,3-triazol-1-yl) analogue of kabiramide C. J Nat Prod 68:157–161
107. Petchprayoon C, Suwanborirux K, Tanaka J, Yan Y, Sakata T, Marriott G (2005) Fluorescent kabiramides: new probes to quantify actin *in vitro* and *in vivo*. Bioconjug Chem 16:1382–1389
108. Sirirak T, Intaraksa N, Kaewsuwan S, Yuenyongsawad S, Suwanborirux K, Plubrukarn A (2011) Intracolonial allocation of trisoxazole macrolides in the sponge *Pachastrissa nux.* Chem Biodivers 8:2238–2246
109. Olatunji O, Brecker L, Plubrukarn A (2020) Metabolomics approach towards the chemical distribution in the sponge *Penares* cf. *nux*. Songklanakarin J Sci Technol 43:696–702
110. Butsuri A, Chanthathamrongsiri N, Sirirak T, Waiyaput W, Srimongkol A, Dittaroj K, Jiso A, Jittorntam P, Plubrukarn A, Tracing trisoxazole macrolide deposition in the tissues of the *Penares nux* sponge with matrix-assisted laser desorption ionization imaging mass spectrometry. Chem Biodivers n/a, e202403215
111. Olatunji O, Plubrukarn A (2015) Localization of kabiramides in the sponge *Pachastrissa nux* and sponge surface-attached bacteria. Planta Med 81:P12
112. Olatunji OO, Petchoubon A, Thawai C, Plubrukarn A (2020) Allocation of trisoxazole macrolides in the sponge *Penares* cf. *nux* and the impact on epiphytic bacterial cohabitants. Nat Prod Res 34:710–713
113. van Soest RW, Beglinger EJ, de Voogd NJ (2010) Skeletons in confusion: a review of astrophorid sponges with (dicho-)calthrops as structural megascleres (Porifera, Demospongiae, Astrophorida). Zookeys 1–88
114. Plubrukarn A, Yuenyongsawad S, Thammasaroj T, Jitsue A (2003) Cytotoxic isoquinoline quinones from the Thai sponge *Cribrochalina.* Pharm Biol 41:439–442
115. Frincke JM, Faulkner DJ (1982) Antimicrobial metabolites of the sponge *Reniera* sp. J Am Chem Soc 104:265–269

116. Fukumi H, Kurihara H, Hata T, Tamura C, Mishima H, Kubo A, Arai T (1977) Mimosamycin, a novel antibiotic produced by *Streptomyces lavendulae* no. 314: structure and synthesis. Tetrahedron Lett 18:3825–3828
117. Guo J (2023) Recent advances in the synthesis and activity of analogues of bistetrahydroisoquinoline alkaloids as antitumor agents. Eur J Med Chem 262:115917
118. Le Cesne A, Martín-Broto J, Grignani G (2022) A review of the efficacy of trabectedin as second-line treatment of advanced soft tissue sarcoma. Future Oncol 18:5–11
119. Manzo A, Sforza V, Carillio G, Palumbo G, Montanino A, Sandomenico C, Costanzo R, Esposito G, Laudato F, Mercadante E, La Manna C, Muto P, Totaro G, De Cecio R, Picone C, Piccirillo MC, Pascarella G, Normanno N, Morabito A (2022) Lurbinectedin in small cell lung cancer. Front Oncol 12:932105
120. Nuñez GS, Diez MM, Guillén MJ, Garrido-Martin EM, Avilés P, Cuevas C (2023) Abstract 1622: ecubectedin is a novel transcriptional inhibitor that displays potent antitumor effects in vitro and in vivo. Can Res 83:1622–1622
121. Fang Y, Li H, Ji B, Cheng K, Wu B, Li Z, Zheng C, Hua H, Li D (2021) Renieramycin-type alkaloids from marine-derived organisms: synthetic chemistry, biological activity and structural modification. Eur J Med Chem 210:113092
122. Arai T, Takahashi K, Kubo A, Nakahara S, Sato S, Aiba K, Tamura C (1979) The structures of novel antibiotics, saframycin B and C. Tetrahedron Lett 20:2355–2358
123. Arai T, Takahashi K, Nakahara S, Kubo A (1980) The structure of a novel antitumor antibiotic, saframycin A. Experientia 36:1025–1027
124. Irschik H, Trowitzsch-Kienast W, Gerth K, Höfle G, Reichenbach H (1988) Saframycin Mx1, a new natural saframycin isolated from a myxobacterium. J Antibiot (Tokyo) 41:993–998
125. Ikeda Y, Matsuki H, Ogawa T, Munakata T (1983) Safracins, new antitumor antibiotics. II. Physicochemical properties and chemical structures. J Antibiot (Tokyo) 36:1284–1289
126. Ikeda Y, Shimada Y, Honjo K, Okumoto T, Munakata T (1983) Safracins, new antitumor antibiotics. III. Biological activity. J Antibiot (Tokyo) 36:1290–1294
127. Rath CM, Janto B, Earl J, Ahmed A, Hu FZ, Hiller L, Dahlgren M, Kreft R, Yu F, Wolff JJ, Kweon HK, Christiansen MA, Håkansson K, Williams RM, Ehrlich GD, Sherman DH (2011) Meta-omic characterization of the marine invertebrate microbial consortium that produces the chemotherapeutic natural product ET-743. ACS Chem Biol 6:1244–1256
128. Schofield MM, Jain S, Porat D, Dick GJ, Sherman DH (2015) Identification and analysis of the bacterial endosymbiont specialized for production of the chemotherapeutic natural product ET-743. Environ Microbiol 17:3964–3975
129. Tianero MD, Balaich JN, Donia MS (2019) Localized production of defence chemicals by intracellular symbionts of *Haliclona* sponges. Nat Microbiol 4:1149–1159
130. Kim AN, Ngamnithiporn A, Du E, Stoltz BM (2023) Recent advances in the total synthesis of the tetrahydroisoquinoline alkaloids (2002–2020). Chem Rev 123:9447–9496
131. Saito N, Hiramatsu A, Hirade H, Kubota M, Toyoshima R, Fujino A, Sirimangkalakitti N, Suwanborirux K, Concepcion G (2017) Chemistry of renieramycins. 16. Structure of 7-desmethylrenieramycin O (= 14α-hydroxyrenieramycin S) from blue sponge, *Xestospongia* sp. Heterocycles 95:748–752
132. Suwanborirux K, Charupant K, Amnuoypol S, Pummangura S, Kubo A, Saito N (2002) Ecteinascidins 770 and 786 from the Thai tunicate *Ecteinascidia thurstoni*. J Nat Prod 65:935–937
133. Daikuhara N, Tada Y, Yamaki S, Charupant K, Amnuoypol S, Suwanborirux K, Saito N (2009) Chemistry of renieramycins. Part 7: Renieramycins T and U, novel renieramycin–ecteinascidin hybrid marine natural products from Thai sponge *Xestospongia* sp. Tetrahedron Lett 50:4276–4278
134. Saito N, Yoshino M, Charupant K, Suwanborirux K (2012) Chemistry of renieramycins. Part 10. structure of renieramycin V, a novel renieramycin marine natural product having a sterol ether at C-14 Position. Heterocycles 84:309–314
135. Charupant K, Suwanborirux K, Amnuoypol S, Saito E, Kubo A, Saito N (2007) Jorunnamycins A-C, new stabilized renieramycin-type bistetrahydroisoquinolines isolated from the Thai Nudibranch *Jorunna funebris*. Chem Pharm Bull 55:81–86

136. Wu Q, Li S-W, de Voogd NJ, Wang H, Yao L-G, Guo Y-W, Li X-W (2021) Marine alkaloids as the chemical marker for the prey–predator relationship of the sponge *Xestospongia* sp. and the nudibranch *Jorunna funebris*. Mar Life Sci Technol 3:375–381
137. Waropastrakul D, Suwanborirux K, De-Eknamkul W, Chuanasa T (2013) Detection of esterase activity converting cytotoxic renieramycin M to Jorunnamycin A in the crude enzyme of the nudibranch *Jorunna funebris*. Thai J Pharm Sci 38:41–44
138. Saito N, Koizumi Y-i, Kubo A, Suwanborirux K (2002) Chemistry of renieramycins. Part 2. Partial reduction and nucleophilic substitution of hexahydro-1,5-imino-4-oxo-3-benzazocine-7,10-dione: promising method to construct renieramycin J from renieramycin G via renieramycin E. Heterocycles 57:2345–2355
139. Arai T, Takahashi K, Ishiguro K, Yazawa K (1980) Increased production of saframycin A and isolation of saframycin S. J Antibiot (Tokyo) 33:951–960
140. Halim H, Chunhacha P, Suwanborirux K, Chanvorachote P (2011) Anticancer and antimetastatic activities of renieramycin M, a marine tetrahydroisoquinoline alkaloid, in human non-small cell lung cancer cells. Anticancer Res 31:193
141. Charupant K, Suwanborirux K, Daikuhara N, Yokoya M, Ushijima-Sugano R, Kawai T, Owa T, Saito N (2009) Microarray-based transcriptional profiling of renieramycin M and jorunnamycin C, isolated from Thai marine organisms. Mar Drugs 7:483–494
142. Tabunoki H, Saito N, Suwanborirux K, Charupant K, Satoh J-i (2012) Molecular network profiling of U373MG human glioblastoma cells following induction of apoptosis by novel marine-derived anti-cancer 1,2,3,4-tetrahydroisoquinoline alkaloids. Cancer Cell Int 12:14
143. Sirimangkalakitti N, Chamni S, Suwanborirux K, Chanvorachote P (2016) Renieramycin M sensitizes anoikis-resistant H460 lung cancer cells to anoikis. Anticancer Res 36:1665–1671
144. Sirimangkalakitti N, Chamni S, Suwanborirux K, Chanvorachote P (2017) Renieramycin M attenuates cancer stem cell-like phenotypes in H460 lung cancer cells. Anticancer Res 37:615–621
145. Tun JO, Salvador-Reyes LA, Velarde MC, Saito N, Suwanborirux K, Concepcion GP (2019) Synergistic cytotoxicity of renieramycin M and doxorubicin in MCF-7 breast cancer cells. Mar Drugs 17:536
146. Petsri K, Chamni S, Suwanborirux K, Saito N, Chanvorachote P (2019) Renieramycin T induces lung cancer cell apoptosis by targeting Mcl-1 degradation: a new insight in the mechanism of action. Mar Drugs 17:301
147. Petsri K, Yokoya M, Tungsukruthai S, Rungrotmongkol T, Nutho B, Vinayanuwattikun C, Saito N, Matsubara T, Sato R, Chanvorachote P (2020) Structure-activity relationships and molecular docking analysis of Mcl-1 targeting renieramycin T analogues in patient-derived lung cancer cells. Cancers 12:875
148. Yu B, Liang J, Li X, Liu L, Yao J, Chen X, Chen R (2022) Renieramycin T inhibits melanoma B16F10 cell metastasis and invasion via regulating Nrf2 and STAT3 signaling pathways. Molecules 27:5337
149. Ecoy GAU, Chamni S, Suwanborirux K, Chanvorachote P, Chaotham C (2019) Jorunnamycin A from *Xestospongia* sp. suppresses epithelial to mesenchymal transition and sensitizes anoikis in human lung cancer cells. J Nat Prod 82:1861–1873
150. Sumkhemthong S, Chamni S, Ecoy GU, Taweecheep P, Suwanborirux K, Prompetchara E, Chanvorachote P, Chaotham C (2021) Jorunnamycin A suppresses stem-like phenotypes and sensitizes cisplatin-induced apoptosis in cancer stem-like cell-enriched spheroids of human lung cancer cells. Mar Drugs 19:261
151. Cheun-Arom T, Chanvorachote P, Sirimangkalakitti N, Chuanasa T, Saito N, Abe I, Suwanborirux K (2013) Replacement of a quinone by a 5-*O*-Acetylhydroquinone abolishes the accidental necrosis inducing effect while preserving the apoptosis-inducing effect of renieramycin M on lung cancer cells. J Nat Prod 76:1468–1474
152. Pinkhien T, Maiuthed A, Chamni S, Suwanborirux K, Saito N, Chanvorachote P (2016) Bishydroquinone renieramycin m induces apoptosis of human lung cancer cells through a mitochondria-dependent pathway. Anticancer Res 36:6327–6333

153. Chamni S, Sirimangkalakitti N, Chanvorachote P, Saito N, Suwanborirux K (2017) Chemistry of renieramycins. 17. A new generation of renieramycins: hydroquinone 5-*O*-Monoester analogues of renieramycin M as potential cytotoxic agents against non-small-cell lung cancer cells. J Nat Prod 80:1541–1547
154. Maiuthed A, Pinkhien T, Chamni S, Suwanborirux K, Saito N, Petpiroon N, Chanvorachote P (2017) Apoptosis-inducing effect of hydroquinone 5-*O*-Cinnamoyl ester analog of renieramycin m on non-small cell lung cancer cells. Anticancer Res 37:6259–6267
155. Hongwiangchan N, Sriratanasak N, Wichadakul D, Aksorn N, Chamni S, Chanvorachote P (2021) Hydroquinone 5-*O*-Cinnamoyl ester of renieramycin M suppresses lung cancer stem cells by targeting Akt and destabilizes c-Myc. Pharmaceuticals 14:1112
156. Song M, Bode AM, Dong Z, Lee M-H (2019) AKT as a therapeutic target for cancer. Can Res 79:1019–1031
157. Madden SK, de Araujo AD, Gerhardt M, Fairlie DP, Mason JM (2021) Taking the Myc out of cancer: toward therapeutic strategies to directly inhibit c-Myc. Mol Cancer 20:3
158. Saito N, Tanaka C, Koizumi Y-i, Suwanborirux K, Amnuoypol S, Pummangura S, Kubo A (2004) Chemistry of renieramycins. Part 6: transformation of renieramycin M into jorumycin and renieramycin J including oxidative degradation products, mimosamycin, renierone, and renierol acetate. Tetrahedron 60:3873–3881
159. Charupant K, Daikuhara N, Saito E, Amnuoypol S, Suwanborirux K, Owa T, Saito N (2009) Chemistry of renieramycins. Part 8: synthesis and cytotoxicity evaluation of renieramycin M–jorunnamycin A analogues. Bioorg Med Chem 17:4548–4558
160. Fontana A, Cavaliere P, Wahidulla S, Naik CG, Cimino G (2000) A new antitumor isoquinoline Alkaloid from the marine Nudibranch *Jorunna funebris*. Tetrahedron 56:7305–7308
161. Sirimangkalakitti N, Chamni S, Charupant K, Chanvorachote P, Mori N, Saito N, Suwanborirux K (2016) Chemistry of renieramycins. 15. synthesis of 22-*O*-Ester derivatives of jorunnamycin A and their cytotoxicity against non-small-cell lung cancer cells. J Nat Prod 79:2089–2093
162. Iksen I, Sinsook S, Wattanathamsan O, Buaban K, Chamni S, Pongrakhananon V (2022) Target identification of 22-(4-Pyridinecarbonyl) Jorunnamycin A, a tetrahydroisoquinoline derivative from the sponge *Xestospongia* sp., in mediating non-Small-cell lung cancer cell apoptosis. Molecules 27:8948
163. Iksen I, Seephan S, Limprasutr V, Sinsook S, Buaban K, Chamni S, Pongrakhananon V (2023) Preclinical characterization of 22-(4′-pyridinecarbonyl) jorunnamycin A against lung cancer cell invasion and angiogenesis via AKT/mTOR signaling. ACS Pharmacol Trans Sci 6:1143–1154
164. Chamni S, Sirimangkalakitti N, Chanvorachote P, Suwanborirux K, Saito N (2020) Chemistry of renieramycins. Part 19: semi-syntheses of 22-*O*-amino ester and hydroquinone 5-*O*-amino ester derivatives of renieramycin M and their cytotoxicity against non-small-cell lung cancer cell lines. Mar Drugs 18:418
165. Oo Y, Nealiga JQL, Suwanborirux K, Chamni S, Ecoy GAU, Pongrakhananon V, Chanvorachote P, Chaotham C (2021) 22-*O*-(*N*-Boc-L-glycine) ester of renieramycin M inhibits migratory activity and suppresses epithelial–mesenchymal transition in human lung cancer cells. J Nat Med 75:949–966
166. Saito N, Yokoya M, Takahashi S (2019) Preparation of tricyclic analog as CDE ring model of renieramycin marine natural product by novel photo-induced transformation of 6-methoxy 1,2,3,4-tetrahydroisoquinoline-5,8-dione. Heterocycles 99:1276–1303
167. Yokoya M, Yamazaki-Nakai M, Nakai K, Sirimangkalakitti N, Chamni S, Suwanborirux K, Saito N (2023) Transformation of renieramycin M into renieramycins T and S by intramolecular photoredox reaction of 7-methoxy-6-methyl-1,2,3,4-tetrahydroisoquinoline-5,8-dione derivatives. J Nat Prod 86:222–231
168. Chantarawong W, Chamni S, Suwanborirux K, Saito N, Chanvorachote P (2019) 5-*O*-Acetyl-renieramycin T from blue Sponge *Xestospongia* sp. Induces Lung Cancer Stem Cell Apoptosis. Mar Drugs 17:109

169. Suksamai D, Racha S, Sriratanasak N, Chaotham C, Aphicho K, Lin ACK, Chansriniyom C, Suwanborirux K, Chamni S, Chanvorachote P (2022) 5-*O*-(*N*-Boc-L-Alanine)-renieramycin T induces cancer stem cell apoptosis via targeting Akt signaling. Mar Drugs 20:235
170. Buaban K, Innets B, Petsri K, Sinsook S, Chanvorachote P, Chansriniyom C, Suwanborirux K, Yokoya M, Saito N, Chamni S (2023) Semisynthesis of 5-*O*-ester derivatives of renieramycin T and their cytotoxicity against non-small-cell lung cancer cell lines. Sci Rep 13:21485
171. Sinsook S, Buaban K, Iksen I, Petsri K, Innets B, Chansriniyom C, Suwanborirux K, Yokoya M, Saito N, Pongrakhananon V, Chanvorachote P, Chamni S (2023) Light-Mediated transformation of renieramycins and semisynthesis of 4′-pyridinecarbonyl-substituted renieramycin-type derivatives as potential cytotoxic agents against non-small-cell lung cancer cells. Mar Drugs 21:400
172. Matsubara T, Yokoya M, Sirimangkalakitti N, Saito N (2019) Asymmetric synthesis and cytotoxicity evaluation of right-half models of antitumor renieramycin marine natural products. Mar Drugs 17:3
173. Nakai K, Kubo K, Yokoya M, Saito N (2014) Preparation of renieramycin left-half model compounds. Tetrahedron 70:6529–6545
174. Willis SN, Chen L, Dewson G, Wei A, Naik E, Fletcher JI, Adams JM, Huang DC (2005) Proapoptotic Bak is sequestered by Mcl-1 and Bcl-xL, but not Bcl-2, until displaced by BH3-only proteins. Genes Dev 19:1294–1305
175. Liao XW, Liu W, Dong WF, Guan BH, Chen SZ, Liu ZZ (2009) Total synthesis of (−)-renieramycin G from L-tyrosine. Tetrahedron 65:5709–5715
176. Anggraeni IG, Ei ZZ, Hotta D, Yokoya M, Chanvorachote P (2023) Renieramycin T derivative DH_22 induces p53-dependent apoptosis in lung cancer cells. In Vivo 37:1960–1966
177. Petsri K, Yokoya M, Racha S, Thongsom S, Thepthanee C, Innets B, Ei ZZ, Hotta D, Zou H, Chanvorachote P (2023) Novel synthetic derivative of renieramycin T right-half analog induces apoptosis and inhibits cancer stem cells via targeting the Akt signal in lung cancer cells. Int J Mol Sci 24:5345
178. Ei ZZ, Racha S, Yokoya M, Hotta D, Zou H, Chanvorachote P (2023) Simplified synthesis of renieramycin T derivatives to target cancer stem cells via β-Catenin proteasomal degradation in human lung cancer. Mar Drugs 21:627
179. Uko NE, Güner OF, Matesic DF, Bowen JP (2020) Akt pathway inhibitors. Curr Top Med Chem 20:883–900
180. Zhang Y, Wang X (2020) Targeting the Wnt/β-catenin signaling pathway in cancer. J Hematol Oncol 13:165
181. Phookphan P, Racha S, Yokoya M, Ei ZZ, Hotta D, Zou H, Chanvorachote P (2024) A new renieramycin T right-half analog as a small molecule degrader of STAT3. Mar Drugs 22:370
182. Darumas U, Chavanich S, Suwanborirux K (2007) Distribution patterns of the renieramycin-producing sponge, *Xestospongia* sp., and Its association with other reef organisms in the gulf of Thailand. Zool Stud 46:695–702
183. Darumas U, Salem GEM, Suwanborirux K, Viyakarn V, Chavanich S (2022) Variation of carbon–nitrogen contents and allelopathic disruption of renieramycin M–producing sponge *Xestospongia* sp. in the gulf of Thailand. Front Mar Sci 9:735702
184. Kieattisak Y, Udomsak D, Wattana WP (2017) Cultivation of a blue marine sponge, *Xestospongia* sp. (c.f. *Neopretrosia* sp.) in hatchery as an ornamental species. Agricult Res Technol Open Access J 12:555838
185. Pedpradab P, Yokseng K, Darumas U (2020) The cultivation of a blue Thai marine sponge, *Xestospongia* sp., under natural conditions to produce anticancer compound, renieramycin M. Wichcha J 39:1–15
186. Santiago VS, Manzano GG, Yu CC, Aliño PM, Salvador-Reyes LA (2019) Mariculture potential of renieramycin-producing Philippine blue sponge *Xestospongia* sp. (Porifera: Haplosclerida). Aquaculture 502:356–364
187. Yokseng K, Darumas U, Pedpradab P (2022) Culture of a blue marine sponge, *Xestospongia* sp., in semi-circulate close system stressed with calcium and magnesium concentration. J Aquaculture Mar Biol 10:230–235

188. Yokseng K, Darumas U, Khawchamnan R, Pedpradab P (2023) Effect of phytoplankton feed on growth and renieramycin M producing of a blue marine sponge (*Xestospongia* sp.). Aquaculture Studies 23:AQUAST1199
189. Abdelaleem ER, Samy MN, Desoukey SY, Liu M, Quinn RJ, Abdelmohsen UR (2020) Marine natural products from sponges (Porifera) of the order Dictyoceratida (2013 to 2019); a promising source for drug discovery. RSC Adv 10:34959–34976

Chapter 8
Biodiversity and Bioprospecting of Benthic Marine Invertebrates from Vietnamese Coastal Waters

Minh-Quang Thai and Lik Tong Tan

1 Introduction

The Vietnamese coastline is extensive with over 3200 km long and covers 15 degrees in latitude, stretching from the Gulf of Siam in the south (8° N) to the Chinese border in the north (23° N). The nearshore territorial water area of Vietnam, including over 3000 islands, is approximately 206,000 km^2. The coverage of the coral reefs is limited to the near-mouth areas of two big rivers, namely the Red River in the north and the Mekong River in the south. In addition, the coral reef ecosystems of Vietnam feature high bioproductivity, with a primary production of up to 30–100 mg C/m^3 per day [1]. Moreover, Vietnam is considered by the International Union for Conservation of Nature (IUCN) as one of the ten marine biodiversity centers, with more than 11,000 species found in more than 20 typical ecosystems at six different marine biodiversity regions.

Until recently, the study of low-molecular weight secondary metabolites of marine organisms obtained from Vietnamese territorial waters was fragmentary. According to marine natural products reviews by Faulkner [2] and Blunt et al. [3], only four new compounds were reported from 1977 to 2003. For instance, neothyonidioside C was the main component of the triterpene glycosides mixture isolated from an alcoholic extract of the holothurian, *Neothyonidium magnum*, collected in South Vietnam [4]. A new brominated triterpene polyether, callicladol, was isolated from red alga, *Laurencia calliclada*, collected at An Thoi, Phu Quoc Island [5]. Another species, *L. pannosa*, from AnThoi, Phu Quoc Island, contained a new C15

M.-Q. Thai
Institute of Oceanography, Vietnam Academy of Science and Technology, Nha Trang, Khanh Hoa, Vietnam

L. T. Tan (✉)
Natural Sciences and Science Education, National Institute of Education, Nanyang Technological University, Singapore, Singapore
e-mail: liktong.tan@nie.edu.sg

L. T. Tan and N. Hanif (eds.), *Marine Natural Products Research in South-East Asia*,
https://doi.org/10.1007/978-3-032-01098-8_8

nonterpenoid bromoallene, pannosallene [6]. Finally, hematoside with 8-*O*-methyl-*N*-glycolylneuraminic acid was isolated from starfish samples of *Linckia laevigata* collected in Nha Trang Bay [7].

Several joint Russian-Vietnamese expeditions aboard the research vessel "Akademik Oparin" made possible the detailed study on the chemical diversity of marine micro- and macroorganisms in Vietnamese waters. These expeditions were organized based on an agreement on scientific collaboration between the Russian Academy of Sciences and the Vietnamese Academy of Science and Technology. The research vessel "Akademik Oparin" is highly equipped with diving facilities, including a compressor and a pressure chamber, as well as chemical and biochemical facilities, including evaporators, a high-pressure liquid chromatograph, a microbiological laboratory and a bioassay laboratory. Such extensive on-board facilities allowed not only the collection and storage of marine samples but also facilitated chemical and biochemical investigations. For instance, from the 34th scientific expedition on board the "Akademik Oparin" in May–June 2007, almost 50% of the extracts from 148 marine samples, including marine invertebrates and algae, showed potential new cancer preventive compounds via their inhibitory effects on the AP-1 transcriptional activity in JB6 Cl41 AP-1 cells [8]. Between 2004 and 2017, a total of five expeditions were made which resulted in more than 250 natural products, including 117 new molecules, reported from diverse marine invertebrates and microbes collected from four regions, including northern (Gulf of Tonkin), central (Ly Son, Con Co, Cu Lao Cham Islands, Van Phong and Nha Trang Bays), south eastern (Ladd Reef, Con Dao and Cu Lao Thu Islands) and south western (Gulf of Siam). Currently, an ongoing expedition on the research vessel "Akademik Oparin" is being carried out for the period 2018–2025 [9].

This chapter focuses on the significant discovery of new specialized molecules from dominant benthic marine invertebrates, including sponges, soft corals and echinoderms, collected from Vietnamese territorial waters. The diversity and distribution patterns of these selected marine benthic invertebrates are discussed to support marine conservation and ongoing bioprospecting efforts in Vietnam. The natural products coverage is not meant to be comprehensive but to highlight the structures of new molecules as well as their biological properties and their impact on drug discovery from marine sources. Selected molecules with significant activities and/or unique structures will be further presented within each taxonomic group. The discussion of these natural products will be classified based on taxonomic groups (e.g., porifera, cnidarians and echinoderms) and within each taxonomic group, the molecules are further presented based on structural classes, including fatty acid derivatives, steroids, peptides, polyketides and alkaloids, or minor taxonomic rank.

2 Overview on the Marine Biodiversity in Vietnam

Vietnam's marine ecosystems are rich in biodiversity, supporting numerous invertebrate species with ecological and economic importance. The unique biodiversity of sponges, octocorals, and echinoderms in Vietnamese waters contributes to ecosystem stability, while their bioactive compounds present significant potential for bioprospecting. This section explores the diversity, distribution, ecological roles, and bioprospecting potential of these invertebrate groups across Vietnam's coastal regions.

2.1 Biodiversity Overview of Vietnamese Sponges

Vietnamese sponge diversity has been documented since 1898, with researchers identifying 291 species from 112 genera across multiple coastal regions. These areas include the Northern, Central, and Southwest coastal regions and the Spratly Islands. Northern Vietnam, including Ha Long Bay, has been extensively studied, with Ha Long alone hosting 61 species across 40 genera [10]. In the North Central Coastal Region, Con Co Island holds the highest documented diversity, with 111 species from 62 genera [11–14]. In the Southeast Coastal Region, Nha Trang has a notable sponge diversity with 66 species from 42 genera, making it a significant area for sponge research in Vietnam [15, 16]. Moreover, 71 species were recorded from Vietnam's coastal waters without specific locations, and these include chemical investigations of Vietnamese sponges [17, 18].

2.1.1 Species Richness and Spatial Distribution of Sponges in Vietnam

A total of 291 sponge species are documented, with 227 identified to specific level (Table 1). These species are grouped into three classes, 20 orders, 56 families and 112 genera. This study adds 150 new records for Vietnam, compared to the list published by Lim et al. in 2016 [19]. Most sponge species in Vietnam belong to three main classes: Calcarea, Demospongiae and Homoscleromorpha. The class Demospongiae dominates with 283 species spread across various orders, while Calcarea (seven species) and Homoscleromorpha (one species) have only a few species each, contributing less to the overall sponge diversity. The orders Haplosclerida and Poecilosclerida contain the highest number of species, representing 25.4% and 16.8% of the total species, respectively. Tetractinellida and Dictyoceratida also display significant diversity, with the largest number of families and genera.

The hierarchical clustering analysis (Fig. 1) and Multidimensional Scaling (MDS) (Fig. 2) were conducted to evaluate the differences in sponge fauna between Vietnam's coastal regions and the Spratly Islands. The results show that the Southwest Coastal Region and the Southeast Coastal Region have the highest similarity in terms

Table 1 Taxonomic distribution of sponge species in Vietnam

Class	Order	Family	Genus	Species	Percent (%)
Calcarea	Clathrinida	2	3	3	1.0
	Leucosolenida	3	3	4	1.4
Demospongiae	Agelasida	2	2	7	2.4
	Axinellida	2	7	11	3.8
	Biemnida	2	4	7	2.4
	Bubarida	2	3	8	2.8
	Chondrillida	1	1	2	0.7
	Chondrosiida	1	1	1	0.3
	Clionaida	3	6	19	6.5
	Dendroceratida	2	2	2	0.7
	Dictyoceratida	4	14	28	9.6
	Haplosclerida	6	14	74	25.4
	Poecilosclerida	11	15	49	16.8
	Scopalinida	1	2	4	1.4
	Spongillida	1	1	2	0.7
	Suberitida	2	12	25	8.6
	Tethyida	1	2	6	2.1
	Tetractinellida	6	16	33	11.3
	Verongiida	3	3	5	1.7
Homoscleromorpha	Homosclerophorida	1	1	1	0.3
Total	20	56	112	291	100

of sponge species composition, as indicated by their close grouping in both analyses. The Northern Coastal Region displays relatively low dissimilarity compared to these two regions, suggesting some shared species but also maintaining distinct characteristics. The Spratly Islands and the Central Coastal Region also exhibit notable similarity to each other, but they are clearly distinct from the other regions. The North Central Coastal Region stands out as having the greatest difference in species composition compared to all other regions, both in the MDS ordination and the hierarchical clustering. These findings suggest that geographical, ecological or environmental factors may be influencing the diversity and distribution of sponge species in these areas, with regions like the North Central Coastal Region being more isolated in terms of species composition.

Three species were recorded in four out of the six regions, including *Acanthella cavernosa*, *Haliclona* (*Gellius*) *cymaeformis* and *Xestospongia testudinaria.* Five species, *Callyspongia* (*Cladochalina*) *subarmigera*, *Chondrilla australiensis*, *Cinachyrella australiensis*, *Gelliodes fibulata* and *Haliclona* (*Gellius*) *varia*, were recorded in three out of the six regions. Additionally, 28 species were recorded in two out of the six regions. Considering 236 out of the 291 recorded species in

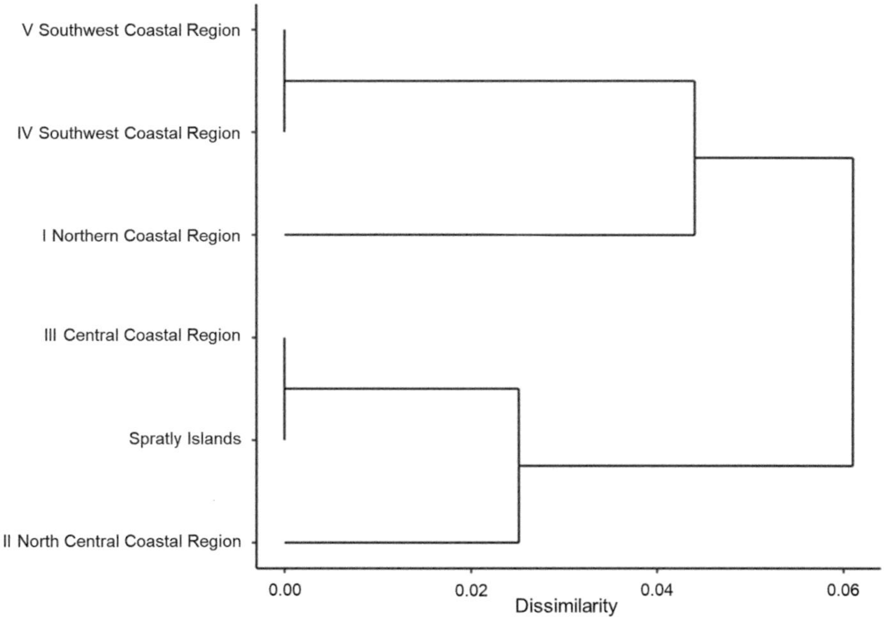

Fig. 1 Dendrogram of sponge fauna in Vietnam based on hierarchical clustering

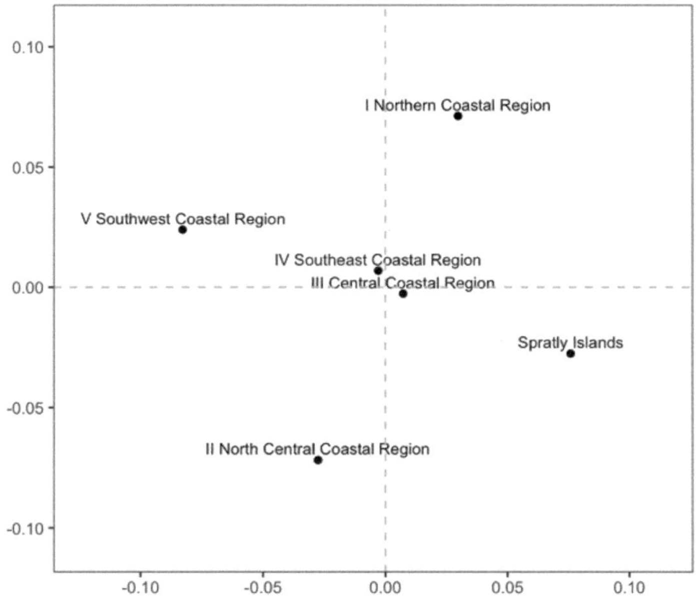

Fig. 2 MDS (multidimensional scaling) ordination based on the Jaccard distance matrix of sponge species

the five coastal regions and the Spratly Islands, 200 species were recorded in only one region. The large number of "restricted" species recorded in this review may not necessarily indicate endemism. MDS analysis of species composition similarity between Vietnam's coastal regions shows that the two northern regions, I. Northern Coastal Region and II. North Central Coastal Region are completely distinct and distant, while there is a transition to III. Central Coastal Region. This could be due to the limited study locations in Region III, as well as potential misidentifications between Regions I and II.

Sponges are a key component of marine life commonly recorded in coral reefs and areas with dead coral intertidal zone. However, collecting specimens and re-encountering sponge species in these areas has become increasingly challenging due to the decline in biodiversity, coral cover, and the effects of coral bleaching in Vietnam's coastal waters. Additionally, sponge identification remains difficult due to the challenges in distinguishing species, while current studies lack detailed visual information and specimen descriptions, leading to frequent misidentifications and reducing the accuracy of species classification. This issue is further complicated by the fact that several species mentioned in this review and others, such as *Aaptos aaptos*, *Acanthella obtusa*, *Agelas dirpar*, *Biemna variantia*, *Callyspongia australis*, *Cliona viridis*, *Dictyonella pelligera*, *Dysidea fragilis*, *Erylus formosus*, *Haliclona oculata*, *Mycale laevis* and *Tethya aurantium*, are primarily cited from studies on natural bioactive compounds, which often lack detailed taxonomic descriptions of the specimens.

2.2 The Octocoral Fauna of Vietnam

The study of Octocorallia (soft corals) in Vietnam has progressed through significant regional surveys, revealing notable diversity across coastal areas. Early research by World Wildlife Fund (WWF) in 1990s at the Northern Coastal Region, specifically around Cat Ba, documented 16 genera and 20 species of soft corals, establishing a foundation for biodiversity knowledge [20]. In the North Central Region, studies by Hung et al. and others further contributed, with surveys in Con Co and Hai Van–Son Cha, recording ten species across four genera, and 27 species across 18 genera, respectively [21]. In the Central Coastal Region, sites like Cu Lao Cham and Ly Son have shown remarkable biodiversity. Research by Hoang and Quang [22] and Hoang et al. [23] identified 14 genera and 44 species in Cu Lao Cham, while Ly Son exhibited even greater diversity, with 60 species across 12 genera recorded by Hoang and Dautova [24] and Hoang [25]. Moreover, in Phu Yen and Cam Ranh, surveys by Hoang noted lower diversity with two genera and two species each [25].

The Southeast Coastal Region has some of the richest coral diversity, especially in Nha Trang, where surveys since Tixier-Durivault in 1970 have recorded 24 genera and 135 species [26]. Contributions by WWF [27, 28], Dautova and Savinkin [29] and Hoang [25] have solidified Nha Trang's significance as a biodiversity hotspot. In Con Dao, research has been limited, identifying only one genus and one species, while Hon Cau recorded nine genera and 26 species. In the Southwest Region, Phu Quoc's soft coral diversity was captured in WWF's 1993 survey [27], documenting eight genera and 17 species. Overall, a comprehensive study by Hoang and Dautova [24] across Vietnam's coastal waters reported 43 genera and 182 species, enhancing the understanding of soft coral diversity nationwide.

2.2.1 Species Richness and Spatial Distribution of Vietnamese Octocoral

Vietnam's coastal regions host a rich diversity of octocorals, encompassing 220 species from 48 genera, 23 families and two orders (Table 2). The dominant order, Malacalcyonacea, contributes significantly to the region's taxonomic diversity, with notable representation from families like Sarcophytidae, Acanthogorgiidae and Nephtheidae. Sarcophytidae stands out with three genera and 116 species, accounting for over half (52.7%) of the total octocoral species, highlighting its crucial role in Vietnam's soft coral ecosystems.

Other families, such as Acanthogorgiidae (six genera, 14 species) and Nephtheidae (four genera, 21 species), represent 6.7% and 9.6% of the total diversity, respectively. Families like Cladiellidae (two genera, 12 species) and Melithaeidae (three genera, 10 species) also contribute meaningfully. Additional families with smaller proportions, such as Anthogorgiidae, Euplexauridae, Isididae and Xeniidae, make up between 1.4% and 2.7%, while Alcyoniidae, Clavulariidae, Eunicellidae and Plexauridae each include a single genus and species, representing 0.5% of the total diversity. This structural composition highlights the prominence of certain families in supporting Vietnam's octocoral biodiversity.

The results from the MDS (Fig. 3) and hierarchical clustering analysis (Fig. 4), based on the presence/absence of 220 soft coral species across the coastal regions of Vietnam, reveal distinct differences between the regions. The MDS analysis shows a clear separation between the North Central Coastal Region (II) and the Southeast Coastal Region (IV) from the other areas, indicating significant differences in species composition. In contrast, the Central Coastal Region (III), Southwest Coastal Region (V), and Northern Coastal Region (I) are more closely positioned, suggesting higher similarity in species presence. Hierarchical clustering supports this by grouping the Northern and Central Coastal Regions first, highlighting their shared species composition. Although the North Central and Southeast Coastal Regions are separated in MDS, they are clustered together in the hierarchical analysis, suggesting some commonalities. The Southwest Coastal Region (V), however, stands out as distinctly different from the others in terms of soft coral species presence.

The distribution of octocoral species in Vietnam's coastal regions reveals both commonalities and regional distinctions. *Junceella fragilis* and *Sinularia brassica* are widespread, found in all five coastal regions, showcasing their adaptability across diverse environments. Other species, particularly from the genus *Sarcophyton* (e.g., *S. ehrenbergi* and *S. elegans*), are prevalent across four regions but are less common overall. Species found in three regions represent a small proportion (6%) of total

Table 2 Taxonomic distribution of soft coral species in Vietnam

Class	Order	Family	Genus	Species	Percent (%)
Octocorallia	Malacalcyonacea	Acanthogorgiidae	6	14	6.4
		Alcyoniidae	1	1	0.5
		Anthogorgiidae	2	4	1.8
		Astrogorgiidae	1	1	0.5
		Carijoidae	1	1	0.5
		Cladiellidae	2	12	5.5
		Clavulariidae	1	1	0.5
		Eunicellidae	1	1	0.5
		Euplexauridae	1	4	1.8
		Gorgoniidae	2	2	0.9
		Isididae	3	4	1.8
		Lemnaliidae	2	3	1.4
		Melithaeidae	3	10	4.6
		Nephtheidae	4	21	9.6
		Plexauridae	1	1	0.5
		Sarcophytidae	3	116	52.7
		Sinulariidae	1	3	1.4
		Subergorgiidae	1	3	1.4
		Tubiporidae	2	2	0.9
		Xeniidae	2	6	2.7
	Scleralcyonacea	Briareidae	1	1	0.5
		Ellisellidae	6	8	3.6
		Parasphaerascleridae	1	1	0.5
Total	2	23	48	220	100

diversity and are primarily concentrated in the North Central, Central and Southeast Coastal Regions. Examples include *Cladiella conifera* and *Lobophytum batarum.* Southern regions, including the Central, Southeast, and Southwest Coastal Regions, host species like *Sclerophytum arctium* and *Sarcophyton glaucum*, indicating a southern distribution tendency.

Some 56 species (25%) are present in only two regions, often in the Central and Southeast Coastal Regions, suggesting localized distribution influenced by specific ecological factors. Species like *Menella flora* are confined to the Northern and North Central Regions, indicating a preference for northern waters. The majority, 116 species (53%), are restricted to a single region, mainly in the Southeast and Central Coastal Regions. Overall, the North Central and Southeast Coastal Regions exhibit unique species compositions, while the Northern and Central Coastal Regions show more similarity in their soft coral diversity.

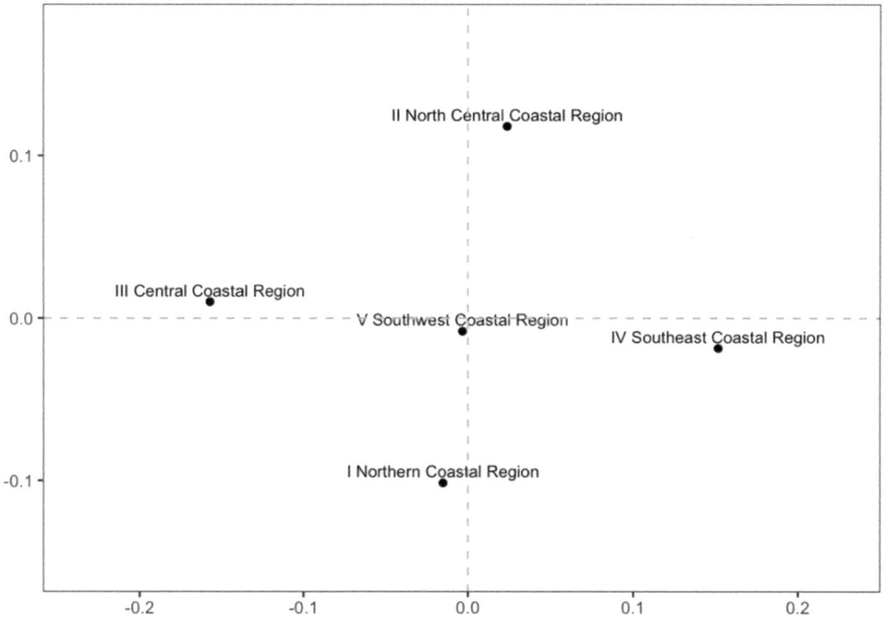

Fig. 3 MDS ordination based on the Jaccard distance matrix of Octocorallia species

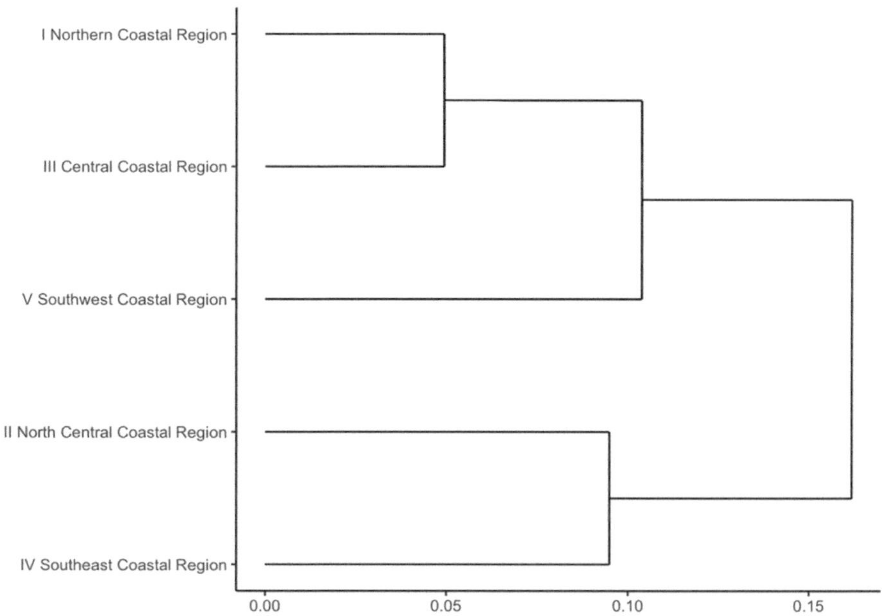

Fig. 4 Dendrogram of Octocorallia fauna in Vietnam based on hierarchical clustering

2.3 *The Echinoderm Fauna of Vietnam*

Echinoderm research in Vietnam has a rich history, beginning with foundational studies by Dao Tan Ho [30–32], which focused on regions like the Southeast Coastal Region, Spratly Islands, and Southwest Coastal Region. His surveys in Con Dao, Phu Quoc, Tho Chu Islands, and the Gulf of Thailand documented a wide range of species, setting a baseline for future studies. Subsequent research broadened to various ecosystems, including coral reefs and seagrass beds, enhancing understanding of Echinoderm diversity in Vietnam.

Species diversity varies significantly across Vietnam's coastal regions. In the Northern Coastal Region, 25 species across 19 genera have been identified, particularly in areas like Bai Tu Long, Co To–Thanh Lan, Ha Long Bay and Cat Ba Islands, with the latter hosting 36 species across 24 genera [33, 34]. In the North Central Coastal Region, fewer species are found, with Con Co and Hai Van–Son Cha recording four to 16 species, though the Paracel Islands stand out with 22 species across 16 genera [31]. The Central Coastal Region also exhibits substantial diversity. Cu Lao Cham has 33 species from 23 genera, and Xuan Dai Bay has 19 species from 14 genera. Ly Son Island adds to this diversity with 11 species from eight genera [35, 36].

The Southeast Coastal Region is the most species-rich, with Khanh Hoa and Nha Trang hosting 168 and 112 species, respectively. Extensive studies by Dao Tan Ho [31] and others underscore the area's biodiversity. Con Dao and Minh Hai (Bac Lieu–Ca Mau) also show high diversity, with 78 and 80 species, respectively, across numerous genera. In the Southwest Coastal Region, notable sites include the Gulf of Thailand and Phu Quoc, with 22 and 37 species, respectively. These findings highlight Vietnam's diverse Echinoderm fauna and the ecological richness of its varied coastal regions.

2.3.1 Species Richness and Spatial Distribution of Vietnamese Echinoderms

Echinoderm diversity in Vietnam is extensive, with 345 species distributed across 192 genera, 81 families, 24 orders, and five main classes, each contributing uniquely to the overall taxonomic structure (Table 3). The first class, Asteroidea (starfish), is primarily represented by the order Valvatida, which encompasses 52 species across 11 families, making up 15.1% of all echinoderm species in Vietnam. Other orders within Asteroidea, such as Forcipulatida, Paxillosida, Spinulosida, and Velatida, are less common, with Paxillosida being the next most diverse, contributing nine species (2.6%). Crinoidea, the class including sea lilies and feather stars, is mainly represented by the order Comatulida, which includes 49 species across ten families, making up 14.2% of the echinoderm species. The order Isocrinida within Crinoidea has minimal diversity, with only three species (0.9%). The third class, Echinoidea (sea urchins), shows moderate diversity, led by the order Camarodonta, which contributes 27 species (7.8%). Additional orders such as Spatangoida (19 species, 5.5%) and Echinolampadacea (13 species, 3.8%) further enhance the diversity within this class, with smaller contributions from orders like Cidaroida, Diadematoida, and Echinothurioida. Holothuroidea, or sea cucumbers, is mainly represented by the order Holothuriida with 25 species (7.3%). Other orders such as Dendrochirotida (14 species, 4.1%) and Apodida (11 species, 3.2%) add to the class's diversity. Lastly, the class Ophiuroidea (brittle stars) shows the highest species diversity, largely due to the order Amphilepidida, which has 53 species (15.4%). Amphilepidida stands out as a major contributor to both this class and the overall taxonomic diversity. Other orders, including Ophiacanthida (26 species, 7.5%) and Ophiurida, also contribute, though Ophiurida has limited diversity. Together, these five classes reveal a rich echinoderm biodiversity in Vietnam, with specific orders playing key roles within each class.

Table 3 Taxonomic distribution of echinoderm species in Vietnam

Class	Order	Family	Genus	Species	Percent (%)
Asteroidea	Forcipulatida	1	1	1	0.3
	Paxillosida	3	4	9	2.6
	Spinulosida	1	2	3	0.9
	Valvatida	11	30	52	15.1
	Velatida	1	1	1	0.3
Crinoidea	Comatulida	10	27	49	14.2
	Isocrinida	2	3	3	0.9
Echinoidea	Camarodonta	6	15	27	7.8
	Cidaroida	2	7	9	2.6
	Clypeasteroida	1	2	4	1.2
	Diadematoida	1	5	8	2.3
	Echinolampadacea	5	10	13	3.8
	Echinoneoida	1	1	1	0.3
	Echinothurioida	1	2	2	0.6
	Spatangoida	7	13	19	5.5
Holothuroidea	Apodida	2	8	11	3.2
	Dendrochirotida	3	11	14	4.1
	Holothuriida	1	4	25	7.3
	Molpadida	1	2	2	0.6
	Synallactida	1	2	8	2.3
Ophiuroidea	Amphilepidida	9	24	53	15.4
	Euryalida	2	3	3	0.9
	Ophiacanthida	8	13	26	7.5
	Ophiurida	2	2	2	0.6
Total	24	81	192	345	100

The results from the MDS (Fig. 5) and hierarchical clustering analyses (Fig. 6) of echinoderm species composition across Vietnam's coastal regions reveal clear patterns of species distribution. The MDS analysis shows a distinct separation between the North Central Coastal Region (II) and the Southeast Coastal Region (IV), indicating significant differences in species composition between these areas. In contrast, the Southwest Coastal Region (V) and the Northern Coastal Region (I) exhibit greater similarity in species composition, with clear overlap shown in both the MDS plot and hierarchical clustering. The Spratly Islands, while relatively isolated, still maintain some degree of species overlap with the Northern and Central Coastal Regions, reflecting a shared species composition. The Central Coastal Region (III) stands out with a clearly distinct species diversity profile compared to other regions.

These results reflect a complex biodiversity pattern, where the Southeast and Southwest Coastal Regions demonstrate strong species linkage, while the North Central Coastal Region exhibits a more distinct species composition. The Central and Northern Coastal Regions form their own cluster, while the Spratly Islands showcase unique biodiversity, yet remain connected to other regions. This combined analysis shows that geographical, ecological, and environmental factors such as water temperature, ocean currents and habitat types have a direct influence on the distribution of echinoderm species across Vietnam's coastal areas.

Several echinoderm species are distributed across different coastal regions. *Diadema setosum* and *Synapta maculata* are the most widespread, found in all seven regions, including the Northern, North Central, Central, Southeast, and Southwest Coastal Regions, the Spratly Islands, and Northern Vietnam. *Bohadschia argus* and

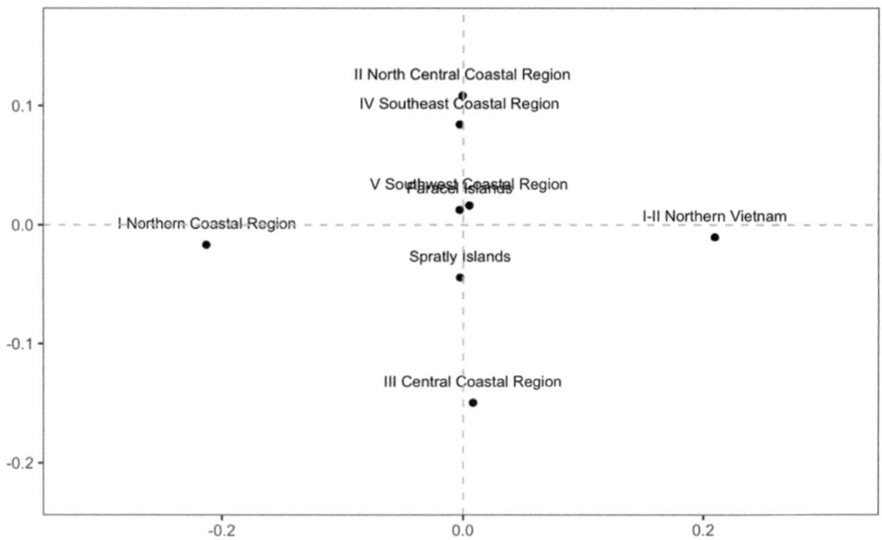

Fig. 5 MDS ordination based on the Jaccard distance matrix of Echinoderm species

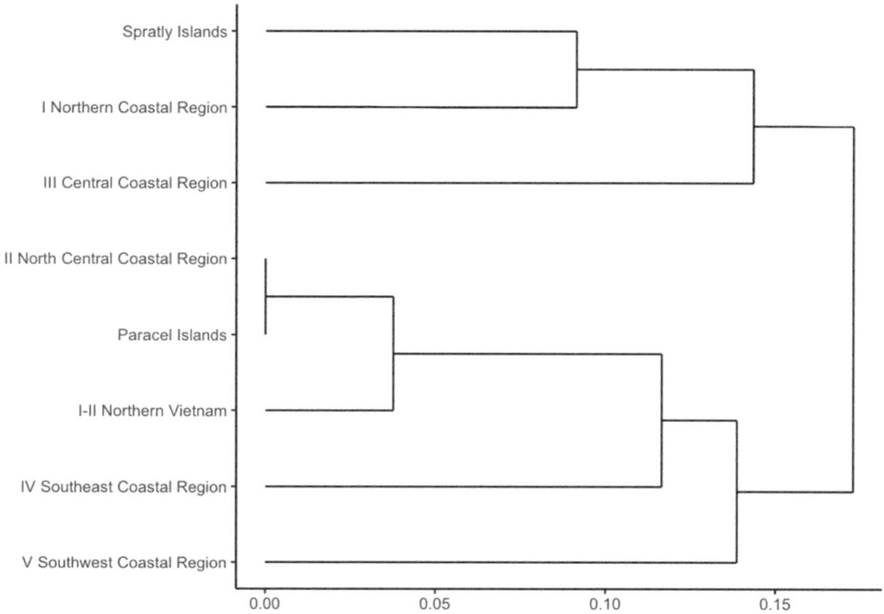

Fig. 6 Dendrogram of echinoderm fauna in Vietnam based on hierarchical clustering

Echinothrix calamaris are present in six regions, contributing to biodiversity in both northern and southern waters. Other species, such as *Acanthaster planci, Amphimetra moelleri, Comaster schlegelii* and *Culcita novaeguineae*, are found in five regions, particularly in the Central and Southeast Coastal Regions and the Spratly Islands. Additionally, species like *Stichopus chloronotus* and *S. horrens* are also distributed across five regions. Most of the widely distributed echinoderm species mentioned above are commonly recorded in coral reef ecosystems.

3 Chemistry and Biological Activities of Natural Products from Vietnamese Marine Invertebrates

3.1 Natural Products from Vietnamese Sponges

More than 25 sponge species from Vietnamese water have been investigated for structurally new bioactive molecules. The structural classes of these compounds range from fatty acid derivatives, polyketides, terpenoids, steroids to alkaloids (Tables 4, 5, 6, 7; Figs. 5, 6, 7, 8, 9, 10, 11, 12, 13, 14, 15, 16, 17, 18, 19 and 20). Highlighted below are selected sponge-derived compounds having unique structures and/or biological properties.

Table 4 List of new fatty acid derivatives/polyketides reported from Vietnamese marine sponges

Classification/Compound[a]	Location	Biological activity	References
Gelliodes sp.			
Gelliodesinic acid (**1**)	Hon Thom, Phu Quoc	–	[37]
Xestospongia testudinaria			
Xestospongiene VN1 (**2**)	Van Phong, Nha Trang	Weak cytotoxicity	[38]
Dysidea cinerea			
Dysideolides A (**3**) and B (**4**)	Lang Co beach	–	[39]
Smenospongia cerebriformis			
Smenocerones A (**5**) and B (**6**)	Sea of Quangtri, Vietnam	Smenocerone B: cytotoxic activity towards LU-1, HepG-2, HL-60, MCF-7 and SK-Mel-2 human cancer cells with IC_{50} values ranging 3.2–5.7 μg/mL	[40]
Arenosclera sp.			
Tribromoiododiphenyl ether (**7**)	Hon Thom, Phu Quoc	Strong antibacterial activities against *B. subtilis*, *S. aureus* and *Klebsiella pneumoniae*, with MIC values ranging between 3.1 and 6.3 μM	[41]
Clathria (*Thalysias*) *reinwardti*			
8-(2′,3′,4′-Trimethylphenyl)-6-methyl-oct-3(*E*),5(*E*),7(*E*)-trien-2-one (**8**)	Van Phong Bay, Nha Trang	–	[42]

[a]Chemical structures in Fig. 7

Table 5 List of new terpenoids reported from Vietnamese marine sponges

Classification/Compound[a]	Location	Biological activity	References
Spongia sp.			
Langcoquinones A (**9**)–F (**14**)	Son Cha, Lang Co, Thua Thien-Hue	Langcoquinones A–D: significant antibacterial activities against *S. aureus* and *B. subtilis* Langcoquinone D: significant cytotoxic activities against the A549, MCF7, HeLa and WI-38 cell lines	[43–45]
Langconols A (**15**)–C (**17**)		Langconols A and C: antibacterial activities against *B. subtilis* and/or *S. aureus* Langconol C: significant cytotoxicity against A549, MCF7, HeLa and WI-38 fibroblast	[45]
Smenospongia cerebriformis			
Smenohaimiens A (**18**)–F (**23**)	Vinh Moc, Quang Tri	Smenohaimiens A, B and E: moderate inhibitory activities on NO production Smenohaimien F: moderate cytotoxic activities	[46, 47]
Dysidea fragilis			
Dysinidins A (**24**)–E (**28**)	Van Don, Quang Ninh	–	[48, 49]
Dysinidin I (**29**)–III (**31**)		–	[50]
D. cinerea			
(4*R*,5*R*)-Muurol-1(6),10(14)-diene-4,5-diol (**32**), (4*R*,5*R*)-muurol-1(6)-ene-4,5-diol (**33**), (4*R*,5*R*,10*R*)-10-methoxymuurol-1(6)-ene-4,5-diol (**34**), (4*S*)-4-hydroxy-1,10-seco-muurol-5-ene-1,10-dione (**35**), (4*R*)-4-hydroxy-1,10-seco-muurol-5-ene-1,10-dione (**36**), (6*S*,10*S*)-6,10-dihydroxy-7,8-seco-2,8-cyclo-muurol-4(5),7(11)-diene-12-oic acid (**37**) and (6*R*,10*S*)-6,10-dihydroxy-7,8-seco-2,8-cyclo-muurol-4(5),7(11)-diene-12-oic acid (**38**)	Lang Co beach, Thua Thien Hue	–	[51]

(continued)

Table 5 (continued)

Classification/Compound[a]	Location	Biological activity	References
Hippospongia fistulosa			
Hippomeroterpenes A (**39**) and B (**40**)	Van Phong Bay, Nha Trang	Weak cytotoxic effects on Hep-G2, MCF-7, SK-LU-1 and SK-Mel-2 human cancer cells	[52]
Hippolactone A (**41**)		Weak/moderate cytotoxic effects on Hep-G2, MCF-7, SK-LU-1 and SK-Mel-2 human cancer cells	[53]
Hippotulosas A (**42**)–D (**45**)		–	[54]
Petrosia nigricans			
Petronigrione (**46**)	Hai Van, Danang	Moderate cytotoxic activities against HepG2, KB, LU-1 and MCF-7 cancer cells	[55]
Ircinia echinata			
8-Hydroxyisovariabilin (**47**)	Coto Island, Quang Ninh	–	[56]
Penares sp.			
3-Oxo-7β,8β-epoxy-5α-lanost-24-en-30,9α-olide (**48**), 29-nor-3-oxo-7β,8β-epoxy-5α-lanost-24-en-30,9α-olide (**49**), 3β-acetoxy-7β,8β-epoxy-5α-lanost-24-en-30,9α-olide (**50**), 3β-hydroxy-7β,8β-epoxy-5α-lanost-24-en-30,9α-olide (**51**), 29-nor-penasterone (**52**) and 3β-hydroxy-5α-lanosta-7,24-dien-30,9α-olide (**53**)	Collected by dredging at a depth of 95 m (16° 07′ N, 114° 47′ E)	3β-hydroxy-5α-lanosta-7,24-dien-30,9α-olide: active against human leukemia HL-60 cells with an IC_{50} of 9.7 μM	[57]
29-nor-24(*R*),25-Dihydroxypenasterone (**54**), 29-nor-25-hydroperoxy-3-oxo-7β,8β-epoxy-5α-lanost-23(*E*)-en-30,9α-olide (**55**), 24(*R*),25-dihydroxypenasterone (**56**), 24(*S*),25-dihydroxypenasterone (**57**), 24(*R*)-hydroxy-3-oxo-7β,8β-epoxy-5α-lanost-25-en-30,9α-olide (**58**), 24(*S*)-hydroxy-3-oxo-7β,8β-epoxy-5α-lanost-25-en-30,9α-olide (**59**), 29-nor-24ε-hydroxy-25-chloro-3-oxo-7β,8β-epoxy-5α-lanost-30,9α-olide (**60**) and 3β,24-dihydroxy-7β,8β-epoxy-5α-lanost-24-en-30,9α-olide (**61**)		-	[58]

(continued)

Table 5 (continued)

Classification/Compound[a]	Location	Biological activity	References
Stelletta sp.			
Stellettins Q (**62**)–V (**67**)	Cham Island	–	[59]
Stellettins W (**68**) and X (**69**)			[60]
Cyclobutastellettolides A (**70**) and B (**71**)		Cyclobutastellettolides A and B: significantly increase in ROS productions in peritoneal macrophages at the concentrations of 10 and 1 μM	[61]
Rhabdastrella providentiae			
Rhabdastrellins G (**72**)–K (**76**)	Depth of 20 m in the coral reef of Con Co, Quang Tri	Rhabdastrellin H: cytotoxic effects against HepG2, LU-1, MCF-7, HL-60 and SK-Mel2	[62]
Rhabdaprovidines A (**77**)–G (**83**)	Con Co, Quang Tri	Rhabdaprovidines A–C: inhibited NO production in LPS stimulated BV2 cells	[63, 64]
R. globostellata			
Rhabdaglostelones A (**84**)–C (**86**)	Van Phong Bay, Khanh Hoa	Rhabdaglostelones A and B: Moderate cytotoxic activities toward LU1, HepG2, MCF7 and SKMel2 cell lines	[65]
Rhabdastrellosides A (**87**) and B (**88**)	Near Cham Island	Rhabdastrellosides A and B: weak cytotoxicity against human neuroblastoma SH-SY5Y cells and normal rat H9c2 cardiomyocytes Rhabdastrelloside B: cytoprotective activity	[66]

[a]Chemical structures in Figs. 8, 9, 10, 11, 12, 13, 14 and 15

Table 6 List of new steroids reported from Vietnamese marine sponges

Classification/Compound[a]	Location	Biological activity	References
Topsentia sp.			
Chlorotopsentiasterol sulfate D (**89**), topsentiasterol sulfate F (**90**), and iodotopsentiasterol D (**91**)	Vang Fong Bay	Chlorotopsentiasterol sulfate D: inhibitor of molluscan *endo*-1,3-β-D-glucanase, having an IC_{50} value of 5.4 μM/L	[67]
Ianthella basta			
Lanthebastanoside A (**92**)	Van Phong, Nha Trang	–	[68]
Ianthella sp.			
Petrosterol-3,6-dione (**93**) and 5*α*,6*α*-epoxy-petrosterol (**94**)	Namyet island, Khanh Hoa Province	Petrosterol-3,6-dione and 5α,6α-epoxy-petrosterol: cytotoxic activities on A549, HL-60, MCF-7, SK-OV-3, and U937 cancer cell lines	[69]
Aragusteroketal B (**95**)		Moderate cytotoxic activity against MCF-7, SK-Hep-1, and HeLa cell lines	[70]
Xestospongia testudinaria			
Aragusterol I (**96**), 21-*O*-octadecanoyl-xestokerol A (**97**) and 7*β*-hydroxypetrosterol (**98**)	Truong Sa Archipelago, Khanh Hoa	21-*O*-octadecanoyl-xestokerol A: exhibited significant antifouling activity	[71]
Langcosterol A (**99**)	Coastal waters of Son Cha, Lang Co, Thua Thien-Hue city	Moderate/weak cytotoxic activities against A549, MCF7, HeLa and WI-38 fibroblast	[72]
Testusterol (**100**)	Phu Quoc Island, Kien Giang	Significant antibacterial activity, IC_{50} values below 12.0 nM and MIC values ranging from 4.70 to 75.23 nM against *S. aureus*, *B. subtilis* and *L. fermentum*	[73]

(continued)

Table 6 (continued)

Classification/Compound[a]	Location	Biological activity	References
Ircinia echinata			
5α,6α-epoxycholesta-7,22(*E*)-dien-3β,9α-diol (**101**), 5α,6α-epoxycholesta-7,24(28)-dien-3β,9α-diol (**102**), (24*R*)-5α,6α-epoxy-24-ethyl-cholesta-7-en-3β,9α-diol (**103**), 5α,6α-epoxycholesta-7-en-3β,9α-diol (**104**), (24*S*)-5α,6α-epoxyergosta-7,22-dien-3β,9α-diol (**105**) amd (24*R*)-5α,6α-epoxy-24-methyl-cholesta-7-en-3β,9α-diol (**106**)	Bai Tu Long, Quang Ninh	(24*R*)-5α,6α-epoxy-24-ethyl-cholesta-7-en-3β,9α-diol and 5α,6α-epoxycholesta-7-en-3β,9α-diol: moderate antiproliferative activity against MCF-7, Hep-G2 and LU-1	[74]
Halichondria vansoesti			
Topsentiasterol sulfate G (**107**), topsentiasterol sulfate I (**108**), topsentiasterol sulfate H (**109**), bromotopsentiasterol sulfate D (**110**), dichlorotopsentiasterol sulfate D (**111**), bromochlorotopsentiasterol sulfate D (**112**) and 4β-hydroxyhalistanol sulfate C (**113**)	Territorial waters of Vietnam, South China Sea (12° 34′ 02″ N; 109° 24′ 26″ E)	Topsentiasterol sulfate H: suppressed PSA expression at a concentration as low as 10 μM and glucose uptake in 22Rv1 cells	[75]
H. panicea			
Halipanasterol (**114**)	Van Phong, Nhatrang	Exhibited cytotoxic effects on Hep-G2, SK-Mel-2, LU-1 and MCF-7 cell lines with IC_{50} values ranging from 9.27 to 15.45 μM	[76]

[a]Chemical structures in Figs. 16, 17, 18 and 19

Table 7 List of new alkaloids reported from Vietnamese marine sponges

Classification/Compound[a]	Location	Biological activity	References
Hippospongia fistulosa			
2-Aminotetradec-13-ene-3,4,5,6,7-pentaol (**115**)	Van Phong, Nha Trang	Moderate cytotoxic effects on Hep-G2, MCF-7, SK-LU-1 and SK-Mel-2 human cancer cells	[53]
Desmapsamma anchorata/*Carijoa riisei*			
Fourteen new melibiosylceramides (**116–129**)	Near Van Don Island, Quang Ninh	Total sum of melibiosylceramides and amphimelibioside C at 20 μM caused a 77% reduction in colony formation and growth of SK-MEL-28 cells using soft agar assay	[77]
Aaptos sp.			
3-*N*-Morpholinyl-9-demethyl(oxy)aaptamine (**130**)	Vang Fong Bay	Inhibit epidermal growth factor (EGF)-induced malignant transformation of mouse epidermal JB6 P^{+}C1 41 cells in soft agar	[78]
2,3-Dihydro-2,3-dioxoaaptamine (**131**), 6-(*N*-morpholinyl)-4,5-dihydro-5-oxodemethyl(oxy)aaptamine (**132**) and 3-(methylamino)demethyl(oxy)aaptamine (**133**)		–	[79]
A. aaptos			
Aaptanone (**134**)	Cu Lao Re Island	–	[80]
N-Demethylaaptanone (**135**)	Depth of 6 m at the Tho Chu Island	–	[81]
9-Methoxy-*N*-demethylaaptanone (**136**), 3,5-dicarbomethoxy-1,6-naphthyridine (**137**), aaptosvanphongs A (**138**) and B (**139**)	Van Phong Bay, Nha Trang	Cytotoxic effects against SKLU-1, MCF-7, HepG2 and SK-Mel-2 cell lines with IC_{50} values ranging from 9.4 to 49.2 μM	[82]

(continued)

Table 7 (continued)

Classification/Compound[a]	Location	Biological activity	References
Neopetrosia sp.			
Neopetrosides A (**143**) and B (**144**)	Con Son Island	Neopetroside A: upregulated mitochondrial functions by enhancing the mitochondrial ATP levels and oxygen consumption rate in C2C12 cells	[83]
Xestospongia muta			
meso-Araguspongine C (**145**)	Coral reef of Vinh Moc, Quang Tri	Cytotoxic effects against HepG-2, HL-60, LU-1, MCF-7 and SK-Mel-2 with IC_{50} values ranging from 0.43 to 0.77 μM	[84]
Haliclona sp.			
Methyl *O*-demethylrenierate (**146**), *O*-demethylrenierol (**147**), 1,6-dimethyl-7-hydroxy-5,8-dihydroisoquinoline-5,8-dione (**148**) and 21-dehydroxyrenieramycin F (**149**)	Vicinity of Hon Thom Island	*O*-Demethylrenierol and 1,6-dimethyl-7-hydroxy-5,8-dihydroisoquinoline-5,8-dione: exhibited moderate anti-inflammatory properties	[85]
Ecionemia acervus			
New stereoisomer of fistularin-3 (**150**)	Collected in Vietnam water (09° 54′ 40.9″ N, 104° 01′ 44.2″ E)	Showed potent anti-inflammatory activity in co-culture system of human epithelial Caco-2 cells and THP-1 macrophages	[86]
Penares sp.			
Two new alkaloids (**151** and **152**)	Collected by dredging at a depth of 95 m (16° 07′ N, 114° 47′ E)	**151**: moderate inhibition of HL-60 and HeLa cell lines with IC_{50} of 16.1 and 33.2 μM, respectively	[87]

[a]Chemical structures in Figs. 20, 21 and 22

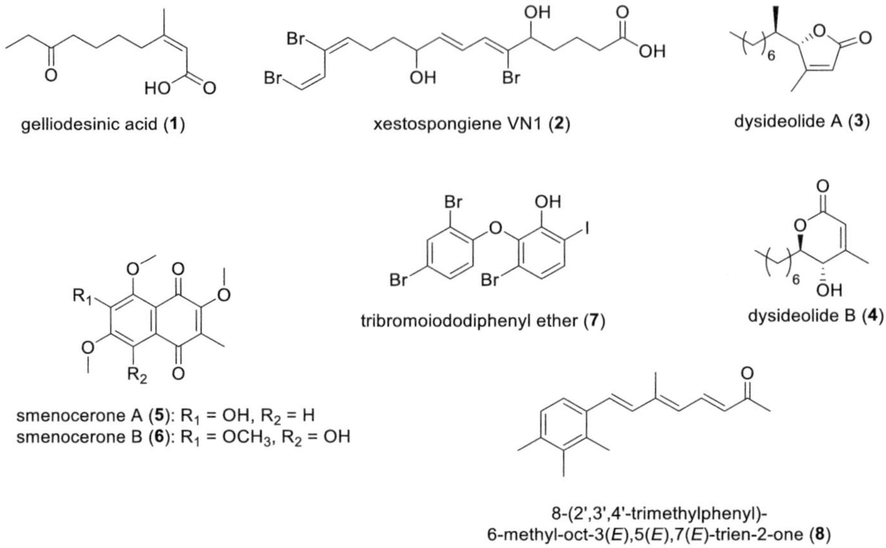

Fig. 7 Chemical structures of new fatty acid derivatives and polyketides reported from Vietnamese sponges

3.1.1 Fatty Acid Derivatives/polyketides

About eight new fatty acid derivatives/polyketide molecules have been reported from at least six sponge species, such as *Arenosclera* sp., *Clathria* (*Thalysias*) *reinwardti*, *Dysidea cinerea*, *Gelliodes* sp., *Smenospongia cerebriformis* and *Xestospongia testudinaria* (Table 4, Fig. 5) [37–42].

Smenospongia cerebriformis

Two new naphtoquinones, smenocerones A (**5**) and B (**6**) (Fig. 7), along with four known sesquiterpene cyclopentenones were isolated from the sponge, *Smenospongia cerebriformis*, collected from the Sea of Quangtri, Vietnam [40]. These molecules were assessed for their cytotoxicity against five human cancer cell lines, including LU-1 (lung carcinoma), HepG-2 (hepatocellular carcinoma), HL-60 (promyelocytic leukemia), MCF-7 (breast carcinoma) and SK-Mel-2 (melanoma). Only smenocerone B (**6**) exhibited significant cytotoxicity against the panel of cancer cell lines with IC_{50} values ranging from 3.2 to 5.7 μg/mL. In addition, when tested at 20 μg/mL, smenocerone B caused over 80% cancer cell death.

3.1.2 Terpenoids

Diverse classes of terpenoids, including sesquiterpenes, diterpenes, sesterterpenes, triterpenes, and rearranged/modified terpenes, have been reported from more than

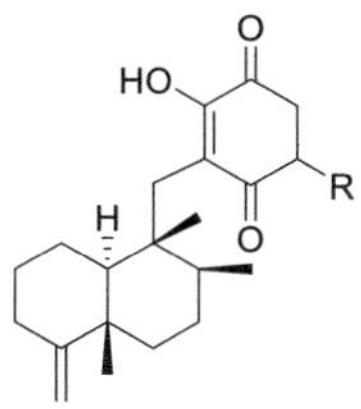

Langcoquinone A (**9**): R = $NHCH_3$
Langcoquinone B (**10**): R = $NHCH_2CH_2CH_2SCH_3$
Langcoquinone C (**11**): R = $OCH_2CH_2CH_2CH_3$
Langcoquinone D (**12**): R = $NHCH_2CH_2CH_2SOCH_3$
Langcoquinone E (**13**): R = $NHCH_2CH_2SO_3H$
Langcoquinone F (**14**): R = $NHCH_2CH_2OH$

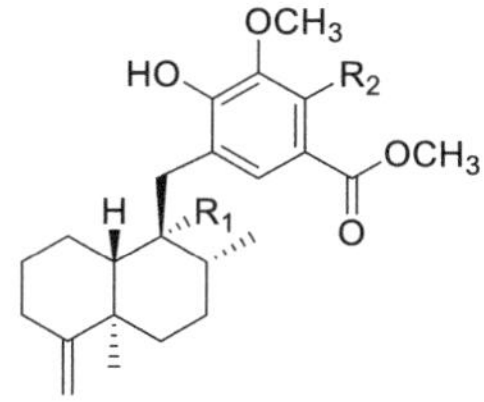

Langconol A (**15**): $R_1 = CH_3$, $R_2 = OH$
Langconol B (**16**): $R_1 = CHO$, $R_2 = H$
Langconol C (**17**): $R_1 = CHO$, $R_2 = OH$

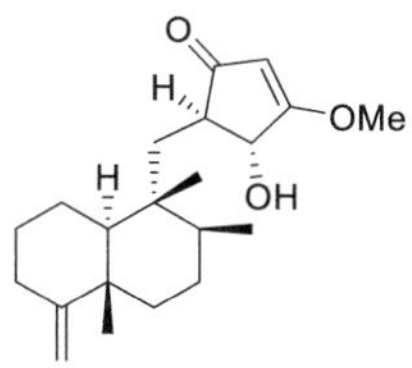

Smenohaimien A (**18**)

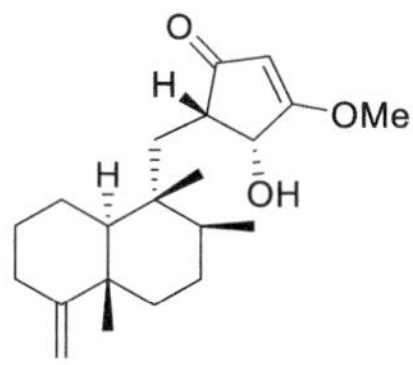

Smenohaimien B (**19**)

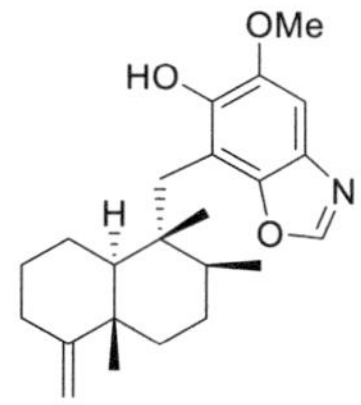

Smenohaimien C (**20**)

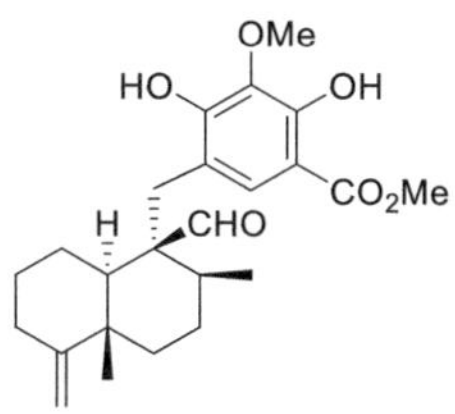

Smenohaimien D (**21**)

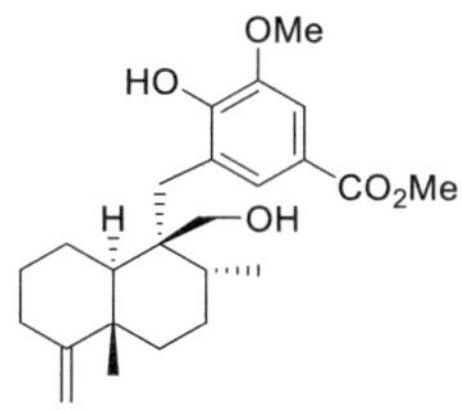

Smenohaimien E (**22**)

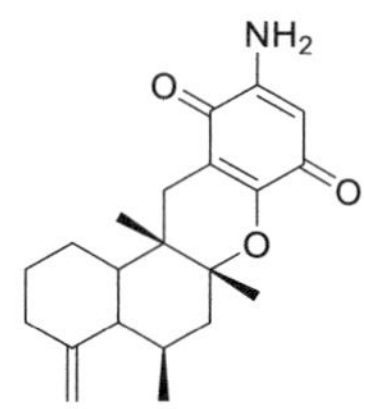

Smenohaimien F (**23**)

Fig. 8 Chemical structures of new terpenoids reported from Vietnamese *Spongia* sp. and *Smenospongia* sp.

10 sponge species (Table 5; Figs. 8, 9, 10, 11, 12, 13, 14 and 15) [43–59, 61–65]. A majority of these molecules exhibit cytotoxicity, antimicrobial as well as anti-inflammatory properties.

Sesquiterpenes

Spongia sp. *Smenospongia cerebriformis* and *Hippospongia fistulosa*

Sesquiterpenoid quinones/merosesquiterpenes with significant anti-inflammatory, antibacterial, antiviral, antitumor, and antiangiogenic properties have reportedly been isolated from several marine sponge genera, including *Dysidea*, *Spongia* and

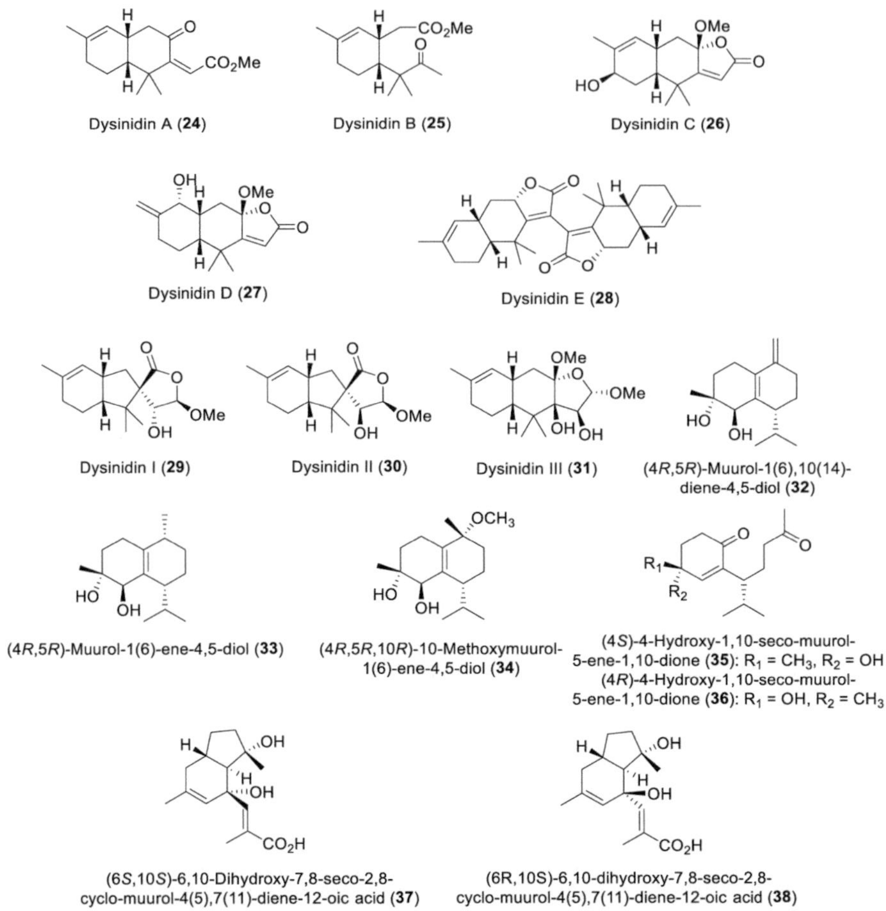

Fig. 9 Chemical structures of new terpenoids reported from Vietnamese *Dysidea* sp.

Dactylospongia. From the Vietnamese sponge, *Spongia* sp., obtained from a depth of 12 m off Son Cha, Lang Co, Thua Thien-Hue city, Vietnam, two new sesquiterpene aminoquinones, langcoquinones A (**9**) and B (**10**) (Fig. 8), were isolated along with seven known meroterpenoids. [43]. The antibacterial activities of these compounds were evaluated against four bacterial strains, including *B. subtilis*, *S. aureus*, *Klebsiella pneumoniae* and *Escherichia coli*. The new sesquiterpene aminoquinones, **9** and **10**, exhibited significant antibacterial activities against *S. aureus* and *B. subtilis*, with MICs values of 12.5 μM. From the same collection that yielded langcoquinones A (**9**) and B (**10**), additional three new related sesquiterpenoid quinones, langcoquinones D (**12**)–F (**14**) (Fig. 8), were subsequently purified from the ethyl acetate extract of *Spongia* sp. Langcoquinone D (**12**) exhibited good antibacterial activities against *B. subtilis* and *S. aureus*, with MIC values of 12.5 and 25.0 μM, respectively [44]. In addition, langcoquinone D showed significant

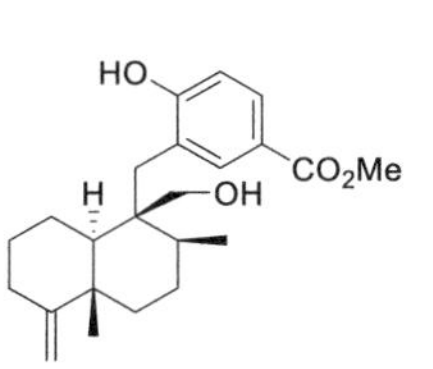

Hippomeroterpene A (**39**)

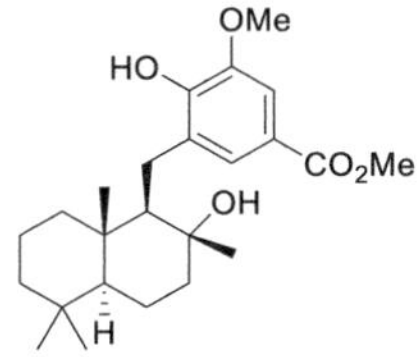

Hippomeroterpene B (**40**)

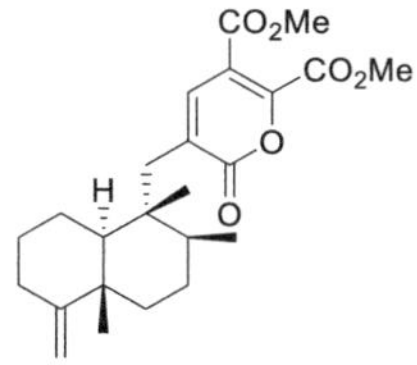

Hippolactone A (**41**)

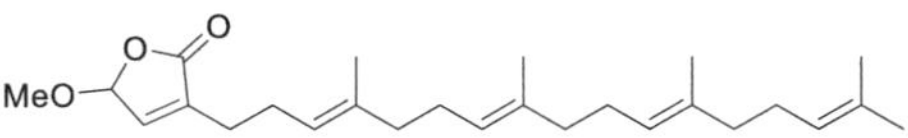

Hippotulosa A (**42**)

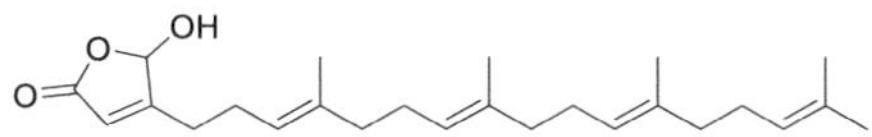

Hippotulosa B (**43**)

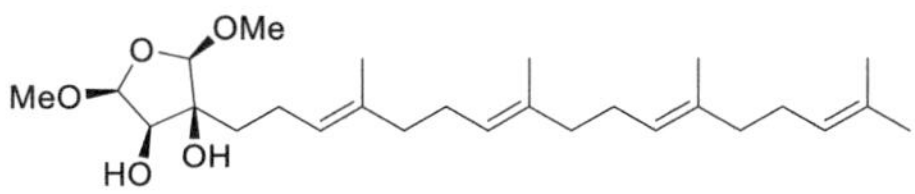

Hippotulosa C (**44**)

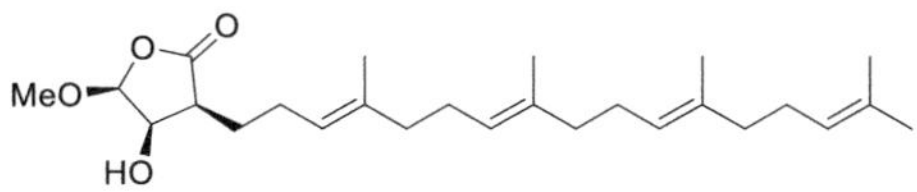

Hippotulosa D (**45**)

Fig. 10 Chemical structures of new terpenoids reported from Vietnamese *Hippospongia* sp.

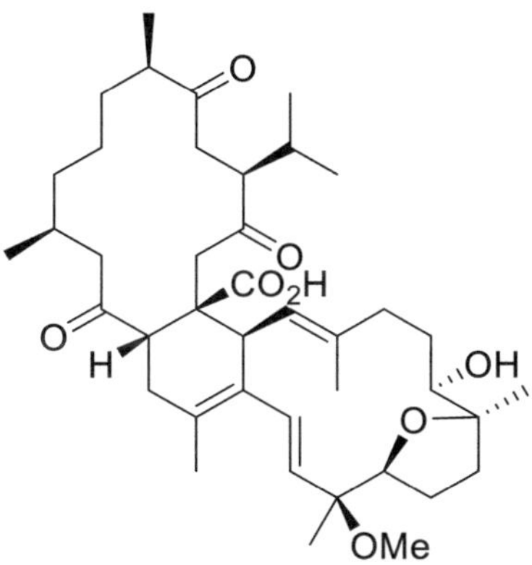

Petronigrione (**46**)

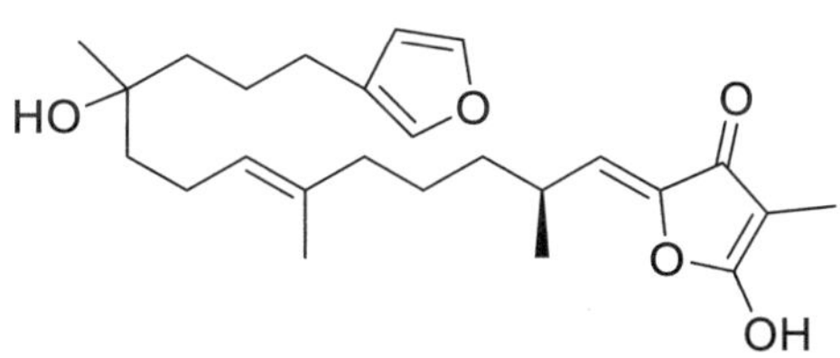

8-Hydroxyisovariabilin (**47**)

Fig. 11 Chemical structures of new terpenoids reported from Vietnamese *Petrosia nigricans* and *Ircinia echinata*

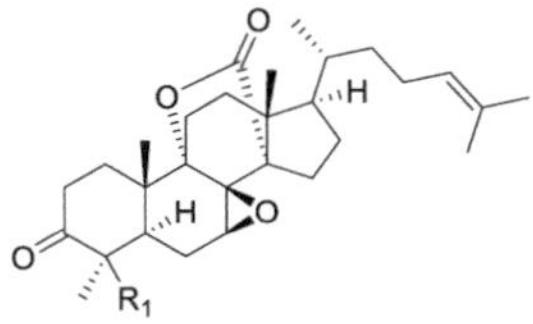

3-Oxo-7β,8β-epoxy-5α-lanost-24-en-30,9α-olide (**48**): R = Me
29-*nor*-3-oxo-7β,8β-epoxy-5α-lanost-24-en-30,9α-olide (**49**): R = H

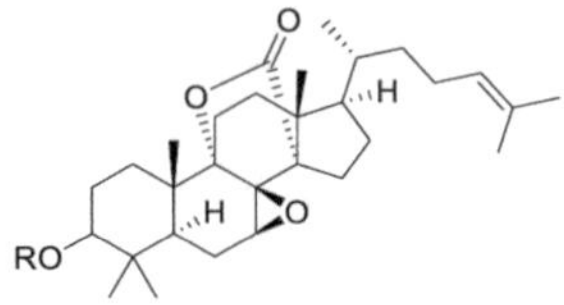

3β-Acetoxy-7β,8β-epoxy-5α-lanost-24-en-30,9α-olide (**50**): R = Ac
3β-Hydroxy-7β,8β-epoxy-5α-lanost-24-en-30,9α-olide (**51**): R = H

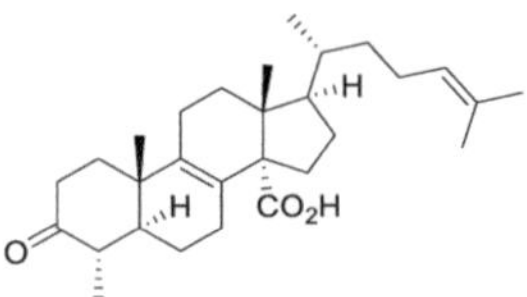

29-*nor*-penasterone (**52**)

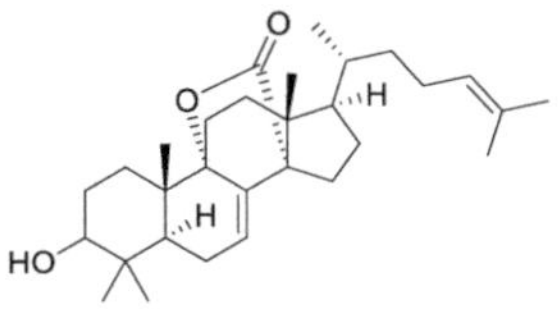

3β-Hydroxy-5α-lanosta-7,24-dien-30,9α-olide (**53**)

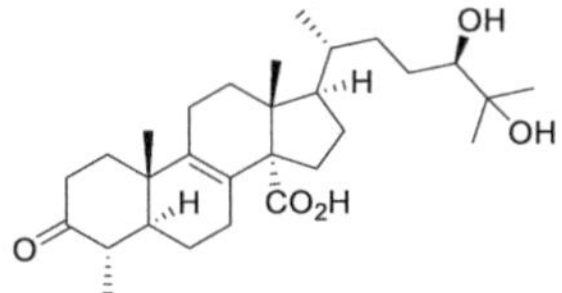

29-*nor*-24(*R*),25-dihydroxypenasterone (**54**)

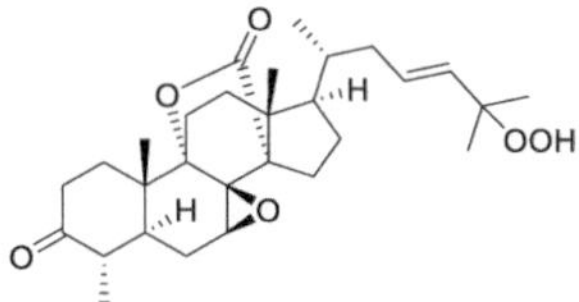

29-*nor*-25-hydroperoxy-3-oxo-7β,8β-epoxy-5α-lanost-23(*E*)-en-30,9α-olide (**55**)

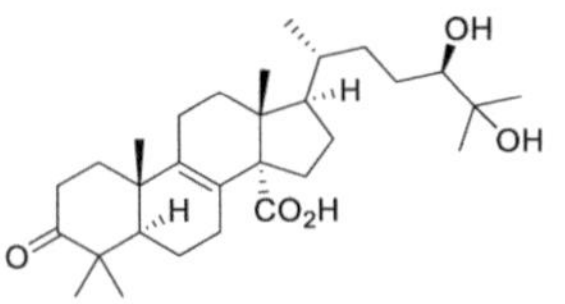

24(*R*),25-Dihydroxypenasterone (**56**)

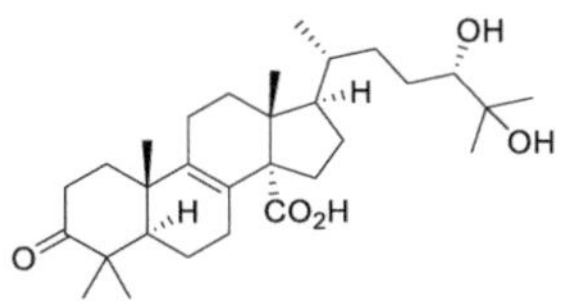

24(*S*),25-Dihydroxypenasterone (**57**)

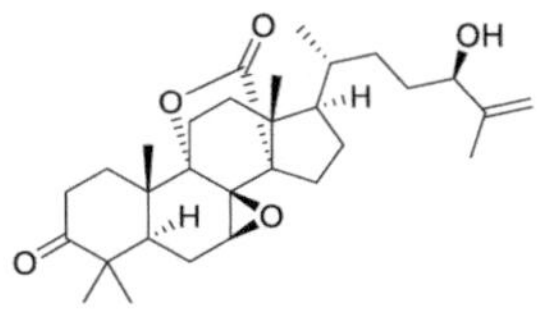

24(*R*)-Hydroxy-3-oxo-7β,8β-epoxy-5α-lanost-25-en-30,9α–olide (**58**)

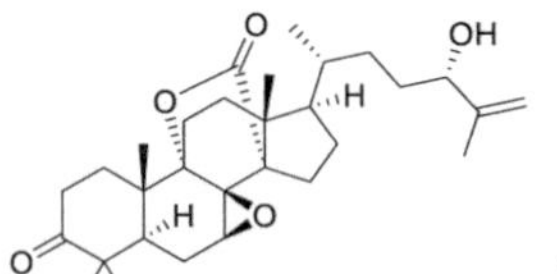

24(*S*)-Hydroxy-3-oxo-7β,8β-epoxy-5α-lanost-25-en-30,9α–olide (**59**)

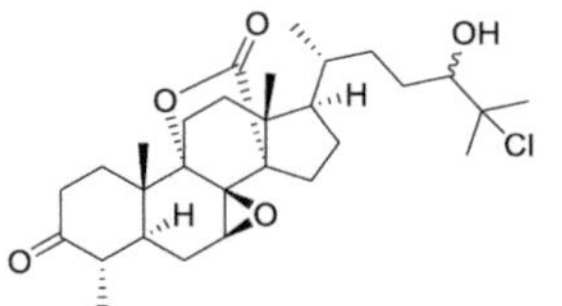

29-*nor*-24ε-hydroxy-25-chloro-3-oxo-7β,8β-epoxy-5α-lanost-30,9α-olide (**60**)

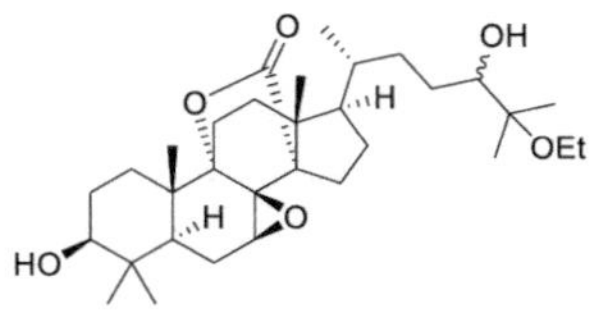

3β,24-Dihydroxy-7β,8β-epoxy-5α-lanost-24-en-30,9α-olide (**61**)

Fig. 12 Chemical structures of new terpenoids reported from Vietnamese *Penares* sp., *Stelletta* sp. and *Rhabdastrella* sp.

cytotoxic activities against the A549, MCF7, HeLa and WI-38 cell lines, with IC_{50} values of 8.9, 5.9, 8.6 and 5.6 μM, respectively.

In their continued search for new antibacterial molecules from *Spongia* sp., the Vietnamese research group reported three new sesquiterpene phenols, namely langconols A (**15**)–C (**17**) (Fig. 8), and one new sesquiterpene hydroxyquinone, langcoquinone C (**11**) (Fig. 8) [45]. The antibacterial assays of langcoquinone C, langconols A and C showed antibacterial activities against *B. subtilis* and/or *S. aureus*, with MICs ranging from 6.25 to 25.0 μM. In addition to the antibacterial activities of langconol C and langcoquinone C, significant cytotoxicity against a panel of cancer

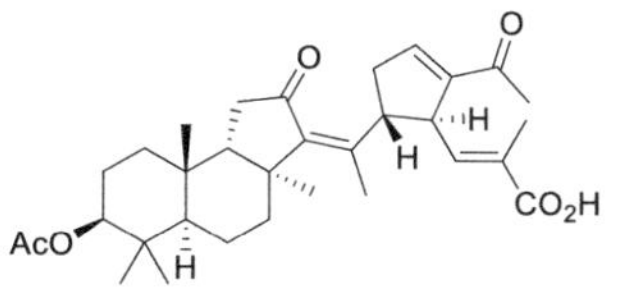

Stelletin Q (**62**)

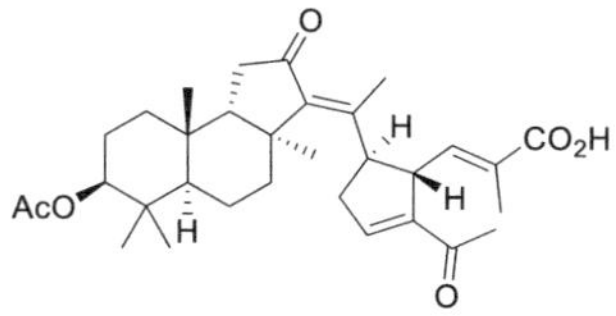

Stelletin R (**63**)

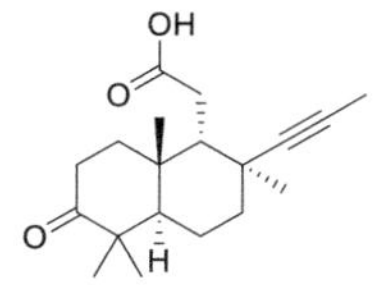

Stelletin S (**64**)

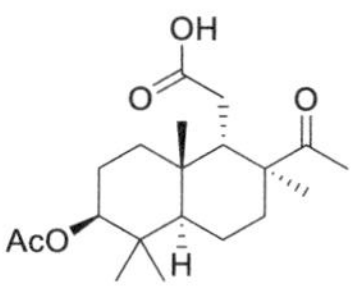

Stelletin T (**65**)

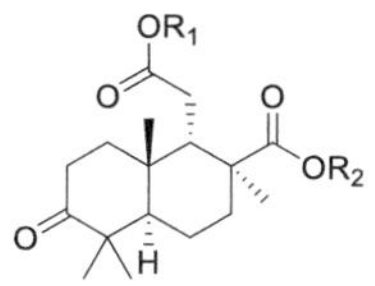

Stelletin U (**66**): R_1 = Et, R_2 = H
Stelletin V (**67**): R_1 = H, R_2 = Et

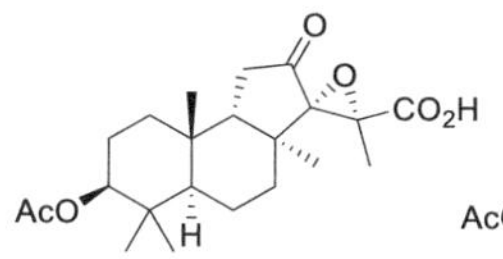

Stelletin W (**68**)

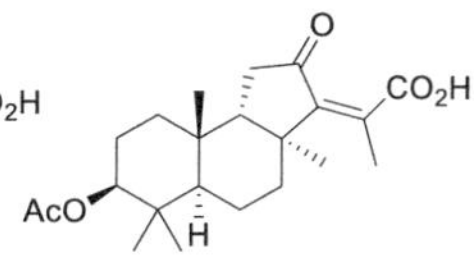

Stelletin X (**69**)

Fig. 13 Chemical structures of new terpenoids reported from Vietnamese *Stelletta* sp.

Fig. 14 Chemical structures of new modified terpenoids reported from Vietnamese *Stelletta* sp.

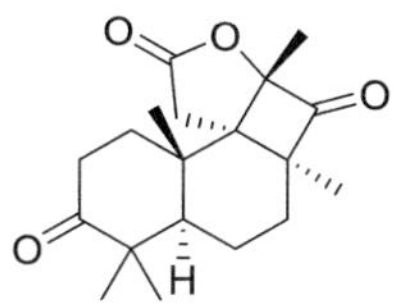

Cyclobutastellettolide A (**70**)

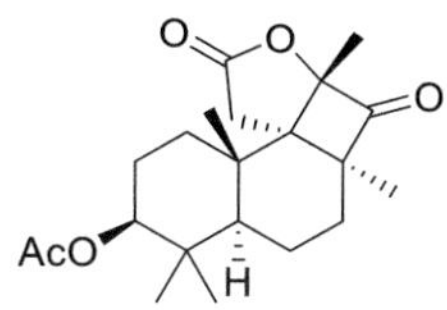

Cyclobutastellettolide B (**71**)

cell lines, including A549, MCF7, HeLa and WI-38 fibroblast, were observed for these molecules, with IC_{50} values ranging from 5.0 to 8.7 μM.

Additional new merosesquiterpenes, namely smenohaimiens A (**18**)–F (**23**) (Fig. 8) as well as hippomeroterpenes A (**39**) and B (**40**) (Fig. 10), have also been reported from at least two Vietnamese sponges, *Smenospongia cerebriformis* and *Hippospongia fistulosa*, respectively [46, 47, 52]. From *S. cerebriformis*, new smenohaimiens A (**18**), B (**19**) and E (**22**) were found to have moderate inhibitory activities on NO production while hippomeroterpenes A (**39**) and B (**40**) from *H. fistulosa* exhibited weak cytotoxicity on Hep-G2, MCF-7, SK-LU-1 and SK-Mel-2 human cancer cells.

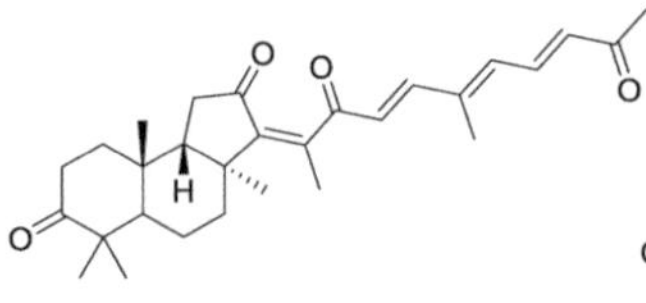

Rhabdastrellin G (**72**)

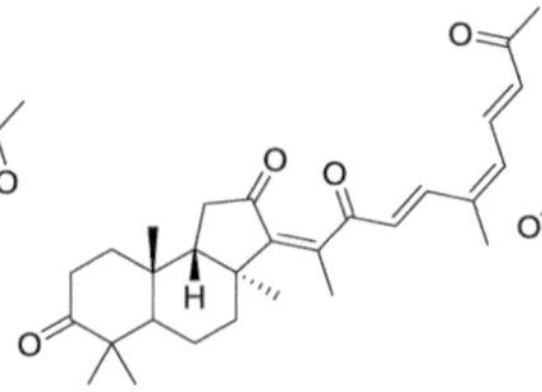

Rhabdastrellin H (**73**)

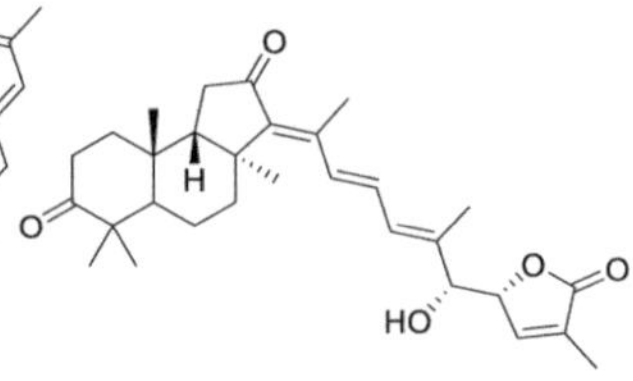

Rhabdastrellin I (**74**)

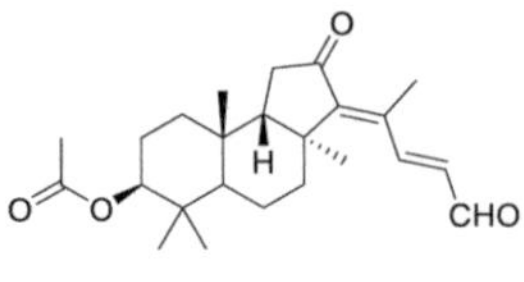

Rhabdastrellin J (**75**)

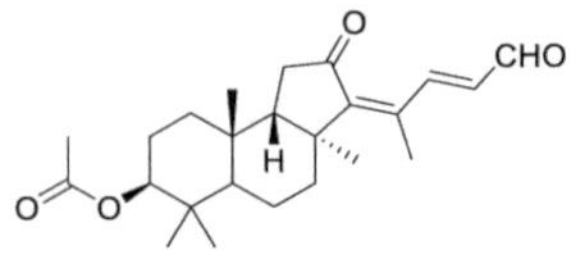

Rhabdastrellin K (**76**)

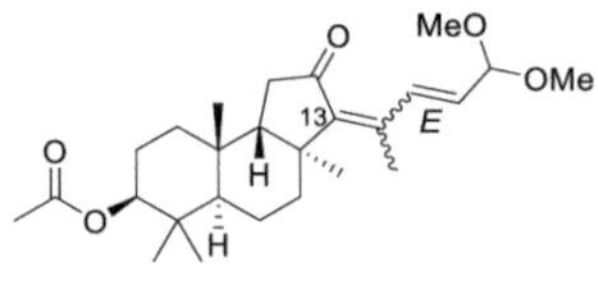

Rhabdaprovidine A (**77**): 13*Z*
Rhabdaprovidine B (**78**): 13*E*

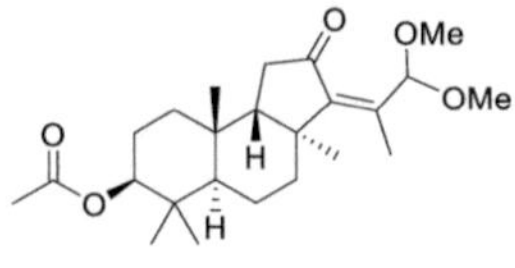

Rhabdaprovidine C (**79**)

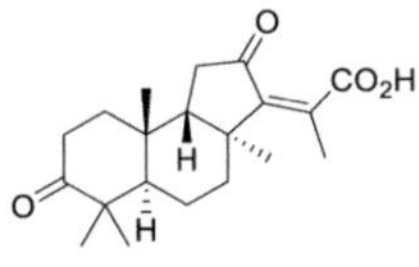

Rhabdaprovidine D (**80**)

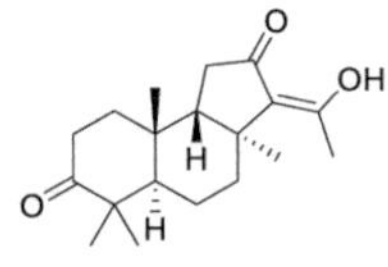

Rhabdaprovidine E (**81**)

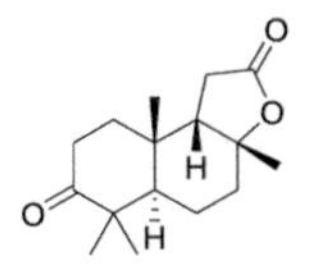

Rhabdaprovidine F (**82**)

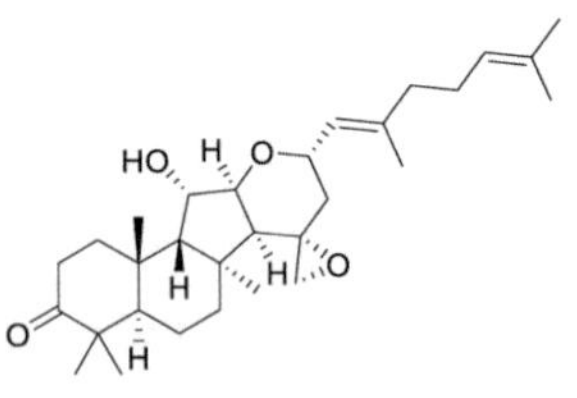

Rhabdaprovidine G (**83**)

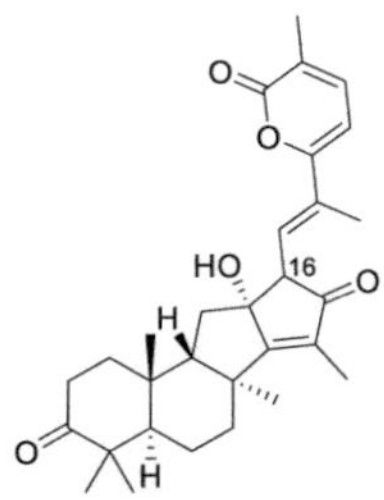

Rhabdaglostelone A (**84**): 16*R*
Rhabdaglostelone B (**85**): 16*S*

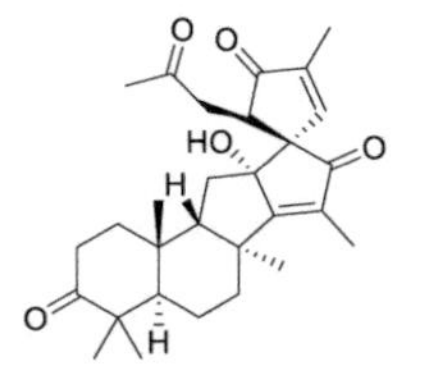

Rhabdaglostelone C (**86**)

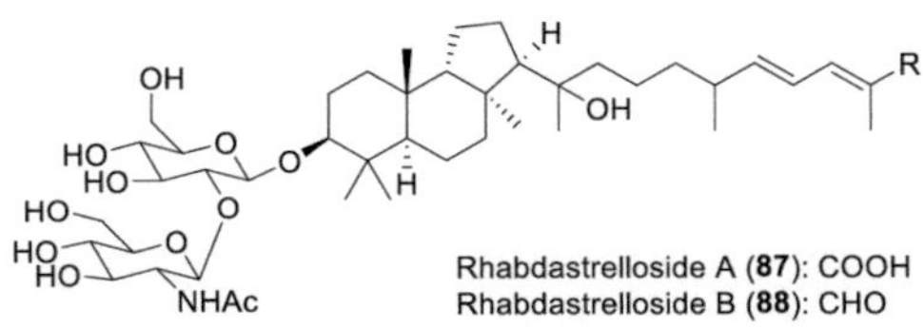

Rhabdastrelloside A (**87**): COOH
Rhabdastrelloside B (**88**): CHO

Fig. 15 Chemical structures of new terpenoids reported from Vietnamese *Rhabdastrella* sp.

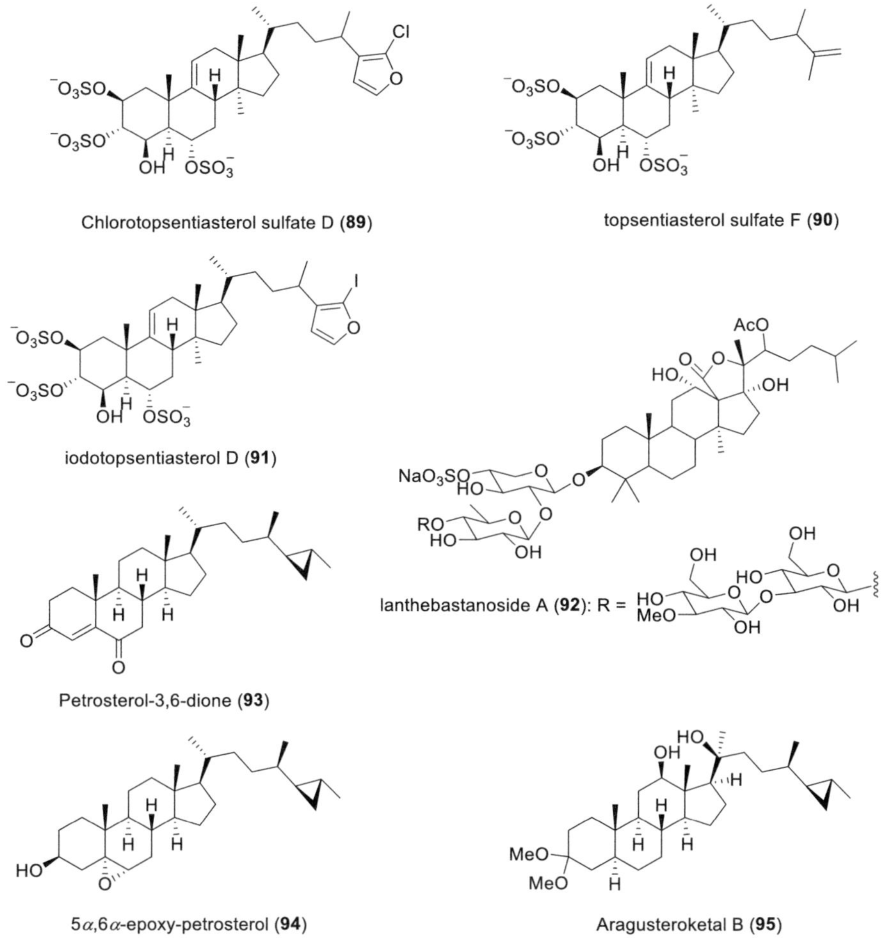

Fig. 16 Chemical structures of new steroids reported from Vietnamese sponges

Triterpenoids

Penares sp.

A dredged sponge sample of *Penares* sp., obtained from a depth of 95 m (16° 07′ N, 114° 47′ E) during the 30th cruise of R/V Akademik Oparin, provided six new lanosterol congeners, including 3-oxo-7*β*,8*β*-epoxy-5*α*-lanost-24-en-30,9*α*-olide (**48**), 29-*nor*-3-oxo-7*β*,8*β*-epoxy-5*α*-lanost-24-en-30,9*α*-olide (**49**), 3*β*-acetoxy-7*β*,8*β*-epoxy-5*α*-lanost-24-en-30,9*α*-olide (**50**), 3*β*-hydroxy-7*β*,8*β*-epoxy-5*α*-lanost-24-en-30,9*α*-olide (**51**), 29-*nor*-penasterone (**52**) and 3*β*-hydroxy-5*α*-lanosta-7,24-dien-30,9*α*-olide (**53**) (Fig. 12) [57]. Configurations of the triterpene epoxy lactones in **48**–**51** were established based on NOESY and CD spectral data as well as calculation of spin coupling constants and confirmation by X-ray crystallographic analysis

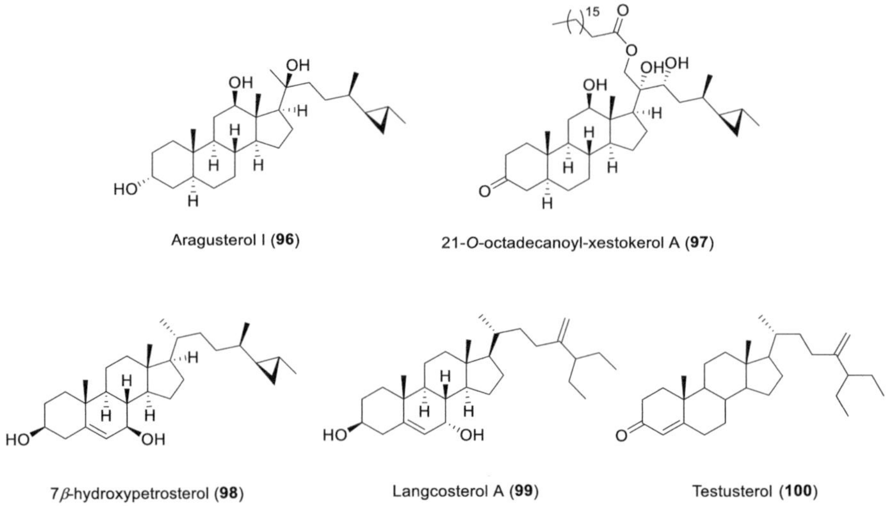

Fig. 17 Chemical structures of new steroids reported from Vietnamese *Xestospongia* sp.

of molecule **49**. Only 3*β*-hydroxy-5*α*-lanosta-7,24-dien-30,9*α*-olide (**53**) significant activity against HL-60 cells, with IC_{50} value of 9.7 μM.

Subsequent chemical investigation of the same sponge sample provided additional eight new oxidized lanostane and *nor*-lanostane derivatives, including 29-nor-24(*R*),25-dihydroxypenasterone (**54**), 29-nor-25-hydroperoxy-3-oxo-7*β*,8*β*-epoxy-5*α*-lanost-23(*E*)-en-30,9*α*-olide (**55**), 24(*R*),25-dihydroxypenasterone (**56**), 24(*S*),25-dihydroxypenasterone (**57**), 24(*R*)-hydroxy-3-oxo-7*β*,8*β*-epoxy-5*α*-lanost-25-en-30,9α-olide (**58**), 24(*S*)-hydroxy-3-oxo-7*β*,8*β*-epoxy-5*α*-lanost-25-en-30,9α-olide (**59**), 29-nor-24ε-hydroxy-25-chloro-3-oxo-7*β*,8*β*-epoxy-5*α*-lanost-30,9*α*-olide (**60**) and 3*β*,24-dihydroxy-7*β*,8*β*-epoxy-5*α*-lanost-24-en-30,9*α*-olide (**61**) (Fig. 12) [58].

Stelletta sp.—Triterpenoids and modified triterpenoids

The malabaricane and isomalabaricane triterpenoids, consisting of rare 6,6,5-tricyclic ring systems with long side chains, have been reported from terrestrial and marine sources. The skeletal systems of malabaricanes and isomalabaricanes are similar to each other but differ mainly in the stereochemistry of the tricyclic core fragments. These tricyclic terpenoids have attracted a lot of attention due to their diverse biological activity, including cytotoxicity against tumor cells and neuroprotective properties. New isomalabaricanes have been isolated from at least three Vietnamese sponge species, including *Stelletta* sp., *Rhabdastrella globostellata* and *R. providentiae.*

Stelletins are a series of isomalabaricane derivatives reported from diverse sponge species collected from many locations. New isomalabaricane triterpenoids, stellettins Q (**62**) and R (**63**) (Fig. 13), and four new isomalabaricane-derived *nor*-terpenoids, stellettins S (**64**)–V (**67**) (Fig. 13), were isolated from *Stelletta* sp., collected in

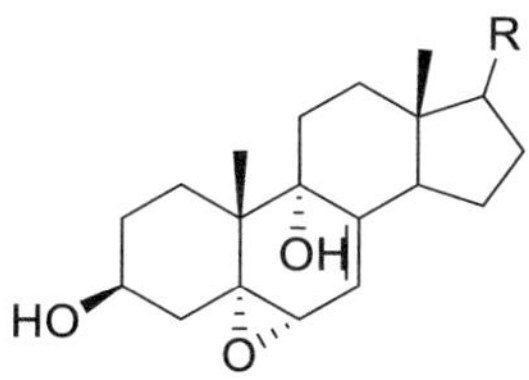

5α,6α-epoxycholesta-7,22(*E*)-dien-3β,9α-diol (**101**): R = 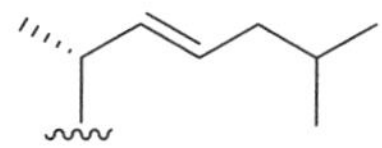

5α,6α-epoxycholesta-7,24(28)-dien-3β,9α-diol (**102**): R = 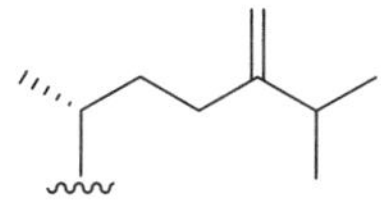

(24*R*)-5α,6α-epoxy-24-ethyl-cholesta-7-en-3β,9α-diol (**103**): R = 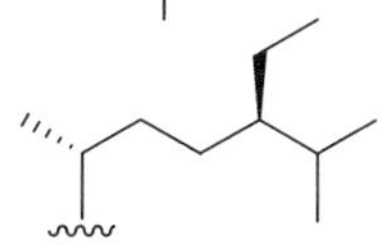

5α,6α-epoxycholesta-7-en-3β,9α-diol (**104**): R = 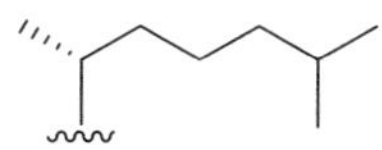

(24*S*)-5α,6α-epoxyergosta-7,22-dien-3β,9α-diol (**105**): R = 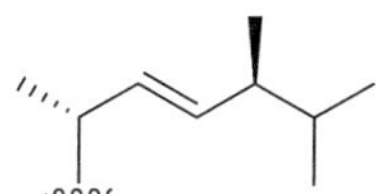

(24*R*)-5α,6α-epoxy-24-methyl-cholesta-7-en-3β,9α-diol (**106**): R = 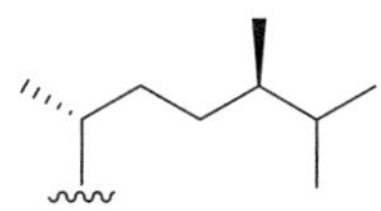

Fig. 18 Chemical structures of new steroids reported from Vietnamese *Ircinia echinata*

the vicinity of Cham Island, Vietnam [59]. The structures and absolute configurations of these new terpenoids were determined by extensive NMR, MS, ECD analyses as well as quantum-chemical modeling. The presence of isomalabaricane-related *nor*-terpenoids, stellettins S (**64**)–V (**67**), in the sponge samples could derive from oxidative degradation of isomalabaricane precursors. In particular, stelletin S (**64**) contains an acetylenic fragment, unprecedented in the isomalabaricane family and extremely rare in other marine sponge terpenoids. Recently, two new *nor*-isomalabaricanic acids, stellettins W (**68**) and X (**69**), were reported from the Vietnamese marine sponge, *Rhabdastrella globostellata*, with the former molecule being the first stellettin with a 13,14-epoxide group [60].

From a sample of the sponge, *Stelletta* sp., collected in the vicinity of Cham Island, two novel C19 terpenoids, cyclobutastellettolides A (**70**) and B (**71**) (Fig. 14), with an unprecedented carbon skeleton, were isolated [61]. Cyclobutastellettolides A and B are the first examples of a new family of norterpenoids having four fused rings, including a cyclobutanone. The biogenesis of cyclobutastellettolide A has

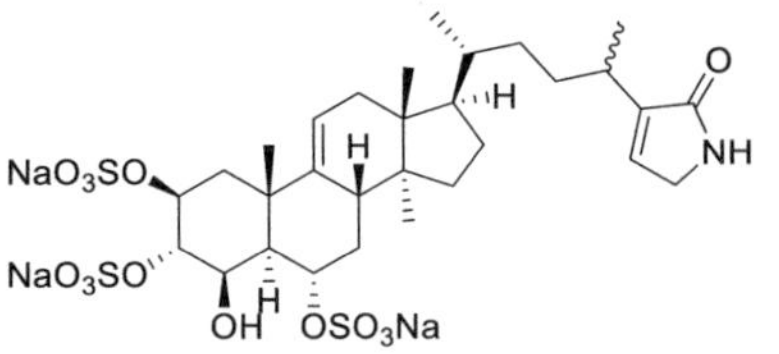

Topsentiasterol sulfate G (**107**)

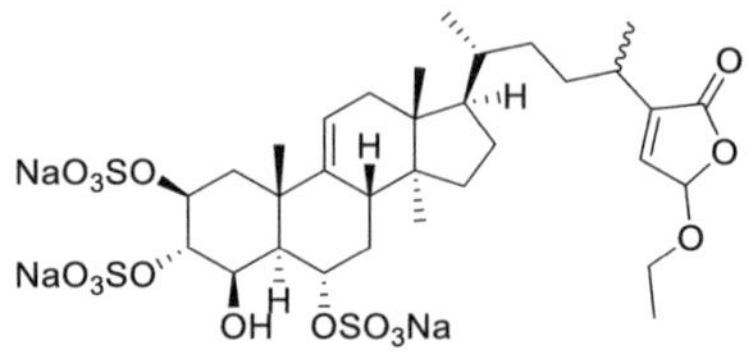

topsentiasterol sulfate I (**108**)

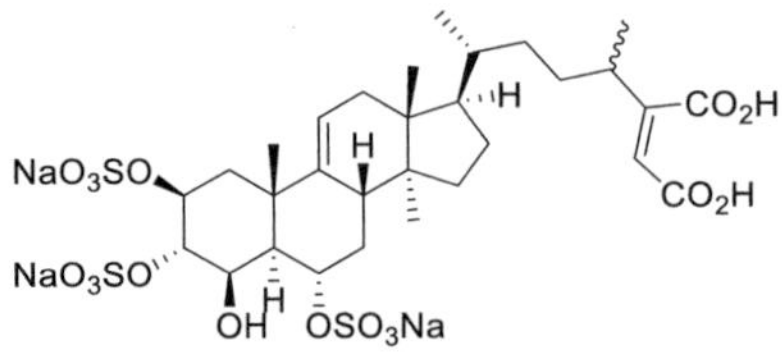

topsentiasterol sulfate H (**109**)

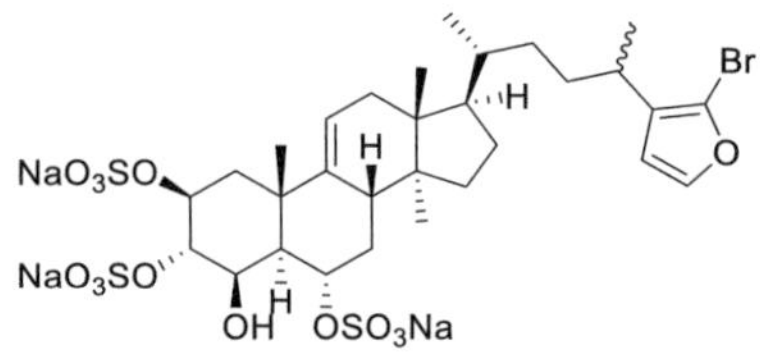

bromotopsentiasterol sulfate D (**110**)

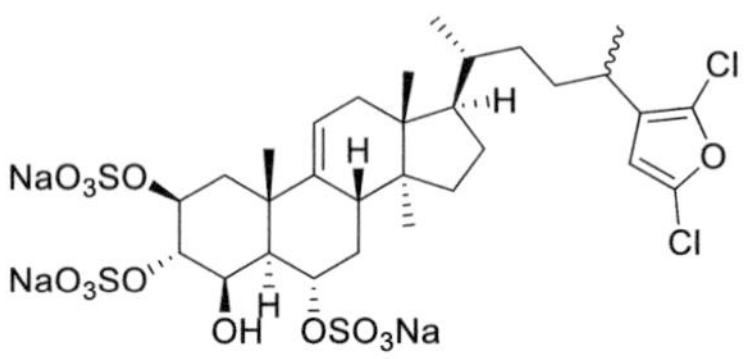

dichlorotopsentiasterol sulfate D (**111**)

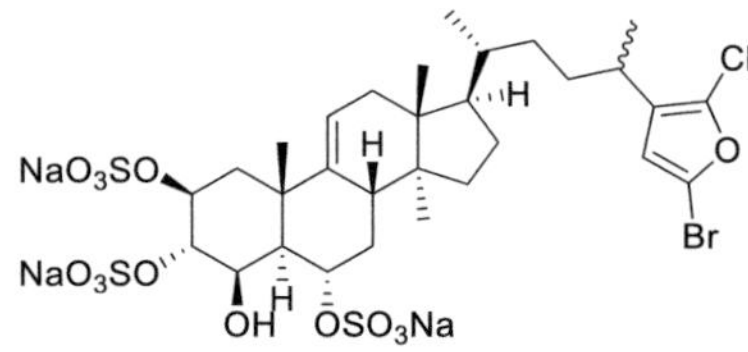

bromochloro topsentiasterol sulfate D (**112**)

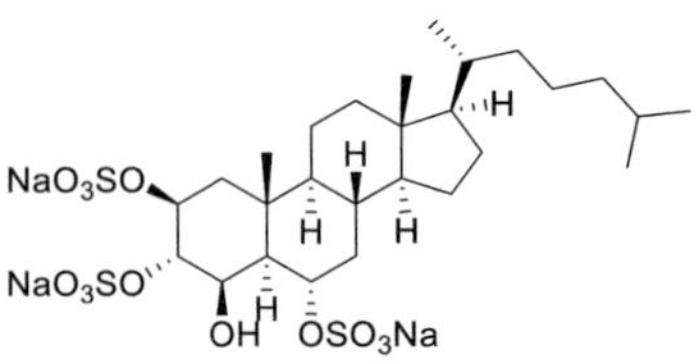

4β-hydroxyhalistanol sulfate C (**113**)

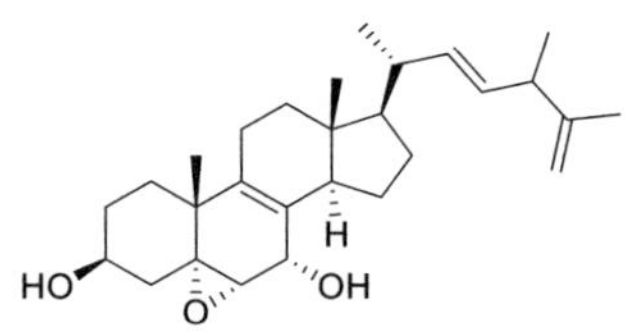

Halipanasterol (**114**)

Fig. 19 Chemical structures of new steroids reported from Vietnamese *Halichondria* sp.

been proposed by the authors to derive from the isomalabricane triterpenoid, (–)-stelletin E, via globostelletins A and B. Both cyclobutastellettolides A and B showed a significant increase in ROS productions by 93% and 90%, respectively, in peritoneal macrophages at the concentrations of 10 μM. The proinflammatory stimulation of cyclobutastellettolides in macrophages suggested that they could be potential immunomodulatory agents.

Due to the unique structural features of cyclobutastellettolide B (**71**) by having a 6/6/4 fused tricyclic core, including a fully substituted cyclobutane, it is an interesting target for total synthesis. Recently, a convenient enantioselective total synthesis of

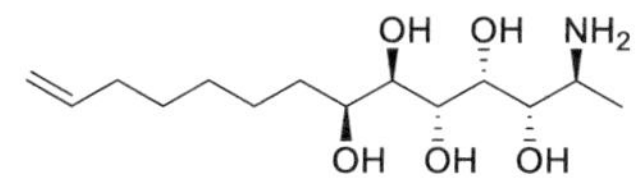

2-Aminotetradec-13-ene-3,4,5,6,7-pentaol (**115**)

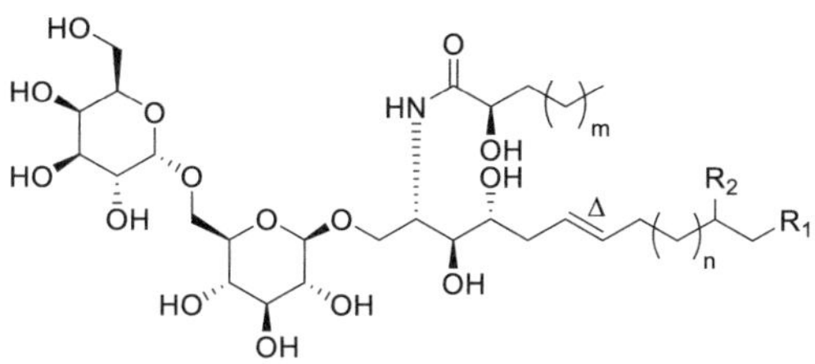

116: m = 18, n = 7, $R_1 = R_2$ = H, Δ sat
117: m = 18, n = 6, R_1 = H, R_2 = Me, Δ sat
118: m = 17, n = 8, $R_1 = R_2$ = H, Δ sat
119: m = 18, n = 8, $R_1 = R_2$ = H, Δ sat
120: m = 19, n = 8, $R_1 = R_2$ = H, Δ sat
121: m = 17, n = 7, R_1 = H, R_2 = Me, Δ sat
122: m = 18, n = 7, R_1 = H, R_2 = Me, Δ sat
123: m = 19, n = 7, R_1 = H, R_2 = Me, Δ sat
124: m = 18, n = 8, R_1 = H, R_2 = Me, Δ sat
125: m = 17, n = 8, R_1 = H, R_2 = Me, Δ
126: m = 18, n = 8, R_1 = H, R_2 = Me, Δ
127: m = 18, n = 7, $R_1 = R_2$ = Me, Δ sat
128: m = 17, n = 7, $R_1 = R_2$ = Me, Δ
129: m = 18, n = 7, $R_1 = R_2$ = Me, Δ

Fig. 20 New nitrogen-containing fatty-acid derivatives reported from Vietnamese *Hippospongia fistulosa* and *Desmapsamma anchorata*/*Carijoa riisei*

(+)-cyclobutastellettolide B was reported, via a strategy involving a diastereoselective Johnson–Claisen rearrangement, a regioselective cyclopropoxytrimethylsilane ring-opening reaction as well as a late stage Norrish–Yang cyclization of α-diketone (Scheme 1) [88]. The overall synthesis of (+)-cyclobutastellettolide B was achieved in 13 steps from the epoxide, with a total yield of 31.5%. The synthesis started from commercially available geranyl acetate, which was converted to an epoxide intermediate in five steps. This epoxide intermediate was subjected to a series of reactions, including polycyclization and hydroxymethylation. A Johnson–Claisen rearrangement then provided the 6/6 ring system with a terminal alkene, which was later converted to an aldehyde. A Wittig reaction and oxidation step were used to transform this aldehyde group to an α-diketone. In the final key step, the diketone was used in a Norrish–Yang photocyclization to form the four-membered ring of cyclobutastellettolide B. Finally, an intramolecular lactonization gave the desired product.

Rhabdastrella sp.

Chemical investigation on a Vietnamese sponge, *Rhabdastrella providentiae*, collected at Con Co, Quang Tri, Vietnam, provided a series of isomalabaricane *nor*-terpenoids, rhabdaprovidines A (**77**)–G (**83**) (Fig. 15) [63, 64]. The anti-inflammatory effects of rhabdaprovidines A (**77**)–C (**79**) were evaluated by measuring NO production in LPS-stimulated microglial BV-2 cells. These molecules inhibited NO production with IC_{50} values ranging from 17.5 to 46.8 μM [63]. Further isolation of the

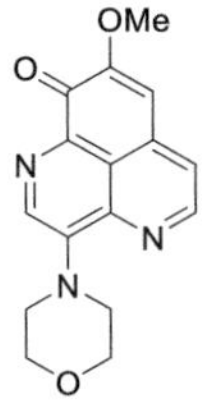

3-*N*-Morpholinyl-9-demethyl(oxy)aaptamine (**130**)

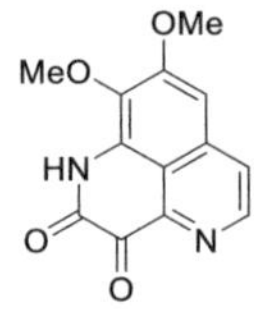

2,3-Dihydro-2,3-dioxoaaptamine (**131**)

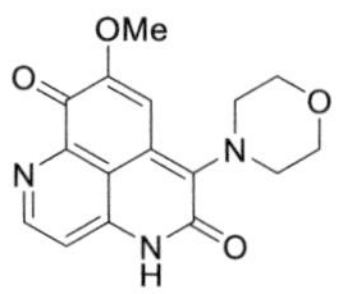

6-(*N*-morpholinyl)-4,5-dihydro-5-oxodemethyl(oxy)aaptamine (**132**)

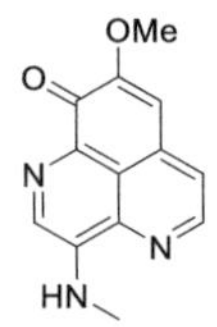

3-(methylamino)demethyl(oxy)aaptamine (**133**)

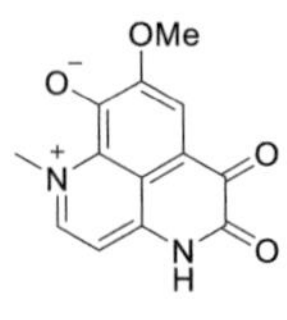

Aaptanone (**134**)

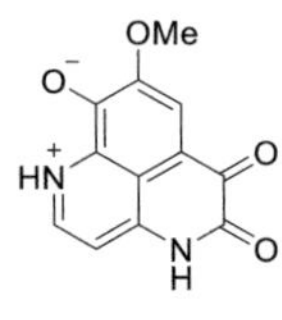

N-Demethylaaptanone (**135**)

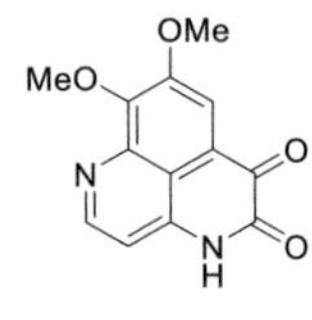

9-Methoxy-*N*-demethylaaptanone (**136**)

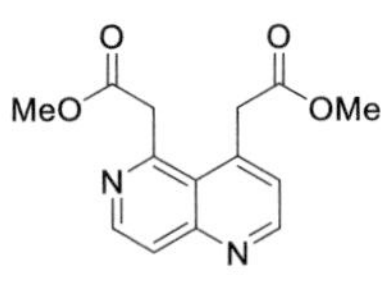

3,5-dicarbomethoxy-1,6-naphthyridine (**137**)

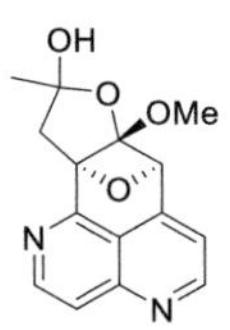

aaptosvanphong A (**138**)

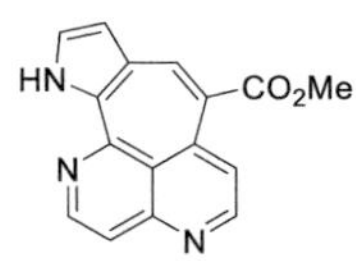

aaptosvanphong B (**139**)

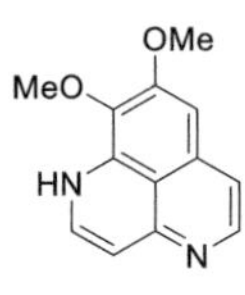

Aaptamine (**140**)

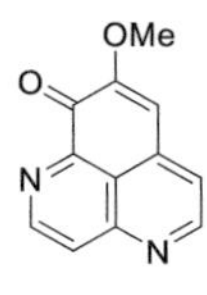

Demethyloxyaaptamine (**141**)

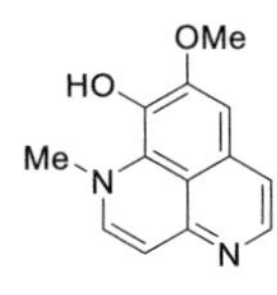

Isoaaptamine (**142**)

Fig. 21 Chemical structures of new and known alkaloids reported from Vietnamese *Aaptos* sp.

constituents of methanol extract of *R. providentiae* yielded additional five new isomalabaricane analogues, rhabdastrellins G (**72**)–K (**76**) (Fig. 15) [62]. They were evaluated for cytotoxicity against HepG2, LU-1, MCF-7, HL-60 and SK-Mel2 human cancer cell lines. Only rhabdastrellin H exhibited cytotoxicity with IC_{50} values ranging from 11.2 to 16.0 μM. It was observed that the 20(22)*Z* stereoisomer in rhabdastrellin H exhibited increased cytotoxicity in comparison with 20(22)*E* isomer in rhabdastrellin G.

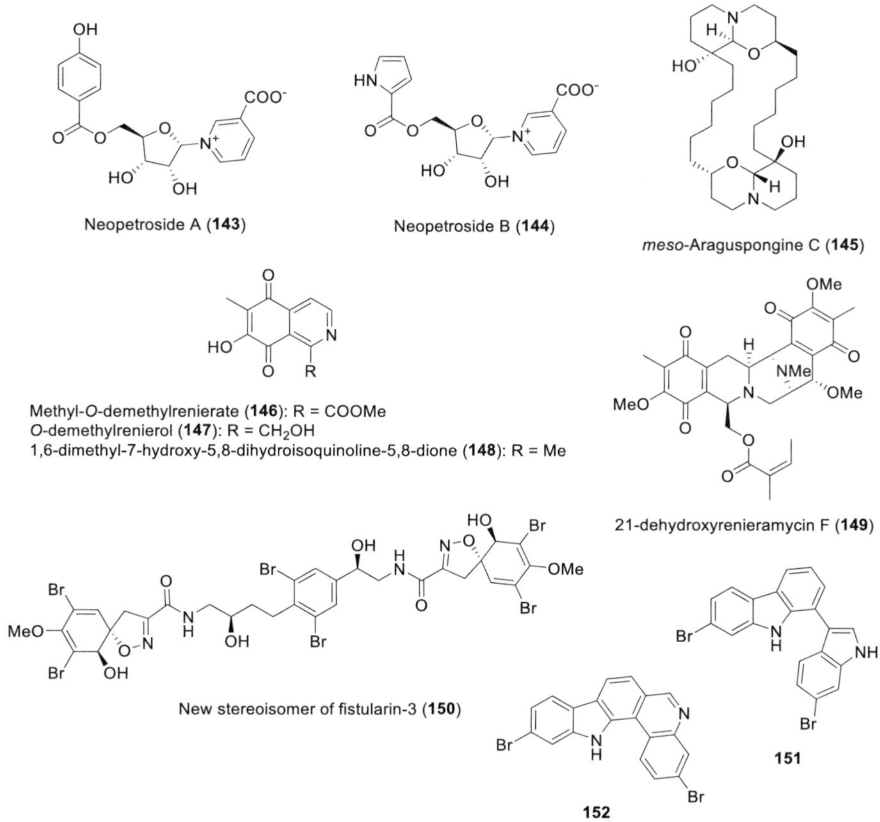

Fig. 22 Chemical structures of new alkaloids reported from Vietnamese *Neopetrosia* sp., *Xestospongia muta*, *Haliclona* sp., *Ecionemia acervus* and *Penares* sp.

Three unprecedented tetracyclic and pentacyclic isomalabaricanes, rhabdaglostelones A (**84**)–C (**86**) (Fig. 15), were recently isolated from another *Rhabdastrella* species, *R. globostellata*, collected from Vanphong Bay, Khanh Hoa, Vietnam [65]. Rhabdaglostelones A and B exhibited cytotoxic activities against LU1, HepG2, MCF7 and SKMel2 cell lines with their IC_{50} values ranging from 15.91 to 23.21 μM. Subsequent investigation of the *R. globostellata*, collected from Cham Island, resulted in the isolation of two new polar isomalabaricanes, rhabdastrellosides A (**87**) and B (**88**) [66]. These molecules possess a 3-*O*-disaccharide moiety in their structures, making them a very rare group of isomalabaricane glycosides. Rhabdastrellosides A and B have weak cytotoxicity when tested on human neuroblastoma SH-SY5Y cells and normal rat H9c2 cardiomyocytes. However, the cytoprotective activity of rhabdastrelloside B, tested at 1 μM, was demonstrated when evaluated using $CoCl_2$-treated SH-SY5Y and H9c2 cells. The presence of the 26-aldehyde unit in the side chain of rhabdastrelloside B could be attributed to its cytoprotective property.

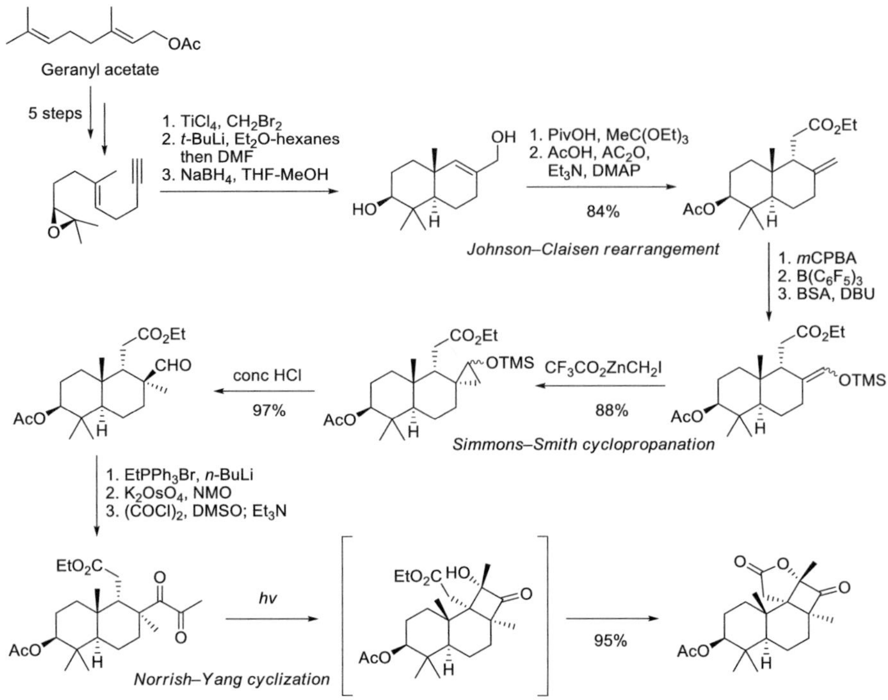

Scheme 1 Summary on the total synthesis of (+)-cyclobutastellettolide B

3.1.3 Steroids

Topsentia sp.

Structurally diverse new bioactive steroids have been isolated from a range of Vietnamese marine sponges including *Halichondria* sp. *Topsentia* sp. *Ianthella* sp., *Ircinia* sp. and *Xestospongia* sp. (Table 6; Figs. 16, 17, 18 and 19) [67–72, 74–76]. For instance, three new polar steroids, topsentiasterol sulfate F (**90**), along with the halogenated chlorotopsentiasterol sulfate D (**89**) and iodotopsentiasterol D (**91**) (Fig. 16), were obtained from samples of *Topsentia* sp. at Vang Fong Bay, Vietnam [67]. Chlorotopsentiasterol sulfate D was revealed to be an inhibitor of molluscan (*Spisula sachalinensis*) *endo*-1,3-*β*-D-glucanase, having an IC_{50} value of 5.4 μM/L. Additionally, chlorotopsentiasterol sulfate D caused a 98% irreversible inhibition of *endo*-1,3-*β*-D-glucanase. Interestingly, it also exhibited reversible inhibition of bifunctional alginate lyase, achieving complete (100%) inhibition, an uncommon characteristic for reversible inhibitors [89].

Xestospongia testudinaria

Three new cyclopropane-containing C_{29} sterols, aragusterol I (**96**), 21-*O*-octadecanoyl-xestokerol A (**97**) and 7*β*-hydroxypetrosterol (**98**) (Fig. 17), were isolated from *Xestospongia testudinaria*, obtained from the Truong Sa Archipelago, Khanh Hoa, Vietnam [71]. When evaluated for their inhibitory activity on the adhesion of isolated bacterial strains, including *Pseudoalteromonas* sp. D41, *Pseudoalteromonas* sp. TC8 and *Polaribacter* sp. TC5, 21-*O*-octadecanoyl-xestokerol A exhibited the significant antifouling activity with EC_{50} values ranging from 10 to 36 μM. These values are close to those reported for the banned antifoulant marine pollutant, tributyltin oxide.

Recently, a new sterol, designated testusterol (**100**), along with known compounds, was isolated from the non-polar extracts of the *X. testudinaria*, collected from Phu Quoc Island. Notably, testusterol demonstrated potent antibacterial activity against both Gram-positive bacteria (*Staphylococcus aureus*, *Bacillus subtilis* and *Lactobacillus fermentum*) and Gram-negative bacteria (*Escherichia coli*, *Salmonella enterica*, *Pseudomonas aeruginosa*), with IC_{50} values below 12.0 nM and MIC values ranging from 4.70 to 75.23 nM, as determined by broth microdilution assays.

Ircinia echinata

Chemical investigation of the extracts of the Vietnamese sponge, *Ircinia echinata*, led to the isolation of several furanosesterterpenes and new hydroxygenated sterols [56, 74]. For instance, six new sterols, namely, 5*α*,6*α*-epoxycholesta-7,22(*E*)-dien-3*β*,9*α*-diol (**101**), 5*α*,6*α*-epoxycholesta-7,24(28)-dien-3*β*,9*α*-diol (**102**), (24*R*)-5*α*,6*α*-epoxy-24-ethyl-cholesta-7-en-3*β*,9*α*-diol (**103**), 5*α*,6*α*-epoxycholesta-7-en-3*β*,9*α*-diol (**104**), (24*S*)-5*α*,6*α*-epoxyergosta-7,22-dien-3*β*,9*α*-diol (**105**) and (24*R*)-5*α*,6*α*-epoxy-24-methyl-cholesta-7-en-3*β*,9*α*-diol (**106**) (Fig. 18), were isolated from *I. echinate* obtained from Bai Tu Long, Quang Ninh, Vietnam [74]. Upon evaluation for cytotoxicity against three human cancer cell lines, including MCF-7, Hep-G2 and LU-1, only sterols **103** and **104** showed moderate antiproliferative activity, with compound **103** exhibiting a selective inhibition towards the MCF-7 human breast cancer cells.

Halichondria vansoesti

Seven new polysulfated steroids, including topsentiasterol sulfate G (**107**), topsentiasterol sulfate I (**108**), topsentiasterol sulfate H (**109**), bromotopsentiasterol sulfate D (**110**), dichlorotopsentiasterol sulfate D (**111**), bromochloro topsentiasterol sulfate D (**112**) and 4β-hydroxyhalistanol sulfate C (**113**) (Fig. 19), were isolated from a sample of *Halichondria vansoesti*, obtained within the territorial waters of Vietnam at the South China Sea [75]. In particular, the unusual side chains in the new topsentiasterol sulfate-related molecules have been hypothesized to be biosynthesized from codisterol by the authors. Several trisulfated steroids were found to suppress PSA (prostate-spectific antigen) expression and glucose uptake in human prostate cancer cells. It is known that the suppression of PSA could indicate the inhibition of androgen receptor (AR)-signalling, which is essential for the growth and survival of several prostate cancer cell types. Topsentiasterol sulfate H was found to suppress PSA expression at a concentration as low as 10 μM as well as suppressed glucose uptake in 22Rv1 cells. Furthermore, mixtures of bromotopsentiasterol sulfate D with known steroids, chlorotopsentiasterol sulfate D and iodotopsentiasterol sulfate D, were found to exert similar activities.

3.1.4 Alkaloids

More than 30 new alkaloids have been isolated from about nine Vietnamese sponge species, with a majority of these molecules having cytotoxic activity (Table 7, Figs. 20, 21 and 22) [53, 77–87]. Selected alkaloids with unique structures and/or biological activities are highlighted below.

Aaptos sp.

Members of the sponge genus, *Aaptos*, are attractive sources of bioactive molecules, especially aaptamine class of compounds with diverse biological properties, ranging from anticancer, antimicrobial to antifouling activities [90, 91]. The first reports of aaptamine-related molecules, 3-*N*-morpholinyl-9-demethyl(oxy)aaptamine (**130**) and aaptanone (**134**) (Fig. 21), were reported by Shubina et al. and Utkina et al., respectively, in 2009 [78, 80]. The new aaptamine-type alkaloid, 3-*N*-morpholinyl-9-demethyl(oxy)aaptamine (**130**), from *Aaptos* sp. collected in Vang Fong Bay, Vietnam, was reported to inhibit epidermal growth factor (EGF)-induced malignant transformation of mouse epidermal JB6 P^+C1 41 cells in soft agar [78]. In the same year, another new zwitterionic aaptamine alkaloid, aaptanone (**134**), having a rare oxygenated 1,6-naphthyridine core, was isolated from the *A. aaptos*, obtained from Cu Lao Re Island, Vietnam [80]. In the following year, three new aaptamines, including 2,3-dihydro-2,3-dioxoaaptamine (**131**), 6-(*N*-morpholinyl)-4,5-dihydro-5-oxodemethyl(oxy)aaptamine (**132**) and 3-(methylamino)demethyl(oxy)aaptamine (**133**) (Fig. 21), with weak apoptotic activity, were purified from *Aaptos* sp. collected in Vang Fong Bay, Vietnam [79]. Additional new aaptamine alkaloids, *N*-demethylaaptanone (**135**), 9-methoxy-*N*-demethylaaptanone (**136**), 3,5-dicarbomethoxy-1,6-naphthyridine (**137**), aaptosvanphongs A (**138**) and B (**139**) (Fig. 21), were subsequently reported from *A. aaptos* from Tho Chu Island and Vanphong Bay, Nha Trang, Vietnam [81, 82]. 9-Methoxy-*N*-demethylaaptanone (**131**), 3,5-dicarbomethoxy-1,6-naphthyridine (**137**), aaptosvanphongs A (**138**) and B (**139**) showed cytotoxic effects against SKLU-1, MCF-7, HepG2 and SK-Mel-2 cell lines with IC_{50} values ranging from 9.4 to 49.2 μM [82].

In addition to the isolation of new aaptamines, known aaptamines, e.g., aaptamine (**140**), demethyloxyaaptamine (**141**) and isoaaptamine (**142**) (Fig. 21), were obtained from Vietnamese sponges and they provided material for further biological evaluation on their anticancer/apoptotic properties [92–94]. In 2012, Dyshlovoy and co-workers isolated about 80 mg of aaptamine from *Aaptos* sp. collected from Vang Fong Bay, Vietnam and proceeded to assess its effects on the proliferation and protein expression of the pluripotent human embryonal carcinoma cell lineNT2 [92]. When tested at lower concentrations, aaptamine arrested the cell cycle at the G2/M phase, whereas at higher concentrations, apoptotic induction was observed. Aaptamine treatment at a concentration of 50 μM resulted in alteration of 10 proteins, of which five each showed up- (e.g., PEA15, TPT1, CTSD, CRABP2 and CFL1) and down-regulation (e.g., eIF5A, ENO1, CSDE1, ACTL6A and SRSF3). Further analysis revealed these proteins are involved in protein biosynthesis, cell metabolism, actin reorganization, cell migration, cell signaling and transcriptional and translational regulation. In particular, the study demonstrated the posttranslational hypusine modification of the eukaryotic initiation factor 5A (eIF5A) as a main target of aaptamine action. In subsequent studies, the same group reported the effects of aaptamine, demethyloxyaaptamine and isoaaptamine on NT2-R, a cisplatin-resistant subline of the human embryonal carcinoma cell line NT2 [94]. At the inhibitory concentration (IC_{50}), aaptamine exhibited an antiproliferative effect, while demethyloxyaaptamine and isoaaptamine showed strong apoptotic inducing activity. In addition, the changes in the proteome of NT2-R cells treated with these compounds were analyzed and revealed that Myc, p53 and tumor necrosis factor were the possible direct/indirect targets/effectors of these alkaloids. Hypusination of eIF5A, a key feature of aaptamine treatment, was not observed upon treatment with either demethyloxyaaptamine or isoaaptamine. Anticancer activity of aaptamine, demethyloxyaaptamine and isoaaptamine were also explored. These three alkaloids demonstrated anticancer activity in THP-1, HeLa, SNU-C4, SK-MEL-28, and MDA-MB-231 human cancer cell lines. Moreover, all molecules were found to prevent the EGF-induced neoplastic transformation of murine JB6 Cl41 cells, and the nuclear factors AP-1, NF-B and p53 were involved in the cellular response following treatment with high and nontoxic concentrations of these aaptamine alkaloids [93].

Neopetrosia sp.

Two highly unusual pyridine nucleosides, neopetrosides A (**143**) and B (**144**) (Fig. 22), with rare α-*N*-glycosidic linkages and residues of either *p*-hydroxybenzoic or pyrrole-2-carboxylic acid attached at C-5’, were reported from *Neopetrosia* sp. from Con Son Island, Vietnam [83]. Confirmation on the unusual structure of neopetroside A was achieved through total synthesis from 2,3-*O*-benzylidene-D-ribose. The noncytotoxic neopetroside A was found to upregulates mitochondrial functions by enhancing the mitochondrial ATP levels and oxygen consumption rate in C2C12 cells treated with 10 μM of the molecule.

Haliclona sp.

Based on bioactivity-guided fractionation, eight isoquinolinequinones, including four new derivatives, including methyl *O*-demethylrenierate (**146**), *O*-demethylrenierol (**147**), 1,6-dimethyl-7-hydroxy-5,8-dihydroisoquinoline-5,8-dione (**148**) and 21-dehydroxyrenieramycin F (**149**) (Fig. 22), were isolated from *Haliclona* sp. obtained from the vicinity of Hon Thom Island, Vietnam [85]. Anti-inflammatory activities of the isolated alkaloids were subsequently tested in a co-culture system of human epithelial Caco-2 and THP-1 macrophages. *O*-Demethylrenierol (**147**) and 1,6-dimethyl-7-hydroxy-5,8-dihydroisoquinoline-5,8-dione (**148**) exhibited moderate anti-inflammatory properties. In addition, they regulated the expression of iNOS, COX-2, the MAPK family (ERK, p38, JNK), nuclear translocation of NF-κB as well as Nrf-2 and HO-1.

3.2 Natural Products from Vietnamese Cnidarians

At least 20 Vietnamese soft coral species have been investigated for new molecules as well as known compounds (Table 8; Figs. 23, 24, 25, 26, 27, 28, 29, 30, 31, 32, 33, 34 and 35). These soft coral species are mainly from four taxonomic families, including Alcyoniidea [95–117], Gorgoniidae [95, 118–120], Nephtheidae [96, 97] and Clavulariidae [98]. The majority of these soft coral species investigated are from the *Sinularia* genus and cembrane-type as the common terpenoid structural class. Many of these compounds exhibit anti-inflammatory and/or cytotoxic properties. Highlighted below are selected new molecules reported from Vietnamese cnidarians having unique structures and/or biological activities.

3.2.1 Family Alcyoniidea

Terpenoids

Sinularia maxima

From a suite of nine new diterpenes, sinumaximols A (**168**)–I (**176**) (Fig. 25), along with known congeners, were isolated from *Sinularia maxima*, collected from Nha trang Bay, Vietnam, sinumaximols B (**169**) and C (**170**) were found to possess anti-inflammatory activities [105]. Specifically, sinumaximol B reduced the production of pro-inflammatory cytokines, IL-12 and IL-6 in lipopolysaccharide (LPS)-stimulated bone marrow dendritic cells, with IC_{50} values of 15.20 and 59.77 μM, respectively. In addition, moderate inhibition of sinumaximol C on TNF-α production was detected with IC_{50} of 62.33 μM. Subsequent antiparasitic screening revealed that sinumaximols A and C have significant inhibitory activity against *P. falciparum* (IC_{50} = 7.20 μM) and *T. b. rhodesiense* (IC_{50} = 9.88 μM), respectively [126].

Subsequent collection of *S. maxima*, made from Nha Trang Bay a year later, led to the isolation of seven norditerpenoids, including two new molecules, 12-hydroxy-scabrolide A (**177**) and 13-*epi*-scabrolide C (**178**) (Fig. 25) [106]. 13-*epi*-scabrolide C (**173**) was found to potently inhibit IL-12 and IL-6 production in LPS-stimulated bone marrow derived dendritic, with IC_{50} values of 5.30 and 13.12 μM, respectively. It has been suggested that the methoxy group at C-13 of 13-*epi*-scabrolide C could play an important role in its anti-inflammatory activity. In addition, further biological screening revealed that 13-*epi*-scabrolide C has inhibitory activity against bloodstream forms of *T. brucei*, with IC_{50} of 11.76 μM.

Table 8 List of new molecules reported from Vietnamese cnidarians

Classification/Compound[a]	Location	Biological activity	References
Family Alcyoniidea			
Sinularia brassica			
Sinubrassione (**153**) and sinubrassioside (**154**)	Near Con Co Island, Quangtri Province	Sinubrassione: cytotoxic activity against PANC-1 cell line	[99]
Sinulaketols A (**155**), B (**156**), (24*R*)-28-chloroergost-5-ene-3β,4α,24-triol (**157**) and 5α,6α -epoxygorgosterol (**158**)	Van Phong bay, Khanh Hoa Province	Chlorinated steroid: antileishmanial activity against amastigote forms of *Leishmania donovani*	[100]
S. conferta			
7α-Methoxygorgosterol (**159**), 7α-methoxy-ergosta-5-ene-3β-ol (**160**), 3β-hydroxyergosta-4-ene-6-one (**161**), 3β-hydroxyergosta-4,24(28)-diene-6-one (**162**) and ergosta-24(28)-ene-3β,5α,6β-triol-3-acetate (**163**)	Near the Con Co Island, Quangtri	–	[101]
S. dissecta			
Dissesterol (**164**)	Hai Van-Son Cha, Hue	Inhibitory effects on LPS-stimulated production of pro-inflammatory cytokines in bone marrow-derived dendritic cells	[102]
S. erecta			
3β,5α-dihydroxyeudesma-4(15),11-diene (**165**)	Cu Lao Cham Island, Quang Nam	Selective cytotoxicity against the A549 cell line	[103]
S. leptoclados			
Leptosteroid (**166**) and 5,6β-epoxygorgosterol (**167**)	Con Co Island, Quangtri	Leptosteroid: significant cytotoxicity against HepG2 and SW480	[104]
S. maxima			

(continued)

Table 8 (continued)

Classification/Compound[a]	Location	Biological activity	References
Sinumaximols A (**168**)–I (**176**)	Nha Trang Bay	Sinumaximols B and C: potent inhibitors of LPS-stimulated interleukin-12 (IL-12) p40; moderate inhibitory activity on IL-6 production Sinumaximols A and C: significant inhibitory activity against *P. falciparum* and *T. b. rhodesiense*, respectively	[105]
12-Hydroxy-scabrolide A (**177**) and 13-*epi*-scabrolide C (**178**)		13-*epi*-scabrolide C: inhibited IL-12 and IL-6 production in LPS-stimulated bone marrow derived dendritic; inhibitory activity against bloodstream forms of *T. brucei*	[106]
S. microspiculata			
7-Oxogorgosterol (**179**) and 16α-hydroxysarcosterol (**180**)	Da Den, Quang Ninh	–	[107]
S. nanolobata			
Nanolobatols A (**181**) and B (**182**)	Lang Co, Hue	–	[108]
3β,4α-Dihydroxyergosta-5,24(28)-diene (**183**) and 24(*S*),28-epoxyergost-5-ene-3β,4α-diol (**184**)	Lang Co, Hue	24(*S*),28-epoxyergost-5-ene-3β,4α-diol: moderate cytotoxicity against the acute leukemia (HL-60) cell line	[109]
Sarcophyton ehrenbergi			
Ehrenbergols F (**185**) and G (**186**)	Van Phong Bay, Khanh Hoa Province	Ehrenbergol F: moderate inhibitory activity on LPS-induced NO production in RAW264.7 cells	[110]

(continued)

Table 8 (continued)

Classification/Compound[a]	Location	Biological activity	References
Ehrensteroids A (**187**)–F (**192**)		Ehrensteroids A, B and F: moderate cytotoxicity against LU-1, HepG2, MCF-7, HL-60 and SW480, IC_{50} values ranging from 15.95 to 38.27 μM Ehrensteroid D: moderate cytotoxicity against LU-1 and HepG2, IC_{50} values of 21.91 and 20.32 μM, respectively	[111]
S. mililatensis			
(–)-7*β*-Hydroxy-8*α*-methoxydeepoxysarcophytoxide (**193**) [previously reported from an Okinawan collection of *Sarcophyton* sp.], (+)-7*β*,8*β*-dihydroxydeepoxysarcophytoxide (**194**), (–)-17-hydroxysarcophytonin A (**195**) and sarcophytol V (**196**)	Baycanh Island, Condao district, Baria-Vungtau Province	(-)-7*β*-hydroxy-8*α*-methoxydeepoxysarcophytoxide: significant elevation of alkaline phosphatase activity, collagen content and nodules mineralization	[112]
Sarcomilasterol (**197**)		Significantly increased the growth of MC3T3-E1 cells and elevation of alkaline phosphatase activity and nodules mineralization	[113]
S. pauciplicatum			
Sarcophytolides M (**198**) and N (**199**)	Hai Phong	Sarcophytolide M: significant cytotoxic effects	[114]
Sarcopanol A (**200**)		Significantly inhibited TNFα/INFγ-induced NF-κB transcriptional activity in human keratinocyte cells in a dose-dependent manner	[115]
Lobophytum sp.			
Lobophytene (**201**)	Conco island, Quang Tri Province	Significant cytotoxic activities against lung (A549) and colon (HT-29) cell lines with IC_{50}s of 8.2 and 5.6 μM, respectively	[116]
L. compactum			

(continued)

Table 8 (continued)

Classification/Compound[a]	Location	Biological activity	References
Lobocompactols A (**202**) and B (**203**)	Bay Canh Island, Truong Sa archipelago, Khanh Hoa	Lobocompactols A and B: moderate peroxyl radical scavenging activity	[117]
L. crissum			
Crassumols D (**204**)–G (**207**)	Conco, Quang Tri	Crassumol E: inhibitory effect on tumor necrosis factor-alpha (TNFα)-induced nuclear factor-kappa B (NF-κB) transcriptional activation in HepG2 cells Crassumols D, E and G: showed inhibitory activity against bloodstream forms of *T. brucei*	[121]
Lobocrasols A (**208**)–D (**211**)		Lobocrasols A and B: significantly inhibited TNFα-induced NF-κB transcriptional activity in HepG2 cells in a dose-dependent manner; decrease in cyclooxygenase-2 (COX-2) and inducible nitric oxide synthase (iNOS) gene expression levels in HepG2 cells Lobocrasols A and C: inhibitory activity against bloodstream forms of *T. brucei* and strong selective activity against *L. donovani*	[122]
L. laevigatum			
Laevigatols A (**212**)–D (**215**)	Khanh Hoa Province	Laevigatols A and B: dose-dependent inhibitory effects on the TNFα-induced NF-κB transcriptional activity in Hep-G2 cells; significantly inhibited the induction of COX-2 and iNOS mRNA dose-dependently Laevigatol A: antiplasmodial activity against *P. falciparum* Laevigatol B: exhibited significant inhibitory effects on *T. brucei*	[123]

(continued)

Table 8 (continued)

Classification/Compound[a]	Location	Biological activity	References
Lobophytosterol (**216**)		Cytotoxic effects on HCT-116, A549 and HL-60 cells via induction of apoptosis, IC_{50}s of 3.2, 4.5 and 5.6 μM, respectively	[124]
Family Gorgoniidae			
Junceella fragilis			
17-*epi*-Junceellolide B (**217**)	Conco Island	–	[118]
Leptogorgia sp.			
Leptogorgins A (**218**)–C (**220**) and leptogorgoid A (**221**)	South China Sea, Vietnamese water	Leptogorgins A and B: moderate cytotoxicity and some selectivity against human drug-resistant prostate cancer cells 22Rv1	[119]
Menella woodin			
Menelloides F (**222**) and G (**223**)	Coto Island, Quang Ninh province	–	[120]
Menellsteroids E (**224**)–H (**227**)		Menellsteroids E and H: decrease production of ROS in murine macrophages of RAW 264.7 line at induction by LPS from *E coli*	[95]
Verrucella corona			
Verrucorosteroids A (**228**)–F (**233**) and verrucorosterone (**234**)	Vinh Moc, Quang Tri Province	Verrucorosterone: cytotoxicity against eight human cancer cell lines as LNCaP, HepG2, KB, MCF-7, SK-Mel2, HL-60, LU-1 and SW480, IC_{50} values ranging from 12.85 to 19.48 μM	[125]
Family Nephtheidae			
Dendronephthya mucronata			
5-α-Pregn-20-en-3,6-dione (**235**)	Phu Quoc Island, Kien Giang	–	[96]
Dendronephthyones A (**236**)–C (**238**)		Moderate cytotoxicity against HeLa cell line	[97]

(continued)

Table 8 (continued)

Classification/Compound[a]	Location	Biological activity	References
Family Clavulariidae			
Carijoa riisei			
15β-Hydroxypregna-4,20-dien-3-one (**239**)	Hon Me, Thanh Hoa	–	[98]

[a]Chemical structures in Figs. 23, 24, 25, 26, 27, 28, 29, 30, 31, 32, 33, 34 and 35

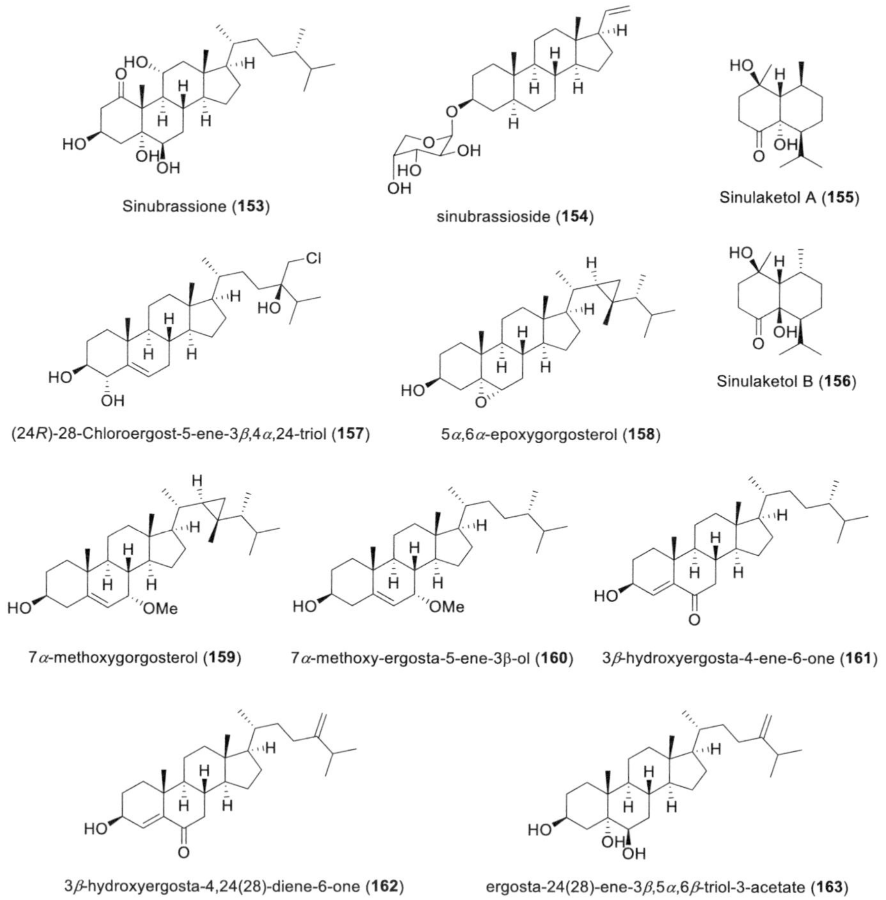

Fig. 23 Chemical structures of new natural products from *Sinularia brassica* and *S. conferta*

Lobophytum sp.

Members of soft corals belonging to the genus *Lobophytum* are a prolific source of steroids and terpenoids. The majority of the reported diterpenes are cembranoid molecules, and are often found in high concentrations, up to 5% dry weight, in soft corals and have possible chemical defense. Two new diterpenes, lobocompactols A (**202**) and B (**203**) (Fig. 29), along with five known compounds, were isolated from *Lobophytum compactum*, collected from Bay Canh Island, Truong Sa archipelago, Khanh Hoa, Vietnam [117]. Subsequent bioassays revealed that known 3*β*,11-dihydroxy-24-methylene-9,11-secocholestan-5-en-9-one exhibited strong cytotoxic activity against the A549 cell line, with an IC_{50} of 4.97 μM. The antioxidant capacity of these diterpenes was evaluated using an oxygen radical absorbance capacity (ORAC) assay. Only lobocompactols A (**202**) and B (**203**), at a concentration of

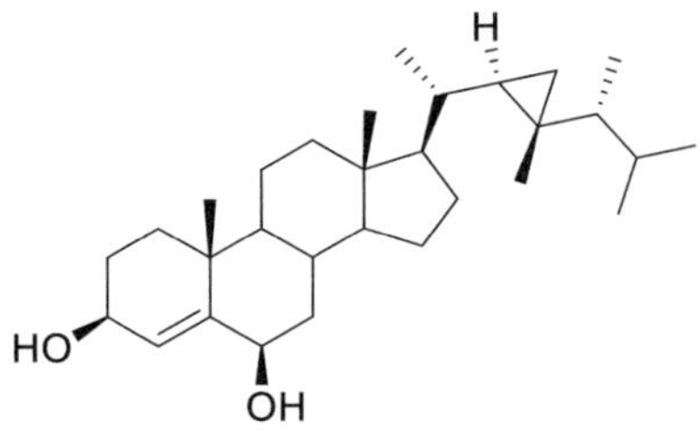

Dissesterol (**164**)

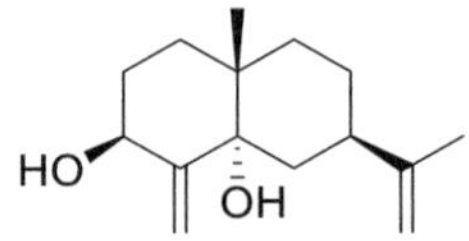

3β,5α-dihydroxyeudesma-4(15),11-diene (**165**)

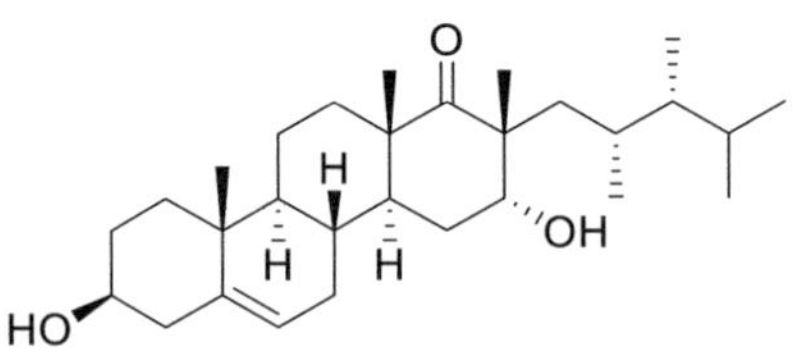

Leptosteroid (**166**)

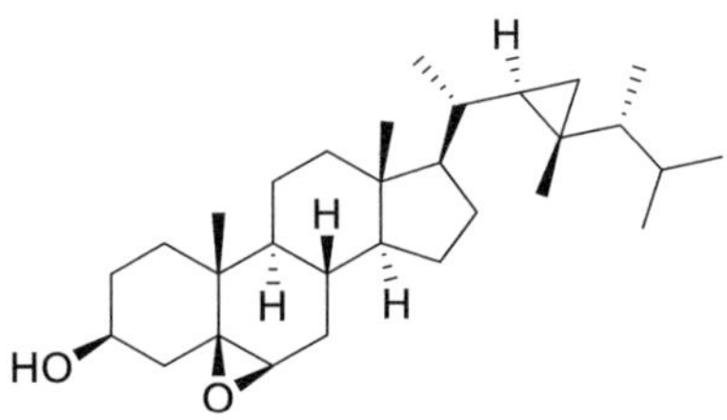

5,6β-epoxygorgosterol (**167**)

Fig. 24 Chemical structures of new natural products from *Sinularia dissecta*, *S. erecta* and *S. leptoclados*

5 μM, showed moderate peroxyl radical-scavenging activities of 1.4 and 1.3 μM Trolox equivalents, respectively.

From the soft coral samples of *Lobophytum crassum*, four new terpenoids, crassumols D (**204**)–G (**207**) (Fig. 30), were purified from its methanol extract [121]. Crassumol E (**205**) exhibited significant inhibitory effect on tumor necrosis factor-alpha (TNFα)-induced nuclear factor-kappa B (NF-κB) transcriptional activation in HepG2 cells, with IC_{50} value of 9.23 μM. Subsequent in vitro antiprotozoal screening of a series of cembrane-type diterpenes revealed that lobocrasols A and C exhibited significant inhibitory effects on *Leishmania donovani* with IC_{50} values < 0.2 μM while crassumols D and E showed the highest level of inhibitory activity against *Trypanosoma brucei rhodesiense*, with IC_{50} values of 0.61 and 0.72 μM, respectively [126].

Additional four new cembranoid diterpenes lobocrasols A (**208**)–D (**211**) (Fig. 30), were isolated from the methanolic extract of the soft coral, *Lobophytum crassum*, obtained from Con Co, Quangtri, Vietnam [122]. The anti-inflammatory effects of these molecules were assessed using NF-κB luciferase and reverse transcription polymerase chain reaction. Lobocrasols A and B significantly inhibited TNFα-induced NF-κB transcriptional activity in HepG2 cells in a dose-dependent manner, having IC_{50} values of 6.30 and 6.63 μM, respectively. Moreover, the transcriptional inhibition of these diterpenes was confirmed by a decrease in cyclooxygenase-2 (COX-2) and inducible nitric oxide synthase (iNOS) expression levels in HepG2 cells. It is known that the activation of NF-κB has been linked

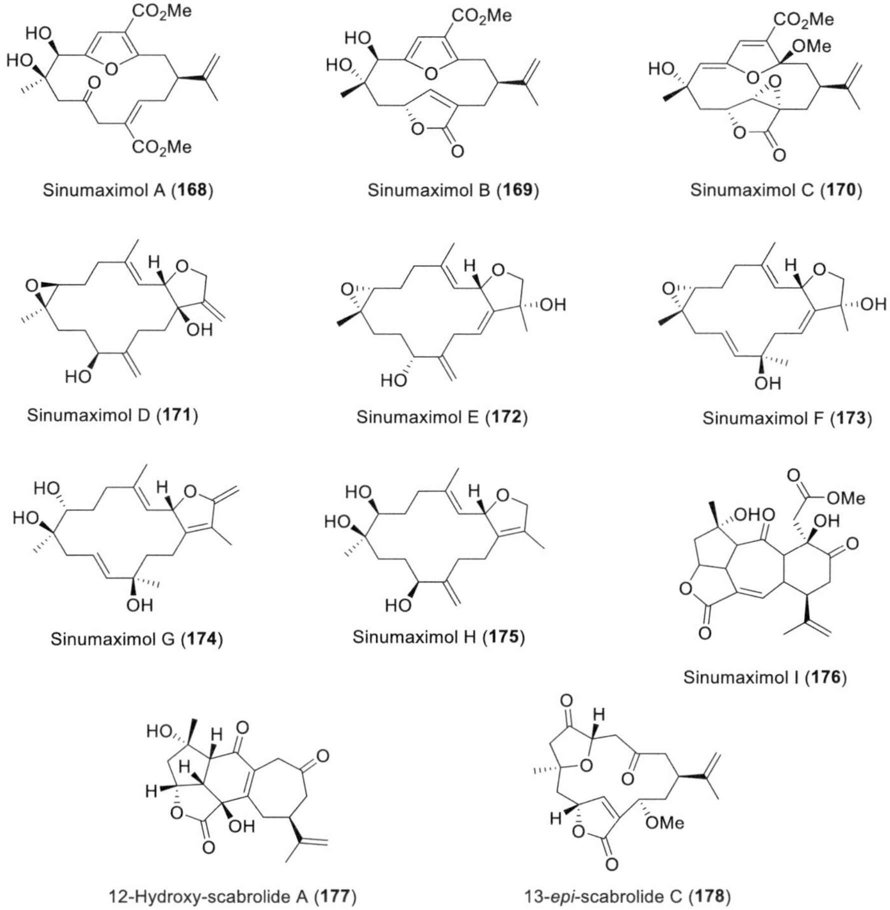

Fig. 25 Chemical structures of new natural products from *Sinularia maxima*

to multiple pathophysiological conditions, such as cancer, arthritis, asthma, inflammatory bowel disease and other inflammatory conditions. The induction of several pro-inflammatory mediators also occurs due to the increase of iNOS and COX-2 activities. As such, suppression of iNOS and COX-2 activities is important for preventing inflammation.

Four new furan-containing analogues, laevigatols A (**212**)–D (**215**) (Fig. 31), in addition to known cembranes were isolated from *Lobophytum laevigatum* collected from Khanh Hoa province, Vietnam [123]. The anti-inflammatory activity of these cembranoids was determined through the inhibition of TNFα-induced NF-κB luciferase reporter, and by attenuation of TNFα-induced pro-inflammatory proteins, namely iNOS and COX-2, expression in Hep-G2 cells. The new molecules, laevigatols A and B, exhibited inhibitory effects on TNFα-induced NF-κB transcriptional activity in Hep-G2 cells in a dose-dependent manner, with IC_{50} values of 9.4 and

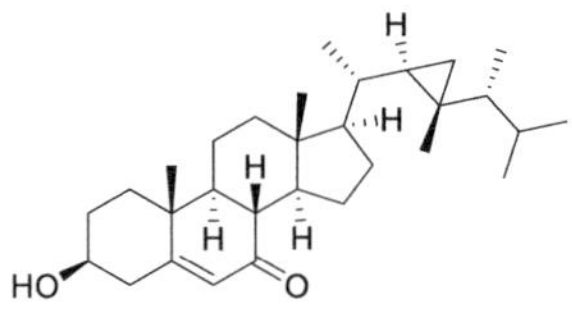

7-Oxogorgosterol (**179**)

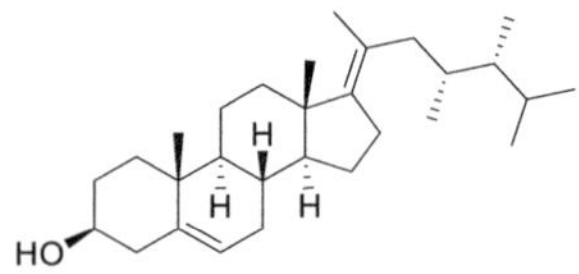

16α-hydroxysarcosterol (**180**)

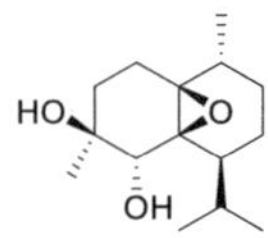

Nanolobatol A (**181**)

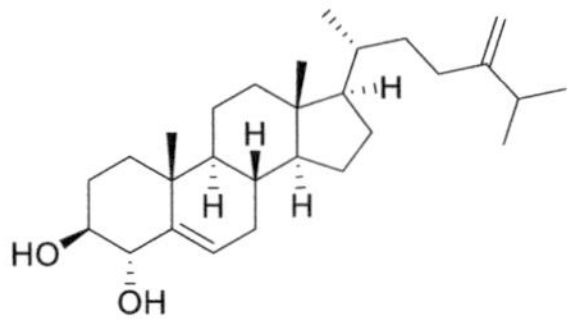

3β,4α-Dihydroxyergosta-5,24(28)-diene (**183**)

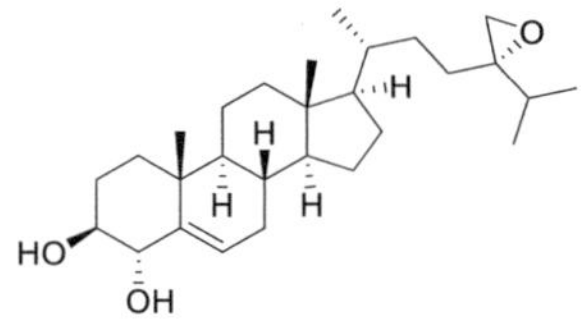

24(*S*),28-epoxyergost-5-ene-3β,4α-diol (**184**)

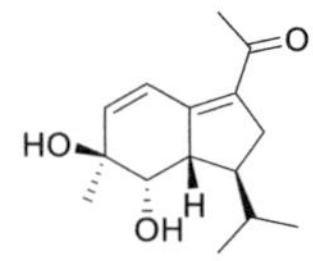

Nanolobatol B (**182**)

Fig. 26 Chemical structures of new natural products from *Sinularia microspiculata* and *S. nanolobata*

9.7 μM, respectively. Subsequent biological testing revealed that laevigatol B (**213**) exhibited significant inhibitory effects on *Trypanosoma brucei*, with EC_{50} of 5.34 μM [127]. Moreover, laevigatol B did not show cytotoxicity against HEK293T and HepG2 cancer cell lines and thus displayed a certain degree of selectivity. Laevigatol A also showed antiplasmodial with an IC_{50} value of 3.02 μM against *P. falciparum* [126].

Steriods

Sinularia sp.

Four new molecules, including two rare rearranged cadinane-type sesquiterpenoids, sinulaketols A (**155**) and B (**156**) as well as one new chlorinated steroid, (24*R*)-28-chloroergost-5-ene-3β,4α,24-triol (**157**) and one new gorgosterol, 5α,6α-epoxygorgosterol (**158**) (Fig. 23), were isolated from the soft coral *Sinularia brassica*, collected from Van Phong bay, Khanh Hoa province, Vietnam [100]. Sinulaketols A and B were the first reports of the cadinane sesquiterpenoids with unprecedented carbon backbone at C-15. The new chlorinated steroid (**157**) was found to suppress the growth of the amastigote forms of *L. donovani* by 58.7% without cytotoxicity when tested at 50 μM.

An ongoing chemical investigation on Vietnamese soft corals led to the isolation of a new steroid, dissesterol (**164**) (Fig. 24), along with nine known molecules from a methanolic extract prepared from *Sinularia dissecta*, obtained from Hai Van-Son Cha, Hue, Vietnam [102]. The new steroid was a strong suppressor of IL-12 p40 cytokine production in LPS-stimulated bone marrow-derived dendritic cells, with

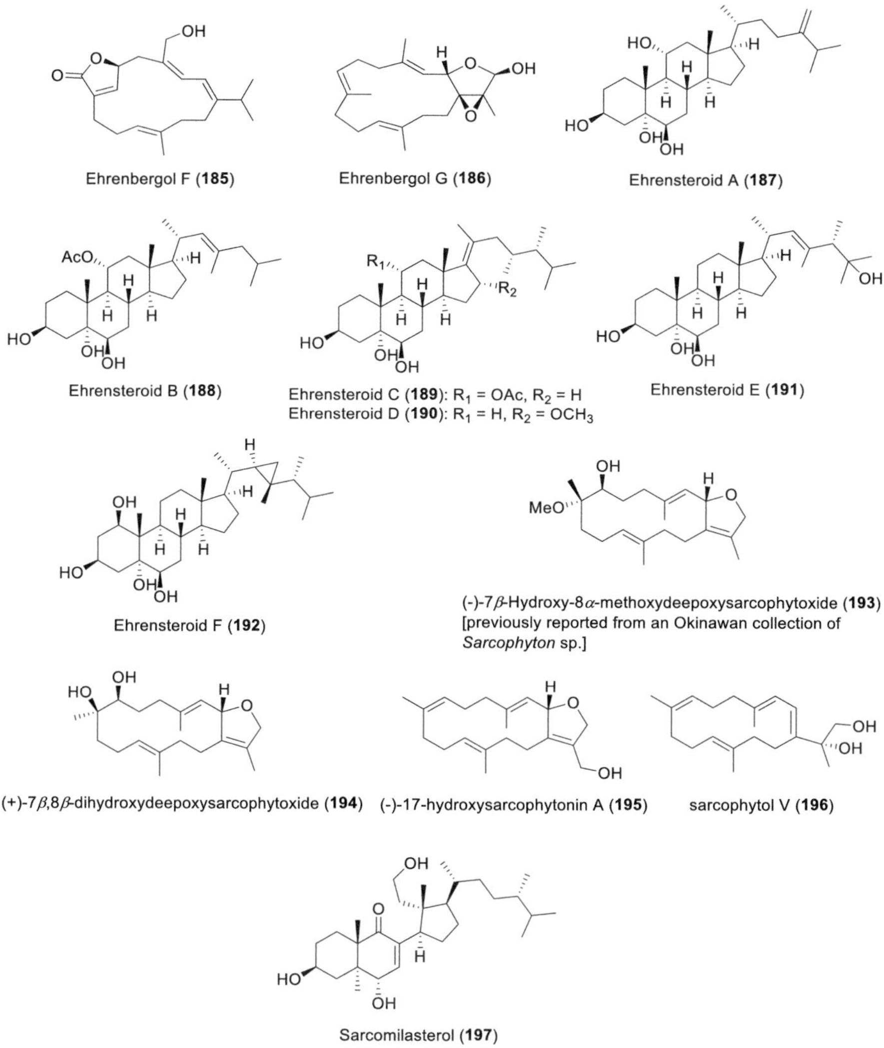

Fig. 27 Chemical structures of new natural products from *Sarcophyton ehrenbergi* and *S. mililatensis*

IC_{50} value of 4.0 μM. The known steroid, (24*S*)-ergost-4-ene-3-one, was subsequently tested to have anti-trypanosomal activity against *T. brucei*, with EC_{50} of 5.47 μM [127].

Sarcophyton sp.

From the Vietnamese soft coral species, *Sarcophyton mililatensis*, collected from Baycanh Island, Baria-Vungtau province, Vietnam, a new 9,11-secosterol, sarcomilasterol (**197**) (Fig. 27), was isolated [113]. This new steroid, when tested at 3 μM,

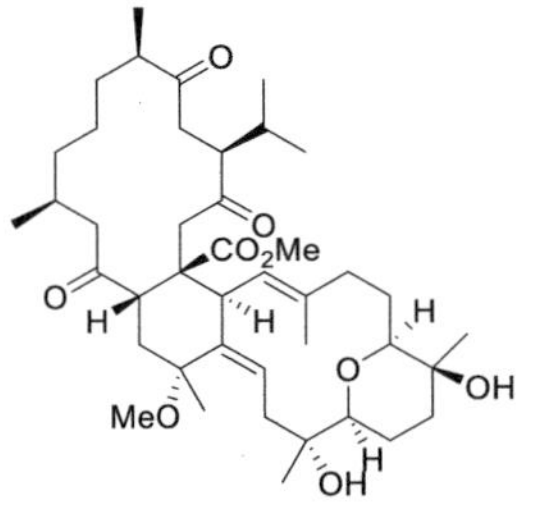

Sarcophytolide M (**198**)

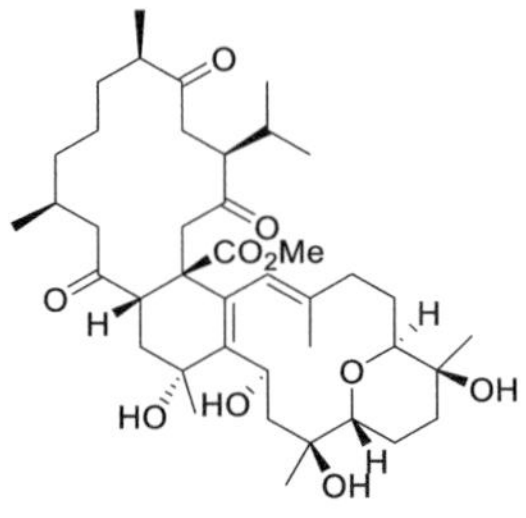

Sarcophytolide N (**199**)

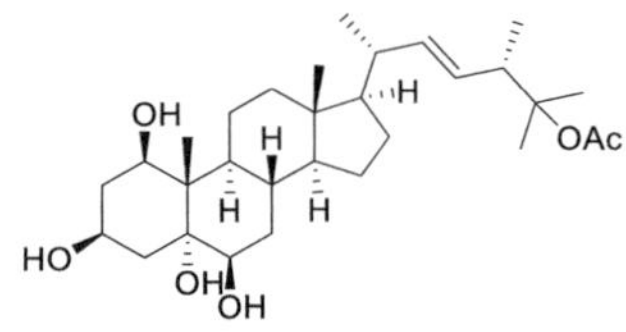

Sarcopanol A (**200**)

Fig. 28 Chemical structures of new natural products from *Sarcophyton pauciplicatum*

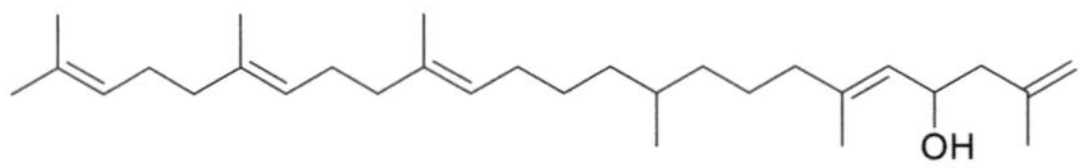

Lobophytene (**201**)

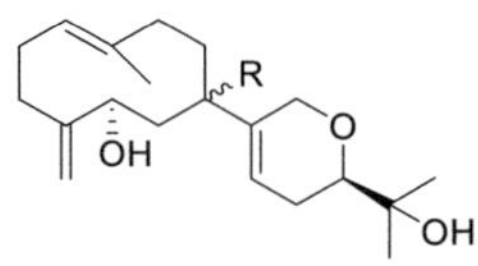

Lobocompactol A (**202**): R = βH
Lobocompactol B (**203**): R = αH

Fig. 29 Chemical structures of new natural products from *Lobophytum* sp. and *L. compactum*

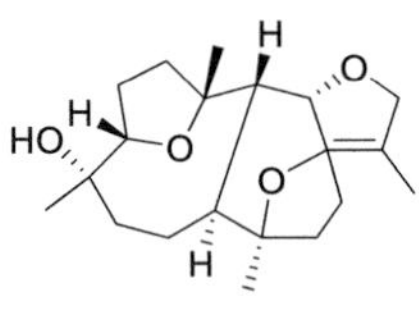

Crassumol D (**204**)

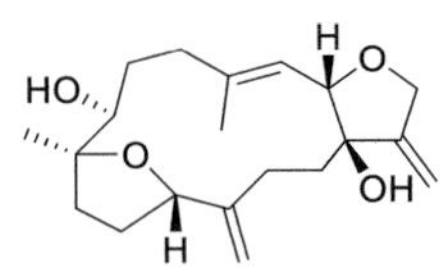

Crassumol E (**205**)

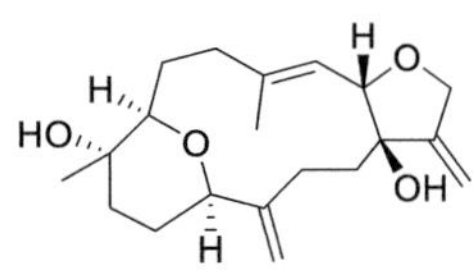

Crassumol F (**206**)

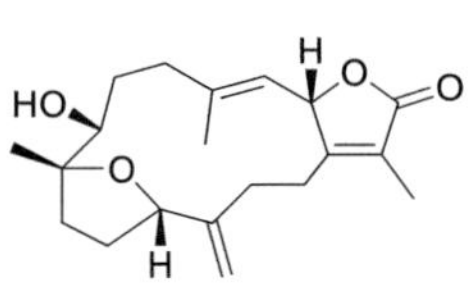

Crassumol G (**207**)

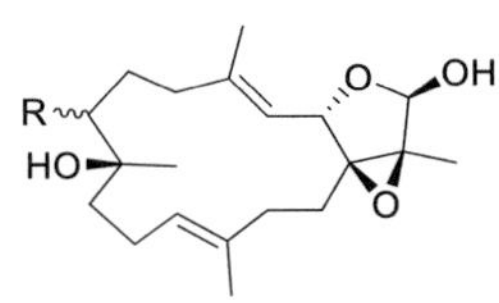

Lobocrasol A (**208**): R = ◄OH
Lobocrasol B (**209**): R = ''''OH

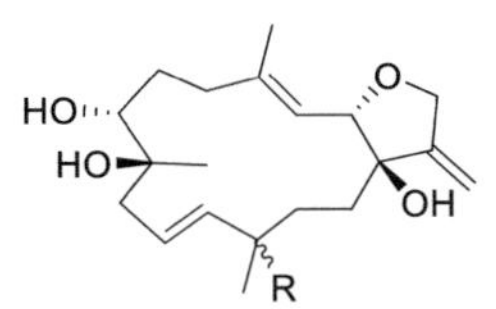

Lobocrasol C (**210**): R = ◄OH
Lobocrasol D (**211**): R = ''''OH

Fig. 30 Chemical structures of new natural products from *Lobophytum crissum*

was found to increase growth of MC3T3-E1 cells and significantly elevated alkaline phosphatase activity and nodules mineralization. These results showed direct effects of the compound on bone formation in vitro and could contribute to the prevention of osteoporosis.

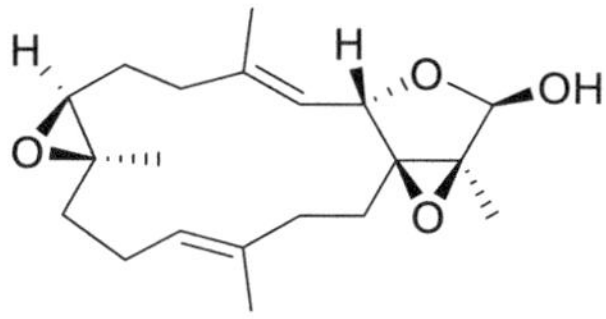

Laevigatol A (**212**)

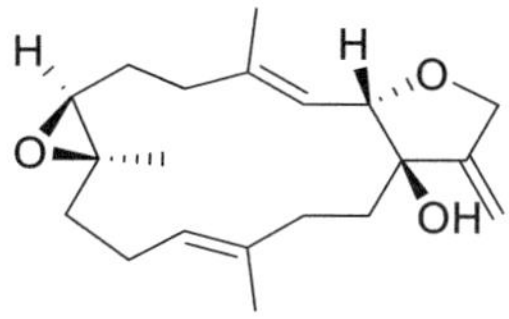

Laevigatol B (**213**)

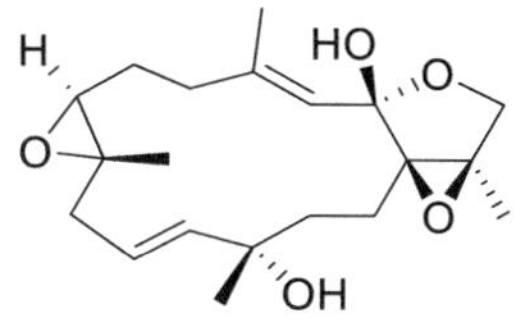

Laevigatol C (**214**)

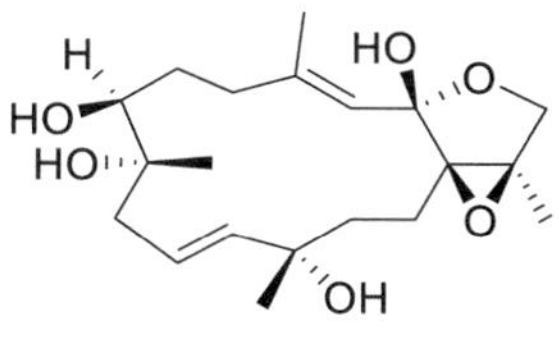

Laevigatol D (**215**)

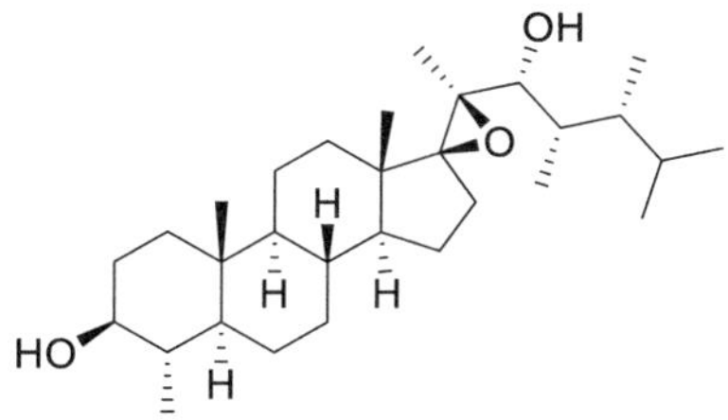

Lobophytosterol (**216**)

Fig. 31 Chemical structures of new natural products from *Lobophytum laevigatum*

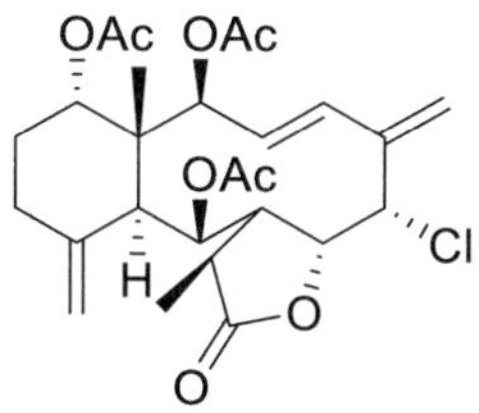

17-*epi*-junceellolide B (**217**)

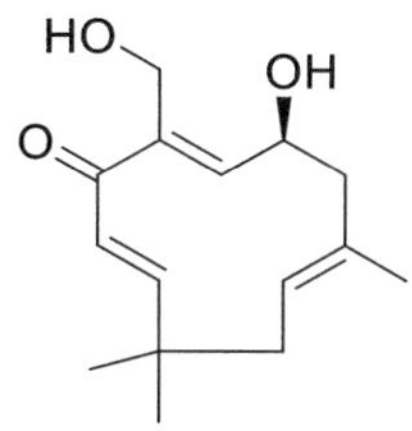

Leptogorgin A (**218**)

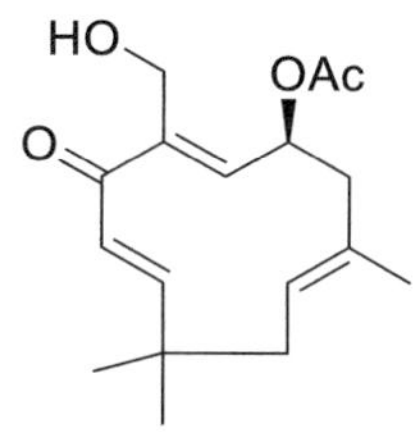

Leptogorgin B (**219**)

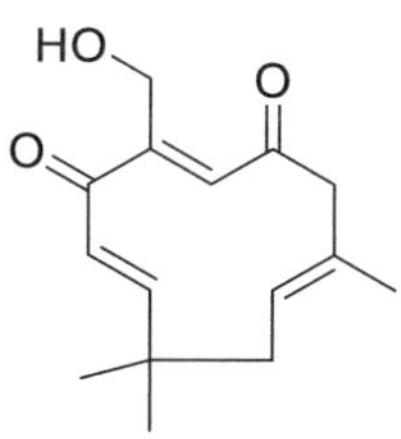

Leptogorgin C (**220**)

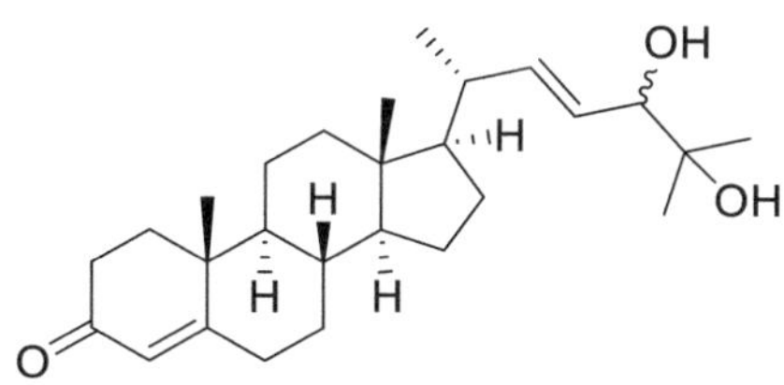

Leptogorgoid A (**221**)

Fig. 32 Chemical structures of new natural products from *Junceella fragilis* and *Leptogorgia* sp.

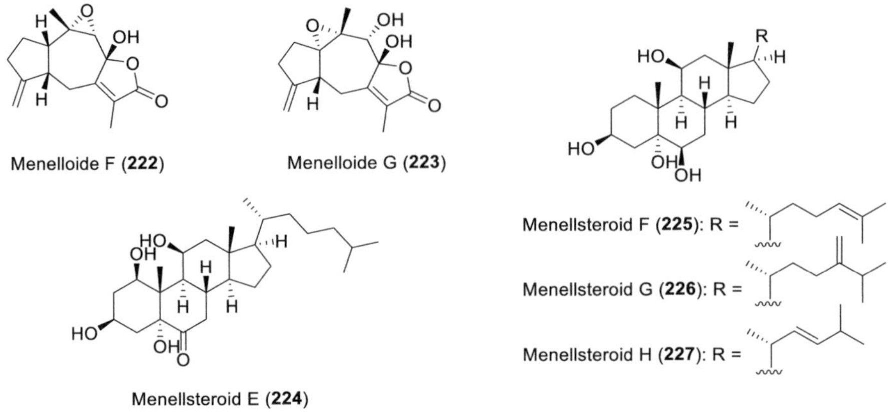

Fig. 33 Chemical structures of new natural products from *Menella woodin*

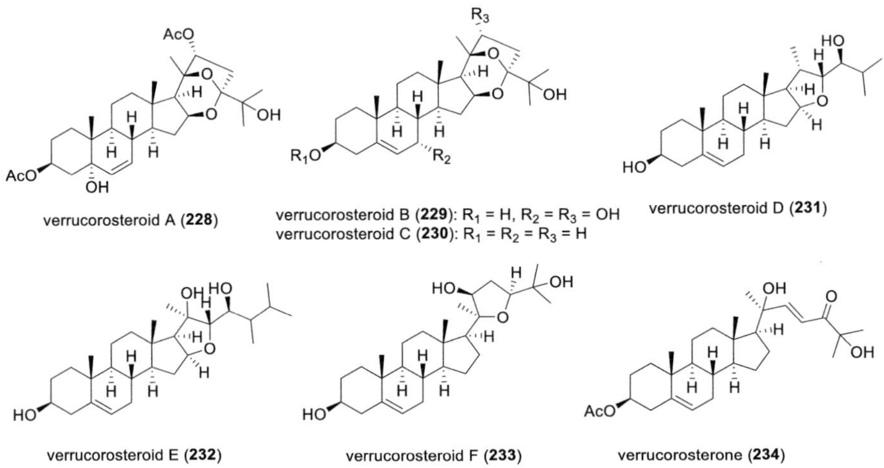

Fig. 34 Chemical structures of new natural products from *Verrucella corona*

From another *Sarcophyton* species, *Sarcophyton pauciplicatum*, a new steroid, sarcopanol A (**200**) (Fig. 28), together with two known compounds, were isolated [115]. Samples of this soft coral species were collected in Hai Phong, Vietnam. Sarcopanol A exhibited significant inhibitory effects on TNFα-induced NF-κB transcriptional activity, with an EC_{50} value of 8.27 μM, while the positive control, apigenin, exhibited activity with EC_{50} of 4.51 μM. In addition, the transcriptional inhibition of sarcopanol A and the known (24*S*)-ergost-1*β*,3*β*,5*α*,6*β*-tetraol-25-monoacetate was confirmed by a decrease in COX-2, iNOS and intercellular adhesion molecule-1 (ICAM-1) gene expression levels in HaCaT cells.

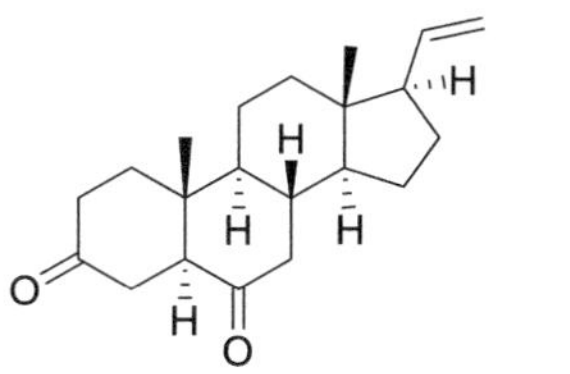

5-α-Pregn-20-en-3,6-dione (**235**)

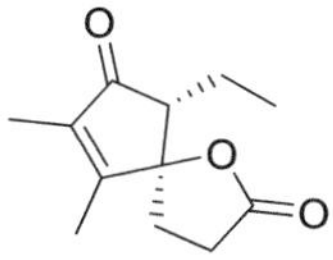

Dendronephthyone A (**236**)

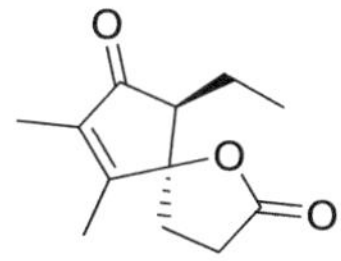

Dendronephthyone B (**237**)

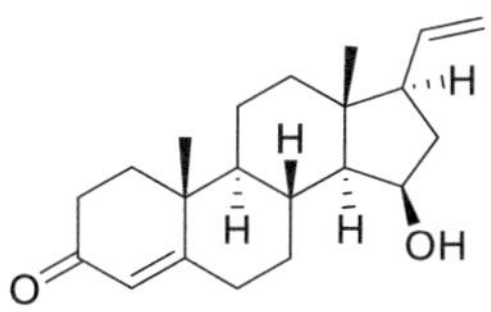

15β-Hydroxypregna-4,20-dien-3-one (**239**)

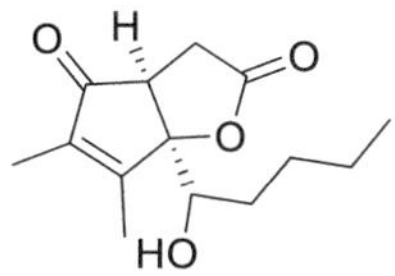

Dendronephthyone C (**238**)

Fig. 35 Chemical structures of new natural products from *Dendronephthya mucronata* and *Carijoa riisei*

Lobophytum laevigatum

Samples of the soft coral *Lobophytum laevigatum*, collected from Khanh Hoa Province, Vietnam, provided an unusual new sterol, lobophytosterol (**216**) (Fig. 31), and five known steroids [124]. The new sterol showed cytotoxicity against three cancer cell lines, including A549, HCT-116 and HL-60, with IC_{50} values ranging from 3.2 to 5.6 μM. The inhibitory activity of lobophytosterol could be due to the induction of apoptosis in these cancer cell lines. Among the known molecules, (22*S*,24*S*)-24-methyl-22,25-epoxyfurost-5-ene-3β,20β-diol, (24*S*)-ergost-5-ene-3β,7α-diol and pregnenolone significantly upregulated PPAR transcriptional activity in Hep-G2 cells in a dose dependent manner. In addition, at a concentration of 1 μM, (22*S*,24*S*)-24-methyl-22,25-epoxyfurost-5-ene-3β,20β-diol, (24*S*)-ergost-5-ene-3β,7α-diol and pregnenolone were found to enhance PPAR transcriptional activity by 1.4-, 1.2- and 1.6-fold, respectively.

3.2.2 Family Gorgoniidae

About 17 new metabolites have been reported from at least four Vietnamese gorgonian species, including *Junceella fragilis*, *Leptogorgia* sp., *Menella woodin* and *Verrucella corona* (Table 8; Figs. 32, 33 and 34).

Terpenoids

Leptogorgia sp.

Three new humulane sesquiterpenoids, leptogorgins A (**218**)–C (**220**) and a new dihydroxyketosteroid, leptogorgoid A (**221**) (Fig. 32), were isolated from the gorgonian, *Leptogorgia* sp., obtained by dredging from a depth of 134 m at the South China Sea in Vietnamese territorial waters [119]. The effects of leptogorgins A (**218**) and B (**219**) on the viability of 22Rv1 cells (human drug-resistant prostate cancer cells) as well as on PNT2 cells (human prostate non-cancer cells) were assessed and found that only leptogorgin A exhibited moderate cytotoxicity to these two cancer cell lines with IC_{50} values of 31.0 and 35.8 μM, respectively. In addition, no inhibition on the colony formation of 22Rv1 prostate cancer cells was found for both leptogorgins A and B.

Steriods

Menella woodin

Four new polyoxygenated sterol derivatives, menellsteroids E (**224**)–H (**227**) (Fig. 33), were isolated from the gorgonian *Menella woodin*, collected from the vicinity of Coto Island, Quang Ninh Province, Vietnam [95]. The potential immunomodulatory properties of isolated polar steroids using murine macrophages of RAW 264.7 cell line were investigated. At the concentration of 10 μM, menellsteroids E and H exhibited significant reduction in reactive oxygen species (ROS) levels in macrophages, pre-stimulated with pro-inflammatory lipopolysaccharide (LPS) from *E. coli*. Specifically, steroids **224** and **227** decreased the ROS formation in the macrophages by 34.0% and 27.9%, respectively. The discovery of new steroids **224** and **227** could be further investigated as potential anti-endotoxic, immunomodulatory and cytoprotective molecules.

3.3 Natural Products from Vietnamese Echinoderms

New natural products reported from almost 30 species of Vietnamese echinoderms, represented by three classes, namely class crinoidea, Asteroidea and Holothuroidea, are being presented in this section (Tables 9, 10 and 11; Figs. 36, 37, 38, 39, 40, 41, 42, 43, 44, 45, 46, 47, 48, 49, 50, 51, 52, 53, 54, 55, 56 and 57). Of these classes, a majority of new chemistry have been reported from classes Asteroidea and Holothuridea, with biological activities ranging from cytotoxicity to immunomodulatory properties. Saponins are characteristic metabolites of starfish and sea cucumbers, and they are also reported from other marine taxonomic groups, such as sponges and soft coral. In addition to having defensive ecological roles against predators,

Table 9 List of new molecules reported from Vietnamese echinoderms belonging to class Crinoidea

Classification/ Compound[a]	Location	Biological activity	References
Capillaster multiradiatus			
Capillasterquinones A (**240**), B (**241**) and capillasterolide (**242**)	Son Tra, Da Nang	Capillasterquinone A: strong NO production inhibitory activity with an IC_{50} of 5.89 μM; reduced the LPS-induced iNOS and COX-2 expressions in a dose-dependent manner	[131]
Comanthus delicata			
Delicapyrons A (**243**)–H (**250**)	Van Phong, Khanhhoa	Delicapyron B: selective cytotoxicity against SK-Mel-2 and MCF7 human cancer cell lines Delicapyron D: selective activity against the SK-Mel-2 with IC_{50} at 8.51 μM Delicapyrons F–H: weak cytotoxic effects on SK-Mel-2	[132, 133]
Delicaquinons A (**251**) and B (**252**)		Delicaquinon B: weak cytotoxic effectson SK-Mel-2	[133]

[a]Chemical structures in Fig. 36

these echinoderm-derived steroid or triterpenoid glycosides often show remarkable pharmacological activities, such as antifungal, antifouling, antitumor and anti-inflammatory activities [128–130]. Many of them can be categorized into three main structural types, including asterosaponins (sea stars), polyhydroxysteroid glycosides (sea stars) and triterpene glycosides (sea cucumbers). As in previous sections, only selected molecules having unique chemical structures and/or biological properties are discussed below.

3.3.1 Class Crinoidea

Capillaster multiradiatus

Within class Crinoidea, the most common molecules isolated are the anthraquinonic pigments, which have been reported in all crinoid species (Table 9). From a collection of crinoids, *Capillaster multiradiatus*, from Son Tra, Da Nang, Vietnam, seven anthraquinones, including two new molecules, capillasterquinones A (**240**) and B (**241**), as well as one new butanolide, capillasterolide (**242**) (Fig. 36) were isolated [131]. The inhibitory effects of all eight compounds were evaluated on LPS-induced NO production in RAW264.7 cells and found that capillasterquinone A exhibited strong NO production inhibitory activity with IC_{50} of 5.9 μM. Subsequent testing of capillasterquinone A on its effects on expression of inducible nitric oxide synthase

Table 10 List of new molecules reported from Vietnamese echinoderms belonging to class Asteroidea

Classification/Compound[a]	Location	Biological activity	References
Acanthaster planci			
Plancisides A (**253**)–F (**258**)	Van Phong Bay, near Nha Trang; Con Co island	Planciside A: cytotoxicity against HCT-116 and RPMI-7951 cancer cells, IC_{50} values of 36 and 58 μM, respectively	[134–136]
Plancipyrrosides A (**259**) and B (**260**)	Con Co Island	Plancipyrrosides A and B: suppressed NO production in RAW264.7 cells, IC_{50} values of 16.61 and 5.94 μM, respectively	[137]
Acanthaglycoside G (**261**)	Van Phong Bay near Nha Trang	Not cytotoxic	[138]
Anthenea aspera			
Anthenosides L (**262**)–X (**274**)	Tu Long Bay, near Khuan Lan Island	Anthenosides T and U: as mixture exhibited hemolytic property with an EC_{50} of 8 μM Anthenoside O: immunomodulatory action by reducing the intracellular ROS content by 24% in RAW 264.7 murine macrophages	[139, 140]
Anthenosides A_1 (**275**) and A_2 (**276**)		–	[141]
A. sibogae			
Anthenosides S1 (**278**)–S6 (**283**)	Bai Tu Long Bay, near Hon Da Den Island	–	[142]
Anthenoides laevigatus			

(continued)

Table 10 (continued)

Classification/Compound[a]	Location	Biological activity	References
(25*S*)-5α-cholestane-3β,5,6β,14,15α,16β,26-heptaol (**284**), (25*S*)-5α-cholestane-3β,6β,14,15α,16β,26-hexaol (**285**), (25*S*)-5β-cholestane-3β,6β,14,15α,16β,26-hexaol (**286**) and (25*S*)-5β-cholestane-3β,4α,6β,14,15α,16β,26-heptaol (**287**)	Coastal waters of the Qui Nhon Province	(25*S*)-5α-cholestane-3β,5,6β,14,15α,16β,26-heptaol: inhibited colony formation of cancer cells and the decrease in the colony number of HT-29 and MDA-MB-231 cells by 18% and 35%, respectively	[143]
Archaster typicus			
(24*R*)-27-*nor*-5α-Cholestane-3β,4β,6α,8,14,15α,24-heptaol, 5α-cholestane-3β,4β,5,6α,8,14,15α,24,26-nonaol, 5α-cholest-25(27)-ene-3β,6α,8,14,15α,24,26-heptaol-15-*O*-sulfate as sodium salt, (23*E*)-27-*nor*-25-oxo-5α-cholest-23-ene-3β,6α,8,14,15α-pentaol-15-*O*-sulfate as sodium salt and typicusoside A	Shallow waters of Quang Ninh Province	All compounds inactive in an urchin embryo sperm-test and embryo development assay	[144]
Archasterosides A (**288**)–C (**290**)		Archasterosides A and B: moderate cytotoxic activities against HeLa and mouse JB6 P^+ Cl41 cell lines Archasteroside B: induced basal AP-1- and p53-, but not NF-κB-transcriptional activations, in JB6 Cl41 cells	[145, 146]
Sodium salts of (24*S*)-5α-cholestane-3β,4β,5,6α,7β,8,14,15α,24-nonaol 6-sulfate, (24*E*)-5α-cholest-24-ene-26-yde-3β,6α,8,14,15α-pentaol-15-sulfate and 5α-cholest-3β,6α,8,14,15α,24,25,26-octaol-15-sulfate	Haiphong	–	[147]
Asterina bather			
Astebatheriosides A (**291**)–D (**294**)	Catba, Haiphong	Astebatheriosides B–D: moderately inhibited IL-12 p40 production in lipopolysaccharide (LPS)-stimulated bone marrow-derived dendritic cells	[148]

(continued)

Table 10 (continued)

Classification/Compound[a]	Location	Biological activity	References
Asteropsis carinifera			
(24*R*,25*S*)-24-Methyl-5*α*-choles-tane-3*β*,6*α*,8,15*β*,16*β*,26-hexaol, (22*E*,24*R*,25*S*)-24-methyl-5*α*-cholest-22-ene-3*β*,6*α*,8,15*β*,16*β*,26-hexaol and (22*E*,24*R*,25*S*)-24-methyl-5*α*-cholest-22-ene-3*β*,4*β*,6*α*,8,15*β*,16*β*,26-heptaol	Van Fong Bay	–	[149]
Cariniferosides A (**295**)–F (**300**)		Cariniferoside A: cytotoxicity against HCT-116, T-47D, and RPMI-7951 cell lines, with IC_{50} values ranging from 32 to 66 μM Cariniferoside F: significant inhibition of RPMI-7951 and T-47D cell colony formation in a soft agar clonogenic assay	[150]
Asteropsiside A		–	[151]
Astropecten monacanthus			
Astrosteriosides A (**301**)–D (**304**)	Cat Ba, Haiphong	Astrosteriosides A and D: exhibited potent anti-inflammatory activity comparable to that of the positive control Astrosterioside D: potent cytotoxic effects against HL-60, PC-3 and SNU-C5 cells, IC_{50} values ranging from 4.31 to 5.21 μM	[152]
A. polyacanthus			
Astropectenols A (**310**)–D (**313**)	Cat Ba, Haiphong	Astropectenols A, C and D: moderate or weak activity against HL-60 and PC-3 cells Astropectenol A: inhibitory activity against *T. brucei* with EC_{50} of 1.57 μM	[153]
Polyacanthoside A	Nhat Le, Quang Binh	–	[154]

(continued)

Table 10 (continued)

Classification/Compound[a]	Location	Biological activity	References
Choriaster granulatus			
Granulatosides C–E	Van Phong Bay	Granulatoside D: increased the intracellular ROS level in peritoneal murine macrophages by 20% and decreased intracellular ROS level by 21% in pre-treated with endotoxic lipopolysaccharide from *E. coli* peritoneal macrophages	[155, 156]
Culcita novaeguineae			
Novaeguinoside E	Quang Ninh	–	[157]
Culcinoside E	Cam Ranh Bay, Khanh Hoa Province	Inhibited growth of *E. faecalis* with MIC value of 200 µM	[158]
Echinaster luzonicus			
Luzonicosides B (**314**)–F (**318**) and known luzonicoside A	Van Phong Bay	Luzonicoside A: potent in lysosomal activity stimulation, intracellular ROS level elevation, and NO synthesis up-regulation in RAW 264.7 murine macrophages	[159]
Linckia laevigata			
Hematoside with 8-*O*-methyi-*N*-glycolylneuraminic acid	Nha Trang Bay	–	[7]
Linckosides L1 (**319**)–L7 (**325**)	Van Phong Bay	Linckosides L1 and L2: able to induce neuronal differentiation and enhance neuritogenic activity of NGF Linckoside L6: cytotoxic to eggs of sea urchin *Strongylocentrotus intermedius*	[160–162]

(continued)

Table 10 (continued)

Classification/Compound[a]	Location	Biological activity	References
Pentaceraster gracilis			
Pentacerosides A and B	Bac Van, Co To, Quangninh	–	[163]
P. regulus			
Regulusosides A (**326**)–D (**329**) and a new polyhydroxylated steroid, (24*S*)-cholestane-3*β*,5,6*β*,8,15*α*,24-hexol	Cham Islands/ Con Co Islands, Quang Tri	Regulusosides A, B and D: potent immunomodulatory activity by inducing intracellular ROS formation in RAW 264.7 murine macrophage cells	[164, 165]
Regulusosides S1 and S2		–	[166]
Pentaregulosides A (**330**)–G (**336**)		Pentareguloside A: cytotoxic activity, IC_{50} value of 6.4 μM against RAW 264.7 murine macrophages Pentaregulosides C–E: immunomodulatory action by reducing reactive oxygen species formation in the RAW 264.7 cells	[167]

[a]Chemical structures in Figs. 37, 38, 39, 40, 41, 42, 43, 44, 45, 46 and 47

Table 11 List of new molecules reported from Vietnamese echinoderms belonging to class Holothuroidea

Classification/ Compound[a]	Location	Biological activity	References
Cercodemas anceps			
Cercodemasoides A (**337**)–E (**341**)	Co To, Quangninh	All compounds showed potent cytotoxicity on five human tumour cell lines with IC_{50} values ranging from 0.03 to 7.36 μM	[168]
Cladolabes schmeltzii			
Cladolosides A_1 (**349**)–A_6	Nha Trang	Cladolosides A_1 and A_2 (**350**): strong cytotoxicity against mouse spleenocytes, IC_{50} values of 2.8 and 1.8 μg/mL, respectively; hemolytic activity, MEC_{100} values of 3.12 μg/mL each	[169]
Cladolosides B_1, B_2, C (**351**), C_1, C_2 (**352**) and D (**343**)		All compounds demonstrated strong cytotoxic and hemolytic properties, with IC_{50} values ranging from 2.16 to 4.47 μg/mL and MIC ranging from 3.1 to 6.26 μg/mL	[170]
Cladolosides C_3, E_1, E_2, F_1, F_2, G (**353**), H_1 (**354**) and H_2(**355**)		All compounds demonstrated strong/moderate cytotoxic against mouse spleenocytes (IC_{50} ranging from 0.3 to 28.4 μg/mL) and hemolytic effects against mouse erythrocytes (MIC ranging from 0.2 to 50.0 μg/mL); hexaosides being more active than the corresponding pentaosides	[171]
Cladolosides C_4 (**342**), D_1 (**344**), D_2 (**345**), M (**356**), M_1 (**358**), M_2, N (**357**) and Q (**346**)		All compounds, except for cladoloside D_2, inhibited the colony formation and growth of HT-29 cells at non-cytotoxic concentrations Cladoloside M_1: highest inhibitory activity on HT-29 cells with $ICCF_{50}$ of 0.05 μM Cladoloside N: showed significant synergism with radioactive irradiation by decreasing the number of colonies of HT-29 cells; increased the inhibitory effect from radiation by 75%	[172]

(continued)

Table 11 (continued)

Classification/ Compound[a]	Location	Biological activity	References
Cladolosides I_1, I_2, J_1 (**348**), K_1, K_2 and L_1		All molecules demonstrated strong/ moderate hemolytic against mouse erythrocytes (ED_{50} of 0.18–11.41 μM) and cytotoxic effects against mouse splenocytes (ED_{50} of 1.02–10.31 μM) and the ascite form of mouse Ehrlich carcinoma cells (ED_{50} of 1.39–10.32 μM)	[173]
Cladolosides O, P (**359**), P_1 (**360**)–P_3 and R (**347**)		All compounds demonstrated strong cytotoxic activities against mouse Ehrlich carcinoma cells (ascite form) (IC_{50} of 0.61–2.41 μM) and mouse erythrocytes (EC_{50} of 0.34–1.99 μM) Cladolosides P_2 and P_3: increased the inhibitory effect of radiation by more than 70% on HT-29 cancer cells	[174]
Colochirus quadrangularis			
Quadrangularisosides A (**361**)–D_4	Coral reefs near the seashore of Vietnam	All compounds strongly hemolytics Quadrangularisosides A (**361**), B and D_1: enhanced the effect of radiation by about 30% Quadrangularisosides B, B_1 (**363**), B_2, D and D_1 (**366**): suppressed the cell viability of HT-29 cells Quadrangularisosides A_1 (**362**), C (**364**), C_1 (**365**) and E (**367**): strong inhibitory activity on colony formation in HT-29 cells	[175]
C. robustus			
Colochirosides A_1 (**368**), A_2, A_3 (**369**) and D (**371**)	Nha Trang	Colochirosides A_1, A_3 and D: strong cytotoxic and hemolytic effects against the ascite form of mouse Ehrlich carcinoma cells (ED_{50} of 1.9–3.9 μM) and mouse erythrocytes (ED_{50} of < 0.8–2.4 μM), respectively	[176]
Colochirosides B_1, B_2, B_3 and C (**370**)		Colochirosides B_3 and C: strong cytotoxic and hemolytic effects against the ascite form of mouse Ehrlich carcinoma cells and mouse erythrocytes (ED_{50} of < 0.8 μM), respectively	[177]
Colochiroside E (**372**)		–	[178]

(continued)

Table 11 (continued)

Classification/ Compound[a]	Location	Biological activity	References
Holothuria edulis			
Holothurin A_5	Cu Lao Cham Island, Quangnam	Weak cytotoxic effect against cancer cell lines	[179]
H. scabra			
Holothurins A_3 (**373**) and A_4 (**374**)	Cat Ba, Haiphong Province	Holothurins A_3 and A_4: strong cytotoxicity against KB and Hep-G2 cell lines with IC_{50} values ranging from 0.32 to 1.12 μg/mL	[180]
Holothurinogenin B		–	[181]
Neothyonidium magnum			
Magnumosides A_1 (**375**)–A_4, B_1, B_2, C_1 (**378**), C_2 and C_4 (**379**)	Near Tam Island, Nha Trang Bay	Magnumosides A_2, B_1 and B_2: moderate hemolytic activity against mouse erythrocytes Magnumosides C_1 and C_4: most hemolytic (ED_{50} of 6.97 and 6.52 μM, respectively) and cytotoxic (against mouse spleenocytes) activity (IC_{50} of 8.97 and 12.23 μM, respectively) Magnumosides A_3 and C_1: inhibited spontaneous colony formation of DLD-1 cells by 22% and 26%, respectively and reduced colonies size of DLD-1 cancer cells by 49% and 43%, respectively Magnumosides A_3 (**376**), C_1 and C_2: enhanced the effect from radiation by more than 30%	[182]
Magnumosides B_3, B_4 (**377**) and C_3		Magnumoside B_4: most hemolytic (ED_{50} of 1.42 μM) and cytotoxic against mouse spleenocytes (ED_{50} of 4.26 μM); decrease the number and size of DLD-1 cells by 67% and 81%, resepectively	[183]
Neothyonidioside C	Shores of South Vietnam	–	[4]
Stichopus chloronotus			
Stichloroside F (**380**)	Cat Ba, Haiphong	–	[184]
S. herrmanni			
Herrmananes A and B	Bai Giau island and Khai Luong island, Van Phong,	Herrmanane A: moderate cytotoxic against SK-Mel-2 cell, IC_{50} of 23.1 μM	[185]

(continued)

Table 11 (continued)

Classification/ Compound[a]	Location	Biological activity	References
Holothurin A_6	Bai Giau island and Khai Luong island, Van Phong	Holothurin A_6: moderate cytotoxic against four cancer cell lines, IC_{50} ranging from 13.9 to 52.6 μM	[186]
S. horrens			
Stichorrenosides A (**382**)–E (**386**)	Hai Van–Son Cha, Hue	Stichorrenoside D: strong cytotoxicity on Hep-G2, KB, LNCaP, MCF7 and SK-Mel2 with IC_{50} ranging from 1.92 to 3.02 μM	[187, 188]
Synapta maculata			
Synaptosides A (**387**) and A_1 (**388**)	Van Phong Bay	Synaptoside A: IC_{50} of 8.6 μg/mL against HeLa human tumor cell line	[189]

[a]Chemical structures in Figs. 48, 49, 50, 51, 52, 53, 54, 55, 56 and 57

(iNOS) and cyclooxygenase 2 (COX-2) proteins in RAW264.7 cells showed reduction of LPS-induced iNOS and COX-2 expressions in a dose-dependent manner. Both iNOS and COX-2 are major inflammatory mediators, and their increased cellular production is found in many chronic inflammatory and autoimmune diseases. Due to the strong inhibitory activity of capillasterquinone A on NO production as well as iNOS and COX-2 expressions, it could be a potential drug lead as anti-inflammatory agent. Interestingly, capillasterquinone B was recently detected in coculture of two Red Sea sponge-derived actinobacterial strains, *Actinokineospora spheciospongiae* strain EG49 and *Rhodococcus* sp. UR59 [190]. In addition, capillasterquinone B exhibited antimalarial effect against *Plasmodium falciparum* with IC_{50} value of 9.2 μg/mL.

Comanthus delicata

An investigation on the chemical constituents of a Vietnamese crinoid, *Comanthus delicata*, obtained from Van Phong, Khanhhoa, provided five new naphthopyrone-related compounds, delicapyrons A (**243**)–E (**247**) (Fig. 36), and nine known molecules [132]. Delicapyrons A and B contain unique bisnaphthopyrone structures consisting of linear and angular naphthopyrone units linked via a methylene group. When tested in a panel of five human cancer cell lines, including Hep-G2, KB, LNCaP, MCF7 and SK-Mel2, delicapyron B exhibited selective cytotoxicity against SK-Mel-2 and MCF7, with IC_{50} values of 11.9 and 14.9 μM, respectively. In addition, delicapyron D showed selective cytotoxicity to only SK-Mel-2 cell lines with IC_{50} value of 8.5 μM. An additional three new naphthopyrone derivatives, delicapyrons F (**248**)–H (**250**) (Fig. 36), and two new sulfated anthraquinone derivatives, delicaquinons A (**251**) and B (**252**) (Fig. 36), were isolated from the water-soluble extract of *C. delicata* [133]. Weak cytotoxicity was observed for these compounds when tested in SK-Mel-2 cancer cell line.

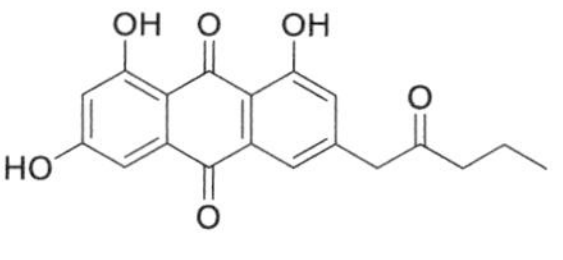

Capillasterquinone A (**240**)

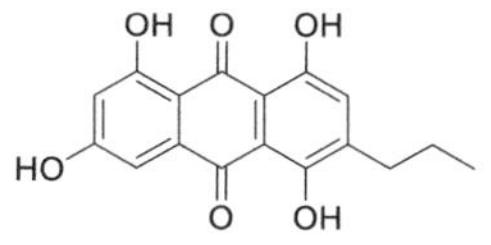

Capillasterquinone B (**241**)

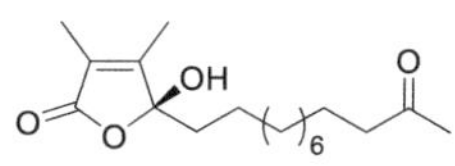

Capillasterolide (**242**)

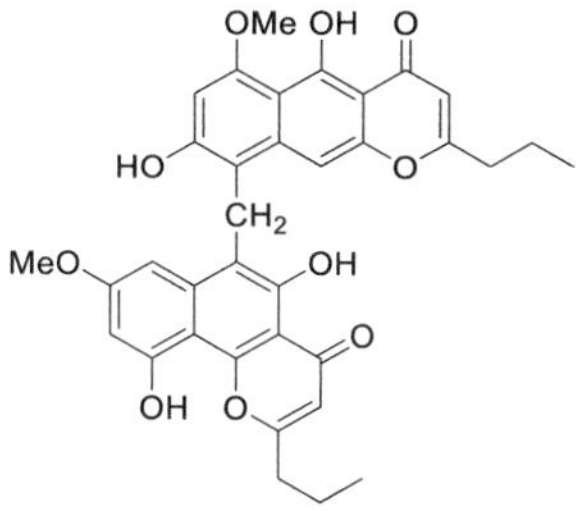

Delicapyron A (**243**)

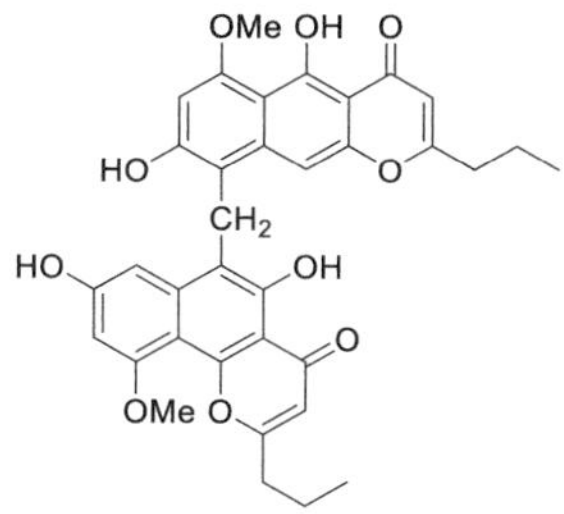

Delicapyron B (**244**)

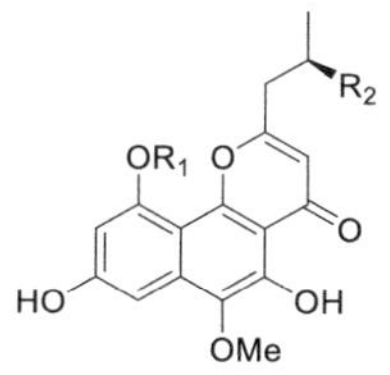

Delicapyron C (**245**): $R_1 = R_2 = H$
Delicapyron D (**246**): $R_1 = Me$, $R_2 = OH$

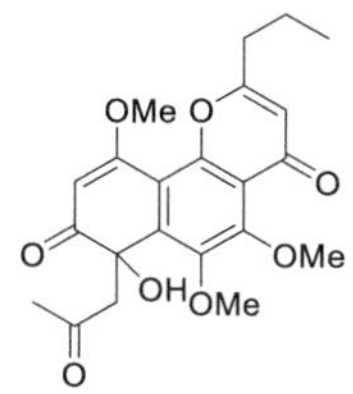

Delicapyron E (**247**)

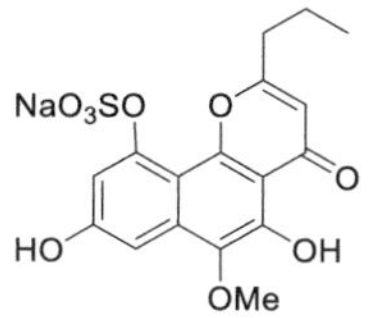

Delicapyron F (**248**)

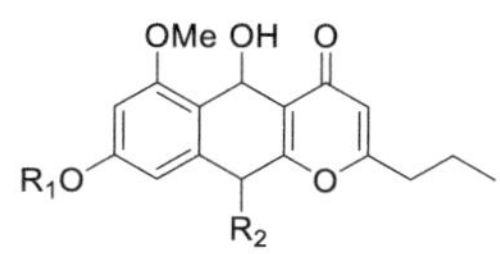

Delicapyron G (**249**): $R_1 = H$, $R_2 = OSO_3Na$
Delicapyron H (**250**): $R_1 = SO_3Na$, $R_2 = H$

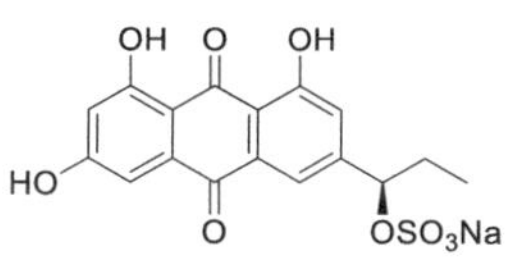

Delicaquinon A (**251**)

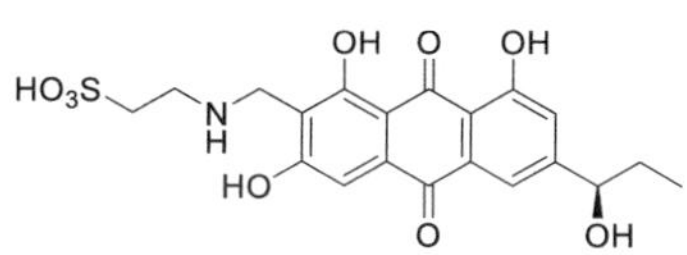

Delicaquinon B (**252**)

Fig. 36 Chemical structures of new natural products from *Capillaster multiradiatus* and *Comanthus delicata*

3.3.2 Class Asteroidea

Structurally new natural products have been reported from more than a dozen Vietnamese starfish species, including members belonging to the genera *Acanthaster*, *Anthenea*, *Anthenoides*, *Archaster*, *Asterina*, *Asteropsis*, *Astropecten*, *Choriaster*, *Culcita*, *Echinaster*, *Linckia* and *Pentaceraster* (Table 10; Figs. 37, 38, 39, 40, 41, 42, 43, 44, 45, 46 and 47). Similar to those of sea cucumbers, several species of sea stars have been historically used as tonic agents in Vietnamese folk medicines and medicinal foods. The predominant class of compounds reported from Vietnamese sea stars

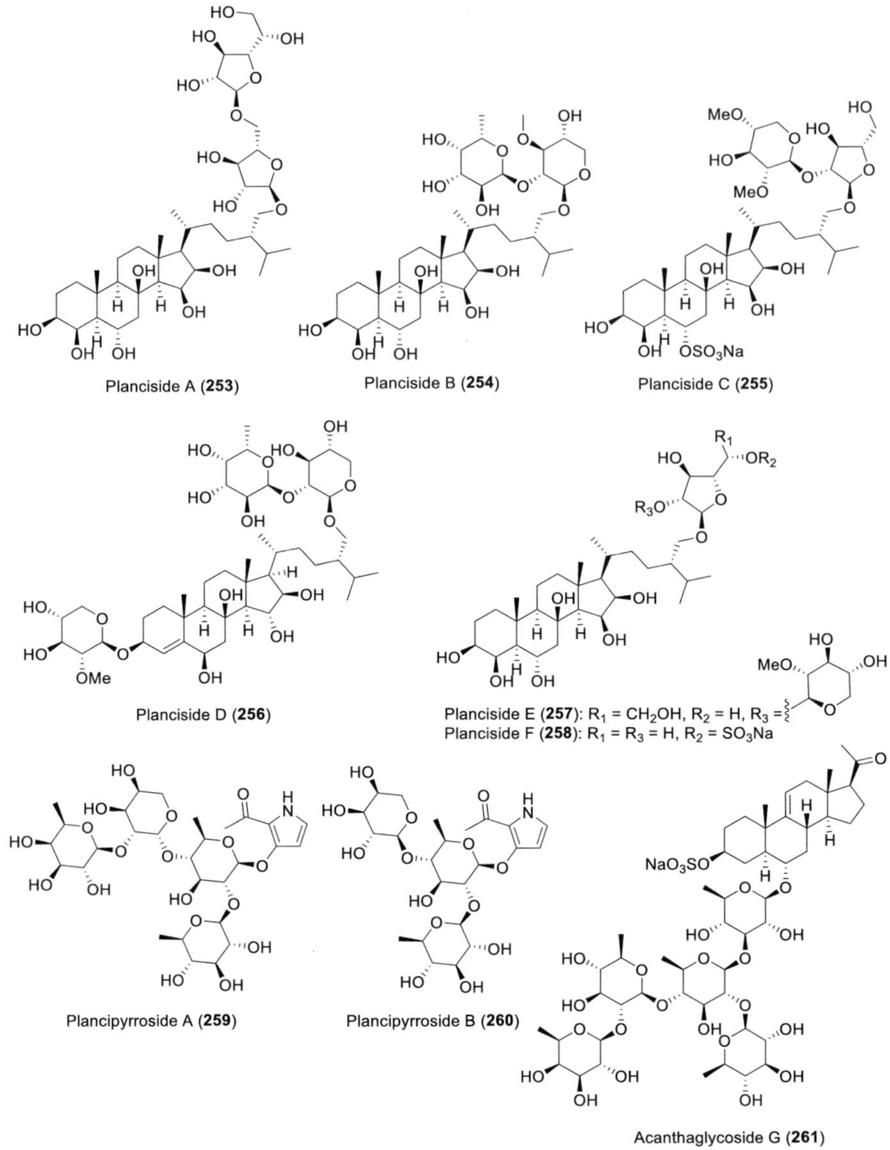

Fig. 37 Chemical structures of new natural products from *Acanthaster planci*

belong to asterosaponins, a class of steroid oligoglycosides, having pharmacological properties, including anti-inflammatory and anticancer activities [128].

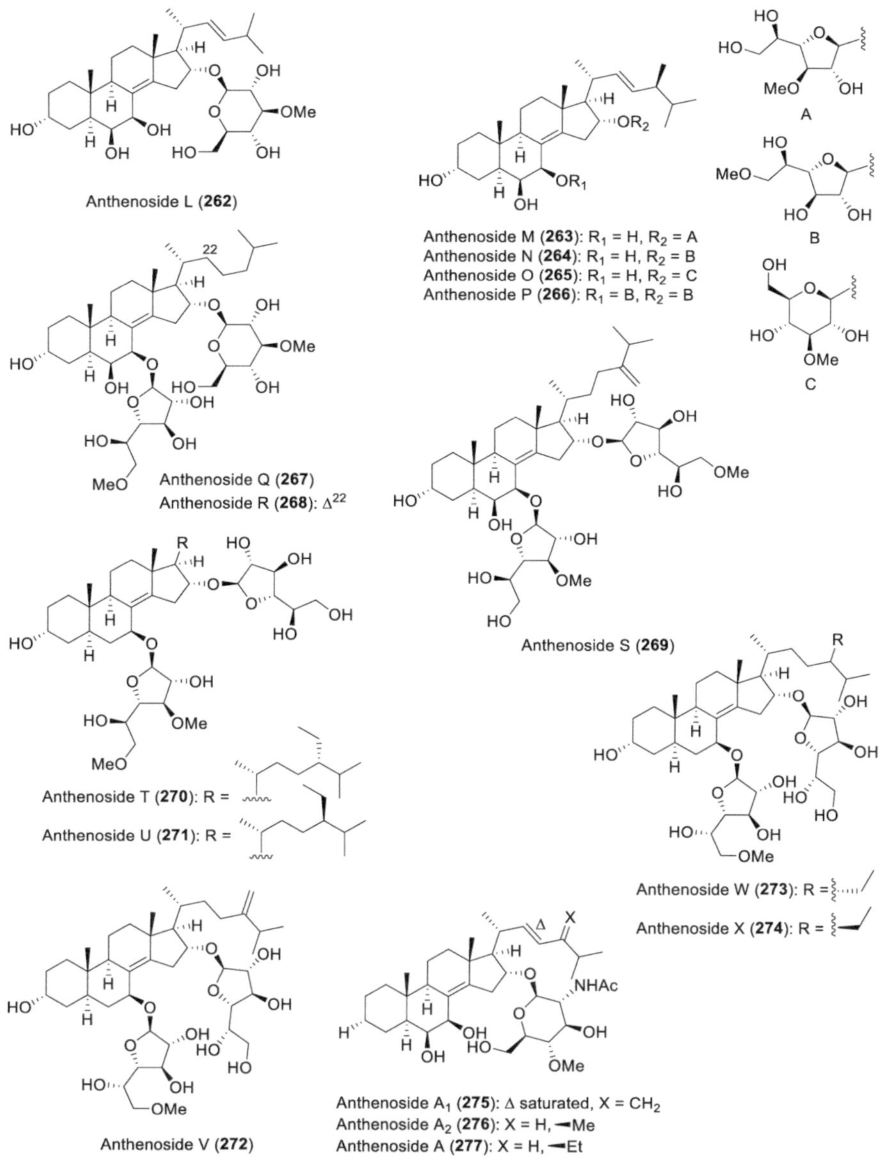

Fig. 38 Chemical structures of new natural products from *Anthenea aspera*

Acanthaster planci

Acanthaster planci, also known as the crown-of-thorns starfish and one of the largest sea stars in the world, is distributed throughout the Indo-Pacific region. The toxins found in the thorn spines consist of polar steroid glycosides, asterosaponins. From

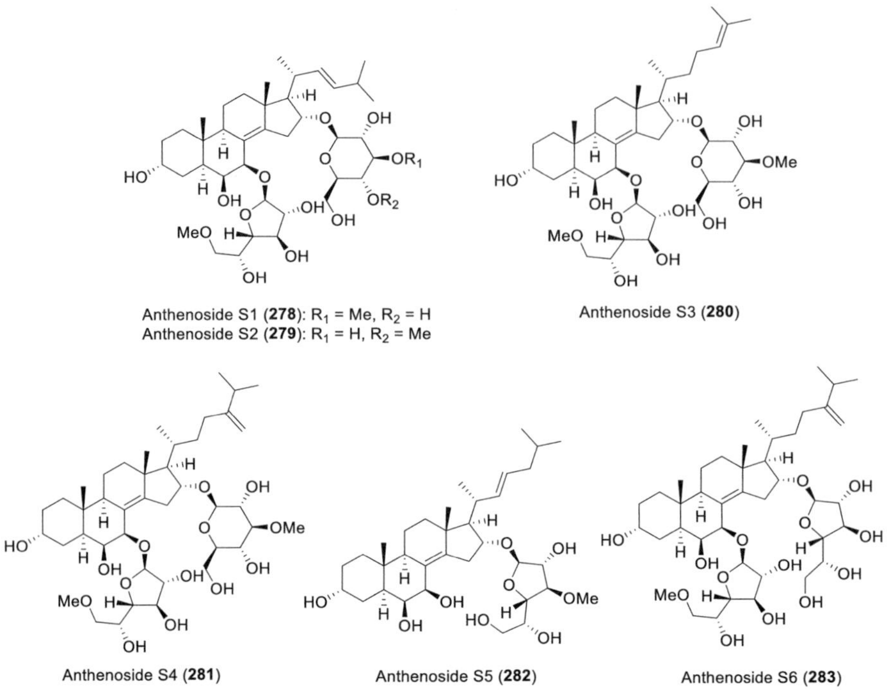

Fig. 39 Chemical structures of new natural products from *Anthenea sibogae*

an ethanolic/mathanolic extracts of *A. planci*, collected from various locations, such as Van Phong Bay, near Nha Trang and Con Co island, a series of new steroidal glycosides, plancisides A (**253**)–F (**258**) and acanthaglycoside G (**261**) (Fig. 37), have been isolated [134–136, 136]. Of these series, the biglycoside, planciside A (**253**) showed cytotoxicity against HCT-116 and RPMI-7951 cancer cells having IC_{50} values of 36 and 58 μM, respectively. In addition, planciside A showed slight antiproliferative property on HCT-116, T-47D and RPMI-7951 cell lines when tested at 15 μM, with no effect on colony formation of these cancer cells in a soft agar clonogenic assay.

From another collection of *A. planci* from the vicinity of Con Co Island, Vietnam, two new pyrrole oligoglycosides, plancipyrrosides A (**259**) and B (**260**) (Fig. 37), were isolated from its methanol extract [137]. When the inhibitory activity of these compounds were being assessed in RAW264.7 cells, plancipyrrosides A and B suppressed NO production in these cells with IC_{50} values of 16.61 and 5.94 μM, respectively. As the inhibitory activity of plancipyrroside B was comparable to that of positive control, cardamonin, it is suggested that plancipyrroside B could be further evaluated on its mode of actions as an anti-inflammatory agent.

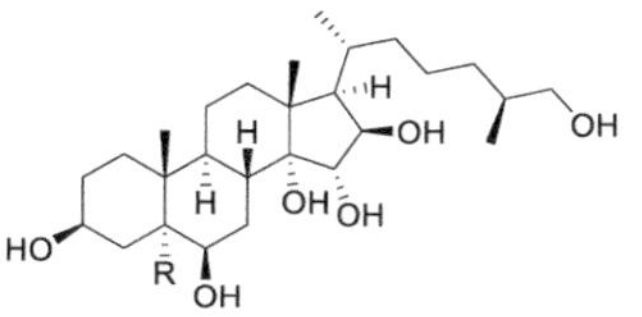

(25*S*)-5α-Cholestane-3β,5,6β,14,15α,16β,26-heptaol (**284**): R = OH
(25*S*)-5α-cholestane-3β,6β,14,15α,16β,26-hexaol (**285**): R = H

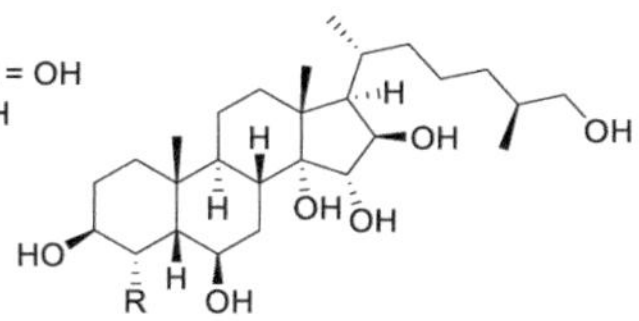

(25*S*)-5β-cholestane-3β,6β,14,15α,16β,26-hexaol (**286**): R = H
(25*S*)-5β-cholestane-3β,4α,6β,14,15α,16β,26-heptaol (**287**): R = OH

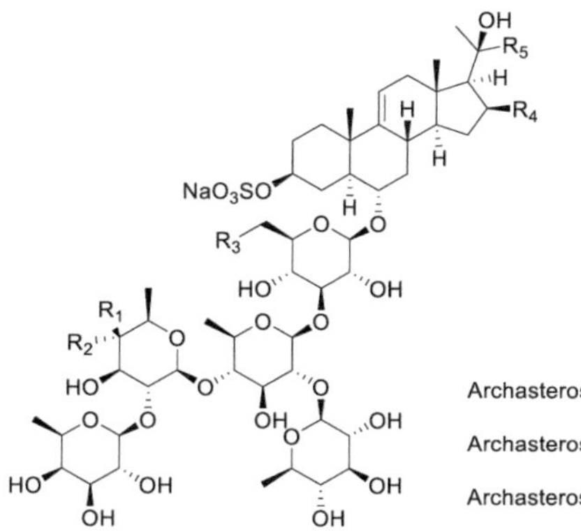

Archasteroside A (**288**): $R_1 = R_4$ = H, $R_2 = R_3$ = OH, R_5 =
Archasteroside B (**289**): $R_1 = R_3$ = H, $R_2 = R_4$ = OH, R_5 =
Archasteroside C (**290**): R_1 = OH, $R_2 = R_3 = R_4$ = H, R_5 =

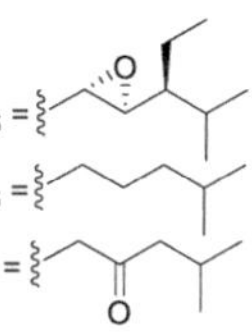

Fig. 40 Chemical structures of new natural products from *Anthenoides laevigatus* and *Archaster typicus*

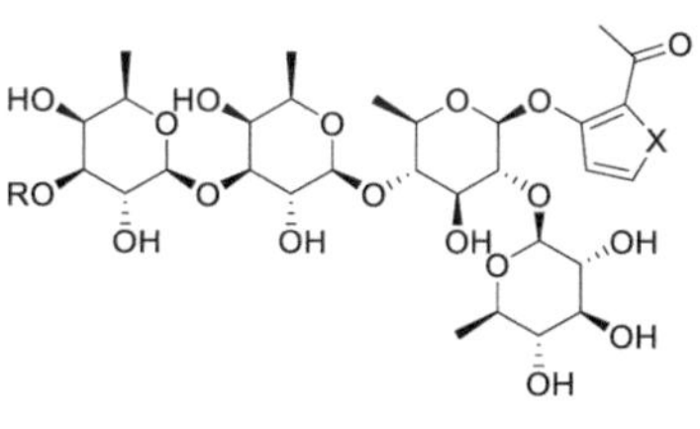

Astebatherioside A (**291**): X = NH, R =

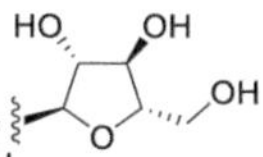

Astebatherioside D (**294**): X = O, R = H

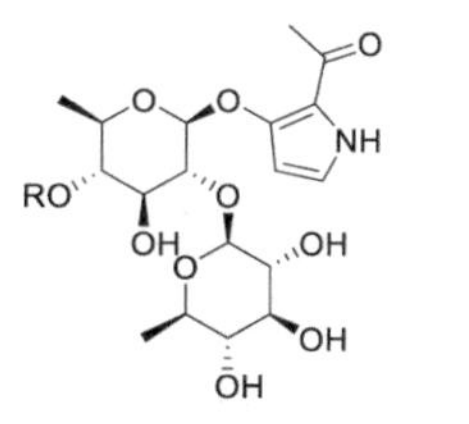

Astebatherioside B (**292**): R =

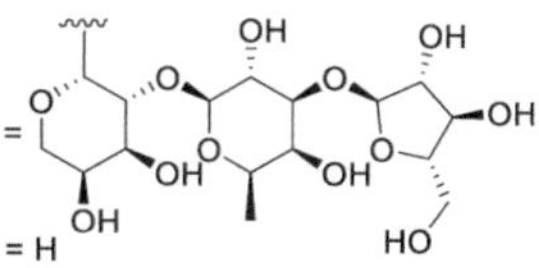

Astebatherioside C (**293**): R = H

Fig. 41 Chemical structures of new natural products from *Asterina batheri*

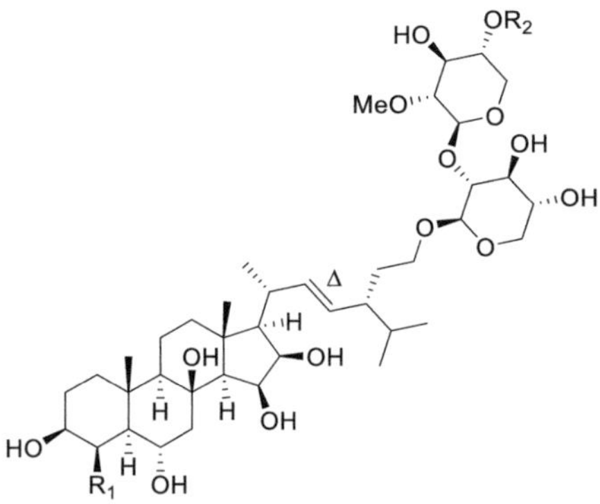

Cariniferoside A (**295**): R_1 = H, R_2 = Me, Δ saturated
Cariniferoside B (**296**): R_1 = R_2 = H
Cariniferoside C (**297**): R_1 = OH, R_2 = Me, Δ saturated

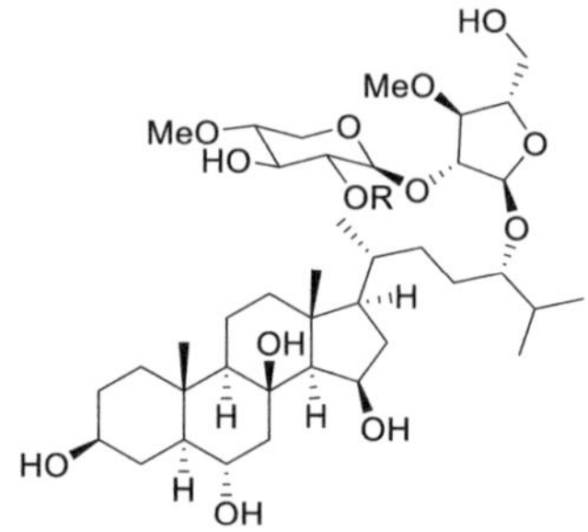

Cariniferoside D (**298**): R = Me
Cariniferoside E (**299**): R = H

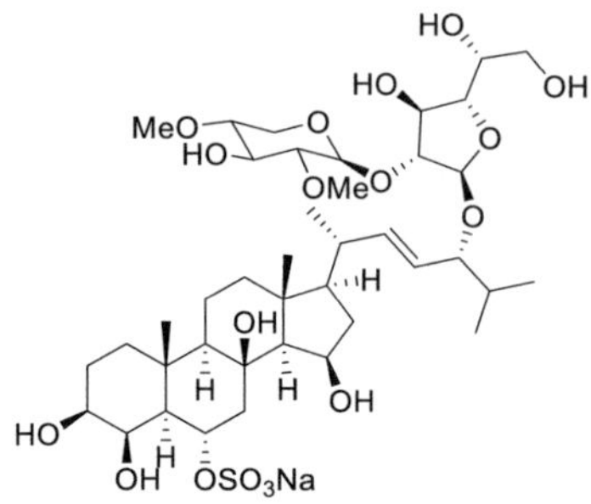

Cariniferoside F (**300**)

Fig. 42 Chemical structures of new natural products from *Asteropsis carinifera*

Anthenea sp.

Ten new polyhydroxysteroidal glycosides related to the anthenoside structural class, anthenosides L (**262**)–U (**271**) (Fig. 38), with rare positions of carbohydrate fragment attachments, were obtained from the starfish, *Anthenea aspera*, collected at Tu Long Bay, near Khuan Lan Island in the South China Sea [139]. The starfish samples were collected during the 34th scientific cruise of the research vessel "Akademik Oparin" in May 2007. These molecules have atypical structure fragments with attachment positions of carbohydrates at only C-16 or both C-16 and C-7 of the steroidal aglycons. This contrasted with the majority of starfish-derived polyhydroxysteroidal glycosides having oxygenated steroidal aglycons with the number of hydroxy groups ranging from three to nine and attachments of one, two and rarely three monosaccharide residues at C-3 in the steroidal nucleus and/or to C-24, C-26, C-28, and C-29 in the side chain. It was reported that a mixture of anthenosides T (**270**) and U (**271**) exhibited hemolytic property with an EC_{50} of 8 μM. In addition, anthenoside O (**265**), when tested at 10 μM, showed potential immunomodulatory action, reducing the intracellular ROS content by 24% in RAW 264.7 murine macrophages, co-stimulated by pro-inflammatory endotoxin, lipopolysaccharide, from *E. coli*.

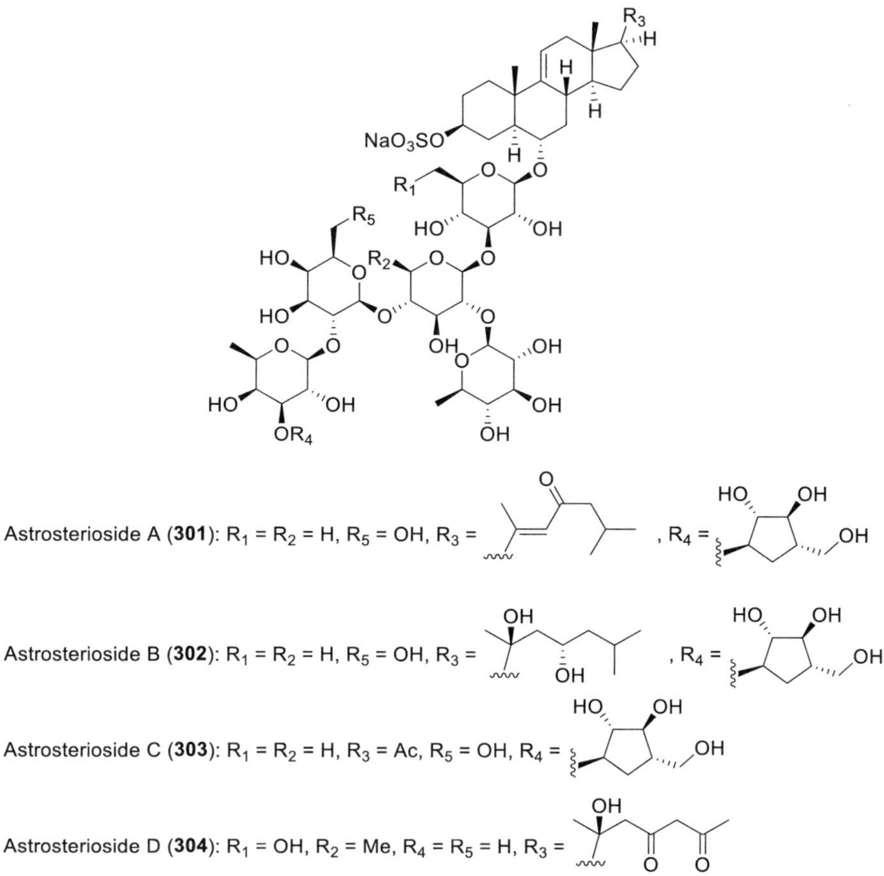

Fig. 43 Chemical structures of new natural products from *Astropecten monacanthus*

Subsequent chemical investigations on the ethanolic extracts of *A. aspera* led to the isolation of new anthenosides A_1 (**275**), A_2 (**276**) (Fig. 38) and V (**272**)–X (**274**) (Fig. 38) [140, 141]. A revision to the structure of anthenoside A (**277**) has also been reported. Anthenosides A_1 and A_2, containing a 2-acetamido-2-deoxy-4-*O*-methyl-β-*D*-glucopyranosyl unit, showed weak inhibition of cell viability in human cancer T-47D cells and no cytotoxic effects against RPMI-7951 cells. In addition, anthenosides A_1 and A_2 inhibited colony formation in human cancer RPMI-7951 and T-47D cells by 16% and 40%, respectively [141].

Chemical investigation on another *Anthenea* species, *A. sibogae*, collected from Bai Tu Long Bay, near Hon Da Den Island, afforded six new polyhydroxysteroidal glycosides, anthenosides S1 (**278**)–S6 (**283**) (Fig. 39) [142]. These molecules consist of a 5α-cholest-8(14)-ene-3α,6β,7β,16α-tetrahydroxysteroidal nucleus as well as rare positions of carbohydrate fragment attached at C7 and C16 or only at C16.

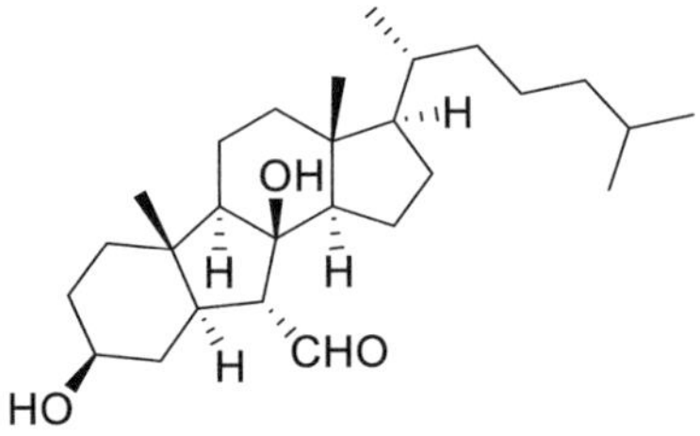

Astropectenol A (**310**)

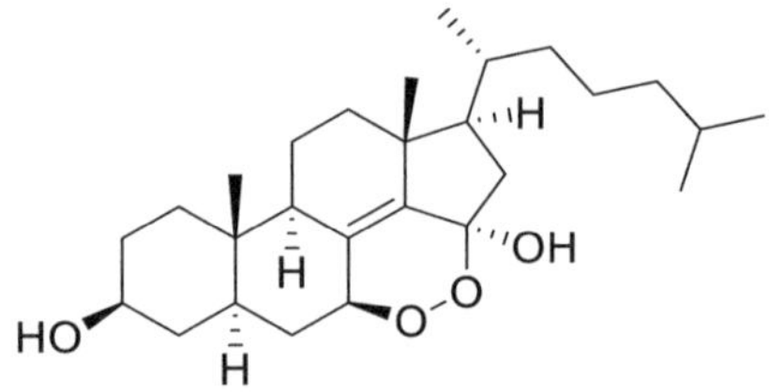

Astropectenol B (**311**)

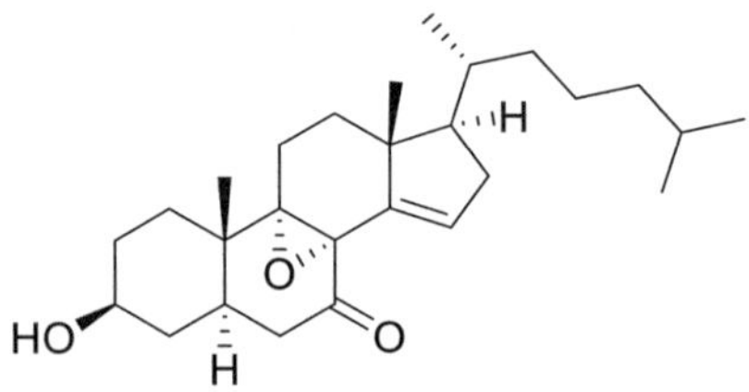

Astropectenol C (**312**)

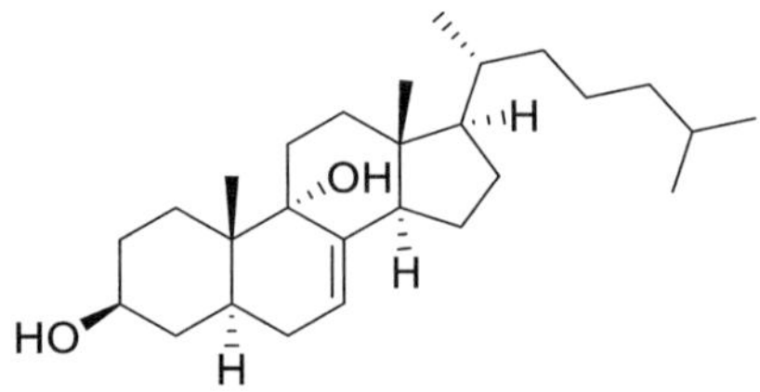

Astropectenol D (**313**)

Fig. 44 Chemical structures of new natural products from *A. polyacanthus*

Anthenosides S1, S2 (**279**) and S4 (**281**) were non-toxic against T-47D cell line at concentrations ranging from 50 to 150 μM.

Anthenoides laevigatus

Four unusual polyhydroxylated steroids, (25*S*)-5*α*-cholestane-3*β*,5,6*β*,14,15*α*,16*β*,26-heptaol (**284**), (25*S*)-5*α*-cholestane-3*β*,6*β*,14,15*α*,16*β*,26-hexaol (**285**), (25*S*)-5*β*-cholestane-3*β*,6*β*,14,15*α*,16*β*,26-hexaol (**286**) and (25*S*)-5*β*-cholestane-3*β*,4*α*,6*β*,14,15*α*,16*β*,26-heptaol (**287**) (Fig. 40), along with two previously known related steroids were isolated from the starfish, *Anthenoides laevigatus* collected off the coastal waters of Vietnam at Qui Nhon Province, at a depth of 20 to 30 m [143]. While heptaol **284** and hexaol **285** contain the common 5*α*-cholestane skeleton, hexaol **286** and heptaol **287** have the rare 5*β*-cholestane skeleton. (25*S*)-5*α*-Cholestane-3*β*,5,6*β*,14,15*α*,16*β*,26-heptaol (**284**) was not cytotoxic against normal JB6 Cl41 and cancer HT-29 and MDA-MB-231 cells, however it inhibited cell proliferation and colony formation of cancer HT-29 and MDA-MB-231 cells by 18% and 35%, respectively.

Archaster typicus

A shallow water collection of the starfish, *Archaster typicus*, obtained from Quang Ninh province, Vietnam, provided a series of four new highly hydroxylated steroids, (24*R*)-27-nor-5*α*-cholestane-3*β*,4*β*,6*α*,8,14,15*α*,24-heptaol, 5*α*-cholestane-3*β*,4*β*,5,6*α*,8,14,15*α*,24,26-nonaol, 5*α*-cholest-25(27)-ene-3*β*,6*α*,8,14,15*α*,24,26-heptaol-15-*O*-sulfate as sodium salt and (23*E*)-27-nor-25-oxo-5*α*-cholest-23-ene-3*β*,6*α*,8,14,15*α*-pentaol-15-*O*-sulfate as sodium salt as well as a

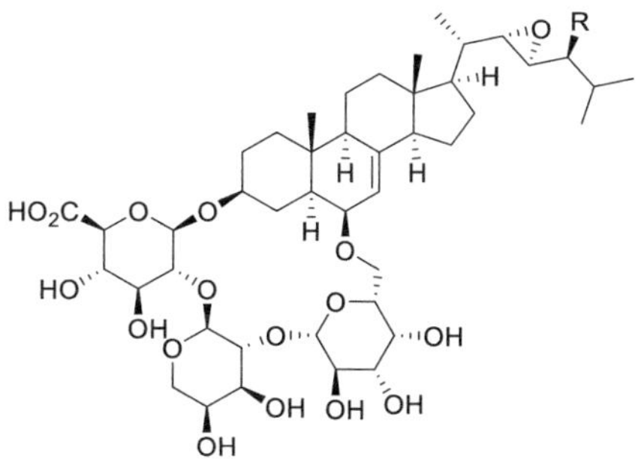

Luzonicoside B (**314**): R = H
Luzonicoside C (**315**): R = Me

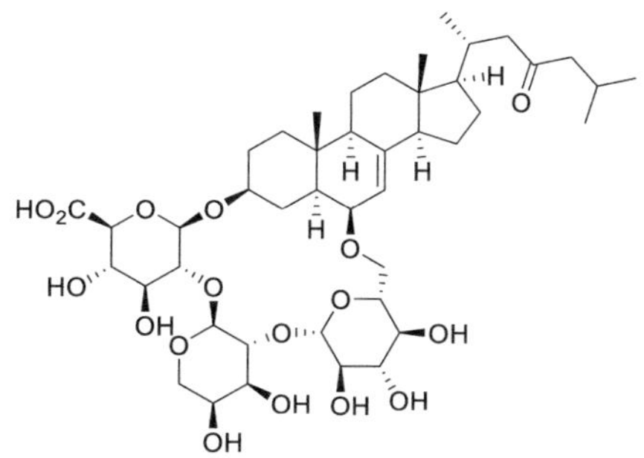

Luzonicoside D (**316**)

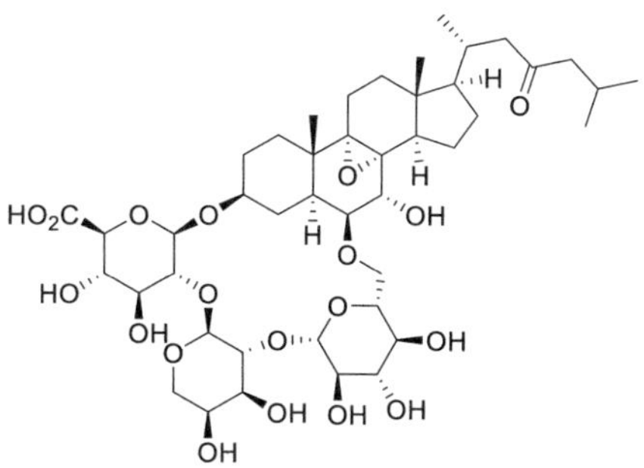

Luzonicoside E (**317**)

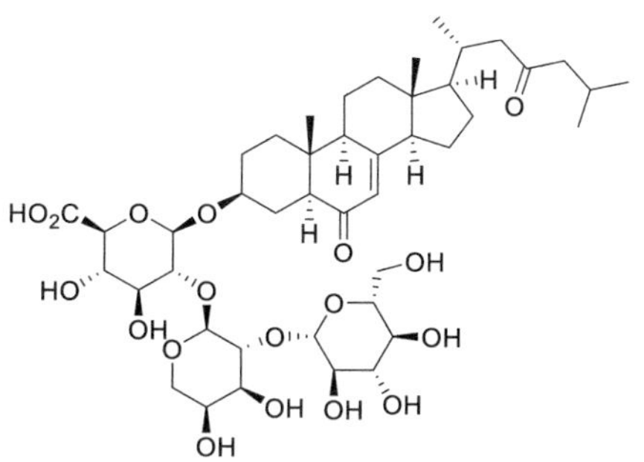

Luzonicoside F (**318**)

Fig. 45 Chemical structures of new natural products from *Echinaster luzonicus*

new glycoside, typicusoside A [144]. This new glycoside is structurally unusual as it represents a rare example of a naturally occurring α-D-glucoside. These compounds were inactive in an urchin embryo sperm-test and embryo development assay.

An investigation on the chemistry and biological activities of an ethanolic extract prepared from a sample of *A. typicus*, collected off Quang Ninh, Vietnam, led to the isolation of new asterosaponins, archasterosides A (**288**)–C (**290**) (Fig. 40) [145, 146]. Archasterosides are asterosaponins are pentasaccharide examples bearing either one (e.g., archasterosides A and B) or two (e.g., archasteroside C) fucose sugar units. Archasterosides A (**288**) and B (**289**) exhibited moderate anticancer activity against human cancer HeLa cells, with IC_{50} values of 24 and 14 μM, respectively. In addition, the activity of archasteroside B on the basal AP-1-, p53- and NF-kB-dependent transcriptional activities was examined based on JB6 Cl41 cells expressing a luciferase reporter gene controlled by an AP-1, p53, or NF-kB DNA binding sequence. Upon 6 h incubation with the cells, the molecule induced basal AP-1- and p53-, but not NF-kB-transcriptional activations in JB6 Cl41 cells.

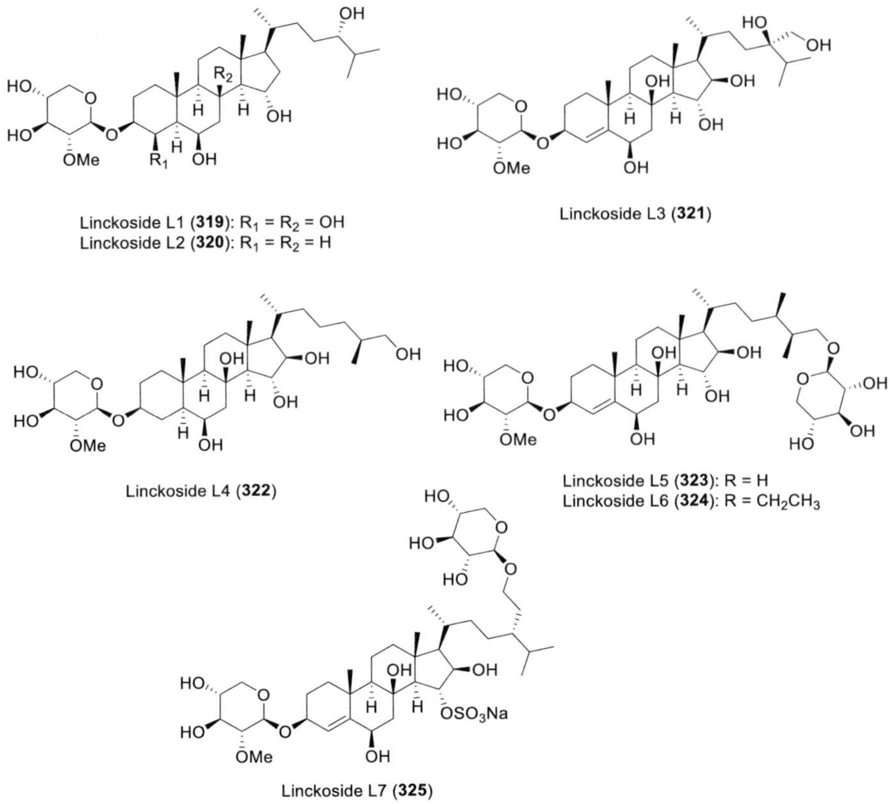

Fig. 46 Chemical structures of new natural products from *Linckia laevigata*

Asterina batheri

Three new pyrrole oligoglycosides, astebatheriosides A (**291**)–C (**293**) (Fig. 41), and one new furan oligoglycoside, astebatherioside D (**294**) (Fig. 41), were characterized from a sample of *Asterina batheri* obtained from Catba, Haiphong, Vietnam [148]. These molecules were evaluated for their biological effects in the inflammatory response of bone marrow-derived dendritic cells. Astebatheriosides B (**292**)–D (**293**) showed moderate inhibitory effects on the interleukin-12 (IL-12) p40 production, with IC_{50} values of 36.4, 31.6 and 22.8 μM, respectively. IL-12 is involved in type-1 helper T-cell-mediated inflammation as well as implicated in inflammatory diseases, such as rheumatoid arthritis, asthma, psoriasis and Crohn's disease. The discovery of astebatheriosides B–D having inhibitory effects on IL-12 p40 production warrants further biological investigation.

Asteropsis carinifera

Six new steroidal biglycosides, cariniferosides A (**295**)–F (**300**) (Fig. 42), were isolated together with six known glycosides from the alcoholic extract of the starfish,

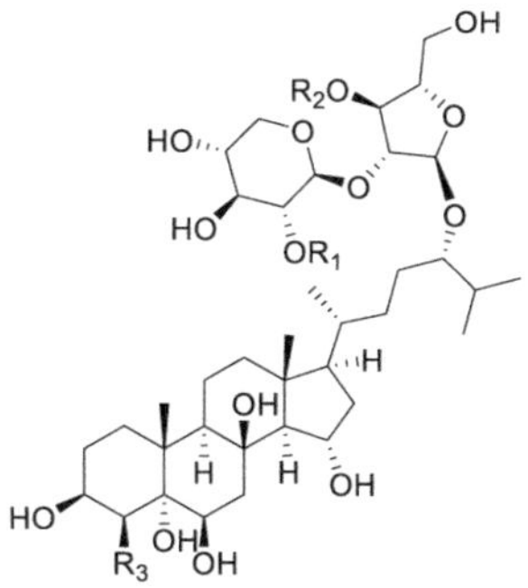

Regulusoside A (**326**): $R_1 = R_3 = H$, $R_2 = Me$
Regulusoside B (**327**): $R_1 = R_2 = R_3 = H$
Regulusoside C (**328**): $R_1 = R_2 = Me$, $R_3 = H$
Regulusoside D (**329**): $R_1 = H$, $R_2 = Me$, $R_3 = OH$

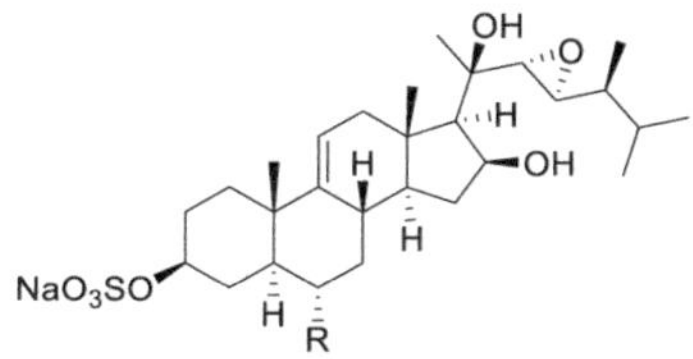

Pentareguloside A (**330**): R = Y

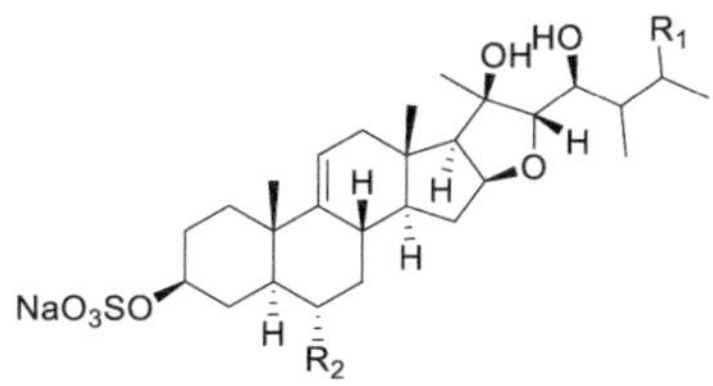

Pentareguloside B (**331**): $R_1 = Me$, $R_2 = Y$
Pentareguloside C (**332**): $R_1 = H$, $R_2 = Y$

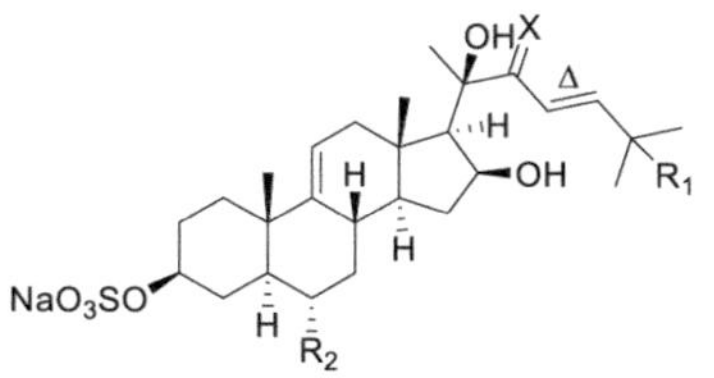

Pentareguloside D (**333**): X = H,–OH , $R_1 = H$, $R_2 = Z$, Δ saturated
Pentareguloside E (**334**): X = O, $R_1 = H$, $R_2 = Z$, Δ saturated
Pentareguloside F (**335**): X = O, $R_1 = OH$, $R_2 = Z$

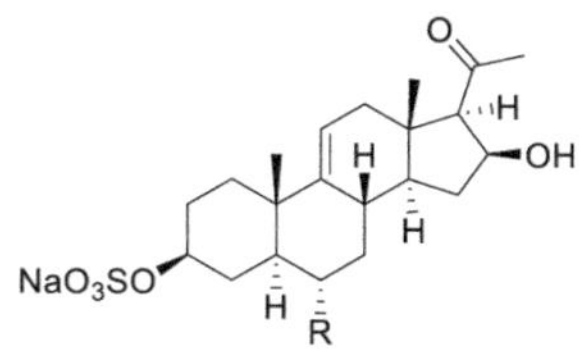

Pentareguloside G (**336**): R = Z

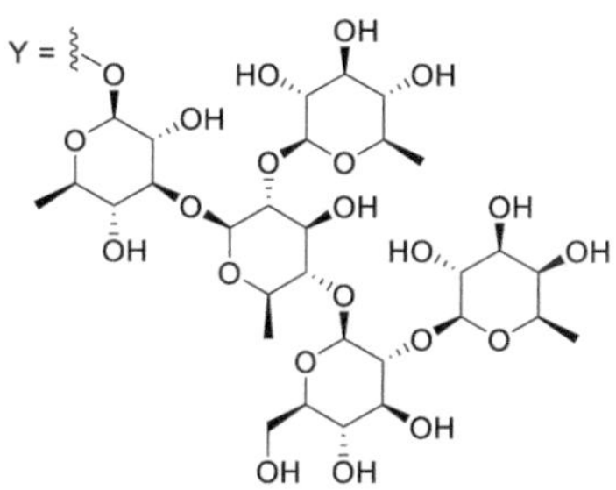

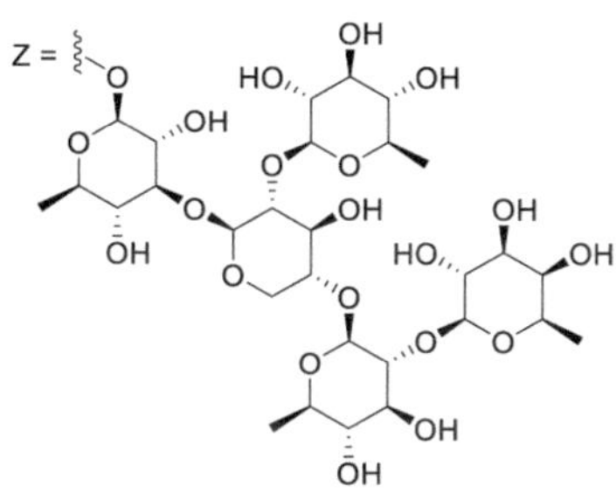

Fig. 47 Chemical structures of new natural products from *Pentaceraster regulus*

Asteropsis carinifera, collected at Van Phong Bay in the South China Sea near the Vietnam coast [150]. When tested for cytotoxicity against a panel of human cancer HCT-116, T-47D and RPMI-7951 cells, only glycoside cariniferoside A displayed moderate cytotoxic activity with IC_{50} values ranging from 32 to 66 μM. A soft agar clonogenic assay using the human breast T-47D and melanoma RPMI-7951 cancer cell lines were conducted on these molecules to evaluate their effect on the colony formation of cancer cells. Cariniferoside F was found to reduce the numbers of colonies of T-47D cells as well as RPMI-7951 cells by 38% and 57%, respectively.

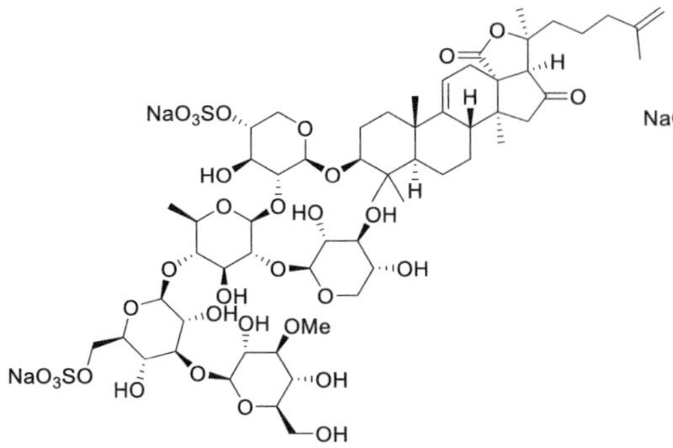

Cercodemasoide A (**337**)

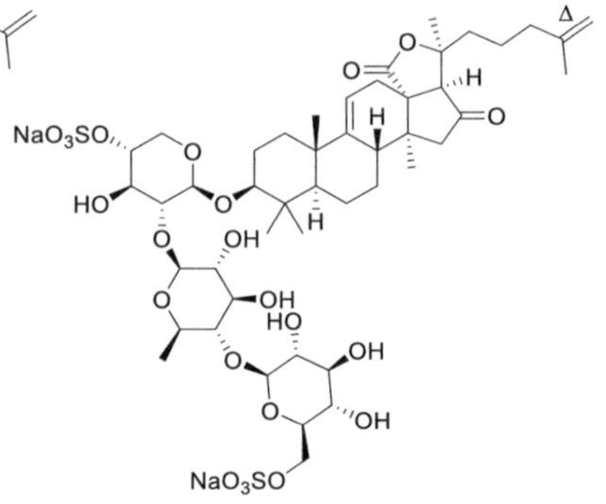

Cercodemasoide B (**338**): Δ saturated
Cercodemasoide C (**339**)

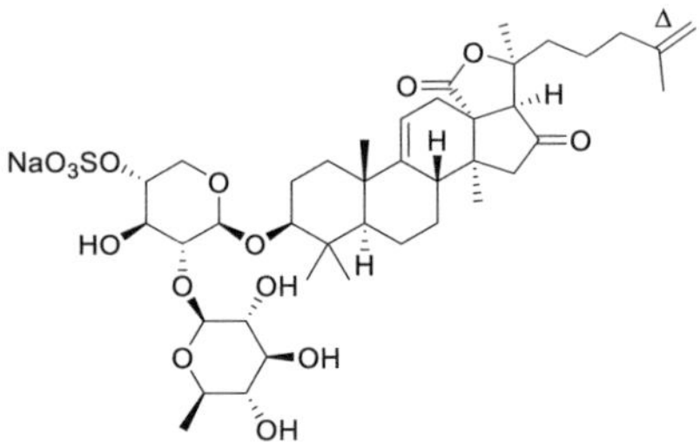

Cercodemasoide D (**340**): Δ saturated
Cercodemasoide E (**341**)

Fig. 48 Chemical structures of new natural products from *Cercodemas anceps*

Astropecten sp.

From the methanolic extract prepared from the edible starfish, *Astropecten monacanthus*, collected from Cat Ba, Haiphong, Vietnam, four new asterosaponins, namely astrosteriosides A (**301**)–D (**304**) (Fig. 43), and two known sulfated steroids, psilasteroside and marthasteroside B, were isolated [152]. The effects of these compounds on pro-inflammatory cytokines were assessed by determining the production of IL-12 p40, IL-6 and TNF-α in LPS-stimulated bone marrow-derived dendritic cells. Of the new compounds, astrosterioside D, with two ketone groups present in the side chain, showed potent inhibitory effects on the production of IL-12 p40, IL-6, and TNF-α with IC_{50} values of 0.60, 3.51 and 1.21 μM, respectively. In addition, astrosterioside A and marthasteroside B exhibited significant inhibitory effects on IL-6 production, with IC_{50} values of 3.17 and 4.37 μM, respectively. The congeners astrosterioside B and C were much less active. These data suggested that the steroidal side chain, such as the presence of α,β-unsaturated carbonyl structure astrosterioside A and two ketone groups in astrosterioside D are important for their inhibitory effects on the production of the pro-inflammatory cytokines. Moreover, astrosterioside D showed potent cytotoxic effects against all three tested human cancer cell lines, including HL-60, PC-3, and SNU-C5 cells, with IC_{50} values ranging from 4.31 to 5.21 μM

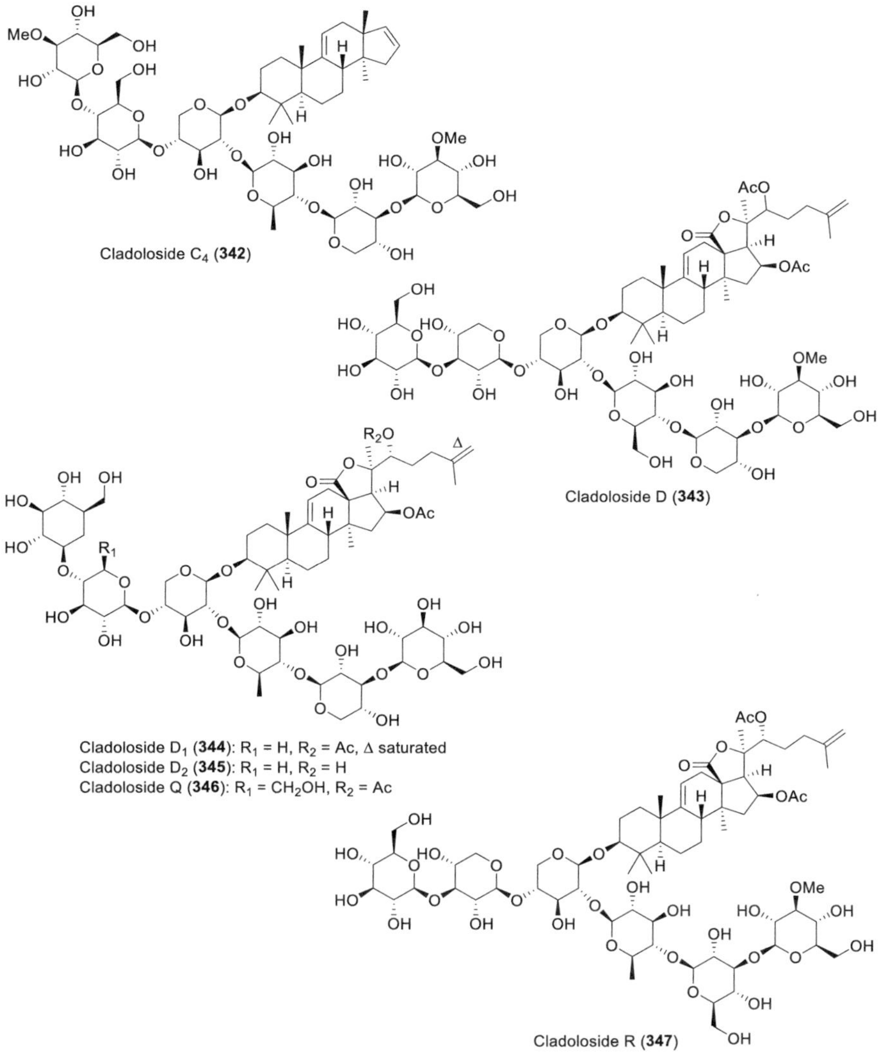

Fig. 49 Chemical structures of new natural products, **342** to **347**, from *Cladolabes schmeltzii*

[191]. Astrosterioside D have showed apoptotic effect that was accompanied by down-regulation of phosphatidyl inositol 3-kinase (PI3K)/AKT signaling and extra-cellular signal-regulated kinase (ERK) 1/2 mitogen-activated protein kinase (MAPK) signaling as well as decrease of c-myc expression.

The total synthesis of astrosterioside A was accomplished in a convergent linear sequence of 24 steps, with 6.8% overall yield from adrenosterone [192]. The key synthetic steps include a stereoselective Horner–Wadsworth–Emmons (HWE) olefination to construct the steroidal 20(22)*E*-ene-23-one side chain, a

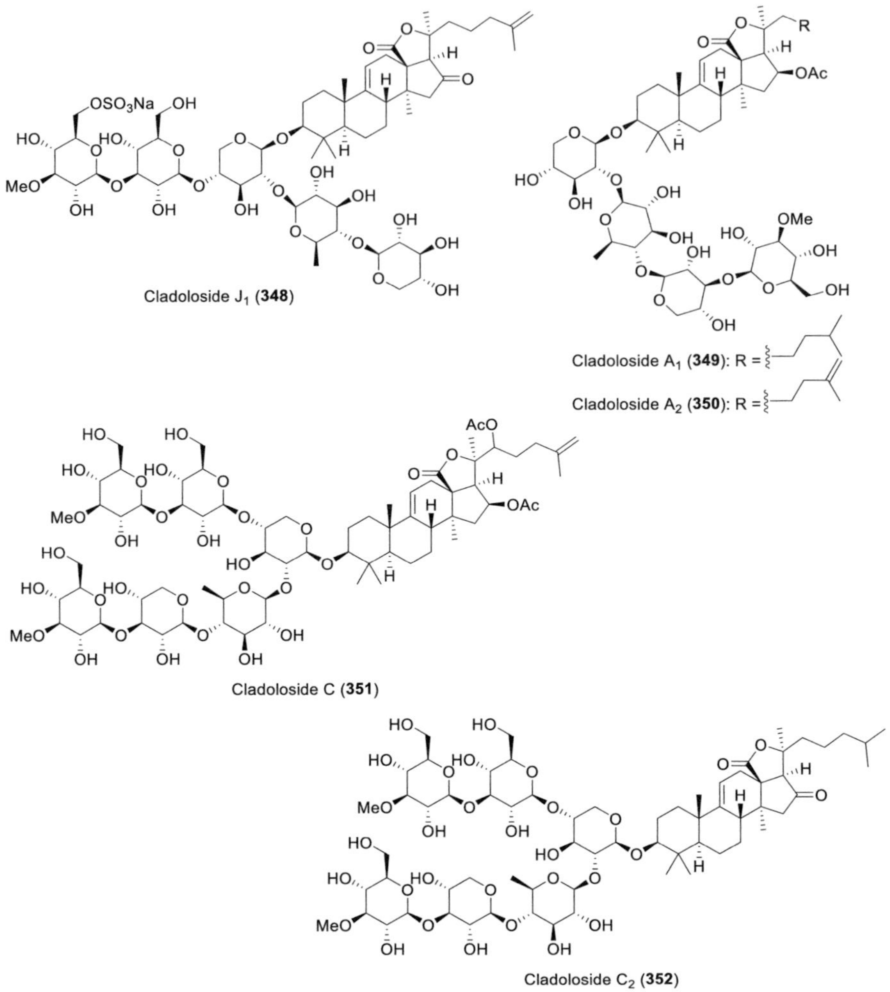

Fig. 50 Chemical structures of new natural products, **348** to **352**, from *Cladolabes schmeltzii*

regioselective glycosylation of the xylopyranoside 4-OH as well as a highly efficient gold(I)-catalyzed coupling of the aglycon with hexasaccharide (cyclopropylethynyl)benzoate donor. Briefly, the starting material, adrenosterone, was converted into ketone **305** in 13 steps (21% overall yield) (Scheme 2). Elaboration of the side chain in **305** to the desired α,β-unsaturated ketone **306** was carried out step wise via HWE reaction. Subsequent removal of the two TBS groups in **306** followed by selective TBS protection resulted in the formation of 6*α*-ol **307**. Glycosylation of **307** with hexasaccharide ortho-cyclopropylethynylbenzoate **308** using catalysis, $Ph_3PAuNTf_2$, gave 6*α*-*O*-β-glycoside **309** (83% yield). Upon removal of the 3-*O*-TBS group in **309**, sulfonation of the resulting C3-OH and global deprotection of the

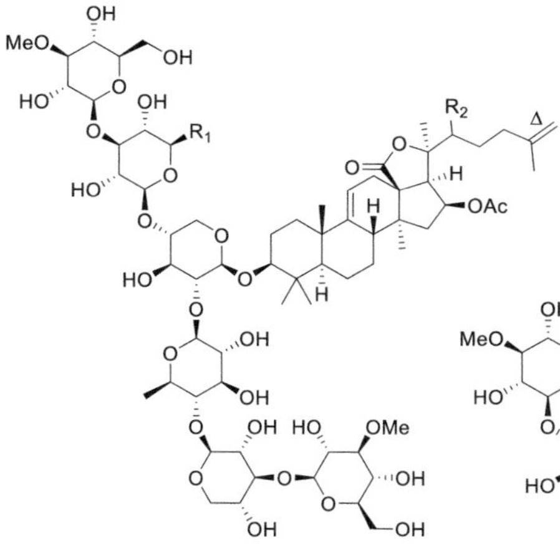

Cladoloside G (**353**): R_1 = H, R_2 = OAc
Cladoloside H_1 (**354**): R_1 = Me, R_2 = OAc, Δ saturated
Cladoloside H_2 (**355**): R_1 = Me, R_2 = OAc

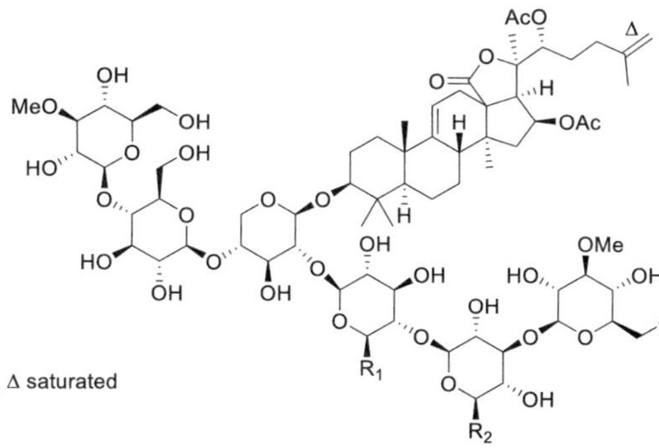

Cladoloside M (**356**): R_1 = Me, R_2 = CH_2OH, Δ saturated
Cladoloside N (**357**): R_1 = R_2 = H

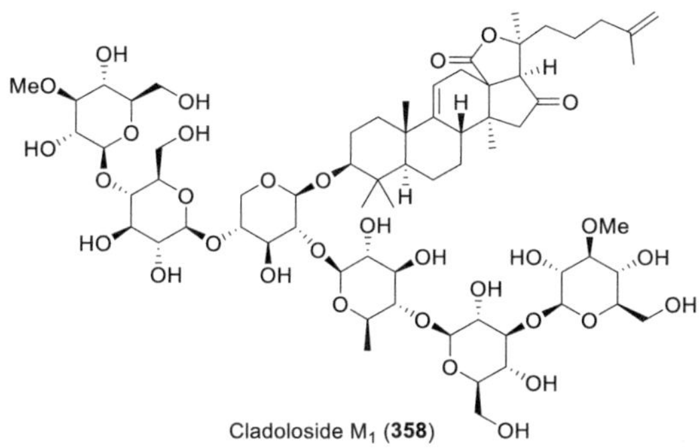

Cladoloside M_1 (**358**)

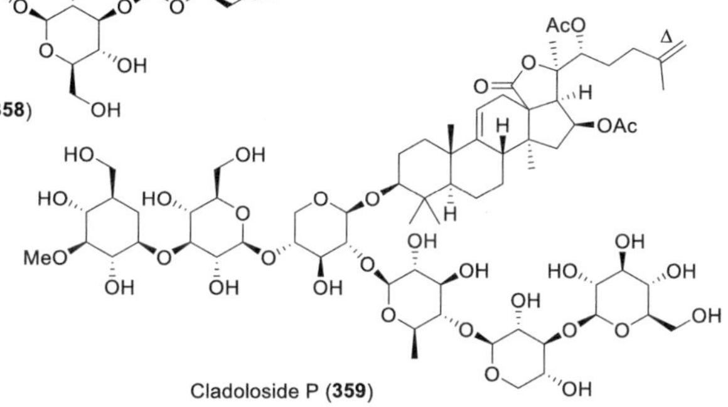

Cladoloside P (**359**)
Cladoloside P_1 (**360**): Δ saturated

Fig. 51 Chemical structures of new natural products, **353** to **360**, from *Cladolabes schmeltzii*

acetyl and benzoyl groups provided astrosteroiside A (**301**) (overall 69% yield). The key synthetic steps involved in the total synthesis of astrosterioside A could facilitate the synthesis of other asterosaponins to allow further studies on their biological and pharmacological activities.

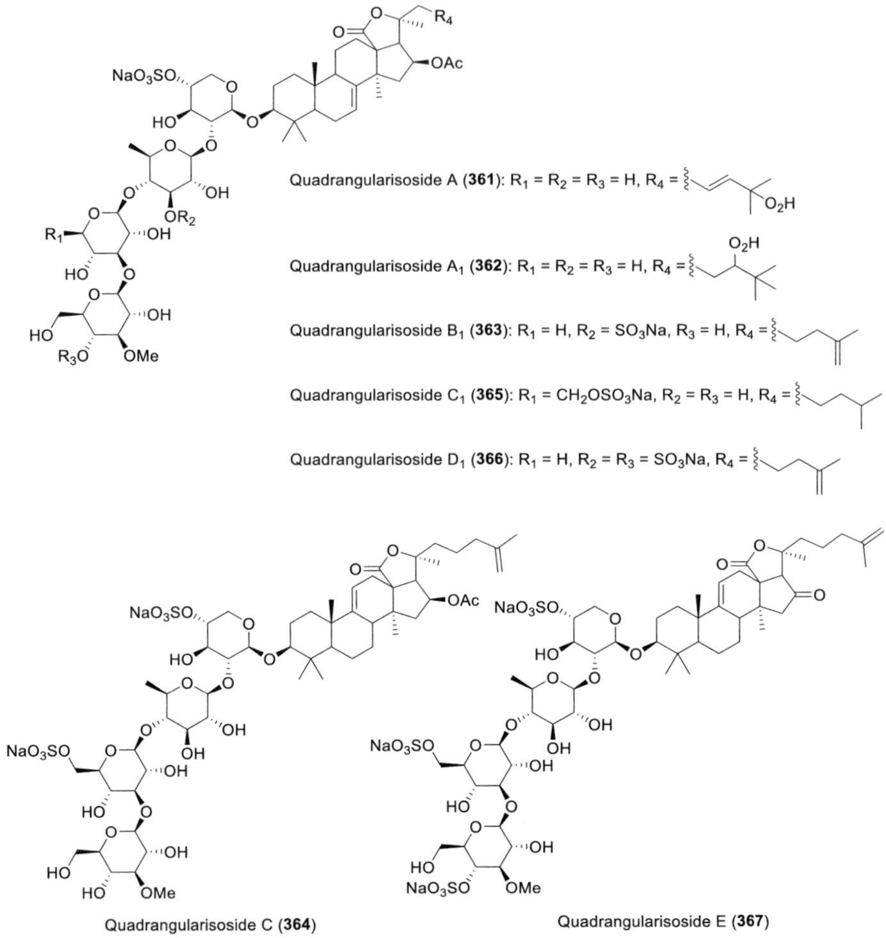

Fig. 52 Chemical structures of new natural products from *Colochirus quadrangularis*

Based on various chromatographic techniques, four new steroids, astropectenols A (**310**)–D (**313**) (Fig. 44) along with known compounds, were isolated from a methanolic extract of the starfish *Astropecten polyacanthus*, collected from Cat Ba, Haiphong, Vietnam [153]. When tested for cytotoxicity activity against HL-60 and PC-3 cells, astropectenols A, C (**312**) and D showed moderate or weak activity while Astropectenol B was inactive. Astropectenol A also exhibited inhibitory activity against *T. brucei* with EC_{50} of 1.57 μM [127].

Echinaster luzonicus

A series of rare structure group of marine glycosides, containing carbohydrate moieties incorporated into a macrocycle, namely luzonicosides B (**314**)–E (**317**)

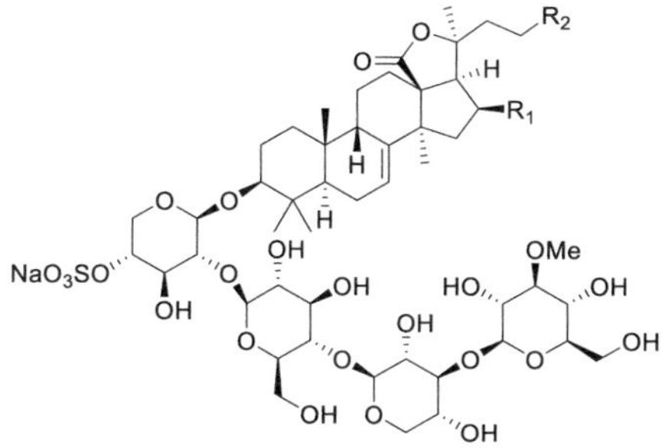

Colochiroside A_1 (**368**): R_1 = OAc, R_2 = 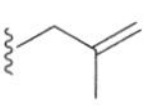

Colochiroside A_3 (**369**): R_1 = H, R_2 = 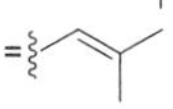

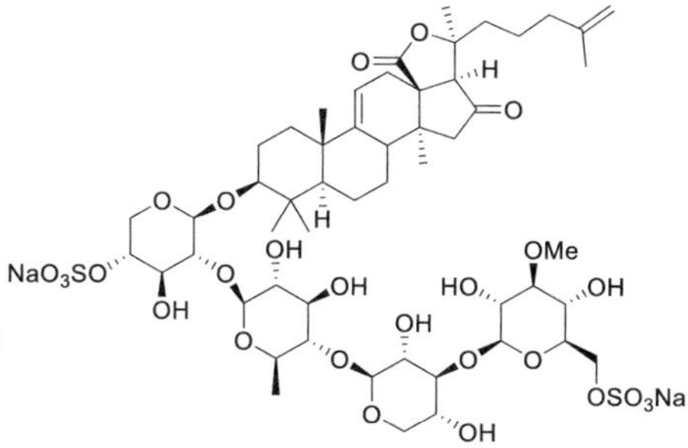

Colochiroside C (**370**)

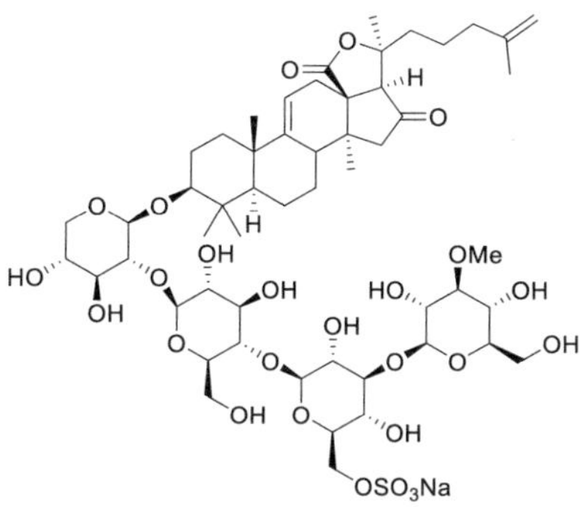

Colochiroside D (**371**)

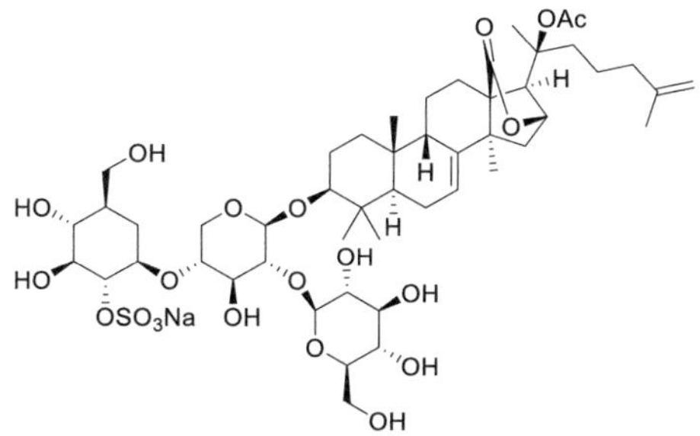

Colochiroside E (**372**)

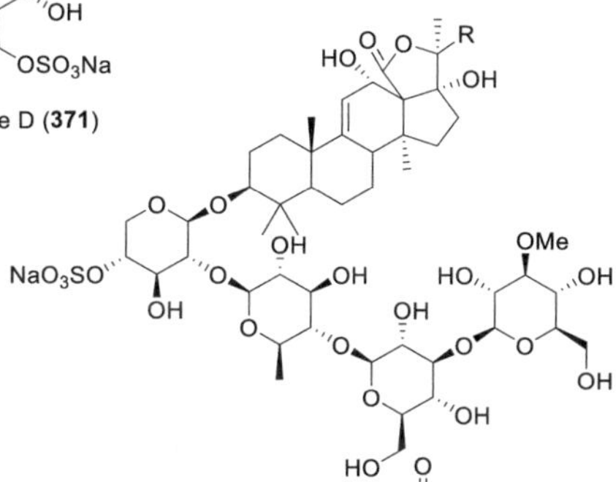

Holothurin A_3 (**373**): R = 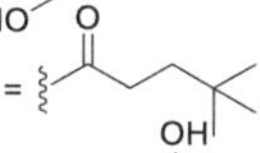

Holothurin A_4 (**374**): R = 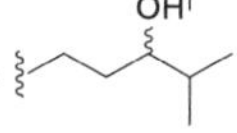

Fig. 53 Chemical structures of new natural products from *Colochirus robustus* and *Holothuria scabra*

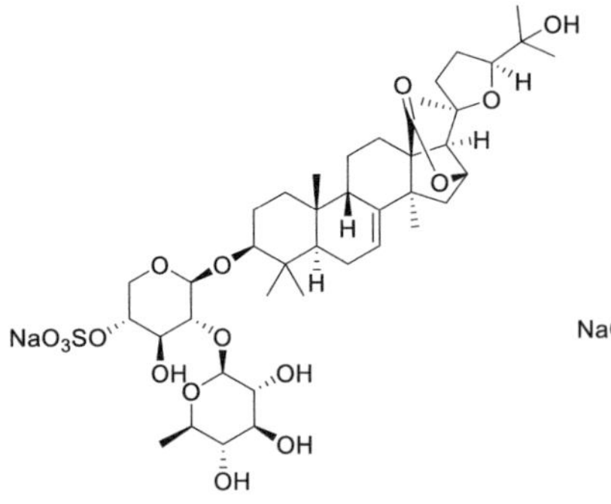

Magnumoside A$_1$ (**375**)

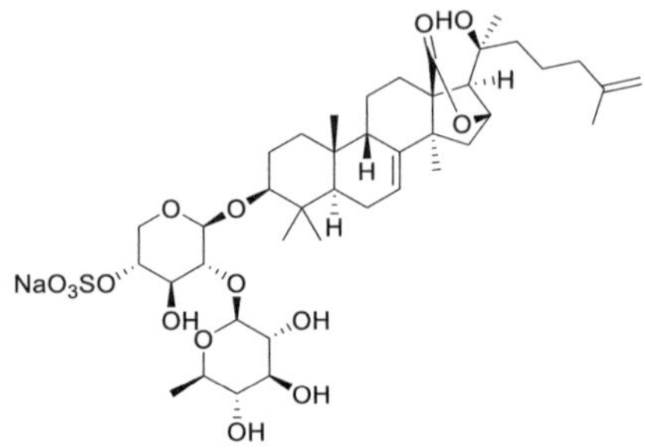

Magnumosides A$_3$ (**376**)

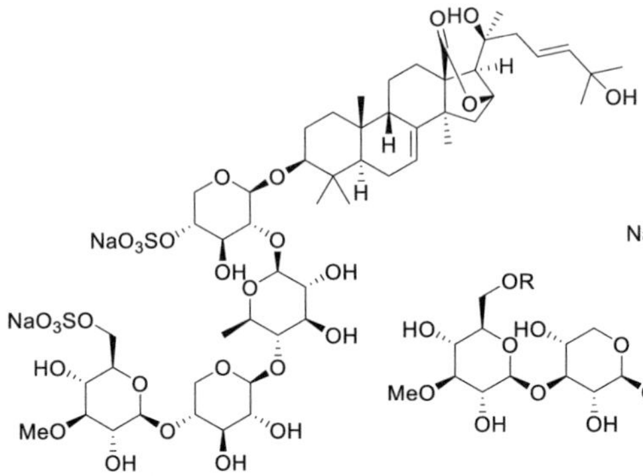

Magnumosides C$_1$ (**378**)

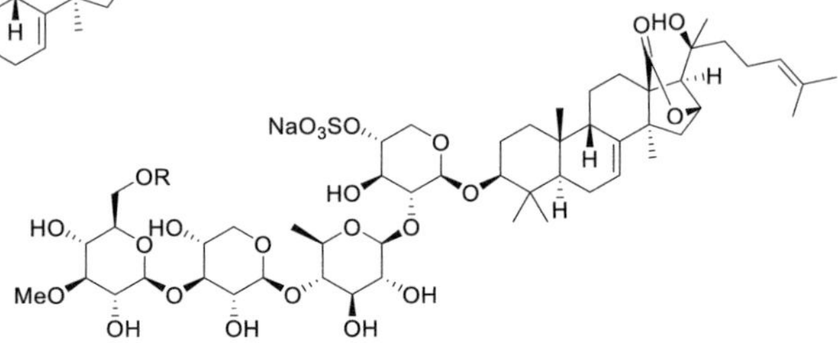

Magnumosides B$_4$ (**377**): R = H
Magnumosides C$_4$ (**379**): R = SO_3Na

Fig. 54 Chemical structures of new natural products from *Neothyonidium* (=*Massinium*) *magnum*

(Fig. 45), were isolated along with a related open carbohydrate chain steroid glycoside, luzonicoside F (**318**) (Fig. 45), from a sample of *Echinaster luzonicus*, collected from Van Phong Bay, Vietnam [159]. These new molecules were isolated together with the known compound, luzonicoside A, previously reported from the Pacific starfish, *E. luzonicus* [193]. Unlike luzonicoside D, luzonicoside A, at concentrations of 0.01–0.1 μM, exhibited potent lysosomal activity stimulation, intracellular ROS level elevation as well as NO synthesis up-regulation in RAW 264.7 murine macrophages. Moreover, luzonicoside A inhibited proliferation, the formation of colonies and the migration of SK-Mel-28 cells significantly more than luzonicoside D (**316**) [194]. Their molecular mode of action appears to be related to the regulation of the activity of cleaved caspase-3 and poly(ADP-ribose) polymerase (PARP), along with Survivin, Bcl-2, p21 and cyclin D1 level.

Linckia laevigata

A chemical investigation by the Russian group on *Linckia laevigata*, collected near the Vietnam coastline at Van Phong Bay, led to a series of related glycosides,

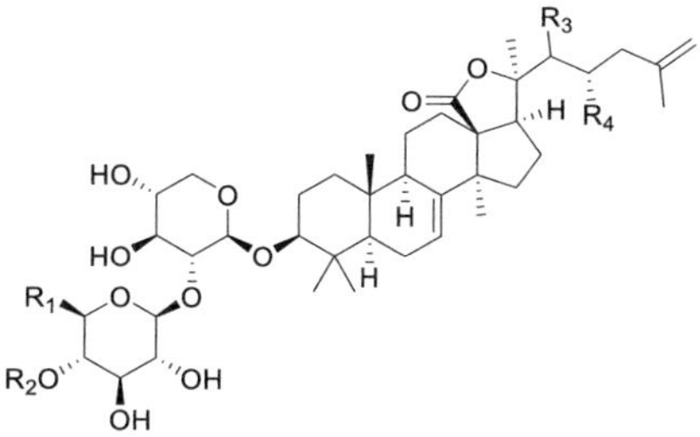

Stichorrenoside A (**382**): $R_1 = CH_2OH$, $R_2 = H$, $R_3 = OH$, $R_4 = H$
Stichorrenoside B (**383**): $R_1 = CH_2OH$, $R_2 = SO_3Na$, $R_3 = OH$, $R_4 = H$
Stichorrenoside C (**384**): $R_1 = CH_2OH$, $R_2 = R_3 = H$, $R_4 = OAc$
Stichorrenoside D (**385**): $R_1 = R_2 = R_3 = H$, $R_4 = OAc$

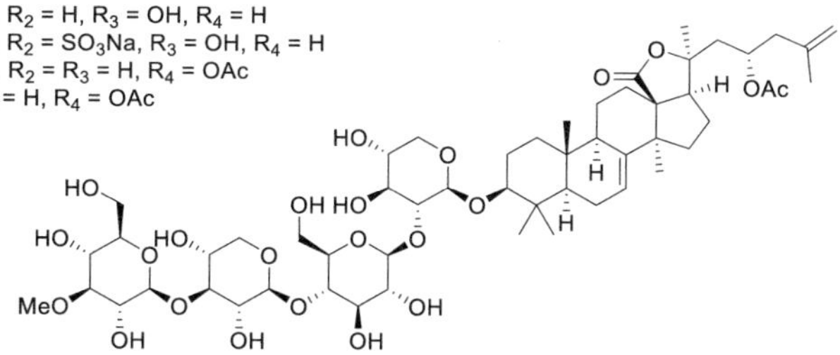

Stichorrenoside E (**386**)

Fig. 55 Chemical structures of new natural products from *Stichopus horrens*

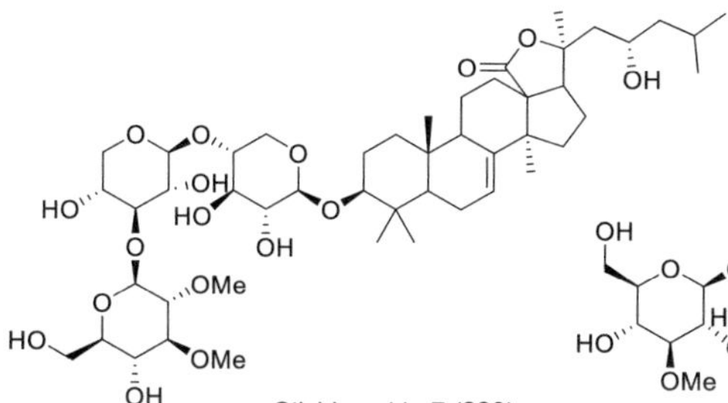

Stichloroside F (**380**)

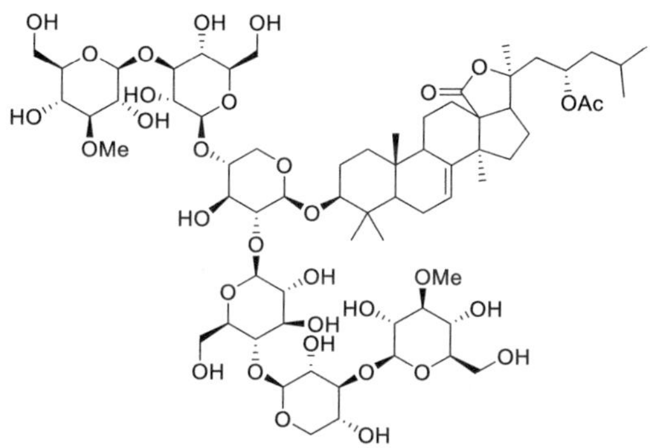

Stichoposide D (**381**)

Fig. 56 Chemical structures of new natural products from *Stichopus chloronotus*

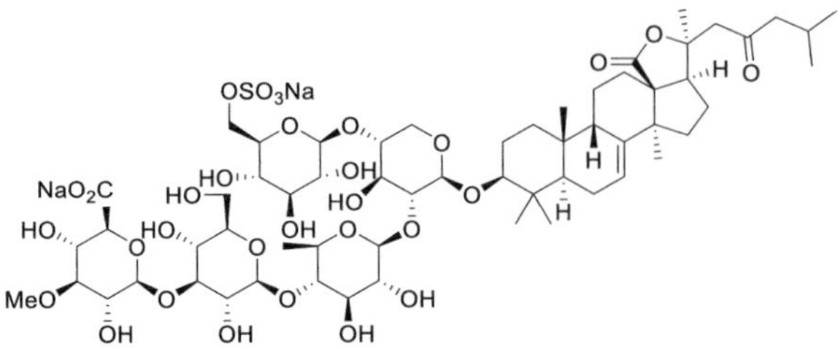

Synaptoside A (**387**)

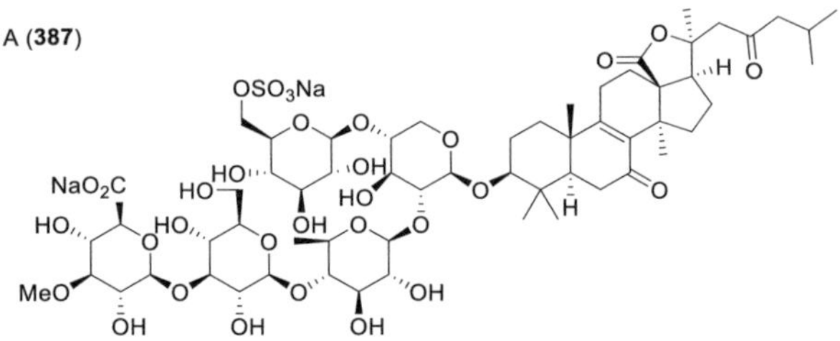

Synaptoside A$_1$ (**388**)

Fig. 57 Chemical structures of new natural products from *Synapta maculata*

including seven new molecules, linckosides L1 (**319**)–L7 (**325**) (Fig. 46) [160–162]. The biological activity of monosides linckosides L1 and L2 (**320**), containing a saturated polyhydroxylated steroid aglycon and 2-*O*-methyl-β-xylopyranose at C-3, were evaluated using mouse neuroblastoma (NB) C-1300 cells. Lifetime observations and analyses of silver impregnation preparations revealed linckosides L1 and L2 to induce neuronal differentiation in the same manner as neurotrophins and to enhance significantly the neuritogenic activity of NGF. In addition, linckoside L3 (**321**) inhibited fertilization and egg cell development in the sea urchin, *Strongylocentrotus intermedius*. Linckoside L6 (**324**), having an additional monosaccharide residue in the side chain, and linckoside L7 (**325**) showed to be more active as cytotoxins with EC_{100} of 50 μg/mL. A recent study on the combined anticancer activity of sulfated laminaran (AaLs) from the brown alga, *Alaria angusta*, and polyhydroxysteroid glycosides, including linckoside L1, on 3D colorectal carcinoma HCT116 cell line was explored [195]. It was revealed that AaLs when combined with linckoside L1 exerted synergistic anticancer effect through the inactivation of protein kinase B (AKT) kinase and, subsequently led to the induction of apoptosis via the regulation of proapoptotic/antiapoptotic proteins balance.

Pentaceraster sp.

The starfish, *Pentaceraster regulus*, also known as the "spotted starfish", is found in tropical waters of the Indian and Pacific Oceans from the northern shores of Australia to the southern coast of Japan. From the Vietnamese starfish, *P. regulus*, collected from either Cham Islands or Con Co Islands, at least 14 new bioactive molecules have been described. An initial series of new polar polyhydroxysteroid biglycosides, regulusosides A (**326**)–C (**328**) (Fig. 47), were isolated from a collection of the

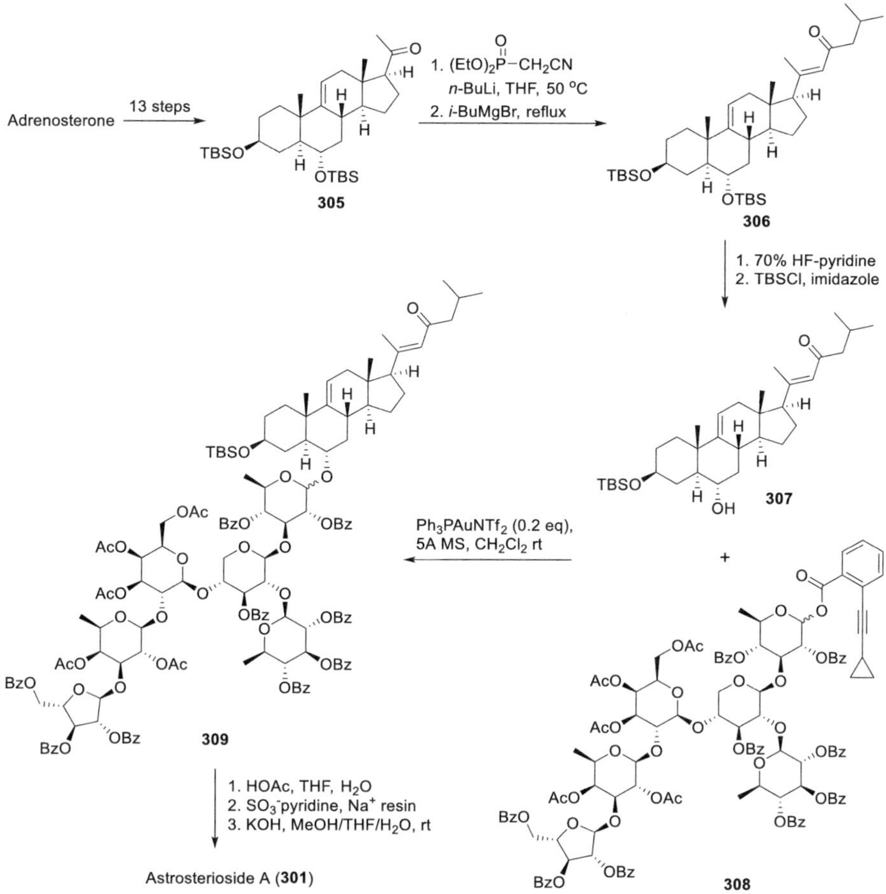

Scheme 2 Summary on the total synthesis of astrosteroiside A

starfish, *Pentaceraster regulus*, obtained from Cham Islands, Vietnam in May 2010 [164]. Subsequent collection of the same species made near Con Co Islands, Quang Tri, Vietnam in May 2026, yielded additional analog, regulusoside D (**329**) (Fig. 47), as well as a new polyhydroxylated steroid, (24*S*)-cholestane-3*β*,5,6*β*,8,15*α*,24-hexol [165]. The regulusosides are structurally related to the previously known nodososide from the starfish, *Protoreaster nodosus* and they differ in the structures of the carbohydrate chains. The cytotoxicity and immunomodulatory properties of regulusosides A, B and D were assessed using murine macrophages of the RAW264.7 cell line. At a non-toxic concentration of 10 μM, regulusosides A and B increased ROS formation in the RAW264.7 cells by 35 and 28%, respectively. In addition, regulusoside D exhibited strong inhibitory effects on LPS-induced NO production in RAW264.7 cells with an IC_{50} value of 9.13 μM.

Seven new asterosaponins, pentaregulosides A (**330**)–G (**336**) (Fig. 47), including two furostane-type steroid oligoglycosides (i.e., pentaregulosides B and C), were purified and characterized from the ethanolic extract of *P. regulus* collected off the coast of Vietnam at Cham Islands [167]. It is known that steroid glycosides having a furostane carbon skeleton are common in terrestrial plants but rare in starfish. Of the new asterosaponins, only pentareguloside A showed cytotoxic activities in RAW 264.7 murine macrophage cells, with IC_{50} value of 6.4 μM. Furthermore, pentaregulosides C (**332**)–E (**334**), when tested at 5 μM, exhibited immunomodulatory action by reducing the ROS content in the RAW 264.7 cells by 40%, 28% and 55%, respectively, upon co-stimulation of the cells with the proinflammatory endotoxin lipopolysaccharide (LPS) from *E. coli*.

3.3.3 Class Holothuroidea

Sea cucumbers are marine invertebrates belonging to the class Holothuroidea of the phylum Echinodermata. During evolution, their skeleton has been lost and reduced to separate ossicles. The toxic triterpenes glycosides belonging to either the common lanostane series or rare products of intramolecular rearrangements are the main defensive chemicals protecting these benthic animals from predation [130]. Most of the triterpene glycosides consist of an 18(20)-lactone as well as carbohydrate chains, mostly containing two to six sugar units, some of which are sulfated. These sugar units are typically belonging to the D-series, including glucose, xylose, quinovose, 3-*O*-methylglucose, 3-*O*-methylxylose, 3-*O*-methylglucuronic acid, 3-*O*-methylquinovose and 6-*O*-acetylglucose, and are in the β-pyranose form [196]. Lanostane derivatives without the lactone ring or containing 18(16)-lactone fragment instead of 18(20)-lactone belong to the non-holostane glycosides [197]. Due to the presence or absence of double bonds, hydroxy groups, acetates, keto- and other functionalities in their aglycones and multitude variants of sulfated or non-sulfated carbohydrate chains, the chemical structures of sea cucumber glycosides are highly diverse. Moreover, these glycosides possess interesting biological activities, such as cytotoxic, membranolytic, ichthyotoxic, antifungal, hemolytic, immunomodulatory and cancer preventive properties [129]. More than a dozen Vietnamese sea cucumber species have been investigated for their natural products and selected molecules as well as their biological activities are being highlighted in the following sections (Table 11; Figs. 48, 49, 50, 51, 52, 53, 54, 55, 56 and 57).

Recently, a comprehensive MS database of sea cucumber triterpene glycosides have been created by research group at the G. B. Elyakov Pacific Institute of Bioorganic Chemistry, Far Eastern Branch of Russian Academy of Sciences [198]. This useful resource will be immensely valuable to researchers in natural products, offering a powerful tool for dereplication and significantly streamlining the isolation of novel bioactive compounds. Additionally, the spectral library will greatly accelerate metabolite annotation in MS-based studies on echinoderms, elucidation of triterpene glycoside biosynthesis and biological functions and chemotaxonomic investigations of newly described or previously unstudied species.

Cercodemas anceps

Five new holostane-type triterpene saponins, cercodemasoides A (**337**)–E (**341**) (Fig. 48), were isolated from the sea cucumber *Cercodemas anceps*, collected from Co To, Quangninh, Vietnam [168]. These molecules showed potent cytotoxicity against a panel of five cancer cell lines, including Hep-G2, KB, LNCaP, MCF7 and SK-Mel2, with IC_{50} values ranging from 0.04 to 7.36 μM. In particular, cercodemasoide A exhibited potent cytotoxicity against KB and Hep-G2 cancer cell lines, having IC_{50} values of 40 and 70 nM, respectively. It is observed that the decrease of sugar units and/or dehydrogenation at C-25/C-26 slightly reduced the cytotoxic property of these molecules. A recent study revealed that cercodemasoide A and its nanoliposomal forms have promising cytotoxic activity against NTERA-2 cancer stem cells (CSC), with IC_{50}s of 1.03 and 0.41 μM, respectively [199]. Moreover, the anti-CSC effects of cercodemasoide A nanoliposomes on NTERA-2 cells were due to their apoptotic induction through enhancing caspase-3 activity and arresting of cell cycle at the S phase.

Cladolabes schmeltzii

It has been reported that the Vietnamese sea cucumber species, *Cladolabes schmeltzii*, contains multicomponent mixture of glycosides [169–174]. At least 40 new cladolosides with carbohydrate units attributed to 19 groups, consisting of nine different aglycones, were isolated from this sea cucumber species [174]. It has been suggested that the structural diversity of the glycosides observed in these cladolosides arises from the active evolution and selection of molecules most effective for biological/ecological function(s) and thus, quick adaptation of the producing benthic organism to changing environmental conditions. Several cladolosides display unique structural features. For instance, cladoloside C_4 (**342**) (Fig. 49) has an uncommon 20,21,22,23,24,25,26,27-octanorlanostane aglycone while cladolosides D_1 (**344**), D_2 (**345**) and Q (**346**) (Fig. 49) were new representatives of the hexaosides having a non-methylated terminal sugar unit in the "upper" half-chain [172]. An unusual pentasaccharide carbohydrate chain is present in cladoloside J_1 (**348**) (Fig. 50) by having a disaccharide moiety of glucose and a sulfated 3-*O*-methylglucose linked to the first xylose residue in a linear trisaccharide fragment [173]. In addition, cladoloside D (**343**) (Fig. 49) contains a new variant of carbohydrate chain with xylose and glucose units as fifth and sixth monosaccharides and cladoloside R (**347**) (Fig. 49) has a novel hexasaccharide carbohydrate chain with different positions of non-methylated terminal sugar units [170, 174].

The aglycones of *Cladodiscus schmeltzii* glycosides, primarily of the holostane type with an 18(20)-lactone ring, show clear biogenetic relationships and derive from a common precursor through two main biosynthetic routes [196]. These pathways involve different timings of oxidation at C-18 and C-22, as well as functional modifications at C-16 and in the side chain. A distinct pathway, seen in cladoloside C4, includes early oxidation at C-22, side chain cleavage, and C17/C20-bond disruption. The frequent presence of hydroxy or acetoxy groups at C-22 across compounds

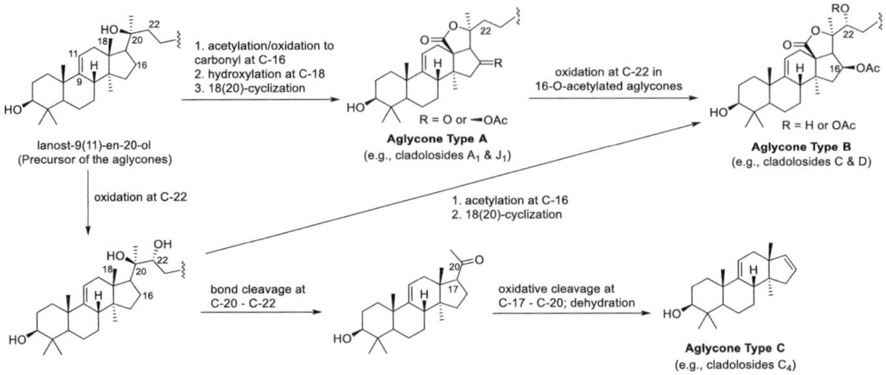

Scheme 3 Proposed biosynthetic pathways of aglycones in *C. schmeltzii*. Modified from Kalinin et al. [196]

supports this. Together, these variations highlight a mosaic biosynthetic process underlying the structural diversity of sea cucumber triterpene glycosides (Scheme 3).

The cytotoxic activities of the cladolosides have been evaluated against various cell types, including HT-29, spleenocytes and the ascite form of mouse Ehrlich carcinoma cells, as well as hemolytic activity against mouse erythrocytes. Several molecules demonstrated strong to potent cytotoxic and hemolytic properties. In the first series of the cladolosides, cladolosides A_1 (**349**), A_2 (**350**), C (**351**) and C_2 (**352**) (Fig. 50) were found to be the most cytotoxic molecules, with IC_{50}/EC_{50} values ranging from 2.8 and 1.8 μg/mL, while the MIC/MEC_{100} values of their hemolytic activity averaging at about 3.11 μg/mL [169, 170]. Cladoloside C_2 was further investigated on its molecular mechanism of induced apoptosis of human leukemia cells and its antitumor effects in a HL-60 and K562 leukemia xenograft model [200]. Results revealed that cladoloside C_2 has in vitro and in vivo anti-leukemic effects via the activation of Fas/CerS6/p38 kinase/JNK/caspase-8 in lipid rafts, supporting its potential therapeutic use in the treatment of human leukemia.

Subsequent investigation of the Vietnamese sea cucumber *C. schmeltzii* led to the isolation of eight new nonsulfated triterpene glycosides, designated as cladolosides C_3, E_1, E_2, F_1, F_2, G (**353**), H_1 (**354**) and H_2 (**355**) [171]. Among these, cladolosides G, H_1 and H_2 exhibited the strongest cytotoxicity against mouse splenocytes, with IC_{50} values ranging from 0.3 to 0.6 μg/mL, and also showed potent hemolytic activity against mouse erythrocytes, with MIC values between 0.2 and 0.4 μg/mL. Interestingly, these three glycosides, which possess six monosaccharide residues and aglycones bearing 16- and 22-O-acetyl groups, are the predominant components in the glycoside mixture of *C. schmeltzii*. These hexaosides likely represent fully biosynthesized carbohydrate chains, wherein further oxidation of the aglycones is blocked by acetylation at C-16 and C-22 hydroxyl groups. It is suggested that the predominance of these structures in the glycoside profile of *C. schmeltzii* is a result of biochemical evolution, favoring the selection of more active defensive natural products.

In a later series of cladolosides, cladolosides M (**356**), M_1 (**358**), P (**359**) and P_1 (**360**) (Fig. 51) were the most cytotoxic and hemolytic against ascite form of Ehrlich carcinoma cells and mouse erythrocytes, respectively, with IC_{50}/EC_{50} values ranging from 0.34 to 0.96 μM [172, 174]. The action of several cladolosides were also investigated for their effects on colony formation of HT-cells by soft agar assay. Cladolosides M_1 and P_1 exhibited strong inhibition of colony formation and growth of HT-29 cells, whereas they were active against the cancer cells at IC_{50} values of 16.9 and 4.4 μM, respectively. Synergism was observed in the decrease on the number of colonies of HT-29 cells using both radioactive irradiation and non-toxic dose of cladolosides N (**357**), P_2 and P_3, where it increased the inhibitory effect from irradiation by more than 70% [172, 174].

Colochirus sp.

A total of thirteen new mono-, di-, and tri-sulfated triterpene glycosides, quadrangularisosides A (**361**), A_1 (**362**), B, B_1 (**363**), B_2, C (**364**), C_1 (**365**), D, D_1 (**366**)–D_4 and E (**367**) (Fig. 52) have been reported from the sea cucumber, *Colochirus quadrangularis*, collected coral reefs near the seashore of Vietnam in the South China Sea [175]. Several unique structural features are observed in these series of sulfated triterpene glycosides. For instance, the novel carbohydrate moieties of quadrangularisosides D–D_4, belonging to the group D [sulfate groups at C(4)Xyl1, C(3)Qui2, and C(4)MeGlc4], and quadrangularisoside E [sulfate groups at C(4)Xyl1, C(6)Glc3, and C(4)MeGlc4] contain three sulfate groups, including one at an unusual position at C(4) of terminal 3-O-methylglucose residue. In addition, quadrangularisosides A (**361**) and D_3 as well as quadrangularisosides A_1 (**362**) and D_4 are characterized by the new aglycones with 25-hydroperoxyl or 24-hydroperoxyl groups in their side chains, respectively. All molecules displayed strong cytotoxicity, but the presence of hydroperoxide groups decreased the activities. Quadrangularisosides A_1 (**362**), C (**364**), C_1 (**365**) and E (**367**) possessed strong inhibitory activity on colony formation in HT-29 cells. The synergic effects of these glycosides (at 0.02 μM) and radioactive irradiation, led to a decreasing number of colonies detected. In particular, the synergistic effects of quadrangularisosides A, B_1 (**363**) and D_1 (**366**) were the highest by enhancing the effect of radiation by about 30%.

A series of new sulfated triterpene glycosides, known as the colochirosides, were isolated from a collection of *Colochirus robustus* obtained from Nha Trang Bay, Vietnam [176–178]. In particular, colochiroside E (**372**) (Fig. 53) belongs to a rare group of glycosylated 9β-H-lanosta-18(16)-lactones and has an unprecedented sulfated trisaccharide carbohydrate chain consisting of two glucose and one xylose residues. Of these series of colochirosides, colochirosides A_1 (**368**), A_3 (**369**), C (**370**) and D (**371**) (Fig. 53) possessed strong cytotoxicity, with ED_{50} values of below 5 μM against the ascite form of Ehrlich mouse carcinoma cells as well as significant hemolytic property against mouse erythrocytes with ED_{50} values of below 2.5 μM.

Neothyonidium (= *Massinium*) *magnum*

A sample of the sea cucumber, *Neothyonidium* (=*Massinium*) *magnum*, collected from Nha Trang bay near Tam Island, afforded nine new sulfated triterpene glycosides, magnumosides A_1, A_2, A_3, A_4, B_1, B_2, C_1, C_2 and C_4 [182]. These new glycosides contain the non-holostane type lanostane aglycones by having the 18(16)-lactone and 7(8)-double bond and they differed from each other by the side chains and carbohydrate units. Magnumoside A_1 (**375**) (Fig. 54), consisting of disaccharide monosulfated carbohydrate moieties, has an unprecedented 20(24)-epoxy-group in the aglycone side chain. When tested for cytotoxicity, the erythrocytes were found to be more sensitive to the glycosides as compared to the spleenocytes and ascites form of mouse Ehrlich carcinoma cells. Of the series of molecules, magnumosides C_1 (**378**) and C_4 (**379**) (Fig. 54) were the most cytotoxic, with IC_{50} values ranging from 6.52 to 35.0 μM. In addition, magnumosides A_3 (**376**) and C_1 (**378**) (Fig. 54) significantly inhibited the colony formation and decreased the size of DLD-1 cancer cells colonies at non-cytotoxic concentrations. Furthermore, Magnumoside C_4 (**379**) (Fig. 54) enhanced the antiproliferative property of radioactive irradiation by increasing the inhibitory effect of radiation on proliferation of DLD-1 cancer cells by 45%. Further chemical investigation of the same sea cucumber sample provided additional three new mono- and disulfated triterpene tetraosides, magnumosides B_3, B_4 and C_3 [183]. Of these three new molecules, magnumosides B_4 (**377**) significantly suppressed the colony formation and decreased the size of colonies of DLD-1 cancer cells. Magnumoside C_4 was able to block the purinergic receptors in vitro [201]. Specifically, the molecule inhibited ATP-dependent Ca^{2+} intake into the Neuro 2A cells, containing P2X7 receptors on the surface, by 47%,

Stichopus sp.

The genus *Stichopus* has at least 13 described species, including *S. chloronotus*, *S. ellipes*, *S. hermanni*, *S. horrens*, *S. ludwigi*, *S. monotuberculatus*, *S. naso*, *S. noctivagus*, *S. ocellatus*, *S. pseudohorrens*, *S. quadrifasciatus*, *S. rubermaculosus* and *S. vastus*, and is an important fishery commodity [202]. In certain parts of Asia, such as Indonesia and Malaysia, *S. hermanni* has long been used for preparation of traditional medicinal product as gamat water and oil. Of the six *Stichopus* species found in Vietnamese waters, new molecules have been reported from about four species, including *S. variegatus*, *S. horrens*, *S. chloronotus* and *S. herrmanni* (Table 8). For instance, new triterpene diglycosides/tetraglycoside, stichorrenosides A (**382**)–E (**386**) (Fig. 55), have been reported from a methanolic extract of *S. horrens* collected at Hai Van–Son Cha, Hue, Vietnam [187, 188]. This species has been found abundantly in coral reefs from Con Co (Quangtri) to Phu Quoc (Kiengiang) at depths of 0–20 m. Of the five new molecules, stichorrenoside D (**385**) exhibited strong cytotoxicity against five tested cancer cell lines, including Hep-G2, KB, LNCaP, MCF7 and SK-Mel2, with IC_{50} values ranging from 1.92 to 3.13 μM.

Chemical investigations of various *Stichopus* sea cucumber resulted in the isolation of previously known molecules in sufficient quantities for further biological evaluation. For instance, the Vietnamese *Stichopus chloronotus*, obtained from Cat

Ba, Haiphong, afforded one new compound, stichloroside F (**380**) (Fig. 56), along with five known triterpene saponins, including stichoposide D (**381**) (Fig. 56) [184]. Stichoposide D was previously isolated from *S. chloronotus* collected at the Great Barrier Reef, Australia by Russian scientists [203]. The isolated of cytotoxic stichoposide D from *S. chloronotus* was subsequently investigated by the Vietnamese scientists for its potential anti-cancer stem cells (CSCs) activities on pluripotent human embryonic carcinomas NTERA-2 cells. The molecule exhibited potent anti-cancer properties on NTERA-2 cancer stem cells with IC_{50} value of 0.26 μM. The mode of action of stichoposide D is suggested to be via apoptotic induction. In addition, treatment of stichoposide D at 1.0 μM induced 76.4% of apoptotic cells and 1.72 relatively fold change of caspase-3 activation. It also exhibited positive effects on reducing the number of highly expressed $CD44^+/CD24^+$ cell surface markers as well as arresting NTERA-2 cells at sub-G1 phase [204].

Synapta maculata

Two new triterpene holostane glycosides, synaptosides A (**387**) and A_1 (**388**) (Fig. 57), were reported from the extract of the sea cucumber, *Synapta maculata*, collected at Van Phong Bay, Vietnam [189]. These glycosides are the first representatives of triterpene glycosides from sea cucumbers consisting of methylated glucuronic acid as one of monosaccharide residues in carbohydrate chains. In addition, synaptoside A contains an unusual polycyclic nucleus having a 7-oxo-8(9)-en system that is unique in sea cucumber triterpene glycosides. Synaptoside A was found to be active against human cervical cancer HeLa cells with an IC_{50} of 8.6 μg/mL.

4 Conclusion

The marine ecosystems of Vietnam, particularly those rich in invertebrate life, showcase impressive biodiversity. Notable among the marine invertebrates are sponges, soft corals, and echinoderms, each group significantly contributing to the ecological diversity across various coastal regions. Sponges are particularly diverse, with 291 recorded species, mainly of the Demospongiae class, while soft corals include 220 species from 48 genera, predominantly within the Southeast Coastal Region. Echinoderms, represented by 345 species across five classes, also demonstrate significant biodiversity. The varying environmental and ecological factors across these regions create unique biogeographic patterns, with the North Central Coastal Region exhibiting a distinct species composition compared to others. This extensive biodiversity is vital not only for the health of marine ecosystems but also for sustaining ongoing research and conservation efforts, which can benefit from the ecological roles and resource potential these organisms provide.

Ongoing international collaborative efforts related to marine bioprospecting have highlighted a wealth of bioactive specialized compounds derived from these marine invertebrates, showing promising biological activities with potential pharmaceutical applications. Vietnamese sponges, for instance, have yielded potent compounds, such as terpenoids and alkaloids, which exhibit cytotoxicity against cancer cell lines, as well as anti-inflammatory and antibacterial properties. Soft corals and echinoderms have also provided a range of biologically active molecules, including specific fatty acid derivatives, polyketides and saponin derivatives with notable antimicrobial and anti-tumor effects. For instance, several potent cytotoxic triterpene saponins from sea cucumbers, such as cercodemasoides and cladolosides, have been uncovered from Vietnamese sea cucumbers. Additionally, selected molecules, such as cladoloside N, cladolosides P_2, P_3 and magnumoside B_4, exhibited significant synergism with radioactive irradiation by decreasing the number of colonies of cancer cells. The diversity of these natural products points to the therapeutic potential within Vietnam's marine biodiversity, with compounds from sponges and echinoderms demonstrating significant cytotoxicity and pro-apoptotic effects on various human cancer cells. Continued bioprospecting of these marine organisms, coupled with conservation, holds promise for discovering novel drug leads and understanding the ecological value of Vietnam's unique marine invertebrates.

References

1. Latypov YY (2005) Reef-building corals of Vietnam as a part of the Indo-Pacific reef ecosystem. Russ J Mar Biol 31:S34–S40
2. Faulkner D (1986) Marine natural products. Nat Prod Rep 3:1–33
3. Blunt JW, Copp BR, Munro MH, Northcote PT, Prinsep MR (2004) Marine natural products. Nat Prod Rep 21:1–49
4. Avilov SA, Kalinovskii AI, Stonik VA (1990) New triterpene glycoside from the holothurian *Neothyonidium magnum*. Chem Nat Compd 26:42–45
5. Suzuki M, Matsuo Y, Takahashi Y, Masuda M (1995) Callicladol, a novel cytotoxic bromotriterpene polyether from a Vietnamese species of the red algal genus *Laurencia*. Chem Lett 24:1045–1046
6. Suzuki M, Takahashi Y, Matsuo Y, Masuda M (1996) Pannosallene, a brominated C15 nonterpenoid from *Laurencia pannosa*. Phytochemistry 41:1101–1103
7. Smirnova GP (2000) Hematoside with 8-*O*-methyl-*N*-glycolylneuraminic acid from the starfish *Linckia laevigata*. Russ Chem Bull 49:165–168
8. Fedorov SN, Krasokhin VB, Shubina LK, Dyshlovoy SA, Nam NH, Minh CV (2013) The extracts of some marine invertebrates and algae collected off the coast waters of Vietnam induce the inhibitory effects on the activator protein-1 transcriptional activity in JB6 Cl41 cells. J Chem 2013:896709
9. Yurchenko EA, Yurchenko AN, Van Minh C, Aminin DL (2019) Achievements in the study of marine low-molecular weight biologically active metabolites from the Vietnamese territorial waters as a result of expeditions aboard the research vessel 'Akademik Oparin' (2004–2017). Chem Biodivers 16:e1800654
10. Calcinai B, Azzini F, Bavestrello G, Cerrano C, Pansini M, Thung D-C (2006) Boring sponges from the Ha Long Bay, Tonkin Gulf, Vietnam. Zool Stud 45:201–212
11. Linh TM, Le Quynh Lien NCM, Tuan PM, Le Thanh Long NKB (2013) DNA extraction method from marine organisms for molecular analysis. In Proceeding of VAST-IRD symposium on marine science. Publishing House for Science and Technology

12. Linh TM, Mai NC, Giang VH, Tuan PM, Van NT, Ban NK (2014) Molecular markers based on 18S ribosome gene fragments of sponges (Porifera: Demospongiae) collected from Con Co island, Quang Tri province. Academia J Biol 36:65–72
13. Linh TM, Mai NC, Le Quynh Lien LQT, Giang VH, MinhTuan P, Tien DD, Van NT, Ban NK, Van Kiem P, Van Minh C (2016) Phylogenetic variation of marine sponges (Porifera: Demospongiae) surrounding Con Co island of Vietnam employing 28S rRNA gene fragments. Int J Adv Res 4:1413–1426
14. Do CT (2022) Biodiversity spongia in the marine waters from Hon Me (Thanh Hoa) to Hai Van-Son Tra (Thua Thien Hue). Vietnam J Marine Sci Technol 22:387–398
15. Lindgren NG (1898) Beitrag zur Kenntniss der Spongienfauna des Malayischen Archipels und der chinesischen Meere. 1898: Verlag von Gustav Fischer
16. Lévi C (1961) Éponges intercoditales de Nhatrang (Vietnam)
17. Dawydoff C (1952) Contribution a l'etude des invertebres de la faune marine benthique de l'Indochine. Contr Inst Ocean Nha Trang 9:155
18. Hooper J, Kennedy J, Van Soest R (2000) Annotated checklist of sponges (Porifera) of the South China Sea region. Raffles Bull Zool Suppl:125–207
19. Lim S-C, Putchakarn S, Thai M-Q, Wang D, Huang YM (2016) Inventory of sponge fauna from the Singapore Strait to Taiwan Strait along the western coastline of the South China Sea. Raffles Bull Zool:104–129
20. WWF (1993) WWF Vietnam marine conservation southern survey team. Survey report on the Biodiversity, Resource Utilization and Conservation Potential of Phu Quoc (An Thoi) Islands, Kien Giang Province, Gulf of Thailand. Institute of Oceanography (Nhatrang, Vietnam) and WWF-International (Gland, Switzerland)
21. Hung TQ, Yet NH, Dautova TN, Van Ken L (2010) Species composition of soft coral family (Alcyoniidae) in Con Co Island, Quang Tri Province, Vietnam. In: Marine biodiversity of East Asian seas: status, challenges and sustainable development
22. Hoang XB, Quang TM (2019) Biodiversity and characteristic of octocoral communities (Octocorallia: Alcyonacea and Gorgonacea) in Cu Lao Cham Marine Protected Area, Quang Nam province. Vietnam J Mar Sci Technol 19:589–599
23. Hoang X-B, Thai M-Q, Nguyen X-V, Dao V-H (2024) Morphological observations reveal new record of *Sarcophyton cherbonnieri* Tixier-Durivault, 1958 (Octocorallia: Alcyonacea: Alcyoniidae) in Vietnam. Natl Acad Sci Lett 47:13–16
24. Hoang XB, Dautova TN (2010) Soft corals (Octocorallia: Alcyonacea) in Ly Son islands, the central of Vietnam. Vietnam J Marine Sci Technol 10:39–49
25. Hoang XB (2024) The family Xeniidae (Octocorallia: Malacalcyonacea) of Vietnam with two new records. Academia J Biol 46:103–111
26. Tixier-Durivault A (1970) Les Octocoralliaires de Nhatrang. Cah du Pacifique 14:115–236
27. WWF (1993) WWF Vietnam marine conservation Southern Survey Team. Survey report on the biodiversity, resource utilization and conservation potential of Hon Mun, Nha Trang, Southern Vietnam. WWF-International, Switzerland
28. WWF (1993) WWF Vietnam marine conservation southern survey team. Survey report on the Biodiversity, Resource Utilization and Conservation Potential of Hon Cau Island, Binh Thuan Province, S. Vietnam. Institute of Oceanography (Nhatrang, Vietnam) and WWF-International (Gland, Switzerland)
29. Dautova TN, Savinkin OV (2009) New data on soft corals (Cnidaria: Octocorallia: Alcyonacea) from Nha Trang Bay, South China Sea. Zootaxa 2027:1–27
30. Đào TH (1991) Preliminary study on the Echinoderms of Truong Sa Archipelago (Viet Nam). Tap chi Sinh Hoc 44–47
31. Ho DT (1994) Check list of Echinoderms in Vietnam. Vol. I. Crinoidea and Echinoidea. Science and Technics Publishing House, Ha Noi
32. Dao T (2002) The echinoderm fauna in the coastal waters of Khanh Hoa Province (Central Vietnam). Vietnam J Marine Sci Technol 2:1–11
33. Ngai ND, Van Thao D, Thung DC, Van Luong P, Trang CT, Luu VT, Tien DD, Van Quan N, Van Chien P (2016) New data on water environment and organism in submerged caves and saltwater lakes in Ha Long and Cat Ba areas. Vietnam J Marine Sci Technol 16:167–175

34. Thung DC, Dong DM (2018) Diversity of echinoderms (Echinodermata) in the North Eastern Islands, Vietnam. Vietnam J Marine Sci Technol 18:296–303
35. Quang TM, Tuyen HT, Khang NA (2018) Species composition and distribution of mollusca and echinodermata on coral reefs in survey on Akademik Oparin RC in 2016–2017. Vietnam J Marine Sci Technol 18:81–92
36. Hong PTK, Khang NA, Hoc DT, Ngan NTM, Tuyen HT (2019) Macrozoobenthos in the intertidal zone of the Ly Son island. Vietnam J Marine Sci Technol 19:287–297
37. Ki D-W, Kodama T, El-Desoky AH, Wong CP, Nguyen HM, Do KM, Thai QM, Ton Nu LH, Morita H (2020) Chemical constituents of the Vietnamese marine sponge *Gelliodes* sp. and their cytotoxic activities. Chem Biodiv 17:e2000303
38. Tai BH, Hang DTT, Thung DC, Trang DT, Yen PH, Yen DTH, Huong PTT, Dung DT, Cuong NC, Thao DT, Nhiem NX, Van Kiem P (2022) Xestospongiene VN1, a new brominated polyunsaturated lipid from marine sponge *Xestospongia testudinaria* (Lamark, 1815). Vietnam J Chem 60:540–545
39. Van Kiem P, Nhiem NX, Van Quang N, Van Minh C, Nam NH, Cuc NT, Le Anh HT, Tai BH, Yen PH, Cuong NX, Thao NP, Hoai NT, Kim NY, Park SJ, Hyun KS (2013) New butenolide and pentenolide from *Dysidea cinerea*. Nat Prod Commun 8:1751–1752
40. Huyen LT, Hang DTT, Nhiem NX, Yen PH, Anh HLT, Quang TH, Tai BH, Dau NV, Kiem PV (2017) Naphtoquinones and sesquiterpene cyclopentenones from the sponge *Smenospongia cerebriformis* with their cytotoxic activity. Chem Pharm Bull 65:589–592
41. Ki D-W, Awouafack MD, Wong CP, Nguyen HM, Thai QM, Ton Nu LH, Morita H (2019) Brominated diphenyl ethers including a new tribromoiododiphenyl ether from the Vietnamese marine sponge *Arenosclera* sp. and their antibacterial activities. Chem Biodiv 16:e1800593
42. Tai BH, Hang DT, Trang DT, Yen PH, Huong PTT, Nhiem NX, Thung DC, Thao DT, Hoai NT, Kiem PV (2021) Conjugated polyene ketones from the marine sponge *Clathria* (*Thalysias*) *reinwardti* (Vosmaer, 1880) and their cytotoxic activity. Nat Prod Commun 16:1934578X211043732
43. Nguyen HM, Ito T, Win NN, Kodama T, Hung VQ, Nguyen HT, Morita H (2016) New antibacterial sesquiterpene aminoquinones from a Vietnamese marine sponge of *Spongia* sp. Phytochem Lett 17:288–292
44. Ito T, Nguyen HM, Win NN, Vo HQ, Nguyen HT, Morita H (2018) Three new sesquiterpene aminoquinones from a Vietnamese *Spongia* sp. and their biological activities. J Nat Med 72:298–303
45. Nguyen HM, Ito T, Kurimoto SI, Ogawa M, Win NN, Hung VQ, Nguyen HT, Kubota T, Kobayashi J, Morita H (2017) New merosesquiterpenes from a Vietnamese marine sponge of *Spongia* sp. and their biological activities. Bioorg Med Chem Lett 27:3043–3047
46. Kiem PV, Huyen LT, Hang DT, Nhiem NX, Tai BH, Anh HLT, Cuong PV, Quang TH, Minh CV, Dau NV, Kim Y-A, Subedi L, Kim SY, Kim SH (2017) Sesquiterpene derivatives from marine sponge *Smenospongia cerebriformis* and their anti-inflammatory activity. Bioorg Med Chem Lett 27:1525–1529
47. Huyen LT, Hang DT, Nhiem NX, Tai BH, Anh HLT, Quang TH, Yen PH, Van Minh C, Van Dau N, Van Kiem P (2017) Sesquiterpene quinones and diterpenes from *Smenospongia cerebriformis* and their cytotoxic activity. Nat Prod Commun 12:1934578X1701200402
48. Cuc NT, Anh HLT, Hang DTT, Nhiem NX, Dang NH, Nam NH, Yen PH, Thung DC, Thu VK, Van Minh C, Van Kiem P (2015) Sesquiterpenes from the Vietnamese marine sponge *Dysidea fragilis*. Nat Prod Commun 10:1934578X1501000804
49. Kiem PV, Nhiem NX, Tai BH, Anh Hle T, Hang DT, Cuc NT, Huyen le T, Nam NH, Yen PH, Thung do C, Minh CV (2016) *Bis*-sesquiterpene from the marine sponge *Dysidea fragilis*. Nat Prod Commun 11:439–441
50. Nguyen XN, Nguyen TC, Dan TT, Do TT, Nguyen HN, Pham HY, Do CT, Vu KT, le Hoang TA, Bui HT, Chau VM, Phan VK (2015) ^{1}H and ^{13}C NMR assignments of sesquiterpenes from *Dysidea fragilis*. Magn Reson Chem 53:1057–1060
51. Kiem PV, Minh CV, Nhiem NX, Cuc NT, Quang NV, Tuan Anh HL, Tai BH, Yen PH, Hoai NT, Ho KY, Kim N, Park S, Kim SH (2014) Muurolane-type sesquiterpenes from marine sponge *Dysidea cinerea*. Magn Reson Chem 52:51–56

52. Thu VK, Trang DT, Hang DTT, Nhiem NX, Quang TH, Yen PH, Tai BH, Kiem PV (2021) New merosesquiterpenes from the Vietnamese sponge *Hippospongia fistulosa* and their cytotoxic activity. Phytochem Lett 44:115–119
53. Hang DTT, Trang DT, Yen PH, Cuc NT, Dung DT, Tai BH, Nhiem NX, Van Kiem P (2021) A new meroterpene lactone and a new alkyl amino alcohol from the Vietnamese marine sponge *Hippospongia fistulosa* Lendenfeld, 1889. Vietnam J Chem 59:522–526
54. Hang DT, Trang DT, Tai BH, Yen PH, Thu VK, Nhiem NX, Van Kiem P (2022) Hippotulosas A-D: four new sesterterpenes from marine sponge *Hippospongia fistulosa* Lendenfeld, 1889. Nat Prod Res 36:5247–5254
55. Nhiem NX, Quang NV, Minh CV, Hang DT, Anh Hle T, Tai BH, Yen PH, Hoai NT, Thung do C, Kiem PV (2013) Biscembranoids from the marine sponge *Petrosia nigricans*. Nat Prod Commun 8:1209–1212
56. Kiem VP, Dung TD, Trang TD, Quang HT, Ngan TTN, Ha MT, Anh LTH, Yen HP, Thi Thao TD, Nhiem XN, Tai HB, Oh H, Kim CY, Minh VC (2017) Constituents from *Ircinia echinata* and their antiproliferative effect on six human cancer cell strains. Lett Org Chem 14:248–253
57. Kolesnikova SA, Lyakhova EG, Kalinovsky AI, Pushilin MA, Afiyatullov S, Yurchenko EA, Dyshlovoy SA, Minh CV, Stonik VA (2013) Isolation, structures, and biological activities of triterpenoids from a *Penares* sp. marine sponge. J Nat Prod 76:1746–1752
58. Lyakhova EG, Kolesnikova SA, Kalinovsky AI, Dmitrenok PS, Nam NH, Stonik VA (2015) Further study on *Penares* sp. from Vietnamese waters: minor lanostane and nor-lanostane triterpenes. Steroids 96:37–43
59. Kolesnikova SA, Lyakhova EG, Kozhushnaya AB, Kalinovsky AI, Berdyshev DV, Popov RS, Stonik VA (2021) New isomalabaricane-derived metabolites from a *Stelletta* sp. marine sponge. Molecules 26:678
60. Kozhushnaya AB, Kolesnikova SA, Kalinovsky AI, Popov RS, Ivanchina NV (2024) New *Nor*-Isomalabaricanic Acids from the Vietnamese marine sponge *Rhabdastrella globostellata*. Chem Nat Compd 60:1056–1060
61. Kolesnikova SA, Lyakhova EG, Kalinovsky AI, Berdyshev DV, Pislyagin EA, Popov RS, Grebnev BB, Makarieva TN, Minh CV, Stonik VA (2019) Cyclobutastellettolides A and B, C_{19} norterpenoids from a *Stelletta* sp. marine sponge. J Nat Prod 82:3196–3200
62. Kiem PV, Dung DT, Yen PH, Nhiem NX, Quang TH, Tai BH, Minh CV (2018) New isomalabaricane analogues from the sponge *Rhabdastrella providentiae* and their cytotoxic activities. Phytochem Lett 26:199–204
63. Dung DT, Yen PH, Nhiem NX, Quang TH, Tai BH, Van Minh C, Kim DC, Oh H, Kim YC, Van Kiem P (2018) New acetylated terpenoids from sponge *Rhabdastrella providentiae* inhibit NO production in LPS stimulated BV2 cells. Nat Prod Commun 13:1934578X1801300602
64. Dung DT, Hang DTT, Nhiem NX, Quang TH, Tai BH, Yen PH, Hoai NT, Thung DC, Van Minh C, Van Kiem P (2018) Rhabdaprovidines D–G, four new 6,6,5-tricyclic terpenoids from the Vietnamese sponge *Rhabdastrella providentiae*. Nat Prod Commun 13:1934578X1801301004
65. Trang DT, Dung DT, Nhiem NX, Cuc NT, Yen PH, Hang DTT, Linh TM, Mai NC, Huong PTT, Tai BH, Kiem PV (2022) New tetracyclic and pentacyclic isomalabaricanes from the marine sponge *Rhabdastrella globostellata* (Carter, 1883). Tetrahedron Lett 89:153607
66. Kozhushnaya AB, Kolesnikova SA, Yurchenko EA, Lyakhova EG, Menshov AS, Kalinovsky AI, Popov RS, Dmitrenok PS, Ivanchina NV (2023) Rhabdastrellosides A and B: two new isomalabaricane glycosides from the marine sponge *Rhabdastrella globostellata*, and their cytotoxic and cytoprotective effects. Mar Drugs 21:554
67. Guzii AG, Makarieva TN, Denisenko VA, Dmitrenok PS, Burtseva YV, Krasokhin VB, Stonik VA (2008) Topsentiasterol sulfates with novel iodinated and chlorinated side chains from the marine sponge *Topsentia* sp. Tetrahedron Lett 49:7191–7193
68. Hang DTT, Thung DC, Trang DT, Yen PH, Cuc NT, Dung DT, Tai BH, Nhiem NX, Van Kiem P (2022) Chemical constituents of marine sponge *Ianthella basta* (Pallas, 1766). Vietnam J Chem 60:238–244

69. Tung NH, Minh CV, Ha TT, Kiem PV, Huong HT, Dat NT, Nhiem NX, Tai BH, Hyun J-H, Kang H-K, Kim YH (2009) C29 sterols with a cyclopropane ring at C-25 and 26 from the Vietnamese marine sponge *Ianthella* sp. and their anticancer properties. Bioorg Med Chem Lett 19:4584–4588
70. Tung NH, Van Minh C, Van Kiem P, Huong HT, Ha TT, Dat NT, Nhiem NX, Cuong NX, Hyun JH, Kang HK, Kim YH (2009) A new C29-sterol with a cyclopropane ring at C-25 and 26 from the Vietnamese marine sponge *Ianthella* sp. Arch Pharm Res 32:1695–1698
71. Nguyen XC, Longeon A, Pham VC, Urvois F, Bressy C, Trinh TT, Nguyen HN, Phan VK, Chau VM, Briand JF, Bourguet-Kondracki ML (2013) Antifouling 26,27-cyclosterols from the Vietnamese marine sponge *Xestospongia testudinaria*. J Nat Prod 76:1313–1318
72. Nguyen HM, Ito T, Win NN, Vo HQ, Nguyen HT, Morita H (2019) A new sterol from the Vietnamese marine sponge *Xestospongia testudinaria* and its biological activities. Nat Prod Res 33:1175–1181
73. Vu Luu P, Hien MN, Phuc MP, An DV, Ton-Nu HL (2024) Testusterol, a new sterol of the sponge species *Xestospongia testudinaria* from Phu Quoc island, Vietnam. Nat Prod Res:1–9
74. Trinh TTV, Truong BN, Longeon A, Doan TMH, Deville A, Chau VM, Pham VC, Bourguet-Kondracki ML (2018) New 9α-hydroxy-5α,6α-epoxyhydroxysterols from the Vietnamese marine sponge *Ircinia echinata*. Marine Drugs 16
75. Tabakmakher KM, Makarieva TN, Denisenko VA, Popov RS, Dmitrenok PS, Dyshlovoy SA, Grebnev BB, Bokemeyer C, von Amsberg G, Cuong NX (2019) New trisulfated steroids from the Vietnamese marine sponge *Halichondria vansoesti* and their PSA expression and glucose uptake inhibitory activities. Marine Drugs 17
76. Van Kiem P, Trang DT, Huu Tai B, Nhiem NX, Hang DTT, Thung DC, Thi Thao D, Yen PH, Huong PTT (2022) Halipanasterol, a new sterol isolated from the marine sponge *Halichondria panicea*. Nat Prod Commun 17:1934578X221088061
77. Santalova EA, Denisenko VA, Dmitrenok PS, Ha DT, Hung NA, Drozdov AL, Malyarenko OS, Thuy TTT, Quan PM, Long PQ (2019) Occurrence of melibiose-containing glycosphingolipids in a sample of a sponge-coral association (*Desmapsamma anchorata*/*Carijoa riisei*). Chem Biodivers 16:e1800401
78. Shubina LK, Kalinovsky AI, Fedorov SN, Radchenko OS, Denisenko VA, Dmitrenok PS, Dyshlovoy SA, Krasokhin VB, Stonik VA (2009) Aaptamine alkaloids from the Vietnamese sponge *Aaptos* sp. Nat Prod Commun 4:1085–1088
79. Shubina LK, Makarieva TN, Dyshlovoy SA, Fedorov SN, Dmitrenok PS, Stonik VA (2010) Three new aaptamines from the marine sponge *Aaptos* sp. and their proapoptotic properties. Nat Prod Commun 5:1881–1884
80. Utkina NK, Denisenko VA, Pushilin MA (2009) Aaptanone, a novel zwitterionic metabolite of the aaptamine class with an oxygenated 1,6-naphthyridine core from the Vietnamese marine sponge *Aaptos aaptos*. Tetrahedron Lett 50:2580–2582
81. Utkina NK, Denisenko VA (2016) N-Demethylaaptanone, a new congener of aaptamine alkaloids from the Vietnamese marine sponge *Aaptos aaptos*. Nat Prod Commun 11:1259–1260
82. Trang DT, Tai BH, Hang DTT, Yen PH, Nhiem NX, Van Kiem P (2022) Four new aaptamine alkaloids from marine sponge *Aaptos aaptos*. Nat Prod Res 36:5022–5031
83. Shubina LK, Makarieva TN, Yashunsky DV, Nifantiev NE, Denisenko VA, Dmitrenok PS, Dyshlovoy SA, Fedorov SN, Krasokhin VB, Jeong SH, Han J, Stonik VA (2015) Pyridine nucleosides neopetrosides A and B from a marine *Neopetrosia* sp. sponge. Synthesis of neopetroside A and its β-riboside analogue. J Nat Prod 78:1383–1389
84. Dung DT, Hang DTT, Yen PH, Quang TH, Nhiem NX, Tai BH, Minh CV, Kim Y-C, Kim DC, Oh H, Kiem PV (2019) Macrocyclic *bis*-quinolizidine alkaloids from *Xestospongia muta*. Nat Prod Res 33:400–406
85. Kim YN, Ji YK, Kim NH, Van Tu N, Rho JR, Jeong EJ (2021) Isoquinolinequinone derivatives from a marine sponge (*Haliclona* sp.) regulate inflammation in in vitro system of intestine. Marine Drugs 19
86. Ji YK, Lee SM, Kim NH, Tu NV, Kim YN, Heo JD, Jeong EJ, Rho JR (2021) Stereochemical determination of fistularins isolated from the marine sponge *Ecionemia acervus* and their regulatory effect on intestinal inflammation. Marine Drugs 19

87. Lyakhova EG, Kolesnikova SA, Kalinovsky AI, Afiyatullov SS, Dyshlovoy SA, Krasokhin VB, Minh CV, Stonik VA (2012) Bromine-containing alkaloids from the marine sponge *Penares* sp. Tetrahedron Lett 53:6119–6122
88. Zhang Z, Chen S, Tang F, Guo K, Liang XT, Huang J, Yang Z (2021) Total synthesis of (+)-cyclobutastellettolide B. J Am Chem Soc 143:18287–18293
89. Belik AA, Tabakmakher KM, Silchenko AS, Makarieva TN, Minh CV, Ermakova SP, Zvyagintseva TN (2019) Sulfated steroids of Halichondriidae family sponges—natural inhibitors of polysaccharide-degrading enzymes of bacterium *Formosa algae*, inhabiting brown alga *Fucus evanescens*. Carbohyd Res 484:107776
90. He Q, Miao S, Ni N, Man Y, Gong K (2020) A review of the secondary metabolites from the marine sponges of the genus *Aaptos*. Nat Prod Commun 15:1934578X20951439
91. Nadar VM, Manivannan S, Chinnaiyan R, Govarthanan M, Ponnuchamy K (2022) Review on marine sponge alkaloid, aaptamine: a potential antibacterial and anticancer drug. Chem Biol Drug Des 99:103–110
92. Dyshlovoy SA, Naeth I, Venz S, Preukschas M, Sievert H, Jacobsen C, Shubina LK, Gesell Salazar M, Scharf C, Walther R, Krepstakies M, Priyadarshini P, Hauber J, Fedorov SN, Bokemeyer C, Stonik VA, Balabanov S, Honecker F (2012) Proteomic profiling of germ cell cancer cells treated with aaptamine, a marine alkaloid with antiproliferative activity. J Proteome Res 11:2316–2330
93. Dyshlovoy SA, Fedorov SN, Shubina LK, Kuzmich AS, Bokemeyer C, Keller-von Amsberg G, Honecker F (2014) Aaptamines from the marine sponge *Aaptos* sp. display anticancer activities in human cancer cell lines and modulate AP-1-, NF-κ-B-, and p53-dependent transcriptional activity in mouse JB6 Cl41 cells. BioMed Res Int 2014:469309
94. Dyshlovoy SA, Venz S, Shubina LK, Fedorov SN, Walther R, Jacobsen C, Stonik VA, Bokemeyer C, Balabanov S, Honecker F (2014) Activity of aaptamine and two derivatives, demethyloxyaaptamine and isoaaptamine, in cisplatin-resistant germ cell cancer. J Proteomics 96:223–239
95. Tu VA, Lyakhova EG, Diep CN, Kalinovsky AI, Dmitrenok PS, Cuong NX, Thanh NV, Menchinskaya ES, Pislyagin EA, Nam NH, Kiem PV, Stonik VA, Minh CV (2015) Polyoxygenated steroids from the gorgonian *Menella woodin* with capabilities to modulate ROS levels in macrophages at response to LPS. Steroids 104:246–251
96. Ngoc NT, Hanh TTH, Cuong NX, Nam NH, Thung DC, Ivanchina NV, Dang NH, Kicha AA, Kiem PV, Minh CV (2019) Steroids from *Dendronephthya mucronata* and their inhibitory effects on lipopolysaccharide-induced no formation in RAW264.7 cells. Chem Nat Compd 55:1090–1093
97. Ngoc NT, Hanh TTH, Nguyen HD, Quang TH, Cuong NX, Nam NH, Thung DC, Ngai ND, Kiem PV, Minh CV (2021) Bicyclic lactones from the octocoral *Dendronephthya mucronata*. Nat Prod Res 35:1134–1138
98. Nguyen Hoai N, Nguyen Thi H, Tran Thi Hong H, Van Nguyen T, Nguyen Xuan C, Do Cong T, Van Phan K, Van Chau M (2017) Pregnane steroids from the Vietnamese octocoral *Carijoa riisei*. Nat Prod Res 31:2435–2440
99. Tran HHT, Nguyen Viet P, Van Nguyen T, Tran HT, Nguyen Xuan C, Nguyen Hoai N, Do Cong T, Van Phan K, Van Chau M (2017) Cytotoxic steroid derivatives from the Vietnamese soft coral *Sinularia brassica*. J Asian Nat Prod Res 19:1183–1190
100. Pham GN, Kang DY, Kim MJ, Han SJ, Lee JH, Na M (2021) Isolation of sesquiterpenoids and steroids from the soft coral *Sinularia brassica* and determination of their absolute configuration. Marine Drugs 19
101. Ngoc NT, Huong PT, Thanh NV, Chi NT, Dang NH, Cuong NX, Nam NH, Thung DC, Kiem PV, Minh CV (2017) Cytotoxic steroids from the Vietnamese soft coral *Sinularia conferta*. Chem Pharm Bull (Tokyo) 65:300–305
102. Nguyen PT, Nguyen HN, Nguyen XC, Bui HT, Tran HQ, Nguyen TTN (2013) Steroidal constituents from the soft coral *Sinularia dissecta* and their inhibitory effects on lipopolysaccharide-stimulated production of pro-inflammatory cytokines in bone marrow-derived dendritic cells. Bull Korean Chem Soc 34:949–952

103. Huong NT, Ngoc NT, Thanh NV, Dang NH, Cuong NX, Nam NH, Thung DC, The HV, Tuan VS, Kiem PV, Minh CV (2018) Eudesmane and aromadendrane sesquiterpenoids from the Vietnamese soft coral *Sinularia erecta*. Nat Prod Res 32:1798–1802
104. Ngoc NT, Hanh TTH, Thanh NV, Thao DT, Cuong NX, Nam NH, Thung DC, Kiem PV, Minh CV (2017) Cytotoxic steroids from the Vietnamese soft coral *Sinularia leptoclados*. Chem Pharm Bull (Tokyo) 65:593–597
105. Thao NP, Nam NH, Cuong NX, Quang TH, Tung PT, Tai BH, Luyen BTT, Chae D, Kim S, Koh Y-S, Kiem PV, Minh CV, Kim YH (2012) Diterpenoids from the soft coral *Sinularia maxima* and their inhibitory effects on lipopolysaccharide-stimulated production of pro-inflammatory cytokines in bone marrow-derived dendritic cells. Chem Pharm Bull 60:1581–1589
106. Thao NP, Nam NH, Cuong NX, Quang TH, Tung PT, le Dat D, Chae D, Kim S, Koh YS, Kiem PV, Minh CV, Kim YH (2013) Anti-inflammatory norditerpenoids from the soft coral *Sinularia maxima*. Bioorg Med Chem Lett 23:228–231
107. Thanh NV, Ngoc NT, Anh Hle T, Thung do C, Thao do T, Cuong NX, Nam NH, Kiem PV, Minh CV (2016) Steroid constituents from the soft coral *Sinularia microspiculata*. J Asian Nat Prod Res 18:938–944
108. Ngoc NT, Huong PTM, Thanh NV, Cuong NX, Nam NH, Thung DC, Kiem PV, Minh CV (2017) Sesquiterpene constituents from the soft coral *Sinularia nanolobata*. Nat Prod Res 31:1799–1804
109. Ngoc NT, Huong PT, Thanh NV, Cuong NX, Nam NH, Thung do C, Kiem PV, Minh CV (2016) Steroid constituents from the soft coral *Sinularia nanolobata*. Chem Pharm Bull (Tokyo) 64:1417–1419.
110. Ninh Thi N, Tran Thi Hong H, Tran Hong Q, Nguyen Xuan C, Nguyen Hoai N, Thi Thao D, Cuong PV, Do Cong T, Van Phan K, Van Minh C (2022) Cembranoids from the Vietnamese soft coral *Sarcophyton ehrenbergi*. Nat Prod Res 36:5517–5523
111. Ngoc NT, Hanh TTH, Quang TH, Cuong NX, Nam NH, Thao DT, Thung DC, Kiem PV, Minh CV (2021) Polyhydroxylated steroids from the Vietnamese soft coral *Sarcophyton ehrenbergi*. Steroids 176:108932
112. Cuong NX, Tuan TA, Kiem PV, Minh CV, Choi EM, Kim YH (2008) New cembranoid diterpenes from the Vietnamese soft coral *Sarcophyton mililatensis* stimulate osteoblastic differentiation in MC3T3-E1 cells. Chem Pharm Bull 56:988–992
113. Van Minh C, Cuong NX, Tuan TA, Choi EM, Kim YH, Van Kiem P (2007) A new 9,11-secosterol from the Vietnamese sea soft coral, *Sarcophyton mililatensis*, increases the function of osteoblastic MC3T3-E1 cells. Nat Prod Commun 2:1934578X0700201109.
114. Nam NH, Tung PT, Ngoc NT, Hanh TT, Thao NP, Thanh NV, Cuong NX, Thao do T, Huong TT, Thung do C, Kiem PV, Kim YH, Minh CV (2015) Cytotoxic biscembranoids from the soft coral *Sarcophyton pauciplicatum*. Chem Pharm Bull (Tokyo) 63:636–640.
115. Thao NP, Luyen BT, Sun YN, Song SB, Thanh NV, Cuong NX, Nam NH, Kiem PV, Kim YH, Minh CV (2014) NF-κB inhibitory activity of polyoxygenated steroids from the Vietnamese soft coral *Sarcophyton pauciplicatum*. Bioorg Med Chem Lett 24:2834–2838
116. Tung NH, Van Minh C, Van Kiem P, Huong HT, Nam NH, Cuong NX, Quang TH, Nhiem NX, Hyun J-H, Kang H-K, Kim YH (2010) Chemical components from the Vietnamese soft coral *Lobophytum* sp. Arch Pharmacal Res 33:503–508
117. Minh CV, Kiem PV, Nhiem NX, Cuong NX, Thao NP, Nam NH, Anh HLT, Thung DC, Thuy DTT, Kang H-K, Jang H-D, Kim YH (2011) Cytotoxic and antioxidant activities of diterpenes and sterols from the Vietnamese soft coral *Lobophytum compactum*. Bioorg Med Chem Lett 21:2155–2159
118. Ninh Thi N, Tran Thi Hong H, Van Nguyen T, Nguyen Xuan C, Nguyen Hoai N, Do Cong T, Van Phan K, Van Chau M (2020) Briarane-type diterpenoids from the Vietnamese gorgonian *Junceella fragilis*. Nat Prod Res 34:385–389
119. Kapustina, II, Makarieva TN, Guzii AG, Kalinovsky AI, Popov RS, Dyshlovoy SA, Grebnev BB, Amsberg GV, Stonik VA (2020) Leptogorgins A-C, humulane sesquiterpenoids from the Vietnamese gorgonian *Leptogorgia* sp. Marine Drugs 18

120. Lyakhova EG, Diep CN, Berdyshev DV, Kolesnikova SA, Kalinovsky AI, Dmitrenok PS, Tu VA, Cuong NX, Van Thanh N, Nam NH, Kiemb PV, Stonik VA, Van Minh C (2016) Guaiane sesquiterpenoids from the gorgonian *Menella woodin*. Nat Prod Commun 11:913–916
121. Cuong NX, Thao NP, Luyen BT, Ngan NT, Thuy DT, Song SB, Nam NH, Kiem PV, Kim YH, Minh CV (2014) Cembranoid diterpenes from the soft coral *Lobophytum crassum* and their anti-inflammatory activities. Chem Pharm Bull (Tokyo) 62:203–208
122. Thao NP, Luyen BT, Ngan NT, Song SB, Cuong NX, Nam NH, Kiem PV, Kim YH, Minh CV (2014) New anti-inflammatory cembranoid diterpenoids from the Vietnamese soft coral *Lobophytum crassum*. Bioorg Med Chem Lett 24:228–232
123. Quang TH, Ha TT, Minh CV, Kiem PV, Huong HT, Ngan NTT, Nhiem NX, Tung NH, Tai BH, Thuy DTT, Song SB, Kang H-K, Kim YH (2011) Cytotoxic and anti-inflammatory cembranoids from the Vietnamese soft coral *Lobophytum laevigatum*. Bioorg Med Chem 19:2625–2632
124. Quang TH, Ha TT, Minh CV, Kiem PV, Huong HT, Ngan NTT, Nhiem NX, Tung NH, Thao NP, Thuy DTT, Song SB, Boo H-J, Kang H-K, Kim YH (2011) Cytotoxic and PPARs transcriptional activities of sterols from the Vietnamese soft coral *Lobophytum laevigatum*. Bioorg Med Chem Lett 21:2845–2849
125. Nam NH, Ngoc NT, Hanh TTH, Cuong NX, Thanh NV, Thao DT, Thung DC, Kiem PV, Minh CV (2018) Cytotoxic steroids from the Vietnamese gorgonian *Verrucella corona*. Steroids 138:57–63
126. Thao NP, Luyen BT, Brun R, Kaiser M, Van Kiem P, Van Minh C, Schmidt TJ, Kang JS, Kim YH (2015) Anti-protozoal activities of cembrane-type diterpenes from Vietnamese soft corals. Molecules 20:12459–12468
127. Thao NP, No JH, Luyen BT, Yang G, Byun SY, Goo J, Kim KT, Cuong NX, Nam NH, Van Minh C, Schmidt TJ, Kang JS, Kim YH (2014) Secondary metabolites from Vietnamese marine invertebrates with activity against *Trypanosoma brucei* and *T. cruzi*. Molecules 19:7869–7880
128. Stonik VA, Kicha AA, Malyarenko TV, Ivanchina NV (2020) Asterosaponins: structures, taxonomic distribution, biogenesis and biological activities. Marine Drugs 18
129. Zhao YC, Xue CH, Zhang TT, Wang YM (2018) Saponins from sea cucumber and their biological activities. J Agric Food Chem 66:7222–7237
130. Thimmappa R, Wang S, Zheng M, Misra RC, Huang AC, Saalbach G, Chang Y, Zhou Z, Hinman V, Bao Z, Osbourn A (2022) Biosynthesis of saponin defensive compounds in sea cucumbers. Nat Chem Biol 18:774–781
131. Vien LT, Hanh TTH, Huong PTT, Dang NH, Thanh NV, Cuong NX, Nam NH, Thung DC, Kiem PV, Minh CV (2018) Anthraquinone and butenolide constituents from the crinoid *Capillaster multiradiatus*. Chem Pharm Bull (Tokyo) 66:1023–1026
132. Vien LT, Hanh TTH, Quang TH, Thao DT, Thung DC, Cuong NX, Nam NH, Cuong PV, Kiem PV, Minh CV (2021) Naphthopyrone and anthraquinone derivatives from *Comanthus delicata*. Phytochem Lett 46:129–135
133. Vien LT, Hanh TTH, Quang TH, Thung DC, Cuong NX, Nam NH, Cuong PV, Kiem PV, Minh CV (2022) Sulfated naphthopyrones and anthraquinones from the Vietnamese crinoid *Comanthus delicata*. Chem Pharm Bull 70:408–412
134. Kicha AA, Dinh TH, Ivanchina NV, Malyarenko TV, Kalinovsky AI, Popov RS, Ermakova SP, Tran TTT, Doan LP (2014) Three new steroid biglycosides, plancisides A, B, and C, from the starfish *Acanthaster planci*. Nat Prod Commun 9:1934578X1400900912
135. Kicha AA, Kalinovskii AI, Ivanchina NV, Malyarenko TV, Popov RS, Long FK, Hung NA (2014) Minor steroidal triglycoside planciside D from the tropical starfish *Acanthaster planci*. Chem Nat Compd 50:1032–1036
136. Vien LT, Hanh TTH, Huong PTT, Tu VA, Thanh NV, Lyakhova EG, Cuong NX, Nam NH, Kiem PV, Minh CV, Kicha AA, Stonik VA (2016) New steroidal glycosides from the starfish *Acanthaster planci*. Chem Nat Compd 52:1056–1060
137. Vien LT, Hanh TT, Huong PT, Dang NH, Thanh NV, Lyakhova E, Cuong NX, Nam NH, Kiem PV, Kicha A, Minh CV (2016) Pyrrole oligoglycosides from the starfish *Acanthaster planci* suppress lipopolysaccharide-induced nitric oxide production in RAW264.7 macrophages. Chem Pharm Bull (Tokyo) 64:1654–1657

138. Ha DT, Kicha AA, Kalinovsky AI, Malyarenko TV, Popov RS, Malyarenko OS, Ermakova SP, Thuy TTT, Long PQ, Ivanchina NV (2021) Asterosaponins from the tropical starfish *Acanthaster planci* and their cytotoxic and anticancer activities in vitro. Nat Prod Res 35:548–555
139. Malyarenko TV, Kharchenko SD, Kicha AA, Ivanchina NV, Dmitrenok PS, Chingizova EA, Pislyagin EA, Evtushenko EV, Antokhina TI, Minh CV, Stonik VA (2016) Anthenosides L-U, steroidal glycosides with unusual structural features from the starfish *Anthenea aspera*. J Nat Prod 79:3047–3056
140. Malyarenko TV, Malyarenko OS, Kicha AA, Ivanchina NV, Kalinovsky AI, Dmitrenok PS, Ermakova SP, Stonik VA (2018) In vitro anticancer and proapoptotic activities of steroidal glycosides from the starfish *Anthenea aspera*. Mar Drugs 16:420
141. Malyarenko TV, Ivanchina NV, Malyarenko OS, Kalinovsky AI, Dmitrenok PS, Evtushenko EV, Minh CV, Kicha AA (2018) Two new steroidal monoglycosides, anthenosides A_1 and A_2, and revision of the structure of known anthenoside A with unusual monosaccharide residue from the starfish *Anthenea aspera*. Molecules 23
142. Kicha AA, Ha DT, Ivanchina NV, Malyarenko TV, Kalinovsky AI, Dmitrenok PS, Ermakova SP, Malyarenko OS, Hung NA, Thuy TTT, Long PQ (2018) Six new polyhydroxysteroidal glycosides, anthenosides S1–S6, from the starfish Anthenea sibogae. Chem Biodivers 15:e1700553
143. Kicha AA, Ha DT, Malyarenko TV, Kalinovsky AI, Popov RS, Malyarenko OS, Thuy TTT, Long PQ, Ha NTT, Ivanchina NV (2020) Unusual polyhydroxylated steroids from the starfish *Anthenoides laevigatus*, collected off the coastal waters of Vietnam. Molecules 25
144. Ivanchina NV, Kicha AA, Huong TT, Kalinovsky AI, Dmitrenok PS, Agafonova IG, Long PQ, Stonik VA (2010) Highly hydroxylated steroids of the starfish *Archaster typicus* from the Vietnamese waters. Steroids 75:897–904
145. Kicha AA, Ivanchina NV, Huong TTT, Kalinovsky AI, Dmitrenok PS, Fedorov SN, Dyshlovoy SA, Long PQ, Stonik VA (2010) Two new asterosaponins, archasterosides A and B, from the Vietnamese starfish *Archaster typicus* and their anticancer properties. Bioorg Med Chem Lett 20:3826–3830
146. Kicha AA, Ivanchina NV, Huong TTT, Kalinovsky AI, Dmitrenok PS, Long PQ (2010) Minor asterosaponin archasteroside C from the starfish *Archaster typicus*. Russ Chem Bull 59:2133–2136
147. Hanh TTH, Vien LT, Vinh LB, Thanh NV, Cuong NX, Nam NH, Thung DC, Kiem PV, Minh CV (2016) Further highly hydroxylated steroids from the Vietnamese starfish *Archaster typicus*. Chem Pharm Bull 64:1523–1527
148. Thao NP, Dat LD, Ngoc NT, Tu VA, Hanh TTH, Huong PTT, Nhiem NX, Tai BH, Cuong NX, Nam NH, Cuong PV, Yang SY, Kim S, Chae D, Koh Y-S, Kiem PV, Minh CV, Kim YH (2013) Pyrrole and furan oligoglycosides from the starfish *Asterina batheri* and their inhibitory effect on the production of pro-inflammatory cytokines in lipopolysaccharide-stimulated bone marrow-derived dendritic cells. Bioorg Med Chem Lett 23:1823–1827
149. Malyarenko TV, Kicha AA, Ivanchina NV, Kalinovsky AI, Dmitrenok PS, Smirnov AV (2010) Three new polyhydroxysteroids from the tropical starfish *Asteropsis carinifera*. Russ J Bioorg Chem 36:755–761
150. Malyarenko TV, Kicha AA, Ivanchina NV, Kalinovsky AI, Dmitrenok PS, Ermakova SP, Stonik VA (2011) Cariniferosides A-F and other steroidal biglycosides from the starfish *Asteropsis carinifera*. Steroids 76:1280–1287
151. Malyarenko TV, Kicha AA, Ivanchina NV, Kalinovskii AI, Dmitrenok PS, Ermakova SP, Minkh CV (2012) Asteropsiside A and other asterosaponins from the starfish *Asteropsis carinifera*. Russ Chem Bull 61:1986–1991
152. Thao NP, Cuong NX, Luyen BTT, Thanh NV, Nhiem NX, Koh Y-S, Ly BM, Nam NH, Kiem PV, Minh CV, Kim YH (2013) Anti-inflammatory asterosaponins from the starfish *Astropecten monacanthus*. J Nat Prod 76:1764–1770
153. Thao NP, Cuong NX, Luyen BTT, Nam NH, Cuong PV, Thanh NV, Nhiem NX, Hanh TTH, Kim E-J, Kang H-K, Kiem PV, Minh CV, Kim YH (2013) Steroidal constituents from the

starfish *Astropecten polyacanthus* and their anticancer effects. Chem Pharm Bull 61:1044–1051
154. Vien LT, Hanh TTH, Hong PT, Thanh NV, Huong TT, Cuong NX, Nam NH, Thung DC, Kiem PV, Minh CV (2018) Polar steroid derivatives from the Vietnamese starfish *Astropecten polyacanthus*. Nat Prod Res 32:54–59
155. Ivanchina NV, Malyarenko TV, Kicha AA, Kalinovsky AI, Dmitrenok PS, Stonik VA (2017) A new steroidal glycoside granulatoside C from the starfish *Choriaster granulatus*, unexpectedly combining structural features of polar steroids from several different marine invertebrate Phyla. Nat Prod Commun 12:1934578X1701201015
156. Ivanchina NV, Kicha AA, Malyarenko TV, Ermolaeva SD, Yurchenko EA, Pislyagin EA, Van Minh C, Dmitrenok PS (2019) Granulatosides D, E and other polar steroid compounds from the starfish *Choriaster granulatus*. Their immunomodulatory activity and cytotoxicity. Nat Prod Res 33:2623–2630
157. Ngoan BT, Hanh TTH, Vien LT, Diep CN, Thao NP, Thao DT, Thanh NV, Cuong NX, Nam NH, Thung DC, Kiem PV, Kim YH, Minh CV (2015) Asterosaponins and glycosylated polyhydroxysteroids from the starfish *Culcita novaeguineae* and their cytotoxic activities. J Asian Nat Prod Res 17:1010–1017
158. Vien LT, Hong HTT, Hong QT, Dang NN, Viet HD, Cuong NX (2024) Culcinoside E, a new sulfated steroidal biglycoside from the starfish *Culcita novaeguineae*. J Asian Nat Prod Res:1–7
159. Kicha AA, Kalinovsky AI, Malyarenko TV, Ivanchina NV, Dmitrenok PS, Menchinskaya ES, Yurchenko EA, Pislyagin EA, Aminin DL, Huong TTT, Long PQ, Stonik VA (2015) Cyclic steroid glycosides from the starfish *Echinaster luzonicus*: structures and immunomodulatory activities. J Nat Prod 78:1397–1405
160. Kicha AA, Ivanchina NV, Kalinovsky AI, Dmitrenok PS, Palyanova NV, Pankova TM, Starostina MV, Gavagnin M, Stonik VA (2007) New neuritogenic steroid glycosides from the Vietnamese starfish *Linckia laevigata*. Nat Prod Commun 2:1934578X0700200108
161. Kicha AA, Ivanchina NV, Kalinovsky AI, Dmitrenok PS, Sokolova EV, Agafonova IG, Morre J, Stonik VA (2007) Four new steroid glycosides from the Vietnamese starfish *Linckia laevigata*. Russ Chem Bull 56:823–830
162. Kicha AA, Ivanchina NV, Kalinovskii AI, Dmitrenok PS, Sokolova EV, Agafonova IG (2007) Sulfated steroid glycosides from the Viet Namese starfish *Linckia laevigata*. Chem Nat Compd 43:76–80
163. Vien LT, Ngoan BT, Hanh TTH, Vinh LB, Thung DC, Thao DT, Thanh NV, Cuong NX, Nam NH, Kiem PV, Minh CV (2017) Steroid glycosides from the starfish *Pentaceraster gracilis*. J Asian Nat Prod Res 19:474–480
164. Kicha AA, Ivanchina NV, Malyarenko TV, Kalinovsky AI, Dmitrenok PS, Pislyagin EA, Yurchenko EA (2016) Regulusosides A, B, and C, three new polyhydroxysteroid glycosides from the starfish *Pentaceraster regulus*. Nat Prod Commun 11:1934578X1601100912
165. Vien LT, Hanh TTH, Quang TH, Cuong NX, Thao DT, Nam NH, Thung DC, Kiem PV, Minh CV (2022) Polyhydroxylated steroid derivatives from the starfish *Pentaceraster regulus*. Nat Prod Res 36:2223–2229
166. Kicha AA, Ivanchina NV, Malyarenko TV, Kalinovsky AI, Dmitrenok PS (2017) Sulfated steroidal glycosides, regulusosides S1 and S2, from the tropical starfish *Pentaceraster regulus*. Chem Nat Compd 53:88–92
167. Kicha AA, Kalinovsky AI, Ivanchina NV, Malyarenko TV, Dmitrenok PS, Kuzmich AS, Sokolova EV, Stonik VA (2017) Furostane series asterosaponins and other unusual steroid oligoglycosides from the tropical starfish *Pentaceraster regulus*. J Nat Prod 80:2761–2770
168. Cuong NX, Vien LT, Hanh TTH, Thao NP, Thao DT, Thanh NV, Nam NH, Thung DC, Kiem PV, Minh CV (2015) Cytotoxic triterpene saponins from *Cercodemas anceps*. Bioorg Med Chem Lett 25:3151–3156
169. Silchenko AS, Kalinovsky AI, Avilov SA, Andryjaschenko PV, Dmitrenok PS, Yurchenko EA, Dolmatov IY, Savchenko AM, Kalinin VI (2014) Triterpene glycosides from the sea cucumber *Cladolabes schmeltzii*. II. Structure and biological action of cladolosides A_1-A_6. Nat Prod Commun 9:1934578X1400901006

170. Silchenko AS, Kalinovsky AI, Avilov SA, Andryjaschenko PV, Dmitrenok PS, Yurchenko EA, Dolmatov IY, Kalinin VI, Stonik VA (2013) Structure and biological action of cladolosides B_1, B_2, C, C_1, C_2 and D, six new triterpene glycosides from the sea cucumber *Cladolabes schmeltzii*. Nat Prod Commun 8:1527–1534
171. Silchenko AS, Kalinovsky AI, Avilov SA, Andryjaschenko PV, Dmitrenok PS, Yurchenko EA, Dolmatov IY, Kalinin VI (2015) Structures and biological activities of cladolosides C_3, E_1, E_2, F_1, F_2, G, H_1 and H_2, eight triterpene glycosides from the sea cucumber *Cladolabes schmeltzii* with one known and four new carbohydrate chains. Carbohyd Res 414:22–31
172. Silchenko AS, Kalinovsky AI, Avilov SA, Andryjaschenko PV, Dmitrenok PS, Yurchenko EA, Ermakova SP, Malyarenko OS, Dolmatov IY, Kalinin VI (2018) Cladolosides C_4, D_1, D_2, M, M_1, M_2, N and Q, new triterpene glycosides with diverse carbohydrate chains from sea cucumber *Cladolabes schmeltzii*. An uncommon 20,21,22,23,24,25,26,27-okta-*nor*-lanostane aglycone. The synergism of inhibitory action of non-toxic dose of the glycosides and radioactive irradiation on colony formation of HT-29 cancer cells. Carbohyd Res 468:36–44
173. Silchenko AS, Kalinovsky AI, Avilov SA, Andryjaschenko PV, Dmitrenok PS, Chingizova EA, Dolmatov IY, Kalinin VI (2017) Cladolosides I_1, I_2, J_1, K_1, K_2 and L_1, monosulfated triterpene glycosides with new carbohydrate chains from the sea cucumber *Cladolabes schmeltzii*. Carbohyd Res 445:80–87
174. Silchenko AS, Kalinovsky AI, Avilov SA, Andryjaschenko PV, Dmitrenok PS, Yurchenko EA, Ermakova SP, Malyarenko OS, Dolmatov IY, Kalinin VI (2018) Cladolosides O, P, P_1–P_3 and R, triterpene glycosides with two novel types of carbohydrate chains from the sea cucumber *Cladolabes schmeltzii*. Inhibition of cancer cells colony formation and its synergy with radioactive irradiation. Carbohyd Res 468:73–79
175. Silchenko AS, Kalinovsky AI, Avilov SA, Andrijaschenko PV, Popov RS, Dmitrenok PS, Chingizova EA, Ermakova SP, Malyarenko OS, Dautov SS, Kalinin VI (2020) Structures and bioactivities of quadrangularisosides A, A_1, B, B_1, B_2, C, C_1, D, D_1-D_4, and E from the sea cucumber *Colochirus quadrangularis*: the first discovery of the glycosides, sulfated by C-4 of the terminal 3-*O*-methylglucose residue. Synergetic effect on colony formation of tumor HT-29 cells of these glycosides with radioactive irradiation. Marine Drugs 18:394
176. Silchenko AS, Kalinovsky AI, Avilov SA, Andryjaschenko PV, Dmitrenok PS, Kalinin VI, Yurchenko EA, Dolmatov IY (2016) Colochirosides A_1, A_2, A_3, and D, four novel sulfated triterpene glycosides from the sea cucumber *Colochirus robustus* (Cucumariidae, Dendrochirotida). Nat Prod Commun 11:381–387
177. Silchenko AS, Kalinovsky AI, Avilov SA, Andryjaschenko PV, Dmitrenok PS, Kalinin VI, Yurchenko EA, Dolmatov IY (2015) Colochirosides B_1, B_2, B_3 and C, novel sulfated triterpene glycosides from the sea cucumber *Colochirus robustus* (Cucumariidae, Dendrochirotida). Nat Prod Commun 10:1934578X1501001014
178. Silchenko AS, Kalinovsky AL, Avilov SA, Andryjaschenko PV, Dmitrenok PS, Yurchenko EA, Dolmatov IY, Dautov SS, Stonik VA, Kalinin VI (2016) Colochiroside E, an unusual non-holostane triterpene sulfated trioside from the sea cucumber *Colochirus robustus* and evidence of the impossibility of a 7(8)-double bond migration in lanostane derivatives having an 18(16)-lactone. Nat Prod Commun 11:1934578X1601100611
179. Hoang L, Le Thi V, Tran Thi Hong H, Van Nguyen T, Nguyen Xuan C, Nguyen Hoai N, Do Cong T, Ivanchina NV, Do Thi T, Dmitrenok PS, Kicha AA, Van Phan K, Van Chau M (2020) Triterpene glycosides from the Vietnamese sea cucumber *Holothuria edulis*. Nat Prod Res 34:1061–1067
180. Dang NH, Van Thanh N, Van Kiem P, Huong LM, Van Minh C, Kim YH (2007) Two new triterpene glycosides from the Vietnamese sea cucumber *Holothuria scabra*. Arch Pharmacal Res 30:1387–1391
181. Van TN, Hai DN, Van KP, Xuan CN, Thanh HH, Van MC (2017) A new triterpene glycoside from the sea cucumber *Holothuria scabra* collected in Vietnam. ASEAN J Sci Technol Dev 23:253–259
182. Silchenko AS, Kalinovsky AI, Avilov SA, Kalinin VI, Andrijaschenko PV, Dmitrenok PS, Chingizova EA, Ermakova SP, Malyarenko OS, Dautova TN (2017) Nine new triterpene glycosides, magnumosides A_1-A_4, B_1, B_2, C_1, C_2 and C_4, from the Vietnamese sea

cucumber *Neothyonidium (=Massinium) magnum*: structures and activities against tumor cells independently and in synergy with radioactive irradiation. Marine Drugs 15
183. Silchenko AS, Kalinovsky AI, Avilov SA, Kalinin VI, Andrijaschenko PV, Dmitrenok PS, Chingizova EA, Ermakova SP, Malyarenko OS, Dautova TN (2017) Magnumosides B3, B4 and C3, mono- and disulfated triterpene tetraosides from the Vietnamese sea cucumber *Neothyonidium (= Massinium) magnum.* Nat Prod Commun 12:1934578X1701201013
184. Thao NP, Luyen BTT, Vien LT, Tai BH, Dat LD, Cuong NX, Nam NH, Van Kiem P, Van Minh C, Kim YH (2014) Triterpene saponins from the sea cucumber *Stichopus chloronotus.* Nat Prod Commun 9:1934578X1400900505
185. Vien LT, Hanh TTH, Quang TH, Van Thanh N, Cuong NX, Nam NH (2022) Herrmananes A and B, two new sulfated hydrocarbons from the sea cumber *Stichopus herrmanni.* Vietnam J Chem 60:685–687
186. Vien LT, Hanh TTH, Quang TH, Thanh NV, Thao DT, Cuong NX, Nam NH, Thung DC, Van Kiem P (2022) Triterpene tetraglycosides from *Stichopus herrmanni* Semper, 1868. Nat Prod Commun 17:1934578X221105369
187. Cuong NX, Vien LT, Hoang L, Hanh TTH, Thao DT, Thanh NV, Nam NH, Thung DC, Kiem PV, Minh CV (2017) Cytotoxic triterpene diglycosides from the sea cucumber *Stichopus horrens.* Bioorg Med Chem Lett 27:2939–2942
188. Vien LT, Hoang L, Hanh TTH, Thanh NV, Cuong NX, Nam NH, Thung DC, Kiem PV, Minh CV (2018) Triterpene tetraglycosides from the sea cucumber *Stichopus horrens.* Nat Prod Res 32:1039–1043
189. Avilov SA, Silchenko AS, Antonov AS, Kalinin VI, Kalinovsky AI, Smirnov AV, Dmitrenok PS, Evtushenko EV, Fedorov SN, Savina AS, Shubina LK, Stonik VA (2008) Synaptosides A and A1, triterpene glycosides from the sea cucumber *Synapta maculata* containing 3-*O*-methylglucuronic acid and their cytotoxic activity against tumor cells. J Nat Prod 71:525–531
190. Alhadrami HA, Thissera B, Hassan MHA, Behery FA, Ngwa CJ, Hassan HM, Pradel G, Abdelmohsen UR, Rateb ME (2021) Bio-guided isolation of antimalarial metabolites from the coculture of two Red Sea sponge-derived *Actinokineospora* and *Rhodococcus* spp. Mar Drugs 19:109
191. Thao NP, Luyen BTT, Kim E-J, Kang H-K, Kim S, Cuong NX, Nam NH, Kiem PV, Minh CV, Kim YH (2014) Asterosaponins from the starfish *Astropecten monacanthus* suppress growth and induce apoptosis in HL-60, PC-3, and SNU-C5 human cancer cell lines. Biol Pharm Bull 37:315–321
192. Dai Y, Yu B (2015) Total synthesis of astrosterioside A, an anti-inflammatory asterosaponin. Chem Commun (Camb) 51:13826–13829
193. Riccio R, Dini A, Minale L, Pizza C, Zollo F, Sevenet T (1982) Starfish saponins VII. Structure of luzonicoside, a further steroidal cyclic glycoside from the pacific starfish *Echinaster luzonicus.* Experientia 38:68–70
194. Malyarenko OS, Dyshlovoy SA, Kicha AA, Ivanchina NV, Malyarenko TV, Carsten B, Gunhild VA, Stonik VA, Ermakova SP (2017) The inhibitory activity of luzonicosides from the starfish *Echinaster luzonicus* against human melanoma cells. Mar Drugs 15:227
195. Malyarenko OS, Malyarenko TV, Usoltseva RV, Surits VV, Kicha AA, Ivanchina NV, Ermakova SP (2021) Combined anticancer effect of sulfated laminaran from the brown alga *Alaria angusta* and polyhydroxysteroid glycosides from the starfish *Protoreaster lincki* on 3D colorectal carcinoma HCT 116 cell line. Mar Drugs 19:540
196. Kalinin VI, Silchenko AS, Avilov SA, Stonik VA (2021) Progress in the studies of triterpene glycosides from sea cucumbers (Holothuroidea, Echinodermata) between 2017 and 2021. Nat Prod Commun 16:1934578X211053934
197. Kalinin VI, Silchenko AS, Avilov SA, Stonik VA (2019) Non-holostane aglycones of sea cucumber triterpene glycosides. Structure, biosynthesis, evolution. Steroids 147:42–51
198. Popov RS, Ivanchina NV, Silchenko AS, Avilov SA, Kalinin VI, Malyarenko TV, Stonik VA, Dmitrenok PS (2023) A mass spectrometry database for sea cucumber triterpene glycosides. Metabolites 13:783

199. Nga NT, Phuong DT, Cuc NT, Phuong TH, Huong PTM, Cuong NX, Huu Tai B, Van Kiem P, Thao DT (2020) Nanoliposomal cercodemasoide A and its improved activities against NTERA-2 cancer stem cells. Nat Prod Commun 15:1934578X20982108
200. Yun SH, Sim EH, Han SH, Kim TR, Ju MH, Han JY, Jeong JS, Kim SH, Silchenko AS, Stonik VA, Park JI (2018) In vitro and in vivo anti-leukemic effects of cladoloside C_2 are mediated by activation of Fas/ceramide synthase 6/p38 kinase/c-Jun NH_2-terminal kinase/caspase-8. Oncotarget 9:495–511
201. Pislyagin EA, Menchinskaya ES, Aminin DL, Avilov SA, Silchenko AS (2018) Sulfated glycosides from the sea cucumbers block Ca^{2+} flow in murine neuroblastoma cells. Nat Prod Commun 13:1934578X1801300808.
202. Van Khanh L, Anh NTN, Dinh TD (2020) Investigating species compositions of sea cucumbers in Nam Du Island, Kien Giang province, Vietnam. Oceanogr Fisheries Open Access J 11:96–102
203. Stonik VA, Mal'tsev II, Kalinovskii AI, Elyakov GB (1982) Glycosides of marine invertebrates. XII. Structure of a new triterpene oligoglycoside from holothurians of family Stichopodidae. Chem Nat Compd 18:182–186
204. Cuc NT, Thi Nga N, Ha Phuong T, Thi Phuong D, Xuan Cuong N, Thi Mai Huong P, Van Minh C, Thi Thao D (2020) The anticancer activities of stichoposide D isolated from the sea cucumber *Stichopus chloronotus* on NTERA-2 cancer stem cells. Vietnam J Biotechnol 18:273–281

Chapter 9
Marine Microbial Natural Products Research in the East Sea Region

Ton That Huu Dat and Lik Tong Tan

1 Introduction

Marine microbes are increasingly recognized as a vital source of potential drugs due to their unique bioactive compounds and diverse pharmacological activities. The marine environment, largely unexplored, offers a rich reservoir of novel molecules that can be used to treat a variety of diseases, including cancer, malaria, tuberculosis and HIV [1]. Marine microorganisms, such as bacteria, cyanobacteria and fungi, produce diverse secondary metabolites, including polyketides, macrolides, terpenoids, alkaloids and nonribosomal peptides, with significant antimicrobial, anticancer and antimalarial properties [2–4]. These microbes are more amenable to culturing compared to macro-organisms, making them a sustainable and renewable source of bioactive compounds. In addition, the symbiotic relationships between marine macroorganisms, such as sponges, corals and macroalgae, and their associated microbes, have led to the discovery of numerous bioactive metabolites with unique properties that are pharmaceutically important [5]. Advances in molecular biology and cultivation technologies have further enhanced the potential of marine microbes in drug discovery, offering an environmentally friendly approach to meet the growing demands of the pharmaceutical industry [6].

The East Sea region, bordered by multiple coastal countries including Vietnam, Southern China, the Philippines, Borneo and the east coast of the Malay Peninsular, presents a particularly rich and diverse marine environment where microbial life

T. T. H. Dat (✉)
Mientrung Institute for Scientific Research, Vietnam National Museum of Nature, Vietnam Academy of Science and Technology, Hue city, Vietnam
e-mail: tthdat@vnmn.vast.vn

L. T. Tan (✉)
Natural Sciences and Science Education, National Institute of Education, Nanyang Technological University, Singapore, Singapore
e-mail: liktong.tan@nie.edu.sg

L. T. Tan and N. Hanif (eds.), *Marine Natural Products Research in South-East Asia*,
https://doi.org/10.1007/978-3-032-01098-8_9

thrives in various niches—from coastal waters to the more isolated archipelagos and deep waters. These microbes, including marine bacteria and fungi, have evolved sophisticated biochemical pathways to survive and adapt to unique pressures such as nutrient scarcity, high salinity and fluctuating temperatures, resulting in the production of secondary metabolites with unique structural diversity and potent bioactivity.

The high biodiversity and unique ecological conditions within the East Sea made it an ideal setting for the discovery of novel bioactive compounds. In recent years, researchers have identified a broad spectrum of natural products from marine microorganisms in this region, including antibiotics, anti-inflammatory agents and anticancer compounds. Coastal ecosystems, characterized by high biological interactions, have been particularly fruitful, with microbial isolates yielding a variety of bioactive metabolites with promising applications in medicine and biotechnology. Additionally, the relatively remote archipelagos in the East Sea, including the Pratas, Paracel and Spratly Islands, present ecosystems less affected by human activities, potentially harboring untapped microbial diversity and novel compounds with therapeutic potential. Beyond coastal and archipelagic environments, deep waters within the East Sea remain a largely uncharted frontier in microbial natural product research. Extreme environmental conditions in these depths, such as high pressure, low light and low temperatures, have driven the evolution of unique microbial communities and bioactive compounds that may be absent in shallower waters.

The chapter highlights natural products from marine microbes across the East Sea region (Fig. 1), including coastal Vietnam, relatively remote archipelagos and associated deep water regions, and organizes them by structural class, specifically terpenoids/steroids, polyketides, alkaloids, peptides and others. Terpenoids and steroids, known for their anti-inflammatory and antiviral properties, are extensively studied within this region, while polyketides and alkaloids exhibit potent cytotoxic and antibacterial activities, underscoring their value in drug discovery. Additionally, marine-derived peptides isolated from marine microbes found in the East Sea have shown considerable bioactivity against a range of pathogens, making them highly relevant for addressing rising antibiotic resistance. The wide range of bioactive compounds identified thus far not only enhances our understanding of marine microbial biochemistry but also points to the potential for developing new therapeutic agents from these resources.

2 Microbial Marine Natural Products from Coastal Vietnam

The discovery of marine microbial natural products from the unique coastal ecosystems of Vietnam has revealed a rich diversity of bioactive compounds with significant potential for pharmaceutical applications. Studies conducted in various regions, such as Nha Trang Bay, Son Tra Peninsula and the coastal waters of Ninh Thuan and Binh

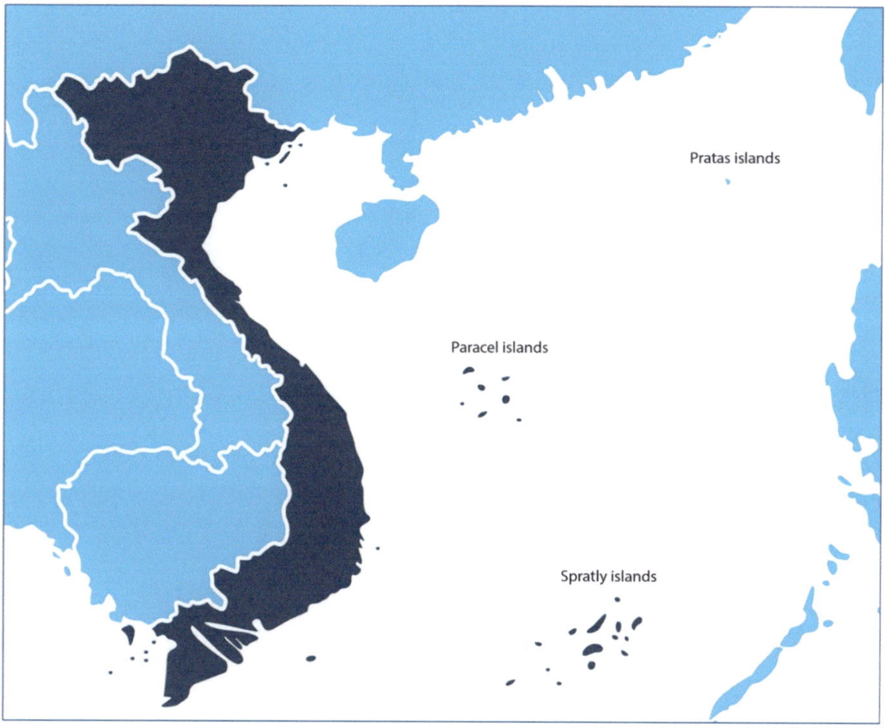

Fig. 1 Map of the east sea region, including coastal Vietnam and remote archipelagos

Thuan, have isolated numerous fungal and bacterial strains exhibiting antimicrobial properties. For instance, research in Nha Trang Bay identified 100 fungal strains from different marine sources, with 59 strains showing antimicrobial activity against multiple human pathogens [7]. Similarly, investigations in the Son Tra Peninsula found that 29 out of 73 fungal isolates exhibited antimicrobial activity against at least two tested pathogens [8]. Additionally, the coastal waters of Ninh Thuan and Binh Thuan yielded 86 microbial strains, with 68 demonstrating antibacterial activity [9]. These findings underscore the untapped potential of Vietnam's coastal ecosystems in contributing to the discovery of novel marine natural products, which could lead to the development of new drugs and therapeutic agents. This section provides an extensive survey of microbial natural products reported diverse marine microbes found at Vietnam coastal regions. A summary of new marine microbial molecules from coastal Vietnam are presented in Table 1.

Table 1 List of new microbial natural products reported from coastal Vietnam

Marine microorganism/Source	Compound*	Biological activity	Refs.
Marine bacteria			
Actinoalloteichus cyanogriseus G631/soft coral *Sarcophyton Infundibuliforme*	Isocyanogramide (**51**)	Antimicrobial	[31]
Bacillus siamensis 168CLC-66.1/ green alga *Caulerpa* sp.	Pityriacitrin D (**1**)	Cytotoxic	[10]
Micromonospora sp. G019/sediment	3,4-dihydroxy-6,7-dimethyl-quinoline-2-carboxylic acid (**3**); 2-[(5-methyl-1,4-dioxan-2-yl)methoxy]ethanol (**288**)	Antimicrobial	[11]
Micromonospora sp./sediment	Isopimara-2-one-3-ol-8,15-diene (**87**)	–	[40]
Micromonospora sp. G047/sediment	1-methyl-6-acetyl-α-D-glucopyranoside (**281**)*	–	[59]
Nocardiopsis sp. G057/sediment	2-[(2*R*-hydroxypropanoyl)amino]benzamide (**4**); 3-acetyl-4-hydroxycinnoline (**5**)*; 3,3′-bis-indole (**6**)*	Antimicrobial; cytotoxic (for **6**)	[13]
Nocardiopsis synnemataformans HT06/sediment	4*βC*-glucopyranosyl-3,5-dihydroxy-2-methoxybenzoic acid (**247**)	Antimicrobial	[17]
Streptomyces sp. G278/sea cucumber *Holothuria edulis*	3-hydroxyl-2-methylpyridine (**17**)*; 2,5-bis(5-*tert*-butyl-2-benzoxazolyl)thiophene (**287**)*	**17**: antimicrobial	[20]
Streptomyces sp. G248/marine sponge *Halichondria panicea*	(2*S*,2″*S*)-6-lavandulyl-7,4′-dimethoxy-5,2′-dihydroxylflavanone (**75**); (2*S*,2″*S*)-6-lavandulyl-5,7,2′,4′-tetrahydroxylflavanone (**76**); (2″*S*)-5′-lavandulyl-2′-methoxy-2,4,4′,6′-tetrahydroxylchalcone (**77**)	Antimicrobial	[15]
Streptomyces sp. G212/sediment	4,5-dihydroxy-7-methylphthalide (**269**); 2,4-dichlorophenyl 2,4-dichloro benzoate (**309**)*	Antimicrobial	[64]
Streptomyces sp. G261/sediment	1,3-benzodioxole-2-one-4-carboxylamide (**308**)	Antimicrobial; cytotoxic	[19]

(continued)

Table 1 (continued)

Marine microorganism/Source	Compound*	Biological activity	Refs.
Tsukamurella pseudospumae IV19-045/sediment	Tsukalipids A (**245**) & B (**246**)	–	[10]
Marine fungi			
Ascomycota sp. VK12/marine sponge	(3*R*)-(3′,5′-dihydroxyphenyl)butan-2-one (**162**)	Cytotoxic; anti-inflammatory	[54]
Aspergillus flocculosus/brown alga *Padina* sp.	14,15-dehydro-6-*epi*-ophiobolin K (**88**); 14,15-dehydro-ophiobolin K (**89**); 14,15-dehydro-6-*epi*-ophiobolin G (**90**); 14,15-dehydro-ophiobolin G (**91**); 14,15-dehydro-(*Z*)-14-ophiobolin G (**92**)	Cytotoxic	[41]
Aspergillus flocculosus 01NT.1.1.5/marine sponge *Stylissa* sp.	Ochraceopone F (**97**)	–	[28]
Aspergillus flocculosus/sediment	12-*epi*-aspertetranone D (**100**); 6β,9α,14-trihydroxycinnamolide (**101**); 6β,7β,14-trihydroxyconfertifolin (**102**); aspilactonol G (**137**)	**101**: cytotoxic	[43]
Aspergillus flocculosus 168ST-16.1/brown alga *Padina* sp.	Deketo-phomaligol A (**237**); phomaligol E (**238**)	–	[52]
Aspergillus fumigatus M580/sea cucumber *Colochirus quadrangularis*	6-methoxyindole-3-carboxylic acid *O*-β-D-glucopyranosyl ester (**280**)	–	[30]
Aspergillus micronesiensis/red alga *Kappaphycus alvarezii*	Aspermicrones A (**146**)–C (**148**)	**147**: cytotoxic **147** & **148**: antimicrobial	[51]
Aspergillus niveoglaucus/sediment	Niveoglaucin A (**140**); niveoglaucin B (**141**)	–	[50]
Aspergillus nidulans M256/marine sponge *Echinodictyum conulosum*	7-dechloronidulin (**151**)	Antimicrobial	[53]
Aspergillus terreus LM.5.2/mangrove	Asterripeptides A (**178**)–C (**180**)	Cytotoxic **179** & **180**: sortase inhibitors	[56]

(continued)

Table 1 (continued)

Marine microorganism/Source	Compound*	Biological activity	Refs.
Hamigera avellanea/marine sponge *Tethya maza*	Hamiavemin A (**37**)	Antimicrobial; cytotoxic	[27]
Lopadostoma pouzarii strain 168CLC-57.3/marine sponge	Lopouzanone A (**127**); lopouzanone B (**128**); 1-*O*-acetyl & 2-*O*-acetyl derivatives of dendrodochol B (**129** & **130**)	–	[49]
Paraconiothyrium sp. VK-13	1-(2,5-dihydroxyphenyl)-3-methoxy-butan-1-one (**210**); 1-(2,5-dihydroxyphenyl)-3-hydroxybutan-1-one (**211**); modiolides D (**256**)–G (**259**)	**211**: anti-inflammatory	[65]
Penicillium janthinellum 168CLC-17.1/sediment	Restricticin B (**166**)	Anti-neuroinflammatory	[25]
Penicillium janthinellum/sediment	Resorcinoside A (**221**); resorcinoside B (**222**)	**221**: cytotoxic	[66]
Penicillium oxalicum M893/brown alga *Spatoglossum* sp.	Oxaliterpenoid (**98**)	Antimicrobial	[42]
Talaromyces cyanescens/green alga *Caulerpa* sp.	Talacyanols A (**277**)–C (**279**)	**277**: cytototxic **278**: anti-neuroinflammatory	[69]
Marine Microalgae			
Aurantiochytrium sp. SC145	*N*-methoxymethyltryptophol (**22**); (2*R*)-2-(4-hydroxyphenyl)ethyl 2-hydroxy-3-phenylpropanoate (**215**)	**215**: antimicrobial	[21]
Lyngbya majuscula	Nhatrangins A (**174**) & B (**175**)	–	[55]

*isolated from natural source for first time

2.1 Alkaloids

Two alkaloids, including a new indole alkaloid, pityriacitrin D (**1**) and known, 1-acetyl-*b*-carboline (**2**), were isolated from the bacterium, *Bacillus siamensis* 168CLC-66.1, derived from the marine seaweed *Caulerpa* sp. (Fig. 2). Cytotoxic assays revealed that compound **1** showed significant cytotoxic activity against all the tested cancer cell lines HCT-15, NUGC-3, NCI-H23, ACHN, PC-3 and MDA-MB-231 with GI_{50} values ranging from 8.0 to 10.9 μM [10]. Additionally, a new quinoline alkaloid, 3,4-dihydroxy-6,7-dimethyl-quinoline-2-carboxylic acid (**3**) was isolated from the marine sediment-derived actinomycete *Micromonospora* sp. G019 (Fig. 2) [11]. Compound **3** was found to exhibit antibacterial activity against *E. coli*, *S. enterica*, *E. faecalis*, *S. aureus*, *B. cereus*, *E. coli* and *P. aeruginosa* with MIC values from 32 to 256 μg/mL [11, 12].

Seven alkaloids, including 2-[(2*R*-hydroxypropanoyl)amino]benzamide (**4**), 3-acetyl-4-hydroxycinnoline (**5**), 3,3′-bis-indole (**6**), 3,3-(2,3-dihydroxypropyl)diindole (**7**), 3-(2*R*-hydroxyl-1-one-propane)-indole (**8**), 3-formylindole (**9**), 3-indolylacetic acid (**10**) were obtained from the marine-derived actinomycete *Nocardiopsis* sp. G057 (Fig. 2). Of these, **4** was identified to be a new enantiomeric isomer while **5** and **6** were isolated from a natural source for the first time. Compound **4** demonstrated selective inhibition against *E. coli* and *P. aeruginosa* with MIC values of 16 and 128 μg/mL, respectively, whereas **5–8** and **10** exhibited antimicrobial activity against both Gram-positive and Gram-negative bacteria, *E. faecalis*, *S. aureus*, *B. cereus*, *E. coli*, *P. aeruginosa*, *S. enterica*, and the yeast *Candida albicans* with MIC values ranging from 64 to 256 μg/mL. Furthermore, **6** exhibited selective inhibition toward KB and LU-1 cancer cell lines with IC_{50} values of 12.5 and 25.6 μg/mL, respectively [13].

Five alkaloids, comprising 3-indolecarbaldehyde (**11**), chrysogine (**12**), and 1*H*-indole-3-acetic acid (**13**) were characterized from the marine sediment *Penicillium* sp.

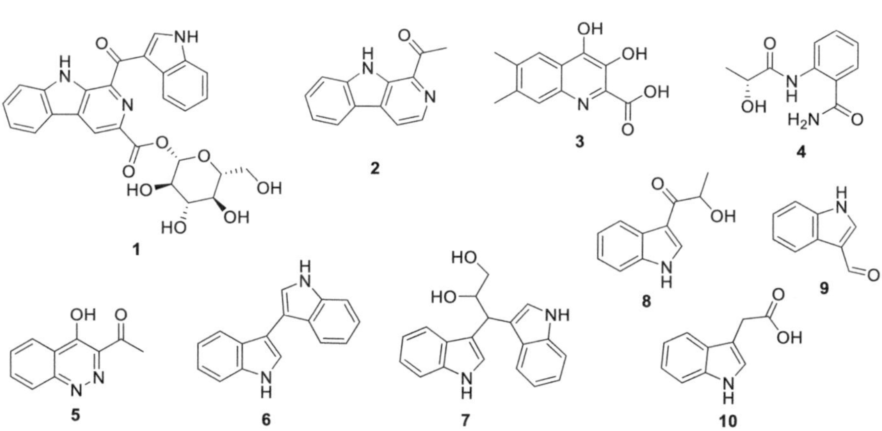

Fig. 2 Chemical structures of **1–10**

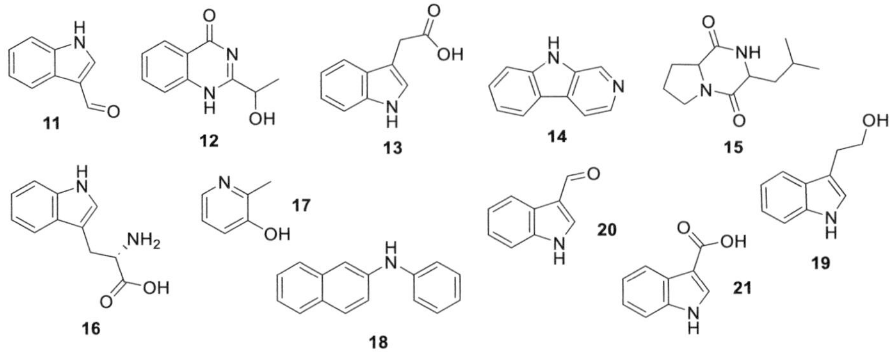

Fig. 3 Chemical structures of **11–21**

(Fig. 3). Of these, **13** exhibited selective antibacterial effect against *E. faecalis* with MIC of 64 μg/mL [14]. In addition, norharman (**14**) was isolated from the sponge-derived *Streptomyces* sp. G248 [15], and 9H-pyrido[3,4-b]indole (**15**), tryptophan (**16**) were isolated from the sponge-derived actinomycete *Streptomyces* sp. G246 (Fig. 3) [16]. Of these, **14** exhibited selective inhibitory activity against *E.* coli, *S. aureus*, and *B. cereus* with MIC of 32–128 μg/mL [17–19] and **15** exhibited selectively antibacterial activity against *E. coli* with 128 μg/mL [16]. Moreover, 3-hydroxyl-2-methylpyridine (**17**), *N*-phenylnaphthalen-2-amine (**18**), tryptophol (**19**), indole-3-carbaldehyde (**20**), and indole-3-carboxylic acid (**21**) were isolated from the echinoderm-derived actinomycete *Streptomyces* sp. G278 (Fig. 3) [20]. Compound **17** was isolated from a natural source for the first time. Compound **17** was found to exhibit antibacterial and antifungal activity against *E. coli, S. enterica, S. aureus, E. faecalis,* and *C. albicans* with MIC values of 64–256 μg/mL, whereas **18** inhibited *E. faecalis, S. aureus, B. cereus, E. coli* and *C. albicans* with MIC values of 128–256 μg/mL, and **19** selectively inhibited *E. coli* with MIC values of 8 μg/mL.

One new indole, *N*-methoxymethyltryptophol (**22**), and three known ones, indole-3-acetic acid (**23**), methyl indole-3-acetate (**24**), were isolated from the marine microalga, *Aurantiochytrium* sp. SC145 (Fig. 4) [21], whereas, oxaline (**25**) was isolated from the sponge-associated *P. oxalicum* CLC-MF05 (Fig. 4) [22]. Bioassays demonstrated that **25** inhibited NO and PGE2 overproduction, as well as iNOS and COX-2 overexpression in both LPS-stimulated BV2 and rat primary microglia with IC_{50} of 8.8–9.2 μM. Compound **25** also repressed mRNA expression of TNF-α, IL-1β, IL-6, and IL-12. Subsequently, mechanistic studies revealed that the inhibitory actions of **25** were regulated by the inactivation of the NF-κB and MAPK signaling pathways. Furthermore, the inactivation of the TLR4/MyD88 pathway contributed to the anti-neuroinfammatory activity of compounds **25** [22].

Notably, seven echinulin-related indolediketopiperazine alkaloids, (+)-cryptoechinuline B (**26a**) and (−)-cryptoechinuline B (**26b**), (+)-cryptoechinuline D (**27a**), (−)-cryptoechinuline D (**27b**), echinulin (**28**), neoechinulins B (**29**), C (**30**) and E (**31**) as well as neoechinulin (**32**) were isolated from the sediment-derived fungus

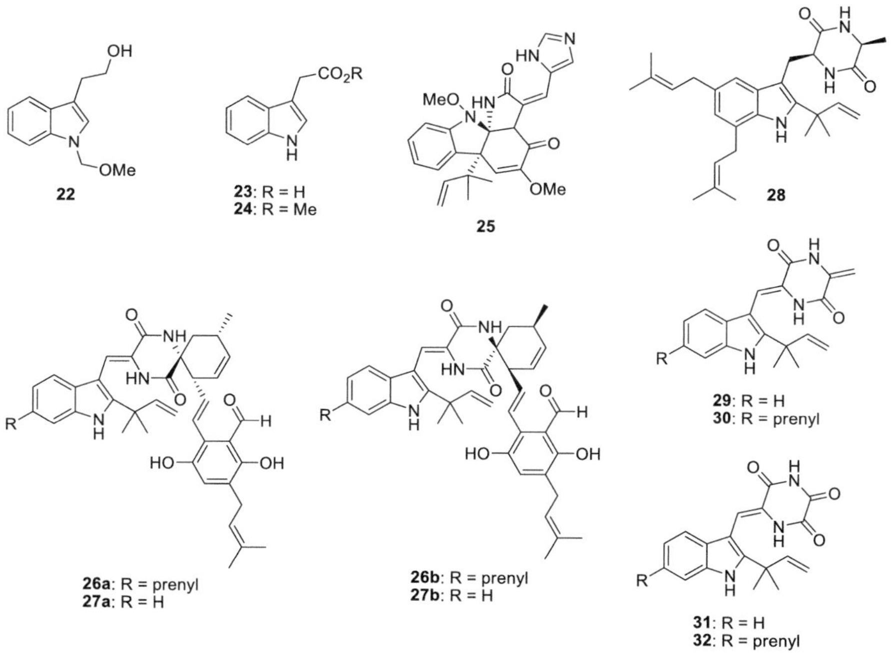

Fig. 4 Chemical structures of **22–32**

Aspergillus niveoglaucus (Fig. 4) [23]. Of these, compounds **29** and **30** showed cytotoxicity with IC_{50} values of 50.9 and 40.6 μM, respectively. This results support the idea that the α,β-unsaturated carbonyl moiety in **29** and **30** control the toxicity of these molecules [24]. Of note, compounds **28** and **32** were cytotoxic against human drug-resistant and drug-sensitive prostate cancer cells. Compound **28** exhibited cytotoxicity towards 22Rv1, PC-3, and LNCaP cells with IC_{50} values of 63.2, 41.7, and 25.9 μM, respectively, and **32** showed similar activities with IC_{50} values of 49.9, 63.8, and 38.9 μM, respectively. Compound **32** also binds DPPH free radicals with IC_{50} of 62.6 μM [23].

Nine alkaloids, including cycloechinulin (**33**) and mactanamide (**34**), were characterized from the marine fungus *Aspergillus flocculosus* 01NT.1.1.5, fellutanine B (**35**) were obtained from the marine sediment-derived fungus *Penicillium janthinellum* 168CLC-17.1 (Fig. 5) [25], (2*S*)-2,3-dihydro-7-hydroxy-6,8-dimethyl-2[(*E*)-prop-1-enyl]-chroman-4-one (**36**) was isolated from the marine sediment-derived fungus *Penicillium citrinum* VM6 [26], and a pair of undescribed alkaloid enantiomers (+)-hamiavemin A (4*S*) (**37a**) and (−)-hamiavemin A (4*R*) (**37b**), together three known metabolites, fumiquinazoline C (**38**), *trans*-*p*-coumaroyl-L-tryptophan (**39**), and 3-(4-hydroxyphenyl)pyrazin-2-ol (**40**), were isolated from marine sponge-associated fungus *Hamigera avellanea* (Fig. 5) [27].

Among them, **34** displayed suppressive activities on receptor activator of nuclear factor κB ligand (RANKL)-induced differentiation of Bone marrow macrophages

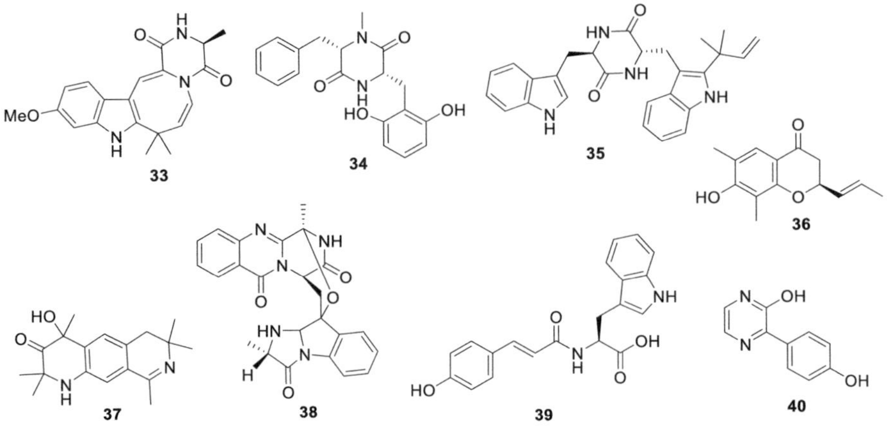

Fig. 5 Chemical structures of **33**–**40**

(BMMs) into osteoclasts without cytotoxic effect on BMMs [28], whereas compound **35** decreased the production of NO in lipopolysaccharide (LPS)-stimulated BV-2 microglial cells without cytotoxicity [25]. Antimicrobial assay indicated that **36** exhibited antimicrobial activity against *E. faecalis, S. aureus, B. cereus, S. enterica* and *C. albicans* with MIC values of 128, 64, 128, 256, 16 μg/mL [26], whereas **37** showed antimicrobial activity against *E. faecalis, S. aureus, B. cereus, E. coli* with MICs of 32–256 μg/mL. Furthermore, **38** showed antimicrobial activity against *E. faecalis* with MIC of 128 μg/mL, whereas **39** showed antibacterial activity against *E. faecalis, S. aureus, B. cereus, E. coli* and *C. albicans* with MICs of 64–256 μg/mL and **40** showed antimicrobial activity against *E. faecalis, S. aureus* with the MICs of 64 μg/mL. In addition, **37** showed cytotoxic activity against three cancer cell lines, HepG2, A549, and MCF-7 with the IC_{50} values ranging from 55.35 ± 1.70 to 63.58 ± 5.77, respectively [27].

From the prickly pen shell-derived *Streptomyces fradiae* G650 collected in Van Phong bay, Vietnam, 15*α*-hydroxyneoline (**41**) was isolated (Fig. 6) [29]. Six other alkaloids, fumiquinazoline D (**42**), fumiquinazoline J (**43**), bisdethiobis(methylthio)gliotoxin (**44**), tryprostatin B (**45**), 12,13-dihydroxyfumitremorgin C (**46**) and 6-methoxyspirotryprostatin B (**47**) (Fig. 6), were extracted from marine sea cucumber-derived *Aspergillus fumigatus* M580 [30]. Bioassays revealed that **42** and **45** exhibited selectively antimicrobial activity against *E. faecalis* with MIC values of 32 and 64 μg/mL, respectively and **42**–**45**, and **47** showed selectively antimicrobial activity on yeast *C. albicans* with all MIC values of 128 μg/mL. In addition, **42**–**44** was found to exhibit cytotoxic activity against Huh-7 and HT-29 cancer cell lines with IC_{50} values ranging from 9.7 to 70.9 μM.

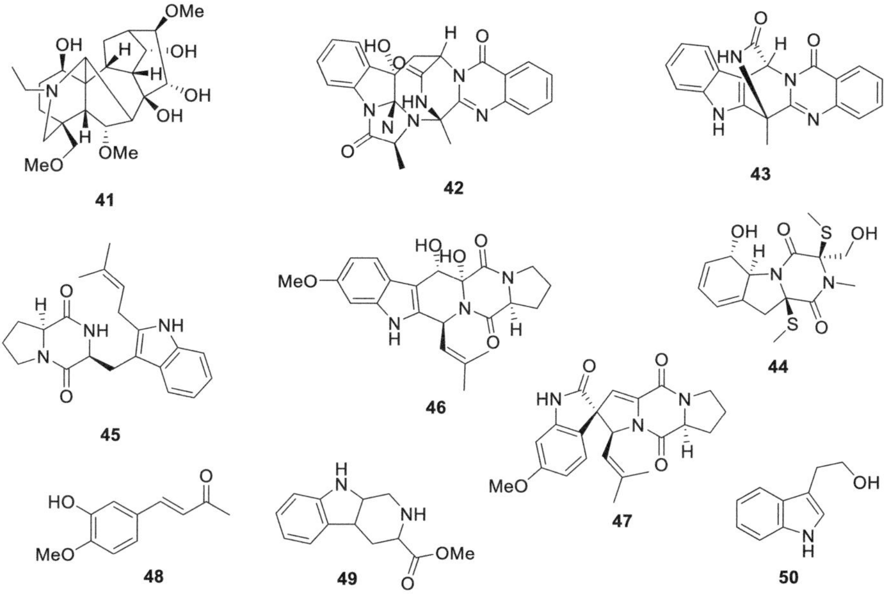

Fig. 6 Chemical structures of **41–50**

Analysis of chemical components from the marine sediment-derived *Nocardiopsis synnemataformans* HT06 led to the isolation of (*E*)-4-(3-hydroxy-4-methoxyphenyl)but-3-en-2-one (**48**), methyl 2,3,4,9-tetrahydro-1*H*-pyrido[3,4-*b*]indole-3-carboxylate (**49**), and 1*H*-indole-3-ethanol (**50**) (Fig. 6) [17]. Of these, **48** displayed selectively antifungal activity against *C. albicans* with MIC of 128 μg/mL, whereas **49** showed antimicrobial activity against *E. faecalis*, *S. aureus*, *B. cereus* and *C. albicans* with MIC values of 64–256 μg/mL and **50** displayed selectively antifungal activity against displayed selectively antibacterial activity against *S. aureus* with MIC of 64 μg/mL.

A new alkaloid, isocyanogramide (**51**) together with five known ones, including cyanogramide (**52**), marinacarboline F (**53**), marinacarboline H (**54**), caerulomycinamide (**55**), cerulomyconitrile (**56**) were isolated from the marine coral-derived actinomycete *Actinoalloteichus cyanogriseus* G631 (Fig. 7) [31]. All compounds **51–56** exhibited significant antimicrobial effects on *E. faecalis*, *S. aureus* and *C. albicans* with MIC values ranging from 64 to 256 μg/mL. Compounds **54** and **55** inhibited *B. cereus* with MIC values of 16 and 128 μg/mL, respectively. In addition, compounds **52** and **54** also showed potent inhibitory activity against *P. aeruginosa* with a MIC value of 64 μg/mL. Compounds **52** and **53** showed selectively inhibitory activity against the *E. coli* with MIC values of 32 and 256 μg/mL, respectively. From another investigation, diketopiperazine dimer WIN 64821 (**57**) (Fig. 7) was isolated from the marine sediment-derived *Penicillium* sp. OM07 [12]. Compound **57** demonstrated selective inhibition against *E. faecalis*, *S. aureus* and *C. albicans* with a MIC value of 64–256 μg/mL.

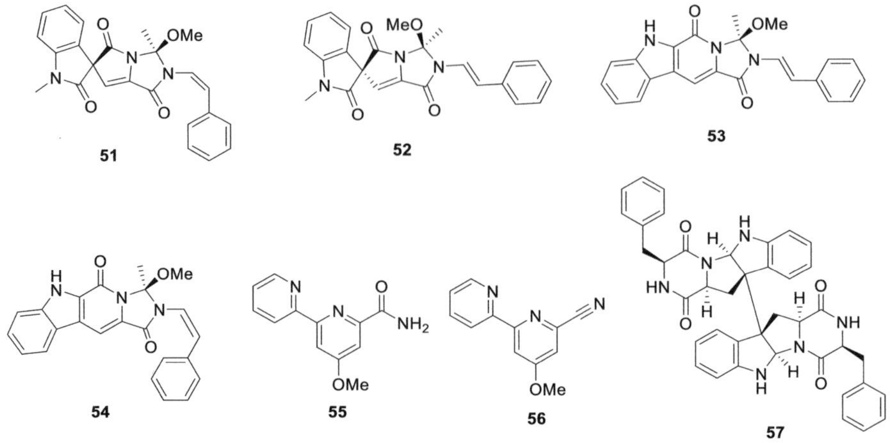

Fig. 7 Chemical structures of **51**–**57**

In addition, numerous other known alkaloids isolated from marine microorganisms in Vietnam have been reported (Fig. 8), including spirotryprostatin A (**58**) from sponge-associated *Streptomyces* sp. G246 [32], terezine D (**59**), cyclotryprostatin A (**60**), pseurotin A (**61**), tryptoquivaline J (**62**) from the marine sediment-fungus *Aspergillus* sp. M28 [33], tryptoline (**63**) from the marine crinoid-derived *Streptomyces* sp. C011 [9], 2-acetamidobenzamide (**64**) from marine sediment-derived *Streptomyces* sp. G261 [19], bacilsubteramide A (**65**), 4,9-dihydroxy-3-isobutyl-2,3,4,5-tetrahydrobenzo-azepin-1-one (**66**), *N*-acetyltryptamine (**67**), *N*-(4-hydroxyphenyl)ethyl-methylbutanamide) (**68**), *N*-phenylacetamide (**69**) from the sponge-derived *Micromonospora* sp. A258 [34], indole 3-acetic acid (**70**), 3-hydroxyacetylindole (**71**) from the marine sediment-derived *Streptomyces* sp. G330 [35], *N*-(4-hydroxyphenylethyl)propionamide (**72**) and *N*-(4-hydroxyphenylethyl)acetamide (**73**) from the marine sediment bacterium *Micromonospora* sp. G019 [36] and from marine sediment-derived *Streptomyces* sp. G039 [37] and (2*S*,4*S*)-4-hydroxyproline (**74**) from the marine sediment-derived bacterium *Oceanisphaera* sp. [38]. However, the biological properties of these compounds were not reported.

2.2 *Flavonoids*

Three new lavandulylated flavonoids, (2*S*,2″*S*)-6-lavandulyl-7,4′-dimethoxy-5,2′-dihydroxylflavanone (**75**), (2*S*,2″*S*)-6-lavandulyl-5,7,2′,4′-tetrahydroxylflavanone (**76**) and (2″*S*)-5′-lavandulyl-2′-methoxy-2,4,4′,6′-tetrahydroxylchalcone (**77**), along with two known flavonoids (2*S*,2″*S*)-6-lavandulyl-7-methoxy-5,2′,4′-trihydroxylflavanone (**78**) and 6-prenyl-4′-methoxy-5,7-dihydroxylflavanone (**79**),

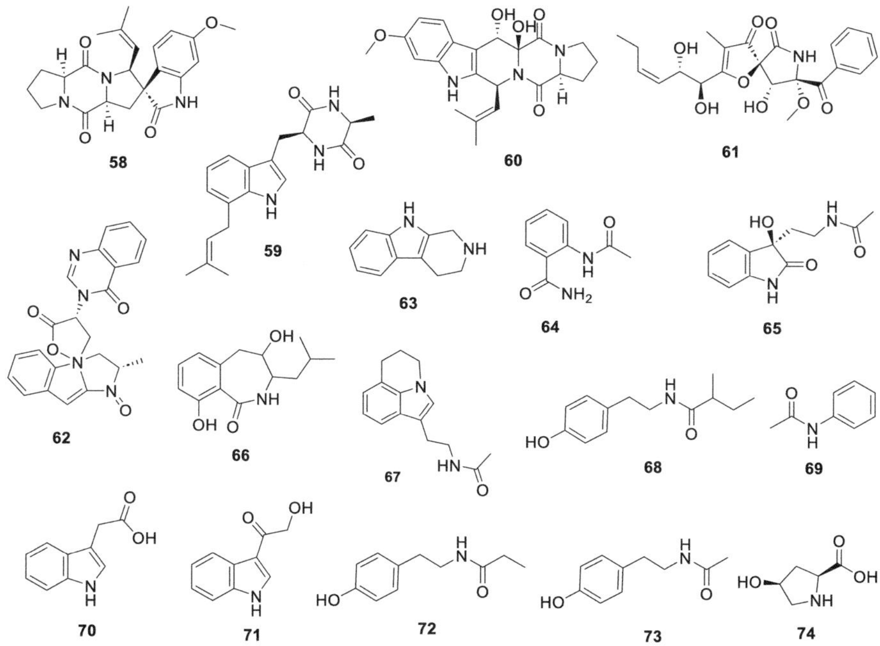

Fig. 8 Chemical structures of **58**–**74**

were isolated from the sponge-derived *Streptomyces* sp. G248 (Fig. 9) [15]. Of these, compounds **75**–**77** exhibited remarkable antimicrobial activity against a broad spectrum of pathogenic microorganisms *E. faecalis*, *S. aureus*, *B. cereus*, *E. coli*, *P. aeruginosa*, *S. enterica*, *C. albicans*, *Mycobacterium tuberculosis* H37Rv with IC_{90} values of 1–48 μg/mL, whereas compounds **78** and **79** exhibited inhibitory activity against *Mycobacterium tuberculosis* H37Rv with MIC values of 6.0 μg/mL and 11.1 μg/mL, respectively. Structure–activity analyses imply that the methoxy group at C-7 and hydroxyl functionality at C-4′ seem to be critical for the anti-TB activity of the lavandulylated flavonoids. Furthermore, cytotoxicity evaluation indicated that compounds **78** and **79** exhibited significant cytotoxic activity against the cancer cell lines KB, Hep-G2, Lu-1 and MCF-7 with IC_{50} values from 2.0 to 14.6 μg/mL.

In another investigation, two new lavandulylated flavonoids, 6-lavandulyl-7-methoxy-5,2′,4′-trihydroxyflavanone (**80**) and 5′-lavandulyl-4′-methoxy-2,4,2′,6′-tetrahydroxylchalcone (**81**) were isolated from the sponge-derived actinomycete *Streptomyces* sp. G246 (Fig. 9) [16]. Notably, both compounds **80** and **81** had a broad spectrum of antimicrobial activity against *E. coli*, *P. aeruginosa*, *S. enterica*, *E. faecalis*, *S. aureus*, *B. cereus* and *C. albicans* strains with MIC values of 1–32 μg/mL. In addition, two other lavandulylated flavonoids, 6-lavandulyl-5,7,2′,4′-tetrahydroxyflavanone (**82**) and 5′-lavandulyl-2′-methoxy-2,4,4′,6′-tetrahydroxychalcone (**83**), were characterized from the marine crinoid-derived *Streptomyces* sp. C011 (Fig. 9) [18]. Compound **82** exhibited antimicrobial

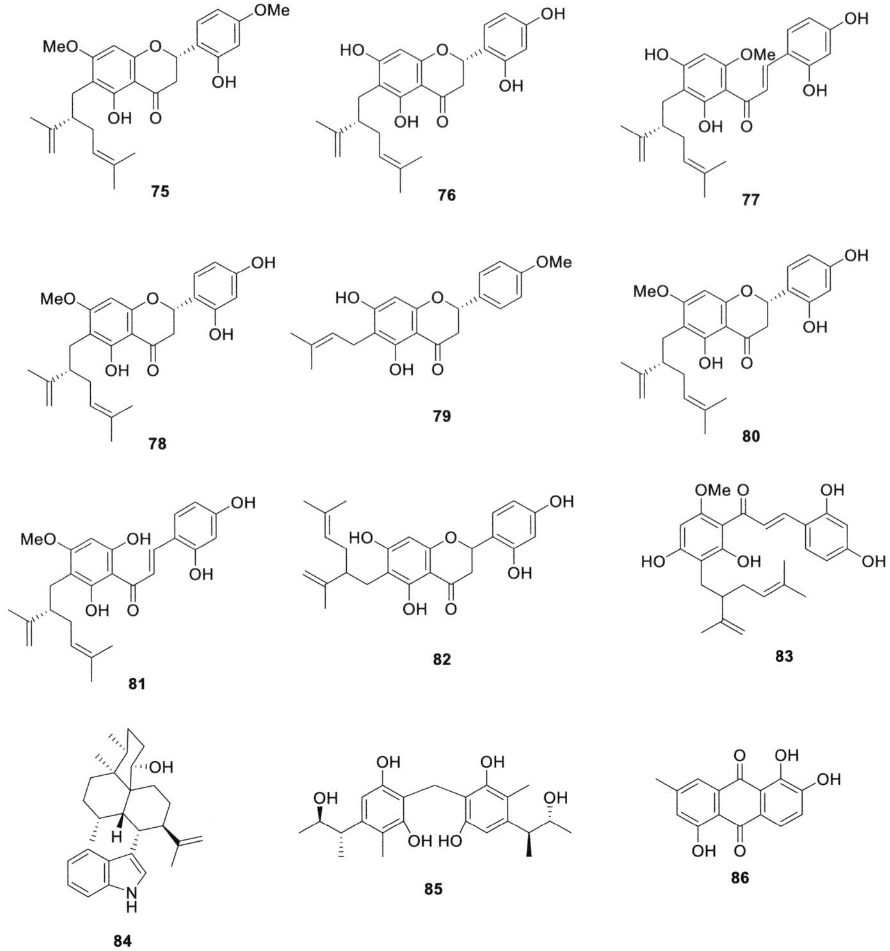

Fig. 9 Chemical structures of **75**–**86**

activity against *E. faecalis* and *S. aureus, B. cereus, P. aeruginosa, S. enterica* and *C. albicans* with MIC values of 1–8 μg/mL, whereas **83** showed inhibitory activity against all tested Gram-positive, Gram-negative and *C. albicans* with MIC values of 4–16 μg/mL.

Several other flavonoids were isolated from marine microorganisms in Vietnam (Fig. 9), including 10,23-dihydro-24,25-dehydroaflavinine (**84**) from the marine sediment-derived fungus *Penicillium janthinellum* 168CLC-17.1 [25], dicitrinone F (**85**) from the marine sediment-derived fungus *Penicillium citrinum* VM6 [26] and 1,2,5-trihydroxy-7-methyl-9,10-antraquinone (**86**) from the marine fungus *Aspergillus terreus* LM.1.5 [39]. Among them, **84** exhibited inhibitory activity on nitric oxide (NO) production in lipopolysaccharide (LPS)-stimulated BV-2 microglial cells. However, **84** also displayed strong toxicity toward BV-2 microglial cells [25]. Compound **85** exhibited antimicrobial activity against *E. faecalis, S. aureus, B. cereus, S. enterica* and *C. albicans* with MIC values of 16–256 μg/mL. Furthermore, **85** displayed cytotoxicity against all tested cell lines A549, MCF7, MDA-MB-231, Hela, and AGS with IC_{50} values of 14.6 ± 1.3 to 29.6 ± 2.2 μg/mL [26].

2.3 Terpenoids

A novel pimarane diterpene, isopimara-2-one-3-ol-8,15-diene (**87**) was isolated from marine sediment-derived *Micromonospora* sp. (Fig. 10) [40] and five new sesterterpenes, 14,15-dehydro-6-*epi*-ophiobolin K (**88**), 14,15-dehydro-ophiobolin K (**89**), 14,15-dehydro-6-*epi*-ophiobolin G (**90**), 14,15-dehydro-ophiobolin G (**91**) and 14,15-dehydro-(Z)-14-ophiobolin G (**92**), together with four known ophiobolins, 6-*epi*-ophiobolins C and N (**93** and **94**) and ophiobolins C and N (**95** and **96**) were isolated from the marine fungus *Aspergillus flocculosus* derived from the seaweed *Padina* sp. (Fig. 10) [41]. All compounds **88–96** showed potent cytotoxicity toward all cancer cell lines HCT-15, NUGC-3, NCI-H23, ACHN, PC-3 and MDA-MB-231 with GI_{50} values ranging from 0.14 to 2.01 μM. The analysis of structure–activity relationships of **92–100** suggests that the geometry of C-14/C-15 might influence their activities, whereas the stereochemistry of C-6 and the hydroxyl group at C-3 might not noticeably affect the cytotoxicity [41].

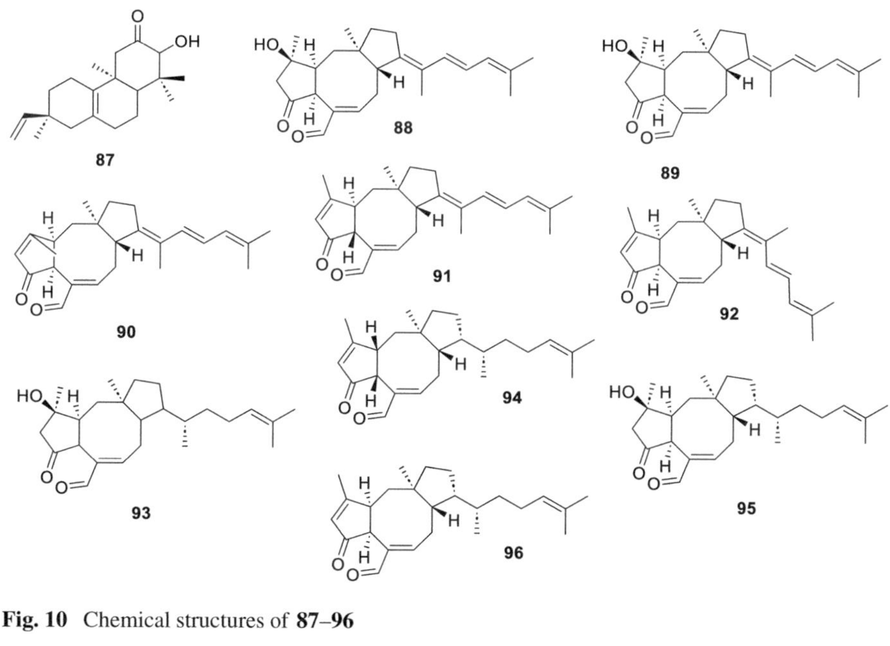

Fig. 10 Chemical structures of **87–96**

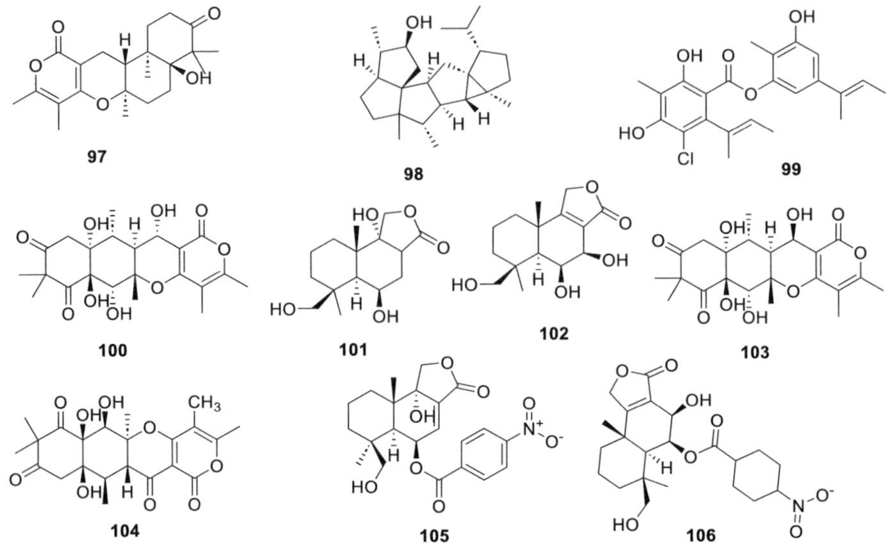

Fig. 11 Chemical structures of **97–106**

A new α-pyrone merosesquiterpenoid with an angular tetracyclic carbon skeleton, ochraceopone F (**97**) was isolated from *Aspergillus flocculosus* 01NT.1.1.5 derived from the marine sponge *Stylissa* sp. [28] and one new sesterterpenoid, oxaliterpenoid (**98**), and known guisinol (**99**) were isolated from the marine algal-derived *Penicillium oxalicum* M893 (Fig. 11) [42]. Compound **98** showed antibacterial activity against *E. faecalis, S. aureus, B. cereus, E. coli* and *C. albicans* with the MICs of 32–128 μg/mL, whereas compound **99** showed antimicrobial activity against *E. faecalis, S. aureus, B. cereus, S. enterica*, and *C. albicans* with the MICs of 4–16 μg/mL.

Seven terpenoids, including three new compounds, 12-*epi*-aspertetranone D (**100**), 6β,9α,14-trihydroxycinnamolide (**101**), 6β,7β,14-trihydroxyconfertifolin (**102**), together with four known compounds, aspertetranone D (**103**), aspertetranone A (**104**), insulicolide A (**105**), and 7α,14-dihydroxy-6β-*p*-nitrobenzoylconfertifolin (**106**) were isolated from marine sediment-derived *Aspergillus flocculosus* (Fig. 11) [43]. Of these, **101** demonstrated cytotoxic activity toward Neuro-2a cell with IC_{50} of 24.1 μM, while compound **105** showed cytotoxic activity toward Neuro-2a cell with IC_{50} of 4.9 μM. Compounds **101** and **105** showed cytotoxic activity toward 22Rv1 cells with IC_{50} values of 31.5 μM and 3.0 μM, respectively. Furthermore, compound **105** exhibited a cytotoxic *ef*fect toward MCF-7 cells with IC50 of 59.6 μM.

The structure–activity analysis of compounds **101–106,** together with literature data, revealed three relevant structural sites. First, a double bond at C7 = C8, which is part of an α,β-unsaturated lactone. It has been demonstrated that the cytotoxicity of such moiety can be explained by a nucleophilic Michael addition reaction with biological nucleophiles [44, 45]. In the case of the non-cytotoxic compounds **102** and **106**, the double bond of the α,β-unsuturated lactone may be rendered inaccessible to a nucleophile attack because of steric obstacles. Second, the presence of a hydroxyl group at C-9 in the drimane core is also crucial for cytotoxicity. Indeed, a recent report of a series of similar compounds indicated that the most pronounced cytotoxicity was observed in compounds containing a 9-OH group [46]. In addition, the structure–activity analysis of compounds **101–106** suggests that the presence of a *p*-nitrobenzoyl moiety substantially enhances the cytotoxicity. The nitrobezoylation of 6-OH resulted in the increase of the cytotoxicity of related compounds against human renal cell carcinoma cells in comparison to 14-OH-derivatives. At the same time, it should be noted that another investigation into the cytotoxicity of 6- and 14-nitrobenzoate derivatives against other cancer cell lines did not support this observation [47].

Other known terpenoids were isolated from marine microorganisms in Vietnam (Fig. 12), including asterriquinone F (**107**), asterriquinone A3 (**108**), asterriquinone B4 (**109**), asterriquinone C1 (**110**), asterriquinone C2 (**111**) and asterriquinone D (**112**) from *Aspergillus terreus* LM.1.5 [39] and chaxine C (**113**) from the marine microalga *Aurantiochytrium* sp. SC145 [21]. Of these, compounds **109** and **110** exhibited cytotoxic activity against Neuro-2a cells with IC_{50} at 91.45 and 42.32 μM, respectively, whereas compound **113** selectively inhibited *E. faecalis* and *C. albicans* with MIC values of 128 and 256 μg/mL.

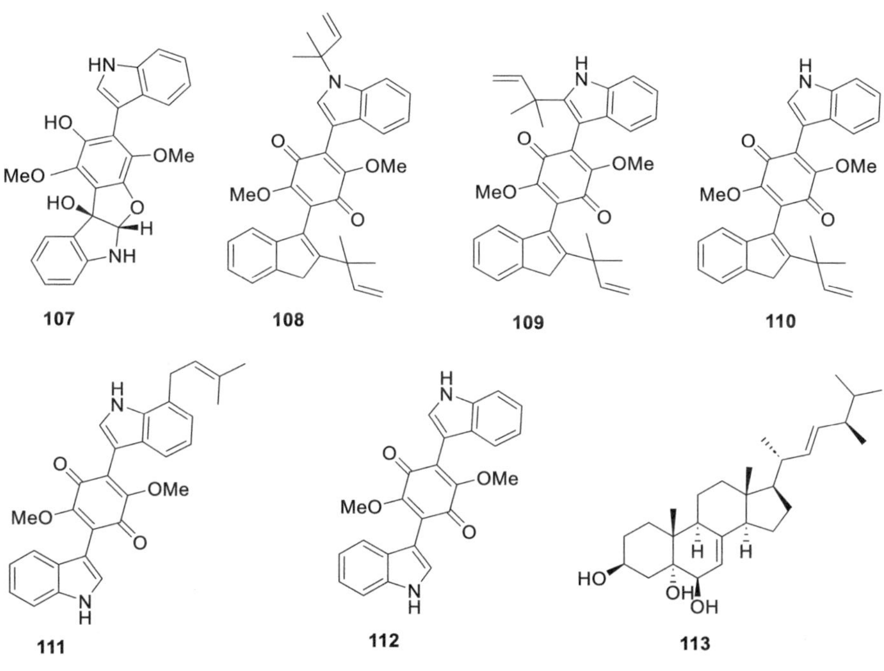

Fig. 12 Chemical structures of **107–113**

2.4 Steroids

Five steroids, (22*E*,24*R*)-5α,8α-*epi*-dioxyergosta-6,22-dien-3β-ol (**114**), (22*E*,24*R*)-5α,8α-epidioxyergosta-6,9(11),22-trien-3β-ol (**115**), cerevisterol (**116**), ergosta-5,7,22-trien-3β-ol (**117**) and fungisterol (**118**), were isolated from the marine microalga *Aurantiochytrium* sp. SC145 (Fig. 13) [21]. Compounds **114–116** significantly inhibited Gram-positive bacteria (*E. faecalis, S. aureus, B. cereus)* and the yeast *C. albicans* with MIC values ranging from 64 and 128 μg/mL, whereas compound **117** selectively inhibited *B. cereus* and *C. albicans* with MIC values of 128 μg/mL and compound **118** inhibited *E. faecalis, B. cereus* and *C. albicans* with MIC values of 32–128 μg/mL. The chemical investigation of the marine sediment-derived fungus *Penicillium citrinum* VM6 also led to the isolation of 3β-hydroxy-5,9-epoxy-(22*E*,24*R*)-ergosta-7,22-dien-6-one (**119**), melithasterol B (**120**) (Fig. 13) [26]. Of these, compound **119** was isolated for the first time from the *Penicillium* genus. Both compounds **119** and **120** selectively inhibited gram-positive bacteria *E. faecalis, S. aureus, B. cereus* and the yeast *C. albicans* with MIC values ranging from 32 to 256 μg/mL. In addition, compound **116** alleviated NO production in LPS-induced murine RAW264.7 macrophages with IC_{50} of 25.45 μg/mL [48] and compound **120** exhibited cytotoxicity against all tested cell lines A549, MCF7, MDA-MB-231, Hela, and AGS with IC_{50} values of 6.7 ± 0.2 to 10.7 ± 1.3 μg/mL [26].

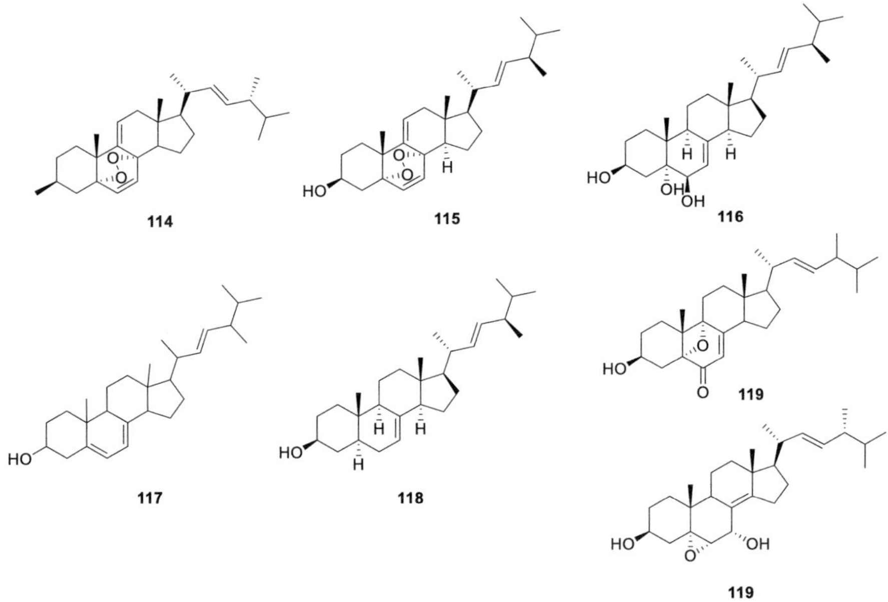

Fig. 13 Chemical structures of **114–120**

Other known steroids were also isolated from marine microorganisms in Vietnam (Fig. 14), including helvolic acid (**121**) and ergosterol (**122**) from marine sponge-associated fungus *Hamigera avellanea* [27], ergosterol peroxide (**123**) and (3*β*,5*α*,22E)-ergosta-6,8(14),22-triene-3,5-diol (**124**) from the seagrass-derived *Penicillium levitum* N33.2. [48], 3*β*,5*α*,9*α*-trihydroxyergosta-7,22-dien-6-one (**125**) from the marine sediment-derived *Penicillium* sp. OM07 and 6*β*-hydroxyergosta-4,7,22-trien-3-on (**126**) from *Aspergillus terreus* LM.1.5 [12]. Of these, **121** showed antimicrobial activity against *E. faecalis*, *S. aureus*, *B. cereus*, *E. coli*, *S. enterica* and *C. albicans* with the MICs of 2–64 μg/mL, whereas **122** showed antimicrobial activity against *E. faecalis*, *S. aureus*, *B. cereus*, *E. coli*, and *C. albicans* with the MICs of 32–128 μg/mL. In addition, compound **122** showed cytotoxic activity against three cancer cell lines, HepG2, A549, and MCF-7 with the IC_{50} values ranging from 70.40 ± 2.57 to 83.02 ± 2.85 μg/mL [27]. Compound **124** exhibited the most potent cytotoxicity against human cancer cells Hep-G2, A549 and MCF-7 with IC_{50} values of 2.89, 18.51 and 16.47 μg/mL, respectively, whereas compound **123** appeared to be less toxic to tested lines with IC_{50} values at 16.22, 22.48 and 27.11 μg/mL, respectively. Structure–activity analysis implies that the presence of the peroxide bridge might be the main cause of cytotoxicity due to its inducibility of apoptosis. In addition, compounds **123** and **124** alleviated NO production in LPS-induced murine RAW264.7 macrophages with IC_{50} of 2.85 and 2.79 μg/mL, respectively [48]. Compounds **123** and **125** selectively inhibited *E. faecalis*, *S. aureus* and *C. albicans* with a MIC value of 64– 256 μg/mL [12].

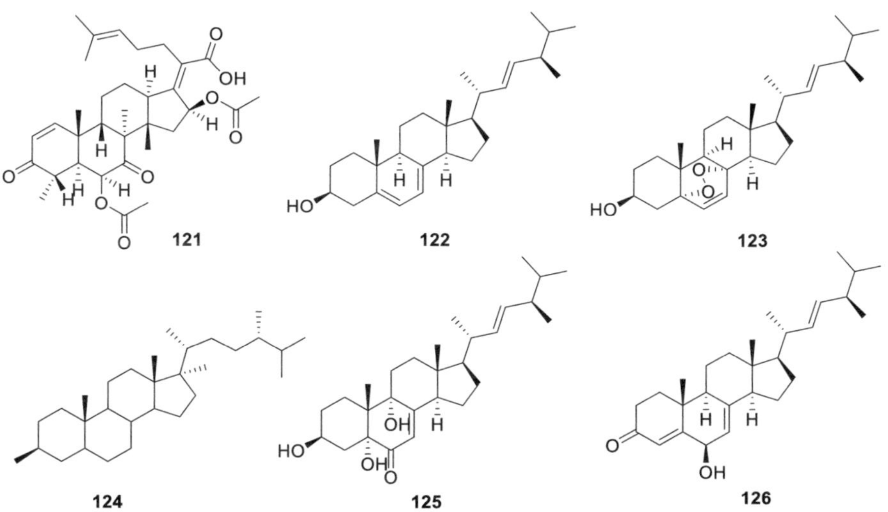

Fig. 14 Chemical structures of **121–126**

2.5 Polyketides

Ten polyketides, including two new polyketides, lopouzanone A (**127**) and lopouzanone B (**128**), two new 1-*O*-acetyl and 2-*O*-acetyl derivatives of dendrodochol B (**129** and **130**), along with six known polyketides, gliorosein (**131**), balticolid (**132**), dendrodolide G (**133**), dihydroisocoumarine (**134**), (–)-5-methylmellein (**135**), and dendrodochol B (**136**) were isolated from the sponge-derived marine fungus *Lopadostoma pouzarii* strain 168CLC-57.3 (Fig. 15) [49]. Of these, compound **131** exhibited DPPH radical-scavenging activity and in vitro cardioprotective effects toward rotenone toxicity and $CoCl_2$-mimic hypoxia.

One new aspyrone-related polyketide, aspilactonol G (**137**), together with two known polyketides, aspilactonol F (**138**) and dihydroaspyrone (**139**), were isolated from marine sediment-derived *Aspergillus flocculosus* (Fig. 16) [43], and two new auroglaucin-derived compounds, niveoglaucin A (**140**) and niveoglaucin B (**141**), together with four known compounds, 5-hydroxy-6-(3-methylbut-2-enyl)-2-(pent-1-enyl)benzofuran-4-carbaldehyde (**142**), flavoglaucin (**143**), aspergin (**144**) and isodihydroauroglaucin (**145**) were isolated from the marine sediment-derived strain of *Aspergillus niveoglaucus* (Fig. 16) [50]. Notably, **143** was cytotoxic activity against LNCaP cells with an IC_{50} of 64.3 μM, whereas **144** exhibited cytotoxicity toward Neuro2a cells and PC-3 cells with IC_{50} of 87.2 and 64.9 μM, respectively without apoptosis induction.

Chemical investigation of the seaweed-derived endophytic fungus *Aspergillus micronesiensis* led to the isolation of three novel dibenzospiroketals, aspermicrones A-C (**146–148**) (Fig. 17). Compound **147** exhibited a selective cytotoxic effect toward HepG2 cell line with $IC_{50} = 9.9$ μM. Additionally, compounds **147** and **148** displayed antibacterial activity against *S. aureus* with MIC = 123.2 μM [51]. Also, compound 2,6-dimethyl-3-*O*-methyl-4-(2-methylbutyryl) phloroglucinol (**149**) isolated from

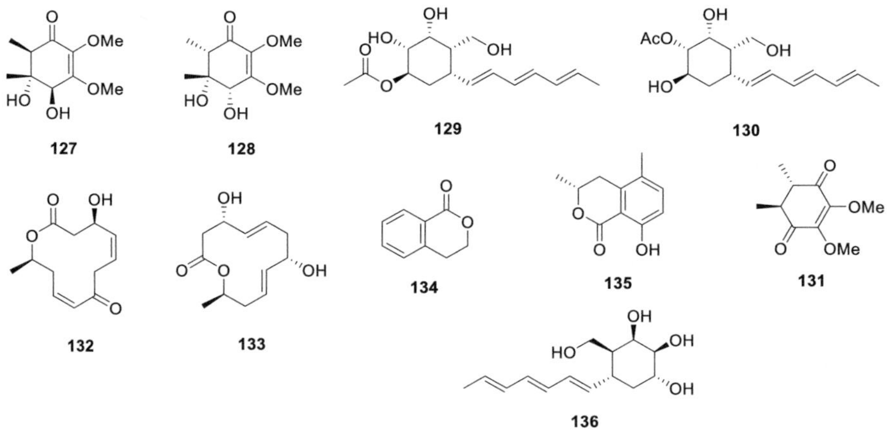

Fig. 15 Chemical structures of **127–136**

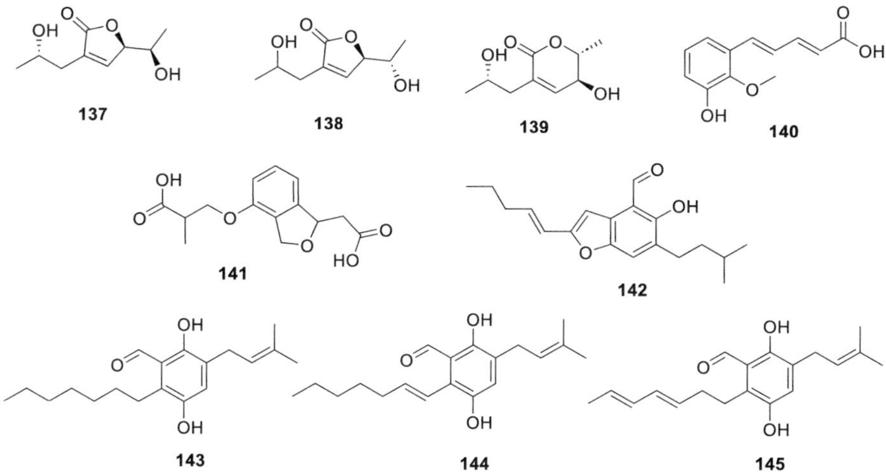

Fig. 16 Chemical structures of **137–145**

the fungus *Aspergillus flocculosus* 168ST-16.1 derived from the marine alga *Padina* sp. showed anti-neuroinflammatory effects with an IC_{50} value of 56.6 μM by suppressing the production of pro-inflammatory mediators in activated microglial cells without cytotoxicity [52], whereas wasabidienone E (**150**) isolated from the marine fungus *Aspergillus flocculosus* 01NT.1.1.5 derived from a sponge *Stylissa* sp. showed suppressive activities on RANKL-induced differentiation of BMMs into osteoclasts without significant cytotoxic effect on BMMs (Fig. 17) [28].

Chemical investigation of marine sponge-derivated fungus *Aspergillus nidulans* resulted in the isolation of a new depsidone, 7-dechloronidulin (**151**), together with six depsidones, nidulin (**152**), 2,4-dichlorounguinol (**153**), unguinol (**154**), 2-chlorounguinol C (**155**), emeguisin A (**156**), emeguisin B (**157**) (Fig. 18) [53]. Of these, chlorinated depsidones **151–153**, and **154** displayed potential antimicrobial activity. Interestingly, undescribed depsidone **151** was selectively bioactive on

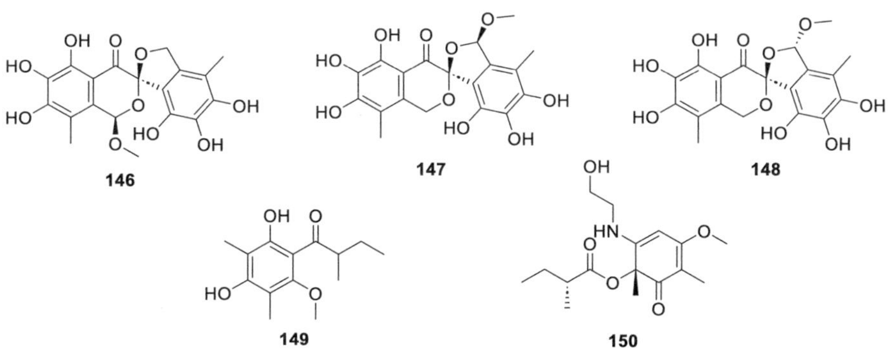

Fig. 17 Chemical structures of **146–150**

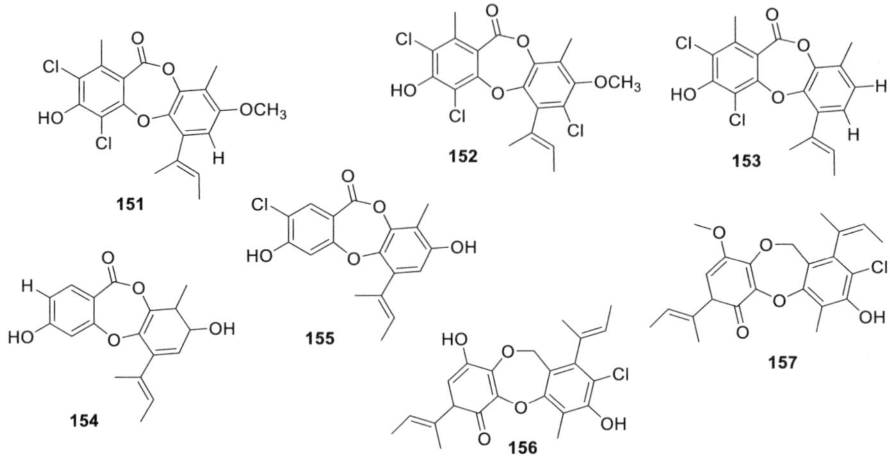

Fig. 18 Chemical structures of **151–157**

Gram-positive bacteria *B. cereus*, *E. faecalis* and *S. aureus* with MICs of 2–4 μg/mL and the yeast *C. albicans* with MIC of 8 μg/mL, whereas compound **152** possessed good activity against *B. cereus*, *E. faecalis*, *E. coli*, *P. aeruginosa*, *S. enterica* and *C. albicans* with MIC values in range of 2–16 μg/mL, and compound **157** showed antimicrobial activity against *E. faecalis*, *S. aureus*, *B. cereus*, *S. enterica*, and *C. albicans* with the MICs of 8–128 μg/mL [42, 53]. The structure–activity analysis suggested that the methyl group at C-1 of depsidone skeleton (in compounds **151–155**) played an important role in antimicrobial activity. The replacement of this methyl group by a 1-methyl-1-propenyl group (in compounds **156** and **157**) significantly reduced activity. Moreover, the presence of chlorine atoms in depsidones (in compounds **151–153**, and **155**) also enhanced antimicrobial activity in comparison with a non-chlorinated depsidone (in compound **154**).

In another investigation, four angucyclines, including lagumycin B (**158**), dehydrorabelomycin (**159**), phenanthroviridone (**160**), and WS-5995 A (**161**) were recovered from marine sediment-derived *Micromonospora* sp. (Fig. 19) [40]. Cytotoxic assays indicated **159** and **160** to be non-specifically cytotoxic. In contrast, **158** exhibited cytotoxic activity against non-cancerous murine cell lines MOSE and MOE with LC_{50} values of 9.80 μM and 10.8 μM, respectively and **161** showed cytotoxic activity toward Kuramochi ovarian cancer cells with an LC_{50} of 18.6 μM.

Four other polyketides, including a new compound, (3*R*)-(3′,5′-dihydroxyphenyl)butan-2-one (**162**) and three known compounds, AGI-7 (**163**), sescandelin (**164**), and sescandelin-B (**165**) were isolated from the sponge-derived fungus *Ascomycota* sp. VK12 (Fig. 19) [54]. Of these, compounds **162** and **163** showed cytotoxicity towards HepG2, MCF-7, and SK-Mel2 carcinoma cells with IC_{50} values ranging from 48.6 to 96.5 μM. Furthermore, compounds **162**, **163**, and **165** displayed NO inhibitory effects in LPS-stimulated BV2 cells, with IC_{50} values in a range from 24.2 to 76.5 μM. Compound **163** further

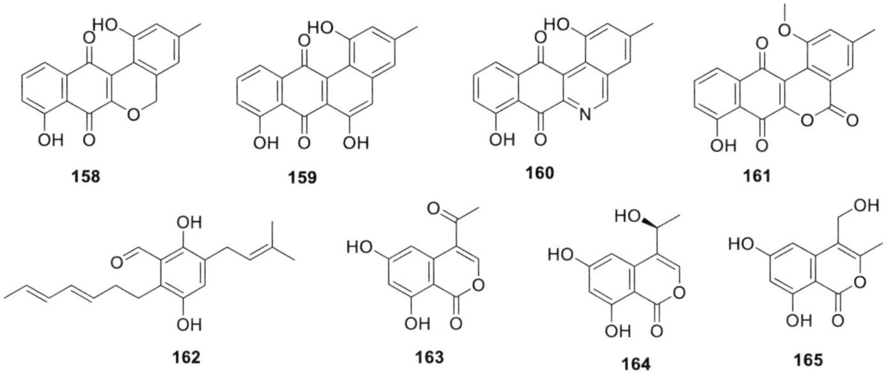

Fig. 19 Chemical structures of **158–165**

inhibited PGE_2 overproduction with an IC_{50} value of 25.3 μM. Notably, the inhibitory effects of **163** towards NO and PGE_2 overproduction were found to relate to its suppression of iNOS and COX-2 protein expression, respectively.

A new compound containing a triene, a tetrahydropyran ring and glycine ester functionalities, restricticin B (**166**), together known *N*-acetyl restricticin (**167**) were obtained from the marine sediment-derived fungus *Penicillium janthinellum* 168CLC-17.1 (Fig. 20) [25]. Both compounds **166** and **167** exhibited anti-neuroinflammatory effects by suppressing the production of pro-inflammatory mediators in activated microglial cells. However, **166** and **167** showed no antibacterial activity and cytotoxicity against cancer cell lines, suggesting that the free amine of the glycine unit is important for antifungal activity.

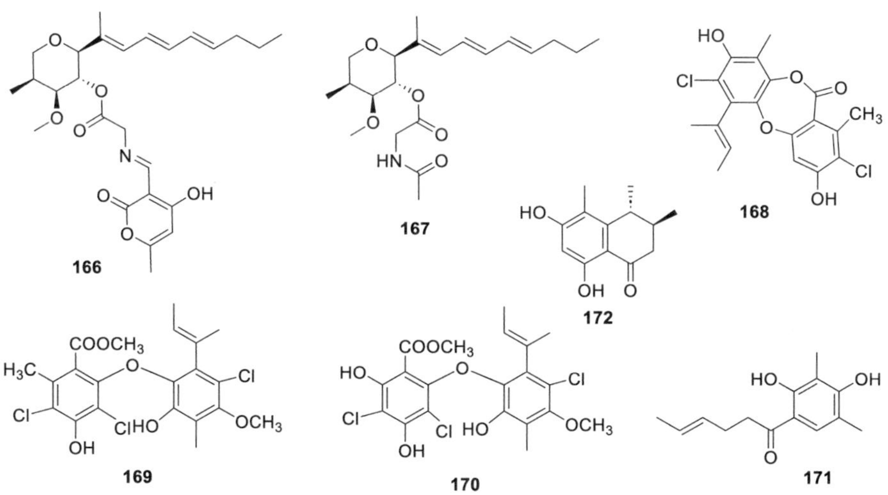

Fig. 20 Chemical structures of **166–172**

One depsidone, aspergillusidone C (**168**), and two diphenyl ethers, aspergillusether A (**179**), aspergillusether J (**170**), were also characterized from the marine algal-derived *Penicillium oxalicum* M893 (Fig. 20) [42]. All compounds **168**–**170** showed antimicrobial activities with MIC values ranging from 2 to 256 μg/mL. Compound **168** showed antimicrobial activity against *E. faecalis, S. aureus, B. cereus, S. enterica*, and *C. albicans* with MICs of 2 μg/mL, whereas **169** showed antimicrobial activity against *E. faecalis, S. aureus, B. cereus, E. coli, P. aeruginosa*, and *C. albicans* with the MICs of 16–64 μg/mL and **170** showed antimicrobial activity against *E. faecalis, S. aureus, B. cereus, E. coli*, and *C. albicans* with the MICs of 4–256 μg/mL. Moreover, two other compounds were isolated from the genus *Penicillium*, including 2′,3′-dihydrosorbicillin (**171**) from the marine sediment *Penicillium* sp. [14] and stoloniferol B (**172**) from the marine sediment-derived fungus *Penicillium citrinum* VM6 (Fig. 20) [26]. Of these, **171** exhibited potential inhibitory activity against α-glucosidase [5], whereas **172** selectively inhibited gram-positive bacteria *E. faecalis, S. aureus, B. cereus* and the yeast *C. albicans* with MIC values ranging from 16 to 256 μg/mL.

Several other new/known polyketides were isolated from marine microorganisms in Vietnam (Fig. 21), including ethylene terephthalate cyclic trimer (**173**) from the echinoderm-derived actinomycete *Streptomyces* sp. G278 [20], new nhatrangins A (**174**) and B (**175**) from marine cyanobacterium *Lyngbya majuscula* [55], monomethylsulochrin 5-sulfonic acid (**176**) and monomethylsulochrin (**177**) from the marine sponge-associated fungus *Hamigera avellanea* [27]. Compound **173** was found to selectively inhibit *E. faecalis* with MIC of 256 μg/mL [20], whereas **176** showed antimicrobial activity against *E. faecalis, S. aureus, B. cereus, E. coli, C. albicans* with MICs of 32–128 μg/mL, and **177** showed antimicrobial activity against *E. faecalis, S. aureus, B.* cereus and C. *albicans* with MICs of 16–256 μg/mL. In addition, compound **176** displayed cytotoxic activity against cancer cell lines HepG2, A549, and MCF-7 with IC_{50} values of 64.79 ± 2.47 to 72.50 ± 2.85 μg/mL.

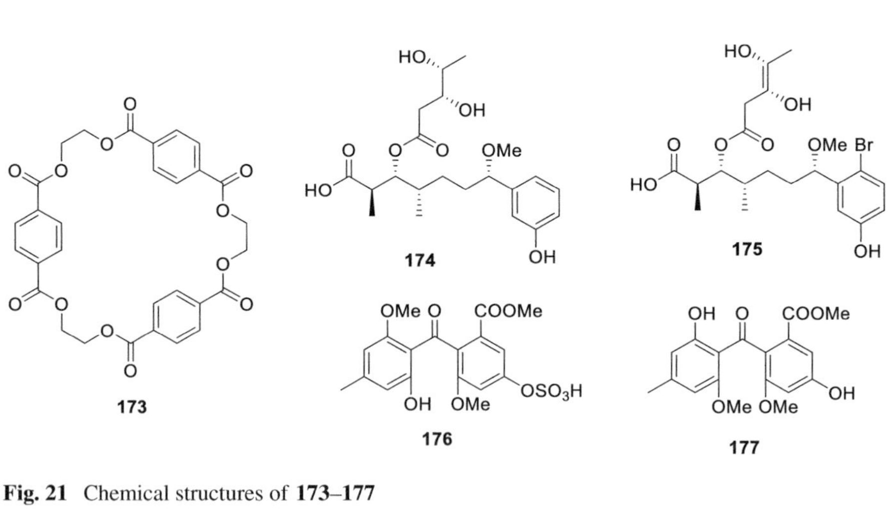

Fig. 21 Chemical structures of **173**–**177**

2.6 Peptides

Three new tripeptide derivatives asterripeptides A–C (**178**–**180**) were isolated from the mangrove-derived fungus *Aspergillus terreus* LM.5.2 (Fig. 22) [56]. Compound **178** showed cytotoxicity towards MCF-7, DLD-1, PC-3 and H9c2 cells with IC_{50} ranging from 64.6 ± 2.4 to 96.8 ± 7.0 μM, whereas **179** showed cytotoxicity towards PC-3 cells with IC_{50} of 75.5 μM and **180** showed cytotoxicity towards MCF-7, DLD-1, PC-3, and H9c2 cells with IC_{50} of 58.3 ± 3.2 to 96.6 ± 1.5 μM. Furthermore, **179** and **180** inhibited sortase A, one of the key enzymes for *S. aureus* virulence, whereas compound **178** was inactive. It is noted that bicyclic peptides with Leu-Pro-Pro motive were identified as potent and selective inhibitors of sortase A via binding to active sites of the enzyme, owing to structural mimicking of the Leu-Pro-Xaa-Thr-Gly region of natural substrate of sortase A [57]. Therefore, the presence of the terminal ethyl group in isoLeu may be the main reason for the inactivity of compound **178**, whereas the presence of Leu-Pro in **179** as well as Val-Pro in **180** may play a crucial role in their sortase A inhibitory effect [56].

A chemical study of the marine sponge-derivated fungus *Aspergillus nidulans* resulted in the isolation of two macrocyclic peptides, unguisin A (**181**) and unguisin B (**182**) (Fig. 23) [53]. Notably, both compounds **181** and **182** displayed selective antibacterial activity against *E. faecalis* with MIC values of 32 and 128 μg/mL, respectively. In another investigation, six lichenysins were isolated from the marine sediment-derived *Streptomyces* G222, including (ante-*iso*)-and (*iso*)-C13 lichenysins (**183**–**184**), (*n*)- and (*iso*)-C14 lichenysins (**185a**–**185b**), and (*ante-iso*)- and (*iso*)-C15 lichenysins (**186**–**187**) (Fig. 23) [58]. Interestingly, all compounds **183**–**187** exhibited anti-biofilm activity with MIC of 100 μM. Furthermore, **183**–**184** and **186**–**187** exhibited selective activity against the MRSA strain without affecting its growth curve and without membranotropic activity.

Several other known peptides were isolated from marine microorganisms in Vietnam (Fig. 24), including 12′,13′-*trans*-14′,15′-dihydrobacillaene B (**188**) from the marine seaweed-derived *Bacillus siamensis* 168CLC-66.1 [10], phenylalanin (**189**) from the sponge-derived actinomycete *Streptomyces* sp. G246 [16], and L-phenyl alanine (**190**) from the marine sediment-derived *Micromospora* sp. G047 [59]. However, their biological properties were not reported.

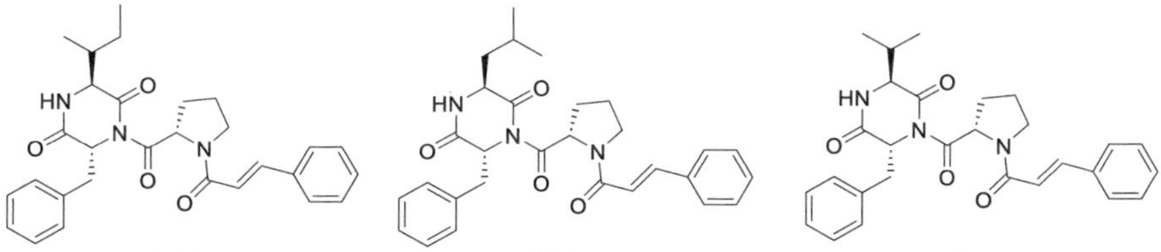

178 **179** **180**

Fig. 22 Chemical structures of **178**–**180**

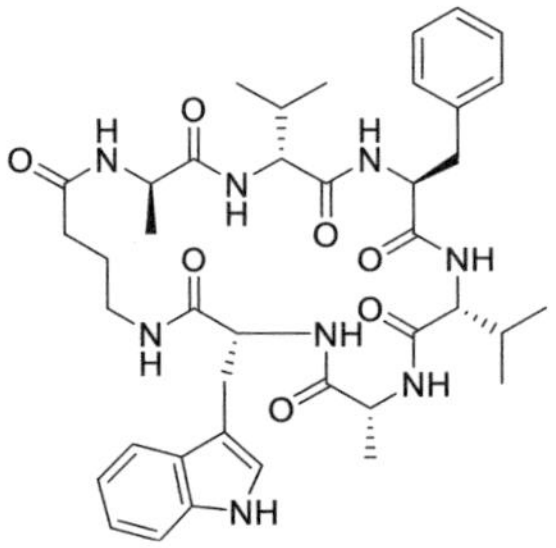

181

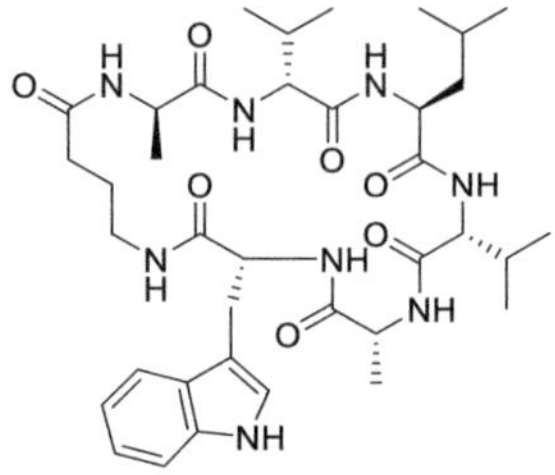

182

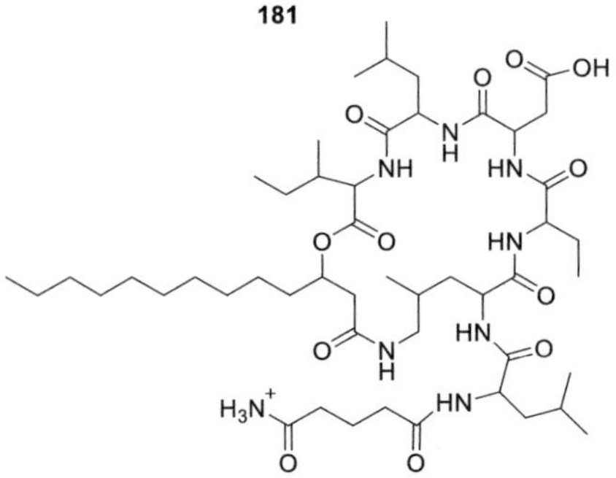

H_2O

183-184

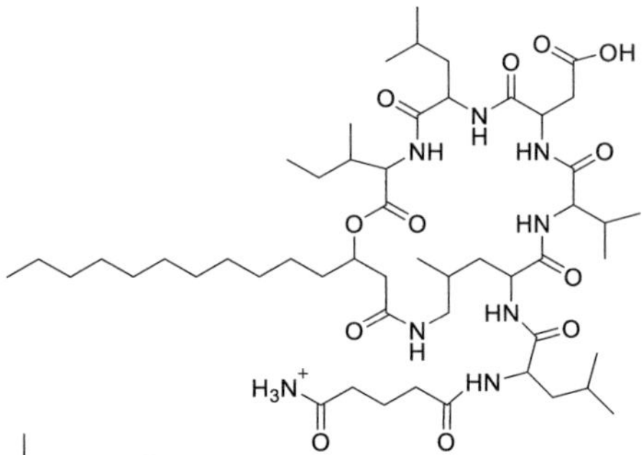

185a-185b

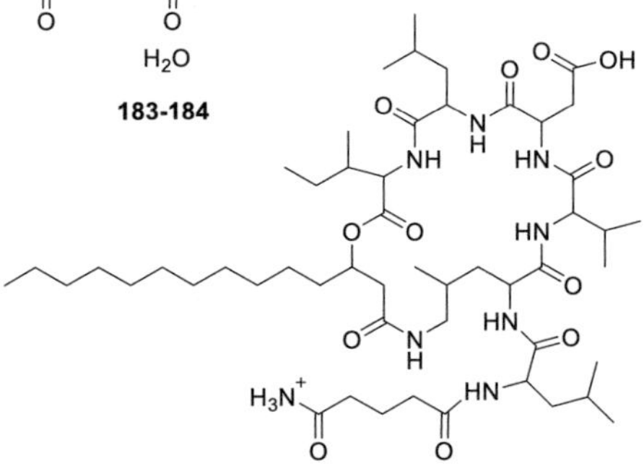

186-187

Fig. 23 Chemical structures of **181–187**

Furthermore, a series of cyclic peptides were isolated from marine microorganisms in Vietnam (Fig. 25), including cyclo-(Pro-Tyr) (**191**) [13, 15, 16, 37, 60–62], cyclo-(Trp-Tyr) (**192**), and cyclo-(Pro-Phe) (**193**) [13, 15, 30, 61], cyclo-(Tyr-Trp) (**194**) [14], cyclo-(Pro-Leu) (**195**) [15, 19, 35, 61, 62], cyclo-(Pro-Ala) (**196**) [16, 19, 63], cyclo-(3-hyroxy-Pro-Tyr) (**197**) [16], cyclo-(Pro-Met) (**198**) [16, 32], cyclo-(Pro-Trp) (**199**) [20, 30, 33, 35, 61, 64], cyclo-(Leu-Tyr) (**200**) [19], cyclo-(Pro-Gly) (**201**) [19, 29, 32, 38, 63], cyclo-(Pro-Val) (**202**) [30, 61], cyclo-(Leu-*trans*-4-hydroxy-Pro) (**203**), cyclo-(Phe-*trans*-4-hydroxy-Pro) (**204**), cyclo-(Phe-Pro) (**205**) [62], cyclo-(Trp-Ser) (**206**), cyclo-(Trp-Ala) (**207**) [35], cyclo-(Leu-Pro) (**208**) [37, 63, 64], cyclo-*trans*-4-OH-(Pro-Phe) (**209**) [19, 61, 63].

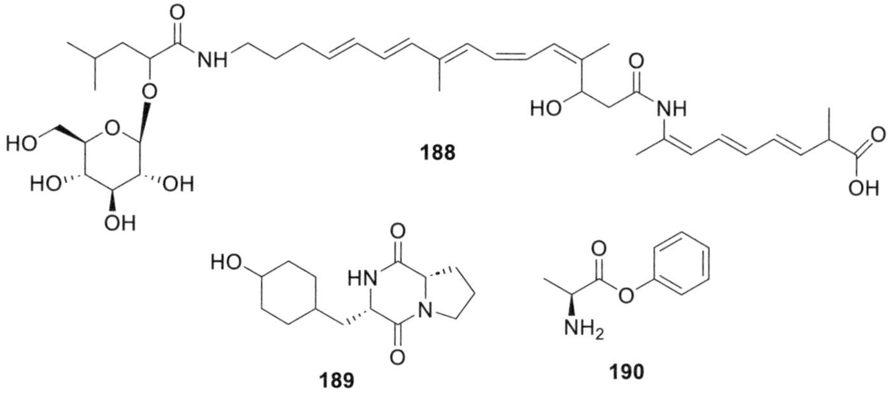

Fig. 24 Chemical structures of **188–190**

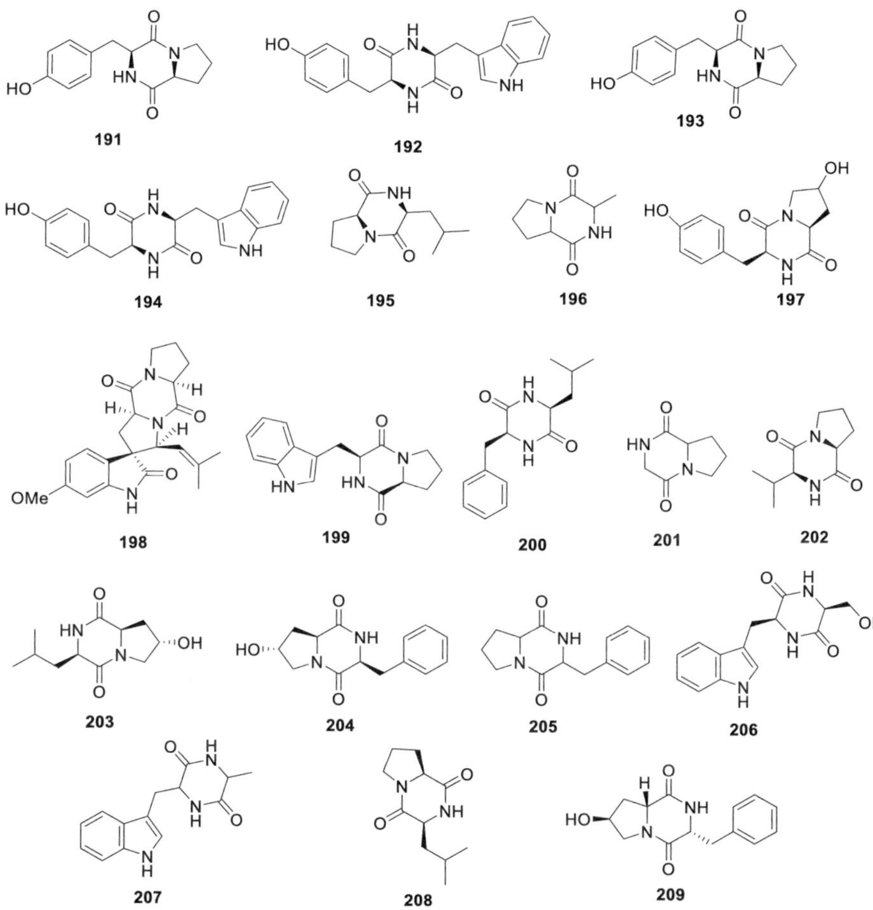

Fig. 25 Chemical structures of **191–209**

Antimicrobial assay indicated that **192** exhibited antimicrobial activity against *E. faecalis*, *S. aureus*, and *P. aeruginosa* with MIC values of 64–256 μg/mL, whereas **193** exhibited selective antimicrobial activity against *E. faecalis*, and *S. aureus* with MIC = 128 μg/mL [13]. Compound **196** exhibited selectively antimicrobial activity against *E. faecalis* and *C. albicans* with MIC values of 256 μg/mL and 2 μg/mL, respectively, whereas **198** exhibited selectively antibacterial activity against *E. coli* with 128 μg/mL [16]. Compound **199** selectively inhibited *E. coli* with MIC values of 64 μg/mL [20], and compounds **209** and **200** had selective inhibition against *S. aureus* and *E. coli* with MIC values of 256 and 32 μg/mL, respectively [19].

2.7 Phenolic Compounds and Derivatives

Two new metabolites, 1-(2,5-dihydroxyphenyl)-3-methoxy-butan-1-one (**210**) and 1-(2,5-dihydroxyphenyl)-3-hydroxybutan-1-one (**211**), along with known 1-(2,5-dihydroxyphenyl)-2-buten-1-one (**212**) were isolated from the marine fungus *Paraconiothyrium* sp. VK-13 (Fig. 26) [65]. Compounds **211** and **212** inhibited the overproduction of proinflammatory mediators NO and PGE2 in LPS-stimulated RAW264.7 cells with IC_{50} values ranging from 3.9 to 12.5 μM. Interestingly, the inhibitory effects of **211** and **212** on the release of NO and PGE2 were correlated with their significant suppression of iNOS and COX-2 protein expression, respectively. Furthermore, both compounds **211** and **212** inhibited the mRNA expression of proinflammatory cytokines, including TNF-α, IL-1β, IL-6, and IL-12, with IC_{50} values in a range of 2.4–12.5 μM. Two other metabolites, sporogen AO-1 (**213**) and 4-hydroxybenzaldehyde (**214**) were isolated from the marine sediment-derived *Penicillium* sp. (Fig. 26). Compound **214** exhibited selectively antibacterial activity against *E. coli* with MIC 8 μg/mL [14].

One new metabolite, (2*R*)-2-(4-hydroxyphenyl)ethyl 2-hydroxy-3-phenylpropanoate (**215**) and five known ones, 3-hydroxy-4-(4-hydrophenyl)-2-butanone (**216**), (erythro)-2-hydroxy-3-methylpentanoic acid (**217**), (2*R*)-2-hydroxy-3-phenylpropanoic acid (**218**), (2*R*)-methyl 2-hydroxy-3-phenylpropanoate (**219**), tyrosol (**220**) were also characterized from the marine microalga *Aurantiochytrium* sp. SC145 (Fig. 27) [21]. Compounds **215**, **216** and **220** significantly exhibited antimicrobial activities on all tested microorganisms (*E. faecalis*, *S. aureus*, *B. cereus*, *E. coli*, *P. aeruginosa*, *S. enterica* and *C. albicans*) with MIC values ranging from 32 to 256 μg/mL. In another investigation, two new glycosylated alkylresorcinols, resorcinoside A (**221**) and resorcinoside B (**222**), were

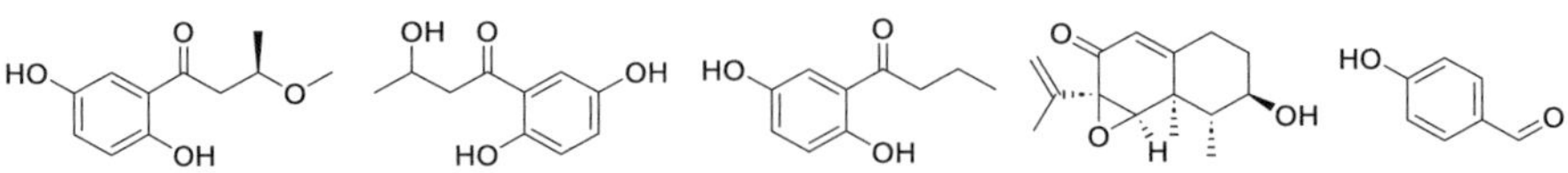

Fig. 26 Chemical structures of **210–214**

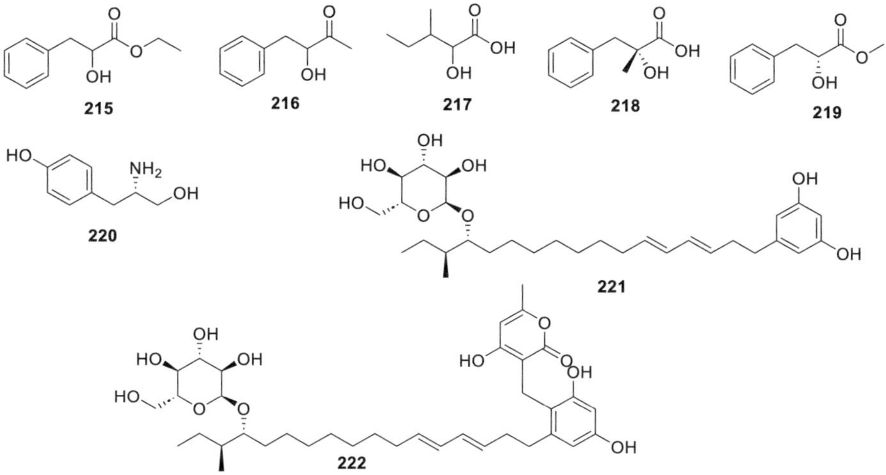

Fig. 27 Chemical structures of **215**–**222**

isolated from the marine sediment-derived *Penicillium janthinellum* (Fig. 27) [66]. Compounds **221** and **222** are new alkylresorcinols with D-glucose, and **222** has an α-pyrone moiety attached to the aromatic ring. Cytotoxic assays revealed that **221** exhibited cytotoxic activity against all tested cancer cell lines HCT-15, NUGC-3, NCI-H23, ACHN, PC-3, and MDA-MB-231 with GI_{50} values of 9.3–31 μM.

Furthermore, numerous other known compounds and derivatives were isolated from marine microorganisms in Vietnam (Fig. 28), including 4-hydroxyscytalone **(223),** 4-hydroxy-6-dehydroxyscytalone **(224)** and demethylcitreoviranol **(225)** from *Penicillium* sp. KMM 4672, 4-hydroxy-3-(3-methylbut-2-enyl)benzaldehyde (**226**) from *Aspergillus terreus* LM.1.5 [39], 1-hydroxy-4-methoxy-2-naphthoic acid (**227**) from the marine-derived actinomycete *Nocardiopsis* sp. G057 mL [13], phenol A acid (**228**) and phenol A (**229**) from the marine sediment-derived fungus *Penicillium citrinum* VM6 [17], isosclerone (**230**) from marine sponge-associated fungus *Hamigera avellanea* [27], 4-hydroxy-3-methoxybenzaldehyde (**231**) and phloretic acid (**232**) from the marine sediment-derived actinomycete *Nocardiopsis synnemataformans* HT06 [17], phenethyl 2-phenyl acetate (**233**), *n*-butyl–isobutyl phthalate (**234**), 4-(2-hydroxyethyl) phenol (**235**), 2-phenylacetic acid (**236**) from the marine sediment-derived *Streptomyces* sp. G039 [37].

Subsequent bioassays revealed that **227** exhibited selectively antibacterial activity against *E. coli* with MIC = 32 μg/ mL, whereas **228** inhibited *E. faecalis, S. aureus, B. cereus* and *C. albicans* with MIC values ranging from 8 to 128 μg/mL [12, 26, 32]. In addition, **228** exhibited cytotoxicity against the MCF7 cell lines with IC_{50} of 98.1 ± 7.8 μg/mL [26]. Compound **230** showed antimicrobial activity against *E. faecalis, S. aureus*, and *C. albicans* with MICs of 128–256 μg/mL [27] and compound **231** inhibited *S. aureus* and *C. albicans*, with MIC values of 64 and 128 μg/mL,

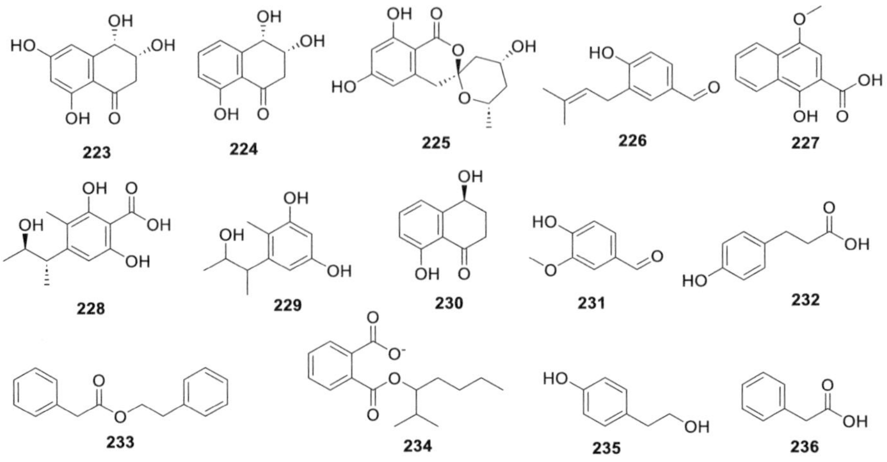

Fig. 28 Chemical structures of **123–236**

respectively, whereas compound **232** inhibited *E.* feacalis, *B. cereus*, and *C. albicans* with MIC values ranging from 64 to 256 μg/mL [17].

2.8 Others

2.8.1 Quinones and Derivatives

Two new phomaligols, deketo-phomaligol A (**237**) and phomaligol E (**238**), together with three known ones, phomaligol A (**239**), phomaligol A1 (**240**), and phomaligol D (**241**) were isolated from the fungus *Aspergillus flocculosus* 168ST-16.1 derived from the marine alga *Padina* sp. (Fig. 29) [52]. Of these, compound **237** was first isolated as a phomaligol derivative possessing a five-membered ring. Also, one known anthraquinone derivative, vismione E (**242**), was obtained from sponge-associated fungus *Aspergillus* sp. 1901NT-1.2.2 (Fig. 29) [67]. Cytotoxic assays indicated that **242** exhibited high cytotoxic activity against MCF-7 cells with IC_{50} of 9.0 μM but exhibited low toxicity for normal human breast MCF-10A cells with IC_{50} of 65.3 μM. It was found that **242** inhibits MCF-7 cell proliferation and arrests the cell cycle in the G1 phase. Moreover, **242** influenced negatively on MCF-7 cell migration.

In another investigation, isorhodoptilometrin (**243**) was isolated from the sponge-associated *P. oxalicum* CLC-MF05 (Fig. 29). Interestingly, **243** inhibited NO and PGE2 overproduction as well as iNOS and COX-2 overexpression in both LPS-stimulated BV2 and rat primary microglia with IC_{50} of 7.5–15.2 μM. Compound **243** also repressed mRNA expression of TNF-α, IL-1β, IL-6, and IL-12. Notably, mechanistic studies revealed that the inhibitory actions of **243** were regulated by

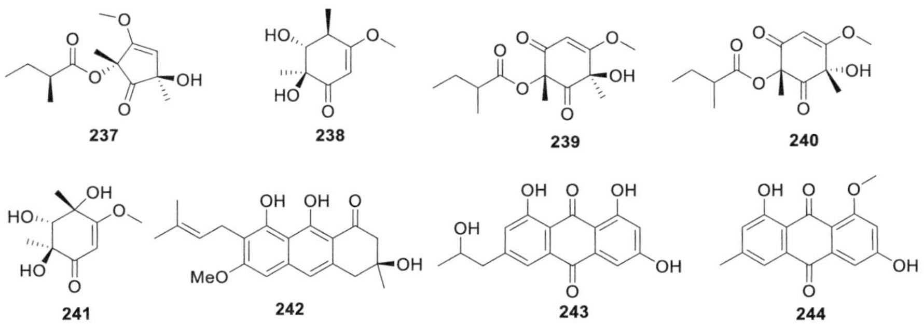

Fig. 29 Chemical structures of **237–244**

the inactivation of the NF-κB and MAPK signaling pathways. Furthermore, the inactivation of the TLR4/MyD88 pathway contributed to the anti-neuroinfammatory activity of compound **243** [22]. From the marine sponge-associated fungus *Hamigera avellanea*, questin (**244**) was isolated (Fig. 29) [27]. Subsequent experiments revealed that **244** exhibited antimicrobial activity against *E. faecalis*, *S. aureus*, and *C. albicans* with MICs of 128–256 μg/mL.

2.8.2 Lipids and Fatty Acids

Several lipids and fatty acids were also isolated from marine microorganisms in coastal Vietnam (Fig. 30), including two new trehalose lipids, tsukalipid A (**245**) and tsukalipid B (**246**), from marine sediment-derived bacterium *Tsukamurella pseudospumae* IV19-045 [10]; a new *C*-glucoside benzoic acid derivative, 4*βC*-glucopyranosyl-3,5-dihydroxy-2-methoxybenzoic acid (**247**), along two known carboxylic acids, 3-phenylpropanoic acid (**248**) and cinnamic acid (**249**), from the marine sediment-derived actinomycete *Nocardiopsis synnemataformans* HT06 [17]; hydroxysydonic acid (**250**) from the sponge-derived fungus *Ascomycota* sp. VK12 [54]; saccharonol A (**251**) from the marine algal-derived fungus *Aspergillus flocculosus* 168ST-16.1 [52]; 1*H*-pyrrole-2-carboxylic acid (**252**) from the marine sediment-derived actinomycete *Streptomyces* sp. G261 [19]; heptadecanoic acid (**253**) from the marine sediment-derived fungus *Neurospora* sp. M45 [68]; as well as acetamide (**254**) and 2-aminopropanoic acid (**255**) from the marine sediment-derived *Oceanisphaera* sp. [38]. Among them, **247** exhibited antimicrobial activity against *S. aureus* and *C. albicans* with MIC values of 32 and 16 μg/mL, respectively, whereas compound **248** inhibited *E. feacalis*, *S. aureus*, and *C. albicans* with MIC values of 8–128 μg/mL, and compound **249** inhibited *S. aureus* and *C. albicans* with MIC values of 256 and 128 μg/mL, respectively [17]. In addition, **250** displayed NO inhibitory effects in LPS-stimulated BV2 cells with IC_{50} values of 41.5 ± 2.1 μM [54].

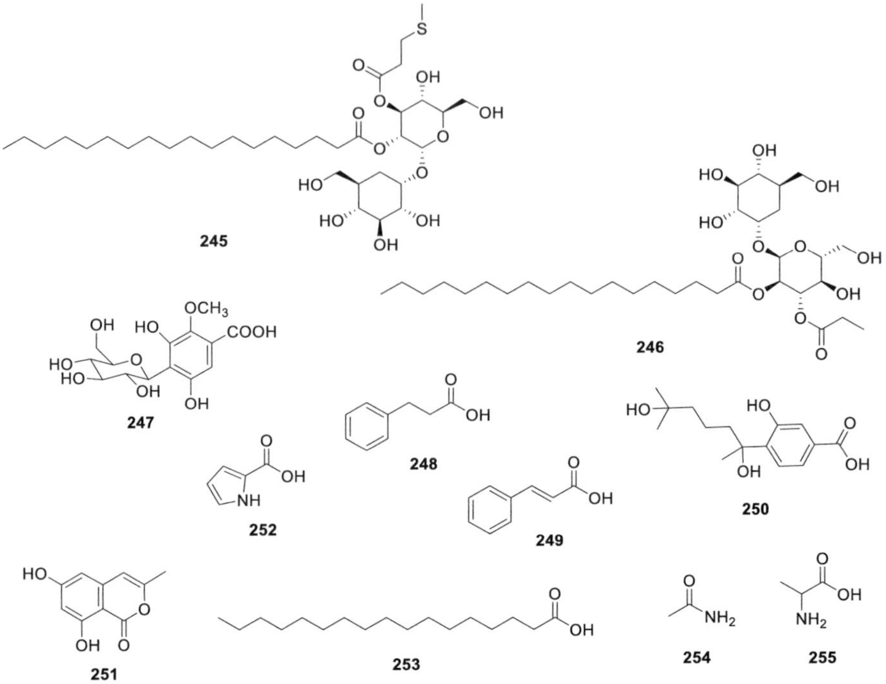

Fig. 30 Chemical structures of **245–255**

2.8.3 Macrolides

Four new macrolides, modiolides D-G (**256–259**) and two known macrolides, modiolide A (**260**) and modiolide B (**261**), were isolated from the marine fungus *Paraconiothyrium* sp. VK-13 (Fig. 31) [65]. Several known macrolides were also isolated obtained other marine microorganisms in Vietnam (Fig. 31): macrolactin B (**262**) from the marine seaweed-derived *Bacillus siamensis* 168CLC-66.1 [10] as well as macrolactin A (**263**), macrolactin H (**264**), and 15,17-epoxy-16-hydroxy macrolactin A (**265**) from sponge-associated *Bacillus* sp. M1_CRV_171 [60]. Of these, compounds **263–265** exhibited antimicrobial activity against a broad spectrum of reference microorganisms, including Gram positive bacteria *B. subtilis*, *S. aureus*, *E. faecalis*, Gram negative bacteria *E. coli*, *P. aeruginosa*, *S. enterica* and the yeast *C. albicans*, with MIC values of 8–256 μg/mL.

2.8.4 Coumarins

Three coumarins were isolated from marine microorganisms in Vietnam, including scopoletin (**266**) from the echinoderm-derived actinomycete *Streptomyces* sp. G278 [20]; ramulosin (**267**) from the marine seaweed-derived fungus *Talaromyces*

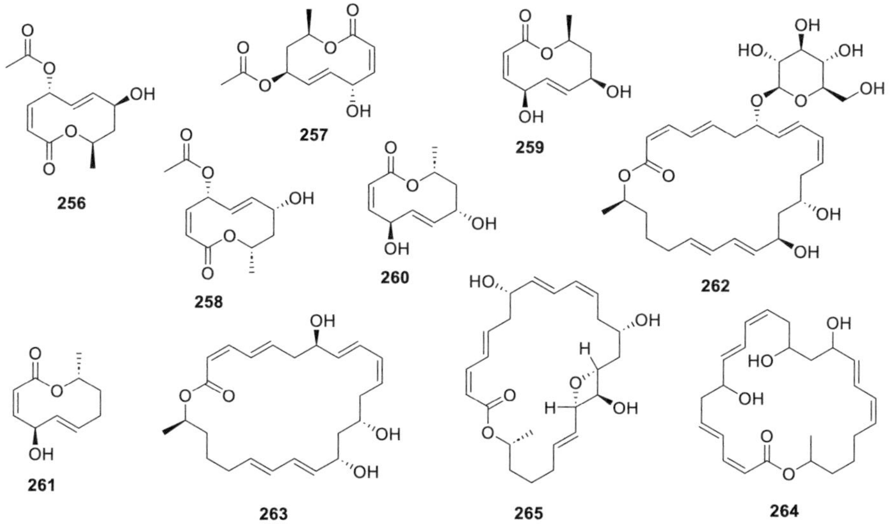

Fig. 31 Chemical structures of **256–264**

cyanescens [69]; and 5-hydroxy-7-(2′-hydroxypropyl)-2-methyl-chromone (**268**) from the sponge-associated fungus *Penicillium oxalicum* CLC-MF05 [22] (Fig. 32). Of these, compound **266** selectively inhibited *E. coli* with MIC values of 64 μg/mL [20], whereas **268** inhibited NO and PGE2 overproduction as well as iNOS and COX-2 overexpression in both LPS-stimulated BV2 and rat primary microglia with IC_{50} of 48.4–75.5 μM. Moreover, **268** also repressed mRNA expression of TNF-α, IL-1β, IL-6, and IL-12. Mechanistic studies revealed that the inhibitory actions of **268** were regulated by the inactivation of the NF-κB and MAPK signaling pathways. Furthermore, the inactivation of the TLR4/MyD88 pathway contributed to the anti-neuroinflammatory activity of **268** [22].

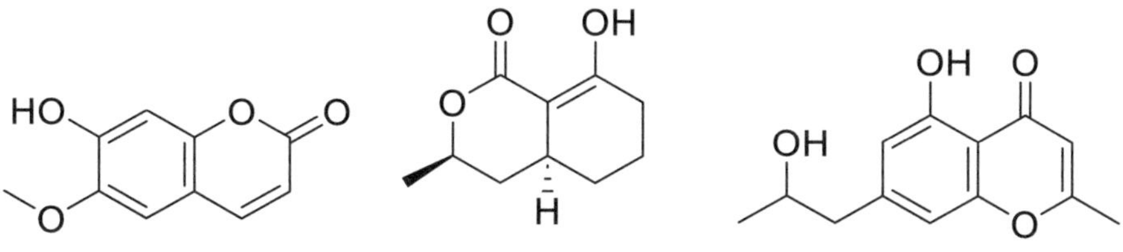

266 **267** **268**

Fig. 32 Chemical structures of **266–268**

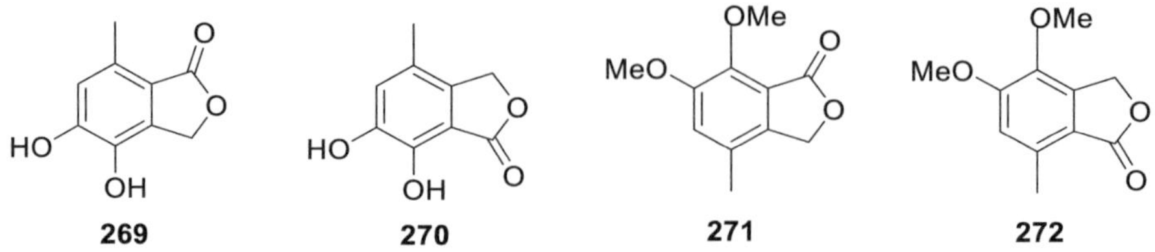

Fig. 33 Chemical structures of **269–272**

2.8.5 Phthalides

A new phthalide, 4,5-dihydroxy-7-methylphthalide (**269**) (Fig. 33), was isolated from marine sediment-derived actinomycete *Streptomyces* sp. G212 [64]. Compound **269** and its isomers, **270–272** (Fig. 33), were subsequently synthesised for biological evaluation. Compound **269** exhibited antimicrobial activity against *E. faecalis*, *S. aureus*, *B. cereus*, and *S. enterica* with MIC values of 64–256 μg/mL and **270** exhibited antimicrobial activity against *E. faecalis*, *S. aureus*, *B. cereus*, *E. coli*, *P. aeruginosa*, and *C. albicans* with MIC values of 32–128 μg/mL, whereas **271** exhibited antimicrobial activity against *E. faecalis*, *S. aureus*, *B. cereus*, *E. coli*, *P. aeruginosa*, and *S. enterica* with MIC values of 64–128 μg/mL and **272** exhibited antimicrobial activity against *S. aureus*, *B. cereus*, *E. coli*, *P. aeruginosa*, and *C. albicans* with MIC values of 32–128 μg/mL. It is noted that **269** exhibited no activity against *C. albicans* whereas its synthetic isomer **270** demonstrated a significant inhibitory effect against *C. albicans* with MIC value of 32 μg/mL. Furthermore, the two synthetic intermediates, **271** and **272**, displayed slightly different inhibitions when compared to their respective demethylated derivatives **269** and **270**. This observation indicates that the methylation of the hydroxyl groups in **269** and **270** has an influence on their antimicrobial properties.

2.8.6 Lactones

Three lactones, including germicidin A (**273**), germicidin B (**274**), and 5-hydroxymethyl-4-hydroxy-2,4-dimethyl-2-cyclopentenone (**275**) (Fig. 34), were isolated from marine sediment-derived actinomycete *Streptomyces* sp. G212 [64]. Of these, **273** exhibited antimicrobial activity against *E. faecalis*, *S. aureus*, *B. cereus*, *S. enterica*, and *C. albicans* with MIC values of 16–128 μg/mL, whereas **274** exhibited antimicrobial activity against *E. faecalis*, *B. cereus*, *S. enterica*, and *C. albicans* with MIC values of 64–128 μg/mL, and **275** exhibited antimicrobial activity against *E. faecalis*, *S. aureus*, *B. cereus*, *E. coli*, *S. enterica*, and *C. albicans* with MIC values of 32–128 μg/mL.

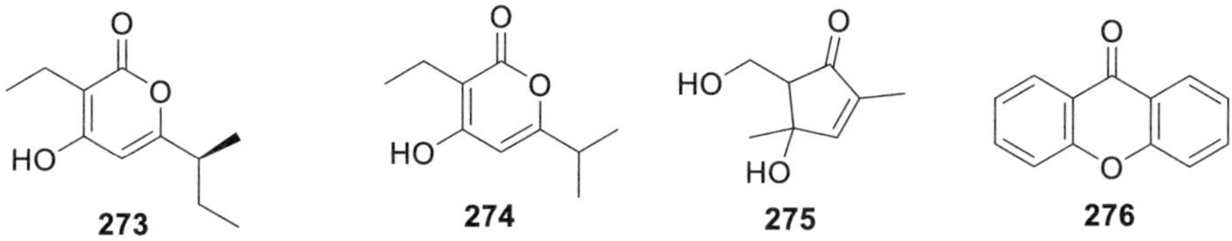

273 274 275 276

Fig. 34 Chemical structures of **273–276**

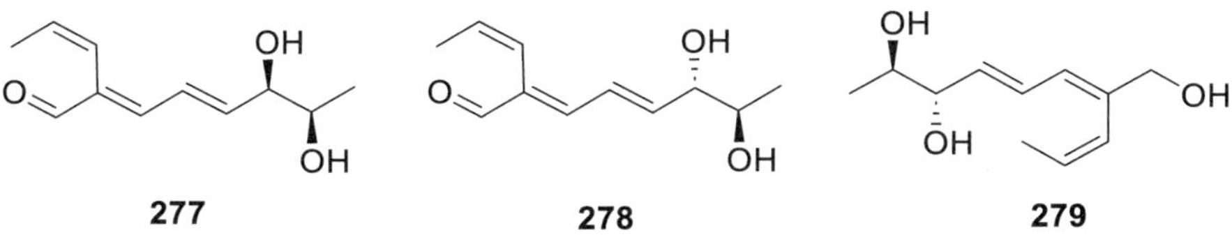

277 278 279

Fig. 35 Chemical structures of **277–279**

2.8.7 Xanthones

Xanthone (**276**) was isolated from the marine-derived actinomycete *Nocardiopsis* sp. G057 [13] and from the marine sediment-derived fungus *Neurospora* sp. M45 (Fig. 34) [68]. Compound **276** exhibited antimicrobial activity against all tested strains of both Gram-positive and Gram-negative bacteria, *E. faecalis*, *S. aureus*, *B. cereus*, *E. coli*, *P. aeruginosa*, *S. enterica*, and the yeast *Candida albicans* with MIC values of 64–256 μg/mL.

2.8.8 Polyenes

Three new polyenes, talacyanols A–C (**277–279**), were isolated from the marine fungus *Talaromyces cyanescens* derived from a seaweed *Caulerpa* sp. (Fig. 35) [69]. Among them, **277** showed moderate cytotoxic activity against a panel of cancer cell lines (HCT-15, NUGC-3, NCI-H23, ACHN, PC-3, and MDA-MB-231) with GI_{50} values ranging from 44.4 to 91.6 μM, whereas **278** exhibited anti-neuroinflammatory effect without cytotoxicity against all the tested cell lines.

2.8.9 Glycosides

Two glycosides, including one new indole glucoside, 6-methoxyindole-3-carboxylic acid, *O*-*β*-D-glucopyranosyl ester (**280**) was isolated from marine sea cucumber-derived *Aspergillus fumigatus* M580 [30] and 1-methyl-6-acetyl-α-D-glucopyranoside (**281**) was isolated from the marine sediment bacterium *Micromonospora* sp. G047 (Fig. 36) [59]. Compound **281** is reported for the first time from a natural source.

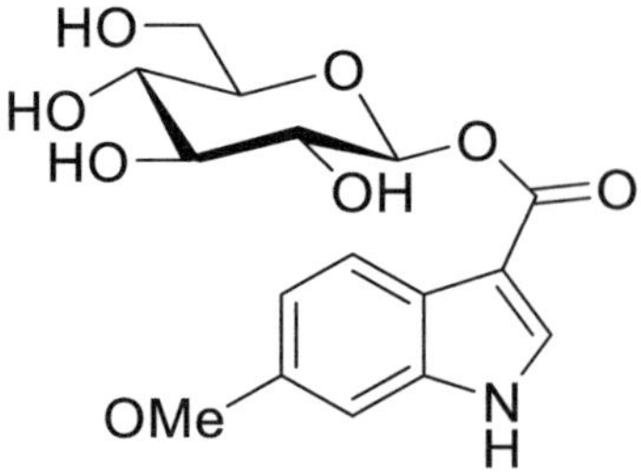

280

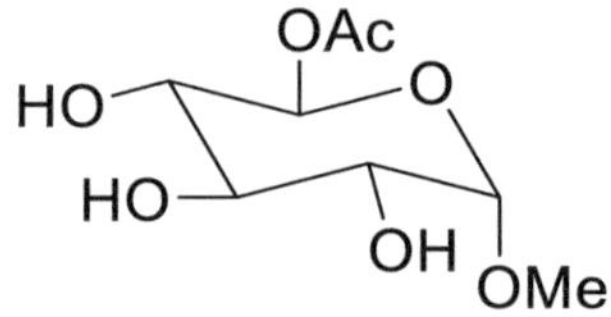

281

Fig. 36 Chemical structures of **280** and **281**

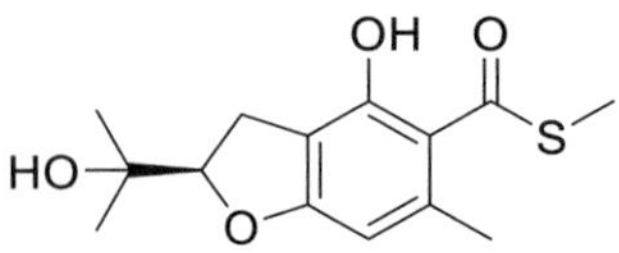

282

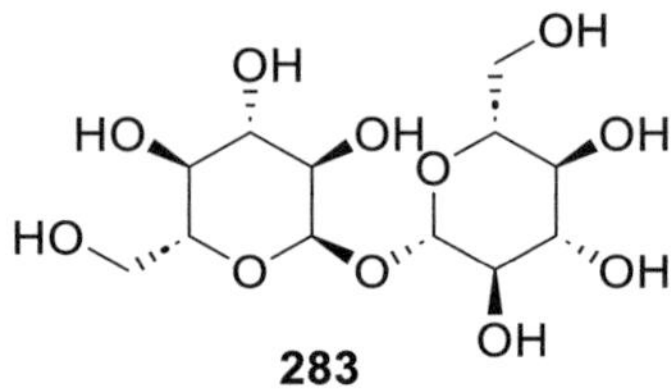

283

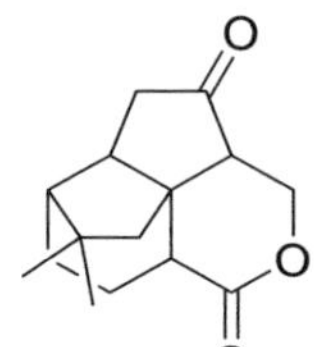

284

Fig. 37 Chemical structures of **282–284**

2.8.10 Benzofurans

Eurothiocin A (**282**) (Fig. 37) was isolated from the marine fungus *Talaromyces cyanescens* derived from a seaweed *Caulerpa* sp. [69]. Compound **282** exhibited anti-neuroinflammatory effect without cytotoxicity against all the tested cell lines HCT-15, NUGC-3, NCI-H23, ACHN, PC-3, and MDA-MB-231.

2.8.11 Disaccharides

α,β-Trehalose (**283**) (Fig. 37) was isolated and characterized from the marine sediment-derived fungus *Neurospora* sp. M45 collected in Co To-Thanh Lan island, Vietnam [68].

2.8.12 Ketones

A diketone, quadrone (**284**) (Fig. 37), was isolated from the marine fungus *Aspergillus terreus* LM.1.5 [39].

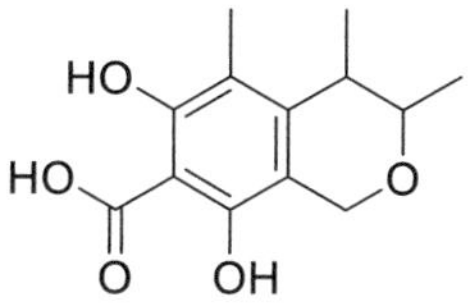

285

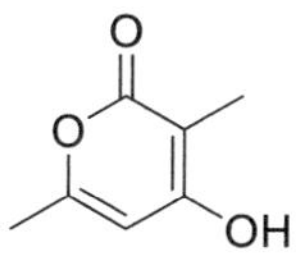

286

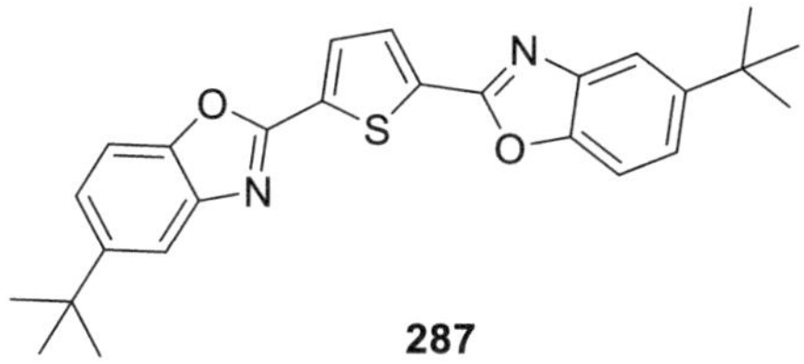

287

Fig. 38 Chemical structures of **285–287**

2.8.13 Citrinins

Dihydrocitrinin (**285**) (Fig. 38) was isolated from the marine sediment-derived *Penicillium* sp. OM07 collected in Son Cha island, Vietnam [12]. Compound **285** inhibited against *E. faecalis*, *S. aureus*, *B. cereus*, and *C. albicans* with MIC values from 8 to 256 μg/mL.

2.8.14 Heterocyclic Compounds

The pyrone, 4-hydroxy-3,6-dimethyl-2-pyrone (**286**) (Fig. 38), was isolated from *Penicillium* sp. KMM 4672 [39]. In addition, 2,5-bis(5-*tert*-butyl-2-benzoxazolyl)thiophene (**287**), was isolated from the echinoderm-derived actinomycete *Streptomyces* sp. G278 (Fig. 38) [20]. Compound **287** was isolated from a natural source for the first time and selectively inhibited *E. faecalis* with MIC of 256 μg/mL.

2.8.15 Alcohols

A new 1,4-dioxane derivative, 2-[(5-methyl-1,4-dioxan-2-yl)methoxy]ethanol (**288**) (Fig. 39) was isolated from the marine sediment-derived actinomycete *Micromonospora* sp. G019 [11]. Compound **288** showed antimicrobial activity against *E. faecalis* and *C. albicans* with MIC values of 32 and 64 μg/mL, respectively. In addition, several known metabolites were isolated from marine microorganisms in Vietnam, including (2*R*,3*S*)-butane-2,3-diol (**289**) from the sponge-derived *Micromonospora* sp. A258 [34]; (2*R*,3*S*)-butanediol (**290**) and (2*S*,3*S*)-butanediol (**291**) from the marine sediment-derived fungus *Neurospora* sp. M45 [68]; and 2,3-butanediol (**292**) from the marine sediment-derived bacterium *Oceanisphaera* sp. (Fig. 39) [38].

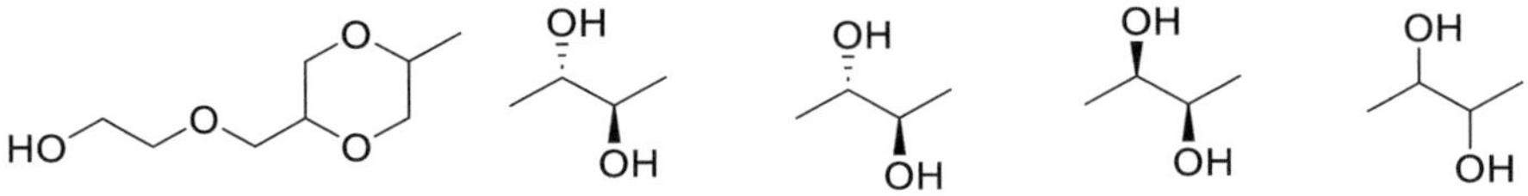

288 289 290 291 292

Fig. 39 Chemical structures of **288–292**

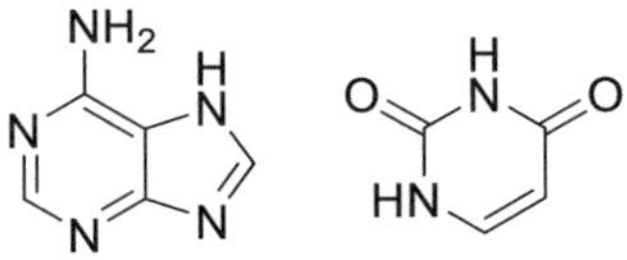

293 294

Fig. 40 Chemical structures of **293** and **294**

2.8.16 Purines and Pyrimidines

The purine, adenine (**293**), and pyrimidine, uracil (**294**), were isolated from various marine microorganisms in Vietnam (Fig. 40). Compound **293** was isolated from the sponge-derived *Streptomyces* sp. G248 [15] and from the prickly pen shell-derived *Streptomyces fradiae* G650 [29], whereas **294** was obtained from several marine sediment-derived microbes, including *Neurospora* sp. M45 [68], *Micromonospora ectrinospora* G017 [61], *Micromonospora* sp. G047 [59], *Streptomyces* sp. G039 [37], and *Oceanisphaera* sp. [38].

2.8.17 Nucleosides

Several nucleosides (Fig. 41) were isolated from marine microorganisms in Vietnam, including guanosine (**295**), 2′-deoxyadenosine (**296**), and adenosine (**297**) from sponge-associated *Streptomyces* sp. G246 [32]; 3-hydroxy-adenosine (**298**) from *Streptomyces* sp. G261 [19]; 2′-deoxyuridine (**299**) and uridine (**300**) from the prickly pen shell-derived *Streptomyces fradiae* G650 [29]; 5′-deoxyuridine (**301**) from the marine coral-derived actinomycete *Actinoalloteichus cyanogriseus* G631 [31]; 2′-deoxythymidine (**302**) from marine sediment-derived actinomycete *Streptomyces* sp. G212; as well as thymine (**303**) from the marine sediment bacterium *Micromospora* sp. G047 [59]. Among them, **299** exhibited antimicrobial activity against *E. faecalis*, *S. aureus*, *B. cereus*, *S. enterica*, and *C. albicans* with MIC values of 16–64 μg/mL [64], whereas **301** inhibited *B. cereus* bacteria with MIC values of 128 μg/mL, respectively, and **302** exhibited antimicrobial activity against *E. faecalis*, *S. aureus*, *B. cereus*, *S. enterica*, and *C. albicans* with MIC values of 32–64 μg/mL.

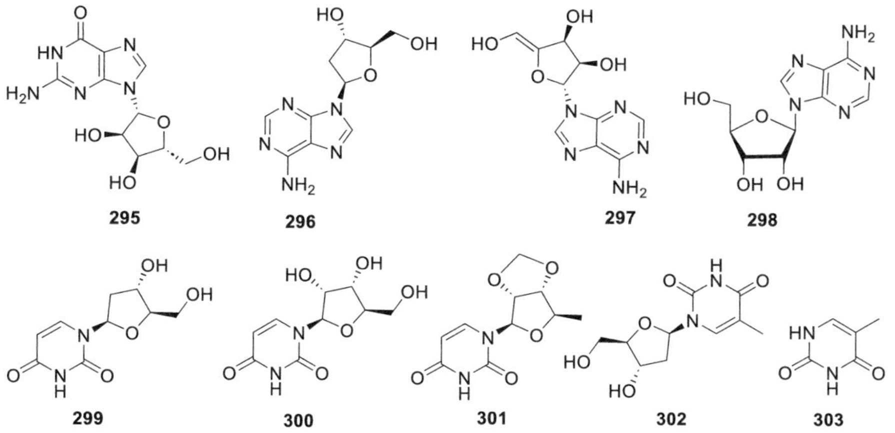

Fig. 41 Chemical structures of **295–303**

2.8.18 Hybrid Compounds

Numerous hybrid compounds (Fig. 42) were isolated from marine microorganisms in Vietnam, including benzyl salicylate (**304**) from the echinoderm-derived actinomycete *Streptomyces* sp. G278 [20]; 1,5-dideoxy-3-C-methyl-arabinitol (**305**) from the marine sediment-derived bacterium *Oceanisphaera* sp. [38]; sydowione A (**306**) from the fungus *Aspergillus flocculosus* 168ST-16.1 derived from the marine alga *Padina* sp. [52]; 3,3″-dihydroxy-6′-*O*-desmethyl terphenyllin (**307**) from the marine sediment-derived fungus *Penicillium janthinellum* 168CLC-17.1 [25]; a new 1,3-benzodioxole derivative, 1,3-benzodioxole-2-one-4-carboxylamide (**308**), from the marine sediment-derived actinomycete *Streptomyces* sp. G261 [19]; a naturally new ester, 2,4-dichlorophenyl 2,4-dichloro benzoate (**309**) and known metabolite, *N*-[2-(1*H*-indol-3-yl)-2-oxo-ethyl] acetamide (**310**), from marine sediment-derived actinomycete *Streptomyces* sp. G212 [64]; as well as betaine (**311**) from the marine sediment-derived bacterium *Oceanisphaera* sp. [38].

Bioassays revealed that **304** selectively inhibited *E. faecalis* with MIC value of 64 μg/mL, whereas **307** exhibited inhibitory activity on nitric oxide (NO) production in lipopolysaccharide (LPS)-stimulated BV-2 microglial cells. However, compound **307** also displayed strong toxicity toward BV-2 microglial cells. Compound **308** exhibited inhibitory activity against *E. faecalis* and *S. aureus* with MIC values of 128 and 256 μg/mL, respectively. Additionally, **308** displayed cytotoxicity against cancer cell lines KB, HepG-2, LU-1 and MCF-7 with IC_{50} values of 71.8 ± 4.1, 73.8 ± 5.3, 52.0 and 82.0 ± 5.9 μg/mL, respectively. Compound **309** was synthesised and antimicrobial assays indicated that **309** exhibited antimicrobial activity against *B. cereus* and *C. albicans* with MIC values of 256 and 64 μg/mL, respectively.

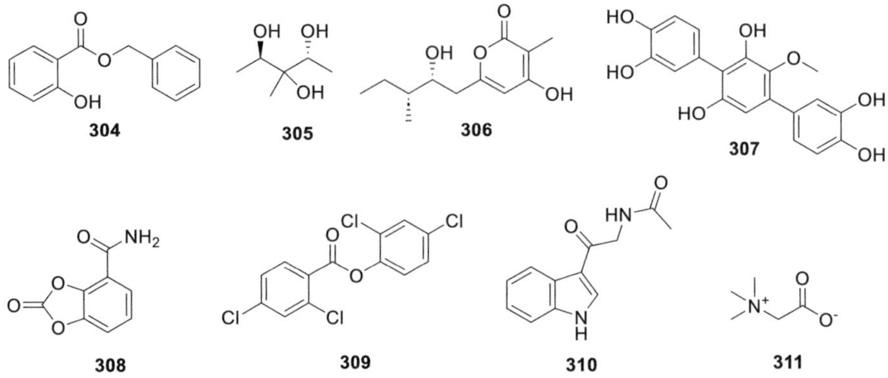

Fig. 42 Chemical structures of **304–311**

3 Microbial Marine Natural Products from Remote Archipelagos Within the East Sea Ecoregion

The East Sea Islands ecoregion consists of over 250 islands, atolls and shoals scattered throughout the oceanic area, and grouped into three archipelagos, including Pratas, Paracel and Spratly Islands. Despite being spread across a vast oceanic area, the actual land area above sea level is quite small, and the important biodiversity remains hidden in the surrounding coral reefs. Marine microbial natural products research within the Paracel Islands, Spratly Islands, Pratas Island and their surrounding deep-water regions is an emerging field marked by significant potential yet limited exploration. These areas harbor unique marine microbial communities adapted to extreme and diverse habitats, such as coral reefs, seamounts and deep-sea hydrothermal environments. Recent investigations into the diversity of fungal communities in nine different deep-sea sediment samples from the East Sea ecoregion using culture-dependent methods have revealed a rich reservoir of deep-sea fungi with promising bioactive potential [70]. Notably, a relatively high proportion (56%) of these isolates demonstrated antimicrobial activity against at least one of four tested marine pathogenic microbes, including *Micrococcus luteus*, *Pseudoalternomonas piscida*, *Aspergillus versicolor* and *A. sydowii*, underscoring the potential of these underexplored deep-sea fungi as valuable sources for future antibiotic discovery. Similarly, the East Sea ecoregion has been emerging as a potentially abundant source of novel species and genera of marine actinomycetes, whose isolated natural products have exhibited a wide range of bioactivities, including antitumor, antimicrobial, antifouling and anti-fibrotic effects [71].

While studies on microbial natural products have yielded novel bioactive compounds with pharmaceutical and biotechnological applications, geopolitical complexities and limited access to these regions have constrained comprehensive research efforts. Recent advancements in metagenomics, bioinformatics and deep-sea sampling techniques are facilitating the discovery of new microbial species and their secondary metabolites, highlighting these archipelagos as promising but under-explored hotspots for natural product discovery. Due to the immense number of marine microbial natural products reported in the literature, this section aims to highlight structural diversity and biological activities of selected new bioactive marine natural products reported within the past 6 years (2019–2024) from marine microbes obtained from the East Sea ecoregion.

At least a dozen marine microbial strains, presented in this review, have been previously investigated before 2019 for new natural products (based on Google Scholar). These microbial strains include: *Micromonospora echinospora* SCSIO 04089, *Streptomyces atratus* SCSIO ZH16, *S. niveus* SCSIO 3406, *S. rochei* MB037, *Acremonium persicinum* SCSIO 115, *Arthrinium arundinis* ZSDS1-F3, *Aspergillus candidus* (CHNSCLM-0393), *A. flocculosus* 16D-1, *A. versicolor* XS-20090066, *Cochliobolus lunatus* SCSIO41401, *Eurotium* sp. SCSIO F452 and *Trichoderma harzianum* XS-20090075. Based on genomic information and/or metabolomic analysis, these strains were reinvestigated for new natural products. A summary of new microbial marine natural products from the remote archipelagos within the East Sea ecoregion are presented in Table 2

3.1 Terpenoids

3.1.1 Sesquiterpenoids

Under MS/MS-based molecular networking guidance, eight odoriferous sesquiterpenes were isolated from the ethyl acetate extract of the marine mesophotic zone sponge (*Spongia* sp.)-associated *Streptomyces* sp. NBU3428, collected near Paracel Island. These compounds include two previously undescribed geosmin-type sesquiterpenoid degradations, named odoripenoids A (**312**) and B (**313**) (Fig. 43), as well as two novel germacrane-type sesquiterpenoids, odoripenoids C (**314**) and D (**315**) (Fig. 43) [72]. Notably, odoripenoids A and B are rare geosmin-related metabolites isolated as natural products from actinomycetes. Both compounds also demonstrated anti-*Candida albicans* activity, with minimum inhibitory concentration (MIC) values of 16 and 32 μg/mL, respectively.

Four new sesquiterpenoids, including nigerin (**316**) and ochracenes J (**317**)–L (**319**) (Fig. 43), were isolated from *Aspergillus niger*, a symbiotic fungus associated with the marine sponge *Dysidea* sp. collected near the Paracel Islands [73]. Nigerin features a unique 1-(3-*n*-pentyl)-2,5,6-trimethyl-cycloheptane carbon skeleton, representing a novel sesquiterpene scaffold. Both nigerin and ochracene J demonstrated significant inhibitory activity on nitric oxide (NO) production in lipopolysaccharide (LPS)-stimulated RAW264.7 macrophages, with IC_{50} values of 8.5 and 4.6 μM, respectively.

Table 2 List of new microbial natural products reported from the remote archipelagos within the East Sea ecoregion

Marine microorganism/ Source	Compound	Biological activity	Refs.
Marine bacteria			
Actinoalloteichus cyanogriseus LHW52806/ marine sponge *Phakellia fusca*	Marinacarboline glucuronide (**584**); cyanogripeptides A (**706**)–C (**708**)	**584** & **708**: antimicrobial	[133, 162]
Micromonospora echinospora SCSIO 04089/deep-sea sediment	Gephysulfuromycin (**533**); homophenanthroviridone (**534**); homophenanthridonamide (**535**); homostealthin D (**536**); nenesfuran (**537**); WS-5995 D (**538**); nenesophanol (**539**)	**534**: cytotoxic; antimicrobial	[120]
Nocardiopsis dassonvillei SCSIO 40065/marine sponge *Petrosia* sp.	Dassonmycins A (**622**) & B (**623**)	Antimicrobial; cytotoxic	[141]
Streptomyces sp. NBU3428/ marine sponge *Spongia* sp.	Odoripenoids A (**312**)–D (**315**)	**312** & **313**: antifungal	[72]
Streptomyces sp. ZYX-F-97/ sediment	1,19-epithio-geldanamycin A (**566**); 17-demethoxylherbimycin H (**567**); herbimycin M (**568**); *seco*-geldanamycin B (**569**)	**569**: antimicrobial	[128]
Streptomyces sp. SCSIO 40070/deep-sea sediment	Aburatubolactams D (**574**)–F (**576**)	**574** & **575**: cytotoxic	[129]
Streptomyces sp. NBU3142/ marine sponge *Dasychalina* sp.	Bresmycins A (**585**) & B (**586**)	Cytotoxic	[134]
Streptomyces sp. OUCMDZ-4539/marine sponge *Dysidea avara*	Pyridapeptides A (**656**)–I (**664**)	**658–660**, **662–664**: cytotoxic	[151, 152]
Streptomyces atratus SCSIO ZH16/deep-sea sediment	Ilamycin congeners (**682–689**, **691–702**); atratumycin (**690**) via *genetic engineering*	antimicrobial	[157, 158]

(continued)

Table 2 (continued)

Marine microorganism/Source	Compound	Biological activity	Refs.
Streptomyces diacarni LHW51701/marine sponge *Diacarnus megaspinorhabdosa*	Chlocarbazomycins A (**580**)–D (**583**)	**582**: antimicrobial	[132]
Streptomyces niveus SCSIO 3406/deep-sea sediment	Aclidinomycins C (**606**)–K (**614**) via *genome-directed discovery*	**607**, **608**, **610**, **613**, **614**: antimicrobial	[139]
Co-culture of *Streptomyces rochei* MB037/marine sponge *Dysidea arenaria* & *Rhinocladiella similis* 35/staghorn gorgonian	Borrelidins J (**428**) & K (**429**); 7-methoxy-2,3-dimethylchromone-4-one (**430**) via *co-culturing method*	**428**: antimicrobial	[98]
Marine fungi			
Acremonium sp. NBUF150/marine sponge *Ciocalypta* sp.	Acremocholone (**418**)	Antimicrobial	[93]
Acremonium persicinum SCSIO 115/deep-sea sediment	Acremonpeptides A (**665**)–E (**669**)	**665**, **666**, **669**: antiviral	[153]
Alternaria sp. ZH-15/soft coral *Lobophytum crassum*	Alternarin A (**383**); territrem F (**384**)	Neuroactive	[86, 87]
Amphichorda felina (syn. *Beauveria felina*) SYSU-MS7908/ascidian *Styela plicata*	Amphichoterpenoid A (**366**)–E (**370**); amphicordins A (**371**)–D (**374**); felinoids A (**376**)–E (**380**); isaridins I (**650**)–N (**655**)	**366–370**: acetylcholinesterase inhibitors **377** & **378**: anti-inflammatory **651**: antifungal	[81–84, 150]
Amphichorda felina/ascidian *Styela plicata*	Amphichopyrones A (**440**) & B (**441**) via *heterologous expression*	Anti-inflammatory	[100]
Arthrinium sp./deep-sea sediment	Arthpyrone L (**595**)	Cytotoxic	[136]

(continued)

Table 2 (continued)

Marine microorganism/ Source	Compound	Biological activity	Refs.
Arthrinium arundinis ZSDS1-F3/marine sponge *Phakellia fusca*	Alternapyrones G (**442**) & H (**443**) via *heterologous expression* Arthpyrones M (**596**)–O (**598**)	**442**: neuroprotective **598**: cytotoxic	[101, 137]
Aspergillus sp. NBUF87/ marine sponge *Hymeniacidon* sp.	Aspergimarins A (**448**)–G (**454**)	**448** & **452**: inhibited root growth of *Arabidopsis thaliana* Columbia-0 **452**: antimicrobial	[102, 103]
Aspergillus sp. ITBBc1	Sanshamycins A (**510**)–E (**514**)	**512**: phosphodiesterase 4D inhibitor	[116]
Aspergillus sp. WHU0154/ deep-sea sediment	18,19-dihydro-18,19-dihydroxyasperophiobolin E (**349**); $\Delta^{16,17}$–8-dehydroxyophiobolin D (**350**); $\Delta^{16,17}$-ophiobolin D (**351**); asperophiobolin L (**352**); (16*E*)-asperophiobolin L (**353**)	**352**: anti-inflammatory	[79]
Aspergillus sp. 151304/ marine sponge *Petrosia* sp.	Petrosamides A (**703**)–C (**705**)	Pancreatic lipase inhibitors	[161]
Aspergillus candidus OUCMDZ-1051/marine sponge	4-*O*-methylcandidusin A (**509**)	Cytotoxic; antimicrobial	[115]
Aspergillus candidus/ gorgonian *Junceella fragilis*	4″-*O*-methyl-prenylterphenyllin B (**515**); phenylcandilides A (**516**) & B (**517**); asperindoles E (**518**)–G (**520**)	**516**, **517**: neuroactive	[117]
Aspergillus candidus (CHNSCLM-0393)/ gorgonian *Juncella fragilis*	Asperazepanones A (**578**) & B (**579**)	**579**: anti-inflammatory	[131]
Aspergillus flavipes 164,013/ marine sponge *Dysidea* sp.	Flavipesides A (**463**)–C (**465**)	Pancreatic lipase inhibitors	[106]
Aspergillus flocculosus 16D-1/marine sponge *Phakellia fusca*	Asperflotone (**423**); asperfloroid (**424**); ochrasperfloroid (**425**); asperfloketals A (**426**) B (**427**)	Anti-inflammatory	[95–97]

(continued)

Table 2 (continued)

Marine microorganism/ Source	Compound	Biological activity	Refs.
Aspergillus hiratsukae SCSIO 7S2001/gorgonian	Chevalones H (**386**)–M (**391**)	Antimicrobial	[88]
Aspergillus niger/marine sponge *Dysidea* sp.	Nigerin (**316**); ochracenes J (**317**)–L (**319**)	**316** & **317**: anti-inflammatory	[73]
Aspergillus puniceus FAHY0085/sediment	Oxepinamide L (**637**); puniceloids E (**638**)–G (**640**)	**639**: transcriptional activator of liver X receptor α	[146]
Aspergillus sydowii SCSIO 41301/marine sponge *Phakellia fusca*	2-hydroxy-1-(hydroxymethyl)-8-methoxy-3-methyl-9H-xanthen-9-one (**455**); 2-hydroxy-1-(hydroxymethyl)-7,8-dimethoxy-3-methyl-9H-xanthen-9-one (**456**); catechol derivatives (**457** and **458**); aspergillusene D (**459**)	**455** & **456**: antiviral	[104]
Aspergillus terreus/soft coral *Sarcophyton subviride*	Asperteramide A (**577**)	Antimicrobial	[130]
Aspergillus versicolor (15XS43ZD-1)/marine sponge	Sterigmatocystins A (**460**)–C (**462**)	Antiviral **460**: angiogenesis-promoting activity	[105]
Aspergillus versicolor XS-20090066/gorgonian *Dichotella gemmacea*	Kipukasins K (**709**) & L (**710**); aspergillusene E (**711**) via *epigenetic method*	**709** & **711**: antimicrobial	[163]
Aspergillus versicolor CHNSCLM-0063/gorgonian *Rumphella aggregata*	Asperversiamides A (**670**)–C (**672**); asperheptatides A (**673**)–D (**676**); asperpyrroindotide A (**681**)	Antimicrobial	[154–156]
Chaetomium globosum HDN151398/deep-sea sediment	*N*-glutarylchaetoviridins A (**599**)–C (**601**)	**601**: cytotoxic	[138]
Cochliobolus lunatus SCSIO41401/marine alga *Coelarthrum* sp.	Sinulariapeptides A (**644**) & B (**645**); phthalide glycerol ether (**646**)	Acetylcholinesterase inhibitors	[147]

(continued)

Table 2 (continued)

Marine microorganism/ Source	Compound	Biological activity	Refs.
Eurotium sp. SCSIO F452/ sediment	Euroticins A (**540**)–I (**548**) eurotinoids A (**633**)–C (**635**)	**541**: antioxidant **542** & **544**: cytotoxic **546** & **547**: α-glucosidase inhibitors **630–632**: antioxidative; cytotoxic	[121, 123, 145]
Fusarium sp. L1/sea star *Acanthaster planci*	Fusaindoterpenes A (**587**) & B (**588**), fusariumindoles A (**589**)–C (**591**); (±)-isoalternatine A (**592**) via L-tryptophan supplementation	**588**: antiviral	[135]
Myrothecium sp. BZO-L062/ deep-sea sediment	Myrothecol (**466**); a methoxy-myrothecol racemate (**467**); myrothin (**468**)	**466**: anti-inflammatory; antioxidant	[107]
Penicillium sp. NBUF154/ marine sponge *Crella* sp.	Guignardones Y (**381**) and Z (**382**)	**381**: antiviral	[85]
Penicillium sp. SCSIO 41512/ soft coral	Penicimeroterpenoids A (**392**)–C (**394**); pyrrospirones K (**615**)–Q (**621**)	**392–394**: enzyme inhibitors **616** & **619**: antimicrobial	[89, 140]
Penicillium sp. SCSIO 41691/ soft coral	1-methyl-12a,12b-epoxyarisugacin M (**395**), 1-methyl-4a,12b-epoxyarisugacin M (**396**), 2,3-dihydroxy-3,4a-epoxy-12a-dehydroxyisoterreulactone A (**397**), 2-hydroxy-12a-dehydroxyisoterreulactone A (**398**), 3′-demethoxyterritrems B′ (**399**), 4a-hydroxyarisugacin P (**400**) and 1-*epi*-arisugacin H (**401**)	**399**: anti-inflammatory	[90]
Penicillium sp. ZYX-Z143/ arthropod *Dardanus scutellatus*	Penpaxilloids A (**411**)–E (**415**); 7-methoxypaxilline-13-ene (**416**); 10-hydroxy-paspaline (**417**)	**411**: noncompetitive inhibitor of protein tyrosine phosphatase 1B **414**: anti-inflammatory	[92]
Penicillium sp. KWF32/ dee-sea sediment	Penitanzacids A (**482**)–J (**491**); hatsusamides C (**492**) & D (**493**)	**492** & **493**: cytotoxic	[112]
Penicillium citrinum SCSIO41402/marine alga *Coelarthrum* sp.	Sorbicillfurans A (**558**) and B (**559**)	**559**: cytotoxic	[125]

(continued)

Table 2 (continued)

Marine microorganism/ Source	Compound	Biological activity	Refs.
Penicillium sclerotiorum E23Y-1A/marine sponge *Holoxea* sp.	Penicilazaphilones I (**494**)–N (**499**); *epi*-geumsanol D (**500**); penidioxolanes C (**501**) & D (**502**)	**499**: anti-inflammatory **501**: cytotoxic	[113]
Pestalotiopsis sp. NBUF145/ marine sponge of a Chalinidae sp.	Pestanoid A (**320**)	anti-osteoclastogenic	[74]
Pestalotiopsis heterocornis XWS03F09/marine sponge *Phakellia fusca*	Heterocornols Q (**549**)–X (**556**); ceramide (**557**) via *OSMAC approach*	**552** & **556**: anti-inflammatory	[124]
Pestalotiopsis neglecta SCSIO41403/red alga *Coelarthrum* sp.	Pestalotioquinols C (**469**)–F (**472**); pestalotiochromenoic acids A (**476**)–D (**479**); pestalotiochromones A (**480**) & B (**481**)	Modulators of liver X receptor alpha	[108]
Pseudallescheria boydii/soft coral *Sinularia sandensis*	**503–505**	Regulatory effects on osteoclastogenesis	[114]
Simplicillium sp. SCSIO 41513/soft coral	Simplifusidic acids A (**355**)–K (**365**)	**362** & **363**: antimicrobial	[80]
Simplicillium lanosoniveum SCSIO41212/soft coral *Sinularia* sp.	Arthriniumsteroids A (**419**)–D (**422**)	anti-inflammatory	[94]
Stagonospora sp. SYSU-MS7888/marine sponge	Stagonospones A (**431**) & B (**432**); stapyrones A (**433**)–G (**439**)	**431**, **432**, **437**, **439**: anti-inflammatory	[99]
Stachybotrys chartarum/soft coral *Sarcophyton subviride*	Stachatranones A (**329**)–C (**331**); stachatranones D (**340**)–F (**342**); atranones Q (**332**)–S (**334**); atranones V (**335**)–Z (**339**); stachybatranones A (**343a/343b**)–F (**348**)	**330** & **332**: antimicrobial **342** & **343**: cardioprotective	[76–78]

(continued)

Table 2 (continued)

Marine microorganism/ Source	Compound	Biological activity	Refs.
Stachybotrys chartarum SCSIO41201/soft coral *Sinularia* sp.	Arthproliferins A (**521**)–E (**525**)	**521**: antimicrobial	[118]
Talaromyces indigoticus FS688/deep-sea sediment	Azaphilone derivatives (**527–530**); isoprenyl-substituted benzoate derivative (**531**)	**530**: cytotoxic	[119]
Talaromyces purpureogenus SCSIO 41517/soft coral	Talaromynoids A (**402**)–I (**410**) (+)-/(−)-talaromyoxaones A (**560/561**) & B (**562/563**); (±)-11-apopyrenulin (**564**); abeopyrenulin (**565**); talaromyines A (**624**)–F (**629**)	**408–410**: lipid-lowering properties **560/561** & **562/563**: inhibitors of phosphatases SHP1, SHP2 & MEG2 **625**, **627**, **628**: inhibitors of phosphatases TCPTP & MEG2 **627** & **628**: cytotoxic	[91, 126, 143]
Trichoderma harzianum XS-20090075/unidentified soft coral	Harzianelactones A (**322**) & B (**323**); harzianones A (**324**)–D (**327**); harziane (**328**)	Phytotoxic	[75]
Marine microalgae			
Oscillatoria sp.	Amantamide B (**647**)	Neuroactive; cytotoxic	[148]

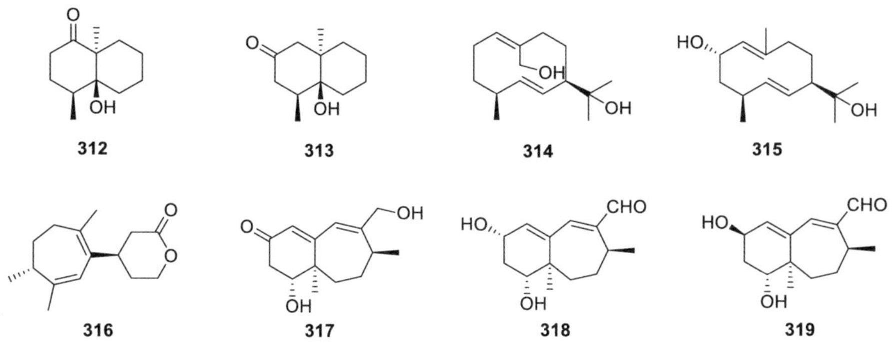

Fig. 43 Terpenoids from *Streptomyces* sp. NBU3428 and *Aspergillus niger*

3.1.2 Diterpenoids

A novel rearranged pimarane diterpenoid, pestanoid A (**320**) (Fig. 44), along with two previously reported compounds, nodulisporenones A (**321**) (Fig. 44) and B, were isolated from *Pestalotiopsis* sp. NBUF145, a fungus associated with a mesophotic zone Chalinidae sponge collected at a depth of 62 m near the Paracel Islands [74]. Pestanoid A and nodulisporenone A effectively inhibited bone marrow monocyte osteoclastogenesis in vitro, displaying similar IC_{50} values of 4.2 μM and 3.0 μM, respectively, without detectable cytotoxicity. In addition, mechanistic studies revealed that both compounds suppressed receptor activator of NF-κB ligand (RANKL)-induced MAPK and NF-κB signaling pathways by inhibiting the phosphorylation of ERK1/2, JNK1/2, p38 MAPKs and the nuclear translocation of NF-κB. As phosphorylation of MAPK signaling and NF-κB nuclear translocation are important for osteoclast differentiation, these molecules may act as potential anti-osteoclastogenic agents.

Two novel harziane diterpene lactones, harzianelactones A (**322**) and B (**323**) (Fig. 44), featuring a distinctive 6/5/7/5-fused carbocyclic core with a lactone ring system, along with five new harziane diterpenes, harzianones A (**324**)–D (**327**) and harziane (**328**) (Fig. 44), were isolated from the fungus *Trichoderma harzianum* XS-20090075 [75] isolated from an unidentified soft coral. This fungal strain was obtained from the inner tissue of an unidentified soft coral collected from the coral reefs of the Paracel Islands. Harzianelactones A and B represent a unique class of harziane diterpene lactones formed through Baeyer–Villiger monooxygenase-catalyzed oxidations of harziane diterpenes. Harzianelactones A and B, harzianones A–C as well as harziane were assessed for phytotoxic and antibacterial activities, with all showing significant phytotoxic effects on the seedling growth of amaranth and lettuce at a concentration of 200 ppm.

From a marine-derived strain of the toxigenic fungus, *Stachybotrys chartarum*, isolated from the soft coral *Sarcophyton subviride* collected at Paracel Island, three new dolabellane-type diterpenoids, stachatranones A (**329**)–C (**331**) (Fig. 45), alongside three new atranones, atranones Q (**332**)–S (**334**) (Fig. 45), were isolated [76].

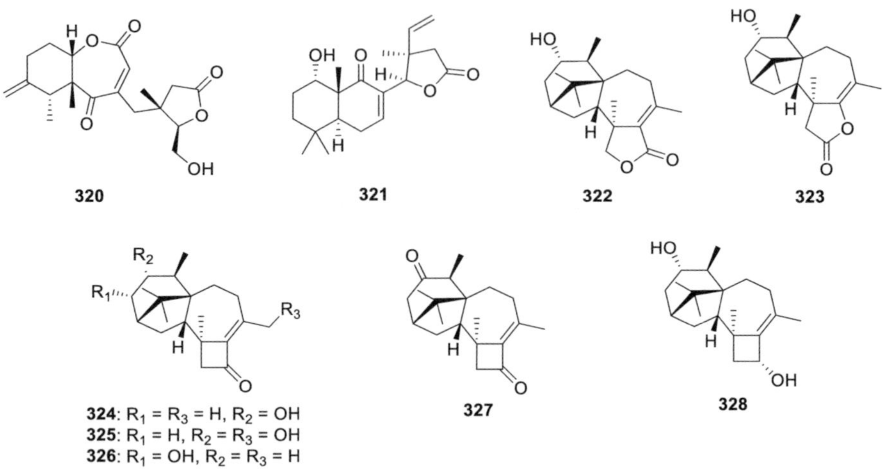

Fig. 44 Natural products from *Pestalotiopsis* sp. NBUF145 and *Trichoderma harzianum* XS-20090075

Stachatranones B and C displayed a unique 1,14-seco dolabellane-type diterpenoid skeleton. Notably, atranone Q is the first C23 atranone featuring a propan-2-one motif linked to a dolabellane-type diterpenoid via a carbon–carbon bond, while atranone R represents the first C24 atranone incorporating a 2-methyltetrahydrofuran-3-carboxylate motif fused to a dolabellane-type diterpenoid at C-5–C-6. In antimicrobial activity assays, stachatranone B exhibited activity against *Acinetobacter baumannii* and *Enterococcus faecalis* with MIC values of 16 and 32 μg/mL, respectively. Additionally, atranone Q demonstrated significant inhibition of *Candida albicans*, *Enterococcus faecalis* and methicillin-resistant *Staphylococcus aureus* (MRSA) with MIC values of 8, 16 and 32 μg/mL, respectively.

Further chemical investigation of a coral-associated strain of the toxigenic fungus *S. chartarum* led to the isolation of five previously undescribed atranones, designated as atranones V (**335**)–Z (**339**) (Fig. 45), along with three novel dolabellane-type diterpenoids, identified as stachatranones D (**340**)–F (**342**) (Fig. 45) [77]. Among these, stachatranone F demonstrated the ability to prevent cold ischemia (CI)-induced dephosphorylation of phosphatidylinositol-3-kinase (PI3K) and RAC-α serine/threonine-protein kinase (AKT) 12 h post-CI in a dose-dependent manner. Notably, stachatranone F significantly protected cardiomyocytes from cold ischemic injury, underscoring its promising potential for development as a novel cardioprotective agent in heart transplant medicine.

Additional new atranone-related molecules, stachybatranones A (**343a/343b**)–F (**348**) (Fig. 45), along with three known analogues, including methylatranones A and B and atranone B, were isolated and characterized from the marine fungus *S. chartarum* [78]. Notably, stachybatranone A represents the first reported example of

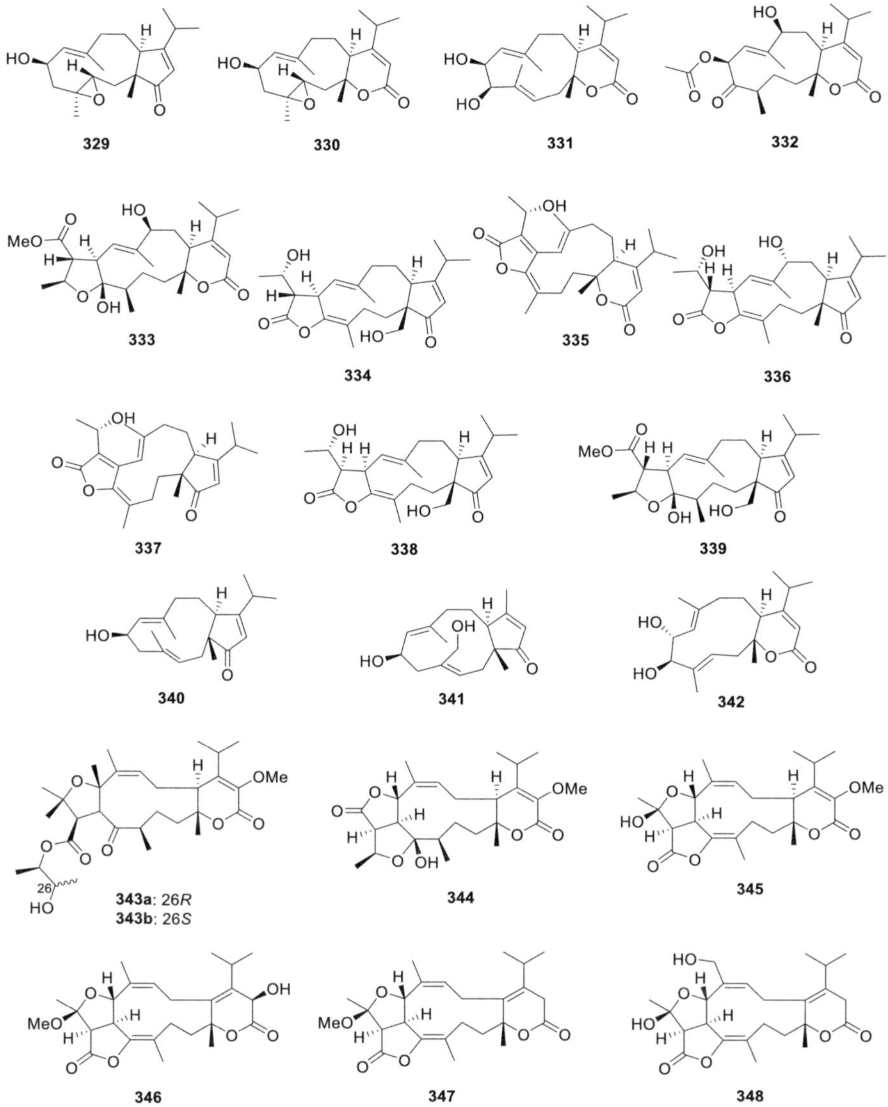

Fig. 45 Natural products from *Stachybotrys chartarum*

a 5–11-6-fused atranone structure featuring a 2,3-butanediol moiety. In a cardiomyocyte protective activity assay, these compounds demonstrated the ability to ameliorate cold ischemic injury (CI) 24 h post-injury. Stachybatranones A and D exhibited dose-dependent efficacy in this assay. Additionally, stachybatranone A was found to prevent cold ischemia-induced dephosphorylation of PI3K and AKT in a dose-dependent manner, suggesting a potential mechanism of action.

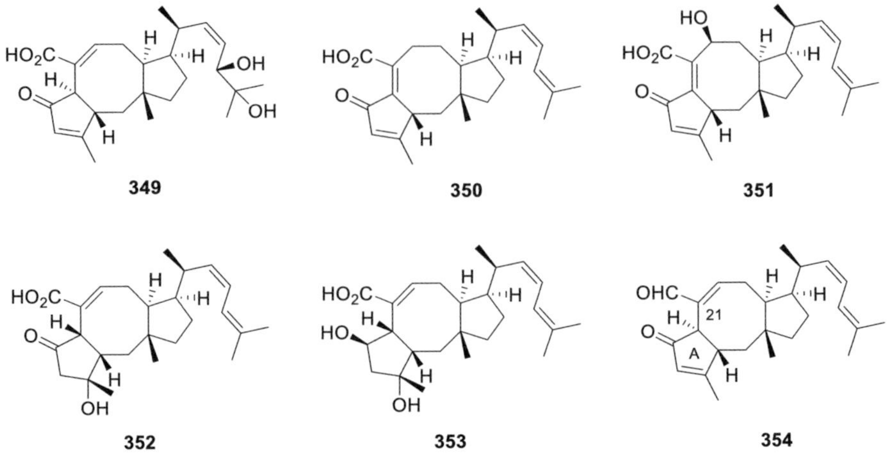

Fig. 46 Natural products from deep-sea-derived fungus WHU0154

3.1.3 Sesterterpenes

Five new ophiobolins, including 18,19-dihydro-18,19-dihydroxyasperophiobolin E (**349**), $\Delta^{16,17}$–8-dehydroxyophiobolin D (**350**), $\Delta^{16,17}$-ophiobolin D (**351**), asperophiobolin L (**352**) and (16*E*)-asperophiobolin L (**353**) (Fig. 46), along with six known analogues, were isolated from the deep-sea-derived fungus, *Aspergillus* sp. WHU0154 [79]. Among these, asperophiobolin L and known 6-*epi*-ophiobolin G (**354**) (Fig. 46), displayed significant inhibitory effects on nitric oxide (NO) production induced by lipopolysaccharide (LPS) in murine macrophage RAW264.7 cells. Preliminary structure–activity relationship (SAR) analysis indicated that the aldehyde group at C-21 and the α,β-unsaturated ketone functionality in the A ring in 6-*epi*-ophiobolin G (**354**) are crucial for the anti-inflammatory activity of ophiobolins.

3.1.4 Triterpenoids

Eleven new fusidane-type nortriterpenoids, named simplifusidic acids A (**355**)–K (**365**) (Fig. 47), were isolated from the marine fungus *Simplicillium* sp. SCSIO 41513, isolated from an unidentified soft coral collected at the Spratly Islands [80]. Notably, simplifusidic acid A featured a unique fusidane triterpene skeleton characterized by a rare 6/6/7/5/5 polycyclic system. Additionally, simplifusidic acids H and I exhibited strong antibacterial activity against *Staphylococcus aureus*, with minimum inhibitory concentration (MIC) values of 2.5 and 0.078 μg/mL, respectively.

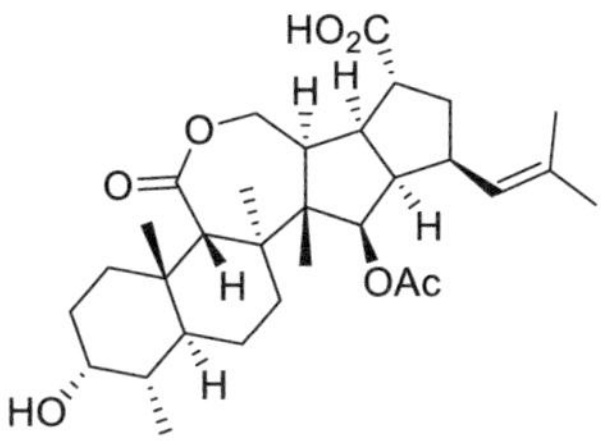

355

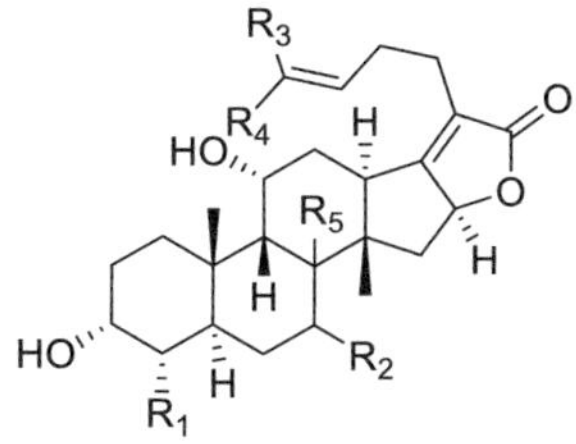

356: $R_1 = R_5 = \alpha CH_3$, $R_2 = \alpha OH$, $R_3 = CH_3$, $R_4 = CO_2H$
357: $R_1 = R_5 = \alpha CH_3$, $R_2 = H$, $R_3 = R_4 = CH_2OH$
358: $R_1 = R_5 = \alpha CH_3$, $R_2 = H$, $R_3 = CH_3$, $R_4 = CH_2OH$
359: $R_1 = \alpha CH_2OH$, $R_2 = H$, $R_3 = CH_3$, $R_4 = CO_2H$, $R_5 = \alpha CH_3$
360: $R_1 = \alpha CH_3$, $R_2 = H$, $R_3 = CH_3$, $R_4 = CO_2H$, $R_5 = \alpha CH_2OH$

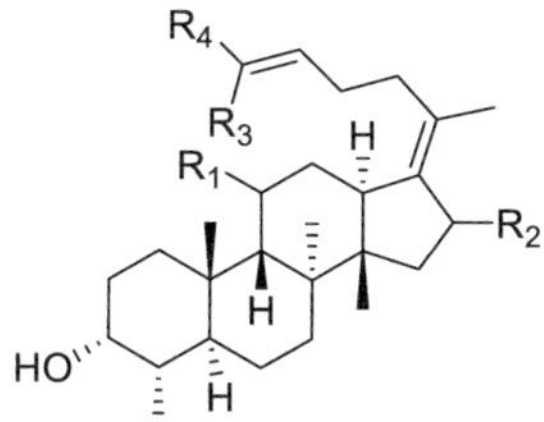

361: $R_1 = \alpha OH$, $R_2 = \beta OAc$, $R_3 = CH_2OH$, $R_4 = CH_3$
362: $R_1 = \alpha OH$, $R_2 = \beta OAc$, $R_3 = CH_3$, $R_4 = CH_2OH$
363: $R_1 = O$, $R_2 = \beta OAc$, $R_3 = R_4 = CH_3$
364: $R_1 = \alpha OH$, $R_2 = O$, $R_3 = CO_2H$, $R_4 = CH_3$
365: $R_1 = R_2 = O$, $R_3 = CO_2H$, $R_4 = CH_3$

Fig. 47 Natural products from *Simplicillium* sp. SCSIO 41513

3.1.5 Meroterpenoids

A pair of enantiomers, amphichoterpenoid A [(+)-**366a** and (−)-**366b**], and two epimers, amphichoterpenoids B (**367**) and C (**368**) (Fig. 48), were identified as novel picoline-derived meroterpenoids featuring a unique pyrano[3,2-c]pyridinyl-γ-pyranone scaffold [81]. These compounds were isolated from the fungus *Amphichorda felina* SYSU-MS7908, which was obtained from the marine ascidian *Styela plicata* collected from the North Atoll of the Paracel Islands. A year later, additional analogues, amphichoterpenoids D (**369**) and E (**370**) (Fig. 48), were isolated from the same fungal strain [82]. All compounds were evaluated for their inhibitory activities against acetylcholinesterase (AChE), which is key therapeutic target for Alzheimer's disease. The compounds (+)-amphichoterpenoid A, (−)-amphichoterpenoid A, amphichoterpenoids B and C demonstrated selective cholinesterase inhibition, exhibiting moderate inhibitory activity against AChE, with IC_{50} values ranging from 18.8 to 53.2 μM. In addition, amphichoterpenoids D and E showed anti-acetylcholinesterase (anti-AChE) activities with IC_{50} values of 12.5 μM and 11.6 μM, respectively.

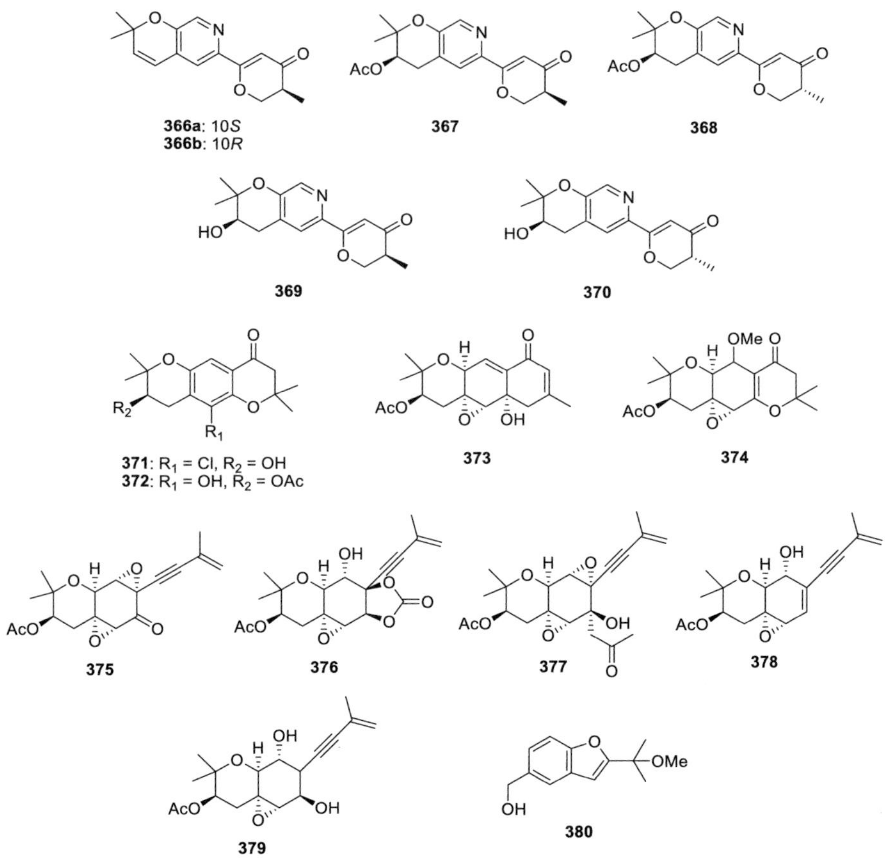

Fig. 48 Natural products from *Amphichorda felina* SYSU-MS7908

Additional chemical investigation of *Amphichorda felina* SYSU-MS7908 yielded four new compounds, amphicordins A (**371**)–D (**374**) (Fig. 48), and five known derivatives, including oxirapentyn A (**375**) (Fig. 48) [83]. Among these, amphicordin C is notable for its unique benzo[g]chromene (6/6/6) skeleton within this meroterpenoid family. Bioassays demonstrated that the known compound, oxirapentyn A, significantly inhibited the proliferation, migration and invasion of glioma cell lines (U87MG, T98G and U251) and induced apoptosis in these cells. In silico studies further suggested that oxirapentyn A has the potential to penetrate the blood–brain barrier, supporting its promise as an anti-glioma agent.

Further chemical work on the fungal strain, *Amphichorda felina* SYSU-MS7908, led to the identification of 13 acetylenic meroterpenoids and four biogenic analogs. Among these, five new compounds were characterized and named felinoids A (**376**)–E (**380**) (Fig. 48) [84]. Notably, felinoid A contained a rare cyclic carbonate, a unique feature in natural acetylenic meroterpenoids. Bioassays revealed that five isolates, including felinoids B (**377**) and C (**378**), exhibited anti-inflammatory activity by inhibiting nitric oxide (NO) production in LPS-induced RAW264.7 cells, with IC_{50} values ranging from 11.6 to 19.5 μM. Additionally, structural modification of the known compound, oxirapentyn B, generated 29 new derivatives, seven of which demonstrated enhanced activity ($IC_{50} < 3$ μM) and higher selectivity indices (SI > 22). Structure–activity relationship analysis highlighted the importance of 7,8-epoxy and 6-acylation groups for bioactivity.

Two new meroterpenoids, guignardones Y (**381**) and Z (**382**) (Fig. 49), along with six known metabolites, including guignardones A–H, gyorgy-isoflavone, daidzein, blumenol A and guignardianone A, were isolated from *Penicillium* sp. NBUF154, a fungus obtained from a *Crella* sponge collected at a depth of 60 m near the Paracel Islands [85]. Antiviral assays against human enterovirus 71 (EV71) revealed that guignardone Y (**381**) and blumenol A exhibited notable inhibitory activities, with EC_{50} values of 1.25 μM and 7.25 μM, respectively, highlighting their potential as antiviral agents.

Alternarin A (**383**) (Fig. 49), a rearranged drimane meroterpenoid with a unique thioglycerate moiety, was isolated alongside two known analogues from the coral-associated fungus *Alternaria* sp. ZH-15 [86]. This fungus was extracted from the internal tissues of the soft coral *Lobophytum crassum*, collected from an atoll, located at Pratas Island, at a depth of 15 m. Alternarin A exhibited potent bioactivity by effectively inhibiting spontaneous synchronous Ca^{2+} oscillations (SCOs)

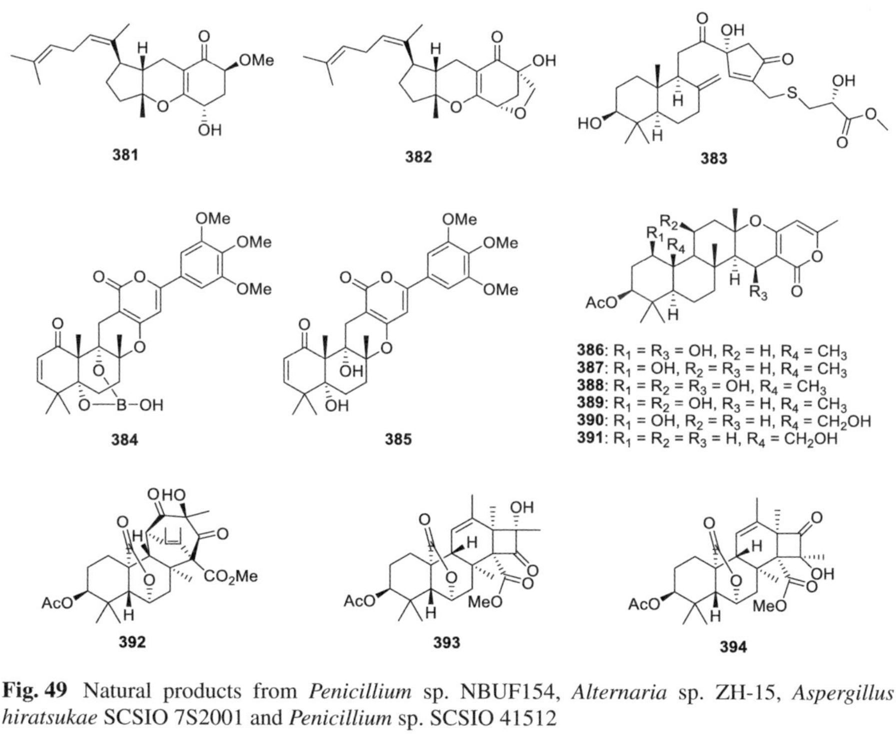

Fig. 49 Natural products from *Penicillium* sp. NBUF154, *Alternaria* sp. ZH-15, *Aspergillus hiratsukae* SCSIO 7S2001 and *Penicillium* sp. SCSIO 41512

and 4-aminopyridine-induced epileptic discharges in cortical neurons at low micromolar concentrations. Specifically, it reduced spontaneous SCO frequency (IC_{50} = 3.2 μM) and amplitude (IC_{50} = 1.8 μM) while suppressing hyperactive SCOs induced by 4-AP, with IC_{50} values of 10.0 μM for frequency and 4.6 μM for amplitude. SCOs play a critical role in neuronal development, and alterations in their patterns serve as indicators for studying the signaling mechanisms of chemical toxins. The inhibitory effects of alternarin A on neuronal excitability underscore its potential significance in neuropharmacological and toxicological research.

Further investigation of the fungal extract of *Alternaria* sp. ZH-15 led to the isolation of the boronic ester territrem F (**384**) (Fig. 49) and its diol precursor, the known drimane meroterpenoid territrem B (**385**) (Fig. 49) [87]. Territrem F, distinguished by its unique borate ring system, demonstrated concentration-dependent inhibition of spontaneous synchronous Ca^{2+} oscillation frequency (IC_{50} = 15 μM) and amplitude (IC_{50} = 16 μM). Additionally, it effectively mitigated the hyperactivity of SCOs induced by the seizurogenic agent 4-aminopyridine, with IC_{50} values of 12 μM for frequency and 34 μM for amplitude. Similarly, territrem B exhibited comparable inhibitory effects on SCO activity in neocortical neurons, further highlighting the bioactivity of these drimane meroterpenoids.

Six new α-pyrone meroterpenoids, chevalones H (**386**)–M (**391**) (Fig. 49), along with six known chevalones, were isolated from the gorgonian coral-derived fungus *Aspergillus hiratsukae* SCSIO 7S2001, collected from the Spratly Islands [88]. Chevalones H–L exhibited varying degrees of antibacterial activity against a panel of bacterial strains, including *Micrococcus lutea*, *Klebsiella pneumoniae*, methicillin-resistant *Staphylococcus aureus* (MRSA) and *Streptococcus faecalis*, with minimum inhibitory concentration (MIC) values ranging from 6.25 to 100 μg/mL. Additionally, chevalones I and L displayed weak cytotoxic activity against the SF-268, MCF-7, HepG-2, and A549 cancer cell lines.

Three novel andrastin-type meroterpenoids, penicimeroterpenoids A (**392**)–C (**394**) (Fig. 49), featuring two unprecedented skeletons with fused 6/5/6/6/7 and 6/5/6/6/4 polycyclic systems, were isolated from the soft coral-derived fungus *Penicillium* sp. SCSIO 41512 [89]. These molecules were evaluated for their enzyme-inhibitory activity against 11 different enzymes, including seven protein tyrosine phosphatases, cathepsin L, xanthine oxidase (XO), human leukocyte elastase (HLE) and trypsin. At a concentration of 40 μg/mL, these compounds showed inhibition ratios of 10–50% against cathepsin L, XO, HLE and trypsin. Notably, penicimeroterpenoid B exhibited mild inhibitory activity against HLE, while penicimeroterpenoids A–C demonstrated moderate inhibitory effects on CDC25B and SHP1. The significant selectivity of their inhibitory activity toward PTPs suggests their potential as candidates for anti-inflammatory and anticancer therapies.

Seven new 4-hydroxy-6-phenyl-2H-pyran-2-one (HPPO)-derived meroterpenoids, namely 1-methyl-12a,12b-epoxyarisugacin M (**395**), 1-methyl-4a,12b-epoxyarisugacin M (**396**), 2,3-dihydroxy-3,4a-epoxy-12a-dehydroxyisoterreulactone A (**397**), 2-hydroxy-12a-dehydroxyisoterreulactone A (**398**), 3′-demethoxyterritrems B′ (**399**), 4a-hydroxyarisugacin P (**400**) and 1-*epi*-arisugacin H (**401**) (Fig. 50), were isolated from the soft coral-derived fungal strain *Penicillium* sp. SCSIO 41691 [90]. 1-Methyl-12a,12b-epoxyarisugacin M and 1-methyl-4a,12b-epoxyarisugacin M featured a unique methyl migration within the basic meroterpenoid skeleton and were characterized by a 12a,12b-epoxy or 4a,12b-epoxy group, while 2,3-dihydroxy-3,4a-epoxy-12a-dehydroxyisoterreulactone A was a highly oxygenated HPPO-derived meroterpenoid with a rare 6/5/6/6/6/6 hexacyclic system containing a 3,4a-epoxy group. Biologically, 3′-demethoxyterritrems B′ demonstrated significant inhibitory activity against lipopolysaccharide-induced nitric oxide production in RAW 264.7 cells, with an IC_{50} value of 21 μM, outperforming the positive control, indomethacin.

Nine new highly oxygenated 3,5-dimethylorsellinic acid-derived meroterpenoids, named talaromynoids A (**402**)–I (**410**) (Fig. 50), were isolated from the soft coral-derived fungus *Talaromyces purpureogenus* SCSIO 41517, which was obtained from a soft coral collected at the Spratly Islands [91]. These compounds exhibited unprecedented polycyclic structures, with talaromynoid A and talaromynoids G-I containing 5/7/6/5/6/6, 6/7/6/6/6/5, 6/7/6/5/6/5/4 and 7/6/5/6/5/4 systems, respectively. Biological evaluation revealed that talaromynoid E selectively inhibited phosphatase CDC25B, with an IC_{50} value of 13 μM. Additionally, talaromynoids G–I demonstrated a dose-dependent reduction of triglyceride levels in 3T3-L1 adipocytes, a well-established in vitro model for evaluating lipid-lowering properties. These compounds also reduced intracellular lipid accumulation in a dose-dependent manner, with IC_{50} values of 21, 37 and 29 μM, respectively.

From the marine-derived fungus *Penicillium* sp. ZYX-Z143, isolated from the arthropod *Dardanus scutellatus* collected the Paracel Islands, researchers identified seven new indole-diterpenoids, namely penpaxilloids A (**411**)–E (**415**), 7-methoxypaxilline-13-ene (**416**) and 10-hydroxy-paspaline (**417**) (Fig. 51), along with 20 known compounds [92]. Notably, penpaxilloid A contained a spiro indole-diterpenoid structure with a 2,3,3a,5-tetrahydro-1*H*-benzo[d]pyrrolo[2,1–*b*][1,3]oxazin-1-one motif, while penpaxilloid B featured a unique heptacyclic system containing a rare 3,6,8-trioxabicyclo[3.2.1]octane unit. Functionally, penpaxilloid A acted as a noncompetitive inhibitor of protein tyrosine phosphatase 1B (PTP1B) with an IC_{50} value of 8.60 μM, and penpaxilloid D showed potent inhibition of nitric oxide production in lipopolysaccharide-stimulated RAW264.7 macrophages, with an IC_{50} value of 7.11 μM.

3.2 Steroids

The marine-derived fungus *Acremonium* sp. NBUF150, isolated from a *Ciocalypta* sponge located at a depth of 70 m near the Paracel Islands, yielded a new steroid, acremocholone (**418**) (Fig. 52), along with three known analogues [93]. The planar structures and absolute configurations of these compounds were elucidated using a

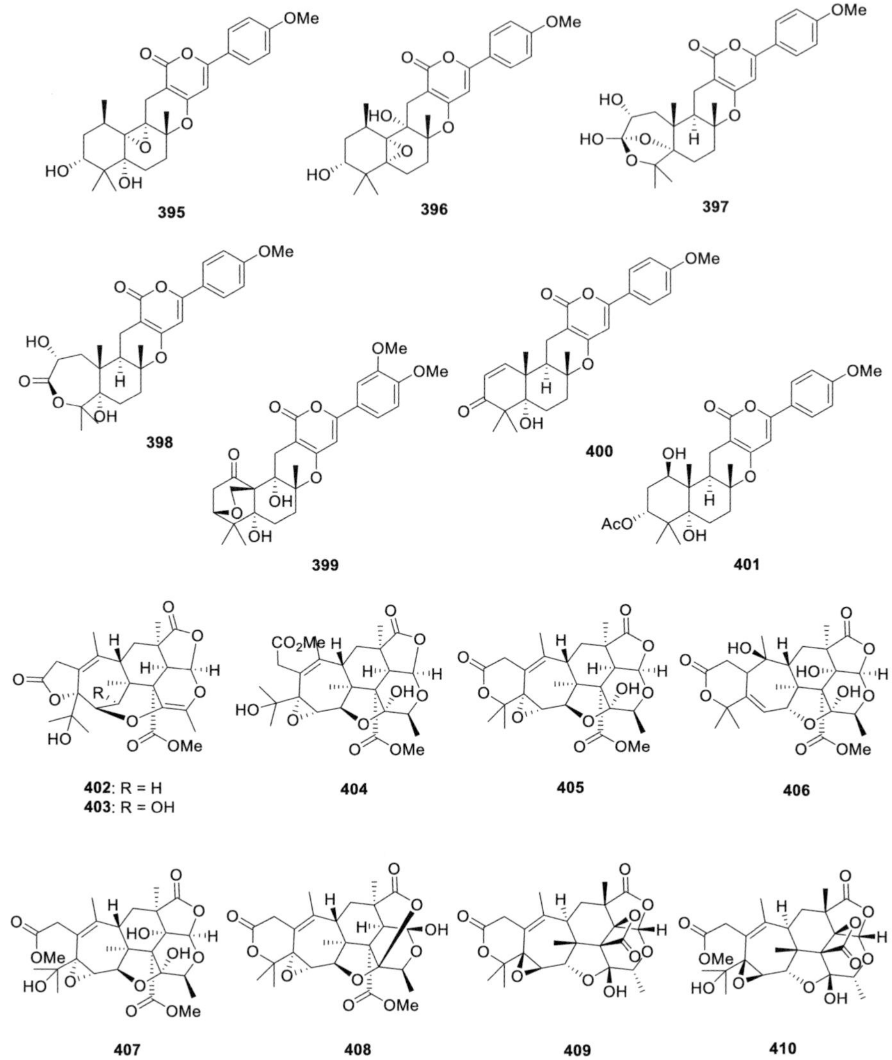

Fig. 50 Natural products from *Penicillium* sp. SCSIO 41691 and *Talaromyces purpureogenus* SCSIO 41517

combination of NMR spectroscopy, HR-ESI–MS and X-ray crystallography. Biologically, acremocholone demonstrated notable antimicrobial activity, exhibiting inhibition against *Vibrio scophthalmi*, *Vibrio shilonii* and *Vibrio brasiliensis* with minimum inhibitory concentrations of 8 μg/mL.

Four new steroid derivatives, arthriniumsteroids A (**419**)–D (**422**) (Fig. 52), along with two known compounds, were isolated from the soft coral-derived fungus *Simplicillium lanosoniveum* SCSIO41212, obtained from a *Sinularia* sp. soft coral collected

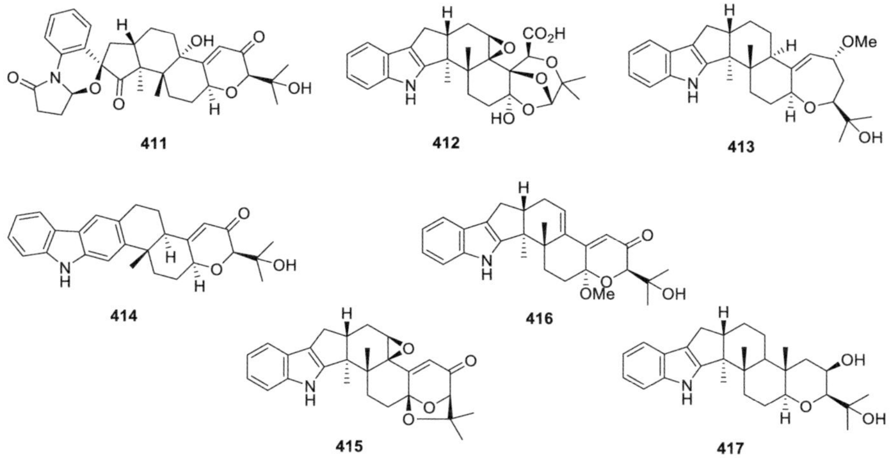

Fig. 51 Natural products from *Penicillium* sp. ZYX-Z143

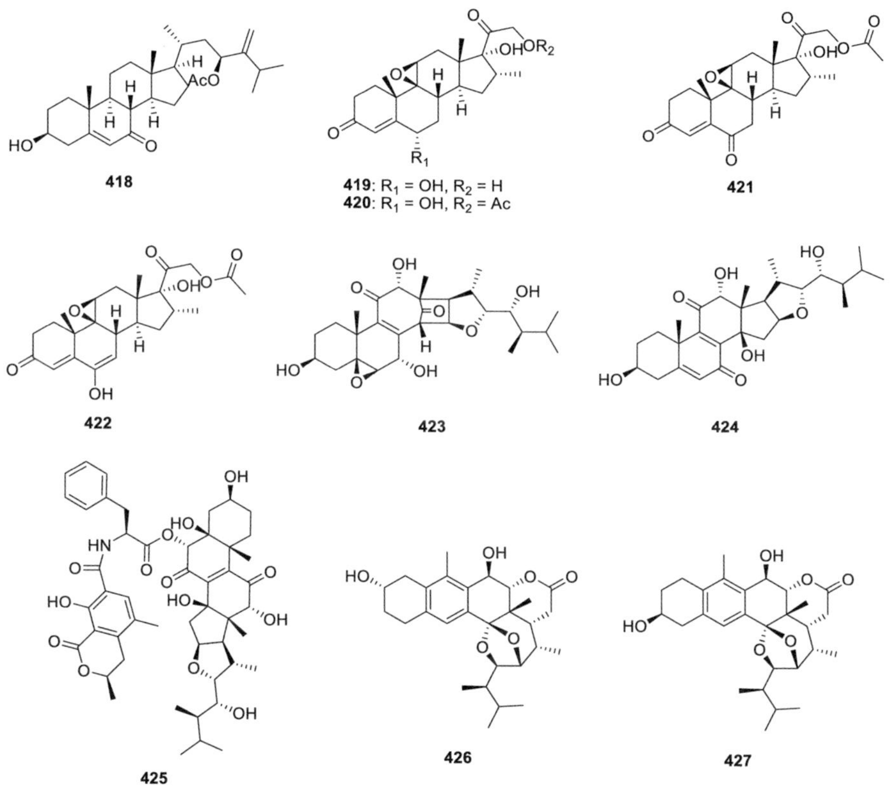

Fig. 52 Steroids from marine microorganisms

at the Paracel Islands [94]. In bioassays, all compounds exhibited weak inhibitory activity against lipopolysaccharide (LPS)-induced nitric oxide (NO) production in RAW 264.7 cells, with inhibitory rates ranging from 21.4% to 44.6%.

Several steroidal molecules have been isolated from the marine fungus, *Aspergillus flocculosus* 16D-1. For instance, from the solid culture of *A. flocculosus* 16D-1, isolated from the inner tissue of the sponge *Phakellia fusca* collected at the Paracel Islands, two steroids were identified: a novel 8(14 → 15)-abeo-ergostane-type steroid named asperflotone (**423**) and an ergostane steroid, asperfloroid (**424**) (Fig. 52) [95]. Asperflotone represents an unprecedented ergosteroid featuring a unique rearranged bicyclo[4.2.1]non-2-ene ring system, likely formed through an α-ketol rearrangement during biosynthesis. Both compounds exhibited significant biological activity, showing inhibitory effects on IL-6 production in LPS-induced THP-1 cell lines, with IC_{50} values of 22 μM.

Additionally, a novel ochratoxin–ergosteroid heterodimer, ochrasperfloroid (**425**) (Fig. 52), along with the known mycotoxin ochratoxin A, was isolated from *A. flocculosus* 16D-1 [96]. Ochrasperfloroid exhibited remarkable biological activity, demonstrating potent inhibitory effects on IL-6 production in lipopolysaccharide (LPS)-induced THP-1 cell lines, with an IC_{50} value of 2.02 μM, and on NO production in LPS-activated RAW264.7 macrophages, with an IC_{50} value of 1.11 μM.

Asperfloketals A (**426**) and B (**427**) (Fig. 52), two 1(10 → 6)-abeo-14,15-secosteroids, featuring a novel trioxahexaheterocyclic ring system, were subsequently isolated from the sponge-associated fungus *A. flocculosus* 16D-1 [97]. Asperfloketals are classified as anthrasteroids which constitute a compound class of steroidal descendents with an aromatic B ring, attached in linear fashion to saturated, six-membered A and C rings, forming the anthracene framework. Both compounds demonstrated strong anti-inflammatory activity in $CuSO_4$-induced transgenic fluorescent zebrafish without exhibiting cytotoxicity against HeLa, HepG2 and SW480 cell lines. Quantitative analysis revealed that asperfloketals A and B significantly alleviated macrophage migration and reduced their numbers around the neuromasts of zebrafish in a dose-dependent manner, showing superior in vivo anti-inflammatory activity compared to the positive control.

3.3 Fatty Acid Derivatives, Polyketides and Macrolides

In a study by Wang and co-workers, a co-culture of *Streptomyces rochei* MB037 and a gorgonian-derived fungus, *Rhinocladiella similis* 35, was conducted to stimulate the production of new metabolites, with pure cultivation serving as a control [98]. The actinomycete *S. rochei* MB037 was derived from the sponge *Dysidea arenaria*, while the fungal strain *R. similis* 35 was isolated from a staghorn gorgonian. Five metabolites were successfully isolated from the co-culture broth, including two new fatty acids with rare nitrile groups, borrelidins J (**428**) and K (**429**) (Fig. 53), a new chromone derivative, 7-methoxy-2,3-dimethylchromone-4-one (**430**) (Fig. 53) as well as two known 18-membered macrolides, borrelidin and borrelidin F. Notably, borrelidins J and K were exclusive to the co-culture, while the production of 7-methoxy-2,3-dimethylchromone-4-one was significantly enhanced in co-culture compared to single cultures. Among these, borrelidin J demonstrated significant antibacterial activity against methicillin-resistant *Staphylococcus aureus* (MRSA), with a MIC value of 0.195 μg/mL.

From the marine fungus *Stagonospora* sp. SYSU-MS7888, isolated from a sponge collected at the North Reef of the Paracel Islands, researchers identified ten metabolites, including two novel cyclopropane derivatives, stagonospones A (**431**) and B (**432**) (Fig. 53), and seven previously undescribed α-pyrone derivatives, stapyrones A (**433**)–G (**439**) (Fig. 53) [99]. Bioactivity assays revealed that stagonospones A and B as well as stapyrones E and G exhibited significant anti-inflammatory activity, with IC_{50} values ranging from 3.6 to 22.8 μM, outperforming the positive control, indomethacin.

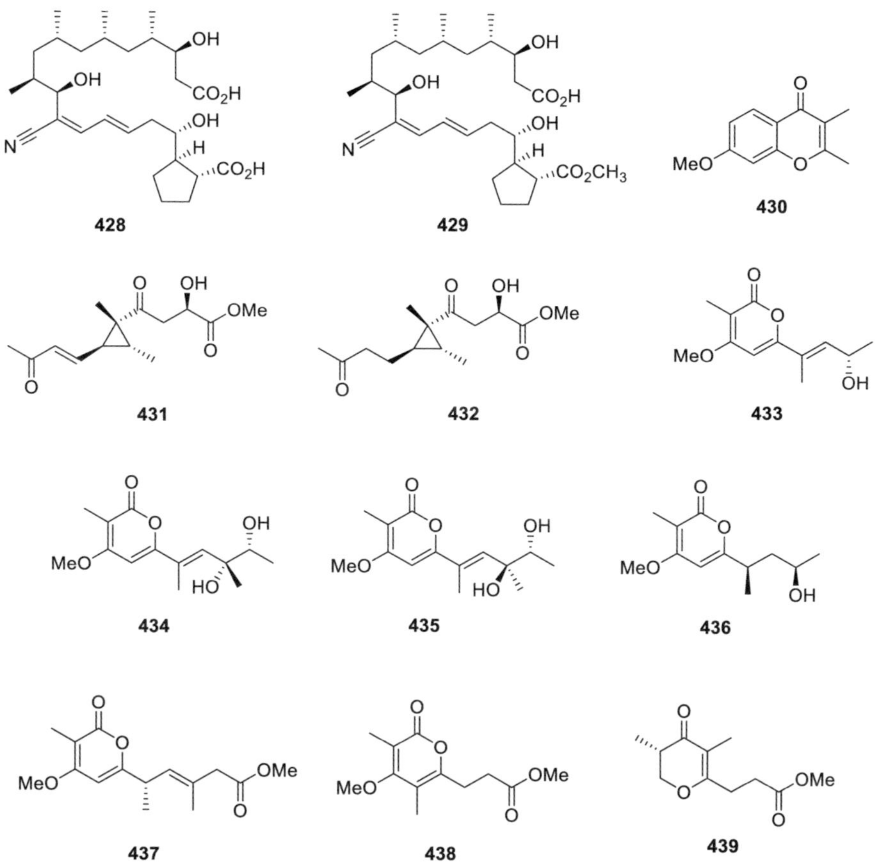

Fig. 53 Natural products from co-culturing of *Streptomyces rochei* MB037 and *Rhinocladiella similis* 35 as well as monoculture of *Stagonospora* sp. SYSU-MS7888

The genome of *Amphichorda felina* SYSU-MS7908, a fungal strain isolated from the marine ascidian *Styela plicata* collected at the north atoll of the Paracel Islands, revealed 56 biosynthetic gene clusters (BGCs) through bioinformatics analysis, most of which appear to be silent. Targeted genome mining of one cryptic BGC using heterologous expression in *Aspergillus oryzae* NSAR1 led to the discovery of two novel α-pyrone derivatives, amphichopyrones A (**440**) and B (**441**) (Fig. 54) [100]. These compounds demonstrated potent anti-inflammatory activity by inhibiting nitric oxide (NO) production in RAW264.7 cells, with IC_{50} values of 18.09 μM and 7.18 μM, respectively.

The marine fungus, *Arthrinium arundinis* ZSDS1-F3, isolated from the marine sponge, *Phakellia fusca* (collected from the at the Paracel Islands in 2012), has

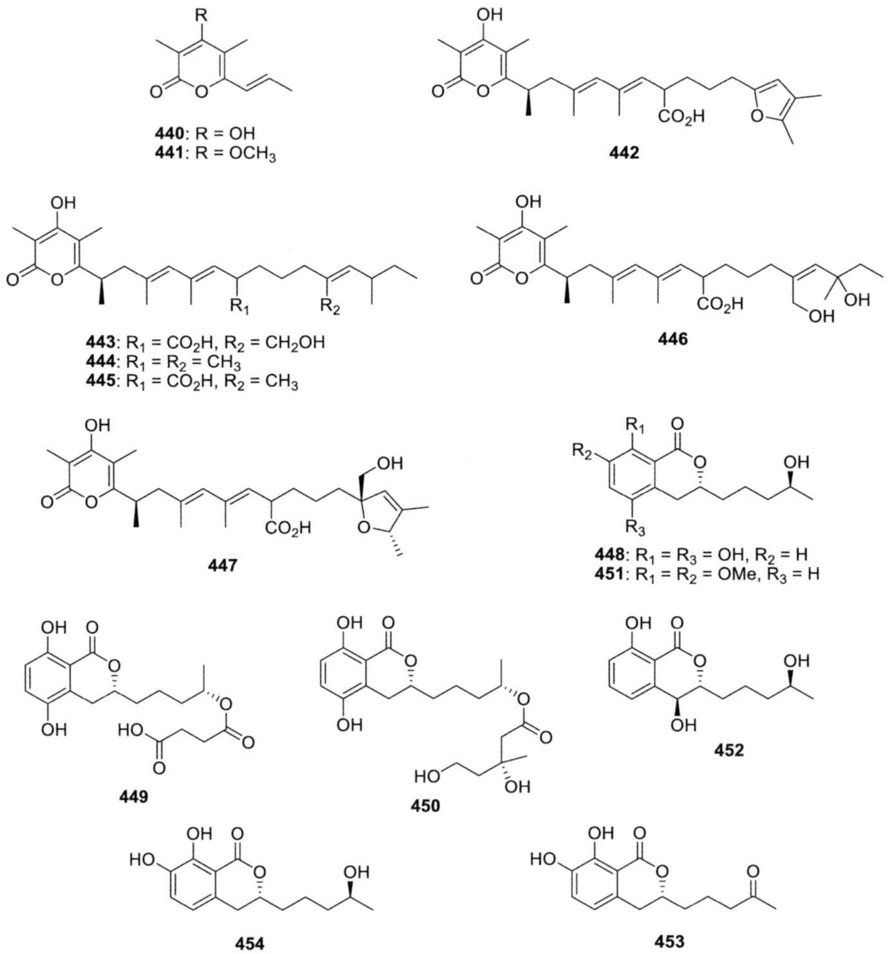

Fig. 54 Natural products from *Amphichorda felina* SYSU-MS7908, *Arthrinium arundinis* ZSDS1-F3 and *Aspergillus* sp. NBUF87

been investigated extensively for new molecules. Further investigation of this fungal strain resulted in the isolation of two new 6-alkenylpyrone polyketides, alternapyrones G (**442**) and H (**443**) (Fig. 54), along with four known compounds, including alternapyrone (**444**), alternapyrones B (**445**), D (**446**) and E (**447**) (Fig. 54) [101]. The biosynthetic gene cluster (*alt*) responsible for alternapyrones was identified and functionally characterized through heterologous expression in *Aspergillus nidulans* A1145 ΔSTΔEM. This analysis revealed the roles of cytochrome P450 monooxygenases Alt2′ and Alt3′ in modifying precursor compounds: Alt2′ converted the 26-CH_3 group to a carboxyl group, generating alternapyrone B from alternapyrone, while Alt3′ catalyzed successive hydroxylation, epoxidation and oxidation steps to produce alternapyrones G, H, D and E from alternapyrone B. Notably, alternapyrone G demonstrated neuroprotective properties, suppressing M1 polarization in lipopolysaccharide (LPS)-stimulated BV2 microglia and promoting dendrite regeneration and neuronal survival after Aβ treatment, positioning it as a promising scaffold for Alzheimer's disease drug discovery.

Six new dihydroisocoumarins, aspergimarins A (**448**)–F (**453**) (Fig. 54), along with five known analogs, were isolated from a monoculture of the sponge-derived fungus *Aspergillus* sp. NBUF87 [102]. This fungus was obtained from the sponge *Hymeniacidon* sp., collected in the Paracel Islands. Among the compounds, aspergimarin A and aspergimarin E exhibited significant inhibitory effects on the lateral root growth of *Arabidopsis thaliana* Columbia-0 (Col-0). While aspergimarin A specifically inhibited lateral root growth, aspergimarin E caused notable inhibition of both lateral and primary root growth, with a particularly strong effect on primary root growth at a concentration of 100 μM. An additional analogue, aspergimarin G (**454**) (Fig. 54), was subsequently isolated from *Aspergillus* sp. NBUF87 [103]. Aspergimarin G demonstrated moderate antibacterial activity against *Staphylococcus aureus* and *Salmonella enteritidis*, with MIC values ranging from 16 to 64 μg/mL.

The marine-derived fungus, *Aspergillus sydowii* SCSIO 41301, was isolated from the sponge *Phakellia fusca*, collected from the Paracel Islands. This fungus produced five new compounds, including two xanthones, identified as 2-hydroxy-1-(hydroxymethyl)-8-methoxy-3-methyl-9H-xanthen-9-one (**455**) and 2-hydroxy-1-(hydroxymethyl)-7,8-dimethoxy-3-methyl-9H-xanthen-9-one (**456**) and two catechol derivatives (**457** and **458**) as well as one bisabolane-type sesquiterpenoid, aspergillusene D (**459**) (Fig. 55) [104]. Notably, the two new xanthones exhibited selective inhibitory activity against two subtypes of the influenza A virus, A/Puerto Rico/8/34 (H1N1) and A/FM-1/1/47 (H1N1), with IC_{50} values ranging from 2.17 to 4.70 μM.

A molecular networking approach was employed to facilitate the targeted isolation of three new sterigmatocystin derivatives, sterigmatocystins A (**460**)–C (**462**) (Fig. 55), from the marine sponge-derived fungus *Aspergillus versicolor* (15XS43ZD-1) [105]. Sterigmatocystin A is distinguished by its rare 6/6/6/6/5 polycyclic structure. Biological evaluations revealed that sterigmatocystin A exhibited superior angiogenesis-promoting activity compared to the positive control at a

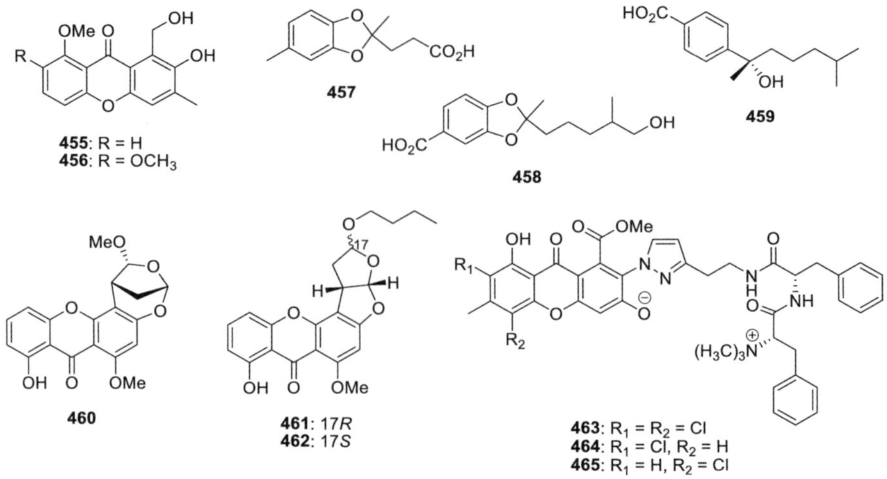

Fig. 55 Natural products from *Aspergillus sydowii* SCSIO 41301, *Aspergillus versicolor* and *Aspergillus flavipes* 164013

concentration of 1.25 μM in transgenic fluorescent zebrafish. Additionally, sterigmatocystins A–C demonstrated moderate antiviral activity, inhibiting HSV-2 with IC_{50} values of 47.11, 39.45, and 38.73 μM, respectively.

Three unusual chlorinated PKS-NRPS hybrid metabolites, flavipesides A (**463**)–C (**465**) (Fig. 55), were isolated from a strain of the marine sponge symbiotic fungus *Aspergillus flavipes* 164013, which was obtained from the marine sponge *Dysidea* sp. collected at the Paracel Islands [106]. Interestingly, the same fungus was also isolated from a cyanobacterium sample within the same area. Flavipesides A–C represent a new structural family of PKS-NRPS hybrid metabolites, characterized by an uncommon assembly of structural scaffolds and functional groups. Each compound consists of a chlorinated, methylated, and hydroxylated xanthone, an aminoethyl-modified pyrazole, and a methylated dipeptide. These metabolites demonstrated potent pancreatic lipase (PL) inhibitory activity, with IC_{50} values ranging from 0.07 to 0.23 μM. Pancreatic lipase is a key target for inhibitors designed to prevent hyperlipidemia and obesity by reducing the absorption of triglycerides, as it hydrolyzes triglycerides for absorption in the human intestine.

From cultures of the deep-sea fungus *Myrothecium* sp. BZO-L062, isolated from sediment at a depth of 2130 m at the Paracel Islands, researchers identified four new compounds: a pair of 2-benzoyl tetrahydrofuran enantiomers, (−)-1*S*-myrothecol (**466a**) and (+)-1*R*-myrothecol (**466b**); a methoxy-myrothecol racemate (**467**); and an azaphilone derivative, myrothin (**468**) (Fig. 56) [107]. Bioactivity studies revealed that (−)-1*S*-myrothecol and (+)-1*R*-myrothecol displayed significant anti-inflammatory activity by inhibiting nitric oxide (NO) production in lipopolysaccharide (LPS)-treated RAW264.7 cells at non-toxic concentrations. Additionally, both enantiomers exhibited strong antioxidant activity, with EC_{50} values of 1.20 and 1.41 μg/mL in the ABTS assay and trolox equivalent antioxidant capacities of 1.41 μM trolox/μM for (−)-1*S*-myrothecol and 1.19 μM trolox/μM for (+)-1*R*-myrothecol. These findings highlight their potential as bioactive compounds with dual anti-inflammatory and antioxidant properties.

Four new ene-yne hydroquinones, pestalotioquinols C (**469**)–F (**472**) (Fig. 56), along with three rare glycosylated derivatives named pestalotioquinosides A (**473**)–C (**475**) (Fig. 56), were isolated from the fungal culture broth of *Pestalotiopsis neglecta* SCSIO41403, a strain obtained from the marine red alga *Coelarthrum* sp., collected at the Paracel Islands [108]. Among these compounds, pestalotioquinoside C demonstrated promising activity as a modulator of liver X receptor alpha (LXRα) based on in silico molecular docking and in vitro surface plasmon resonance studies. This compound significantly upregulated the canonical LXRα

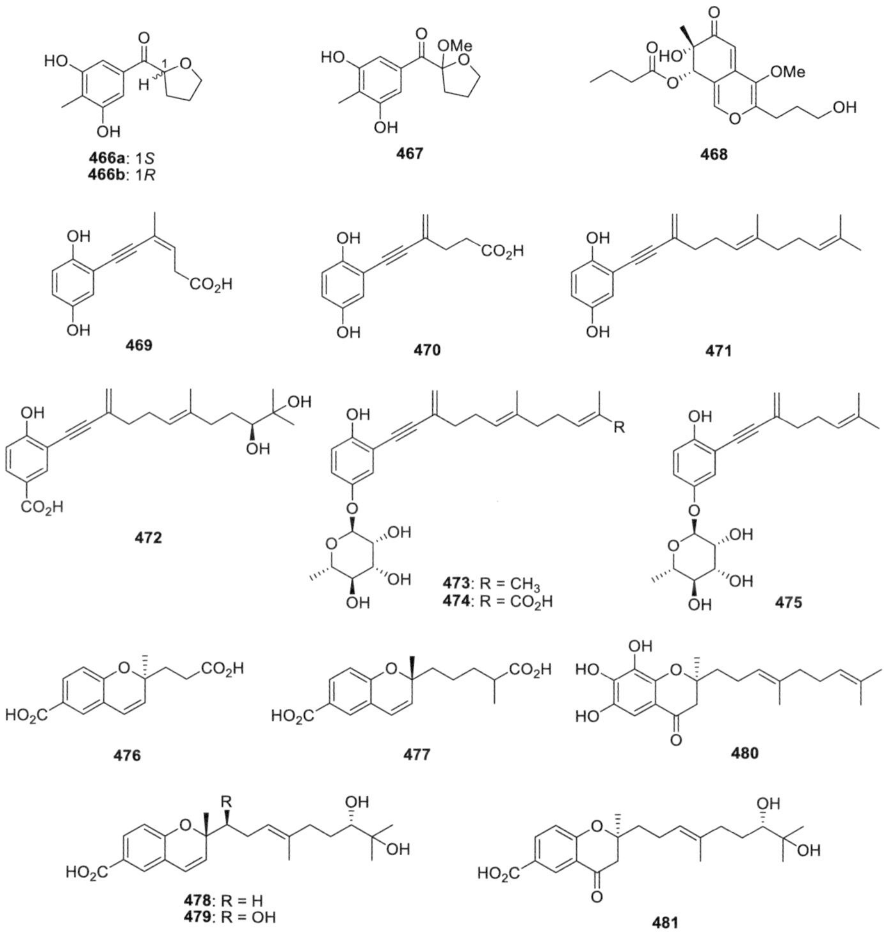

Fig. 56 Natural products from *Myrothecium* sp. BZO-L062 and *Pestalotiopsis neglecta* SCSIO41403

target gene *ABCA1*, identifying it as a potential LXRα agonist. LXRα, a nuclear receptor and transcription factor, plays a critical role in regulating cholesterol homeostasis, lipid metabolism, and inflammation. Molecules that modulate LXRα are of great biomedical interest for their potential applications in treating cardiovascular, neurodegenerative, inflammatory and autoimmune diseases [109, 110].

Further investigation of *Pestalotiopsis neglecta* SCSIO41403 led to the discovery of four new chromene derivatives, pestalotiochromenoic acids A (**476**)–D (**479**) (Fig. 56), and two new chromone derivatives, pestalotiochromones A (**480**) and B (**481**) (Fig. 56) [111]. While these compounds exhibited weak cytotoxicity, they demonstrated notable liver X receptor (LXR) modulatory activities in in vitro assays assessing the expression of LXRα, LXRβ, and their target gene *ABCA1* mRNA using RT-qPCR, supported by in silico docking analysis. Among these, pestalotiochromenoic acid B upregulated LXRα and *ABCA1* expression, whereas pestalotiochromenoic acid A downregulated LXRβ and *ABCA1*. Pestalotiochromenoic acid D and pestalotiochromone A, both containing a geranyl group, significantly upregulated *ABCA1* and LXRβ expression without altering LXRα levels. Notably, pestalotiochromone A exhibited a high binding affinity for LXRα ($K_D = 6.2$ μM), identifying it as a promising potential LXR agonist.

Twelve new tanzawaic acid derivatives, penitanzacids A (**482**)–J (**491**) (Fig. 57), along with hatsusamides C (**492**) and D (**493**) (Fig. 57), were isolated from the deep-sea-derived fungus *Penicillium* sp. KWF32 [112]. This fungus was obtained from the tissues of *Bathymodiolus* sp., collected from a cold spring area in the East Sea at a depth of 1352 m. The study also led to the revision of the structures of tanzawaic acids I and J. Penitanzacids H (**489**)–J (**491**) are the first members of this family featuring a C3 side chain, supporting a proposed biosynthetic pathway in which the side chain attaches to the decalin backbone. Hatsusamides C and D possess a hybrid skeleton formed by linking a tanzawaic acid and a diketopiperazine through an ester bond. Among the bioactivities observed, hatsusamides C and D exhibited weak cytotoxicity against A549 cells, with inhibition rates of 51.7% and 90.7%, respectively, at a concentration of 13 μM.

Nine new azaphilones, including penicilazaphilones I (**494**)–N (**499**), *epi*-geumsanol D (**500**) and penidioxolanes C (**501**) and D (**502**) (Fig. 57), were isolated from the marine-derived fungus *Penicillium sclerotiorum* E23Y-1A, which was obtained from the marine sponge *Holoxea* sp. collected at the Paracel Islands [113]. These compounds share a common azaphilone framework characterized by a highly oxygenated pyranoquinone bicyclic core and a quaternary carbon center, suggesting their origin from a shared biogenetic pathway. Among the isolated compounds, penicilazaphilone N demonstrated moderate anti-inflammatory activity with an IC_{50} value of 22.63 μM, while penidioxolane C exhibited moderate cytotoxicity against various human cancer cell lines, including K562 (23.94 μM), BEL-7402 (60.66 μM), SGC-7901 (46.17 μM), A549 (60.16 μM) and HeLa (59.30 μM).

Three new compounds (**503–505**) (Fig. 58), along with eight known analogues, including pseurotin A (**506**) (Fig. 58), (−)-ovalicin, chlovalicin, dihydroxybergamotene, AM6898B (**507**) (Fig. 58), aspergiketone and two (−)-ovalicin derivatives, were isolated from the fungus *Pseudallescheria boydii*, which was associated with

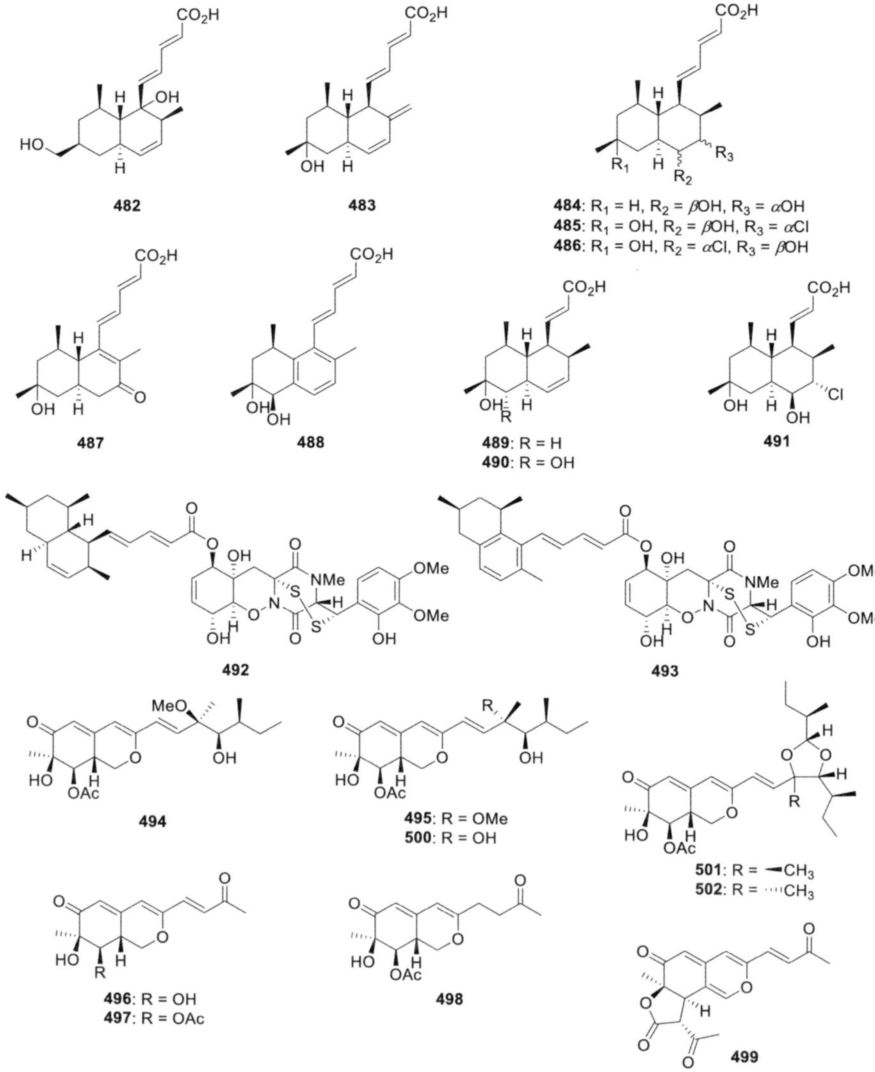

Fig. 57 Natural products from *Penicillium* sp. KWF32 and *P. sclerotiorum* E23Y-1A

the soft coral *Sinularia sandensis* collected from an atoll located in the Pratas Island [114]. These 11 compounds were evaluated in vitro for their regulatory effects on osteoclastogenesis at concentrations of 0.1 and 1.0 μM, demonstrating no significant cytotoxicity below 10 μM. Known pseurotin A, AM6898B, (−)-ovalicin derivatives (**508**) and **504** (Fig. 58) exhibited stimulatory activity on osteoclastogenesis, with AM6898B and **508** significantly increasing osteoclast numbers at both 0.1 and 1.0 μM and enhancing the osteoclast area at 0.1 μM. Pseurotin A increased osteoclast

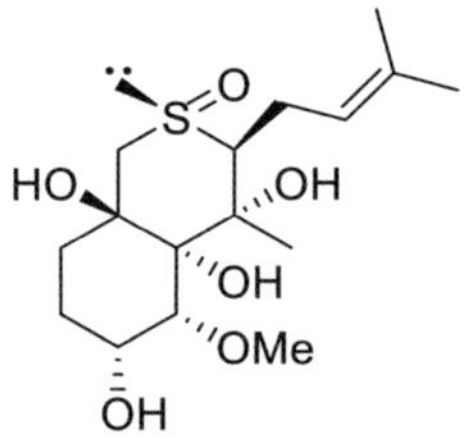

503

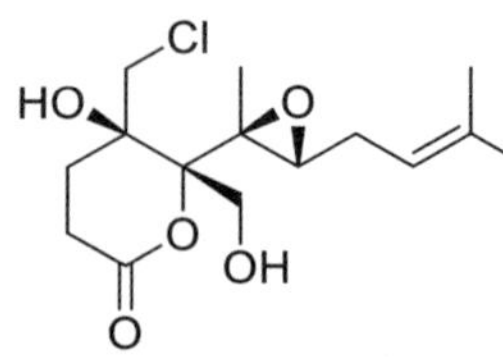

504

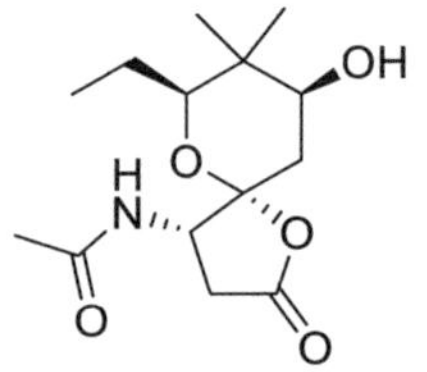

505

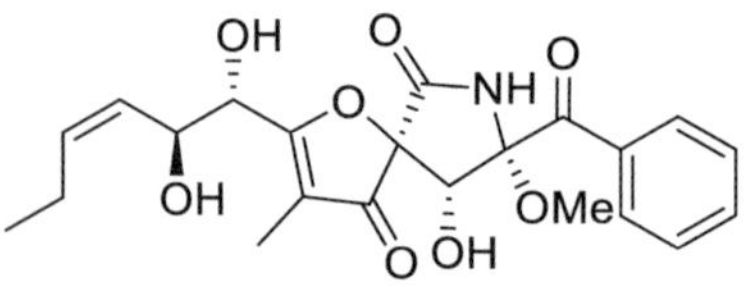

506

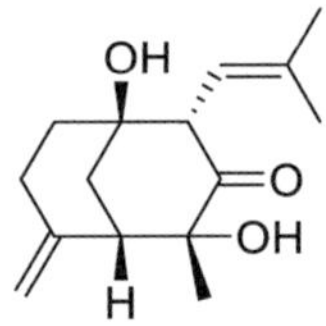

507

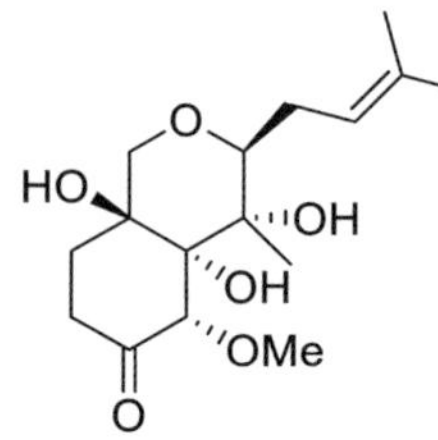

508

Fig. 58 Natural products from *Pseudallescheria boydii*

numbers at 0.1 μM, while **504** elevated osteoclast numbers at 1.0 μM, highlighting their potential roles in bone remodeling processes.

A new *p*-terphenyl alcohol, 4-*O*-methylcandidusin A (**509**) (Fig. 59), along with nine known analogs, was isolated from the marine sponge-derived fungus *Aspergillus candidus* OUCMDZ-1051 [115]. This fungal strain was obtained from an unidentified sponge sample (No. XS-3) collected from the Paracel Islands. 4-*O*-Methylcandidusin A exhibited selective anticancer activity, inhibiting the growth of human cancer cell lines MDA-MB-468, BT474 and A431 with IC_{50} values of 1.84, 6.05 and 0.98 μM, respectively. Additionally, 4-*O*-methylcandidusin A demonstrated selective antimicrobial activity against *Escherichia coli*, with an MIC value of 27.3 μM.

The fermentation extract of the coral-associated fungus *Aspergillus* sp. ITBBc1 yielded five novel *p*-terphenyl derivatives, sanshamycins A (**510**)–E (**514**) (Fig. 59) [116]. Among these, sanshamycin A is the first reported *p*-terphenyl with an aldehyde substitution on the benzene ring, while sanshamycins B–D feature diverse methoxyl and isopentenyl substitutions. Sanshamycin E is distinguished by a five-membered lactone linked to a biphenyl structure. Of these derivatives, sanshamycin C exhibited the most potent activity, inhibiting phosphodiesterase 4D (PDE4D) with an IC_{50} value of 5.54 μM. PDE4 enzymes, which hydrolyze the secondary messenger cyclic adenosine 3′,5′-monophosphate (cAMP), consist of four subtypes (PDE4A, PDE4B, PDE4C, and PDE4D) with highly conserved catalytic domains. These enzymes play critical roles in various physiological processes and are implicated in the pathogenesis of inflammatory, respiratory, autoimmune, and neurological diseases, as well as certain cancers.

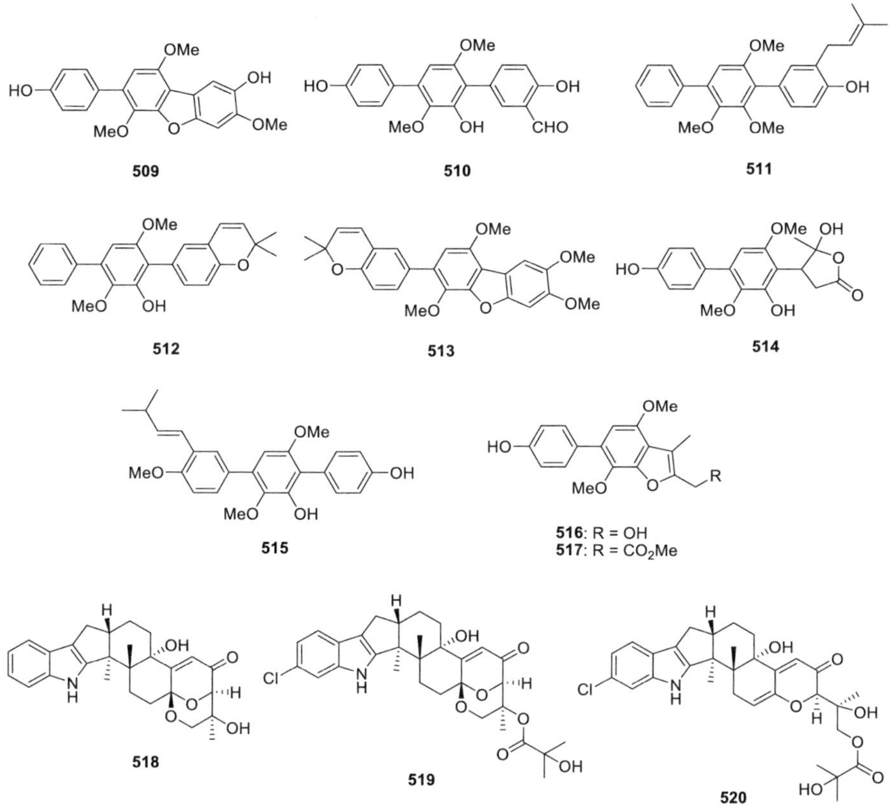

Fig. 59 Natural products from *Aspergillus candidus* OUCMDZ-1051, *Aspergillus* sp. ITBBc1 and *Aspergillus candidus*

From the fungal strain, *Aspergillus candidus*, associated with the gorgonian *Junceella fragilis* collected in the Paracel Islands, researchers isolated three new *p*-terphenyl derivatives, 4″-*O*-methyl-prenylterphenyllin B (**515**) (Fig. 59) as well as phenylcandilides A (**516**) and B (**517**) (Fig. 59), along with three novel indole-diterpene alkaloids, asperindoles E (**518**)–G (**520**) (Fig. 59), and eighteen known analogues [117]. Biological evaluation revealed that phenylcandilides A, B and asperindole G exhibited modulatory effects on spontaneous Ca^{2+} oscillations and 4-aminopyridine-induced hyperexcited neuronal activity in a primary cultured cortical neuronal network.

Four previously undescribed polyketide derivatives, arthproliferins A (**521**)–D (**524**) (Fig. 60), and a novel phenylspirodrimane derivative, arthproliferin E (**525**) (Fig. 60), along with 11 known metabolites, including satratoxin F (**526**) (Fig. 60), were isolated from the soft coral-associated fungus *Stachybotrys chartarum* SCSIO41201 [118]. This fungal strain was originally derived from the soft coral *Sinularia* sp. collected in from the Paracel Islands. Among the isolated compounds, arthproliferin A and satratoxin F demonstrated moderate inhibitory activity against methicillin-resistant *Staphylococcus aureus* (MRSA) ATCC 29213, with minimum inhibitory concentration (MIC) values of 78 μg/mL and 39 μg/mL, respectively.

Four new azaphilone derivatives (**527–530**) (Fig. 60) and a new isoprenyl-substituted benzoate derivative (**531**) (Fig. 60), along with a known analogue (**532**) (Fig. 60), were isolated from the deep-sea-derived fungus *Talaromyces indigoticus*

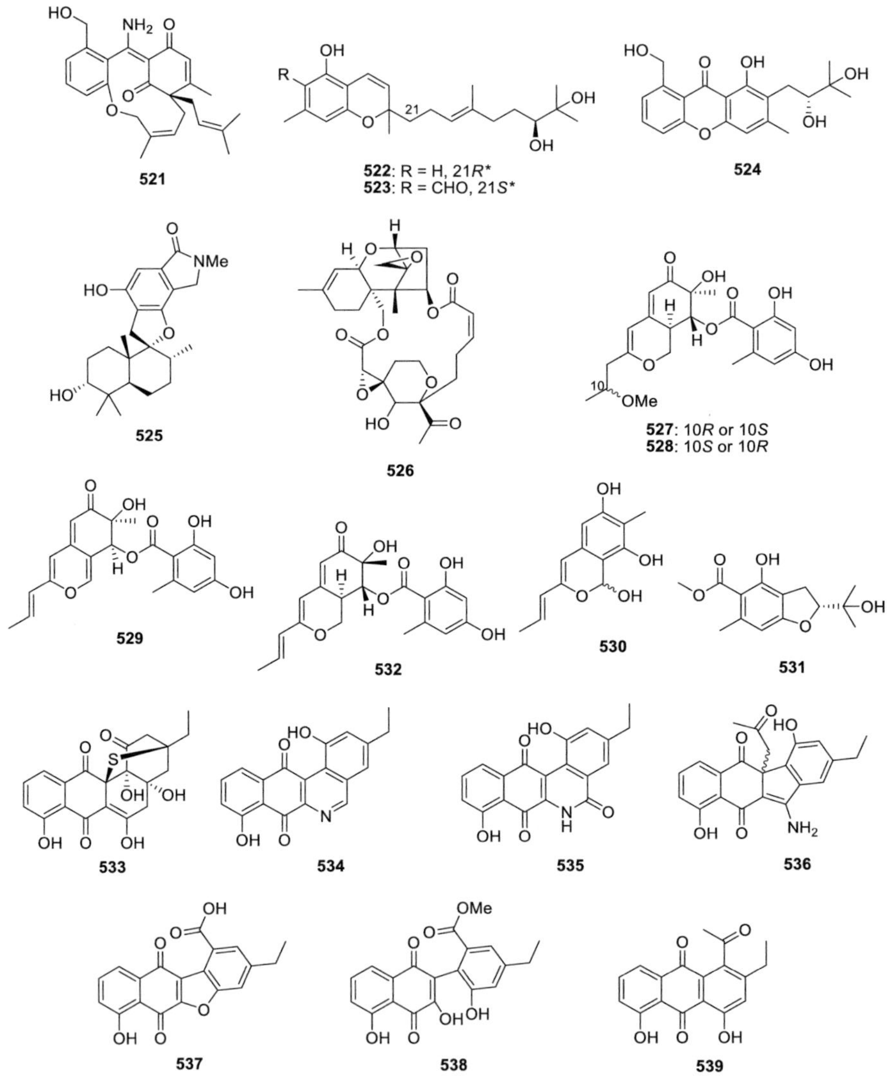

Fig. 60 Natural products from *Stachybotrys chartarum* SCSIO41201, *Talaromyces indigoticus* FS688 and *Micromonospora echinospora* SCSIO 04089

FS688 [119]. Among the compounds, **532** and **530** demonstrated moderate cytotoxicity against four human cancer cell lines, with IC_{50} values ranging from 30 to 60 μM. Bioactivity tests further revealed that compound **532** exhibited significant nitric oxide (NO) production and α-glucosidase inhibitory activities, with IC_{50} values of 31.1 μM and 22.7 μM, respectively.

A bacterial strain of *Micromonospora echinospora*, identified based on its 16S rDNA gene sequence, was isolated from a depth of 3025 m in the northern East Sea (E 119° 31.949, N 18° 2.441). This strain, *M. echinospora* SCSIO 04089, was cultured in eight different media to explore its potential for producing secondary metabolites. Subsequent culturing in N4 medium with addition of prewashed HP20 resin led to the isolation of various angucyclinone derivatives and anthracene compounds derived from this strain. Notable metabolites include the new thioether gephysulfuromycin (**533**), two new benzo[*b*]phenanthridines, homophenanthroviridone (**534**) and homophenanthridonamide (**535**), a new benzo[*b*]fluorene, homostealthin D (**536**), a new naphtho[2,3-*b*]benzofuran, nenesfuran (**537**), a new naphthoquinone, WS-5995 D (**538**) and a new anthracene, nenesophanol (**539**) (Fig. 60) [120]. Gephysulfuromycin is characterized by a rare single *S*-bridged 3,12a-epithiotetraphene skeleton. Homophenanthroviridone exhibited notable cytotoxicity against SF-268, MCF-7 and HepG2 cell lines, with IC_{50} values of 5.4, 6.8 and 1.4 μM, respectively, and demonstrated antibacterial activity against Gram-positive bacteria, with MIC values ranging from 2 to 4 μg/mL.

Two pairs of salicylaldehyde derivative enantiomers, euroticins A (**540**) and B (**541**) (Fig. 61), were isolated from the marine-derived fungus *Eurotium* sp. SCSIO F452, which was obtained from a sediment sample (17°29.804′N, 110°0.292′E) at a depth of 158 m [121]. Euroticin A features a highly intricate 6/6/6/5/7 pentacyclic structure with an unprecedented 2,11-dioxatricyclo[5.3.1.$0^{4,8}$]undecane core, while euroticin B represents the first reported example of a 6/6/6/6 tetracyclic salicylaldehyde derivative. Although both enantiomers of euroticin B showed no cytotoxic activity against HepG-2, MCF-7, SF-268 and A549 cell lines at a concentration of 40 μM, they exhibited remarkable DPPH radical scavenging activity, with EC_{50} values of 37.5 [(+)-euroticin B (**541a**)] and 21.6 μM [(−)-euroticin B (**541b**)], comparable to or better than the positive control ascorbic acid. Interestingly, (−)-euroticin B demonstrated superior antioxidative activity compared to its enantiomer, suggesting that stereochemistry plays a crucial role in the biological activities of these compounds.

Additional chemical analysis of the crude extract of *Eurotium* sp. SCSIO F452 led to the isolation of three pairs of salicylaldehyde derivative enantiomers, euroticins C (**542**)–E (**544**) (Fig. 61) [122]. Euroticins C and E were evaluated for their in vitro cytotoxicity against four human cancer cell lines: SF-268 (glioblastoma carcinoma), MCF-7 (breast cancer), HepG-2 (liver cancer) and A549 (lung cancer). They exhibited moderate to weak cytotoxic activities, with IC_{50} values ranging from 11.77 to 94.66 μM, except for (−)-euroticin E, which showed no activity against HepG-2. Additionally, both enantiomers of euroticin C demonstrated significant DPPH radical scavenging activity, with EC_{50} values of 27.0 [(+)-euroticin C] and 30.3 μM [(−)-euroticin C].

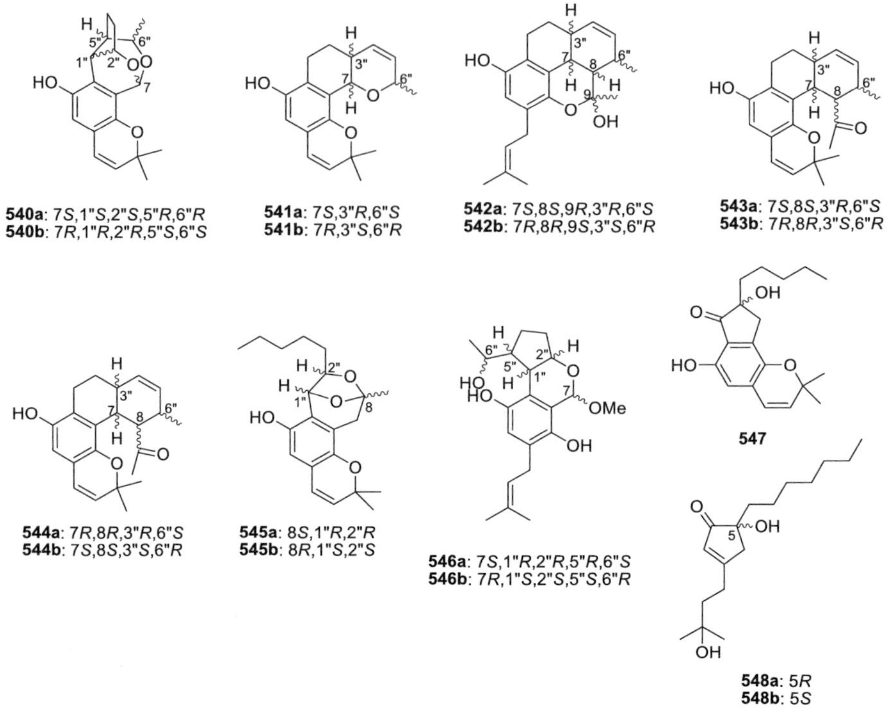

Fig. 61 Natural products from *Eurotium* sp. SCSIO F452

Further investigations of *Eurotium* sp. SCSIO F452 led to the isolation and characterization of additional related compounds, including four pairs of new salicylaldehyde derivative enantiomers, euroticins F (**545**)–I (**548**) (Fig. 61), along with a known compound, eurotirumin [123]. Euroticin F features an unprecedented 6/6/6/5 tetracyclic structure, while euroticins G and H represent two novel types of 6/6/5 scaffolds. Notably, (±)-euroticin H and (+)-euroticin G exhibited significant inhibitory activity against α-glucosidase, with IC_{50} values of 16.31 μM and 38.04 μM, respectively, outperforming or being comparable to the positive control, acarbose.

The one strain-many compounds (OSMAC) approach applied to the sponge-derived fungus *Pestalotiopsis heterocornis* XWS03F09 led to the discovery of several novel secondary metabolites. The fungal strain XWS03F09 was isolated from the sponge *Phakellia fusca*, collected from the Paracel Islands. This manipulation yielded twelve compounds, including eight new polyketide derivatives, heterocornols Q (**549**)–X (**556**) (Fig. 62), and one new ceramide (**557**) (Fig. 62) [124]. Among these, heterocornols Q (**549**) and R (**550**) represent novel 6/5/7/5 tetracyclic polyketide derivatives characterized by a dihydroisobenzofuran and a benzo-fused dioxabicyclo[4.2.1]nonane system, likely biosynthesized via acetyl-CoA-mediated epoxidation, polyene cyclization and skeletal rearrangement. Notably, heterocornols T (**552**) and X (**556**) demonstrated significant inhibition of LPS-induced nitric oxide

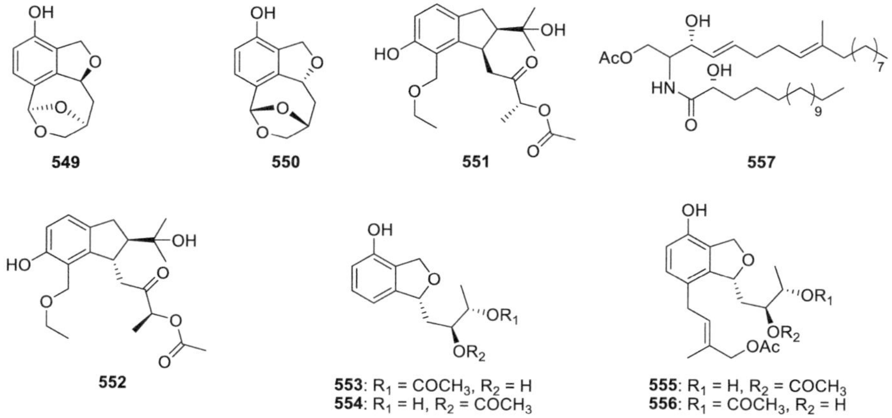

Fig. 62 Natural products from *Pestalotiopsis heterocornis* XWS03F09 using one strain-many compounds (OSMAC) approach

(NO) production, comparable to dexamethasone, and further Western blot analysis confirmed their dose-dependent suppression of inducible nitric oxide synthase (iNOS) protein expression in LPS-stimulated RAW 264.7 cells.

Two novel sorbicillinoid adducts, sorbicillfurans A (**558**) and B (**559**) (Fig. 63), featuring bicyclo[2.2.2]octane and tetrahydrofuran moieties, were identified from the static culture of the marine-derived fungus *Penicillium citrinum* SCSIO41402 [125]. This fungal strain was isolated from the marine alga *Coelarthrum* sp., collected from the Paracel Islands. Among the compounds, sorbicillfuran B exhibited weak cytotoxicity against HL-60 cells, with an IC_{50} value of 9.6 μM. Further analysis suggested that the citrinin moiety in sorbicillfuran B might play a significant role in its cytotoxic activity against HL-60 cells.

Two new oxaphenalenone derivatives, (+)- and (−)-talaromyoxaones A (**560/561**) and B (**562/563**) (Fig. 63), were isolated from the marine-derived fungus *Talaromyces purpureogenus* SCSIO 41517, which was obtained from a soft coral collected at the Spratly Islands [126]. These compounds feature a unique hemiacetal framework and an unprecedented spirolactone structure, specifically a 2′H,3H,4′H-spiro[isobenzofuran-1,3′-pyran]-3-one unit, exhibiting biosynthetic enantiodivergence. Additionally, two new oxaphenalenone analogues, (±)-11-apopyrenulin (**564**) and (+)- or (−)-abeopyrenulin (**565**) (Fig. 63), were identified. Talaromyoxaones A and B demonstrated selective inhibitory activity against phosphatases SHP1, SHP2 and MEG2, with IC_{50} values ranging from 1.3 to 3.4 μM.

Ansamycins are metabolites belonging to the macrolactam family, primarily produced by actinomycetes. This important class includes several clinically significant compounds, such as rifamycins, maytansinoids, ansamitocin P-3 and geldanamycins [127]. Their structures are characterized by a macrocyclic framework, which features an aromatic "core" linked by an aliphatic "ansa bridge" that connects two nonadjacent positions on the core.

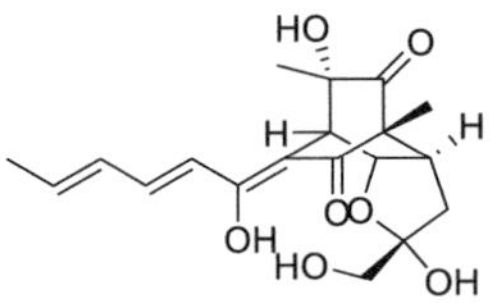

558

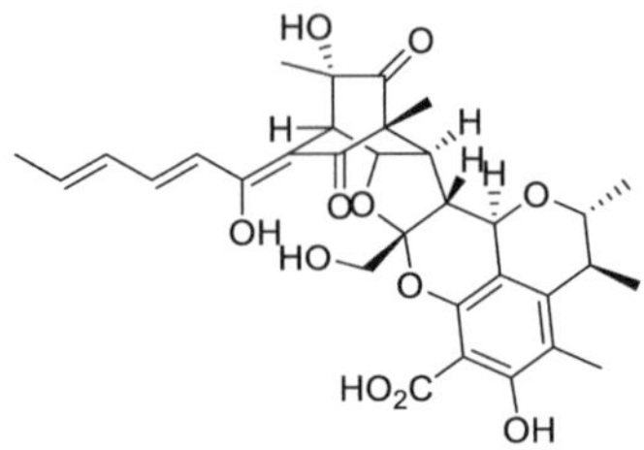

559

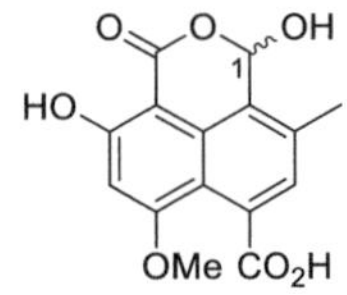

564

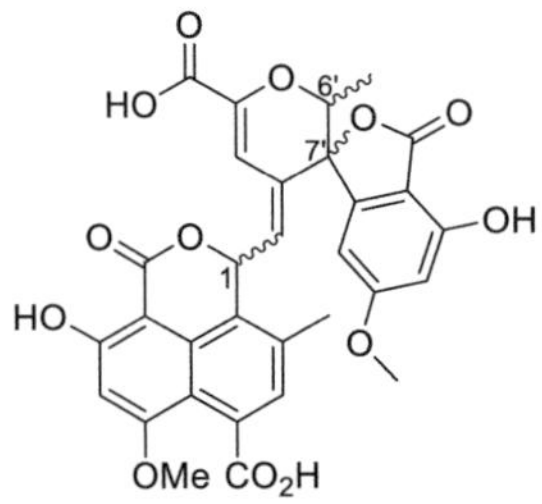

560a: 1*S*,6'*S*,7'*R*
561a: 1*S*,6'*R*,7'*R*
560b: 1*R*,6'*R*,7'*S*
561b: 1*R*,6'*S*,7'*S*

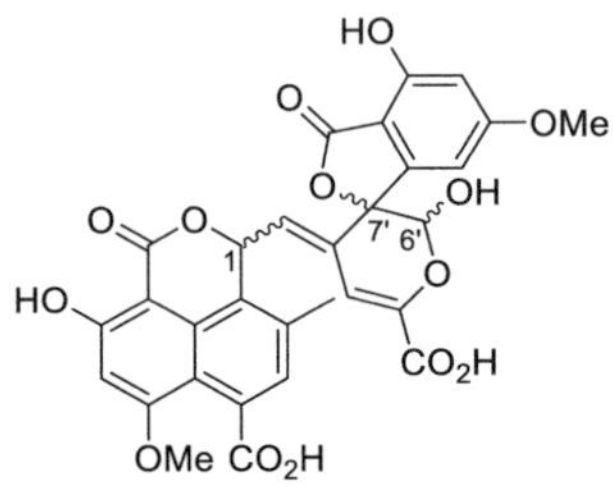

562a: 1*R*,6'*S*,7'*R*
563a: 1*R*,6'*R*,7'*R*
562b: 1*S*,6'*R*,7'*S*
563b: 1*S*,6'*S*,7'*S*

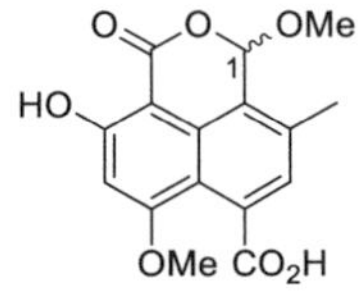

565a: 1*S*
565b: 1*R*

Fig. 63 Natural products from *Penicillium citrinum* SCSIO41402 and *Talaromyces purpureogenus* SCSIO 41517

Four new ansamycin derivatives, 1,19-epithio-geldanamycin A (**566**), 17-demethoxylherbimycin H (**567**), herbimycin M (**568**) and *seco*-geldanamycin B (**569**) (Fig. 64), along with eight known ansamycin analogues [(e.g., 4,5-dihydro-17-*O*-demethylgeldanamycin (**570**), 4,5-dihydro-19-*S*-methylgeldanamycin (**571**) and *seco*-geldanamycin A (**572**) (Fig. 64)] were isolated from the solid fermentation of the marine-derived actinomycete *Streptomyces* sp. ZYX-F-97 [128]. Among the isolated compounds, *seco*-geldanamycin B, 4,5-dihydro-17-*O*-demethylgeldanamycin and *seco*-geldanamycin A displayed notable antibacterial activity against *Listeria monocytogenes* with minimum inhibitory concentrations (MIC) ranging from 8 to 64 μg/mL and exhibited moderate inhibition against methicillin-resistant *Staphylococcus aureus* (MRSA) with an MIC of 64 $\mu g \cdot mL^{-1}$. Additionally, *seco*-geldanamycin B, 4,5-dihydro-17-*O*-demethylgeldanamycin, 4,5-dihydro-19-*S*-methylgeldanamycin and *seco*-geldanamycin A demonstrated moderate inhibitory effects against both *Staphylococcus aureus* and *Bacillus subtilis* with MIC values between 32 and 128 μg/mL.

A known polycyclic tetramate macrolactam (PoTeM), aburatubolactam C (**573**), and three new derivatives, aburatubolactams D (**574**)–F (**576**) (Fig. 64), were isolated from the marine-derived *Streptomyces* sp. SCSIO 40070 [129]. This strain was obtained from sediment collected at a depth of 645 m in the East Sea (E 115°48.7902′,

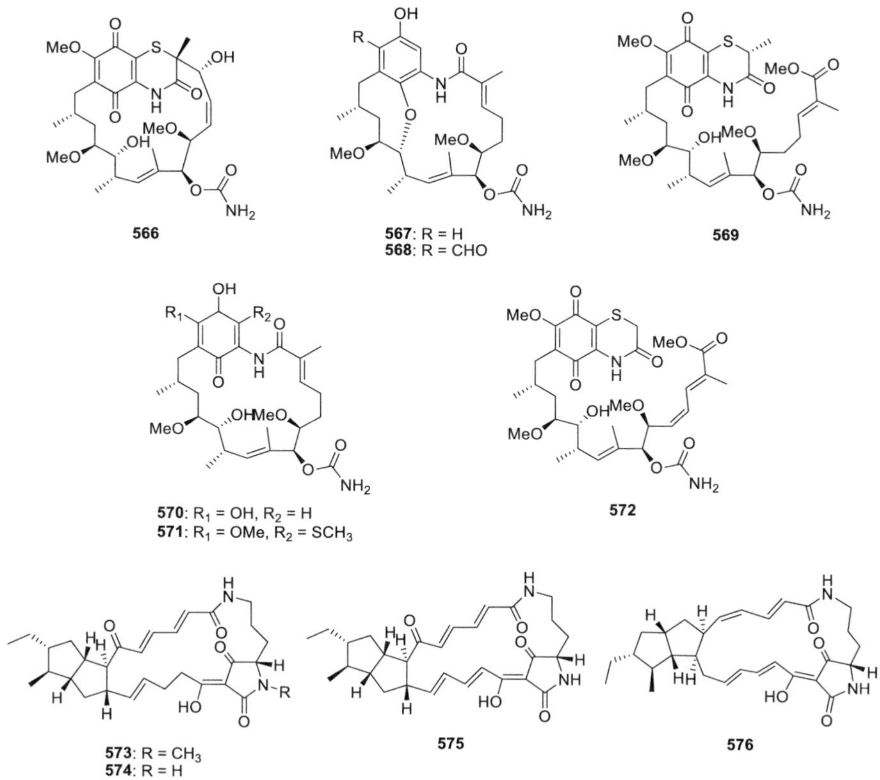

Fig. 64 Natural products from *Streptomyces* sp. ZYX-F-97 and *Streptomyces* sp. SCSIO 40070

N 20°19.4258′). PoTeMs are a family of pharmacologically promising natural products characterized by a tetramate-containing macrolactam fused to diverse carbocyclic ring systems, including 5/6, 5/5 (C6–C13 or C7–C14), 5/6/5, or 5/5/6 configurations. Aburatubolactams C–E demonstrated moderate cytotoxic activity against the C4–2B, A549 and HepG2 cell lines, with IC_{50} values ranging from 2.43 to 16.16 μM.

3.4 Alkaloids and Other Nitrogen-Containing Molecules

Asperteramide A (**577**) (Fig. 65), an unusual *N*-phenyl-carbamic acid methyl ester trimer, was isolated from the coral-derived fungus *Aspergillus terreus*. This fungus was obtained from the inner tissue of the soft coral *Sarcophyton subviride*, collected from the Paracel Islands [130]. Asperteramide A displayed remarkable antibacterial activity against *Candida albicans*, with a minimum inhibitory concentration (MIC) value of 2 μg/mL. Additionally, it exhibited significant activity against six drug-resistant microbial pathogens, including *Escherichia coli*, *Acinetobacter baumannii*,

Pseudomonas aeruginosa, *Klebsiella pneumoniae*, methicillin-resistant *Staphylococcus aureus* (MRSA) and *Enterococcus faecalis*, with MIC values of 8, 8, 16, 64, 64 and 8 μg/mL, respectively.

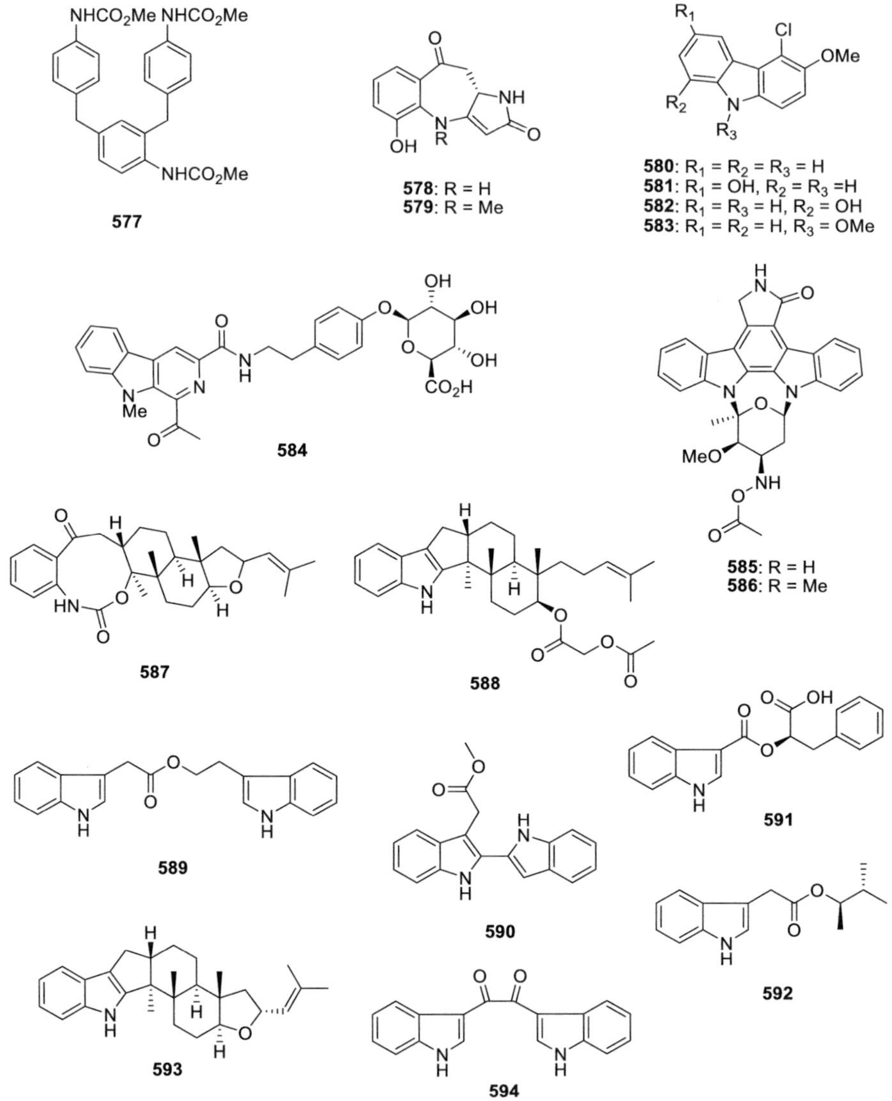

Fig. 65 Chemical structures of nitrogen-containing molecules from *Aspergillus terreus*, *Aspergillus candidus* (CHNSCLM-0393), *Streptomyces diacarni* LHW51701, *Actinoalloteichus cyanogriseus* LHW52806, *Streptomyces* sp. NBU3142 and *Fusarium* sp. L1

Two pyrrolinone-fused benzoazepine alkaloids, (+)-asperazepanones A (**578**) and B (**579**) (Fig. 65), featuring a 6/7/5 ring system, along with the artifact (−)-asperazepanone A, were isolated from the coral-derived fungus *Aspergillus candidus* (CHNSCLM-0393) [131]. This fungal strain was obtained from fresh tissue within the gorgonian coral *Juncella fragilis* (CHNNS-2016-01). The total syntheses of (±)-asperazepanone A and (±)-asperazepanone B were accomplished in 7 and 8 steps, respectively, using commercially available L-aspartic acid diethyl ester hydrochloride and monoethyl malonate as starting materials. A key step in the synthesis involved an intramolecular Friedel–Crafts reaction to construct the unique tricyclic skeleton. Biological evaluation revealed that (+)-asperazepanone B exhibited notable inhibitory activity against nitric oxide (NO) production and suppressed lipopolysaccharide (LPS)-induced expression of pro-inflammatory cytokines TNF-α and IL-6 at a concentration of 0.1 μM. Enzyme-linked immunosorbent assay (ELISA) further demonstrated that (+)-asperazepanone B reduced TNF-α and IL-6 release levels by 40% and 77%, respectively, at this concentration, without compromising cell viability.

Four new chlorinated carbazole alkaloids, chlocarbazomycins A (**580**)–D (**583**) (Fig. 65), were isolated from the sponge-associated bacterium *Streptomyces diacarni* LHW51701 [132]. This bacterium was obtained from the marine sponge, *Diacarnus megaspinorhabdosa.* Structurally, the 4-chloro group coupled with the 3-methoxyl substituents in the tricyclic nucleus of chlocarbazomycins A–D, as well as the *N*-methoxyl group of chlocarbazomycin D, represent rare features among naturally occurring tricyclic carbazole alkaloids. Biological evaluation revealed that chlocarbazomycin C (**582**) exhibited inhibitory activity against pathogenic microorganisms, including methicillin-resistant *Staphylococcus aureus* (MRSA), *Mycobacterium smegmatis*, *Bacillus mycoides*, and *Candida albicans*, with a minimum inhibitory concentration (MIC) of 32 μg/mL.

A novel β-carboline alkaloid, marinacarboline glucuronide (**584**) (Fig. 65), along with nine previously known compounds, was obtained from the static liquid fermentation extracts of the marine actinomycete *Actinoalloteichus cyanogriseus* LHW52806 [133]. This strain was isolated from the tissue of the marine sponge *Phakellia fusca*, collected from the Paracel Islands. In an anti-inflammatory assay using mouse macrophage RAW 264.7 cells, marinacarboline glucuronide demonstrated significant activity by effectively reducing the expression of IL-6 at a concentration of 20 μM.

Two previously unreported indolocarbazole alkaloids, bresmycins A (**585**) and B (**586**) (Fig. 65), were isolated from the culture extract of the marine mesophotic zone sponge-associated actinomycete *Streptomyces* sp. NBU3142 [134]. This strain was obtained from the tissue of the sponge *Dasychalina* sp., collected at a depth of 60 m in the mesophotic zone near the Paracel Islands. Both molecules feature a rare *O*-acetylhydroxylamine substituent at C-4″, distinguishing them from the aminomethyl acetate group commonly found in other indolocarbazole alkaloids. Biological evaluations of bresmycins A and B revealed significant cytotoxicity against the breast cancer cell line MCF-7, with IC_{50} values of 5.01 and 2.01 μM, respectively. Notably, bresmycin B demonstrated a strong inhibitory effect ($IC_{50} = 4.47$ μM) against adriamycin-resistant MCF-7/ADR cells. Further mechanistic studies indicated that bresmycin B exerts its anti-cancer activity by inducing DNA damage and promoting apoptosis in MCF-7 cells.

The effects of L-tryptophan supplementation on secondary metabolite production in the marine-derived fungus *Fusarium* sp. L1 were investigated by culturing the fungus in GPY medium with and without the amino acid. *Fusarium* sp. L1 was isolated from the inner tissue of the sea star *Acanthaster planci* collected from the Paracel Islands. This study led to the isolation of 23 distinct indole alkaloids, including six new compounds: fusaindoterpenes A (**587**) and B (**588**), fusariumindoles A (**589**)–C (**591**) and (±)-isoalternatine A (**592**), along with 17 known compounds (e.g., **593** and **594**) (Fig. 65) [135]. Biological evaluation revealed that fusaindoterpene B, **593** and **594** exhibited inhibitory activity against the Zika virus (ZIKV) in a standard plaque assay, with EC_{50} values of 7.5, 4.2 and 5.0 μM, respectively.

Fractionation of the ethanol extract of a deep-sea sediment-derived fungus, *Arthrinium* sp., provided a new pyridone alkaloid, arthpyrone L (**595**) (Fig. 66) [136]. In vitro cell viability assays revealed that arthpyrone L exhibited antiproliferative effects against human A549 (lung), MG63, U2OS (bone), MCF-7, and MDA-MB-231 (breast) cancer cells. Arthpyrone L notably inhibited the proliferation of all tested cell lines, with the lowest IC_{50} value observed against MG63 cells, ranging from 6.10 to 7.14 μM. Mechanistic studies indicated that the growth inhibition of MG63 cells by arthpyrone L occurred through the activation of a caspase-modulated apoptotic pathway and the inhibition of the PI3K/Akt signaling pathway.

Three undescribed pyridone alkaloids, arthpyrones M (**596**)–O (**598**) (Fig. 66), along with two known pyridone derivatives, arthpyrones C and G, were isolated from the reinvestigation of the sponge (*Phakellia fusca*)-derived fungus, *Arthrinium arundinis* ZSDS1-F3, guided by molecular networking [137]. All molecules exhibited cytotoxicity against one or more of five cancer cell lines, with IC_{50} values ranging from 0.26 to 6.43 μM. Notably, arthpyrone O demonstrated potent efficacy in inhibiting the proliferative activity of small cell lung cancer (SCLC) cells and induced apoptosis in vitro. In vivo, it also significantly inhibited the growth of xenograft tumors derived from SCLC cells. These findings suggest that 4-hydroxy-2-pyridone alkaloids may serve as promising scaffolds for drug discovery.

Three new azaphilone alkaloids containing glutamine residues, *N*-glutarylchaetoviridins A (**599**)–C (**601**) (Fig. 66), along with two known related compounds, chaetomugilins A (**602**) and C (**603**) (Fig. 66), were isolated from the extract of a deep-sea sediment-derived fungus *Chaetomium globosum* HDN151398 [138]. *N*-glutarylchaetoviridins A–C represent the first class of chaetoviridins characterized by embedded glutamate residues. Biological evaluation revealed that *N*-glutarylchaetoviridin C, chaetomugilins A and C exhibited a broad spectrum of cytotoxicity. Notably, *N*-glutarylchaetoviridin C demonstrated significant activity against MGC-803 and HO8910 cancer cell lines, with IC_{50} values of 6.6 and 9.7 μM, respectively.

A genome-directed discovery strategy was employed to identify novel tetrahydroisoquinolines (THIQs) from the deep-sea-derived *Streptomyces niveus* SCSIO

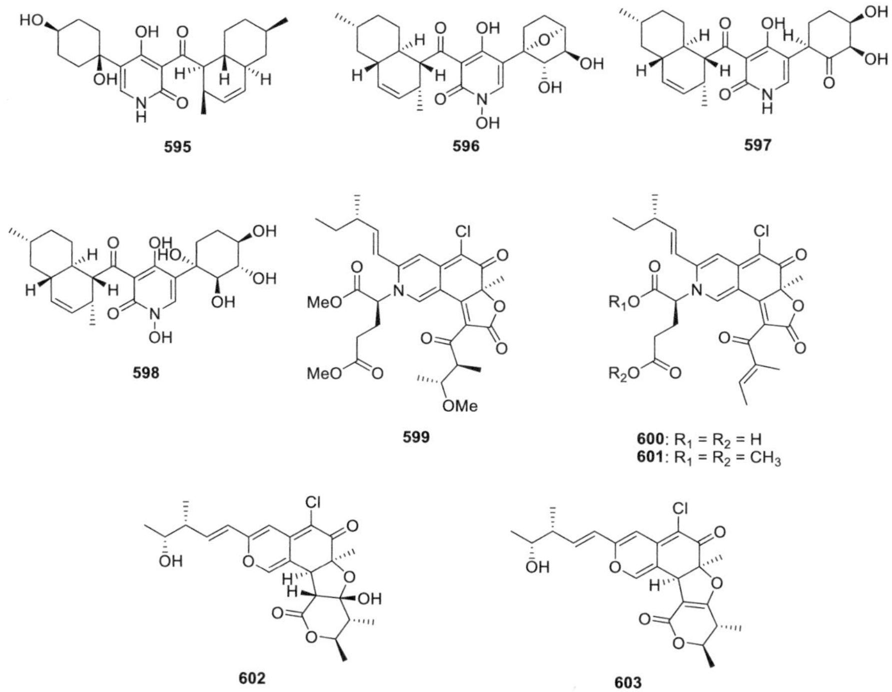

Fig. 66 Natural products from *Arthrinium* sp., *Arthrinium arundinis* ZSDS1-F3 and *Chaetomium globosum* HDN151398

3406, leading to the isolation of 11 THIQs representing three structural classes. This included the known compounds aclidinomycins A (**604**) and B (**605**), along with nine newly identified derivatives, aclidinomycins C (**606**)–K (**614**) (Fig. 67) [139]. Among these, aclidinomycins F (**609**)–I (**612**) (Fig. 67) were notable for their unique 6/6/6/6/5/5 polycyclic skeleton incorporating a fused tetrahydropyran (THP) ring. The strain SCSIO 3406 was isolated from a sediment sample collected in the East Sea at a depth of 3536 m. Antibacterial activities of all compounds were assessed using disk diffusion assays against Gram-positive bacteria, including *Bacillus subtilis* BS01, *Micrococcus luteus* ML01, *Staphylococcus aureus* ATCC 29213 and methicillin-resistant *S. aureus* (MRSA) strains from human and pig sources. Most compounds, except aclidinomycins B, C, F, H and I, demonstrated antimicrobial activities with minimum inhibitory concentrations (MICs) ranging from 4 to 64 μg/mL. Further structure–activity relationship analysis revealed that C-9′ hydroxylation and C-5′ methylation negatively impacted the antibacterial properties of the THIQs.

Seven new decahydrofluorene-class alkaloids, pyrrospirones K (**615**)–Q (**621**) (Fig. 67), along with six known analogues, were isolated from the marine-derived fungal strain *Penicillium* sp. SCSIO 41512, which was obtained from a soft coral collected in the Spratly Islands [140]. Pyrrospirones K (**615**) and M (**617**) feature novel decahydrofluorene-class alkaloid skeletons, with a 6/5/6/8/5/6/13 and a 6/

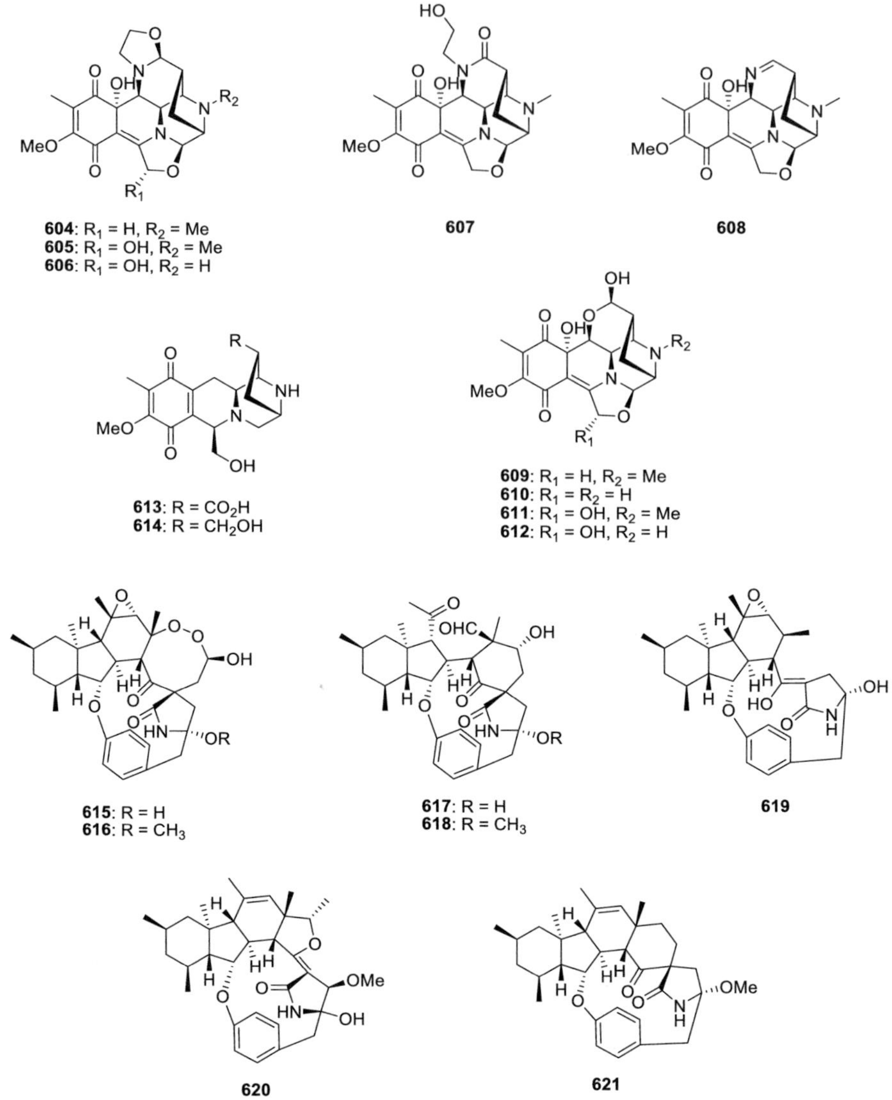

Fig. 67 Natural products from *Streptomyces niveus* SCSIO 3406 and *Penicillium* sp. SCSIO 41512

5/6/5/6/13 polycyclic system, respectively. Biological evaluations revealed that pyrrospirones K, L (**616**) and O (**619**) exhibited moderate antibacterial activity against six pathogens, including *Bacillus amyloliquefaciens*, *Bacillus subtilis*, *Escherichia coli*, *Staphylococcus aureus*, MRSA and *Streptococcus agalactiae*, with IC_{50} values ranging from 20 to 50 μg/mL.

Two polycyclic thioalkaloids, dassonmycins A (**622**) and B (**623**) (Fig. 68), were isolated from *Nocardiopsis dassonvillei* SCSIO 40065, an actinomycete strain associated with the marine sponge *Petrosia* sp. collected from the seabed at the Paracel Islands [141]. The structures of these compounds were elucidated through comprehensive spectroscopic analysis, confirming a 6/6/6/6-fused tetracyclic ring system with a naphthoquinone[2,3-*e*]piperazine[1,2-*c*]thiomorpholine scaffold. Dassonmycins A and B demonstrated antibacterial activity against *Bacillus subtilis* 1064, *Micrococcus luteus* SCSIO ML01, *Staphylococcus aureus* ATCC 29213 and MRSA shhs-A1, with minimum inhibitory concentration (MIC) values ranging from 8 to 64 μg/mL. Additionally, both dassonmycins exhibited moderate cytotoxicity against four human cancer cell lines, with half inhibitory concentration (IC_{50}) values between 12 and 34 μM. Recently, the synthesis of dassonmycins A and B was reported, involving a key chemoselective reduction of a diketopiperazine, yielding a 2-piperazinone intermediate [142]. This intermediate undergoes heteroannulation with naphthoquinone to produce dassonmycin A. At physiological pH, dassonmycin A slowly cyclizes to form dassonmycin B, supporting the biosynthetic hypothesis that this transformation could naturally occur in the marine bacterial cytosol.

Six undescribed prenylated indole diketopiperazine alkaloids, talaromyines A (**624**)–F (**629**) (Fig. 68), were isolated from the marine-derived fungus *Talaromyces*

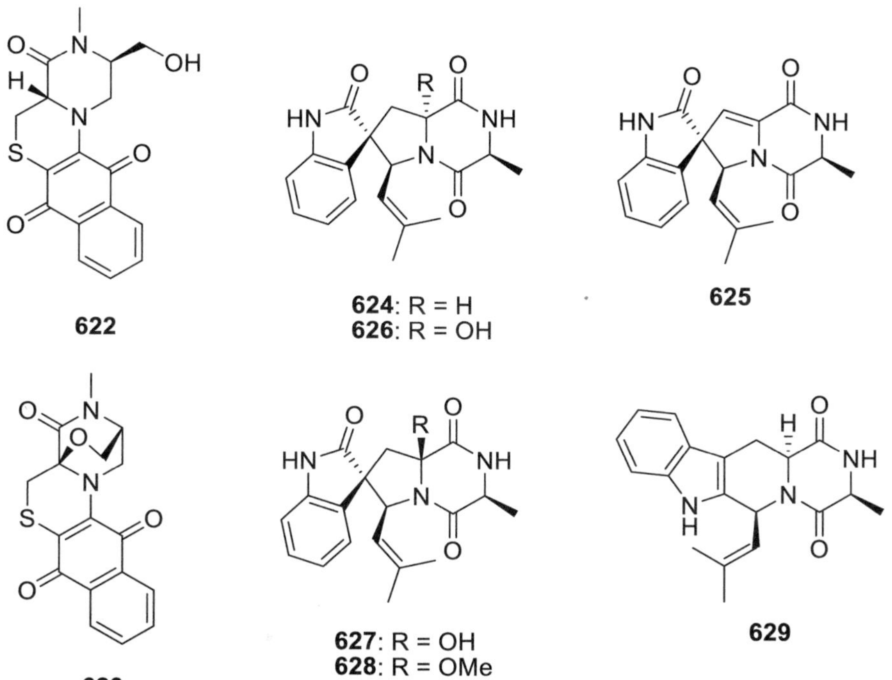

Fig. 68 Natural products from *Nocardiopsis dassonvillei* SCSIO 40065 and *Talaromyces purpureogenus* SCSIO 41517

purpureogenus SCSIO 41517, which was obtained from a soft coral collected at the Spratly Islands [143]. Among these compounds, talaromyines B (**625**) as well as D (**627**) and E (**628**) exhibited selective inhibitory activities against phosphatases TCPTP and MEG2, with IC_{50} values ranging from 17.9 to 29.7 μM. Additionally, talaromyines D and E demonstrated mild cytotoxic activities against two human cancer cell lines, H1975 and HepG-2, with IC_{50} values of 18.6 and 17.9 μM, respectively.

In addition to salicylaldehyde derivatives isolated from *Eurotium* sp. SCSIO F452, this marine fungal strain is a source of many unique diketopiperazines, such as variecolortins A (**630**)–C (**632**) (Fig. 69) previously reported in 2018 [144]. Recently, three pairs of new spirocyclic alkaloid enantiomers, eurotinoids A (**633**)–C (**635**) (Fig. 70), along with a known biogenetically related racemate, dihydrocryptoechinulin D (**636**) (Fig. 70), were isolated from *Eurotium* sp. SCSIO F452 [145]. The enantiomers of these compounds exhibited distinct antioxidative and cytotoxic activities. Specifically, all compounds demonstrated significant radical scavenging activity against DPPH, with IC_{50} values ranging from 3.7 to 24.9 μM. Additionally, compound (+)-dihydrocryptoechinulin D showed weak cytotoxicities against SF-268 and HepG2 cell lines, with IC_{50} values of 51.7 and 49.9 μM, respectively. Molecular docking studies suggested that lipoxygenase (LOX) could be a potential antioxidative target for (±)-eurotinoids and (±)-dihydrocryptoechinulin D, while epidermal growth factor receptor (EGFR) may be a potential cytotoxic target for compound (±)-dihydrocryptoechinulin D.

Four previously undescribed diketopiperazine-type alkaloids, including one oxepin-containing diketopiperazine alkaloid, oxepinamide L (**637**), and three 4-quinazolinone alkaloids, puniceloids E (**638**)–G (**640**) (Fig. 70), were isolated from the culture of the marine-derived fungus *Aspergillus puniceus* FAHY0085 [146]. The fungal strain *A. puniceus* FAHY0085 was isolated from corals collected from the East Sea at a depth of 20 m. In addition to new alkaloids, 12 known analogues were identified, including protuboxepin D, oxepinamides D–G, J–K and I, puniceloids B (**641**)–D (**643**) (Fig. 69) as well as protubonine B. Puniceloid F (**639**), along with known puniceloids B–D, exhibited significant transcriptional activation of liver X receptor α, with EC_{50} values ranging from 2 to 50 μM.

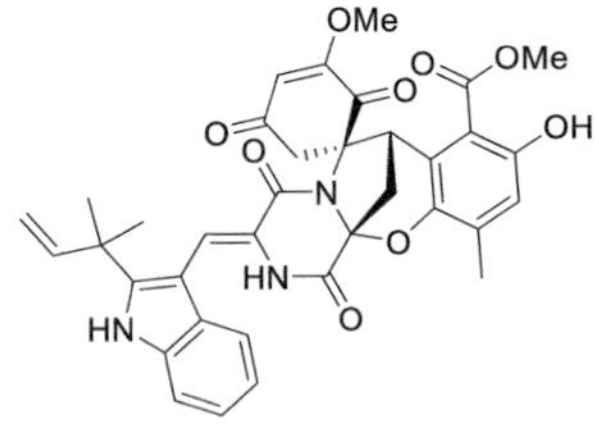

630

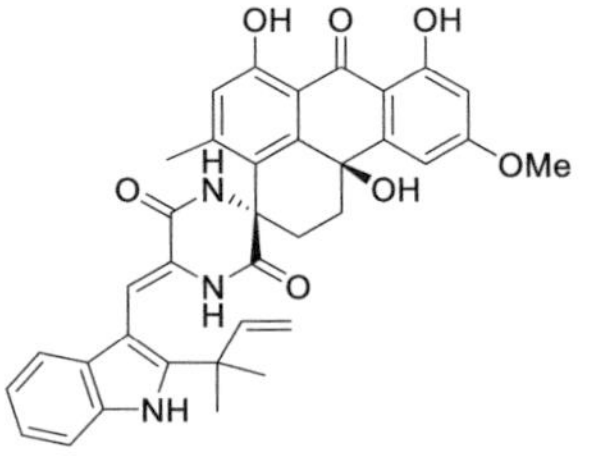

631

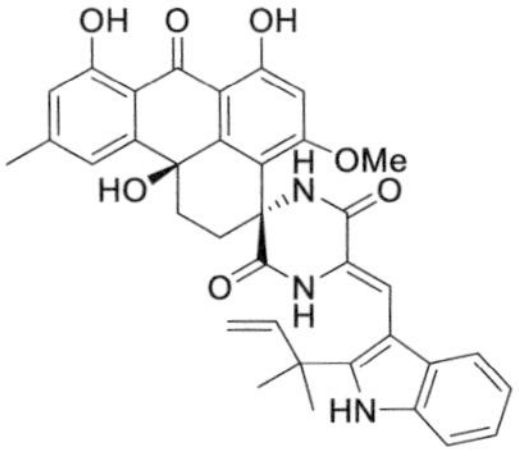

633

Fig. 69 Variecolortins A–C from *Eurotium* sp. SCSIO F452

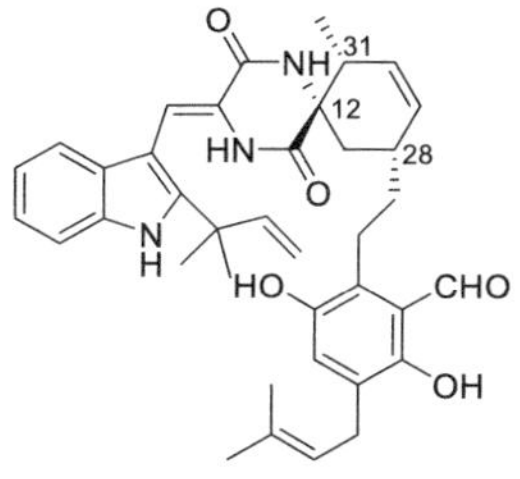

633a: 12*S*,28*R*,31*R*
633b: 12*R*,28*S*,31*S*

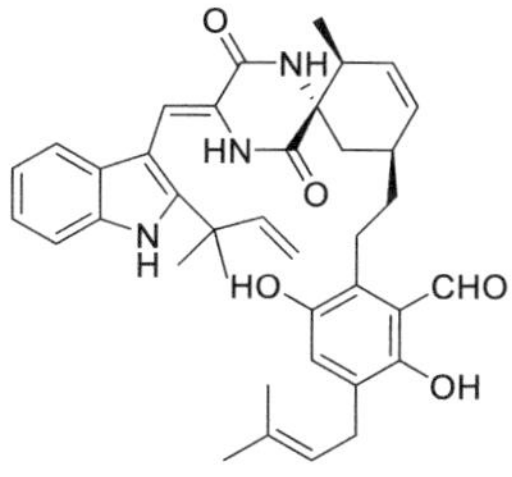

634a: 12*R*,28*R*,31*R*
634b: 12*S*,28*S*,31*S*

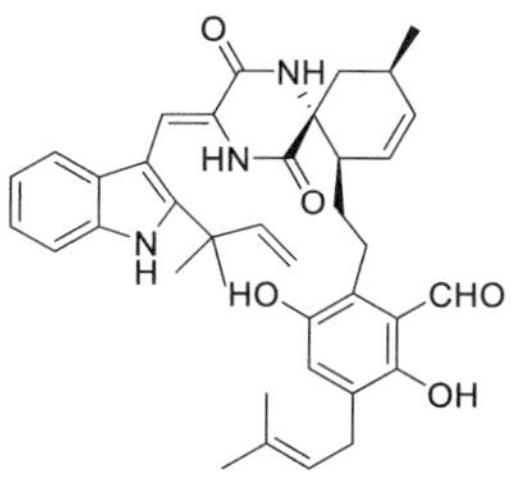

635a: 12*R*,28*R*,31*R*
635b: 12*S*,28*S*,31*S*

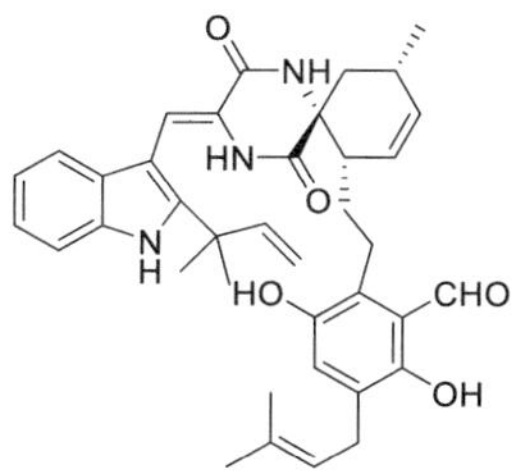

636a: 12*R*,28*S*,31*S*
636b: 12*S*,28*R*,31*R*

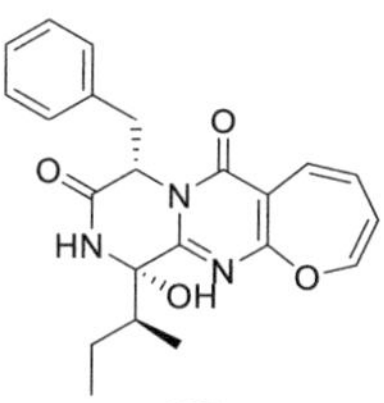

637

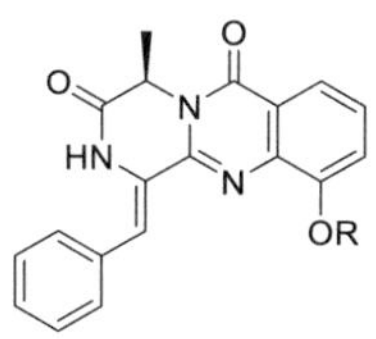

638: R = Me
641: R = H

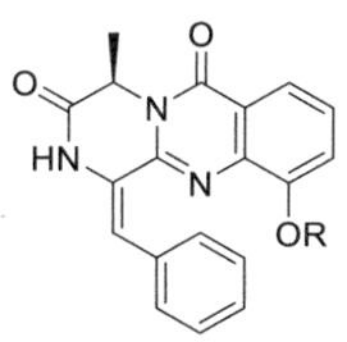

639: R = H
640: R = Me

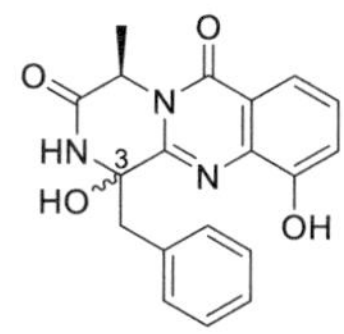

642: 3*S*
643: 3*R*

Fig. 70 Natural products from *Eurotium* sp. SCSIO F452 and *Aspergillus puniceus* FAHY0085

3.5 Linear and Cyclic Depsipeptides/Peptides

A pair of novel lipopeptide epimers, sinulariapeptides A (**644**) and B (**645**), along with a new phthalide glycerol ether (**646**) (Fig. 71), were isolated from the marine algal (*Coelarthrum* sp.)-associated fungus *Cochliobolus lunatus* SCSIO41401 [147]. Sinulariapeptides A, B and phthalide glycerol ether demonstrated moderate inhibitory activity against acetylcholinesterase (AChE), with IC_{50} values of 1.8, 1.3 and 2.5 μM, respectively. Molecular docking studies revealed that these compounds could bind to the active pocket of AChE, centered around TYR124. Specifically, the carbonyl group at C-8 connected to the Pro residue in sinulariapeptides A and B formed a hydrogen bond interaction with TYR124. Additionally, the CONH2-19 group in sinulariapeptides A and B interacted with different residues, SER125

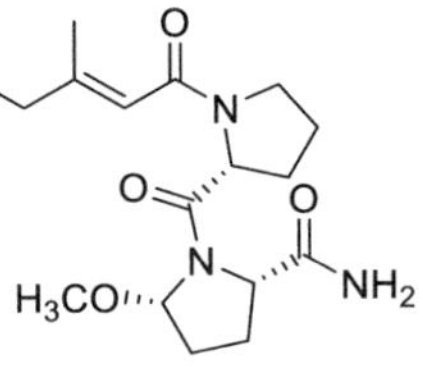

644

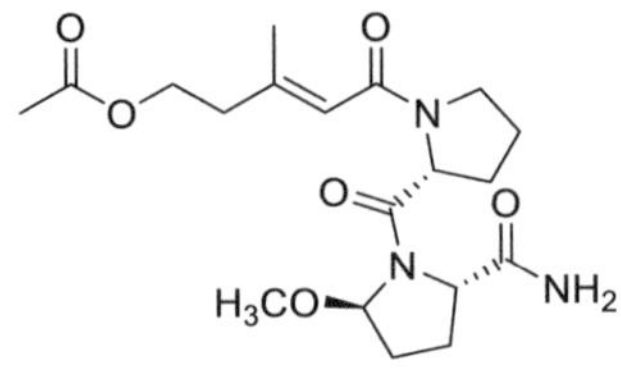

645

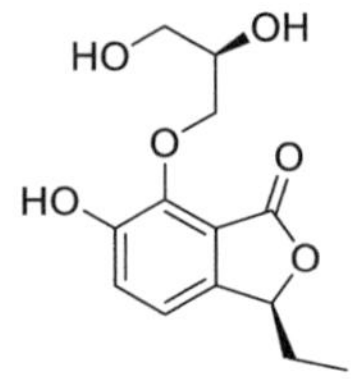

646

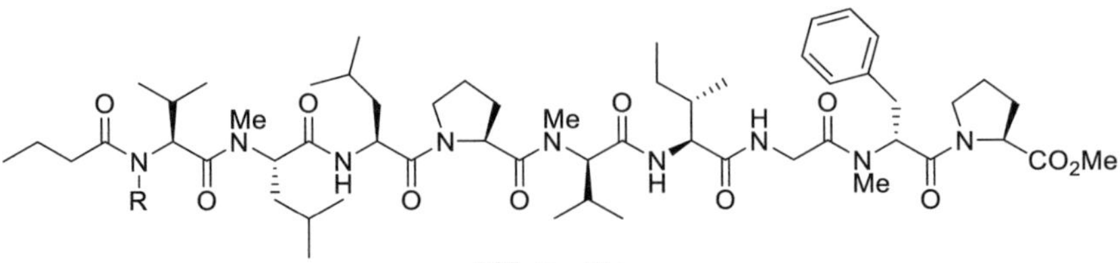

647: R = CH_3
648: R = H

Fig. 71 Natural products from *Cochliobolus lunatus* SCSIO41401 and *Oscillatoria* sp

and GLU202, respectively, though no significant differences in docking effects were observed between the two epimers.

Amantamide B (**647**) (Fig. 71), a newly identified linear nonapeptide analogue of the cyanobacterial natural product amantamide A (**648**) (Fig. 71), shares structural features including methyl ester and butanamide termini [148]. These compounds were isolated from the organic extract of the tropical marine filamentous cyanobacterium *Oscillatoria* sp., collected near the Paracel Islands. Amantamides A and B modulated spontaneous calcium oscillations in primary cultured neocortical neurons, a model for assessing neuronal excitability and Ca^{2+} signaling, without significant cytotoxicity at 10 μM during short-duration in vitro testing. However, in longer-duration bioassays, both compounds exhibited moderate cytotoxicity, with IC_{50} values ranging from 1 to 10 μM against CCRF-CEM human T lymphoblastoid cells and U937 human histiocytic lymphoma cells.

Isaridins are cyclodepsipeptides isolated from the marine-derived fungus, *Amphichorda felina* (syn. *Beauveria felina*) SYSU-MS7908, which was isolated from a marine ascidian, *Styela plicata*, collected in the North Atoll of the Paracel Islands. This marine fungal strain is also a rich source of other natural product classes, such as meroterpenoids and polyketide (see above sections). In 2022, isaridin E (**649**) (Fig. 72), obtained in gram quantity from the fungal strain, was demonstrated to possess anti-inflammatory and insecticidal activities [149]. Isaridin E was found to inhibit ADP-induced platelet aggregation, activation and secretion in vitro in a concentration-dependent manner, while having no effect on platelet aggregation induced by collagen or thrombin. In vivo, isaridin E demonstrated dose-dependent antithrombotic activity in an $FeCl_3$-induced mouse carotid thrombosis model, notably without prolonging bleeding time. Mechanistic studies revealed that isaridin E significantly reduced ADP-mediated phosphorylation of PI3K and Akt. These findings indicate that isaridin E exerts potent antithrombotic effects by selectively inhibiting ADP-induced platelet activation, secretion and aggregation through modulation of the PI3K/Akt signaling pathway, without increasing bleeding risk.

Additional six new molecules, isaridins I (**650**)–N (**655**) (Fig. 72), were identified from *Beauveria felina* SYSU-MS7908 in 2023 [150]. Among these, isaridin K (**652**) is notable for its peptide backbone containing an *N*-methyl-2-aminobutyric acid residue, a rare feature in natural cyclopeptides. Bioassays revealed that isaridin

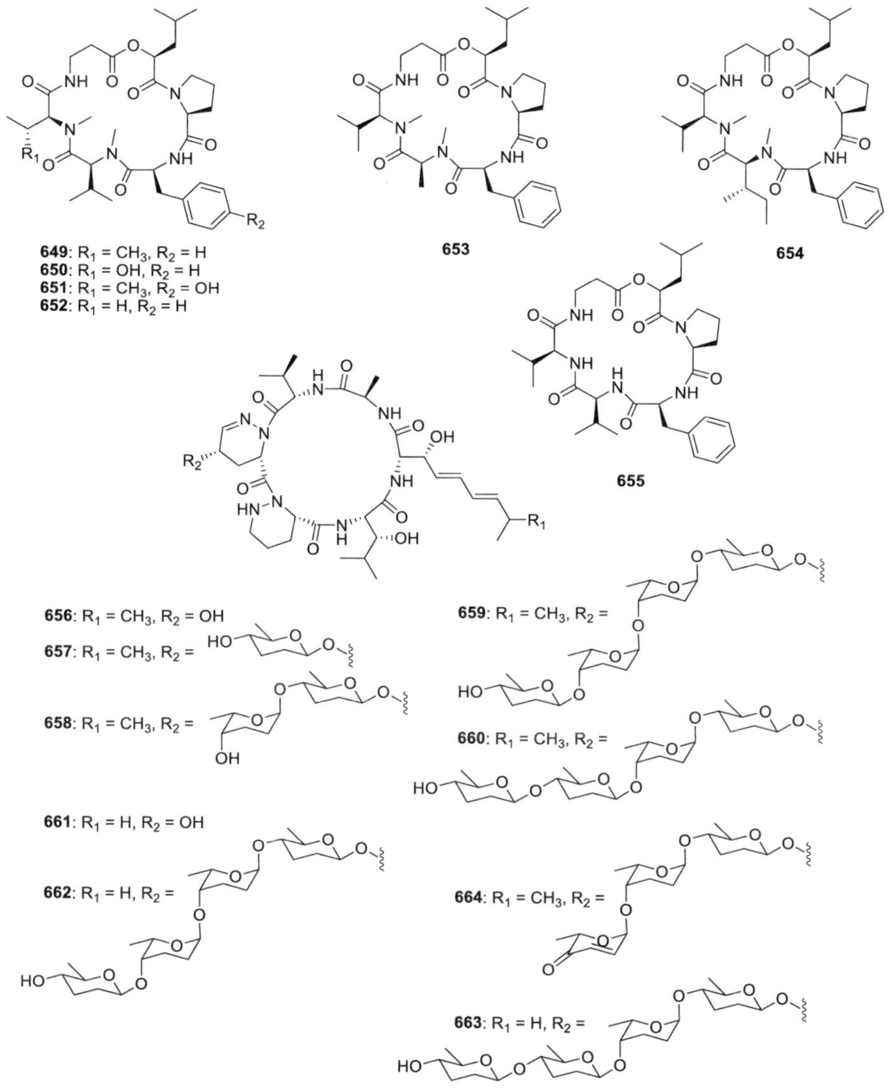

Fig. 72 Chemical structures of isaridins and pyridapeptides

J (**651**) effectively inhibited the mycelial growth of *Geotrichum citri-aurantii* by disrupting the fungal cell membrane. Given that sour rot caused by *G. citri-aurantii* is a major postharvest disease in citrus, leading to significant economic losses, these findings highlight the potential of isaridins as a basis for developing new agrochemical fungicides and underscore the value of marine fungal peptides in agricultural applications.

Pyridapeptide A (**656**) (Fig. 72), a cyclohexapeptide incorporating unique residues, such as hexahydropyridazine-3-carboxylic acid (HPDA), 5-hydroxytetrahydropyridazine-3-carboxylic acid (γ-OH-TPDA) and (2*S*,3*R*,4*E*,6*E*)-2-amino-3-hydroxy-8-methylnona-4,6-dienoic acid, along with its four glycopeptides, pyridapeptides B (**657**)–E (**660**) (Fig. 72), were isolated from the fermentation broth of *Streptomyces* sp. OUCMDZ-4539, an actinobacterium derived from the marine sponge *Dysidea avara* [151]. Pyridapeptides D (**659**) and E (**660**), featuring linear tetrasaccharide chains, demonstrated broad-spectrum antiproliferative activity against six human cancer cell lines, including PC9, MKN45, HepG2, HCT-116 and K562, with IC_{50} values ranging from 0.45 to 2.11 μM. In contrast, pyridapeptide C (**658**), bearing a disaccharide chain, exhibited selective antiproliferative activity against the human colon cancer cell line HCT-116, with an IC_{50} value of 1.28 μM. Pyridapeptide A and its monoglycoside, pyridapeptide B, showed no significant antiproliferative activity against the six tested cell lines, highlighting the critical role of oligosaccharide chains in conferring antiproliferative properties.

Re-fermentation of *Streptomyces* sp. OUCMDZ-4539 in a large volume of liquid media led to the isolation of four additional cyclohexapeptides, pyridapeptides F (**661**)–I (**664**) (Fig. 72) [152]. Notably, Pyridapeptide I was found to contain a highly rare glycose residue, aculose. Among the isolated compounds, pyridapeptides G (**662**) and H (**663**), characterized by linear tetrasaccharide chains, demonstrated broad-spectrum antiproliferative activity against five human cell lines (PC9, MKN45, HepG2, K562 and L-02 cell lines), with IC_{50} values ranging from 1.33 to 3.33 μM. In contrast, pyridapeptide I (**664**), which possesses a trisaccharide chain, exhibited selective antiproliferative effects against the human lung cancer cell line (PC9) and human hepatoma cell line (HepG2), with IC_{50} values of 3.39 and 3.04 μM, respectively.

Four new hydroxamate-containing cyclopeptides, designated acremonpeptides A (**665**)–D (**668**) (Fig. 73), along with an Al(III)–acremonpeptide D complex, acremonpeptide E (**669**) (Fig. 73), were identified from the marine fungus *Acremonium persicinum* SCSIO 115 [153]. This fungal strain was isolated from marine sediment in the Paracel Islands at a depth of 205 m. Acremonpeptides A–D share a common structural feature of three 2-amino-5-(*N*-hydroxyacetamido)pentanoic acid (N^5-hydroxy-N^5-acetyl-L-ornithine) moieties, which contribute to their metal ion-chelating properties. In vitro bioassays revealed that acremonpeptides A (**665**), B (**666**) and acremonpeptide E (**669**) exhibit moderate antiviral activity against herpes simplex virus 1, with EC_{50} values of 16, 8.7 and 14 μM, respectively.

By integrating molecular networking and ^1H NMR techniques, three novel cycloheptapeptides, asperversiamides A (**670**)–C (**672**) (Fig. 74), were discovered from the coral-derived fungus *Aspergillus versicolor* CHNSCLM-0063 [154]. The structures

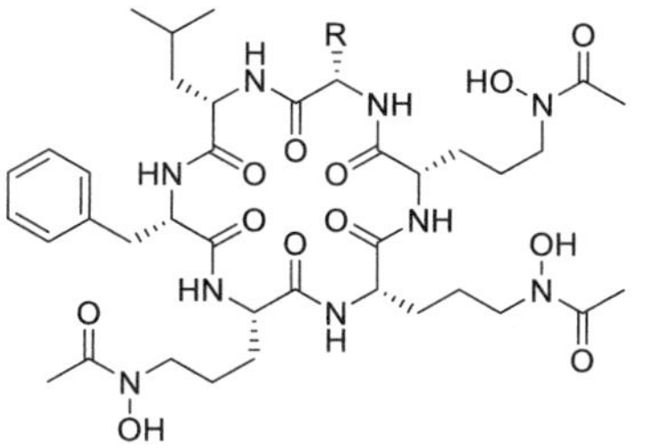

665: R = CH_2OH
666: R = Me

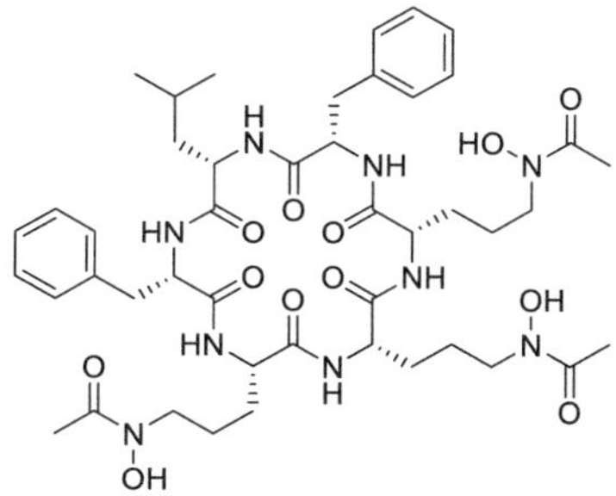

667

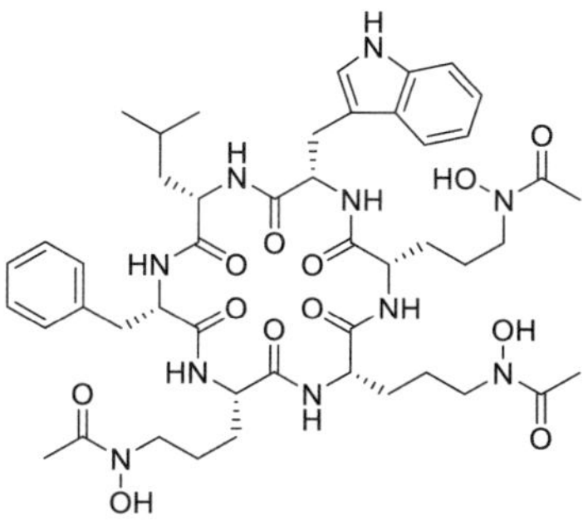

668

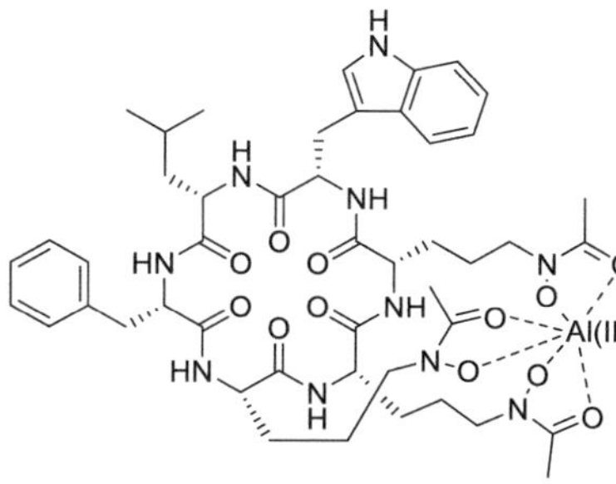

669

Fig. 73 Chemical structures of acremonpeptides

of these compounds were confirmed through total synthesis. This fungal strain was isolated from the gorgonian coral *Rumphella aggregata* (NSM02). All three asperversiamides exhibited potent inhibitory activity against *Mycobacterium marinum*, with minimum inhibitory concentrations (MICs) of 23.4, 81.2 and 87.5 μM for asperversiamides A–C, respectively. These activities are comparable to those of positive controls rifampin (19.0 μM), streptomycin (20.1 μM), and isoniazid (88.5 μM), underscoring asperversiamides potential as antimicrobial agents.

Guided by MS/MS-based molecular networking, additional four new cycloheptapeptides, asperheptatides A (**673**)–D (**676**) (Fig. 74), along with three known analogues, asperversiamides A–C, were isolated from *Aspergillus versicolor* CHNSCLM-0063 [155]. Furthermore, a series of 19 new derivatives of asperversiamide A (**670**) were semi-synthesized and evaluated for antitubercular activity against *Mycobacterium tuberculosis* H37Ra. Among these, four semi-synthetic compounds **677**–**680** (Fig. 74) exhibited moderate activity with minimum inhibitory concentration (MIC) values of 12.5 μM, suggesting their potential as a new class of antitubercular agents.

Further investigation of the secondary metabolites produced by *Aspergillus versicolor* CHNSCLM-0063, guided by MS/MS-based molecular networking, resulted

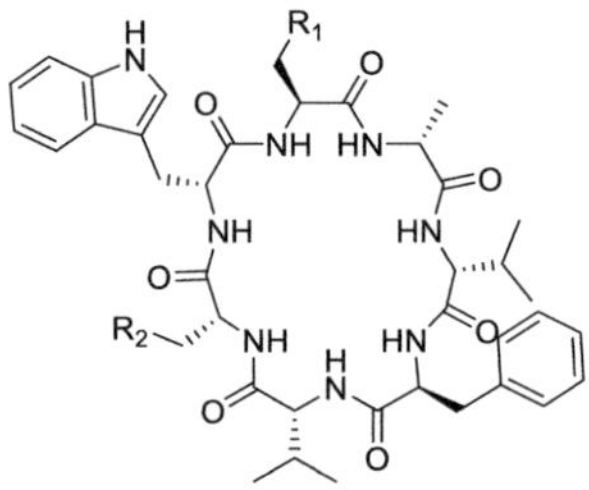

670: R_1 = OH, R_2 = OH
671: R_1 = H, R_2 = OH
672: R_1 = OH, R_2 = H

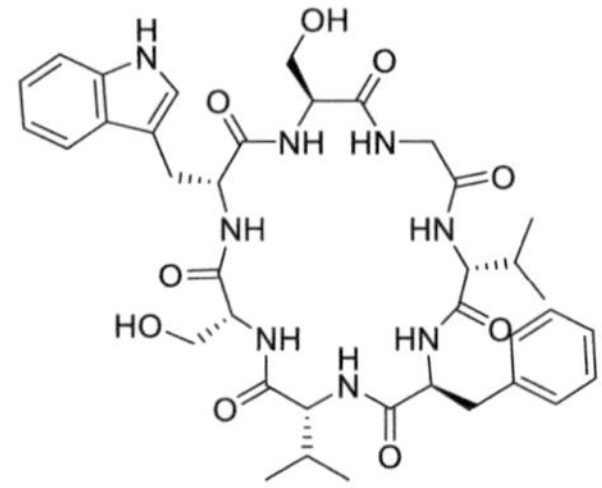

673

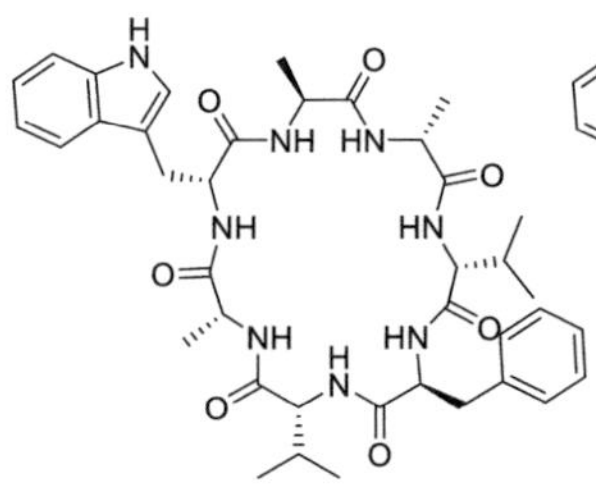

674

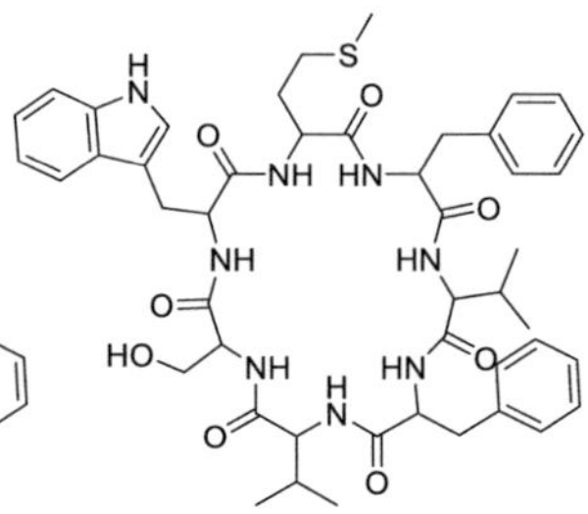

675

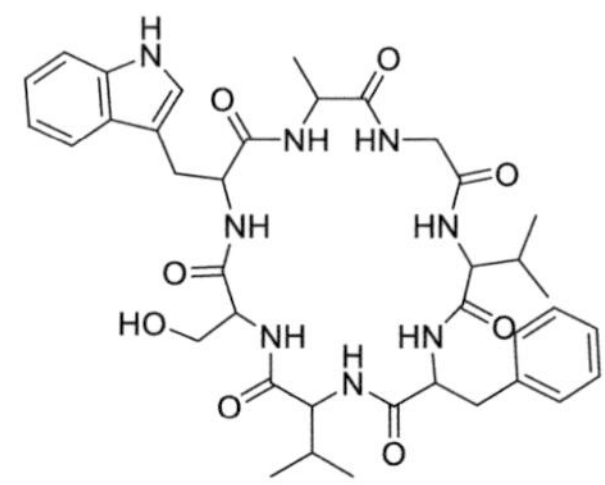

676

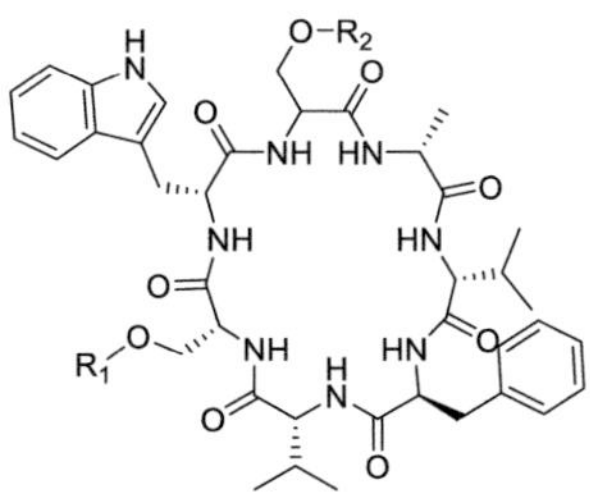

677: R_1 = R_2 = 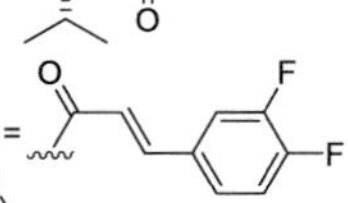

678: R_1 =  R_2 = H

679: R_1 = R_2 = 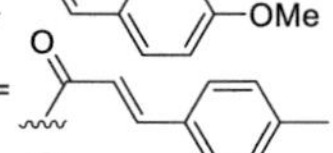

680: R_1 = R_2 = 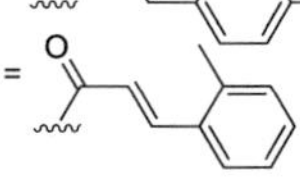

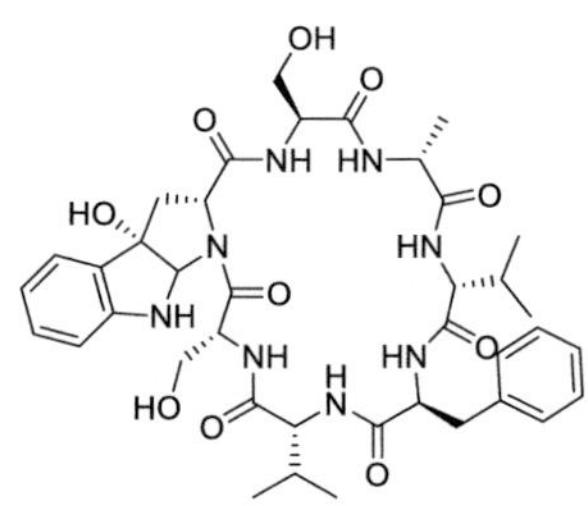

681

Fig. 74 Chemical structures of cyclopeptides from *Aspergillus versicolor* CHNSCLM-0063 and synthetic analogues

in the discovery of an unusual pyrroloindoline-containing cycloheptapeptide, asperpyrroindotide A (**681**) (Fig. 74) [156]. The semisynthetic transformation of asperversiamides A into asperpyrroindotide A was successfully achieved with optimized reaction conditions. Furthermore, a series of new derivatives of asperversiamide A were semi-synthesized, and their antitubercular activities were evaluated. Preliminary structure–activity relationship studies revealed that the serine hydroxyl groups and the tryptophan residue within this class of cycloheptapeptides play a crucial role in their antitubercular activity.

Streptomyces atratus SCSIO ZH16, a bacterium isolated from sediment samples collected at a depth of 3536 m east of the Pratas Islands, has demonstrated significant biosynthetic potential. In 2017, fermentations of the wild-type strain yielded a series of ilamycin congeners, ilamycins B_1 (**682**), B_2 (**683**), C_1 (**684**), C_2 (**685**), D (**686**) and E_1 (**687**) (Fig. 75), with antituberculosis activity, with the biosynthetic gene cluster (BGC) responsible for ilamycin production subsequently elucidated [157]. In addition, the biosynthesis of the rare L-3-nitrotyrosine and L-2-amino-4-hexenoic acid units, along with three pre-tailoring and two post-tailoring steps, was investigated within the ilamycin biosynthetic machinery. This was achieved through a combination of gene inactivation, precursor chemical complementation, isotope-labeled precursor feeding experiments and eventual structural elucidation of intermediates, such as ilamycins E_2 (**688**) and F (**689**) (Fig. 75), derived from the respective mutants. Genome sequencing and annotation revealed the presence of 26 BGCs, highlighting its capacity to produce diverse specialized natural products. Through genome mining, researchers activated a cyclodepsipeptide gene cluster, leading to the isolation of atratumycin (**690**) (Fig. 75), a cinnamic acid-bearing compound [158]. Atratumycin exhibits notable activity against *Mycobacterium tuberculosis* H37Ra and H37Rv, with minimum inhibitory concentrations (MICs) of 3.8 and 14.6 μM, respectively.

Further genetic engineering of wild type *Streptomyces atratus* SCSIO ZH16 resulted in the creation of the Δ*ilaR* mutant strain, which facilitated the discovery of new ilamycin analogues. From a 200 L culture of this marine-derived mutant actinomycete, 12 novel cyclopeptides, ilamycins G (**691**)–R (**702**) (Fig. 76) containing various nonproteinogenic amino acids were isolated, along with ilamycins E_1 (**687**) and F (**689**) (Fig. 76) [159]. Bioassays against *Mycobacterium tuberculosis* H37Rv demonstrated that all 12 new compounds exhibited antitubercular activity, with minimum inhibitory concentration (MIC) values ranging from 0.0096 to 10 μM. Structure–activity relationship (SAR) analyses revealed that: (i) reduction and cyclization at C-33 significantly enhanced antitubercular activity; (ii) oxidation at C-15 did not impact the activity of ilamycin E_1, while oxidation at C-32 decreased its activity; (iii) for ilamycin F, C-15 oxidation reduced activity, whereas C-32 oxidation increased it; and (iv) the presence of a C-43 nitro group played a critical role in determining antitubercular efficacy.

Previously it was reported by Ma and co-workers in 2017 that ilamycins E_1/E_2 exhibited potent and selective anti-tuberculosis activity against *Mycobacterium tuberculosis* H37Rv, with a MIC of 9.8 nM. However, initial fermentation conditions for the strain producing ilamycins E_1/E_2 yielded a production level of only 13.51 mg/L. To overcome this limitation, a combinatorial strategy was developed recently by

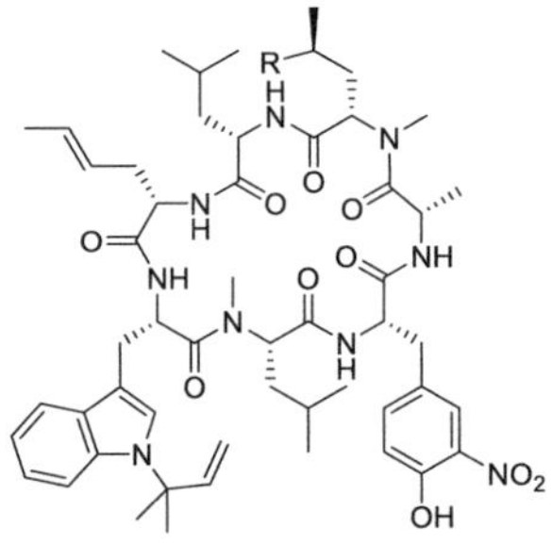

682: R = CH_3
689: R = CO_2H

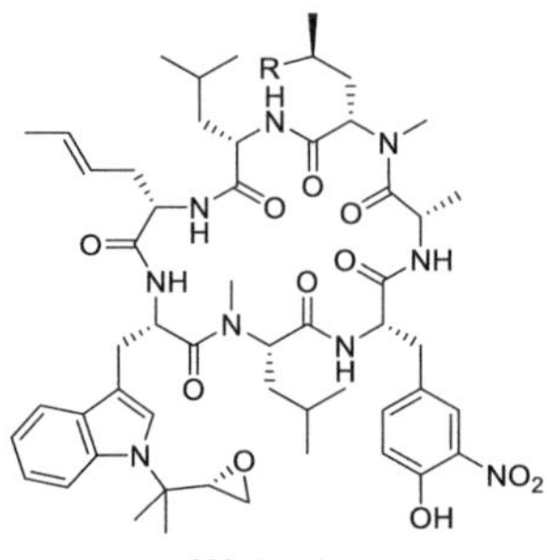

683: R = CH_3
686: R = CO_2H

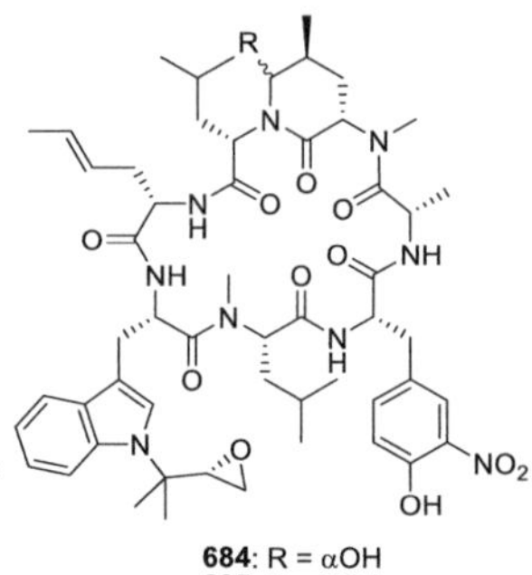

684: R = αOH
685: R = βOH

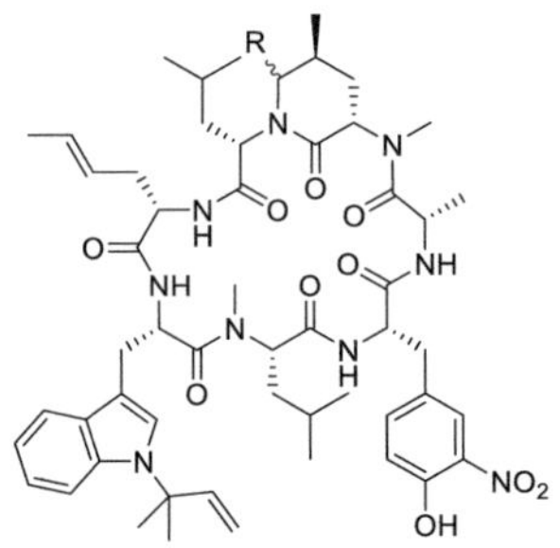

687: R = αOH
688: R = βOH

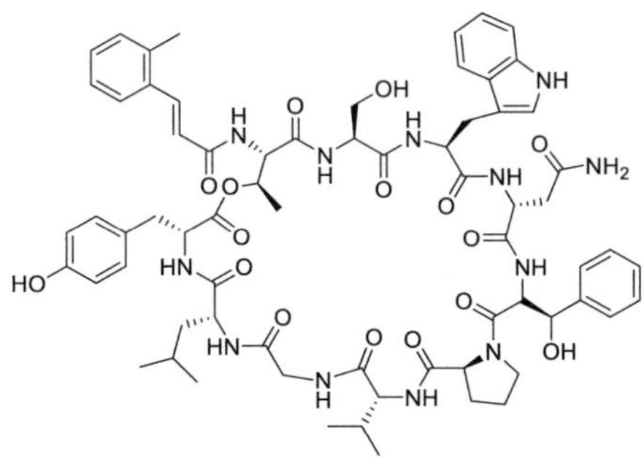

690

Fig. 75 Chemical structures of cyclopeptides from *Streptomyces atratus* SCSIO ZH16 and related mutant strains

Zhu and co-workers, incorporating medium optimization, adjustments to fermentative parameters and the exogenous addition of metal ions, precursors and surfactants. Specific interventions, such as the addition of 1 mM $ZnCl_2$ at 0 h, 1 g/L tyrosine at 96 h and 2 g/L shikimic acid at 48 h, significantly enhanced average production by 53 times [160]. This optimized production of ilamycins E_1/E_2 in *S. atratus* Δ*ilaR* establishes a robust foundation for their further clinical development.

Three new cyclohexapeptides, petrosamides A (**703**)–C (**705**) (Fig. 77), were isolated from *Aspergillus* sp. 151304, a sponge-derived fungus obtained from the inner tissue of a *Petrosia* sp. marine sponge [161]. Petrosamides A–C exhibited significant, dose-dependent pancreatic lipase (PL) inhibitory activities, with IC_{50} values of 7.6, 1.8 and 0.5 μM, respectively. Notably, petrosamide C inhibited PL in a noncompetitive manner, and molecular dynamics simulations indicated that it binds to PL at the entrance of the catalytic pocket.

Cyanogripeptides A (**706**)–C (**708**) (Fig. 77), three novel cyclolipopeptides containing unusual β-methyl-leucine residues, were discovered from a rare bacterium, *Actinoalloteichus cyanogriseus* LHW52806, through an LC–MS-guided approach [162]. The strain *A. cyanogriseus* LHW52806 was isolated from the sponge

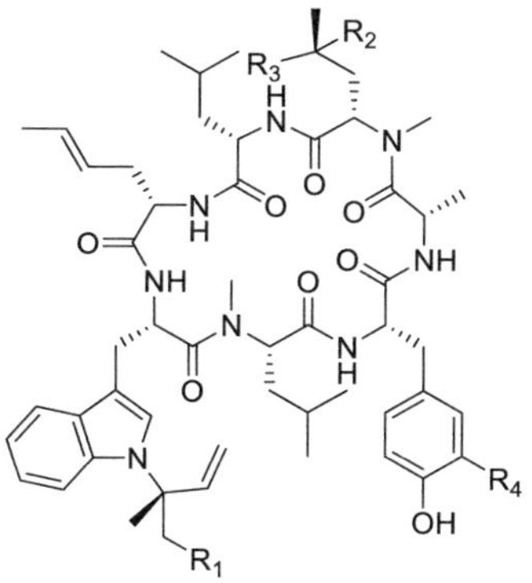

691: R_1 = OH, R_2 = H, R_3 = CO_2H, R_4 = NO_2
696: R_1 = R_2 = OH, R_3 = CO_2H, R_4 = NO_2
697: R_1 = H, R_2 = OH, R_3 = CH_2OH, R_4 = NO_2
698: R_1 = R_2 = H, R_3 = CO_2H, R_4 = H
699: R_1 = R_2 = H, R_3 = CO_2H, R_4 = NO_2

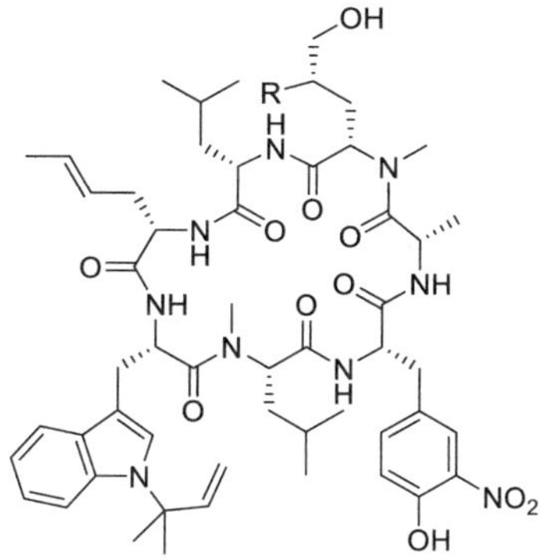

692: R = CO_2H
693: R = CH_2OH

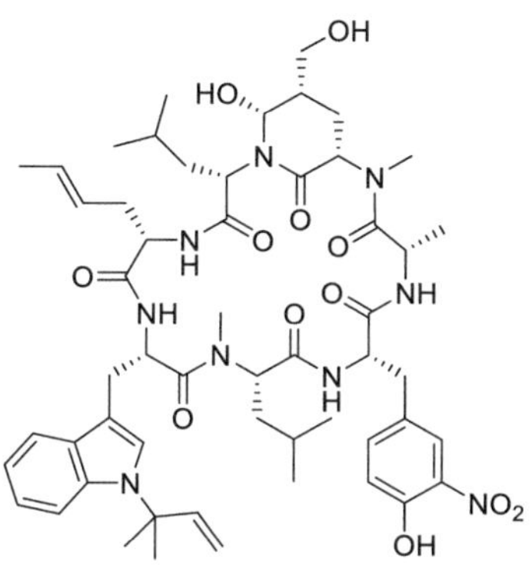

700

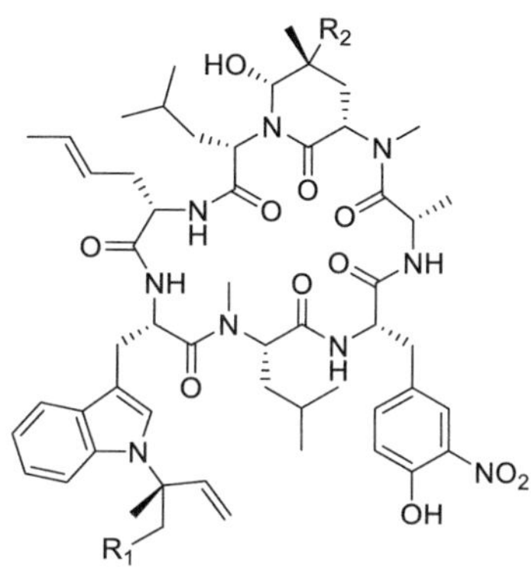

694: R_1 = OH, R_2 = H
695: R_1 = R_2 = OH

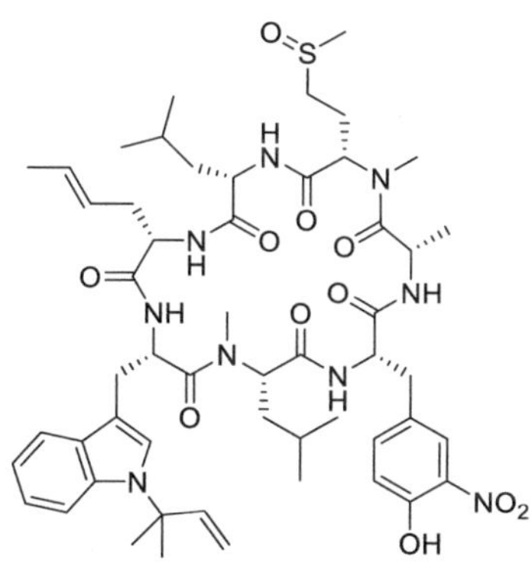

701

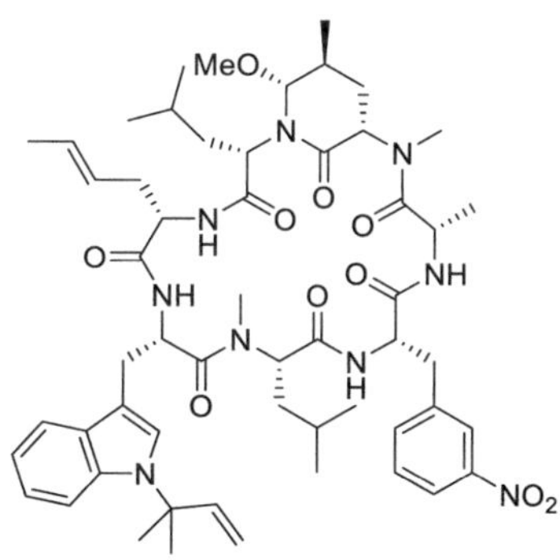

702

Fig. 76 Chemical structures of cyclopeptides from *Streptomyces atratus* SCSIO ZH16 and related mutant strains

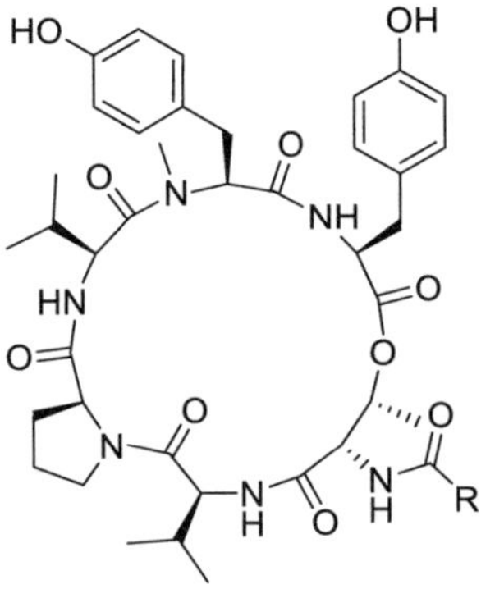

703: R = CH_3
704: R = CH_2CH_3

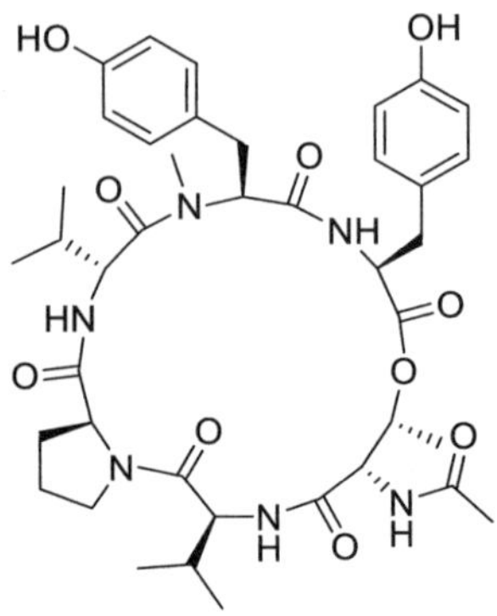

705

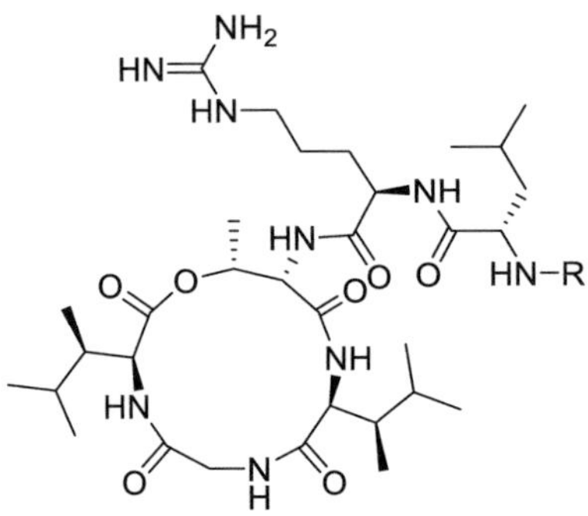

706: R =
707: R =
708: R =

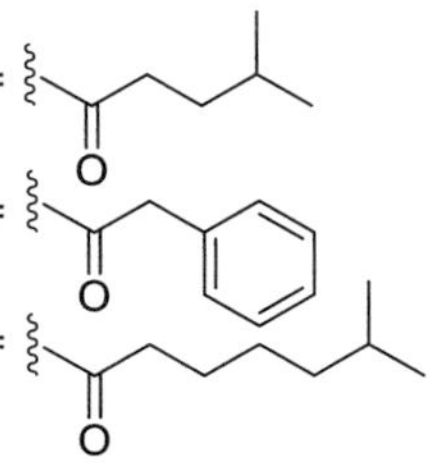

Fig. 77 Chemical structures of cyclopeptides from *Aspergillus* sp. 151304 and *Actinoalloteichus cyanogriseus* LHW52806

Phakellia fusca. Among the compounds, cyanogripeptide C demonstrated antibacterial activity against *Helicobacter pylori* strains G27 and 26,695, as well as *Mycolicibacterium smegmatis* ATCC607, with minimum inhibitory concentration (MIC) values of 32 μg/mL.

3.6 Nucleosides

Epigenetic agents, specifically the histone deacetylase inhibitor suberoylanilide hydroxamic acid and the DNA methyltransferase inhibitor 5-azacytidine, were introduced into Czapek-Dox medium to enhance the chemical diversity of the marine-derived fungus *Aspergillus versicolor* XS-20090066 [163]. This fungus was isolated from the inner tissue of the gorgonian *Dichotella gemmacea*, collected from the Paracel Islands coral reef. The study led to the discovery of two new nucleoside

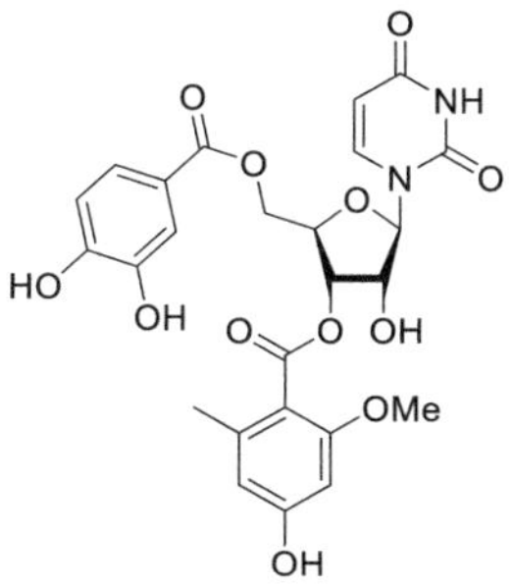
709

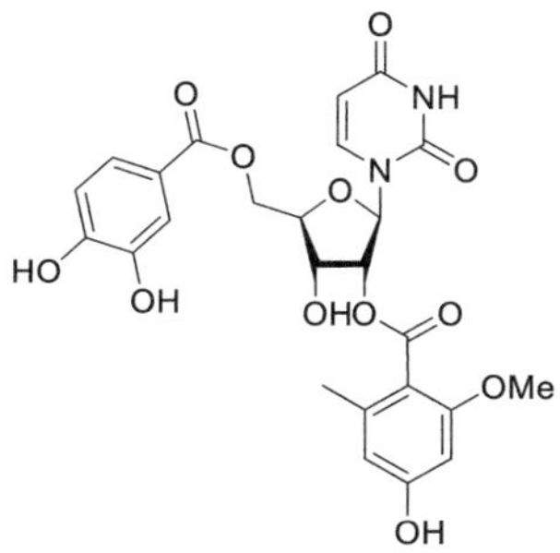
710

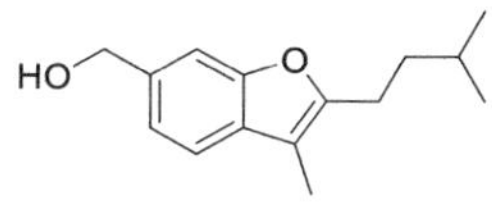
711

Fig. 78 Nucleoside derivatives and aspergillusene E isolated from *Aspergillus versicolor* XS-20090066

derivatives, kipukasins K (**709**) and L (**710**) (Fig. 78), alongside a significant increase in the yields of four known nucleoside derivatives. Additionally, one new bisabolane sesquiterpene, aspergillusene E (**711**) (Fig. 78), and ten known derivatives were identified. Kipukasin K and aspergillusene E exhibited antibacterial activity against *Staphylococcus epidermidis* and *Staphylococcus aureus*, with minimum inhibitory concentration (MIC) values ranging from 8 to 16 μg/mL.

4 Conclusions

The research on marine microbial natural products in the East Sea region underscores the immense potential of this area for drug discovery and biotechnology. The region's diverse ecosystems, ranging from coastal areas to deep-sea environments, host a wealth of microbial life that has evolved to produce unique bioactive compounds. Advances in cultivation techniques and molecular biology, including epigenetics regulation, co-culture, precursor feeding, heterologous expression and changing fermentation parameters to activate silent BGCs, have been instrumental in isolating and characterizing these natural products, particularly from marine bacteria and fungi, highlighting their structural novelty and pharmacological relevance [164]. Moreover, certain activation techniques, such as epigenetic regulation, co-culture, precursor feeding and varying fermentation parameters, are relatively easy to perform and independent of genomic information.

This chapter also emphasizes the distinct biochemistry of microorganisms associated with the East Sea's marine habitats. These microbes produce diverse classes of compounds, such as alkaloids, peptides, terpenoids and polyketides, with significant bioactivity against pathogens, cancer cells and modulation of inflammatory processes. The ecological conditions of remote archipelagos and deep-sea areas, less impacted by human activity, within the East Sea ecoregion offer an untapped source of microbial diversity. In addition, findings from Vietnamese coastal regions reveal numerous fungal and bacterial strains with remarkable antimicrobial and cytotoxic properties, validating the region's importance for natural product research. Furthermore, marine fauna and flora as well as their symbiotic or associated microbial communities represent a rich and largely untapped reservoir of novel bioactive compounds. The mutualistic interactions between invertebrates, such as sponges, soft corals and echinoderms, and diverse marine bacteria and fungi have led to the isolation of structurally unique secondary metabolites with significant pharmacological potential. The capacity of these marine microbes to biosynthesize such diverse molecules is driven by evolutionary pressures in their unique ecological niches.

Overall, the East Sea region represents a frontier for marine natural product exploration, with promising implications for pharmaceutical innovation. The research highlights the need for sustainable and systematic bioprospecting efforts, leveraging cutting-edge technologies to harness the therapeutic potential of marine microorganisms. This approach not only enriches our understanding of marine biochemistry but also aligns with global efforts to address challenges, such as antibiotic resistance and the need for novel anticancer treatments.

Acknowledgements Ton That Huu Dat was funded by the Postdoctoral Scholarship Programme of Vingroup Innovation Foundation (VINIF), code VINIF.2024.STS.58.

References

1. Romano G, Costantini M, Sansone C, Lauritano C, Ruocco N, Ianora A (2017) Marine microorganisms as a promising and sustainable source of bioactive molecules. Mar Environ Res 128:58–69
2. Agrawal S, Acharya D, Adholeya A, Barrow CJ, Deshmukh SK (2017) Nonribosomal peptides from marine microbes and their antimicrobial and anticancer potential. Front Pharmacol 8:828
3. McCarthy PJ, Roberts BF, Carbonell A, Roberts J, Wright AE, Chakrabarti D (2019) Marine microbiome as a source of antimalarials. Trop Med Infect Dis 4:103
4. Mondal A, Bose S, Banerjee S, Patra JK, Malik J, Mandal SK, Kilpatrick KL, Das G, Kerry RG, Fimognari C (2020) Marine cyanobacteria and microalgae metabolites—a rich source of potential anticancer drugs. Mar Drugs 18:476
5. Tan LT (2023) Impact of marine chemical ecology research on the discovery and development of new pharmaceuticals. Mar Drugs 21:174
6. Maghembe R, Damian D, Makaranga A, Nyandoro SS, Lyantagaye SL, Kusari S, Hatti-Kaul R (2020) Omics for bioprospecting and drug discovery from bacteria and microalgae. Antibiotics 9:229
7. Trinh PTH, Tien PQ, Ngoc NTD, Ly BM, Van TTT (2018) Isolation and screening marine fungi with antimicrobial activity from samples collected in Nha Trang Bay, Vietanam. Vietnam J Biotechnol 16:181–187
8. Trinh PTH, Ngoc NTD, Trang VTD, Tien PQ, Ly BM, Van TTT (2017) Antimicrobial activity of marine fungi isolated from the Son Tra peninsula, Da Nang, Vietnam. Academia J Biol 39:457–462
9. Vu TTH, Le THM, Nguyen MA, Vu TQ, Doan TMH, Pham VC (2022) Screening of actinomycetes from Ninh Thuan and Binh Thuan sea for antimicrobial producers. Vietnam J Mar Sci Technol 22:423–432

10. Anh CV, Kang JS, Lee H-S, Trinh PTH, Heo C-S, Shin HJ (2022) New glycosylated secondary metabolites from marine-derived bacteria. Mar Drugs 20:464
11. Thi QV, Tran VH, Mai HDT, Le CV, Hong MLT, Murphy BT, Chau VM, Pham VC (2016) Antimicrobial metabolites from a marine-derived actinomycete in Vietnam's East Sea. Nat Prod Commun 11, 1934578X1601100116
12. Phi TD, Nguyen TL, Vu TQ, Nguyen MA, Vu TTH, Le THM, Pham VC, Doan TMH (2024) Antimicrobial secondary metabolites from a marine-derived fungus *Penicillium* sp. OM07. Vietnam J Mar Sci Technol 24:175–184
13. Thi QV, Tran VH, Mai HDT, Le CV, Hong MLT, Murphy BT, Chau VM, Pham VC (2016) Secondary metabolites from an Actinomycete from Vietnam's East Sea. Nat Prod Commun 11,1934578X1601100320.
14. Le HMT, Do QT, Doan MHT, Vu QT, Nguyen MA, Vu THT, Nguyen HD, Duong NTT, Tran MH, Chau VM (2019) Chemical composition and biological activities of metabolites from the marine fungi *Penicillium* sp. isolated from sediments of Co To Island, Vietnam. Molecules 24:3830
15. Cao DD, Trinh TTV, Doan Thi Mai H, Vu VN, Le HM, Vu Thi Q, Nguyen MA, Duong TT, Tran DT, Chau VM (2019) Antimicrobial lavandulylated flavonoids from a sponge-derived *Streptomyces* sp. G248 in East Vietnam Sea. Marine Drugs 17:529
16. Cao DD, Do TQ, Doan Thi Mai H, Vu Thi Q, Nguyen MA, Le Thi HM, Tran DT, Chau VM, Cong Thung D, Pham VC (2020) Antimicrobial lavandulylated flavonoids from a sponge-derived actinomycete. Nat Prod Res 34:413–420
17. Quyen VT, Minh LTH, Huyen VTT, Anh NM, Hue NT, Dao PT, Linh NT, Van Cuong P, Huong DTM (2022) Antimicrobial secondary metabolites from the marine-derived actinomycete *Nocardiopsis synnemataformans* HT06. Chem Nat Compd 58:704–707
18. Hoa TTT, Van Nam V, Huong DTM, Dao PT, Phuong NN, Minh LTH, Quyen VT, Anh NM, Murphy BT, Ly HTM (2020) Flavonoids and alkaloids from marine-derived actinomycete *Streptomyces* sp. C011. Vietnam J Chem 58:610–614
19. Phi Thi D, Doan Thi Mai H, Cao DD, Vu Thi Q, Nguyen MA, Le Thi HM, Tran DT, Chau VM, Pham VC (2020) Novel 1,3-benzodioxole from marine-derived actinomycete in East Vietnam Sea. Nat Prod Commun 15, 1934578X20920042
20. Cao DT, Tran VH, Vu VN, Mai HDT, Le THM, Vu TQ, Nguyen HH, Chau VM, Pham VC (2019) Antimicrobial metabolites from a marine-derived Actinomycete *Streptomyces* sp. G278. Nat Prod Res 33:3223–3230
21. Thuy NTT, Hien HTM, Ha NC, Thom LT, Hong DD, Thinh NV, Dan NT, Hoi ND, Loan VT, Quang HD (2024) Chemical constituents from the heterotrophic marine microalgae *Aurantiochytrium* sp. SC145 and their antimicrobial activities. Nat Prod Res 38:789–795
22. Kim D-C, Quang TH, Tien NT, Kim K-W, Kim Y-C, Ngan NTT, Cuong NX, Nam NH, Oh H (2022) Anti-neuroinflammatory effect of oxaline, isorhodoptilometrin, and 5-hydroxy-7-(2′-hydroxypropyl)-2-methyl-chromone obtained from the marine fungal strain *Penicillium oxalicum* CLC-MF05. Arch Pharmacal Res 45:90–104
23. Smetanina OF, Yurchenko AN, Girich EV, Trinh PTH, Antonov AS, Dyshlovoy SA, von Amsberg G, Kim NY, Chingizova EA, Pislyagin EA (2019) Biologically active echinulin-related indolediketopiperazines from the marine sediment-derived fungus *Aspergillus niveoglaucus*. Molecules 25:61
24. Koleva YK, Madden JC, Cronin MT (2008) Formation of categories from structure−activity relationships to allow read-across for risk assessment: toxicity of α, β-unsaturated carbonyl compound*s*. Chem Res Toxicol 21:2300–2312
25. Choi B-K, Jo S-H, Choi D-K, Trinh PTH, Lee H-S, Anh CV, Van TTT, Shin HJ (2020) Anti-neuroinflammatory agent, restricticin B, from the marine-derived fungus *Penicillium janthinellum* and its inhibitory activity on the NO production in BV-2 microglia cells. Mar Drugs 18:465
26. Anh NM, Huyen VTT, Quyen VT, Dao PT, Quynh DT, Huong DTM, Van Cuong P, Dat TTH, Minh LTH (2024) Antimicrobial and cytotoxic secondary metabolites from a marine-derived fungus *Penicillium citrinum* VM6. Curr Microbiol 81:32

27. Minh LTH, Anh NM, Huyen VTT, Quyen VT, Dao PT, Huong DTM, Cuong PV, Tuan CD, Kiem PV, Tai BH (2023) A pair of undescribed alkaloid enantiomers from marine sponge-derived fungus *Hamigera avellanea* and their antimicrobial and cytotoxic activities. Chem Biodivers 20:e202301425
28. Shin HJ, Choi B-K, Trinh PTH, Lee H-S, Kang JS, Van TTT, Lee H-S, Lee JS, Lee Y-J, Lee J (2018) Suppression of RANKL-induced osteoclastogenesis by the metabolites from the marine fungus *Aspergillus flocculosus* isolated from a sponge *Stylissa* sp. Mar Drugs 16:14
29. Hue NT, Linh NT, Huyen VTT, Huong DTM, Van Cuong P (2022) Chemical constituents of marine-derived actinomycete *Streptomyces fradiae* G650. Vietnam J Sci Technol 60:974–981
30. Cao DT, Van Hung N, Lien HTH, Jung-Woo C, Yun H-y, Kim YH, Van Cuong P, Huong DTM (2022) A new indole glucoside and other constituents from the sea cucumber-derived *Aspergillus fumigatus* M580 and their antimicrobial, cytotoxic, and inhibitory α-glucosidase activities. Rec Nat Prod 16:638
31. Do TQ, Van TTT, Linh NT, Minh LTH, Quyen VT, Murphy B, Huong DTM, Van Cuong P (2023) A new alkaloid from marine-derived actinomycete *Actinoalloteichus cyanogriseus* G631. Rec Nat Prod 1024–1030
32. Huong DTM, Van Hieu T, Ngan TB, Quyen VT, Anh NTH, Murphy BT, Van Cuong P (2021) Secondary metabolites produced by marine actinomycete *Streptomyces* sp. G246. Vietnam J Sci Technol 59:1–8
33. Minh LTH, Hue NT, Linh NT, Ngan TB, Quyen VT, Anh NM, Tuan CD, Van Minh C, Van Cuong P, Huong DTM (2020) Antimicrobial secondary metabolites from the marine-derived fungus *Aspergillus* sp. M28. Chem Nat Compd 56:1173–1175
34. Van Nam V, Huong DTM, Dao PT, Anh NM, Minh LTH, Van Cuong P (2019) Secondary metabolites produced by marine actinomycete *Micromonospora* sp. A258. Vietnam J Chem 57:707–711
35. Van Hieu T, Ngan TB, Huong DTM, Quyen VT, Minh LTH, Murphy BT, Van Cuong P (2019) Secondary metabolites from marine actinomycete *Streptomyces* sp. G330. Vietnam J Chem 57:480–484
36. Quyen VT, Van Hieu T, Huong DTM, Murphy B, Van Minh C, Van Cuong P (2015) Secondary metabolites produced by marine bacterirum *Micromonospora* sp. G019. Vietnam J Chem 53:150–153
37. Quyen VT, Van Nam V, Van Hieu T, Huong DTM, Murphy B, Van Minh C, Van Cuong P (2015) Secondary metabolites from marine bacterium *Streptomyces* sp. G039. Vietnam J Chem 53:142–145
38. Dao PT, Huong DTM, Murphy B, Van Minh C, Van Cuong P (2015) Compounds from culture broth of marine bacterium *Oceanisphaera* sp. Vietnam J Chem 53:120–123
39. Girich EV, Yurchenko AN, Smetanina OF, Trinh PTH, Ngoc NTD, Pivkin MV, Popov RS, Pislyagin EA, Menchinskaya ES, Chingizova EA (2020) Neuroprotective metabolites from vietnamese marine derived fungi of *Aspergillus* and *Penicillium* genera. Mar Drugs 18:608
40. Mullowney MW, Ó hAinmhire E, Tanouye U, Burdette JE, Pham VC, Murphy BT (2015) A pimarane diterpene and cytotoxic angucyclines from a marine-derived *Micromonospora* sp. in Vietnam's east sea. Marine Drugs 13:5815–5827
41. Choi B-K, Trinh PTH, Lee H-S, Choi B-W, Kang JS, Ngoc NTD, Van TTT, Shin HJ (2019) New ophiobolin derivatives from the marine fungus *Aspergillus flocculosus* and their cytotoxicities against cancer cells. Mar Drugs 17:346
42. Nguyen THA, Do TQ, Nguyen TL, Le Thi HM, Nguyen MA, Murphy BT, Dam TX, Huong DTM, Cuong PV (2023) New sesterterpenoid from the marine fungus *Penicillium oxalicum* M893. Nat Prod Commun 18, 1934578X231191636
43. Yurchenko AN, Trinh PTH, Girich EV, Smetanina OF, Rasin AB, Popov RS, Dyshlovoy SA, von Amsberg G, Menchinskaya ES, Van Thanh TT (2019) Biologically active metabolites from the marine sediment-derived fungus *Aspergillus flocculosus*. Mar Drugs 17:579
44. Beekman AC, Woerdenbag HJ, van Uden W, Pras N, Konings AW, Wikström HV, Schmidt TJ (1997) Structure−cytotoxicity relationships of some helenanolide-type sesquiterpene lactones. J Nat Prod 60:252–257

45. Belofsky GN, Jensen PR, Renner MK, Fenical W (1998) New cytotoxic sesquiterpenoid nitrobenzoyl esters from a marine isolate of the fungus *Aspergillus versicolor*. Tetrahedron 54:1715–1724
46. Tan Y, Yang B, Lin X, Luo X, Pang X, Tang L, Liu Y, Li X, Zhou X (2018) Nitrobenzoyl sesquiterpenoids with cytotoxic activities from a marine-derived *Aspergillus ochraceus* fungus. J Nat Prod 81:92–97
47. Fang W, Lin X, Zhou X, Wan J, Lu X, Yang B, Ai W, Lin J, Zhang T, Tu Z (2014) Cytotoxic and antiviral nitrobenzoyl sesquiterpenoids from the marine-derived fungus *Aspergillus ochraceus* Jcma1F17. MedChemComm 5:701–705
48. Hoang CK, Le CH, Nguyen DT, Tran HTN, Luu CV, Le HM, Tran HTH (2023) Steroid components of marine-derived fungal strain *Penicillium levitum* N33.2 and their biological activities. Mycobiology 51:246–255
49. Trinh PTH, Yurchenko AN, Khmel OO, Dieu TVT, Ngoc NTD, Girich EV, Menshov AS, Kim NY, Chingizova EA, Van TTT, Lee JS, Lee H-S, Yurchenko EA (2022) Cytoprotective polyketides from sponge-derived fungus *Lopadostoma pouzarii*. Molecules 27:7650
50. Yurchenko AN, Smetanina OF, Ivanets EV, Phan TTH, Ngo NTD, Zhuravleva OI, Rasin AB, Dyshlovoy SA, Menchinskaya ES, Pislyagin EA, von Amsberg G, Afiyatullov SS, Yurchenko EA (2020) Auroglaucin-related neuroprotective compounds from Vietnamese marine sediment-derived fungus *Aspergillus niveoglaucus*. Nat Prod Res 34:2589–2594
51. Luyen ND, Huong LM, Thi Hong Ha T, Cuong LH, Thi Hai Yen D, Nhiem NX, Tai BH, Gardes A, Kopprio G, Van Kiem P (2019) Aspermicrones A-C, novel dibenzospiroketals from the seaweed-derived endophytic fungus *Aspergillus micronesiensis*. J Antibiot 72:843–847
52. Choi B-K, Cho D-Y, Choi D-K, Trinh PTH, Shin HJ (2021) Two new phomaligols from the marine-derived fungus *Aspergillus flocculosus* and their anti-neuroinflammatory activity in BV-2 microglial cells. Mar Drugs 19:65
53. Thi Hoang Anh N, Mai Anh N, Thi Thu Huyen V, Thi Dao P, Thi Mai Huong D, Van Cuong P, Thanh Xuan D, Huu Tai B, Thi Hong Minh L, Van Kiem P (2023) Antimicrobial activity of depsidones and macrocyclic peptides isolated from marine sponge-derived fungus *Aspergillus nidulans* M256. Chem Biodivers 20:e202301660
54. Quang TH, Phong NV, Hanh TTH, Cuong NX, Ngan NTT, Oh H, Nam NH, Van Minh C (2021) Cytotoxic and immunomodulatory phenol derivatives from a marine sponge-derived fungus *Ascomycota* sp. VK12. Nat Prod Res 35:5153–5159
55. Chlipala GE, Tri PH, Hung NV, Krunic A, Shim SH, Soejarto DD, Orjala J (2010) Nhatrangins A and B, aplysiatoxin-related metabolites from the marine cyanobacterium *Lyngbya majuscula* from Vietnam. J Nat Prod 73:784–787
56. Girich EV, Rasin AB, Popov RS, Yurchenko EA, Chingizova EA, Trinh PTH, Ngoc NTD, Pivkin MV, Zhuravleva OI, Yurchenko AN (2022) New tripeptide derivatives asterripeptides A–C from Vietnamese mangrove-derived fungus *Aspergillus terreus* LM.5.2. Marine Drugs 20:77
57. Rentero Rebollo I, McCallin S, Bertoldo D, Entenza JM, Moreillon P, Heinis C (2016) Development of potent and selective *S. aureus* sortase A inhibitors based on peptide macrocycles. ACS Med Chem Lett 7:606–611
58. Castaldi A, Truong BN, Vu QT, Le THM, Marie A, Le Pennec G, Rouvier F, Brunel J-M, Longeon A, Pham VC, Doan TMH, Bourguet-Kondracki M-L (2024) Computational methods reveal a series of cyclic and linear lichenysins and surfactins from the Vietnamese marine sediment-derived *Streptomyces* strain G222. Molecules 29:1458
59. Nam VV, Mai Huong DT, Minh CV, Cuong PV (2017) Compounds from culture broth of marine bacterium *Micromonospora* sp. (G047). Vietnam J Chem 55:207–210
60. Dat TTH, Cuc NTK, Cuong PV, Smidt H, Sipkema D (2021) Diversity and antimicrobial activity of vietnamese sponge-associated bacteria. Mar Drugs 19:353
61. Quyen VT, Ngan TB, Mai Huong DT, Bourguet Kondracki M-L, Longeon A, Murphy BM, Minh CV, Cuong PV (2015) Secondary metabolites from *Micromonospora ectrinospora* G017. Vietnam J Chem 53:146–149

62. Quynh DT, Tuan CD, Chuc DV, Lien HTH, Huyen LK, Minh LTH, Cuong PV, Hung NV, Kim YH, Huong DTM (2020) Alkaloids and cyclopeptides from the marine-derived fungus *Aspergillus* sp. M512. Vietnam J Sci Technol 58:541–548
63. Danh CD, Dao PT, Huong DTM, Thach TD, Anh NM, Minh LTH, Anh TT, Van Cuong P (2018) Cyclopeptides from marine actinomycete *Streptomyces* sp. G261. Vietnam J Chem 56:570–573
64. Cao D-T, Nguyen T-L, Tran V-H, Huong DTM, Quyen VT, Nguyen M-A, Minh LTH, Chau VM, Pham VC (2019) Synthesis, structure and antimicrobial activity of novel metabolites from a marine actinomycete in Vietnam's East Sea. Nat Prod Commun 14:121–124
65. Quang TH, Kim DC, Van Kiem P, Van Minh C, Nhiem NX, Tai BH, Yen PH, Thi Thanh Ngan N, Kim HJ, Oh H (2018) Macrolide and phenolic metabolites from the marine-derived fungus *Paraconiothyrium* sp. VK-13 with anti-inflammatory activity. J Antibiot 71:826–830
66. Choi B-K, Phan THT, Hwang S, Oh D-C, Kang JS, Lee H-S, Ngo TDN, Tran TTV, Shin HJ (2019) Resorcinosides A and B, glycosylated alkylresorcinols from a marine-derived strain of the fungus *Penicillium janthinellum*. J Nat Prod 82:3186–3190
67. Girich EV, Trinh PTH, Nesterenko LE, Popov RS, Kim NY, Rasin AB, Menchinskaya ES, Kuzmich AS, Chingizova EA, Minin AS, Ngoc NTD, Van TTT, Yurchenko EA, Yurchenko AN, Berdyshev DV (2023) Absolute stereochemistry and cytotoxic effects of vismione E from marine sponge-derived fungus *Aspergillus* sp. 1901NT-1.2.2. Int J Molecul Sci 24:8150
68. Anh TT, Van Nam V, Le Thi Hong Minh VT, Quyen NMA, Huyen VTT, Huong DTM, Vuong NQ, Van Cuong P (2017) Secondary metabolites from marine fungus *Neurospora* sp. M45. Vietnam J Chem 55:131–134
69. Shin HJ, Anh CV, Cho D-Y, Choi D-K, Kang JS, Trinh PTH, Choi B-K, Lee H-S (2021) New polyenes from the marine-derived fungus *Talaromyces cyanescens* with anti-neuroinflammatory and cytotoxic activities. Molecules 26:836
70. Zhang X-Y, Zhang Y, Xu X-Y, Qi S-H (2013) Diverse deep-sea fungi from the South China Sea and their antimicrobial activity. Curr Microbiol 67:525–530
71. Kamjam M, Sivalingam P, Deng Z, Hong K (2017) Deep sea actinomycetes and their secondary metabolites. Front Microbiol 8:760
72. Wen H, Zhang D, Zhao H, Zhang Y, Yan X, Lin W, He S, Ding L (2023) Molecular networking-guided isolation of undescribed antifungal odoriferous sesquiterpenoids from a marine mesophotic zone sponge-associated *Streptomyces* sp. NBU3428. Phytochemistry 213:113779
73. Shang R-Y, Cui J, Li J-X, Miao X-X, Zhang L, Xie D-D, Zhang L, Lin H-W, Jiao W-H (2022) Nigerin and ochracenes J–L, new sesquiterpenoids from the marine sponge symbiotic fungus *Aspergillus niger*. Tetrahedron 104:132599
74. Wang T, Feng Y, Huang J, Wu S, Hu K, Wu J, Naman CB, Wang H, Lin W, He S (2024) Pestanoid A, a rearranged pimarane diterpenoid osteoclastogenesis inhibitor from a marine mesophotic zone chalinidae sponge-associated fungus, *Pestalotiopsis* sp. NBUF145. J Nat Prod 87:160–165
75. Zhao D-L, Yang L-J, Shi T, Wang C-Y, Shao C-L, Wang C-Y (2019) Potent phytotoxic harziane diterpenes from a soft coral-derived strain of the fungus *Trichoderma harzianum* XS-20090075. Sci Rep 9:13345
76. Yang B, He Y, Lin S, Zhang J, Li H, Wang J, Hu Z, Zhang Y (2019) Antimicrobial dolabellanes and atranones from a marine-derived strain of the toxigenic fungus *Stachybotrys chartarum*. J Nat Prod 82:1923–1929
77. Lin S, Chai Z, Zeng H, Yang B, Chi J, Zhang Y, Hu Z (2024) Atranones and dolabellanes with cardiomyocyte protective activity against cold ischemic injury from a coral-associated fungus *Stachybotrys chartarum*. Phytochemistry 225:114199
78. Lin S, Zeng H, Wang C, Chai Z, Zhang X, Yang B, Chi J, Zhang Y, Hu Z (2024) Discovery of novel natural cardiomyocyte protectants from a toxigenic fungus *Stachybotrys chartarum*. Bioorg Chem 148:107461
79. Ding W, Uvarani C, Wang F, Xue Y, Wu N, He L, Tian D, Chen M, Zhang Y, Hong K, Tang J (2020) New Ophiobolins from the deep-sea derived fungus *Aspergillus* sp. WHU0154 and their anti-inflammatory effects. Marine Drugs 18:575

80. Cheng X, Liang X, Yao F-H, Liu X-B, Qi S-H (2021) Fusidane-type antibiotics from the marine-derived fungus *Simplicillium* sp. SCSIO 41513. J Nat Prod 84:2945–2952
81. Jiang M, Wu Z, Wu Q, Yin H, Guo H, Yuan S, Liu Z, Chen S, Liu L (2021) Amphichoterpenoids A-C, unprecedented picoline-derived meroterpenoids from the ascidian-derived fungus *Amphichorda felina* SYSU-MS7908. Chin Chem Lett 32:1893–1896
82. Jiang M, Guo H, Wu Q, Yuan S, Liu L (2022) Two new picoline-derived meroterpenoids with anti-acetylcholinesterase activity from ascidian-derived fungus *Amphichorda felina*. Molecules 27:5076
83. Jiang M, Wu Q, Guo H, Lu X, Chen S, Liu L, Chen S (2023) Shikimate-Derived meroterpenoids from the ascidian-derived fungus *amphichorda felina* SYSU-MS7908 and their anti-glioma activity. J Nat Prod 86:2651–2660
84. Jiang M, Guo H, Wu Q, Lu X, Zou Y, Fu Q, Chen S, Liu L, Peng B, Chen S (2023) Anti-inflammatory acetylenic meroterpenoids from the ascidian-derived fungus *Amphichorda felina* SYSU-MS7908. Bioorg Chem 139:106715
85. Zou J, Wu J, Ding L, Wang W, Liu Y, Feng Y, Lai Q, Lin W, Wang T, He S (2022) Guignardones Y–Z, antiviral meroterpenes from *Penicillium* sp. NBUF154 associated with a crella sponge from the marine mesophotic zone. Chem Biodivers 19:e202200475
86. Wang H-L, Li R, Li J, He J, Cao Z-Y, Kurtán T, Mándi A, Zheng G-L, Zhang W (2020) Alternarin A, a drimane meroterpenoid, suppresses neuronal excitability from the coral-associated fungi *Alternaria* sp. ZH-15. Org Lett 22:2995–2998
87. Wang H-L, Li R, Zhao M, Wang Z-Y, Tang H, Cao Z-Y, Zheng G-L, Zhang W (2023) A drimane meroterpenoid borate as a synchronous Ca^+ oscillation inhibitor from the coral-associated fungus *Alternaria* sp. ZH-15. J Nat Prod 86:429–433
88. Chen X-Y, Zeng Q, Chen Y-C, Zhong W-M, Xiang Y, Wang J-F, Shi X-F, Zhang W-M, Zhang S, Wang F-Z (2022) Chevalones H-M: six new α-pyrone meroterpenoids from the gorgonian coral-derived fungus *Aspergillus hiratsukae* SCSIO 7S2001. Mar Drugs 20:71
89. Cheng X, Liang X, Zheng Z-H, Zhang X-X, Lu X-H, Yao F-H, Qi S-H (2020) Penicimeroterpenoids A-C, meroterpenoids with rearrangement skeletons from the marine-derived fungus *Penicillium* sp. SCSIO 41512. Org Lett 22:6330–6333
90. Tang Y-Q, Liang X, Zou Q-H, Cui H, Luo L-X, Qi S-H (2024) HPPO-Derived meroterpenoids from the marine-derived fungus *Penicillium* sp. SCSIO 41691. J Nat Prod 87:1209–1216
91. Huang Z-H, Liang X, Li C-J, Gu Q, Ma X, Qi S-H (2021) Talaromynoids A-I, highly oxygenated meroterpenoids from the marine-derived fungus *Talaromyces purpureogenus* SCSIO 41517 and their lipid accumulation inhibitory activities. J Nat Prod 84:2727–2737
92. Dai L-T, Yang L, Guo J-C, Ma Q-Y, Xie Q-Y, Jiang L, Yu Z-F, Dai H-F, Zhao Y-X (2024) Anti-diabetic and anti-inflammatory indole diterpenes from the marine-derived fungus *Penicillium* sp. ZYX-Z-143. Bioorg Chem 145:107205
93. Feng Y-P, Wang H-K, Wu J-L, Shao P, Zhou W-L, Lai Q-L, Lin H-W, Naman CB, Wang T-T, He S (2022) Acremocholone, an anti-*Vibrio* steroid from the marine Mesophotic Zone *Ciocalypta* sponge-associated fungus *Acremonium* sp. NBUF150. Chem Biodivers 19:e202200028
94. Li J, Tao H, Lei X-x, Zhang H, Zhou X, Liu Y, Li Y, Yang B (2021) Arthriniumsteroids A-D, four new steroids from the soft coral-derived fungus *Simplicillium lanosoniveum* SCSIO41212. Steroids 171:108831
95. Gu B-B, Wu W, Jiao F-R, Jiao W-h, Li L, Sun F, Wang S-P, Yang F, Lin H-W (2019) Asperflotone, an 8(14→15)-abeo-ergostane from the sponge-derived fungus *Aspergillus flocculosus* 16D–1. J Org Chem 84:300–306
96. Gu B-B, Jiao F-R, Wu W, Liu L, Jiao W-H, Sun F, Wang S-P, Yang F, Lin H-W (2019) Ochrasperfloroid, an ochratoxin–ergosteroid heterodimer with inhibition of IL-6 and NO production from *Aspergillus flocculosus* 16D–1. RSC Adv 9:7251–7256
97. Jiao F-R, Gu B-B, Zhu H-R, Zhang Y, Liu K-C, Zhang W, Han H, Xu S-H, Lin H-W (2021) Asperfloketals A and B, the first two ergostanes with rearranged A and D Rings: from the sponge-associated *Aspergillus flocculosus* 16D–1. J Org Chem 86:10954–10961
98. Yu M, Li Y, Banakar SP, Liu L, Shao C, Li Z, Wang C (2019) New metabolites from the co-culture of marine-derived actinomycete *Streptomyces rochei* MB037 and fungus *Rhinocladiella similis* 35. Front Microbiol 10:915

99. Wu Z, Guo H, Wu Q, Jiang M, Chen J, Chen B, Li H, Liu L, Chen S (2023) Absolute configuration of cyclopropanes and the structural revision of pyrones from marine-derived fungus *Stagonospora* sp. SYSU-MS7888. Bioorg Chem 136:106542
100. Yuan S, Chen L, Wu Q, Jiang M, Guo H, Hu Z, Chen S, Liu L, Gao Z (2022) Genome mining of α-pyrone natural products from ascidian-derived fungus *Amphichorda felina* SYSU-MS7908. Mar Drugs 20:294
101. Hu Y, Zhao X, Song Y, Jiang J, Long T, Cong M, Miao Y, Liu Y, Yang Z, Zhu Y, Wang J (2024) Anti-inflammatory and neuroprotective α-pyrones from a marine-derived strain of the fungus *Arthrinium arundinis* and their heterologous expression. J Nat Prod 87:1975–1982
102. Huang L, Ding L, Li X, Wang N, Cui W, Wang X, Naman CB, Lazaro JEH, Yan X, He S (2019) New dihydroisocoumarin root growth inhibitors from the sponge-derived fungus *Aspergillus* sp. NBUF87. Front Microbiol 10
103. Lin S, Li J, Chen W, He J, Shi Y, Jin H, Ding L (2023) A new antibacterial dihydroisocoumarin from the marine sponge-associated fungus *Aspergillus* sp. Chem Nat Compd 59:246–248
104. Liu N, Peng S, Yang J, Cong Z, Lin X, Liao S, Yang B, Zhou X, Zhou X, Liu Y, Wang J (2019) Structurally diverse sesquiterpenoids and polyketides from a sponge-associated fungus *Aspergillus sydowii* SCSIO41301. Fitoterapia 135:27–32
105. Han X, Tang X, Luo X, Sun C, Liu K, Zhang Y, Li P, Li G (2020) Isolation and identification of three new sterigmatocystin derivatives from the fungus *Aspergillus versicolor* guided by molecular networking approach. Chem Biodivers 17:e2000208
106. Jiao W-H, Xu Q-H, Ge G-B, Shang R-Y, Zhu H-R, Liu H-Y, Cui J, Sun F, Lin H-W (2020) Flavipesides A-C, PKS-NRPS hybrids as pancreatic lipase inhibitors from a marine sponge symbiotic fungus *Aspergillus flavipes* 164013. Org Lett 22:1825–1829
107. Lu X, He J, Wu Y, Du N, Li X, Ju J, Hu Z, Umezawa K, Wang L (2020) Isolation and characterization of new anti-inflammatory and antioxidant components from deep marine-derived fungus *Myrothecium* sp. Bzo-l062. Mar Drugs 18:597
108. Wang J, Liang Z, Li K, Yang B, Liu Y, Fang W, Tang L, Zhou X (2020) Ene-yne hydroquinones from a marine-derived strain of the fungus *Pestalotiopsis neglecta* with effects on liver X receptor alpha. J Nat Prod 83:1258–1264
109. Wang B, Tontonoz P (2018) Liver X receptors in lipid signalling and membrane homeostasis. Nat Rev Endocrinol 14:452–463
110. Zhao L, Lei W, Deng C, Wu Z, Sun M, Jin Z, Song Y, Yang Z, Jiang S, Shen M, Yang Y (2021) The roles of liver X receptor α in inflammation and inflammation-associated diseases. J Cell Physiol 236:4807–4828
111. Liang Z, Gu T, Wang J, She J, Ye Y, Cao W, Luo X, Xiao J, Liu Y, Tang L, Zhou X (2021) Chromene and chromone derivatives as liver X receptors modulators from a marine-derived *Pestalotiopsis neglecta* fungus. Bioorg Chem 112:104927
112. Wang J, Li T, Wang P, Ding W, Xu J (2022) Tanzawaic acids from a deep-sea derived *Penicillium* species. J Nat Prod 85:1218–1228
113. Zeng Y, Wang Z, Chang W, Zhao W, Wang H, Chen H, Dai H, Lv F (2023) New azaphilones from the marine-derived fungus *Penicillium sclerotiorum* E23Y–1A with their anti-inflammatory and antitumor activities. Mar Drugs 21:75
114. Liu D-H, Sun Y-Z, Kurtán T, Mándi A, Tang H, Li J, Su L, Zhuang C-L, Liu Z-Y, Zhang W (2019) Osteoclastogenesis regulation metabolites from the coral-associated fungus *Pseudallescheria boydii* TW-1024-3. J Nat Prod 82:1274–1282
115. Wang D, Qu P, Zhou J, Wang Y, Wang L, Zhu W (2020) P-Terphenyl alcohols from a marine sponge-derived fungus, *Aspergillus candidus* OUCMDZ-1051. Mar Life Sci Technol 2:262–267
116. Guo Z, Abulaizi A, Huang L, Xiong Z, Zhang S, Liu T, Wang R (2022) Discovery of p-terphenyl metabolites as potential phosphodiesterase PDE4D inhibitors from the coral-associated fungus *Aspergillus* sp. ITBBc1. Mar Drugs 20:679
117. Peng G-Y, Kurtán T, Mándi A, He J, Cao Z-Y, Tang H, Mao S-C, Zhang W (2021) Neuronal modulators from the coral-associated fungi *Aspergillus candidus*. Mar Drugs 19:281

118. Yang B, Long J, Pang X, Lin X, Liao S, Wang J, Zhou X, Li Y, Liu Y (2021) Structurally diverse polyketides and phenylspirodrimanes from the soft coral-associated fungus *Stachybotrys chartarum* SCSIO41201. J Antibiot 74:190–198
119. Liu Z, Kou J, Li M, Chen Y, Li S, Li H, Zhang W (2023) Molecular networking-based discovery of new polyketides from the deep-sea-derived fungus *Talaromyces indigoticus* FS688. Tetrahedron 138:133410
120. Fang Z, Jiang X, Zhang Q, Zhang L, Zhang W, Yang C, Zhang H, Zhu Y, Zhang C (2020) S-Bridged thioether and structure-diversified angucyclinone derivatives from the South China Sea-Derived *Micromonospora echinospora* SCSIO 04089. J Nat Prod 83:3122–3130
121. Zhong W, Chen Y, Mai Z, Wei X, Wang J, Zeng Q, Chen X, Tian X, Zhang W, Wang F, Zhang S (2020) Euroticins A and B, two pairs of highly constructed salicylaldehyde derivative enantiomers from a marine-derived fungus *Eurotium* sp. SCSIO F452. J Org Chem 85:12754–12759
122. Zhong W, Chen Y, Wei X, Wang J, Zeng Q, Tian X, Zhang W, Wang F, Zhang S (2021) Euroticins C-E, three pairs of polycyclic salicylaldehyde derivative enantiomers from a marine-derived fungus *Eurotium* sp. SCSIO F452. Organic Chem Front 8:1466–1473
123. Zhong W-M, Wei X-Y, Chen Y-C, Zeng Q, Wang J-F, Shi X-F, Tian X-P, Zhang W-M, Wang F-Z, Zhang S (2021) Structurally diverse polycyclic salicylaldehyde derivative enantiomers from a marine-derived fungus *Eurotium* sp. SCSIO F452. Mar Drugs 19:543
124. Lei H, Bi X, Lin X, She J, Luo X, Niu H, Zhang D, Yang B (2021) Heterocornols from the sponge-derived fungus *Pestalotiopsis heterocornis* with anti-inflammatory activity. Mar Drugs 19:585
125. Wang J, Li K, Luo X, Wu Z, Gu T, Liao S, Lin X, Yang B, Liu Y, Fang W, Zhou X (2019) Sorbicillfurans A and B, two novel sorbicillinoid adducts from the fungus *Penicillium citrinum* SCSIO41402. Org Biomol Chem 17:8721–8725
126. Liang X, Huang Z-H, Shen W-B, Lu X-H, Zhang X-X, Ma X, Qi S-H (2021) Talaromyoxaones A and B: unusual oxaphenalenone spirolactones as phosphatase inhibitors from the marine-derived fungus *Talaromyces purpureogenus* SCSIO 41517. J Org Chem 86:12831–12839
127. Bryskier A (2005) Ansamycins. In: Antimicrobial agents, pp 906–924
128. Yi K-X, Xie Q-Y, Ma Q-Y, Yang L, Dai H-F, Zhao Y-X, Hao Y-E (2024) Diverse ansamycin derivatives from the marine-derived *Streptomyces* sp. ZYX-F-97 and their antibacterial activities. Fitoterapia 173:105814
129. Chen Y, Jin H, Xiong W, Fang Z, Sun W, Zhu Y, Zhang L, Zhang Y, Zhang W, Zhang C (2024) Discovery of aburatubolactams reveals biosynthetic logic for distinct 5/5-type polycyclic tetramate macrolactams. Org Lett 26:1677–1682
130. Liu M, He Y, Shen L, Anbari WHA, Li H, Wang J, Qi C, Hu Z, Zhang Y (2019) Asperteramide A, an unusual N-phenyl-carbamic acid methyl ester trimer isolated from the coral-derived fungus *Aspergillus terreus*. Eur J Org Chem 2019:2928–2932
131. Xu L, Guo F-W, Zhang X-Q, Zhou T-Y, Wang C-J, Wei M-Y, Gu Y-C, Wang C-Y, Shao C-L (2022) Discovery, total syntheses and potent anti-inflammatory activity of pyrrolinone-fused benzoazepine alkaloids Asperazepanones A and B from *Aspergillus candidus*. Commun Chem 5:80
132. Cheng Y, Chen N, Li J, Su J-C, Yang J, Zhang C-X, Lin H-W, Zhou Y (2021) Antimicrobial chlorinated carbazole alkaloids from the sponge-associated Actinomycete *Streptomyces diacarni* LHW51701. Chin J Chem 39:1188–1192
133. Zhang D, Xu J, Qin Q, An F, Wang S, Li L, Lin H (2022) Marinacarboline glucuronide, a new member of β-carboline alkaloids from sponge-derived actinomycete *Actinoalloteichus cyanogriseus* LHW52806. J Antibiot 75:523–525
134. Ding L, Li W, Zhong X, Feng F, Xin Y, Yan X, Lazaro JEH, Zhang B, Shi Y, Yang G-J, He S (2023) Bresmycins A and B, potent anti-breast cancer indolocarbazole alkaloids from the sponge-associated *Streptomyces* sp. NBU3142. J Mol Struct 1290:135809
135. Guo Y-W, Liu X-J, Yuan J, Li H-J, Mahmud T, Hong M-J, Yu J-C, Lan W-J (2020) L-Tryptophan induces a marine-derived *Fusarium* sp. to produce indole alkaloids with activity against the zika virus. J Nat Prod 83:3372–3380

136. Zhang Y, Li M, Zhang Q, Wang Z, Li X, Bao J, Zhang H (2021) Arthpyrone L, a new pyridone alkaloid from a Deep-Sea *Arthrinium* sp., inhibits proliferation of MG63 osteosarcoma cells by inducing G0/G1 arrest and apoptosis. Chem Biodivers 18:e2000639
137. Hu Y, Ma S, Pang X, Cong M, Liu Q, Han F, Wang J, Feng W, Liu Y, Wang J (2023) Cytotoxic pyridine alkaloids from a marine-derived fungus *Arthrinium arundinis* exhibiting apoptosis-inducing activities against small cell lung cancer. Phytochemistry 213:113765
138. Sun C, Ge X, Mudassir S, Zhou L, Yu G, Che Q, Zhang G, Peng J, Gu Q, Zhu T, Li D (2019) New glutamine-containing azaphilone alkaloids from deep-sea-derived Fungus *Chaetomium globosum* HDN151398. Mar Drugs 17:253
139. Yang J, Song Y, Tang M-C, Li M, Deng J, Wong N-K, Ju J (2021) Genome-directed discovery of tetrahydroisoquinolines from deep-sea derived *Streptomyces niveus* SCSIO 3406. J Org Chem 86:11107–11116
140. Yao F-H, Liang X, Lu X-H, Cheng X, Luo L-X, Qi S-H (2022) Pyrrospirones K-Q, decahydrofluorene-class alkaloids from the marine-derived fungus *Penicillium* sp. SCSIO 41512. J Nat Prod 85:2071–2081
141. Zhang X, Chen S, Zhang L, Zhang Q, Zhang W, Chen Y, Zhang W, Zhang H, Zhang C (2021) Dassonmycins A and B, Polycyclic thioalkaloids from a marine sponge-derived *Nocardiopsis dassonvillei* SCSIO 40065. Org Lett 23:2858–2862
142. Murgatroyd TC, Söhnel T, Sperry J (2024) Biomimetic synthesis of the marine-derived thioalkaloids dassonmycins A and B. Org Lett 26:10041–10045
143. Liang X, Huang Z-H, Shen W-B, Lu X-H, Zhang X-X, Ma X, Qi S-H (2024) Prenylated indole diketopiperazine alkaloids as phosphatase inhibitors from the marine-derived fungus *Talaromyces purpureogenus*. Phytochemistry 223:114119
144. Zhong W, Wang J, Wei X, Chen Y, Fu T, Xiang Y, Huang X, Tian X, Xiao Z, Zhang W, Zhang S, Long L, Wang F (2018) Variecolortins A-C, three pairs of spirocyclic diketopiperazine enantiomers from the marine-derived fungus *Eurotium* sp. SCSIO F452. Org Lett 20:4593–4596
145. Zhong W, Wang J, Wei X, Fu T, Chen Y, Zeng Q, Huang Z, Huang X, Zhang W, Zhang S, Long L, Wang F (2019) Three pairs of new spirocyclic alkaloid enantiomers from the marine-derived fungus *Eurotium* sp. SCSIO F452. Front Chem 7
146. Hong Q, Guo M-M, Yang J, Wei X, Liao L, Xin X-J, Zhang D, An F-L (2023) Four previously undescribed diketopiperazines from marine fungus *Aspergillus puniceus* FAHY0085 and their effects on liver X receptor α. Phytochemistry 214:113816
147. Dai Y, Li K, She J, Zeng Y, Wang H, Liao S, Lin X, Yang B, Wang J, Tao H, Dai H, Zhou X, Liu Y (2020) Lipopeptide epimers and a phthalide glycerol ether with AChE inhibitory activities from the marine-derived fungus *Cochliobolus lunatus* SCSIO41401. Mar Drugs 18:547
148. Li T, Xi C, Yu Y, Wang N, Wang X, Iwasaki A, Fang F, Ding L, Li S, Zhang W, Yuan Y, Wang T, Yan X, He S, Cao Z, Naman CB (2022) Targeted discovery of amantamide B, an Ion channel modulating nonapeptide from a South China Sea *Oscillatoria* Cyanobacterium. J Nat Prod 85:493–500
149. Pan N, Li Z-C, Li Z-H, Chen S-H, Jiang M-H, Yang H-Y, Liu Y-S, Hu R, Zeng Y-W, Dai L-H, Liu L, Wang G-L (2022) Antiplatelet and antithrombotic effects of Isaridin E isolated from the marine-derived fungus via downregulating the PI3K/Akt signaling pathway. Mar Drugs 20:23
150. Jiang M, Chen S, Lu X, Guo H, Chen S, Yin X, Li H, Dai G, Liu L (2023) Integrating genomics and metabolomics for the targeted discovery of new cyclopeptides with antifungal activity from a marine-derived fungus *Beauveria felina*. J Agric Food Chem 71:9782–9795
151. Zhao S, Xia Y, Liu H, Cui T, Fu P, Zhu W (2022) A Cyclohexapeptide and its rare glycosides from marine sponge-derived *Streptomyces* sp. OUCMDZ-4539. Org Lett 24:6750–6754
152. Zhao S, Yan P, Liu P, Liu H, Li N, Fu P, Zhu W (2024) Pyridapeptides F–I, cyclohexapeptides from marine sponge-derived *Streptomyces* sp. OUCMDZ-4539. Chinese Chem Lett 35:108950
153. Luo M, Zang R, Wang X, Chen Z, Song X, Ju J, Huang H (2019) Natural hydroxamate-containing siderophore acremonpeptides A-D and an aluminum complex of acremonpeptide D from the marine-derived *Acremonium persicinum* SCSIO 115. J Nat Prod 82:2594–2600

154. Hou X-M, Liang T-M, Guo Z-Y, Wang C-Y, Shao C-L (2019) Discovery, absolute assignments, and total synthesis of asperversiamides A-C and their potent activity against *Mycobacterium marinum*. Chem Commun 55:1104–1107
155. Chao R, Hou X-M, Xu W-F, Hai Y, Wei M-Y, Wang C-Y, Gu Y-C, Shao C-L (2021) Targeted isolation of asperheptatides from a coral-derived fungus using LC-MS/MS-based molecular networking and antitubercular activities of modified cinnamate derivatives. J Nat Prod 84:11–19
156. Han Y-Q, Zhang Q, Xu W-F, Hai Y, Chao R, Wang C-F, Hou X-M, Wei M-Y, Gu Y-C, Wang C-Y, Shao C-L (2023) Targeted isolation of antitubercular cycloheptapeptides and an unusual pyrroloindoline-containing new analog, asperpyrroindotide A, using LC–MS/MS-based molecular networking. Mar Life Sci Technol 5:85–93
157. Ma J, Huang H, Xie Y, Liu Z, Zhao J, Zhang C, Jia Y, Zhang Y, Zhang H, Zhang T, Ju J (2017) Biosynthesis of ilamycins featuring unusual building blocks and engineered production of enhanced anti-tuberculosis agents. Nat Commun 8:391
158. Sun C, Yang Z, Zhang C, Liu Z, He J, Liu Q, Zhang T, Ju J, Ma J (2019) Genome mining of *Streptomyces atratus* SCSIO ZH16: Discovery of atratumycin and identification of its biosynthetic gene cluster. Org Lett 21:1453–1457
159. Sun C, Liu Z, Zhu X, Fan Z, Huang X, Wu Q, Zheng X, Qin X, Zhang T, Zhang H, Ju J, Ma J (2020) Antitubercular Ilamycins from marine-derived *Streptomyces atratus* SCSIO ZH16 ΔilaR. J Nat Prod 83:1646–1657
160. Zhu Y, Zheng G, Xin X, Ma J, Ju J, An F (2022) Combinatorial strategies for production improvement of anti-tuberculosis antibiotics ilamycins E1/E2 from deep sea-derived *Streptomyces atratus* SCSIO ZH16 ΔilaR. Bioresour Bioprocess 9:111
161. Tang W-Z, Liu J-T, Hu Q, He R-J, Guan X-Q, Ge G-B, Han H, Yang F, Lin H-W (2020) Pancreatic lipase inhibitory cyclohexapeptides from the marine sponge-derived fungus *Aspergillus* sp. 151304. J Nat Prod 83:2287–2293
162. Zhang S-S, Li L, Wu Y, Wu Z-M, Kong C, Hong L-L, Zhang S, Lin X-L, Lin H-W, Wang S-P (2023) LC-MS-guided isolation of cyanogripeptides A-C, cyclolipopeptides with β-methylleucine residues, from an *Actinoalloteichus cyanogriseus* LHW52806. J Nat Prod 86:1708–1714
163. Wu J-S, Yao G-S, Shi X-H, Rehman SU, Xu Y, Fu X-M, Zhang X-L, Liu Y, Wang C-Y (2020) Epigenetic agents trigger the production of bioactive nucleoside derivatives and bisabolane sesquiterpenes from the marine-derived fungus *Aspergillus versicolor*. Front Microbiol 11:85
164. Zou R, Chen B, Sun J, Guo Y-W, Xu B (2023) Recent advances of activation techniques-based discovery of new compounds from marine fungi. Fitoterapia 167:105503

Chapter 10
Impact of Papua New Guinean Marine Biodiversity on the Discovery and Development of New Pharmaceuticals

Lik Tong Tan

1 Introduction

Papua New Guinea (PNG), part of maritime Southeast Asia, is situated within the Coral Triangle, encompassing the oceans between the Philippines to the north and Indonesia to the west. The waters of PNG represent one of the few megadiverse marine ecosystems globally, with many areas still unexplored for natural-product discovery [1]. PNG has an estimated total coastline of 17,110 km. Its coral reefs and associated marine habitats host approximately 2800 species of marine fish, accounting for about 10% of the global total. Most reef types in Papua New Guinea are either fringing or barrier reefs, covering an estimated area of 40,000 km^2. Additionally, the country boasts some of the largest unpolluted tropical freshwater systems in the Asia–Pacific region. Mangrove swamps cover 51.6 million hectares, about 2% of the forest estate. Papua New Guinea is home to 33 species of mangrove trees, representing the highest mangrove diversity in the world. Studies found that the extent of mangroves in the Gulf of Papua has remained relatively stable for nearly 40 years, with expansion in some areas offset by regression in others [2].

Papua New Guinea's marine biodiversity is largely due to its strategic location. Positioned at the eastern edge of the extensive Southeast Asian archipelago, the country is influenced by the South Equatorial Current and the powerful Indonesian Throughflow. The South Equatorial Current flows westward and splits at the Papua New Guinea mainland, forming the East Australian Current to the south [3]. During the northwest monsoon season, which lasts from December to March each year, Papua New Guinea is influenced by the Indonesian Throughflow. This current brings

L. T. Tan (✉)
Natural Sciences and Science Education, National Institute of Education, Nanyang Technological University, Singapore, Singapore
e-mail: liktong.tan@nie.edu.sg

L. T. Tan and N. Hanif (eds.), *Marine Natural Products Research in South-East Asia*,
https://doi.org/10.1007/978-3-032-01098-8_10

nutrient-rich waters along the northeast coast of the main island of New Guinea. Coastal air temperatures remain relatively stable throughout the year, ranging from 25 to 30 °C. The region experiences a monsoonal climate, characterized by a high average annual rainfall of approximately 3000 mm [3]. These vast water currents, combined with abundant tropical sunlight and warm regional temperatures, create the ideal conditions that have led to Papua New Guinea's remarkable marine diversity.

Research on marine organisms for drug discovery in Papua New Guinea (PNG) began intensifying in the 1980s, with several research groups conducting Scuba-diving expeditions to find new and unexplored species. Over the last 30 years, efforts to discover natural products from PNG's marine biota have escalated, largely driven by the establishment of the PNG International Cooperative Biodiversity Group (ICBG) program coordinated by the Fogarty International Center of the U.S. NIH [4]. Led by Principal Investigator Lou Barrows from Utah, the ICBG aimed to enhance scientific activities, conserve biodiversity and improve human health in PNG [4]. Similar initiative to explore PNG marine biodiversity for anticancer drugs was provided by the National Cooperative Natural Products Drug Discovery Group (NCNPDDG) through the NCI NCDDG Program in 1995 [5]. As a result of these programs, several research groups, including Ireland (University of Utah), Andersen (University of British Columbia), Crews (University of California at Santa Cruz) and Gerwick (Oregon State University/Scripps Institution of Oceanography), in collaboration with Matainaho (University of Papua New Guinea), have discovered many new bioactive marine natural products, many of which are detailed in this review. Marine sponges, known for their relative abundance and potential to produce diverse natural products, have been a primary focus of intense investigations by these researchers. Moreover, molecular networking has quantitatively demonstrated that marine cyanobacteria and algae from PNG harbors a wealth of unique natural products, surpassing those found in other geographical locations [6]. For instance, around 4% of all MS/MS features contributing to the overall diversity originate from Panama-Bocas del Toro, while approximately 7–8% come from both Panama-Portobelo and Palmyra Atoll. In contrast, Papua New Guinea accounts for 20% of these features. More importantly, the initial research efforts on the discoveries of marine natural products derived from Papua New Guinean marine organisms led to several drug candidates that have progressed to clinical trials for the treatment of inflammation and cancer [7].

In this review, the focus is on new natural products discovered from two most studied groups of marine organisms from PNG, including marine sponges and filamentous cyanobacteria, with emphasis on the bioactive metabolites of diverse structural types, including polyketides and linear/cyclic peptides. Additionally, an overview on the biodiversity of marine sponges and filamentous cyanobacteria at PNG will be presented. Most of the sample collections for natural products investigation have been made from the waters around Loloata, Milne Bay Province, Madang Province and the islands of New Britain and New Ireland. These curated molecules discussed in this chapter are of particular interest due to their structural novelty, and/or biological activities as well as impact on drug discovery and development.

2 Biodiversity of Papua New Guinean Sponges and Marine Cyanobacteria

Only a few studies have documented the Porifera species in Papua New Guinea [8]. Kelly-Borges and Valentine (1995) reported 184 species ranging from shallow to deep waters of unspecified depth in PNG [9]. However, this list is derived from collections in limited areas and is not considered geographically representative. A recent study by Gonzalez-Murcia and co-workers found that the nearshore and island reefs of Kimbe Bay, Papua New Guinea are known for their high sponge abundance and diversity [10]. On the composition of sponge assemblages on a coral reef in Kimbe Bay, their study found sponges ranked 3rd, with an average of 13.1% cover, including 63 recognizable species. Based on morphology, there were 38 encrusting, 21 erect and four massive sponge species, with the encrusting sponges *Lamellodysidea* cf. *chlorea* and *Dysidea* sp. showing the highest cover.

During the US National Cancer Institute's natural product screening program conducted over 10 years from 1994 to 2014, more than 500 sponge species were recorded in Papua New Guinea [11]. In addition, about 433 species were found exclusively in shallow waters, making up 86% of the total species. At mesophotic depths (> 45 m), 73 (14% of the total) sponge species were recorded, with 51 (10%) species occurring in both shallow and mesophotic depths, and 22 (4%) species found only at mesophotic depths. Sponge collections from the neighboring central Solomon Islands showed somewhat lower species diversity compared to PNG, with several species common in both shallow and mesophotic depths in PNG absent in the Solomon Islands. Current findings suggest a trend of decreasing sponge diversity toward the margins of the Coral Triangle [11].

Another Papua New Guinean marine-derived taxonomic group that has been studied for its production of bioactive natural products are the filamentous cyanobacteria. Marine cyanobacteria are a prolific source of unique bioactive molecules, and it is not surprising that numerous structurally unique cyanobacterial compounds have been reported from PNG. These marine microbes thrive in diverse habitats within tropical marine ecosystems, ranging from coastal areas to the open ocean. Their highest biodiversity is found in littoral zones, where they form intertidal and infralittoral mats, establish symbiotic relationships with sponges and ascidians, and live as endoliths in carbonate substrates [12]. A series of detailed floristic surveys of marine cyanobacteria in PNG was initiated by Hoffman's research group in the mid-1980s till 1997 [12–20]. These early surveys resulted in the uncovering of at least 77 marine cyanobacterial species belonging to four taxonomic orders, such as Chroococcales, Oscillatoriales, Nostocales and Stigonematales. In addition, at least 25 new taxa were also described, including the chemically rich *L. bouillonii*, originally collected from Madang Province (Hansa Bay, Laing Island) and Central province (Barrier Reef), which proved to be a rich source of novel macrolides (e.g., madangolide and laingolides) and modified peptides (e.g., alotamide A, lyngbyapeptin A and apratoxins) [15, 21–26]. Based on additional phylogenetic analysis, the taxonomy of *L. bouillonii* has been placed in a new genus *Moorea* by Engene and co-workers [27]. However,

Moorea was subsequently renamed to *Moorena* as the name *Moorea* has been used in botanical code previously [28].

3 Porifera (Sponges)

3.1 Terpenoids and Steroids

3.1.1 Terpenoids as Lipoxygenase Inhibitors

Several sponge-derived terpenoids having potent lipoxygenase inhibitory activities have been reported by Crew's research group at University of California Santa Cruz. A human 15-lipoxygenase (15-hLO) assay was utilized by them to identify new bioactive compounds derived from marine sponges. Among the extracts tested, one potent extract originated from a Papua New Guinea sponge, *Suberea* sp., showed an IC_{50} of 0.1 μg/mL. The *Suberea* sp. sponge specimen was collected by SCUBA at a depth of 20 m in the Madang region of Papua New Guinea. Subsequent bioassay-guided approach using the 15-hLO assay led to the isolation of two terpenoids having strong 15-hLO inhibitory activities, including jaspaquinol (**1**, $IC_{50} = 0.3$ μM) and (–)-jaspic acid (**2a**, $IC_{50} = 1.4$ μM) (Fig. 1) [29]. Two other new compounds, (+)-subersin (**3**) and (–)-subersic acid (**4**) (Fig. 1) were also isolated from *Suberea* sp, though they exhibited significantly lower potency. In reevaluation of jaspaquinol and jaspic acid against a panel of hLOs in the presence of detergent, Triton X100, they showed IC_{50} values of 19 μM (15-hLO-1) and 0.45 μM (5-hLO) for the former compound and 7.9 μM (12-hLO), 4.7 μM (15-hLO-1) and 14 μM (5-hLO) for the latter molecule [30]. The major compounds, jaspaquinol (**1**) and (–)-jaspic acid (**2a**), along with jasplakinolide, were previously isolated by Crew's group in 1996 from a different sponge, *Jaspis* cf. *johnstoni*, also obtained from Madang, PNG [31]. No biological activities were reported for the original isolation of jaspaquinol and jaspic acid in 1996.

A new meroditerpene, (+)-isojaspic acid (**2b**), was subsequently isolated from a marine sponge *Cacospongia* sp. collected from Milne Bay, Papua New Guinea [32]. This molecule showed significant antimicrobial activity against *S. epidermidis* with MIC of 2.5 μg/mL. Jaspaquinol and (–)-jaspic acid were also tested and showed significant antimicrobial activity against S. epidermidis with MIC values of 5.0 and 2.5 μg/mL, respectively. However, no activity against *E. coli* and *B. subtilus* were detected (MIC > 400 μg/mL). In addition, evaluation of (+)-isojaspic acid against a panel of hLOs in the presence of detergent, Triton X100, showed IC_{50} values of 31 μM (12-hLO), 16 μM (15-hLO-1) and 18 μM (5-hLO) [30].

A screening for selective human 15-lipoxygenase inhibitors from an extract of the Papua New Guinean sponge *Psammocinia* sp. (order Dictyoceratida, family Irciniidae), collected from Madang region, afforded chromarols A (**5**) to E (**9**) (Fig. 2) [33]. The IC_{50} values for chromarols A to D were found to be 0.6, 4.0, 0.7 and 1.1 μM,

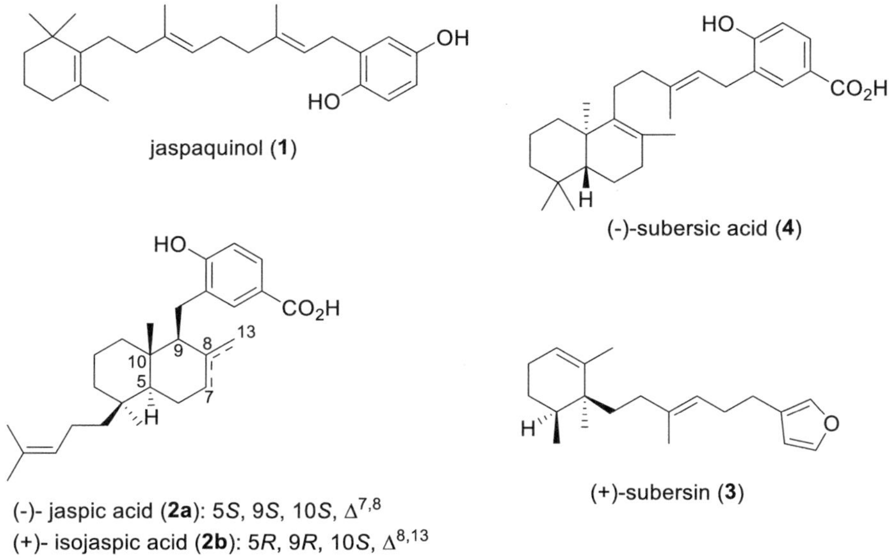

Fig. 1 Molecules reported from PNG *Suberea* sp. and *Cacospongia* sp.

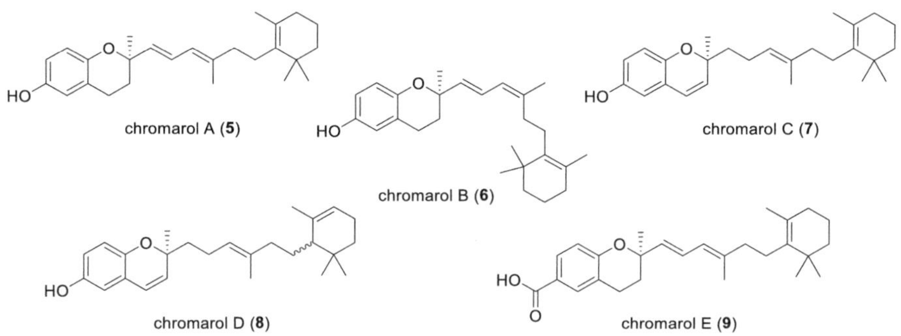

Fig. 2 Chromarols from Papua New Guinean sponge *Psammocinia* sp.

respectively. The 6-hydroxychromane moiety present in the chromarols is crucial for the selective redox inhibition of 15-hLO but not human platelet 12-lipoxygenase (12-hLO). In addition, the oxidized form of 6-hydroxychromane can be reduced by ascorbate, indicating a potential regeneration pathway for these inhibitors in the body. This pharmacophore represents a promising moiety for developing a unique class of recyclable 15-hLO redox inhibitors for treating atherosclerosis. Moreover, the biosynthesis of the cyclohexene ring system in the side chain of chromarols is presumably derived from an acid-catalyzed cyclization.

From a 1999 collection of *Hyrtios* sponges obtained at Gulf of Papua, Papua New Guinea, new compound, hyrtenone A (**10**), a member of the puupehenone family, along with known puupehenone (**11**) and chloropuupehenone (**12**) (Fig. 3) [34]. The lipoxygenase inhibitory screening was conducted on these molecules as well as ten additional sponge-derived compounds maintained at Crew's research lab. Based on the lipoxygenase screening panel, consisting of 12-human lipoxygenase (12-hLO), 15-human lipoxygenase (15-hLO) and soybean 15-lipoxygenase (15-SLO), it was revealed that puupehenone (**11**) demonstrated potency comparable to or exceeding that of the known redox inhibitor nordihydroguarierate acid. However, none of the terpenoid inhibitors matched the selectivity ratio of nordihydroguarierate acid. Several potent non-redox inhibitors were also identified, with one in particular, 15,20-dimethoxypuupehenol (**13**) (Fig. 3), showing notable selectivity.

Enzyme screening of crude extracts from a 2005 Papua New Guinea (at Pocklington Reef) collection of another sample of *Hyrtios* sponges led to the prioritization of the MeOH extract for further investigation [30]. This extract contained puupehenone and four related puupehenone analogues, including

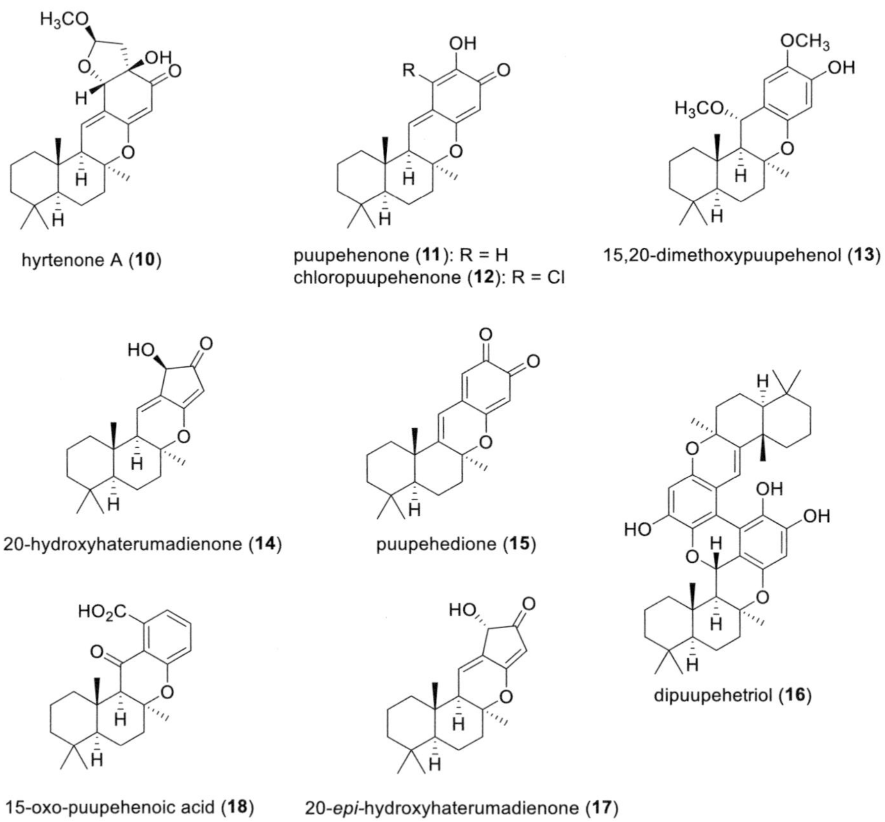

Fig. 3 Terpenoids from the Papua New Guinean sponge *Hyrtios* sp. and related molecules

15,20-dimethoxypuupehenol (**13**), 20-hydroxyhaterumadienone (**14**), puupehedione (**15**) and dipuupehetriol (**16**), along with a new diastereomer, 20-*epi*-hydroxyhaterumadienone (**17**), and a novel analogue, 15-oxo-puupehenoic acid (**18**) (Fig. 3). These compounds, along with repository natural products, including jaspaquinol, maintained at Crew's research lab, were evaluated for their activity against three human lipoxygenase isozymes: 5-, 12- and 15-lipoxygenases. In the human 5-lipoxygenase (5-hLO) assay, both puupehenone and jaspaquinol demonstrated significant potency and selectivity. Jaspaquinol showed exceptional selectivity for 5-hLO over 12-hLO (with a selectivity ratio greater than 111-fold) and strong selectivity over 15-hLO (selectivity ratio of 42). Puupehenone exhibited a high selectivity for 5-hLO over 12-hLO (selectivity ratio of 32), though it displayed lower selectivity against 15-hLO (selectivity ratio of 9).

3.1.2 Peroxiterpenes

Several sponge *Diacarnus* species from Papua New Guinea have been investigated for bioactive peroxiterpenes. Three samples of the marine sponge *Diacarnus* cf. *spinopoculum*, collected from PNG, have yielded a series of norterpenes, including five new compounds (**19**–**23**), two new *ent*-compounds [(–)-**24a** and (+)-**24b**] as well as two known compounds, including **25** (Fig. 4) [35]. Further confirmation that nuapapuin B (**19a**) was the exocylic double-bond isomer of known nuapapuin A was achieved via chemical synthesis involving carboxylic acid methylation and double-bond migration. All compounds were assessed for their cytotoxic properties using a soft agar assay system and the NCI's 60-cell line assays. Generally, terpenes lacking peroxide functionality showed no activity while norsesterterpene peroxides were less selective as cytotoxins compared to norditerpene peroxide analogues. The norsesterterpene peroxide **25**, in particular, demonstrated a GI_{50} at submicromolar concentrations across most cell types tested, including leukemic, non-small cell lung, colon, melanoma, ovarian, renal and breast cancer cells. Conversely, the norditerpene peroxides **19a** and **19b** also gave submicromolar GI_{50}s, but selectively, excluding leukemic cell lines. Subsequently, nuapapuin A methyl ester and nuapapuin B were selected for further in vivo evaluation using mouse model but no antitumor activity were observed.

As part of an ongoing research program aimed at discovering natural products that inhibit hypoxia-activated tumor survival pathways, researchers found that the organic extract of the Papua New Guinea sponge *Diacarnus levii* suppressed hypoxia-induced HIF-1 activation and hypoxic tumor cell survival. Bioassay-guided isolation of the extract afforded four new norsesterterpene peroxides, diacarnoxides A (**26**)–D (**29**) (Fig. 4) [36]. Against a panel of human prostate (DU145 and PC-3) and breast (MCF-7, MDA-MB-231 and T47D) carcinoma cell lines, the most active diacarnoxide B, inhibited cell proliferation and viability under both normoxic and hypoxic conditions, with IC_{50} values ranging from 5.3 to 16.7 μM. Using the same concentration of 12.7 μM, diacarnoxides A, C and D showed three times lower inhibition of both normoxic and hypoxic conditions compared to the percentage inhibition induced by

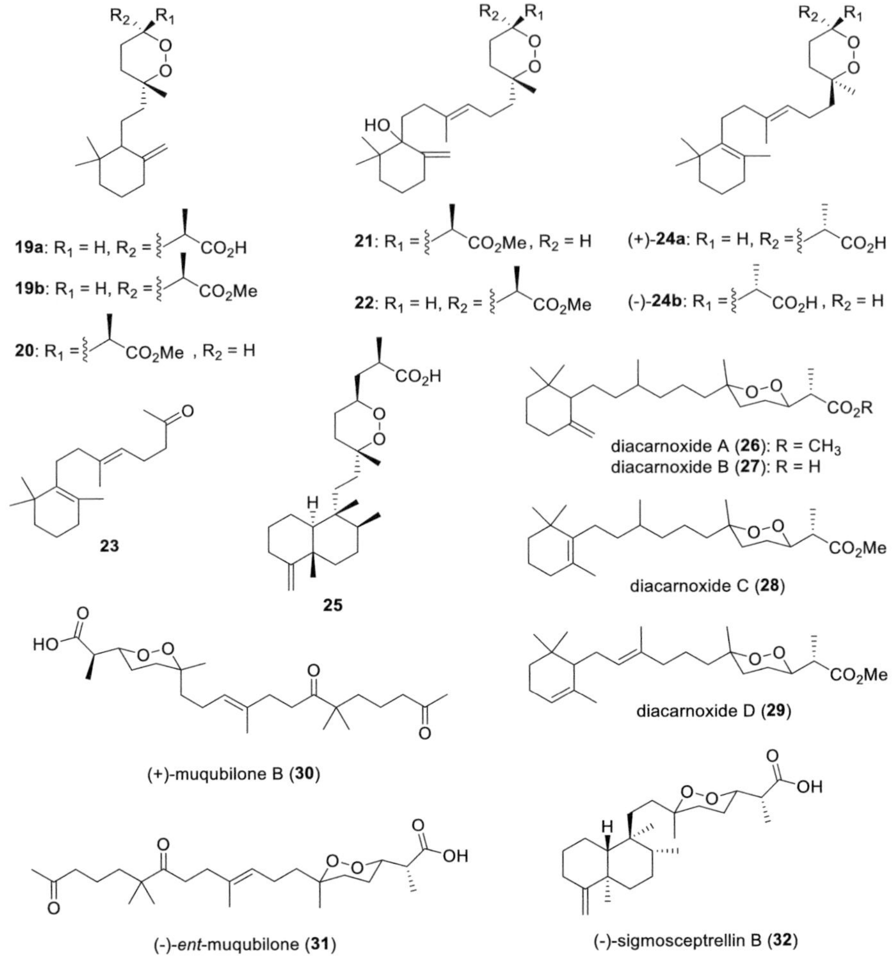

Fig. 4 Molecules from various PNG-derived sponge *Diacarnus* sp.

diacarnoxide B. In addition, at lower concentrations, diacarnoxide B exhibited an unusual enhanced growth and viability inhibitory activity in MCF-7 and MDA-MB-231 cells under hypoxic conditions.

A collaborative lead discovery project, utilizing a repository of extracts and natural product compounds to screen for growth inhibition against *Trypanosoma brucei*, led to the discovery of peroxiterpene inhibitors of the parasite from the sponge *Diacarnus bismarckensis* collected near Sanaroa, Papua New Guinea. These inhibitors include two new peroxiterpenes: (+)-muqubilone B (**30**) and (–)-*ent*-muqubilone (**31**) as well as five known compounds, such as (+)-sigmosceptrellin A, (+)-sigmosceptrellin A methyl ester, (–)-sigmosceptrellin B (**32**), (+)-*epi*-muqubillin A and (–)-*epi*-nuapapuin B methyl ester (Fig. 4) [37]. The isolated peroxiterpenes exhibited anti-parasitic activities with IC_{50} values ranging from 0.2 to 2 μg/mL. It was suggested that the structural features of (–)-sigmosceptrellin B (**32**) (*T. brucei* IC_{50} of 0.2 μg/mL) would be a good starting point for further SAR modifications, as it demonstrated similar potencies with a current HAT therapeutic, pentamidine (*T. b. rhodesiense* $IC_{50} = 0.4$ μg/mL).

3.1.3 Thorectidiols

Thorectidiols, a family of symmetrical and unsymmetrical dimeric biphenyl meroterpenoid stereoisomers, were isolated from the marine sponge, *Dactylospongia elegans*, collected from reefs near Keviang, Papua New Guinea [38]. A member of this family, thorectidiol A (**33**) (Fig. 5), has been isolated in its natural form, with its structure determined through extensive NMR, MS and electronic circular dichroism (ECD) data analysis. Subsequent acetylation of the sponge extract, obtained from a large-scale extraction using 40 g of sponge material, facilitated the isolation of additional thorectidiol diacetate stereoisomers, thorectidiol A diacetate (**34**), thorectidiol B diacetate (**35**), as well as the racemic monomer thorectidol acetate (**36**) (Fig. 5). It has been proposed that the formation of thorectidiol A involves phenol coupling between two molecules of the co-occurring racemate of thorectidol (**37**) (Fig. 5), which can potentially yield six thorectidiol stereoisomers. Racemic thorectidiol A demonstrated selective inhibition of the SARS-CoV-2 spike receptor binding domain (RBD) interaction with the host angiotensin-converting enzyme II (ACE2) receptor, with an IC_{50} of 1.0 μM. SARS-CoV-2 enters host cells primarily by the interaction between its spike protein's receptor binding domain (RBD) and the host's angiotensin-converting enzyme II (ACE2) receptor [39]. Hence, small molecules, such as thorectidiol A, that can selectively disrupt this RBD-ACE2 binding could serve as promising structural leads for developing antiviral agents to prevent SARS-CoV-2 infection.

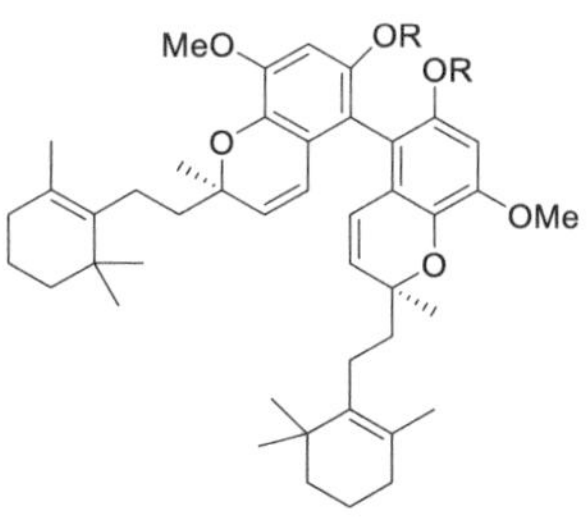

thorectidiol A (**33**): R = H
thorectidiol A diacetate (**34**): R = Ac

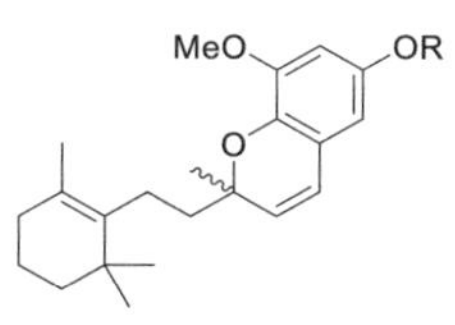

37: R = H
thorectidol acetate (**36**): R = Ac

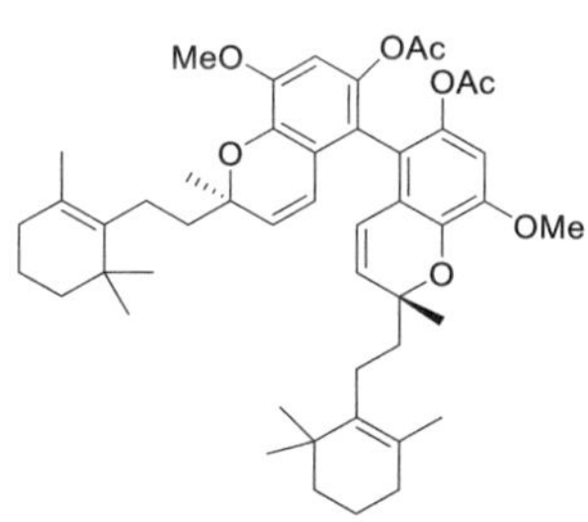

thorectidiol B diacetate (**35**)

Fig. 5 Thorectidiols from *Dactylospongia elegans*

3.1.4 Alotaketals

In 2009, Kieffer and Andersen isolated alotaketals A (**38**) and B (**39**) (Fig. 6) from extracts of the marine sponge *Hamigera* sp. in Milne Bay, Papua New Guinea [40]. Their congeners, alotaketals C (**40**) to E (**42**) (Fig. 6), were later isolated in 2013 and 2016 from the sponge *Phorbas* sp. collected in British Columbia [41, 42]. These molecules belong to a new class of sesterterpenoid natural products featuring unique didehydro[6.6]spiroketal structures. Due to their unique structural characteristics and intriguing biological activities, several total syntheses and synthetic strategies for these natural products have been reported [43]. Recently, a gold-catalyzed tandem spiroketalization of epoxyalkynes, along with the rearrangement of epoxides into allylic alcohols, was developed by Lee and co-workers for the synthesis of functionalized spiroketals [44]. This novel synthetic approach for creating unsaturated spiroketals facilitated the efficient total synthesis of alotaketal A, starting from carvone through an epoxyalkyne intermediate.

Alotaketals A and B were reported to activate the cAMP cell signaling pathway in HEK293T (human embryonic kidney) cells transfected with the pHTS-CRE plasmid, without hormone binding [40]. These sesterterpenoids demonstrated EC_{50} values of 18 nM and 240 nM, respectively, making them more potent than the commonly used

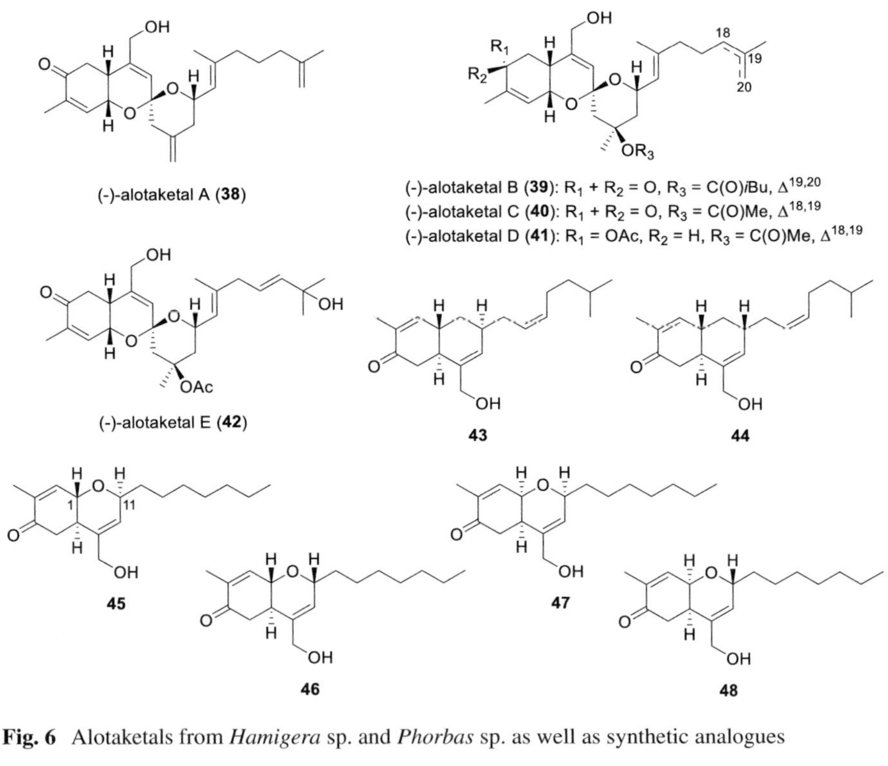

Fig. 6 Alotaketals from *Hamigera* sp. and *Phorbas* sp. as well as synthetic analogues

forskolin, which has an EC_{50} of 3 μM. Further research by Zhang and Yang confirmed that alotaketal A can enhance protein kinase A (PKA) activity by increasing intracellular levels of cAMP [45, 46]. A structure–activity relationship (SAR) analysis indicated that the functionality at the C22 position is essential for the cAMP-activating pharmacophore of the alotaketals [42].

In another development by Irie and co-workers, alotaketals were predicted to have a protein kinase C (PKC) binding affinity based on screening of PubChem compounds using machine learning [47]. This led to the synthesizing of new PKC ligands **43** and **44** (Fig. 6) based on the alotaketals, with **43** displaying isozyme selectivity for cPKCs. The PKC family comprises a large group of related serine/threonine kinases involved in cellular signal transduction. In humans, this family consists of 11 isozymes classified based on their cofactor requirements and PKC isozymes are known to play key roles in cell proliferation, differentiation and apoptosis. Given that most cancer-associated PKC mutations result in loss-of-function, ligands that activate PKCs present promising potential as drug leads [48]. More recently, Irie and co-workers designed four diastereomers (**45** − **48**) (Fig. 6) as simplified analogs of alotaketals using Rh-catalyzed O − H insertion and ring-closing metathesis as key reactions [49]. Evaluation of their PKC binding affinities revealed that the stereochemistry at C11 of the simplified analogs, which influences the orientation of the alkyl side chain, was crucial for binding affinity. In addition, the stereochemistry at C1 in these analogs played a role in determining PKC isozyme selectivity.

The proposed biogenesis of the tricyclic spiroketal skeleton in alotaketals begins with the cyclization of geranylfarnesyl pyrophosphate [42]. The dissociation of pyrophosphate generates an allylic cation, which is then attacked by a nearby olefin, leading to the formation of the alotane skeleton. Oxidation at the C1, C4 and C9 positions produces phorbin A, a precursor for the synthesis of alotaketals. Subsequent oxidation and spiroketalization can yield phorbaketal A (**49**) and alotaketal A (**38**), which exist in equilibrium through double bond migration (Scheme 1).

3.1.5 Aignopsanes

Chemo-centric investigations of the Indo-Pacific marine sponge *Cacospongia mycofijiensis*, collected from the coast of New Britain (Kimbe Bay), Papua New Guinea, led to the discovery of novel sesquiterpenes: aignopsanoic acid A (**50**), methyl aignopsanoate A (**51**) and isoaignopsanoic acid A (**52**) (Fig. 7) [50]. Two of these metabolites, aignopsanoic acid A (**50**) and methyl aignopsanoate A (**51**), showed moderate activity against *Trypanosoma brucei*, with IC_{50} values of 6 and 16 μg/mL, respectively. The total synthesis of (+)-aignopsanoic acid A, (−)-methyl aignopsanoate A and (−)-isoaignopsanoic acid A has been accomplished, confirming their absolute configurations [51]. Interestingly, (+)-microcionin-1 (**53**) (Fig. 7) can be converted to aignopsanoic acid A through a simple oxidation process, supporting the hypothesis that (+)-microcionin-1 may serve as an intermediate in the biosynthesis of aignopsanes. Using a high-throughput approach that generated LC-MS-UV-ELSD-based natural product libraries, additional four new compounds were

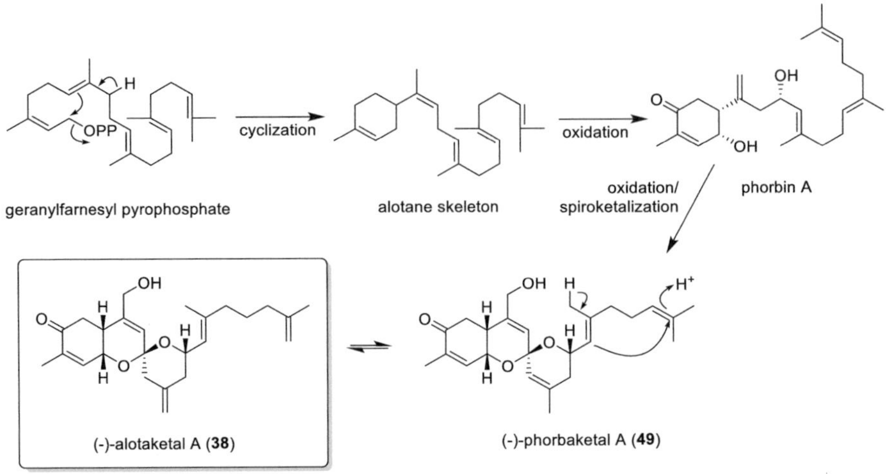

Scheme 1 Proposed biosynthesis of alotaketal A

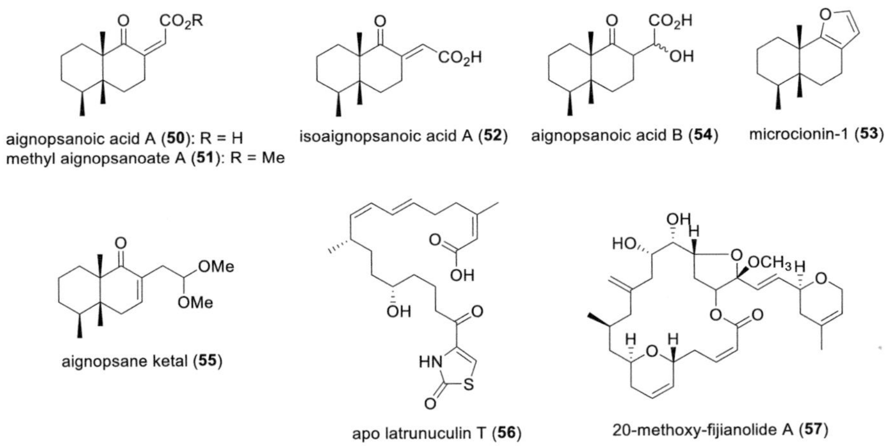

Fig. 7 Aignopsane-related compounds, apo latrunculin T and 20-methoxy-fijinolide A

discovered from a Papua New Guinea collection of *C. mycofijiensis*. These four new molecules include two new derivatives of the aignopsane class, namely aignopsanoic acid B (**54**) and aignopsane ketal (**55**) as well as apo-latrunculin T (**56**) and 20-methoxy-fijianolide A (**57**) (Fig. 7) [52]. apo-Latrunculin T (**56**) exhibited inhibitory activity against *T. brucei* without disrupting microfilament assembly, while 20-methoxy-fijianolide A demonstrated modest microtubule-stabilizing effects.

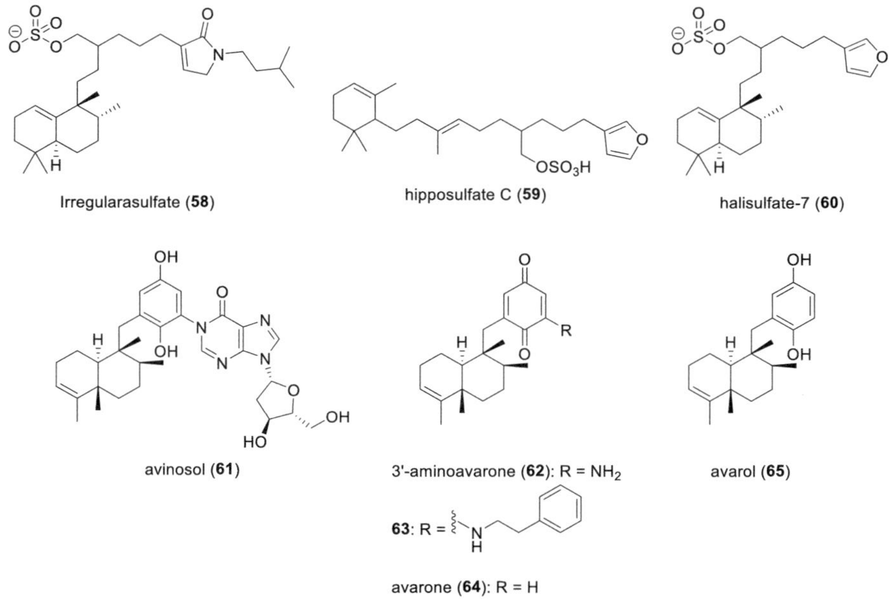

Fig. 8 Irregularasulfate, hipposulfate C, halisulfate-7, avinosol and related compounds from Papua New Guinean sponges

3.1.6 Irregularasulfate

Irregularasulfate (**58**), a new nitrogen-containing sesterterpenoid, along with the known sesterterpenoids, hipposulfate C (**59**), halisulfate-7 (**60**) and igernellin, were isolated from the marine sponge *Spongia irregularis* collected in Keviang, Papua New Guinea (Fig. 8) [53]. Irregularasulfate, hipposulfate C and halisulfate-7 demonstrated moderate inhibition of the catalytic subunits of mammalian Ser/Thr protein phosphatases, including calcineurin, PP-1 and PP-2A. In vitro testing revealed that irregularasulfate (**58**) inhibited calcineurin with an IC_{50} of 59 μM, while hipposulfate C and halisulfate-7 exhibited calcineurin inhibition with IC_{50} values of 66 μM and 69 μM, respectively. Calcineurin inhibitors are crucial in medical treatments, primarily for their immunosuppressive effects. Calcineurin is a calcium-dependent enzyme involved in activating T-cells of the immune system, and inhibiting it can suppress immune responses effectively [54].

3.1.7 Avinosol and Related Compounds

Crude extracts of the sponge *Dysidea* sp. collected near Milne Bay, Papua New Guinea demonstrated promising activity in the assay that detects molecules capable of inhibiting invasion of the extracellular matrix. Subsequent bioassay-guided fractionation gave a new meroterpenoid-nucleoside conjugate, avinosol (**61**) (Fig. 8), as the anti-invasive compound in the extract [55]. In addition, two new meroterpenoids were discovered, including the selective cytotoxin 3′-aminoavarone (**62**) and the phenethylamine analogue (**63**) as well as known meroterpenoids avarone (**64**) and avarol (**65**) (Fig. 8). These meroterpenoids were tested in an anti-invasion assay against two human tumor cell lines with distinct modes of invasion, including MDA-MB-231 breast cancer cells that utilize the mesenchymal mode of invasion (path-generating) and LS174T colon carcinoma cells that invade in an amoeboid manner (path-finding). Avinosol exhibited an IC_{50} of approximately 50 μg/mL in the anti-invasion assay against both cell lines.

3.1.8 Strongylophorine-26

Strongylophorines, a class of meroditerpenoids containing either a quinol or a quinone moiety, are the most prevalent meroditerpenoids found in *Strongylophora* sponges. The first compounds of this series, strongylophorines 1 (**66**)–3 (**68**) (Fig. 9), were initially reported from *Strongylophora durissima* specimens collected near Laing Island, Papua New Guinea in 1978 [56]. Related strongylophorines have also been reported from sponges collected in the Philippines, Okinawa, North Sulawesi, Indonesia and Taiwan. Subsequently, strongylophorine 26 (**69**) was isolated from *P. corticata* collected in Papua New Guinea, along with known strongylophorine 8 (**70**) (Fig. 9) [57]. In the anti-invasion assay using MDA-231 cells, strongylophorine 26 exhibited an IC_{50} value of approximately 1.0 μg/mL, while strongylophorine 8 was significantly less potent, with an IC_{50} value of around 7.0 μg/mL. In a follow up study, strongylophorine 26 was revealed to be a Rho-dependent inhibitor of tumor cell invasion that reduces actin stress fibers and induces nonpolarized lamellipodial extensions [58]. Synthetic analogues (+)- and (−)-**71** (Fig. 9) were prepared to investigate the structure–activity relationship of the anti-invasion pharmacophore in strongylophorine-26 and found that the lactone ring moiety plays a crucial role in the anti-invasion activity of strongylophorine-26 [59]. The first synthesis of strongylophorine-26, using catalytic oxidative quinone heterofunctionalization method, was achieved by Poulsen and co-workers [60].

Additional research by Ireland research groups uncovered strongylophorines 2 (**67**), 3 (**68**) and 8 (**70**), purified from *Petrosia* (*Strongylophora*) *strongylata* sponge collected from Milne Bay, Papua New Guinea, to be responsible for the hypoxia inducible factor-1 (HIF-1) inhibition with EC_{50} values of 8, 13 and 6 μM, respectively, and also effectively decrease expression of VEGF, a downstream HIF-1 target gene [61]. The pursuit of inhibitors targeting the HIF-1 transcriptional pathway has largely been driven by evidence suggesting that selective HIF-1 inhibitors could be developed as molecular-targeted antitumor agents, potentially offering fewer side effects. The availability of strongylophorine 8, obtained from the PNG sponge, facilitated its discovery as a first example of a neuroprotective pro-electrophilic molecule from marine organisms [62]. Specifically, strongylophorine 8 activated the Nrf2/ARE pathway, induced phase 2 enzymes and increased glutathione, thus protecting neuronal cells from oxidative stress.

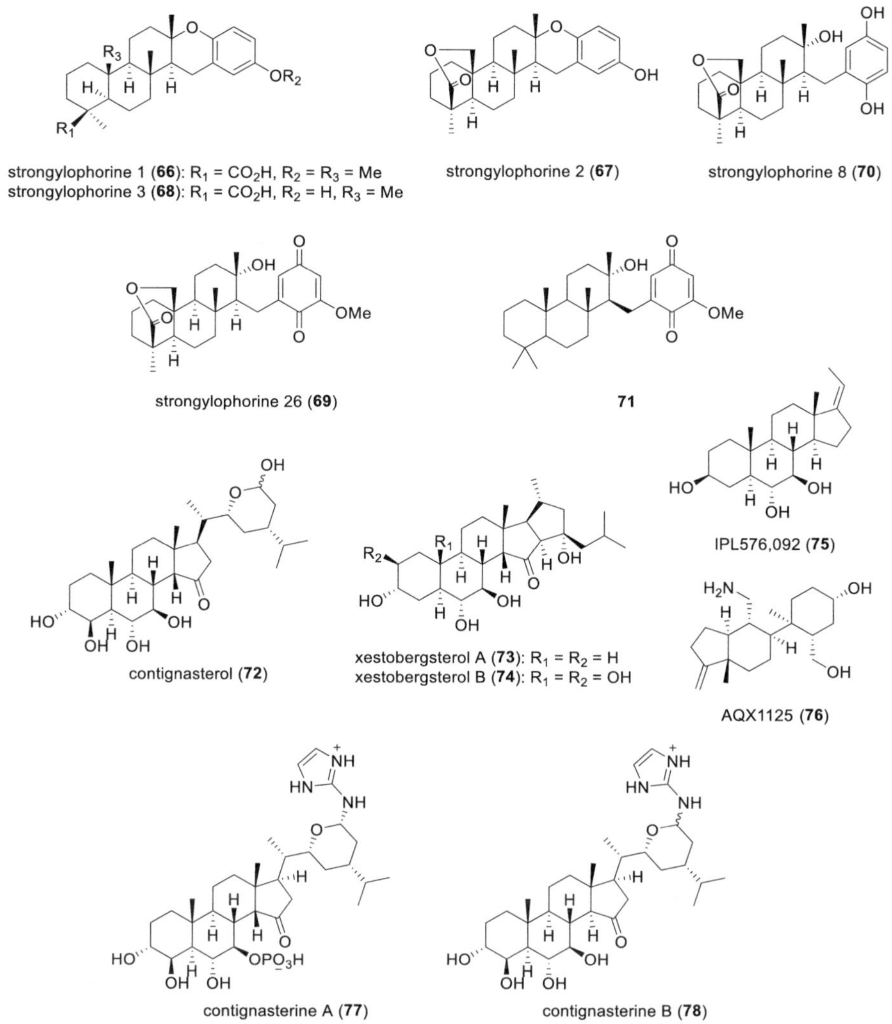

Fig. 9 Strongylophorines, contignasterol and related compounds

3.1.9 Contignasterol and Contignasterines

One of the inactive fractions obtained during the chromatographic separation of the sponge, *Petrosia contignata* (from Madang, PNG), extract contained contignasterol (**72**) (Fig. 9), a novel, highly oxygenated steroid with a *cis* CD ring junction substructure [63]. Shortly after the structure of contignasterol was reported in 1992, Shoji and co-workers published the isolation of closely related steroids, xestobergsterols A (**73**) and B (**74**) (Fig. 9), from the sponge, *Xestospongia bergquista*, collected in Okinawa [64]. The xestobergsterols were found to potently inhibit histamine release from rat mast cells stimulated with anti-IgE in a standard cell-based screening assay for anti-asthma drug leads. A sample of contignasterol was subsequently sent to Masao Takai, who had conducted the histamine release bioassay on the xestobergsterols. He demonstrated that contignasterol exhibited the same inhibitory effect on histamine release in his assay, with similar potency [65].

A series of synthetic efforts ensued and led to a drug candidate, IPL576,092 (**75**) (Fig. 9), was greater than tenfold more active than contignasterol in a rodent model of allergen- induced lung inflammation and it could be synthesized from 5-androsten-3β-ol-17-one in nine steps with an overall yield of 25% [66]. IPL576,092 was eventually evaluated in Phase I and II human trials by Inflazyme Pharmaceuticals as a new anti-asthma agent based on the molecule high oral bioavailability, removing the need for the use of an inhaler, and having a biological activity profile distinct from the widely used glucocorticoids. Unfortunately, further development of this drug candidate was halted due to subsequent closure of the company. Another contignasterol-inspired drug candidate, AQX-1125 (**76**) (Fig. 9), a Src homology 2 domain-containing inositol-5-phosphatase 1 (SHIP1) activating anti-inflammatory molecule, was developed by Aquinox Pharmaceuticals for the treatment of interstitial cystitis/bladder pain syndrome. SHIP1 is an enzyme that plays a pivotal role in the regulation of immune cell function and signaling pathways. However, AQX-1125 development was also discontinued in 2018 following the failure of its LEADERSHIP 301 Phase III clinical trial.

Recently, related analogs, contignasterines A (**77**) and B (**78**), along with contignasterol, were isolated as major metabolites from the sponge *Neopetrosia* cf. *rava*, collected in the Bismarck Sea during the Tara Pacific expedition [67]. Structurally, the contignasterines are oxidized steroids featuring a 2-aminoimidazole moiety branched on their side chains; notably, contignasterine A also possesses an unusual phosphate group at the C-7 position. The anti-inflammatory potential of these compounds was evaluated using the murine macrophage cell line RAW 264.7, motivated by previous findings that contignasterol exhibits both antiasthmatic and anti-inflammatory properties. All three compounds inhibited nitric oxide (NO) release induced by lipopolysaccharide (LPS) stimulation, showing efficacy comparable to that of cyclosporin A, a benchmark anti-inflammatory drug. Contignasterine B significantly reduced NO levels at concentrations ranging from 0.01 to 1 μM, while contignasterol was effective at even lower concentrations (0.001 to 0.1 μM). Furthermore, reactive oxygen species (ROS) production was significantly suppressed, particularly by contignasterines A and B, indicating their potent redox-related anti-inflammatory activity. While contignasterol's mechanism includes modulation of transient receptor potential vanilloid 1 (TRPV1) channels, the anti-inflammatory effects of the contignasterines appear to act independently of TRPV1, relying instead on pathways linked to oxidative stress [67]. TRPV1 channels are a type of ion channel found in the membranes of cells, particularly in sensory neurons, and play crucial roles in the perception of pain, thermoregulation and inflammation [68].

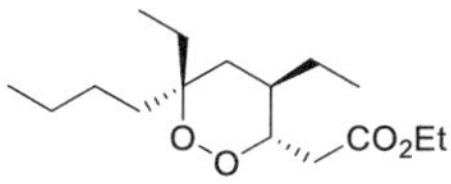

ethyl plakortide Z (**79**),

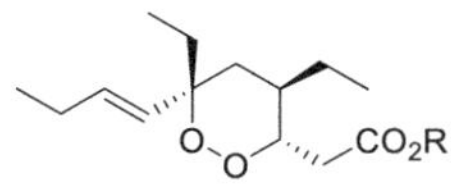

ethyl didehydroplakortide Z (**80**): R = CH_2CH_3
methyl didehydroplakortide Z (**81**): R = CH_3

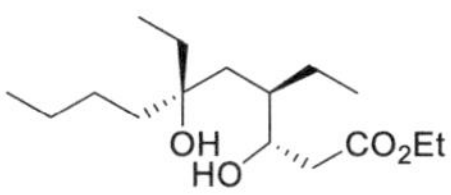

ethyl *seco*-plakortide Z (**82**)

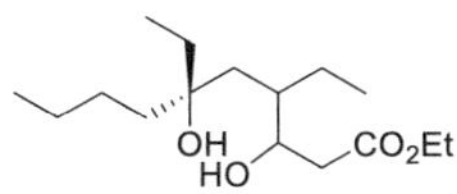

epi-ethyl *seco*-plakortide Z (**83**)

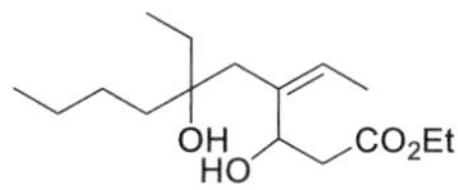

ethyl didehydro-*seco*-plakortide Z (**84**)

Fig. 10 Cyclic polyketide peroxides and acyclic diol analogues from the PNG sponge *Plakortis lita*

3.2 Polyketides

3.2.1 Plakortide-Related Molecules

Plakortides are stable cyclic polyketide peroxides, along with acyclic diol analogues, typically found from *Plakortis* marine sponges. The sponge *Plakortis lita* from Wongat Island, Papua New Guinea provided three cyclic peroxides, including ethyl plakortide Z (**79**), ethyl didehydroplakortide Z (**80**) and methyl didehydroplakortide Z (**81**) as well as three acyclic diol analogues, such as ethyl *seco*-plakortide Z (**82**), *epi*-ethyl *seco*-plakortide Z (**83**) and ethyl didehydro-*seco*-plakortide Z (**82**) (Fig. 10) [69]. Ethyl plakortide Z demonstrated equal activity in vitro against solid tumor and L-1210 leukemia cell lines. In contrast, ethyl didehydroplakortide Z showed moderate selectivity for solid tumors in vitro but did not exhibit activity against solid tumors in vivo in mice.

3.2.2 Psymberin

The marine polyketide, psymberin (**85**) (Fig. 11) (also known as irciniastatin A) is a biologically active natural product belonging to the pederin structural class. It was independently discovered in 2004 by the research teams of Crews and Pettit, who isolated it from the sponges *Psammocinia* sp. and *Ircinia ramosa*, respectively [70, 71]. The isolation of psymberin by the Crews research group was accomplished from pooling of extracts from small collections of the sponge samples originated from multiple proveniences in the waters of Papua New Guinea during the period of 1990 to 2001. Psymberin demonstrated potent cytotoxicity against melanoma, breast and colon cancer cell lines ($LC_{50} < 2.5 \times 10^{-9}$ M) in tests involving 60 cancer cell lines. Due to its exquisite anticancer activities, it has garnered significant interest as a marine natural product and several synthetic efforts have been carried out [72].

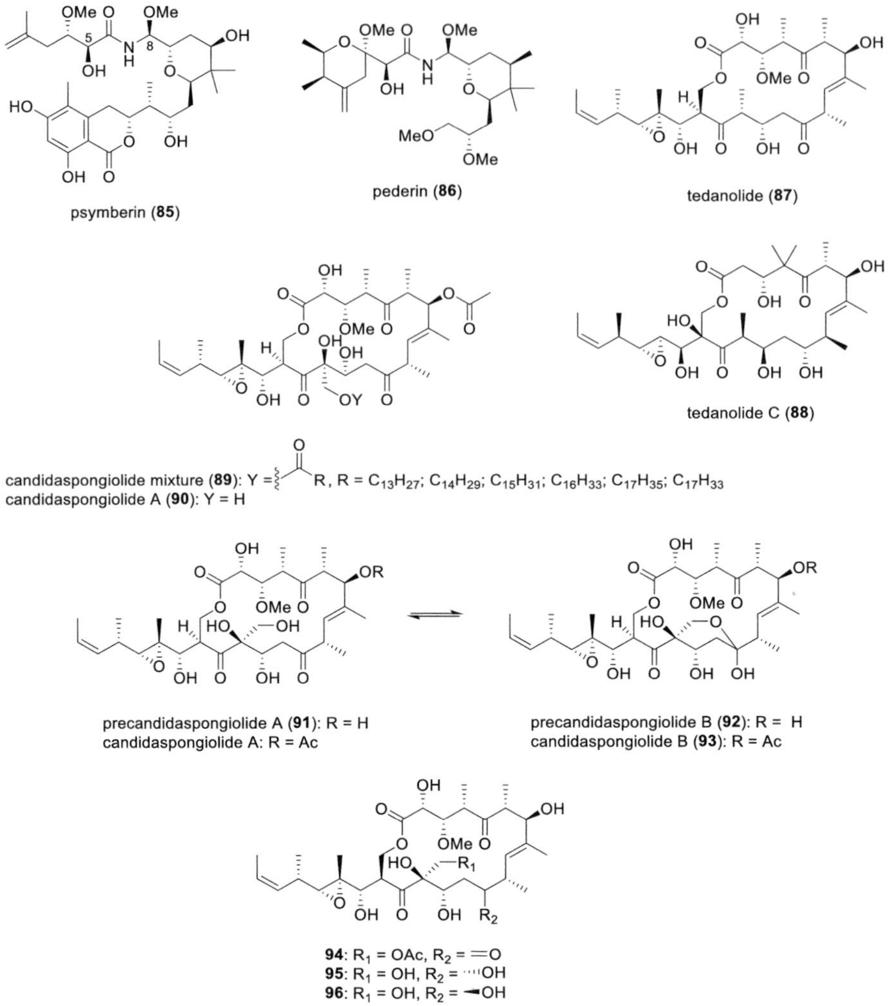

Fig. 11 Chemical structures of psymberin, tedanolide and related compounds

In 2004, researchers identified the biosynthetic gene cluster for a related polyketide, pederin (**86**) (Fig. 11), from the metagenome of the beetle, *Paederus fuscipes*, tracing it back to an uncultured *Pseudomonas*-like bacterium [73]. This breakthrough enabled researchers to investigate the metagenomes of *Theonella swinhoei* and *Psammocinia* aff. *bulbosa* for similar polyketide synthase domains. By 2009, the nearly complete biosynthetic gene cluster for psymberin was identified from an unknown bacterium in *P.* aff. *bulbosa* collected at Milne Bay, Papua New Guinea [74]. Starting from acetic acid thioester, the psymberic acid unit is initially constructed through typical polyketide synthesis reactions, such as keto reduction, crotylation and dehydration. The chain then undergoes further transformations, leading to elongation until cyclization and aromatization take place. The details of the final steps remain unclear. However, it is believed that chemo- and stereoselective oxidations occur at the 5- and 8-positions, followed by *O*-methylation at the 8-position to complete the synthesis. Unfortunately, the identity of the bacterium producing psymberin remains unknown.

Due to psymberin extremely potent and selective cytotoxicity against certain human cancer cell lines, several researchers investigated the mode of action of this molecule. Studies by Japanese researchers showed that psymberin potently inhibits protein synthesis as well as promotes the ectodomain shedding of TNF receptor 1 by TACE and/or other metalloproteinases [75, 76]. In addition, genetic studies with *Caenorhabditis elegans* show that irciniastatin A primarily targets the ribosome [77]. In a recent work by Hsu and co-workers on the use of new pipeline based on patient-derived organoids, psymberin was identified as a potent protein translation inhibitor with anti-cancer properties in colorectal cancer [78].

3.2.3 Tedanolide-Related Compounds: Candidaspongiolides and Tedanolide C

Tedanolide (**87**) (Fig. 11) is a polyketide originally reported from the sponge *Tedania ignis*, commonly known as the fire sponge [79] which is widely distributed across the greater Caribbean, from Mexico to Bermuda. The structure of tedanolide features an 18-membered macrocyclic lactone, originating from a combination of propionate and acetate biosynthesis. Its complex framework includes 13 stereogenic centers and multiple potentially unstable β-hydroxyketone units. A distinctive aspect of tedanolide is its epoxide-containing sidechain, which adds to its structural uniqueness. Ireland and colleagues at the University of Utah discovered tedanolide C (**88**) (Fig. 11) in methanol extracts of *Ircinia* sp., a marine sponge collected from Milne Bay, Papua New Guinea [80]. Although tedanolide C exhibits distinct methylation, oxidation and stereochemical patterns compared to tedanolide and 13-deoxytedanolide, it still demonstrates notable cytotoxicity. Tedanolide C was evaluated against the HCT-116 wild-type colorectal cancer cell line, showing an IC_{50} of 0.057 μg/mL. Cell cycle analysis revealed that treating HCT-116 cells with 0.2 μg/mL of tedanolide C caused significant accumulation of cells in the S-phase after 24 h, which persisted through 48 h. These potent effects suggest that tedanolide C may function as a protein synthesis inhibitor.

Extracts from another PNG sponge genus *Candidaspongia* sp. have been identified as highly cytotoxic, with a specific selectivity towards melanoma cells in the NCI 60-cell-line screen. Subsequent analysis of these *Candidaspongia* extracts afforded a complex mixture of acyl esters of a macrolide structurally related to tedanolide, termed candidaspongiolides (**89**) (Fig. 11) [81]. Lipase-catalyzed hydrolysis of this candidaspongiolide complex mixture resulted in the isolation and identification of the macrolide core, candidaspongiolide A (**90**) (Fig. 11). In the NCI 60-cell-line screen, the candidaspongiolide acyl ester mixture and candidaspongiolide A gave GI_{50} values of approximately 14 nM and less than 4 nM, respectively, against seven melanoma cell lines, including LOX IMVI, MALME-3 M, M14, M19-MEL, SK-MEL-2, SK-MEL-28 and UACC-257. Trisciuoglio and co-workers showed that treating HCT-116 and human glioblastoma (U251) cells with 50 μM of candidaspongiolide A induced hallmarks of apoptosis, including PARP cleavage and the activation of caspase-3 and -12 [82]. Moreover, they demonstrated that candidaspongiolide A triggered a non-classic apoptotic pathway dependent on PKR/eIF2alpha/caspase-12 and inhibited protein synthesis in both HCT-116 and U251 cells.

Further investigation of *Candidaspongia* sp. extracts from Papua New Guinea resulted in the isolation of three new compounds, such as precandidaspongiolides A (**91**), B (**92**) and candidaspongiolide B (**93**), along with candidaspongiolide A (**90**) and (+)-tedanolide (**87**) (Fig. 11) [83]. Precandidaspongiolides A and B were isolated as an inseparable mixture of two isomers in equilibrium, with the major isomer, precandidaspongiolide A, containing a primary alcohol, while the minor isomer, precandidaspongiolide B, contains a hemiketal. Similarly, candidaspongiolides A and B were also isolated as an equilibrium mixture [ratio 1.7 (candidaspongiolide A):1 (candidaspongiolide B)].

Semisynthetic derivatives (**94–96**) (Fig. 11) based on precandidaspongiolide A were prepared to evaluate the importance of the hemiketal/hydroxy ketone for melanoma inhibitory activity. Precandidaspongiolides A and B exhibited excellent selectivity against melanoma cell lines in the NCI 60-cell-line screen. The LC_{50} values for **91/92** against melanoma cell lines were significantly lower than for other tumor cell lines, with seven of the nine melanoma cell lines in the panel showing nanomolar LC_{50} values (19–174 nM), while most other tumor cell lines had LC_{50} values greater than 100 μM. Comparing the activities of **91/92** and the semisynthetic derivatives (**94–96**) indicated that the hemiketal is not essential for activity, as **95** retains the same level of potency as **91/92**. However, potency is reduced when the primary alcohol is substituted, as in **94**, which is approximately 5–15 times less potent than **91/92**.

3.2.4 Swinhoeiamide A

A new calyculinamide-related compound, swinhoeiamide A (**97**) (Fig. 12), was isolated from *Theonella swinhoei* collected off the coast of Karkar Island, Papua New Guinea [84]. Swinhoeiamide A differs from calyculinamide A by replacing the side chain attached to C-29 in the oxazole ring system with a methyl substituent and hydrogenation of the double bond at C-4 and C-5. Swinhoeiamide A exhibited insecticidal activity against the neonate larvae of *Spodoptera littoralis* in a chronic feeding bioassay, with an ED_{50} of 2.11 ppm and a LD_{50} of 2.98 ppm. In addition, it inhibited the growth of fungal *Candida albicans* and *Aspergillus fumigatus* with MIC values of 1.2 and 1.0 μg/mL, respectively. Furthermore, swinhoeiamide A demonstrated dose-dependent cytotoxicity against various undisclosed cell lines and tissues, with IC_{50} values ranging from 20 to 90 ng/mL. Swinhoeiamide A, along with other structurally related molecules, are probably produced by sponge-derived

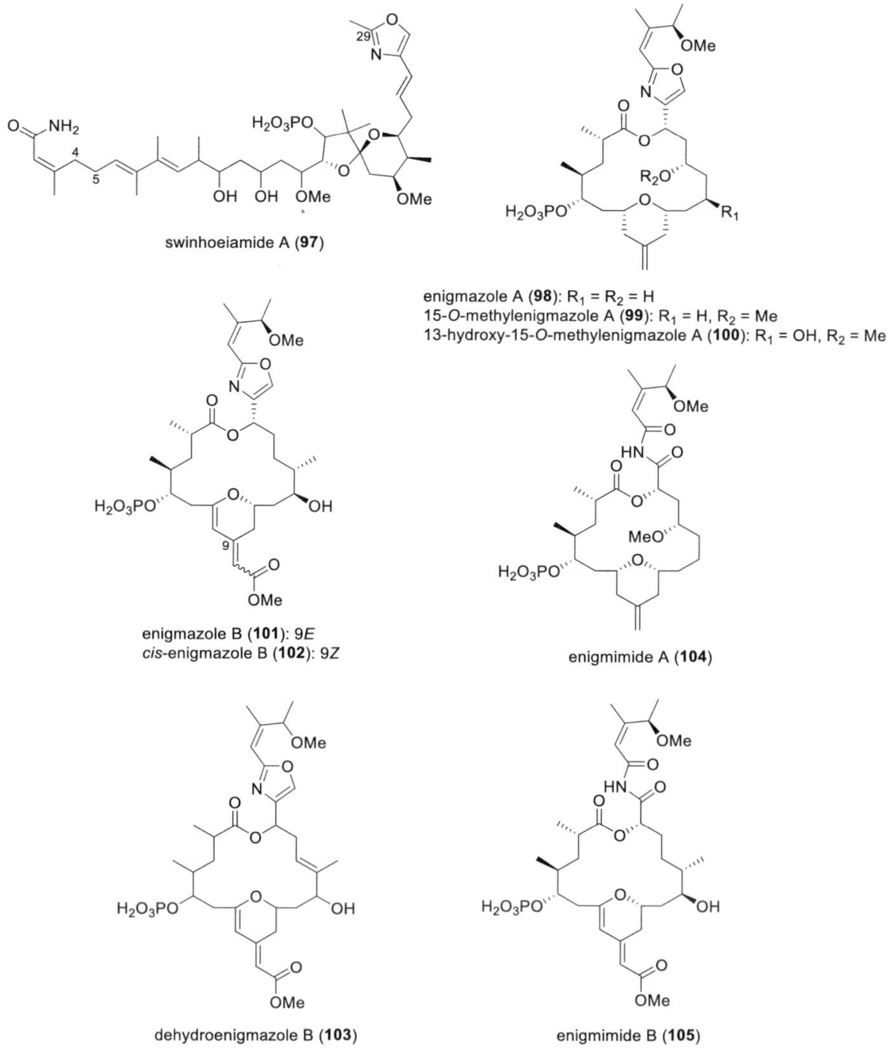

Fig. 12 Chemical structures of swinhoeiamide A, enigmazole A and related compounds

microbial symbionts due to the discovery of calyculin A, being biosynthesized by the microbial symbiont, '*Candidatus* Entotheonella' sp. [85].

3.2.5 Enigmazole A

Enigmazole A (**98**) (Fig. 12), a novel phosphate-containing macrolide, along with two analogues, 15-*O*-methylenigmazole A (**99**) and 13-hydroxy-15-*O*-methylenigmazole

A (**100**) (Fig. 12), were isolated from a marine sponge, *Cinachyrella enigmatica*, collected along the south coast of Papua New Guinea [86]. Enigmazole A features an 18-membered phosphomacrolide structure with an embedded exomethylene-substituted tetrahydropyran ring and an acyclic portion containing an oxazole moiety. The enigmazoles are the first phosphomacrolides discovered from a marine source, with enigmazole A showing significant cytotoxicity in the NCI 60-cell line antitumor screen, achieving a mean GI_{50} of 1.7 μM.

Since the discovery of enigmazole A, at least seven total and formal syntheses of (−)-enigmazole A have been reported. Each synthesis showcases the unique and creative synthetic strategies developed by different research groups, including Molinski, Smith, Fürstner, Fuwa and Kadota research teams [87]. The first total synthesis of (−)-enigmazole A, reported by Molinski, successfully established its absolute configuration [88]. This was followed by six additional total syntheses by Smith [89], Fürstner [90] and Fuwa [91–93], as well as a formal synthesis by Kadota [94]. These synthetic efforts employed distinct synthetic strategies, requiring between 18 and 29 steps to provide enigmazole A. For instance, the Corey–Fuchs alkyne synthesis reaction was one of the key synthetic strategies used by Sakurai et al. in the total synthesis of (−)-enigmazole A and (−)-15-*O*-methylenigmazole A [91]. A recent report on the 18-step total synthesis of enigmazole A by Masuda et al. provided a concise access to the macrocyclic framework of enigmazole A for further structure–activity relationship (SAR) studies [92]. Their work is based on a modular assembly of three similarly complex building blocks, followed by a macrocyclic ring-closing metathesis to provide the 18-membered macrolactone core and a desilylative transannular oxa-Michael addition to stereoselectively form the 2,6-*cis*-substituted tetrahydropyran ring.

Further study of the organic-soluble portion of an aqueous extract from *C. enigmatica* yielded a new macrolide, enigmazole B (**101**), along with two enigmazole B derivatives, *cis*-enigmazole B (**102**) and dehydroenigmazole B (**103**) as well as two imide-type analogues named enigmimides A (**104**) and B (**105**) (Fig. 12) [95]. These natural products were tested for cytotoxic activity against IC-2 murine mast cells and found that enigmazole B, *cis*-enigmazole B, dehydroenigmazole B and compounds **98**–**100** exhibited cytotoxic activity against IC-2 cells with IC_{50} values ranging from 2.0 to 7.0 μM. Additional enigmazole analogues, enigmazoles C–E, were isolated as cytotoxic components from an Indonesian *Homophymia* sponge [96].

3.3 Alkaloids

3.3.1 Motuporamines

The macrocyclic alkaloids, motuporamines A (**106**) – C (**108**) (Fig. 13), were isolated from the marine sponge, *Xestospongia exigua*, collected from outer reef off Motupore Island and Madang, Papua New Guinea [97]. The mixture of motuporamines A–C exhibited moderate in vitro cytotoxicity, with a mean IC_{50} of 0.6 μg/mL against a panel of human solid tumor cancer cell lines. Subsequent work by Roskelley and co-workers identified motuporamines A and C as micromolar inhibitors of basement membrane gel invasion by MDA-231 breast carcinoma, PC-3 prostate carcinoma and U-87 and U-251 glioma cells [98]. In addition, motuporamine C inhibited cell migration in monolayer cultures and impaired actin-mediated membrane ruffling at the leading edge of lamellae. Moreover, motuporamine C reduced β1-integrin activation, suggesting it may interfere with "inside-out" signaling to integrins. The anti-invasive compound, motuporamine C, was later shown to be a robust stimulator of neuronal growth cone collapse [99].

Further investigation of the same sponge species led to the isolation of three new motuporamines D (**109**)–F (**111**) (Fig. 13), a mixture of motuporamines G–I (**112**) (Fig. 13), along with motuporamines A–C [100]. Motuporamines A to C, along with the mixture of G, H and I, were responsible for the anti-invasion activity of the crude extract. A series of motuporamine analogues have been synthesized and evaluated for anti-invasive activity. These SAR results demonstrated that a saturated 15-membered cyclic amine fused to the natural motuporamine diamine side chain represented the optimal structure for anti-invasive activity. Subsequent SAR data from both natural and synthetic motuporamines were combined with conformational analyses to outline the functionality and shape required for activity and resulted in the design of a new carbazole analogue, **113** (Fig. 13), which has good anti-invasion activity. Further synthetic work by Muth and co-workers generated new dihydromotuporamine C

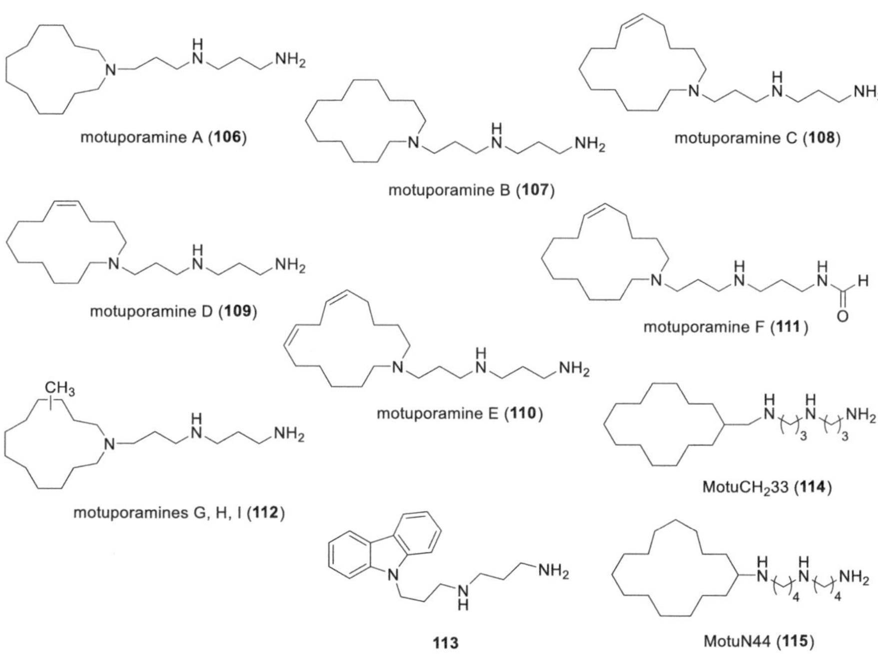

Fig. 13 Chemical structures of motuporamines and synthetic analogues

analogue, **114** (Fig. 13), by placing the polyamine chain outside the ring that doubled the antimigration potency and reduced compound toxicity by 133-fold. Moreover, the antimigration properties of **114** were demonstrated both in vitro and in vivo [101]. In another development by Borselli and co-workers, dihydromotuporamine C and its derivatives were assessed for their in vitro antimicrobial activities and their ability to enhance antibiotic effectiveness against Gram-negative bacteria and clinical isolates [102]. Using fluorescent dyes to measure outer-membrane depolarization and permeabilization, they found that a derivative, MOTU-N44 (**115**) (Fig. 13), inhibited antibiotic transporters by de-energizing the efflux pump rather than by directly interacting with the pump. Their study showed that targeting the efflux pump's energy source could be an alternative strategy to overcome antibiotic resistance.

3.3.2 Papuamine

Papuamine (**116**) (Fig. 14) is a pentacyclic diamine originally reported by Scheuer and co-workers from the marine sponge *Haliclona* sp., collected at South Lion Island, Papua New Guinea [103]. Since this initial report in 1988, papuamine and related analogue, haliclonadiamine (**117**) (Fig. 14), have been reported from *Haliclona* sp from different locations [104, 105] as well as other sponge species, including *Halichondria panicea* and *Neopetrosia* cf *exigua* [106, 107]. Papuamine and haliclonadiamine are unsymmetrical diastereoisomers, having all configurations of the two alkaloids to be opposite except C22 [108]. Research revealed that papuamine exhibits excellent bioactivities against bacteria, e.g., *Bacillus subtilis*, *Staphylococcus aureus* and *Mycobacterium* sp. [104, 109], fungi, e.g., *Trichophyton mentagrophytes* and *Candida albicans* [103, 104] as well as human cancer cells, including breast cancer cell MCF-7, prostate cancer cell LNCap and colon cancer cell lines Caco-2 and HCT-15, having IC_{50} values in the nanomolar range [105, 110]. In addition, papuamine showed potent inhibitory activity against human non-small cell lung cancer cell lines (e.g., H1299, H226B and A549) and glioblastoma cells SF-295 via induction of mitochondrial dysfunction [111]. Moreover, papuamine displays synergistic anticancer effects with doxorubicin, suggesting that the alkoid may represent a novel type of modulator for doxorubicin chemotherapy [112].

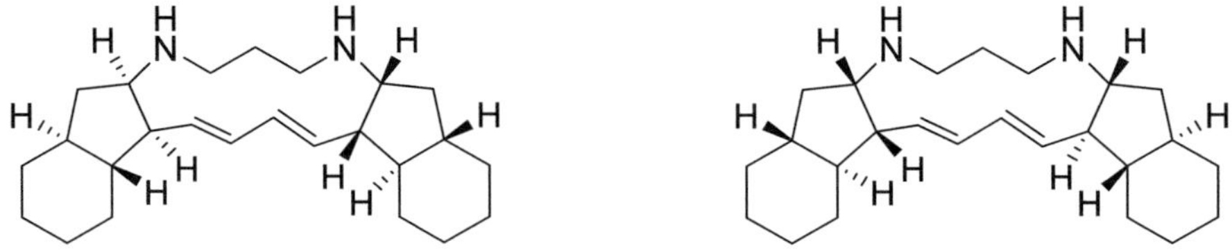

papuamine (**116**) haliclonadiamine (**117**)

Fig. 14 Chemical structures of papuamine and haliclonadiamine

3.3.3 3-Alkylpiperidine Alkaloids

Due to the immense number of sponge-derived alkaloids, it is possible to recognize major categories of compounds that are related by their putative biogenetic origins. An example of such group of compounds is the structurally diverse collection of alkaloids that are related to each other by the presence of a 3-alkylpiperdine moiety having a common biogenesis [113]. The alkyl portion of the common structural motif is typically a linear chain, either saturated or unsaturated, with a length ranging from eight to sixteen carbon atoms. The six-membered nitrogen heterocycle found in these alkaloids can exist in different oxidation states, including pyridine, tetrahydropyridine or piperidine. About 30% of the known compounds in this group, which include both simple monomers and high molecular weight oligomers or polymers, consist exclusively of pyridine rings. The majority of the remaining two-thirds, which feature more complex cyclic structures and are thought to originate biogenetically from macrocyclic dimers, contain piperidine or tetrahydropyridine heterocycles.

Xestocyclamines, Ingenamines and Ingamines

A series of unique 3-alkylpiperidine alkaloids, including xestocyclamines A (**118**) and B (**119**) [114, 115], ingamines A (**120**) and B (**121**) as well as ingenamines A (**122**)–F (**127**) (Fig. 15) [116–118], were all isolated from samples of *Xestospongia* sp./*X. ingens* collected from various locations, including Milne Bay province and Sek Point off Madang, Papua New Guinea. The isolation of these alkaloids provided further evidence for Baldwin and Whitehead's biogenetic pathway on proposed intermediates involved in manzamine biosynthesis [119]. Xestocyclamine A (**118**) was isolated from a *Xestospongia* sp. collected in the Milne Bay region of Papua New Guinea [114]. The original structure of xestocyclamine A was later revised as **118** and was an isomer of ingenamine, differing only in the placement of the double bonds with xestocyclamine A having a Δ^{22} while ingenamine has a Δ^{23} unit [115]. However, recent total synthesis of xestocyclamine A by provided strong evidence that this molecule is in fact the enantiomer of ingenamine A and not a pseudo-enantiomer [120]. It was proposed that the biogenesis of the unprecedented pentacyclic skeleton of xestocylamine A involved an intermolecular [4 + 2] cycloadditon reaction between two monomeric 3-alkyldehydropiperidine macrocycles [114]. In addition, xestocyclamine A demonstrated moderate inhibitory activity against protein kinase C, with an IC_{50} value of 4 μg/mL. Notably, xestocyclamine A exhibited selectivity against IL-1 (interleukin) while showing no activity against other cancer-relevant targets.

Ingamines A (**120**) and B (**121**) [118], ingenamine A (**122**) [116] as well as ingenamines B (**123**)–F (**127**) [117] were all isolated from *X. ingens*, collected from Sek Point off Madang, PNG. The biogenesis of ingenamine was proposed to involve an intramolecular [4 + 2] cycloaddition reaction between two dehydropiperidine rings in a *bis*-3-alkylpiperidine macrocyclic precursor. When tested against murine leukemia P388 in vitro, ingenamine A (**122**) exhibited cytotoxic activity with an ED_{50} of 1 μg/mL while ingamines A (**120**) and B (**121**) both exhibited cytotoxicity with ED_{50} values of 1.5 μg/mL. Interestingly, ingamine A and another analog,

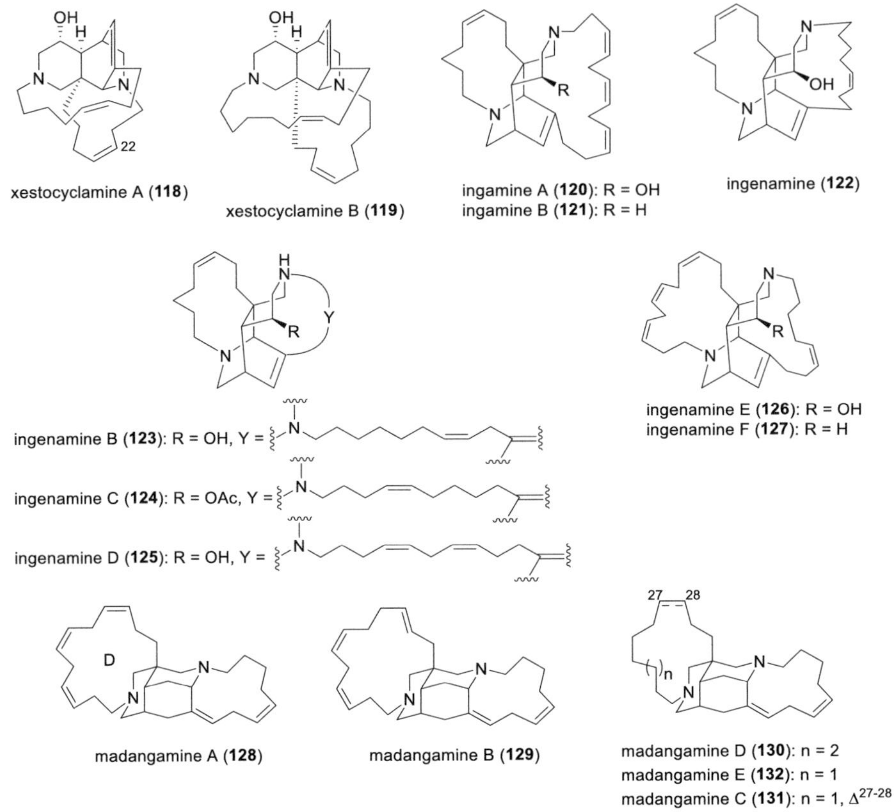

Fig. 15 Chemical structures of xestocyclamines, ingenamines, ingamines and madangamines

dihydroingenamine D, exhibited potent antiplasmodial activity against chloroquine-sensitive (D6) and -resistant (W2) strains of *Plasmodium falciparum* with IC_{50} values of 90 and 78 ng/mL as well as 72 and 57 ng/mL, respectively [121]. Furthermore, ingamine A showed weak antimicrobial activity and moderate antileishmanial activity against *Leishmania donovani* promastigotes. These polycyclic ingamine alkaloids are notable as the first antiplasmodial leads that lack a β-carboline ring, which is typically associated with the cytotoxicity found in manzamine class of alkaloids related to the ingamines.

Madangamines

Madangamines form a distinct and intricate class of 3-alkylpiperidine alkaloids, resulting from the dimerization of tetrahydropyridine. The first compound from this series, madangamine A (**128**) (Fig. 15), was reported from the marine sponge, *Xestospongia ingens*, collected near Sek Point off Madang, Papua New Guinea [122]. This was followed by the discovery of madangamines B (**129**)–E (**132**) (Fig. 15) also from *X. ingens* from Papua New Guinea [123]. As a small unique class of complex 3-alkylpiperidine molecules, madangamine alkaloids have garnered significant interest from the synthetic chemistry community [124]. This is largely due to their fascinating polycyclic structures, biological activities and their biogenetic connections to other alkaloids, such as the manzamines. The madangamines showed significant cytotoxicity to a variety of cancer cell lines. For instance, madangamine A exhibits good cytotoxicity in vitro to the human leukaemia cell line HL-60 with IC_{50} of 16.7 μg/mL and was found to be a novel lysosomotropic agent to inhibit autophagy [125]. In addition, the olefin groups in the D-ring of madangamines play an important role in the cytotoxic activity [126].

Manzamine-related Alkaloids

Manzamine alkaloids are distinguished by a unique polycyclic ring structure, which is thought to originate biogenetically from ammonia, C_{10} and C_3 units and tryptamine. In most manzamine alkaloids, these multicyclic structures are fused with tryptamine to form a β-carboline core. The first compound of this alkaloid class, manzamine A, was isolated as a hydrochloride salt from the Okinawan sponge *Haliclona* sp. by Higa and colleagues in 1986 [127]. A sample of *Petrosia contignata* collected in Milne Bay, Papua New Guinea, was found to contain 1,2,3,4-tetrahydro-2-*N*-methyl-8-hydroxymanzamine A (**133**) and 1,2,3,4-tetrahydro-8-hydroxymanzamine A (**134**) (Fig. 16) [128]. The desmethyl variant, compound **134**, was also isolated from a *Cribochalina* sp. specimen obtained near Madang, Papua New Guinea. It was reported that 1,2,3,4-tetrahydro-2-*N*-methyl-8-hydroxymanzamine A exhibited cytotoxic activity against murine leukemia P388 cells in vitro, with an IC_{50} value of 0.8 μg/mL.

Petrosines

Extracts from the sponge *Petrosia seriata* collected near Laing Island, Papua New Guinea, have shown toxicity to the tropical fish *Lebistes reticulatus*. The toxic activity is linked to three bisquinolizidine alkaloids present in the extracts, namley petrosin (**135**) (LD_{100} = 10 mg/L), petrosins A (**136**) and B (**137**) (Fig. 16) [129–131]. Petrosin, the major component, was isolated as optically inactive crystals displaying a two-fold symmetry in the NMR data. Its structure was confirmed through X-ray diffraction analysis, revealing it as a naturally occurring racemate in *P. seriata*. Interestingly, in petrosin the two quinolizidine units share the same absolute configuration, while petrosin A, a meso compound, has opposite absolute configurations for the two quinolizidine units. In petrosin B, the units have different relative configurations. These petrosins represent the first known examples of naturally occurring compounds with a bisquinolizidine macrocyclic structure.

3.3.4 Hyrtiocarboline

A β-carboline alkaloid, hyrtiocarboline (**138**) (Fig. 17), was isolated from *Hyrtios reticulatus* collected at depths of 13–20 m in Bismarck Sea, Papua New Guinea [132].

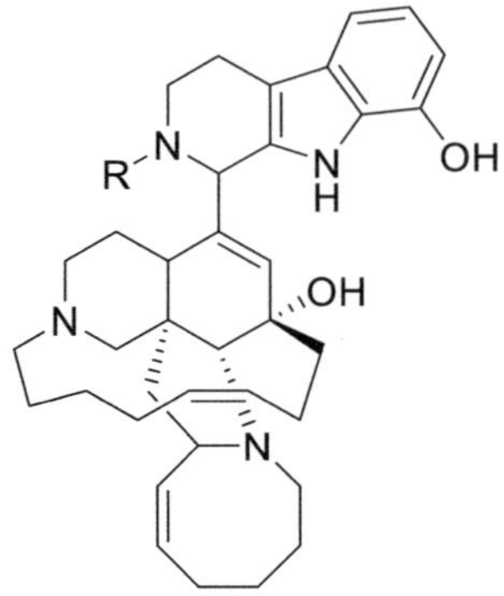

1,2,3,4-tetrahydro-2-*N*-methyl-8-hydroxymanzamine A (**133**): R = Me
1,2,3,4-tetrahydro-8-hydroxymanzamine A (**134**): R = H

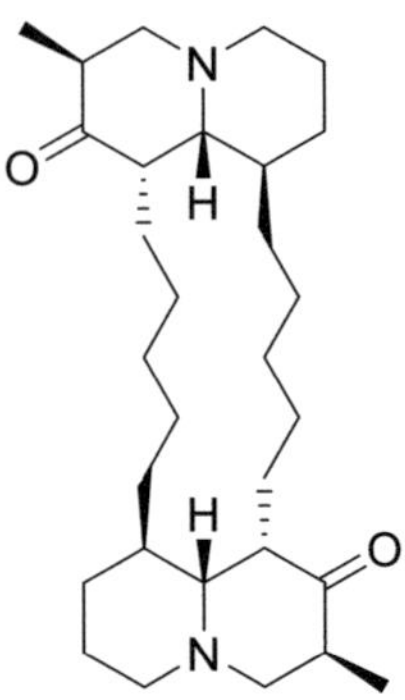

petrosin (**135**)

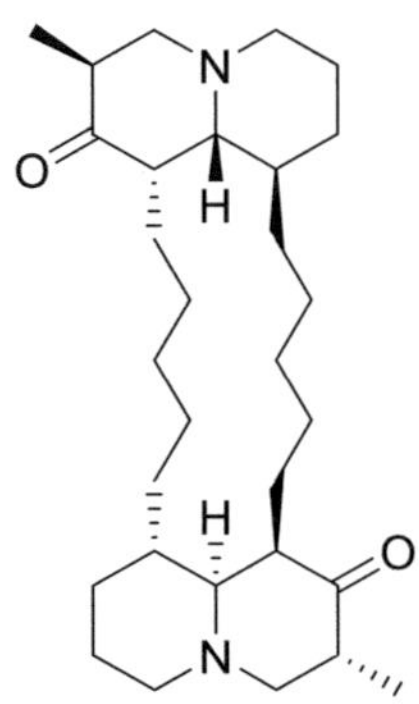

petrosin A (**136**)

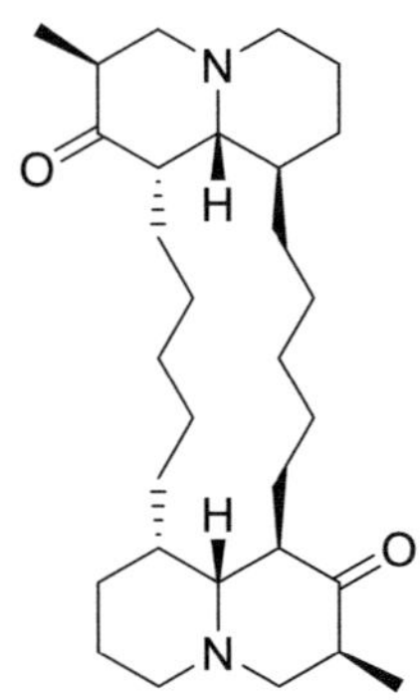

petrosin B (**137**)

Fig. 16 Chemical structures of manzamine-related molecules and petrosins

Hyrtiocarboline demonstrated selective anticancer activity against three cancer cell lines: H522-T1 non-small cell lung cancer, MDA-MB-435 melanoma and U937 lymphoma cell lines, with IC_{50} values of 1.2, 3.0 and 1.5 μg/mL, respectively. Hyrtiocarboline also demonstrated 57% inhibition of HeLa cells at a concentration of 230 μM after 24 h of incubation, leading to induced cytotoxicity, cytoskeletal disruption and nuclear blebbing.

3.3.5 Alkaloids from Agelas Sponges

Marine sponges of the genus *Agelas*, commonly found in tropical and subtropical oceans, have emerged as promising yet underexplored sources of natural products. These sponges exhibit remarkable molecular diversity and a wide range of

Fig. 17 Chemical structure of hyrtiocarboline

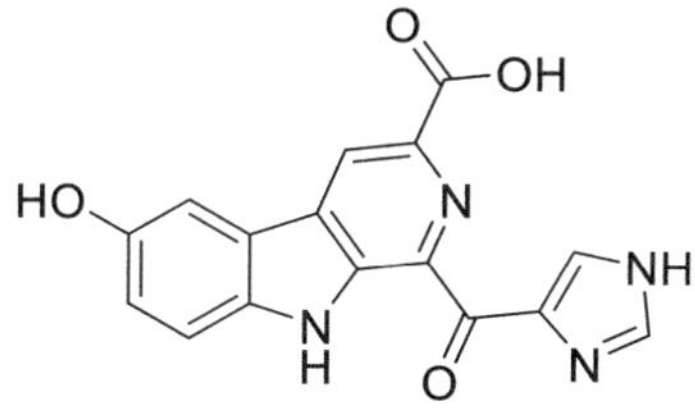

hyrtiocarboline (**138**)

intriguing biological activities. Many of their metabolites are of mixed biogenetic origin, including alkaloids (notably pyrrole and terpenoid alkaloids), glycosphingolipids, sterols and carotenoids. Notably, several unique alkaloids have been isolated from *Agelas* species in Papua New Guinea. For instance, two new bromopyrrole alkaloids, 5-bromopyrrole-2-carboamide (**139**) and 5-bromopyrrole-2-(*N*-methoxymethyl) carboxamide (**140**) as well as a novel diterpene featuring a 9-methyladenium moiety (**141**) (Fig. 18) were isolated from the Papua New Guinean sponge *Agelas nakamurai*, collected from the Fly Islands [133]. 5-Bromopyrrole-2-carboamide, demonstrated antimicrobial activity against a panel of Gram-positive bacteria (*Staphylococcus epidermidis*, *S. aureus*, MRSA, *Enterococcus faecalis* and *E. faecium*) and fungi (*Candida albicans* and *C. neoformans*) with IC_{50} values ranging from 0.78 to 6.25 μg/mL. Another *Agelas* sp. samples from Papua New Guinea yielded two additional agelasines with clerodane diterpenoid structures, agelasine M (**142**) and 2-oxo-agelasine B (**143**), along with the rare norditerpenoid-agelasines, agelasines A (**144**) and B (**145**) (Fig. 18) [134]. Among these, agelasine M showed notable inhibitory activity against *Trypanosoma brucei* (IC_{50}, 3.0 μg/mL) and exhibited mild cytotoxicity against Jurkat cells (IC_{50}, 25 μg/mL).

Two novel *N*-methylated phakellin derivatives, (−)-7-*N*-methyldibromophakellin (**146**) and 7-*N*-methylmonobromophakellin (**147**) (Fig. 18), were isolated from the Papua New Guinean *Agelas* sp. (collected near the coast of Wewak) through bioassay- and LC–MS-guided fractionation [135]. Compound **146** exhibited moderate inhibitory activity against 12-human lipoxygenase isozymes, 12-hLO and 15-hLO, with an IC_{50} values of 10.7 and 48 μM, respectively. Lastly, a human cancer cell line bioassay-directed investigation of the Papua New Guinean *Agelas* sp. led to isolation of agelagalastatin (**148**) (Fig. 18), which is the first example of a natural product containing a digalactofuranosyl unit [136]. Agelagalastatin displayed significant in vitro activity against panel of cancer cell lines with GI_{50} values ranging from 0.77 μg mL^{-1} for lung NCI-H460 to 2.8 μg mL^{-1} for the ovarian OVCAR-3.

3.3.6 Bromotyrosine-Containing Alkaloids: Ceratamines, Bastadins, Hemibastadinols and Psammaplins

The heterocyclic alkaloids, ceratamines A (**149**) and B (**150**) (Fig. 19) were isolated from *Pseudoceratina* sp. obtained on the outer reefs near Motupore Island, Papua

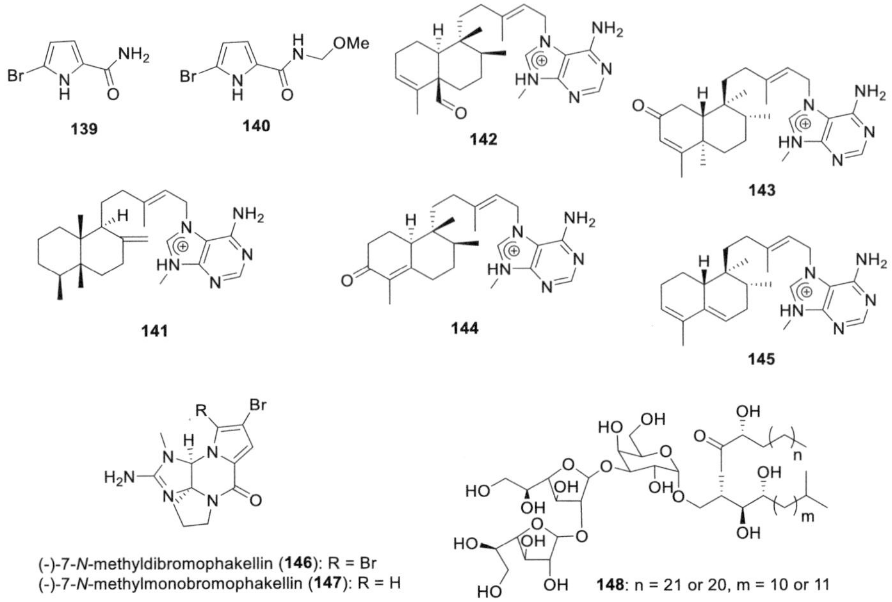

Fig. 18 Alkaloids from Papua New Guinean *Agelas* sponges

New Guinea [137]. Both alkaloids exhibited IC_{50} values of 10 μg/mL in a cell-based antimitotic assay, marking them as the first members of a novel family of antimitotic heterocyclic alkaloids. In vitro studies with purified tubulin revealed that ceratamines directly stimulate microtubule polymerization in the absence of microtubule-associated proteins [138]. In addition, cells treated with ceratamines displayed a dense perinuclear microtubule network during interphase and multiple pillar-like tubulin structures in mitotic cells. Notably, ceratamines did not compete with paclitaxel for microtubule binding in vitro [138]. Unlike other microtubule-stabilizing agents, ceratamines possess relatively simple structures without chiral centers, making them attractive drug candidates.

To develop novel substances with enhanced antiproliferative properties, syntheses of ceratamine analogs have been performed. For instance, Nodwell and co-workers in 2010 reported a derivative, compound **151**, by replacing the bromine atoms with methyl groups [139]. In MCF-7 cells, compound **151** showed an IC_{50} value of 3 μM. Subsequent research by Tao et al. (2017) highlighted the significance of the methyl groups at the C-14 and C-16 positions. Another derivative **152** (Fig. 19) demonstrated relatively good anticancer activity, with an IC_{50} of 23 μM in lung carcinoma cells (A549 cell line) [140]. In addition, a significant increase in antiproliferative activity was observed with derivative **153** (Fig. 19), which had an IC_{50} of 8.56 μM in A549 cells; this compound featured a benzyl group at the N-7 position and three methoxy groups on the benzene ring [141].

Several sponges from Papua New Guinea are sources of bioactive bromotyrosine alkaloids. One such classes are the macrocyclic bromotyrosine derivatives, bastadins,

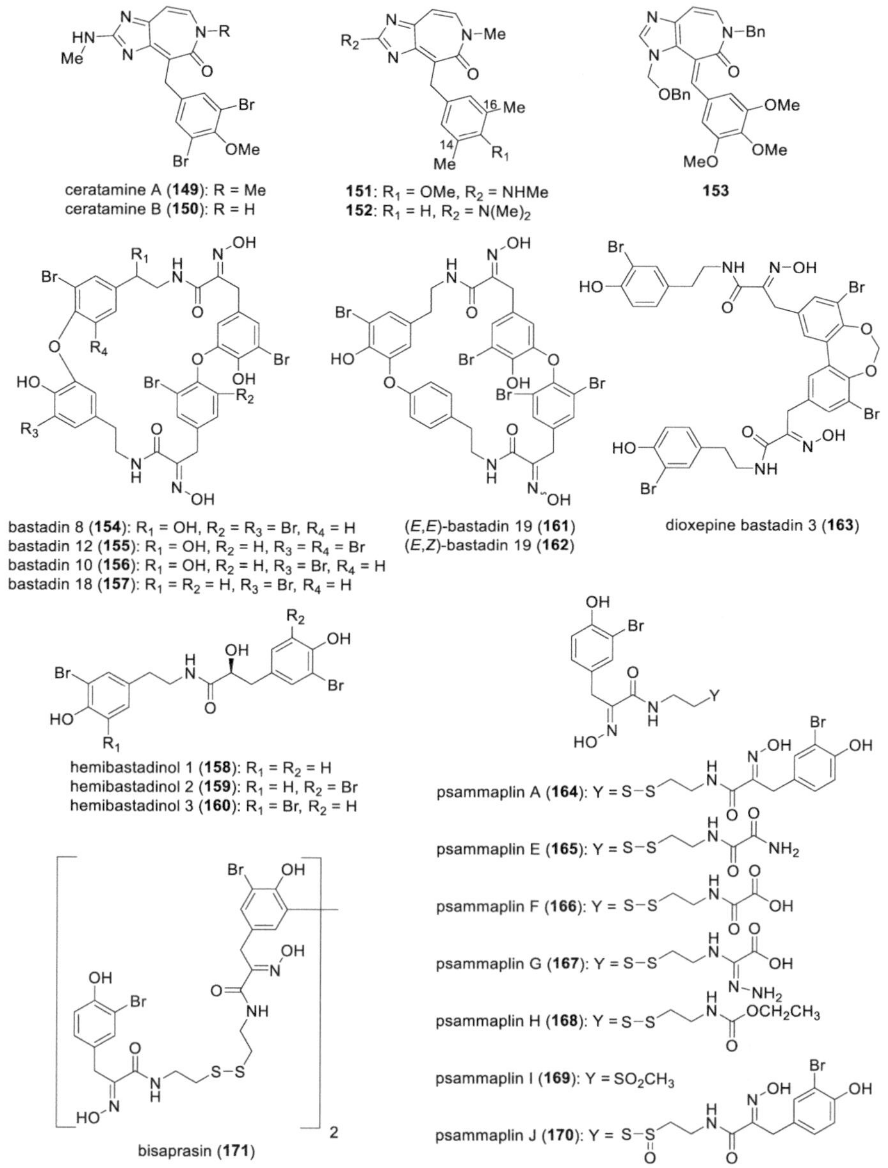

Fig. 19 Bromotyrosine-related molecules isolated from various PNG sponges

which are biogenetically derived from four bromotyrosines via oxidative phenolic coupling of two tyramine–tyrosine units connected through an amide bond. Two new cytotoxic metabolites, bastadins 8 (**154**) and 12 (**155**) (Fig. 19), were isolated from *Ianthella basta*, collected from Madang, Papua New Guinea [142]. These bastadins exhibited in vitro L-1210 cytotoxicity with ED_{50} value of 5 μg/mL. Reisolation of bastidins, by Pettit's research groups, from *I. basta* collected at Idler Bay Point, PNG, revealed inhibitory activity of bastidins 8, 10 and 12 against *Staphylococcus aureus* and *Enterococcus faecalis* [143].

From another collection of *I. basta*, six known bastidins and one new bastidin 18 (**157**) (Fig. 19) were isolated [144]. Bastadins 8 (**154**) and 10 (**156**) were found to be moderate inhibitors of inosine 5'-phosphate dehydrogenase. Subsequent investigation of the Bismarck Archipelago *I. basta* led to the isolation of known hemibastadins 1 - 3 and the new brominated tyrosine derivatives, hemibastadinols 1 (**158**) - 3 (**160**) (Fig. 19) [145]. The new hemibastadinols inhibited growth of the Gram-negative pathogen *Neisseria gonorrhoeae.*

Chemical investigations on another species of *Ianthella*, *I.* cf. *reticulata*, collected from Milne Bay, PNG, resulted in the isolation of two new bastadins, (*E*,*Z*)-bastadin 19 (**162**), a diastereoisomer of the known (*E*,*E*)-bastadin 19 (**161**), and dioxepine bastadin 3 (**163**) (Fig. 19) [146]. Dioxepine bastadin-3 is particularly notable for featuring an uncommon dibenzo-1,3-dioxepine functionality, a rare scaffold in natural products chemistry. In addition to the two new bastadins, 10 known related molecules, including bastadins 2–9, 13 and 19 were purified. Based on chemical investigations, it has been suggested that all bastadins and related molecules are originally produced with the (*Z*)-oxime configuration and subsequently isomerize to the more thermodynamically favored *E* isomer in solution during extraction process. It was revealed that bastadins 13 and 19 (**161**) exhibited moderate inhibition against *Trypanosoma brucei* while bastadin 4 was cytotoxic to HCT-116 colon cancer cells.

New disulfide bromotyrosine derivatives, psammaplins E (**165**)–J (**170**) (Fig. 19) were isolated from the sponge *Psammaplysilla purpurea* collected from Papua New Guinea by Crews' group in 2003 [147]. Interestingly, the *N*-substituted oxalamide functionality of psammaplin E is rare among marine natural products, while psammaplin F (**166**) is the first marine natural product containing an oxalamic functionality. In addition, psammaplins A (**164**), F (**166**) and bisaprasin (**171**) (Fig. 19) are potent histone deacetylase inhibitors, with IC_{50} ranging from 2.1 to 9.0 nM, while psammaplins A, G (**167**) and bisaprasin are potent DNA methyltransferase inhibitors.

3.3.7 Leucosolenamines

Bioassay-guided fractionation of *Leucosolenia* sp. samples collected from a various sites off Milne Bay Papua New Guinea led to the isolation of two novel compounds, leucosolenamines A (**172**) and B (**173**) (Fig. 20), along with the known compound thymidine [148]. Leucosolenamine A features a 2-aminoimidazole core with substitutions at C-4 and C-5 by an *N*,*N*-dimethyl-5,6-diaminopyrimidine-2,4-dione group and a benzyl group, respectively. Similarly, leucosolenamine B shares the same core structure, but with a distinct substitution at C-4 by a 5,6-diamino-1,3-dimethyl-4-(methylimino)-3,4-dihydropyrimidin-2(1H)-one moiety. This unique substitution pattern has not been previously reported in calcareous imidazole alkaloid chemistry. Based on standard disk diffusion assay, it was revealed that leucosolenamines A, at 180 μg/disk, has mild cytotoxicity and inhibited murine colon adenocarcinoma C-38 cells with a 10.5 mm zone as compared to 0 mm against CFU-GM cells.

3.3.8 Acridine Alkaloids

In 1990, a collaborative program was launched between Dr. Phillip Crews' natural product chemistry laboratory at the University of California, Santa Cruz and Dr. Fred Valeriote's experimental therapeutics laboratory at Henry Ford Hospital in Detroit. The initiative focused on discovering and developing anticancer drugs derived from sponge extracts using a novel in vitro disk diffusion assay, selective for solid tumors. This bioassay-guided fractionation process resulted in the identification of active pure compounds from many of these sponges, samples collected from Indo-Pacific region [149].

The investigation of various sponge samples from the genus *Xestospongia*, collected in Indonesia (*X.* cf. *carbonaria*) and Papua New Guinea (*X.* cf. *exigua*), led to the isolation of several alkaloids, including one known pyridoacridine, neoamphimedine (**174**), and three new analogues, 5-methoxyneoamphimedine (**175**), neoamphimedine Y (**176**) and neoamphimedine Z (**177**) as well as a new acridine, alpkinidine (**178**) (Fig. 20) [150]. Several of these bisannulated acridines were evaluated for their differential cytotoxicity (solid tumor selectivity) based on a disk diffusion soft agar assay, using a panel of five cancer cell lines (solid tumors and leukemias) and two normal cells. Results showed that neoamphimedine, 5-methoxyneoamphimedine and alpkinidine were solid tumor selective with neoamphimedine and 5-methoxyneoamphimedine exhibiting the most potent and highest selective activity for solid tumors, respectively.

3.3.9 Exiguamines

Exiguamine A (**179**) (Fig. 20) was isolated as deep-red optically-inactive crystals from the marine sponge, *Neopetrosia exigua*, collected in Milne Bay, Papua New Guinea [151]. It possesses a complex hexacyclic alkaloid structure, incorporating both indole and quinone elements. Exiguamine A exhibited a Ki of 41 nM for the inhibition of indoleamine-2,3-dioxygenase (IDO), making it the most potent inhibitor known at the time of its discovery. IDO is an enzyme that catalyzes the initial step in the degradation of the amino acid tryptophan into kynurenine. This pathway has significant immunoregulatory effects, particularly in the tumor microenvironment. Tumors often upregulate IDO expression as a mechanism to suppress the immune response. By depleting tryptophan in the local environment and increasing levels of

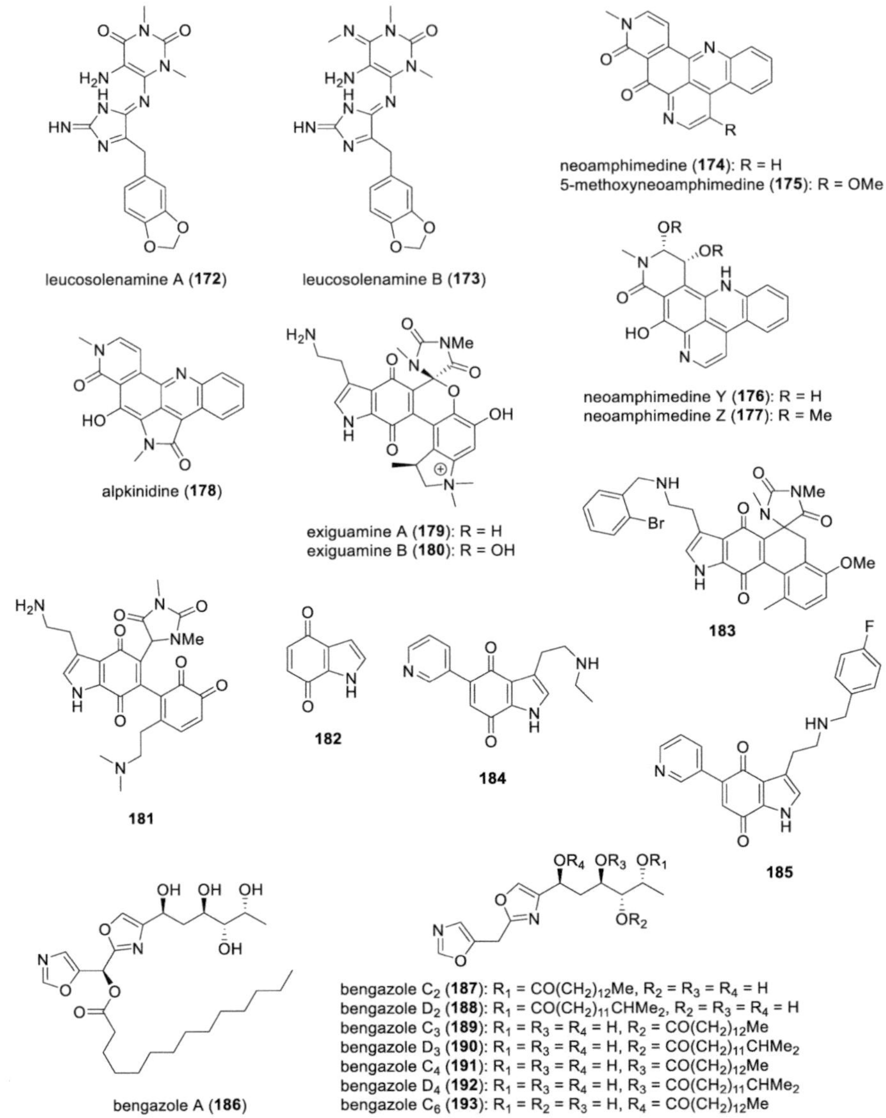

Fig. 20 Chemical structure of leucosolenamines, bisannulated acridines, exiguamines, bengazoles and synthetic analogues

kynurenine, IDO can inhibit the proliferation and function of effector T cells, which are crucial for the immune system's ability to target and destroy cancer cells. Hence, IDO inhibitors play an important role in cancer treatment by targeting the immune evasion mechanisms employed by tumors [152]. Subsequently, the hydroxylated analog, exiguamine B (**180**) (Fig. 20), from the same sponge extract was discovered [153].

Research efforts have been made by several synthetic groups on the total synthesis of the exiguamines and derivatives. For instance, Trauner successfully synthesized both exiguamines A and B, along with the putative biosynthetic intermediate **181** (Fig. 20), and demonstrated that exiguamine B and **181** had *Ki* values of approximately 80 nM for IDO inhibition [153]. Carr and co-workers synthesized a simplified pharmacophore analog, compound **182** (Fig. 20), which is more accessible than the natural alkaloids and still exhibits a potent K_i of approximately 200 nM [154]. A series of exiguamine A analogs were designed and synthesized by Dong and co-workers through a 15-step process. Among them, compound **183** (Fig. 20) emerged as the most potent IDO1 inhibitor, with an IC_{50} value of 65.3 nM, closely matching that of the positive control drug epacadostat ($IC_{50} = 46$ nM). Additionally, compound **183** demonstrated greater selectivity for IDO1 over tryptophan 2,3-dioxygenase and exhibited no cytotoxicity at its effective concentration, making it a strong candidate for further optimization and evaluation [155].

A series of 1*H*-indole-4,7-dione derivatives were designed and synthesized as IDO1 inhibitors, drawing inspiration from the structure and bioactivity of exiguamine A and tsitsikammamines [156]. Initial SAR studies indicated that incorporating a 3-pyridinyl group at the 5-position and a smaller *N*-alkyl substituent in the side chain at the 3-position of the 1*H*-indole-4,7-dione scaffold enhances inhibitory activity. The 5-(3-pyridinyl)-1*H*-indole-4,7-dione scaffold emerged as a promising pharmacophore for IDO1 inhibition, with compound **184** (Fig. 20) identified as a hit, exhibiting an IC_{50} value of 0.16 μM in enzymatic assays. Based on this initial work, compounds featuring different side-chain substituents and linkers of varying lengths at the 3-position were subsequently synthesized to incorporate an aryl motif capable of occupying pocket B of the IDO1 protein [157]. Among them, 3-(2-((4-fluorobenzyl)amino)ethyl)-5-(3-pyridyl)-1*H*-indole-4,7-dione (**185**) (Fig. 20) exhibited the most potent inhibitory activity, with an IC_{50} of 0.125 μM in an enzymatic assay and an EC_{50} of 0.821 μM in a cellular assay. Compound **185** also demonstrated greater selectivity for IDO1 over indoleamine 2,3-dioxygenase 2 (IDO2) and tryptophan 2,3-dioxygenase (TDO) at the effective concentration.

3.3.10 Bengazoles

Bengazoles are among the most common secondary metabolites identified in *Jaspis* species. In 1988, Crews and colleagues isolated two unusual oxazole compounds, bengazoles A (**186**) (Fig. 20) and B, from a large quantity of Jaspidae sponge collected from new sites in the Benga Lagoon, Fiji [158]. Later, in 1993, Crews et al. reported the discovery of an inseparable mixture of seven new analogues, namely bengazoles C_2 (**187**), D_2 (**188**), C_3 (**189**), D_3 (**190**), C_4 (**191**), D_4 (**192**) and C_6 (**193**) (Fig. 20), in a specimen of *Jaspis* cf. *corticea* collected from Boi-Boi Waga Island, Papua New Guinea [159]. This mixture also contained bengazole A, as well as bengamides A and B. Cytotoxicity screening by the U.S. National Cancer Institute across its panel of 60 human cancer cell lines revealed that bengazole A displayed notable in vitro activity, with GI_{50} values of 0.181 μM and 1.13 μM against COLO-205 and SK-MEL-5 cell lines, respectively.

3.4 Linear and Cyclic Peptides/Depsipeptides

3.4.1 Hemiasterlins, Milnamides and Criamides

Hemiasterlins and milnamides are two closely related families of small cytotoxic peptides derived from several marine sponge species. These peptides feature a dipeptide side chain composed of two unusual amino acid residues, namely a *tert*-leucine and a *N*-methylvinylogous valine. In hemiasterlins, this dipeptide fragment is combined with another atypical amino acid, either tri- or tetra-methylated tryptophan. In milnamides, the dipeptide fragment is linked to a tetrahydro-β-carboline residue. Criamides A and B share structural similarities with hemiasterlins and milnamides, with the key difference being the inclusion of an arginine unit. The fact that these related peptides have been reported from different sponge species point to the microbial symbionts as the true origin of the natural products.

In an effort to re-isolate additional cyclodepsipeptides, geodiamolides, from the sponge *Cymbastela* sp. (previously identified as *Pseudaxinyssa* sp.), obtained from Madang, Papua New Guinea, for in vivo evaluation, Andersen and co-workers obtained novel peptides, hemiasterlins A (**195**) and B (**196**), criamides A (**198**) and B (**199**) as well as known hemiasterlin (**194**) (Fig. 21) [160]. In contrast to criamide A, criamide B demonstrated activity against various cell lines, including murine leukemia, human breast cancer, human glioblastoma/astrocytoma, human ovarian carcinoma, human colon and lung cancer cells. Hemiasterlin A is structurally related to the known cytotoxic compound hemiasterlin, which was first isolated from a South African sponge *Hemiasterella minor* by Kashman's group [161]. Further chemical and biological investigations of extracts from the sponge genus *Auletta* and two collections of *Siphonochalina* sp., collected off the coast of PNG by the Coral Reef Foundation (under contract to the NCI), afforded a new analogue, hemiasterlin C (**197**) (Fig. 21) [162]. Bioactivity testing confirmed that hemiasterlin A showed activity against human glioblastoma/astrocytoma and ovarian carcinoma cell lines, while hemiasterlin B was cytotoxic toward P388, MCF7 and human ovarian carcinoma cells. It was also reported that these pharmacologic effects are caused by hemiasterlins binding to tubulin and inhibiting spindle microtubule formation, which leads to mitotic arrest [163].

Fortuitously, the PNG collection of *Cymbastella* sp. by Andersen and co-workers provided enough pure hemiasterlin to accurately determine IC_{50} values and conduct the first in vivo mouse experiments. Hemiasterlin inhibited P388 in vitro with an astonishing IC_{50} of 87 pM, more than two orders of magnitude lower than previously reported by Kashman [160]. Additionally, the compound demonstrated significant anti-leukemic activity in a mouse model, with a 308% increase in life extension

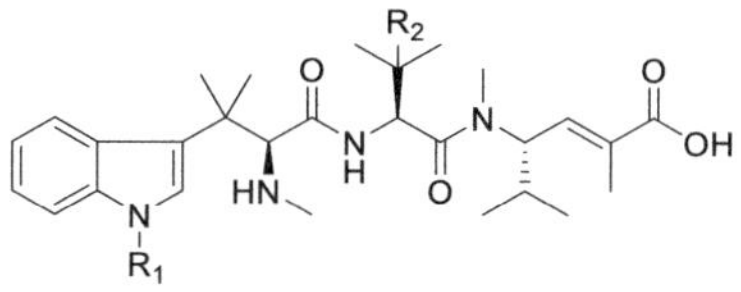

hemiasterlin (**194**): $R_1 = R_2 = Me$
hemiasterlin A (**195**): $R_1 = H$, $R_2 = Me$
hemiasterlin B (**196**): $R_1 = R_2 = H$
hemiasterlin C (**197**): $R_1 = Me$, $R_2 = H$

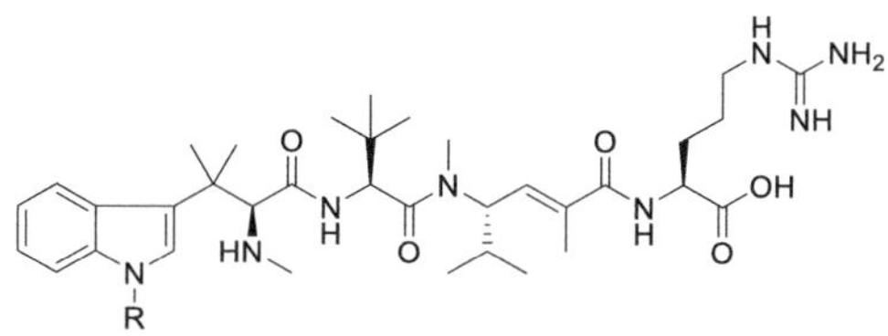

criamide A (**198**): R = H
criamide B (**199**): R = Me

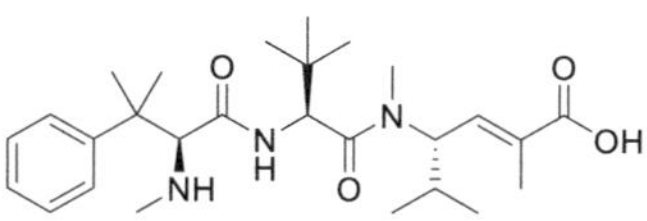

HTI-286 (taltobulin) (**200**)

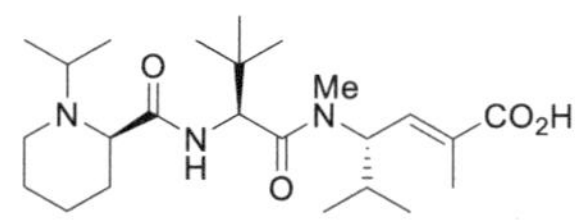

E7974 (**201**)

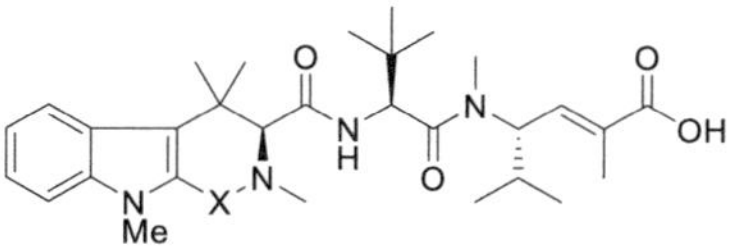

milnamide A (**202**): X = CH_2
milnamide C (**203**): X = C=O

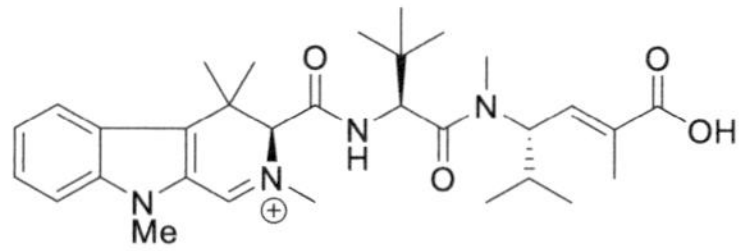

milnamide D (**204**)

Fig. 21 Chemical structures of hemiasterlins, milnamides and synthetic derivatives

(Test/Control) after five intraperitoneal doses of 0.45 μg per mouse. In addition, hemiasterlin demonstrated greater potency as an in vitro cytotoxin and antimitotic agent compared to the anticancer drugs taxol and vincristine [163]. This made it a particularly promising natural product lead for developing new anticancer drugs and numerous synthetic methodologies have been established to create a wide array of hemiasterlin analogues [164]. Structure–activity relationship studies revealed that the *tert*-leucine and *N*-methylvinylogous valine residues are crucial for its bioactivity.

The outcome of these synthetic efforts led to the development of two hemiasterlin analogues, including HTI-286 (taltobulin) (**200**) and E7974 (**201**) (Fig. 21), who have gone through clinical trials as anticancer drugs. One of the significant characteristics of HTI compounds is their ability to overcome P-glycoprotein-mediated chemoresistance [165]. In addition, HTI-286 has demonstrated exceptional activity against paclitaxel- and vincristine-resistant solid tumors both in vitro and in vivo and is also orally available. Through detailed binding kinetics studies using photoaffinity analogs of HTI-286, it was found that HTI-286 may interact with alpha-tubulin. Both HTI-286 and E7974 are currently being formulated as antibody–drug conjugate for potential treatment of cancer diseases [166].

The structure of the highly methylated tripeptide milnamide A (**202**) (Fig. 21) was first reported by Crews and co-workers in 1993 [167]. This molecule, which closely resembles hemiasterlin A, was isolated from samples of an orange sponge, *Auletta* cf. *constricta* obtained from Milne Bay, PNG. Initial bioactivity screening showed that milnamide A was broadly cytotoxic as it showed in vitro IC_{50}'s µg/mL values of 4.1 (A549), 2.8 (HT-29), 3.3 (B16/F10) and 0.74 (P388). Subsequently, additional analogue, milnamide C (**203**) (Fig. 21), was obtained during a re-investigation of an *Auletta* sp., along with known milnamides A and B (= hemiasterlin), jasplakinolide (**236**) and geodiamolides A, D, E and G [168]. Milnamide C was found to be significantly less potent than hemiasterlin when tested against MDA-MB-435 cancer cells with IC_{50} values of 3.2×10^{-1} and 1.48×10^{-4} µg/mL, respectively. Ireland and co-workers reported another derivative, milnamide D (**204**) (Fig. 21), from a Papua New Guinea *Cymbastela* sp., which, like milnamide A, exhibited cytotoxic properties [169]. When tested against two colorectal cancer cell lines, including HCT-116 wild type and p53-deficient mutant, milnamide D (IC_{50} 66.8 nM) was significantly more potent than milnamide A (IC_{50} 1652.6 nM) but less potent than hemiasterlin (IC_{50} 6.8 nM). Milnamides A, C and D were also found to inhibit tubulin polymerization [168, 169].

3.4.2 Stylostatin 1 and Stylopeptides

Based on samples of *Stylotella* sp. collected in 1983 and 1986 from the Bismarck Archipelago, Papua New Guinea, several cyclic peptides, including stylostatin 1 (**205**), stylopeptides 1 (**206**) and 2 (**207**) (Fig. 22), have been reported [170–172]. Stylostatin 1 exhibited cell growth inhibitory activity against P388 lymphocytic leukemia cell lines with ED_{50} of 0.8 µg/mL. In addition, stylopeptide 2 inhibited two breast cancer cell lines, namely BT-549 displayed growth at 77% of the mean and HS 578 T at 56% of the mean, in the National Cancer Institute's one-dose 60-cell line assay.

3.4.3 Stylissatin A

The cyclic heptapeptide stylissatin A (**208**) (Fig. 22), with the sequence cyclo-(Tyr1-Phe7-Pro6-Ile5-Pro4-Phe3-Ile2), was isolated from the orange marine sponge, *Stylissa massa*, collected from Loloata Island, Papua New Guinea. Stylissatin A inhibited nitric oxide (NO) generation in RAW264.7 macrophages activated by lipopolysaccharide, with an IC_{50} of 87 µM [173]. Since its discovery, several groups have reported the synthesis and biological activities of stylissatin A and its derivatives [174–176]. Among synthetic stylissatin A, the *tert*-butyl ether analog, *D*-Tyr^1-*t*BuSA (**209**) (Fig. 22) stands out for its remarkable ability to inhibit NO production with minimal cytotoxicity [174, 177]. Additionally, **209** suppresses the production of the proinflammatory cytokines Interleukin-6 (IL-6) and tumor necrosis factor alpha (TNF-α), as well as the expression of inducible nitric oxide synthase (iNOS) [178]. Using a stylissatin A biotin probe (3-BP) (**210**) (Fig. 22) derived from propargyl stylissatin A, acyl-CoA dehydrogenase long chain (ACADL) was identified as a

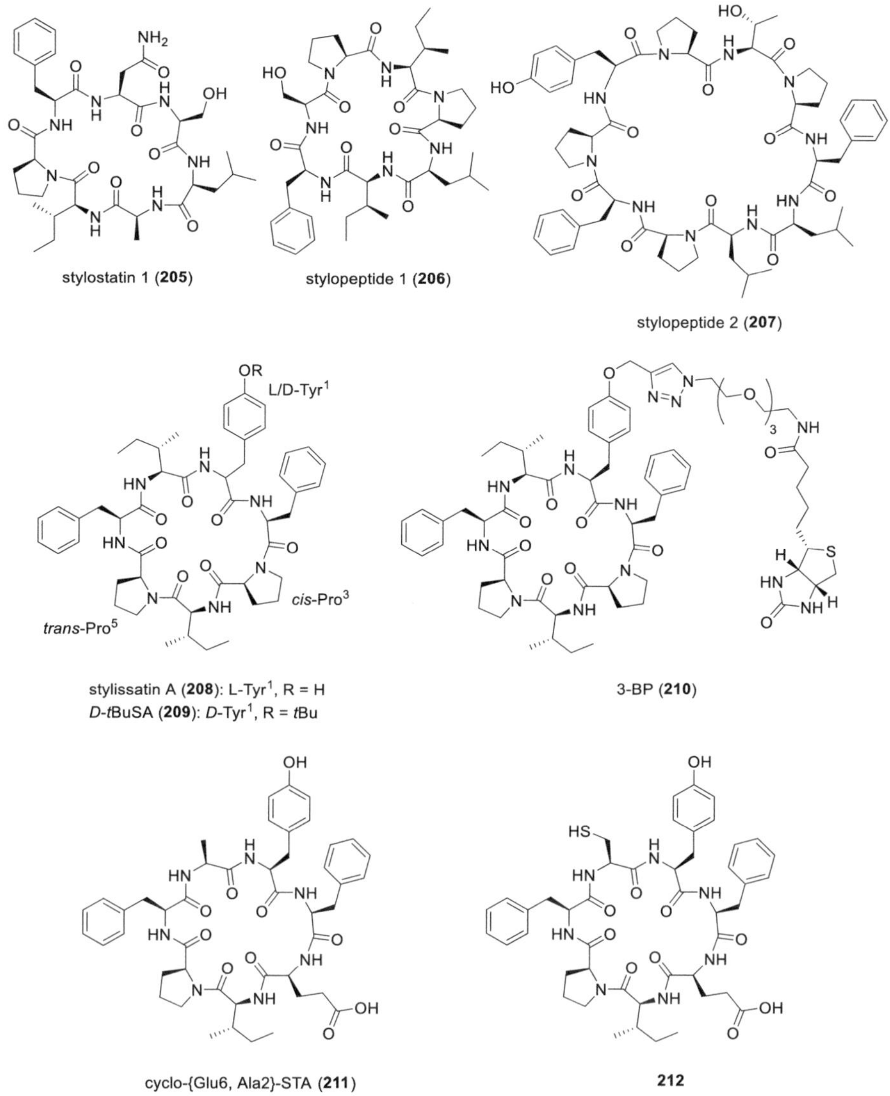

Fig. 22 Chemical structures of stylostatin 1, stylopeptides, stylissatin A and synthetic analogues

target protein in macrophage lysates. ACADL is involved in the β-oxidation of long-chain fatty acids, playing a crucial role in mitochondrial fatty acid metabolism [179]. Inhibition of ACADL leads to the accumulation of fatty acids, which in turn activates peroxisome proliferator-activated receptors (PPAR). Activation of PPARα has been linked to anti-inflammatory effects through the inhibition of nuclear factor-kappa B (NF-κB) transcription in cells [180]. Therefore, it has been proposed that stylissatin

A may inhibit ACADL, leading to fatty acid accumulation and subsequent inhibition of the NF-κB signaling pathway and iNOS expression, ultimately displaying anti-inflammatory activity.

In another study led by Shaheen and co-workers, several synthetic analogs of stylissatin A have been reported to exhibit significant inhibitory activity against IL-2 cytokine production and T-cell proliferation, surpassing that of stylissatin A itself [175]. However, none of these analogs demonstrated inhibition of nitric oxide (NO) production. Among them, the analog cyclo-[Glu6, Ala2]-STA (**211**) (Fig. 22) showed excellent inhibition of interleukin-2 and T-cell proliferation but was ineffective in inhibiting NO production. A second-generation of analogs was then carried out to develop new non-cytotoxic compounds with potent anti-inflammatory activity [181]. Analog **212** (Fig. 22), in particular, demonstrated potent suppression of NO production with IC_{50} of 46.0 μM, surpassing the activity of stylissatin A. Additionally, analog **212** downregulated the transcription of pro-inflammatory molecules, including tumor necrosis factor-α, interleukin-1β, caspase-1 and ASC, further emphasizing its anti-inflammatory and potential therapeutic properties. Notably, analog **212** was non-toxic to BJ and Vero cell lines.

As chronic inflammation in adipose tissue is closely associated with metabolic disorders, including obesity, insulin resistance and type 2 diabetes, stylissatin A and its analogs have been explored for their possible modulatory activity on adipocyte differentiation and fat accumulation. Kita and co-workers demonstrated that stylissatin A and analogs effectively inhibit the differentiation of murine 3T3-L1 preadipocytes [177]. Among these analogs, *D*-Tyr^1 analog **209** exhibited the strongest inhibitory effect on fat accumulation in preadipocytes, with an EC_{50} of 0.44 μM and minimal cytotoxicity. Additionally, stylissatin A, **209** and its *L*-Tyr^1 isomer, *t*BuSA, significantly reduced triglyceride accumulation in mature adipocytes, with EC_{50} values of 6.1, 7.0 and 1.9 μM, respectively. Recently, stylissatin A and its analogs were shown to inhibit membranous neuraminidase 1 (Neu1) function by interacting with lysosomal protective protein cathepsin A (PPCA) [182]. Stylissatin A analogs do not directly bind to Neu1 but modulate the molecular chaperone activity of PPCA, resulting in degradation of perilipin 1 around lipid droplets and inhibited fat accumulation, along with decrease in membranous Neu1. These results demonstrated that stylissatin A and analogs play critical role in protein–protein interaction between PPCA and Neu1, at least in adipocytes. On going studies are in progress on the effects of stylissatin A and its analogs on cellular signaling, structural modification for bioavailability as well as in vivo experiments using mouse models of inflammation and obesity.

3.4.4 Cyclocinamide A

Cyclocinamide A (**213**) (Fig. 23) is a minor cyclohexapeptide isolated by Crews and co-workers from the marine sponge, *Psammocinia* sp., collected from the Milne Bay Province, Papua New Guinea [183]. Its tetrapeptide core consists of a 14-membered ring, along with the dipeptide side chain terminating in the proline-derived

N-methylchloropyrrole. Cyclocinamide A exhibited robust tumor selectivity against C38 with the following cytotoxic profile: [μg/disk, L1210/C38/M17-Adr/CX1] 50, 0/500/0/0. When cyclocinamide A was first reported in 1997 by Crews' group, they were unable to fully determine its absolute structure using chiral TLC analysis of its hydrolysate. Only two of the four chiral centers were identified, with stereochemistries of 7*S* and 14*S* for 5-bromotryptophan and asparagine, respectively. Subsequent efforts to determine the complete structure included the successful synthesis of nominal (4*R*, 11*R*)-cyclocinamide A by Grieco and Reilly, as well as nominal (4*R*, 11*S*)-cyclocinamide A by Postema and Liu [184]. However, both synthetic studies concluded that the natural product differed from the synthetic versions. In 2008, a recollection of the cyclocinamide-containing sponge led to the isolation of more of the cyclopeptide [185]. The acid hydrolysate of this sample was derivatized using Marfey's method, which supposedly resulted in the complete assignment of the absolute structure of cyclocinamide A as all-*S*. However, in 2012, Konopelski et al., who synthesized the remaining two nominal cyclocinamide A diastereomers, found that the natural product did not match the NMR spectra of either the 4*S*, 11*S*- or 4*S*, 11*R*-diastereomer [186]. Another analogue, cyclocinamide B, characterized by Ireland and co-workers in 2007 from the Fijian sponge *Corticium* sp., was assigned a 4*S*, 7*R*, 11*S*, 14*R* stereochemistry using Marfey's and modified Marfey's analyses [187]. Despite these efforts, the complete structures of the cyclocinamides remain uncertain due to conflicting results.

3.4.5 Scleritodermin A

Samples of *Scleritoderma nodosum*, collected from the northwest side of Olango Island, Cebu, Philippines and from Milne Bay, Papua New Guinea, yielded the cytotoxic cyclic peptide, scleritodermin A (**214**) (Fig. 23) [188]. Scleritodermin A was tested against various cancer cell lines, including HCT-116, HCT-116/VM46 multidrug-resistant colon cancer, A2780 human ovarian carcinoma and SKBR3 breast carcinoma, with significant IC_{50} values of 1.9, 5.6, 0.940 and 0.670 μM, respectively. In A2780 cells treated with scleritodermin A at 1.3 μM for 24 h, a G2/M cell cycle block was observed, which is characteristic of molecules targeting tubulin. Further studies revealed that scleritodermin A inhibited GTP-induced tubulin polymerization by 50% at a concentration of 10 μM. Additionally, a 5.5-fold increase in apoptosis induction was observed after 24 h of compound exposure at a concentration near its cytotoxic IC_{50}. Given its significant in vitro cytotoxicity and its unique structural features, including an *O*-methyl-*N*-sulfoserine, a conjugated thiazole moiety and an α-ketoamide unit, the total synthesis of scleritodermin A was undertaken [189]. Initially, NOESY data suggested a 2*Z*, 4*E* configuration for the conjugated thiazole moiety, but this was revised to a 2*E*, 4*E* configuration after its total synthesis, based on the observation that the methine protons, CH-3 and CH-5, appeared 1.0 ppm further upfield than in the natural product. Additionally, differences in the chemical shifts of the two methyl groups in the keto-Ile moiety led to the revision of the 14*R* assignment to a 14*S* configuration.

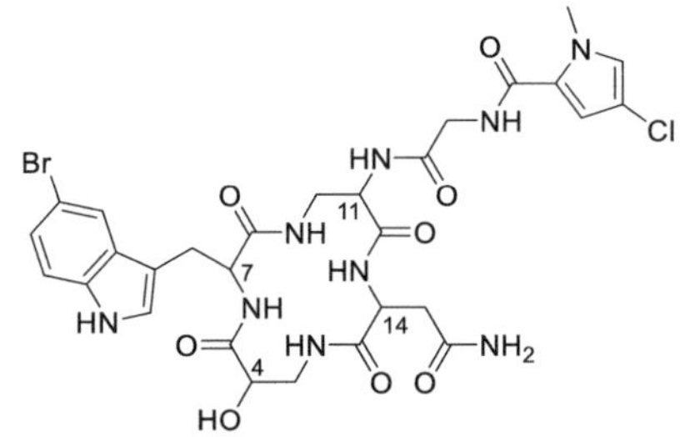

cyclocinamide A (**213**)

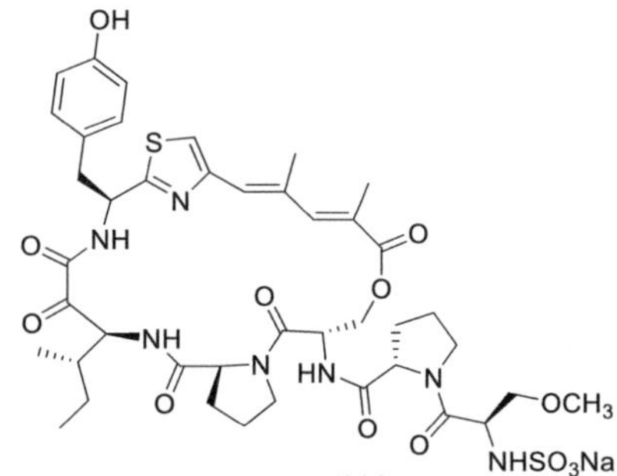

scleritodermin A (**214**)

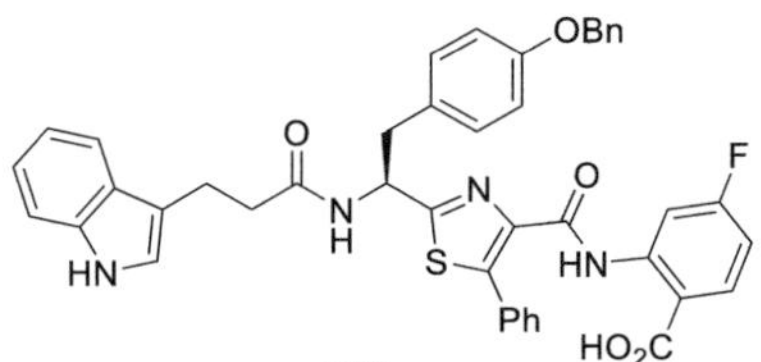

215

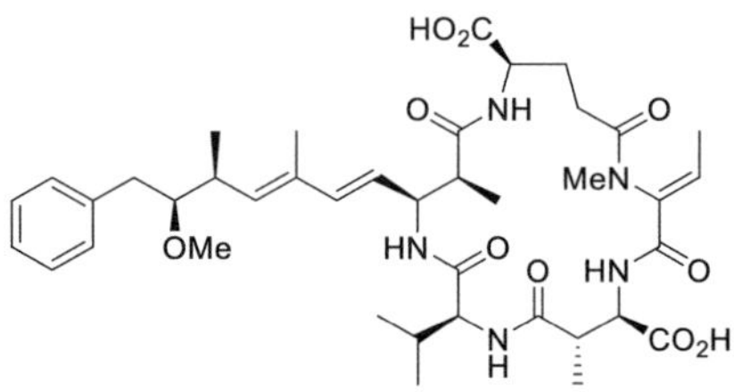

Motuporin (**216**)

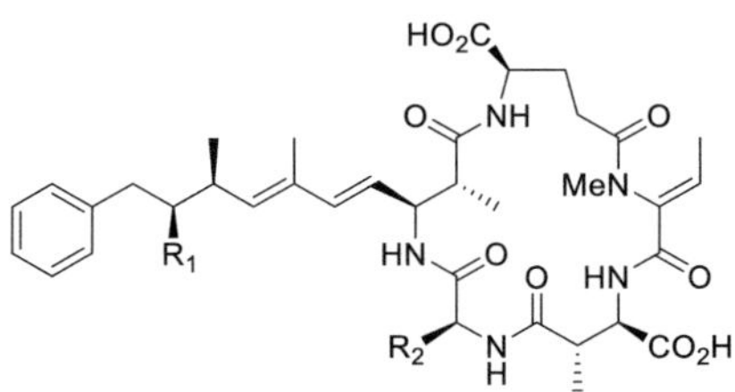

*iso*motuporin (**217**): $R_1 = OCH_3$, $R_2 = CH(CH_3)_2$
*iso*motuporin B (**218**): $R_1 = OCH_3$, $R_2 = CH(CH_3)CH_2CH_3$
*iso*motuporin C (**219**): $R_1 = OCH_3$, $R_2 = CH_2CH_3$
*iso*motuporin D (**220**): $R_1 = H$, $R_2 = CH(CH_3)_2$

Fig. 23 Chemical structures of cyclocinamide A, scleritodermin A, motuporin and related molecules

Inspired by the key structure of scleritodermin A, synthetic chemists reported the discovery of novel protein tyrosine phosphatase 1B (PTP1B) inhibitors featuring an ACT [2-(1-amino-2-p-hydroxyphenylethane)-4-(4-carboxy-2,4-dimethyl-2Z,4E-propadiene)-thiazole] skeleton [190, 191]. PTP1B is recognized as a key negative regulator of insulin and leptin signal transduction pathways. As such, PTP1B inhibitors have become attractive and potent pharmaceutical candidates for treating type 2 diabetes and obesity. One such synthetic derivative, **215** (Fig. 23), demonstrated the ability to inhibit intracellular PTP1B, thereby activating the insulin signaling pathway. Treatment of cells with **215** significantly increased the phosphorylation levels of IRβ and Akt, as well as the rate of glucose uptake.

3.4.6 Motuporins

Motuporin (**216**) (Fig. 23) is a cyclic pentapeptide derived from the sponge *Theonella swinhoei* collected from Motupore Island, Papua New Guinea [192]. It is structurally related to nodularin and microcystin LR, both of which are cyclic peptides associated with toxicity in cyanobacteria *Nodularia* and *Microcystis*, respectively [193]. The most unique structural feature of motuporin is the double unsaturated β-amino acid, (2*S*,3*S*,8*S*,9*S*)-3-amino-9-methoxy-2,6,8-trimethyl-10-phenyl-4,6-decadienoic acid (Adda), which is also found in the cyanobacterial compound, microcystin. Motuporin is also termed nodularin-V due to the valine in its variable position. Motuporin was found to inhibit protein phosphatase 1 (PP1) at concentrations below 1 nM [192]. Additionally, motuporin exhibited activity against murine leukemia as well as human lung, ovarian, colon, breast and brain cancer cell lines [192]. Structural comparisons between motuporin and the peptides nodularin and microcystin LR revealed that the polar arginine residue is not essential for PP1 inhibition, as motuporin retained potent activity even when arginine was substituted with the non-polar valine [194]. Subsequent screening of 14 collections of *T. swinhoei* from various Indo-Pacific locations using LCMS led to the identification of motuporin analogues, including isomotuporin (**217**) and isomotuporins B (**218**) to D (**220**) (Fig. 23) in two sponge collections from Normanby, Papua New Guinea [195]. These analogues featured a novel 2*R* configuration for the Adda amino acid, with one analogue containing a unique non-oxygenated Adda amino acid.

3.4.7 Geodiamolides and Jaspamides (= Jasplakinolide)

Two closely related families of cyclic depsipeptides, known as jaspamides and geodiamolides, have been isolated from various tropical marine sponges. Jaspamide (also called jasplakinolide) (**236**) (Fig. 24) was the first reported member of this group, independently discovered in samples of *Jaspis* sp. collected from Palau and Fiji [196, 197]. Soon after, geodiamolides A (**221**) and B (**222**) (Fig. 24) were identified from the Caribbean sponge *Geodia* sp. [198]

Geodiamolides are notable for their macrocyclic structures that include both peptide and polyketide-derived fragments. The known peptides geodiamolides A and B, along with the newly discovered geodiamolides C (**223**)–F (**226**) (Fig. 24), were isolated from a sample of *Pseudaxinyssa* sp. (now classified as *Cymbastela* sp.) collected at reefs off Matupore and Madang, Papua New Guinea [198, 199]. These peptides were found to be cytotoxic in the nanomolar range. Subsequently, geodiamolide G (**227**) (Fig. 24) was reported along with hemiasterlins A and B as well as criamides A and B from *Cymbastela* sp. recollected at Motupore and Madang, Papua New Guinea [160]. Further analogs, geodiamolides J (**228**)–P (**234**) and R (**235**) (Fig. 24), were subsequently isolated from another *Cymbastela* sp. sample from Papua New Guinea [200]. Geodiamolides A–F have very similar structures, with only slight variations in their macrocyclic peptide backbones. Additional structural diversity comes from the halide substituent on the aromatic ring. Unlike other

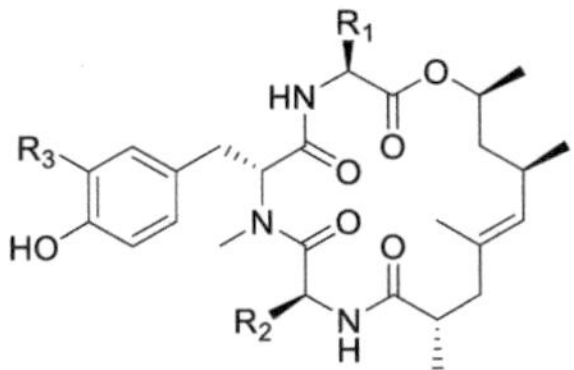

geodiamolide A (**221**): $R_1 = R_2 = Me$, $R_3 = I$
geodiamolide B (**222**): $R_1 = R_2 = Me$, $R_3 = Br$
geodiamolide C (**223**): $R_1 = R_2 = Me$, $R_3 = Cl$
geodiamolide D (**224**): $R_1 = Me$, $R_2 = H$, $R_3 = I$
geodiamolide E (**225**): $R_1 = Me$, $R_2 = H$, $R_3 = Br$
geodiamolide F (**226**): $R_1 = Me$, $R_2 = H$, $R_3 = Cl$
geodiamolide L (**230**): $R_1 = CH_2OH$, $R_2 = Me$, $R_3 = I$
geodiamolide M (**231**): $R_1 = CH_2OH$, $R_2 = Me$, $R_3 = Br$
geodiamolide N (**232**): $R_1 = CH_2OH$, $R_2 = Me$, $R_3 = Cl$
geodiamolide O (**233**): $R_1 = Me$, $R_2 = CH_2OH$, $R_3 = I$
geodiamolide P (**234**): $R_1 = Me$, $R_2 = CH_2OH$, $R_3 = Br$
geodiamolide R (**235**): $R_1 = CH_2OH$, $R_2 = H$, $R_3 = I$

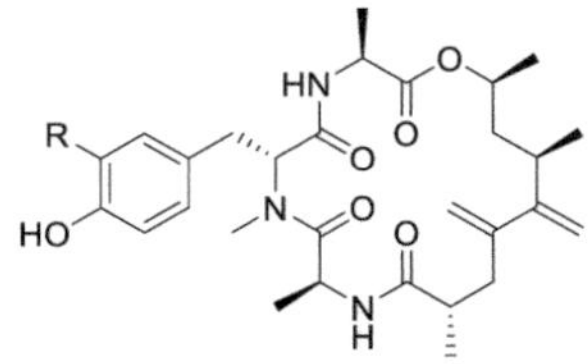

geodiamolide G (**227**): R = I
geodiamolide J (**228**): R = Br
geodiamolide K (**229**): R = Cl

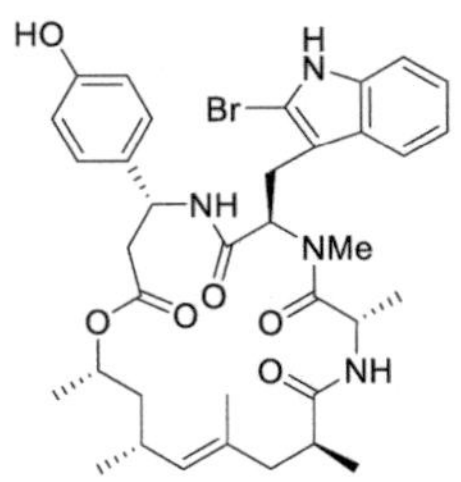

jasplakinolide (**236**)

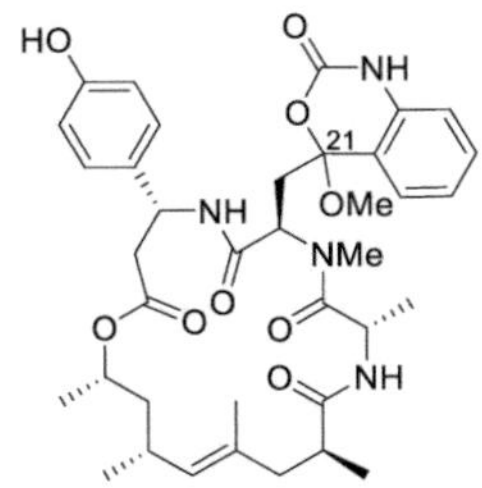

jasplakinolide P/
21-*epi*-jasplakinolide P (**237**)

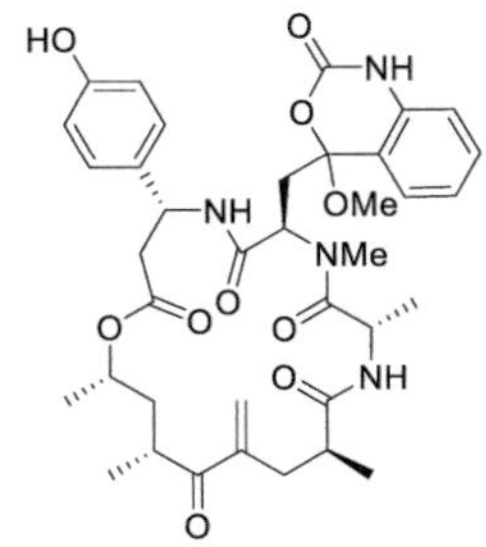

jasplakinolide U (**241**)

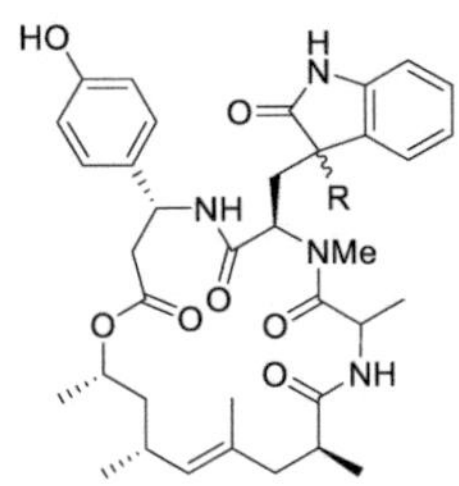

jasplakinolide S (**238**): R = ◄OH
21-*epi*-jasplakinolide S (**239**): R = ···OH

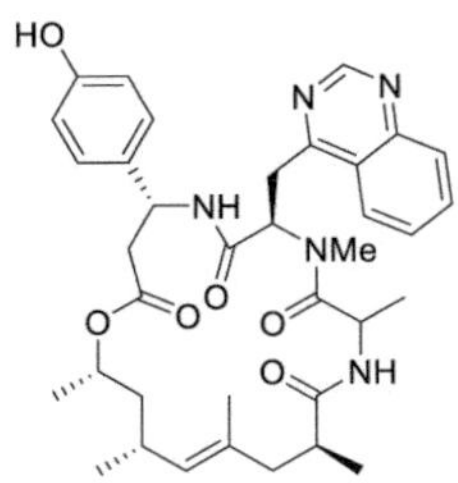

jasplakinolide T (**240**)

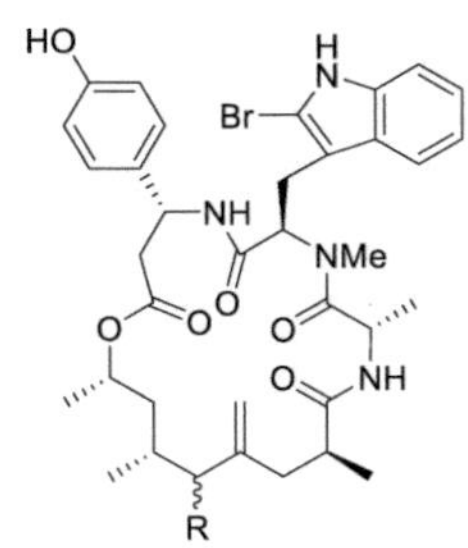

jasplakinolide B (**242**): R = =O
jasplakinolide C_a (**243**): R = ···OH
jasplakinolide C_b (**244**): R = ◄OH

Fig. 24 Chemical structures of geodiamolides and jasplakinolides

compounds in this class, geodiamolides L–P and R contain serine-derived fragments in their macrocycles, adding to the range of structural variants. The fact that these cyclic depsipeptides have been isolated from taxonomically distant sponge species obtained from distinct geographical location, suggest that microbial symbionts may be involved in their biosyntheses.

In their study on structure–activity relationships, Andersen and colleagues observed that geodiamolide G (**227**) has a slight difference in the polyketide fragment compared to geodiamolides A–F. This specific modification was found to reduce the cytotoxicity levels of the geodiamolides in vitro against human glioblastoma/astrocytoma and human ovarian carcinoma cells. However, variations in the amino acid residues in geodiamolides C–F did not result in significant changes in toxicity levels in the same bioassays.

In 1986, two publications were released detailing the structure of a unique cyclic depsipeptide-polyketide compound named jasplakinolide by Crew's lab, or jaspamide, by the Ireland-Faulkner-Clardy research consortium [196, 197]. This molecule has been reported to exhibit cytotoxic effects against P388 murine leukemia cells [161], as well as insecticidal activity against *Heliothis virescens* and antimicrobial activity against *Candida albicans* [196, 197, 201]. It also demonstrated antihelminthic properties, along with ichthyotoxicity [196, 202]. In addition, cytotoxicity studies conducted at the US National Cancer Institute revealed that the compound showed significant activity against renal, prostate and CNS tumor cell lines, prompting an investigation of two samples of *Auletta* sp. from Papua New Guinea and a Fijian *Jaspis splendens* (formerly *Jaspis johnstoni*) for additional analogs [167, 203]. This work resulted in the isolation of jasplakinolide (**236**) and eleven jasplakinolide analogues, including seven new analogues, namely compounds 21-*epi*-jasplakinolide P (**237**), jasplakinolide S (**238**), 21-*epi*-jasplakinolide S (**239**), jasplakinolide T (**240**), jasplakinolide U (**241**), jasplakinolide C_a (**243**) and jasplakinolide C_b (**244**) (Fig. 24). Among the new compounds, five are oxidized tryptophan derivatives of jasplakinolide, including a unique quinazoline derivative jasplakinolide T which was the only isolated from the Fijian *J. splendens*. Several selected compounds were assessed in the NCI 60 cell line screen and tested in a microfilament disruption assay. In particular, jasplakinolide B (**242**) (Fig. 24) exhibited potent cytotoxicity (GI_{50} < 1 nM against human colorectal adenocarcinoma (HCT-116) cells) but did not show microfilament-disrupting activity at 80 nM.

3.4.8 Papuamides

The papuamides A (**245**)–D (**248**) (Fig. 25) are cyclic depsipeptides isolated from the sponges *Theonella mirabilis* and *T. swinhoei* collected from Madang Harbour, Papua New Guinea [204]. *Theonella* species are known for producing many significant peptide natural products, including the cyclotheonamides, pseudotheonamides, theonellapeptolides, theonellamides and microsclerodermins [205]. These peptides exhibit a range of biological activities such as cytotoxic, immunosuppressive, antifungal, antitumor and serine protease inhibition. Some of these compounds are

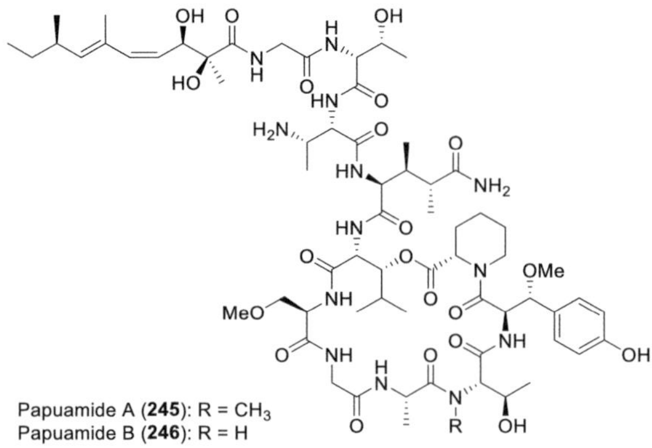

Papuamide A (**245**): R = CH_3
Papuamide B (**246**): R = H

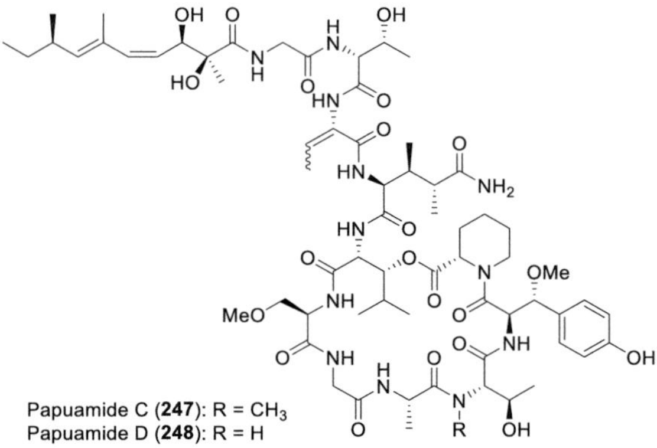

Papuamide C (**247**): R = CH_3
Papuamide D (**248**): R = H

Fig. 25 Chemical structures of papuamides

believed to be produced by bacterial symbionts. The papuamides feature a cyclic depsipeptide ring composed of seven amino acids, to which a linear tetrapeptide is attached through an amide bond. The *N*-terminus of the tetrapeptide chain is capped by a polyketide chain, specifically 2,3-dihydroxy-2,6,8-trimethyldeca-(4*Z*,6*E*)-dienoic acid. Subsequent total synthesis by Ma and colleagues confirmed the stereochemistry of these cyclodepsipeptides [206].

In vitro screening of papuamides A and B (**246**) revealed that they inhibited HIV-1 RF infection of human T-lymphoblastoid cells with an EC_{50} value of approximately 4 ng/mL. Studies conducted by Barrows and co-workers found that papuamide A has a direct virucidal effect on HIV-1 by inhibiting viral entry into host cells, thereby preventing the transmission of the virus [207]. They also discovered that the observed response was not due to papuamide A binding to CD4 or HIV gp120, the proteins responsible for cell-virus recognition and binding. Papuamides B–D also demonstrated similar viral inhibition properties, sparking interest in these peptides as synthetic targets. In addition, papuamides A and B demonstrated potent cytotoxicity against human cancer cells, with IC_{50} values of less than 74 ng/mL. A screening panel of human cancer cell lines indicated that multi-drug-resistant cell lines were relatively sensitive to papuamide A, while leukemia cell lines exhibited relative resistance. The specific target molecule responsible for the anti-HIV activity and cytotoxicity of these compounds has yet to be identified.

Papuamides A and B exhibit growth inhibitory effects against the yeasts *Saccharomyces cerevisiae* and *Candida albicans*. The target molecule of papuamides in yeast was identified as phosphatidylserine (PS) through a combination of budding yeast chemical genomic analysis and traditional genetic strategies [208]. After it was determined that papuamide B recognizes PS, this natural product became an important research tool: the amount of PS in the outer leaflet of the plasma membrane of living cells can be assessed by testing cellular sensitivity to papuamides. This strategy enabled biologists to investigate the cellular function of aminophospholipid translocases and to explore new factors involved in the regulation of phospholipid asymmetry [209–212]. In another study, papuamide A was used in high-throughput phenotypic drug screening targeting fungal phosphatidylserine (PS) synthase [213]. Compounds that counteract the growth inhibitory activity of papuamide A against fungi are considered potential drug candidates, as PS synthase is essential for virulence in *C. albicans* [214].

3.4.9 Swinhopeptolides

Two new cyclic depsipeptides, swinhopeptolides A (**249**) and B (**250**) (Fig. 26), have been isolated from the marine sponge *Theonella swinhoei* cf. *verrucosa*, collected from Papua New Guinea [215]. These compounds each contain 11 diverse amino acid residues and 13-carbon polyketide moieties attached at the *N*-terminus. In addition, swinhopeptolide B features a previously undescribed 2,6,8-trimethyldeca-(2*E*,4*E*,6*E*)-trienoic acid moiety *N*-linked to a terminal serine residue. Both swinhopeptolides A and B demonstrated significant inhibition of the Ras/Raf signaling pathway, with IC_{50} values of 5.8 μM and 8.5 μM, respectively. Small molecules that can effectively inhibit protein–protein interactions are relatively rare, largely due to the flat and large nature of these interaction domains. The activity of swinhopeptolides A and B in the cell-based Bioluminescence Resonance Energy Transfer (BRET) assay for Ras–Raf interactions suggests that they can suppress or destabilize the Ras–Raf complex. As such, swinhopeptolides may represent a new structural class of protein–protein interaction modulators.

3.4.10 Neamphamide A

A new HIV-inhibitory cyclic depsipeptide, neamphamide A (**251**) (Fig. 27), was isolated from a marine sponge, *Neamphius huxleyi*, collected at Duke of York Islands, Papua New Guinea [216]. Its complete absolute stereochemistry was established a year after its initial report [217]. Neamphamide A consists of 11 amino acid residues and an amide-linked 3-hydroxy-2,4,6-trimethylheptanoic acid moiety. This molecule exhibits a high degree of structural similarity with another antiviral depsipeptide,

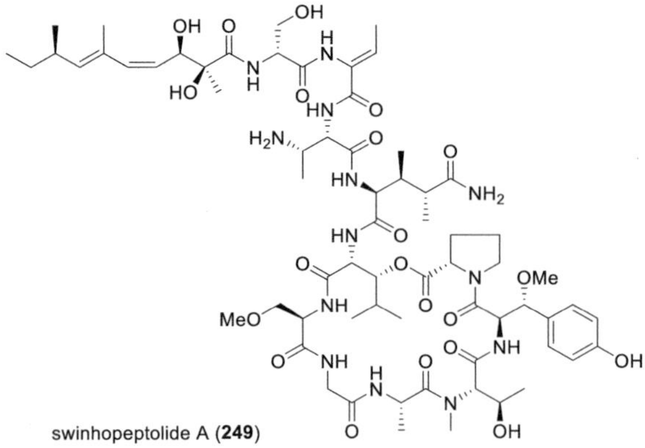

swinhopeptolide A (**249**)

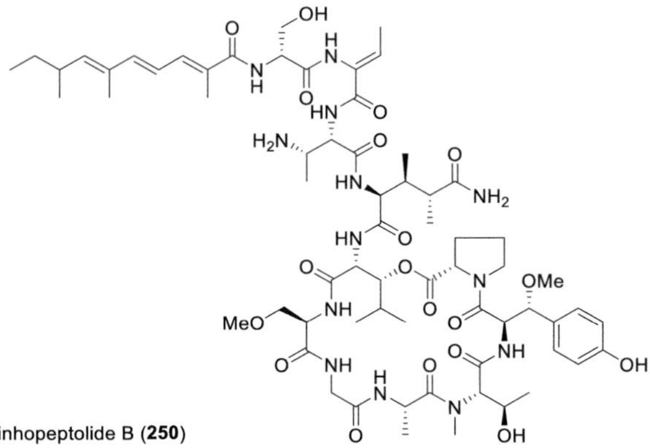

swinhopeptolide B (**250**)

Fig. 26 Chemical structures of swinhopeptolides

callipeltin A initially isolated from the New Caledonian sponge *Callipelta* sp. and later discovered in a *Latrunculia* sp. collected in Vanuatu. The primary difference between neamphamide A and callipeltin A is the expansion of the macrocyclic ring in neamphamide A, due to the incorporation of a Hpr residue at the C-terminus and the presence of two Asn residues. Both neamphamide A (2) and the callipeltins, as well as the papuamides, contain the unusual 3,4-diMeGln and β-*O*MeTyr residues. In a cell-based XTT assay, neamphamide A demonstrated potent cytoprotective activity against HIV-1 infection, with an EC_{50} of approximately 28 nM.

3.4.11 Theopapuamides

Theopapuamide A (**252**) (Fig. 27) is a cytotoxic undecapeptide isolated from *Theonella swinhoei*, collected from Milne Bay, Papua New Guinea [218]. It is notable as the first natural peptide containing β-methoxyasparagine and 4-amino-5-methyl-2,3,5-trihydroxyhexanoic acid residues. Theopapuamide A was tested against the

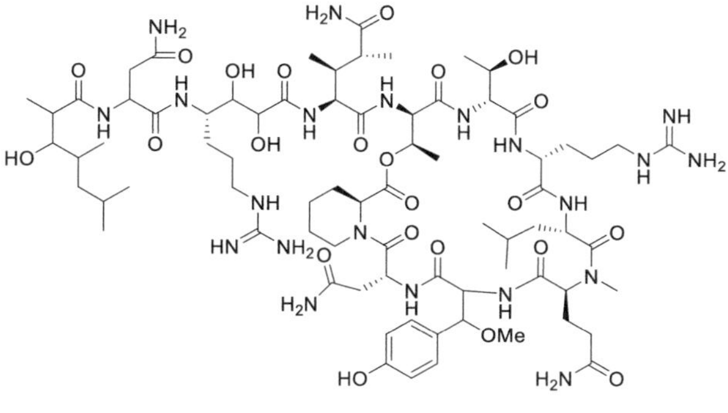

neamphamide A (**251**)

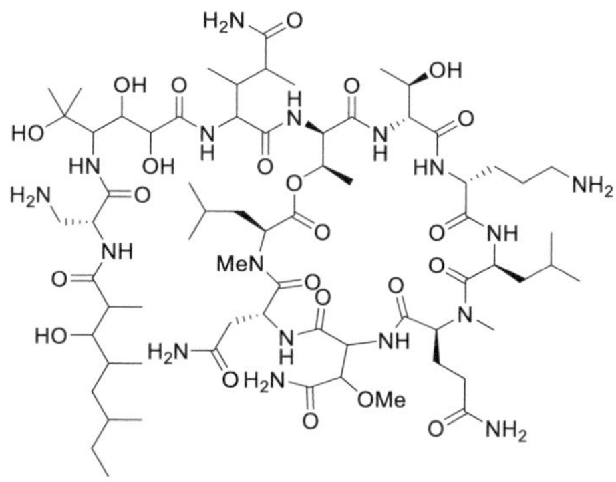

theopapuamide A (**252**)

Fig. 27 Chemical structures of neamphamide A and theopapuamide

CEM-TART cell line, which are T-cells expressing both HIV-1 tat and rev, and the HCT-116 colorectal carcinoma cell line, showing IC_{50} values of 0.5 and 0.9 μM, respectively. Additional new analogs, theopapuamides B–D were reported from an extract of *Siliquariaspongia mirabilis*, collected off Sulawesi Island, Indonesia [219]. The ability to inhibit HIV-1 entry was also evaluated for theopapuamide B, which exhibited IC_{50} value of 0.5 μM. Theopapuamides A - C were also assessed for their ability to inhibit the growth of both wild-type and amphotericin B-resistant strains of *Candida albicans*. Theopapuamide A inhibited the growth of both fungal strains, with zones of inhibition of 8 mm at 1 μg/disk, while theopapuamides B and C displayed zones of 10 mm against both strains at 5 μg/disk.

3.4.12 Neopetrosiamides

Neopetrosiamides A (**253**) and B (**254**) (Fig. 28) are two diastereomeric 28-amino acid tricyclic peptides that inhibit amoeboid invasion of human tumor cells at a concentration of 6 μg/mL [220]. These peptides were isolated from the marine sponge *Neopetrosia* sp., collected near Milne Bay, Papua New Guinea. The two

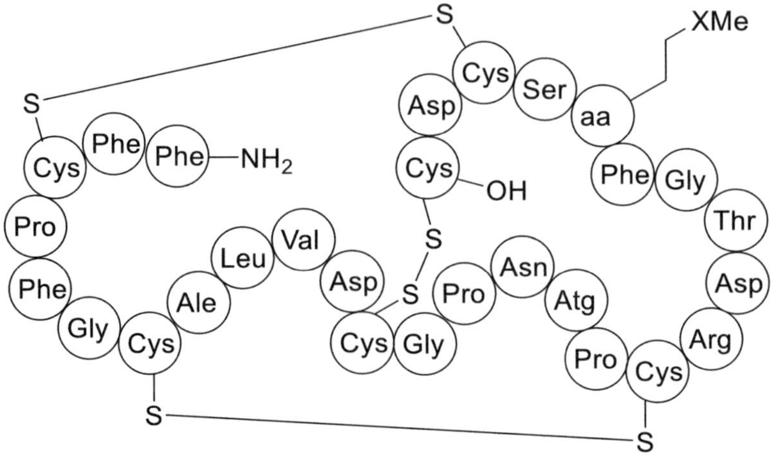

neopetrosiamides A (**253**) and B (**254**): X = S-O

Fig. 28 Chemical structure of neopetrosiamides

diastereomers differ only in the stereochemistry of the methionine sulfoxide at position 24 and contain three intramolecular crosslinking disulfides. They are present in the sponge extract at very high concentrations (0.55% dry weight), which is typical of many sponge peptides, such as motuporin and the papuamides. Subsequent synthetic work by Vederas and co-workers demonstrated that the disulfide bond connectivity in the neopetrosiamides required reassignment, and the correct connectivity is shown in structures **253** and **254** [221]. In addition, neopetrosiamide A was found to be an anti-adhesive peptide that decreases cell surface integrin levels through a novel mechanism, yet to be fully elucidated, involving the release of adhesion molecule-containing vesicles from the tumor cell surface [222].

Based on orthogonal sulfur protection strategy, 15 analogues of neopetrosiamides A and B were prepared using Fmoc solid-phase peptide chemistry [223]. These included six analogues with selected disulfide bond replacements and nine with diastereomeric methionine sulfoxide replacements. Disulfide replacement analogues were shown to lose activity, while only one methionine sulfoxide analogue retained full bioactivity in morphological studies. In another study, synthetic replacement of single disulfide bridges with methylene bridges resulted in active analogues [224]. Moreover, pre-stapling one ring enhances the correct formation of the remaining disulfides by reducing isomeric possibilities and potentially initiating the correct 3D fold.

4 Marine Cyanobacteria

4.1 *Fatty Acid Amides: Molecules Targeting Cannabinoid Receptors*

The class of marine metabolites most commonly active on cannabinoid receptors is predominantly composed of fatty acid amides produced by marine cyanobacteria. Their activity is likely attributed to their chemical resemblance to the endogenous agonist anandamide. Several fatty acid amides have been reported from Papua New Guinean marine cyanobacteria. For instance, semiplenamides A (**255**)–G (**261**) (Fig. 29), a series of newly discovered anandamide-like fatty acid amides, were isolated from a sample of the marine cyanobacterium, *Lyngbya semiplena* in Wewak Bay, Papua New Guinea [225]. Semiplenamides A, B and G bear a clear resemblance to anandamide in their polar head groups, but their apolar tails are less unsaturated and, notably, lack the critical cis double bond located in the middle of the chain. This structural feature, which imparts a U-shape to the alkyl chain, is known to be important for CB_1 receptor ligands. As such, these fatty acid amides exhibited only weak affinity for the CB_1 receptor (K_i around 20 μM). In addition, only semiplenamide A acted as a moderate inhibitor of anandamide cellular uptake ($IC_{50} = 18.1$ μM). Semiplenamides have been synthesized by several research groups using stereoselective methods [226–228].

Another fatty acid amide, serinolamide A (**262**) (Fig. 29) was obtained from two samples of marine cyanobacteria, *Lyngbya majuscula* and *Oscillatoria* sp., collected from New Ireland, Papua New Guinea and sla Canales de Afuera, Coiba National Park, Panama, respectively [229]. Serinolamide A exhibited good agonist effects and selectivity for CB_1 cannabinoid receptors ($K_i = 1.3$ μM). Several research groups have reported on the total synthesis, including enantioselective synthesis, of serinolamide A [230–233]. Recently, several serinolamide A derivatives were synthesized and tested in sphingosine-1-phosphate-1 ($S1P_1$) receptor internalization assay to evaluate $S1P_1$ receptor agonist activity [234]. Sphingosine-1-phosphate-1 (S1P1) receptor agonists can be used for treating multiple sclerosis (MS) caused by autoreactive lymphocytes that attack the myelin sheath [235]. Compounds **263** and **264** (Fig. 29) were found to have superior $S1P_1$ agonistic effects to serinolamide A. In addition, compound **263** effectively inhibited lymphocyte outflow in peripheral lymphocyte count (PLC) assay.

Mooreamide A (**265**) (Fig. 29), a new alkyl amide isolated from the marine cyanobacterium, *Moorea bouillonii*, found growing on the tropical coral *Stylophora pistillata* at Pigeon Island, PNG, shows structural similarity to anandamide and 2-arachidonoylglycerol [236]. Notably, its aminoglycerol moiety, which is also present in serinolamides, is unusual and likely derives from the reduction of the serine. Mooreamide A has demonstrated potent and selective binding activity to the CB_1 neuroreceptor ($K_i = 0.47$ μM for CB_1 vs. $K_i > 25$ μM for CB_2), making it the most active member of this class of marine metabolites. The enhanced affinity of mooreamide A compared to serinolamides may be attributed to the presence of two

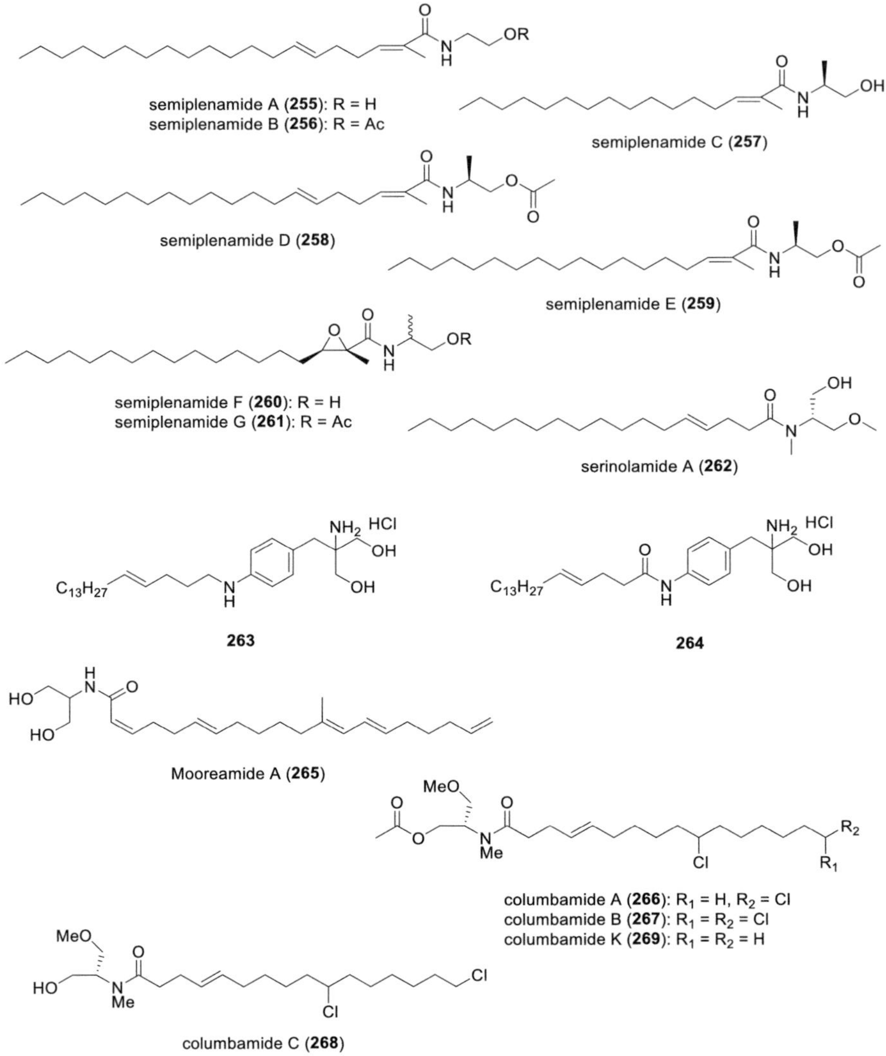

Fig. 29 Cyanobacterial cannabinomimetic fatty acid amides and synthetic analogues

free hydroxyl groups in the amine head group and a more unsaturated, longer and branched lipid chain.

Using a combination of orthologous techniques, including genome mining and mass spectrometric networking, researchers isolated and characterized the structurally unique acyl amides, columbamides A (**266**)–C (**268**) (Fig. 29), from cultured *M. bouillonii* PNG05-198, collected off Pigeon Island, Papua New Guinea [237]. Columbamides A and B were identified as potent ligands for the CB_1 and CB_2 receptors. For instance, for CB_1 receptor, the K_i values for columbamides A and

B are 0.59 μM and 0.41 μM, respectively, while for CB_2 receptor, K_i values for columbamides A and B are 1.03 μM and 0.86 μM, respectively. Columbamide B displayed slightly higher activity than columbamide A. Both compounds exhibited nearly twice the affinity for CB_1 compared to CB_2, although they were less selective for the CB_1 receptor compared to mooreamide A.

The availability of cultured *M. bouillonii* PNG05-198 facilitated identification of the columbamide biosynthetic gene cluster as well as heterologous expression of its 28.5 kb biosynthetic gene cluster (BGC) in *Anabaena* (*Nostoc*) PCC 7120 [238]. The production of columbamides in *Anabaena* was examined under various culture conditions, leading to the identification of several new columbamide analogues using liquid chromatography–tandem mass spectrometry and NMR. Along with the previously known columbamides A, B and C, the experiments yielded new columbamides I–M. Among them, the structure of the most abundant monochlorinated analogue, columbamide K (**269**) (Fig. 29), was fully characterized.

4.2 Credneramides

Two metabolites containing vinyl chloride, credneramides A (**270**) and B (**271**) (Fig. 30), were isolated from a Papua New Guinea collection of cf. *Trichodesmium* sp. nov. [239]. The precursor fatty acid, credneric acid (**272**) (Fig. 30), was also isolated from the same source in the Credner Islands of Papua New Guinea. Credneramides A and B were found to inhibit spontaneous calcium oscillations in murine cerebrocortical neurons at low micromolar concentrations, with IC_{50} values of 4.0 μM and 3.8 μM, respectively. Shortly after their discovery, Erver and Hilt reported their total synthesis via Alder–ene reactions between a terminal alkene and an internal alkyne [240].

4.3 Janthielamide A and Kimbeamides A–C

Structurally related to the credneramides are a series of chlorinated lipoamides, janthielamide A (**273**) as well as kimbeamides A (**274**)–C (**276**) and kimbelactone A (**277**) (Fig. 30), isolated from two geographically distant samples of marine cyanobacteria obtained from Curacao and Papua New Guinea [241]. Janthielamide A, a major constituent from the Curacao sample, possessed sodium channel blocking activity with IC_{50} value of 11.5 μM when tested on murine Neuro-2a cells. It was also found to be an agonist of veratridine-induced sodium influx with an IC_{50} of 5.2 μM in cerebrocortical neurons. From the Papua New Guinea collections, kimbeamide A was identified as a sodium channel blocker in murine Neuro-2a cells at a concentration of 20 μg/mL. However, further biological evaluation of this molecule and related lipoamides was hindered by their rapid decomposition during storage, whether in dry conditions or in organic solvents.

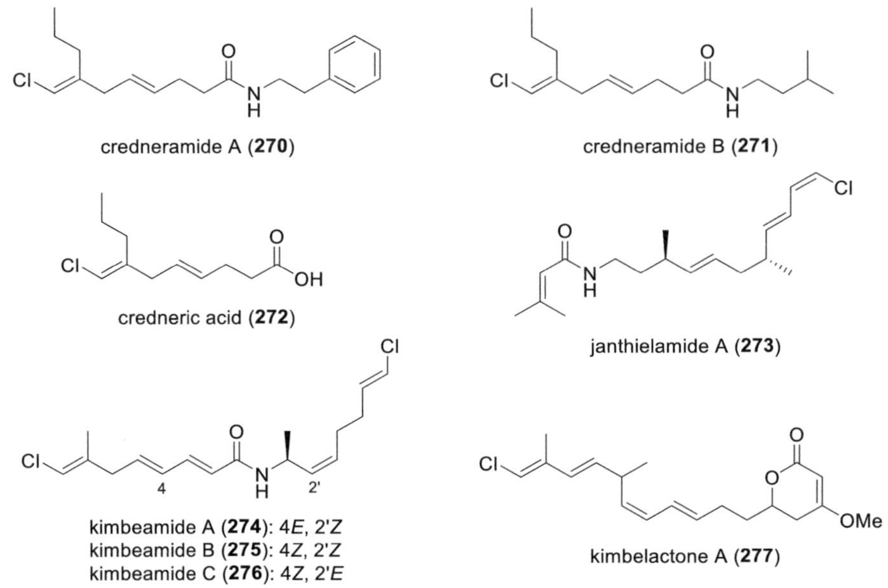

Fig. 30 Chemical structures of credneramides, janthielamide A and kimbeamides

4.4 Malyngamide-Related Molecules

The marine cyanobacterial malyngamide class of molecules consists of over 30 different compounds, each distinguished by various *N*-substitution groups of amides. These compounds have demonstrated a broad spectrum of biological activities, including antifeedant properties, ichthyotoxicity, toxicity to other marine organisms, cytotoxicity against cancer cells and potential anti-HIV, antileukemic and antitumor effects. Several bioactive malyngamide related molecules have been reported from cyanobacterial strains collected from Papua New Guinea. For instance, hermitamides A (**278**) and B (**279**) (Fig. 31) were isolated from the marine cyanobacterium, *Lyngbya majuscula*, collected from Hermit Village Islands, Papua New Guinea [242]. These compounds were tested for their biological activity across various systems. In a brine shrimp (*Artemia salina*) toxicity assay, hermitamides A and B exhibited LD_{50} values of 5 μM and 18 μM, respectively. Additionally, they demonstrated IC_{50} values of 2.2 μM and 5.5 μM, respectively, against Neuro-2a neuroblastoma cells in tissue culture. In a later study by Paige and co-workers, hermitamides were found to displace [^{3}H]-batrachotoxin at 10 μM more effectively than phenytoin, a clinically used sodium channel blocker [243]. Additionally, they proposed a potential binding mode for (*S*)-hermitamide B at the BTX-binding site and provided electrophysiological evidence demonstrating that these compounds are potent blockers of the hNav1.2 voltage-gated sodium channel. There are currently at least four reports on the total synthesis of the hermitamides [243–247].

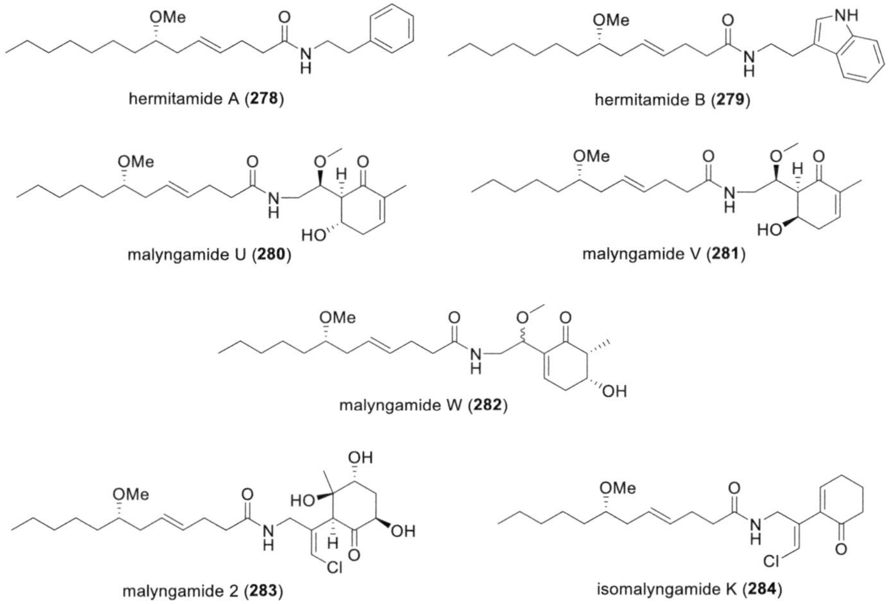

Fig. 31 Chemical structures of malyngamide-related compounds

Other malyngamide-related molecules, such as malyngamides U (**280**)–W (**282**), malyngamide 2 (**283**) and isomalyngamide K (**284**) (Fig. 31), have also been reported from Papua New Guinean marine cyanobacteria collected from Crown Island, Dutchess Island and Alotau Bay, respectively [248–250]. Malyngamide 2, from *L. sordida*, demonstrated anti-inflammatory activity in LPS-induced RAW macrophage cells, with an IC_{50} of 8.0 μM, while showing only modest cytotoxicity to mammalian cells [249].

4.5 Tumonoic Acids

The marine cyanobacterium, *Blennothrix cantharidosmum*, was a source of six new acyl proline derivatives, tumonoic acids D (**286**)–I (**291**), along with the known tumonoic acid A (**285**) (Fig. 32) and lyngbyastatins as active metabolites [251]. *B. cantharidomum* is a ubiquitous mat forming benthic cyanobacterium and the isolation of these new acyl proline derivatives represents the first report of secondary metabolites from this species. The new compounds were evaluated in various assays, including anticancer, antimalarial, anti-Chagas, antileishmanial and antimicrobial activity, but only tumonoic acid I exhibited antimalarial activity, with an IC_{50} of 2 μM, while none of the other naturally occurring analogues showed any activity in this assay at concentrations up to 10 μg/mL. Due to their structural similarities to

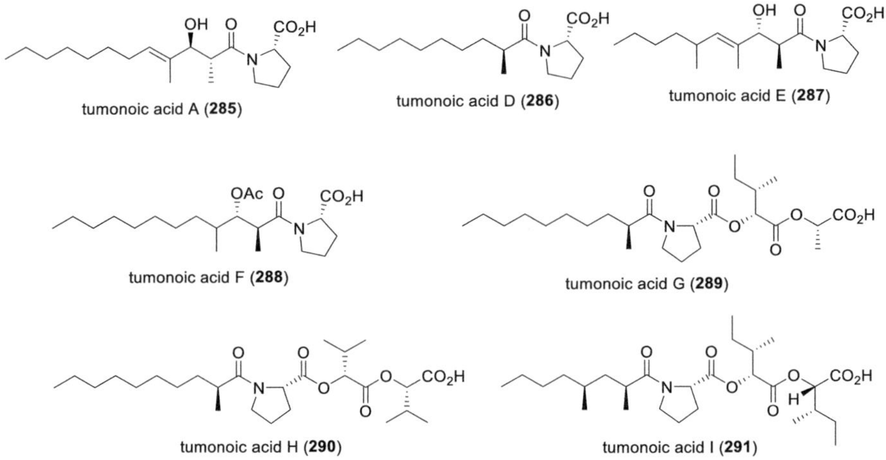

Fig. 32 Chemical structures of tumonoic acids

the bacterial signaling molecules, homoserine lactones, tumonoic acids were tested for their effects on bacterial quorum sensing. Based on an assay using a wild-type strain of *Vibrio harveyi*, which regulates bioluminescence through quorum sensing, tumonoic acids E–H demonstrated modest inhibition of bioluminescence without impacting bacterial growth. Among them, tumonoic acid F (**288**) was the most active, with an IC_{50} of 62 μM.

4.6 *Lyngbyaloside and Related Compounds*

Among the variety of marine cyanobacteria, *Lyngbya bouillonii*, first described by Hoffmann and Demoulin in 1991, has emerged as an exceptional source of new natural products [15]. The first 16-membered glycosidic macrolide, lyngbyaloside (**292**) (Fig. 33), was reported from *L. bouillonii* collected from Laing Island, Papua New Guinea [23]. In 2002, Gerwick and co-workers isolated lyngbouilloside (**293**) (Fig. 33) from a subtidal sample of *L. bouillonii*, collected off the North Coast of Papua New Guinea [252]. Due to the extreme scarcity of lyngbouilloside, a complete biological profile could not be established. However, Gerwick et al. showed that it exhibited moderate cytotoxicity against neuro-2a neuroblastoma cells, with an IC_{50} of 17 μM. These molecules have attracted significant attention related to their synthesis due to their intriguing architecture, natural scarcity as well as potential biological activities [253].

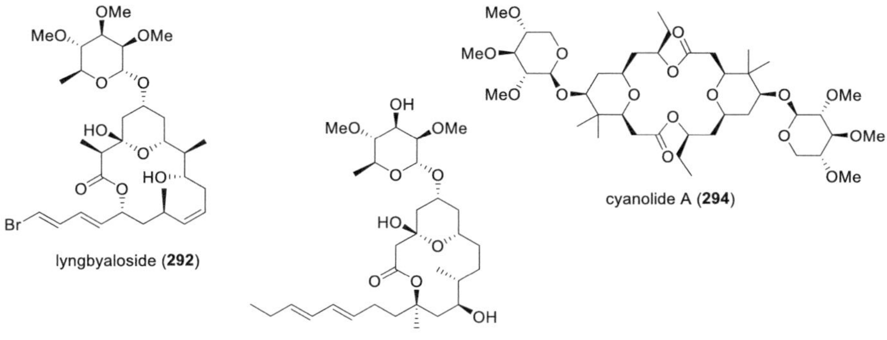

lyngbyaloside (**292**)

lyngbouilloside (**293**)

cyanolide A (**294**)

Fig. 33 Chemical structures of lyngbyaloside, lyngbouilloside and cyanolide A

4.7 Cyanolide A

Cyanolide A (**294**) (Fig. 33), a glycosidic macrolide, isolated from *Lyngbya bouillonii* obtained from Pigeon Island, Papua New Guinea, possess highly potent molluscicidal agent against *B. glabrata*, with LC_{50} value of 1.2 μM [254]. Cyanolide A is a C_2-symmetric macrodiolide having a central 20-membered macrocycle fused to two tetrahydropyran units, each bearing a pendant xylose residue. Subsequent biological evaluation revealed that the glycosidic macrolide was relatively noncytotoxic against H-460 human lung adenocarcinoma and Neuro-2a mouse neuroblastoma cell lines. Its total synthesis and its aglycon have been successfully carried by several researchers [255–264].

4.8 Tasiamides C–E

A collection of the tropical marine cyanobacterium *Symploca* sp., gathered near Kimbe Bay, Papua New Guinea, previously led to the discovery of several new metabolites, including kimbeamides, kimbelactone A and tasihalide C. Further investigation into a more polar cytotoxic fraction resulted in the identification of three new lipopeptides, tasiamides C (**295**)–E (**297**) (Fig. 34) [265]. Only tasiamides C and D were evaluated for their cytotoxicity against the HCT-116 colon cancer cell line but were found to be inactive. However, previously reported analogs, tasiamide and tasiamide B, exhibited moderate cytotoxicity against KB cells, with IC_{50} values of 0.48 μM and 0.8 μM, respectively.

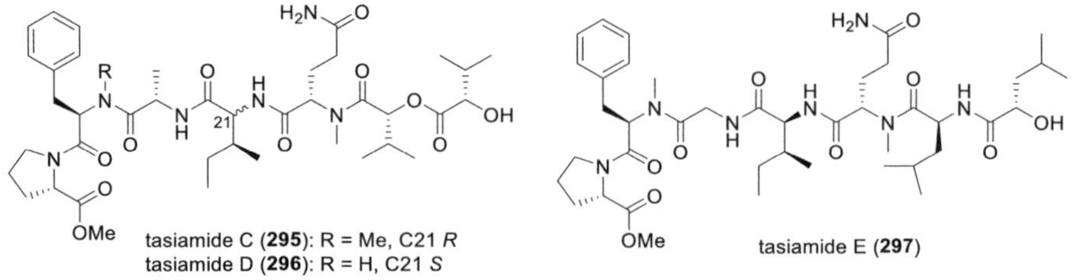

Fig. 34 Chemical structures of tasiamides C–E

4.9 Beta-Amino and Beta-Hydroxy-Containing Cyclic Depsipeptides

The presence of β-amino or β-hydroxy acid-containing depsipeptides is a distinguishing feature of marine cyanobacterial natural products. Several of such bioactive cyclic depsipeptides have been isolated from Papua New Guinean marine cyanobacteria, including guineamides, wewakamide A, dudawalamides, wewakpeptins and lyngbyabellins.

A screening program aimed at identifying cytotoxic molecules from the marine cyanobacterium *Lyngbya majuscula* (now *Moorena producens*) from reefs of Seamount, Papua New Guinea led to the discovery of novel cyclic depsipeptides known as guineamides A (**298**) to F (**303**) (Fig. 35) [266]. The cytotoxic activities of these guineamides were evaluated using a bioassay on the Neuro-2a mouse neuroblastoma cell line. Among them, only guineamides B (**299**) and C (**300**) demonstrated moderate cytotoxic activity, with IC_{50} values of 15 μM and 16 μM, respectively. Additional new analog, guineamide G (**304**), was isolated along with wewakamide A (**305**) (Fig. 35) from *L. majuscula* and *L. semiplena*, respectively collected from Wewak Bay, Papua New Guinea [267]. Guineamide G was cytotoxic to mouse neuroblastoma cell line, with an LC_{50} value of 2.7 μM.

The Dhoya-containing cyclic depsipeptides, the dudawalamides A (**306**)–D (**309**) (Fig. 36), were isolated from *Moorea producens* collected in the shallow marine habitat of Dudawali Bay, Papua New Guinea [268]. Both dudawalamides A and D exhibited moderate activity against *Plasmodium falciparum* (IC_{50} 3.5–3.6 μM), with dudawalamide A showing less activity against *Leishmania donovani* and *Trypanosoma cruzi* compared to dudawalamide D. Another series of cyclic depsipeptides containing Dhoya or its fully reduced form, Dhoaa, residues, was found in wewakpeptins A (**310**)–D (**313**) (Fig. 36), isolated from *L. semiplena* collected from Wewak Bay, Papua New Guinea [269]. Interestingly, wewakpeptins A and B were about 10 times more toxic than wewakpeptins C and D, with an LC_{50} of approximately 0.4 μM against NCI H-460 human lung tumor cells and mouse neuroblastoma cells.

Five new lyngbyabellin analogs, lyngbyabellins E (**314**)–I (**318**) (Fig. 36), along with the known dolabellin, were isolated from *Lyngbya majuscula* collected from

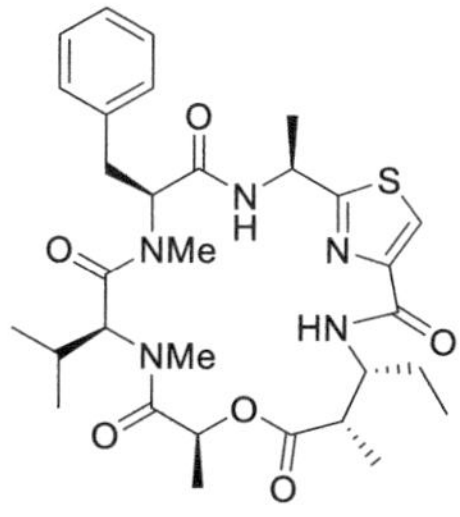
guineamide A (**298**)

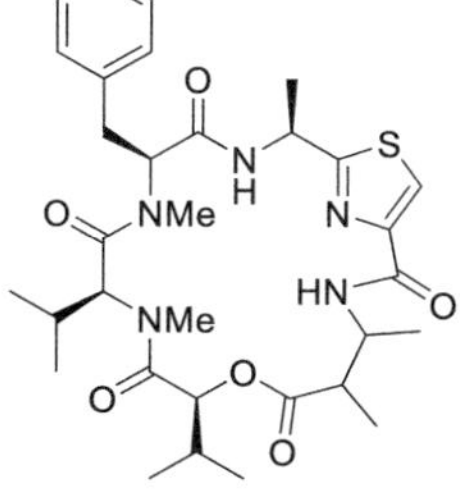
guineamide B (**299**)

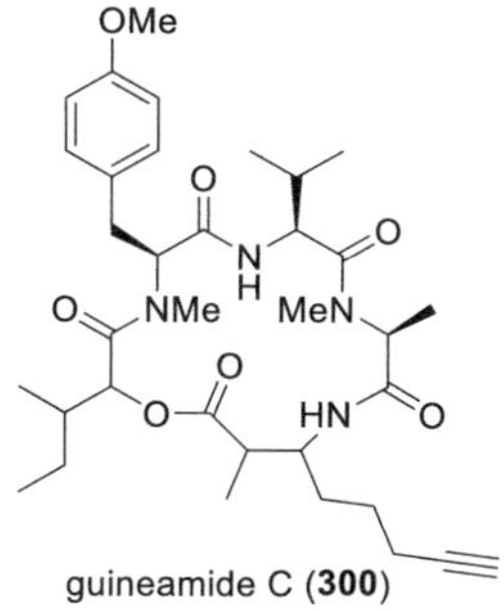
guineamide C (**300**)

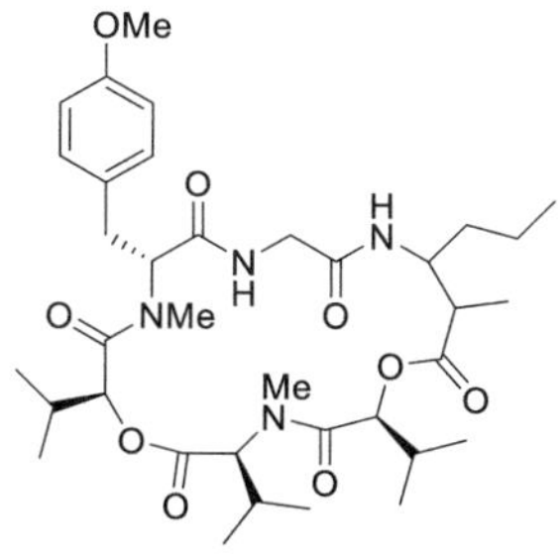
guineamide D (**301**)

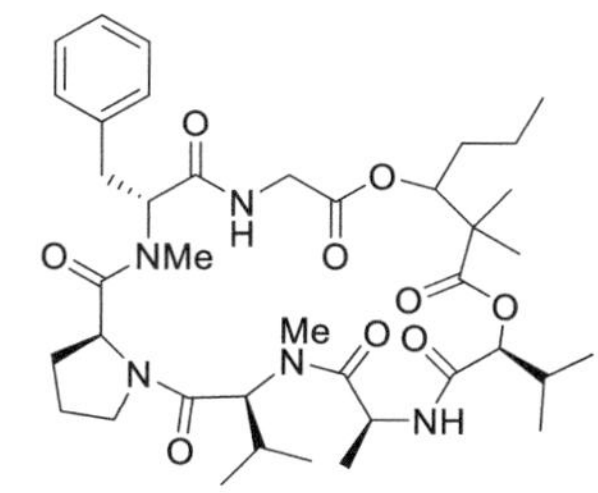
guineamide E (**302**)

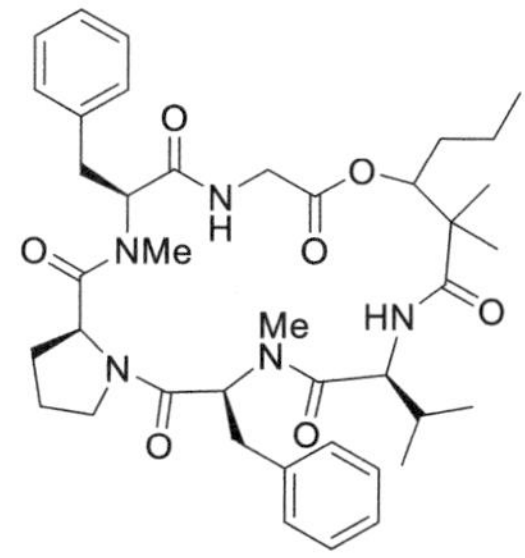
guineamide F (**303**)

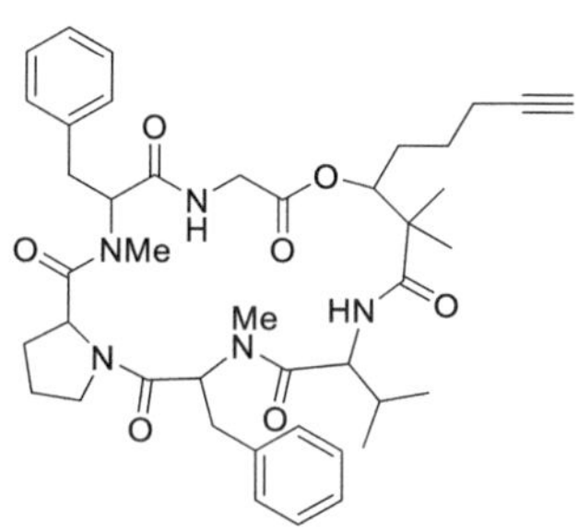
guineamide G (**304**)

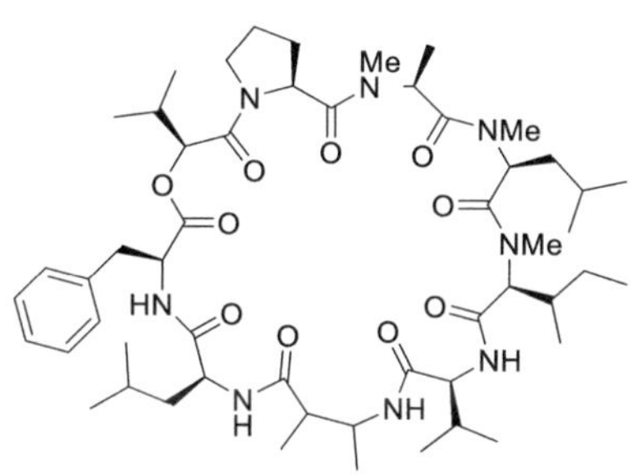
wewakamide A (**305**)

Fig. 35 Beta-amino and beta-hydroxy-containing cyclic depsipeptides

Alotau Bay, Papua New Guinea [270]. All five lyngbyabellins demonstrated cytotoxicity against NCI-H460 human lung tumor and Neuro-2a mouse neuroblastoma cell lines, with LC_{50} values ranging from 0.2 to 4.8 μM. Notably, lyngbyabellin E disrupted the cellular microfilament network in A-10 cells at concentrations of 0.01–6.0 μM. Moreover, at higher concentrations, many cells exhibited two nuclei, suggesting inhibition of cytokinesis, a common result of microfilament network disruption.

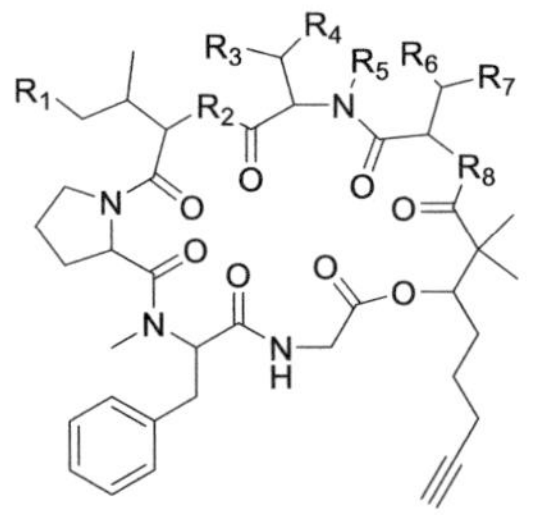

Dudawalamide A (**306**): R_1 = Me, R_2 = CH_3N, R_3 to R_7 = H, R_8 = O
Dudawalamide C (**308**): R_1 = H, R_2 = O, R_3 to R_7 = CH_3, R_8 = NH
Dudawalamide D (**309**): R_1 = CH_3, R_2 = O, R_3 to R_7 = CH_3, R_8 = NH

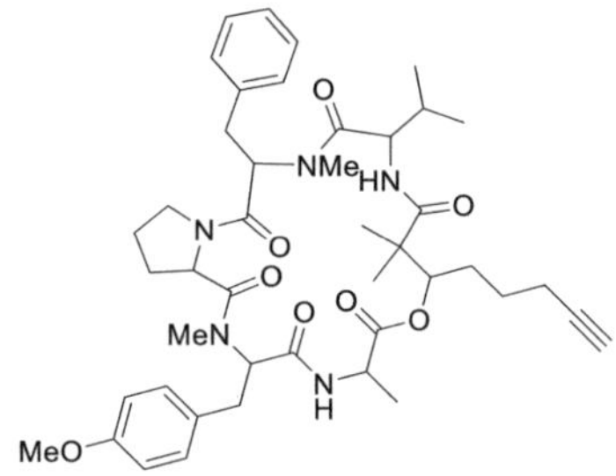

Dudawalamide B (**307**)

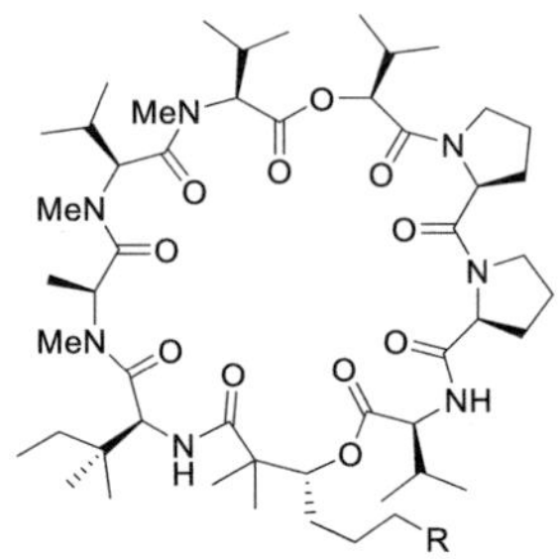

wewakpeptin A (**310**): R =

wewakpeptin B (**311**): R =

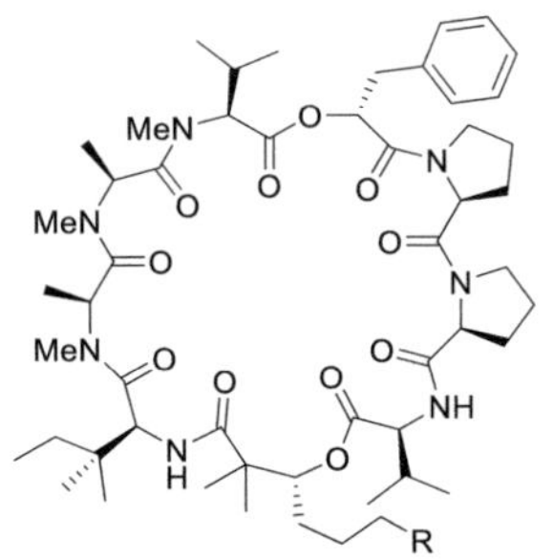

wewakpeptin C (**312**): R =

wewakpeptin D (**313**): R =

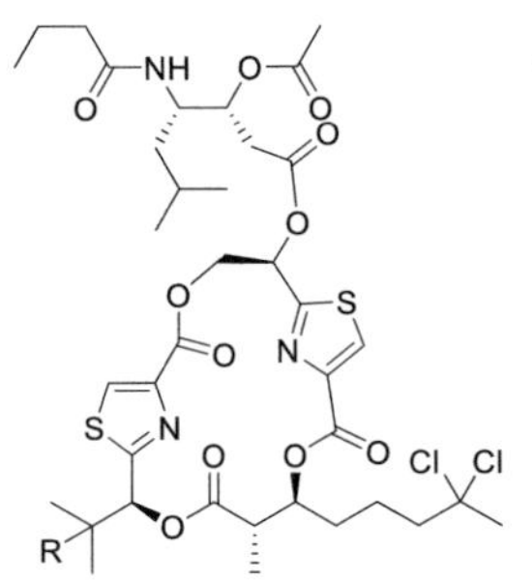

lyngbyabellin E (**314**): R = OH
lyngbyabellin H (**317**): R = H

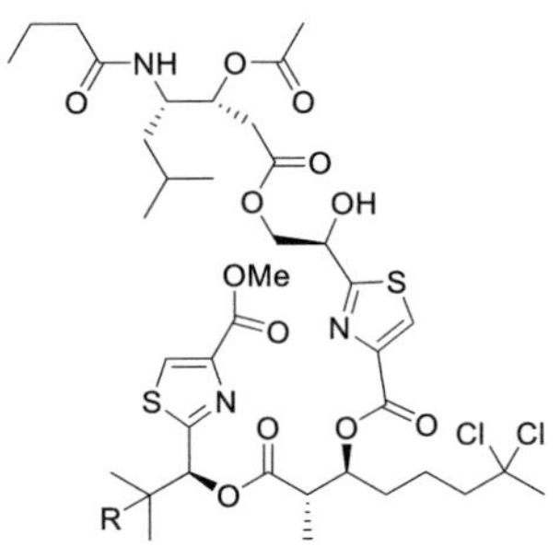

lyngbyabellin F (**315**): R = OH
lyngbyabellin I (**318**): R = H

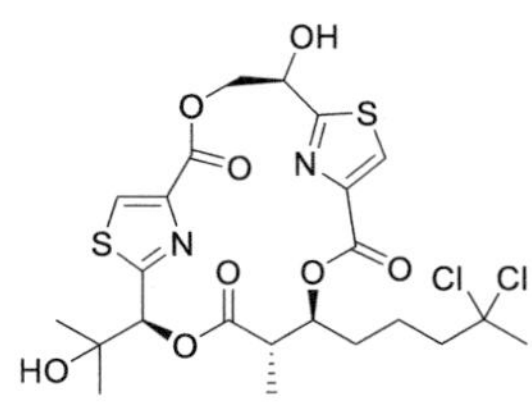

lyngbyabellin G (**316**)

Fig. 36 Beta-hydroxy-containing cyclic depsipeptides

4.10 *Symplocamide A*

Symplocamide A (**319**) (Fig. 37) is a potent cytotoxin and chymotrypsin inhibitor isolated from *Symploca* sp. collected from Sunday Island, Papua New Guinea [271].

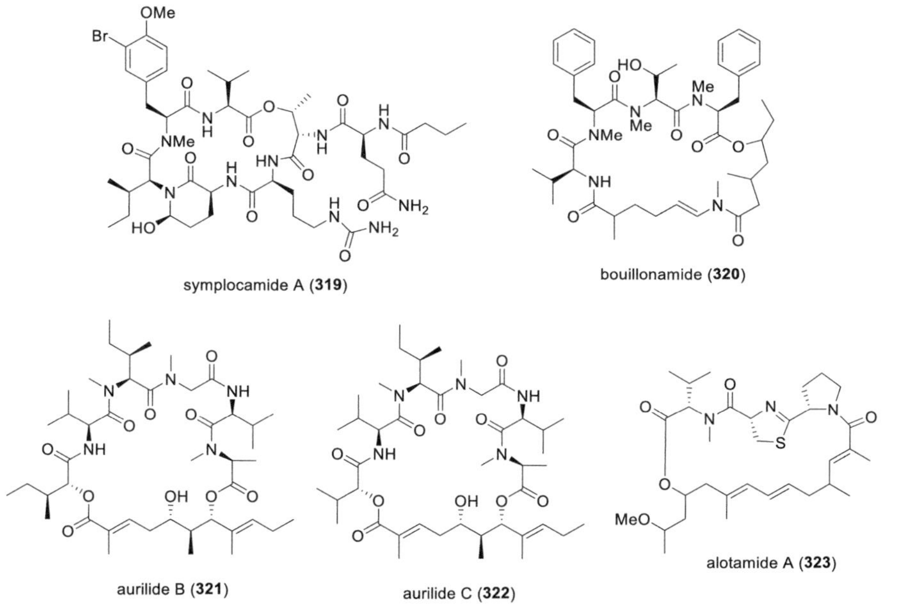

Fig. 37 Chemical structures of symplocamide A, bouillonamide, aurilides and alotamide A

This compound is unique among Ahp-containing cyclodepsipeptides due to the presence of citrulline and *N*,*O*-diMe-Br-Tyr units in its cyclic core structure. Symplocamide A inhibits chymotrypsin and trypsin with IC_{50} values of 0.38 μM and 80.2 μM, respectively. Additionally, it exhibits high potency against NCI-H460 non-small cell lung cancer cells and Neuro-2a mouse neuroblastoma cells, with IC_{50} values of 40 nM and 29 nM, respectively. Furthermore, when tested against three tropical parasites, symplocamide A was found to be particularly active against W2 *Plasmodium falciparum*, with an IC_{50} of 0.95 μM. Two years after its initial report of the natural symplocamide A, its total synthesis, using solid phase method, was accomplished by Stolze and co-workers [272].

4.11 Bouillonamide

Further chemical investigation of *M. bouillonii*, collected from New Britain, Papua New Guinea, led to the discovery of a novel cytotoxic cyclic depsipeptide, bouillonamide (**320**) (Fig. 37), along with known compounds, including ulongamide A and apratoxin A [273]. In addition to α-amino acid residues, bouillonamide features two unique polyketide-derived moieties: a 2-methyl-6-methylamino-hex-5-enoic acid

(Mmaha) residue and a 3-methyl-5-hydroxy-heptanoic acid (Mhha) unit. Bouillonamide exhibited mild toxicity, with an IC_{50} of 6.0 μM against Neuro-2a mouse neuroblastoma cells.

4.12 Aurilides

Aurilides are a group of depsipeptides primarily found in marine cyanobacteria. These compounds are known for their potent cytotoxic effects against a range of cancer cell lines [274]. Two aurilide analogues, aurilides B (**321**) and C (**322**) (Fig. 37), have been isolated from *L. majuscula* collected at Alotau Bay, Papua New Guinea [275]. Aurilide was previously identified in the sea hare *Dolabella auricularia*, and the discovery of these new analogues further supports their cyanobacterial origin. Among them, aurilide B was found to be the more potent compound, exhibiting LC_{50} values of 10 nM and 40 nM in NCI-H460 human lung tumor and Neuro-2a mouse neuroblastoma cell lines, respectively. Additionally, aurilide B demonstrated significant growth inhibitory properties in the NCI panel of 60 cell lines, particularly against leukemia, renal and prostate cancer cells. Furthermore, aurilide B exhibited tumor cell-killing activity in the NCI's in vivo hollow fiber assay model. Moreover, a quantitative short-hairpin RNA screen identifies *ATP1A1* as a gene that regulates cytotoxicity by aurilide B [276]. The *ATP1A1* gene provides instructions for making one part (the α-1 subunit) of a protein pump known as a Na^+/K^+ ATPase. In fact, combined treatment of aurilide B with ouabain, which inhibits Na+/K+ ATPase by targeting α-subunits, potentiated sensitivity to aurilide B, suggesting that ATP1A1 regulates mitochondria-mediated apoptosis.

4.13 Alotamide A

Alotamide A (**323**) (Fig. 37) is a structurally intriguing cyclic depsipeptide isolated from *L. bouillonii* collected in Milne Bay, Papua New Guinea [26]. The polyketide portion of alotamide A consists of seven acetate units as part of the macrocyclic ring structure. Three other peptidic residues, an *N*-Me-Val, a cysteine-derived thiazoline unit and Pro, are linked to the polyketide unit completing the overall macrocyclic structure of the molecule. When tested on murine cerebrocortical neurons, alotamide A showed unusual Ca^{2+} influx activation profile with EC_{50} of 4.18 μM.

4.14 Hoiamides

Hoiamides A (**324**)–D (**327**) (Fig. 38) represent a newly identified family of marine secondary metabolites. The hoiamides consist of a peptidic segment that includes an

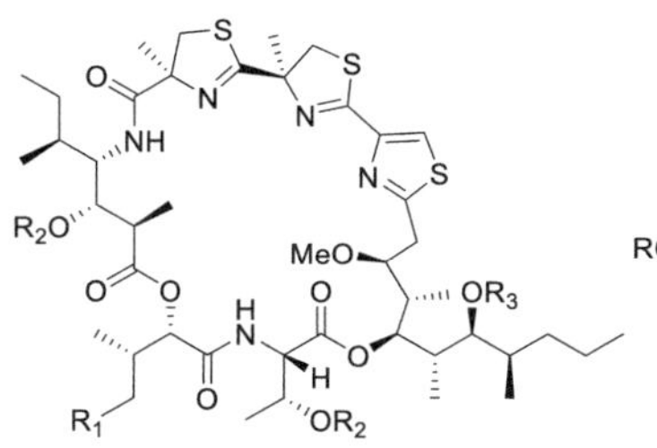

hoiamide A (**324**): $R_1 = R_2 = R_3 = H$
hoiamide B (**325**): $R_1 = Me$, $R_2 = R_3 = H$

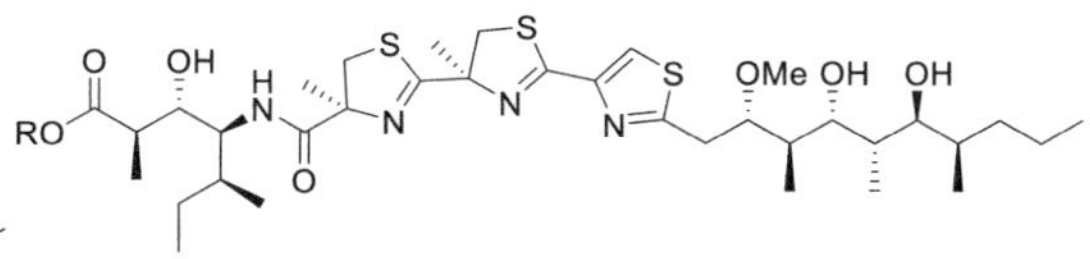

hoiamide C (**326**): R = H
hoiamide D (**327**): $R = CH_2CH_3$

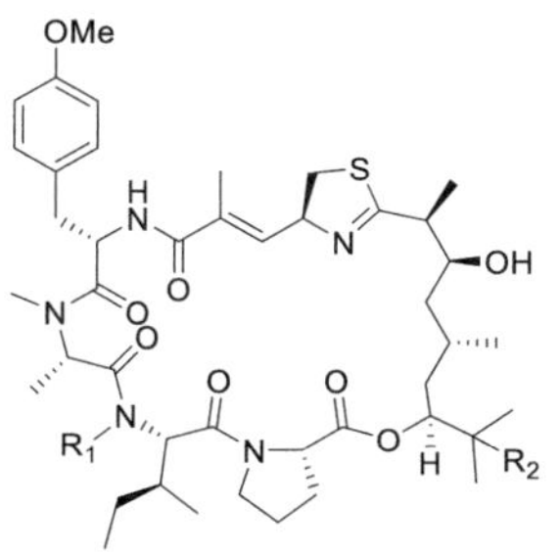

apratoxin A (**328**): $R_1 = R_2 = Me$
apratoxin B (**329**): $R_1 = H$, $R_2 = Me$
apratoxin C (**330**): $R_1 = Me$, $R_2 = H$

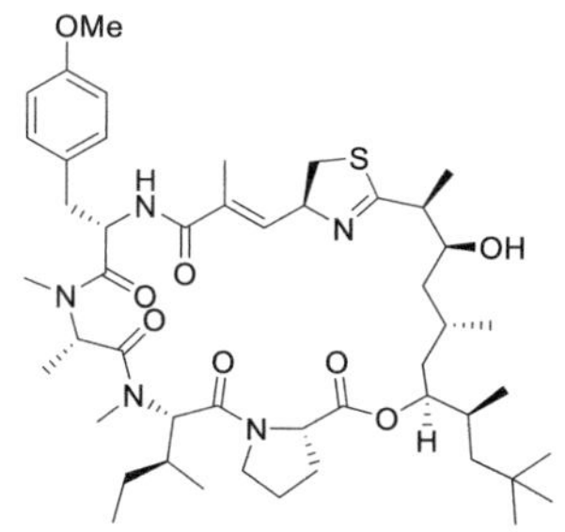

apratoxin D (**331**)

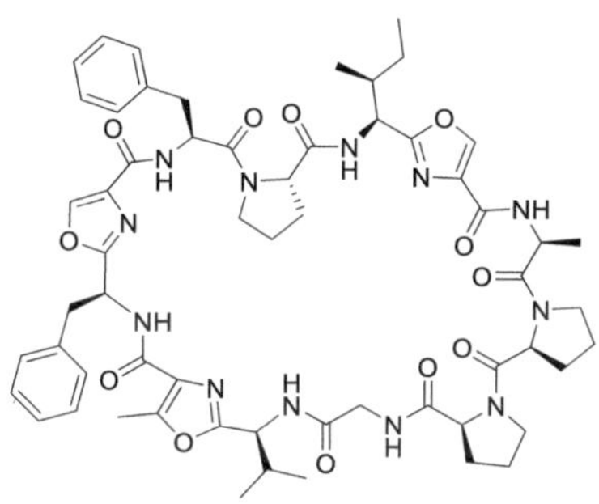

wewakazole A (**332**)

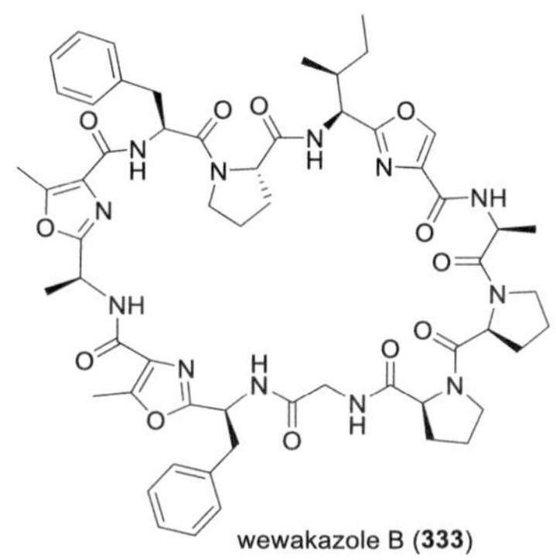

wewakazole B (**333**)

Fig. 38 Chemical structures of hoiamides, apratoxins and wewakazoles

acetylated and *S*-adenosyl methionine-modified isoleucine moiety, a central triheterocyclic core featuring two α-methylated thiazolines and one thiazole, and a highly oxygenated as well as a methylated C-15 polyketide substructure. The first member, hoiamide A was reported in 2009 from the marine cyanobacteria *Lyngbya majuscula* and *Phormidium gracile* collected in Hoia Bay, Papua New Guinea [277]. Hoiamide A is a novel cyclic depsipeptide with a highly unusual structure, demonstrating potent inhibition of [^{3}H]batrachotoxin binding to voltage-gated sodium channels

($IC_{50} = 92.8$ nM), activated sodium influx ($EC_{50} = 2.31$ μM) in mouse neocortical neurons and showed moderate cytotoxicity against cancer cells. In addition, nanomolar concentrations of hoiamide A induced a distinctive neuronal death profile characterized by both necrotic and apoptotic mechanisms, likely stemming from an interaction with a common molecular target [278]. Pharmacological studies revealed that hoiamide A-induced neuronal death depends on both JNK and caspase signaling pathways. The potent neurotoxicity and unique neuronal death profile of hoiamide A establish it as a novel neurotoxic chemotype derived from marine cyanobacteria. Its total synthesis was eventually accomplished by Ye synthetic group in 2019 using Evans-Tishchenko reaction as a key step in coupling between a five-stereogenic-center-containing β-hydroxyketone and a chiral aldehyde derived from threonine [279]. The synthesis was carried out in 15 steps with a 7% overall yield.

A year later in 2010, Gerwick and co-workers reported two additional members of the hoiamide family, namely hoiamide B (**325**) and the linear lipopeptide hoiamide C (**326**), both isolated from different collections of marine cyanobacteria obtained in Gallows Reef and Pigeon Island, Papua New Guinea, respectively [280]. In neocortical neurons, hoiamide B induced sodium influx and inhibited spontaneous Ca^{2+} oscillations with EC_{50} values of 3.9 μM and 79.8 nM, respectively, while hoiamide C did not show significant effects in these assays. The total synthesis of hoiamide C was again reported by Ye research group in 16 steps with an overall yield of 1.8% in 2011 [281]. Hoiamide D (**327**) was later reported in 2012 from the cyanobacterium *Symploca* sp. collected at Kape Point, Papua New Guinea [282]. Hoiamide D exhibited inhibitory activity against p53/MDM2 interaction with EC_{50} value of 4.5 μM. Additionally, based on in silico analysis, both hoiamides A and D have been identified as possible new activators of phosphoglycerate kinase 1, an enzyme implicated in neurodegenerative diseases, such as stroke, amyotrophic lateral sclerosis (ALS) and Parkinson's disease [283].

4.15 Apratoxins

Apratoxins are a fascinating group of cyclodepsipeptides of mixed biosynthetic origin possessing both peptide and polyketide moieties. Members of this compound family exhibit potent in vitro cytotoxicity against the KB (0.52–21.3 nM) and LoVo cell lines. The parent compound, apratoxin A (**328**) (Fig. 38), exhibits subnanomolar cytotoxicity against various cancer cell lines in vitro [284]. However, when apratoxin A was evaluated for in vivo anticancer activity, it demonstrated limited or no efficacy against colon and breast tumors at sublethal doses, and higher doses led to animal toxicity and weight loss. In 2008, Gerwick and co-workers reported the discovery of apratoxin D (**331**) (Fig. 38) from collections of *Lyngbya sordida* and *L. majuscula*, along with apratoxins A (**328**)–C (**330**) (Fig. 38) from a *L. bouillonii* collection made at several locations in Papua New Guinea [25]. Apratoxin D, features a novel carbon skeleton, likely resulting from an additional polyketide synthase (PKS) element in its biosynthetic pathway. Additionally, apratoxin D demonstrated potent in vitro cytotoxicity against H-460 human lung cancer cells with an IC_{50} value of 2.6 nM. Its total synthesis was accomplished by Coltart research group [285].

A collection of apratoxin-producing *L. bouillonii* from PNG was eventually established as unicyanobacterial cultures in the laboratory of Gerwick research group, which facilitated biosynthetic studies on apratoxin A. Subsequently, the biosynthetic gene cluster of apratoxin A was identified via a single-cell genome-amplification approach and reveals a PKS-type loading module and nine extension modules, including four polyketide synthases (PKS) and five nonribosomal peptide synthetases (NRPS) [286]. Moreover, it was recently reported that the *tert*-butyl group in apratoxin A is formed as a pivaloyl acyl carrier protein by AprA, the PKS loading module of the apratoxin A biosynthetic pathway.

The apratoxin-producing marine cyanobacterium facilitated studies on the mechanism of action of apratoxin A. Studies by Paavilainen and co-workers showed apratoxin A inhibits protein translocation by directly targeting Sec61α, the core component of the protein translocation channel [287]. In addition, mutagenesis and competitive photocrosslinking experiments suggest that apratoxin A binds to the Sec61 lateral gate in a unique manner, distinct from the binding mechanism of cotransin, a substrate-selective Sec61 inhibitor.

4.16 Wewakazoles

Wewakazole (**332**) (Fig. 38) is a novel trioxazole-containing cyclic dodecapeptide isolated from *L. majuscula* found off the coast of Papua New Guinea [288]. A second analog, wewakazole B (**333**) (Fig. 38), was eventually isolated from *M. producens* in the Red Sea [289]. Wewakazole B has been shown to exhibit toxicity against several cancer cell lines, including MCF7 (breast cancer) and H460 (lung cancer) cell lines, with IC_{50} values of 0.58 μM and 1.0 μM, respectively. However, there is no available data on the anticancer bioactivity of wewakazole. Their total syntheses were successfully carried out by at least two research groups, including a gram scale synthesis of wewakazole [290, 291]. Biological evaluation indicated that synthetic wewakazole significantly inhibited cancer cell proliferation in a dose-dependent manner [291].

4.17 Divamides

Symbiotic relationships between marine microbes and its hosts could pose a challenge that many researchers face when characterizing the discovery of natural products, albeit the potential that these small molecules possess. One example of which the symbiotic relationship has been studied is that of tunicates and cyanobacteria of the genus *Prochloron*, and even after more than 40 years of study, it still remains uncultivated [292]. Investigating these symbionts are interesting as they are known

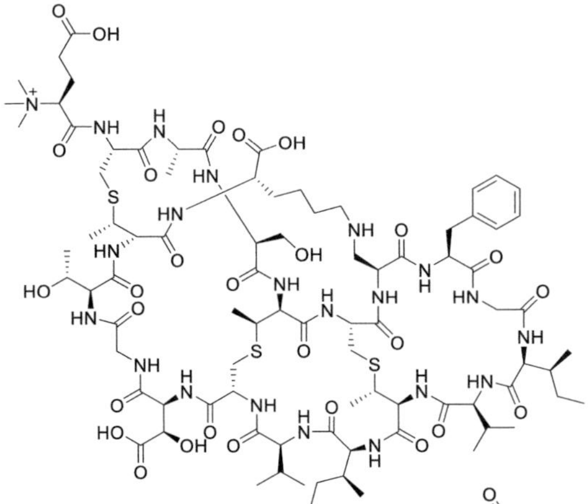

divamide A (**334**)

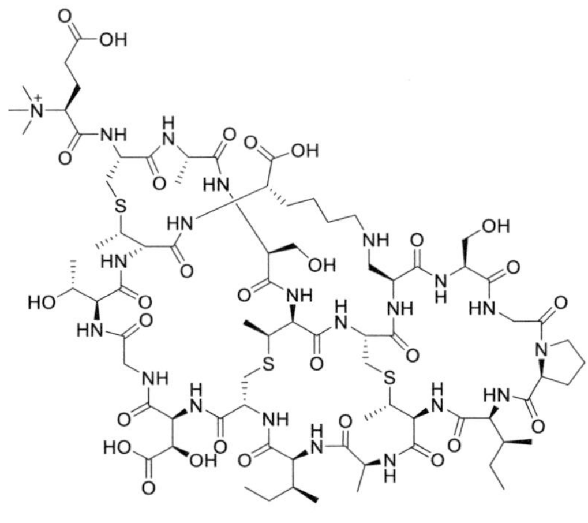

divamide B (**335**)

Fig. 39 Structures of divamides A and B

to be producers of active secondary metabolites and toxins, and able to function as chemical defenses to protect itself and the host from predation.

In a study by Smith and co-workers, additional RiPP molecules has been discovered within *Prochloron didemni* cyanobionts found in two neighbouring *Didemnum molle* tunicates, collected in Eastern Papua New Guinea [293]. Two anti-HIV compounds from novel family of lanthipeptides named divamides A (**334**) and B (**335**) (Fig. 39) were isolated and investigated with the help of metagenomics, chemistry and synthetic biology. Divamide A was found to contain three methyllanthionines, one lysinoalanine, a β-hydroxy aspartic acid and a *N*-terminal trimethylation, which is a naturally rare post-translational modification. It was found from the biosynthetic clusters that these compounds were synthesized from the symbiont cyanobacteria *Prochloron didemni*. The strategy implemented in the study was the first to overcome the limitations of supply when it comes to studying the secondary metabolites produced by the symbiont marine cyanobacteria. It proved to be an efficient strategy when working with a limited resource, as not only was it able to analyze the structure of the compound, but it also allowed for structure–activity relationships investigations with biological testing.

4.18 Dolastatins and Majusculamide C

A series of cytotoxic dolastatin-related molecules, including dolastatins 16 (**336**)–18 (**338**) (Fig. 40) [294–296] as well as majusculamide C (**339**) (Fig. 40) [297] have been isolated from the sea hare *Dolabella auricularia* collected from Papua New Guinea by the late Prof George Pettit's research group. These compounds are structurally related to marine cyanobacterial molecules and are now thought to be sequestered by the sea hare through its cyanobacterial diet. Further evidence of this is provided from reports of actual re-isolation of dolastatin compounds, such as dolastatins 10 and 16, from various marine cyanobacterial strains [298]. Dolastatin 16, a unique cyclodepsipeptide having novel β-amino acid, dolamethylleucine, and β-branched valine, dolaphenvaline, was originally isolated from *D. auricularia* obtained from Bismarck archipelago, Papua New Guinea [294]. The X-ray structure of dolastatin 16 as well as the syntheses of dolamethylleucine and dolaphenvaline were eventually accomplished, pathing the way to full-scale preclinical development as an anticancer drug [299]. In addition to its potent anticancer activity, dolastatin 16 was reported to exhibit potent antifouling activity as well as its potential use as diabetic wound healing agent as revealed by network pharmacology [300, 301]. Similar to dolastatin 16, dolastatin 17 was isolated from the Papua New Guinea sea hare *D. auricularia* and found to contain a new acetylenic β-amino acid designated dolayne [295]. Dolastatin 17 was found to exhibit significant human cancer cell growth inhibitory activity with GI_{50} values ranging from 0.45 to 0.74 μg/mL. Lastly, the cytotoxic dolastatin 18 complete structure was elucidated via total synthesis and X-ray crystallography [302]. Dolastatin 18 was found to inhibit a selection of cancer cell lines, including non-small cell lung cancer NCI-H460 with GI_{50} value of 0.39 μg/mL.

5 Conclusion

The waters around Papua New Guinea, located within the ecologically rich Coral Triangle, harbor a vast array of marine species, especially sponges and cyanobacteria, which are sources of structurally unique and biologically active compounds. This review on selected new bioactive sponge- and cyanobacterial-derived natural products has revealed tremendous chemical diversity obtained from rich marine resources found at Papua New Guinea. If the wide array of previously known natural products from Papua New Guinea were also considered, the overall impact of its

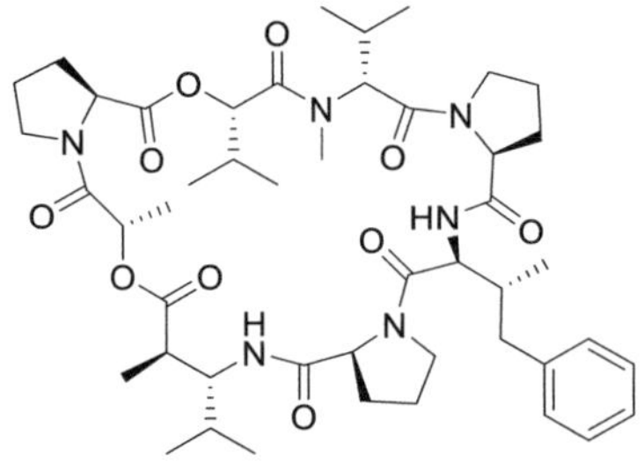

dolastatin 16 (**336**)

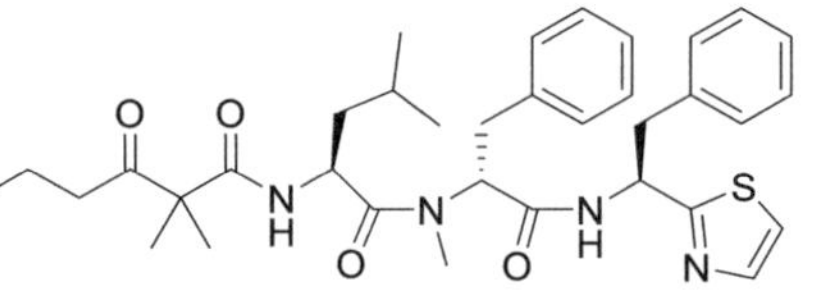

dolastatin 18 (**338**)

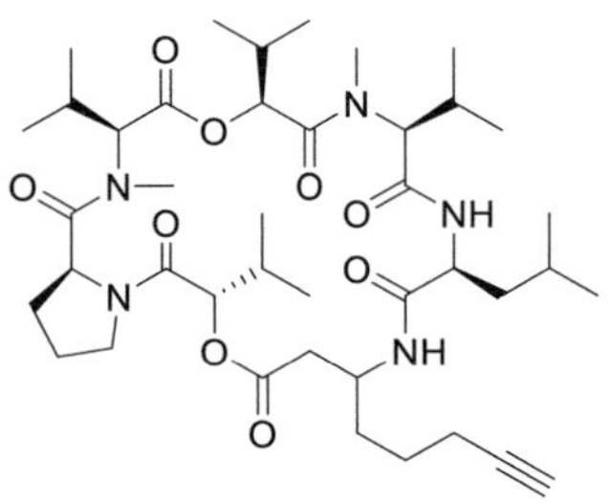

dolastatin 17 (**337**)

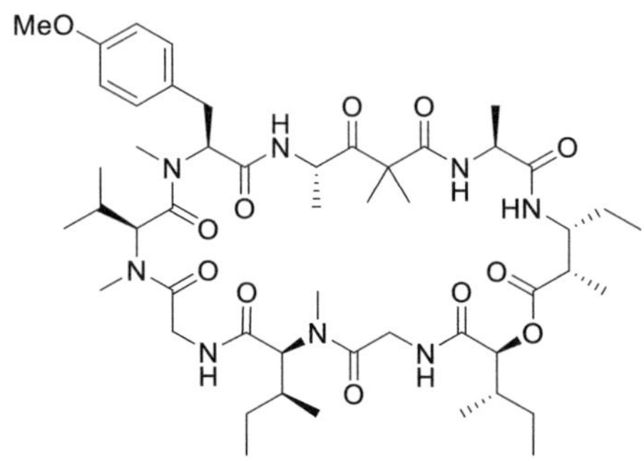

majusculamide C (**339**)

Fig. 40 Dolastatin-related compounds and majusculamide C isolated from PNG *Dolabella auricularia*

marine biodiversity on drug discovery and development would be even more significant. For instance, the sponge *Dactylospongia elegans*, collected in Papua New Guinea, yielded pelorol (Fig. 41), a compound originally identified through bioassay-guided screening for SHIP1 activation, a promising therapeutic target in cancer, by Andersen's group [303]. While further biological studies were in progress, the molecule was reported by Konig's and Schmitz's groups, from *D. elegans* (Pelorus Island, Australia) [304] and *Petrosaspongia metachromia* (Yap Archipelago, Federated States of Micronesia) [305], respectively. The structure of pelorol later inspired the synthesis of AQX-016A, AQX-MN100 and AQX-151 (Fig. 41), which demonstrated greater SHIP1 activation than pelorol, underscoring the enduring relevance of previously discovered marine metabolites in drug development pipelines [7].

These PNG-derived molecules were uncovered through the collaborative efforts of several international marine natural products research groups led mainly by Profs George Pettit, Raymond Andersen, Phillip Crews and William Gerwick, with fundings from diverse agencies, including the PNG International Cooperative Biodiversity Group, the NCI National Coorperative Natural Products Drug Discovery group programmes, Natural Sciences and Engineering Research Council of Canada, the Canadian Cancer Society and other sources. Productive collaboration forged in 1989 by Professor Crews with Dr. Fred Valeriote of the Ford Cancer in Detroit, Michigan resulted in screening thousands of crude extracts from marine sources in

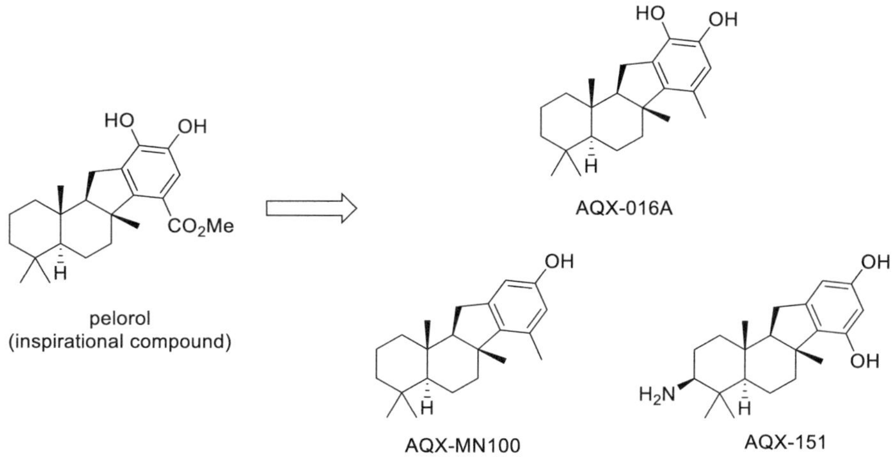

Fig. 41 The marine sponge-derived compound, pelorol, as the structural inspiration for the synthesis of selective SHIP1 activators: AQX-016A, AQX-MN100 and AQX-151

a soft agar colony-based disk diffusion assay using a panel of murine and human cancer and normal cell lines. This led to the discovery of several solid tumor selective compounds, including milnamides and psymberin. Another important collaboration between Profs Raymond Andersen and several researchers, such as Teatulohi Matainaho (University of Papua New Guinea) and Theresa Allen (Department of Pharmacology, University of Alberta) resulted in several potent sponge-derived lead compounds, including contignasterol and hemiasterlins, for development of therapeutic agents for clinical trials. Equally productive are research efforts by Prof. Gerwick on the discovery of potent molecules, such as apratoxins and hoiamides, from Papua New Guinean filamentous marine cyanobacteria.

Through the review of sponge metabolites from PNG samples, it becomes clear that most bioactive molecules originate from peptide and/or polyketide biosynthesis pathways. This suggests that these compounds are more likely produced by symbiotic microbes than by the host sponges. The current challenge lies in proving that these bioactive compounds are produced by microbial symbionts or commensals within the sponge rather than the sponge itself. If it can be conclusively shown that microorganisms are the primary or sole producers of bioactive natural products with pharmaceutical potential, the next step would be to focus on optimizing their large-scale production. One such method using CARD-FISH analysis was demonstrated on a Papua New Guinean sponge, *Dysidea* (*Lamellodysidea*) *herbacea*, by Flatt et al. [306]. Using genetic information available for the marine cyanobacterial barbamide biosynthetic gene cluster, a *barB1* homolog (*dysB1*) from *D. herbacea* samples was amplified by PCR. In addition, CARD-FISH analysis revealed hybridization of the *dysB1* oligonucleotide probes to sequences in the filamentous cyanobacterial symbiont *O. spongeliae*.

In summary, Papua New Guinea is renowned for its rich marine biodiversity, making it a vital region for drug discovery and development efforts. Its unique marine habitats, ranging from the shallow waters of coastal mangroves, estuaries and reefs to the abyssal plains, sea mounts and hydrothermal vents, host a wide array of organisms that produce bioactive compounds with potential pharmaceutical applications. Despite the successes, the biodiversity of Papua New Guinea remains largely unexplored, suggesting a significant potential for future discoveries. Continued investment in scientific research, sustainable practices and collaborative programs could further unlock the therapeutic value of the region's marine natural products while supporting biodiversity conservation efforts.

References

1. Bowen BW, Rocha LA, Toonen RJ, Karl SA (2013) The origins of tropical marine biodiversity. Trends Ecol Evol 28:359–366
2. Shearman PL (2010) Recent change in the extent of mangroves in the northern Gulf of Papua, Papua New Guinea. Ambio 39:181–189
3. Haywood M, Dennis D, Thomson D, Pillans R (2016) Mine waste disposal leads to lower coral cover, reduced species richness and a predominance of simple coral growth forms on a fringing coral reef in Papua New Guinea. Mar Environ Res 115:36–48
4. Barrows L, Matainaho T, Ireland C, Miller S, Carter G, Bugni T, Rai P, Gideon O, Manoka B, Piskaut P (2009) Making the most of Papua New Guinea's biodiversity: establishment of an integrated set of programs that link botanical survey with pharmacological assessment in "The Land of the Unexpected." Pharm Biol 47:795–808
5. Gupta R, Gabrielsen B, Ferguson SM (2005) Nature's medicines: traditional knowledge and intellectual property management. Case studies from the National Institutes of Health (NIH), USA. Curr Drug Discov Technol 2:203–219
6. Luzzatto-Knaan T, Garg N, Wang M, Glukhov E, Peng Y, Ackermann G, Amir A, Duggan BM, Ryazanov S, Gerwick L, Knight R, Alexandrov T, Bandeira N, Gerwick WH, Dorrestein PC (2017) Digitizing mass spectrometry data to explore the chemical diversity and distribution of marine cyanobacteria and algae. eLife 6:e24214
7. Andersen RJ (2017) Sponging off nature for new drug leads. Biochem Pharmacol 139:3–14
8. Pulitzer-Finali G, Pronzato R (1999) Horny sponges from the north-eastern coast of Papua New Guinea, Bismark Sea. J Mar Biol Assoc UK 79:593–607
9. Kelly-Borges M, Valentine C (1995) The sponges of the tropical island region of Oceania: a taxonomic status review. Mar Coastal Biodiv Trop Island Pacific Region 1:83–120
10. Coppock AG, Kingsford MJ, Jones GP (2024) Importance of complex sponges as habitat and feeding substrata for coral reef fishes. Mar Biol 171:154
11. Longenecker K, Roberts TE, Colin PL (2019) Papua New Guinea. In: Loya Y, Puglise KA, Bridge TCL (eds) Mesophotic coral ecosystems. Springer International Publishing, Cham, pp 321–336
12. Hoffmann L (1999) Marine cyanobacteria in tropical regions: diversity and ecology. Eur J Phycol 34:371–379
13. Hoffmann L (1988) Macroscopic benthic blue-green algal vegetation of Laing island (northern coast of Papua New Guinea). Bulletin de la Société Royale de Botanique de Belgique/Bulletin van de Koninklijke Belgische Botanische Vereniging 121:3–11
14. Hoffmann L (1991) Marine cyanophyceae of Papua New Guinea. I. The genera *Arthrospira* and *Spirulina*. Nova Hedwigia

15. Hoffmann L, Demoulin V (1991) Marine cyanophyceae of Papua New Guinea. II. *Lyngbya bouillonii* sp. nov., a remarkable tropical reef-inhabiting blue-green alga. Belgian J Bot:82–88
16. Hoffmann L, Demoulin V (1993) Marine cyanophyceae of Papua New Guinea. III. The genera *Borzia* and *Oscillatoria*. Bot Mar 36:451–460
17. Hoffmann L (1992) Marine cyanophyceae of Papua New Guinea IV New Chroococcales. Belgian J Bot 125:29–32
18. Hoffmann L (1993) Marine cyanophyceae of Papua New Guinea. V: Nostocales (except Rivulariaceae) and Stigonemateles. Nova Hedwigia 57:393–408
19. Hoffmann L (1994) Marine cyanophyceae of Papua New Guinea. VI. The genus Lyngbya S.L. Belg J Bot 127:79–86
20. Taton A, Hoffmann L (2003) Marine cyanophyceae of Papua New Guinea. VII. Endoliths. Archiv für Hydrobiologie. Supplementband, Algological Studies 148:537–554
21. Klein D, Braekman J-C, Daloze D, Hoffmann L, Demoulin V (1996) Laingolide, a novel 15-membered macrolide from *Lyngbya bouillonii* (Cyanophyceae). Tetrahedron Lett 37:7519–7520
22. Klein D, Braekman JC, Daloze D, Hoffmann L, Castillo G, Demoulin V (1999) Madangolide and laingolide A, two novel macrolides from *Lyngbya bouillonii* (Cyanobacteria). J Nat Prod 62:934–936
23. Klein D, Braekman JC, Daloze D, Hoffmann L, Demoulin V (1997) Lyngbyaloside, a novel 2,3,4-tri-*O*-methyl-6-deoxy-α-mannopyranoside macrolide from *Lyngbya bouillonii* (Cyanobacteria). J Nat Prod 60:1057–1059
24. Klein D, Braekman J-C, Daloze D, Hoffmann L, Castillo G, Demoulin V (1999) Lyngbyapeptin A, a modified tetrapeptide from *Lyngbya bouillonii* (Cyanophyceae). Tetrahedron Lett 40:695–696
25. Gutiérrez M, Suyama TL, Engene N, Wingerd JS, Matainaho T, Gerwick WH (2008) Apratoxin D, a potent cytotoxic cyclodepsipeptide from Papua New Guinea collections of the marine cyanobacteria *Lyngbya majuscula* and *Lyngbya sordida*. J Nat Prod 71:1099–1103
26. Soria-Mercado IE, Pereira A, Cao Z, Murray TF, Gerwick WH (2009) Alotamide A, a novel neuropharmacological agent from the marine cyanobacterium *Lyngbya bouillonii*. Org Lett 11:4704–4707
27. Engene N, Rottacker EC, Kaštovský J, Byrum T, Choi H, Ellisman MH, Komárek J, Gerwick WH (2012) *Moorea producens* gen. nov., sp. nov. and *Moorea bouillonii* comb. nov., tropical marine cyanobacteria rich in bioactive secondary metabolites. Int J Syst Evol Microbiol 62:1171–1178
28. Tronholm A (2019) *Moorena* gen. nov., a valid name for "*Moorea* Engene & al." nom. inval.(Oscillatoriaceae, Cyanobacteria). Notulae algarum 1
29. Carroll J, Jonsson EN, Ebel R, Hartman MS, Holman TR, Crews P (2001) Probing sponge-derived terpenoids for human 15-lipoxygenase inhibitors. J Org Chem 66:6847–6851
30. Robinson SJ, Hoobler EK, Riener M, Loveridge ST, Tenney K, Valeriote FA, Holman TR, Crews P (2009) Using enzyme assays to evaluate the structure and bioactivity of sponge-derived meroterpenes. J Nat Prod 72:1857–1863
31. Murray LM, Johnson A, Diaz MC, Crews P (1997) Geographic variation in the tropical marine sponge Jaspis cf. johnstoni: an unexpected source of new terpene-benzenoids. J Org Chem 62:5638–5641
32. Rubio BK, van Soest RW, Crews P (2007) Extending the record of meroditerpenes from *Cacospongia* marine sponges. J Nat Prod 70:628–631
33. Cichewicz RH, Kenyon VA, Whitman S, Morales NM, Arguello JF, Holman TR, Crews P (2004) Redox inactivation of human 15-lipoxygenase by marine-derived meroditerpenes and synthetic chromanes: archetypes for a unique class of selective and recyclable inhibitors. J Am Chem Soc 126:14910–14920
34. Amagata T, Whitman S, Johnson TA, Stessman CC, Loo CP, Lobkovsky E, Clardy J, Crews P, Holman TR (2003) Exploring sponge-derived terpenoids for their potency and selectivity against 12-human, 15-human, and 15-soybean lipoxygenases. J Nat Prod 66:230–235

35. Sperry S, Valeriote FA, Corbett TH, Crews P (1998) Isolation and cytotoxic evaluation of marine sponge-derived norterpene peroxides. J Nat Prod 61:241–247
36. Dai J, Liu Y, Zhou Y-D, Nagle DG (2007) Hypoxia-selective antitumor agents: norsesterterpene peroxides from the marine sponge *Diacarnus levii* preferentially suppress the growth of tumor cells under hypoxic conditions. J Nat Prod 70:130–133
37. Rubio BK, Tenney K, Ang K-H, Abdulla M, Arkin M, McKerrow JH, Crews P (2009) The marine sponge *Diacarnus bismarckensis* as a source of peroxiterpene inhibitors of *Trypanosoma brucei*, the causative agent of sleeping sickness. J Nat Prod 72:218–222
38. Williams DE, Cassel J, Zhu J-L, Yang J-X, de Voogd NJ, Matainaho T, Salvino JM, Wang YA, Montaner LJ, Tietjen I (2023) Thorectidiol A isolated from the marine sponge *Dactylospongia elegans* disrupts interactions of the SARS-CoV-2 spike receptor binding domain with the host ACE2 receptor. J Nat Prod 86:582–588
39. Wrapp D, Wang N, Corbett KS, Goldsmith JA, Hsieh C-L, Abiona O, Graham BS, McLellan JS (2020) Cryo-EM structure of the 2019-nCoV spike in the prefusion conformation. Science 367:1260–1263
40. Forestieri R, Merchant CE, de Voogd NJ, Matainaho T, Kieffer TJ, Andersen RJ (2009) Alotaketals A and B, sesterterpenoids from the marine sponge *Hamigera* species that activate the cAMP cell signaling pathway. Org Lett 11:5166–5169
41. Daoust J, Chen M, Wang M, Williams DE, Chavez MAG, Wang YA, Merchant CE, Fontana A, Kieffer TJ, Andersen RJ (2013) Sesterterpenoids isolated from a northeastern Pacific *Phorbas* sp. J Org Chem 78:8267–8273
42. Wang M, Tietjen I, Chen M, Williams DE, Daoust J, Brockman MA, Andersen RJ (2016) Sesterterpenoids isolated from the sponge *Phorbas* sp. activate latent HIV-1 provirus expression. J Org Chem 81:11324–11334
43. Yao H, Zhou N, Zhang Z, Guan W, Wang H, Cheng H (2020) Recent developments in syntheses of alotaketals and phorbaketals. Tetrahedron Lett 61:151480
44. Lee J, Kim J, Lee H-Y (2020) Au (I)-catalyzed cyclization of epoxyalkynes to allylic alcohol containing spiroketals and application to the total synthesis of (−)-alotaketal A. Org Lett 22:4073–4077
45. Huang J, Yang JR, Zhang J, Yang J (2012) Total synthesis of the potent cAMP signaling agonist (−)-alotaketal A. J Am Chem Soc 134:8806–8809
46. Huang J, Yang JR, Zhang J, Yang J (2013) Total synthesis and structure–activity relationship study of the potent cAMP signaling agonist (−)-alotaketal A. Org Biomol Chem 11:3212–3222
47. Maki J, Oshimura A, Tsukano C, Yanagita RC, Saito Y, Sakakibara Y, Irie K (2022) AI and computational chemistry-accelerated development of an alotaketal analogue with conventional PKC selectivity. Chem Commun 58:6693–6696
48. Antal CE, Hudson AM, Kang E, Zanca C, Wirth C, Stephenson NL, Trotter EW, Gallegos LL, Miller CJ, Furnari FB (2015) Cancer-associated protein kinase C mutations reveal kinase's role as tumor suppressor. Cell 160:489–502
49. Okuda S, Tsukano C, Irie K (2023) Synthesis of stereoisomeric simplified analogs of alotaketals toward the elucidation of the structural requirements of protein kinase C isozyme-selective binding. Org Lett 25:805–809
50. Johnson TA, Amagata T, Sashidhara KV, Oliver AG, Tenney K, Matainaho T, Ang KK-H, McKerrow JH, Crews P (2009) The aignopsanes, a new class of sesquiterpenes from selected chemotypes of the sponge *Cacospongia mycofijiensis*. Org Lett 11:1975–1978
51. Bürki C, Bonjoch J, Bradshaw B, Villa G, Renaud P (2015) Total synthesis of aignopsanes, a class of sesquiterpenes: (+)-aignopsanoic acid A, (−)-methyl aignopsanoate A, and (−)-isoaignopsanoic A. Chem Eur J 21:395–401
52. Johnson TA, Sohn J, Inman WD, Estee SA, Loveridge ST, Vervoort HC, Tenney K, Liu J, Ang KK-H, Ratnam J (2011) Natural product libraries to accelerate the high-throughput discovery of therapeutic leads. J Nat Prod 74:2545–2555
53. Carr G, Raszek M, Soest RV, Matainaho T, Shopik M, Holmes CF, Andersen RJ (2007) Protein phosphatase inhibitors isolated from *Spongia irregularis* collected in Papua New Guinea. J Nat Prod 70:1812–1815

54. Rusnak F, Mertz P (2000) Calcineurin: form and function. Physiol Rev 80:1483–1521
55. Diaz-Marrero AR, Austin P, Van Soest R, Matainaho T, Roskelley CD, Roberge M, Andersen RJ (2006) Avinosol, a meroterpenoid-nucleoside conjugate with antiinvasion activity isolated from the marine sponge *Dysidea* sp. Org Lett 8:3749–3752
56. Braekman JC, Daloze D, Hulot G, Tursch B, Declercq J-P, Germain G, Van Meerssche M (1978) Chemical studies of marine invertebrates. XXXVII (1). Three novel meroditerpenoids from the sponge *Strongylophora durissima.* Bulletin des Sociétés Chimiques Belges 87:917–926
57. Warabi K, McHardy LM, Matainaho L, Van Soest R, Roskelley CD, Roberge M, Andersen RJ (2004) Strongylophorine-26, a new meroditerpenoid isolated from the marine sponge *Petrosia (Strongylophora) corticata* that exhibits anti-invasion activity. J Nat Prod 67:1387–1389
58. McHardy LM, Warabi K, Andersen RJ, Roskelley CD, Roberge M (2005) Strongylophorine-26, a Rho-dependent inhibitor of tumor cell invasion that reduces actin stress fibers and induces nonpolarized lamellipodial extensions. Mol Cancer Ther 4:772–778
59. Warabi K, Patrick BO, Austin P, Roskelley CD, Roberge M, Andersen RJ (2007) Strongylophorine-26, an inhibitor of cancer cell invasion: SAR revealed by synthesis of analogues. J Nat Prod 70:736–740
60. Yu W, Hjerrild P, Jacobsen KM, Tobiesen HN, Clemmensen L, Poulsen TB (2018) A catalytic oxidative quinone heterofunctionalization method: synthesis of strongylophorine-26. Angew Chem 130:9953–9957
61. Mohammed KA, Jadulco RC, Bugni TS, Harper MK, Sturdy M, Ireland CM (2008) Strongylophorines: natural product inhibitors of hypoxia-inducible factor-1 transcriptional pathway. J Med Chem 51:1402–1405
62. Sasaki S, Tozawa T, Van Wagoner RM, Ireland CM, Harper MK, Satoh T (2011) Strongylophorine-8, a pro-electrophilic compound from the marine sponge *Petrosia* (*Strongylophora*) *corticata*, provides neuroprotection through Nrf2/ARE pathway. Biochem Biophys Res Commun 415:6–10
63. Burgoyne DL, Andersen RJ, Allen TM (1992) Contignasterol, a highly oxygenated steroid with the unnatural 14. beta. configuration from the marine sponge *Petrosia contignata* Thiele, 1899. J Org Chem 57:525–528
64. Shoji N, Umeyama A, Shin K, Takeda K, Arihara S, Kobayashi J, Takei M (1992) Two unique pentacyclic steroids with cis C/D ring junction from Xestospongia bergquistia Fromont, powerful inhibitors of histamine release. J Org Chem 57:2996–2997
65. Takei M, Burgoyne DL, Andersen RJ (1994) Effect of contignasterol on histamine release induced by anti-immunoglobuline from rat peritoneal mast cells. J Pharm Sci 83:1234–1235
66. Shen Y, Burgoyne DL (2002) Efficient synthesis of IPL576,092: a novel anti-asthma agent. J Org Chem 67:3908–3910
67. Ortega-Vidal J, Reddy MM, Pérez-Fuentes N, Alvariño R, Burth S, Alfonso A, Vale C, Virués-Segovia JR, Mungkaje A, Botana LM, Thomas OP (2025) Contignasterines, anti-inflammatory 2-aminoimidazole steroids from the sponge *Neopetrosia* cf. *rava* collected in the Bismarck sea. J Nat Prod
68. Brito R, Sheth S, Mukherjea D, Rybak LP, Ramkumar V (2014) TRPV1: A Potential Drug Target for Treating Various Diseases. Cells 3:517–545
69. Harrison B, Crews P (1998) Cyclic polyketide peroxides and acyclic diol analogues from the sponge *Plakortis lita.* J Nat Prod 61:1033–1037
70. Cichewicz RH, Valeriote FA, Crews P (2004) Psymberin, a potent sponge-derived cytotoxin from Psammocinia distantly related to the pederin family. Org Lett 6:1951–1954
71. Pettit GR, Xu J-P, Chapuis J-C, Pettit RK, Tackett LP, Doubek DL, Hooper JN, Schmidt JM (2004) Antineoplastic Agents. 520. Isolation and structure of irciniastatins A and B from the Indo-Pacific marine sponge *Ircinia ramosa.* J Med Chem 47:1149–1152
72. Bielitza M, Pietruszka J (2013) The psymberin story—biological properties and approaches towards total and analogue syntheses. Angew Chem Int Ed 52:10960–10985
73. Piel Jr, Höfer I, Hui D (2004) Evidence for a symbiosis island involved in horizontal acquisition of pederin biosynthetic capabilities by the bacterial symbiont of *Paederus fuscipes* beetles. J Bacteriol 186:1280–1286

74. Fisch KM, Gurgui C, Heycke N, Van Der Sar SA, Anderson SA, Webb VL, Taudien S, Platzer M, Rubio BK, Robinson SJ (2009) Polyketide assembly lines of uncultivated sponge symbionts from structure-based gene targeting. Nat Chem Biol 5:494
75. Chinen T, Nagumo Y, Watanabe T, Imaizumi T, Shibuya M, Kataoka T, Kanoh N, Iwabuchi Y, Usui T (2010) Irciniastatin A induces JNK activation that is involved in caspase-8-dependent apoptosis via the mitochondrial pathway. Toxicol Lett 199:341–346
76. Hirano S, Quach HT, Watanabe T, Kanoh N, Iwabuchi Y, Usui T, Kataoka T (2015) Irciniastatin A, a pederin-type translation inhibitor, promotes ectodomain shedding of cell-surface tumor necrosis factor receptor 1. J Antibiot 68:417–420
77. Wu C-Y, Feng Y, Cardenas ER, Williams N, Floreancig PE, De Brabander JK, Roth MG (2012) Studies toward the unique pederin family member psymberin: structure–activity relationships, biochemical studies, and genetics identify the mode-of-action of psymberin. J Am Chem Soc 134:18998–19003
78. Dayanidhi DL, Somarelli JA, Mantyh JB, Rupprecht G, Roghani RS, Vincoff S, Shin I, Zhao Y, Kim SY, McCall S (2022) Psymberin, a marine-derived natural product, induces cancer cell growth arrest and protein translation inhibition. Front Med 9:999004
79. Schmitz FJ, Gunasekera SP, Yalamanchili G, Hossain MB, Van der Helm D (1984) Tedanolide: a potent cytotoxic macrolide from the Caribbean sponge *Tedania ignis.* J Am Chem Soc 106:7251–7252
80. Chevallier C, Bugni TS, Feng X, Harper MK, Orendt AM, Ireland CM (2006) Tedanolide C: a potent new 18-membered-ring cytotoxic macrolide isolated from the Papua New Guinea marine sponge *Ircinia* sp. J Org Chem 71:2510–2513
81. Meragelman TL, Willis RH, Woldemichael GM, Heaton A, Murphy PT, Snader KM, Newman DJ, Van Soest R, Boyd MR, Cardellina JH (2007) Candidaspongiolides, distinctive analogues of tedanolide from sponges of the genus *Candidaspongia.* J Nat Prod 70:1133–1138
82. Trisciuoglio D, Uranchimeg B, Cardellina JH, Meragelman TL, Matsunaga S, Fusetani N, Del Bufalo D, Shoemaker RH, Melillo G (2008) Induction of apoptosis in human cancer cells by candidaspongiolide, a novel sponge polyketide. J Natl Cancer Inst 100:1233–1246
83. Whitson EL, Pluchino KM, Hall MD, McMahon JB, McKee TC (2011) New candidaspongiolides, tedanolide analogues that selectively inhibit melanoma cell growth. Org Lett 13:3518–3521
84. Edrada RA, Ebel R, Supriyono A, Wray V, Schupp P, Steube K, van Soest R, Proksch P (2002) Swinhoeiamide A, a new highly active calyculin derivative from the marine sponge *Theonella swinhoei.* J Nat Prod 65:1168–1172
85. Wakimoto T, Egami Y, Nakashima Y, Wakimoto Y, Mori T, Awakawa T, Ito T, Kenmoku H, Asakawa Y, Piel J (2014) Calyculin biogenesis from a pyrophosphate protoxin produced by a sponge symbiont. Nat Chem Biol 10:648–655
86. Oku N, Takada K, Fuller RW, Wilson JA, Peach ML, Pannell LK, McMahon JB, Gustafson KR (2010) Isolation, structural elucidation, and absolute stereochemistry of enigmazole A, a cytotoxic phosphomacrolide from the Papua New Guinea marine sponge *Cinachyrella enigmatica.* J Am Chem Soc 132:10278–10285
87. Sugai T, Fuwa H (2020) Total Synthesis of (−)-enigmazole A. In: Marine natural products. Springer, p 361–386
88. Skepper CK, Quach T, Molinski TF (2010) Total synthesis of enigmazole A from *Cinachyrella enigmatica.* Bidirectional bond constructions with an ambident 2,4-disubstituted oxazole synthon. J Am Chem Soc 132:10286–10292
89. Ai Y, Kozytska MV, Zou Y, Khartulyari AS, Maio WA, Smith AB III (2018) Total synthesis of the marine phosphomacrolide, (−)-Enigmazole A, exploiting multicomponent type I anion relay chemistry (ARC) in conjunction with a late-stage petasis-ferrier union/rearrangement. J Org Chem 83:6110–6126
90. Ahlers A, de Haro T, Gabor B, Fürstner A (2016) Concise total synthesis of enigmazole A. Angew Chem 128:1428–1433
91. Sakurai K, Sakamoto K, Sasaki M, Fuwa H (2020) Unified total synthesis of (−)-enigmazole A and (−)-15-*O*-methylenigmazole A. Chem Asian J 15:3494–3502

92. Masuda T, Ohyama K, Yoshimura A, Fuwa H (2024) Total synthesis of (−)-enigmazole A by the macrocyclization/transannular pyran cyclization strategy. Org Lett 26:2045–2050
93. Sakurai K, Sasaki M, Fuwa H (2018) Total Synthesis of (−)-Enigmazole A. Angew Chem Int Ed 57:5143–5146
94. Meissner A, Kishi T, Fujisawa Y, Murai Y, Takamura H, Kadota I (2018) Formal total synthesis of enigmazole A. Tetrahedron Lett 59:4492–4495
95. Takada K, Oku N, Peach ML, Ransom TT, Henrich CJ, Gustafson KR (2023) Enigmazole phosphomacrolides from the marine sponge *Cinachyrella enigmatica*. J Org Chem 88:10996–11002
96. Tarazona G, Fernández R, Pérez M, Millán RE, Jiménez C, Rodríguez J, Cuevas C (2022) Enigmazole C: a cytotoxic macrocyclic lactone and its ring-opened derivatives from a new species of homophymia sponge. J Nat Prod 85:1059–1066
97. Williams DE, Lassota P, Andersen RJ (1998) Motuporamines A−C, cytotoxic alkaloids isolated from the marine sponge *Xestospongia exigua* (Kirkpatrick). J Org Chem 63:4838–4841
98. Roskelley CD, Williams DE, McHardy LM, Leong KG, Troussard A, Karsan A, Andersen RJ, Dedhar S, Roberge M (2001) Inhibition of tumor cell invasion and angiogenesis by motuporamines. Can Res 61:6788–6794
99. To K, Loh K, Roskelley C, Andersen R, O'Connor T (2006) The anti-invasive compound motuporamine C is a robust stimulator of neuronal growth cone collapse. Neuroscience 139:1263–1274
100. Williams DE, Craig KS, Patrick B, McHardy LM, van Soest R, Roberge M, Andersen RJ (2002) Motuporamines, anti-invasion and anti-angiogenic alkaloids from the marine sponge *Xestospongia exigua* (Kirkpatrick): isolation, structure elucidation, analogue synthesis, and conformational analysis. J Org Chem 67:245–258
101. Muth A, Pandey V, Kaur N, Wason M, Baker C, Han X, Johnson TR, Altomare DA, Phanstiel O IV (2014) Synthesis and biological evaluation of antimetastatic agents predicated upon dihydromotuporamine C and its carbocyclic derivatives. J Med Chem 57:4023–4034
102. Borselli D, Blanchet M, Bolla JM, Muth A, Skruber K, Phanstiel O IV, Brunel JM (2017) Motuporamine derivatives as antimicrobial agents and antibiotic enhancers against resistant Gram-negative bacteria. ChemBioChem 18:276–283
103. Baker BJ, Scheuer PJ, Shoolery JN (1988) Papuamine, an antifungal pentacyclic alkaloid from a marine sponge, *Haliclona* sp. J Am Chem Soc 110:965–966
104. Fahy E, Molinski TF, Harper MK, Sullivan BW, Faulkner DJ, Parkanyi L, Clardy J (1988) Haliclonadiamine, an antimicrobial alkaloid from the sponge *Haliclona* sp. Tetrahedron Lett 29:3427–3428
105. Yamazaki H, Wewengkang DS, Kanno S-i, Ishikawa M, Rotinsulu H, Mangindaan RE, Namikoshi M (2013) Papuamine and haliclonadiamine, obtained from an Indonesian sponge *Haliclona* sp., inhibited cell proliferation of human cancer cell lines. Nat Prod Res 27:1012–1015
106. Liang Z, Sulzmaier FJ, Yoshida WY, Kelly M, Ramos JW, Williams PG (2015) Neopetrocyclamines A and B, polycyclic diamine alkaloids from the sponge *Neopetrosia* cf. *exigua*. J Nat Prod 78:543–547
107. Abdjul DB, Yamazaki H, Kanno S-i, Takahashi O, Kirikoshi R, Ukai K, Namikoshi M (2016) Haliclonadiamine derivatives and 6-*epi*-monanchorin from the marine sponge *Halichondria panicea* collected at Iriomote Island. J Nat Prod 79:1149–1154
108. Liu H-B, Imler GH, Baldridge KK, O'Connor RD, Siegel JS, Deschamps JR, Bewley CA (2020) X-Ray crystallography and unexpected chiroptical properties reassign the configuration of haliclonadiamine. J Am Chem Soc 142:2755–2759
109. Abdjul DB, Yagi A, Yamazaki H, Kirikoshi R, Takahashi O, Namikoshi M, Uchida R (2018) Anti-mycobacterial haliclonadiamine alkaloids from the Okinawan marine sponge *Haliclona* sp. collected at Iriomote Island. Phytochem Lett 26:130–133
110. Kanno S-I, Yomogida S, Tomizawa A, Yamazaki H, Ukai K, Mangindaan RE, Namikoshi M, Ishikawa M (2013) Papuamine causes autophagy following the reduction of cell survival

through mitochondrial damage and JNK activation in MCF-7 human breast cancer cells. Int J Oncol 43:1413–1419
111. Min H-Y, Jung Y, Park KH, Lee H-Y (2020) Papuamine inhibits viability of non-small cell lung cancer cells by inducing mitochondrial dysfunction. Anticancer Res 40:323–333
112. Kanno SI, Yomogida S, Tomizawa A, Yamazaki H, Ukai K, Mangindaan RE, Namikoshi M, Ishikawa M (2014) Combined effect of papuamine and doxorubicin in human breast cancer MCF-7 cells. Oncol Lett 8:547–550
113. Andersen RJ, Van Soest RWM, Kong F (1996) Chapter Three 3-Alkylpiperidine alkaloids isolated from marine sponges in the order haplosclerida. In: Pelletier SW (ed) Alkaloids: chemical and biological perspectives. Pergamon, pp 301–355
114. Rodriguez J, Peters BM, Kurz L, Schatzman RC, McCarley D, Lou L, Crews P (1993) An alkaloid protein kinase C inhibitor, xestocyclamine A, from the marine sponge *Xestospongia* sp. J Am Chem Soc 115:10436–10437
115. Rodríguez J, Crews P (1994) Revised structure of xestocyclamine A and description of a new analogue. Tetrahedron Lett 35:4719–4722
116. Kong F, Andersen RJ, Allen TM (1994) Ingenamine, a novel pentacyclic alkaloid from the marine sponge *Xestospongia ingens*. Tetrahedron Lett 35:1643–1646
117. Kong F, Andersen RJ (1995) Ingenamine alkaloids isolated from the sponge *Xestospongia ingens*: structures and absolute configurations. Tetrahedron 51:2895–2906
118. Kong F, Andersen RJ, Allen TM (1994) Ingamines A and B, new cytotoxic alkaloids from the marine sponge *Xestospongia ingens*. Tetrahedron 50:6137–6144
119. Baldwin J, Whitehead R (1992) On the biosynthesis of manzamines. Tetrahedron Lett 33:2059–2062
120. Meng Z, Fürstner A (2020) Total synthesis provides strong evidence: xestocyclamine A is the enantiomer of ingenamine. J Am Chem Soc 142:11703–11708
121. Ilias M, Ibrahim MA, Khan SI, Jacob MR, Tekwani BL, Walker LA, Samoylenko V (2012) Pentacyclic ingamine alkaloids, a new antiplasmodial pharmacophore from the marine sponge *Petrosid Ng5 Sp5*. Planta Med 78:1690–1697
122. Kong F, Andersen RJ, Allen TM (1994) Madangamine A, a novel cytotoxic alkaloid from the marine sponge *Xestospongia ingens*. J Am Chem Soc 116:6007–6008
123. Kong F, Graziani EI, Andersen RJ (1998) Madangamines B– E, pentacyclic alkaloids from the marine sponge *Xestospongia ingens*. J Nat Prod 61:267–271
124. Tang Y, Zhu L, Hong R (2022) Madangamine alkaloids: Madness and tranquility. Tetrahedron Chem 3:100025
125. Miura K, Kawano S, Suto T, Sato T, Chida N, Simizu S (2021) Identification of madangamine A as a novel lysosomotropic agent to inhibit autophagy. Bioorg Med Chem 34:116041
126. Suto T, Yanagita Y, Nagashima Y, Takikawa S, Kurosu Y, Matsuo N, Miura K, Simizu S, Sato T, Chida N (2019) Unified total synthesis of madangamine alkaloids. Bull Chem Soc Jpn 92:545–571
127. Sakai R, Higa T, Jefford CW, Bernardinelli G (1986) Manzamine A, a novel antitumor alkaloid from a sponge. J Am Chem Soc 108:6404–6405
128. Crews P, Cheng X-C, Adamczeski M, Rodríguez J, Jaspars M, Schmitz FJ, Traeger SC, Pordesimo EO (1994) 1, 2, 3, 4-tetrahydro-8-hydroxymanzamines, alkaloids from two different haplosclerid sponges. Tetrahedron 50:13567–13574
129. Braekman JC, Daloze D, de Abreu PM, Piccinni-Leopardi C, Germain G, Van Meerssche M (1982) A novel type of *bis*-quinolizidine alkaloid from the sponge: *Petrosia seriata*. Tetrahedron Lett 23:4277–4280
130. Braekman JC, Daloze D, Defay N, Zimmermann D (1984) Petrosin-A and-B, two new *bis*-quinolizidine alkaloids from the sponge *Petrosia seriata*. Bulletin des Sociétés Chimiques Belges 93:941–944
131. Braekman JC, Daloze D, Cimino G, Trivellone E (1988) 2D-NMR study of petrosins: revised structure for petrosin-A. Bulletin des Sociétés Chimiques Belges 97:519–524
132. Inman WD, Bray WM, Gassner NC, Lokey RS, Tenney K, Shen YY, TenDyke K, Suh T, Crews P (2010) A β-carboline alkaloid from the Papua New Guinea marine sponge *Hyrtios reticulatus*. J Nat Prod 73:255–257

133. Iwagawa T, Kaneko M, Okamura H, Nakatani M, van Soest RW (1998) New alkaloids from the Papua New Guinean sponge *Agelas nakamurai*. J Nat Prod 61:1310–1312
134. Calcul L, Tenney K, Ratnam J, McKerrow JH, Crews P (2010) Structural variations to the 9-*N*-methyladeninium diterpenoid hybrid commonly isolated from *Agelas* sponges. Aust J Chem 63:915–921
135. Gautschi JT, Whitman S, Holman TR, Crews P (2004) An analysis of phakellin and oroidin structures stimulated by further study of an *Agelas* sponge. J Nat Prod 67:1256–1261
136. Pettit GR, Xu J-p, Gingrich DE, Williams MD, Doubek DL, Chapuis J-C, Schmidt JM (1999) Antineoplastic agents. Part 395. 1 Isolation and structure of agelagalastatin from the Papua New Guinea marine sponge *Agelas* sp. Chem Commun:915–916
137. Manzo E, Van Soest R, Matainaho L, Roberge M, Andersen RJ (2003) Ceratamines A and B, antimitotic heterocyclic alkaloids isolated from the marine sponge *Pseudoceratina* sp. collected in Papua New Guinea. Org Lett 5:4591–4594
138. Karjala G, Chan Q, Manzo E, Andersen RJ, Roberge M (2005) Ceratamines, structurally simple microtubule-stabilizing antimitotic agents with unusual cellular effects. Can Res 65:3040–3043
139. Nodwell M, Zimmerman C, Roberge M, Andersen RJ (2010) Synthetic analogues of the microtubule-stabilizing sponge alkaloid ceratamine A are more active than the natural product. J Med Chem 53:7843–7851
140. Tao L, Pan X, Ji M, Chen X, Liu Z (2017) Efficient synthesis and cytotoxicity of novel microtubule-stabilizing agent ceratamine A analogues. Tetrahedron 73:2159–2171
141. Pan X, Tao L, Ji M, Chen X, Liu Z (2018) Synthesis and cytotoxicity of novel imidazo[4,5-d]azepine compounds derived from marine natural product ceratamine A. Bioorg Med Chem Lett 28:866–868
142. Miao S, Andersen RJ, Allen TM (1990) Cytotoxic metabolites from the sponge *Lanthella basta* collected in Papua New Guinea. J Nat Prod 53:1441–1446
143. Pettit GR, Butler MS, Bass CG, Doubek DL, Williams MD, Schmidt JM, Patit RK, Hooper JN, Tackett LP, Filiatrault MJ (1995) Antineoplastic agents, 326. The stereochemistry of bastadins 8, 10, and 12 from the Bismarck archipelago marine sponge *Lanthella basta*. J Nat Prod 58:680–688
144. Jaspars M, Rali T, Laney M, Schatzman RC, Diaz MC, Schmitz FJ, Pordesimo EO, Crews P (1994) The search for inosine 5′-phosphate dehydrogenase (IMPDH) inhibitors from marine sponges. Evaluation of the bastadin alkaloids. Tetrahedron 50:7367–7374
145. Pettit GR, Butler MS, Williams MD, Filiatrault MJ, Pettit RK (1996) Isolation and structure of hemibastadinols 1–3 from the Papua New Guinea marine sponge Ianthella basta. J Nat Prod 59:927–934
146. Calcul L, Inman WD, Morris AA, Tenney K, Ratnam J, McKerrow JH, Valeriote FA, Crews P (2010) Additional insights on the bastadins: isolation of analogues from the sponge *Ianthella* cf. *reticulata* and exploration of the oxime configurations. J Nat Prod 73:365–372
147. Pina IC, Gautschi JT, Wang G-Y-S, Sanders ML, Schmitz FJ, France D, Cornell-Kennon S, Sambucetti LC, Remiszewski SW, Perez LB (2003) Psammaplins from the sponge *Pseudoceratina purpurea*: inhibition of both histone deacetylase and DNA methyltransferase. J Org Chem 68:3866–3873
148. Ralifo P, Tenney K, Valeriote FA, Crews P (2007) A distinctive structural twist in the aminoimidazole alkaloids from a calcareous marine sponge: isolation and characterization of leucosolenamines A and B. J Nat Prod 70:33–38
149. Valeriote FA, Tenney K, Pietraszkiewicz H, Edelstein M, Johnson TA, Amagata T, Crews P (2012) Discovery and development of anticancer agents from marine sponges: perspectives based on a chemistry-experimental therapeutics collaborative program. J Exp Therapeutics Oncol 10
150. Thale Z, Johnson T, Tenney K, Wenzel PJ, Lobkovsky E, Clardy J, Media J, Pietraszkiewicz H, Valeriote FA, Crews P (2002) Structures and cytotoxic properties of sponge-derived bisannulated acridines. J Org Chem 67:9384–9391

151. Brastianos HC, Vottero E, Patrick BO, Van Soest R, Matainaho T, Mauk AG, Andersen RJ (2006) Exiguamine A, an indoleamine-2, 3-dioxygenase (IDO) inhibitor isolated from the marine sponge *Neopetrosia exigua.* J Am Chem Soc 128:16046–16047
152. Fujiwara Y, Kato S, Nesline MK, Conroy JM, DePietro P, Pabla S, Kurzrock R (2022) Indoleamine 2,3-dioxygenase (IDO) inhibitors and cancer immunotherapy. Cancer Treat Rev 110:102461
153. Volgraf M, Lumb J-P, Brastianos HC, Carr G, Chung MK, Münzel M, Mauk AG, Andersen RJ, Trauner D (2008) Biomimetic synthesis of the IDO inhibitors exiguamine A and B. Nat Chem Biol 4:535–537
154. Carr G, Chung MK, Mauk AG, Andersen RJ (2008) Synthesis of indoleamine 2, 3-dioxygenase inhibitory analogues of the sponge alkaloid exiguamine A. J Med Chem 51:2634–2637
155. Dong J, Pan X, Yang Y, Zhang G, Xiao Z, Liu Z (2021) Design, synthesis and biological evaluation of exiguamine A analogues as IDO1 inhibitors. Eur J Med Chem 223:113631
156. Kong K-m, Zhang J-w, Liu B-z, Meng G-r, Zhang Q (2020) Discovery of 5-(pyridin-3-yl)-1H-indole-4, 7-diones as indoleamine 2, 3-dioxygenase 1 (IDO1) inhibitors. Bioorg Med Chem Lett 30:126901
157. Zhang J, Liu J, Kong K, Li X, Zhang Q (2024) Structure–activity relationship studies of 5-(Pyridin-3-yl)-1H-indole-4,7-diones as indoleamine 2,3-dioxygenase 1 inhibitors. Results Chem 7:101285
158. Adamczeski M, Quinoa E, Crews P (1988) Unusual anthelminthic oxazoles from a marine sponge. J Am Chem Soc 110:1598–1602
159. Rodríguez J, Nieto RM, Crews P (1993) New structures and bioactivity patterns of bengazole alkaloids from a Choristid marine sponge. J Nat Prod 56:2034–2040
160. Coleman JE, de Silva ED, Kong F, Andersen RJ, Allen TM (1995) Cytotoxic peptides from the marine sponge *Cymbastela* sp. Tetrahedron 51:10653–10662
161. Talpir R, Benayahu Y, Kashman Y, Pannell L, Schleyer M (1994) Hemiasterlin and geodiamolide TA; two new cytotoxic peptides from the marine sponge *Hemiasterella minor* (Kirkpatrick). Tetrahedron Lett 35:4453–4456
162. Gamble WR, Durso NA, Fuller RW, Westergaard CK, Johnson TR, Sackett DL, Hamel E, Cardellina JH II, Boyd MR (1999) Cytotoxic and tubulin-interactive hemiasterlins from *Auletta* sp. and *Siphonochalina* spp. sponges. Bioorg Med Chem 7:1611–1615
163. Anderson HJ, Coleman JE, Andersen RJ, Roberge M (1996) Cytotoxic peptides hemiasterlin, hemiasterlin A and hemiasterlin B induce mitotic arrest and abnormal spindle formation. Cancer Chemother Pharmacol 39:223–226
164. Nieman JA, Coleman JE, Wallace DJ, Piers E, Lim LY, Roberge M, Andersen RJ (2003) Synthesis and antimitotic/cytotoxic activity of hemiasterlin analogues. J Nat Prod 66:183–199
165. Loganzo F, Discafani CM, Annable T, Beyer C, Musto S, Hari M, Tan X, Hardy C, Hernandez R, Baxter M (2003) HTI-286, a synthetic analogue of the tripeptide hemiasterlin, is a potent antimicrotubule agent that circumvents P-glycoprotein-mediated resistance *in vitro* and *in vivo.* Can Res 63:1838–1845
166. Newman DJ (2021) Natural product based antibody drug conjugates: clinical status as of November 9, 2020. J Nat Prod 84:917–931
167. Crews P, Farias JJ, Emrich R, Keifer PA (1994) Milnamide A, an unusual cytotoxic tripeptide from the marine sponge *Auletta* cf. *constricta.* J Org Chem 59:2932–2934
168. Sonnenschein RN, Farias JJ, Tenney K, Mooberry SL, Lobkovsky E, Clardy J, Crews P (2004) A further study of the cytotoxic constituents of a milnamide-producing sponge. Org Lett 6:779–782
169. Chevallier C, Richardson AD, Edler MC, Hamel E, Harper MK, Ireland CM (2003) A new cytotoxic and tubulin-interactive milnamide derivative from a marine sponge *Cymbastela* sp. Org Lett 5:3737–3739
170. Pettit GR, Srirangam JK, Herald DL, Erickson KL, Doubek DL, Schmidt JM, Tackett LP, Bakus GJ (1992) Antineoplastic agents. 251. Isolation and structure of stylostatin 1 from the Papua New Guinea marine sponge *Stylotella aurantium.* J Org Chem 57:7217–7220

171. Pettit GR, Srirangam JK, Herald DL, Xu J-p, Boyd MR, Cichacz Z, Kamano Y, Schmidt JM, Erickson KL (1995) Isolation and crystal structure of stylopeptide 1, a new marine porifera cycloheptapeptide. J Org Chem 60:8257–8261
172. Brennan MR, Costello CE, Maleknia SD, Pettit GR, Erickson KL (2008) Stylopeptide 2, a proline-rich cyclodecapeptide from the sponge *Stylotella* sp. J Nat Prod 71:453–456
173. Kita M, Gise B, Kawamura A, Kigoshi H (2013) Stylissatin A, a cyclic peptide that inhibits nitric oxide production from the marine sponge *Stylissa massa*. Tetrahedron Lett 54:6826–6828
174. Akindele T, Gise B, Sunaba T, Kita M, Kigoshi H (2015) Total synthesis of stylissatin A, a cyclic peptide that inhibits nitric oxide production. Bull Chem Soc Jpn 88:600–609
175. Shaheen F, Jabeen A, Ashraf S, Nadeem-ul-Haque M, Shah ZA, Ziaee MA, Dastagir N, Ganesan A (2016) Total synthesis, structural, and biological evaluation of stylissatin A and related analogs. J Pept Sci 22:607–617
176. Li Y-l, Bao X-c, Gao Y-y, Liu C, Zou Y, Yan F (2016) Solid-phase synthesis of cyclic heptapeptide stylissatin A. Chem Nat Compd 52:1069–1072
177. Zhang M, Sunaba T, Sun Y, Sasaki K, Isoda H, Kigoshi H, Kita M (2019) Anti-inflammatory marine cyclic peptide stylissatin A and its derivatives inhibit differentiation of murine preadipocytes. Chem Commun 55:5471–5474
178. Zhang M, Sunaba T, Sun Y, Shibata T, Sasaki K, Isoda H, Kigoshi H, Kita M (2020) Acyl-CoA dehydrogenase long chain (ACADL) is a target protein of stylissatin A, an anti-inflammatory cyclic heptapeptide. J Antibiot 73:589–592
179. Houten SM, Wanders RJ (2010) A general introduction to the biochemistry of mitochondrial fatty acid β-oxidation. J Inherit Metab Dis 33:469–477
180. Poynter ME, Daynes RA (1998) Peroxisome proliferator-activated receptor α activation modulates cellular redox status, represses nuclear factor-κB signaling, and reduces inflammatory cytokine production in aging. J Biol Chem 273:32833–32841
181. Ullah A, Shams S, Khan AI, Ahmad A, Jabeen A, Shaheen F (2022) Discovery of stylissatin A analogs exhibiting potent nitric oxide inhibition. Bioorg Chem 124:105841
182. Kita M, Sun Y, Dakiiwa A, Zhang M, Shibata T (2024) Inhibition of lysosomal cathepsin A and neuraminidase 1 interaction by anti-obesity cyclic peptide. Chem Eur J:e202402049
183. Clark WD, Corbett T, Valeriote F, Crews P (1997) Cyclocinamide A. An unusual cytotoxic halogenated hexapeptide from the marine sponge *Psammocinia*. J Am Chem Soc 119:9285–9286
184. Grieco PA, Reilly M (1998) Studies related to the absolute configuration of cyclocinamide A: total synthesis of 4(*R*), 11(*R*)-cyclocinamide A. Tetrahedron Lett 39:8925–8928
185. Rubio BK, Robinson SJ, Avalos CE, Valeriote FA, De Voogd NJ, Crews P (2008) Revisiting the sponge sources, stereostructure, and biological activity of cyclocinamide A. J Nat Prod 71:1475–1478
186. Garcia JM, Curzon SS, Watts KR, Konopelski JP (2012) Total synthesis of nominal (11*S*)-and (11*R*)-cyclocinamide A. Org Lett 14:2054–2057
187. Laird DW, LaBarbera DV, Feng X, Bugni TS, Harper MK, Ireland CM (2007) Halogenated cyclic peptides isolated from the sponge *Corticium* sp. J Nat Prod 70:741–746
188. Schmidt EW, Raventos-Suarez C, Bifano M, Menendez AT, Fairchild CR, Faulkner DJ (2004) Scleritodermin A, a cytotoxic cyclic peptide from the lithistid sponge *Scleritoderma nodosum*. J Nat Prod 67:475–478
189. Liu S, Cui Y-M, Nan F-J (2008) Total synthesis of the originally proposed and revised structures of scleritodermin A. Org Lett 10:3765–3768
190. Wei Y, Chen Y-T, Shi L, Gao L-X, Liu S, Cui Y-M, Zhang W, Shen Q, Li J, Nan F-J (2011) Discovery and structural modification of novel inhibitors of PTP1B inspired by the ACT fragment of scleritodermin A. MedChemComm 2:1104–1109
191. Chen YT, Tang CL, Ma WP, Gao LX, Wei Y, Zhang W, Li JY, Li J, Nan FJ (2013) Design, synthesis, and biological evaluation of novel 2-ethyl-5-phenylthiazole-4-carboxamide derivatives as protein tyrosine phosphatase 1B inhibitors with improved cellular efficacy. Eur J Med Chem 69:399–412

192. De Silva ED, Williams DE, Andersen RJ, Klix H, Holmes CF, Allen TM (1992) Motuporin, a potent protein phosphatase inhibitor isolated from the Papua New Guinea sponge *Theonella swinhoei* Gray. Tetrahedron Lett 33:1561–1564
193. Melaram R, Newton AR, Lee A, Herber S, El-Khouri A, Chafin J (2024) A review of microcystin and nodularin toxins derived from freshwater cyanobacterial harmful algal blooms and their impact on human health. Toxicol Environ Health Sci
194. Bagu JR, Sönnichsen FD, Williams D, Andersen RJ, Sykes BD, Holmes CF (1995) Comparison of the solution structures of microcystin-LR and motuporin. Nat Struct Biol 2:114–116
195. Wegerski CJ, Hammond J, Tenney K, Matainaho T, Crews P (2007) A serendipitous discovery of isomotuporin-containing sponge populations of *Theonella swinhoei*. J Nat Prod 70:89–94
196. Crews P, Manes LV, Boehler M (1986) Jasplakinolide, a cyclodepsipeptide from the marine sponge, *Jaspis* sp. Tetrahedron Lett 27:2797–2800
197. Zabriskie TM, Klocke JA, Ireland CM, Marcus AH, Molinski TF, Faulkner DJ, Xu C, Clardy J (1986) Jaspamide, a modified peptide from a *Jaspis* sponge, with insecticidal and antifungal activity. J Am Chem Soc 108:3123–3124
198. Chan WR, Tinto WF, Manchand PS, Todaro LJ (1987) Stereostructures of geodiamolides A and B, novel cyclodepsipeptides from the marine sponge *Geodia* sp. J Org Chem 52:3091–3093
199. de Silva ED, Andersen RJ, Allen TM (1990) Geodiamolides C to F, new cytotoxic cyclodepsipeptides from the marine sponge *Pseudaxinyssa* sp. Tetrahedron Lett 31:489–492
200. Coleman JE, Van Soest R, Andersen RJ (1999) New geodiamolides from the sponge *Cymbastela* sp. collected in Papua New Guinea. J Nat Prod 62:1137–1141
201. Scott V, Boehme R, Matthews T (1988) New class of antifungal agents: jasplakinolide, a cyclodepsipeptide from the marine sponge, *Jaspis* species. Antimicrob Agents Chemother 32:1154–1157
202. Braekman JC, Daloze D, Moussiaux B, Riccio R (1987) Jaspamide from the marine sponge *Jaspis johnstoni*. J Nat Prod 50:994–995
203. Robinson SJ, Morinaka BI, Amagata T, Tenney K, Bray WM, Gassner NC, Lokey RS, Crews P (2010) New structures and bioactivity properties of jasplakinolide (jaspamide) analogues from marine sponges. J Med Chem 53:1651–1661
204. Ford PW, Gustafson KR, McKee TC, Shigematsu N, Maurizi LK, Pannell LK, Williams DE, Dilip de Silva E, Lassota P, Allen TM (1999) Papuamides A– D, HIV-inhibitory and cytotoxic depsipeptides from the Sponges *Theonella mirabilis* and *Theonella swinhoei* collected in Papua New Guinea. J Am Chem Soc 121:5899–5909
205. Festa C, De Marino S, Zampella A, Fiorucci S (2023) *Theonella*: a treasure trove of structurally unique and biologically active sterols. Marine Drugs 21
206. Xie W, Ding D, Zi W, Li G, Ma D (2008) Total synthesis and structure assignment of papuamide B, a potent marine cyclodepsipeptide with anti-HIV properties. Angew Chem 120:2886–2890
207. Andjelic CD, Planelles V, Barrows LR (2008) Characterizing the anti-HIV activity of papuamide A. Mar Drugs 6:528–549
208. Parsons AB, Lopez A, Givoni IE, Williams DE, Gray CA, Porter J, Chua G, Sopko R, Brost RL, Ho C-H (2006) Exploring the mode-of-action of bioactive compounds by chemical-genetic profiling in yeast. Cell 126:611–625
209. Takar M, Wu Y, Graham TR (2016) The essential Neo1 protein from budding yeast plays a role in establishing aminophospholipid asymmetry of the plasma membrane. J Biol Chem 291:15727–15739
210. Chen S, Wang J, Muthusamy BP, Liu K, Zare S, Andersen RJ, Graham TR (2006) Roles for the Drs2p–Cdc50p complex in protein transport and phosphatidylserine asymmetry of the yeast plasma membrane. Traffic 7:1503–1517
211. Mioka T, Fujimura-Kamada K, Mizugaki N, Kishimoto T, Sano T, Nunome H, Williams DE, Andersen RJ, Tanaka K (2018) Phospholipid flippases and Sfk1p, a novel regulator of phospholipid asymmetry, contribute to low permeability of the plasma membrane. Mol Biol Cell 29:1203–1218

212. Curto M-Á, Sharifmoghadam MR, Calpena E, De León N, Hoya M, Doncel C, Leatherwood J, Valdivieso M-H (2014) Membrane organization and cell fusion during mating in fission yeast requires multipass membrane protein Prm1. Genetics 196:1059–1076
213. Cassilly CD, Maddox MM, Cherian PT, Bowling JJ, Hamann MT, Lee RE, Reynolds TB (2016) SB-224289 antagonizes the antifungal mechanism of the marine depsipeptide papuamide A. PLoS ONE 11:e0154932
214. Chen YL, Montedonico AE, Kauffman S, Dunlap JR, Menn FM, Reynolds TB (2010) Phosphatidylserine synthase and phosphatidylserine decarboxylase are essential for cell wall integrity and virulence in Candida albicans. Mol Microbiol 75:1112–1132
215. Kim C-K, Wang D, Bokesch HR, Fuller RW, Smith E, Henrich CJ, Durrant DE, Morrison DK, Bewley CA, Gustafson KR (2020) Swinhopeptolides A and B: cyclic depsipeptides from the sponge *Theonella swinhoei* that inhibit Ras/Raf interaction. J Nat Prod 83:1288–1294
216. Oku N, Gustafson KR, Cartner LK, Wilson JA, Shigematsu N, Hess S, Pannell LK, Boyd MR, McMahon JB (2004) Neamphamide A, a new HIV-inhibitory depsipeptide from the Papua New Guinea marine sponge *Neamphius huxleyi*. J Nat Prod 67:1407–1411
217. Oku N, Krishnamoorthy R, Benson AG, Ferguson RL, Lipton MA, Phillips LR, Gustafson KR, McMahon JB (2005) Complete stereochemistry of neamphamide A and absolute configuration of the β-methoxytyrosine residue in papuamide B. J Org Chem 70:6842–6847
218. Ratnayake AS, Bugni TS, Feng X, Harper MK, Skalicky JJ, Mohammed KA, Andjelic CD, Barrows LR, Ireland CM (2006) Theopapuamide, a cyclic depsipeptide from a Papua New Guinea lithistid sponge *Theonella swinhoei*. J Nat Prod 69:1582–1586
219. Plaza A, Bifulco G, Keffer JL, Lloyd JR, Baker HL, Bewley CA (2009) Celebesides A– C and theopapuamides B– D, depsipeptides from an Indonesian sponge that inhibit HIV-1 entry. J Org Chem 74:504–512
220. Williams DE, Austin P, Diaz-Marrero AR, Soest RV, Matainaho T, Roskelley CD, Roberge M, Andersen RJ (2005) Neopetrosiamides, peptides from the marine sponge *Neopetrosia* sp. that inhibit amoeboid invasion by human tumor cells. Org Lett 7:4173–4176
221. Liu H, Boudreau MA, Zheng J, Whittal RM, Austin P, Roskelley CD, Roberge M, Andersen RJ, Vederas JC (2010) Chemical synthesis and biological activity of the neopetrosiamides and their analogues: revision of disulfide bond connectivity. J Am Chem Soc 132:1486–1487
222. Austin P, Heller M, Williams DE, McIntosh LP, Vogl AW, Foster LJ, Andersen RJ, Roberge M, Roskelley CD (2010) Release of membrane-bound vesicles and inhibition of tumor cell adhesion by the peptide neopetrosiamide A. PLoS ONE 5:e10836
223. Towle KM, Chaytor JL, Liu H, Austin P, Roberge M, Roskelley CD, Vederas JC (2013) Synthesis and biological studies of neopetrosiamides as inhibitors of cancer cell invasion. Org Biomol Chem 11:1476–1481
224. Pascoe CA, Engelhardt DB, Rosana ARR, van Belkum MJ, Vederas JC (2021) Methylene analogues of neopetrosiamide as potential antimetastatic agents: solid-supported syntheses using diamino diacids for pre-stapling of peptides with multiple disulfides. Org Lett 23:9216–9220
225. Han B, McPhail KL, Ligresti A, Di Marzo V, Gerwick WH (2003) Semiplenamides A– G, fatty acid amides from a Papua New Guinea collection of the marine cyanobacterium *Lyngbya semiplena*. J Nat Prod 66:1364–1368
226. Davies IR, Cheeseman M, Niyadurupola DG, Bull SD (2005) An efficient synthesis of semiplenamide C. Tetrahedron Lett 46:5547–5549
227. Das B, Damodar K, Bhunia N, Shashikanth B (2009) Mild and practical stereoselective synthesis of (*Z*)- and (*E*)-allyl bromides from Baylis-Hillman adducts using Appel agents (PPh3/CBr 4): a facile synthesis of semiplenamides C and *E*. Tetrahedron Lett 50:2072–2074
228. Das B, Srinivas Y, Sudhakar C, Damodar K, Reddy PR (2010) A rapid and efficient synthesis of allyl iodides from Baylis-Hillman adducts: a facile synthesis of semiplenamide D. Chem Lett 39:246–247
229. Gutiérrez M, Pereira AR, Debonsi HM, Ligresti A, Di Marzo V, Gerwick WH (2011) Cannabinomimetic lipid from a marine cyanobacterium. J Nat Prod 74:2313–2317

230. Gao Y-R, Guo S-H, Zhang Z-X, Mao S, Zhang Y-L, Wang Y-Q (2013) Concise synthesis of (+)-serinolamide A. Tetrahedron Lett 54:6511–6513
231. Gahalawat S, Pandey SK (2015) Enantioselective total synthesis of (+)-serinolamide A. RSC Adv 5:41013–41016
232. Ghotekar GS, Mujahid M, Muthukrishnan M (2019) Total synthesis of marine natural products serinolamide A and columbamide D. ACS Omega 4:1322–1328
233. Jadhao AR, Waghmode SB (2024) Organocatalytic approach to the enantioselective total synthesis of (+)-serinolamides A and B and (+)-lacosamide. Synthesis 56:1407–1414
234. Park SJ, Kim J, Kim J, Kim Y, Lee EH, Kim HJ, Kim S, Kim B, Kim R, Choi JW (2022) Synthesis and evaluation of serinolamide derivatives as sphingosine-1-phosphate-1 (S1P1) receptor agonists. Molecules 27:2818
235. Park SJ, Yeon SK, Kim Y, Kim HJ, Kim S, Kim J, Choi JW, Kim B, Lee EH, Kim R (2022) Discovery of novel sphingosine-1-phosphate-1 receptor agonists for the treatment of multiple sclerosis. J Med Chem 65:3539–3562
236. Mevers E, Matainaho T, Allara' M, Di Marzo V, Gerwick WH (2014) Mooreamide A: a cannabinomimetic lipid from the marine cyanobacterium *Moorea bouillonii*. Lipids 49:1127–1132
237. Kleigrewe K, Almaliti J, Tian IY, Kinnel RB, Korobeynikov A, Monroe EA, Duggan BM, Di Marzo V, Sherman DH, Dorrestein PC (2015) Combining mass spectrometric metabolic profiling with genomic analysis: a powerful approach for discovering natural products from cyanobacteria. J Nat Prod 78:1671–1682
238. Taton A, Rohrer S, Diaz B, Reher R, Caraballo Rodriguez AM, Pierce ML, Dorrestein PC, Gerwick L, Gerwick WH, Golden JW (2022) Heterologous expression in *Anabaena* of the columbamide pathway from the cyanobacterium *Moorena bouillonii* and production of new analogs. ACS Chem Biol 17:1910–1923
239. Malloy KL, Suyama TL, Engene N, Debonsi H, Cao Z, Matainaho T, Spadafora C, Murray TF, Gerwick WH (2012) Credneramides A and B: neuromodulatory phenethylamine and isopentylamine derivatives of a vinyl chloride-containing fatty acid from cf. *Trichodesmium* sp. nov. J Nat Prod 75:60–66
240. Erver F, Hilt G (2012) Cobalt- versus ruthenium-catalyzed Alder-Ene reaction for the synthesis of credneramide A and B. J Org Chem 77:5215–5219
241. Nunnery JK, Engene N, Byrum T, Cao Z, Jabba SV, Pereira AR, Matainaho T, Murray TF, Gerwick WH (2012) Biosynthetically intriguing chlorinated lipophilic metabolites from geographically distant tropical marine cyanobacteria. J Org Chem 77:4198–4208
242. Tan LT, Okino T, Gerwick WH (2000) Hermitamides A and B, toxic malyngamide-type natural products from the marine cyanobacterium *Lyngbya majuscula*. J Nat Prod 63:952–955
243. De Oliveira EO, Graf KM, Patel MK, Baheti A, Kong H-S, MacArthur LH, Dakshanamurthy S, Wang K, Brown ML, Paige M (2011) Synthesis and evaluation of hermitamides A and B as human voltage-gated sodium channel blockers. Bioorg Med Chem 19:4322–4329
244. Virolleaud M-A, Menant C, Fenet B, Piva O (2006) Total and formal enantioselective synthesis of lyngbic acid and hermitamides A and B. Tetrahedron Lett 47:5127–5130
245. Frost CG, Penrose SD, Gleave R (2008) Rhodium catalysed conjugate addition of a chiral alkenyltrifluoroborate salt: the enantioselective synthesis of hermitamides A and B. Org Biomol Chem 6:4340–4347
246. Satyanarayana S, Reddy BS, Narender R (2014) A concise total synthesis of lyngbic acid, hermitamides A and B. Tetrahedron Lett 55:6027–6029
247. Knouse KW, Wuest WM (2016) The enantioselective synthesis and biological evaluation of chimeric promysalin analogs facilitated by diverted total synthesis. J Antibiot 69:337–339
248. McPhail KL, Gerwick WH (2003) Three new malyngamides from a Papua New Guinea collection of the marine cyanobacterium *Lyngbya majuscula*. J Nat Prod 66:132–135
249. Malloy KL, Villa FA, Engene N, Matainaho T, Gerwick L, Gerwick WH (2011) Malyngamide 2, an oxidized lipopeptide with nitric oxide inhibiting activity from a Papua New Guinea marine cyanobacterium. J Nat Prod 74:95–98

250. Han B, Reinscheid UM, Gerwick WH, Gross H (2011) The structure elucidation of isomalyngamide K from the marine cyanobacterium *Lyngbya majuscula* by experimental and DFT computational methods. J Mol Struct 989:109–113
251. Clark BR, Engene N, Teasdale ME, Rowley DC, Matainaho T, Valeriote FA, Gerwick WH (2008) Natural products chemistry and taxonomy of the marine cyanobacterium *Blennothrix cantharidosmum*. J Nat Prod 71:1530–1537
252. Tan LT, Márquez BL, Gerwick WH (2002) Lyngbouilloside, a novel glycosidic macrolide from the marine cyanobacterium *Lyngbya bouillonii*. J Nat Prod 65:925–928
253. ElMarrouni A, Kolleth A, Lebeuf R, Gebauer J, Prevost S, Heras M, Arseniyadis S, Cossy J (2013) Lyngbouilloside and related macrolides from marine cyanobacteria. Nat Prod Commun 8:1934578X1300800723
254. Pereira AR, McCue CF, Gerwick WH (2010) Cyanolide A, a glycosidic macrolide with potent molluscicidal activity from the Papua New Guinea cyanobacterium *Lyngbya bouillonii*. J Nat Prod 73:217–220
255. Kim H, Hong J (2010) Total synthesis of cyanolide A and confirmation of its absolute configuration. Org Lett 12:2880–2883
256. Hajare AK, Ravikumar V, Khaleel S, Bhuniya D, Reddy DS (2011) Synthesis of molluscicidal agent cyanolide A macrolactone from D-(−)-pantolactone. J Org Chem 76:963–966
257. Gesinski MR, Rychnovsky SD (2011) Total synthesis of the cyanolide A aglycon. J Am Chem Soc 133:9727–9729
258. Sharpe RJ, Jennings MP (2011) A formal synthesis of (−)-cyanolide A featuring a stereoselective mukaiyama aldol reaction and oxocarbenium reduction. J Org Chem 76:8027–8032
259. Pabbaraja S, Satyanarayana K, Ganganna B, Yadav J (2011) Formal total synthesis of cyanolide A. J Org Chem 76:1922–1925
260. Yang Z, Xie X, Jing P, Zhao G, Zheng J, Zhao C, She X (2011) Total synthesis of cyanolide A. Org Biomol Chem 9:984–986
261. Waldeck AR, Krische MJ (2013) Total synthesis of cyanolide A in the absence of protecting groups, chiral auxiliaries, or premetalated carbon nucleophiles. Angewandte Chemie (International ed. in English) 52
262. Tay GC, Gesinski MR, Rychnovsky SD (2013) Formation of highly substituted tetrahydropyranones: application to the total synthesis of cyanolide A. Org Lett 15:4536–4539
263. Bates RW, Lek TG (2014) A synthesis of cyanolide A by intramolecular oxa-michael addition. Synthesis 46:1731–1738
264. Che W, Li Y-Z, Liu J-C, Zhu S-F, Xie J-H, Zhou Q-L (2019) Stereodiverse iterative synthesis of 1,3-polyol arrays through asymmetric catalytic hydrogenation. formal total synthesis of (−)-cyanolide A. Org Lett 21:2369–2373
265. Mevers E, Haeckl FJ, Boudreau PD, Byrum T, Dorrestein PC, Valeriote FA, Gerwick WH (2014) Lipopeptides from the tropical marine cyanobacterium *Symploca* sp. J Nat Prod 77:969–975
266. Tan LT, Sitachitta N, Gerwick WH (2003) The guineamides, novel cyclic depsipeptides from a Papua New Guinea collection of the marine cyanobacterium *Lyngbya majuscula*. J Nat Prod 66:764–771
267. Han B, Gross H, McPhail KL, Goeger D, Maier CS, Gerwick WH (2011) Wewakamide A and guineamide G, cyclic depsipeptides from the marine cyanobacteria *Lyngbya semiplena* and *Lyngbya majuscula*. J Microbiol Biotechnol 21:930–936
268. Almaliti J, Malloy KL, Glukhov E, Spadafora C, Gutiérrez M, Gerwick WH (2017) Dudawalamides A-D, antiparasitic cyclic depsipeptides from the marine cyanobacterium *Moorea producens*. J Nat Prod 80:1827–1836
269. Han B, Goeger D, Maier CS, Gerwick WH (2005) The wewakpeptins, cyclic depsipeptides from a Papua New Guinea Collection of the marine cyanobacterium *Lyngbya semiplena*. J Org Chem 70:3133–3139
270. Han B, McPhail KL, Gross H, Goeger DE, Mooberry SL, Gerwick WH (2005) Isolation and structure of five lyngbyabellin derivatives from a Papua New Guinea collection of the marine cyanobacterium *Lyngbya majuscula*. Tetrahedron 61:11723–11729

271. Linington RG, Edwards DJ, Shuman CF, McPhail KL, Matainaho T, Gerwick WH (2008) Symplocamide A, a potent cytotoxin and chymotrypsin inhibitor from the marine Cyanobacterium *Symploca* sp. J Nat Prod 71:22–27
272. Stolze SC, Meltzer M, Ehrmann M, Kaiser M (2010) Solid phase total synthesis of the 3-amino-6-hydroxy-2-piperidone (Ahp) cyclodepsipeptide and protease inhibitor symplocamide A. Chem Commun (Camb) 46:8857–8859
273. Tan LT, Okino T, Gerwick WH (2013) Bouillonamide: A mixed polyketide–peptide cytotoxin from the marine cyanobacterium *Moorea bouillonii*. Mar Drugs 11:3015–3024
274. Michon S, Cavelier F, Salom-Roig XJ (2021) Synthesis and biological activities of cyclodepsipeptides of aurilide family from marine origin. Mar Drugs 19:55
275. Han B, Gross H, Goeger DE, Mooberry SL, Gerwick WH (2006) Aurilides B and C, cancer cell toxins from a Papua New Guinea collection of the marine cyanobacterium *Lyngbya majuscula*. J Nat Prod 69:572–575
276. Takase S, Kurokawa R, Arai D, Kanemoto Kanto K, Okino T, Nakao Y, Kushiro T, Yoshida M, Matsumoto K (2017) A quantitative shRNA screen identifies ATP1A1 as a gene that regulates cytotoxicity by aurilide B. Sci Rep 7:2002
277. Pereira A, Cao Z, Murray TF, Gerwick WH (2009) Hoiamide a, a sodium channel activator of unusual architecture from a consortium of two Papua New Guinea cyanobacteria. Chem Biol 16:893–906
278. Cao Z, Li X, Zou X, Greenwood M, Gerwick WH, Murray TF (2015) Involvement of JNK and caspase activation in hoiamide A-induced neurotoxicity in neocortical neurons. Mar Drugs 13:903–919
279. Guo Y, Zhou J, Gao B, Zhao M, Yan J-L, Xu Z, Choi S, Ye T (2019) Total synthesis of hoiamide A using an Evans-Tishchenko reaction as a key step. Org Lett 21:5471–5474
280. Choi H, Pereira AR, Cao Z, Shuman CF, Engene N, Byrum T, Matainaho T, Murray TF, Mangoni A, Gerwick WH (2010) The hoiamides, structurally intriguing neurotoxic lipopeptides from Papua New Guinea marine cyanobacteria. J Nat Prod 73:1411–1421
281. Wang L, Xu Z, Ye T (2011) Total synthesis of hoiamide C. Org Lett 13:2506–2509
282. Malloy KL, Choi H, Fiorilla C, Valeriote FA, Matainaho T, Gerwick WH (2012) Hoiamide D, a marine cyanobacteria-derived inhibitor of p53/MDM2 interaction. Bioorg Med Chem Lett 22:683–688
283. Prabhu S, Kalaimathi K, Jayasree S, Ayyanar M, Kadaikunnan S, Thiruvengadam M, Amalraj S, Ceasar SA, Alharbi NS, Sanjeevi B, Priya SP (2024) Cyanobacterial metabolites as promising neuroprotective agents by targeting phosphoglycerate kinase 1: dynamic *In Silico* approaches. J Comput Biophys Chem 23:691–708
284. Luesch H, Yoshida WY, Moore RE, Paul VJ, Corbett TH (2001) Total structure determination of apratoxin A, a potent novel cytotoxin from the marine cyanobacterium *Lyngbya majuscula*. J Am Chem Soc 123:5418–5423
285. Robertson BD, Wengryniuk SE, Coltart DM (2012) Asymmetric total synthesis of apratoxin D. Org Lett 14:5192–5195
286. Grindberg RV, Ishoey T, Brinza D, Esquenazi E, Coates RC, Liu W-t, Gerwick L, Dorrestein PC, Pevzner P, Lasken R (2011) Single cell genome amplification accelerates identification of the apratoxin biosynthetic pathway from a complex microbial assemblage. PLoS ONE 6:e18565
287. Paatero AO, Kellosalo J, Dunyak BM, Almaliti J, Gestwicki JE, Gerwick WH, Taunton J, Paavilainen VO (2016) Apratoxin kills cells by direct blockade of the Sec61 protein translocation channel. Cell Chem Biol 23:561–566
288. Nogle LM, Marquez BL, Gerwick WH (2003) Wewakazole, a novel cyclic dodecapeptide from a Papua New Guinea *Lyngbya majuscula*. Org Lett 5:3–6
289. Lopez JAV, Al-Lihaibi SS, Alarif WM, Abdel-Lateff A, Nogata Y, Washio K, Morikawa M, Okino T (2016) Wewakazole B, a cytotoxic cyanobactin from the cyanobacterium *Moorea producens* collected in the Red Sea. J Nat Prod 79:1213–1218
290. Inman M, Dexter HL, Moody CJ (2017) Total synthesis of the cyclic dodecapeptides wewakazole and wewakazole B. Org Lett 19:3454–3457

291. Long B, Zhang J, Wang X, Tang X, Wu Z (2017) Total synthesis and biological evaluation of wewakazole. Chem Res Chin Univ 33:890–894
292. Lin Z, Torres JP, Tianero MD, Kwan JC, Schmidt EW (2016) Origin of chemical diversity in *Prochloron*-tunicate symbiosis. Appl Environ Microbiol 82:3450–3460
293. Smith TE, Pond CD, Pierce E, Harmer ZP, Kwan J, Zachariah MM, Harper MK, Wyche TP, Matainaho TK, Bugni TS (2018) Accessing chemical diversity from the uncultivated symbionts of small marine animals. Nat Chem Biol 14:179–185
294. Pettit GR, Xu J-p, Hogan F, Williams MD, Doubek DL, Schmidt JM, Cerny RL, Boyd MR (1997) Isolation and structure of the human cancer cell growth inhibitory cyclodepsipeptide dolastatin 16. J Nat Prod 60:752–754
295. Pettit GR, Xu J-P, Hogan F, Cerny RL (1998) Isolation and structure of dolastatin 17. Heterocycles 47:491–496
296. Pettit GR, Xu J-p, Williams MD, Hogan F, Schmidt JM, Cerny RL (1997) Antineoplastic agents 370. Isolation and structure of dolastatin 18. Bioorg Med Chem Lett 7:827–832
297. Pettit GR, Hogan F, Xu J-P, Tan R, Nogawa T, Cichacz Z, Pettit RK, Du J, Ye Q-H, Cragg GM (2008) Antineoplastic agents. 536. New sources of naturally occurring cancer cell growth inhibitors from marine organisms, terrestrial plants, and microorganisms. J Nat Prod 71:438–444
298. Luesch H, Harrigan G, Goetz G, Horgen F (2002) The cyanobacterial origin of potent anticancer agents originally isolated from sea hares. Curr Med Chem 9:1791–1806
299. Pettit GR, Smith TH, Xu J-P, Herald DL, Flahive EJ, Anderson CR, Belcher PE, Knight JC (2011) Antineoplastic Agents. 590. X-ray crystal structure of dolastatin 16 and syntheses of the dolamethylleuine and dolaphenvaline units. J Nat Prod 74:1003–1008
300. Tan LT, Goh BP, Tripathi A, Lim MG, Dickinson GH, Lee SS, Teo SL (2010) Natural antifoulants from the marine cyanobacterium *Lyngbya majuscula*. Biofouling 26:685–695
301. Luthfiana D, Utomo DH (2023) Network pharmacology reveals the potential of dolastatin 16 as a diabetic wound healing agent. In Silico Pharmacology 11:23
302. Pettit GR, Hogan F, Herald DL (2004) Synthesis and X-ray crystal structure of the *Dolabella auricularia* peptide dolastatin 18. J Org Chem 69:4019–4022
303. Yang L, Williams DE, Mui A, Ong C, Krystal G, van Soest R, Andersen RJ (2005) Synthesis of pelorol and analogues: activators of the inositol 5-phosphatase SHIP. Org Lett 7:1073–1076
304. Goclik E, König GM, Wright AD, Kaminsky R (2000) Pelorol from the tropical marine sponge *Dactylospongia elegans*. J Nat Prod 63:1150–1152
305. Kwak JH, Schmitz FJ, Kelly M (2000) Sesquiterpene quinols/quinones from the micronesian sponge *Petrosaspongia metachromia*. J Nat Prod 63:1153–1156
306. Flatt PM, Gautschi JT, Thacker RW, Musafija-Girt M, Crews P, Gerwick WH (2005) Identification of the cellular site of polychlorinated peptide biosynthesis in the marine sponge *Dysidea* (*Lamellodysidea*) *herbacea* and symbiotic cyanobacterium *Oscillatoria spongeliae* by CARD-FISH analysis. Mar Biol 147:761–774

MIX
Papier aus verantwortungsvollen Quellen
Paper from responsible sources
FSC® C105338

If you have any concerns about our products, you can contact us on
ProductSafety@springernature.com

In case Publisher is established outside the EU, the EU authorized representative is:
Springer Nature Customer Service Center GmbH
Europaplatz 3, 69115 Heidelberg, Germany

Printed by Libri Plureos GmbH
in Hamburg, Germany